J. Hansen · J. Pospiech · K. Lücke

Tables for Texture Analysis of Cubic Crystals

Springer-Verlag
Berlin Heidelberg GmbH 1978

Dr. rer. nat. Jörn Hansen

Member of the Institute for Materials Research of the German Aero-Space Research Establishment (Institut für Werkstoff-Forschung der Deutschen Forschungs- und Versuchsanstalt für Luft- und Raumfahrt e.V.),

Köln, West-Germany.

Dr. Jan Pospiech

Member of the Institute for Metal Research of the Polish Academy of Sciences (Polska Akademia Nauk, Zaklad Podstaw Metalurgii),

Kraków, Poland.

Prof. Dr. rer. nat. Kurt Lücke

Director of the Institute for Physical Metallurgy and Metal Physics at the Rhenian Westphalian Institute of Technology (Institut für Allgemeine Metallkunde und Metallphysik der Rheinisch-Westfälischen Technischen Hochschule),

Aachen, West-Germany.

ISBN 978-3-540-08689-5 ISBN 978-3-662-01630-5 (eBook)
DOI 10.1007/978-3-662-01630-5

Library of Congress Cataloging in Publication Data. Hansen, Jörn, 1940- Tables for texture analysis of cubic crystals. Bibliography: p. Includes index. 1. Texture (Crystallography) -Tables. I. Pospiech, Jan, 1936- joint author. II. Lücke, Kurt, joint author. III. Title. QD921.H33 547'.8 78-2014

Originally published by Springer-Verlag Berlin Heidelberg New York in 1978

PREFACE

Recent progress in the material sciences has led to an increasing amount of interest in the role of textures for the behaviour of materials and in the mechanisms controlling the texture formation. This development was supported by a rather powerfull development taking place in the area of texture studies itself: Besides the usual, more qualitative, characterization of a texture by pole figures a fully quantitative description by a three-dimensional orientation distribution function (ODF) has been increasingly applied.

There are two sides to the problem of quantitative representation of textures. One involves the mathematical technique associated with the acquisition of an ODF, and its transforms, from the experimental data, whereas the other concerns the methods of a rational description and interpretation of an ODF.

The first side can be considered from the practical point of view as experimental-data processing which is accomplished by a computer and is sort of a continuation of the measurement itself. Much attention has been paid to this problem, particulary by Bunge, who has written up achievements in this field in this extensive monograph /1/ and in conference proceedings /2/. There is also avaiable a rather detailed presentation of a system of subroutines written in Fortran /3/ which allows standard computations to be made without having to go into the mathematical details of the method.

The second side of the problem is the concern of the present study which contains information facilitating the analysis of an ODF obtained by calculative methods. The ODF defines the frequency of occurrence of a given orientation in a sample and is presented in a three-dimensional space formed by the three parameters describing an orientation, usually by the three Euler angles. For the purpose of approximate description and interpretation of the ODF, "ideal orientations" or "components" are often identified and crystallographic relationships between the components determined (e. g. twin relationships). This allows simple comparison to mathematical or physical models.

The present "Tables" give the most important notions and auxiliary data used in texture analysis on the basis of ODF's and of pole figures. The largest part of this work is tabulated data presenting numerical relationships between Miller indeces, Euler angles and pole figure positions. Such data is very important and helpful in almost all texture investigations. The lack of a generally accessible collection of such data has frequently made it necessary to calculate appropriate tabular values separately as appendices of publications.

The tables are limited to cubic crystal symmetry. They are further limited to orientations characterized by Miller indices with $0 \leqslant h, k, l, u, v, w \leqslant 15$ or $\leqslant 12$, respectively. But even then nearly 15000 different orientations had to be considered. The authors regret that the use of the tables is not as simple as e.g. the use of logarithm tables. They hope, however, that – with the aid of the explanations before each table – they will quickly become a useful tool for researchers in the area of texture analysis.

Especially emphasized shall also be the first part of this book. In a rather complete but also easily accessible form it contains a detailed review of the different ways of representing orientation distributions and orientation relationships including the most important mathematical derivations in this filed. Particularly the symmetry relations in the Euler angle space – hitherto a little known although rather important area – have been thoroughly discussed. In order to achieve an optimum understandig also original contributions not published elsewhere are included into this review.

The authors are deeply indebted to Dip.-Ing. K.H. Virnich for his continuous assistance and for numerous valuable discussions and contributions. They acknowledge the understanding shown for this work by Prof. W. Bunk, Köln, and Prof. Truszkowski, Krakow. They like to express their gratitude to Mrs. V. Boldin and to the staff of the computer center of the RWTH Aachen and of the DFVLR, Köln, for valuable aid in calculating and printing the tables.

CONTENTS

A. REVIEW OF THE REPRESENTATION OF ORIENTATIONS AND ORIENTATION DISTRIBUTIONS

1. INTRODUCTION

One of the most important quantities describing the internal structure of a polycrystalline material is the distrubtion of the orientations of its crystallites. This orientation distribution is commonly denoted as texture. The accepted practice is to speak of a texture when the orientation distribution is not a random one.

Texture analysis is based on a simple geometrical model in which the polycrystalline aggregate is represented by rectangular right-handed reference frames associated with the sample and with the crystallographic lattice of the crystallites. The axes of the sample reference system are chosen mostly in accordance with the external shape of the sample or, if its orientation distribution is symmetrical, in accordance with this symmetry. E.g. in the case of a rolled sheet, usually the rolling direction (RD), the transverse direction (TD) and the normal direction (ND) are used. The axes of the crystal reference systems are chosen parallel to (mostly low indiced) crystallographic directions, e.g. in the case of cubic symmetry parallel to the three edges of the cubic cell [100], [010] and [001], respectively. These frames are thought to be brought to a common origin at which also the center of point symmetry is located.

The fundamental notion when describing textures is the orientation of a crystallite. It is defined as the position of the reference frame of the crystallite relative to that of the sample, and thus can be expressed by the rotation of one frame into the other. Thus the notions and relationships used for describing textures are based on the properties of rotations and can be established by employing vectorial and matrix calculus. Since for the determination of a rotation 3 parameters are needed, an orientation can be represented by a point in a 3-dimensional "orientation space" formed by the 3 orientation parameters as coordinates. In quantitative texture analysis mostly the 3 Euler angles φ_1, ϕ, φ_2 are chosen as orientation parameters.

Another, rather illustrative way of representing an orientation is to consider a unit sphere fixed with respect to the sample frame, and to indicate on its surface the directions normal to a set of symmetrically equivalent low-indiced crystallographic planes ("poles"). The common practice is to consider the stereographic projection of this unit sphere ("pole figure") and to describe the positions of the poles by means of the spherical coordinates α, β. Frequently employed in analyzing and interpreting textures is also a description of orientations by assigning crystallographic indices (HKL) [UVW] to a certain plane of the sample and to a certain direction within this plane. E.g. in the case of a rolled sheed, (HKL) [UVW] commonly denotes the rolling plane and the rolling direction.

A texture is quantitatively described by its orientation distribution function ("ODF"). This is a density function in the three-dimensional orientation space and represents the frequency of a certain orientation as function of the 3 orientation parameters. It is obtained by numerical techniques from experimental data, which usually are pole figures, i.e. two-dimensional distribution functions of the poles of specific lattice planes. The reason for chosing this type of data is that the pole of a reflecting plane (in contrast to the angle of rotation around this pole) can be obtained rather easily by simple Debye-Scherrer X-ray technique. In practice, often the pole figures themselves (i.e. without calculating an ODF) are used to characterize a texture. They do not allow a complete recognition of the orientation distribution, but give some information about its main features, e.g. allow the identification of the orientations of the maxima of the ODF.

The following derivations are limited to cubic crystals. Furthermore, the sample geometry is mostly considered to be orthorhombic. In order to have something specific in mind, the samples will be then reffered to as rolled sheets. To a certain extent, however, also monoclinic samples (e.g. sheets after uni-directional rolling) or triclinic samples (e.g. rolled single crystals) are considered. Fiber textures are not especially discussed since they are thoroughly described in /1/ and can be presented unequivocally and mostly more simply by inverse pole figures.

2. REPRESENTATION OF AN ORIENTATION

2.1 Definition of Orientation

In all what follows, the base vectors $\vec{s}_1$, $\vec{s}_2$, $\vec{s}_3$ of the reference frame S associated with the sample are chosen parallel to RD, TD and ND and the base vectors $\vec{c}_1$, $\vec{c}_2$, $\vec{c}_3$ of the frame C associated with a crystal lattice are chosen parallel to the crystallographic directions [100], [010] and [001]. The crystallite orientation is defined as the rotation which transforms the sample reference frame S into that of the crystallite C.

The base vectors of the reference frames are related through the linear relationships

$$\begin{aligned} \vec{c}_1 &= g_{11}\vec{s}_1 + g_{12}\vec{s}_2 + g_{13}\vec{s}_3 \\ \vec{c}_2 &= g_{21}\vec{s}_1 + g_{22}\vec{s}_2 + g_{23}\vec{s}_3 \\ \vec{c}_3 &= g_{31}\vec{s}_1 + g_{32}\vec{s}_2 + g_{33}\vec{s}_3 \end{aligned} \tag{1}$$

In matrix notation the transformation (1) has the form

$$\begin{pmatrix} \vec{c}_1 \\ \vec{c}_2 \\ \vec{c}_3 \end{pmatrix} = g \begin{pmatrix} \vec{s}_1 \\ \vec{s}_2 \\ \vec{s}_3 \end{pmatrix} \quad \text{with } g = \begin{pmatrix} g_{11} & g_{12} & g_{13} \\ g_{21} & g_{22} & g_{23} \\ g_{31} & g_{32} & g_{33} \end{pmatrix} \tag{2}$$

or abbreviated $\{C\} = g \cdot \{S\}$. The relationship (1) and thus the matrix g describe the rotation of the frame S into frame C. Hence, according to the above definition of orientation, defines the orientation in matrix representation.

Since the 3 vectors $\vec{s}_i$ as well as, the 3 vectors $\vec{c}_i$ are orthogonal to each other and since, in addition, the $\vec{s}_i$, $\vec{c}_i$ are supposed to be unity vectors, one has for the scalar products

$$\vec{s}_i \cdot \vec{s}_j \text{ or } \vec{c}_i \cdot \vec{c}_j = \delta_{ij} \equiv \begin{cases} 1 & \text{for } i = j \\ 0 & \text{for } i \neq j \end{cases} \tag{3}$$

With this, one obtains for the scalar products

$$\vec{c}_i \cdot \vec{s}_k = g_{ik}. \tag{4}$$

The matrix elements g_{ik} are the cosines of the angles between the base vectors $\vec{c}_i$ and $\vec{s}_k$. The elements in the rows g_{ik} $(k = 1, 2, 3)$ are the direction cosines of the $\vec{c}_i$ vectors in the S system, whereas the elements in the columns g_{ik} $(i = 1, 2, 3)$ are the direction cosines of the $\vec{s}_i$ vectors in the C system.

The orthogonality of the frames expressed by Eq. (3) results in six independent conditions for elements g_{ik}. They can be written in the form

$$\sum_{k=1}^{3} g_{ik}\, g_{jk} = \delta_{ij} \tag{5}$$

and are obtained from Eqs. (1) and (3) by taking the appropriate scalar products $\vec{c}_i\, \vec{c}_j$. The left-hand side of this equation represents the matrix element of the product of the matrix g and the transposed matrix g^T (which is defined by $g^T_{mn} = g_{nm}$):

$$(g \cdot g^T)_{ij} = \sum_{k=1}^{3} g_{ik} g^T_{kj} = \sum_{k=1}^{3} g_{ik}\, g_{jk}. \tag{6}$$

The right hand side represents the unity matrix E so that Eq. (5) can be written as

$$g \cdot g^T = E = \begin{pmatrix} 1 & 0 & 0 \\ 0 & 1 & 0 \\ 0 & 0 & 1 \end{pmatrix} \tag{7}$$

Since the inverse matrix g^{-1} is defined by the equation $g \cdot g^{-1} = E$, the orthogonality conditions are identical with statement that here the inverse matrix g^{-1} (which describes the transformation $\{S\} = g^{-1}\{C\}$) is equal to the transposed matrix g^T.

The coordinates (x_S, y_S, z_S) and (x_C, y_C, z_C) of any vector $\vec{R}$ in the two reference frames S and C are transformed by the same matrices g and g^{-1} as the base vectors. With

$$\vec{R} = x_C\vec{c}_1 + y_C\vec{c}_2 + z_C\vec{c}_3 = x_S\vec{s}_1 + y_S\vec{s}_2 + z_S\vec{s}_3 \tag{8}$$

one obtains by scalar multiplication in succession by $\vec{c}_1, \vec{c}_2, \vec{c}_3$ and considering (4):

$$\begin{pmatrix} x_C \\ y_C \\ z_C \end{pmatrix} = g \begin{pmatrix} x_S \\ y_S \\ z_S \end{pmatrix} ; \quad \begin{pmatrix} x_S \\ y_S \\ z_S \end{pmatrix} = g^{-1} \begin{pmatrix} x_C \\ y_C \\ z_C \end{pmatrix} \tag{9}$$

Because of the six orthogonality conditions (5) between the nine matrix elements g_{ik}, the number of independent angles defining a rotation g becomes reduced from nine to three. Hence, the description of an orientation requires only 3 angles which are called orientation parameters. They may be chosen in different ways some of them will now be discussed.

2.2. Description of an Orientation by the Miller Indices (HKL) [UVW]

A method rather frequently used for describing an orientation is to indicate rolling plane and rolling direction by the Miller indices (HKL) [UVW]. They have the advantage of directly giving an insight into the crystallographic nature of the orientation.

The indices (HKL) are assigned to the normal direction $\vec{s}_3$ of the sheet plane and [UVW] to the rolling direction $\vec{s}_1$. They define the direction cosines of the $\vec{s}_3$ and $\vec{s}_1$ vectors in the crystallite system C according to

$$\vec{s}_3 = \frac{H}{M}\vec{c}_1 + \frac{K}{M}\vec{c}_2 + \frac{L}{M}\vec{c}_3 \tag{10}$$

and $$\vec{s}_1 = \frac{U}{N}\vec{c}_1 + \frac{V}{N}\vec{c}_2 + \frac{W}{N}\vec{c}_3 \tag{11}$$

where $M = \sqrt{H^2 + K^2 + L^2}$ and $N = \sqrt{U^2 + V^2 + W^2}$.

The vector $\vec{s}_2$ in TD follows from these equations according to $\vec{s}_2 = (\vec{s}_1 \times \vec{s}_2)$ or

$$\vec{s}_2 = \frac{KW - LV}{MN}\vec{c}_1 + \frac{LU - HW}{MN}\vec{c}_2 + \frac{HV - KU}{MN}\vec{c}_3 \tag{12}$$

i.e. the indices [QRS] of TD are given by

$$Q = KW - LV, \qquad R = LU - HW, \qquad S = HV - KU.$$

Scalar multiplication of Eqs. (10), (11) and (12) by $\vec{c}_1$, $\vec{c}_2$ and $\vec{c}_3$ and consideration of (4) yields the matrix of rotations defined by the indices (HKL) [UVW]

$$g\,((HKL)\,[UVW]) = \begin{pmatrix} \frac{U}{N} & \frac{KW - LV}{MN} & \frac{H}{M} \\ \frac{V}{N} & \frac{LU - HW}{MN} & \frac{K}{M} \\ \frac{W}{N} & \frac{HV - KU}{MN} & \frac{L}{M} \end{pmatrix} \tag{13}$$

The introduction of the Miller indices reduces the number of nine quantities q_{ik} for the description of an orientation to six and, at the same time, reduces the number of the six orthogonality conditions between the g_{ik} (Eq. (5)) to the following three relationships between the H, K, L, U, V, W:

$$\begin{aligned} &(H/M)^2 + (K/M)^2 + (L/M)^2 = 1 \\ &(U/N)^2 + (V/N)^2 + (W/N)^2 = 1 \\ &HU + KV + LW = 0. \end{aligned} \tag{14}$$

2.3 Description of an Orientation by the Euler Angles φ_1 ϕ φ_2

For the purpose of quantitative texture analysis the orientations are mostly described by the three Euler angles φ_1, ϕ, φ_2 which lead to a simpler mathematical formalism than the other orientation parameters being in use. Using Euler angles, the transformation of the sample frame S into the crystallite frame C occurs by a set of three consecutive rotations (Fig. 1):

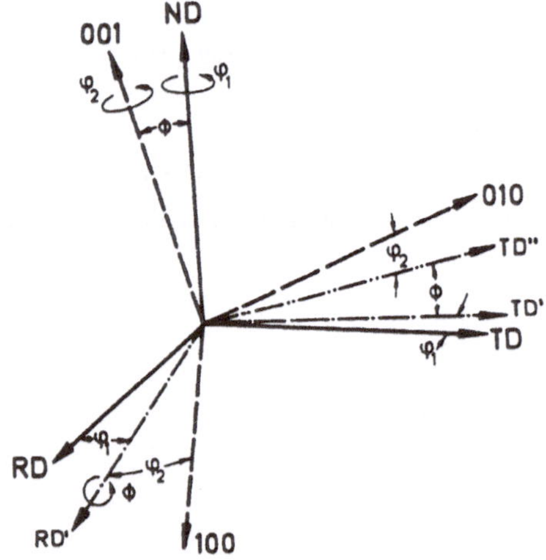

Fig. 1:
Definition of the Euler angles φ_1, ϕ, φ_2.

1. A first rotation φ_1 around the normal direction ND transforming the transverse direction TD and the rolling direction RD into the new directions TD′ and RD′, respectively. φ_1 has to have such a value that RD′ will be perpendicular to the plane formed by ND and [001].

2. A second rotation ϕ around the new direction RD′ with ϕ having such a value that ND is transformed into [001] (= ND′) (and TD′ into TD″).

3. A third rotation φ_2 around [001] (= ND′) with φ_2 having such a value that RD′ is transformed into [100] (and TD″ into [010]).

These 3 rotations can be expressed mathematically in the following way:

1. Rotation about $\vec{s}_3$ by the angle φ_1 ($\vec{s}_3 = \vec{s}\,'_3, \vec{s}_2 \to \vec{s}\,'_2, \vec{s}_1 \to \vec{s}\,'_1$) which corresponds to the transformation

$$\{S'\} = g(\varphi_1)\{S\} \tag{15}$$

with the rotation matrix $g(\varphi_1) = \begin{pmatrix} \cos\varphi_1 & \sin\varphi_1 & 0 \\ -\sin\varphi_1 & \cos\varphi_1 & 0 \\ 0 & 0 & 1 \end{pmatrix}$

2. about $\vec{s}\,'_1$ by the angle ϕ ($\vec{s}\,'_3 \to \vec{s}\,''_3, \vec{s}\,'_2 \to \vec{s}\,''_2, \vec{s}\,'_1 = \vec{s}\,''_1$) i.e.

$$\{S''\} = g(\phi)\cdot\{S'\} \tag{16}$$

where

$$g(\phi) = \begin{pmatrix} 1 & 0 & 0 \\ 0 & \cos\phi & \sin\phi \\ 0 & -\sin\phi & \cos\phi \end{pmatrix}$$

3. about $\vec{s}\,''_3$ by the angle φ_2 ($\vec{s}\,''_3 = \vec{c}_3, \vec{s}\,''_2 \to \vec{c}_2, \vec{s}\,''_1, \to \vec{c}_1$)

i.e. $$\{C\} = g(\varphi_2)\ \{S''\} \tag{17}$$

where

$$g(\varphi_2) = \begin{pmatrix} \cos\varphi_2 & \sin\varphi_2 & 0 \\ -\sin\varphi_2 & \cos\varphi_2 & 0 \\ 0 & 0 & 1 \end{pmatrix}$$

Successive elimination $\{S'\}$ and $\{S''\}$ from formulae (15), (16) and (17) gives the rotation matrix defined by Euler angles,

$$\{C\} = g(\varphi_2)\cdot g(\phi)\cdot g(\varphi_1)\ \{S\} = g(\varphi_1\ \phi\ \varphi_2)\ \{S\} \tag{18}$$

which has the form:

$$g(\varphi_1\,\phi\,\varphi_2) = \begin{pmatrix} \cos\varphi_1\cos\varphi_2 - \sin\varphi_1\sin\varphi_2\cos\phi & \sin\varphi_1\cos\varphi_2 + \cos\varphi_1\sin\varphi_2\cos\phi & \sin\varphi_2\sin\phi \\ -\cos\varphi_1\sin\varphi_2 - \sin\varphi_1\cos\varphi_2\cos\phi & -\sin\varphi_1\sin\varphi_2 + \cos\varphi_1\cos\varphi_2\cos\phi & \cos\varphi_2\sin\phi \\ \sin\varphi_1\sin\phi & -\cos\varphi_1\sin\phi & \cos\phi \end{pmatrix} \tag{19}$$

This matrix does not change if integer multiples of $\pm\,2\pi$ are added to the angles φ_1, ϕ, φ_2 or when this set of angles is replaced by

$$\varphi_1^e = \pi + \varphi_1,\quad \phi^e = -\phi,\quad \varphi_2^e = \pi + \varphi_2. \tag{20}$$

All orientations resulting from this transformations are called identically equivalent (c.f. Sec. 3.3).

2.4 Description of an Orientation by the Euler Angles $\psi\ \theta\ \phi$

In the literature also somewhat differently defined Euler angles are encountered /4/ which are denoted ψ, θ, ϕ. Their definition differs from that given above in that the second rotation (by the angle θ) takes place about the $\vec{s}_2$-axis instead of about the $\vec{s}_1$-axis (Fig. 2).

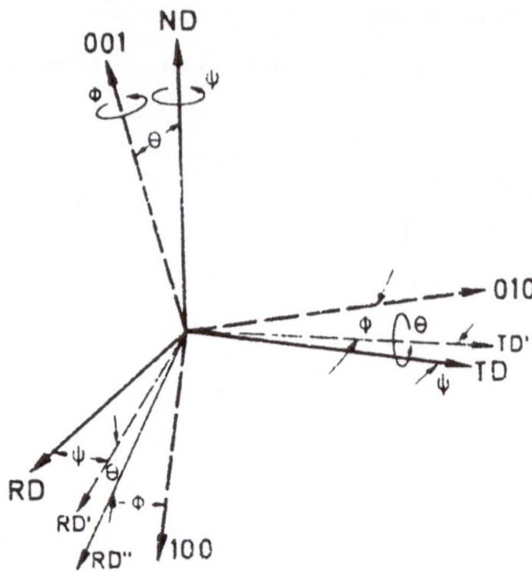

Fig. 2:
Definition of the Euler angles ψ, θ, ϕ.

The $g(\theta)$ matrix in Eq. (16) then assumes a different form,

$$g(\theta) = \begin{pmatrix} \cos\theta & 0 & -\sin\theta \\ 0 & 1 & 0 \\ \sin\theta & 0 & \cos\theta \end{pmatrix} \tag{21}$$

After the change of notation for the angles in matrices $g(\varphi_1)$ and $g(\varphi_2)$, and considering (18), we get

$$g(\phi) \cdot g(\phi) \cdot g(\psi) = g(\psi, \theta, \phi) \tag{22}$$

where

$$g(\psi, \theta, \phi) = \begin{pmatrix} \cos\phi\,\cos\theta\,\cos\psi - \sin\phi\,\sin\psi & \cos\phi\,\cos\theta\,\cos\psi + \sin\phi\,\cos\psi & -\cos\phi\,\sin\theta \\ -\sin\phi\,\cos\theta\,\cos\psi - \cos\phi\,\sin\psi & -\sin\phi\,\cos\theta\,\sin\psi + \cos\phi\,\cos\psi & \sin\phi\,\sin\theta \\ \sin\theta\,\cos\psi & \sin\theta\,\sin\psi & \cos\phi \end{pmatrix} \tag{23}$$

There exist the following general relationships between the two types of Euler angles.

$$\varphi_1 = \psi + \frac{\pi}{2}; \quad \phi = \theta; \quad \varphi_2 = \phi - \frac{\pi}{2}. \tag{24}$$

If one additionally considers cubic crystal symmetry and orthorhombic sample symmetry (Sec. 3.4) one obtains

$$\varphi_1 = \frac{\pi}{2} - \psi; \quad \phi = \theta; \quad \varphi_2 = \frac{\pi}{2} - \phi. \tag{25}$$

In Fig. 3 these relationships are illustrated in a section ψ, θ = constant. One recognizes that these two sets of parameters can be transformed into each other by a two-fold axis parallel to ϕ und θ, respectively.

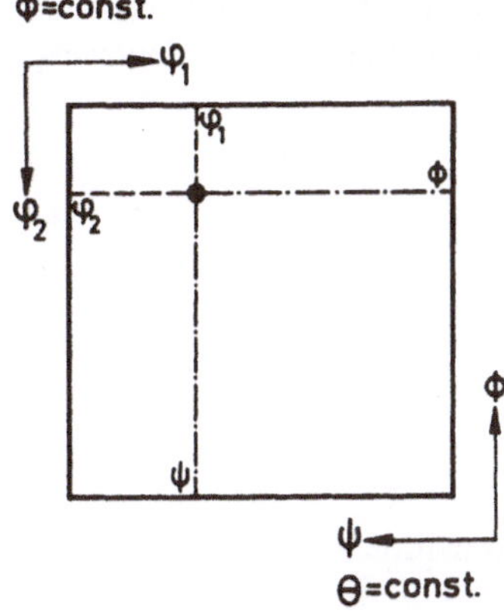

Fig. 3:
Relationships between the angles $\varphi_1, \phi, \varphi_2$ and ψ, θ, ϕ.

2.5 Description of an Orientation by Rotational Coordinates $\vec{v}$, ω

The rotational coordinates characterize an orientation by a single axis and angle of a rotation which transforms the S frame into C frame (Fig. 4). These coordinates have the advantage of being easy to visualize, much easier than e.g. the Euler-coordinates which describe a set of three consecutive rotations. The derivations of the expressions given in the following section can be found in /5/ and /6/.

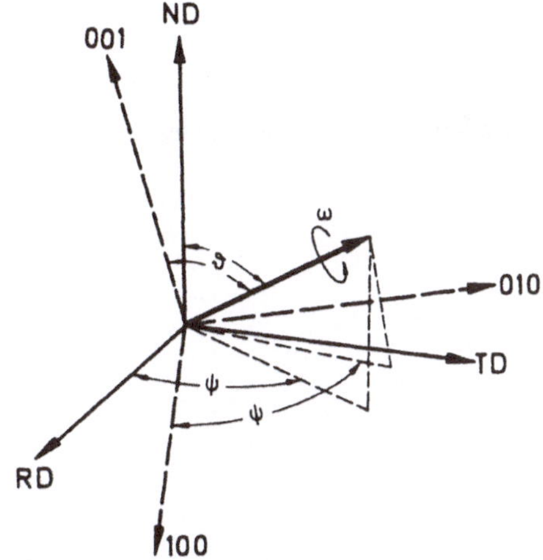

Fig. 4:
Definition of the rotational coordinates (axis of rotation $\vec{v} = \{\vartheta, \psi\}$, angle of rotation ω).

In the following ϑ and ψ shall denote the spherical coordinates of the unit vector $\vec{v}$ which indicates the axis of rotation and ω shall be the angle of rotation around $\vec{v}$. As can be recognized from Fig. 4, the values of the spherical coordinates ϑ and ψ are equal in both reference frames S and C. That means that the direction cosines v_x, v_y, v_z of the rotation axis $\vec{v}$ satisfy the condition

$$\vec{v} = v_x\,\vec{s}_1 + v_y\,\vec{s}_2 + v_z\,\vec{s}_3 = v_x\,\vec{c}_1 + v_y\,\vec{c}_2 + v_z\,\vec{c}_3$$

by scalar multiplication in succession by $\vec{s}_1$, $\vec{s}_2$ or $\vec{s}_3$ (c.f. Eq. (4)) one obtains

$$\begin{pmatrix} v_x \\ v_y \\ v_z \end{pmatrix} = g \cdot \begin{pmatrix} v_x \\ v_y \\ v_z \end{pmatrix} \qquad (26)$$

i.e. the rotation axes v are eigen-vectors of the matrix g. The angle of rotation ω and the direction cosines of the rotation axis $\vec{v}$ can be calculated from the elements g_{ik} of the matrix g in the following way

$$\begin{aligned} g_{11} + g_{22} + g_{33} &= 1 + 2\cos\omega \\ \frac{1}{2}(g_{23} - g_{32}) &= v_x \sin\omega \\ \frac{1}{2}(g_{31} - g_{13}) &= v_y \sin\omega \\ \frac{1}{2}(g_{12} - g_{21}) &= v_z \sin\omega \end{aligned} \qquad (27)$$

The matrix expressed in terms of parameters ϑ, ψ, ω in the transformation

$$\{C\} = g(\vartheta, \psi, \omega) \cdot \{S\} = g(\vec{v}, \omega) \cdot \{S\} \qquad (28)$$

possesses the following form:

$$g(\vec{v}, \omega) = \begin{pmatrix} (1 - v_x^2)\cos\omega + v_x^2 & v_x v_y (1 - \cos\omega) + v_z \sin\omega & v_x v_z (1 - \cos\omega) - v_y \sin\omega \\ v_x v_y (1 - \cos\omega) - v_z \sin\omega & (1 - v_y^2)\cos\omega + v_y^2 & v_g v_z (1 - \cos\omega) + v_x \sin\omega \\ v_x v_y (1 - \cos\omega) + v_z \sin\omega & v_y v_z (1 - \cos\omega) - v_x \sin\omega & (1 - v_z^2)\cos\omega + v_z^2 \end{pmatrix} \qquad (29)$$

where

$$v_x = \cos\psi \sin\vartheta\,; \quad v_y = \sin\psi \sin\varphi\,; \quad v_z = \cos\vartheta\,.$$

The matrix (29) does not change if integer multiples of $\pm 2\pi$ are added to the angles ϑ, ψ, ω or when this set of angles is replaced by

$$\begin{aligned} \vartheta' &= -\vartheta & \qquad & \vartheta' = \pi - \vartheta \\ \psi' &= \pi + \psi & \text{or by} \qquad & \psi' = \pi + \psi \\ \omega' &= \omega & \qquad & \omega' = -\omega, \end{aligned} \qquad (30)$$

respectively.

2.6 Description of an Orientation by Pole Figures

In a pole figure an orientation is defined by the positions of the poles $(X_i Y_i Z_i)$ of the symmetrically equivalent lattice planes $\{XYZ\}$. These positions can be visualized as the intersection points of the normals to these planes with the surface of a unit sphere which is bound to the S-frame (sample). It can graphically be represented on the stereographic projection of the unit sphere („pole figure") and numerically be described by the spherical coordinates α_i, β_i. Fig. 5 a gives an example for the $\{001\}$ poles of an arbitrary oriented crystal. Fig. 5 b indicates the projection of the intersection points into the equatorial plane and Fig. 5 c shows for one pole the angles α_i, β_i within this plane.

If R_i is the unit vector of the pole $(X_i Y_i Z_i)$ then, according to Eq. (8), it has the following components in the sample's frame S (i.e. in the pole figure) and in the crystallite's frame C (i.f. also Fig. 5):

$$\begin{aligned} \vec{R}_i &= (\sin\alpha_i \cos\beta_i)\vec{s}_1 + (\sin\alpha_i \sin\beta_i)\vec{s}_2 + (\cos\alpha_i)\vec{s}_3 \\ &= \frac{1}{P}(X_i \vec{c}_1 + Y_i \vec{c}_2 + Z_i \vec{c}_3). \end{aligned} \qquad (31)$$

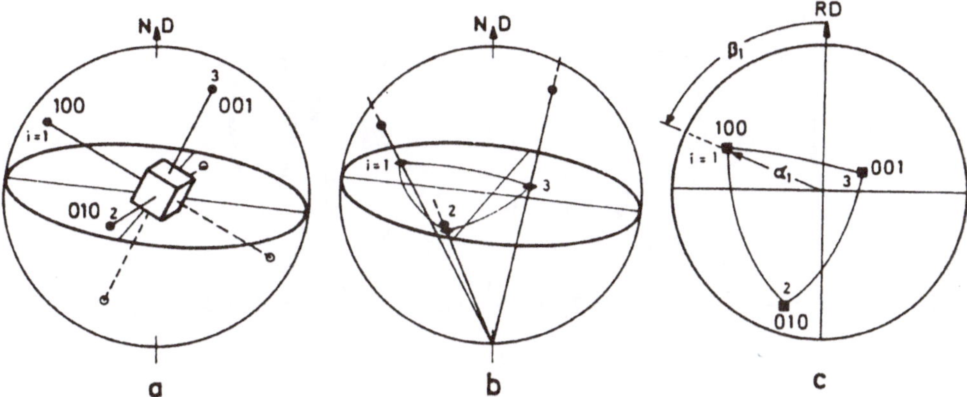

Fig. 5:
Presentation of a $\{100\}$-pole figure. a) Position of a crystal in the center of the orientation sphere; b) Projection of the cube half axes on the equatorial plane, c) $\{100\}$-pole figure and definition of the spherical coordinates α_i, β_i of the pole i.

Here the quantity $P = \sqrt{X_i^2 + Y_i^2 + Z_i^2}$ has the same value for all symmetrically equivalent poles $(X_iY_iZ_i)$, i.e. it is independent of i. Scalar multiplications of Eq. (31) in succession by the three vectors $\vec{s}_i$ and taking into consideration Eq. (4) yields the following transformation:

$$\begin{pmatrix} \sin\alpha_i \cos\beta_i \\ \sin\alpha_i \sin\beta_i \\ \cos\alpha_i \end{pmatrix} = \frac{1}{P} \begin{pmatrix} g_{11} & g_{21} & g_{31} \\ g_{12} & g_{22} & g_{32} \\ g_{13} & g_{23} & g_{33} \end{pmatrix} \begin{pmatrix} x_i \\ y_i \\ z_i \end{pmatrix} \tag{32}$$

It is to be recognized that here matrix $g^T = g^{-1}$ appears.

This equation allows the matrix elements g_{ik} to be determined from the poles $(X_iY_iZ_i)$ given by the angles α_1, β_i. If, for example, for the three $\{001\}$-poles

$X_1 = 1,\ Y_1 = 0,\ Z_1 = 0$ the coordinates α_1, β_1

$X_2 = 0,\ Y_2 = 1,\ Z_2 = 0$ the coordinates α_1, β_2

$X_3 = 0,\ Y_3 = 0,\ Z_3 = 1$ the coordinates α_3, β_3

have been found, Eq. (32) yields the following nine equations for the g_{ik}:

$$g_{i1} = \sin\alpha_i \cos\beta_i; \quad g_{i2} = \sin\alpha_i \sin\beta_i; \quad g_{i3} = \cos\alpha_i. \tag{33}$$

Because of the six orthogonality conditions between the nine g_{ik}, it would suffice to consider only 3 angles (two α_i and one β_i or vice versa), i.e. two poles only. In practice, however, it is better to use all poles and to apply the orthogonality conditions as a control. Then, by equating the nine calculated g_{ik} with the elements of the matrix Eq. (19) (or Eqs. (23), (29) or (13)}, the three Euler angles (or any other three orientation parameters) can be derived. Also here not all nine equations have to be solved, three independent equations would suffice. However, because of the multivalency of the trigonometric functions, sometimes five equations are necessary in order to fix also the signs of the angles *.

Conversely, if the orientation g is given, the positions α_i, β_i of the poles $(X_iY_iZ_i)$ in the pole figure can be calculated from the g_{ik}. For example, the position α_i, β_i of the pole (111) (i.e. $X_1 = Y_1 = Z_1 = 1$) of the orientation (101) $[\bar{1}\bar{2}1]$ are found from Eqs. (13) and (32):

* Often is it faster and more practical to derive the poles $(X_iY_iZ_i)$ in a graphical way. Such a method for obtaining the poles from the Euler angles is described in Appendix I.

$$\begin{pmatrix} \sin\alpha_1 \cos\beta_1 \\ \sin\alpha_1 \sin\beta_1 \\ \cos\alpha_1 \end{pmatrix} = \frac{1}{\sqrt{3}} \begin{pmatrix} -\frac{1}{\sqrt{6}} & -\frac{2}{\sqrt{6}} & \frac{1}{\sqrt{6}} \\ \frac{1}{\sqrt{3}} & -\frac{1}{\sqrt{3}} & -\frac{1}{\sqrt{3}} \\ \frac{1}{\sqrt{2}} & 0 & \frac{1}{\sqrt{2}} \end{pmatrix} \begin{pmatrix} 1 \\ 1 \\ 1 \end{pmatrix}$$

This leads to $\sin\alpha_1 \cos\beta_1 = -\frac{\sqrt{2}}{3}$; $\sin\alpha_1 \sin\beta_1 = -\frac{1}{3}$; $\cos\alpha_1 = \frac{\sqrt{2}}{\sqrt{3}}$, and thus to $\alpha_1 = 35°, 3$, $\beta_1 = 215°, 3$.

2.7 Description of an Orientation by Inverse Pole Figures

The reference system of an inverse pole figure is the frame C associated with the crystal and the orientation is defined by the directions of axes connected with the specimen. These axes are usually the axes $\vec{s}_1, \vec{s}_2, \vec{s}_3$ parallel to RD, TD, ND. In analogy to Eq. (31) where the spherical coordinates α_i, β_i of a unit vector R_i parallel to the crystallographic axis $(X_i, Y_i Z_i)$ have been considered in the system S, here the spherical coordinates γ_i, δ_i of a unit vector $\vec{R}_i$ parallel to a sample axis $\vec{s}_i$ in the coordinate system C must be introduced

$$\vec{R}_i = \vec{s}_i = \sin\gamma_i \cos\delta_i \vec{c}_1 + \sin\gamma_i \sin\delta_i c_2 + \cos\delta_i c_3. \tag{34}$$

According to Eq. (4) one obtains by scalar multiplication in succession by $\vec{c}_1, \vec{c}_2, \vec{c}_3$

$$g_{1i} = \sin\gamma_i \cos\delta_i; \quad g_{2i} = \sin\gamma_i \sin\delta_i; \quad g_{3i} = \cos\delta_i. \tag{35}$$

These expressions describe the elements of the orientation matrix g in terms of the positions of ND, TD, RD in the inverse pole figure.

2.8 Relationships Between the Different Types of Orientation Parameters

Since all the above matrix representations of an orientation (Eqs. (13), (19), (23), (29)) are equivalent, it is possible to establish the relationships between the different types of orientation parameters by comparing the different types of matrix elements. The most important ones of the resulting relationships are the following:

For $(\varphi_1\ \phi\ \varphi_2) \rightleftarrows (\vartheta\ \psi\ \omega)$ (36)

$$\sin\frac{\omega}{2} \sin\vartheta = \sin\frac{\phi}{2}$$

$$\cos\frac{\omega}{2} = \cos\frac{\phi}{2} \cos\frac{\varphi_1 + \varphi_2}{2}$$

$$\psi = \tfrac{1}{2}(\varphi_1 - \varphi_2).$$

For (HKL) [UVW] → $(\varphi_1\ \phi\ \varphi_2)$ (37)

$$\operatorname{tg}\phi\ \cos\varphi_2 = \frac{K}{L}$$

$$\operatorname{tg}\varphi_2 = \frac{H}{K}$$

$$\cos\phi\ \operatorname{tg}\varphi_1 = \frac{LW}{KU - HV}$$

For (HKL) [UVW] → $(\vartheta\ \psi\ \omega)$ (38)

$$\operatorname{tg}\psi = \frac{WM - HN}{KN - HV + KU}$$

$$\cos\omega = \frac{1}{2NM}(UM + LU - HW + LN - NM)$$

$$\cos\vartheta\ \sin\omega = \frac{1}{2NM}(KW - LV - VM).$$

For (HKL) [UVW] → $\alpha_i\ \beta_i$ (39)

$$\cos\alpha_i = \frac{1}{PM}(HX_i + KY_i + LZ_i)$$

$$\sin\alpha_i\ \cos\beta_i = \frac{1}{PN}(UX_i + VY_i + WZ_i)$$

$$\sin\alpha_i\ \sin\beta_i = \frac{1}{PMN}[(KW - LV)X_i + (LU - HW)Y_i + (HV - KU)Z_i]$$

For $(\varphi_1\ \phi\ \varphi_2)$ → $\alpha_i\ \beta_i$ (40)

$$\cos\alpha_i = \frac{1}{P}(X_i\ \sin\varphi_2\ \sin\phi + Y_i\ \cos\varphi_2\ \sin\phi + Z_i\ \cos\phi)$$

$$\sin\alpha_i\ \cos\beta_i = \frac{1}{P}[X_i(\cos\varphi_1\ \cos\varphi_2 - \sin\varphi_1\ \sin\varphi_2\ \cos\phi) - Y_i(\cos\varphi_1\ \sin\varphi_2 + \sin\varphi_1\ \cos\varphi_2\ \cos\phi) + Z_i\ \sin\varphi_1\ \sin\phi]$$

$$\sin\alpha_i\ \sin\beta_i = \frac{1}{P}[X_i(\sin\varphi_1\ \cos\varphi_2 + \cos\varphi_1\ \sin\varphi_2\ \cos\phi) - Y_i(\sin\varphi_1\ \sin\varphi_2 - \cos\varphi_1\ \cos\varphi_2\ \cos\phi) - Z_i\ \cos\varphi_1\ \sin\phi)]$$

For $(\psi\ \theta\ \phi) \rightleftarrows (\varphi_1\ \phi\ \varphi_2)$ (41)

$$\varphi_1 = \frac{\pi}{2} - \psi$$

$$\phi = \theta$$

$$\varphi_2 = \frac{\pi}{2} - \phi.$$

3. THE ORIENTATION SPACE

3.1 Examples for Various Types of Orientation Spaces

The three parameters describing an orientation can be used as coordinates of a three dimensional orientation space in which each point represents an orientation. Applying different types of orientation parameters and coordinate systems, the orientation space can be formed in many ways.

In Fig. 6 use is made of the rotational coordinates ϑ, ψ, ω. The direction of a radius vector R characterizes the axis of rotation $\vec{v}$ and its length the angle of rotation ω. If one further defines that radius vectors with opposite direction characterize rotations with opposite sense (e.g. that vectors above the equatorial plane characterize counter-clockwise and those below clockwise rotations), a sphere with a radius $R_0 = \pi$ will contain all possible orientations.

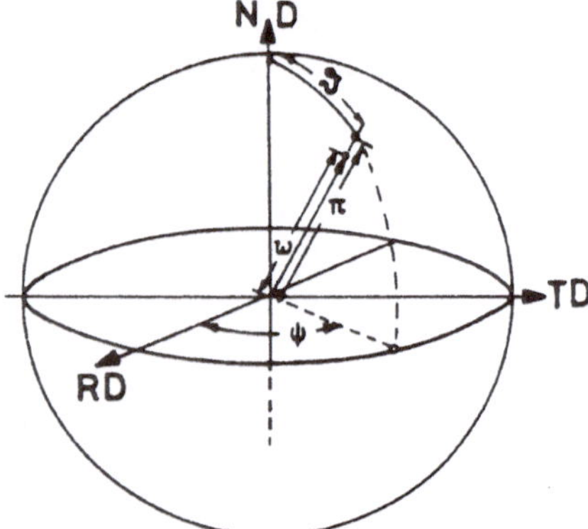

Fig. 6:
Presentation of a spherical orientation space with rotational coordinates $\vec{v}$, ω.

In Fig. 7 the same parameters ϑ, ψ, ω are applied in another form. The ω-axis (from $-\pi$ to $+\pi$) is chosen as cylinder-axis and the sections ω = constant represent the stereographic projection of the axes of rotation $\vec{v}$. This cylindrical orientation space has proved very useful for describing orientation relationships.

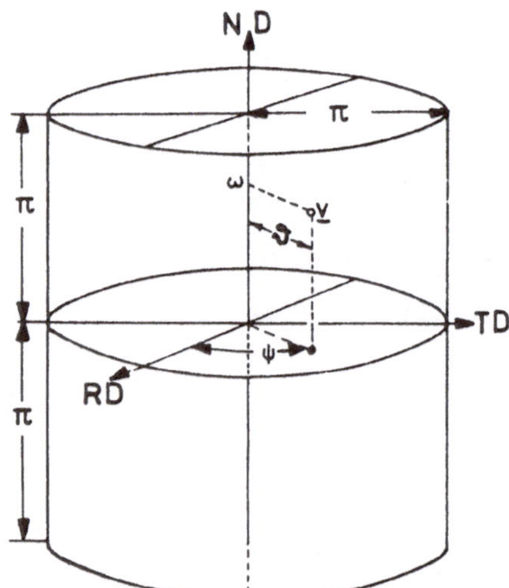

Fig. 7:
Presentation of a cylindrical orientation space will the rotational coordinates $\vec{v}$, ω.

Most widespread, however, is the use of the orientation space in which the Euler angles φ_1, ϕ, φ_2 form an Cartesian coordinate system. To present an orientation in this Euler angle space it suffices to consider the range

$$H = \{0 \leqslant \varphi_1 \leqslant 2\pi, \quad 0 \leqslant \phi \leqslant \pi, \quad 0 \leqslant \varphi_2 \leqslant 2\pi\} \tag{42}$$

the so-called asymmetric unit (Fig. 8). This can be recognized in Fig. 5 a where an orientation is characterized by a set of 3 points on the surface of the sphere. By the rotation φ_1 around ND (from 0 to 2π) and by the angle ϕ between ND and [001] (from 0 to π) the position of [001] is determined, whereas φ_2 gives the rotation around [001] (from 0 to 2π).

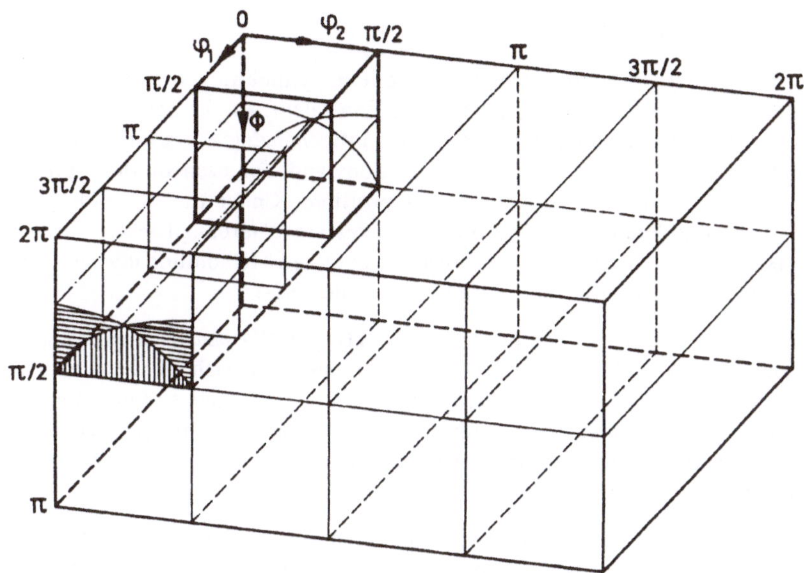

Fig. 8:
Asymmetric unit H of the Euler angle space divided into the subspaces H′ and H°.

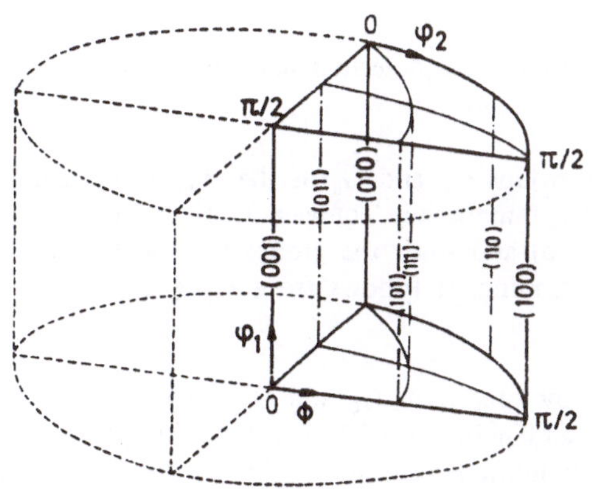

Fig. 9
Presentation of a cylindrical Euler angle space with φ_2 as cylinder axis. The base corresponds to a pole figure.

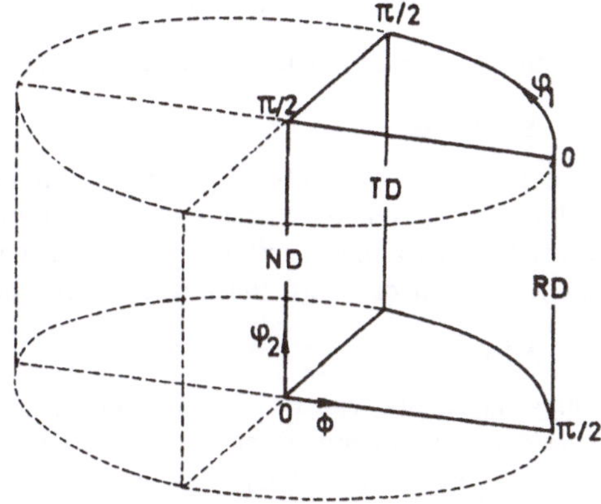

Fig. 10:
Presentation of a cylindrical Euler angle space with φ_2 as cylinder axis. The base corresponds to a pole figure.

Fig. 9 shows a cylindrical type of orientation space based also on the Euler angles φ_1, ϕ, φ_2. (For reasons given in Sec. 3.3 only the section $0 \leqslant \varphi_1, \phi, \varphi_2 \leqslant \pi/2$ is represented in the figure.) The cylinder axis is put into the direction [001] and the base plane represents a stereographic projection of the ND's thus corresponding to an inverse pole figure. As follows directly from the definition of the Euler angles, ϕ then gives the pole distance and φ_2 the azimuth of the ND, whereas φ_1, the rotation around ND, is plotted parallel to the axis. It is often convenient to use this type of Euler angle space together with pole figure considerations. It is possible, of course, to exchange the coordinates φ_1 and φ_2 (Fig. 10). Then an Euler angle space is obtained which is fixed to the sample coordinate system and corresponds to the usual pole figure.

3.2 Symmetrically Equivalent Orientations

If the crystal lattice and/or the specimen (i.e. the pole figure) contain symmetries, orientations which would be different without these symmetries will become symmetrically equivalent. In the case of cubic crystal symmetry, the symmetrically equivalent orientations are found by interchanging the six (equivalent) half axes of the cube. A given half axis can assume six different positions and, for each of these positions by rotation around this half axis, a second half axes four different positions. Under the supposition that only right hand coordinate systems are considered, also the other half axes are determined when these two are fixed so that 6 · 4 = 24 symmetrically equivalent orientations are obtained (If additionally also left hand systems would be taken into account, alltogether 48 possibilities would arise).

In the case of a rolled polycrystalline sheet, the sample axes ND, TD and RD can be interchanged according to the (orthorhombic) symmetry of rolling. E.g. + RD and – RD and, for each of these two positions, also + ND and – ND can be interchanged. Since, considering again only right hand systems, the two TD-half axes are then fixed so that here 4 symmetrically equivalent orientations are obtained. This shows up also the symmetry of the 4 quadrants of the pole figure. Hence, for a rolled sheet of a cubic material, all together 24 · 4 = 96 orientations are symmetrically equivalent.

These relationships shall now be looked at somewhat closer. Beside the unity element, the elements of point symmetry for crystal and specimen are given by rotation axes and by an inversion. In the following the symbols E and I shall denote the unity and inversion element, L_α^i an i-fold rotation axis parallel to the direction α and T_α a translation in direction of α. The here considered crystal symmetry given by the G_c = m3m group of the cubic system has the symmetry elements I, L_{001}^4, L_{100}^2, L_{111}^3 *. For a sample symmetry given by the G_s = mmm group of the orthorhombic system, one has as symmetry elements the axes L_{ND}^2, L_{RD}^2 and again the inversion I.

The orientations g^e symmetrically equivalent to an orientation g are determined by the relation

$$g^e = g_c \cdot g \cdot g_s \tag{43}$$

where g_c and g_s are the elements of the point symmetry groups G_c and G_s of the crystal and sample. Since the application of an inversion element does not generate symmetrically equivalent orientations (it only converts a left-hand coordinate system into a right-hand one) and since the groups G_c and G_s both cointain an inversion I, one can disregard the inversions for our reasoning. It follows from Eq. (43):

$$g^e = I \cdot g \cdot I = g \cdot I \cdot I = g \cdot E = g \tag{44}$$

i.e. the groups G_c and G_s can be thought to be replaced by subgroups G_c' and G_s' not containing I. Thus the here considered group mmm can be replaced by 222 and m3m by 432. The generating elements of the group 432 are the rotation axes L_{001}^4 and L_{111}^3. Their combination leads to all the 24 symmetry elements ** listed in Table 3.2.

---

* The L_{100}^2 element is not necessary for the formation of the m3m group, but its consideration is convenient in the deliberations. The three-fold axis can be replaced by another four-fold axis according to $L_{111}^3 = L_{100}^4 \cdot L_{001}^4$. The product of a two-fold axis and the inversion gives a mirror plane $P = IL^2$. From this it follows also that two two-fold axes result in two mirror planes: $L_Z^2 \cdot L_X^2 = P_Z \cdot P_X$.

** This remark see on page 16.

Table 3.2: The 24 possible symmetry elements of the group 432. Column 1 contains the here applied symbols for the symmetries and column 2 the corresponding matrix representation. Column 3 contains the positions for the orthogonal coordinate axes x, y, z after the symmetry operation has been carried out.

	1	2	3
1	E	$\begin{matrix} 1 & 0 & 0 \\ 0 & 1 & 0 \\ 0 & 0 & 1 \end{matrix}$	x y z
2	L^2_{010}	$\begin{matrix} -1 & 0 & 0 \\ 0 & 1 & 0 \\ 0 & 0 & -1 \end{matrix}$	$\bar{x}$ y $\bar{z}$
3	L^2_{001}	$\begin{matrix} -1 & 0 & 0 \\ 0 & -1 & 0 \\ 0 & 0 & 1 \end{matrix}$	$\bar{x}$ $\bar{y}$ z
4	L^2_{100}	$\begin{matrix} 1 & 0 & 0 \\ 0 & -1 & 0 \\ 0 & 0 & -1 \end{matrix}$	x $\bar{y}$ $\bar{z}$
5	L^2_{111}	$\begin{matrix} 0 & 1 & 0 \\ 0 & 0 & 1 \\ 1 & 0 & 0 \end{matrix}$	y z x
6	$L^2_{1\bar{1}\bar{1}}$	$\begin{matrix} 0 & -1 & 0 \\ 0 & 0 & 1 \\ -1 & 0 & 0 \end{matrix}$	$\bar{y}$ z $\bar{x}$
7	$L^3_{\bar{1}1\bar{1}}$	$\begin{matrix} 0 & -1 & 0 \\ 0 & 0 & -1 \\ 1 & 0 & 0 \end{matrix}$	$\bar{y}$ $\bar{z}$ x
8	$L^3_{\bar{1}\bar{1}1}$	$\begin{matrix} 0 & 1 & 1 \\ 0 & 0 & -1 \\ -1 & 0 & 0 \end{matrix}$	y $\bar{z}$ $\bar{x}$
9	$L^3_{\bar{1}\bar{1}\bar{1}}$	$\begin{matrix} 0 & 0 & 1 \\ 1 & 0 & 0 \\ 0 & 1 & 0 \end{matrix}$	z x y
10	$L^3_{11\bar{1}}$	$\begin{matrix} 0 & 0 & -1 \\ 1 & 0 & 0 \\ 0 & -1 & 0 \end{matrix}$	$\bar{z}$ x $\bar{y}$
11	$L^3_{\bar{1}11}$	$\begin{matrix} 0 & 0 & -1 \\ -1 & 0 & 0 \\ 0 & 1 & 1 \end{matrix}$	$\bar{z}$ $\bar{x}$ y
12	$L^3_{1\bar{1}1}$	$\begin{matrix} 0 & 0 & 1 \\ -1 & 0 & 0 \\ 0 & -1 & 0 \end{matrix}$	z $\bar{x}$ $\bar{y}$
13	$L^2_{10\bar{1}}$	$\begin{matrix} 0 & 0 & -1 \\ 0 & -1 & 0 \\ -1 & 0 & 0 \end{matrix}$	$\bar{z}$ $\bar{y}$ $\bar{x}$
14	L^2_{101}	$\begin{matrix} 0 & 0 & 1 \\ 0 & -1 & 0 \\ 1 & 0 & 0 \end{matrix}$	z $\bar{y}$ x
15	$L^4_{0\bar{1}0}$	$\begin{matrix} 0 & 0 & 1 \\ 0 & 1 & 0 \\ -1 & 0 & 0 \end{matrix}$	z y $\bar{x}$
16	L^4_{010}	$\begin{matrix} 0 & 0 & -1 \\ 0 & 1 & 0 \\ 1 & 0 & 0 \end{matrix}$	$\bar{z}$ y x
17	$L^2_{01\bar{1}}$	$\begin{matrix} -1 & 0 & 0 \\ 0 & 0 & -1 \\ 0 & -1 & 0 \end{matrix}$	$\bar{x}$ $\bar{z}$ $\bar{y}$
18	$L^4_{\bar{1}00}$	$\begin{matrix} 1 & 0 & 0 \\ 0 & 0 & -1 \\ 0 & 1 & 0 \end{matrix}$	x $\bar{z}$ y
19	L^4_{100}	$\begin{matrix} 1 & 0 & 0 \\ 0 & 0 & 1 \\ 0 & -1 & 0 \end{matrix}$	x z $\bar{y}$
20	L^2_{011}	$\begin{matrix} -1 & 0 & 0 \\ 0 & 0 & 1 \\ 0 & 1 & 0 \end{matrix}$	$\bar{x}$ z y
21	$L^2_{\bar{1}10}$	$\begin{matrix} 0 & -1 & 0 \\ -1 & 0 & 1 \\ 0 & 0 & -1 \end{matrix}$	$\bar{y}$ $\bar{x}$ $\bar{z}$
22	L^4_{001}	$\begin{matrix} 0 & 1 & 0 \\ -1 & 0 & 0 \\ 0 & 0 & 1 \end{matrix}$	y $\bar{x}$ z
23	L^2_{110}	$\begin{matrix} 0 & 1 & 0 \\ 1 & 0 & 0 \\ 0 & 0 & -1 \end{matrix}$	y x $\bar{z}$
24	$L^4_{00\bar{1}}$	$\begin{matrix} 0 & -1 & 0 \\ 1 & 0 & 0 \\ 0 & 0 & 1 \end{matrix}$	$\bar{y}$ x z

The point symmetry elements g_c and g_s for crystal and sample which then can be considered as pure rotation axes induce symmetries into the orientation space. These symmetries are especially pronounced if the Euler angle parametrisation is used for which case Eq. (43) assumes the form

$$g\,(\varphi_1^e\ \phi^e\ \varphi_2^e) = g_c \cdot g(\varphi_1\ \phi\ \varphi_2) \cdot g_s \tag{45}$$

By introducing into Eq. (45) the matrix Eq. (19) for g and g^e and the corresponding matrices of Table I for g_c and g_s, the relationships between the φ_1^e, ϕ^e, φ_2^e and φ_1, ϕ, φ_2 can be calculated. Exept for the case that the inducing element is the three-fold axis (or any other which can be written as a product containing such an axis) they can be expressed by simple symmetries in the Euler angle space. As shown in /7/, one then obtains the linear transformation

$$\epsilon^e = T + M \cdot \epsilon \tag{46}$$

with

$$\epsilon = \begin{pmatrix} \varphi_1 \\ \phi \\ \varphi_2 \end{pmatrix}; \quad T = \begin{pmatrix} \Delta\varphi_1 \\ \Delta\phi \\ \Delta\varphi_2 \end{pmatrix}; \quad M = \begin{pmatrix} \pm 1 & 0 & 0 \\ 0 & \pm 1 & 0 \\ 0 & 0 & \pm 1 \end{pmatrix}. \tag{47}$$

Here is ϵ a vector in the Euler angle space and T and M are elements of space symmetry in this space. T represents a translation and M a matrix the signs of which can assume all possible combinations, i.e. M may describe two-fold axes, mirror planes, the inversion and the unity element (c.f. Table I). For the case that g_c contains a three-fold axis, non linear (trigonometric) relationships between ϵ^e and ϵ result.

In expressing the effect of a pair of point symmetry elements on the symmetries of the Euler angle space by an operator $[g_c, g_s]$, Eq. (46) will be written as

$$[g_c, g_s] = T + M. \tag{48}$$

Also if g_c and g_s are products of symmetry elements, such a linear relationship holds as long as no three-fold axis is involved. According to /7/ one then obtains

$$[g_c^{(2)} \cdot g_c^{(1)}, g_s^{(1)} \cdot g_s^{(2)}] \equiv [g_c^{(2)}, g_s^{(2)}] \cdot [g_c^{(1)}, g_s^{(1)}] = T + M \tag{49}$$

where T and M are found to be

$$T = T^{(2)} + M^{(2)} \cdot T^{(1)} \text{ and } M = M^{(2)} \cdot M^{(1)}. \tag{50}$$

(The writing of the left hand side of Eq. (49) as a product of two symmetry operators of the type $[g_c^{(i)}, g_s^{(i)}]$ is only a more convenient way of notation).

Summarizing the main results of this section, it has been shown (i) that only the rotational symmetry elements of crystal and sample have to be considered for describing the relations between symmetrically equivalent orientations; (ii) that these relations imply a division of the region H of the Euler angle space into symmetrically equivalent regions; and (iii) that, except for the 3-fold axis L^3_{111}, they lead to symmetries in the Euler angle space (Eq. (47)) which can be described by the linear transformation (Eqs. (47, 48)):

---

* Here some examples:

$$L^4_{001} \cdot L^4_{001} = \begin{pmatrix} 0 & 1 & 0 \\ -1 & 0 & 0 \\ 0 & 0 & 1 \end{pmatrix} \begin{pmatrix} 0 & 1 & 0 \\ -1 & 0 & 0 \\ 0 & 0 & 1 \end{pmatrix} = \begin{pmatrix} -1 & 0 & 0 \\ 0 & -1 & 0 \\ 0 & 0 & 1 \end{pmatrix} = L^2_{001}$$

$L^3_{111}\ L^3_{111} = L^3_{\bar{1}\bar{1}\bar{1}}$; $\quad L^4_{001} \cdot L^4_{001} \cdot L^4_{001} \cdot L^3_{111} = L^4_{010}$

$L^3_{111}\ L^4_{001} = L^2_{011}$; $\quad L^3_{111}\ L^4_{001}\ L^4_{001} \cdot L^4_{001} = L^4_{100}$.

$$[L_c^i\, L_s^j] = \begin{pmatrix} \Delta\varphi_1 \\ \Delta\phi \\ \Delta\varphi_2 \end{pmatrix} + \begin{pmatrix} \pm 1 & 0 & 0 \\ 0 & \pm 1 & 0 \\ 0 & 0 & \pm 1 \end{pmatrix}. \tag{51}$$

Before considering the effect of the different symmetry elements of crystal and sample in detail (Sec. 3.4 to 3.6), the symmetries of the Euler angle space induced by identical equivalencies will be briefly discussed.

3.3 Symmetries of the Euler Angle Space due to Identical Equivalencies

As shown in Sec. 2.3 the matrix elements g_{ik} may assume the same values for different sets of Euler angles $\varphi_1\ \phi\ \varphi_2$. Since the rotation between the two frames C and S is completely defined by the matrix g, different sets of Euler angles leading to the same matrix are called identically equivalent orientations.

Such identically equivalent orientations are defined by the equation

$$g\,(\varphi_1{}^e\ \phi^e\ \varphi_2{}^e) = g(\varphi_1\ \phi\ \varphi_2). \tag{52}$$

It is easy to recognize that the solution of this equation which is given by Eq. (20) can be expressed as a linear transformation of the form of Eq. (46)

$$\epsilon^e = \begin{pmatrix} \varphi_1{}^e \\ \phi^e \\ \varphi_2^e \end{pmatrix} = \begin{pmatrix} \pi \\ 0 \\ \pi \end{pmatrix} + \begin{pmatrix} 1 & 0 & 0 \\ 0 & -1 & 0 \\ 0 & 0 & 1 \end{pmatrix} \begin{pmatrix} \varphi_1 \\ \phi \\ \varphi_2 \end{pmatrix} = \begin{pmatrix} \pi \\ 0 \\ \pi \end{pmatrix} + P_\phi \cdot \epsilon. \tag{53}$$

This means one has in the Euler angle space a mirror plane P_ϕ perpendicular to the axis ϕ in $\phi = 0$ superimposed by a translation $\Delta\varphi_1 = \pi$, $\Delta\phi = 0$ and $\Delta\varphi_2 = \pi$, i.e. one has a glide-mirror plane. Since Eq. (52) can be derived from Eq. (45) by setting $g_c = g_s = E$, Eq. (53) can also be written in an operator notation of Eq. (48) as

$$[E, E] = \begin{pmatrix} \pi \\ 0 \\ \pi \end{pmatrix} + P_\phi. \tag{54}$$

As also pointed out in Sec. 2.3, further solutions of Eq. (52), i.e. further identical equivalencies, are obtained by translations by multiples of 2π parallel to each of the axes $\varphi_1, \phi, \varphi_2$. As can easily be recognized and is indicated in Fig. 11, this leads to a family of glide-mirror planes perpendicular to the ϕ-axis in $\phi = k\pi$ with translation components $(2k_1 + 1)\pi$ and $(2k_2 + 1)\pi$ parallel to the axes φ_1 and φ_2 with k, k_1, k_2 being integers. Thus the symmetries induced by the identical equivalencies imply a division of the Euler angle space into equivalent regions. These can be chosen in different ways. Here the range

$$H = \{\, 0 \leqslant \varphi_1 \leqslant 2\pi,\ \ 0 \leqslant \phi \leqslant \pi,\ \ 0 \leqslant \varphi_2 \leqslant 2\pi \}. \tag{55}$$

(Eq. (42) and Fig. 11) is chosen as the basic range and will be refered to as asymmetric unit (c.f. /7/).

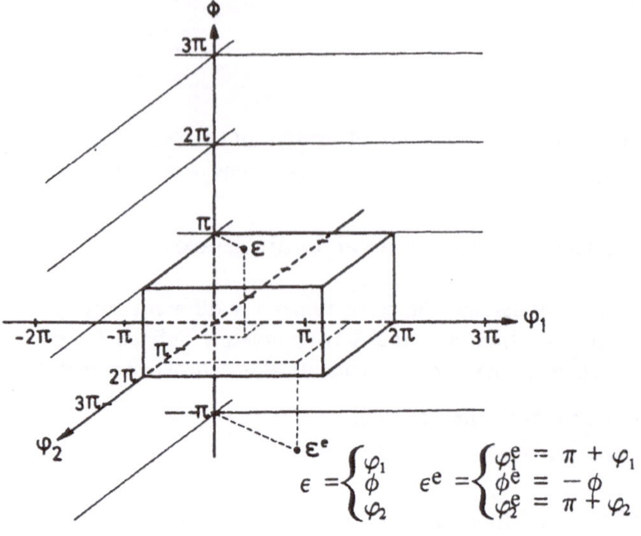

Fig. 11:
Glide mirror planes in the infinite Euler angle space induced by identical equivalences. They divide the space into identically equivalent subspaces, one of them being the asymmetric unit H.

3.4 Symmetries of the Euler Angle Space Induced by the Cubic Crystal Symmetry

Now let us consider the effect of the cubic crystal symmetry. Using the form of writing given by Eq. (51) the elements L^2_{100} and L^4_{001} give /7/,

$$\text{(I)} \quad [L^2_{100}, E] = \begin{pmatrix} \pi \\ \pi \\ \pi \end{pmatrix} + \begin{pmatrix} 1 & 0 & 0 \\ 0 & -1 & 0 \\ 0 & 0 & -1 \end{pmatrix} = \begin{pmatrix} \pi \\ \pi \\ \pi \end{pmatrix} + L^2_{\varphi_1} \tag{56}$$

$$\text{(II)} \quad [L^4_{001}, E] = \begin{pmatrix} \pi 0 \\ \pi 0 \\ \pi/2 \end{pmatrix} + \begin{pmatrix} 1 & 0 & 0 \\ 0 & 1 & 0 \\ 0 & 0 & 1 \end{pmatrix} = \begin{pmatrix} 0 \\ 0 \\ \pi/2 \end{pmatrix} + E$$

Combinig I and II on the basis of formulae (49): leads to

$$\text{(III)} \quad [L^2_{100} \cdot L^4_{001}, E] = [L^2_{100}, E] \cdot [L^4_{001}, E] = \begin{pmatrix} \pi \\ \pi \\ \pi/2 \end{pmatrix} + L^2_{\varphi_1} . \tag{57}$$

The resulting symmetry elements for the range H of the Euler angle space are shown in Fig. 12 a–c. One recognizes that L^2_{100} induces two-fold screw axes $L^2_{\varphi_1} + (\Delta\varphi_1 = \pi)$ parallel to φ_1 with a translation vector of $\Delta\varphi_1 = \pi$ and the coordinates

$$\phi^\circ = \frac{\Delta\phi}{2} = \frac{\pi}{2};\ \varphi_2^0 = \frac{\Delta\varphi_2}{2} = \frac{\pi}{2} \quad \text{and} \quad \phi^\circ = \frac{\Delta\phi}{2} = \frac{\pi}{2};\ \varphi_2^0 = \frac{\Delta\varphi_2 + 2\pi}{2} = \frac{3\pi}{2} .$$

By these axes a point a is brought into the position b (Fig. 12 d).

The L^4_{001} axis induces a translation vector $\Delta\varphi_2 = \pi/2$, that is, a periodicity of $\pi/2$ for the angle φ_2 (Fig. (Fig. 12 b). This translation converts a point a into the position a′ and b into b′.
Combined, the two elements L^2_{100} and L^4_{001} introduce a family of two-fold screw axes $L^2_{\varphi_2} + (\Delta\varphi_1 = \pi)$ having the coordinates $\phi^\circ = \pi/2$, $\varphi_2^0 = 0, \pi/4, 2\pi/4 \ldots 8\pi/4$, which transform the point b′ back into the original position a. Thus these symmetries $[L^2_{100}, E]$ and $[L^4_{001}, E]$ allow the region H to be divided into $2 \cdot 4 = 8$ symmetrically equivalent subregions. Here mainly the subregion

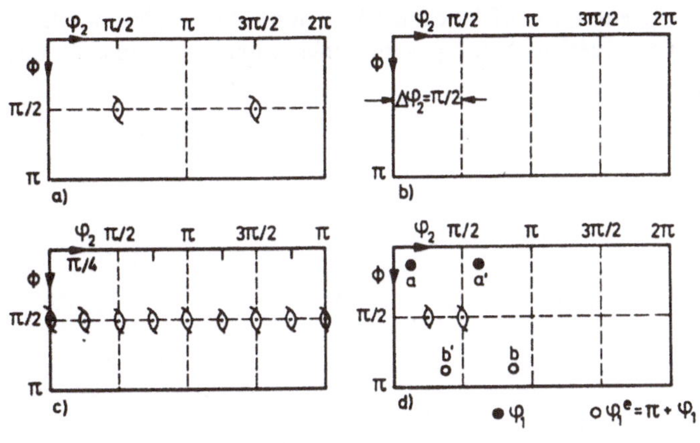

Fig. 12:
Symmetry elements in the Euler angle space induced by the cubic crystal symmetry. Fig. 12 d illustrates the transformation effected by these elements.

$$H' = \{0 \leqslant \varphi_1 \leqslant 2\pi,\ \ 0 \leqslant \phi \leqslant \pi/2,\ \ 0 \leqslant \varphi_2 \leqslant \pi/2\} \tag{58}$$

(Fig. 8) will be considered.

The remaining three-fold axis L^3_{111} causes a non-linear transformation which cannot be expressed by elements of space symmetry in the Euler angle space. It causes a quasi three-fold screw axis parallel to φ_1 having the coordinates ϕ° = arc cos $(1/\sqrt{3})$, $\varphi_2^0 = \pi/4$ (Q^3 in Fig. 13).

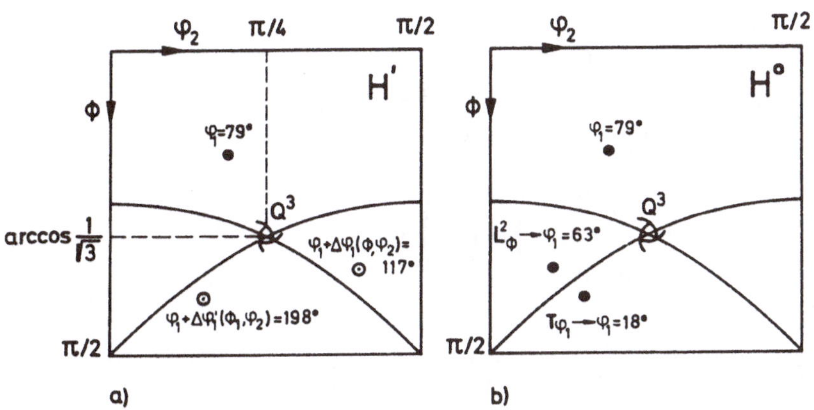

Fig. 13:
Position and effect of the quasi-screw axis Q^3 induced by the three-fold crystal axis: a) in the subspace H'; b) in the subspace H°.

It effects that a point lying on this axis is shifted along this axis by $\Delta\varphi_1' = 120^\circ$ or $\Delta\varphi_1'' = 240^\circ$ and that a point lying beside it (i.e. having the coordinates $\phi, \varphi_2 \neq \phi^\circ, \varphi_2^0$ is displaced screw-wise into two other positions $\varphi_1', \phi', \varphi_2'$ and $\varphi_1'', \phi'', \varphi_2''$. The new coordinates $\phi', \varphi_2', \phi'', \varphi_2''$ as well as the shifts $\Delta\varphi_1 = \varphi_1' - \varphi_1$ and $\Delta\varphi_1' = \varphi_1'' - \varphi_1$ depend only on the coordinates ϕ, φ_2, but not on φ_1. Thus a segment parallel to φ_1 remains parallel to φ_1 and conserves its length at this transformation (Fig. 13).

This transformation described by the quasi-three-fold axis divides the range H' into three symmetrically equivalent parts. It thus follows that the cubic crystal symmetry induce into the orientation space H $2 \cdot 4 \cdot 3 = 24$ symmetrical equivalences for orientations, thereby dividing the range H into 24 basic ranges.

There are many possible choices of surface limiting the basic regions which all are plane except those which follow from the three-fold axis. A simple possibility is to divide the range H' into three right prisms the three basic areas are indicated by different types of hatching. Thus for cubic single crystals one of the 3 prismatic regions would be needed for representing every possible orientation once and only once.

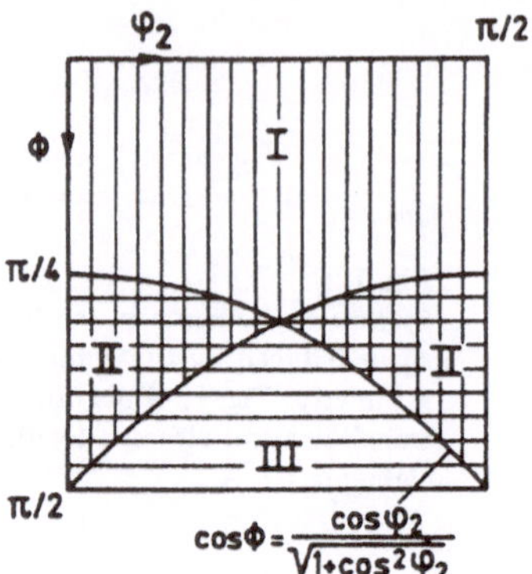

Fig. 14:
Division of the subspaces H′ and H° into three symmetrically equivalent subspaces by the three-fold crystal symmetry.

3.5 Symmetries of the Euler Angle Space in the Case of Orthorhombic Sample Symmetry

Now let us consider orthorhombic symmetry for the sample and thus the effect of the elements L^2_{ND} and L^2_{RD}. They induce symmetries into the Euler angle space which are indicated in Fig. 15 for the range H /7/:

$$\text{(IV)}\quad [E, L^2_{ND}] = \begin{pmatrix}\pi\\0\\0\end{pmatrix} + \begin{pmatrix}1&0&0\\0&1&0\\0&0&1\end{pmatrix} = \begin{pmatrix}\pi\\0\\0\end{pmatrix} + E \tag{59}$$

$$\text{(V)}\quad [E, L^2_{RD}] = \begin{pmatrix}\pi\\\pi\\\pi\end{pmatrix} + \begin{pmatrix}-1&0&0\\0&-1&0\\0&0&1\end{pmatrix} = \begin{pmatrix}\pi\\\pi\\\pi\end{pmatrix} + L^2_{\varphi_2} \tag{60}$$

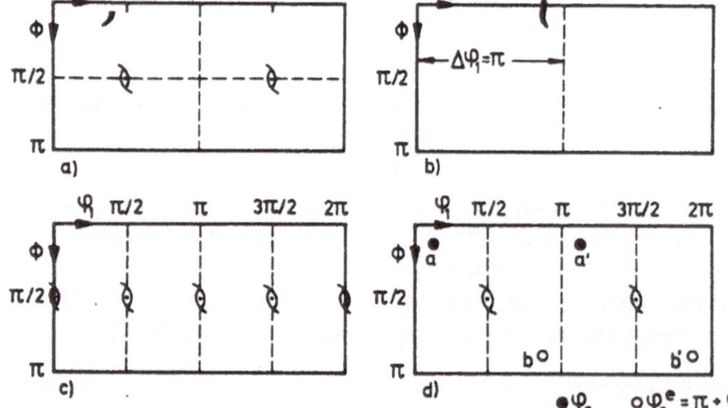

Fig. 15:
Symmetry elements in the Euler angle space induced by the orthorhombic sample symmetry. Fig. 15 d illustrates the transformation effected by these elements.

One recognizes that the axis L^2_{ND} induces into the orientation space the translation $\Delta\varphi_1 = \pi$, i.e. the periodicity π for φ_1 which transforms a point a into the position a′. The element L^2_{RD} induces a two-fold screw axis $L^2_{\varphi_2} + (\Delta\varphi_2 = \pi)$ having the coordinates $\phi^\circ = \pi/2$, $\varphi_1 = \pi/2, 3\pi/2$. It transforms a into the position b and a′ into b′. It can easily be seen that the elements L^2_{ND} and L^2_{RD} introduce a division of the range H′ into yet another $2 \cdot 2 = 4$ symmetrically equivalent subregions with intersection planes perpendicular to φ_1 at the points $\varphi_1 = \pi/2, \pi, 3\pi/2$.

Interesting is the combination with the crystal symmetry elements. The translation $\Delta\varphi_1 = \pi$ generated by the axis L^2_{ND} transforms the two-fold screw axis $L^2_{\varphi_1} + (\Delta\varphi_1 = \pi)$ introduced by the element L^2_{100} into an ordinary two-fold axis $L^2_{\varphi_1}$. This also follows from Eqs. (49) and (57):

$$\text{(VI)}\ [L^2_{100} \cdot L^4_{001}, L^2_{ND}] = [L^2_{100} \cdot L^4_{001}, E] \cdot [E, L^2_{ND}] = \begin{pmatrix} 0 \\ \pi \\ \pi/2 \end{pmatrix} + L^2_{\varphi_1}. \tag{61}$$

Similarly, with Eqs. (57) and (60), L^2_{RD} leads to a translation in the φ_2-direction and a two-fold axis parallel ϕ:

$$\text{(VII)}\ [L^2_{100}\, L^4_{001}, L^2_{RD}] = [L^2_{100} \cdot L^4_{001}, E] \cdot [E, L^2_{RD}] = \begin{pmatrix} 0 \\ 0 \\ \pi/2 \end{pmatrix} + L^2_{\phi}. \tag{62}$$

Finally, the combination of L^2_{ND} and L^2_{RD} results in

$$\text{(VIII)}\ [L^2_{100} \cdot L^4_{001}, L^2_{ND} \cdot L^2_{RD}] = [L^2_{100} \cdot L^4_{001}, L^2_{ND}] \cdot [E, L^2_{RD}] = \begin{pmatrix} \pi \\ 0 \\ \pi/2 \end{pmatrix} + L^2_{\phi}. \tag{63}$$

This gives a family of two-fold axes parallel to ϕ having the coordinates $\varphi_1^0 = 0, \pi/2, 2\pi/2 \ldots$; $\varphi_2^0 = 0, \pi/4, 2\pi/4 \ldots$

As already pointed out, the transformation introduced by the three-fold axis L^3_{111} of the crystal symmetry cannot be described by space symmetry elements in the Euler angle space. It introduces, however, some two- and one-dimensional symmetries within special planes and along special lines. One finds in the planes $\varphi_2 = n\pi/2$ $(n = 0, \pm 1, \pm 2 \ldots)$

$$[L^4_{001} \cdot L^3_{111}, L^2_{ND}]_{\varphi_2 = n\pi/2} = \begin{cases} \varphi_1^e = \varphi_1 \\ \phi^e = \pi/2 - \phi \end{cases} = \begin{pmatrix} 0 \\ \pi/2 \end{pmatrix} + L^M. \tag{64}$$

i.e. in these planes there is a line parallel to φ_1 at $\phi = \pi/4$, which can be called a mirror line L^M. Furthermore, along the lines (called l) parallel to φ_1 at $\phi = 0$, $\varphi_2 = 0, \pi/2$ and $\phi = \pi/2$, $\varphi_2 = 0, \pi/2$ one has

$$[L^3_{111} \cdot L^4_{001}, L^2_{ND} \cdot L^2_{RD}]_l = \left\{\varphi_1{}^e = \frac{\pi}{2} - \varphi_1\right\} = \left(\frac{\pi}{2}\right) + L^P \tag{65}$$

i.e. along these lines at $\varphi_1 = \pi/2$ points are situated which can be called mirror points L^P. Further mirror points along the axis Q^3 $(\phi = \arccos(1/\sqrt{3}),\ \varphi_2 = \pi/4)$ follow from

$$[L^4_{001} \cdot L^2_{100} \cdot L^3_{111}, L^2_{ND} \cdot L^2_{RD}]_{Q^3} = \left\{\varphi_1^e = k\frac{\pi}{3} - \varphi_1\right\} = \left(\frac{\pi}{3}\right) + L^P \tag{66}$$

and are situated at $\varphi_1 = \frac{1}{2}\gamma\,\frac{\pi}{3}$ $(\gamma = 0, \pm 1, \pm 2 \ldots)$ *. In the above, the operators L^M and L^P have the meaning

* The points on Q^3 at $\varphi_1 = \gamma\pi/2$ are situated on two-fold axes and cause the same symmetry operations along the line Q^3 as these axes. Thus, in contrast to the other mirror points along Q^3, these ones are no true additional symmetry elements.

$$L^M = \begin{pmatrix} 1 & 0 \\ 0 & -1 \end{pmatrix}; \quad L^P = (-1).$$

The description of symmetries in the Euler angle space given in this section is complete within the range H. These symmetries, however, give no possibility of transforming of orientations from the range $0 \leqslant \phi \leqslant \pi$ considered here to the neighbouring ranges $-\pi \leqslant \phi \leqslant 0$ or $\pi \leqslant \phi \leqslant 2\pi$. This possibility is obtained only by combining the symmetries induced by the rotation axes with those induced by the glide-mirror plane and given by Eq. (54). Using Eq. (54) and /7/ it can be shown that in the present case the translation $\Delta\varphi_1 = \pi$ and $\Delta\varphi_2 = \pi$ introduced by the axis L^2_{ND} or L^4_{001}, respectively, reduce the glide-mirror plane into a simple mirror plane:

$$\begin{pmatrix} \pi \\ 0 \\ \pi \end{pmatrix} + P_\phi \Rightarrow \begin{pmatrix} 0 \\ 0 \\ 0 \end{pmatrix} + P_\phi .$$

This mirror plane is perpendicular to the axis ϕ in $\phi = 0$. Considering also the translations 2π parallel to the axis ϕ leads to a family of mirror planes perpendicular to the axis ϕ in $\phi = k\pi$ ($k = 0, \pm 1, \pm 2 \ldots$).

Recapitulating, we find that the symmetries of the cubic system of the crystal and the orthorhombic system of the sample induces into Euler angle space symmetries corresponding to linear transformations which are listed in Table 3.5. The symmetries 1 to 4 divide the H region into $2 \cdot 4 \cdot 2 \cdot 2 = 32$ symmetrically equivalent subregions. One of them,

$$H^\circ = \{ 0 \leqslant \varphi_1 \leqslant \pi/2,\ 0 \leqslant \phi \leqslant \pi/2,\ 0 \leqslant \varphi_2 \leqslant \pi/2 \} \tag{67}$$

having the form of a cube with edge length 90° is depicted in Fig. 16 with the outlined symmetry elements. Due to the three-fold axis L^3_{111} in the crystal's symmetry, the subregion H° becomes divided further into three symmetrically equivalent parts denoted in Figs. 14 by the numerals I, II, III.

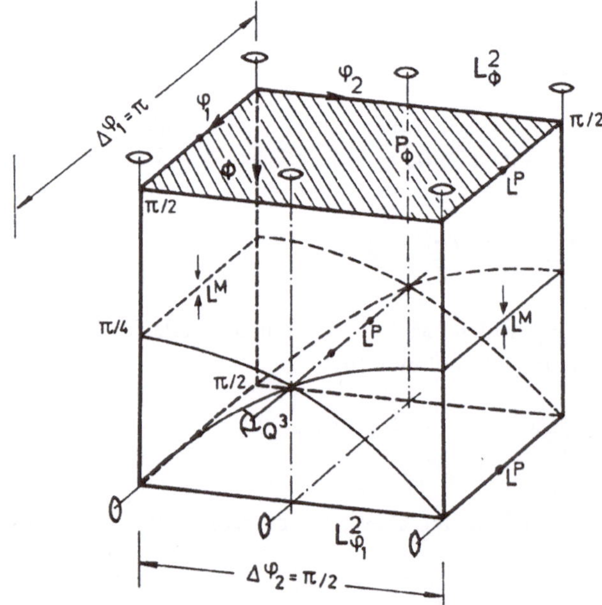

Fig. 16:
Presentation of the symmetry elements in the subspace H° of the Euler angle space.

Table 3.5: Symmetry elements in the Eulers angle space for different sample symmetries (in the triclinic and monoclinic case only the true space symmetries are listed, i.e. Mirror lines and mirror points are omitted).

Triclinic	Orthorhombic
1. translation $\Delta\varphi_2 = \pi/2$ $\begin{pmatrix}0\\0\\\pi/2\end{pmatrix} + E$ $\varphi_1^e = \varphi_1$ $\phi^e = \phi$ $\varphi_2^e = \varphi_2 + \pi/2$	1. translation $\Delta\varphi_1 = \pi$ $\begin{pmatrix}\pi\\0\\0\end{pmatrix} + E$ $\varphi_1^e = \varphi_1 + \pi$ $\phi^e = \phi$ $\varphi_2^e = \varphi_2$
2. two-fold screw axis $\begin{pmatrix}\pi\\\pi\\\pi/2\end{pmatrix} + L_{\varphi_1}^2$ $\varphi_1^e = \pi + \varphi_1$ $\phi^e = \pi - \phi$ $\varphi_2^e = \pi/2 - \varphi_2$	2. translation $\Delta\varphi_2 = \pi/2$ $\begin{pmatrix}0\\0\\\pi/2\end{pmatrix} + E$ $\varphi_1^e = \varphi_1$ $\phi^e = \phi$ $\varphi_2^e = \varphi_2 + \pi/2$
3. glide-mirror plane $\begin{pmatrix}\pi\\0\\0\end{pmatrix} + P_\phi$ $\varphi_1^e = \pi + \varphi_1$ $\phi^e = -\phi$ $\varphi_2^e = \varphi_2$	3. two-fold axis $\begin{pmatrix}0\\\pi\\\pi/2\end{pmatrix} + L_{\Phi_1}^2$ $\varphi_1^e = \varphi_1$ $\phi^e = \pi - \phi$ $\varphi_2^e = \pi/2 - \varphi_2$
	4. two-fold axis $\begin{pmatrix}\pi\\0\\\pi/2\end{pmatrix} + L_\phi^2$ $\varphi_1^e = \pi - \varphi_1$ $\phi^e = \phi$ $\varphi_2^e = \pi/2 - \varphi_2$
Monoclinic	
1. translation $\Delta\varphi_2 = \pi/2$ $\begin{pmatrix}0\\0\\\pi/2\end{pmatrix} + E$ $\varphi_1^e = \varphi_1$ $\phi^e = \phi$ $\varphi_2^e = \varphi_2 + \pi/2$	5. mirror plane $\begin{pmatrix}0\\0\\0\end{pmatrix} + P_\phi$ $\varphi_1^e = \varphi_1$ $\phi^e = -\phi$ $\varphi_2^e = \varphi_2$
2. two-fold screw axis $\begin{pmatrix}\pi\\\pi\\\pi/2\end{pmatrix} + L_{\varphi_1}^2$ $\varphi_1^e = \pi + \varphi_1$ $\phi^e = \pi - \phi$ $\varphi_2^e = \pi/2 - \varphi_2$	6. mirror line $\begin{pmatrix}0\\\pi/2\end{pmatrix} + L^M$ $\varphi_1^e = \varphi_1$ $\phi^e = \pi/2 - \phi$ $\varphi_2^e = 0$
3. two-fold axis $\begin{pmatrix}0\\0\\\pi/2\end{pmatrix} + L_\phi^2$ $\varphi_1^e = -\varphi_1$ $\phi^e = \phi$ $\varphi_2^e = \pi/2 - \varphi_2$	7. mirror point $(\pi/2) + L^P$ $\varphi_1^e = \pi/2 - \varphi_1$ $\phi^e = 0, \pi/2$ $\varphi_2^e = 0, \pi/2$
4. glide-mirror plane $\begin{pmatrix}\pi\\0\\0\end{pmatrix} + P_\phi$ $\varphi_1^e = \pi + \varphi_2$ $\phi^e = -\phi$ $\varphi_2^e = \varphi_2$	8. mirror point $(\pi/3) + L^P$ $\varphi_1^e = k\pi/3 - \varphi_1$ $\phi^e = \text{arc cos } 1/\sqrt{3}$ $\varphi_2^e = \pi/4$ $(k = 0, 1, 2 \ldots)$

3.6 Symmetries of the Euler Angle Space in the Case of Lower Sample Symmetry

In the following some cases shall be briefly discussed for which the cubic crystal symmetry is retained, but the sample symmetry is less than orthorhombic. If, for example, the sample is a single crystal, its symmetry (i.e. also that of the pole figure) is given in general by the triclinic group $\bar{1}$ containing only the inversion I. Only if the crystal is oriented in such a way that symmetry axes of the crystal are parallel to ND, RD oder TD, a higher sample symmetry is obtained. A crystal in cube orientation $\{001\}$ $\langle 100\rangle$ or Goss orientation $\{011\}$ $\langle 100\rangle$ possesses the orthorhombic, or a crystal in the brass rolling orientation $\{011\}$ $\langle 211\rangle$ the monoclinic sample symmetry.

For triclinic sample symmetry it is nessessary to use the subregion H′ (Eq. (58), Fig. 8) instead of H°. Since for each single crystal orientation H′ contains three symmetrically equivalent positions (which have distances $> \pi/2$ in φ_1 /7/) and since H′ can be divided into 4 regions of the size H°, only 3 of these 4 regions would contain such a position. This means not all crystal orientations would be contained in the range H° used for polycrystals. If, however, it is allowed to rotate the crystal by 180° around RD, TD or ND, one can always manage to have one position in H°. In these cases one can use H° also for the presentation of single crystal orientations. In other cases, e.g. if the orientation relationship with respect to a second crystal is considered, it might not be allowed to think the first crystal being rotated. Then the larger region H′ must be used.

In the case of uni-directional rolling (e.g. of sheets or tubes) monoclinic sample symmetry is obtained. One has here the symmetry group m characterized by the inversion I and the two fold-axis L_{RD}. As shown in Sec. 3.5, this axis causes 2-fold screw axes parallel φ_2 in the Euler angle space situated at $\varphi_1 = \pi/2, 3\pi/2$ and $\phi = \pi/2$ (Eq. (60)). This leads to reduction of the range H′ into two parts of witch one is given by

$$H'' = \left\{0 \leqslant \varphi_1 \leqslant \pi, \quad 0 \leqslant \phi, \quad \varphi_2 \leqslant \frac{\pi}{2}\right\}. \tag{68}$$

In combination with the elements induced by the cubic crystal symmetry, this axis gives ordinary two-fold axes parallel ϕ situated at $\varphi_1 = 0, 2\pi$ and $\varphi_2 = \pi/4$ (Eq. (62)).

4. REPRESENTATION OF ORIENTATION DISTRIBUTIONS IN THE EULER ANGLE SPACE

4.1 The Orientation Distribution Function (ODF)

For a random orientation distribution the density of orientation points in the (linear) Euler angle space would come out not to be constant but would change proportional to $1/\sin\phi$ /8/. For the purpose of representing orientation distributions, however, a function f(g) must be defined in such a manner that f(g) dg is the volume fraction of orientations in the element dg and that in the random case f(g) = const. = 1. This is fulfilled by introducing for the volume element *

$$dg = \frac{1}{8\pi^2}\sin\phi\, d\varphi_1\, d\phi\, d\varphi_2 \tag{69}$$

with
$$\int f(g)\,dg = \int_0^{2\pi}\int_0^{\pi}\int_0^{2\pi} f(g)\,\frac{1}{8\pi^2}\sin\phi\, d\varphi_1\, d\phi\, d\varphi_2 = 1 \tag{70}$$

f(g) is commonly denoted as orientation distribution function (ODF). The density $P_{XYZ}(\alpha, \beta)$ of the poles {XYZ} at the point α, β of the pole figure and the density $R_{s_i}(\gamma, \delta)$ of the sample axis s_i at the point γ, δ of the inverse pole figure can be obtained from the ODF by integration:

$$P_{XYZ}(\alpha, \beta) = \frac{1}{2\pi}\int_0^{2\pi} f(g)\, d\omega_{XYZ} \tag{71}$$

$$R_{si}(\gamma, \delta) = \frac{1}{2\pi}\int_0^{2\pi} f(g)\, d\omega_{s_i}. \tag{72}$$

Here is ω_{XYZ} and ω_{s_i} the angle of rotation around the axis {XYZ} or s_i, respectively.

For representation of the ODF in the case of orthorhombic sample symmetry mostly the subregion H° (Eq. (67)) is used although this region generally contains each orientation 3 times. The use of the region has the advantage that it has no curved surfaces. The 3 subranges denoted as I, II and III which contain each orientation only once can be recognized in Fig. 14 and 16. In the radial cuts φ_2 = constant (which are the radial cuts in Fig. 9), their boundaries appear as lines ϕ = constant which are shown in Fig. 17. It can be seen that best suited for studying the details of the orientation distribution is range I, since it is the largest range, since it is limited by only one curved surface and since it changes its dimension only little with changing φ_2. For $\varphi_2 = 0°$ and 90° it extends over $0 \leqslant \phi \leqslant 45°$ and for $\varphi_2 = 45°$ over $0 \leqslant \phi \leqslant 54{,}7°$.

The ODF is commonly displayed by contour lines in sections φ_2 = const or φ_1 = const. The sections are positioned mostly in distances of 5° from each other. Examples of ODF's displayed in this way are given by Fig. 17 ({236} ⟨385⟩ texture) and Fig. 18 (cube texture {001} ⟨100⟩). Both types of ODF's are observed as recrystallization texture of f.c.c. metals. Fig. 17 also demonstrates the appearance of a maximum of the ODF in the three basic regions of H°!

* This follows directly from the condition for the function $I(\varphi_1, \phi, \varphi_2)$

$$\iiint f(g)\, I\, d\varphi_1\, d\phi\, d\varphi_2 = \iiint f(g\, g_0)\, I\, d\varphi_1\, d\phi\, d\varphi_2$$

with g_0 being any rotation. Togehter with Eq. (70) this condition leads to $I = \sin\phi/8\pi^2$.

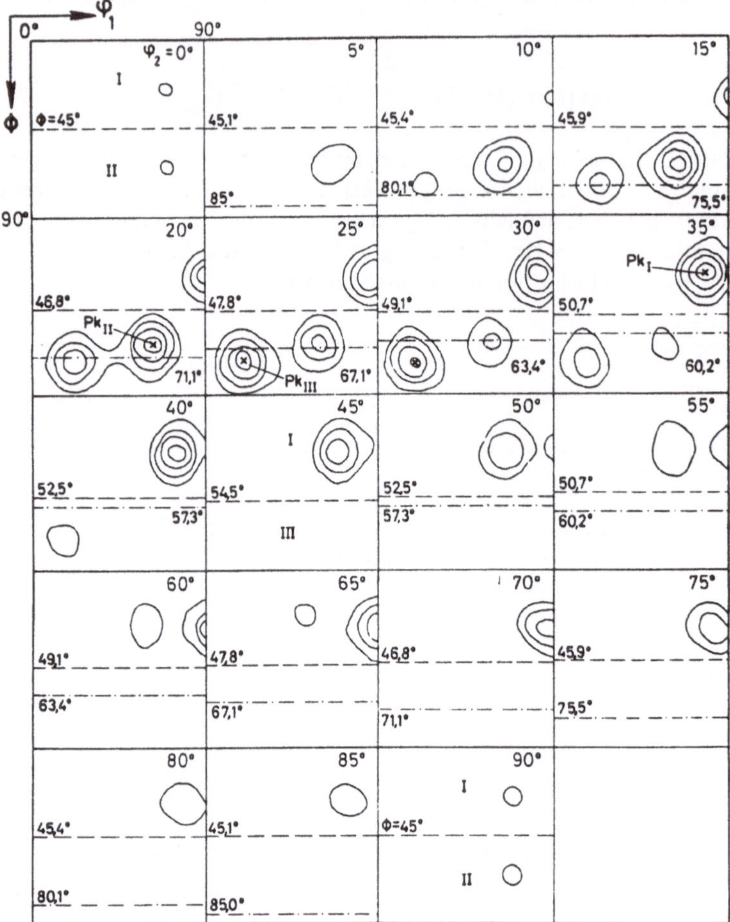

Fig. 17:
Presentation of an ODF consisting of a single component around the ideal orientation $\{236\}\langle 385\rangle$ in the range H° of the Euler angle space. The orientation density is presented by contour lines in sections φ_2 = constant taken at $\varphi_2 = 0°, 5°, 10°$ etc. The maximum appears three times (Pk_I, Pk_{II}, Pk_{III}) corresponding to the three symmetrically equivalent subspaces indicated here by the dashed and dashed-dotted lines.

4.2 Distortion of the Euler Angle Space Near $\phi = 0$; Texture Components

Let us here consider the plane $\phi = 0$ of the Euler angle space. For the point $\varphi_1 = \phi = \varphi_2 = 0$ all three crystal axes [001], [010] and [100] are parallel to the sample axes ND, TD and RD, i.e. this point describes the cube orientation (001) [100]. Since for $\phi = 0$ always sample coordinate ND and crystal coordinate are identical, the orientations in this plane are given by rotations of the cube orientations around ND.

In the plane $\phi = 0$ of the Euler angle space a peculiar degeneracy occurs. While, in general, an orientation is represented in the Euler angle space by a point, in the case $\phi = 0$ it is represented by a line. This follows from the matrix Eq. (19) which for $\phi = 0$ simplifies to

$$g(\varphi_1\, \phi\, \varphi_2)_{\phi=0} = \begin{pmatrix} \cos(\varphi_1 + \varphi_2) & \sin(\varphi_1 + \varphi_2) & 0 \\ -\sin(\varphi_1 + \varphi_2) & \cos(\varphi_1 + \varphi_2) & 0 \\ 0 & 0 & 1 \end{pmatrix}. \tag{73}$$

One recognizes that in this plane an orientation is described not by one pair of angles φ_1, φ_2, but by all pairs for which the sum $\varphi_1 + \varphi_2$ = constant, i.e. by -45° lines in the plane $\phi = 0$. This can be seen also directly from the definition of Euler angles. Since for $\phi = 0$ ND and [001] are identical, the two rotations φ_1 and φ_2 can no longer be separated and only the total rotation $\varphi_1 + \varphi_2$ determines the orientation.

As example let us consider the cube orientation $\varphi_1 = \phi = \varphi_2 = 0$. By application of Eq. (19) one finds the symmetrically equivalent orientations which turn out to be the seven other corners of the cube of Fig. 16. Additionally, however, the cube orientation is also given by the -45°-lines in the plane $\phi = 0$ which are running through the cube corners situated in this plane, i.e. in particular by the line $\varphi_1 + \varphi_2 = \pi/2$ which runs between the corners $\varphi_2 = 0$, $\varphi_1 = \pi/2$ and $\varphi_2 = \pi/2$, $\varphi_1 = 0$. For this reason, the intensity along this line is always constant. This can be seen in Fig. 18 where a cube texture is shown. With increasing φ_2, the maximum moves along the line $\phi = 0$ from $\varphi_1 = \pi/2$ to $\varphi_1 = 0$ without changing its height. In contrast to this, the symmetrically equivalent positions in the plane $\phi = \pi/2$ are not degenerated, i.e. with changing φ_2 the maxima situated at the corners at $\phi = 90^\circ$ do not shift but decrease in height, as also can be seen in Fig. 18.

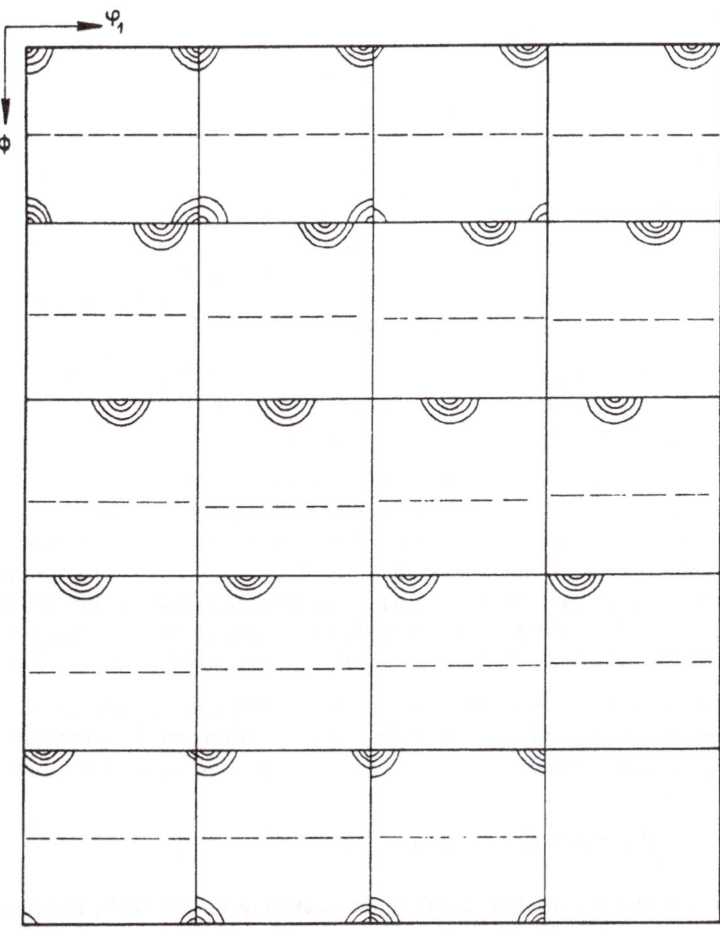

Fig. 18:
Presentation of the ODF of a cube texture $\{001\}\langle 100\rangle$ in the manner described at Fig. 17.

Obviously such a degeneracy is obtained for all orientations in the plane $\phi = 0$, i.e. for those formed by rotation of the cube orientation around ND. E.g. for the orientation (001) [110] one has $\varphi_1 + \varphi_2 = \pi/4$ which means that it is presented by all points on the -45°-line going from $\varphi_2 = 0$, $\varphi_1 = \pi/4$ to $\varphi_2 = \pi/4$, $\varphi_1 = 0$. (The other symmetrically equivalent orientations $\{001\}\langle 110\rangle$ are given by the $\{\varphi_1 \phi \varphi_2\}$-values $\{45, 90, 0\}$ and $\{45, 90, 90\}$).

This degeneracy at $\phi = 0$ is the reason for a strong distortion of the Euler angle space also for values $\phi \neq 0$. This can be seen best by considering all points in the Euler angle space which from a given point have a constant desorientation. (This is the smallest angle of rotation ω by which, independent of the axis of rotation, two orientations can be transformed into each other.) These points form a surface which, however, does not assume a shape of a sphere but, in the general case, a shape similar to that of an ellipsoide. Particularly in the plane $\phi = 0$ where an orientation is given by a $-45°$-line, one has only two other orientations with a given desorientation which form two lines parallel to the first one. At $\phi \neq 0$ an orientation is given again by a point, but this degeneracy still has the effect that for small values of ϕ ellipsoids are obtained which are strongly stretched out in the $-45°$ direction.

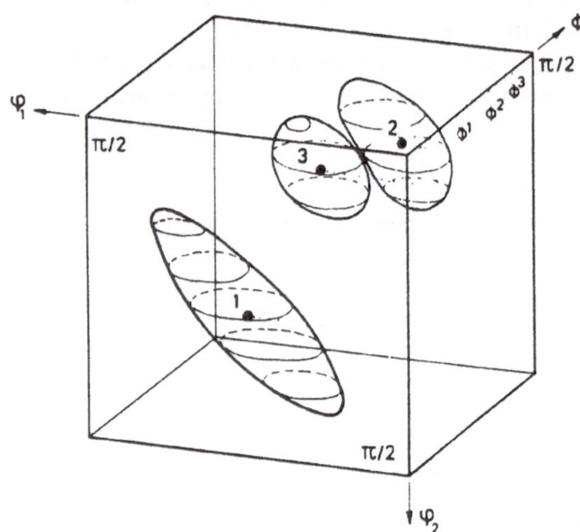

Fig. 19:
3-dimensional presentation of the three symmetrically equivalent components of the orientation {123} ⟨634⟩ in the range H_0. The surfaces around the ideal orientation characterize a constant desorientation of 10°. The coordinates of the ideal orientations 1, 2 and 3 are $\{\varphi_1, \phi, \varphi_2\} = \{59.0, 36.7, 63.4\}$, $\{27.0, 57.7, 18.4\}$ and $\{52.9, 74.5, 33.7\}$, respectively.

Fig. 19 gives an example for the surfaces corresponding to a 10° desorientation from the three symmetrically equivalent orientations {123} ⟨634⟩. The ideal orientations 1, 2 and 3 are situated at ϕ = 36,7°, 57,7° and 74,5°, and one recognizes that the ellipsoids are stretched the more the smaller ϕ.

Often, in good approximation, the ODF can be considered as being superimposed by „isotropic components", i.e. by orientation accumulations the density of which decreases from a maximum at an ideal orientation depending only of the desorientation (e.g. this accummulation may form of a three-dimensional Gauß type distribution in the desorientations /9, 10/). In such a case the contour lines correspond to surfaces of constant desorientation and thus have the same shape as above discussed for the ellipsoids. This shows up e.g. in Fig. 19 and also 17. It leads there to the fact that the maxima in the three symmetrically equivalent ranges of H° look very different although they represent the same orientation distribution.

The degeneracy in $\phi = 0$ is not the only reason for the deviation of the surfaces of constant desorientation from that of a sphere. For example the metric of the orientation space resulting in the factor $\sin\phi$ in Eq. (69) leads to a distortion in ϕ-direction. These questions, however, shall here not be discussed further.

4.3 Meaning of Special Lines and Planes in the Euler Angle Space

In order to be able to better visualize the orientation relationships occurring in the ODF, the meaning of some simple features of the Euler angle space should be kept in mind. For this purpose it is often useful to consider the cylindrical type of orientation space (Fig. 9).
The range H° is given there by the quadrant indicated by full lines. The sections φ_2 = const. used in Figs. 17 and 18 appear here as radial sections parallel to the cylinder axis. With the definition of the Euler angles (Sec. 2.3) in mind, the meaning of some special lines and planes can immediately be realized.

It directly follows from the definition of the Euler angles that, with respect to the sample system S, the direction [001] is determined, only by the angles φ_1, ϕ. This means the lines parallel to the φ_2-axis at fixed φ_1, ϕ indicate the different rotations around a fixed [001]-axis. For $\{\varphi_1, \phi\} = \{0 \text{ to } 90°, 0\}$ this [001] axis is parallel to ND, for $\{0, 90°\}$ parallel to TD and for $\{90°, 90°\}$ parallel to RD. (For $\phi = 0°$, of course, the rotations are around [001] as well as ND.)

Vice versa, with respect to the crystal system C, the direction ND is determined only by the angles ϕ, φ_2 so that lines parallel to φ_1 at fixed ϕ, φ_2 indicate the rotations around a fixed ND. Since ND is given by the Miller indices $\{HKL\}$, these lines indicate also a rotation around a fixed crystallographic direction [HKL]. Here the values $\phi, \varphi_2 = \{0{,}0 \text{ to } 90°\}$ correspond to [HKL] = [001], $\{90°, 0\}$ to [010], $\{90°, 90°\}$ to [100], $\{45°, 0\}$ to [011], $\{54.7°, 45°\}$ to [111] etc. (c.f. Fig. 9). In Fig. 20 the lines for all values HKL $\leqslant 3$ are compiled and on these lines the orientations corresponding to values UVW $\leqslant 3$ are marked /11/.

One further recognizes that the lines running parallel the ϕ-axis at fixed φ_1, φ_2 indicate the different rotations around a fixed axis lying in the rolling plane forming the angle φ_1 to RD as well as in the plane $\{001\}$ forming the angle φ_2 to [100]. Thus for $\varphi_1 = 0$ one has rotations around a fixed RD, for $\varphi_1 = 90°$ around a fixed TD, for $\varphi_2 = 0$ around [100] and for $\varphi_2 = 90°$ around [010]. If RD is given by the Miller indices [UVW], for $\varphi_1 = 0$ one has rotations around the crystallographic axis [UVW] (which is determined by φ_2). The line determined by the values $\{\varphi_1, \varphi_2\} = \{0, 0\}$ corresponds to rotations of the cube orientation around RD or [100], $\{0, 90°\}$ to such around RD or $[0\,\bar{1}\,0]$, $\{90°, 0\}$ to such around TD or [1 0 0] and $\{90°, 90°\}$ to such around TD or $[0\,\bar{1}\,0]$.

4.4 The Multiplicity of Orientations

The fact that, due to symmetry of crystal and sample, a given orientation possesses symmetrically equivalent orientations in the asymmetric unit H of the Euler angle space (Eq. (55), Fig. 8) is denoted as „multiplicity" of this orientation. As shown above, in the case of cubic crystal and orthorhombic sample symmetry this multiplicity is 96. This, however, is correct only for a general orientation. If by symmetry elements of the sample a certain orientation is transformed within itself, some (in general i) of these equivalent positions fall togehter so that for such a special orientation the multiplicity is reduced to 96/i.

In the here mainly considered case of cubic/orthorhombic specimen/sample symmetry, always two symmetrically equivalent orientations fall togehter if they are situated on the rotation axes $L^2_{\varphi_1}$ or L^2_{ϕ}, the mirror plane P_ϕ, the mirror lines L^M or the mirror points L^P. As can be seen from Fig. 16 and from Table 3.5 where these symmetry elements are listed, all these special orientations are characterized by a ⟨100⟩- or ⟨110⟩-axis lying either in ND, TD or RD. This means that the multiplicity is reduced to 96/2 = 48 for all orientations for which one of the 3 sets of Miller indices $\{HKL\}$, ⟨UVW⟩ or ⟨QRS⟩ is given by ⟨100⟩ or ⟨110⟩.

In the intersection points of two of these symmetry elements the multiplicity is reduced again by a factor of 2 resulting in a multiplicity of 24. This is the case also for the mirror points on the edges parallel to φ_1 of the cube (Fig. 16), since, at the same time, these points lie on a mirror plane for $\phi = 0$ or on a two-fold axis (for $\phi = 90°$). (The mirror points on the line Q^3, also the ones being situated on two-fold axes, have only the multiplicity 48, since along the line Q^3 the symmetry operations of the mirror points and these axes are identical.) As can be seen in Fig. 20, these points with a multiplicity of 24 are the orientations $\{001\}$ ⟨100⟩ ⟨100⟩, $\{001\}$ ⟨110⟩ ⟨110⟩, $\{011\}$ ⟨100⟩ ⟨110⟩ and $\{011\}$ ⟨110⟩ ⟨100⟩. (It is not possible to have only two of the sets $\{HKL\}$, ⟨UVW⟩ or ⟨QRS⟩ given by ⟨100⟩ or ⟨110⟩.)

The multiplicity is rather important for the interpretation of textures. For example, let us consider a texture consisting of „components" in form of pronounced maxima of the ODF. Supposing that their volume fraction and scattering width would be equal, a maximum situated at such a special orientation would possess i times the height of a maximum situated at a general orientation. In such a case i different

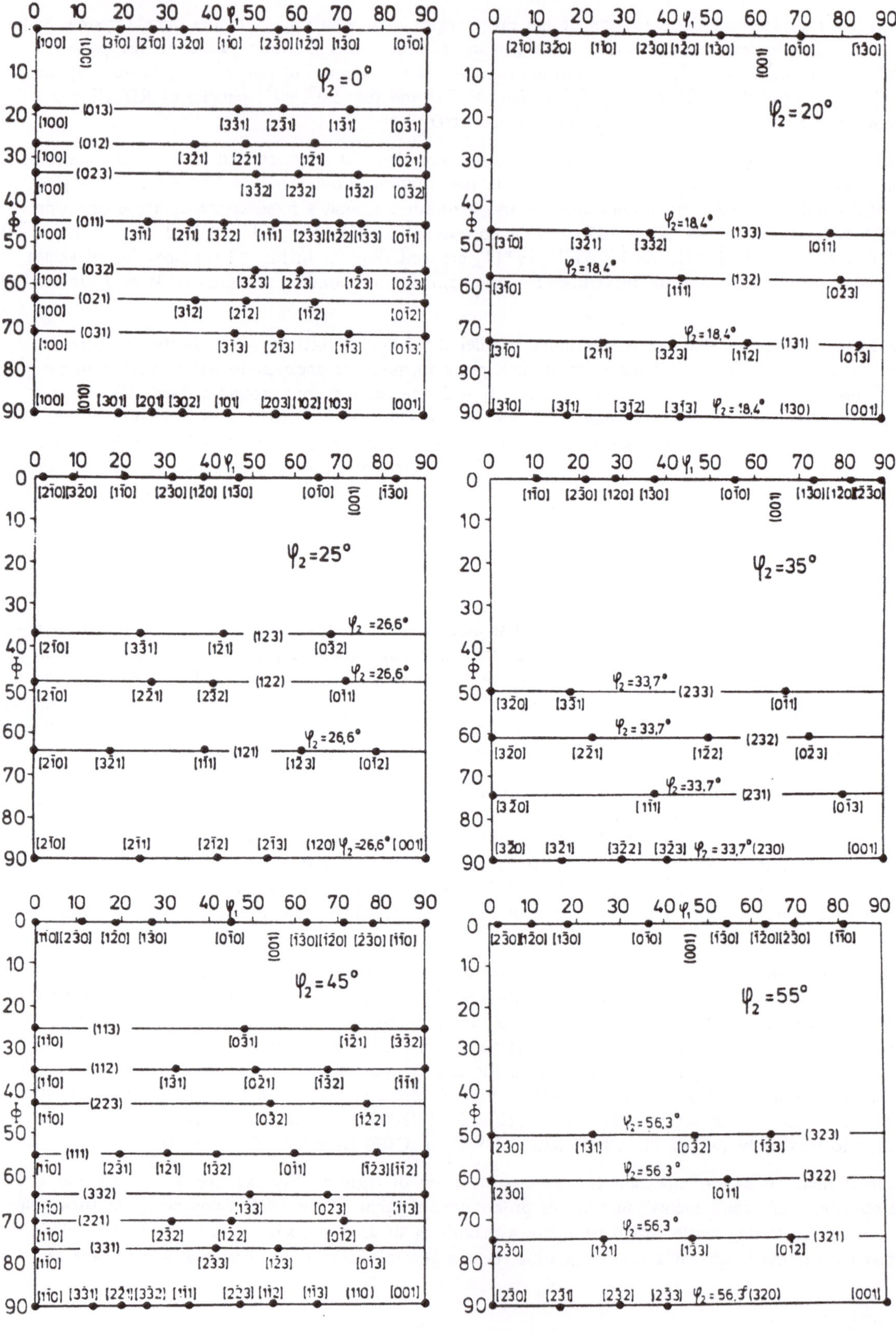
φ2 = 0°
φ2 = 20°
φ2 = 25°
φ2 = 35°
φ2 = 45°
φ2 = 55°
φ1
Φ
φ2 = 18,4°
φ2 = 26,6°
φ2 = 33,7°
φ2 = 56,3°

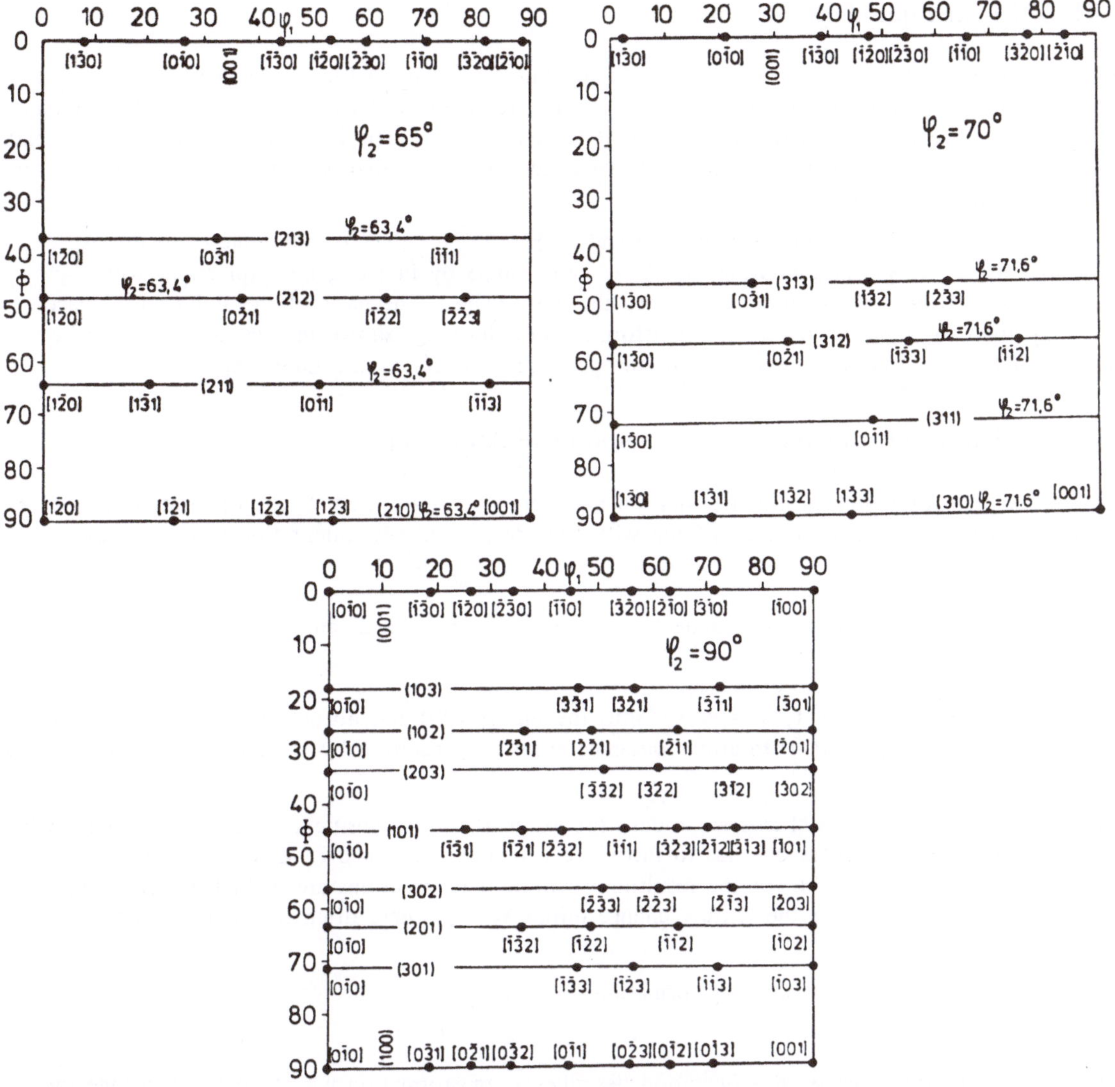

Fig. 20:
Positions of the orientations (HKL) [UVW] with |H, K, L, U, V, W| ⩽ 3 in sections φ_2 = constant through the range H° of the Euler angle space. In the case that the φ_2-value of such an orientation does not exactly coincide with that of the φ_2-values of the selected sections, the points are inserted in the section with a φ_2-value closest to the exact one.

maxima which, in the general case, would occur at i different positions (not necessarily all of them inside H°) will fall on top of each other. Fig. 17 gives an example of a position of the multiplicity of 96 ({236} ⟨385⟩) and Fig. 18 of the multiplicity of only 24 (cube positions {001} ⟨100⟩).

In the case of cubic single crystals one has triclinic sample symmetry and thus the multiplicity 24. In this case no special orientations with reduced multiplicity exist. (The symmetry elements are either translations or screw axes, i.e. elements which do not transform points into themselves.) At monoclinic sample symmetry the multiplicity is 48 in the general case, but can be reduced to 24 for special orientations.

4.5 The ODF in Equivalent Regions

As pointed out above, for representing the ODF in the Euler angle space for samples with triclinic, monoclinic or orthorhombic symmetry the ranges H', H'' or H°, respectively, have to be applied. They contain each orientation three times. Sometimes, however, it is useful to consider the ODF also outside these basic ranges. These values can be obtained from the values inside by transformations which directly follow from the symmetry relationships listed in Table 3.5.

In Fig. 21 (for the triclinic and monoclinic case) and in Fig. 22 (for the orthorhombic case) this is demonstrated for the ranges neighbouring the above basic ranges by inserting the expressions which give the transformation from each of these ranges into the basic range. The parts of the Euler angle space further away from the basic ranges can be transformed into the range shown in Figs. 21 and 22 a) by making use of the identical equivalency due to the periodicity in $2/\pi$ of the Euler angles $\varphi_1\ \phi\ \varphi_2$.

4.6 Orientation Transformations and Orientation Relationships

For orientation transformation mostly the rotation parameters $\vec{v}$, ω are used. If the orientation g is transformed by a rotation around an axis connected with the crystal, the new orientation g' is given by

$$g' = g(\vec{v}, \omega) \cdot g. \tag{74}$$

Transformation with respect to an axis connected with the specimen yields an orientation g'' given by

$$g'' = g \cdot g(\vec{v}, \omega). \tag{75}$$

For example, if one inserts $v_x = 1$, $v_y = v_z = 0$ into the matrix (29) and introduces this matrix into Eq. (74) or (75), g' and g'' represent orientations obtained from g by rotation by the angle ω around [100] or RD, respectively.

Rotations with respect to a crystal axis are found, for example, at martensitic transformations or at twinning. For twinning in f.c.c. crystals one has to introduce $\vec{v} = \langle 111 \rangle$, $\omega = 180^\circ$ and for b.c.c. metals $\vec{v} = \langle 112 \rangle$, $\omega = 180^\circ$. In Table 4.6 for the resulting orientation relationships are compiled. An example for rotations around a sample axis is the texture change during rolling where often rotations around TD or RD are observed.

Another application makes use of the transformation

$$g(\vec{v}', \omega) = h \cdot g(\vec{v}, \omega) \cdot h^{-1}. \tag{76}$$

With h being any rotation matrix, this operation describes a transformation which does not change the trace of the matrix g, i.e. the sum $g_{11} + g_{22} + g_{33}$. This means that this transformation alters only the axis of rotation (from $\vec{v}$ to $\vec{v}'$) but not the angle ω (Eq. (29)). If h is a symmetry element, then $\vec{v}'$ means an axis of rotation symmetrically equivalent to $\vec{v}$, and, correspondingly, $g(\vec{v}', \omega)$ a transformation symmetrically equivalent to $g(\vec{v}, \omega)$. If, for example, $g(\vec{v}, \omega)$ describes the Kurdjumov-Sachs relationship 90° $[11\bar{2}]$, then the 90° rotations around all other ⟨112⟩-axis are obtained simply by applying for h the cubic symmetry elements given in Table 3.2.

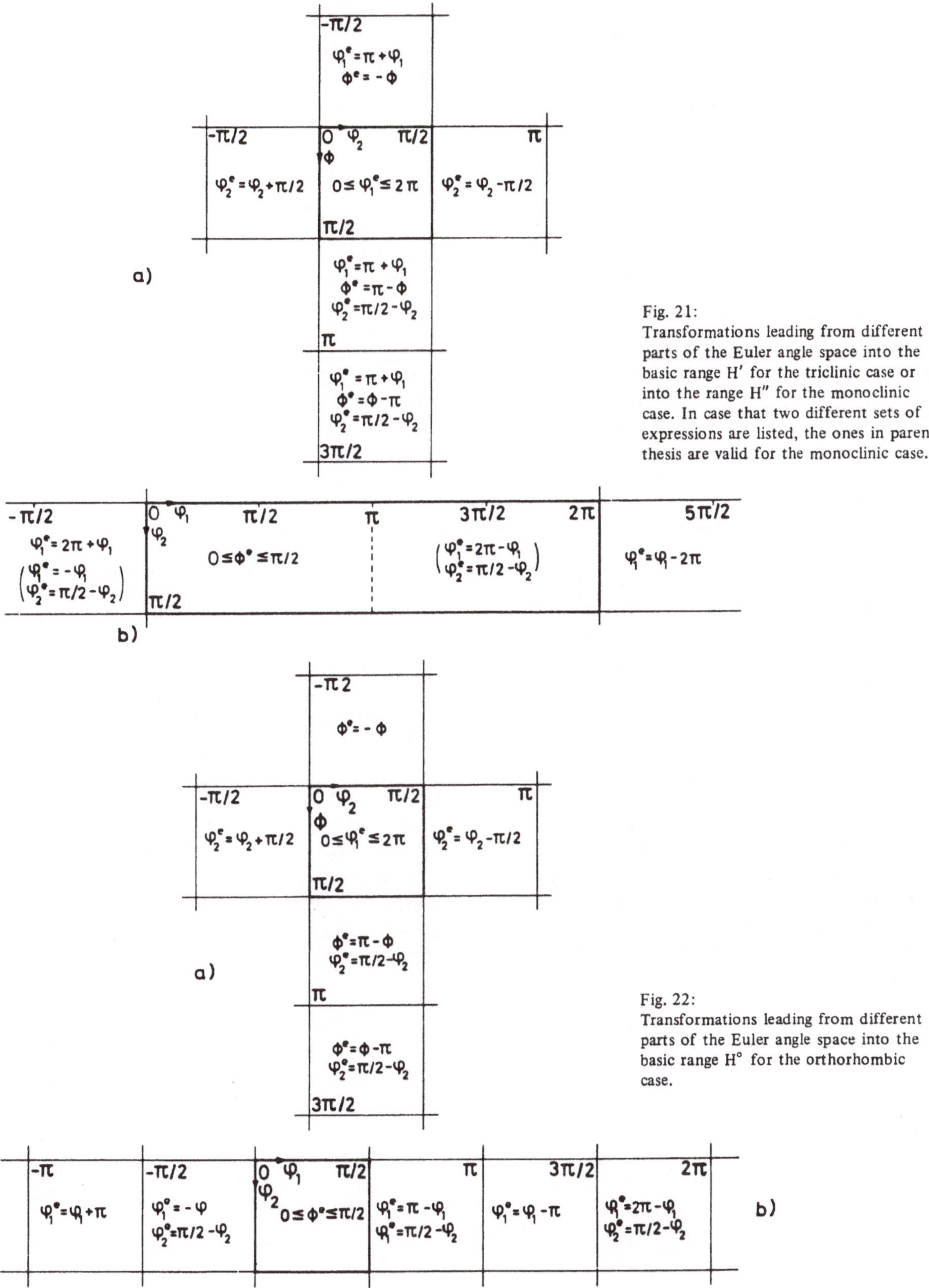

Fig. 21:
Transformations leading from different parts of the Euler angle space into the basic range H′ for the triclinic case or into the range H″ for the monoclinic case. In case that two different sets of expressions are listed, the ones in parenthesis are valid for the monoclinic case.

Fig. 22:
Transformations leading from different parts of the Euler angle space into the basic range H° for the orthorhombic case.

Table 4.6: The Millers indices (H'K'L') [U'V'W'] for the four {111}-twins and the twelve {112}-twins of a given orientation (HKL) [UVW]

	111	$\bar{1}11$	$1\bar{1}1$	$\bar{1}\bar{1}1$
H'	- H + 2K + 2L	- H - 2K - 2L	- H - 2K + 2L	- H + 2K - 2L
K'	2H - K + 2L	- 2H - K + 2L	- 2H - K - 2L	2H - K - 2L
L'	2H + 2K - L	- 2H + 2K - L	2H - 2K - L	- 2H - 2K - L
U'	- U + 2V + 2W	- U - 2V - 2W	- U - 2V + 2W	- U + 2V - 2W
V'	2U - V + 2W	- 2U - V + 2W	- 2U - V - 2W	2U - V - 2W
W'	2U + 2V - W	- 2U + 2V - W	2U - 2V - W	- 2U - 2V - W
	112	$\bar{1}12$	$1\bar{1}2$	$\bar{1}\bar{1}2$
H'	- 2H + K + 2L	- 2H - K - 2L	- 2H - K + 2L	- 2H + K - 2L
K'	H - 2K + 2L	- H - 2K + 2L	- H - 2K - 2L	H - 2K - 2L
L'	2H + 2K + L	- 2H + 2K + L	2H - 2K + L	- 2H - 2K + L
U'	- 2U + V + 2W	- 2U - V - 2W	- 2U - V + 2W	- 2U + V - 2W
V'	U - 2V + 2W	- U - 2V + 2W	- U - 2V - 2W	U - 2V - 2W
W'	2W + 2V + W	- 2U + 2V + W	2U - 2V + W	- 2U - 2V + W
	121	$\bar{1}21$	$1\bar{2}1$	$\bar{1}\bar{2}1$
H'	- 2H + 2K + L	- 2H - 2K - L	- 2H - 2K + L	- 2H + 2K - L
K'	2H + K + 2L	- 2H + K + 2L	- 2H + K - 2L	2H + K - 2L
L'	H + 2K - 2L	- H + 2K - 2L	- H - 2K - 2L	- H - 2K - 2L
U'	- 2U + 2V + W	- 2U - 2V - W	- 2U - 2V + W	- 2U + 2V - W
V'	2U + V + 2W	- 2U + V + 2W	- 2U + V - 2W	2U + V - 2W
W'	U + 2V - 2W	- U + 2V - 2W	- U - 2V - 2W	- U - 2V - 2W
	211	$\bar{2}11$	$2\bar{1}1$	$\bar{2}\bar{1}1$
H'	H + 2K + 2L	H - 2K - 2L	H - 2K + 2L	H + 2K - 2L
K'	2H - 2K + L	- 2H - 2K + L	- 2H - 2K - L	2H - 2K - L
L'	2H + K - 2L	- 2H + K - 2L	2H - K - 2L	- 2H - K - 2L
U'	U + 2V + 2W	U - 2V - 2W	U - 2V + 2W	U + 2V - 2W
V'	2U - 2V + W	- 2U - 2V + W	- 2U - 2V - W	2U - 2V - W
W'	2U + V - 2W	- 2U + V - 2W	2U - V - 2W	- 2U - V - 2W

APPENDIX

GRAPHICAL CONSTRUCTION OF A POLE FIGURE FROM EULER ANGLES

As shown in Sec. 2.3 the Euler angles transform the sample system S into the crystal system C by a sequence of three independent rotations. These rotations can easily be carried out graphically in such a way that only by the usual application of Wulff net (and a sheet of transparent paper) a pole figure is obtained. The idea of the method is to exchange the sequence of the partial rotations in the total rotation

$$g(\varphi_1\ \phi\ \varphi_2) = g(\varphi_2) \cdot g(\phi) \cdot g(\varphi_1)$$

and correct this by the proper choice of the rotation axes.

First the direction RD will be laid in the direction N („north") of the Wulff net. Then, by a rotation by $(\varphi_1 + \varphi_2)$ around the center, RD is brought into the direction R (Fig. 23 a) and the poles in question (in Fig. 23 a the $\{111\}$-poles) are drawn into the sheet in standard position with rexpect to the original position (filled symbols in Fig. 23 a). Now by a reversed rotation by φ_2 a sample axis RD' is brought into N, and, by this rotation, the poles (here $\{111\}$) are taken into a new position (open symbols in Fig. 23 b). Subsequently by a rotation by ϕ around N = RD' the poles are shifted again (into the filled symbols of Fig. 23 b). Finally by reversed rotation by φ_1, RD is brought back into N and thus the poles into their final positions (filled symbols in Fig. 23 c).

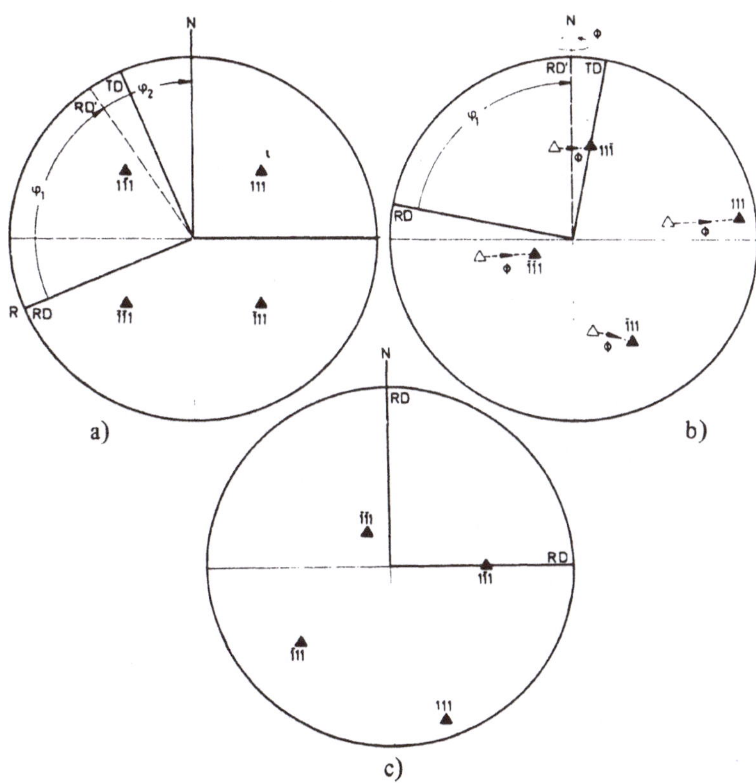

REFERENCES

/1/ H.J. Bunge, Mathematische Methoden der Texturanalyse, Akademie-Verlag, Berlin 1961.

/2/ Proceedings of the Internal Seminar, Cracow 1971.

/3/ J. Jura, J. Pospiech, H.J. Bunge, Papers of Commission of Metallurgy and Foundry, Metalurgia *24* (1976), 111.

/4/ R.J. Roe, Journ. Appl. Phys. *36* (1965), 2024.

/5/ J. Pospiech, Kristall u. Technik *7* (1972), 1074.

/6/ G. Ibe, K. Lücke, Texture *1* (1972), 87.

/7/ J. Pospiech, A. Gnatek, K. Fichtner, Kristall u. Technik *9* (1974), 729.

/8/ I.M. Gelfand, R.A. Minlos, Z.YA. Shapiro, Representations of the Rotation and Lorentz Groups an their Applications, Pergamon Press, Oxford–London–New York–Paris 1963.

/9/ W. Truszkowski, J. Pospiech, J. Jura, B. Major, 3e colloque européén sur les textur de deformation et de recrystallisation des métaux et leurs applications industrielles, Pont-à-Mousson 1973.

/10/ K.H. Virnich, J. Pospiech, A. Flemmer, K. Lücke, (paper 1.1); K.H. Virnich, K. Lücke, (paper 5.2); K.H. Virnich, G. Köhlhoff, K. Lücke, J. Pospiech, (paper 5.9); Proceedings of the fifth International Conferences of Textures of Materials, Springer-Verlag, Berlin–Heidelberg–New York, in press.

/11/ D. Schläfer, H.J. Bunge, VII Konferencja Naukowa-Techniczna, Gliwice 1974.

TABLE I

The General Miller-Indices {HKL} ⟨UVW⟩ for H, K, L, U, V, W ⩽ 15

Since the Miller indices (HKL) [UVW] have to fulfill the orthogonality condition

$$HU + KV + LW = 0,$$

they cannot be chosen independently of each other. This table contains all possible general Miller indices {HKL} ⟨UVW⟩ with

$$H, K, L, U, V, W \leqslant 15.$$

The indices are chosen to be positive and ordered according to

$$H \leqslant K \leqslant L \quad \text{and} \quad U \leqslant V \leqslant W.$$

The table is made up of blocks of constant HKL with the HKL values unterlined on top of each block. The bocks are ordered according to increasing HKL. Within each block the orientations are ordered according to increasing UVW. From these general orientations the special orientations, i.e. the signs and sequence of the digits, can be found either by trying to fulfill the orthogonality condition or from Table V where for the case H, K, L, U, V, W ⩽ 12 also special orientations are listed.

TAB. I THE TYPES OF HKL-UVW FOR H,K,L,U,V,W ≤15 1 - 2

Column 1

H U	K V	L W
0	**0**	**1**
0	0	1
0	1	1
0	1	2
0	1	3
0	1	4
0	1	5
0	1	6
0	1	7
0	1	8
0	1	9
0	1	10
0	1	11
0	1	12
0	1	13
0	1	14
0	1	15
0	2	3
0	2	5
0	2	7
0	2	9
0	2	11
0	2	13
0	2	15
0	3	4
0	3	5
0	3	7
0	3	8
0	3	10
0	3	11
0	3	13
0	3	14
0	4	5
0	4	7
0	4	9
0	4	11
0	4	13
0	4	15
0	5	6
0	5	7
0	5	8
0	5	9
0	5	11
0	5	12
0	5	13
0	5	14

Column 2

H U	K V	L W
0	**0**	**1**
0	6	7
0	6	11
0	6	13
0	7	8
0	7	9
0	7	10
0	7	11
0	7	12
0	7	13
0	7	15
0	8	9
0	8	11
0	8	13
0	8	15
0	9	10
0	9	11
0	9	13
0	9	14
0	10	11
0	10	13
0	11	12
0	11	13
0	11	14
0	11	15
0	12	13
0	13	14
0	13	15
0	14	15
0	**1**	**1**
0	0	1
0	1	1
1	1	1
1	1	2
1	1	3
1	1	4
1	1	5
1	1	6
1	1	7
1	1	8
1	1	9
1	1	10
1	1	11
1	1	12
1	1	13

Column 3

H U	K V	L W
0	**1**	**1**
1	1	14
1	1	15
1	2	2
1	3	3
1	4	4
1	5	5
1	6	6
1	7	7
1	8	8
1	9	9
1	10	10
1	11	11
1	12	12
1	13	13
1	14	14
1	15	15
2	2	3
2	2	5
2	2	7
2	2	9
2	2	11
2	2	13
2	2	15
2	3	3
2	5	5
2	7	7
2	9	9
2	11	11
2	13	13
2	15	15
3	3	4
3	3	5
3	3	7
3	3	8
3	3	10
3	3	11
3	3	13
3	3	14
3	4	4
3	5	5
3	7	7
3	8	8
3	10	10
3	11	11
3	13	13
3	14	14

Column 4

H U	K V	L W
0	**1**	**1**
4	4	5
4	4	7
4	4	9
4	4	11
4	4	13
4	4	15
4	5	5
4	7	7
4	9	9
4	11	11
4	13	13
4	15	15
5	5	6
5	5	7
5	5	8
5	5	9
5	5	11
5	5	12
5	5	13
5	5	14
5	6	6
5	7	7
5	8	8
5	9	9
5	11	11
5	12	12
5	13	13
5	14	14
6	6	7
6	6	11
6	6	13
6	7	7
6	11	11
6	13	13
7	7	8
7	7	9
7	7	10
7	7	11
7	7	12
7	7	13
7	7	15
7	8	8
7	9	9
7	10	10
7	11	11
7	12	12

Column 5

H U	K V	L W
0	**1**	**1**
7	13	13
7	15	15
8	8	9
8	8	11
8	8	13
8	8	15
8	9	9
8	11	11
8	13	13
8	15	15
9	9	10
9	9	11
9	9	13
9	9	14
9	10	10
9	11	11
9	13	13
9	14	14
10	10	11
10	10	13
10	11	11
10	13	13
11	11	12
11	11	13
11	11	14
11	11	15
11	12	12
11	13	13
11	14	14
11	15	15
12	12	13
12	13	13
13	13	14
13	13	15
13	14	14
13	15	15
14	14	15
14	15	15
0	**1**	**2**
0	0	1
0	1	2
1	1	2
1	2	2
1	2	3

Column 6

H U	K V	L W
0	**1**	**2**
1	2	4
1	2	5
1	2	6
1	2	7
1	2	8
1	2	9
1	2	10
1	2	11
1	2	12
1	2	13
1	2	14
1	2	15
1	3	6
1	4	8
1	5	10
1	6	12
1	7	14
2	3	4
2	3	6
2	4	5
2	4	7
2	4	9
2	4	11
2	4	13
2	4	15
2	5	10
2	7	14
3	4	6
3	4	8
3	5	6
3	5	10
3	6	7
3	6	8
3	6	10
3	6	11
3	6	13
3	6	14
3	7	14
4	5	8
4	5	10
4	7	8
4	7	14
4	8	9
4	8	11
4	8	13
4	8	15

Column 7

H U	K V	L W
0	**1**	**2**
5	6	10
5	6	12
5	7	10
5	7	14
5	8	10
5	9	10
5	10	11
5	10	12
5	10	13
5	10	14
6	7	12
6	7	14
6	11	12
6	12	13
7	8	14
7	9	14
7	10	14
7	11	14
7	12	14
7	13	14
7	14	15
0	**1**	**3**
0	0	1
0	1	3
1	1	3
1	2	3
1	2	6
1	3	3
1	3	4
1	3	5
1	3	6
1	3	7
1	3	8
1	3	9
1	3	10
1	3	11
1	3	12
1	3	13
1	3	14
1	3	15
1	4	12
1	5	15
2	3	6
2	3	9

Column 8

H U	K V	L W
0	**1**	**3**
2	5	6
2	5	15
2	6	7
2	6	9
2	6	11
2	6	13
2	6	15
3	4	9
3	4	12
3	5	9
3	5	15
3	7	9
3	8	9
3	9	10
3	9	11
3	9	13
3	9	14
4	5	12
4	5	15
4	7	12
4	9	12
4	11	12
4	12	13
4	12	15
5	6	15
5	7	15
5	8	15
5	9	15
5	11	15
5	12	15
5	13	15
5	14	15
0	**1**	**4**
0	0	1
0	1	4
1	1	4
1	2	4
1	2	8
1	3	4
1	3	12
1	4	4
1	4	5
1	4	6
1	4	7

Column 9

H U	K V	L W
0	**1**	**4**
1	4	8
1	4	9
1	4	10
1	4	11
1	4	12
1	4	13
1	4	14
1	4	15
2	3	8
2	3	12
2	5	8
2	7	8
2	8	9
2	8	11
2	8	13
2	8	15
3	4	12
3	5	12
3	7	12
3	8	12
3	10	12
3	11	12
3	12	13
3	12	14
0	**1**	**5**
0	0	1
0	1	5
1	1	5
1	2	5
1	2	10
1	3	5
1	3	15
1	4	5
1	5	5
1	5	6
1	5	7
1	5	8
1	5	9
1	5	10
1	5	11
1	5	12
1	5	13
1	5	14
1	5	15

Column 10

H U	K V	L W
0	**1**	**5**
2	3	10
2	3	15
2	5	10
2	7	10
2	9	10
2	10	11
2	10	13
2	10	15
3	4	15
3	5	15
3	7	15
3	8	15
3	10	15
3	11	15
3	13	15
3	14	15
0	**1**	**6**
0	0	1
0	1	6
1	1	6
1	2	6
1	2	12
1	3	6
1	4	6
1	5	6
1	6	6
1	6	7
1	6	8
1	6	9
1	6	10
1	6	11
1	6	12
1	6	13
1	6	14
1	6	15
2	3	12
2	5	12
2	7	12
2	9	12
2	11	12
2	12	13
2	12	15

Column 11

H U	K V	L W
0	**1**	**7**
0	0	1
0	1	7
1	1	7
1	2	7
1	2	14
1	3	7
1	4	7
1	5	7
1	6	7
1	7	7
1	7	8
1	7	9
1	7	10
1	7	11
1	7	12
1	7	13
1	7	14
1	7	15
2	3	14
2	5	14
2	7	14
2	9	14
2	11	14
2	13	14
2	14	15
0	**1**	**8**
0	0	1
0	1	8
1	1	8
1	2	8
1	3	8
1	4	8
1	5	8
1	6	8
1	7	8
1	8	8
1	8	9
1	8	10
1	8	11
1	8	12
1	8	13
1	8	14
1	8	15

Column 12

H U	K V	L W
0	**1**	**9**
0	0	1
0	1	9
1	1	9
1	2	9
1	3	9
1	4	9
1	5	9
1	6	9
1	7	9
1	8	9
1	9	9
1	9	10
1	9	11
1	9	12
1	9	13
1	9	14
1	9	15
0	**1**	**10**
0	0	1
0	1	10
1	1	10
1	2	10
1	3	10
1	4	10
1	5	10
1	6	10
1	7	10
1	8	10
1	9	10
1	10	10
1	10	11
1	10	12
1	10	13
1	10	14
1	10	15
0	**1**	**11**
0	0	1
0	1	11
1	1	11
1	2	11
1	3	11

TAB. I

THE TYPES OF HKL-UVW FOR H,K,L,U,V,W ≤15


```
 H  K  L
 U  V  W
--------
 0  1 11
--------
 1  4 11
 1  5 11
 1  6 11
 1  7 11
 1  8 11
 1  9 11
 1 10 11
 1 11 11
 1 11 12
 1 11 13
 1 11 14
 1 11 15

 0  1 12
========
 0  0  1
 0  1 12
 1  1 12
 1  2 12
 1  3 12
 1  4 12
 1  5 12
 1  6 12
 1  7 12
 1  8 12
 1  9 12
 1 10 12
 1 11 12
 1 12 12
 1 12 13
 1 12 14
 1 12 15

 0  1 13
========
 0  0  1
 0  1 13
 1  1 13
 1  2 13
 1  3 13
 1  4 13
 1  5 13
 1  6 13
 1  7 13
 1  8 13
 1  9 13
```

```
 H  K  L
 U  V  W
--------
 0  1 13
--------
 1 10 13
 1 11 13
 1 12 13
 1 13 13
 1 13 14
 1 13 15

 0  1 14
========
 0  0  1
 0  1 14
 1  1 14
 1  2 14
 1  3 14
 1  4 14
 1  5 14
 1  6 14
 1  7 14
 1  8 14
 1  9 14
 1 10 14
 1 11 14
 1 12 14
 1 13 14
 1 14 14
 1 14 15

 0  1 15
========
 0  0  1
 0  1 15
 1  1 15
 1  2 15
 1  3 15
 1  4 15
 1  5 15
 1  6 15
 1  7 15
 1  8 15
 1  9 15
 1 10 15
 1 11 15
 1 12 15
 1 13 15
 1 14 15
 1 15 15
```

```
 H  K  L
 U  V  W
--------
 0  2  3
========
 0  0  1
 0  2  3
 1  2  3
 1  4  6
 1  6  9
 1  8 12
 1 10 15
 2  2  3
 2  3  3
 2  3  4
 2  3  5
 2  3  6
 2  3  7
 2  3  8
 2  3  9
 2  3 10
 2  3 11
 2  3 12
 2  3 13
 2  3 14
 2  3 15
 2  6  9
 2 10 15
 3  4  6
 3  8 12
 3 10 15
 4  5  6
 4  6  7
 4  6  9
 4  6 11
 4  6 13
 4  6 15
 4 10 15
 5  6  9
 5  8 12
 6  7  9
 6  8  9
 6  9 10
 6  9 11
 6  9 13
 6  9 14
 6 10 15
 7  8 12
 7 10 15
 8  9 12
```

```
 H  K  L
 U  V  W
--------
 0  2  3
--------
 8 10 15
 8 11 12
 8 12 13
 8 12 15
 9 10 15
10 11 15
10 12 15
10 13 15
10 14 15

 0  2  5
========
 0  0  1
 0  2  5
 1  2  5
 1  4 10
 1  6 15
 2  2  5
 2  3  5
 2  4  5
 2  5  5
 2  5  6
 2  5  7
 2  5  8
 2  5  9
 2  5 10
 2  5 11
 2  5 12
 2  5 13
 2  5 14
 2  5 15
 2  6 15
 3  4 10
 4  5 10
 4  6 15
 4  7 10
 4  9 10
 4 10 11
 4 10 13
 4 10 15
 5  6 15
 6  7 15
 6  8 15
 6 10 15
 6 11 15
 6 13 15
```

```
 H  K  L
 U  V  W
--------
 0  2  5
--------
 6 14 15

 0  2  7
========
 0  0  1
 0  2  7
 1  2  7
 1  4 14
 2  2  7
 2  3  7
 2  4  7
 2  5  7
 2  6  7
 2  7  7
 2  7  8
 2  7  9
 2  7 10
 2  7 11
 2  7 12
 2  7 13
 2  7 14
 2  7 15
 3  4 14
 4  5 14
 4  7 14
 4  9 14
 4 11 14
 4 13 14
 4 14 15

 0  2  9
========
 0  0  1
 0  2  9
 1  2  9
 2  2  9
 2  3  9
 2  4  9
 2  5  9
 2  6  9
 2  7  9
 2  8  9
 2  9  9
 2  9 10
 2  9 11
 2  9 12
```

```
 H  K  L
 U  V  W
--------
 0  2  9
--------
 2  9 13
 2  9 14
 2  9 15

 0  2 11
========
 0  0  1
 0  2 11
 1  2 11
 2  2 11
 2  3 11
 2  4 11
 2  5 11
 2  6 11
 2  7 11
 2  8 11
 2  9 11
 2 10 11
 2 11 11
 2 11 12
 2 11 13
 2 11 14
 2 11 15

 0  2 13
========
 0  0  1
 0  2 13
 1  2 13
 2  2 13
 2  3 13
 2  4 13
 2  5 13
 2  6 13
 2  7 13
 2  8 13
 2  9 13
 2 10 13
 2 11 13
 2 12 13
 2 13 13
 2 13 14
 2 13 15
```

```
 H  K  L
 U  V  W
--------
 0  2 15
========
 0  0  1
 0  2 15
 1  2 15
 2  2 15
 2  3 15
 2  4 15
 2  5 15
 2  6 15
 2  7 15
 2  8 15
 2  9 15
 2 10 15
 2 11 15
 2 12 15
 2 13 15
 2 14 15
 2 15 15

 0  3  4
========
 0  0  1
 0  3  4
 1  3  4
 1  6  8
 1  9 12
 2  3  4
 2  9 12
 3  3  4
 3  4  4
 3  4  5
 3  4  6
 3  4  7
 3  4  8
 3  4  9
 3  4 10
 3  4 11
 3  4 12
 3  4 13
 3  4 14
 3  4 15
 3  6  8
 4  9 12
 5  6  8
 5  9 12
 6  7  8
```

```
 H  K  L
 U  V  W
--------
 0  3  4
--------
 6  6  9
 6  8 11
 6  8 13
 6  8 15
 7  9 12
 8  9 12
 9 10 12
 9 11 12
 9 12 13
 9 12 14

 0  3  5
========
 0  0  1
 0  3  5
 1  3  5
 1  6 10
 1  9 15
 2  3  5
 2  9 15
 3  3  5
 3  4  5
 3  5  5
 3  5  6
 3  5  7
 3  5  8
 3  5  9
 3  5 10
 3  5 11
 3  5 12
 3  5 13
 3  5 14
 3  5 15
 3  6 10
 4  9 15
 5  6 10
 5  9 15
 6  7 10
 6  9 10
 6 10 11
 6 10 13
 6 10 15
 7  9 15
 8  9 15
 9 10 15
 9 11 15
```

```
 H  K  L
 U  V  W
--------
 0  3  5
--------
 9 13 15
 9 14 15

 0  3  7
========
 0  0  1
 0  3  7
 1  3  7
 1  6 14
 2  3  7
 3  3  7
 3  4  7
 3  5  7
 3  6  7
 3  6 14
 3  7  7
 3  7  8
 3  7  9
 3  7 10
 3  7 11
 3  7 12
 3  7 13
 3  7 14
 3  7 15
 5  6 14
 6  7 14
 6  9 14
 6 11 14
 6 13 14
 6 14 15

 0  3  8
========
 0  0  1
 0  3  8
 1  3  8
 2  3  8
 3  3  8
 3  4  8
 3  5  8
 3  6  8
 3  7  8
 3  8  8
 3  8  9
 3  8 10
 3  8 11
```

```
 H  K  L
 U  V  W
--------
 0  3  8
--------
 3  8 12
 3  8 13
 3  8 14
 3  8 15

 0  3 10
========
 0  0  1
 0  3 10
 1  3 10
 2  3 10
 3  3 10
 3  4 10
 3  5 10
 3  6 10
 3  7 10
 3  8 10
 3  9 10
 3 10 10
 3 10 11
 3 10 12
 3 10 13
 3 10 14
 3 10 15

 0  3 11
========
 0  0  1
 0  3 11
 1  3 11
 2  3 11
 3  3 11
 3  4 11
 3  5 11
 3  6 11
 3  7 11
 3  8 11
 3  9 11
 3 10 11
 3 11 11
 3 11 12
 3 11 13
 3 11 14
 3 11 15
```

```
 H  K  L
 U  V  W
--------
 0  3 13
========
 0  0  1
 0  3 13
 1  3 13
 2  3 13
 3  3 13
 3  4 13
 3  5 13
 3  6 13
 3  7 13
 3  8 13
 3  9 13
 3 10 13
 3 11 13
 3 12 13
 3 13 13
 3 13 14
 3 13 15

 0  3 14
========
 0  0  1
 0  3 14
 1  3 14
 2  3 14
 3  3 14
 3  4 14
 3  5 14
 3  6 14
 3  7 14
 3  8 14
 3  9 14
 3 10 14
 3 11 14
 3 12 14
 3 13 14
 3 14 14
 3 14 15

 0  4  5
========
 0  0  1
 0  4  5
 1  4  5
 1  8 10
 1 12 15
```

```
 H  K  L
 U  V  W
--------
 0  4  5
--------
 2  4  5
 2 12 15
 3  4  5
 3  8 10
 4  4  5
 4  5  5
 4  5  6
 4  5  7
 4  5  8
 4  5  9
 4  5 10
 4  5 11
 4  5 12
 4  5 13
 4  5 14
 4  5 15
 4 12 15
 5  8 10
 5 12 15
 7  8 10
 7 12 15
 8  9 10
 8 10 11
 8 10 13
 8 10 15
 8 12 15
10 12 15
11 12 15
12 13 15
12 14 15

 0  4  7
========
 0  0  1
 0  4  7
 1  4  7
 1  8 14
 2  4  7
 3  4  7
 3  8 14
 4  4  7
 4  5  7
 4  6  7
 4  7  7
 4  7  8
 4  7  9
```

TAB. I THE TYPES OF HKL-UVW FOR H,K,L,U,V,W ≤15

H K L	H K L	H K L	H K L	H K L	H K L	H K L	H K L	H K L	H K L	H K L	H K L
U V W	U V W	U V W	U V W	U V W	U V W	U V W	U V W	U V W	U V W	U V W	U V W
0 4 7	0 4 11		0 5 7	0 5 9		0 6 7	0 6 13	0 7 9	0 7 11		0 8 11
--------	--------	0 5 6	--------	--------	0 5 13	--------	--------	--------	--------	0 7 15	--------
4 7 10	4 10 11	========	5 7 14	5 9 14	========	3 6 7	2 6 13	7 8 9	7 11 14	========	4 8 11
4 7 11	4 11 11	0 0 1	5 7 15	5 9 15	0 0 1	3 12 14	3 6 13	7 9 9	7 11 15	0 0 1	5 8 11
4 7 12	4 11 12	0 5 6	5 10 14		0 5 13	4 6 7	4 6 13	7 9 10		0 7 15	6 8 11
4 7 13	4 11 13	1 5 6	7 10 14	0 5 11	1 5 13	5 6 7	5 6 13	7 9 11	0 7 12	1 7 15	7 8 11
4 7 14	4 11 14	1 10 12	9 10 14	========	2 5 13	5 12 14	6 6 13	7 9 12	========	2 7 15	8 8 11
4 7 15	4 11 15	2 5 6	10 11 14	0 0 1	3 5 13	6 6 7	6 7 13	7 9 13	0 0 1	3 7 15	8 9 11
5 8 14		3 5 6	10 13 14	0 5 11	4 5 13	6 7 7	6 8 13	7 9 14	0 7 12	4 7 15	8 10 11
7 8 14	0 4 13	3 10 12	10 14 15	1 5 11	5 5 13	6 7 8	6 9 13	7 9 15	1 7 12	5 7 15	8 11 11
8 9 14	========	4 5 6		2 5 11	5 6 13	6 7 9	6 10 13		2 7 12	6 7 15	8 11 12
6 11 14	0 0 1	5 5 6	0 5 8	3 5 11	5 7 13	6 7 10	6 11 13	0 7 10	3 7 12	7 7 15	8 11 13
8 13 14	0 4 13	5 6 6	========	4 5 11	5 8 13	6 7 11	6 12 13	========	4 7 12	7 8 15	8 11 14
8 14 15	1 4 13	5 6 7	0 0 1	5 5 11	5 9 13	6 7 12	6 13 13	0 0 1	5 7 12	7 9 15	8 11 15
	2 4 13	5 6 8	0 5 8	5 6 11	5 10 13	6 7 13	6 13 14	0 7 10	6 7 12	7 10 15	
0 4 9	3 4 13	5 6 9	1 5 8	5 7 11	5 11 13	6 7 14	6 13 15	1 7 10	7 7 12	7 11 15	0 8 13
========	4 4 13	5 6 10	2 5 8	5 8 11	5 12 13	6 7 15		2 7 10	7 8 12	7 12 15	========
0 0 1	4 5 13	5 6 11	3 5 8	5 9 11	5 13 13	7 12 14	0 7 8	3 7 10	7 9 12	7 13 15	0 0 1
0 4 9	4 6 13	5 6 12	4 5 8	5 10 11	5 13 14	9 12 14	========	4 7 10	7 10 12	7 14 15	0 8 13
1 4 9	4 7 13	5 6 13	5 5 8	5 11 11	5 13 15	11 12 14	0 0 1	5 7 10	7 11 12	7 15 15	1 8 13
2 4 9	4 8 13	5 6 14	5 6 8	5 11 12		12 13 14	0 7 8	6 7 10	7 12 12		2 8 13
3 4 9	4 9 13	5 6 15	5 7 8	5 11 13	0 5 14	12 14 15	1 7 8	7 7 10	7 12 13	0 8 9	3 8 13
4 4 9	4 10 13	5 10 12	5 8 8	5 11 14	========		2 7 8	7 8 10	7 12 14	========	4 8 13
4 5 9	4 11 13	7 10 12	5 8 9	5 11 15	0 0 1	0 6 11	3 7 8	7 9 10	7 12 15	0 0 1	5 8 13
4 6 9	4 12 13	9 10 12	5 8 10		0 5 14	========	4 7 8	7 10 10		0 8 9	6 8 13
4 7 9	4 13 13	10 11 12	5 8 11	0 5 12	1 5 14	0 0 1	5 7 8	7 10 11	0 7 13	1 8 9	7 8 13
4 8 9	4 13 14	10 12 13	5 8 12	========	2 5 14	0 6 11	6 7 8	7 10 12	========	2 8 9	8 8 13
4 9 9	4 13 15	10 12 15	5 8 13	0 0 1	3 5 14	1 6 11	7 7 8	7 10 13	0 0 1	3 8 9	8 9 13
4 9 10			5 8 14	0 5 12	4 5 14	2 6 11	7 8 8	7 10 14	0 7 13	4 8 9	8 10 13
4 9 11	0 4 15	0 5 7	5 8 15	1 5 12	5 5 14	3 6 11	7 8 9	7 10 15	1 7 13	5 8 9	8 11 13
4 9 12	========	========		2 5 12	5 6 14	4 6 11	7 8 10		2 7 13	6 8 9	8 12 13
4 9 13	0 0 1	0 0 1	0 5 9	3 5 12	5 7 14	5 6 11	7 8 11	0 7 11	3 7 13	7 8 9	8 13 13
4 9 14	0 4 15	0 5 7	========	4 5 12	5 8 14	6 6 11	7 8 12	========	4 7 13	8 8 9	8 13 14
4 9 15	1 4 15	1 5 7	0 0 1	5 5 12	5 9 14	6 7 11	7 8 13	0 0 1	5 7 13	8 9 9	8 13 15
	2 4 15	1 10 14	0 5 9	5 6 12	5 10 14	6 8 11	7 8 14	0 7 11	6 7 13	8 9 10	
0 4 11	3 4 15	2 5 7	1 5 9	5 7 12	5 11 14	6 9 11	7 8 15	1 7 11	7 7 13	8 9 11	0 8 15
========	4 4 15	3 5 7	2 5 9	5 8 12	5 12 14	6 10 11		2 7 11	7 8 13	8 9 12	========
0 0 1	4 5 15	3 10 14	3 5 9	5 9 12	5 13 14	6 11 11	0 7 9	3 7 11	7 9 13	8 9 13	0 0 1
0 4 11	4 6 15	4 5 7	4 5 9	5 10 12	5 14 14	6 11 12	========	4 7 11	7 10 13	8 9 14	0 8 15
1 4 11	4 7 15	5 5 7	5 5 9	5 11 12	5 14 15	6 11 13	0 0 1	5 7 11	7 11 13	8 9 15	1 8 15
2 4 11	4 8 15	5 6 7	5 6 9	5 12 12		6 11 14	0 7 9	6 7 11	7 12 13		2 8 15
3 4 11	4 9 15	5 7 7	5 7 9	5 12 13	0 6 7	6 11 15	1 7 9	7 7 11	7 13 13	0 8 11	3 8 15
4 4 11	4 10 15	5 7 8	5 8 9	5 12 14	========		2 7 9	7 8 11	7 13 14	========	4 8 15
4 5 11	4 11 15	5 7 9	5 9 9	5 12 15	0 0 1	0 6 13	3 7 9	7 9 11	7 13 15	0 0 1	5 8 15
4 6 11	4 12 15	5 7 10	5 9 10		0 6 7	========	4 7 9	7 10 11		0 8 11	6 8 15
4 7 11	4 13 15	5 7 11	5 9 11		1 6 7	0 0 1	5 7 9	7 11 11		1 8 11	7 8 15
4 8 11	4 14 15	5 7 12	5 9 12		1 12 14	0 6 13	6 7 9	7 11 12		2 8 11	8 8 15
4 9 11	4 15 15	5 7 13	5 9 13		2 6 7	1 6 13	7 7 9	7 11 13		3 8 11	8 9 15

TAB. I THE TYPES OF HKL-UVW FUR H,K,L,U,V,W ≤15

H K L	H K L	H K L	H K L	H K L	H K L	H K L	H K L	H K L	H K L	H K L	H K L
U V W	U V W	U V W	U V W	U V W	U V W	U V W	U V W	U V W	U V W	U V W	U V W
0 8 15		0 10 11	0 11 12		0 13 14	0 14 15		1 1 2	1 1 3	1 1 4	1 1 5
--------	0 9 13	--------	--------	0 11 15	--------	--------	1 1 2	--------	--------	--------	--------
6 10 15	========	4 10 11	10 11 12	========	4 13 14	10 14 15	========	3 7 13	1 8 11	1 4 8	2 3 7
8 11 15	0 0 1	5 10 11	11 11 12	0 0 1	5 13 14	11 14 15	0 1 1	3 8 13	1 9 12	1 4 15	2 3 13
8 12 15	0 9 13	6 10 11	11 12 12	0 11 15	6 13 14	12 14 15	0 1 2	3 8 14	1 10 13	1 5 9	2 5 5
8 13 15	1 9 13	7 10 11	11 12 13	1 11 15	7 13 14	13 14 15	1 1 1	4 5 6	1 11 14	1 6 10	2 5 15
8 14 15	2 9 13	8 10 11	11 12 14	2 11 15	8 13 14	14 14 15	1 1 3	4 5 13	1 12 15	1 7 11	3 4 11
8 15 15	3 9 13	9 10 11	11 12 15	3 11 15	9 13 14	14 15 15	1 2 3	4 5 14	2 3 3	1 8 12	3 5 10
	4 9 13	10 10 11		4 11 15	10 13 14		1 2 4	4 7 10	2 3 7	1 9 13	3 7 8
0 9 10	5 9 13	10 11 11	0 11 13	5 11 15	11 13 14	1 1 1	1 2 5	4 7 15	2 3 9	1 10 14	4 5 15
========	6 9 13	10 11 12	========	6 11 15	12 13 14	========	1 3 5	4 9 14	2 3 11	1 11 15	4 7 13
0 0 1	7 9 13	10 11 13	0 0 1	7 11 15	13 13 14	0 1 1	1 3 7	5 6 7	2 5 11	2 3 5	4 9 11
0 9 10	8 9 13	10 11 14	0 11 13	8 11 15	13 14 14	1 1 2	1 4 6	5 7 9	2 5 13	2 3 10	5 11 14
1 9 10	9 9 13	10 11 15	1 11 13	9 11 15	13 14 15	1 2 3	1 4 7	5 8 11	2 7 13	2 3 11	5 12 13
2 9 10	9 10 13		2 11 13	10 11 15		1 3 4	1 4 9	5 9 13	2 9 15	2 3 14	
3 9 10	9 11 13	0 10 13	3 11 13	11 11 15	0 13 15	1 4 5	1 5 7	6 7 8	3 4 5	2 5 13	1 1 6
4 9 10	9 12 13	========	4 11 13	11 12 15	========	1 5 6	1 5 9	7 8 9	3 4 9	2 7 15	========
5 9 10	9 13 13	0 0 1	5 11 13	11 13 15	0 0 1	1 6 7	1 5 11	7 9 11	3 4 13	3 4 8	0 1 1
6 9 10	9 13 14	0 10 13	6 11 13	11 14 15	0 13 15	1 7 8	1 6 8	7 10 13	3 4 15	3 4 13	0 1 6
7 9 10	9 13 15	1 10 13	7 11 13	11 15 15	1 13 15	1 8 9	1 6 11	7 11 15	3 5 12	3 5 7	1 1 5
8 9 10		2 10 13	8 11 13		2 13 15	1 9 10	1 6 13	8 9 10	3 5 14	4 5 11	1 1 7
9 9 10	0 9 14	3 10 13	9 11 13	0 12 13	3 13 15	1 10 11	1 7 9	8 11 14	4 5 7	4 7 9	1 2 4
9 10 10	========	4 10 13	10 11 13	========	4 13 15	1 11 12	1 7 13	9 10 11	4 5 11	5 6 14	1 2 8
9 10 11	0 0 1	5 10 13	11 11 13	0 0 1	5 13 15	1 12 13	1 7 15	9 11 13	5 6 9	5 7 13	1 2 11
9 10 12	0 9 14	6 10 13	11 12 13	0 12 13	6 13 15	1 13 14	1 8 10	10 11 12	5 6 13	5 8 12	1 2 13
9 10 13	1 9 14	7 10 13	11 13 13	1 12 13	7 13 15	1 14 15	1 8 15	11 12 13	5 7 8	5 9 11	1 3 3
9 10 14	2 9 14	8 10 13	11 13 14	2 12 13	8 13 15	2 3 5	1 9 11	11 13 15	6 7 11	6 11 13	1 3 9
9 10 15	3 9 14	9 10 13	11 13 15	3 12 13	9 13 15	2 5 7	1 10 12	12 13 14	6 7 15	7 13 15	1 4 10
	4 9 14	10 10 13		4 12 13	10 13 15	2 7 9	1 11 13	13 14 15	7 8 13		1 5 11
0 9 11	5 9 14	10 11 13	0 11 14	5 12 13	11 13 15	2 9 11	1 12 14		7 9 12	1 1 5	1 6 12
========	6 9 14	10 12 13	========	6 12 13	12 13 15	2 11 13	1 13 15	1 1 3	7 10 11	========	1 7 13
0 0 1	7 9 14	10 13 13	0 0 1	7 12 13	13 13 15	2 13 15	2 3 4	========	8 9 15	0 1 1	1 8 14
0 9 11	8 9 14	10 13 14	0 11 14	8 12 13	13 14 15	3 4 7	2 3 7	0 1 1	8 11 13	0 1 5	1 9 15
1 9 11	9 9 14	10 13 15	1 11 14	9 12 13	13 15 15	3 5 8	2 3 8	0 1 3	9 13 14	1 1 4	2 3 9
2 9 11	9 10 14		2 11 14	10 12 13		3 7 10	2 5 8	1 1 2		1 1 6	2 3 15
3 9 11	9 11 14	0 11 12	3 11 14	11 12 13	0 14 15	3 8 11	2 5 9	1 1 4	1 1 4	1 2 3	2 5 7
4 9 11	9 12 14	========	4 11 14	12 12 13	========	3 10 13	2 5 12	1 2 5	========	1 2 7	3 4 14
5 9 11	9 13 14	0 0 1	5 11 14	12 13 13	0 0 1	3 11 14	2 7 11	1 2 7	0 1 1	1 2 9	3 5 13
6 9 11	9 14 14	0 11 12	6 11 14	12 13 14	0 14 15	4 5 9	2 7 12	1 3 6	0 1 4	1 2 11	3 7 11
7 9 11	9 14 15	1 11 12	7 11 14	12 13 15	1 14 15	4 7 11	2 9 13	1 3 8	1 1 3	1 3 8	3 8 10
8 9 11		2 11 12	8 11 14		2 14 15	4 9 13	2 11 15	1 3 10	1 1 5	1 3 14	4 9 15
9 9 11	0 10 11	3 11 12	9 11 14	0 13 14	3 14 15	4 11 15	3 4 5	1 4 7	1 2 2	1 4 9	4 11 13
9 10 11	========	4 11 12	10 11 14	========	4 14 15	5 6 11	3 4 10	1 4 11	1 2 6	1 5 10	
9 11 11	0 0 1	5 11 12	11 11 14	0 0 1	5 14 15	5 7 12	3 4 11	1 4 13	1 2 7	1 6 11	1 1 7
9 11 12	0 10 11	6 11 12	11 12 14	0 13 14	6 14 15	5 8 13	3 5 7	1 5 8	1 2 9	1 7 12	========
9 11 13	1 10 11	7 11 12	11 13 14	1 13 14	7 14 15	5 9 14	3 5 11	1 5 14	1 3 7	1 8 13	0 1 1
9 11 14	2 10 11	8 11 12	11 14 14	2 13 14	8 14 15	6 7 13	3 5 13	1 6 9	1 3 11	1 9 14	0 1 7
9 11 15	3 10 11	9 11 12	11 14 15	3 13 14	9 14 15	7 8 15	3 7 11	1 7 10	1 3 13	1 10 15	1 1 6

TAB. I　　THE TYPES OF HKL-UVW FOR H,K,L,U,V,W ≤15　　I - 6

H K L	H K L	H K L	H K L	H K L	H K L	H K L	H K L	H K L	H K L	H K L	H K L
U V W	U V W	U V W	U V W	U V W	U V W	U V W	U V W	U V W	U V W	U V W	U V W
1 1 7	1 1 9	1 1 11	1 1 14	1 2 2	1 2 3	1 2 3	1 2 4	1 2 4	1 2 5	1 2 6	1 2 7
1 1 8	1 1 8	1 5 6	1 6 8	2 10 11	1 2 7	4 6 9	1 2 8	5 6 8	2 4 9	1 8 10	1 3 5
1 2 5	1 1 10	2 7 15	1 7 7	2 11 12	1 2 8	4 6 11	1 2 10	5 6 13	2 4 11	1 9 12	1 3 10
1 2 9	1 2 7	2 9 13	2 13 15	2 12 13	1 3 3	4 7 13	1 3 10	5 6 14	2 8 9	1 10 14	1 3 11
1 2 13	1 2 11	2 11 11		2 13 14	1 3 4	4 10 11	1 3 14	5 8 14	2 11 12	2 2 5	1 3 13
1 2 15	1 3 6		1 1 15	2 14 15	1 3 5	4 10 13	1 4 4	6 7 10	3 4 7	2 2 7	1 4 15
1 3 4	1 3 12	1 1 12	========	3 4 5	1 3 7	4 13 14	1 4 12	6 7 11	3 4 14	2 3 6	1 5 6
1 3 10	1 4 5	========	0 1 1	3 4 14	1 3 9	5 6 8	1 4 14	6 8 13	3 5 5	2 3 8	1 7 7
1 4 11	1 4 13	0 1 1	0 1 15	3 7 8	1 3 11	5 7 8	1 5 6	7 8 10	3 5 10	2 3 10	1 8 9
1 5 12	1 5 14	0 1 12	1 1 14	3 8 10	1 4 5	5 7 11	1 5 14	7 6 12	3 5 11	2 3 14	1 9 11
1 6 13	1 6 15	1 1 11	1 2 13	3 10 14	1 4 10	5 7 13	1 6 8	7 9 10	3 5 14	2 4 11	1 10 13
1 7 14	2 3 15	1 1 13	1 3 12	4 5 7	1 4 11	5 8 14	1 7 10	8 9 14	3 7 11	2 4 13	1 11 15
1 8 15	2 5 13	1 2 10	1 4 11	4 6 7	1 4 14	5 9 11	1 8 12	8 10 11	3 11 13	2 5 14	2 3 8
2 3 11	2 7 11	1 2 14	1 5 10	4 7 9	1 5 7	5 9 12	1 9 14	9 10 13	3 13 14	2 6 9	2 4 5
2 5 9	2 9 9	1 3 9	1 6 9	4 9 10	1 5 8	5 11 13	2 2 3	9 11 14	4 5 10	2 10 11	2 4 9
2 7 7	3 13 14	1 3 15	1 7 8	4 9 11	1 5 13	5 11 14	2 2 5	10 13 14	4 6 7	2 13 14	2 4 13
3 7 14		1 4 8	2 15 15	4 11 13	1 6 9	5 13 14	2 3 5	11 14 15	4 6 13	3 4 7	2 4 15
3 8 13	1 1 10	1 5 7		4 11 14	1 6 15	6 7 9	2 3 7		4 10 15	3 4 10	2 8 11
3 10 11	========	1 6 6	1 2 2	4 13 15	1 7 10	6 8 9	2 3 8	1 2 5	5 6 13	3 4 11	2 12 13
4 13 15	0 1 1	2 9 15	========	5 6 8	1 7 11	6 8 13	2 3 14	========	5 7 9	3 5 8	3 4 13
	0 1 10	2 11 13	0 1 1	5 8 9	1 8 13	6 8 15	2 4 7	0 1 2	5 7 11	3 6 13	3 5 8
1 1 8	1 1 9		0 1 2	5 11 12	1 9 13	7 8 10	2 4 9	0 1 5	5 7 15	3 10 14	3 5 11
========	1 1 11	1 1 13	1 1 4	5 12 14	1 9 14	7 9 10	2 5 9	0 2 5	5 8 9	4 5 13	3 5 13
0 1 1	1 2 8	========	1 2 2	6 7 10	1 9 15	7 9 13	2 5 11	1 1 2	6 8 11	4 5 14	3 7 7
0 1 8	1 2 12	0 1 1	1 2 6	6 8 11	2 3 5	7 9 15	2 6 7	1 1 3	7 9 13	4 6 9	3 7 14
1 1 7	1 3 7	0 1 13	1 3 4	6 10 11	2 3 12	7 11 13	2 6 11	1 1 7	7 10 15	4 6 15	4 6 11
1 1 9	1 3 13	1 1 12	1 3 8	6 10 13	2 3 13	8 9 11	2 6 13	1 2 8	7 11 12	4 7 10	4 7 14
1 2 6	1 4 6	1 1 14	1 4 6	6 11 14	2 4 5	8 10 11	2 7 13	1 2 9	7 11 13	5 6 12	4 9 10
1 2 10	1 4 14	1 2 11	1 4 10	6 13 14	2 4 7	9 10 12	2 7 15	1 2 12	8 10 15	5 8 11	5 7 14
1 2 15	1 5 5	1 2 15	1 5 8	7 8 11	2 5 11	9 11 12	2 8 15	1 3 4	8 13 14	5 6 14	5 9 13
1 3 5	1 5 15	1 3 10	1 5 12	7 10 12	2 7 8	9 11 15	2 9 10	1 3 7		5 9 12	5 11 12
1 3 11	2 5 15	1 4 9	1 6 10	7 12 13	2 8 11	9 13 15	2 11 14	1 3 8	1 2 6	6 10 13	5 11 13
1 4 4	2 7 13	1 5 8	1 6 14	6 9 13	2 8 13	10 11 13	3 4 4	1 3 11	========	6 11 14	
1 4 12	2 9 11	1 6 7	1 7 12	8 10 13	2 9 12	10 12 13	3 4 8	1 3 13	0 1 2	7 12 15	1 2 8
1 5 13		2 11 15	1 8 14	8 11 15	3 4 6	11 12 14	3 4 10	1 4 13	0 1 3		========
1 6 14	1 1 11	2 13 13	2 2 3	6 14 15	3 5 6	11 13 14	3 5 14	1 5 5	0 1 6	1 2 7	0 1 2
1 7 15	========		2 3 4	9 10 14	3 5 7	12 13 15	3 6 10	1 5 12	1 1 4	========	0 1 4
2 3 13	0 1 1	1 1 14	2 3 10		3 5 9	12 14 15	3 10 11	1 5 13	1 1 8	0 1 2	0 1 6
2 5 11	0 1 11	========	2 4 5	1 2 3	3 7 8		3 13 14	1 5 15	1 2 2	0 1 7	1 1 6
2 7 9	1 1 10	0 1 1	2 5 6	========	3 7 9	1 2 4	4 5 6	1 6 7	1 2 4	0 2 7	1 1 10
3 10 14	1 1 12	0 1 14	2 5 14	0 1 2	3 7 12	========	4 5 6	1 7 9	1 2 10	1 1 3	1 2 3
3 11 13	1 2 9	1 1 13	2 6 7	0 1 3	3 7 15	0 1 2	4 5 12	1 8 11	1 2 14	1 1 4	1 2 4
	1 2 13	1 1 15	2 7 8	0 2 3	3 10 11	0 1 4	4 6 11	1 9 13	1 3 12	1 1 5	1 2 5
1 1 9	1 3 8	1 2 12	2 7 10	1 1 1	3 11 13	1 1 2	4 7 12	1 10 15	1 4 5	1 1 9	1 2 12
========	1 3 14	1 3 11	2 8 9	1 1 2	3 11 15	1 1 6	4 10 13	2 3 4	1 4 14	1 2 3	1 2 14
0 1 1	1 4 7	1 4 10	2 9 10	1 1 5	4 5 7	1 2 3	4 14 15	2 3 11	1 6 6	1 2 11	1 3 14
0 1 9	1 4 15	1 5 9	2 9 14	1 2 4	4 6 7	1 2 6	5 6 7	2 4 7	1 7 8	1 2 12	1 4 6

TAB. I

THE TYPES OF HKL-UVW FOR H,K,L,U,V,W ≤15

I - 7

H	K	L
1	2	8
1	6	7
1	8	8
1	9	10
1	10	12
1	11	14
2	2	7
2	2	9
2	3	10
2	3	11
2	3	13
2	4	15
2	5	6
2	6	11
2	10	13
2	14	15
3	4	10
3	4	14
3	5	14
3	7	10
3	8	8
4	6	13
4	9	14
4	10	11
5	12	14
5	13	14

H	K	L
1	2	9
0	1	2
0	1	9
0	2	9
1	1	4
1	1	5
1	1	7
1	1	11
1	2	5
1	2	13
1	3	3
1	3	6
1	3	13
1	3	14
1	3	15
1	5	7
1	7	8
1	9	9
1	10	11
1	11	13
1	12	15
2	3	12
2	4	7
2	4	11
2	5	8
2	8	13
2	12	15
3	5	11
3	7	10
3	7	13
3	8	11
4	6	15
4	10	13
4	11	14

H	K	L
1	2	10
0	1	2
0	1	5
0	1	10
1	1	8
1	1	12
1	2	4
1	2	6
1	2	14
1	3	4
1	4	7
1	6	8
1	8	9
1	10	10
1	11	12
1	12	14
2	2	9
2	2	11
2	3	14
2	5	10
2	6	7
2	6	13
2	10	15
3	4	13
3	8	11
3	8	14
3	10	10
4	10	15
4	13	14

H	K	L
1	2	11
0	1	2
0	1	11
0	2	11
1	1	5
1	1	6
1	1	9
1	1	13
1	2	7
1	2	15
1	3	4
1	3	5
1	3	7
1	5	8
1	7	9
1	9	10
1	11	11
1	12	13
1	13	15
2	4	9
2	4	13
2	5	12
2	7	8
2	8	15
3	5	14
3	7	13
3	10	13
3	11	11
4	14	15

H	K	L
1	2	12
0	1	2
0	1	6
0	1	12
1	1	10
1	1	14
1	2	5
1	2	7
1	2	8
1	3	6
1	4	4
1	4	8
1	6	9
1	8	10
1	10	11
1	12	12
1	13	14
2	2	11
2	2	13
2	5	14
2	6	9
2	6	15
2	7	10
3	8	14
3	10	13
3	11	14

H	K	L
1	2	13
0	1	2
0	1	13
0	2	13
1	1	6
1	1	7
1	1	11
1	1	15
1	2	9
1	3	5
1	3	7
1	3	8
1	4	5
1	5	9
1	7	10
1	9	11
1	11	12
1	13	13
1	14	15
2	4	11
2	4	15
2	7	12
2	8	9
3	11	14
3	13	13

H	K	L
1	2	14
0	1	2
0	1	7
0	1	14
1	1	12
1	2	6
1	2	8
1	2	10
1	3	8
1	4	5
1	4	6
1	4	9
1	6	10
1	8	11
1	10	12
1	12	13
1	14	14
2	2	13
2	2	15
2	6	11
2	7	14
2	9	10
3	14	14

H	K	L
1	2	15
0	1	2
0	1	15
0	2	15
1	1	7
1	1	8
1	1	13
1	2	11
1	3	6
1	3	9
1	4	7
1	5	5
1	5	10
1	7	11
1	9	12
1	11	13
1	13	14
1	15	15
2	4	13
2	8	11
2	9	12

H	K	L
1	3	3
0	1	1
0	1	3
1	1	6
1	2	3
1	2	9
1	3	6
1	3	12
1	4	9
1	4	15
1	5	12
1	6	15
2	3	3
2	3	15
2	5	9
2	7	15
3	3	4
3	4	5
3	5	6
3	6	7
3	7	8
3	7	12
3	8	9
3	8	15
3	9	10
3	10	11
3	11	12
3	12	13
3	13	14
3	14	15
4	7	9
4	9	15
5	6	7
5	8	9
5	9	12
6	7	9
6	9	11
6	11	13
6	11	15
6	13	15
7	9	10
7	11	12
7	12	15
8	9	11
8	13	15
9	10	13
9	11	14
9	12	13
9	14	15
11	12	15

H	K	L
1	3	4
0	1	3
0	1	4
0	3	4
1	1	1
1	1	7
1	2	2
1	2	3
1	2	5
1	2	10
1	2	11
1	3	5
1	3	9
1	3	13
1	3	15
1	4	5
1	4	8
1	4	13
1	5	7
1	5	11
1	6	14
1	7	9
1	8	11
1	10	13
1	11	15
2	3	6
2	5	6
2	5	7
2	5	14
2	7	10
2	7	13
2	11	14
3	3	5
3	4	7
3	5	11
3	6	7
3	7	9
3	8	12
3	9	11
3	9	13
4	5	7
4	5	8
4	7	8
4	9	11
5	6	9
5	7	9
5	7	11
5	7	13
5	8	9
5	11	13
6	7	10
6	9	10
6	9	11
6	9	14
6	13	15
7	8	11
7	8	12
7	8	13
7	9	15
7	10	11
7	11	13
7	13	15
8	9	12
8	11	12
8	13	15
9	10	13
9	12	13
10	11	14
10	13	14
11	12	15
11	14	15

H	K	L
1	3	5
0	1	3
0	1	5
0	3	5
1	1	2
1	1	8
1	2	3
1	2	7
1	2	11
1	2	13
1	3	4
1	3	12
1	3	14
1	4	7
1	5	8
1	5	10
1	6	13
1	7	12
1	8	13
2	3	9
2	5	5
2	5	9
2	7	11
3	3	4
3	4	11
3	6	7
3	6	11
3	8	9
3	9	14
3	10	15
4	5	5
4	5	7
4	5	13
4	7	9
4	7	13
4	11	13
5	6	7
5	6	15
5	7	10
5	8	11
5	8	15
5	11	12
5	13	14
6	7	9
6	9	13
7	8	9
7	8	11
7	10	15
8	9	13
9	10	15
9	11	12
10	11	13
10	11	15
11	13	14

H	K	L
1	3	6
0	1	2
0	1	3
0	1	6
1	1	3
1	1	9
1	2	9
1	2	12
1	2	15
1	3	3
1	3	15
1	4	6
1	5	9
1	6	12
1	7	15
2	3	3
2	3	5
2	3	12
2	7	9
2	9	15
3	3	5
3	3	7
3	4	6
3	4	7
3	4	9
3	4	15
3	5	9
3	5	11
3	6	8
3	6	11
3	6	13
3	7	13
3	7	15
3	8	15
3	10	12
3	11	15
4	5	9
4	9	11
4	13	15
5	6	8
5	6	12
5	7	9
5	9	13
5	12	14
6	7	12
6	7	15
7	9	11
7	9	15
7	10	12
8	9	13
8	11	15
9	12	14
9	13	15

H	K	L
1	3	7
0	1	3
0	1	7
0	3	7
1	1	2
1	1	4
1	1	10
1	2	3
1	2	5
1	2	11
1	2	13
1	4	5
1	4	9
1	5	8
1	5	12
1	6	11
1	7	14
2	3	5
2	3	15
2	5	11
2	7	7
2	9	13
3	3	8
3	4	9
3	5	6
3	6	13
3	9	10
3	11	12
4	5	11
4	5	13
4	7	7
4	7	15
4	11	13
5	7	14
5	8	9
5	8	11
6	9	11
7	10	13
7	12	13
8	11	15

H	K	L
1	3	8
0	1	3
0	1	8
0	3	8
1	1	3
1	1	5
1	1	11
1	2	2
1	2	5
1	2	13
1	2	14
1	4	4
1	4	11
1	5	7
1	5	13
1	6	10
1	7	13
2	3	7
2	5	7
2	5	14
2	9	11
3	3	7
3	4	12
3	5	9
3	6	10
3	9	11
3	13	15
4	5	9
4	5	12
4	7	11
4	7	13
4	13	15
5	7	11
5	9	13
6	9	13
6	11	15
7	11	15

H	K	L
1	3	9
0	1	3
0	1	9
1	1	6
1	1	12
1	2	3
1	2	15
1	3	4
1	5	6
1	6	9
1	7	12

TAB. I THE TYPES OF HKL-UVW FOR H,K,L,U,V,W ≤15 I - 8

H U	K V	L W
1	**3**	**9**
1	8	15
2	3	5
2	3	7
2	3	9
2	9	9
2	11	15
3	3	8
3	3	10
3	4	11
3	4	13
3	4	15
3	5	12
3	5	14
3	6	7
3	6	11
3	12	13
3	14	15
4	7	15
4	9	9
4	9	15
5	6	13
5	9	12
5	11	12
6	13	15
1	**3**	**10**
0	1	3
0	1	10
0	3	10
1	1	3
1	1	7
1	1	13
1	2	4
1	2	7
1	4	13
1	5	5
1	6	8
1	7	11
1	8	14
2	3	11
2	5	5
2	7	9
2	11	13
3	3	11
3	5	15

H U	K V	L W
1	**3**	**10**
3	6	8
3	7	9
3	9	13
3	12	14
4	5	15
4	7	11
4	9	13
5	8	14
5	11	13
5	12	14
1	**3**	**11**
0	1	3
0	1	11
0	3	11
1	1	4
1	1	8
1	1	14
1	2	3
1	2	5
1	2	7
1	4	5
1	4	15
1	6	7
1	7	10
1	8	13
2	3	13
2	5	7
2	5	9
2	11	11
3	3	10
3	6	13
3	7	12
3	8	9
3	9	14
4	5	13
4	11	11
5	13	14
1	**3**	**12**
0	1	3
0	1	4
0	1	12
1	1	9

H U	K V	L W
1	**3**	**12**
1	1	15
1	2	6
1	3	3
1	3	5
1	6	6
1	7	9
1	8	12
1	9	15
2	3	7
2	3	9
2	3	15
2	5	9
2	9	11
2	13	15
3	3	11
3	3	13
3	4	15
3	6	10
3	6	14
3	7	15
3	8	12
4	9	13
4	11	15
1	**3**	**13**
0	1	3
0	1	13
0	3	13
1	1	4
1	1	10
1	2	5
1	2	7
1	2	9
1	3	4
1	5	6
1	7	8
1	8	11
1	9	14
2	5	7
2	5	11
2	7	11
2	13	13
3	3	14
3	6	11
3	8	15

H U	K V	L W
1	**3**	**13**
3	9	10
4	7	15
4	13	13
1	**3**	**14**
0	1	3
0	1	14
0	3	14
1	1	5
1	1	11
1	2	4
1	2	8
1	2	9
1	3	5
1	4	6
1	7	7
1	8	10
1	9	13
2	5	11
2	5	13
2	7	7
2	11	13
3	3	13
3	9	11
3	10	12
4	11	15
1	**3**	**15**
0	1	3
0	1	5
0	1	15
1	1	12
1	2	9
1	3	4
1	3	6
1	6	7
1	8	9
1	9	12
1	10	15
2	3	9
2	3	11
2	5	15
2	7	9
2	9	13

H U	K V	L W
1	**3**	**15**
2	15	15
3	3	14
3	6	13
3	10	15
3	11	12
4	15	15
1	**4**	**4**
0	1	1
0	1	4
1	1	8
1	2	4
1	2	12
1	3	8
1	4	12
2	3	4
2	5	12
3	4	4
3	5	8
4	4	5
4	5	6
4	6	7
4	7	8
4	7	12
4	8	9
4	9	10
4	10	11
4	11	12
4	12	13
4	13	14
4	14	15
5	7	8
5	8	12
7	8	9
7	10	12
8	9	11
8	11	12
8	11	13
8	13	15
10	12	13
11	12	14

H U	K V	L W
1	**4**	**5**
0	1	4
0	1	5
0	4	5
1	1	1
1	1	9
1	2	3
1	2	6
1	2	13
1	2	14
1	3	4
1	3	7
1	3	11
1	4	11
1	5	6
1	5	15
1	7	9
1	9	11
1	11	14
2	2	3
2	3	7
2	5	10
2	6	7
2	10	13
3	4	8
3	5	5
3	5	7
3	5	13
3	7	8
3	7	13
3	11	13
4	5	9
4	7	8
4	8	9
4	8	11
5	6	10
5	7	8
5	7	10
5	7	15
5	9	10
5	9	11
5	11	15
5	13	15
6	7	11
6	10	11
6	11	14

H U	K V	L W
1	**4**	**5**
7	8	12
7	9	11
7	9	13
7	11	12
7	12	13
8	9	13
8	12	13
9	10	14
9	11	14
9	13	14
10	11	15
10	14	15
1	**4**	**6**
0	1	4
0	1	6
0	2	3
1	1	2
1	1	10
1	2	2
1	2	8
1	2	14
1	3	6
1	3	14
1	4	10
1	5	14
2	3	4
2	3	5
2	3	10
2	5	7
2	5	8
2	7	10
2	7	11
2	9	13
2	9	14
3	4	12
3	7	10
3	8	14
4	4	5
4	4	7
4	5	14
4	8	11
4	8	13
4	9	12
5	6	6

H U	K V	L W
1	**4**	**6**
5	6	9
5	11	14
6	7	8
6	7	9
6	7	12
6	8	11
6	11	15
7	8	10
7	10	13
8	9	12
8	10	13
8	12	15
9	10	11
9	10	14
10	11	14
11	13	14
1	**4**	**7**
0	1	4
0	1	7
0	4	7
1	1	2
1	1	3
1	1	11
1	2	10
1	2	15
1	3	5
1	4	9
1	5	9
1	5	13
1	7	12
2	2	3
2	3	13
2	5	6
2	6	11
2	7	14
3	4	5
3	5	8
3	7	7
3	7	13
3	8	11
4	5	8
4	8	9
4	8	13
4	8	15

H U	K V	L W
1	**4**	**7**
5	6	11
5	7	7
5	7	11
5	9	11
5	12	13
6	7	14
6	10	13
7	9	10
7	9	13
7	9	14
7	11	15
8	11	12
9	11	13
1	**4**	**8**
0	1	2
0	1	4
0	1	8
1	1	4
1	1	12
1	2	12
1	3	4
1	4	8
1	5	12
2	3	4
2	4	5
2	7	12
3	4	5
3	4	7
3	4	8
3	8	8
4	4	7
4	4	9
4	5	9
4	5	11
4	5	12
4	6	11
4	6	13
4	7	13
4	7	15
4	8	15
4	11	12
5	7	12
5	8	8
5	8	12

H U	K V	L W
1	**4**	**8**
5	12	13
7	8	12
7	11	12
8	12	13
1	**4**	**9**
0	1	4
0	1	9
0	4	9
1	1	2
1	1	5
1	1	13
1	2	14
1	3	3
1	3	7
1	4	7
1	5	11
1	6	15
2	2	5
2	3	6
2	6	13
2	7	10
3	4	11
3	5	7
3	5	8
3	5	12
3	7	15
3	10	13
4	7	8
4	8	11
5	7	13
5	8	13
5	9	9
5	11	14
6	10	11
7	11	13
7	12	15
8	12	15
1	**4**	**10**
0	1	4
0	1	10
0	2	5
1	1	6

H U	K V	L W
1	**4**	**10**
1	1	14
1	2	2
1	2	3
1	4	6
1	5	10
1	6	14
2	3	7
2	3	8
2	5	12
2	5	13
2	7	8
3	4	14
3	5	10
3	10	10
3	11	14
4	4	9
4	4	11
4	7	12
4	12	13
5	6	11
5	6	14
5	9	14
6	8	13
7	10	15
1	**4**	**11**
0	1	4
0	1	11
0	4	11
1	1	3
1	1	7
1	1	15
1	2	3
1	3	8
1	4	5
1	5	9
1	5	14
1	6	13
2	2	5
2	3	10
2	6	7
2	9	14
3	5	7
3	5	13
3	7	10

H U	K V	L W
1	**4**	**11**
3	11	11
4	8	9
4	8	13
5	7	12
5	11	11
6	13	14
1	**4**	**12**
0	1	3
0	1	4
0	1	12
1	1	6
1	2	4
1	4	4
1	5	6
1	6	12
2	3	12
2	4	5
2	4	7
2	9	12
3	4	6
3	4	10
3	7	8
3	8	11
4	4	11
4	4	13
4	5	14
4	9	12
4	12	15
5	8	13
5	12	12
1	**4**	**13**
0	1	4
0	1	13
0	4	13
1	1	3
1	1	9
1	2	5
1	3	4
1	3	10
1	5	7
1	6	11
1	7	15

TAB. I

THE TYPES OF HKL-UVW FOR H,K,L,U,V,W ≤15

H / U	K / V	L / W
1	4	13
2	2	7
2	3	14
2	5	6
2	9	10
3	5	11
3	7	8
3	7	11
3	13	13
4	8	11
4	8	15
5	9	14
5	13	13
1	4	14
========		
0	1	4
0	1	14
0	2	7
1	1	10
1	2	3
1	2	4
1	2	6
1	5	6
1	6	10
1	7	14
2	3	10
2	3	11
2	5	8
2	8	9
3	7	14
3	8	10
3	10	13
3	14	14
4	4	13
4	4	15
4	11	12
5	14	14
1	4	15
========		
0	1	4
0	1	15
0	4	15
1	1	4
1	1	11
1	2	7

H / U	K / V	L / W
1	4	15
1	3	3
1	3	11
1	5	5
1	6	9
1	7	13
2	2	7
2	5	10
2	6	9
2	11	14
3	5	10
3	7	13
3	8	13
3	11	14
4	8	13
1	5	5
========		
0	1	1
0	1	5
1	1	10
1	2	5
1	2	15
1	3	10
1	4	15
2	3	5
2	5	15
3	4	5
3	5	10
4	5	5
4	7	15
5	5	6
5	6	7
5	7	8
5	7	10
5	8	9
5	8	15
5	9	10
5	10	11
5	11	12
5	12	13
5	13	14
5	14	15
7	9	10
7	10	15
8	11	15
9	10	11

H / U	K / V	L / W
1	5	5
10	11	13
10	13	15
11	14	15
1	5	6
========		
0	1	5
0	1	6
0	5	6
1	1	1
1	1	11
1	2	4
1	2	7
1	3	9
1	3	13
1	4	5
1	4	14
1	6	7
1	9	11
1	11	13
2	3	3
2	3	4
2	3	8
2	5	13
2	7	8
3	4	9
3	5	7
3	5	15
3	7	9
3	8	9
3	13	15
4	5	10
4	7	11
4	9	10
4	11	14
5	5	7
5	6	11
5	8	10
5	9	15
5	10	11
5	10	13
6	7	12
6	11	12
7	8	10
7	8	11
7	8	13

H / U	K / V	L / W
1	5	6
7	11	13
7	12	13
8	9	14
8	13	14
9	10	15
9	11	13
9	11	15
9	14	15
1	5	7
========		
0	1	5
0	1	7
0	5	7
1	1	2
1	1	12
1	2	3
1	2	9
1	3	4
1	3	8
1	4	13
1	7	10
1	8	11
2	3	11
2	5	11
2	9	13
2	11	15
3	4	5
3	4	13
3	7	14
4	5	15
4	7	7
4	7	9
5	5	6
5	5	8
5	9	10
5	10	13
6	7	7
6	7	11
6	11	13
7	8	9
7	9	14
7	11	12
7	11	14
8	9	11
8	9	13

H / U	K / V	L / W
1	5	7
9	10	13
10	11	15
11	12	13
1	5	8
========		
0	1	5
0	1	8
0	5	8
1	1	3
1	1	13
1	2	2
1	2	3
1	2	11
1	3	5
1	3	7
1	4	12
1	7	11
1	8	13
2	3	14
2	5	9
2	9	14
3	4	4
3	4	7
3	7	11
4	5	7
4	7	12
4	9	13
5	5	7
5	5	9
5	6	10
5	11	15
6	7	10
6	7	11
6	7	13
7	9	11
7	9	13
8	9	11
8	11	15
10	13	15
11	13	15
1	5	9
========		
0	1	5
0	1	9

H / U	K / V	L / W
1	5	9
0	5	9
1	1	2
1	1	4
1	1	14
1	2	13
1	3	6
1	4	7
1	4	11
1	6	11
2	3	3
2	3	5
2	5	7
2	7	13
3	4	7
3	7	8
3	7	12
3	8	13
3	8	15
4	5	11
4	9	9
5	5	8
5	6	15
5	7	10
5	10	11
5	12	15
6	11	13
7	8	11
7	8	13
7	10	13
7	12	15
9	11	14
1	5	10
========		
0	1	2
0	1	5
0	1	10
1	1	5
1	1	15
1	2	15
1	3	5
1	4	10
1	5	15
2	3	5
2	5	5
2	7	15

H / U	K / V	L / W
1	5	10
3	4	10
3	5	5
3	5	7
3	8	10
4	5	7
4	5	9
4	5	15
4	11	15
5	5	9
5	5	11
5	6	11
5	6	13
5	7	13
5	7	15
5	8	10
5	8	15
5	10	12
5	13	15
7	10	12
7	11	15
8	13	15
1	5	11
========		
0	1	5
0	1	11
0	5	11
1	1	2
1	1	6
1	3	4
1	4	9
1	5	14
1	6	13
2	3	5
2	3	7
2	7	13
2	7	15
3	4	13
3	5	8
3	7	8
4	7	9
4	11	11
5	5	12
5	8	15
5	9	10
5	10	13

H / U	K / V	L / W
1	5	11
5	14	15
6	11	11
7	12	13
8	13	15
1	5	12
========		
0	1	5
0	1	12
0	5	12
1	1	7
1	2	2
1	2	5
1	3	3
1	3	7
1	4	8
1	5	13
2	3	9
2	7	11
3	4	8
3	4	9
3	5	11
3	10	14
4	7	11
4	7	13
5	5	11
5	5	13
5	9	15
5	10	14
6	7	13
7	9	15
1	5	13
========		
0	1	5
0	1	13
0	5	13
1	1	8
1	2	3
1	2	5
1	3	8
1	4	7
1	5	12
2	3	11
2	7	9
3	4	7

H / U	K / V	L / W
1	5	13
3	4	11
3	5	14
3	10	11
4	7	9
4	13	13
5	5	12
5	5	14
5	10	11
6	13	13
1	5	14
========		
0	1	5
0	1	14
0	5	14
1	1	3
1	1	9
1	2	4
1	4	6
1	4	11
1	5	11
2	3	5
2	3	8
2	3	13
2	7	7
3	7	7
3	8	10
3	11	13
4	9	11
4	9	13
5	5	13
5	10	12
5	11	15
1	5	15
========		
0	1	3
0	1	5
0	1	15
1	1	10
1	2	5
1	4	5
1	5	10
1	6	15
2	3	15
2	5	5

H / U	K / V	L / W
1	5	15
2	5	7
2	9	15
3	5	8
3	5	10
3	7	10
3	10	11
4	5	11
4	5	13
4	9	15
4	15	15
5	5	14
5	10	13
5	12	15
1	6	6
========		
0	1	1
0	1	6
1	1	12
1	2	6
1	3	12
2	3	6
3	4	6
3	5	12
4	5	6
5	6	6
5	7	12
6	6	7
6	7	8
6	8	9
6	9	10
6	10	11
6	11	12
6	12	13
6	13	14
6	14	15
7	9	12
9	11	12
11	12	13
12	13	15
1	6	7
========		
0	1	6
0	1	7
0	6	7

H / U	K / V	L / W
1	6	7
1	1	1
1	1	13
1	2	5
1	2	8
1	3	11
1	3	15
1	5	6
1	7	8
1	11	13
1	13	15
2	3	4
2	3	9
2	4	5
2	8	9
3	3	4
3	4	10
3	5	9
3	9	10
3	9	11
4	5	11
4	7	14
4	10	11
5	6	12
5	7	7
5	7	9
5	8	13
5	11	12
6	7	13
6	12	13
7	8	14
7	10	11
7	11	14
7	13	14
8	9	15
8	10	11
8	10	13
8	14	15
9	13	15
11	13	15
1	6	8
========		
0	1	6
0	1	8
0	3	4
1	1	2

H / U	K / V	L / W
1	6	8
1	1	14
1	2	4
1	2	10
1	3	10
2	2	3
2	3	12
2	4	5
2	5	7
2	5	14
2	7	9
2	8	11
2	10	13
2	11	15
3	4	14
3	5	6
4	5	6
4	7	10
4	11	14
5	8	8
5	9	14
6	6	7
6	9	11
6	9	13
7	8	8
7	10	11
8	9	10
8	13	14
9	10	12
9	12	14
10	11	13
10	11	14
13	14	15
1	6	9
========		
0	1	6
0	1	9
0	2	3
1	1	3
1	1	15
1	2	3
1	2	12
1	3	9
1	4	15
2	3	15
2	5	12

TAB. I

THE TYPES OF HKL-UVW FOR H,K,L,U,V,W ≤15

H K L / U V W	H K L / U V W	H K L / U V W	H K L / U V W	H K L / U V W	H K L / U V W	H K L / U V W	H K L / U V W	H K L / U V W	H K L / U V W	H K L / U V W	H K L / U V W
1 6 9	1 6 10	1 6 12	1 6 13	1 6 15	1 7 8	1 7 9	1 7 10	1 7 12	1 7 14	1 8 8	1 8 9
3 3 4	4 9 14	0 1 12	6 11 12	4 6 9	1 13 15	4 7 13	11 12 14	4 5 13	2 5 7	2 3 8	10 13 14
3 3 5	5 6 14	1 1 6	7 13 13	4 6 11	2 3 5	4 9 11		4 9 15	3 4 14	3 4 8	11 13 14
3 5 7	5 7 8	1 3 6	8 14 15	5 9 11	2 3 10	5 6 7	1 7 11	5 6 11	3 5 7	4 5 8	
3 5 8	6 6 11	1 4 12		5 9 14	2 5 6	5 8 11	========	5 7 11	3 7 7	5 6 8	1 8 10
3 7 10	6 9 14	2 3 6	1 6 14	6 12 13	2 9 10	6 11 15	0 1 7	5 8 13	3 8 14	6 7 8	========
3 7 11	7 8 13	2 5 6	========	7 15 15	3 4 4	7 7 8	0 1 11	7 7 11	4 7 7	7 8 8	0 1 8
3 7 15	7 10 10	3 4 12	0 1 6		3 4 5	7 7 10	0 7 11	7 7 13	4 7 9	8 8 9	0 1 10
3 9 13	8 10 15	3 5 6	0 1 14	1 7 7	3 4 11	7 11 14	1 1 4	7 10 14	5 7 9	8 9 10	0 4 5
3 9 14	8 11 14	3 6 7	0 3 7	========	3 5 11	8 9 9	1 2 3	8 9 15	5 7 11	8 10 11	1 1 2
4 5 6	9 12 13	3 8 12	1 1 8	0 1 1	3 10 11	8 11 13	1 2 15	9 10 14	5 8 14	8 11 12	1 2 6
4 6 7		4 6 7	1 2 2	0 1 7	3 11 13	9 10 11	1 3 10		5 12 14	8 12 13	1 2 12
5 6 9	1 6 11	4 6 9	1 3 4	1 1 14	4 5 12	9 10 13	1 5 8	1 7 13	6 7 11	8 13 14	1 3 14
5 9 9	========	5 6 9	1 4 10	1 2 7	4 11 12	10 11 13	1 9 14	========	6 7 13	8 14 15	2 3 4
6 7 12	0 1 6	5 6 11	2 2 5	1 3 14	5 6 13	11 12 15	2 3 5	0 1 7	7 7 13		2 3 14
6 8 11	0 1 11	5 8 12	2 3 10	2 3 7	5 7 9	12 13 15	2 5 13	0 1 13	7 7 15	1 8 9	2 5 6
6 8 13	0 6 11	5 12 12	2 4 9	3 4 7	5 9 11		3 4 5	0 7 13	7 8 15	========	2 7 9
6 11 12	1 1 2	6 6 11	2 5 12	3 5 14	5 12 13	1 7 10	3 8 13	1 1 2	7 12 14	0 1 8	2 9 11
7 8 15	1 1 5	6 6 13	2 7 14	4 5 7	6 7 14	========	4 5 7	1 1 6		0 1 9	2 11 14
7 9 9	1 3 7	6 7 13	3 5 12	5 6 7	6 13 14	0 1 7	4 5 9	1 3 8	1 7 15	0 8 9	3 5 10
7 9 12	1 4 13	6 7 15	3 6 8	5 7 14	7 7 9	0 1 10	5 6 7	1 4 15	========	1 1 1	4 6 7
7 13 15	1 5 9	6 8 15	4 5 11	6 7 7	7 8 15	0 7 10	5 6 13	1 6 11	0 1 7	1 2 7	4 10 13
9 11 15	1 7 13	7 12 12	4 7 14	7 7 8	7 10 14	1 1 3	5 8 9	1 8 15	0 1 15	1 2 10	5 6 7
10 12 13	2 3 4		4 10 11	7 8 9	7 14 15	1 2 3	6 11 11	2 3 5	0 7 15	1 3 15	5 8 14
11 14 15	2 4 7	1 6 13	5 8 13	7 9 10	8 11 13	1 2 4	7 7 10	2 5 9	1 1 2	1 7 8	6 7 8
	2 5 8	========	5 14 14	7 9 14	10 11 13	1 2 13	7 7 12	3 4 5	1 1 8	1 9 10	6 11 13
1 6 10	2 8 15	0 1 6	6 6 13	7 10 11	10 11 14	1 3 11	7 9 14	3 4 7	1 3 6	2 3 6	7 10 10
========	3 3 5	0 1 13	7 8 15	7 11 12		1 5 7	7 13 14	3 4 11	1 4 13	2 3 11	8 8 9
0 1 6	3 4 9	0 6 13		7 12 13	1 7 9	1 9 13	8 9 13	3 7 10	1 6 13	2 6 7	8 8 11
0 1 10	3 7 9	1 1 2	1 6 15	7 13 14	========	2 3 4	8 11 11	4 9 11	2 3 9	2 10 11	9 10 10
0 3 5	3 8 15	1 1 7	========	7 14 15	0 1 7	2 5 15		5 8 9	2 5 5	3 4 5	9 13 14
1 1 4	4 5 14	1 3 5	0 1 6	9 11 14	0 1 9	3 5 5	1 7 12	5 11 12	2 5 11	3 4 12	10 11 12
1 2 2	4 9 10	1 4 11	0 1 15	11 13 14	0 7 9	3 8 11	========	6 13 13	3 4 7	3 5 6	10 11 15
1 2 14	5 7 8	1 5 11	0 2 5	13 14 15	1 1 2	4 5 5	0 1 7	7 7 12	3 4 9	3 5 13	10 13 15
1 3 8	5 7 12	1 7 15	1 1 9		1 2 5	4 7 9	0 1 12	7 11 14	3 5 10	3 11 12	11 12 14
1 4 14	5 7 13	2 3 8	1 2 3	1 7 8	1 2 11	5 6 8	0 7 12	7 14 15	3 6 11	3 13 15	
2 2 3	5 11 11	2 4 5	1 3 3	========	1 3 4	5 9 13	1 1 5	8 13 13	4 7 11	4 4 5	1 8 11
2 4 7	6 12 13	2 4 9	1 4 9	0 1 7	1 3 12	5 11 15	1 2 2		5 9 12	4 5 13	========
2 5 8	7 11 11	3 3 7	1 5 15	0 1 8	1 4 5	6 7 11	1 3 5	1 7 14	6 11 13	4 12 13	0 1 8
2 5 10	7 13 15	3 4 15	2 3 12	0 7 8	1 10 13	7 7 9	1 3 9	========	7 13 14	5 6 14	0 1 11
2 7 12	8 10 13	3 5 9	2 7 12	1 1 1	1 11 14	7 7 11	1 4 7	0 1 2	8 15 15	5 7 11	0 8 11
2 8 13	9 14 15	3 8 9	3 3 7	1 1 15	2 3 3	7 8 14	2 3 3	0 1 7		5 11 13	1 1 3
3 4 6		4 7 10	3 3 8	1 2 6	2 3 13	7 12 14	2 5 11	0 1 14	1 8 8	5 13 14	1 2 5
3 7 12	1 6 12	4 11 14	3 5 12	1 2 9	3 4 15	6 9 14	3 4 8	1 1 7	========	6 7 15	1 2 14
4 5 9	========	5 7 12	3 5 13	1 3 13	3 5 6	8 11 13	3 5 9	1 3 7	0 1 1	6 14 15	1 3 4
4 5 10	0 1 2	5 9 11	3 5 15	1 6 7	3 5 8	9 10 13	3 7 13	1 4 14	0 1 8	7 9 9	1 3 13
4 7 11	0 1 6	5 13 13	3 10 15	1 8 9	4 5 7	9 11 13	4 5 8	2 3 7	1 2 8	7 9 11	1 5 7

TAB. I

THE TYPES OF HKL-UVW FOR H,K,L,U,V,W ≤15

H/U	K/V	L/W
1	8	11
1	11	15
2	2	3
3	5	7
3	7	10
4	4	5
4	7	12
5	6	7
5	7	9
5	8	9
5	9	13
6	7	10
6	10	13
7	8	13
7	11	11
9	11	11
9	11	14
9	13	14

H/U	K/V	L/W
1	8	12
0	1	8
0	1	12
0	2	3
1	1	4
1	2	4
1	3	12
3	4	4
3	4	5
4	5	7
4	5	8
4	7	8
4	7	10
4	7	11
4	9	13
4	9	14
5	6	12
5	9	12
7	9	12
7	12	12
8	8	11
8	8	13
11	12	15

H/U	K/V	L/W
1	8	13
0	1	8
0	1	13
0	8	13
1	1	5
1	2	3
1	3	5
1	3	11
1	5	8
2	2	3
2	5	14
3	4	7
3	7	11
4	4	7
4	5	12
5	7	9
5	7	12
5	9	14
6	10	11
7	10	11
7	13	13
9	13	13

H/U	K/V	L/W
1	8	14
0	1	8
0	1	14
0	4	7
1	1	6
1	2	2
1	3	10
2	3	4
2	3	5
2	5	9
2	5	12
2	7	12
3	4	10
3	7	14
4	6	11
5	6	8
5	7	14
6	7	13
7	10	11
8	8	13
8	8	15
9	14	14

H/U	K/V	L/W
1	8	15
0	1	8
0	1	15
0	8	15
1	1	2
1	1	7
1	3	9
1	7	13
2	3	6
2	5	10
2	6	11
3	4	13
3	5	5
3	5	9
3	7	11
4	4	7
4	9	12
5	8	11
5	11	13
6	13	14
7	9	12
7	15	15

H/U	K/V	L/W
1	9	9
0	1	1
0	1	9
1	2	9
2	3	9
3	4	9
4	5	9
5	6	9
6	7	9
7	8	9
8	9	9
9	9	10
9	10	11
9	11	12
9	12	13
9	13	14
9	14	15

H/U	K/V	L/W
1	9	10
0	1	9
0	1	10

H/U	K/V	L/W
1	9	10
0	9	10
1	1	1
1	2	8
1	2	11
1	8	9
1	10	11
2	3	7
2	3	12
2	7	8
2	11	12
3	4	6
3	4	13
3	5	15
3	6	7
3	12	13
4	5	5
4	5	6
4	5	14
4	13	14
5	6	15
5	7	13
5	13	15
5	14	15
7	9	11
7	11	13
9	9	11

H/U	K/V	L/W
1	9	11
0	1	9
0	1	11
0	9	11
1	1	2
1	2	7
1	2	13
1	4	5
1	5	6
2	3	5
2	3	15
3	3	4
3	5	12
3	6	7
4	7	9
4	11	13
5	7	8
5	11	14

H/U	K/V	L/W
1	9	11
7	8	9
7	10	13
8	11	11
8	13	15
9	9	10
10	11	11
11	12	13
11	14	15
12	13	15

H/U	K/V	L/W
1	9	12
0	1	9
0	1	12
0	3	4
1	1	3
1	2	6
1	2	15
1	3	15
2	3	3
3	4	5
3	5	7
3	5	9
3	7	9
3	8	11
3	10	13
3	11	15
4	7	15
5	6	6
5	8	12
6	7	9
6	7	10
6	11	14
7	8	12
7	11	15
8	9	15
9	9	11
9	9	13
11	13	15

H/U	K/V	L/W
1	9	13
0	1	9
0	1	13
0	9	13
1	1	4

H/U	K/V	L/W
1	9	13
1	2	3
1	2	5
1	3	14
1	7	10
2	5	7
3	3	4
3	5	6
4	5	7
4	7	11
5	6	11
5	7	8
5	8	11
6	7	15
7	8	11
7	9	10
8	11	15
8	13	13
9	9	14
10	13	13

H/U	K/V	L/W
1	9	14
0	1	9
0	1	14
0	9	14
1	1	5
1	2	3
1	2	4
1	3	13
1	7	11
2	5	8
3	3	5
3	4	6
4	5	11
4	7	7
5	7	7
5	8	13
5	9	11
6	11	15
7	8	10
7	10	12
8	11	13
9	9	13

H/U	K/V	L/W
1	9	15
0	1	9
0	1	15
0	3	5
1	1	6
1	2	3
1	3	12
2	3	3
2	5	15
3	4	7
3	4	9
3	5	8
3	7	12
3	8	13
4	5	15
5	6	9
5	7	12
6	7	11
6	9	11
7	10	15
7	12	13
8	15	15
9	9	14

H/U	K/V	L/W
1	10	10
0	1	1
0	1	10
1	2	10
2	3	10
3	4	10
4	5	10
5	6	10
6	7	10
7	8	10
8	9	10
9	10	10
10	10	11
10	11	12
10	12	13
10	13	14
10	14	15

H/U	K/V	L/W
1	10	11
0	1	10
0	1	11
0	10	11
1	1	1
1	2	9
1	2	12
1	9	10
1	11	12
2	3	8
2	3	13
2	8	9
2	12	13
3	4	7
3	4	14
3	7	8
3	13	14
4	5	6
4	5	15
4	6	7
4	14	15
5	5	6
5	7	15
7	9	13
7	13	15
9	11	11
9	11	13

H/U	K/V	L/W
1	10	12
0	1	10
0	1	12
0	5	6
1	1	2
1	2	8
1	2	14
2	3	6
2	4	5
2	6	7
2	9	11
2	11	13
3	4	4
3	5	14
4	7	8
5	7	10
6	7	9

H/U	K/V	L/W
1	10	12
6	6	9
6	13	15
8	9	10
8	11	14
10	10	11
10	10	13
11	12	12
12	13	14

H/U	K/V	L/W
1	10	13
0	1	10
0	1	13
0	10	13
1	1	3
1	2	7
1	3	4
1	7	9
2	3	4
2	4	5
3	5	11
3	11	14
5	5	6
5	5	7
5	8	15
6	7	8
6	8	11
7	8	11
7	9	10
7	9	11
7	11	15
8	9	14
9	13	13
11	13	13

H/U	K/V	L/W
1	10	14
0	1	10
0	1	14
0	5	7
1	1	4
1	2	6
2	2	3
2	3	4
2	7	10
2	8	11

H/U	K/V	L/W
1	10	14
3	5	6
4	5	6
4	7	14
4	9	13
4	11	15
5	6	10
5	8	10
6	7	14
7	8	9
7	11	12
9	14	14
10	10	13
11	14	14

H/U	K/V	L/W
1	10	15
0	1	10
0	1	15
0	2	3
1	1	5
1	2	5
1	3	15
3	4	5
3	5	5
4	5	10
4	7	10
5	5	7
5	5	8
5	6	15
5	7	10
5	7	11
5	9	13
5	9	14
5	9	15
7	9	15
7	12	15
8	10	11
8	10	13
11	15	15

H/U	K/V	L/W
1	11	11
0	1	1
0	1	11
1	2	11
2	3	11

H/U	K/V	L/W
1	11	11
3	4	11
4	5	11
5	6	11
6	7	11
7	8	11
8	9	11
9	10	11
10	11	11
11	11	12
11	12	13
11	13	14
11	14	15

H/U	K/V	L/W
1	11	12
0	1	11
0	1	12
0	11	12
1	1	1
1	2	10
1	2	13
1	10	11
1	12	13
2	3	9
2	3	14
2	9	10
2	13	14
3	4	8
3	4	15
3	6	9
3	14	15
4	5	7
4	7	8
5	6	6
5	6	7
7	9	15
9	11	13
9	13	15
11	11	13

H/U	K/V	L/W
1	11	13
0	1	11
0	1	13
0	11	13
1	1	2

H/U	K/V	L/W
1	11	13
1	2	9
1	2	15
1	5	6
1	6	7
2	3	7
3	4	5
3	7	8
4	9	11
4	13	15
5	7	12
5	8	9
7	8	9
7	9	10
9	10	11
10	13	13
11	11	12
11	11	14
12	13	13
13	14	15

H/U	K/V	L/W
1	11	14
0	1	11
0	1	14
0	11	14
1	1	3
1	2	8
1	4	5
1	7	9
2	3	4
2	3	5
3	5	13
3	10	13
4	5	6
5	7	7
5	9	11
6	7	7
7	8	10
7	10	12
8	9	11
8	9	13
11	11	13
11	11	15

TAB. I

THE TYPES OF HKL-UVW FOR H,K,L,U,V,W ≤15

H K L
U V W

1 11 15
0 1 11
0 1 15
0 11 15
1 1 4
1 2 7
1 3 4
1 8 11
2 3 3
2 5 7
3 5 10
4 5 5
5 6 9
5 7 10
5 8 13
6 7 9
6 7 13
7 10 13
8 9 13
9 11 14
11 11 14

1 12 12
0 1 1
0 1 12
1 2 12
2 3 12
3 4 12
4 5 12
5 6 12
6 7 12
7 8 12
8 9 12
9 10 12
10 11 12
11 12 12
12 12 13
12 13 14
12 14 15

1 12 13
0 1 12
0 1 13
0 12 13
1 1 1
1 2 11
1 2 14
1 11 12
1 13 14
2 3 10
2 3 15
2 10 11
2 14 15
3 4 9
3 9 10
4 5 8
4 8 9
5 6 7
5 7 8
6 6 7
9 11 15
11 13 13
11 13 15

1 12 14
0 1 12
0 1 14
0 6 7
1 1 2
1 2 10
2 3 8
2 5 6
2 7 8
2 11 13
2 13 15
3 4 6
4 4 5
4 8 9
5 7 14
6 9 10
6 9 11
7 9 10
8 10 11
10 11 12
11 14 14
12 12 13
13 14 14

1 12 15
0 1 12
0 1 15
0 4 5
1 1 3
1 2 9
2 3 6
3 3 4
3 5 6
3 5 15
3 7 9
3 9 11
3 11 14
5 7 9
6 6 7
6 10 13
7 8 9
7 10 15
8 9 12
8 11 12
9 10 15
9 11 13
11 15 15
13 15 15

1 13 13
0 1 1
0 1 13
1 2 13
2 3 13
3 4 13
4 5 13
5 6 13
6 7 13
7 8 13
8 9 13
9 10 13
10 11 13
11 12 13
12 13 13
13 13 14
13 14 15

1 13 14
0 1 13
0 1 14
0 13 14
1 1 1
1 2 12
1 2 15
1 12 13
1 14 15
2 3 11
2 11 12
3 4 10
3 10 11
4 5 9
4 9 10
5 6 8
5 8 9
6 7 7
6 7 8
11 13 15
13 13 15

1 13 15
0 1 13
0 1 15
0 13 15
1 1 2
1 2 11
1 6 7
1 7 8
2 3 9
3 4 7
3 5 6
3 8 9
4 5 5
4 11 13
5 9 10
7 9 12
7 10 11
8 9 11
9 11 12
11 12 13
13 13 14
14 15 15

1 14 14
0 1 1
0 1 14
1 2 14
2 3 14
3 4 14
4 5 14
5 6 14
6 7 14
7 8 14
8 9 14
9 10 14
10 11 14
11 12 14
12 13 14
13 14 14
14 14 15

1 14 15
0 1 14
0 1 15
0 14 15
1 1 1
1 2 13
1 13 14
2 3 12
2 12 13
3 4 11
3 11 12
4 5 10
4 10 11
5 6 9
5 9 10
6 7 8
6 8 9
7 7 8
13 15 15

1 15 15
0 1 1
0 1 15
1 2 15
2 3 15
3 4 15
4 5 15
5 6 15
6 7 15
7 8 15
8 9 15
9 10 15
10 11 15
11 12 15
12 13 15
13 14 15
14 15 15

2 2 3
0 1 1
0 2 3
1 2 2
1 2 4
1 4 5
1 4 7
1 6 8
1 6 10
1 8 11
1 8 13
1 10 14
2 2 5
2 3 6
2 4 7
2 5 8
2 6 7
2 6 9
2 6 11
2 7 10
2 8 11
2 9 12
2 10 13
2 11 14
2 12 15
3 3 4
3 4 9
3 8 9
3 8 15
3 10 12
4 5 6
4 5 11
4 6 13
4 7 13
4 9 15
4 10 11
5 6 14
5 7 8
5 12 13
6 9 10
6 14 15
7 8 10
7 11 12
8 13 14
9 12 14
10 11 14

2 2 5
0 1 1
0 2 5
1 2 4
1 2 6
1 4 9
1 4 11
1 6 14
2 2 3
2 2 7
2 3 8
2 4 9
2 5 10
2 6 11
2 6 13
2 7 12
2 8 13
2 9 14
2 10 15
3 4 7
3 4 13
4 5 5
4 5 15
4 6 11
5 6 10
5 8 15
6 7 8
7 8 13
8 9 11
10 11 14
10 12 13

2 2 7
0 1 1
0 2 7
1 2 6
1 2 8
1 4 13
1 4 15
2 2 5
2 2 9
2 3 4
2 3 10
2 4 11
2 5 12
2 6 13
2 7 14
2 8 15
3 4 11
4 5 9
4 7 7
6 7 14
6 8 13
6 10 11
8 13 15

2 2 9
0 1 1
0 2 9
1 2 8
1 2 10
2 2 7
2 2 11
2 3 6
2 3 12
2 4 5
2 4 13
2 5 14
2 6 15
3 4 15
4 5 13
4 7 11
4 9 9
6 13 14

2 2 11
0 1 1
0 2 11
1 2 10
1 2 12
2 2 9
2 2 13
2 3 8
2 3 14
2 4 7
2 4 15
2 5 6
4 7 15
4 9 13
4 11 11

2 2 13
0 1 1
0 2 13
1 2 12
1 2 14
2 2 11
2 2 15
2 3 10
2 4 9
2 5 8
2 6 7
4 11 15
4 13 13

2 2 15
0 1 1
0 2 15
1 2 14
2 2 13
2 3 12
2 4 11
2 5 10
2 6 9
2 7 8
4 15 15

2 3 3
0 1 1
0 2 3
1 1 3
1 3 3
1 3 6
1 5 6
1 5 9
1 7 9
1 7 12
1 9 12
1 9 15
1 11 15
2 3 4
2 4 9
2 8 9
2 8 15
2 12 15
3 3 5
3 4 6
3 5 7
3 5 12
3 6 7
3 6 8
3 7 9
3 7 15
3 8 10
3 9 11
3 10 12
3 11 12
3 11 13
3 12 14
3 13 15
4 6 15
4 9 10
4 14 15
5 6 9
5 9 11
5 12 13
6 7 11
6 9 13
6 11 15
7 9 13
7 12 15
8 9 14

2 3 4
0 1 2
0 2 3
0 3 4
1 1 2
1 2 2
1 2 5
1 4 4
1 4 6
1 4 8
1 5 6
1 6 7
1 6 11
1 7 10
1 8 10
1 8 14
1 10 13
1 10 14
1 11 14
2 2 7
2 3 3
2 3 9
2 4 5
2 4 11
2 5 6
2 5 7
2 5 8
2 5 13
2 6 9
2 6 13
2 6 15
2 7 8
2 7 11
2 8 13
2 9 15
2 10 11
3 4 12
3 6 8
3 6 10
3 9 10
3 9 14
3 12 14
3 14 15
4 4 5
4 7 8
4 7 10
4 7 12
4 9 12
4 11 12
4 13 14
5 6 7
5 8 12
5 8 14
5 11 14
5 13 14
6 7 8
6 8 9
6 9 14
6 10 11
6 11 13
7 10 11
8 10 11
8 12 13
8 13 14
9 10 12
9 14 15
10 11 13
10 13 14

2 3 5
0 2 3
0 2 5
0 3 5
1 1 1
1 1 4
1 2 3
1 2 4
1 3 6
1 3 7
1 3 9
1 4 6
1 5 5
1 5 9
1 5 10
1 5 11
1 5 14
1 7 8
1 7 11
1 7 13
1 8 14
1 9 11

TAB. I THE TYPES OF HKL-UVW FOR H,K,L,U,V,W ≤15

H U	K V	L W
2	**3**	**5**
1	11	14
2	4	7
2	4	11
2	4	13
2	5	7
2	7	8
2	7	13
2	12	13
3	3	7
3	5	6
3	5	15
3	6	8
3	7	13
3	9	11
3	9	13
3	12	13
4	5	10
4	5	11
4	6	9
4	7	9
4	11	14
5	7	10
5	7	13
5	7	15
5	8	10
5	9	15
5	11	15
6	8	11
6	9	11
7	9	12
7	10	12
8	10	13
8	11	13
9	11	14
9	12	14
10	12	15
10	13	15
2	**3**	**6**
0	1	2
0	1	3
0	2	3
1	2	6
1	3	4
1	4	9

H U	K V	L W
2	**3**	**6**
1	6	6
1	6	12
1	8	9
1	8	15
1	10	12
1	12	15
2	2	3
2	2	9
2	3	6
2	3	12
2	4	9
2	4	15
2	5	12
2	6	15
2	9	10
2	14	15
3	3	4
3	3	8
3	4	6
3	4	10
3	4	15
3	5	8
3	5	12
3	6	10
3	6	14
3	7	12
3	8	14
3	12	14
4	5	9
4	7	15
4	9	14
5	6	6
5	6	14
6	7	10
6	7	12
6	8	15
6	9	14
7	8	9
8	9	10
8	9	15
9	10	12
9	10	14
11	12	14
11	12	15
12	14	15

H U	K V	L W
2	**3**	**7**
0	2	3
0	2	7
0	3	7
1	1	2
1	1	3
1	1	5
1	2	4
1	3	8
1	3	9
1	3	12
1	4	5
1	4	10
1	5	11
1	7	7
1	7	14
1	9	10
1	11	13
2	4	11
2	4	13
2	5	8
2	5	13
2	7	15
2	11	12
3	3	5
3	5	13
3	6	11
3	9	13
4	5	6
4	5	9
4	6	15
4	7	14
5	7	7
5	7	13
6	8	9
7	8	11
7	8	14
7	9	11
8	10	13
9	11	15
9	12	13
2	**3**	**8**
0	1	4
0	2	3

H U	K V	L W
2	**3**	**8**
0	3	8
1	1	2
1	2	4
1	2	6
1	2	7
1	4	10
1	5	6
1	5	14
1	6	13
1	8	8
1	10	11
1	12	14
2	2	5
2	2	11
2	3	9
2	3	15
2	4	13
2	5	12
2	7	10
2	13	14
3	3	10
3	4	6
3	6	14
3	9	14
4	5	14
4	6	7
5	6	11
5	8	8
6	9	10
7	10	12
7	10	13
8	11	14
9	14	15
2	**3**	**9**
0	1	3
0	2	3
0	2	9
1	1	3
1	1	6
1	3	5
1	3	9
1	3	12
1	3	15
1	5	12

H U	K V	L W
2	**3**	**9**
1	6	7
1	7	15
1	9	9
1	11	12
1	13	15
2	3	4
2	3	8
2	4	15
2	9	12
3	3	7
3	3	11
3	4	10
3	4	14
3	5	6
3	5	13
3	6	13
4	6	9
5	6	11
5	7	12
5	9	9
6	8	15
7	9	15
7	11	15
7	12	13
8	14	15
2	**3**	**10**
0	1	5
0	2	3
0	3	10
1	1	4
1	2	2
1	2	6
1	2	8
1	4	6
1	4	11
1	4	14
1	6	14
1	7	8
1	10	10
1	12	13
2	2	7
2	2	13
2	3	12
2	5	10

H U	K V	L W
2	**3**	**10**
2	11	14
3	3	8
3	4	9
3	6	14
4	5	10
4	5	14
4	6	11
5	7	12
5	8	13
6	9	14
6	10	15
2	**3**	**11**
0	2	3
0	2	11
0	3	11
1	1	3
1	1	4
1	1	7
1	2	5
1	2	8
1	3	10
1	3	15
1	4	14
1	5	7
1	5	13
1	8	9
1	11	11
1	13	14
2	4	5
2	7	12
3	3	13
3	5	9
3	6	7
4	6	13
4	7	10
5	8	13
5	9	14
5	11	11
2	**3**	**12**
0	1	4
0	1	6
0	2	3

H U	K V	L W
2	**3**	**12**
1	2	3
1	2	9
1	3	6
1	4	12
1	6	8
1	6	15
1	9	10
1	12	12
1	14	15
2	2	9
2	2	15
2	3	6
2	3	10
2	3	15
2	9	14
3	3	10
3	3	14
3	4	12
3	4	14
3	6	8
4	6	15
4	9	10
5	9	14
5	12	12
6	14	15
2	**3**	**13**
0	2	3
0	2	13
0	3	13
1	1	5
1	1	8
1	2	3
1	2	8
1	3	11
1	4	7
1	5	14
1	7	9
1	10	11
1	13	13
2	4	7
2	5	12
3	3	11
3	5	12
3	7	9

H U	K V	L W
2	**3**	**13**
4	5	14
4	10	11
5	13	13
2	**3**	**14**
0	1	7
0	2	3
0	3	14
1	1	4
1	2	4
1	2	6
1	2	10
1	4	13
1	5	8
1	8	10
1	11	12
1	14	14
2	2	11
2	5	6
2	7	14
3	4	15
3	6	10
3	8	9
4	7	14
4	10	13
5	14	14
2	**3**	**15**
0	1	5
0	2	3
0	2	15
1	1	6
1	1	9
1	3	3
1	3	7
1	3	12
1	5	15
1	6	9
1	9	11
1	12	13
1	15	15
2	3	8
2	3	12
2	4	9

H U	K V	L W
2	**3**	**15**
3	3	13
3	5	15
3	6	11
3	7	12
4	9	14
4	10	15
2	**4**	**5**
0	1	2
0	2	5
0	4	5
1	2	2
1	2	3
1	2	7
1	4	8
1	4	12
1	6	7
1	6	8
1	6	13
1	10	12
1	10	13
2	2	9
2	3	4
2	3	11
2	4	13
2	5	5
2	5	15
2	6	7
2	6	11
2	7	9
2	8	9
2	8	11
2	9	13
2	10	15
3	4	4
3	6	14
3	10	11
3	10	14
4	4	7
4	6	7
4	8	9
4	11	12
4	12	13
5	5	6
5	6	10

H U	K V	L W
2	**4**	**5**
5	8	10
5	14	15
6	7	6
6	7	11
6	8	13
6	10	13
6	11	13
6	13	14
7	9	10
7	10	11
8	9	10
8	10	15
8	11	12
9	13	14
10	14	15
11	12	14
11	13	14
2	**4**	**7**
0	1	2
0	2	7
0	4	7
1	2	3
1	2	4
1	2	5
1	2	9
1	4	12
1	6	10
1	6	11
2	2	3
2	2	11
2	3	5
2	3	13
2	4	15
2	5	6
2	7	7
2	8	9
2	8	13
2	8	15
2	9	11
2	10	13
2	11	15
3	4	8
4	4	5
4	4	9

H U	K V	L W
2	**4**	**7**
4	6	13
4	8	11
4	12	[illegible]
5	6	[illegible]
5	6	11
5	6	13
6	7	[illegible]
6	7	14
6	6	11
7	8	14
7	10	14
8	9	10
9	10	13
10	11	12
10	11	13
2	**4**	**9**
0	1	2
0	2	9
0	4	9
1	2	4
1	2	5
1	2	7
1	2	11
1	6	13
1	6	14
2	2	5
2	2	13
2	3	3
2	3	6
2	3	15
2	5	7
2	7	8
2	9	9
2	10	11
2	11	13
2	12	15
3	4	12
4	4	7
4	4	11
4	5	8
4	8	13
4	12	15
5	6	11
6	7	10

H U	K V	L W
2	**4**	**9**
6	7	13
6	8	11
6	6	15
8	10	13
6	11	14
2	**4**	**11**
0	1	2
0	2	11
0	4	11
1	2	5
1	2	6
1	2	9
1	2	13
2	2	7
2	2	15
2	3	4
2	3	5
2	3	7
2	5	8
2	7	9
2	9	10
2	11	11
2	12	13
2	13	15
4	4	9
4	4	13
4	5	12
4	7	8
4	8	15
5	6	14
6	7	13
6	10	13
6	11	11
8	14	15
2	**4**	**13**
0	1	2
0	2	13
0	4	13
1	2	6
1	2	7
1	2	11
1	2	15

TAB. I

THE TYPES OF HKL-UVW FOR H,K,L,U,V,W ≤15

H U	K V	L W
2	**4**	**13**
2	2	9
2	3	5
2	3	7
2	3	8
2	4	5
2	5	9
2	7	10
2	9	11
2	11	12
2	13	13
2	14	15
4	4	11
4	4	15
4	7	12
4	8	9
6	11	14
6	13	13
2	**4**	**15**
0	1	2
0	2	15
0	4	15
1	2	7
1	2	8
1	2	13
2	2	11
2	3	6
2	3	9
2	4	7
2	5	5
2	5	10
2	7	11
2	9	12
2	11	13
2	13	14
2	15	15
4	4	13
4	8	11
4	9	12
2	**5**	**5**
0	1	1
0	2	5
1	1	5

H U	K V	L W
2	**5**	**5**
1	3	5
1	3	10
1	5	10
1	5	15
1	7	15
2	4	5
2	4	15
2	8	15
3	5	5
3	7	10
4	5	6
4	10	15
5	5	7
5	6	8
5	7	9
5	8	10
5	9	10
5	9	11
5	10	12
5	11	13
5	11	15
5	12	14
5	13	15
7	10	11
7	13	15
8	14	15
9	10	13
10	11	15
2	**5**	**6**
0	1	3
0	2	5
0	5	6
1	2	2
1	2	8
1	3	4
1	4	7
1	4	13
1	6	12
1	7	8
1	8	10
1	12	14
2	2	11
2	3	4
2	3	14

H U	K V	L W
2	**5**	**6**
2	4	7
2	5	10
2	6	9
2	6	13
2	9	10
2	11	14
3	4	6
3	8	11
3	11	12
4	5	5
4	7	10
4	7	11
4	10	13
4	13	14
5	5	8
5	10	14
5	12	15
6	6	7
6	7	10
6	8	9
6	10	15
7	8	9
7	9	12
8	9	14
8	10	11
8	11	13
8	11	14
9	12	13
10	11	12
10	13	14
10	14	15
2	**5**	**7**
0	2	5
0	2	7
0	5	7
1	1	1
1	1	6
1	3	4
1	3	8
1	3	11
1	3	13
1	4	6
1	5	9
1	5	15

H U	K V	L W
2	**5**	**7**
1	6	8
1	7	14
1	9	13
1	11	15
2	3	4
2	3	5
2	4	9
2	7	9
2	8	13
3	5	10
3	7	7
3	7	11
3	8	10
4	6	11
4	9	11
4	10	11
5	5	9
5	7	12
5	10	11
5	10	12
5	13	15
6	8	13
6	11	13
7	8	13
7	8	14
7	9	14
7	12	14
7	13	15
8	10	15
8	13	15
2	**5**	**8**
0	1	4
0	2	5
0	5	8
1	1	2
1	2	9
1	4	6
1	4	14
1	6	10
1	6	11
1	9	14
2	2	3
2	2	13
2	3	4

H U	K V	L W
2	**5**	**8**
2	3	7
2	4	11
2	5	15
2	6	7
2	7	12
3	7	10
3	8	8
3	8	14
3	10	13
4	5	10
4	9	10
5	5	6
5	10	12
5	10	14
5	14	15
6	7	13
6	11	14
7	8	8
7	8	10
9	10	11
9	11	14
11	12	14
12	13	14
2	**5**	**9**
0	2	5
0	2	9
0	5	9
1	1	2
1	1	7
1	2	4
1	3	3
1	3	5
1	3	11
1	3	12
1	5	8
1	7	13
1	8	14
2	4	13
2	8	11
3	4	6
3	4	7
3	5	15
3	11	14
4	7	10

H U	K V	L W
2	**5**	**9**
4	7	11
5	5	7
5	5	11
5	10	13
6	7	8
6	7	15
7	9	9
8	9	13
8	10	11
9	11	13
10	12	15
11	12	15
2	**5**	**10**
0	1	2
0	1	5
0	2	5
1	2	10
1	4	5
1	4	15
1	6	10
1	8	15
2	2	5
2	2	15
2	3	10
2	4	15
2	5	6
2	10	15
3	4	5
3	5	8
3	10	10
4	5	6
4	5	10
4	5	15
4	14	15
5	5	8
5	5	12
5	6	10
5	6	14
5	7	12
5	8	14
5	10	14
7	8	15
7	10	10
8	10	15

H U	K V	L W
2	**5**	**10**
9	10	14
10	14	15
2	**5**	**11**
0	2	5
0	2	11
0	5	11
1	1	3
1	1	8
1	2	3
1	2	4
1	3	7
1	3	13
1	3	14
1	5	7
1	7	12
1	7	15
2	8	9
3	4	5
3	4	10
3	11	11
4	6	7
4	13	14
5	5	9
5	5	13
5	7	10
6	7	13
6	8	13
7	11	11
8	9	14
2	**5**	**12**
0	1	6
0	2	5
0	5	12
1	1	2
1	2	11
1	4	4
1	4	10
1	6	9
1	6	14
1	8	14
2	2	7
2	3	6

H U	K V	L W
2	**5**	**12**
2	3	8
2	3	13
2	10	13
3	4	8
3	7	10
4	6	9
4	11	14
5	5	14
5	6	15
5	8	10
6	10	11
7	12	12
8	13	14
2	**5**	**13**
0	2	5
0	2	13
0	5	13
1	1	3
1	1	4
1	1	9
1	3	14
1	4	10
1	5	6
1	7	11
2	3	4
2	3	7
2	7	8
3	5	7
3	8	11
3	13	13
4	6	11
4	9	14
5	5	11
5	7	15
5	9	10
7	8	15
7	13	13
2	**5**	**14**
0	1	7
0	2	5
0	5	14
1	2	2

H U	K V	L W
2	**5**	**14**
1	2	6
1	2	12
1	3	4
1	3	8
1	6	8
1	8	13
2	2	9
2	10	11
3	4	10
3	4	11
3	14	14
4	6	13
4	7	14
5	5	12
5	8	15
6	7	14
2	**5**	**15**
0	1	3
0	2	5
0	2	15
1	1	5
1	1	10
1	3	15
1	5	5
1	7	10
1	9	15
2	4	5
2	5	8
2	12	15
3	5	7
3	5	10
3	5	11
3	10	13
4	5	10
4	5	14
4	6	15
5	5	13
5	9	15
5	10	11
7	15	15

H U	K V	L W
2	**6**	**7**
0	1	3
0	2	7
0	6	7
1	2	2
1	2	4
1	2	10
1	4	5
1	4	11
1	8	9
1	10	12
2	2	3
2	2	13
2	4	5
2	5	8
2	6	11
2	6	15
2	7	14
2	10	11
3	4	5
3	6	8
3	12	13
4	6	9
4	7	7
4	9	13
4	10	13
4	14	15
5	6	6
5	8	11
5	8	13
6	9	10
6	11	12
7	7	8
7	10	14
8	9	10
8	10	11
8	11	13
9	11	12
10	13	14
12	13	14
2	**6**	**9**
0	1	3
0	2	3
0	2	9

H U	K V	L W
2	**6**	**9**
1	2	6
1	2	12
1	4	15
2	2	3
2	2	15
2	3	4
2	5	6
2	6	9
2	7	12
2	8	15
3	4	5
3	4	7
3	4	9
3	6	8
3	6	10
3	8	11
3	8	13
3	10	14
4	6	15
4	9	9
4	11	15
5	6	12
6	6	7
6	6	11
6	10	13
6	12	13
6	14	15
7	8	15
8	9	9
8	9	15
9	10	12
10	11	12
12	13	15
2	**6**	**11**
0	1	3
0	2	11
0	6	11
1	2	4
1	2	8
1	2	14
1	4	7
1	8	15
2	2	3
2	2	5

H U	K V	L W
2	**6**	**11**
2	4	5
2	6	7
2	7	10
2	8	13
3	4	13
3	6	10
4	5	7
4	5	9
4	11	11
5	8	13
6	6	13
6	7	12
6	8	9
6	9	14
8	11	11
10	13	14
2	**6**	**13**
0	1	3
0	2	13
0	6	13
1	2	4
1	2	10
1	4	9
2	2	5
2	2	7
2	3	4
2	5	6
2	7	8
2	8	11
2	9	14
3	6	14
4	5	7
4	5	11
4	7	11
4	13	13
6	6	11
6	8	15
6	9	10
7	8	15
8	13	13

TAB. I

THE TYPES OF HKL-UVW FOR H,K,L,U,V,W ≤15


```
 H  K  L
 U  V  W
--------

 2  6 15
========
 0  1  3
 0  2  5
 0  2 15
 1  2 12
 2  2  9
 2  3  4
 2  3  6
 2  6  7
 2  8  9
 2  9 12
 2 10 15
 3  4  9
 3  4 11
 3  6 14
 4  5 15
 4  7  9
 4  9 13
 4 15 15
 6  6 13
 6 10 15
 6 11 12
 8 15 15

 2  7  7
========
 0  1  1
 0  2  7
 1  1  7
 1  3  7
 1  3 14
 1  5 14
 2  4  7
 3  5  7
 3  7 14
 4  6  7
 5  7  7
 5  9 14
 6  7  8
 7  7  9
 7  8 10
 7  9 11
 7 10 12
 7 11 13
 7 11 14
 7 12 14
```

```
 H  K  L
 U  V  W
--------
 2  7  7
--------
 7 13 15
 9 13 14
11 14 15

 2  7  8
========
 0  1  4
 0  2  7
 0  7  8
 1  2  2
 1  2  3
 1  2 11
 1  4 10
 1  5  6
 1  9 10
 1 12 14
 2  2 15
 2  3  4
 2  3  5
 2  4  9
 2  5 13
 2  6 13
 2 11 12
 3  8 10
 3 13 14
 4  5  6
 4  7 14
 5  8  8
 5 11 14
 6  7  7
 6  8 11
 6 10 11
 7  7 10
 7 12 14
 8  8  9
 9 10 11
 9 10 14
 9 13 14
10 12 13
11 14 15

 2  7  9
========
 0  2  7
 0  2  9
 0  7  9
```

```
 H  K  L
 U  V  W
--------
 2  7  9
--------
 1  1  1
 1  1  8
 1  3  6
 1  3 10
 1  3 15
 1  5 13
 1  6  8
 1  8 10
 2  4  5
 2  4 11
 2  5  7
 2  9 11
 3  4  5
 3  4  6
 3  5 12
 3  7 11
 3 10 12
 3 11 15
 4  6 13
 4 11 13
 5  7 14
 5  9  9
 5  9 13
 5 12 14
 6  8 15
 6 13 15
 7  7 11
 7 13 14
 8 13 14

 2  7 10
========
 0  1  5
 0  2  7
 0  7 10
 1  2  2
 1  2 12
 1  3  4
 1  4  6
 1  4  9
 1 10 14
 2  2  3
 2  3  8
 2  4 13
 2  6 11
 3  5  8
```

```
 H  K  L
 U  V  W
--------
 2  7 10
--------
 3  8 13
 4  5  6
 4  5 11
 4  9 14
 4 10 15
 5  6 10
 5  8 10
 6  7 14
 6 11 14
 7  7  8
 7  7 12
 8  9 14
 9 10 10
11 12 13
11 12 14

 2  7 11
========
 0  2  7
 0  2 11
 0  7 11
 1  1  2
 1  1  9
 1  3  5
 1  3 13
 1  4  6
 1  5 12
 2  3  4
 2  4 15
 3  5  7
 3  7  8
 4  5  9
 4  9 13
 4 10 13
 5  6  8
 5  8 14
 5 11 11
 7  7  9
 7  7 13
 7 14 15
 8  9 10
 8 11 15
 9 11 11
10 12 13
11 13 15
```

```
 H  K  L
 U  V  W
--------

 2  7 12
========
 0  1  6
 0  2  7
 0  7 12
 1  1  2
 1  2 13
 1  4  8
 1  6 10
 1  6 15
 1  6 14
 2  2  5
 2  3 11
 2  5  8
 2  6  9
 3  4  4
 3  4  6
 3  8 10
 4 10 11
 5  6  9
 5  9 14
 5 12 12
 6 13 14
 7  7 10
 7  8 14
 8 10 13
10 11 14

 2  7 13
========
 0  2  7
 0  2 13
 0  7 13
 1  1  3
 1  1 10
 1  2  4
 1  3  4
 1  5  9
 1  5 11
 2  3  5
 2  8 15
 3  5  7
 3  8 14
 4  5  6
 4  9 10
 5  6 11
 5  7  8
```

```
 H  K  L
 U  V  W
--------
 2  7 13
--------
 5 13 13
 6  8 11
 7  7 11
 7  7 15
 7  9 14
 9 13 13

 2  7 14
========
 0  1  2
 0  1  7
 0  2  7
 1  2 14
 1  4  7
 1  6 14
 2  2  7
 2  3 14
 2  6  7
 3  4  7
 3  7  8
 3 10 14
 4  6  7
 4  7 10
 5  6 14
 5  7  8
 5  7 12
 5 14 14
 6  7 10
 6  7 14
 7  7 12
 7  8 14
 7 10 14
 9 14 14

 2  7 15
========
 0  2  7
 0  2 15
 0  7 15
 1  1  4
 1  1 11
 1  2  4
 1  3  3
 1  5 10
 1  5 11
 2  3  7
```

```
 H  K  L
 U  V  W
--------
 2  7 15
--------
 2  8 13
 3  5  5
 4  5 10
 4  6  9
 5  6  9
 5  7 13
 5  8 13
 7  7 13
 7 11 14
 8 11 14

 2  8  9
========
 0  1  4
 0  2  9
 0  8  9
 1  2  2
 1  2  5
 1  2 13
 1  4 14
 1  6  7
 1 10 11
 2  3  3
 2  4  5
 2  4  7
 2  5 11
 2  6 15
 2 12 13
 3  4  6
 3 10 12
 3 14 15
 4  6 11
 4  7 10
 5  6  7
 5  6  8
 6 10 13
 7  8  8
 7 10 13
 8  6 11
 8 10 13
 9  9 10
10 11 12
10 11 14
11 13 14
12 14 15
```

```
 H  K  L
 U  V  W
--------

 2  6 11
========
 0  1  4
 0  2 11
 0  8 11
 1  2  3
 1  2  7
 1  2 15
 1  6  8
 1 10 14
 2  2  3
 2  4  5
 2  5  9
 2  6 13
 3  4 10
 3 10 13
 4  6  7
 4  9 14
 5  6  7
 5  6 13
 6  7 10
 6 11 11
 7 10 12
 8  6  9
 8  8 13
10 11 11
12 13 14

 2  8 13
========
 0  1  4
 0  2 13
 0  8 13
 1  2  3
 1  2  9
 1  6 10
 2  2  5
 2  3  4
 2  4  7
 2  5  7
 2  6 11
 2  7 15
 3  4 14
 4  5  6
 4  9 10
 5  6 11
 6  7  8
```

```
 H  K  L
 U  V  W
--------
 2  8 13
--------
 6  7 11
 6 13 13
 8  8 11
 8  8 15
 9 10 14
10 13 13

 2  8 15
========
 0  1  4
 0  2 15
 0  8 15
 1  2  4
 1  2 11
 1  6 11
 2  2  7
 2  3  3
 2  4  7
 2  5  5
 2  6  9
 2  7 13
 4  5 10
 4  6  9
 4 11 14
 5  6 10
 6  7 13
 6  8 13
 6 11 14
 8  8 13

 2  9  9
========
 0  1  1
 0  2  9
 1  1  9
 1  3  9
 2  4  9
 3  5  9
 4  6  9
 5  7  9
 6  8  9
 7  9  9
 8  9 10
 9  9 11
 9 10 12
 9 11 13
```

```
 H  K  L
 U  V  W
--------
 2  9  9
--------
 9 12 14
 9 13 15

 2  9 10
========
 0  1  5
 0  2  9
 0  9 10
 1  2  2
 1  2  4
 1  2 14
 1  4 13
 1  7  8
 1 11 12
 2  3  6
 2  4 11
 2  5  6
 2 13 14
 3  3  4
 3 12 14
 4  5  7
 4  6  7
 5  8 10
 5  8 11
 6  7  8
 6  8 13
 6 10 15
 7 10 10
 8  9  9
 8 11 14
10 10 11
10 12 15
10 13 14
11 12 13
12 14 15

 2  9 11
========
 0  2  9
 0  2 11
 0  9 11
 1  1  1
 1  1 10
 1  3  8
 1  3 12
 1  8 10
```

```
 H  K  L
 U  V  W
--------
 2  9 11
--------
 1 10 12
 2  4  7
 2  4 13
 2  7  9
 2 11 13
 3  5  6
 3  5 14
 3  6  8
 3  7 15
 3 12 14
 4  5  6
 4  5  7
 4  6 15
 4 13 15
 5  9 13
 7 11 11
 7 11 15
 9  9 13

 2  9 12
========
 0  1  6
 0  2  9
 0  3  4
 1  2  3
 1  2 15
 1  4 12
 2  2  3
 2  3  9
 2  4 15
 2  6 15
 3  4  6
 3  5  6
 3  7 10
 3  8 10
 3 10 14
 3 11 14
 4  5 12
 5  6  8
 6  7  8
 6  7 15
 6  9 10
 7 12 12
 8 14 15
 9  9 10
 9  9 14
```

```
 H  K  L
 U  V  W
--------
 2  9 12
--------
 9 12 14
11 12 12
13 14 15

 2  9 13
========
 0  2  9
 0  2 13
 0  9 13
 1  1  2
 1  1 11
 1  3  7
 1  3 15
 1  4  6
 1  5  7
 2  4  5
 3  3  5
 3  6  8
 3  7 12
 4  7 11
 4 11 15
 5  7  9
 5  8  9
 7  8 10
 7 13 13
 8 11 14
 9  9 11
10 11 12
11 13 13
12 14 15

 2  9 14
========
 0  1  7
 0  2  9
 0  9 14
 1  2  2
 1  4  6
 1  4 11
 1  5  6
 2  2  5
 2  3 12
 2  6 13
 3  3  4
 3  6 10
 3  8 15
```

TAB. I

THE TYPES OF HKL-UVW FOR H,K,L,U,V,W ≤15

H U	K V	L W
2	**9**	**14**
4	7	10
5	6	8
5	7	12
7	8	13
7	8	14
7	10	14
8	10	11
11	14	14
2	**9**	**15**
0	2	9
0	2	15
0	3	5
1	1	3
1	1	12
1	3	6
1	5	15
2	3	4
3	4	6
3	5	9
3	7	9
3	7	11
3	8	14
4	10	15
5	6	7
5	11	12
6	7	13
6	8	9
7	9	12
7	15	15
8	10	15
9	9	13
11	15	15
2	**10**	**11**
0	1	5
0	2	11
0	10	11
1	2	2
1	2	6
1	8	9
1	12	13
2	3	4
2	4	9

H U	K V	L W
2	**10**	**11**
2	5	14
2	6	7
2	14	15
3	4	5
3	4	7
4	6	13
4	7	13
5	6	8
5	10	12
6	7	8
7	8	9
8	10	15
8	11	11
9	10	10
10	10	13
10	14	15
11	11	12
12	13	14
2	**10**	**13**
0	1	5
0	2	13
0	10	13
1	2	8
1	4	5
1	6	8
2	2	3
2	4	7
2	5	12
3	4	11
3	8	11
4	6	7
4	7	9
5	6	14
5	10	12
5	10	14
6	10	11
7	8	9
8	13	13
10	10	11
12	13	13

H U	K V	L W
2	**10**	**15**
0	1	5
0	2	3
0	2	15
1	2	10
2	2	5
2	4	5
2	5	10
2	6	15
3	4	15
4	5	5
4	5	7
4	9	15
5	6	8
5	6	10
5	8	11
5	8	13
5	10	14
6	7	10
6	10	11
8	9	15
8	15	15
10	10	13
10	12	15
2	**11**	**11**
0	1	1
0	2	11
1	1	11
1	3	11
2	4	11
3	5	11
4	6	11
5	7	11
6	8	11
7	9	11
8	10	11
9	11	11
10	11	12
11	11	13
11	12	14
11	13	15

H U	K V	L W
2	**11**	**12**
0	1	6
0	2	11
0	11	12
1	2	2
1	2	5
1	9	10
1	13	14
2	3	7
2	4	13
2	7	8
3	4	4
3	5	6
4	5	8
4	6	9
5	8	14
5	12	14
6	7	9
6	8	15
7	8	10
7	10	13
8	9	10
10	11	11
11	11	14
12	12	13
13	14	15
2	**11**	**13**
0	2	11
0	2	13
0	11	13
1	1	1
1	1	12
1	3	10
1	3	14
1	10	12
1	12	14
2	4	9
2	4	15
2	9	11
2	13	15
3	5	8
3	8	10
4	6	7
4	7	9

H U	K V	L W
2	**11**	**13**
5	6	7
5	6	8
7	11	15
9	13	13
11	11	15
2	**11**	**14**
0	1	7
0	2	11
0	11	14
1	2	4
1	3	4
1	4	15
1	8	10
2	2	3
2	3	10
2	5	6
4	5	6
4	5	13
5	7	8
5	8	9
6	7	10
6	10	13
7	10	14
7	12	14
8	9	10
9	14	14
11	11	12
13	14	14
2	**11**	**15**
0	2	11
0	2	15
0	11	15
1	1	2
1	1	13
1	3	9
1	5	7
1	6	8
2	4	7
3	4	6
3	5	5
3	7	9
4	9	13

H U	K V	L W
2	**11**	**15**
5	8	10
5	9	12
7	8	11
7	9	11
9	10	12
11	11	13
12	13	14
13	15	15
2	**12**	**13**
0	1	6
0	2	13
0	12	13
1	2	2
1	2	7
1	10	11
1	14	15
2	3	5
2	4	11
2	6	9
3	4	8
3	6	7
4	4	5
4	6	15
5	6	9
6	8	9
7	8	10
7	10	12
8	11	14
9	10	11
10	13	13
11	12	12
13	13	14
2	**12**	**15**
0	1	6
0	2	15
0	4	5
1	2	9
2	2	3
2	3	3
2	4	9
2	5	15
3	4	12

H U	K V	L W
2	**12**	**15**
3	6	7
3	6	8
3	10	12
3	10	13
4	7	12
5	6	15
6	8	9
6	8	11
6	10	15
9	10	11
9	10	14
12	12	13
14	15	15
2	**13**	**13**
0	1	1
0	2	13
1	1	13
1	3	13
2	4	13
3	5	13
4	6	13
5	7	13
6	8	13
7	9	13
8	10	13
9	11	13
10	12	13
11	13	13
12	13	14
13	13	15
2	**13**	**14**
0	1	7
0	2	13
0	13	14
1	2	2
1	2	6
1	11	12
2	3	8
2	4	15
2	9	10
3	4	5
3	7	8

H U	K V	L W
2	**13**	**14**
4	5	6
4	5	9
4	6	11
6	7	10
7	12	14
8	9	10
8	9	11
10	11	12
11	14	14
12	13	13
14	14	15
2	**13**	**15**
0	2	13
0	2	15
0	13	15
1	1	1
1	1	14
1	3	12
1	12	14
2	4	11
2	11	13
3	5	10
3	10	12
4	6	9
4	9	11
5	7	8
5	8	10
6	7	8
6	7	9
11	15	15
2	**14**	**15**
0	1	7
0	2	15
0	14	15
1	2	2
1	2	8
1	12	13
2	3	6
2	4	13
2	10	11
3	4	9
3	8	9

H U	K V	L W
2	**14**	**15**
4	5	5
4	6	7
5	6	10
6	8	11
7	8	11
9	10	12
11	12	13
13	14	14
2	**15**	**15**
0	1	1
0	2	15
1	1	15
1	3	15
2	4	15
3	5	15
4	6	15
5	7	15
6	8	15
7	9	15
8	10	15
9	11	15
10	12	15
11	13	15
12	14	15
13	15	15
3	**3**	**4**
0	1	1
0	3	4
1	3	3
1	3	5
1	6	7
1	6	9
1	9	11
1	9	13
1	12	15
2	2	3
2	3	6
2	9	10
2	9	14
3	3	7
3	4	8
3	5	6

H U	K V	L W
3	**3**	**4**
3	5	9
3	6	10
3	6	11
3	7	11
3	8	12
3	9	13
3	10	14
3	11	15
3	12	13
4	8	9
5	6	13
5	7	9
5	11	12
6	7	15
6	14	15
7	9	12
7	13	15
8	12	15
9	11	15
3	**3**	**5**
0	1	1
0	3	5
1	3	4
1	3	6
1	6	9
1	6	11
1	9	14
2	3	3
2	3	7
2	9	13
3	3	8
3	4	9
3	5	10
3	6	7
3	6	11
3	6	13
3	7	12
3	8	13
3	9	14
3	10	15
4	9	11
5	5	6
5	6	15
5	9	10

H U	K V	L W
3	**3**	**5**
5	12	15
7	8	9
7	12	13
9	11	12
11	14	15
12	13	15
3	**3**	**7**
0	1	1
0	3	7
1	3	6
1	3	8
1	6	13
1	6	15
2	3	5
2	3	9
3	3	4
3	3	10
3	4	11
3	5	12
3	6	11
3	6	13
3	7	14
3	8	15
5	6	9
6	7	7
7	9	14
8	9	13
9	10	11
12	13	15
3	**3**	**8**
0	1	1
0	3	8
1	3	7
1	3	9
1	6	15
2	3	6
2	3	10
3	3	5
3	3	11
3	4	4
3	4	12
3	5	13

H U	K V	L W
3	**3**	**8**
3	6	13
3	6	14
3	7	15
5	6	11
6	7	9
9	10	14
9	11	13
3	**3**	**10**
0	1	1
0	3	10
1	3	9
1	3	11
2	3	8
2	3	12
3	3	7
3	3	13
3	4	6
3	4	14
3	5	5
3	5	15
5	6	15
6	7	13
6	9	11
3	**3**	**11**
0	1	1
0	3	11
1	3	10
1	3	12
2	3	9
2	3	13
3	3	8
3	3	14
3	4	7
3	4	15
3	5	6
6	7	15
6	9	13
6	11	11

TAB. I   THE TYPES OF HKL-UVW FOR H,K,L,U,V,W ≤15

```
H  K  L
U  V  W
--------

 3  3 13
========
 0  1  1
 0  3 13
 1  3 12
 1  3 14
 2  3 11
 2  3 15
 3  3 10
 3  4  9
 3  5  8
 3  6  7
 6 11 15
 6 13 13

 3  3 14
========
 0  1  1
 0  3 14
 1  3 13
 1  3 15
 2  3 12
 3  3 11
 3  4 10
 3  5  9
 3  6  8
 3  7  7
 6 13 15

 3  4  4
========
 0  1  1
 0  3  4
 1  2  4
 1  4  4
 1  5  8
 1  7  8
 1  8 12
 1 10 12
 2  4  5
 2  7 12
 2 11 12
 3  3  8
 3  4  6
 3  8  9
 4  4  7
 4  5  8
```

```
H  K  L
U  V  W
--------
 3  4  4
--------
 4  5 12
 4  6  9
 4  7 10
 4  8 11
 4  9 12
 4 10 13
 4 11 14
 4 12 13
 4 12 15
 5  8 11
 5 12 14
 7  8 13
 8  9 15

 3  4  5
========
 0  3  4
 0  3  5
 0  4  5
 1  1  2
 1  1  3
 1  2  2
 1  3  3
 1  4  7
 1  4  8
 1  5  5
 1  5  7
 1  7  8
 1  7 11
 1  7 13
 1  8  9
 1  8 12
 1  9 12
 1 10 15
 1 11 13
 2  5 10
 2  5 11
 2  6  7
 2  6  9
 2  7  9
 2 10 11
 2 13 14
 3  4  5
 3  6 13
 3  6 14
 3  7  9
```

```
H  K  L
U  V  W
--------
 3  4  5
--------
 3  7 11
 3  9 11
 3 11 12
 3 13 14
 4  7  8
 4  8 11
 4  8 13
 4 11 13
 5  5  7
 5  9 13
 5  9 15
 5 10 11
 5 13 15
 6  7  9
 6 10 15
 6 13 14
 7  9 11
 7 11 13
 8  9 12
 8 11 13
 9 13 15

 3  4  6
========
 0  1  2
 0  2  3
 0  3  4
 1  2  3
 1  3  6
 1  6  6
 1  6 10
 1  9 10
 1  9 14
 1 12 14
 2  3  3
 2  3  6
 2  3  8
 2  5  6
 2  5  9
 2  7  9
 2  7 12
 2  8  9
 2  9 12
 2  9 15
 2 11 15
 3  3 10
```

```
H  K  L
U  V  W
--------
 3  4  6
--------
 3  4  4
 3  4 12
 3  5  6
 3  5 14
 3  6  8
 3  6 14
 3  7 10
 3  8 12
 3  9 14
 3 10 12
 3 14 15
 4  4  9
 4  8  9
 4  8 15
 4 12 15
 5  6 12
 6  6  7
 6  7 15
 6  8 15
 6  9 10
 6 11 12
 6 11 14
 6 13 15
 8  9 10
 8 14 15
 9 10 11
 9 13 14
10 12 13
12 14 15

 3  4  7
========
 0  3  4
 0  3  7
 0  4  7
 1  1  1
 1  2  5
 1  2  6
 1  3  4
 1  3  6
 1  4  8
 1  5  8
 1  5  9
 1  5 13
 1  7  7
 1  7 13
```

```
H  K  L
U  V  W
--------
 3  4  7
--------
 1  7 15
 1  8 13
 1  9 15
 1 10 11
 1 13 15
 2  2  5
 2  5  9
 2  6  9
 2  7 14
 2 10 11
 3  3 11
 3  5  9
 3  5 11
 3  6 10
 3  7 10
 3 13 15
 4  7 11
 4  8 11
 4  8 13
 5  7 14
 5  8 12
 5  9 12
 6  9 13
 6 10 13
 7 10 14
 7 11 14
 8 11 15
 8 12 15

 3  4  8
========
 0  1  2
 0  3  4
 0  3  8
 1  1  4
 1  2  4
 1  4  5
 1  4  8
 1  4 12
 1  5 12
 1  7 12
 1  8  8
 1 11 12
 2  4  7
 2  5 12
 2 12 13
```

```
H  K  L
U  V  W
--------
 3  4  8
--------
 3  3  4
 3  4  9
 3  8 12
 4  4  5
 4  4 11
 4  5  7
 4  5  8
 4  5 13
 4  6  9
 4  6 15
 4  7 11
 4  8 13
 4  9 15
 5  7 12
 7  8  8
 7  8 12
 8  9 12
10 11 12
11 12 13

 3  4  9
========
 0  1  3
 0  3  4
 0  4  9
 1  1  3
 1  3  6
 1  3  7
 1  5  6
 1  5 12
 1  6 11
 1  7 15
 1  9  9
 1  9 15
 1 12 13
 2  2  3
 2  3 10
 2  6  9
 2  6 15
 2 14 15
 3  3  5
 3  3 13
 3  4  8
 3  5  9
 3  5 11
 3  6 14
```

```
H  K  L
U  V  W
--------
 3  4  9
--------
 3  7 12
 3 11 15
 4  8 15
 5  6  7
 5  7 12
 6  7 13
 7  9  9
 7  9 15
 8  9 12
 9 11 12
11 13 15

 3  4 10
========
 0  2  5
 0  3  4
 0  3 10
 1  1  2
 1  2  4
 1  2  6
 1  2  8
 1  4 12
 1  5 10
 1  6  7
 1  8 14
 1 10 10
 1 13 14
 2  3  6
 2  3  9
 2  5 11
 2  5 14
 2  8 11
 3  3 14
 3 10 15
 4  4  7
 4  4 13
 5  6  8
 6  7 13
 6  8  9
 7 10 10
 8 11 12
 9 10 15
 9 12 14
```

```
H  K  L
U  V  W
--------

 3  4 11
========
 0  3  4
 0  3 11
 0  4 11
 1  1  2
 1  1  5
 1  2  6
 1  3  5
 1  3  9
 1  4  9
 1  5 13
 1  7  8
 1  7 13
 1 11 11
 1 14 15
 2  2  7
 2  5 14
 2  6 15
 2 10 13
 3  3  7
 3  5 12
 4  5  8
 4  5 13
 5  7  9
 6  9 10
 7  8 15
 7 11 11
 8 12 13
 9 13 15

 3  4 12
========
 0  1  3
 0  1  4
 0  3  4
 1  3  8
 1  4  6
 1  6 12
 1  8  9
 1 12 12
 2  3  4
 2  3 12
 2  4  9
 2 12 15
 3  3  8
 3  4  6
```

```
H  K  L
U  V  W
--------
 3  4 12
--------
 3  4 12
 3  8 15
 4  4  9
 4  4 15
 4  5 12
 4  6 15
 5  6 12
 5  8  9
 7  8 15
 7 12 12
 8 12 15

 3  4 13
========
 0  3  4
 0  3 13
 0  4 13
 1  1  3
 1  1  4
 1  2  7
 1  2 10
 1  3  9
 1  5  7
 1  5 11
 1  6 15
 1  9 10
 1 13 13
 2  2  5
 2  6 11
 3  5  6
 3  5 14
 3  7 15
 4  7  8
 5  7 11
 5  8 11
 6  9 14
 7 13 13

 3  4 14
========
 0  2  7
 0  3  4
 0  3 14
 1  1  6
 1  2  2
 1  2  5
```

```
H  K  L
U  V  W
--------
 3  4 14
--------
 1  2  8
 1  4 10
 1  6  8
 1  7 14
 1 10 11
 1 14 14
 2  3  9
 2  3 12
 2  8 13
 3  3 10
 4  4 11
 4  5 12
 5  6 13
 5  7 14
 6  8 15

 3  4 15
========
 0  1  5
 0  3  4
 0  4 15
 1  1  3
 1  3  6
 1  3  9
 1  3 12
 1  6 13
 1  7  9
 1 11 12
 1 15 15
 2  2  9
 2  3  6
 2  3 14
 2 10 15
 3  3 11
 3  5 15
 3  6  7
 4  8  9
 5  9 12
 5  9 13
 6 10 15
 7 15 15

 3  5  5
========
 0  1  1
 0  3  5
```

```
H  K  L
U  V  W
--------
 3  5  5
--------
 1  2  5
 1  4  5
 1  5 10
 1  7 10
 1  8 15
 1 10 15
 2  5  5
 2  7 15
 2 11 15
 3  3 10
 3  5  6
 3  9 10
 4  5  7
 4  5 15
 4 13 15
 5  5  8
 5  6  9
 5  7 10
 5  8 11
 5  9 12
 5 10 11
 5 10 13
 5 11 14
 5 12 15
 5 14 15
 7 10 13
 9 10 15

 3  5  6
========
 0  1  2
 0  3  5
 0  5  6
 1  2  3
 1  3  3
 1  3  7
 1  6  8
 1  6 12
 1  7  9
 1  8  9
 1  9 13
 1 12 15
 1 13 15
 2  3  9
 2  9 11
 2  9 12
```

```
H  K  L
U  V  W
--------
 3  5  6
--------
 2 11 12
 3  3  4
 3  3 11
 3  4  6
 3  4 13
 3  5  5
 3  5 15
 3  6  7
 3  7  9
 3  8 11
 3  9 13
 3 10 15
 3 11 15
 3 12 14
 3 14 15
 4  6  7
 4  7  9
 5  5  9
 5  9 10
 5 10 12
 6  7 12
 6  8  9
 6 11 12
 6 12 13
 6 13 15
 7  9 11
 7  9 15
 7 11 15
 8  9 15
 9 11 13
 9 13 14
10 12 15

 3  5  7
========
 0  3  5
 0  3  7
 0  5  7
 1  1  2
 1  1  4
 1  2  3
 1  4  5
 1  4  9
 1  4 11
 1  5  6
 1  5 10
```

TAB. I THE TYPES OF HKL-UVW FOR H,K,L,U,V,W ≤15

H U	K V	L W
3	**5**	**7**
1	6	9
1	7	14
1	8	11
1	9	12
2	3	3
2	5	13
2	5	15
2	7	7
2	7	11
2	7	13
3	7	8
3	8	9
3	8	13
3	12	13
4	5	5
4	9	11
4	9	15
4	11	13
5	7	8
5	10	11
5	10	13
6	9	11
6	11	15
7	7	8
7	11	14
7	13	14
9	10	11
9	12	13
11	12	13
11	13	14
3	**5**	**8**
0	3	5
0	3	8
0	5	8
1	1	1
1	2	6
1	2	7
1	4	4
1	4	7
1	4	9
1	5	11
1	5	15
1	6	9
1	7	9

H U	K V	L W
3	**5**	**8**
1	9	15
1	10	14
2	3	5
2	3	6
2	5	10
2	7	10
2	11	13
3	3	13
3	6	11
3	7	9
3	7	13
3	8	11
4	7	12
4	9	12
4	11	13
5	5	11
5	8	13
5	10	13
5	10	14
6	9	14
6	11	14
7	10	15
7	12	15
3	**5**	**9**
0	1	3
0	3	5
0	5	9
1	2	3
1	3	6
1	3	8
1	6	7
1	6	13
1	7	12
1	8	15
1	9	12
2	3	11
2	9	9
2	9	15
3	3	4
3	3	14
3	4	7
3	4	9
3	5	10
3	6	13

H U	K V	L W
3	**5**	**9**
3	11	12
4	7	15
4	13	15
5	5	6
5	5	12
6	7	9
6	7	11
6	7	15
7	9	12
8	9	9
8	11	15
9	10	15
11	12	13
11	12	15
13	14	15
3	**5**	**10**
0	1	2
0	3	5
0	3	10
1	1	5
1	2	5
1	4	10
1	4	15
1	5	5
1	5	15
1	7	15
1	8	10
1	11	15
2	5	7
2	5	15
2	13	15
3	3	5
3	5	9
3	10	12
4	5	5
4	5	10
4	5	11
5	5	7
5	5	13
5	6	9
5	6	15
5	7	11
5	7	15
5	8	13

H U	K V	L W
3	**5**	**10**
5	9	15
7	8	10
7	8	15
9	10	12
10	11	15
11	13	15
3	**5**	**11**
0	3	5
0	3	11
0	5	11
1	1	2
1	2	5
1	2	7
1	2	9
1	3	4
1	3	6
1	4	13
1	5	12
1	7	8
1	10	13
2	5	15
2	11	11
3	4	7
3	4	9
3	14	15
4	7	13
5	5	8
5	5	14
6	7	9
7	9	10
8	9	15
8	11	11
9	12	13
10	13	15
3	**5**	**12**
0	1	4
0	3	5
0	5	12
1	1	3
1	3	9
1	4	9
1	6	6

H U	K V	L W
3	**5**	**12**
1	6	14
1	6	15
1	9	11
2	3	3
2	3	6
2	3	13
2	7	9
3	3	7
3	4	11
3	5	15
3	7	15
3	8	12
3	13	15
4	9	15
5	5	9
5	6	10
7	8	12
7	8	15
7	9	13
9	11	15
3	**5**	**13**
0	3	5
0	3	13
0	5	13
1	1	2
1	1	6
1	2	7
1	4	5
1	4	11
1	6	15
1	8	9
1	11	14
2	3	7
2	3	9
2	13	13
3	3	8
3	7	12
4	5	9
4	5	15
5	7	10
6	9	11
7	11	12
8	13	13
9	14	15

H U	K V	L W
3	**5**	**14**
0	3	5
0	3	14
0	5	14
1	1	3
1	2	4
1	2	5
1	2	8
1	2	11
1	3	9
1	5	13
1	7	7
1	10	12
2	9	11
3	4	6
3	4	13
3	11	15
4	7	7
5	5	11
5	8	10
6	9	13
7	11	13
3	**5**	**15**
0	1	3
0	1	5
0	3	5
1	3	10
1	5	6
1	6	15
1	9	10
1	12	15
2	3	5
2	3	15
2	5	9
2	15	15
3	3	10
3	4	15
3	5	6
3	5	12
3	10	15
4	5	9
4	5	15
5	5	12
5	6	15

H U	K V	L W
3	**5**	**15**
5	9	10
7	10	15
7	12	15
8	15	15
3	**6**	**7**
0	1	2
0	3	7
0	6	7
1	3	3
1	3	4
1	3	5
1	3	9
1	6	12
1	9	10
1	9	11
2	3	3
2	3	11
2	12	13
2	12	15
3	3	5
3	3	13
3	4	15
3	5	6
3	6	8
3	7	7
3	8	9
3	9	11
3	10	13
3	11	15
4	5	6
4	6	9
4	9	13
5	6	9
5	9	11
5	9	13
6	8	11
6	11	12
6	12	13
7	7	9
7	9	14
7	12	14
7	14	15
9	10	12
9	13	15

H U	K V	L W
3	**6**	**7**
11	12	15
11	13	15
3	**6**	**8**
0	1	2
0	3	4
0	3	8
1	3	6
1	3	10
1	6	14
2	3	3
2	3	4
2	3	5
2	3	12
2	6	7
2	6	9
2	9	11
2	9	13
2	12	15
3	3	14
3	4	6
3	6	7
3	6	10
3	6	8
3	9	10
3	10	12
3	11	14
4	9	10
4	9	14
5	6	6
5	9	14
6	6	11
6	10	13
6	12	13
6	14	15
7	9	10
8	8	9
9	12	14
10	11	12
12	14	15
13	14	15

H U	K V	L W
3	**6**	**10**
0	1	2
0	3	5
0	3	10
1	3	8
1	3	12
2	3	4
2	3	6
2	3	14
2	6	9
2	6	11
2	9	14
3	3	4
3	4	7
3	6	8
3	6	14
3	8	9
3	10	10
3	11	12
3	12	14
4	9	13
5	6	10
6	6	7
6	6	13
6	10	15
8	9	11
8	9	14
9	10	10
10	12	15
12	13	14
3	**6**	**11**
0	1	2
0	3	11
0	6	11
1	3	5
1	3	6
1	3	9
1	3	13
2	3	7
2	3	15
3	3	4
3	3	5
3	3	7
3	5	8

H U	K V	L W
3	**6**	**11**
3	7	9
3	9	10
3	11	11
3	12	13
3	13	15
4	6	9
4	6	13
5	6	12
5	9	14
6	7	8
6	8	15
7	9	13
9	10	13
9	11	11
12	14	15
3	**6**	**13**
0	1	2
0	3	13
0	6	13
1	3	6
1	3	7
1	3	11
1	3	15
2	3	9
3	3	5
3	3	7
3	3	8
3	4	5
3	5	9
3	7	10
3	9	11
3	11	12
3	13	13
3	14	15
4	6	11
4	6	15
6	7	12
6	8	9
9	11	14
9	13	13

H U	K V	L W
3	**6**	**14**
0	1	2
0	3	7
0	3	14
1	3	12
2	3	6
2	3	8
2	3	10
2	6	13
2	6	15
3	3	8
3	4	5
3	4	6
3	4	9
3	6	10
3	8	11
3	10	12
3	12	13
3	14	14
6	6	11
6	7	14
6	9	10
9	14	14
3	**7**	**7**
0	1	1
0	3	7
1	2	7
1	4	7
1	5	14
1	7	14
2	5	7
3	3	14
3	6	7
3	9	14
4	7	7
5	7	8
5	11	14
6	7	9
7	7	10
7	8	11
7	9	12
7	10	13
7	11	14
7	12	15

H U	K V	L W
3	**7**	**7**
7	13	14
9	14	15
3	**7**	**8**
0	3	7
0	3	8
0	7	8
1	1	5
1	2	2
1	2	3
1	3	3
1	4	5
1	4	12
1	4	13
1	5	9
1	5	11
1	10	11
1	11	13
2	7	11
2	7	14
2	13	14
3	5	7
3	9	13
4	4	5
4	9	12
4	11	15
5	6	9
5	7	7
5	7	8
6	9	10
6	10	11
7	7	11
7	11	13
7	13	14
8	9	13
9	11	12
9	11	13
9	11	15
3	**7**	**9**
0	1	3
0	3	7
0	7	9
1	2	3

TAB. I — THE TYPES OF HKL-UVW FOR H,K,L,U,V,W ≤15

H K L	U V W
3 7 9	1 3 4
	1 3 10
	1 5 6
	1 6 11
	1 9 12
	1 12 15
	2 3 3
	2 3 13
	2 9 15
	2 11 15
	3 4 5
	3 5 6
	3 5 8
	3 6 11
	3 7 14
	3 8 9
	4 9 9
	4 13 15
	5 6 9
	5 11 12
	5 12 13
	6 7 7
	6 9 13
	7 7 12
	7 14 15
	8 9 15
	8 11 15
	9 9 10
	9 11 12
	11 12 13
3 7 10	0 3 7
	0 3 10
	0 7 10
	1 1 1
	1 2 8
	1 2 9
	1 4 6
	1 4 11
	1 5 15
	1 6 9
	1 7 13
	1 8 11
	2 5 5

H K L	U V W
3 7 10	2 5 8
	2 5 12
	2 9 12
	3 4 6
	3 4 7
	3 6 13
	3 9 11
	3 10 13
	4 7 14
	4 11 14
	5 8 15
	5 9 11
	5 9 15
	5 12 15
	7 7 13
3 7 11	0 3 7
	0 3 11
	0 7 11
	1 1 2
	1 1 6
	1 2 5
	1 4 13
	1 5 8
	1 6 9
	1 8 13
	1 8 15
	2 5 7
	2 7 9
	2 9 15
	3 3 4
	3 4 5
	3 9 10
	4 5 9
	4 11 11
	5 7 14
	5 9 12
	5 12 13
	6 13 15
	7 7 8
	7 8 13
	10 11 11
	10 11 13
	13 14 15

H K L	U V W
3 7 12	0 1 4
	0 3 7
	0 7 12
	1 2 3
	1 3 3
	1 3 11
	1 5 9
	1 6 10
	1 9 15
	2 3 6
	2 3 15
	2 9 13
	3 3 5
	3 4 9
	3 5 13
	3 8 15
	4 5 9
	5 6 6
	5 8 12
	6 7 14
	6 11 15
	7 7 9
	7 7 15
	8 9 11
	8 9 12
	9 11 15
	12 13 15
3 7 13	0 3 7
	0 3 13
	0 7 13
	1 1 2
	1 2 9
	1 2 11
	1 3 6
	1 4 5
	1 4 7
	1 4 15
	1 7 12
	2 3 5
	3 5 8
	3 8 9
	4 5 11

H K L	U V W
3 7 13	4 9 15
	4 13 13
	5 6 9
	5 11 14
	7 7 10
	8 9 11
	9 11 12
	10 13 13
	11 14 15
3 7 14	0 1 2
	0 3 7
	0 3 14
	1 1 7
	1 2 7
	1 4 14
	1 5 7
	1 8 14
	2 7 7
	3 3 7
	3 7 9
	3 12 14
	4 5 7
	4 5 14
	4 7 11
	5 7 7
	5 7 13
	6 7 9
	6 7 15
	7 7 11
	7 8 13
	7 8 14
	7 9 15
	9 12 14
3 7 15	0 1 5
	0 3 7
	0 7 15
	1 2 3
	1 3 6
	1 3 12
	1 4 9

H K L	U V W
3 7 15	1 6 9
	2 3 3
	2 9 11
	3 3 8
	3 4 13
	3 5 12
	3 12 13
	4 5 9
	4 15 15
	5 6 11
	5 6 15
	5 8 15
	7 7 12
	7 9 14
	9 10 15
3 8 8	0 1 1
	0 3 8
	1 2 8
	1 4 8
	2 5 8
	3 6 8
	4 7 8
	5 8 8
	6 8 9
	7 8 10
	8 8 11
	8 9 12
	8 10 13
	8 11 14
	8 12 15
3 8 9	0 1 3
	0 3 8
	0 8 9
	1 3 3
	1 3 5
	1 3 11
	1 5 6
	1 6 13
	1 11 12
	1 13 15

H K L	U V W
3 8 9	2 2 3
	2 3 14
	2 14 15
	3 4 4
	3 5 7
	3 6 7
	3 6 10
	3 7 9
	3 7 13
	4 9 12
	4 12 15
	5 6 7
	5 9 9
	5 9 12
	6 9 10
	6 9 11
	7 11 12
	7 11 15
	7 12 13
	9 9 11
	9 13 15
	12 13 15
3 8 10	0 3 8
	0 3 10
	0 4 5
	1 1 6
	1 2 2
	1 2 4
	1 4 14
	1 5 10
	1 5 14
	2 3 3
	2 5 7
	2 7 8
	2 7 12
	2 9 12
	2 11 13
	3 9 14
	4 5 6
	4 6 9
	4 10 11
	5 7 10
	6 7 11

H K L	U V W
3 8 10	6 9 14
	6 13 14
	7 8 8
	8 8 13
	9 10 15
	10 10 11
	11 13 14
	12 13 14
3 8 11	0 3 8
	0 3 11
	0 8 11
	1 1 1
	1 2 9
	1 2 10
	1 4 7
	1 4 12
	1 7 10
	1 7 15
	1 9 12
	2 5 6
	2 5 13
	2 6 9
	2 10 13
	3 5 6
	3 5 8
	3 6 14
	3 9 13
	3 11 14
	4 4 7
	4 7 15
	4 12 15
	5 11 11
	7 9 13
	7 9 15
3 8 12	0 1 4
	0 2 3
	0 3 8
	1 3 4
	1 3 12
	1 6 12

H K L	U V W
3 8 12	3 3 4
	3 4 6
	3 4 8
	3 5 12
	4 5 6
	4 5 9
	4 7 9
	4 7 12
	4 8 9
	4 9 12
	4 9 15
	4 11 15
	5 12 12
	6 7 12
	7 12 15
	8 8 9
	8 8 15
	8 12 15
	11 12 12
	12 13 15
3 8 13	0 3 8
	0 3 13
	0 8 13
	1 1 2
	1 1 7
	1 2 6
	1 4 15
	1 5 9
	1 7 11
	1 9 15
	2 6 9
	2 7 10
	3 3 5
	3 5 7
	3 9 11
	4 4 5
	4 5 11
	4 11 12
	5 13 13
	6 14 15
	7 8 9
	7 9 12
	10 11 14

H K L	U V W
3 8 13	11 13 13
3 8 14	0 3 8
	0 3 14
	0 4 7
	1 1 2
	1 2 10
	1 2 12
	1 4 6
	1 7 14
	2 3 6
	2 4 5
	2 5 8
	2 7 13
	2 9 15
	3 9 10
	4 6 9
	5 6 11
	5 14 14
	6 7 10
	6 8 11
	9 10 12
	10 12 13
	11 14 14
3 8 15	0 1 5
	0 3 8
	0 8 15
	1 3 3
	1 3 6
	1 3 13
	1 5 9
	1 6 11
	2 2 3
	2 9 14
	3 3 7
	3 4 12
	3 7 12
	4 4 9
	5 6 9
	5 7 12
	5 7 15

H K L	U V W
3 8 15	5 9 15
	6 10 15
	7 9 11
	9 12 13
	11 15 15
3 9 10	0 1 3
	0 3 10
	0 9 10
	1 3 3
	1 3 7
	1 3 13
	1 6 7
	1 12 13
	2 3 4
	3 5 5
	3 6 8
	3 6 11
	3 7 11
	3 8 14
	3 9 11
	5 5 6
	5 9 15
	5 12 15
	6 7 9
	6 8 9
	6 11 13
	7 9 9
	7 11 12
	8 14 15
	9 9 13
	9 12 13
	9 12 14
	11 13 15
	12 14 15
3 9 11	0 1 3
	0 3 11
	0 9 11
	1 3 4
	1 3 8
	1 3 14

H K L	U V W
3 9 11	1 6 7
	1 12 15
	2 3 3
	2 3 5
	3 4 5
	3 6 7
	3 6 13
	3 7 10
	3 8 13
	3 9 10
	5 6 7
	5 6 9
	5 12 13
	6 9 13
	6 11 11
	7 9 12
	8 9 9
	9 9 14
	11 11 12
	13 14 15
3 9 13	0 1 3
	0 3 13
	0 9 13
	1 3 4
	1 3 10
	1 6 9
	2 3 5
	2 3 7
	3 3 4
	3 5 6
	3 7 8
	3 8 11
	3 9 14
	5 6 7
	5 6 11
	6 7 11
	6 9 11
	6 13 13
	7 12 15
	8 9 15
	9 9 10
	12 13 13

H K L	U V W
3 9 14	0 1 3
	0 3 14
	0 9 14
	1 3 5
	1 3 11
	1 6 9
	2 3 4
	2 3 8
	3 3 5
	3 4 6
	3 7 7
	3 8 10
	3 9 13
	5 6 11
	5 6 13
	6 7 7
	6 11 13
	9 9 11
	9 10 12
	11 12 15
3 10 10	0 1 1
	0 3 10
	1 2 10
	1 4 10
	2 5 10
	3 6 10
	4 7 10
	5 8 10
	6 9 10
	7 10 10
	8 10 11
	9 10 12
	10 10 13
	10 11 14
	10 12 15
3 10 11	0 3 10
	0 3 11
	0 10 11
	1 1 7

TAB. I — THE TYPES OF HKL-UVW FOR H,K,L,U,V,W ≤15 — I - 20

```
H K L
U V W
--------
3 10 11
--------
1 2 3
1 2 4
1 3 3
1 5 13
1 5 15
1 7 8
1 13 14
2 4 5
3 11 13
4 5 5
5 5 7
5 7 9
6 7 8
6 8 9
6 9 13
7 9 12
7 10 11
7 11 11
8 13 14
9 12 14
11 11 13

3 10 12
========
0 1 4
0 3 10
0 5 6
1 3 6
1 3 14
2 2 3
2 3 3
2 7 9
2 8 9
2 12 15
2 13 15
3 4 6
3 5 10
3 6 8
3 6 14
4 9 12
4 9 14
4 11 12
5 9 10
6 7 8
6 11 15
6 14 15
```

```
H K L
U V W
--------
3 10 12
--------
7 12 12
9 10 10
9 11 14
9 14 15
12 12 13

3 10 13
========
0 3 10
0 3 13
0 10 13
1 1 1
1 2 11
1 2 12
1 4 9
1 4 14
1 9 12
1 11 14
2 5 8
2 5 15
2 8 11
2 12 15
3 6 7
3 7 10
4 6 7
4 6 9
5 5 8
5 11 15
7 13 13
9 11 15

3 10 14
========
0 3 10
0 3 14
0 5 7
1 1 8
1 2 2
1 2 6
1 5 12
2 4 5
2 6 9
2 7 14
2 9 12
3 3 4
4 5 10
```

```
H K L
U V W
--------
3 10 14
--------
4 7 11
6 7 8
6 8 13
7 8 14
8 9 15
8 11 13
10 10 11
13 14 14

3 10 15
========
0 1 5
0 2 3
0 3 10
1 3 5
1 3 15
1 6 15
3 3 5
3 4 10
3 5 6
3 5 15
4 9 10
5 5 6
5 5 9
5 7 9
5 7 12
5 9 12
5 9 15
5 11 15
5 12 15
6 7 15
7 15 15
8 9 10
8 10 15
10 12 15
11 12 15
13 15 15

3 11 11
========
0 1 1
0 3 11
1 2 11
1 4 11
2 5 11
3 6 11
```

```
H K L
U V W
--------
3 11 11
--------
4 7 11
5 8 11
6 9 11
7 10 11
8 11 11
9 11 12
10 11 13
11 11 14
11 12 15

3 11 12
========
0 1 4
0 3 11
0 11 12
1 3 3
1 3 7
1 3 15
1 8 9
1 14 15
2 3 3
2 5 6
3 4 5
3 5 9
3 6 10
3 6 13
3 13 15
5 7 9
5 9 13
6 6 7
6 9 14
7 9 10
7 12 15
8 9 12
8 12 13
9 11 11
11 11 15

3 11 13
========
0 3 11
0 3 13
0 11 13
1 1 8
1 2 3
1 2 5
```

```
H K L
U V W
--------
3 11 13
--------
1 4 5
1 5 14
1 7 8
2 3 3
3 10 11
4 5 7
5 7 12
5 9 12
5 13 14
6 7 9
7 8 9
7 10 11
8 13 13
10 11 11
13 13 14

3 11 14
========
0 3 11
0 3 14
0 11 14
1 1 1
1 2 12
1 2 13
1 4 10
1 4 15
1 10 13
1 12 15
2 5 9
2 9 12
3 6 8
3 8 11
4 7 7
4 7 10
5 6 8
5 6 9
7 13 15
9 13 15

3 11 15
========
0 1 5
0 3 11
0 11 15
1 2 3
1 3 6
```

```
H K L
U V W
--------
3 11 15
--------
1 9 12
2 7 9
3 3 4
3 4 9
3 5 6
3 5 14
3 6 7
4 9 13
5 8 9
5 12 15
6 7 13
7 8 9
7 9 12
8 15 15
9 10 15
10 13 15
11 11 12
14 15 15

3 12 13
========
0 1 4
0 3 13
0 12 13
1 3 3
1 3 9
1 9 10
2 3 5
2 6 7
3 3 4
3 5 7
3 6 11
3 6 14
3 7 15
5 6 6
5 9 11
6 9 10
7 8 9
7 9 11
8 11 12
8 12 15
9 13 13
9 14 15
13 13 15
```

```
H K L
U V W
--------

3 12 14
========
0 1 4
0 3 14
0 6 7
1 3 10
2 3 3
2 3 4
2 3 6
2 9 10
2 9 11
3 5 6
3 6 10
3 7 14
4 12 13
4 12 15
5 6 8
6 8 9
7 9 14
8 9 10
9 10 13
9 14 14
11 12 12
14 14 15

3 13 13
========
0 1 1
0 3 13
1 2 13
1 4 13
2 5 13
3 6 13
4 7 13
5 8 13
6 9 13
7 10 13
8 11 13
9 12 13
10 13 13
11 13 14
12 13 15

3 13 14
========
0 3 13
0 3 14
```

```
H K L
U V W
--------
3 13 14
--------
0 13 14
1 1 9
1 2 4
1 2 5
1 3 3
1 10 11
2 7 8
3 4 5
4 5 6
5 7 7
5 7 11
5 11 13
6 9 11
7 7 8
7 11 15
8 10 11
9 10 12
9 13 14
11 13 13

3 13 15
========
0 1 5
0 3 13
0 13 15
1 3 8
1 5 6
1 8 9
2 3 3
3 4 7
3 5 12
3 6 11
3 11 12
4 7 9
5 9 14
5 12 15
5 14 15
6 7 9
7 9 12
9 10 11
10 11 15
12 13 13
```

```
H K L
U V W
--------

3 14 14
========
0 1 1
0 3 14
1 2 14
1 4 14
2 5 14
3 6 14
4 7 14
5 8 14
6 9 14
7 10 14
8 11 14
9 12 14
10 13 14
11 14 14
12 14 15

3 14 15
========
0 1 5
0 3 14
0 14 15
1 3 3
1 3 9
1 11 12
2 3 4
2 8 9
3 4 6
3 5 6
3 5 11
3 6 13
5 13 15
6 7 7
7 7 9
8 9 10
9 11 12
9 11 13
9 12 13
10 12 15
11 15 15

4 4 5
========
0 1 1
0 4 5
1 4 4
```

```
H K L
U V W
--------
4 4 5
--------
1 4 6
1 8 9
1 8 11
1 12 14
2 3 4
2 4 7
2 12 13
3 4 8
3 7 8
3 8 13
4 4 9
4 5 10
4 6 11
4 7 12
4 8 13
4 9 14
4 10 15
4 11 12
5 5 8
5 8 15
5 10 12
7 8 12

4 4 7
========
0 1 1
0 4 7
1 4 6
1 4 8
1 8 13
1 8 15
2 4 5
2 4 9
3 4 4
3 4 10
3 8 11
4 4 11
4 5 12
4 6 13
4 7 14
4 8 15
5 8 9
7 7 8
7 12 14
8 12 13
10 11 12
```

```
H K L
U V W
--------

4 4 9
========
0 1 1
0 4 9
1 4 8
1 4 10
2 4 7
2 4 11
3 4 6
3 4 12
3 8 15
4 4 5
4 4 13
4 5 14
4 6 15
5 8 13
7 8 11
8 9 9
12 13 14

4 4 11
========
0 1 1
0 4 11
1 4 10
1 4 12
2 4 9
2 4 13
3 4 8
3 4 14
4 4 7
4 4 15
4 5 6
7 8 15
8 9 13
8 11 11

4 4 13
========
0 1 1
0 4 13
1 4 12
1 4 14
2 4 11
2 4 15
3 4 10
4 4 9
```

```
H K L
U V W
--------
4 4 13
--------
4 5 8
4 6 7
8 11 15
8 13 13

4 4 15
========
0 1 1
0 4 15
1 4 14
2 4 13
3 4 12
4 4 11
4 5 10
4 6 9
4 7 8
6 15 15

4 5 5
========
0 1 1
0 4 5
1 3 5
1 5 5
1 7 10
1 9 10
1 11 15
1 13 15
2 2 5
2 5 6
2 10 15
2 14 15
3 5 7
3 5 10
3 10 11
4 5 8
4 8 15
5 5 9
5 6 10
5 7 11
5 7 15
5 8 12
5 9 13
5 10 13
5 10 14
5 11 15
```

TAB. I THE TYPES OF HKL-UVW FOR H,K,L,U,V,W ≤15

H U	K V	L W
4	5	5
7	10	15
4	5	6
0	2	3
0	4	5
0	5	6
1	1	2
1	2	4
1	4	4
1	6	6
1	6	8
1	6	9
1	9	10
1	10	11
1	10	14
1	11	14
2	2	3
2	3	7
2	5	5
2	5	10
2	7	8
2	7	10
2	7	13
2	8	13
2	9	11
2	11	14
2	13	14
3	6	7
3	8	10
3	8	12
3	13	14
4	4	11
4	7	8
4	9	12
4	9	14
4	10	11
4	12	13
5	6	15
5	8	10
5	10	14
5	14	15
6	9	11
6	12	13
7	10	13
8	11	14

H U	K V	L W
4	5	7
0	4	5
0	4	7
0	5	7
1	1	3
1	2	2
1	2	3
1	3	4
1	3	5
1	5	8
1	5	10
1	7	7
1	7	9
1	7	11
1	8	12
1	9	13
1	11	12
2	6	11
2	6	13
2	9	10
2	9	11
2	10	15
3	4	8
3	5	5
3	7	14
3	11	13
4	7	13
4	8	9
4	13	15
5	5	11
5	6	7
5	11	13
6	7	14
7	7	9
7	13	14
8	9	11
8	11	12
9	11	13
10	13	15
4	5	8
0	1	2
0	4	5
0	5	8
1	2	4

H U	K V	L W
4	5	8
1	3	4
1	4	7
1	7	12
1	8	8
1	8	12
1	12	13
2	4	9
2	11	12
3	4	4
3	4	8
3	4	11
4	4	13
4	5	5
4	5	15
4	6	7
4	7	8
4	7	9
4	7	12
4	8	11
4	9	13
4	10	15
5	5	12
5	10	12
7	11	12
8	8	9
8	11	12
11	12	13
12	13	14
4	5	9
0	4	5
0	4	9
0	5	9
1	1	1
1	3	6
1	3	8
1	4	5
1	4	8
1	5	10
1	6	10
1	7	11
1	9	9
1	13	14
2	2	7
2	3	6

H U	K V	L W
4	5	9
2	3	7
2	6	11
2	7	11
2	13	14
3	5	13
3	5	15
3	7	11
3	7	12
3	7	15
3	8	12
4	8	13
4	9	13
5	5	13
5	9	14
5	10	14
6	10	15
6	11	15
4	5	10
0	1	2
0	2	5
0	4	5
1	2	5
1	5	6
1	6	10
1	10	10
1	10	15
1	14	15
2	3	5
2	3	10
2	5	8
2	5	10
2	5	15
2	7	15
2	8	15
3	5	10
3	10	14
4	4	5
4	4	15
4	5	12
4	8	15
5	5	6
5	5	14
5	6	8
5	7	10

H U	K V	L W
4	5	10
5	8	12
5	9	14
6	7	10
8	10	15
9	10	10
10	11	14
10	11	15
13	14	15
4	5	11
0	4	5
0	4	11
0	5	11
1	1	3
1	1	4
1	3	7
1	4	8
1	5	9
1	5	15
1	6	7
1	6	14
1	9	14
1	11	11
2	2	3
2	3	5
2	6	13
2	7	10
3	5	10
3	7	13
3	8	13
5	5	7
5	7	13
7	8	9
7	8	12
9	11	11
11	13	14
4	5	12
0	1	3
0	4	5
0	5	12
1	2	4
1	3	8
1	4	8

H U	K V	L W
4	5	12
1	7	8
1	8	13
1	12	12
2	4	11
2	9	12
3	4	4
3	4	12
3	4	14
4	4	7
4	5	10
4	6	13
5	5	8
6	7	12
7	8	11
8	9	12
10	12	15
4	5	13
0	4	5
0	4	13
0	5	13
1	1	2
1	2	6
1	3	5
1	3	7
1	3	11
1	5	15
1	7	12
1	8	9
1	13	13
2	2	9
2	11	14
3	4	8
3	4	11
4	7	15
5	5	9
6	7	10
7	9	11
8	11	12
9	13	13
10	14	15

H U	K V	L W
4	5	14
0	2	7
0	4	5
0	5	14
1	1	2
1	2	6
1	4	6
1	4	12
1	6	11
1	9	10
1	14	14
2	3	8
2	3	10
2	3	13
2	5	15
3	7	14
4	4	9
5	6	10
6	8	11
7	10	12
8	12	13
9	14	14
4	5	15
0	1	3
0	4	5
0	4	15
1	1	5
1	3	10
1	5	7
1	5	10
1	9	15
1	10	11
1	15	15
2	2	5
2	5	10
2	6	15
3	5	5
3	5	13
3	5	15
4	5	8
5	5	11
5	6	14
5	7	10
7	9	15

H U	K V	L W
4	5	15
7	10	13
8	12	15
4	6	7
0	2	3
0	4	7
0	6	7
1	2	2
1	2	3
1	2	5
1	4	4
1	6	9
1	6	12
1	8	10
1	10	11
2	3	8
2	4	5
2	5	11
2	7	7
2	7	14
2	8	11
2	9	10
2	10	13
2	11	13
2	14	15
3	5	6
3	10	13
4	4	13
4	5	8
4	9	10
4	11	12
4	12	15
5	6	8
5	13	14
6	6	11
6	9	13
7	7	10
7	8	14
8	9	12
8	11	14
9	11	14

H U	K V	L W
4	6	9
0	2	3
0	4	9
1	2	3
1	2	6
1	6	12
1	6	15
2	3	5
2	3	9
2	5	12
2	6	7
2	7	15
2	8	15
2	9	9
2	11	12
2	13	15
3	4	4
3	4	8
3	6	7
3	6	11
3	8	10
3	8	14
3	10	13
4	4	15
4	9	12
5	6	6
6	6	13
6	8	9
6	10	11
7	10	12
8	12	15
9	9	10
9	14	15
11	14	15
12	13	14
4	6	11
0	2	3
0	4	11
0	6	11
1	2	3
1	2	4
1	2	7
1	4	8
1	6	15

H U	K V	L W
4	6	11
1	8	14
2	2	5
2	3	10
2	5	7
2	5	13
2	8	9
2	11	11
2	13	14
3	6	13
4	4	5
4	7	12
5	6	9
6	6	7
6	8	13
7	8	10
8	10	13
9	10	14
10	11	11
4	6	13
0	2	3
0	4	13
0	6	13
1	2	5
1	2	8
1	4	8
2	2	3
2	3	11
2	4	7
2	5	14
2	7	9
2	10	11
2	13	13
3	6	11
4	4	7
4	5	12
5	6	12
5	8	14
6	7	9
8	10	11
10	13	13

H U	K V	L W
4	6	15
0	2	3
0	2	5
0	4	15
1	2	6
1	2	9
2	3	3
2	3	7
2	3	12
2	5	15
2	6	9
2	9	11
2	12	13
2	15	15
3	4	8
3	4	12
3	6	13
4	4	9
5	6	15
6	6	11
6	7	12
8	9	14
8	10	15
4	7	7
0	1	1
0	4	7
1	3	7
1	5	7
1	7	14
1	9	14
2	2	7
2	6	7
3	5	14
3	7	7
3	11	14
4	7	8
5	7	9
5	13	14
6	7	10
7	7	11
7	8	12
7	9	13
7	10	14
7	11	15

H U	K V	L W
4	7	7
7	14	15
4	7	8
0	1	2
0	4	7
0	7	8
1	3	4
1	4	4
1	4	5
1	4	9
1	8	12
1	10	12
1	11	12
2	3	4
2	4	11
3	4	5
3	4	13
3	8	8
4	4	15
4	5	6
4	5	8
4	7	7
4	8	9
4	9	11
4	10	13
4	11	15
4	12	13
5	8	12
5	11	12
5	12	13
7	7	12
7	12	14
8	8	11
8	12	13
4	7	9
0	4	7
0	4	9
0	7	9
1	1	4
1	2	2
1	3	3
1	3	5
1	5	7

TAB. I THE TYPES OF HKL-UVW FOR H,K,L,U,V,W ≤15

H K L U V W	H K L U V W	H K L U V W	H K L U V W	H K L U V W	H K L U V W	H K L U V W	H K L U V W	H K L U V W	H K L U V W	H K L U V W	H K L U V W
4 7 9	4 7 10	4 7 12		4 7 15	4 8 11		4 9 10	4 9 11	4 9 13	4 9 15	4 10 11
			4 7 14			4 8 15					
1 5 11	4 11 12	1 4 10	========	9 11 13	2 4 15	========	1 2 7	6 14 15	1 5 14	1 1 6	11 11 14
1 5 13	5 9 10	1 5 8	0 1 2	10 13 14	3 4 4	0 1 2	1 4 4	7 9 9	1 8 12	1 3 3	
1 7 10	5 12 14	1 8 11	0 2 7	11 15 15	3 4 5	0 4 15	1 5 6	7 10 13	1 10 14	1 3 9	4 10 13
1 7 14	6 7 7	2 3 4	0 4 7		3 4 7	0 8 15	1 6 11	8 12 15	2 2 11	1 5 15	========
1 9 11	8 10 11	2 4 13	1 2 7	4 8 9	4 5 8	1 4 7	1 13 14	9 10 11	2 6 7	1 7 12	0 2 5
2 3 5	8 13 14	2 12 15	1 6 7	========	4 7 9	1 4 8	2 3 3	11 11 13	2 6 15	2 2 3	0 4 13
2 6 15	10 10 11	3 5 8	1 6 14	0 1 2	4 8 9	1 4 13	2 3 12	12 13 15	2 7 11	2 10 15	0 10 13
2 10 13	10 13 15	3 8 12	1 10 14	0 4 9	4 8 13	2 4 11	2 5 8		2 11 15	3 4 8	1 2 3
2 11 13	13 14 15	4 4 5	2 3 7	0 8 9	4 9 10	3 4 6	2 7 13	4 9 12	3 6 7	3 5 7	1 2 4
3 5 6		4 5 8	2 3 14	1 4 4	4 11 11	3 4 9	2 8 13	========	3 6 10	3 5 12	1 2 9
3 8 12	4 7 11	4 6 11	2 5 14	1 4 5	4 12 13	4 4 7	3 6 8	0 1 3	3 11 15	3 7 13	1 8 10
3 13 15	========	4 7 14	2 7 8	1 4 7	4 13 15	4 5 5	3 10 15	0 3 4	4 5 8	3 8 12	2 3 14
4 5 8	0 4 7	4 9 12	3 7 10	1 4 11	5 8 12	4 5 10	4 6 7	0 4 9	4 5 9	5 6 11	2 5 6
4 8 11	0 4 11	5 6 12	3 14 14	1 12 13	5 12 14	4 7 11	4 8 11	1 4 6	5 13 13	5 7 15	2 6 7
5 7 7	0 7 11	7 7 8	4 4 7	1 12 14	7 8 8	4 8 13	4 11 14	1 4 12	7 11 15	6 7 9	2 7 11
5 8 11	1 1 1	8 9 13	4 7 12	2 4 5	7 12 13	4 9 12	5 7 10	1 8 15		6 9 14	3 4 4
6 11 14	1 3 8	9 10 12	5 6 7	2 4 13	8 8 15	4 11 13	5 10 11	2 3 4	4 9 14	9 9 11	4 7 8
7 7 13	1 3 10	11 12 12	5 7 14	3 3 4	10 12 13	4 13 14	6 11 14	2 4 15	========	11 12 13	5 6 7
7 8 9	1 5 6		6 7 8	3 4 6	11 11 12	4 15 15	6 14 15	3 4 4	0 2 7	13 15 15	5 10 11
8 12 13	1 5 12	4 7 13	6 7 14	3 4 15		8 8 11	7 8 10	3 5 8	0 4 9		6 8 11
9 9 11	1 6 10	========	7 7 10	3 8 12	4 8 13	8 9 12	7 8 12	3 7 8	0 9 14	4 10 11	6 13 13
10 11 13	1 7 15	0 4 7	7 8 12	4 5 7	========		9 9 14	3 8 9	1 1 2	========	7 10 15
11 13 15	1 8 12	0 4 13	7 9 14	4 7 8	0 1 2	4 9 9	10 10 13	3 8 12	1 2 8	0 2 5	8 9 14
	1 9 13	0 7 13	9 10 14	4 8 11	0 4 13	========		3 10 12	1 6 10	0 4 11	8 14 15
4 7 10	2 2 9	1 1 5	11 14 14	4 9 9	0 8 13	0 1 1	4 9 11	4 5 6	1 8 12	0 10 11	9 10 10
========	2 5 6	1 2 3		4 10 11	1 4 6	0 4 9	========	4 6 9	2 3 6	1 2 3	13 13 14
0 2 5	2 5 9	1 3 5	4 7 15	4 11 13	1 4 7	1 3 9	0 4 9	4 7 12	2 3 15	1 2 8	
0 4 7	2 6 13	1 3 8	========	4 12 15	1 4 11	1 5 9	0 4 11	4 8 15	2 5 13	1 4 4	4 10 15
0 7 10	2 9 13	1 4 8	0 4 7	5 8 8	1 4 15	2 2 9	0 9 11	4 12 15	2 7 10	1 6 7	========
1 1 2	3 4 7	1 5 12	0 4 15	5 11 12	2 4 9	2 6 9	1 1 5	5 12 12	2 8 11	1 6 14	0 2 3
1 2 6	3 4 8	1 7 9	0 7 15	7 10 12	3 4 5	3 7 9	1 2 2	6 7 12	3 6 6	1 14 15	0 2 5
1 6 8	3 7 14	2 2 3	1 1 2	7 12 13	3 4 7	4 8 9	1 3 4	6 11 12	3 10 12	2 2 3	0 4 15
1 6 13	3 10 14	2 10 11	1 3 7	8 8 13	3 4 8	5 9 9	1 3 6	8 9 9	4 4 5	2 3 13	1 2 5
1 8 12	3 11 11	3 5 11	1 3 9	8 11 12	4 4 5	6 9 10	1 5 14	8 11 15	4 5 6	2 5 7	1 2 10
1 10 15	4 8 15	3 13 13	1 3 13	8 12 15	4 5 9	7 9 11	1 7 9	12 12 13	4 12 13	2 7 12	2 3 15
2 3 4	4 11 15	4 5 7	1 4 8		4 7 10	8 9 12	1 7 13	12 14 15	5 14 14	3 8 10	2 5 5
2 3 11	5 9 13	5 6 14	1 5 5	4 8 11	4 8 11	9 9 13	1 11 13		6 7 11	4 5 6	2 7 10
2 5 6	7 7 15	5 7 14	1 9 12	========	4 8 15	9 10 14	2 5 7	4 9 13	7 11 14	4 8 9	2 9 15
2 5 9		5 8 12	2 2 11	0 1 2	4 9 11	9 11 15	2 13 15	========	9 9 10	5 9 10	4 4 5
2 7 14	4 7 12	7 7 9	2 3 6	0 4 11	4 11 12		3 3 5	0 4 9	13 14 14	5 10 13	4 5 6
2 8 9	========	9 10 11	3 5 9	0 8 11	4 13 13	4 9 10	3 5 7	0 4 13		6 7 8	4 12 15
2 9 14	0 1 3	11 13 13	3 8 11	1 4 5	4 14 15	========	4 7 8	0 9 13	4 9 15	6 11 11	5 6 7
3 4 4	0 4 7		4 5 13	1 4 6	7 8 12	0 2 5	4 8 13	1 1 1	========	6 13 14	5 6 10
3 10 10	0 7 12		5 6 10	1 4 9	8 8 9	0 4 9	5 7 8	1 3 10	0 3 5	7 10 10	5 6 11
3 11 14	1 2 4		7 7 11	1 4 13	11 12 14	0 9 10	5 6 12	1 3 12	0 4 9	8 12 13	5 8 10
4 8 13	1 4 4		8 9 12	2 4 7	12 13 13	1 2 2	5 11 11	1 5 8	0 4 15	8 13 14	5 8 14

H	K	L
U	V	W
4	10	15
---	---	---
5	10	13
6	8	15
6	10	13
9	10	15
10	10	11
14	15	15
4	11	11
===	===	===
0	1	1
0	4	11
1	3	11
1	5	11
2	2	11
2	6	11
3	7	11
4	8	11
5	9	11
6	10	11
7	11	11
8	11	12
9	11	13
10	11	14
11	11	15
4	11	12
■■■	■■■	■■■
0	1	3
0	4	11
0	11	12
1	4	4
1	4	8
1	4	14
1	7	8
2	3	4
2	4	5
3	8	13
3	10	12
4	4	5
4	6	7
4	7	10
4	8	13
5	7	8
5	8	9
6	12	13
7	12	12

H	K	L
U	V	W
4	11	12
---	---	---
8	9	12
8	11	11
9	12	14
4	11	13
■■■	■■■	■■■
0	4	11
0	4	13
0	11	13
1	1	6
1	2	2
1	3	5
1	3	7
1	9	11
1	13	15
2	7	9
3	4	5
3	5	7
3	5	8
4	8	9
4	8	15
5	7	9
5	11	14
7	8	12
7	9	10
7	13	13
9	11	11
11	12	13
13	13	15
4	11	14
■■■	■■■	■■■
0	2	7
0	4	11
0	11	14
1	2	2
1	2	9
1	5	6
1	6	8
1	6	13
2	3	5
2	5	12
2	8	15
3	4	4
3	7	10
4	9	10

H	K	L
U	V	W
4	11	14
---	---	---
5	8	12
6	7	8
7	9	14
7	13	14
8	10	13
10	11	11
14	14	15
4	11	15
■■■	■■■	■■■
0	4	11
0	4	15
0	11	15
1	1	1
1	3	12
1	3	14
1	5	10
1	10	14
2	2	13
2	6	9
2	9	13
3	7	8
3	8	12
4	7	8
4	7	11
5	6	9
5	6	10
7	15	15
4	12	13
■■■	■■■	■■■
0	1	3
0	4	13
0	12	13
1	4	4
1	4	10
1	8	9
2	4	5
2	4	7
3	4	4
3	12	14
4	5	6
4	7	8
4	8	11
4	9	14
5	7	8

H	K	L
U	V	W
4	12	13
---	---	---
5	8	11
6	11	12
7	8	11
8	12	15
8	13	13
9	10	12
4	12	15
===	===	===
0	1	3
0	4	5
0	4	15
1	4	12
2	4	9
3	4	4
3	4	6
3	8	9
3	8	11
3	12	14
4	6	7
4	8	9
4	9	12
4	10	15
5	8	15
6	12	13
7	8	9
8	9	13
8	15	15
10	12	15
11	12	12
4	13	13
===	===	===
0	1	1
0	4	13
1	3	13
1	5	13
2	2	13
2	6	13
3	7	13
4	8	13
5	9	13
6	10	13
7	11	13
8	12	13
9	13	13

H	K	L
U	V	W
4	13	13
---	---	---
10	13	14
11	13	15
4	13	14
===	===	===
0	2	7
0	4	13
0	13	14
1	2	3
1	2	10
1	4	4
1	9	10
2	3	4
2	5	6
2	5	11
4	8	15
5	6	7
5	6	8
6	10	11
7	8	10
7	11	14
7	14	15
8	9	10
8	11	12
9	14	14
10	13	13
4	13	15
===	===	===
0	4	13
0	4	15
0	13	15
1	1	7
1	2	2
1	3	6
1	3	8
1	11	13
2	9	11
3	5	5
3	5	9
3	7	9
4	5	7
4	8	11
5	7	10
7	9	11
8	9	12

H	K	L
U	V	W
4	13	15
---	---	---
9	11	12
11	13	13
13	14	15
4	14	15
■■■	■■■	■■■
0	2	7
0	4	15
0	14	15
1	2	4
1	2	11
1	4	4
1	10	11
2	3	3
2	5	10
2	6	7
4	8	13
5	5	6
5	8	10
6	8	9
6	9	10
7	10	13
7	13	14
8	10	13
11	14	14
4	15	15
■■■	■■■	■■■
0	1	1
0	4	15
1	3	15
1	5	15
2	2	15
2	6	15
3	7	15
4	8	15
5	9	15
6	10	15
7	11	15
8	12	15
9	13	15
10	14	15
11	15	15

H	K	L
U	V	W
---	---	---
5	5	6
■■■	■■■	■■■
0	1	1
0	5	6
1	5	5
1	5	7
1	10	11
1	10	13
2	4	5
2	5	8
3	3	5
3	5	9
3	9	10
3	10	15
4	5	10
4	14	15
5	5	11
5	6	12
5	7	10
5	7	13
5	8	14
5	9	15
5	13	15
7	11	15
8	10	15
5	5	7
■■■	■■■	■■■
0	1	1
0	5	7
1	5	6
1	5	8
1	10	13
1	10	15
2	5	5
2	5	9
3	4	5
3	5	10
3	10	11
4	5	11
5	5	12
5	6	13
5	7	14
5	8	15
5	9	10
7	7	10
7	14	15

H	K	L
U	V	W
5	5	7
---	---	---
8	13	15
10	11	15
5	5	8
■■■	■■■	■■■
0	1	1
0	5	6
1	5	7
1	5	9
1	10	15
2	5	6
2	5	10
3	5	5
3	5	11
3	10	13
4	4	5
4	5	12
5	5	13
5	6	14
5	7	15
5	10	11
7	9	10
10	14	15
11	13	15
5	5	9
■■■	■■■	■■■
0	1	1
0	5	9
1	5	8
1	5	10
2	5	7
2	5	11
3	5	6
3	5	12
3	10	15
4	5	5
4	5	13
5	5	14
5	6	15
5	10	13
7	10	11
9	9	10
13	14	15

H	K	L
U	V	W
---	---	---
5	5	11
■■■	■■■	■■■
0	1	1
0	5	11
1	5	10
1	5	12
2	5	9
2	5	13
3	5	8
3	5	14
4	5	7
4	5	15
5	5	6
7	10	15
9	10	13
10	11	11
5	5	12
===	===	===
0	1	1
0	5	12
1	5	11
1	5	13
2	5	10
2	5	14
3	5	9
3	5	15
4	5	8
5	5	7
5	6	6
9	10	15
10	11	13
5	5	13
■■■	■■■	■■■
0	1	1
0	5	13
1	5	12
1	5	14
2	5	11
2	5	15
3	5	10
4	5	9
5	5	8
5	6	7
10	11	15
10	13	13

H	K	L
U	V	W
---	---	---
5	5	14
■■■	■■■	■■■
0	1	1
0	5	14
1	5	13
1	5	15
2	5	12
3	5	11
4	5	10
5	5	9
5	6	8
5	7	7
10	13	15
5	6	6
■■■	■■■	■■■
0	1	1
0	5	6
1	4	6
1	6	6
1	9	12
1	11	12
2	3	6
2	6	7
3	6	8
3	7	12
3	12	13
4	6	9
5	5	12
5	6	10
5	12	15
6	6	11
6	7	12
6	8	13
6	9	14
6	10	15
5	6	7
■■■	■■■	■■■
0	5	6
0	5	7
0	6	7
1	1	2
1	2	4
1	3	3
1	3	5
1	5	5

H	K	L
U	V	W
5	6	7
---	---	---
1	7	7
1	7	9
1	7	11
1	8	10
1	8	11
1	11	12
1	12	13
2	3	4
2	6	9
2	8	11
2	9	15
2	11	13
3	4	9
3	8	9
3	9	11
3	9	13
4	5	7
4	7	14
4	10	13
4	10	15
4	13	14
5	5	13
5	8	10
5	11	13
5	11	15
6	11	13
7	7	11
9	10	15
5	6	8
■■■	■■■	■■■
0	3	4
0	5	6
0	5	8
1	2	2
1	2	3
1	2	4
1	3	6
1	7	10
1	8	8
1	8	14
1	13	14
2	4	7
2	5	5
2	7	11
2	8	9

H	K	L
U	V	W
5	6	8
---	---	---
2	9	12
2	9	14
2	10	11
2	10	15
2	11	13
3	4	10
3	11	14
3	12	14
4	5	10
4	6	7
4	13	14
5	5	14
6	6	13
6	7	9
6	11	12
8	8	11
9	10	13
10	12	15
5	6	9
■■■	■■■	■■■
0	2	3
0	5	6
0	5	9
1	1	3
1	3	4
1	4	6
1	6	9
1	6	12
1	9	9
1	9	15
1	11	15
1	14	15
2	3	3
2	7	12
2	7	15
2	12	13
3	3	7
3	5	5
3	5	10
3	7	8
3	7	9
3	7	13
3	8	15
3	9	11
3	11	14

H K L
U V W
----
5 6 9
----
4 6 11
4 11 15
5 9 15
5 10 12
6 7 8
6 12 13
7 13 15
9 9 11
9 12 13
11 12 14

5 6 10
========
0 1 2
0 3 5
0 5 6
1 4 5
1 5 8
1 10 10
1 10 14
2 2 5
2 4 5
2 5 10
2 7 10
2 8 15
2 10 15
2 14 15
3 5 12
3 6 10
4 5 5
4 5 14
4 7 15
4 10 15
4 11 15
5 6 6
5 7 8
5 8 10
5 8 15
5 9 12
5 10 14
6 9 10
8 13 15
10 10 11
10 13 14
10 14 15

H K L
U V W
----
5 6 11
========
0 5 6
0 5 11
0 6 11
1 1 1
1 4 7
1 4 10
1 5 6
1 5 10
1 6 12
1 7 12
1 9 13
1 11 11
2 3 8
2 3 9
2 4 7
2 4 9
2 7 13
2 8 13
3 3 8
3 7 15
3 8 14
3 9 13
3 9 14
4 9 15
4 10 15
5 7 15

5 6 12
========
0 1 2
0 5 6
0 5 12
1 2 6
1 3 6
1 6 7
1 8 12
1 12 12
2 6 9
3 4 6
3 4 12
3 6 11
4 6 13
4 7 12
5 5 6
5 6 15

H K L
U V W
----
5 6 12
----
6 6 7
6 7 9
6 8 11
6 9 13
6 10 15
8 9 12
11 12 12

5 6 13
========
0 5 6
0 5 13
0 6 13
1 1 3
1 2 4
1 2 5
1 3 9
1 5 10
1 7 8
1 7 11
1 13 13
2 4 7
2 8 11
3 3 4
3 4 14
3 9 14
5 5 7
5 7 11
6 7 11
7 9 15
8 9 10
8 10 15
11 13 13

5 6 14
========
0 3 7
0 5 6
0 5 14
1 1 4
1 2 4
1 4 10
1 6 10
1 8 9
1 14 14
2 2 3

H K L
U V W
----
5 6 14
----
2 3 6
2 4 11
2 5 10
2 7 14
2 10 13
4 5 15
4 7 13
5 5 8
7 8 10
7 8 14
9 11 12
11 14 14

5 6 15
========
0 1 3
0 2 5
0 5 6
1 3 5
1 5 9
1 9 10
1 10 15
1 15 15
2 5 12
2 12 15
3 3 5
3 3 10
3 5 10
3 5 15
3 7 15
4 5 6
4 6 15
5 5 9
5 6 12
5 7 15
6 8 15
7 9 10
9 10 15
10 12 15
11 15 15

5 7 7
========
0 1 1
0 5 7
1 4 7

H K L
U V W
----
5 7 7
----
1 6 7
1 9 14
1 11 14
2 3 7
2 7 7
3 7 8
3 7 14
3 13 14
4 7 9
5 5 14
5 7 10
5 14 15
6 7 11
7 7 12
7 8 13
7 9 14
7 10 15

5 7 8
========
0 5 7
0 5 8
0 7 8
1 1 3
1 2 3
1 4 4
1 4 5
1 5 5
1 6 10
1 6 11
1 9 11
1 9 13
1 12 13
2 2 3
2 7 13
2 7 14
2 11 14
2 13 15
3 5 7
3 7 7
3 7 8
4 9 11
4 11 12
4 12 13
5 6 10
5 9 10

H K L
U V W
----
5 7 8
----
5 9 11
5 13 15
7 7 13
7 11 13
7 11 14

5 7 9
========
0 5 7
0 5 9
0 7 9
1 1 2
1 2 5
1 3 4
1 3 6
1 6 7
1 7 8
1 7 14
1 8 11
1 8 13
1 12 15
2 5 5
2 9 9
2 9 13
3 3 4
3 4 11
3 10 11
3 10 15
3 11 12
4 7 7
4 11 13
5 8 9
5 10 11
5 12 15
6 13 15
7 10 11
7 13 14
8 11 13
10 13 15

5 7 10
========
0 1 2
0 5 7
0 7 10
1 3 5

H K L
U V W
----
5 7 10
----
1 4 5
1 5 5
1 5 9
1 10 12
1 10 15
1 11 15
2 3 5
2 5 11
3 5 5
3 5 13
3 6 10
4 5 10
4 5 15
4 9 10
4 13 15
5 5 6
5 7 7
5 8 9
5 8 15
5 9 11
5 10 13
5 11 15
5 13 15
7 7 15
7 14 15
8 10 11
10 12 13

5 7 11
========
0 5 7
0 5 11
0 7 11
1 2 3
1 2 5
1 3 4
1 3 6
1 4 7
1 7 12
1 7 14
1 8 9
1 10 15
2 11 11
3 4 13
3 5 10
3 13 14

H K L
U V W
----
5 7 11
----
4 5 5
5 6 13
5 8 9
6 7 7
6 11 13
7 9 10
8 11 13
9 13 14
11 11 12

5 7 12
========
0 5 7
0 5 12
0 7 12
1 1 1
1 4 6
1 4 11
1 6 6
1 6 11
1 6 13
1 8 13
1 9 15
1 11 13
2 3 9
2 3 10
2 5 7
2 5 10
2 7 14
2 9 14
3 4 6
3 4 9
3 8 15
3 10 15
3 11 13
5 9 15

5 7 13
========
0 5 7
0 5 13
0 7 13
1 1 4
1 2 3
1 3 4
1 4 9

H K L
U V W
----
5 7 13
----
1 5 10
1 6 11
1 9 10
2 3 5
2 3 7
2 7 15
2 13 13
3 7 14
4 5 11
5 5 6
5 8 15
7 7 8
7 8 11
8 9 11
9 11 14
12 13 13

5 7 14
========
0 1 2
0 5 7
0 5 14
1 2 7
1 3 7
1 7 7
1 8 14
1 12 14
2 7 9
3 4 7
3 4 14
3 7 11
4 7 13
4 7 14
5 5 7
5 7 15
6 7 7
7 7 9
7 8 11
7 9 13
7 10 15
8 9 14
11 12 14

H K L
U V W
----
5 7 15
========
0 1 3
0 5 7
0 7 15
1 2 5
1 4 5
1 5 10
1 10 11
2 3 5
2 5 13
2 15 15
3 5 10
3 8 15
4 5 5
4 9 15
5 5 8
5 6 11
5 6 15
5 7 14
7 7 10
9 10 13
9 10 15
11 12 15

5 8 8
========
0 1 1
0 5 8
1 4 8
1 6 6
2 3 6
2 7 8
3 8 8
4 8 9
5 8 10
6 8 11
7 8 12
8 8 13
8 9 14
8 10 15

5 8 9
========
0 5 8
0 5 9
0 8 9

H K L
U V W
----
5 8 9
----
1 2 2
1 2 5
1 3 3
1 3 4
1 3 7
1 5 5
1 7 11
1 7 13
1 8 11
1 11 13
1 13 14
2 9 13
2 11 14
3 4 12
3 6 7
3 11 15
4 4 7
4 11 12
5 7 9
5 7 10
5 7 11
6 10 13
6 10 15
7 8 11
9 9 13
10 11 15

5 8 10
========
0 1 2
0 4 5
0 5 8
1 5 6
1 5 10
1 10 14
2 3 5
2 4 5
2 5 5
2 5 12
2 7 10
2 9 10
2 11 15
2 13 15
3 5 14
3 10 10
4 5 6

H K L
U V W
----
5 8 10
----
4 10 15
4 14 15
5 6 7
5 6 10
5 8 8
5 9 10
5 10 12
5 11 14
5 14 15
6 10 11
7 10 15
8 8 15
10 10 13
10 14 15

5 8 11
========
0 5 8
0 5 11
0 8 11
1 1 2
1 2 6
1 3 5
1 3 7
1 7 9
1 8 15
1 9 13
2 6 7
2 9 10
2 10 15
3 4 4
3 4 13
3 5 5
3 11 11
4 7 9
4 12 13
5 10 14
5 11 12
5 13 15
7 8 9
7 11 13
10 13 14
11 11 13

TAB. I

THE TYPES OF HKL-UVW FOR H,K,L,U,V,W ≤15


```
 H  K  L
 U  V  W
--------

 5  8 12
========
 0  2  3
 0  5  8
 0  5 12
 1  1  4
 1  4  4
 1  4  8
 1  6 12
 1  9 12
 2  3  4
 3  4  7
 3  7 12
 4  5  5
 4  5 10
 4  7  8
 4  7 13
 4  8 11
 4  9 11
 4 11 14
 5 12 15
 7  8  8
 8 12 13
 9 11 12
12 12 13

 5  8 13
========
 0  5  8
 0  5 13
 0  8 13
 1  1  1
 1  4  9
 1  4 12
 1  6  7
 1  6 14
 1  7 12
 1  9 14
 1 11 15
 2  3 10
 2  3 11
 2  6  7
 2  6 11
 2  7 15
 2 10 15
 3  5  8
 3  5 10
```

```
 H  K  L
 U  V  W
--------
 5  8 13
--------
 3 13 13
 4  4  9
 5 11 15

 5  8 14
========
 0  4  7
 0  5  8
 0  5 14
 1  2  3
 1  2  4
 1  2  6
 1  3 10
 1  7 14
 1  8 10
 2  3  4
 2  5 10
 2  7 11
 2 11 12
 3 14 14
 4  6 13
 4 10 15
 5  5  6
 6  7 10
 7  9 14
 8  8  9
10 11 13
13 14 14

 5  8 15
========
 0  1  3
 0  5  8
 0  8 15
 1  3  5
 1  5  5
 1  5 10
 1  5 11
 1 10 13
 2  2  5
 2  5 14
 3  7 10
 3  7 15
 4  4  5
 4 12 15
 5  5  7
```

```
 H  K  L
 U  V  W
--------
 5  8 15
--------
 5  6 10
 5  7 10
 5  7 13
 5  9 15
 6 10 15
 9 10 11
 9 11 15
13 15 15

 5  9  9
========
 0  1  1
 0  5  9
 1  4  9
 1  6  9
 2  3  9
 2  7  9
 3  8  9
 4  9  9
 5  9 10
 6  9 11
 7  9 12
 8  9 13
 9  9 14
 9 10 15

 5  9 10
========
 0  1  2
 0  5  9
 0  9 10
 1  4  5
 1  5  5
 1  5  7
 1  5 11
 1 13 15
 1 14 15
 2  5  5
 2  5 13
 3  3  5
 3  5  6
 3  5 15
 3 10 12
 4  7 10
 4 10 11
 5  5  7
```

```
 H  K  L
 U  V  W
--------
 5  9 10
--------
 5  7  8
 5  8 10
 5  9  9
 5 10 11
 5 11 13
 5 11 15
 5 12 15
 7 10 15
 7 13 15
 8 10 13
 8 11 15
10 12 15

 5  9 11
========
 0  5  9
 0  5 11
 0  9 11
 1  1  4
 1  2  3
 1  4  5
 1  4  7
 1  6 13
 1  6 15
 1  7  8
 1  9 14
 1 11 14
 2  3  3
 2  5  5
 3  6  7
 3  7 10
 3 12 13
 4 11 11
 5  7  8
 5  7 10
 5 10 13
 7  8 13
 7 12 13
 8  9 11
11 11 14

 5  9 12
========
 0  3  4
 0  5  9
 0  5 12
```

```
 H  K  L
 U  V  W
--------
 5  9 12
--------
 1  2  3
 1  2  6
 1  3  3
 1  3  9
 1  7 15
 1  8 12
 2 11 15
 3  4  7
 3  4 15
 3  5  5
 3  7 11
 3  8  9
 3 10 15
 3 11 13
 5  6 10
 6  6  7
 6  9 13
 6 13 14
 7  9  9
 8 11 12
 9 11 12
 9 13 15

 5  9 13
========
 0  5  9
 0  5 13
 0  9 13
 1  1  2
 1  2  7
 1  3  6
 1  3  8
 1  7 10
 1  8 11
 1 10 15
 2  7  9
 2  9 11
 3  4  5
 3  4 15
 3  6  7
 3 11 12
 4  5  5
 4  7 11
 4 13 13
 5 14 15
 7 11 12
```

```
 H  K  L
 U  V  W
--------
 5  9 13
--------
 8  9  9
11 13 14
13 13 14

 5  9 14
========
 0  5  9
 0  5 14
 0  9 14
 1  1  1
 1  4 10
 1  4 13
 1  6  8
 1  6 15
 1  8 13
 1 10 15
 2  3 11
 2  3 12
 2  7  7
 2  7 12
 3  6  8
 3  6 11
 3 13 15
 4  5  9
 4  5 10
 5 13 15

 5  9 15
========
 0  1  3
 0  3  5
 0  5  9
 1  5  6
 1  5 12
 1 10 15
 2  3  5
 2  5 15
 3  4  5
 3  5 10
 3  7 10
 3  8 15
 3  9 10
 3 10 15
 4 15 15
 5  5  6
 5  6  9
```

```
 H  K  L
 U  V  W
--------
 5  9 15
--------
 5  7 12
 5  8 15
 5 12 15
 6  7 15
 6 11 15
 9  9 10
10 11 15
12 13 15
14 15 15

 5 10 11
========
 0  1  2
 0  5 11
 0 10 11
 1  5  5
 1  5  6
 1  5  9
 1  5 13
 2  5  7
 2  5 15
 3  4  5
 3  5  5
 3  5  7
 4  9 10
 4 10 13
 5  5  8
 5  7  9
 5  9 10
 5 10 12
 5 11 11
 5 12 13
 5 13 15
 5 14 15
 7  8 10
 7 13 15
 8 10 15
10 13 15
11 11 15

 5 10 12
========
 0  1  2
 0  5  6
 0  5 12
 1  5 10
```

```
 H  K  L
 U  V  W
--------
 5 10 12
--------
 1  5 14
 2  5  5
 2  5  7
 2  5  8
 2 10 11
 2 10 13
 3  5  6
 4  4  5
 4  5  8
 5  6  9
 5  8 10
 5 10 11
 5 10 14
 5 12 12
 5 13 14
 6  9 10
 6 10 15
 7 10 10
 8 14 15
10 13 15
11 14 15

 5 10 13
========
 0  1  2
 0  5 13
 0 10 13
 1  5  6
 1  5  7
 1  5 11
 1  5 15
 2  5  9
 3  5  5
 3  5  7
 3  5  8
 4  5  5
 4 10 11
 4 10 15
 5  5  9
 5  7 10
 5  9 11
 5 11 12
 5 13 13
 5 14 15
 7 10 12
 8  9 10
```

```
 H  K  L
 U  V  W
--------
 5 10 13
--------
11 14 15
13 13 15

 5 10 14
========
 0  1  2
 0  5  7
 0  5 14
 1  5 12
 2  5  6
 2  5  8
 2  5 10
 2 10 13
 2 10 15
 3  5  8
 4  5  5
 4  5  6
 4  5  9
 5  6 10
 5  8 11
 5 10 12
 5 12 13
 5 14 14
 6 10 11
 7 10 14
 9 10 10
14 14 15

 5 11 11
========
 0  1  1
 0  5 11
 1  4 11
 1  6 11
 2  3 11
 2  7 11
 3  8 11
 4  9 11
 5 10 11
 6 11 11
 7 11 12
 8 11 13
 9 11 14
10 11 15
```

```
 H  K  L
 U  V  W
--------

 5 11 12
========
 0  5 11
 0  5 12
 0 11 12
 1  2  2
 1  2  7
 1  3  5
 1  3  9
 1  5  5
 1  6  7
 1  7 13
 2  9 15
 3  3  4
 3  7  9
 4  7  8
 4  8  9
 5  8 11
 5 10 13
 5 13 14
 7  9 13
 7 11 11
 7 11 13
 9 10 15
10 14 15

 5 11 13
========
 0  5 11
 0  5 13
 0 11 13
 1  2  3
 1  2  7
 1  3  4
 1  3 10
 1  8  9
 1  8 15
 2  5  5
 3 13 14
 4  5  7
 5  6  7
 5  9 10
 6  7 11
 6 13 13
 7  8  9
 7 10 15
 7 11 12
```

```
 H  K  L
 U  V  W
--------
 5 11 13
--------
 8 11 11
 9 13 14

 5 11 14
========
 0  5 11
 0  5 14
 0 11 14
 1  1  5
 1  2  3
 1  4  6
 1  4  9
 1  9 11
 2  3  4
 2  7  8
 3  5  5
 3  7  7
 3  7 13
 4 11 13
 5  8 10
 5 12 13
 7  7  8
 7  9 13
 9 11 11
10 12 13
11 14 15

 5 11 15
========
 0  1  3
 0  5 11
 0 11 15
 1  4  5
 1  5  8
 1  5 14
 1  7 10
 2  3  5
 2  5  5
 3 10 13
 3 10 15
 4  5  5
 5  6  7
 5  7 10
 5  8 13
 5  9 10
 6 13 15
```

```
 H  K  L
 U  V  W
--------
 5 11 15
--------
 7 12 15
 6  9 15
 9 14 15
10 11 11

 5 12 12
========
 0  1  1
 0  5 12
 1  4 12
 1  6 12
 2  3 12
 2  7 12
 3  8 12
 4  9 12
 5 10 12
 6 11 12
 7 12 12
 8 12 13
 9 12 14
10 12 15

 5 12 13
========
 0  5 12
 0  5 13
 0 12 13
 1  1  5
 1  4  7
 1  4  8
 1  5  5
 1  7  8
 2  2  3
 2  3  3
 3  7  9
 3  7 11
 3  9 11
 4  7  8
 5 10 11
 5 10 14
 5 11 14
 5 12 13
 7  9 11
 7 13 13
10 11 14
```

TAB. I THE TYPES OF HKL-UVW FOR H,K,L,U,V,W ≤15 I - 26

H K L U V W	H K L U V W	H K L U V W	H K L U V W	H K L U V W	H K L U V W	H K L U V W	H K L U V W	H K L U V W	H K L U V W	H K L U V W	H K L U V W
	5 12 15		5 14 15	6 6 11		6 7 9	6 7 10	6 7 12	6 7 14	6 7 15	6 8 11
5 12 14	--------	5 13 15	--------	--------	6 7 8	--------	--------	--------	--------	--------	--------
========	9 10 11	========	1 5 11	3 6 8	========	1 11 15	5 8 14	2 6 11	1 4 7	13 15 15	2 3 5
0 5 12	10 13 15	0 1 3	1 9 10	3 6 14	0 3 4	2 11 12	6 8 9	3 5 6	1 7 8		2 4 7
0 5 14		0 5 13	2 4 5	4 6 7	0 6 7	2 13 15	6 12 13	3 6 13	1 10 14	6 8 9	2 4 9
0 6 7	5 13 13	0 13 15	2 5 8	4 6 15	0 7 8	3 3 6	8 9 11	3 8 12	1 14 14	========	2 7 6
1 2 2	========	1 4 5	3 5 5	5 6 6	1 1 2	3 4 5	10 10 13	4 5 12	2 2 7	0 2 3	2 7 13
1 2 8	0 1 1	1 5 10	3 13 15	7 12 15	1 4 6	3 5 9	11 12 15	4 6 15	2 4 7	0 3 4	2 11 11
1 3 6	0 5 13	1 9 10	4 5 6	9 12 13	1 6 6	3 5 11		4 9 12	2 5 14	0 8 9	2 12 15
1 3 10	1 4 13	2 5 5	5 7 7	11 11 12	1 8 8	3 7 7	6 7 11	5 6 6	2 7 10	1 2 3	2 14 15
1 7 14	1 6 13	2 5 7	5 8 10		1 8 10	3 7 14	========	6 7 7	2 7 14	1 6 6	3 6 7
2 5 5	2 3 13	3 4 5	5 9 13	6 6 13	1 10 13	3 9 10	0 6 7	6 8 9	3 6 14	1 10 12	4 5 14
2 7 9	2 7 13	3 14 15	5 10 11	========	1 13 14	3 11 13	0 6 11	6 9 11	3 7 12	1 14 15	4 10 13
2 11 12	3 8 13	5 5 6	5 10 13	0 1 1	1 14 15	3 13 15	0 7 11	6 10 13	4 5 7	2 3 4	5 6 12
3 4 4	4 9 13	5 7 8	7 7 10	0 6 13	2 2 5	4 6 13	1 1 3	6 11 15	4 7 14	2 3 7	5 8 8
4 7 10	5 10 13	5 7 10	9 11 15	1 6 12	2 3 4	4 9 15	1 2 4	8 11 12	6 6 7	2 5 6	7 9 10
4 8 11	6 11 13	5 8 11	10 11 13	1 6 14	2 4 5	5 6 8	1 4 5	12 12 13	6 9 14	2 6 11	8 14 15
6 8 9	7 12 13	5 9 14	10 12 15	2 6 11	2 5 9	5 6 12	1 5 7		7 7 8	2 9 9	9 10 12
6 9 13	8 13 13	5 10 11		2 6 15	2 7 7	6 9 11	1 5 8	6 7 13	7 8 10	2 9 15	11 11 14
7 11 14	9 13 14	6 11 15	6 6 7	3 6 10	2 8 13	6 11 12	1 7 10	========	7 9 12	2 12 13	
8 10 15	10 13 15	7 10 11	========	4 6 9	2 9 10	7 7 15	1 7 14	0 6 7	7 10 14	2 12 15	6 8 13
8 13 14		8 15 15	0 1 1	5 6 8	2 9 12	7 12 14	1 9 15	0 6 13	10 11 14	3 4 10	========
10 11 13	5 13 14	9 10 15	0 6 7	6 6 7	2 10 11	9 9 13	1 11 11	0 7 13	13 14 14	3 5 6	0 3 4
	========	10 13 13	1 6 6	11 12 15	2 13 15		2 3 3	1 1 1		3 6 13	0 6 13
5 12 15	0 5 13		1 6 8	12 13 13	3 6 11	6 7 10	2 4 5	1 5 8	6 7 15	3 8 8	0 8 13
========	0 5 14	5 14 14	1 12 13		3 10 11	========	2 8 13	1 5 12	========	3 9 10	1 2 3
0 1 3	0 13 14	========	1 12 15	6 7 7	3 10 12	0 3 5	3 5 9	1 6 7	0 2 5	3 10 11	1 2 4
0 4 5	1 2 3	0 1 1	2 5 6	========	3 10 14	0 6 7	3 8 10	1 6 12	0 6 7	3 12 14	1 4 10
0 5 12	1 2 8	0 5 14	2 6 9	0 1 1	4 7 14	0 7 10	3 9 13	1 7 14	0 7 15	4 6 9	1 6 9
1 5 9	1 3 5	1 4 14	3 4 6	0 6 7	4 10 11	1 2 2	4 9 14	1 8 14	1 1 3	4 14 15	2 2 7
1 5 15	1 3 11	1 6 14	3 6 10	1 5 7	4 11 14	1 2 4	5 6 13	1 11 15	1 3 6	5 12 14	2 4 5
2 5 6	1 5 5	2 3 14	3 11 12	1 7 7	5 6 9	1 3 4	5 7 7	1 13 13	1 6 12	6 7 10	2 5 7
3 3 5	1 8 9	2 7 14	4 6 11	1 11 14	6 13 15	1 5 8	5 11 13	2 4 9	1 8 9	6 7 12	2 5 11
3 5 5	2 3 4	3 8 14	5 6 12	1 13 14	7 10 14	1 8 11	7 11 12	2 4 11	1 9 13	6 11 15	2 8 15
3 7 10	3 11 13	4 9 14	5 9 12	2 4 7	8 8 13	1 9 12	8 9 10	2 5 8	1 15 15	6 13 14	2 9 10
3 9 10	4 6 7	5 10 14	6 6 13	2 7 8		1 10 10	11 11 13	2 5 11	2 5 9	8 8 15	2 13 13
3 10 15	4 7 7	6 11 14	6 7 14	3 3 7	6 7 9	2 3 6		2 8 15	2 9 12	9 9 14	3 10 14
3 11 15	5 10 12	7 12 14	6 8 15	3 7 9	========	2 4 9	6 7 12	2 9 15	3 3 4		4 6 11
3 13 15	5 11 12	8 13 14	7 7 12	3 9 14	0 2 3	2 5 6	========	3 3 10	3 3 11	6 8 11	5 6 6
5 6 6	7 7 9	9 14 14		3 14 15	0 6 7	2 7 14	0 1 2	3 4 9	3 4 6	========	6 7 15
5 7 9	7 9 11	10 14 15	6 6 11	4 7 10	0 7 9	2 8 11	0 6 7	3 4 10	3 5 9	0 3 4	7 8 8
5 8 12	8 10 11		========	5 7 11	1 2 3	2 10 15	0 7 12	3 11 15	3 7 14	0 6 11	7 10 11
5 9 10	9 13 13	5 14 15	0 1 1	5 7 14	1 3 3	2 11 14	1 3 6		3 10 15	0 8 11	8 10 11
5 9 15	10 11 15	========	0 6 11	6 7 12	1 3 5	2 13 14	1 4 6	6 7 14	5 9 15	1 2 2	9 12 14
6 10 15		0 1 3	1 6 10	7 7 13	1 4 6	3 8 14	1 5 6	========	6 8 13	1 2 5	13 13 14
6 14 15		0 5 14	1 6 12	7 8 14	1 9 9	4 5 10	1 6 9	0 1 2	7 7 9	1 4 6	
7 15 15		0 14 15	2 6 9	7 9 15	1 9 12	4 5 13	1 12 12	0 3 7	10 11 12	1 6 9	
8 12 15		1 5 5	2 6 13		1 10 12	4 7 7	2 3 6	0 6 7	10 11 15	1 10 13	

TAB. I THE TYPES OF HKL-UVW FOR H,K,L,U,V,W ≤15

H U	K V	L W
6	**8**	**15**
0	2	5
0	3	4
0	8	15
1	2	3
1	6	12
2	3	6
2	3	9
2	4	9
2	6	13
2	7	9
2	11	12
2	15	15
3	4	6
3	4	14
3	6	11
4	10	15
5	6	15
6	6	7
8	8	9
9	10	12
9	10	13
10	12	15
14	15	15
6	**9**	**10**
0	2	3
0	3	5
0	9	10
1	3	4
1	6	6
1	12	14
2	2	3
2	3	8
2	6	7
2	6	13
2	9	12
3	4	6
3	4	11
3	6	14
3	7	8
3	8	9
3	10	10
3	12	13
4	9	9

H U	K V	L W
6	**9**	**10**
4	14	15
5	6	10
5	10	12
6	9	14
6	11	12
6	11	14
7	12	15
8	13	15
6	**9**	**11**
0	2	3
0	6	11
0	9	11
1	3	3
1	3	4
1	3	7
1	6	8
1	9	15
1	12	14
2	3	5
3	3	10
3	5	7
3	5	13
3	8	9
3	9	13
3	11	11
3	13	14
4	5	6
5	9	9
6	7	9
6	7	12
6	12	13
7	10	12
8	13	15
9	14	15
11	11	15
6	**9**	**13**
0	2	3
0	6	13
0	9	13
1	3	5
1	3	8
1	6	8

H U	K V	L W
6	**9**	**13**
2	3	3
3	3	11
3	4	7
3	5	14
3	7	9
3	9	11
3	10	11
3	13	13
4	6	7
5	6	12
5	9	12
5	12	14
7	9	9
10	11	12
13	13	15
6	**9**	**14**
0	2	3
0	3	7
0	9	14
1	3	4
1	6	10
2	3	4
2	3	6
2	3	10
2	6	11
3	4	13
3	5	8
3	8	10
3	11	12
3	14	14
4	9	15
5	6	6
6	7	14
6	9	10
7	12	14
8	9	9
10	12	13
14	14	15
6	**10**	**11**
0	3	5
0	6	11
0	10	11

H U	K V	L W
6	**10**	**11**
1	2	2
1	4	5
1	4	9
1	6	6
1	8	13
2	2	7
2	3	4
2	5	12
2	7	8
2	10	13
2	10	15
3	7	8
4	6	9
4	11	11
4	13	14
5	8	10
5	10	14
6	14	15
7	8	13
7	9	12
9	10	14
6	**10**	**13**
0	3	5
0	6	13
0	10	13
1	2	2
1	2	6
1	4	7
1	8	11
1	12	15
2	4	5
2	4	11
2	6	9
2	8	9
2	11	14
3	4	7
3	6	8
4	10	15
4	13	13
5	8	9
6	7	12
7	10	10
9	11	12
11	12	14

H U	K V	L W
6	**10**	**15**
0	2	3
0	2	5
0	3	5
2	3	10
2	5	6
2	6	15
2	9	10
2	12	15
3	4	5
3	4	15
3	6	10
3	8	15
4	5	9
4	15	15
5	6	6
5	6	12
5	8	9
5	8	15
5	10	12
5	12	15
6	10	15
9	10	10
10	14	15
12	14	15
6	**11**	**11**
0	1	1
0	6	11
1	5	11
1	7	11
2	4	11
2	8	11
3	3	11
3	9	11
4	10	11
5	11	11
6	11	12
7	11	13
8	11	14
9	11	15

H U	K V	L W
6	**11**	**12**
0	1	2
0	6	11
0	11	12
1	5	6
1	6	6
1	6	9
1	6	13
2	6	7
2	6	15
3	4	6
3	5	6
3	6	7
4	9	12
4	12	13
5	6	8
5	12	12
6	7	9
6	9	10
6	11	11
6	12	13
6	13	15
7	8	12
8	12	15
6	**11**	**13**
0	6	11
0	6	13
0	11	13
1	1	4
1	3	3
1	5	7
1	5	9
1	7	15
1	8	10
2	3	4
2	4	5
2	5	7
3	9	10
3	9	14
5	6	7
5	7	11
5	13	13
7	8	10
7	11	11

H U	K V	L W
6	**11**	**13**
8	9	15
8	11	13
6	**11**	**14**
0	3	7
0	6	11
0	11	14
1	2	2
1	2	6
1	4	5
1	9	12
2	4	13
2	5	8
2	7	10
2	8	15
3	4	6
3	5	8
4	7	9
4	9	10
5	14	14
6	9	10
7	8	14
8	11	11
10	12	13
12	13	15
6	**11**	**15**
0	2	5
0	6	11
0	11	15
1	3	3
1	3	8
1	4	6
1	7	9
1	9	14
2	3	3
3	3	13
3	5	7
3	7	12
3	10	12
4	5	9
5	9	15
5	13	15
6	7	12

H U	K V	L W
6	**11**	**15**
6	8	9
7	10	15
9	11	11
12	13	14
6	**12**	**13**
0	1	2
0	6	13
0	12	13
1	6	6
1	6	7
1	6	11
1	6	15
2	6	9
3	5	6
3	6	7
3	6	8
4	5	6
4	11	12
4	12	15
5	6	9
6	7	10
6	9	11
6	11	12
6	13	13
6	14	15
7	12	12
8	9	12
6	**13**	**13**
0	1	1
0	6	13
1	5	13
1	7	13
2	4	13
2	8	13
3	3	13
3	9	13
4	10	13
5	11	13
6	12	13
7	13	13
8	13	14
9	13	15

H U	K V	L W
6	**13**	**14**
0	3	7
0	6	13
0	13	14
1	2	2
1	4	11
1	6	6
1	7	8
1	8	15
2	2	9
2	4	5
2	7	12
3	4	5
3	8	10
4	10	11
5	9	12
6	8	9
7	10	14
8	13	13
10	11	14
6	**13**	**15**
0	2	5
0	6	13
0	13	15
1	3	3
1	3	4
1	3	9
1	10	12
2	7	9
3	3	14
3	4	6
3	5	6
3	7	11
5	7	9
5	11	15
6	7	8
6	11	12
7	15	15
8	9	11
9	10	15
9	12	14
9	13	13

H U	K V	L W
6	**14**	**15**
0	2	5
0	3	7
0	14	15
1	6	6
1	8	9
2	2	3
2	3	12
2	6	9
3	3	4
3	6	8
3	8	13
3	10	12
4	9	10
4	9	11
5	12	15
6	10	11
6	12	13
7	12	14
8	15	15
9	14	14
7	**7**	**8**
0	1	1
0	7	8
1	7	7
1	7	9
1	14	15
2	6	7
2	7	10
3	5	7
3	7	11
3	13	14
4	4	7
4	7	12
5	7	13
5	11	14
6	7	14
7	7	15
7	9	14
7	**7**	**9**
0	1	1
0	7	9

H U	K V	L W
7	**7**	**9**
1	7	8
1	7	10
2	7	7
2	7	11
3	6	7
3	7	12
3	14	15
4	5	7
4	7	13
5	7	14
5	13	14
6	7	15
7	11	14
9	9	14
7	**7**	**10**
0	1	1
0	7	10
1	7	9
1	7	11
2	7	8
2	7	12
3	7	7
3	7	13
4	6	7
4	7	14
5	5	7
5	7	15
5	14	15
7	13	14
9	11	14
7	**7**	**11**
0	1	1
0	7	11
1	7	10
1	7	12
2	7	9
2	7	13
3	7	8
3	7	14
4	7	7
4	7	15
5	6	7

H U	K V	L W
7	**7**	**11**
7	14	15
9	13	14
11	11	14
7	**7**	**12**
0	1	1
0	7	12
1	7	11
1	7	13
2	7	10
2	7	14
3	7	9
3	7	15
4	7	8
5	7	7
6	6	7
9	14	15
11	13	14
7	**7**	**13**
0	1	1
0	7	13
1	7	12
1	7	14
2	7	11
2	7	15
3	7	10
4	7	9
5	7	8
6	7	7
11	14	15
13	13	14
7	**7**	**15**
0	1	1
0	7	15
1	7	14
2	7	13
3	7	12
4	7	11
5	7	10
6	7	9
7	7	8

TAB. I

THE TYPES OF HKL-UVW FOR H,K,L,U,V,W ≤15

(Rows in bold are H K L; the rows following each are U V W.)

H U	K V	L W
7	**7**	**15**
14	15	15
7	**8**	**8**
0	1	1
0	7	8
1	6	8
1	8	8
2	5	8
2	8	9
3	4	8
3	8	10
4	8	11
5	8	12
6	8	13
7	8	14
8	8	15
7	**8**	**9**
0	7	8
0	7	9
0	8	9
1	1	2
1	3	5
1	4	4
1	5	7
1	7	7
1	9	9
1	9	11
1	10	14
1	11	13
1	12	15
2	3	6
2	5	6
2	10	11
2	10	13
3	3	5
3	8	13
3	11	13
3	11	15
3	12	13
4	5	11
4	7	9
4	12	15
5	8	11

H U	K V	L W
7	**8**	**9**
5	11	13
7	10	14
7	**8**	**10**
0	4	5
0	7	8
0	7	10
1	2	3
1	2	4
1	4	6
1	5	6
1	9	14
1	10	10
1	11	14
2	2	3
2	5	8
2	7	7
2	9	13
2	11	12
2	12	13
3	5	10
3	8	8
4	6	11
4	9	10
4	13	14
5	6	14
5	10	11
6	7	14
6	11	13
7	12	14
7	**8**	**11**
0	7	8
0	7	11
0	8	11
1	2	2
1	3	4
1	3	5
1	5	6
1	5	9
1	9	13
1	10	13
1	11	11
2	3	7

H U	K V	L W
7	**8**	**11**
2	11	15
2	14	15
3	7	7
4	4	9
4	5	12
4	12	13
5	7	13
5	7	14
5	8	9
7	9	11
9	10	13
11	11	15
7	**8**	**12**
0	2	3
0	7	8
0	7	12
1	2	4
1	3	4
1	4	5
1	4	8
1	9	12
1	12	12
3	4	8
3	5	12
4	4	5
4	5	11
4	7	7
4	7	14
4	8	13
4	9	10
4	11	13
5	8	8
6	11	12
8	11	12
9	12	13
7	**8**	**13**
0	7	8
0	7	13
0	8	13
1	1	3
1	3	4
1	5	6

H U	K V	L W
7	**8**	**13**
1	5	9
1	6	10
1	8	11
1	13	13
2	2	5
2	5	7
2	9	14
3	4	4
3	7	11
3	7	14
5	7	7
5	9	11
6	10	11
7	13	15
8	13	15
9	10	11
13	13	15
7	**8**	**14**
0	1	2
0	4	7
0	7	8
1	6	7
1	7	10
1	14	14
2	3	7
2	4	7
2	5	7
2	7	12
2	7	14
2	9	14
3	7	14
3	10	14
4	6	7
5	6	14
6	7	7
6	11	14
7	8	8
7	9	10
7	10	12
7	11	14
10	13	14
14	14	15

H U	K V	L W
7	**8**	**15**
0	7	8
0	7	15
0	8	15
1	1	1
1	6	9
1	6	14
1	7	8
1	7	14
1	15	15
2	5	10
2	5	13
2	6	9
2	6	13
3	4	11
3	4	12
3	5	10
3	5	12
4	4	11
7	**9**	**9**
0	1	1
0	7	9
1	6	9
1	8	9
2	5	9
2	9	9
3	4	9
3	9	10
4	9	11
5	9	12
6	9	13
7	9	14
8	9	15
7	**9**	**10**
0	7	9
0	7	10
0	9	10
1	2	3
1	2	4
1	3	3
1	4	7
1	5	5

H U	K V	L W
7	**9**	**10**
1	7	7
1	9	13
1	10	13
1	11	13
1	12	14
2	4	5
3	5	11
3	6	8
3	11	12
4	11	13
5	5	8
5	6	15
5	7	11
6	8	11
7	8	14
7	9	13
7	11	14
7	**9**	**11**
0	7	9
0	7	11
0	9	11
1	1	2
1	3	6
1	4	5
1	5	8
1	8	9
1	9	10
1	10	13
2	3	7
2	7	7
2	11	11
2	11	15
3	4	5
3	5	6
3	8	15
3	12	13
4	5	13
4	9	9
4	13	15
5	12	13
5	13	14
7	8	11
7	13	14

H U	K V	L W
7	**9**	**12**
0	3	4
0	7	9
0	7	12
1	1	3
1	6	6
1	6	9
1	8	12
1	8	15
1	13	15
2	3	5
2	5	6
2	9	15
3	3	4
3	5	9
3	7	7
3	8	13
3	9	11
3	10	11
3	11	15
3	13	15
5	9	9
6	7	14
6	10	11
7	14	15
8	12	13
9	13	15
7	**9**	**13**
0	7	9
0	7	13
0	9	13
1	2	3
1	2	5
1	4	5
1	4	7
1	5	8
1	10	11
2	3	3
2	13	13
3	5	12
3	6	11
3	8	11
4	7	7
5	7	14

H U	K V	L W
7	**9**	**13**
5	11	12
5	11	14
7	9	10
10	11	13
7	**9**	**14**
0	1	2
0	7	9
0	9	14
1	4	7
1	5	7
1	7	7
1	7	11
2	5	7
2	7	13
3	3	7
3	6	7
3	7	15
3	12	14
4	7	14
4	11	14
5	7	7
5	8	14
7	7	8
7	9	9
7	10	11
7	11	13
7	12	15
8	13	14
12	14	15
7	**9**	**15**
0	3	5
0	7	9
0	7	15
1	2	3
1	3	4
1	3	6
1	5	12
1	10	15
1	11	12
2	3	9
2	15	15
3	4	9

H U	K V	L W
7	**9**	**15**
3	5	6
3	7	14
3	8	11
4	5	15
5	6	13
6	7	7
8	9	9
9	11	12
9	12	13
10	13	15
7	**10**	**10**
0	1	1
0	7	10
1	6	10
1	8	10
2	5	10
2	9	10
3	4	10
3	10	10
4	10	11
5	10	12
6	10	13
7	10	14
8	10	15
7	**10**	**11**
0	7	10
0	7	11
0	10	11
1	1	3
1	3	4
1	6	7
1	6	8
1	7	7
1	8	13
1	8	14
1	13	15
2	3	4
2	4	5
2	5	5
3	10	11
3	11	11
3	11	13

H U	K V	L W
7	**10**	**11**
5	5	9
5	7	9
5	13	15
6	8	13
7	9	14
7	12	14
8	9	13
7	**10**	**12**
0	5	6
0	7	10
0	7	12
1	2	2
1	3	6
1	4	4
1	5	10
1	9	14
1	11	14
2	3	5
2	3	8
2	7	7
2	8	11
2	12	13
4	5	14
4	6	9
4	8	9
5	10	13
6	9	11
7	8	14
8	10	13
10	11	15
7	**10**	**13**
0	7	10
0	7	13
0	10	13
1	1	2
1	3	7
1	4	6
1	5	9
1	9	11
1	11	15
2	3	6
2	8	9

H U	K V	L W
7	**10**	**13**
2	11	12
3	5	5
3	7	7
3	13	13
4	5	6
4	5	15
4	9	11
4	14	15
7	12	13
8	9	11
7	**10**	**14**
0	1	2
0	5	7
0	7	10
1	7	8
1	7	12
2	4	7
2	6	7
2	7	14
2	9	14
2	11	14
3	4	7
3	14	14
4	7	7
5	10	14
6	7	8
6	7	14
6	13	14
7	8	9
7	10	10
7	11	12
7	12	14
10	14	15
7	**10**	**15**
0	2	3
0	7	10
0	7	15
1	2	5
1	3	5
1	4	10
1	5	5
1	9	15

H U	K V	L W
7	**10**	**15**
1	12	15
3	5	8
3	5	15
4	5	5
4	10	13
5	5	11
5	7	7
5	7	14
5	8	10
5	9	10
5	11	13
6	11	15
9	13	15
10	11	12
7	**11**	**11**
0	1	1
0	7	11
1	6	11
1	8	11
2	5	11
2	9	11
3	4	11
3	10	11
4	11	11
5	11	12
6	11	13
7	11	14
8	11	15
7	**11**	**12**
0	7	11
0	7	12
0	11	12
1	2	5
1	3	3
1	4	5
1	4	8
1	5	7
1	7	7
1	10	14
2	2	3
3	5	13
3	8	9

TAB. I — THE TYPES OF HKL-UVW FOR H,K,L,U,V,W ≤15

H K L = 7 11 12

U	V	W
3	9	10
4	5	8
5	9	13
5	11	11
5	11	13
6	7	11
7	10	14
7	13	14

H K L = 7 11 13

U	V	W
0	7	11
0	7	13
0	11	13
1	2	3
1	2	5
1	3	4
1	4	9
1	5	6
1	9	10
2	7	7
3	4	5
3	5	14
3	7	8
4	13	13
5	7	8
5	8	11
5	11	12
6	11	11
7	9	14
7	14	15
9	10	13

H K L = 7 11 14

U	V	W
0	1	2
0	7	11
0	11	14
1	5	7
1	6	7
1	7	9
1	7	13
2	7	7
2	7	15
3	4	7
3	5	7

H K L = 7 11 14

U	V	W
3	7	7
4	9	14
4	13	14
5	7	8
5	12	14
7	7	9
7	8	14
7	9	10
7	11	11
7	12	13
7	13	15
8	14	15

H K L = 7 11 15

U	V	W
0	7	11
0	7	15
0	11	15
1	1	2
1	3	8
1	4	7
1	5	10
1	9	12
1	10	13
2	3	9
2	9	11
2	11	13
3	5	6
3	8	9
3	13	14
4	7	7
4	9	13
4	15	15
5	5	6
8	11	11
9	12	13

H K L = 7 12 12

U	V	W
0	1	1
0	7	12
1	6	12
1	8	12
2	5	12
2	9	12
3	4	12

H K L = 7 12 12

U	V	W
3	10	12
4	11	12
5	12	12
6	12	13
7	12	14
8	12	15

H K L = 7 12 13

U	V	W
0	7	12
0	7	13
0	12	13
1	2	2
1	3	7
1	4	5
1	5	6
1	5	11
1	7	7
1	9	15
2	3	9
3	3	5
3	8	9
4	8	9
4	8	11
5	9	11
5	13	13
7	10	13
7	11	14
9	11	15

H K L = 7 12 14

U	V	W
0	1	2
0	6	7
0	7	12
1	7	10
1	7	14
2	5	7
2	7	7
2	7	8
2	11	14
2	13	14
3	6	7
4	4	7
4	7	8
5	14	14

H K L = 7 12 14

U	V	W
6	7	9
6	9	14
6	14	15
7	8	10
7	10	11
7	10	14
7	12	12
7	13	14

H K L = 7 12 15

U	V	W
0	4	5
0	7	12
0	7	15
1	2	6
1	3	3
1	4	9
1	5	9
1	10	15
2	3	3
3	5	8
3	5	15
3	7	7
3	8	12
3	9	13
3	11	12
5	11	15
6	6	11
6	9	10
7	9	14
9	11	13

H K L = 7 13 13

U	V	W
0	1	1
0	7	13
1	6	13
1	8	13
2	5	13
2	9	13
3	4	13
3	10	13
4	11	13
5	12	13
6	13	13
7	13	14

H K L = 7 13 13

U	V	W
8	13	15

H K L = 7 13 14

U	V	W
0	1	2
0	7	13
0	13	14
1	6	7
1	7	7
1	7	11
1	7	15
2	7	9
3	5	7
3	7	7
3	7	8
4	5	7
4	11	14
4	14	15
5	7	9
7	7	10
7	9	11
7	11	12
7	12	14
7	13	13
7	14	15
8	9	14

H K L = 7 13 15

U	V	W
0	7	13
0	7	15
0	13	15
1	1	4
1	3	4
1	6	9
1	6	11
1	10	11
2	5	5
2	5	7
2	7	7
3	3	4
3	11	14
5	6	9
5	9	10
5	10	11
7	8	13

H K L = 7 13 15

U	V	W
7	11	14
8	13	13
8	13	15

H K L = 7 14 15

U	V	W
0	1	2
0	7	15
0	14	15
1	7	7
1	7	8
1	7	13
2	7	11
3	6	7
3	7	9
4	7	7
4	13	14
5	5	7
5	7	10
7	7	11
7	9	12
7	11	13
7	13	14
7	15	15
8	11	14
9	12	14

H K L = 7 15 15

U	V	W
0	1	1
0	7	15
1	6	15
1	8	15
2	5	15
2	9	15
3	4	15
3	10	15
4	11	15
5	12	15
6	13	15
7	14	15
8	15	15

H K L = 8 8 9

U	V	W
0	1	1
0	8	9
1	8	8
1	8	10
2	7	8
2	8	11
3	6	8
3	8	12
4	5	8
4	8	13
5	8	14
6	8	15

H K L = 8 8 11

U	V	W
0	1	1
0	8	11
1	8	10
1	8	12
2	8	9
2	8	13
3	8	8
3	8	14
4	7	8
4	8	15
5	6	8

H K L = 8 8 13

U	V	W
0	1	1
0	8	13
1	8	12
1	8	14
2	8	11
2	8	15
3	8	10
4	8	9
5	8	8
6	7	8

H K L = 8 8 15

U	V	W
0	1	1
0	8	15
1	8	14

H K L = 8 8 15

U	V	W
2	8	13
3	8	12
4	8	11
5	8	10
6	8	9
7	8	8

H K L = 8 9 9

U	V	W
0	1	1
0	8	9
1	7	9
1	9	9
2	6	9
2	9	10
3	5	9
3	9	11
4	4	9
4	9	12
5	9	13
6	9	14
7	9	15

H K L = 8 9 10

U	V	W
0	4	5
0	8	9
0	9	10
1	1	2
1	6	8
1	8	8
1	10	10
1	10	12
2	3	6
2	4	5
2	4	7
2	6	7
2	7	11
2	9	9
2	11	12
2	11	14
2	13	14
3	4	6
3	10	15
3	12	14
3	14	15

H K L = 8 9 10

U	V	W
4	13	14
5	6	13
5	10	13
6	7	11

H K L = 8 9 11

U	V	W
0	8	9
0	8	11
0	9	11
1	2	3
1	3	3
1	4	4
1	5	7
1	5	8
1	11	11
1	11	14
1	13	15
2	2	5
2	13	14
3	6	10
3	7	12
3	7	13
4	5	7
5	7	13
5	9	11
6	7	10
6	13	15
7	10	13

H K L = 8 9 12

U	V	W
0	2	3
0	3	4
0	8	9
1	3	4
1	4	6
1	12	12
1	12	15
3	4	5
3	4	8
3	4	9
3	7	12
3	8	8
3	11	12
4	5	12

H K L = 8 9 12

U	V	W
4	6	7
4	7	15
4	8	15
4	9	9
4	11	12
4	13	15
5	6	12
6	12	13
8	9	12

H K L = 8 9 13

U	V	W
0	8	9
0	8	13
0	9	13
1	2	2
1	3	5
1	3	6
1	4	5
1	5	7
1	7	11
1	11	14
1	11	15
1	13	13
2	5	9
3	3	7
3	7	8
4	4	11
4	7	12
4	12	15
5	9	9
7	10	11
9	11	13
11	12	15

H K L = 8 9 14

U	V	W
0	4	7
0	8	9
0	9	14
1	2	4
1	5	6
1	7	10
1	10	13
1	14	14
2	3	3

H K L = 8 9 14

U	V	W
2	5	6
2	5	11
2	7	10
3	6	10
4	6	15
4	10	13
5	7	14
5	8	8
7	13	14
8	10	11
11	12	14

H K L = 8 9 15

U	V	W
0	3	5
0	8	9
0	8	15
1	1	3
1	6	7
1	7	12
1	9	12
1	15	15
2	3	6
2	6	9
2	10	15
3	4	4
3	5	6
3	5	11
3	7	9
3	9	13
3	10	14
5	11	15
6	11	13
7	9	9
10	14	15
11	12	13

H K L = 8 10 11

U	V	W
0	4	5
0	8	11
0	10	11
1	2	3
1	2	4
1	6	7
1	8	8

H K L = 8 10 11

U	V	W
1	10	15
1	12	14
2	3	4
2	5	6
2	5	9
2	6	7
2	9	14
2	11	11
3	10	10
3	13	14
4	6	13
4	7	10
5	7	10
5	13	14
6	8	13
8	9	14

H K L = 8 10 13

U	V	W
0	4	5
0	8	13
0	10	13
1	2	2
1	4	6
1	6	7
1	6	11
1	10	15
2	3	5
2	3	7
2	4	9
2	7	12
2	8	9
2	13	13
3	8	6
4	6	11
4	11	14
4	14	15
5	9	10
7	10	12
9	11	14

H K L = 8 10 15

U	V	W
0	2	3
0	4	5
0	8	15

TAB. I THE TYPES OF HKL-UVW FOR H,K,L,U,V,W ≤15

H U	K V	L W
8	10	15

1	2	5
1	6	10
2	4	5
2	5	7
2	5	10
2	9	15
2	10	11
2	15	15
3	10	15
4	5	10
4	6	15
5	5	6
5	6	13
5	8	8
5	10	11
5	12	14
7	10	10
9	14	15
10	13	14
8	11	11
========		
0	1	1
0	8	11
1	7	11
1	9	11
2	6	11
2	10	11
3	5	11
3	11	11
4	4	11
4	11	12
5	11	13
6	11	14
7	11	15
8	11	12
========		
0	2	3
0	8	11
0	11	12
1	3	4
1	4	4
1	4	7
1	8	8
1	12	15

H U	K V	L W
8	11	12

2	4	5
3	4	10
3	12	13
4	5	7
4	5	8
4	5	13
4	8	9
4	11	11
4	13	14
5	9	12
6	7	12
7	8	12
8	11	13
========		
0	8	11
0	8	13
0	11	13
1	1	3
1	4	4
1	7	8
1	7	9
1	7	10
1	9	14
2	3	5
2	5	6
2	6	7
3	4	5
3	10	14
3	13	13
5	7	9
5	7	11
5	11	11
6	11	13
9	10	14
8	11	14
========		
0	4	7
0	8	11
0	11	14
1	1	2
1	6	10
1	10	12
2	3	8
2	4	9

H U	K V	L W
8	11	14

2	5	6
2	7	15
2	9	10
2	9	13
2	12	13
3	8	8
3	14	14
4	5	6
4	6	7
6	11	11
7	14	15
9	10	13
8	11	15
========		
0	8	11
0	8	15
0	11	15
1	2	2
1	3	6
1	3	7
1	5	5
1	5	8
1	9	13
2	7	11
3	4	7
3	5	9
3	7	9
4	4	13
4	9	12
5	9	10
7	11	11
9	12	13
11	13	15
8	12	13
========		
0	2	3
0	8	13
0	12	13
1	4	5
1	4	8
1	8	8
2	3	4
3	4	11
3	11	12

H U	K V	L W
8	12	13

4	4	7
4	5	14
4	7	8
4	7	9
4	10	11
4	13	13
5	8	12
5	12	12
7	9	12
8	12	15
========		
0	2	3
0	4	5
0	8	15
1	4	6
1	4	9
3	3	4
3	4	7
3	4	12
3	8	8
3	8	12
3	12	13
4	5	15
4	6	9
4	8	9
4	9	11
4	12	13
4	15	15
5	12	15
6	11	12
7	12	12
8	13	13
========		
0	1	1
0	8	13
1	7	13
1	9	13
2	6	13
2	10	13
3	5	13
3	11	13
4	4	13
4	12	13
5	13	13

H U	K V	L W
8	13	13

6	13	14
7	13	15
8	13	14
========		
0	4	7
0	8	13
0	13	14
1	2	5
1	5	6
1	6	8
1	8	8
2	2	3
2	3	4
2	5	12
2	7	9
3	10	11
4	7	10
4	10	11
5	12	14
5	14	14
6	13	13
7	9	14
8	13	15
========		
0	8	13
0	8	15
0	13	15
1	3	3
1	3	4
1	4	4
1	5	10
1	5	11
1	9	11
2	2	7
2	5	7
5	5	7
5	6	10
6	9	10
6	9	11
7	8	13
7	13	13
7	13	15

H U	K V	L W
8	14	15
========		
0	4	7
0	8	15
0	14	15
1	2	2
1	6	7
1	6	13
1	8	8
2	3	9
2	4	11
2	5	5
2	9	12
3	4	6
3	9	10
4	10	13
5	10	12
6	8	11
7	11	14
8	15	15
========		
0	1	1
0	8	15
1	7	15
1	9	15
2	6	15
2	10	15
3	5	15
3	11	15
4	4	15
4	12	15
5	13	15
6	14	15
7	15	15
9	9	10
========		
0	1	1
0	9	10
1	9	9
1	9	11
2	8	9
2	9	12
3	7	9
3	9	13
4	6	9

H U	K V	L W
9	9	10

4	9	14
5	5	9
5	9	15
9	9	11
========		
0	1	1
0	9	11
1	9	10
1	9	12
2	9	9
2	9	13
3	8	9
3	9	14
4	7	9
4	9	15
5	6	9
9	9	13
========		
0	1	1
0	9	13
1	9	12
1	9	14
2	9	11
2	9	15
3	9	10
4	9	9
5	8	9
6	7	9
9	9	14
========		
0	1	1
0	9	14
1	9	13
1	9	15
2	9	12
3	9	11
4	9	10
5	9	9
6	6	9
7	7	9

H U	K V	L W
9	10	10
========		
0	1	1
0	9	10
1	8	10
1	10	10
2	7	10
2	10	11
3	6	10
3	10	12
4	5	10
4	10	13
5	10	14
6	10	15
9	10	11
========		
0	9	10
0	9	11
0	10	11
1	1	2
1	4	6
1	5	5
1	7	9
1	9	9
1	11	11
1	11	13
2	5	6
2	7	8
2	12	13
2	12	15
3	3	7
3	4	6
3	5	7
3	13	15
4	7	13
4	9	11
5	8	15
5	12	15
7	8	13
9	10	12
========		
0	3	4
0	5	6
0	9	10
1	2	3

H U	K V	L W
9	10	12

1	6	8
1	12	12
2	3	4
2	3	6
2	6	9
2	9	9
2	11	15
2	14	15
3	5	10
3	6	7
3	8	14
3	9	14
3	10	10
3	13	14
4	7	12
4	12	13
6	8	11
6	8	15
9	10	15
9	10	13
========		
0	9	10
0	9	13
0	10	13
1	2	4
1	3	3
1	3	4
1	5	7
1	7	9
1	7	10
1	13	13
2	4	7
2	5	5
3	6	11
3	12	14
5	5	11
5	6	8
5	7	15
6	8	15
7	8	11
7	11	13
8	11	14

H U	K V	L W
9	10	14
========		
0	5	7
0	9	10
0	9	14
1	2	2
1	4	5
1	4	6
1	7	12
1	14	14
2	3	6
2	5	10
2	7	8
2	8	13
2	12	15
3	3	8
4	6	11
4	7	14
4	9	9
6	10	11
8	11	13
10	12	15
9	10	15
========		
0	2	3
0	3	5
0	9	10
1	3	5
1	5	6
1	12	15
1	15	15
3	4	10
3	5	5
3	5	9
3	7	15
3	8	10
3	11	15
4	10	15
5	5	12
5	6	7
5	6	15
5	7	15
5	9	9
5	11	12
5	13	15
6	13	15

H U	K V	L W
9	10	15

9	10	12
9	11	11
========		
0	1	1
0	9	11
1	8	11
1	10	11
2	7	11
2	11	11
3	6	11
3	11	12
4	5	11
4	11	13
5	11	14
6	11	15
9	11	12
========		
0	3	4
0	9	11
0	11	12
1	2	3
1	3	5
1	6	6
1	9	9
1	13	15
2	6	7
3	3	5
3	4	9
3	7	8
3	7	9
3	7	13
3	11	11
3	14	15
4	13	15
5	6	14
5	8	12
5	9	12
6	10	13
7	9	15
9	11	13
========		
0	9	11
0	9	13

H U	K V	L W
9	11	13

0	11	13
1	1	2
1	4	7
1	5	6
1	7	10
1	10	11
1	11	12
1	12	15
2	5	9
2	9	9
2	13	13
3	3	8
3	5	6
3	7	8
3	14	15
4	5	7
4	7	15
4	11	11
7	12	15
6	9	13
9	11	14
========		
0	9	11
0	9	14
0	11	14
1	2	4
1	3	3
1	4	5
1	5	9
1	7	7
1	8	11
1	11	15
2	3	5
2	5	8
3	6	13
3	10	12
4	6	7
5	7	13
5	11	11
7	7	10
8	10	13

H K L / U V W
9 11 15
0 3 5
0 9 11
0 11 15
1 2 3
1 5 6
1 6 9
1 12 13
2 3 7
2 15 15
3 3 4
3 5 12
3 7 8
3 7 12
3 10 13
4 9 9
5 8 15
5 14 15
6 11 11
7 12 13
9 14 15

H K L / U V W
9 12 13
0 3 4
0 9 13
0 12 13
1 3 3
1 3 4
1 6 10
1 9 9
2 3 7
2 5 6
3 5 7
3 5 11
3 8 15
3 9 10
3 13 13
3 14 15
5 6 9
6 6 11
7 8 12
7 9 15
7 11 15
8 11 15

H K L / U V W
9 12 14
0 3 4
0 6 7
0 9 14
1 3 6
1 6 8
2 2 3
2 3 5
2 9 9
2 9 12
3 4 10
3 6 8
3 7 14
3 9 10
3 10 11
3 14 14
4 11 12
5 12 12
6 8 13
6 13 15
7 14 15

H K L / U V W
9 13 13
0 1 1
0 9 13
1 8 13
1 10 13
2 7 13
2 11 13
3 6 13
3 12 13
4 5 13
4 13 13
5 13 14
6 13 15

H K L / U V W
9 13 14
0 9 13
0 9 14
0 13 14
1 1 3
1 4 5
1 8 10
1 8 11
1 9 9
2 4 5
2 7 7
2 7 8
3 4 6
3 5 6
3 13 14
5 7 11
5 11 13
5 13 13
7 7 11

H K L / U V W
9 13 15
0 3 5
0 9 13
0 13 15
1 2 3
1 3 6
1 6 7
1 11 12
2 9 9
3 4 5
3 4 11
3 6 7
3 8 9
3 11 14
4 15 15
5 9 12
6 13 13
7 9 12
7 10 15

H K L / U V W
9 14 14
0 1 1
0 9 14
1 8 14
1 10 14
2 7 14
2 11 14
3 6 14
3 12 14
4 5 14
4 13 14
5 14 14
6 14 15

H K L / U V W
9 14 15
0 3 5
0 9 14
0 14 15
1 3 3
1 5 6
1 6 11
1 9 9
2 3 4
2 3 8
3 5 13
3 7 7
3 10 12
3 12 13
4 6 9
5 11 15
6 9 11
7 7 12
8 10 15
9 11 15

H K L / U V W
10 10 11
0 1 1
0 10 11
1 10 10
1 10 12
2 9 10
2 10 13
3 8 10
3 10 14
4 7 10
4 10 15
5 6 10

H K L / U V W
10 10 13
0 1 1
0 10 13
1 10 12
1 10 14
2 10 11
2 10 15
3 10 10
4 9 10
5 8 10
6 7 10

H K L / U V W
10 11 11
0 1 1
0 10 11
1 9 11
1 11 11
2 8 11
2 11 12
3 7 11
3 11 13
4 6 11
4 11 14
5 5 11
5 11 15

H K L / U V W
10 11 12
0 5 6
0 10 11
0 11 12
1 1 2
1 8 10
1 10 10
1 12 12
1 12 14
2 4 7
2 5 6
2 6 9
2 8 9
2 9 13
2 11 11
2 13 14
3 4 8
3 6 8
4 4 7
6 7 15
6 9 13
7 10 15

H K L / U V W
10 11 13
0 10 11
0 10 13
0 11 13
1 2 3
1 3 5
1 5 5
1 6 8
1 7 8
1 7 9
1 13 13
2 3 4
2 4 7
3 7 11
3 11 11
4 7 9
5 5 12
5 8 14
5 12 14
5 14 15
7 9 13

H K L / U V W
10 11 14
0 5 7
0 10 11
0 11 14
1 3 4
1 4 6
1 6 8
1 7 8
1 14 14
2 2 3
2 2 5
2 7 12
2 8 9
3 8 13
3 10 10
4 5 10
4 11 11
5 12 13
6 7 14
6 13 14

H K L / U V W
10 11 15
0 2 3
0 10 11
0 11 15
1 3 5
1 4 5
1 5 7
1 8 10
1 15 15
2 5 5
3 5 10
3 13 15
4 5 10
5 5 7
5 5 13
5 8 9
5 9 15
5 11 11
5 13 14
6 7 15
7 10 12

H K L / U V W
10 12 13
0 5 6
0 10 13
0 12 13
1 2 3
1 4 4
1 6 9
1 10 10
2 2 5
2 7 8
2 7 11
2 13 13
3 4 6
3 8 14
4 8 11
5 7 10
5 11 14
6 9 14
6 11 14

H K L / U V W
10 12 15
0 2 3
0 4 5
0 5 6
2 3 5
2 5 9
2 9 10
2 10 15
2 15 15
3 5 6
3 6 10
3 10 10
3 10 15
3 14 15
4 5 12
4 12 15
5 6 8
5 6 15
5 9 10
5 12 12
5 14 15
6 8 15
9 10 14

H K L / U V W
10 13 13
0 1 1
0 10 13
1 9 13
1 11 13
2 8 13
2 12 13
3 7 13
3 13 13
4 6 13
4 13 14
5 5 13
5 13 15

H K L / U V W
10 13 14
0 5 7
0 10 13
0 13 14
1 2 4
1 3 4
1 8 9
1 10 10
2 3 4
2 5 6
2 6 7
2 6 11
2 9 10
3 14 14
4 7 15
4 13 13
5 6 10
5 8 11
7 8 14
8 10 15

H K L / U V W
10 13 15
0 2 3
0 10 13
0 13 15
1 5 5
1 5 8
1 8 10
2 3 5
3 5 11
3 11 15
4 5 7
4 7 10
5 5 14
5 7 9
5 10 11
5 10 12
5 12 15
5 13 13
7 9 15

H K L / U V W
10 14 15
0 2 3
0 5 7
0 14 15
1 4 5
1 10 10
2 4 5
2 5 6
2 5 10
2 10 11
4 5 13
4 15 15
5 5 8
5 6 10
5 8 10
5 11 12
5 14 14
6 10 15
7 10 14
8 9 15

H K L / U V W
11 11 12
0 1 1
0 11 12
1 11 11
1 11 13
2 10 11
2 11 14
3 9 11
3 11 15
4 8 11
5 7 11
6 6 11

H K L / U V W
11 11 13
0 1 1
0 11 13
1 11 12
1 11 14
2 11 11
2 11 15
3 10 11
4 9 11
5 8 11
6 7 11

H K L / U V W
11 11 14
0 1 1
0 11 14
1 11 13
1 11 15
2 11 12
3 11 11
4 10 11
5 9 11
6 8 11
7 7 11

H K L / U V W
11 11 15
0 1 1
0 11 15
1 11 14
2 11 13
3 11 12
4 11 11
5 10 11
6 9 11
7 8 11

H K L / U V W
11 12 12
0 1 1
0 11 12
1 10 12
1 12 12
2 9 12
2 12 13
3 8 12
3 12 14
4 7 12
4 12 15
5 6 12

H K L / U V W
11 12 13
0 11 12
0 11 13
0 12 13
1 1 2
1 5 7
1 6 6
1 9 11
1 11 11
1 13 13
1 13 15
2 7 10
2 9 10
2 14 15
3 4 8
3 5 7
3 5 9
3 7 9
4 5 8
4 9 15
4 11 13
8 9 15

H K L / U V W
11 12 14
0 6 7
0 11 12
0 11 14
1 2 3
1 4 4
1 7 10
1 8 10
1 14 14
2 3 6
2 4 5
2 5 8
2 7 10
2 11 11
4 8 13
5 6 9
5 7 14
6 10 13
8 9 14

H K L / U V W
11 12 15
0 4 5
0 11 12
0 11 15
1 3 3
1 3 4
1 7 9
1 8 12
1 15 15
2 3 6
2 5 9
3 5 9
3 6 7
3 9 14

TAB. I

THE TYPES OF HKL-UVW FOR H,K,L,U,V,W ≤15

H K L = 11 12 15

U	V	W
3	10	15
3	11	11
5	7	15
6	6	13
6	7	10
8	9	13

H K L = 11 13 13

U	V	W
0	1	1
0	11	13
1	10	13
1	12	13
2	9	13
2	13	13
3	8	13
3	13	14
4	7	13
4	13	15
5	6	13

H K L = 11 13 14

U	V	W
0	11	13
0	11	14
0	13	14
1	2	3
1	3	5
1	4	6
1	7	7
1	8	9
1	11	11
2	4	5
2	8	9
3	5	7
3	8	10
3	13	13
4	5	11
5	9	13
7	7	12

H K L = 11 13 15

U	V	W
0	11	13
0	11	15
0	13	15
1	1	2
1	5	8
1	6	7
1	9	12
1	12	13
1	13	14
2	7	11
2	11	11
2	15	15
3	4	9
3	5	10
3	6	7
3	9	10
4	7	9
4	13	13
5	5	8
8	11	15

H K L = 11 14 14

U	V	W
0	1	1
0	11	14
1	10	14
1	12	14
2	9	14
2	13	14
3	8	14
3	14	14
4	7	14
4	14	15
5	6	14

H K L = 11 14 15

U	V	W
0	11	14
0	11	15
0	14	15
1	2	4
1	3	4
1	5	5
1	6	9
1	9	11
1	11	11
2	7	7
2	7	8
3	3	5
3	7	13
4	6	9
5	10	12
5	10	13
5	11	14
7	7	13

H K L = 11 15 15

U	V	W
0	1	1
0	11	15
1	10	15
1	12	15
2	9	15
2	13	15
3	8	15
3	14	15
4	7	15
4	15	15
5	6	15

H K L = 12 12 13

U	V	W
0	1	1
0	12	13
1	12	12
1	12	14
2	11	12
2	12	15
3	10	12
4	9	12
5	8	12
6	7	12

H K L = 12 13 13

U	V	W
0	1	1
0	12	13
1	11	13
1	13	13
2	10	13
2	13	14
3	9	13
3	13	15
4	6	13
5	7	13
6	6	13

H K L = 12 13 14

U	V	W
0	6	7
0	12	13
0	13	14
1	1	2
1	10	12
1	12	12
1	14	14
2	5	8
2	6	7
2	8	11
2	10	11
2	11	15
2	13	13
3	6	10
3	8	10
4	4	9
4	5	8
4	6	9
6	11	15

H K L = 12 13 15

U	V	W
0	4	5
0	12	13
0	13	15
1	2	3
1	6	6
1	7	9
1	9	11
1	15	15
2	6	9
3	3	5
3	3	7
3	7	12
3	8	9
3	8	12
3	13	13
4	9	11
5	9	15
6	11	14

H K L = 12 14 15

U	V	W
0	4	5
0	6	7
0	14	15
1	2	3
1	12	12
2	3	6
2	8	9
2	9	10
2	9	13
2	15	15
3	4	6
3	6	8
3	6	11
3	9	10
3	14	14
4	9	12
6	10	15
7	9	14

H K L = 13 13 14

U	V	W
0	1	1
0	13	14
1	13	13
1	13	15
2	12	13
3	11	13
4	10	13
5	9	13
6	8	13
7	7	13

H K L = 13 13 15

U	V	W
0	1	1
0	13	15
1	13	14
2	13	13
3	12	13
4	11	13
5	10	13
6	9	13
7	8	13

H K L = 13 14 14

U	V	W
0	1	1
0	13	14
1	12	14
1	14	14
2	11	14
2	14	15
3	10	14
4	9	14
5	8	14
6	7	14

H K L = 13 14 15

U	V	W
0	13	14
0	13	15
0	14	15
1	1	2
1	6	8
1	7	7
1	11	13
1	13	13
1	15	15
2	9	12
2	11	12
3	5	9
3	6	8
3	7	11
3	9	11
4	5	10
4	7	10
4	13	15
5	5	9

H K L = 13 15 15

U	V	W
0	1	1
0	13	15
1	12	15
1	14	15
2	11	15
2	15	15
3	10	15
4	9	15
5	8	15
6	7	15

H K L = 14 14 15

U	V	W
0	1	1
0	14	15
1	14	14
2	13	14
3	12	14
4	11	14
5	10	14
6	9	14
7	8	14

H K L = 14 15 15

U	V	W
0	1	1
0	14	15
1	13	15
1	15	15
2	12	15
3	11	15
4	10	15
5	9	15
6	8	15
7	7	15

TABLE II

From the {001} Pole Figure to Miller Indices

The angles α_i and α_i' which are formed by a {001}-axis ($A_iB_iC_i$) and the normal direction (HKL) or rolling direction [UVW], respectively, are given by the relationships

$$\cos \alpha_i = \frac{(A_iB_iC_i)\,(HKL)}{\sqrt{(A_i^2 + B_i^2 + C_i^2)\,(H^2 + K^2 + L^2)}}$$

$$\cos \alpha_i' = \frac{(A_iB_iC_i)\,[UVW]}{\sqrt{(A_i^2 + B_i^2 + C_i^2)\,(U^2 + V^2 + W^2)}} .$$

The index i characterizes the different equivalent axes of the {001} type, i.e. (100), (010) and (001). (This means i runs here from 1 to 3).

If the α_i and α_i' can be taken from a pole figure, the 6 quantities H, K, L, U, V, W can be calculated from these equations, if a sufficient number of α_i and α_i' is available (c.f. Sec. 2.2). More practical, however, is to use the following Table.

In the table possible sets of the α_1 α_2 α_3 are listed, ordered in blocks with increasing α_1. The sets XYZ listed on the right of the α_i give the corresponding sets of (HKL) which thus can be obtained if a set of α_1 α_2 α_3 is known. If instead of α_1 α_2 α_3 the angles α_1' α_2' α_3' are introduced into the table, the XYZ give the corresponding [UVW].

Since the table only contains values $|H, K, L, U, V, W| \leqslant 15$, the exact measured angles α_i and α_i' will generally not be listed in the table. Thus one has to pick that of the listed sets of α_i which is closest to the given one.

TAB. II

FROM ALPHA ANGLES OF (100) - POLES INTO HKL

α1	α2	α3	X	Y	Z
0	80				
4	86	90	0	1	13
4	86	90	0	1	14
4	86	90	0	1	15
5	85	90	0	1	11
5	85	90	0	1	12
5	86	86	1	1	15
6	84	90	0	1	9
6	84	90	0	1	10
6	86	86	1	1	13
6	86	86	1	1	14
7	83	90	0	1	8
7	85	85	1	1	11
7	85	85	1	1	12
8	82	86	1	2	15
8	82	90	0	2	15
8	82	90	0	1	7
8	84	84	1	1	10
9	81	90	0	2	13
9	81	90	0	1	6
9	82	86	1	2	14
9	84	84	1	1	9
0	90				
0	90	90	0	0	1
10	70				
11	79	90	0	1	5
12	78	90	0	3	14
12	79	86	1	3	15
13	77	90	0	2	9
13	77	90	0	3	13
13	78	86	1	3	14
13	79	84	1	2	10
14	76	90	0	1	4
14	77	86	1	3	13
14	78	82	2	3	14
14	78	84	1	2	9
14	79	83	2	3	15
15	75	86	1	4	15
15	75	90	0	4	15
15	75	90	0	3	11
15	76	85	1	3	12
16	74	86	1	4	14
16	74	90	0	2	7
16	75	85	1	3	11
16	76	83	1	2	8

α1	α2	α3	X	Y	Z
10	70				
16	77	81	2	3	13
16	79	79	1	1	5
17	73	90	0	4	13
17	73	90	0	3	10
17	75	83	2	4	15
17	76	81	2	3	12
17	78	78	3	3	14
17	78	78	2	2	9
18	72	90	0	1	3
18	73	85	1	3	10
18	73	86	1	4	13
18	74	82	1	2	7
18	75	79	3	4	15
18	75	80	2	3	11
18	77	77	3	3	13
19	72	84	1	3	9
19	72	85	1	4	12
19	72	86	1	5	15
19	73	82	2	4	13
19	76	76	1	1	4
10	80				
10	80	90	0	2	11
10	81	86	1	2	13
10	83	83	1	1	8
11	80	85	1	2	11
11	81	85	1	2	12
11	82	82	2	2	15
11	82	82	1	1	7
12	81	81	2	2	13
13	81	81	1	1	6
14	80	60	2	2	11
20	60				
21	69	86	1	5	13
21	69	90	0	5	13
21	69	90	0	3	8
22	68	85	1	4	10
22	68	86	1	6	15
22	68	90	0	2	5
23	67	86	1	5	12
23	67	86	1	6	14
23	67	90	0	5	12
23	67	90	0	3	7
23	68	83	2	6	15
23	69	82	2	5	13

α1	α2	α3	X	Y	Z
20	60				
24	66	90	0	4	9
24	66	90	0	5	11
24	67	83	1	3	7
24	68	81	2	5	12
24	69	78	3	5	13
24	69	79	1	2	5
25	65	86	1	6	13
25	65	87	1	7	15
25	65	90	0	7	15
25	65	90	0	6	13
25	66	84	1	4	9
25	66	85	1	5	11
26	65	82	2	6	13
26	65	83	2	7	15
26	66	81	2	5	11
26	67	79	3	6	14
26	67	79	2	4	9
26	68	77	3	5	12
26	69	76	4	6	15
27	63	86	1	7	14
27	63	90	0	1	2
27	64	83	2	7	14
27	64	84	1	4	8
27	64	85	1	5	10
27	64	86	1	6	12
27	65	80	3	7	15
27	66	78	3	6	13
27	68	75	2	3	7
27	69	74	3	4	10
28	62	87	1	8	15
28	62	90	0	8	15
28	62	90	0	7	13
28	64	80	2	5	10
26	64	82	1	3	6
28	66	76	4	7	15
28	66	76	3	5	11
28	68	73	4	5	12
28	69	73	5	6	15
29	61	85	1	6	11
29	61	90	0	5	9
29	61	90	0	6	11
29	62	82	2	7	13
29	62	83	2	8	15
29	62	86	1	7	13
29	64	77	1	2	4
29	64	79	3	7	14
29	66	74	4	6	13
29	67	73	3	4	9

α1	α2	α3	X	Y	Z
20	60				
29	68	72	5	6	14
20	70				
20	70	86	1	5	14
20	70	90	0	4	11
20	70	90	0	5	14
20	72	81	1	2	6
20	72	83	2	5	15
20	74	78	3	4	14
20	74	79	2	3	10
21	70	85	1	4	11
21	71	82	2	5	14
21	72	79	3	5	15
21	73	78	3	4	13
21	75	75	3	3	11
21	76	76	4	4	15
22	70	80	2	4	11
22	70	83	1	3	8
22	72	78	2	3	9
22	75	75	2	2	7
23	71	79	3	5	14
23	72	76	4	5	15
23	72	77	3	4	12
23	74	74	3	3	10
24	70	77	2	3	8
24	71	76	3	4	11
24	74	74	4	4	13
25	71	75	4	5	14
25	72	72	1	1	3
26	70	74	4	5	13
27	71	71	5	5	14
27	71	71	4	4	11
28	71	71	3	3	8
29	70	70	2	2	5
29	70	70	5	5	13
30	50				
31	59	85	1	6	10
31	59	87	1	9	15
31	59	90	0	3	5
32	58	86	1	8	13
32	58	90	0	8	13
32	58	90	0	5	8
32	58	90	0	7	11
32	59	83	2	9	15
32	59	83	2	8	13

α1	α2	α3	X	Y	Z
30	50				
33	57	87	1	9	14
33	57	90	0	9	14
33	58	81	2	7	11
33	58	83	2	9	14
33	58	84	1	5	8
33	58	86	1	7	11
33	59	79	3	8	13
34	56	85	1	6	9
34	56	86	1	8	12
34	56	87	1	10	15
34	56	90	0	2	3
34	57	82	1	4	6
34	57	84	2	10	15
34	58	80	3	9	14
34	59	76	2	5	8
35	55	85	1	7	10
35	55	86	1	9	13
35	55	90	0	7	10
35	55	90	0	9	13
35	56	83	2	9	13
35	57	78	3	8	12
35	57	80	2	6	9
35	57	81	3	10	15
35	58	76	4	9	14
35	56	77	3	7	11
36	54	86	1	8	11
36	54	87	1	11	15
36	54	90	0	11	15
36	54	90	0	5	7
36	54	90	0	8	11
36	55	83	1	5	7
36	55	87	1	10	14
36	56	79	3	9	13
36	56	81	2	7	10
36	57	77	4	10	15
36	59	73	4	7	11
36	59	73	5	9	14
37	53	84	1	6	8
37	53	86	1	9	12
37	53	90	0	3	4
37	54	81	3	11	15
37	54	82	2	8	11
37	54	84	2	11	15
37	55	80	3	10	14
37	56	76	3	7	10
37	57	76	4	9	13
37	58	74	1	2	3
38	52	85	1	7	9

α1	α2	α3	X	Y	Z
30	50				
38	52	87	1	11	14
38	52	90	0	7	9
38	52	90	0	10	13
38	52	90	0	11	14
38	53	63	2	10	13
38	53	87	1	10	13
38	54	79	1	3	4
38	54	82	2	9	12
38	55	78	4	11	15
38	55	78	3	8	11
38	56	77	2	5	7
38	57	72	5	9	13
38	58	71	5	8	12
38	58	72	6	10	15
38	59	70	6	9	14
39	51	86	1	8	10
39	51	86	1	9	11
39	51	87	1	12	15
39	51	90	0	4	5
39	51	90	0	9	11
39	52	80	3	11	14
39	52	64	2	12	15
39	52	84	2	11	14
39	53	80	2	7	9
39	53	80	3	10	13
39	55	75	4	9	12
39	55	75	5	11	15
39	56	74	4	8	11
39	56	74	5	10	14
39	57	72	4	7	10
39	59	69	7	10	15
39	59	70	4	6	9
30	60				
30	60	66	1	8	14
30	60	90	0	4	7
30	60	90	0	7	12
30	61	83	1	4	7
30	61	84	1	5	9
30	62	79	3	7	13
30	62	80	3	8	15
30	62	61	2	6	11
30	64	75	3	5	10
30	64	76	4	7	14
30	66	73	5	7	15
30	67	72	4	5	11
31	60	62	2	7	12

TAB. II

FROM HKL INTO ALPHA ANGLES

MILL IND X	Y	Z	(100)- α1	α2	α3	(111)-POLES α1	α2	α3	α4
2	2	9	78	78	17	37	72	57	57
2	2	11	80	80	14	40	69	56	56
2	2	13	81	81	12	42	67	56	56
2	2	15	82	82	11	44	65	55	55
2	3	3	65	50	50	10	76	76	61
2	3	4	68	56	42	15	84	71	58
2	3	5	71	61	36	21	90	68	56
2	3	6	73	65	31	25	85	66	55
2	3	7	75	68	27	28	82	64	54
2	3	8	77	70	24	31	79	63	54
2	3	9	78	72	22	34	76	62	53
2	3	10	79	74	20	35	74	61	53
2	3	11	80	75	18	37	73	60	53
2	3	12	81	76	17	38	71	60	53
2	3	13	81	77	16	40	70	59	53
2	3	14	82	78	14	41	69	59	53
2	3	15	83	79	14	42	68	58	53
2	4	5	73	53	42	19	85	75	53
2	4	7	76	61	33	25	86	70	51
2	4	9	79	67	26	30	80	66	51
2	4	11	80	70	22	34	76	64	51
2	4	13	82	73	19	37	73	62	51
2	4	15	83	75	17	39	71	61	51
2	5	5	74	47	47	19	81	81	51
2	5	6	76	52	42	21	86	78	50
2	5	7	77	56	38	24	90	75	49
2	5	8	78	59	34	26	87	73	49
2	5	9	79	62	31	28	84	71	49
2	5	10	80	64	28	30	81	69	49
2	5	11	81	66	26	32	79	68	49
2	5	12	81	68	24	33	77	67	49
2	5	13	82	69	23	35	76	66	49
2	5	14	82	71	21	36	74	65	49
2	5	15	83	72	20	37	73	64	49
2	6	7	78	51	42	23	86	79	48
2	6	9	80	57	35	27	87	75	47
2	6	11	81	62	30	30	82	71	47
2	6	13	82	65	26	33	78	69	47
2	6	15	83	68	23	35	76	67	48
2	7	7	79	46	46	24	83	83	47
2	7	8	79	50	42	25	87	81	46
2	7	9	80	53	39	26	90	78	46
2	7	10	81	56	36	28	87	77	46
2	7	11	81	58	33	29	85	75	46
2	7	12	82	60	31	30	83	73	46
2	7	13	82	62	29	32	81	72	46
2	7	14	83	64	27	33	79	71	46
2	7	15	83	65	26	34	78	70	46
2	8	9	81	49	42	26	87	82	45

MILL IND X	Y	Z	(100)- α1	α2	α3	(111)-POLES α1	α2	α3	α4
2	8	11	82	54	37	28	88	78	44
2	8	13	83	59	32	30	84	75	45
2	8	15	83	62	29	33	80	72	45
2	9	9	81	46	46	26	85	85	44
2	9	10	82	49	43	27	88	83	44
2	9	11	82	51	40	28	90	81	44
2	9	12	82	54	38	29	88	79	44
2	9	13	83	56	35	30	86	77	44
2	9	14	83	58	33	31	84	76	44
2	9	15	83	59	32	32	82	75	44
2	10	11	82	48	43	28	88	83	43
2	10	13	83	53	38	29	88	80	43
2	10	15	84	57	34	31	85	77	43
2	11	11	83	45	45	28	86	86	43
2	11	12	83	48	43	28	88	84	42
2	11	13	83	50	41	29	90	82	42
2	11	14	84	52	39	30	88	81	42
2	11	15	84	54	37	30	86	79	42
2	12	13	84	48	43	29	88	84	42
2	12	15	84	52	39	30	88	81	42
2	13	13	84	45	45	29	86	86	41
2	13	14	84	47	43	29	88	85	41
2	13	15	84	49	41	30	90	83	41
2	14	15	84	47	43	30	88	85	41
2	15	15	85	45	45	30	87	87	41
3	3	4	59	59	47	8	79	67	67
3	3	5	63	63	40	14	85	64	64
3	3	7	68	68	31	24	86	60	60
3	3	8	71	71	28	27	83	59	59
3	3	10	74	74	23	32	78	58	58
3	3	11	75	75	21	34	76	57	57
3	3	13	77	77	18	37	73	57	57
3	3	14	78	78	17	38	72	56	56
3	4	4	62	51	51	7	74	74	63
3	4	5	65	56	45	12	81	71	61
3	4	6	67	59	40	16	86	68	59
3	4	7	70	62	36	20	90	66	58
3	4	8	71	65	32	23	86	65	57
3	4	9	73	67	29	26	84	63	56
3	4	10	74	69	27	29	81	62	55
3	4	11	76	71	24	31	79	61	55
3	4	12	77	72	23	32	77	61	55
3	4	13	78	73	21	34	76	60	55
3	4	14	78	74	20	35	74	60	54
3	4	15	79	75	18	37	73	59	54
3	5	5	67	49	49	12	77	77	58
3	5	6	69	53	44	15	82	74	56
3	5	7	71	57	40	18	86	72	55
3	5	8	72	60	36	21	90	70	54

MILL IND X	Y	Z	(100)- α1	α2	α3	(111)-POLES α1	α2	α3	α4
3	5	9	74	62	33	24	87	68	54
3	5	10	75	64	30	26	84	66	53
3	5	11	76	66	28	28	82	65	53
3	5	12	77	68	26	30	80	64	53
3	5	13	78	69	24	32	78	64	53
3	5	14	79	71	23	33	77	63	52
3	5	15	79	72	21	34	75	62	52
3	6	7	72	52	44	18	83	76	53
3	6	8	73	55	40	20	87	74	53
3	6	10	76	60	34	24	87	70	51
3	6	11	77	62	31	26	85	69	51
3	6	13	78	66	27	30	81	67	51
3	6	14	79	67	26	31	79	66	51
3	7	7	73	47	47	18	80	80	52
3	7	8	74	51	44	20	84	78	51
3	7	9	75	54	40	21	87	76	50
3	7	10	76	56	37	23	90	74	50
3	7	11	77	58	35	25	88	72	50
3	7	12	78	60	32	27	85	71	49
3	7	13	79	62	30	28	83	70	49
3	7	14	79	64	29	30	82	69	49
3	7	15	80	65	27	31	80	68	49
3	8	8	75	47	47	20	81	81	50
3	8	9	76	50	44	21	85	79	49
3	8	10	77	53	41	23	87	77	49
3	8	11	78	55	38	24	90	76	48
3	8	12	78	57	35	26	88	74	48
3	8	13	79	59	33	27	86	73	48
3	8	14	79	61	31	28	84	72	48
3	8	15	80	62	30	30	82	70	48
3	9	10	77	49	43	23	85	80	48
3	9	11	78	52	41	24	88	79	47
3	9	13	79	56	36	26	88	75	47
3	9	14	80	58	34	27	86	74	47
3	10	10	78	46	46	23	83	83	47
3	10	11	79	49	44	24	86	81	47
3	10	12	79	51	41	25	88	80	46
3	10	13	80	53	39	26	90	78	46
3	10	14	80	55	37	27	88	77	46
3	10	15	81	57	35	28	86	75	46
3	11	11	79	46	46	24	84	84	46
3	11	12	80	48	44	25	86	82	46
3	11	13	80	50	41	26	88	80	45
3	11	14	80	52	39	26	90	79	45
3	11	15	81	54	37	27	88	78	45
3	12	13	80	48	44	26	86	83	45
3	12	14	81	50	41	26	88	81	45
3	13	13	81	46	46	26	85	85	45
3	13	14	81	48	44	26	87	83	44

MILL IND X	Y	Z	(100)- α1	α2	α3	(111)-POLES α1	α2	α3	α4
3	13	15	81	50	42	27	88	82	44
3	14	14	81	46	46	27	85	85	44
3	14	15	82	48	44	27	87	84	44
4	4	5	58	58	49	6	77	68	68
4	4	7	64	64	39	16	86	63	63
4	4	9	68	68	32	23	87	61	61
4	4	11	71	71	27	28	82	59	59
4	4	13	74	74	24	31	78	58	58
4	4	15	76	76	21	34	75	57	57
4	5	5	61	52	52	6	73	73	65
4	5	6	63	55	47	9	79	71	63
4	5	7	65	56	42	13	83	69	61
4	5	8	67	61	39	17	87	67	60
4	5	9	69	63	35	20	90	65	58
4	5	10	70	65	33	23	87	64	58
4	5	11	72	67	30	25	85	63	57
4	5	12	73	68	28	27	83	62	57
4	5	13	74	70	26	29	81	61	56
4	5	14	75	71	25	30	79	61	56
4	5	15	76	72	23	32	76	60	56
4	6	7	67	53	46	12	80	73	59
4	6	9	70	59	39	18	87	69	57
4	6	11	72	63	33	23	87	67	55
4	6	13	74	66	29	27	83	65	54
4	6	15	76	69	26	30	80	63	54
4	7	7	68	49	49	13	78	78	57
4	7	8	69	52	45	15	81	75	56
4	7	9	71	55	42	17	85	73	55
4	7	10	72	57	39	19	87	72	54
4	7	11	73	59	36	21	90	70	54
4	7	12	74	61	34	23	88	69	53
4	7	13	75	63	32	25	86	68	53
4	7	14	76	64	30	27	84	67	53
4	7	15	76	66	28	28	82	66	52
4	8	9	72	51	45	17	82	77	54
4	8	11	74	56	39	21	88	73	52
4	8	13	75	60	35	24	88	71	52
4	8	15	77	63	31	27	84	69	51
4	9	9	73	48	48	18	80	80	53
4	9	10	73	50	45	19	83	78	52
4	9	11	74	52	42	20	86	76	51
4	9	12	75	55	39	22	88	75	51
4	9	13	76	57	37	23	90	74	50
4	9	14	76	58	35	24	88	72	50
4	9	15	77	60	33	26	86	71	50
4	10	11	75	49	44	20	84	79	50
4	10	13	76	54	40	23	88	76	49
4	10	15	77	57	36	25	88	74	49
4	11	11	76	47	47	21	82	82	50

TAB. II

FROM HKL INTO ALPHA ANGLES

MILL IND X	Y	Z	(100)- α1	α2	α3	(111)-POLES α1	α2	α3	α4
4	11	12	76	49	44	22	84	80	49
4	11	13	77	51	42	22	86	79	49
4	11	14	77	53	40	23	88	77	48
4	11	15	78	55	38	24	90	76	48
4	12	13	77	49	44	23	85	81	48
4	12	15	78	52	40	24	88	78	47
4	13	13	78	46	46	23	83	83	48
4	13	14	78	48	44	24	85	81	47
4	13	15	79	50	42	24	87	80	47
4	14	15	79	48	44	24	85	82	46
4	15	15	79	46	46	25	84	84	46
5	5	6	57	57	50	5	76	68	68
5	5	7	60	60	45	9	80	66	66
5	5	8	62	62	41	13	84	64	64
5	5	9	64	64	38	17	87	63	63
5	5	11	68	68	33	22	87	61	61
5	5	12	69	69	31	24	85	60	60
5	5	13	70	70	29	26	83	60	60
5	5	14	71	71	27	28	82	59	59
5	6	6	59	52	52	5	73	73	66
5	6	7	62	55	48	8	77	71	64
5	6	8	63	58	44	11	81	69	62
5	6	9	65	60	41	14	84	67	61
5	6	10	67	62	38	17	87	66	60
5	6	11	68	64	35	20	90	65	59
5	6	12	70	65	33	22	88	64	58
5	6	13	71	67	31	24	86	63	58
5	6	14	72	68	29	26	84	62	57
5	6	15	73	69	28	27	82	61	57
5	7	7	63	51	51	8	75	75	62
5	7	8	65	53	47	11	79	73	61
5	7	9	66	56	44	13	82	71	59
5	7	10	68	58	41	16	85	70	58
5	7	11	69	60	38	18	88	68	57
5	7	12	70	62	36	20	90	67	57
5	7	13	71	63	33	22	88	66	56
5	7	14	72	65	32	24	86	65	56
5	7	15	73	66	30	26	84	64	55
5	8	8	66	50	50	11	77	77	59
5	8	9	67	52	46	13	80	75	58
5	8	10	69	54	43	15	83	73	57
5	8	11	70	56	41	17	85	71	56
5	8	12	71	58	38	19	88	70	55
5	8	13	72	60	36	21	90	69	55
5	8	14	73	62	34	23	88	68	54
5	8	15	74	63	32	24	86	67	54
5	9	9	69	49	49	14	78	78	57
5	9	10	70	51	46	15	81	76	56
5	9	11	71	53	43	17	83	74	55

MILL IND X	Y	Z	(100)- α1	α2	α3	(111)-POLES α1	α2	α3	α4
5	9	12	72	55	41	18	86	73	54
5	9	13	72	57	38	20	88	72	54
5	9	14	73	59	36	22	90	71	53
5	9	15	74	60	34	23	88	70	53
5	10	11	71	50	45	17	82	77	54
5	10	12	72	52	43	18	84	76	53
5	10	13	73	54	41	19	86	74	53
5	10	14	74	56	39	21	88	73	52
5	11	11	72	48	48	17	80	80	53
5	11	12	73	50	45	18	82	78	52
5	11	13	74	52	43	19	84	77	52
5	11	14	74	54	41	21	86	76	51
5	11	15	75	55	39	22	88	74	51
5	12	12	74	47	47	19	81	81	52
5	12	13	74	49	45	20	83	79	51
5	12	14	75	51	43	20	85	78	51
5	12	15	75	53	41	21	87	77	50
5	13	13	75	47	47	20	81	81	50
5	13	14	75	49	45	21	83	80	50
5	13	15	76	51	43	21	85	79	50
5	14	14	76	47	47	21	82	82	49
5	14	15	76	48	45	22	84	81	49
6	6	7	57	57	50	4	75	68	68
6	6	11	64	64	38	17	88	63	63
6	6	13	67	67	33	22	88	61	61
6	7	7	59	53	53	4	73	73	66
6	7	8	61	55	49	7	76	71	65
6	7	9	62	57	46	10	80	69	63
6	7	10	64	59	43	13	83	68	62
6	7	11	65	61	40	15	85	66	61
6	7	12	67	62	38	17	88	65	60
6	7	13	68	64	35	20	90	64	60
6	7	14	69	65	33	22	88	63	59
6	7	15	70	67	32	23	86	63	58
6	8	9	64	54	48	9	78	73	62
6	8	11	66	57	42	14	83	70	60
6	8	13	69	61	38	18	88	67	58
6	8	15	71	64	34	22	88	65	57
6	9	10	66	52	47	12	79	74	59
6	9	11	67	54	45	13	81	73	58
6	9	13	69	58	40	17	86	70	57
6	9	14	70	59	38	19	88	69	56
6	10	11	68	51	47	13	80	75	57
6	10	13	70	55	42	17	84	73	56
6	10	15	72	58	38	20	88	70	55
6	11	11	69	49	49	14	78	78	56
6	11	12	70	51	46	15	80	77	56
6	11	13	71	52	44	16	83	75	55
6	11	14	71	54	42	18	85	74	54

MILL IND X	Y	Z	(100)- α1	α2	α3	(111)-POLES α1	α2	α3	α4
6	11	15	72	56	40	19	87	73	54
6	12	13	71	50	46	17	81	78	54
6	13	13	72	48	48	17	80	80	53
6	13	14	73	50	46	18	82	78	53
6	13	15	73	51	44	19	84	77	52
6	14	15	74	49	45	19	82	79	52
7	7	8	57	57	51	4	74	69	69
7	7	9	58	58	48	7	78	67	67
7	7	10	60	60	45	10	81	66	66
7	7	11	62	62	42	13	83	65	65
7	7	12	63	63	40	15	86	64	64
7	7	13	65	65	37	17	88	63	63
7	7	15	67	67	33	21	86	61	61
7	8	8	58	53	53	4	72	72	67
7	8	9	60	55	50	6	76	71	66
7	8	10	61	57	47	9	79	69	64
7	8	11	63	58	44	11	81	68	63
7	8	12	64	60	42	13	84	67	62
7	8	13	65	62	39	16	86	66	61
7	8	14	67	63	37	18	88	65	60
7	8	15	68	64	35	20	90	64	60
7	9	9	61	52	52	6	74	74	64
7	9	10	63	54	49	8	77	72	63
7	9	11	64	55	46	10	80	71	62
7	9	12	65	57	44	12	82	70	61
7	9	13	66	59	41	14	84	68	60
7	9	14	67	60	39	16	86	67	59
7	9	15	68	61	37	18	88	67	59
7	10	10	64	51	51	9	75	75	62
7	10	11	65	53	48	10	78	74	61
7	10	12	66	54	45	12	80	72	60
7	10	13	67	56	43	14	83	71	59
7	10	14	68	57	41	16	85	70	58
7	10	15	69	59	39	17	87	69	57
7	11	11	66	50	50	11	76	76	59
7	11	12	67	52	47	12	79	75	59
7	11	13	68	53	45	14	81	74	58
7	11	14	69	55	43	15	83	72	57
7	11	15	69	56	41	17	85	71	56
7	12	12	68	49	49	13	77	77	58
7	12	13	68	51	47	14	80	76	57
7	12	14	69	53	45	15	82	75	56
7	12	15	70	54	43	16	84	74	56
7	13	13	69	49	49	14	78	78	56
7	13	14	70	50	47	15	80	77	55
7	13	15	71	52	45	16	82	76	55
7	14	15	71	50	46	17	81	78	54
7	15	15	72	48	48	17	80	80	54
8	8	9	56	56	51	3	74	69	69

MILL IND X	Y	Z	(100)- α1	α2	α3	(111)-POLES α1	α2	α3	α4
8	8	11	60	60	46	9	79	66	66
8	8	13	62	62	41	14	84	64	64
8	8	15	65	65	37	18	88	63	63
8	9	9	58	53	53	3	72	72	67
8	9	10	59	55	50	5	75	71	66
8	9	11	61	57	48	8	78	69	65
8	9	12	62	58	45	10	80	68	64
8	9	13	63	59	43	12	83	67	63
8	9	14	64	61	41	14	85	66	62
8	9	15	65	62	39	16	87	65	61
8	10	11	62	54	49	7	76	72	64
8	10	13	64	57	45	11	81	70	62
8	10	15	66	60	40	15	85	68	60
8	11	11	63	51	51	8	75	75	62
8	11	12	64	53	49	9	77	73	61
8	11	13	65	54	46	11	79	72	61
8	11	14	66	56	44	13	81	71	60
8	11	15	67	57	42	14	83	70	59
8	12	13	66	52	48	11	78	74	60
8	12	15	67	55	44	14	82	72	58
8	13	13	66	50	50	12	77	77	59
8	13	14	67	51	47	13	79	75	58
8	13	15	68	53	46	14	81	74	57
8	14	15	69	51	47	14	79	76	57
8	15	15	69	49	49	15	78	78	56
9	9	10	56	56	52	3	73	69	69
9	9	11	58	58	49	6	76	68	68
9	9	13	60	60	44	10	81	66	66
9	9	14	62	62	42	12	83	65	65
9	10	10	58	53	53	3	72	72	68
9	10	11	59	55	51	5	75	71	67
9	10	12	60	56	48	7	77	69	65
9	10	13	61	58	46	9	79	68	64
9	10	14	62	59	44	11	81	67	64
9	10	15	63	60	42	13	83	66	63
9	11	11	60	52	52	5	73	73	65
9	11	12	61	54	50	7	76	72	64
9	11	13	62	55	48	8	78	71	63
9	11	14	63	57	45	10	80	70	62
9	11	15	64	58	43	12	82	69	62
9	12	13	63	53	49	9	77	73	62
9	12	14	64	54	47	10	79	72	61
9	13	13	64	51	51	9	75	75	61
9	13	14	65	52	48	10	77	74	61
9	13	15	66	53	47	11	79	73	60
9	14	14	66	50	50	11	76	76	60
9	14	15	66	51	48	12	78	75	59
10	10	11	56	56	52	3	73	69	69
10	10	13	59	59	47	7	78	67	67

TAB. II FROM HKL INTO ALPHA ANGLES

MILL IND X	Y	Z	(100)- α_1	α_2	α_3	(111)-POLES α_1	α_2	α_3	α_4
10	11	11	57	54	54	3	72	72	68
10	11	12	58	55	51	4	74	71	67
10	11	13	60	56	49	6	76	69	66
10	11	14	61	57	47	8	79	68	65
10	11	15	62	59	45	10	81	67	64
10	12	13	61	54	50	6	75	72	65
10	12	15	62	56	46	9	79	70	63
10	13	13	61	52	52	7	74	74	64
10	13	14	62	53	50	8	76	73	63
10	13	15	63	54	48	9	78	72	62
10	14	15	64	52	49	9	77	74	61
11	11	12	56	56	52	2	73	69	69
11	11	13	57	57	50	5	75	68	68
11	11	14	58	58	48	7	77	67	67
11	11	15	59	59	46	9	79	66	66
11	12	12	57	54	54	2	72	72	68
11	12	13	58	55	51	4	74	71	67
11	12	14	59	56	49	6	76	70	66
11	12	15	60	57	47	8	78	69	65
11	13	13	59	53	53	4	73	73	66
11	13	14	60	54	51	6	75	72	65
11	13	15	61	55	49	7	77	71	64
11	14	14	61	52	52	6	74	74	64
11	14	15	62	53	50	7	76	73	63
11	15	15	63	51	51	8	75	75	63
12	12	13	56	56	53	2	73	69	69
12	13	13	57	54	54	2	72	72	68
12	13	14	58	55	52	4	74	71	67
12	13	15	59	56	50	5	76	70	67
12	14	15	60	54	51	5	75	72	66
13	13	14	56	56	53	2	73	70	70
13	13	15	57	57	51	4	74	69	69
13	14	14	57	54	54	2	72	72	69
13	14	15	58	55	52	3	73	71	68
13	15	15	58	53	53	4	72	72	67
14	14	15	56	56	53	2	72	70	70
14	15	15	57	54	54	2	71	71	69

MILL IND X	Y	Z	(100)- α_1	α_2	α_3	(111)-POLES α_1	α_2	α_3	α_4

MILL IND X	Y	Z	(100)- α_1	α_2	α_3	(111)-POLES α_1	α_2	α_3	α_4

MILL IND X	Y	Z	(100)- α_1	α_2	α_3	(111)-POLES α_1	α_2	α_3	α_4

TABLE III

From $\{011\}$ Pole Figure to Miller Indices

The angles α_i and α_i' which are formed by a $\{011\}$-axis $(A_iB_iC_i)$ and the normal direction (HKL) or rolling direction [UVW], respectively, are given by the relationships

$$\cos \alpha_i = \frac{(A_iB_iC_i)\ (HKL)}{\sqrt{(A_i^2 + B_i^2 + C_i^2)\ (H^2 + K^2 + L^2)}}$$

$$\cos \alpha_i' = \frac{(A_iB_iC_i)\ [UVW]}{\sqrt{(A_i^2 + B_i^2 + C_i^2)\ (U^2 + V^2 + W^2)}}$$

The index i characterizes the different equivalent axes of the $\{011\}$ type, i.e. (110), (101), (011), $(\bar{1}10)$, $(\bar{1}01)$ and $(0\bar{1}1)$. (This means i runs here from 1 to 6).

If the α_i and α_i' can be taken from a pole figure, the 6 quantities H, K, L, U, V, W can be calculated from these equations, if a sufficient number of α_i and α_i' is available (c.f. Sec. 2.2). More practical, however, is to use the following Table.

In the table possible sets of the $\alpha_1 \ldots \alpha_6$ are listed, ordered in blocks with increasing α_1. The sets XYZ listed on the right of the α_i give the corresponding sets of (HKL) which thus can be obtained if a set the $\alpha_1 \ldots \alpha_6$ is known. If instaed of $\alpha_1 \ldots \alpha_6$ the angles $\alpha_1' \ldots \alpha_6'$ are introduced into the table, the XYZ give the corresponding [UVW].

Since the table only contains values (H, K, L, U, V, W) $\leqslant$ 15, the exact measured angles α_i and α_i' will generally not be listed in the table and one has to pick that of the listed sets of α_i which is closest to the given one.

TAB. III FROM ALPHA ANGLES OF (110) – POLES INTO HK_L

α1	α2	α3	α4	α5	α6	X	Y	Z
0	50							
2	59	59	61	61	88	0	11	12
2	59	59	61	61	88	0	12	13
2	59	59	61	61	88	0	13	14
2	59	59	61	61	88	0	14	15
3	57	57	63	63	90	1	12	12
3	57	57	63	63	90	1	13	13
3	57	59	61	63	88	1	14	15
3	58	58	62	62	87	0	9	10
3	58	58	62	62	87	0	10	11
3	58	58	62	62	90	1	14	14
3	58	58	62	62	87	0	8	9
3	58	58	62	62	90	1	15	15
4	56	56	64	64	90	1	9	9
4	56	59	61	64	88	1	12	13
4	56	59	62	64	88	1	11	12
4	56	59	61	64	88	1	13	14
4	57	57	63	63	90	1	10	10
4	57	57	63	63	90	1	11	11
4	58	58	63	63	86	0	6	7
4	58	58	62	62	86	0	7	8
4	58	58	62	62	86	0	13	13
5	55	58	62	65	87	1	9	10
5	55	59	62	65	87	1	10	11
5	55	60	60	65	86	1	13	15
5	55	60	60	65	86	1	12	14
5	56	56	64	64	90	1	8	8
5	56	56	64	64	90	2	15	15
5	57	57	63	63	85	0	11	11
5	57	57	63	63	85	0	5	6
6	54	56	64	66	88	2	13	14
6	54	57	64	66	88	2	14	15
6	54	58	62	66	87	1	8	9
6	54	60	60	66	85	1	10	12
6	55	55	65	65	90	2	13	13
6	55	55	65	65	90	1	7	7
6	55	60	60	66	85	1	11	13
6	56	56	64	64	84	0	4	5
6	57	57	63	63	84	0	9	11
7	53	56	64	67	88	2	11	12
7	53	56	64	67	88	2	12	13
7	53	58	62	67	86	1	7	8
7	53	58	63	67	86	2	13	15
7	53	60	60	67	84	1	9	11
7	54	54	66	66	90	2	11	11
7	54	59	61	66	84	1	12	15
7	55	55	66	66	90	1	6	5
7	56	56	64	64	83	0	7	9
7	56	56	64	64	83	0	10	13

α1	α2	α3	α4	α5	α6	X	Y	Z
0	50							
7	56	56	64	64	83	0	11	14
8	52	56	65	68	87	2	10	11
8	52	58	63	68	86	1	6	7
8	52	58	63	68	85	2	11	13
8	53	59	62	67	83	1	10	13
8	53	60	60	67	84	1	8	10
8	54	54	67	67	90	1	5	5
8	54	59	62	67	83	1	11	14
8	56	56	65	65	82	0	3	4
9	51	55	65	69	87	2	9	10
9	51	57	63	69	85	1	5	6
9	51	59	62	69	83	2	11	14
9	52	54	66	69	88	3	13	14
9	52	55	66	68	88	3	14	15
9	52	59	62	69	84	2	12	15
9	52	59	62	68	82	1	9	12
9	52	60	60	68	83	1	7	9
9	53	53	67	67	90	2	9	9
9	53	53	67	67	90	3	14	14
9	53	53	68	68	90	3	13	13
9	53	58	63	68	81	1	11	15
9	55	55	66	66	81	0	5	7
9	55	55	65	65	81	0	11	15
9	55	55	65	65	81	0	8	11
0	60							
0	60	60	60	60	90	0	1	1
10	40							
11	49	54	67	71	86	2	7	6
11	49	55	66	71	85	3	11	13
11	49	57	64	71	84	1	4	5
11	49	59	62	71	82	2	9	12
12	48	54	67	72	86	4	13	15
12	48	55	66	72	85	3	10	12
12	48	57	64	72	83	3	11	14
12	48	57	65	72	83	2	7	9
12	48	59	62	72	81	2	8	11
12	49	52	69	71	88	4	13	14
12	49	53	68	71	87	3	10	11
12	49	59	63	71	79	1	6	9
12	49	61	61	71	81	1	5	7
13	47	54	67	73	84	3	9	11
13	47	55	67	73	84	4	12	15
13	47	57	65	73	83	3	10	13
13	48	51	69	72	88	4	12	13

α1	α2	α3	α4	α5	α6	X	Y	Z
10	40							
13	48	52	69	72	87	3	9	10
13	48	53	68	73	86	2	6	7
13	48	58	63	73	81	3	11	15
13	48	60	62	72	79	2	10	15
13	48	61	61	72	80	2	9	13
14	46	54	67	74	83	4	11	14
14	46	56	65	74	82	1	3	4
14	47	51	70	73	87	3	8	9
14	47	53	69	74	85	4	11	13
14	47	58	64	74	81	3	10	14
14	47	59	63	73	80	2	7	10
14	47	61	61	73	79	1	4	6
14	48	50	70	72	88	5	14	15
14	48	51	70	73	88	4	11	12
14	48	59	63	73	77	1	5	8
14	48	60	62	73	78	2	9	14
14	49	49	72	72	90	5	14	14
14	49	49	72	72	90	4	11	11
15	45	52	69	75	85	2	5	6
15	45	56	66	75	81	3	8	11
15	45	56	66	75	81	4	11	15
15	45	58	64	75	80	3	9	13
15	45	59	63	75	79	2	6	9
15	46	50	71	74	87	4	10	11
15	46	52	70	74	86	5	13	15
15	46	54	68	74	84	3	8	10
15	46	60	63	74	77	2	8	13
15	46	60	62	74	79	3	10	15
15	46	61	61	74	78	2	7	11
15	47	50	71	73	88	5	13	14
15	47	59	64	74	76	2	9	15
15	48	48	72	72	90	3	8	8
15	48	48	73	73	90	5	13	13
15	48	57	65	72	76	1	6	10
15	49	56	66	72	75	1	7	12
16	44	53	69	76	83	3	7	9
16	44	56	66	76	81	2	5	7
16	44	58	64	76	79	3	8	12
16	45	50	71	75	86	3	7	8
16	45	51	71	75	86	5	12	14
16	45	53	69	76	84	5	12	15
16	45	54	68	75	83	4	10	13
16	45	60	63	75	78	3	9	14
16	46	49	72	74	88	5	12	13
16	46	49	73	75	88	6	14	15
16	47	47	74	74	90	5	12	12
16	48	48	73	73	90	2	5	5
16	49	55	67	72	75	1	8	14

α1	α2	α3	α4	α5	α6	X		Z
10	40							
17	43	52	70	77	83	5	11	14
17	43	54	69	77	82	4	9	12
17	43	54	68	77	82	5	11	15
17	43	56	67	77	80	4	9	13
17	43	56	67	77	80	3	7	10
17	43	58	65	77	79	4	10	15
17	43	60	63	77	77	3	8	13
17	44	50	72	76	86	6	13	15
17	44	50	71	76	85	5	11	13
17	44	51	70	76	85	4	9	11
17	44	61	61	76	76	1	3	5
17	45	48	73	76	88	5	11	12
17	45	49	72	75	87	4	9	10
17	45	60	63	75	75	2	7	12
17	46	46	75	75	90	4	9	9
17	46	59	64	75	75	1	4	7
17	47	47	74	74	90	3	7	7
17	47	57	66	74	74	1	5	9
17	48	54	68	73	73	1	8	15
17	48	55	68	73	73	1	7	13
17	48	56	67	74	74	1	6	11
18	42	51	72	78	84	2	4	5
18	42	52	71	78	83	5	10	13
18	42	58	65	78	78	3	7	11
18	42	58	66	78	78	4	9	14
18	43	50	72	78	85	5	10	12
18	43	59	64	77	77	2	5	8
18	43	62	62	75	78	3	8	14
18	45	48	74	76	88	6	13	14
18	45	58	66	73	76	2	8	15
18	45	59	65	73	76	2	7	13
18	46	46	75	75	90	6	13	13
18	46	46	75	75	90	5	11	11
18	46	46	75	75	90	7	15	15
19	41	54	69	79	81	5	10	14
19	41	55	68	79	79	1	2	3
19	42	52	70	78	82	3	6	8
19	42	53	70	78	81	4	8	11
19	42	59	64	76	79	4	9	15
19	42	60	63	76	79	3	7	12
19	42	62	62	74	78	2	5	9
19	43	49	73	77	86	3	6	7
19	44	47	75	77	88	6	12	13
19	44	47	75	77	88	7	14	15
19	44	47	74	77	87	5	10	11
19	44	48	74	77	87	4	8	9
19	44	60	64	74	77	2	6	11
19	45	45	76	76	90	1	2	2

TAB. III FROM ALPHA ANGLES OF (110) - POLES INTO HKL

α1	α2	α3	α4	α5	α6	X	Y	Z
10	40							
19	45	57	67	72	76	1	4	8
19	46	55	68	72	75	1	5	10
19	47	54	69	72	74	1	7	14
19	47	55	68	72	75	1	6	12
10	50							
10	50	55	66	70	87	2	8	9
10	50	55	65	70	86	3	12	14
10	50	57	64	70	84	2	9	11
10	50	59	62	70	83	2	10	13
10	51	54	67	69	88	3	12	13
10	51	56	65	69	86	3	13	15
10	51	60	60	69	82	1	6	8
10	52	52	68	68	90	1	4	4
10	52	58	63	68	81	1	10	14
10	52	59	62	69	81	1	8	11
10	54	54	66	66	80	0	9	13
10	55	55	66	66	80	0	7	10
11	50	52	68	70	88	4	14	15
11	50	53	67	70	88	3	11	12
11	50	61	61	70	81	2	11	15
11	51	51	70	70	90	2	7	7
11	51	51	69	69	90	3	11	11
11	51	58	63	69	80	1	9	13
11	51	59	62	70	80	1	7	10
11	52	52	69	69	90	4	15	15
11	54	54	67	67	79	0	2	3
12	50	50	70	70	90	4	13	13
12	51	51	70	70	90	3	10	10
12	51	57	64	69	79	1	10	15
12	51	57	64	70	79	1	8	12
12	54	54	68	68	78	0	9	14
13	50	50	71	71	90	1	3	3
13	50	57	65	70	78	1	9	14
13	50	57	64	71	78	1	7	11
13	53	53	68	68	77	0	5	8
13	53	53	68	68	77	0	7	11
13	53	53	68	68	77	0	8	13
14	50	56	66	71	76	1	9	15
14	50	56	65	71	77	1	8	13
14	53	53	69	69	76	0	3	5
15	52	52	69	69	75	0	4	7
15	52	52	69	69	75	0	7	12
16	52	52	70	70	74	0	6	11
16	52	52	70	70	74	0	5	9
17	51	51	71	71	73	0	8	15
17	51	51	70	70	73	0	7	13

α1	α2	α3	α4	α5	α6	X	Y	Z
10	50							
18	51	51	72	72	72	0	1	2
20	30							
21	39	52	72	81	81	3	5	7
21	39	55	69	78	81	5	9	14
21	39	55	69	78	81	4	7	11
21	39	57	67	77	81	5	9	15
22	38	50	73	82	82	7	11	15
22	38	53	71	79	82	5	8	12
22	38	55	69	78	82	3	5	8
22	38	55	69	77	82	5	8	13
22	38	59	65	74	82	4	7	13
22	39	48	75	81	84	7	11	14
22	39	51	73	82	82	5	8	11
22	39	53	70	79	81	6	10	15
22	39	57	67	76	82	4	7	12
23	37	51	73	80	83	6	9	13
23	37	52	72	79	83	4	6	9
23	37	53	71	78	83	6	9	14
23	37	55	70	77	83	2	3	5
23	37	57	68	75	83	5	8	14
23	37	59	66	74	83	5	8	15
23	38	49	75	82	82	2	3	4
23	38	58	67	75	82	3	5	9
23	38	61	64	72	82	4	7	14
23	39	47	76	82	84	8	12	15
23	39	61	64	70	82	3	6	13
24	36	50	74	81	84	5	7	10
24	36	52	73	79	84	7	10	15
24	36	53	72	78	84	5	7	11
24	36	57	68	74	84	4	6	11
24	37	50	75	81	83	7	10	14
24	37	61	65	72	83	3	5	10
24	38	48	76	83	83	7	10	13
24	38	63	63	71	83	4	7	15
24	39	45	78	82	86	4	6	7
24	39	47	77	82	85	6	9	11
24	39	60	66	69	82	3	6	14
24	39	60	65	69	82	2	4	9
25	35	51	74	80	85	3	4	6
25	35	53	72	78	85	6	8	13
25	35	55	70	76	85	5	7	12
25	35	55	71	76	85	3	4	7
25	35	57	69	74	85	5	7	13
25	35	59	67	72	85	5	7	14
25	36	48	76	82	85	6	8	11
25	36	60	66	72	84	2	3	6

α1	α2	α3	α4	α5	α6	X	Y	Z
20	30							
25	36	62	65	71	85	4	6	13
25	37	47	77	84	84	8	11	14
25	37	47	77	83	83	5	7	9
25	37	48	76	82	84	8	11	15
25	37	63	63	70	83	3	5	11
25	38	45	78	83	85	7	10	12
25	39	43	80	83	88	7	10	11
25	39	44	80	83	87	5	7	8
25	39	44	79	83	86	9	13	15
25	39	59	67	67	83	1	2	5
26	34	50	75	80	86	8	10	15
26	34	52	74	78	86	4	5	8
26	34	53	73	77	86	7	9	15
26	34	55	71	75	86	4	5	9
26	34	58	68	73	86	3	4	8
26	35	49	76	81	85	7	9	13
26	35	51	74	79	86	7	9	14
26	35	57	69	74	86	6	8	15
26	35	61	66	71	85	5	7	15
26	36	47	78	83	85	7	9	12
26	36	63	63	69	85	2	3	7
26	37	46	78	84	84	3	4	5
26	37	62	65	68	84	3	5	12
26	38	44	80	84	86	9	12	14
26	38	44	79	84	86	8	11	13
26	39	42	81	83	88	8	11	12
26	39	42	81	83	88	10	14	15
27	33	52	74	77	87	5	6	10
27	33	53	74	76	87	6	7	12
27	33	55	72	75	87	5	6	11
27	33	55	72	75	87	6	7	13
27	33	57	70	73	87	5	6	12
27	34	49	76	80	87	5	6	9
27	34	58	69	73	87	4	5	10
27	34	60	67	71	87	4	5	11
27	34	61	66	70	86	3	4	9
27	35	47	78	82	86	9	11	15
27	35	48	77	81	86	4	5	7
27	36	46	79	83	86	8	10	13
27	36	62	65	68	85	4	6	15
27	37	43	81	85	86	10	13	15
27	37	44	80	85	85	7	9	11
27	37	60	67	67	84	3	5	13
27	38	42	82	84	88	9	12	13
27	38	43	81	84	87	6	8	9
27	39	58	65	69	83	2	4	11
28	32	51	75	77	88	8	9	15
28	32	53	74	76	88	7	8	14

α1	α2	α3	α4	α5	α6	X	Y	Z
20	30							
28	32	55	72	74	88	7	8	15
28	32	57	70	73	88	6	7	14
28	33	49	77	80	87	7	8	12
28	33	49	77	79	88	8	9	14
28	33	50	76	79	87	6	7	11
28	33	51	75	78	88	7	8	13
28	33	59	68	71	87	5	6	13
28	33	59	69	71	88	6	7	15
28	33	61	67	69	87	5	6	14
28	34	47	78	81	87	6	7	10
28	34	62	65	69	87	4	5	12
28	35	44	81	84	86	10	12	15
28	35	45	80	84	86	9	11	14
28	35	46	79	83	86	5	6	8
28	35	64	64	68	86	3	4	10
28	36	43	82	86	86	9	11	13
28	36	44	81	85	85	4	5	6
28	36	61	66	66	85	2	3	8
28	37	41	83	85	88	8	10	11
28	38	41	82	84	88	10	13	14
28	38	41	83	85	88	11	14	15
28	38	42	82	85	87	7	9	10
28	38	59	65	68	85	3	5	14
28	39	39	84	84	90	3	4	4
28	39	56	64	71	84	1	2	6
29	33	45	80	82	88	10	11	15
29	33	46	80	82	88	9	10	14
29	33	47	78	81	88	8	9	13
29	33	48	78	80	88	9	10	15
29	33	63	65	68	88	5	6	15
29	34	45	80	83	88	8	9	12
29	34	46	79	82	87	7	8	11
29	34	63	66	66	87	4	5	14
29	34	64	64	67	87	4	5	13
29	35	43	82	84	87	7	8	10
29	35	44	81	84	87	6	7	9
29	35	62	66	66	87	3	4	11
29	36	42	83	86	86	11	13	15
29	36	42	82	86	86	5	6	7
29	37	41	83	86	88	9	11	12
29	37	59	64	69	86	2	3	9
29	38	38	85	85	90	4	5	5
29	38	58	64	69	85	3	5	15
29	39	39	84	84	90	10	13	13
29	39	39	85	85	90	11	14	14
29	39	39	84	84	90	7	9	9

TAB. III — FROM ALPHA ANGLES OF (110) - POLES INTO HKL — III - 4

α1	α2	α3	α4	α5	α6	X	Y	Z
20	40							
20	40	53	70	80	80	5	9	13
20	40	57	66	77	80	4	8	13
20	40	58	66	76	80	3	6	10
20	41	49	74	79	85	5	9	11
20	41	50	72	79	84	6	11	14
20	41	51	72	80	82	5	9	12
20	41	52	71	80	82	6	11	15
20	41	62	62	74	79	3	7	13
20	42	48	74	79	86	6	11	13
20	42	48	74	78	86	7	13	15
20	42	61	63	73	78	3	8	15
20	43	46	76	78	88	7	13	14
20	43	46	76	78	88	6	11	12
20	43	59	65	72	78	1	3	3
20	44	57	66	72	77	2	7	14
20	47	53	70	70	75	1	7	15
21	40	49	74	80	84	7	12	15
21	40	50	74	81	83	6	10	13
21	40	50	73	80	83	4	7	9
21	40	53	71	80	80	4	7	10
21	40	59	65	75	80	2	4	7
21	40	60	64	74	81	3	6	11
21	40	61	64	74	81	4	8	15
21	41	47	75	80	86	7	12	14
21	41	61	64	72	80	3	7	14
21	42	45	77	79	88	8	14	15
21	42	46	76	79	87	5	9	10
21	42	47	76	79	86	4	7	8
21	42	60	64	72	79	2	5	10
21	44	44	78	78	90	5	9	9
21	44	44	77	77	90	8	15	15
21	44	44	78	78	90	7	13	13
21	44	44	78	78	90	6	11	11
21	44	57	68	70	78	2	7	15
21	46	54	70	70	76	1	6	13
21	46	54	70	70	77	1	5	11
22	40	47	75	80	85	3	5	6
22	40	48	75	81	84	5	8	10
22	40	62	62	72	81	1	2	4
22	41	45	77	80	87	6	10	11
22	41	46	77	80	86	8	13	15
22	41	60	65	70	80	3	7	15
22	42	45	77	79	88	7	12	13
22	43	43	79	79	90	7	12	12
22	43	43	79	79	90	4	7	7
22	43	57	67	70	79	2	6	13
22	44	55	69	69	78	1	4	9
22	46	53	68	71	77	1	6	14

α1	α2	α3	α4	α5	α6	X	Y	Z
20	40							
22	49	49	68	74	74	0	5	12
22	49	49	68	74	74	0	3	7
23	40	46	77	81	86	7	11	13
23	41	44	78	80	88	8	13	14
23	41	44	78	81	88	7	11	12
23	41	45	77	81	87	5	8	9
23	41	59	66	70	80	2	5	11
23	43	43	79	79	90	3	5	5
23	43	56	68	68	79	1	3	7
23	45	53	68	71	77	1	5	12
23	46	52	67	72	77	1	6	15
23	49	49	67	75	75	0	2	5
24	40	43	80	82	88	8	12	13
24	40	44	79	82	87	6	9	10
24	41	43	79	81	88	9	14	15
24	41	57	68	68	81	2	5	12
24	42	42	80	80	90	5	8	8
24	42	42	81	81	90	9	14	14
24	42	42	80	80	90	7	11	11
24	42	42	80	80	90	8	13	13
24	42	56	67	70	80	2	6	15
24	44	54	67	71	79	1	4	10
24	45	53	66	72	78	1	5	13
24	49	49	66	75	75	0	5	13
24	49	49	66	76	76	0	3	8
25	40	43	80	82	88	9	13	14
25	41	41	81	81	90	2	3	3
25	41	56	66	69	81	2	5	13
25	42	55	66	71	81	1	3	8
25	44	53	65	72	80	1	4	11
25	48	48	65	76	76	0	4	11
25	48	48	65	76	76	0	5	14
26	40	40	82	82	90	7	10	10
26	41	41	82	82	90	9	13	13
26	41	56	65	71	82	2	5	14
26	45	52	65	73	79	1	5	14
27	40	40	83	83	90	8	11	11
27	40	40	83	83	90	5	7	7
27	40	40	83	83	90	11	15	15
27	41	55	64	72	82	2	5	15
27	42	54	64	73	81	1	3	9
27	44	51	63	74	80	1	5	15
27	44	52	64	74	80	1	4	12
27	48	48	63	77	77	0	1	3
28	43	52	62	73	81	1	4	13
28	47	47	62	73	78	0	4	13
28	47	47	62	78	78	0	3	10
29	40	56	62	72	84	2	4	13

α1	α2	α3	α4	α5	α6	X	Y	Z
20	40							
29	42	53	62	74	82	1	3	10
29	43	51	61	76	82	1	4	14
29	47	47	61	79	79	0	2	7
20	50							
20	50	50	70	73	73	0	7	15
20	50	50	70	73	73	0	6	13
21	50	50	69	73	73	0	4	9
21	50	50	69	73	73	0	5	11
30	30							
30	30	49	78	78	90	8	8	13
30	30	50	78	78	90	3	3	5
30	30	50	77	77	90	7	7	12
30	30	51	76	76	90	4	4	7
30	30	52	75	75	90	6	6	11
30	30	52	76	76	90	5	5	9
30	30	53	75	75	90	7	7	13
30	30	53	75	75	90	8	8	15
30	30	55	73	73	90	1	1	2
30	30	57	72	72	90	7	7	15
30	30	57	71	71	90	5	5	11
30	30	57	71	71	90	6	6	13
30	30	58	71	71	90	4	4	9
30	30	59	69	69	90	5	5	12
30	30	59	70	70	90	3	3	7
30	30	61	68	68	90	2	2	5
30	34	43	82	84	88	10	11	14
30	34	43	83	85	88	11	12	15
30	34	44	81	83	88	9	10	13
30	35	42	83	86	88	9	10	12
30	35	43	83	85	88	8	9	11
30	35	61	64	68	87	3	4	12
30	35	62	64	67	88	4	5	15
30	36	40	84	87	87	7	8	9
30	36	41	83	87	87	6	7	8
30	36	59	63	69	87	3	4	13
30	37	39	85	87	88	12	14	15
30	37	40	84	86	88	11	13	14
30	37	40	84	86	88	10	12	13
30	37	58	62	71	86	2	3	10
30	38	38	85	85	90	9	11	11
31	31	45	82	82	90	5	5	7
31	31	45	81	81	90	7	7	10
31	31	45	81	81	90	9	9	13
31	31	47	80	80	90	2	2	3

α1	α2	α3	α4	α5	α6	X	Y	Z
30	30							
31	31	48	79	79	90	9	9	14
31	31	48	79	79	90	7	7	11
31	31	49	79	79	90	5	5	8
31	31	61	68	68	90	5	5	13
31	31	62	67	67	90	3	3	8
31	31	63	66	66	90	4	4	11
31	31	63	66	66	90	5	5	14
31	31	65	65	65	90	1	1	3
31	35	40	85	87	88	12	13	15
31	35	41	84	86	88	10	11	13
31	35	41	84	86	88	11	12	14
31	36	39	85	88	88	9	10	11
31	36	40	85	87	87	8	9	10
31	36	58	62	71	87	3	4	14
31	37	37	86	86	90	6	7	7
31	37	57	61	72	86	2	3	11
31	38	38	86	86	90	11	13	13
31	38	38	86	86	90	5	6	6
32	32	42	84	84	90	7	7	9
32	32	42	84	84	90	11	11	14
32	32	43	83	83	90	3	3	4
32	32	43	84	84	90	10	10	13
32	32	44	82	82	90	8	8	11
32	32	44	82	82	90	11	11	15
32	32	63	63	66	90	4	4	13
32	32	63	63	67	90	3	3	10
32	35	38	86	88	88	12	13	14
32	35	38	87	88	88	13	14	15
32	35	39	86	88	88	10	11	12
32	35	39	85	88	88	11	12	13
32	36	58	61	72	87	3	4	15
32	37	37	87	87	90	7	8	8
32	37	37	88	88	90	9	10	10
32	37	37	87	87	90	8	9	9
32	37	37	87	87	90	13	15	15
32	38	56	59	74	87	2	3	12
33	33	39	87	87	90	13	13	15
33	33	40	86	86	90	11	11	13
33	33	40	86	86	90	6	6	7
33	33	40	66	86	90	5	5	6
33	33	41	85	85	90	9	9	11
33	33	41	85	85	90	4	4	5
33	33	61	61	69	90	3	3	11
33	33	61	61	69	90	4	4	15
33	33	62	62	68	90	2	2	7
33	36	36	88	88	90	13	14	14
33	36	36	88	88	90	11	12	12
33	36	36	88	88	90	12	13	13

TAB. III FROM ALPHA ANGLES OF (110) - POLES INTO HKL

30 30

α1	α2	α3	α4	α5	α6	X	Y	Z
33	36	36	88	88	90	14	15	15
33	37	37	88	88	90	10	11	11
33	38	55	58	75	87	2	3	13
34	34	37	88	88	90	13	13	14
34	34	37	88	88	90	14	14	15
34	34	37	88	88	90	12	12	13
34	34	38	88	88	90	10	10	11
34	34	38	88	88	90	11	11	12
34	34	38	87	87	90	9	9	10
34	34	39	87	87	90	7	7	8
34	34	39	87	87	90	8	8	9
34	34	58	58	73	90	2	2	9
34	34	59	59	72	90	3	3	13
34	34	60	60	71	90	1	1	4
34	39	53	57	77	87	2	3	15
34	39	54	57	76	87	2	3	14
35	35	35	90	90	90	1	1	1
35	35	57	57	74	90	1	1	5
35	35	58	58	73	90	3	3	14
36	36	56	56	76	90	2	2	11
37	37	54	54	78	90	2	2	13
37	37	55	55	77	90	1	1	6
38	38	52	52	80	90	1	1	8
38	38	53	53	79	90	2	2	15
38	38	54	54	79	90	1	1	7
39	39	52	52	81	90	1	1	9

30 40

α1	α2	α3	α4	α5	α6	X	Y	Z
30	40	55	61	73	84	1	2	7
30	42	52	60	76	83	1	3	11
30	43	50	60	77	82	1	4	15
30	47	47	60	80	80	0	4	15
30	47	47	60	79	79	0	3	11
31	40	54	60	74	85	2	4	15
31	42	51	59	77	83	1	3	12
31	47	47	59	80	80	0	1	4
32	40	53	59	75	85	1	2	8
32	42	51	58	78	84	1	3	13
32	46	46	58	81	81	0	3	13
32	46	46	58	81	81	0	2	9
33	40	52	58	77	86	1	2	9
33	42	50	57	79	84	1	3	14
33	46	46	57	81	81	0	3	14
34	41	52	56	78	86	1	2	10
34	42	50	56	79	85	1	3	15
34	46	46	56	82	82	0	1	5
35	41	51	55	79	86	1	2	11
35	46	46	55	83	83	0	2	11
36	41	50	54	81	87	1	2	13
36	41	50	55	80	87	1	2	12
36	46	46	54	83	83	0	1	6
36	46	46	54	84	84	0	2	13
37	42	50	53	81	87	1	2	14
37	46	46	53	85	85	0	2	15
37	46	46	53	84	84	0	1	7
38	42	49	53	82	87	1	2	15
38	45	45	52	85	85	0	1	8
39	45	45	51	86	86	0	1	10
39	45	45	51	86	86	0	1	9

40 40

α1	α2	α3	α4	α5	α6	X	Y	Z
40	40	50	50	83	90	1	1	12
40	40	50	50	83	90	1	1	11
40	40	51	51	82	90	1	1	10
40	45	45	50	86	86	0	1	11
40	45	45	50	87	87	0	1	12
41	41	49	49	85	90	1	1	15
41	41	49	49	84	90	1	1	14
41	41	50	50	84	90	1	1	13
41	45	45	49	87	87	0	1	13
41	45	45	49	87	87	0	1	14
41	45	45	49	87	87	0	1	15
45	45	45	45	90	90	0	0	1

TABLE IV

From {111} Pole Figure to Miller Indices

The angles α_i and α_i' which are formed by a {111}-axis $(A_iB_iC_i)$ and the normal direction (HKL) or rolling direction [UVW], respectively, are given by the relationships

$$\cos \alpha_i = \frac{(A_iB_iC_i)\,(HKL)}{\sqrt{(A_i^2 + B_i^2 + C_i^2)\,(H^2 + K^2 + L^2)}}$$

$$\cos \alpha_i' \quad \frac{(A_iB_iC_i)\,[UVW]}{\sqrt{(A_i^2 + B_i^2 + C_i^2)\,(U^2 + V^2 + W^2}}$$

The index i characterizes the different equivalent axes of the {111} type, i.e. (111), $(\bar{1}11)$, $(\bar{1}\bar{1}1)$ and $(1\bar{1}1)$. (This means i runs here from 1 to 4).

If the α_i and α_i' can be taken from a pole figure, the 6 quantities H, K, L, U, V, W can be calculated from these equations, if a sufficient number of α_i und α_i' is available (c.f. Sec. 2.2). More practical, however, is to use the following Table.

In the table possible sets of the $\alpha_1 \ldots \alpha_4$ are listed, ordered in blocks with increasing α_1. The sets XYZ listed on the right of the α_i give the corresponding sets of (HKL) which thus can be obtained if the set $\alpha_1 \ldots \alpha_4$ is known. If instead of $\alpha_1 \ldots \alpha_4$ the angles $\alpha_1' \ldots \alpha_4'$ are introduced into the table, the XYZ give the corresponding [UVW].

Since the table only contains values (H, K, L, U, V, W) $\leqslant$ 15, the exact measured angles α_i and α_i' will generally not be listed in the table and one has to pick that of the listed sets of α_i which is closest to the given one.

TAB. IV FROM ALPHA ANGLES OF (111) - POLES INTO HKL IV - 2

α1	α2	α3	α4	X	Y	Z
0	60					
2	68	72	72	11	12	12
2	68	72	72	12	13	13
2	69	69	73	11	11	12
2	69	69	73	12	12	13
2	69	71	71	14	15	15
2	69	72	72	13	14	14
3	67	72	72	8	9	9
3	68	71	73	13	14	15
3	68	72	72	9	10	10
3	68	72	72	10	11	11
3	69	69	73	9	9	10
3	69	69	73	10	10	11
3	69	69	74	8	8	9
4	66	73	73	6	7	7
4	66	73	73	11	13	13
4	67	71	74	12	13	14
4	67	71	74	10	11	12
4	67	71	74	11	12	13
4	67	72	72	7	8	8
4	67	72	72	13	15	15
4	68	68	75	6	6	7
4	69	69	74	13	13	15
4	69	69	74	7	7	8
5	65	73	73	9	11	11
5	66	71	75	8	9	10
5	66	72	75	12	14	15
5	66	73	73	5	6	6
5	67	70	76	12	13	15
5	67	71	75	9	10	11
5	68	68	75	11	11	13
5	68	68	76	5	5	6
6	64	74	74	11	14	14
6	64	74	74	7	9	9
6	65	72	75	10	12	13
6	65	72	75	11	13	14
6	65	73	73	4	5	5
6	66	69	76	10	11	13
6	66	70	76	11	12	14
6	66	71	76	7	8	9
6	68	68	76	9	9	11
6	68	68	77	4	4	5
7	63	73	76	11	14	15
7	63	74	74	3	4	4
7	64	71	77	11	13	15
7	64	72	76	8	10	11
7	64	72	76	9	11	12
7	64	74	74	10	13	13
7	65	69	77	9	10	12

α1	α2	α3	α4	X	Y	Z
0	60					
7	65	71	76	6	7	8
7	67	67	77	11	11	14
7	67	67	78	7	7	9
7	67	67	78	10	10	13
8	62	75	75	5	7	7
8	62	75	75	8	11	11
8	63	71	78	9	11	13
8	63	72	77	7	9	10
8	63	73	76	10	13	14
8	63	75	75	11	15	15
8	64	71	77	5	6	7
8	65	68	79	10	11	14
8	65	69	78	8	9	11
8	65	69	78	11	12	15
8	67	67	79	3	3	4
9	61	73	77	8	11	12
9	61	74	77	10	14	15
9	61	75	75	9	13	13
9	62	72	78	10	13	15
9	62	73	77	9	12	13
9	62	73	78	6	8	9
9	62	75	75	7	10	10
9	63	70	79	10	12	15
9	63	71	79	4	5	6
9	64	68	79	9	10	13
9	64	69	79	7	8	10
9	66	66	79	11	11	15
9	66	66	79	8	8	11
9	66	66	80	5	5	7
0	70					
0	71	71	71	1	1	1
2	70	70	72	14	14	15
2	70	70	73	13	13	14
10	50					
11	59	76	76	7	11	11
11	59	77	77	5	8	8
12	58	77	77	3	5	5
12	59	73	80	4	6	7
12	59	74	79	6	9	10
12	59	75	78	9	14	15
12	59	75	79	7	11	12
12	59	77	77	8	13	13
13	57	75	80	6	10	11
13	57	78	78	4	7	7

α1	α2	α3	α4	X	Y	Z
10	50					
13	58	73	81	6	9	11
13	58	75	79	8	13	14
13	58	75	80	5	8	9
13	58	77	77	7	12	12
13	59	71	82	5	7	9
14	56	78	78	7	13	13
14	56	78	78	6	11	11
14	57	74	81	8	13	15
14	57	76	79	8	14	15
14	57	76	80	7	12	13
14	57	78	78	5	9	9
14	58	72	82	8	12	15
14	56	74	81	7	11	13
14	59	70	83	6	11	15
14	59	71	83	7	10	13
15	55	77	80	7	13	14
15	56	74	82	3	5	6
15	56	75	81	4	7	8
15	56	75	82	7	12	14
15	56	76	81	5	9	10
15	56	77	80	6	11	12
15	56	78	78	8	15	15
15	57	72	83	7	11	14
15	57	73	83	5	8	10
15	58	71	84	2	3	4
16	55	75	83	6	11	13
16	55	76	82	7	13	15
16	55	79	79	1	2	2
16	56	74	84	7	12	15
16	58	70	85	5	7	10
16	58	70	85	7	10	14
16	59	67	86	7	9	14
16	59	68	86	3	4	6
17	53	80	80	5	11	11
17	53	80	80	6	13	13
17	54	77	82	4	8	9
17	54	77	82	5	10	11
17	54	78	81	7	14	15
17	54	78	81	6	12	13
17	54	80	80	7	15	15
17	55	73	85	4	7	9
17	55	74	83	5	9	11
17	56	71	85	7	11	15
17	56	71	85	5	8	11
17	56	73	84	6	10	13
17	57	69	87	7	10	15
17	57	70	86	6	9	13
18	52	78	82	5	11	12

α1	α2	α3	α4	X	Y	Z
10	50					
18	52	80	80	3	7	7
18	53	76	83	3	6	7
18	53	76	84	5	10	12
18	53	78	82	6	13	14
18	53	80	80	4	9	9
18	54	73	86	5	9	12
18	54	74	85	6	11	14
18	55	72	86	3	5	7
18	57	68	88	5	7	11
18	57	69	87	4	6	9
18	58	67	88	6	8	13
18	59	67	88	7	9	15
19	51	81	81	2	5	5
19	52	77	84	6	13	15
19	52	77	84	5	11	13
19	52	78	83	4	9	10
19	52	79	82	6	14	15
19	52	81	81	5	12	12
19	53	74	86	5	10	13
19	53	75	85	2	4	5
19	54	72	87	4	7	10
19	54	73	87	6	11	15
19	55	70	88	5	8	12
19	56	69	88	6	9	14
10	60					
10	61	72	79	9	12	14
10	61	74	77	9	13	14
10	61	74	78	7	10	11
10	61	76	76	2	3	3
10	62	70	80	9	11	14
10	62	71	80	7	9	11
10	63	69	80	6	7	9
10	64	67	81	10	11	15
10	64	68	80	8	9	12
10	66	66	81	7	7	10
10	66	66	81	9	9	13
11	60	73	79	9	13	15
11	60	74	78	8	12	13
11	60	76	76	9	14	14
11	61	72	79	8	11	13
11	61	73	79	5	7	8
11	62	69	81	5	6	8
11	62	70	81	6	10	13
11	63	68	81	7	8	11
11	64	67	81	9	10	14
11	65	65	82	2	2	3

α1	α2	α3	α4	X	Y	Z
10	60					
12	60	72	80	7	10	1
12	61	70	82	7	9	1
12	61	71	81	3	4	
12	62	69	82	9	11	1
12	63	67	83	6	9	1
12	65	65	83	9	9	1
13	60	71	81	8	11	1
13	61	69	83	4	5	
13	62	67	84	7	8	1
13	62	68	83	6	7	1
13	63	66	83	9	10	1
13	64	64	84	5	5	
13	65	65	83	7	7	1
14	60	68	84	7	9	1
14	60	70	83	6	8	1
14	61	67	84	5	6	
14	62	66	85	8	9	1
14	64	64	84	8	8	1
14	64	64	85	3	3	
15	60	68	85	8	10	1
15	61	66	85	6	7	1
15	64	64	86	7	7	1
16	61	65	87	6	9	1
16	61	66	86	7	8	1
16	63	63	86	4	4	
17	60	65	88	6	7	1
17	60	66	87	5	6	1
17	60	67	87	4	5	
17	63	63	87	5	5	
17	63	63	88	7	7	1
17	63	63	86	6	6	1
18	60	65	88	7	[illegible]	1
18	63	63	88	6	[illegible]	[illegible]
19	62	62	90	1	[illegible]	
20	40					
21	49	79	85	3	8	
21	49	82	82	5	14	[illegible]
22	49	79	86	4	11	[illegible]
22	49	80	84	4	11	1
22	49	81	84	5	14	1
22	49	82	82	1	3	
23	47	83	83	3	10	1
23	48	77	88	4	11	1
23	48	79	86	2	6	
23	48	80	85	3	9	1
23	48	81	85	4	12	1

TAB. IV FROM ALPHA ANGLES OF (111) - POLES INTO HKL

α1	α2	α3	α4	X	Y	Z
20	40					
23	48	83	83	4	13	13
23	49	76	88	4	10	13
23	49	77	87	3	8	10
24	46	82	85	4	14	15
24	46	84	84	3	11	11
24	47	78	88	4	12	15
24	47	79	88	3	9	11
24	47	80	87	4	13	15
24	47	81	85	4	13	14
24	47	81	86	3	10	11
24	47	83	83	2	7	7
24	48	76	90	4	11	15
24	48	76	90	3	8	11
24	49	75	90	2	5	7
25	45	84	84	1	4	4
25	46	80	88	3	10	12
25	46	81	87	2	7	8
25	46	82	86	3	11	12
25	46	84	84	4	15	15
25	47	77	90	1	3	4
25	49	74	88	4	10	15
26	44	83	87	3	13	14
26	44	85	85	2	9	9
26	45	79	90	3	11	14
26	45	80	88	3	11	13
26	45	81	88	3	12	14
26	45	82	87	2	8	9
26	45	83	86	3	12	13
26	45	85	85	3	13	13
26	46	78	90	2	7	9
26	46	78	90	3	10	13
26	47	75	88	3	9	13
26	48	74	88	3	8	12
26	49	73	87	2	5	8
27	43	85	85	1	5	5
27	44	82	88	3	13	15
27	44	83	88	2	9	10
27	44	84	87	3	14	15
27	44	85	85	3	14	14
27	45	78	88	3	11	15
27	45	80	90	1	4	5
27	46	77	88	3	10	14
27	47	74	86	3	9	14
27	47	75	87	2	6	9
27	48	73	86	3	8	13
27	49	71	85	3	7	12
28	42	84	88	2	11	12
28	43	82	90	1	5	6

α1	α2	α3	α4	X	Y	Z
20	40					
28	43	83	88	2	10	11
28	43	86	86	2	11	11
28	44	78	88	2	8	11
28	44	81	90	2	9	11
28	46	75	86	3	10	15
28	46	77	87	2	7	10
28	48	72	84	3	8	14
28	49	70	83	3	7	13
28	49	71	84	2	5	9
29	41	85	88	2	13	14
29	41	86	86	2	13	13
29	41	87	87	1	7	7
29	42	82	90	2	11	13
29	42	83	90	1	6	7
29	42	84	88	2	12	13
29	42	86	86	1	6	6
29	43	80	88	2	10	13
29	44	76	85	1	4	6
29	44	79	88	2	9	12
29	46	75	85	2	7	11
29	47	73	84	1	3	5
20	50					
20	50	79	84	4	10	11
20	50	81	81	3	8	8
20	50	81	81	5	13	13
20	51	76	86	4	9	11
20	51	78	84	3	7	8
20	51	78	85	5	12	14
20	51	79	83	5	12	13
20	53	74	87	3	6	8
20	54	72	88	5	9	13
20	55	70	88	6	10	15
20	57	67	90	5	7	12
20	58	65	90	4	5	9
20	58	66	90	3	4	7
20	59	65	90	5	6	11
21	50	76	87	3	7	9
21	50	77	87	5	12	15
21	50	78	86	2	5	6
21	50	79	85	5	13	15
21	50	80	83	5	13	14
21	50	82	82	4	11	11
21	51	76	86	5	11	14
21	52	73	88	4	8	11
21	52	73	88	5	10	14
21	54	70	90	3	5	8

α1	α2	α3	α4	X	Y	Z
20	50					
21	54	70	90	4	7	11
21	55	69	90	5	8	13
21	56	68	90	2	3	5
22	51	74	86	5	11	15
22	51	75	86	4	9	12
22	52	72	90	1	2	3
22	53	71	90	5	9	14
22	56	66	88	5	7	13
22	57	65	88	6	8	15
22	58	64	88	5	6	12
22	59	63	88	6	7	14
23	50	74	90	3	7	10
23	50	74	90	4	9	13
23	53	69	88	4	7	12
23	53	70	88	5	9	15
23	54	68	88	5	8	14
23	55	67	87	4	6	11
23	57	65	86	3	4	8
23	58	63	86	6	7	15
23	58	64	87	4	5	10
24	50	72	88	4	9	14
24	51	70	87	3	6	10
24	52	71	88	4	8	13
24	54	67	86	5	8	15
24	54	68	87	3	5	9
24	56	65	86	5	7	14
24	58	63	86	5	6	13
25	50	72	88	3	7	11
25	51	70	86	2	4	7
25	53	68	86	4	7	13
25	55	66	85	2	3	6
25	57	63	85	4	5	11
26	50	71	86	4	9	15
26	51	69	85	3	6	11
26	53	66	84	3	5	10
26	55	64	84	5	7	15
26	56	63	84	3	4	9
26	57	62	84	5	6	14
27	51	69	84	4	8	15
27	53	67	84	4	7	14
27	54	65	83	4	6	13
27	57	61	82	5	6	15
27	57	62	83	4	5	12
27	59	59	83	3	3	8
28	51	68	83	1	2	4
28	52	66	82	4	7	15
28	53	65	82	3	5	11
28	54	64	82	2	3	7

α1	α2	α3	α4	X	Y	Z
20	50					
28	59	59	82	5	5	14
28	59	59	82	4	4	11
29	55	62	81	3	4	10
29	56	61	81	4	5	13
29	59	59	80	1	1	3
20	60					
20	60	64	90	7	8	15
20	60	64	90	6	7	13
21	61	61	88	7	7	15
22	61	61	87	5	5	11
22	61	61	88	6	6	13
23	61	61	87	4	4	9
24	60	60	85	5	5	12
24	60	60	86	3	3	7
25	60	60	84	2	2	5
26	60	60	83	5	5	13
30	30					
31	39	88	88	1	10	10
32	38	87	90	1	13	14
32	38	88	88	1	13	13
32	38	88	88	1	14	14
32	39	84	88	1	10	12
32	39	84	88	1	11	13
32	39	85	88	1	12	14
32	39	86	90	1	10	11
32	39	86	90	1	11	12
32	39	86	90	1	12	13
32	39	88	88	1	12	12
32	39	88	88	1	11	11
33	38	85	88	1	13	15
33	38	87	90	1	14	15
33	38	88	88	1	15	15
33	39	81	85	1	11	15
33	39	82	86	1	10	13
33	39	83	86	1	11	14
33	39	83	87	1	12	15
35	35	87	87	0	7	8
35	35	87	87	0	13	15
35	35	87	87	0	8	9
35	35	88	88	0	14	15
35	35	88	88	0	9	10
35	35	88	88	0	11	12
35	35	68	88	0	10	11
35	35	88	88	0	12	13

α1	α2	α3	α4	X	Y	Z
30	30					
35	35	88	88	0	13	1
35	35	90	90	0	1	
36	36	82	62	0	5	
36	36	82	62	0	7	1
36	36	83	83	0	11	1
36	36	83	83	0	3	
36	36	83	83	0	8	1
36	36	84	84	0	11	1
36	36	84	84	0	7	
36	36	84	84	0	10	1
36	36	85	85	0	4	
36	36	85	65	0	9	1
36	36	86	86	0	11	1
36	36	86	86	0	6	
36	36	86	86	0	5	
37	37	79	79	0	8	1
37	37	79	79	0	5	
37	37	80	60	0	7	1
37	37	80	80	0	9	1
37	37	81	81	0	2	
37	37	82	62	0	9	1
38	38	77	77	0	5	
38	38	77	77	0	6	1
38	38	78	78	0	7	1
38	38	78	78	0	4	
38	38	79	79	0	3	
39	39	75	75	0	1	
39	39	76	76	0	8	1
39	39	76	76	0	7	1
30	40					
30	40	87	87	1	8	
30	41	83	90	2	13	1
30	41	84	90	1	7	
30	41	85	88	2	14	1
30	41	87	87	2	15	1
30	42	79	86	2	11	1
30	42	80	87	1	6	
30	42	81	88	2	12	1
30	42	81	88	2	11	1
30	43	78	86	1	5	
30	44	77	86	2	9	1
30	45	75	84	2	6	1
30	46	73	83	2	7	1
30	47	71	82	2	6	1
30	48	70	82	3	8	1
30	49	69	81	2	5	1

TAB. IV

FROM ALPHA ANGLES OF (111) - POLES INTO HKL

α_1	α_2	α_3	α_4	X	Y	Z
30 40						
30	49	69	82	3	7	14
31	40	82	87	1	8	10
31	40	85	90	1	9	10
31	40	85	90	1	8	9
31	40	87	87	1	9	9
31	41	81	87	1	7	9
31	43	77	85	2	10	15
31	44	76	84	2	9	14
31	45	73	82	1	4	7
31	49	68	80	3	7	15
32	40	80	85	1	8	11
32	40	81	86	1	9	12
32	40	83	88	1	9	11
32	41	79	85	1	7	10
32	42	78	84	1	6	9
32	43	76	83	1	5	8
32	44	75	82	2	9	15
32	46	72	81	2	7	13
32	47	70	80	1	3	6
32	49	68	79	2	5	11
33	40	80	84	1	10	14
33	40	80	84	1	9	13
33	41	77	82	1	7	11
33	41	78	83	1	8	12
33	42	76	81	1	6	10
33	43	74	80	1	5	9
33	45	71	79	1	4	8
33	45	72	80	2	8	15
33	46	71	79	2	7	14
33	47	69	78	2	6	13
33	49	67	77	2	5	12
34	40	78	82	1	9	14
34	40	79	83	1	10	15
34	41	77	81	1	8	13
34	42	76	80	1	7	12
34	43	74	79	1	6	11
34	46	70	78	2	7	15
34	47	68	77	1	3	7
35	41	76	80	1	8	14
35	41	77	81	1	9	15
35	42	74	79	1	7	13
35	43	73	78	1	6	12
35	44	72	78	1	5	10
35	46	70	77	1	4	9
35	48	67	76	2	6	15
35	49	66	76	2	5	13
36	42	74	78	1	8	15
36	43	73	77	1	7	14

α_1	α_2	α_3	α_4	X	Y	Z
30 40						
36	44	71	76	1	5	11
36	44	71	76	1	6	13
36	48	66	74	1	3	8
36	49	65	74	2	5	14
37	43	72	76	1	7	15
37	44	70	75	1	6	14
37	45	69	75	1	5	12
37	46	68	75	1	4	10
37	49	64	73	2	5	15
38	44	69	73	1	6	15
38	45	68	73	1	5	13
38	47	67	73	1	4	11
38	48	65	72	1	3	9
39	46	67	72	1	5	14
39	47	66	71	1	4	12
30 50						
30	51	66	80	2	4	9
30	51	67	81	3	6	13
30	53	64	80	3	5	12
30	54	63	80	4	6	15
30	56	61	79	4	5	14
31	51	66	79	3	6	14
31	54	63	79	2	3	8
31	55	61	79	3	4	11
31	58	58	78	4	4	13
32	53	64	78	3	5	13
32	55	61	77	3	4	12
32	56	60	78	4	5	15
32	58	58	78	3	3	10
33	51	65	78	1	2	5
33	52	63	77	3	5	14
33	58	58	77	2	2	7
34	51	64	76	2	4	11
34	52	62	75	3	5	15
34	53	62	76	2	3	9
34	55	60	76	3	4	13
34	57	57	75	4	4	15
34	57	57	76	3	3	11
35	53	61	74	2	3	10
35	54	60	74	3	4	14
35	57	57	74	1	1	4
36	51	63	74	1	2	6
37	51	62	73	2	4	13
37	53	60	73	2	3	11
37	54	59	73	3	4	15
37	57	57	72	2	2	9

α_1	α_2	α_3	α_4	X	Y	Z
30 50						
37	57	57	73	3	3	13
38	51	62	72	1	2	7
38	53	60	71	2	3	12
38	56	56	72	3	3	14
39	51	61	71	2	4	15
39	56	56	71	1	1	5
40 40						
40	40	73	73	0	5	11
40	40	73	73	0	4	9
40	40	74	74	0	7	15
40	40	74	74	0	6	13
40	46	66	71	1	5	15
40	47	65	70	1	4	13
40	49	64	71	1	3	10
41	41	71	71	0	2	5
41	41	72	72	0	5	12
41	41	72	72	0	3	7
41	48	64	69	1	4	14
41	49	63	69	1	3	11
42	42	70	70	0	4	11
42	42	70	70	0	3	8
42	42	70	70	0	5	14
42	42	71	71	0	5	13
42	48	64	68	1	4	15
42	49	62	68	1	3	12
43	43	69	69	0	1	3
44	44	67	67	0	2	7
44	44	67	67	0	3	10
44	44	68	68	0	4	13
45	45	66	66	0	3	11
45	45	66	66	0	4	15
46	46	64	64	0	2	9
46	46	64	64	0	3	13
46	46	65	65	0	1	4
47	47	63	63	0	1	5
47	47	64	64	0	3	14
48	48	62	62	0	1	6
48	48	62	62	0	2	11
49	49	61	61	0	1	7
49	49	61	61	0	2	13
40 50						
40	51	61	70	1	2	8
40	53	59	70	2	3	13
40	56	56	69	2	2	11

α_1	α_2	α_3	α_4	X	Y	Z
40 50						
41	53	59	69	2	3	14
41	56	56	68	1	1	6
42	51	60	68	1	2	9
42	53	58	68	2	3	15
42	56	56	67	2	2	13
43	50	62	67	1	3	13
43	52	60	67	1	2	10
43	56	56	66	1	1	7
44	50	61	66	1	3	15
44	50	61	66	1	3	14
44	52	59	66	1	2	11
44	55	55	65	2	2	15
45	52	59	65	1	2	12
45	55	55	65	1	1	8
46	52	58	63	1	2	14
46	52	58	64	1	2	13
46	55	55	64	1	1	9
47	52	58	63	1	2	15
47	55	55	62	1	1	11
47	55	55	63	1	1	10
48	55	55	61	1	1	12
49	55	55	60	1	1	15
49	55	55	61	1	1	13
49	55	55	61	1	1	14
50 50						
50	50	59	59	0	1	9
50	50	60	60	0	2	15
50	50	60	60	0	1	8
51	51	58	58	0	1	12
51	51	58	58	0	1	11
51	51	59	59	0	1	10
52	52	57	57	0	1	15
52	52	58	58	0	1	13
52	52	58	58	0	1	14
55	55	55	55	0	0	1

α_1	α_2	α_3	α_4	X	Y	Z

TABLE V

From Miller Indices into Euler Angles

The relationships between Miller indices and Euler angles are given by Eq. (37). The present table allows to numerically find the Euler angles $\varphi_1, \phi, \varphi_2$ from a given set of Miller indices (HKL) [UVW] for $|H, K, L, U, V, W| \leqslant 12$. Since for the cubic crystal and orthgonal sample symmetrie Euler angles need to be considered only in the range $H^\circ = 0 \leqslant \varphi_1 \, \phi \, \varphi_2 \leqslant \pi/2$, one obtains the following limits on the indices

$$0 \leqslant H, K, L, \qquad V \leqslant 0, \qquad 0 \leqslant W, \qquad (HV - KU) \leqslant 0.$$

The table is made up of blocks of constant $\{HKL\}$ with this HKL-values doubly underlined on top of each block. The blocks are ordered according to increasing HKL with $H \leqslant K \leqslant L$. Each of the blocks is divided into subblocks for the special (HKL)-values (singly underlined) which fulfill the above conditions. Within each subblock the orientations are ordered according to increasing U-Values.

For finding a special orientation (HKL) [UVW], one permutates the HKL until a combination $H'K'L'$ with $H' \leqslant K' \leqslant L'$ is obtained. Then in the block $H'K'L'$ the subblock corresponding to the special combination (HKL) and in this subblock the special combination [UVW] is looked up. On the right of [UVW] the angle φ_1 is listed, on the right of the special (HKL) the angles ϕ and φ_2.

If one wants to find the orientations symmetrically equivalent to an orientation $\varphi_1 \, \phi \, \varphi_2$, it is easiest to first determine the (HKL) [UVW] corresponding to $\varphi_1 \, \phi \, \varphi_2$ with Table VI and to derive from there the equivalent orientations as described at Table VI.

TAB. V — FROM HKL-UVW INTO EULER ANGLES — V-2

H	K	L	Φ	φ_2
	U	V	W	φ_1
0	0	1	0	--
=====	=====	=====	=====	=====
φ_2=CONST. = ↓				
	-12	-11	0	137
	-12	-7	0	150
	-12	-5	0	157
	-12	-1	0	175
	-11	-12	0	133
	-11	-10	0	138
	-11	-9	0	141
	-11	-8	0	144
	-11	-7	0	148
	-11	-6	0	151
	-11	-5	0	156
	-11	-4	0	160
	-11	-3	0	165
	-11	-2	0	170
	-11	-1	0	175
	-10	-11	0	132
	-10	-9	0	138
	-10	-7	0	145
	-10	-3	0	163
	-10	-1	0	174
	-9	-11	0	129
	-9	-10	0	132
	-9	-8	0	138
	-9	-7	0	142
	-9	-5	0	151
	-9	-4	0	156
	-9	-2	0	167
	-9	-1	0	174
	-8	-11	0	126
	-8	-9	0	132
	-8	-7	0	139
	-8	-5	0	148
	-8	-3	0	159
	-8	-1	0	173
	-7	-12	0	120
	-7	-11	0	122
	-7	-10	0	125
	-7	-9	0	128
	-7	-8	0	131
	-7	-6	0	139
	-7	-5	0	144
	-7	-4	0	150
	-7	-3	0	157
	-7	-2	0	164

H	K	L	Φ	φ_2
	U	V	W	φ_1
	-7	-1	0	172
	-6	-11	0	119
	-6	-7	0	131
	-6	-5	0	140
	-6	-1	0	171
	-5	-12	0	113
	-5	-11	0	114
	-5	-9	0	119
	-5	-8	0	122
	-5	-7	0	126
	-5	-6	0	130
	-5	-4	0	141
	-5	-3	0	149
	-5	-2	0	158
	-5	-1	0	169
	-4	-11	0	110
	-4	-9	0	114
	-4	-7	0	120
	-4	-5	0	129
	-4	-3	0	143
	-4	-1	0	166
	-3	-11	0	105
	-3	-10	0	107
	-3	-8	0	111
	-3	-7	0	113
	-3	-5	0	121
	-3	-4	0	127
	-3	-2	0	146
	-3	-1	0	162
	-2	-11	0	100
	-2	-9	0	103
	-2	-7	0	106
	-2	-5	0	112
	-2	-3	0	124
	-2	-1	0	153
	-1	-12	0	95
	-1	-11	0	95
	-1	-10	0	96
	-1	-9	0	96
	-1	-8	0	97
	-1	-7	0	98
	-1	-6	0	99
	-1	-5	0	101
	-1	-4	0	104
	-1	-3	0	108
	-1	-2	0	117
	-1	-1	0	135
	-1	0	0	180
	0	-1	0	90

H	K	L	Φ	φ_2
	U	V	W	φ_1
	1	-12	0	85
	1	-11	0	85
	1	-10	0	84
	1	-9	0	84
	1	-8	0	83
	1	-7	0	82
	1	-6	0	81
	1	-5	0	79
	1	-4	0	76
	1	-3	0	72
	1	-2	0	63
	1	-1	0	45
	1	0	0	0
	2	-11	0	80
	2	-9	0	77
	2	-7	0	74
	2	-5	0	68
	2	-3	0	56
	2	-1	0	27
	3	-11	0	75
	3	-10	0	73
	3	-8	0	69
	3	-7	0	67
	3	-5	0	59
	3	-4	0	53
	3	-2	0	34
	3	-1	0	18
	4	-11	0	70
	4	-9	0	66
	4	-7	0	60
	4	-5	0	51
	4	-3	0	37
	4	-1	0	14
	5	-12	0	67
	5	-11	0	66
	5	-9	0	61
	5	-8	0	58
	5	-7	0	54
	5	-6	0	50
	5	-4	0	39
	5	-3	0	31
	5	-2	0	22
	5	-1	0	11
	6	-11	0	61
	6	-7	0	49
	6	-5	0	40
	6	-1	0	9
	7	-12	0	60
	7	-11	0	58

H	K	L	Φ	φ_2
	U	V	W	φ_1
	7	-10	0	55
	7	-9	0	52
	7	-8	0	49
	7	-6	0	41
	7	-5	0	36
	7	-4	0	30
	7	-3	0	23
	7	-2	0	16
	7	-1	0	8
	8	-11	0	54
	8	-9	0	48
	8	-7	0	41
	8	-5	0	32
	8	-3	0	21
	8	-1	0	7
	9	-11	0	51
	9	-10	0	48
	9	-8	0	42
	9	-7	0	38
	9	-5	0	29
	9	-4	0	24
	9	-2	0	13
	9	-1	0	6
	10	-11	0	46
	10	-9	0	42
	10	-7	0	35
	10	-3	0	17
	10	-1	0	6
	11	-12	0	47
	11	-10	0	42
	11	-9	0	39
	11	-8	0	36
	11	-7	0	32
	11	-6	0	29
	11	-5	0	24
	11	-4	0	20
	11	-3	0	15
	11	-2	0	10
	11	-1	0	5
	12	-11	0	43
	12	-7	0	30
	12	-5	0	23
	12	-1	0	5
0	1	0	90	0
-----	-----	-----	-----	-----
	0	0	1	90
	1	0	0	0
	1	0	1	45
	1	0	2	63

H	K	L	Φ	φ_2
	U	V	W	φ_1
	1	0	3	72
	1	0	4	76
	1	0	5	79
	1	0	6	81
	1	0	7	82
	1	0	8	83
	1	0	9	84
	1	0	10	84
	1	0	11	85
	1	0	12	85
	2	0	1	27
	2	0	3	56
	2	0	5	68
	2	0	7	74
	2	0	9	77
	2	0	11	80
	3	0	1	18
	3	0	2	34
	3	0	4	53
	3	0	5	59
	3	0	7	67
	3	0	8	69
	3	0	10	73
	3	0	11	75
	4	0	1	14
	4	0	3	37
	4	0	5	51
	4	0	7	60
	4	0	9	66
	4	0	11	70
	5	0	1	11
	5	0	2	22
	5	0	3	31
	5	0	4	39
	5	0	6	50
	5	0	7	54
	5	0	8	58
	5	0	9	61
	5	0	11	66
	5	0	12	67
	6	0	1	9
	6	0	5	40
	6	0	7	49
	6	0	11	61
	7	0	1	8
	7	0	2	16
	7	0	3	23
	7	0	4	30
	7	0	5	36

H	K	L	Φ	φ_2
	U	V	W	φ_1
	7	0	6	41
	7	0	8	49
	7	0	9	52
	7	0	10	55
	7	0	11	58
	7	0	12	60
	8	0	1	7
	8	0	3	21
	8	0	5	32
	8	0	7	41
	8	0	9	48
	8	0	11	54
	9	0	1	6
	9	0	2	13
	9	0	4	24
	9	0	5	29
	9	0	7	38
	9	0	8	42
	9	0	10	48
	9	0	11	51
	10	0	1	6
	10	0	3	17
	10	0	7	35
	10	0	9	42
	10	0	11	48
	11	0	1	5
	11	0	2	10
	11	0	3	15
	11	0	4	20
	11	0	5	24
	11	0	6	29
	11	0	7	32
	11	0	8	36
	11	0	9	39
	11	0	10	42
	11	0	12	47
	12	0	1	5
	12	0	5	23
	12	0	7	30
	12	0	11	43
1	0	0	90	90
-----	-----	-----	-----	-----
	0	-12	1	5
	0	-12	5	23
	0	-12	7	30
	0	-12	11	43
	0	-11	1	5
	0	-11	2	10
	0	-11	3	15

H	K	L	Φ	φ_2
	U	V	W	φ_1
	0	-11	4	20
	0	-11	5	24
	0	-11	6	29
	0	-11	7	32
	0	-11	8	36
	0	-11	9	39
	0	-11	10	42
	0	-11	12	47
	0	-10	1	6
	0	-10	3	17
	0	-10	7	35
	0	-10	9	42
	0	-10	11	48
	0	-9	1	6
	0	-9	2	13
	0	-9	4	24
	0	-9	5	29
	0	-9	7	38
	0	-9	8	42
	0	-9	10	48
	0	-9	11	51
	0	-8	1	7
	0	-8	3	21
	0	-8	5	32
	0	-8	7	41
	0	-8	9	48
	0	-8	11	54
	0	-7	1	8
	0	-7	2	16
	0	-7	3	23
	0	-7	4	30
	0	-7	5	36
	0	-7	6	41
	0	-7	8	49
	0	-7	9	52
	0	-7	10	55
	0	-7	11	58
	0	-7	12	60
	0	-6	1	9
	0	-6	5	40
	0	-6	7	49
	0	-6	11	61
	0	-5	1	11
	0	-5	2	22
	0	-5	3	31
	0	-5	4	39
	0	-5	6	50
	0	-5	7	54
	0	-5	8	58

H	K	L	Φ	φ_2
	U	V	W	φ_1
	0	-5	9	61
	0	-5	11	66
	0	-5	12	67
	0	-4	1	14
	0	-4	3	37
	0	-4	5	51
	0	-4	7	60
	0	-4	9	66
	0	-4	11	70
	0	-3	1	18
	0	-3	2	34
	0	-3	4	53
	0	-3	5	59
	0	-3	7	67
	0	-3	8	69
	0	-3	10	73
	0	-3	11	75
	0	-2	1	27
	0	-2	3	56
	0	-2	5	68
	0	-2	7	74
	0	-2	9	77
	0	-2	11	80
	0	-1	0	0
	0	-1	1	45
	0	-1	2	63
	0	-1	3	72
	0	-1	4	76
	0	-1	5	79
	0	-1	6	81
	0	-1	7	82
	0	-1	8	83
	0	-1	9	84
	0	-1	10	84
	0	-1	11	85
	0	-1	12	85
	0	0	1	90
0	1	1	45	0
=====	=====	=====	=====	=====
	0	-1	1	90
	1	-12	12	87
	1	-11	11	86
	1	-10	10	86
	1	-9	9	86
	1	-8	8	85
	1	-7	7	84
	1	-6	6	83
	1	-5	5	82

AB. V FROM HKL-UVW INTO EULER ANGLES V - 3

H	K U	L V	Φ W	φ2 φ1
0	1	1	45	0
	1	-4	4	80
	1	-3	3	77
	1	-2	2	71
	1	-1	1	55
	1	0	0	0
	2	-11	11	83
	2	-9	9	81
	2	-7	7	79
	2	-5	5	74
	2	-3	3	65
	2	-1	1	35
	3	-11	11	79
	3	-10	10	78
	3	-8	8	75
	3	-7	7	73
	3	-5	5	67
	3	-4	4	62
	3	-2	2	43
	3	-1	1	25
	4	-11	11	76
	4	-9	9	73
	4	-7	7	68
	4	-5	5	61
	4	-3	3	47
	4	-1	1	19
	5	-12	12	74
	5	-11	11	72
	5	-9	9	69
	5	-8	8	66
	5	-7	7	63
	5	-6	6	59
	5	-4	4	49
	5	-3	3	40
	5	-2	2	29
	5	-1	1	16
	6	-11	11	69
	6	-7	7	59
	6	-5	5	50
	6	-1	1	13
	7	-12	12	68
	7	-11	11	66
	7	-10	10	64
	7	-9	9	61
	7	-8	8	58
	7	-6	6	50
	7	-5	5	45
	7	-4	4	39

H	K U	L V	Φ W	φ2 φ1
	7	-3	3	31
	7	-2	2	22
	7	-1	1	11
	8	-11	11	63
	8	-9	9	58
	8	-7	7	51
	8	-5	5	41
	8	-3	3	28
	8	-1	1	10
	9	-11	11	60
	9	-10	10	58
	9	-8	8	51
	9	-7	7	48
	9	-5	5	38
	9	-4	4	32
	9	-2	2	17
	9	-1	1	9
	10	-11	11	57
	10	-9	9	52
	10	-7	7	45
	10	-3	3	23
	10	-1	1	8
	11	-12	12	57
	11	-10	10	52
	11	-9	9	49
	11	-8	8	46
	11	-7	7	42
	11	-6	6	38
	11	-5	5	33
	11	-4	4	27
	11	-3	3	21
	11	-2	2	14
	11	-1	1	7
	12	-11	11	52
	12	-7	7	40
	12	-5	5	31
	12	-1	1	7
1	0	1	45	90
	-12	-11	12	57
	-12	-7	12	68
	-12	-5	12	74
	-12	-1	12	87
	-11	-12	11	52
	-11	-10	11	57
	-11	-9	11	60
	-11	-8	11	63
	-11	-7	11	66
	-11	-6	11	69

H	K U	L V	Φ W	φ2 φ1
	-11	-5	11	72
	-11	-4	11	76
	-11	-3	11	79
	-11	-2	11	83
	-11	-1	11	86
	-10	-11	10	52
	-10	-9	10	58
	-10	-7	10	64
	-10	-3	10	78
	-10	-1	10	86
	-9	-11	9	49
	-9	-10	9	52
	-9	-8	9	58
	-9	-7	9	61
	-9	-5	9	69
	-9	-4	9	73
	-9	-2	9	81
	-9	-1	9	86
	-8	-11	8	46
	-8	-9	8	51
	-8	-7	8	58
	-8	-5	8	66
	-8	-3	8	75
	-8	-1	8	85
	-7	-12	7	40
	-7	-11	7	42
	-7	-10	7	45
	-7	-9	7	48
	-7	-8	7	51
	-7	-6	7	59
	-7	-5	7	63
	-7	-4	7	68
	-7	-3	7	73
	-7	-2	7	79
	-7	-1	7	84
	-6	-11	6	38
	-6	-7	6	50
	-6	-5	6	59
	-6	-1	6	83
	-5	-12	5	31
	-5	-11	5	33
	-5	-9	5	38
	-5	-8	5	41
	-5	-7	5	45
	-5	-6	5	50
	-5	-4	5	61
	-5	-3	5	67
	-5	-2	5	74
	-5	-1	5	82

H	K U	L V	Φ W	φ2 φ1
	-4	-11	4	27
	-4	-9	4	32
	-4	-7	4	39
	-4	-5	4	49
	-4	-3	4	62
	-4	-1	4	80
	-3	-11	3	21
	-3	-10	3	23
	-3	-8	3	28
	-3	-7	3	31
	-3	-5	3	40
	-3	-4	3	47
	-3	-2	3	65
	-3	-1	3	77
	-2	-11	2	14
	-2	-9	2	17
	-2	-7	2	22
	-2	-5	2	29
	-2	-3	2	43
	-2	-1	2	71
	-1	-12	1	7
	-1	-11	1	7
	-1	-10	1	8
	-1	-9	1	9
	-1	-8	1	10
	-1	-7	1	11
	-1	-6	1	13
	-1	-5	1	16
	-1	-4	1	19
	-1	-3	1	25
	-1	-2	1	35
	-1	-1	1	55
	-1	0	1	90
	0	-1	0	0
1	1	0	90	45
	0	0	1	90
	1	-1	0	0
	1	-1	1	35
	1	-1	2	55
	1	-1	3	65
	1	-1	4	71
	1	-1	5	74
	1	-1	6	77
	1	-1	7	79
	1	-1	8	80
	1	-1	9	81
	1	-1	10	82
	1	-1	11	83

H	K U	L V	Φ W	φ2 φ1
	1	-1	12	83
	2	-2	1	19
	2	-2	3	47
	2	-2	5	61
	2	-2	7	68
	2	-2	9	73
	2	-2	11	76
	3	-3	1	13
	3	-3	2	25
	3	-3	4	43
	3	-3	5	50
	3	-3	7	59
	3	-3	8	62
	3	-3	10	67
	3	-3	11	69
	4	-4	1	10
	4	-4	3	28
	4	-4	5	41
	4	-4	7	51
	4	-4	9	58
	4	-4	11	63
	5	-5	1	8
	5	-5	2	16
	5	-5	3	23
	5	-5	4	29
	5	-5	6	40
	5	-5	7	45
	5	-5	8	49
	5	-5	9	52
	5	-5	11	57
	5	-5	12	59
	6	-6	1	7
	6	-6	5	31
	6	-6	7	40
	6	-6	11	52
	7	-7	1	6
	7	-7	2	11
	7	-7	3	17
	7	-7	4	22
	7	-7	5	27
	7	-7	6	31
	7	-7	8	39
	7	-7	9	42
	7	-7	10	45
	7	-7	11	48
	7	-7	12	50
	8	-8	1	5
	8	-8	3	15
	8	-8	5	24

H	K U	L V	Φ W	φ2 φ1
	8	-8	7	32
	8	-8	9	39
	8	-8	11	44
	9	-9	1	4
	9	-9	2	9
	9	-9	4	17
	9	-9	5	21
	9	-9	7	29
	9	-9	8	32
	9	-9	10	38
	9	-9	11	41
	10	-10	1	4
	10	-10	3	12
	10	-10	7	26
	10	-10	9	32
	10	-10	11	38
	11	-11	1	4
	11	-11	2	7
	11	-11	3	11
	11	-11	4	14
	11	-11	5	18
	11	-11	6	21
	11	-11	7	24
	11	-11	8	27
	11	-11	9	30
	11	-11	10	33
	11	-11	12	38
	12	-12	1	3
	12	-12	5	16
	12	-12	7	22
	12	-12	11	33
0	1	2	27	0
	0	-2	1	90
	1	-12	6	86
	1	-10	5	85
	1	-8	4	84
	1	-6	3	82
	1	-4	2	77
	1	-2	1	66
	1	0	0	0
	2	-10	5	80
	2	-6	3	73
	2	-2	1	48
	3	-10	5	75
	3	-8	4	71
	3	-4	2	56
	3	-2	1	37

H	K U	L V	Φ W	φ2 φ1
	4	-10	5	70
	4	-6	3	59
	4	-2	1	29
	5	-12	6	70
	5	-8	4	61
	5	-6	3	53
	5	-4	2	42
	5	-2	1	24
	6	-10	5	62
	6	-2	1	20
	7	-12	6	62
	7	-10	5	56
	7	-8	4	52
	7	-6	3	44
	7	-4	2	33
	7	-2	1	18
	8	-10	5	54
	8	-6	3	40
	8	-2	1	16
	9	-10	5	51
	9	-8	4	45
	9	-4	2	26
	9	-2	1	14
	10	-6	3	34
	10	-2	1	13
	11	-12	6	51
	11	-10	5	45
	11	-8	4	39
	11	-6	3	31
	11	-4	2	22
	11	-2	1	11
	12	-10	5	43
	12	-2	1	11
1	0	2	27	90
	-12	-11	6	51
	-12	-7	6	62
	-12	-5	6	70
	-12	-1	6	86
	-10	-12	5	43
	-10	-11	5	45
	-10	-9	5	51
	-10	-8	5	54
	-10	-7	5	56
	-10	-6	5	62
	-10	-4	5	70
	-10	-3	5	75
	-10	-2	5	80
	-10	-1	5	85

H	K U	L V	Φ W	φ2 φ1
	-8	-11	4	39
	-8	-9	4	45
	-8	-7	4	52
	-8	-5	4	61
	-8	-3	4	71
	-8	-1	4	84
	-6	-11	3	31
	-6	-10	3	34
	-6	-8	3	40
	-6	-7	3	44
	-6	-5	3	53
	-6	-4	3	59
	-6	-2	3	73
	-6	-1	3	82
	-4	-11	2	22
	-4	-9	2	26
	-4	-7	2	33
	-4	-5	2	42
	-4	-3	2	56
	-4	-1	2	77
	-2	-12	1	11
	-2	-11	1	11
	-2	-10	1	13
	-2	-9	1	14
	-2	-8	1	16
	-2	-7	1	18
	-2	-6	1	20
	-2	-5	1	24
	-2	-4	1	29
	-2	-3	1	37
	-2	-2	1	48
	-2	-1	1	66
	-2	0	1	90
	0	-1	0	0
0	2	1	63	0
	0	-1	2	90
	1	-6	12	86
	1	-5	10	85
	1	-4	8	84
	1	-3	6	82
	1	-2	4	77
	1	-1	2	66
	1	0	0	0
	2	-5	10	80
	2	-3	6	79
	2	-1	2	48
	3	-5	10	73
	3	-4	8	71

AB. V

FROM HKL-UVW INTO EULER ANGLES

V - 4

H	K U	L V	Φ W	φ2 φ1
0	1	2		
0	2	1	63	0
	3	-2	4	56
	3	-1	2	37
	4	-5	10	70
	4	-3	6	59
	4	-1	2	29
	5	-6	12	70
	5	-4	8	61
	5	-3	6	53
	5	-2	4	42
	5	-1	2	24
	6	-5	10	62
	6	-1	2	20
	7	-6	12	62
	7	-5	10	58
	7	-4	8	52
	7	-3	6	44
	7	-2	4	33
	7	-1	2	18
	8	-5	10	54
	8	-3	6	40
	8	-1	2	16
	9	-5	10	51
	9	-4	8	45
	9	-2	4	26
	9	-1	2	14
	10	-3	6	34
	10	-1	2	13
	11	-6	12	51
	11	-5	10	45
	11	-4	8	39
	11	-3	6	31
	11	-2	4	22
	11	-1	2	11
	12	-5	10	43
	12	-1	2	11
2	0	1	63	90
	-6	-11	12	51
	-6	-7	12	62
	-6	-5	12	70
	-6	-1	12	86
	-5	-12	10	43
	-5	-11	10	45
	-5	-9	10	51
	-5	-8	10	54

H	K U	L V	Φ W	φ2 φ1
	-5	-7	10	58
	-5	-6	10	62
	-5	-4	10	70
	-5	-3	10	75
	-5	-2	10	80
	-5	-1	10	85
	-4	-11	8	39
	-4	-9	8	45
	-4	-7	8	52
	-4	-5	8	61
	-4	-3	8	71
	-4	-1	8	84
	-3	-11	6	31
	-3	-10	6	34
	-3	-8	6	40
	-3	-7	6	44
	-3	-5	6	53
	-3	-4	6	59
	-3	-2	6	73
	-3	-1	6	82
	-2	-11	4	22
	-2	-9	4	26
	-2	-7	4	33
	-2	-5	4	42
	-2	-3	4	56
	-2	-1	4	77
	-1	-12	2	11
	-1	-11	2	11
	-1	-10	2	13
	-1	-9	2	14
	-1	-8	2	16
	-1	-7	2	18
	-1	-6	2	20
	-1	-5	2	24
	-1	-4	2	29
	-1	-3	2	37
	-1	-2	2	48
	-1	-1	2	66
	-1	0	2	90
	0	-1	0	0
1	2	0	90	27
	0	0	1	90
	2	-1	0	0
	2	-1	1	24
	2	-1	2	42
	2	-1	3	53
	2	-1	4	61
	2	-1	5	66

H	K U	L V	Φ W	φ2 φ1
	2	-1	6	70
	2	-1	7	72
	2	-1	8	74
	2	-1	9	76
	2	-1	10	77
	2	-1	11	79
	2	-1	12	79
	4	-2	1	13
	4	-2	3	34
	4	-2	5	48
	4	-2	7	57
	4	-2	9	64
	4	-2	11	68
	6	-3	1	8
	6	-3	2	17
	6	-3	4	31
	6	-3	5	37
	6	-3	7	46
	6	-3	8	50
	6	-3	10	56
	6	-3	11	59
	8	-4	1	6
	8	-4	3	19
	8	-4	5	29
	8	-4	7	38
	8	-4	9	45
	8	-4	11	51
	10	-5	1	5
	10	-5	2	10
	10	-5	3	15
	10	-5	4	20
	10	-5	6	28
	10	-5	7	32
	10	-5	8	36
	10	-5	9	39
	10	-5	11	45
	10	-5	12	47
	12	-6	1	4
	12	-6	5	20
	12	-6	7	28
	12	-6	11	39
2	1	0	90	63
	0	0	1	90
	1	-2	0	0
	1	-2	1	24
	1	-2	2	42
	1	-2	3	53
	1	-2	4	61

H	K U	L V	Φ W	φ2 φ1
	1	-2	5	66
	1	-2	6	70
	1	-2	7	72
	1	-2	8	74
	1	-2	9	76
	1	-2	10	77
	1	-2	11	79
	1	-2	12	79
	2	-4	1	13
	2	-4	3	34
	2	-4	5	48
	2	-4	7	57
	2	-4	9	64
	2	-4	11	68
	3	-6	1	8
	3	-6	2	17
	3	-6	4	31
	3	-6	5	37
	3	-6	7	46
	3	-6	8	50
	3	-6	10	56
	3	-6	11	59
	4	-8	1	6
	4	-8	3	19
	4	-8	5	29
	4	-8	7	38
	4	-8	9	45
	4	-8	11	51
	5	-10	1	5
	5	-10	2	10
	5	-10	3	15
	5	-10	4	20
	5	-10	6	26
	5	-10	7	32
	5	-10	8	36
	5	-10	9	39
	5	-10	11	45
	5	-10	12	47
	6	-12	1	4
	6	-12	5	20
	6	-12	7	28
	6	-12	11	39
0	1	3	18	0
	0	-3	1	90
	1	-12	4	65
	1	-9	3	84
	1	-6	2	81

H	K U	L V	Φ W	φ2 φ1
	1	-3	1	72
	1	0	0	0
	2	-9	3	78
	2	-3	1	58
	3	-12	4	77
	3	-6	2	65
	3	-3	1	47
	4	-9	3	67
	4	-3	1	38
	5	-12	4	68
	5	-9	3	62
	5	-6	2	52
	5	-3	1	32
	6	-3	1	28
	7	-12	4	61
	7	-9	3	54
	7	-6	2	42
	7	-3	1	24
	8	-9	3	50
	8	-3	1	22
	9	-12	4	55
	9	-6	2	35
	9	-3	1	19
	10	-9	3	43
	10	-3	1	18
	11	-12	4	49
	11	-9	3	41
	11	-6	2	30
	11	-3	1	16
	12	-3	1	15
1	0	3	18	90
	-12	-11	4	49
	-12	-9	4	55
	-12	-7	4	61
	-12	-5	4	68
	-12	-3	4	77
	-12	-1	4	85
	-9	-11	3	41
	-9	-10	3	43
	-9	-8	3	50
	-9	-7	3	54
	-9	-5	3	62
	-9	-4	3	67
	-9	-2	3	78
	-9	-1	3	84
	-6	-11	2	30
	-6	-9	2	35
	-6	-7	2	42

H	K U	L V	Φ W	φ2 φ1
	-6	-5	2	52
	-6	-3	2	65
	-6	-1	2	81
	-3	-12	1	15
	-3	-11	1	16
	-3	-10	1	18
	-3	-9	1	19
	-3	-8	1	22
	-3	-7	1	24
	-3	-6	1	28
	-3	-5	1	32
	-3	-4	1	38
	-3	-3	1	47
	-3	-2	1	58
	-3	-1	1	72
	-3	0	1	90
	0	-1	0	0
0	3	1	72	0
	0	-1	3	90
	1	-4	12	85
	1	-3	9	84
	1	-2	6	81
	1	-1	3	72
	1	0	0	0
	2	-3	9	78
	2	-1	3	58
	3	-4	12	77
	3	-2	6	65
	3	-1	3	47
	4	-3	9	67
	4	-1	3	38
	5	-4	12	68
	5	-3	9	62
	5	-2	6	52
	5	-1	3	32
	6	-1	3	28
	7	-4	12	61
	7	-3	9	54
	7	-2	6	42
	7	-1	3	24
	8	-3	9	50
	8	-1	3	22
	9	-4	12	55
	9	-2	6	35
	9	-1	3	19
	10	-3	9	43
	10	-1	3	18
	11	-4	12	49

H	K U	L V	Φ W	φ2 φ1
	11	-3	9	41
	11	-2	6	30
	11	-1	3	16
	12	-1	3	15
3	0	1	72	90
	-4	-11	12	49
	-4	-9	12	55
	-4	-7	12	61
	-4	-5	12	68
	-4	-3	12	77
	-4	-1	12	85
	-3	-11	9	41
	-3	-10	9	43
	-3	-8	9	50
	-3	-7	9	54
	-3	-5	9	62
	-3	-4	9	67
	-3	-2	9	78
	-3	-1	9	84
	-2	-11	6	30
	-2	-9	6	35
	-2	-7	6	42
	-2	-5	6	52
	-2	-3	6	65
	-2	-1	6	81
	-1	-12	3	15
	-1	-11	3	16
	-1	-10	3	18
	-1	-9	3	19
	-1	-8	3	22
	-1	-7	3	24
	-1	-6	3	28
	-1	-5	3	32
	-1	-4	3	38
	-1	-3	3	47
	-1	-2	3	58
	-1	-1	3	72
	-1	0	3	90
	0	-1	0	0
1	3	0	90	18
	0	0	1	90
	3	-1	0	0
	3	-1	1	18
	3	-1	2	32
	3	-1	3	43
	3	-1	4	52
	3	-1	5	58

H	K U	L V	Φ W	φ2 φ1
	3	-1	6	62
	3	-1	7	66
	3	-1	8	68
	3	-1	9	71
	3	-1	10	72
	3	-1	11	74
	3	-1	12	75
	6	-2	1	9
	6	-2	3	25
	6	-2	5	38
	6	-2	7	48
	6	-2	9	55
	6	-2	11	60
	9	-3	1	6
	9	-3	2	12
	9	-3	4	23
	9	-3	5	28
	9	-3	7	36
	9	-3	8	40
	9	-3	10	47
	9	-3	11	49
	12	-4	1	5
	12	-4	3	13
	12	-4	5	22
	12	-4	7	29
	12	-4	9	35
	12	-4	11	41
3	1	0	90	72
	0	0	1	90
	1	-3	0	0
	1	-3	1	18
	1	-3	2	32
	1	-3	3	43
	1	-3	4	52
	1	-3	5	58
	1	-3	6	62
	1	-3	7	66
	1	-3	8	68
	1	-3	9	71
	1	-3	10	72
	1	-3	11	74
	1	-3	12	75
	2	-6	1	9
	2	-6	3	25
	2	-6	5	38
	2	-6	7	48
	2	-6	9	55
	2	-6	11	60

TAB. V

FROM HKL-UVW INTO EULER ANGLES

H	K U	L V	Φ W	φ2 φ1
0	1	3		
3	1	0	90	72
	3	-9	1	6
	3	-9	2	12
	3	-9	4	23
	3	-9	5	28
	3	-9	7	36
	3	-9	8	40
	3	-9	10	47
	3	-9	11	49
	4	-12	1	5
	4	-12	3	13
	4	-12	5	22
	4	-12	7	29
	4	-12	9	35
	4	-12	11	41
0	1	4	14	0
	0	-4	1	90
	1	-12	3	85
	1	-8	2	83
	1	-4	1	76
	1	0	0	0
	2	-12	3	81
	2	-4	1	64
	3	-8	2	70
	3	-4	1	54
	4	-12	3	72
	4	-4	1	46
	5	-12	3	68
	5	-8	2	59
	5	-4	1	40
	6	-4	1	34
	7	-12	3	60
	7	-8	2	50
	7	-4	1	30
	8	-12	3	57
	8	-4	1	27
	9	-8	2	42
	9	-4	1	25
	10	-12	3	51
	10	-4	1	22
	11	-12	3	48
	11	-8	2	37
	11	-4	1	21
	12	-4	1	19
1	0	4	14	90
	-12	-11	3	48
	-12	-10	3	51
	-12	-8	3	57
	-12	-7	3	60
	-12	-5	3	68
	-12	-4	3	72
	-12	-2	3	81
	-12	-1	3	85
	-8	-11	2	37
	-8	-9	2	42
	-8	-7	2	50
	-8	-5	2	59
	-8	-3	2	70
	-8	-1	2	83
	-4	-12	1	19
	-4	-11	1	21
	-4	-10	1	22
	-4	-9	1	25
	-4	-8	1	27
	-4	-7	1	30
	-4	-6	1	34
	-4	-5	1	40
	-4	-4	1	46
	-4	-3	1	54
	-4	-2	1	64
	-4	-1	1	76
	-4	0	1	90
	0	-1	0	0
0	4	1	76	0
	0	-1	4	90
	1	-3	12	85
	1	-2	8	83
	1	-1	4	76
	1	0	0	0
	2	-3	12	81
	2	-1	4	64
	3	-2	8	70
	3	-1	4	54
	4	-3	12	72
	4	-1	4	46
	5	-3	12	68
	5	-2	8	59
	5	-1	4	40
	6	-1	4	34
	7	-3	12	60
	7	-2	8	50
	7	-1	4	30
	8	-3	12	57
	8	-1	4	27
	9	-2	8	42
	9	-1	4	25
	10	-3	12	51
	10	-1	4	22
	11	-3	12	48
	11	-2	8	37
	11	-1	4	21
	12	-1	4	19
4	0	1	76	90
	-3	-11	12	48
	-3	-10	12	51
	-3	-8	12	57
	-3	-7	12	60
	-3	-5	12	68
	-3	-4	12	72
	-3	-2	12	81
	-3	-1	12	85
	-2	-11	8	37
	-2	-9	8	42
	-2	-7	8	50
	-2	-5	8	59
	-2	-3	8	70
	-2	-1	8	83
	-1	-12	4	19
	-1	-11	4	21
	-1	-10	4	22
	-1	-9	4	25
	-1	-8	4	27
	-1	-7	4	30
	-1	-6	4	34
	-1	-5	4	40
	-1	-4	4	46
	-1	-3	4	54
	-1	-2	4	64
	-1	-1	4	76
	-1	0	4	90
	0	-1	0	0
1	4	0	90	14
	0	0	1	90
	4	-1	0	0
	4	-1	1	14
	4	-1	2	26
	4	-1	3	36
	4	-1	4	44
	4	-1	5	50
	4	-1	6	56
	4	-1	7	60
	4	-1	8	63
	4	-1	9	65
	4	-1	10	68
	4	-1	11	69
	4	-1	12	71
	8	-2	1	7
	8	-2	3	20
	8	-2	5	31
	8	-2	7	40
	8	-2	9	48
	8	-2	11	53
	12	-3	1	5
	12	-3	2	9
	12	-3	4	18
	12	-3	5	22
	12	-3	7	30
	12	-3	8	33
	12	-3	10	39
	12	-3	11	42
4	1	0	90	76
	0	0	1	90
	1	-4	0	0
	1	-4	1	14
	1	-4	2	26
	1	-4	3	36
	1	-4	4	44
	1	-4	5	50
	1	-4	6	56
	1	-4	7	60
	1	-4	8	63
	1	-4	9	65
	1	-4	10	68
	1	-4	11	69
	1	-4	12	71
	2	-8	1	7
	2	-8	3	20
	2	-8	5	31
	2	-8	7	40
	2	-8	9	48
	2	-8	11	53
	3	-12	1	5
	3	-12	2	9
	3	-12	4	18
	3	-12	5	22
	3	-12	7	30
	3	-12	8	33
	3	-12	10	39
	3	-12	11	42
0	1	5	11	0
	0	-5	1	90
	1	-10	2	84
	1	-5	1	79
	1	0	0	0
	2	-5	1	69
	3	-10	2	74
	3	-5	1	60
	4	-5	1	52
	5	-10	2	64
	5	-5	1	46
	6	-5	1	40
	7	-10	2	56
	7	-5	1	36
	8	-5	1	33
	9	-10	2	49
	9	-5	1	30
	10	-5	1	27
	11	-10	2	43
	11	-5	1	25
	12	-5	1	23
1	0	5	11	90
	-10	-11	2	43
	-10	-9	2	49
	-10	-7	2	56
	-10	-5	2	64
	-10	-3	2	74
	-10	-1	2	84
	-5	-12	1	23
	-5	-11	1	25
	-5	-10	1	27
	-5	-9	1	30
	-5	-8	1	33
	-5	-7	1	36
	-5	-6	1	40
	-5	-5	1	46
	-5	-4	1	52
	-5	-3	1	60
	-5	-2	1	69
	-5	-1	1	79
	-5	0	1	90
	0	-1	0	0
0	5	1	79	0
	0	-1	5	90
	1	-2	10	84
	1	-1	5	79
	1	0	0	0
	2	-1	5	69
	3	-2	10	74
	3	-1	5	60
	4	-1	5	52
	5	-2	10	64
	5	-1	5	46
	6	-1	5	40
	7	-2	10	56
	7	-1	5	36
	8	-1	5	33
	9	-2	10	49
	9	-1	5	30
	10	-1	5	27
	11	-2	10	43
	11	-1	5	25
	12	-1	5	23
5	0	1	79	90
	-2	-11	10	43
	-2	-9	10	49
	-2	-7	10	56
	-2	-5	10	64
	-2	-3	10	74
	-2	-1	10	84
	-1	-12	5	23
	-1	-11	5	25
	-1	-10	5	27
	-1	-9	5	30
	-1	-8	5	33
	-1	-7	5	36
	-1	-6	5	40
	-1	-5	5	46
	-1	-4	5	52
	-1	-3	5	60
	-1	-2	5	69
	-1	-1	5	79
	-1	0	5	90
	0	-1	0	0
1	5	0	90	11
	0	0	1	90
	5	-1	0	0
	5	-1	1	11
	5	-1	2	21
	5	-1	3	30
	5	-1	4	38
	5	-1	5	44
	5	-1	6	50
	5	-1	7	54
	5	-1	8	57
	5	-1	9	60
	5	-1	10	63
	5	-1	11	65
	5	-1	12	67
	10	-2	1	6
	10	-2	3	16
	10	-2	5	26
	10	-2	7	34
	10	-2	9	41
	10	-2	11	47
5	1	0	90	79
	0	0	1	90
	1	-5	0	0
	1	-5	1	11
	1	-5	2	21
	1	-5	3	30
	1	-5	4	36
	1	-5	5	44
	1	-5	6	50
	1	-5	7	54
	1	-5	8	57
	1	-5	9	60
	1	-5	10	63
	1	-5	11	65
	1	-5	12	67
	2	-10	1	6
	2	-10	3	16
	2	-10	5	26
	2	-10	7	34
	2	-10	9	41
	2	-10	11	47
0	1	6	9	0
	0	-6	1	90
	1	-12	2	85
	1	-6	1	81
	1	0	0	0
	2	-6	1	72
	3	-12	2	76
	3	-6	1	64
	4	-6	1	57
	5	-12	2	68
	5	-6	1	51
	6	-6	1	45
	7	-12	2	60
	7	-6	1	41
	8	-6	1	37
	9	-12	2	54
	9	-6	1	34
	10	-6	1	31
	11	-12	2	48
	11	-6	1	29
	12	-6	1	27
1	0	6	9	90
	-12	-11	2	48
	-12	-9	2	54
	-12	-7	2	60
	-12	-5	2	68
	-12	-3	2	76
	-12	-1	2	85
	-6	-12	1	27
	-6	-11	1	29
	-6	-10	1	31
	-6	-9	1	34
	-6	-8	1	37
	-6	-7	1	41
	-6	-6	1	45
	-6	-5	1	51
	-6	-4	1	57
	-6	-3	1	64
	-6	-2	1	72
	-6	-1	1	81
	-6	0	1	90
	0	-1	0	0
0	6	1	81	0
	0	-1	6	90
	1	-2	12	85
	1	-1	6	81
	1	0	0	0
	2	-1	6	72
	3	-2	12	76
	3	-1	6	64
	4	-1	6	57
	5	-2	12	68
	5	-1	6	51
	6	-1	6	45
	7	-2	12	60

TAB. V FROM HKL-UVW INTO EULER ANGLES

H	K U	L V	Φ W	φ2 φ1
0	1	6		
0	6	1	81	0
	7	-1	6	41
	8	-1	6	37
	9	-2	12	54
	9	-1	6	34
	10	-1	6	31
	11	-2	12	48
	11	-1	6	29
	12	-1	6	27
6	0	1	81	90
	-2	-11	12	48
	-2	-9	12	54
	-2	-7	12	60
	-2	-5	12	68
	-2	-3	12	76
	-2	-1	12	85
	-1	-12	6	27
	-1	-11	6	29
	-1	-10	6	31
	-1	-9	6	34
	-1	-8	6	37
	-1	-7	6	41
	-1	-6	6	45
	-1	-5	6	51
	-1	-4	6	57
	-1	-3	6	64
	-1	-2	6	72
	-1	-1	6	81
	-1	0	6	90
	0	-1	0	0
1	6	0	90	9
	0	0	1	90
	6	-1	0	0
	6	-1	1	9
	6	-1	2	18
	6	-1	3	26
	6	-1	4	33
	6	-1	5	39
	6	-1	6	45
	6	-1	7	49
	6	-1	8	53
	6	-1	9	56
	6	-1	10	59
	6	-1	11	61

H	K U	L V	Φ W	φ2 φ1
	6	-1	12	63
	12	-2	1	5
	12	-2	3	14
	12	-2	5	22
	12	-2	7	30
	12	-2	9	36
	12	-2	11	42
6	1	0	90	81
	0	0	1	90
	1	-6	0	0
	1	-6	1	9
	1	-6	2	18
	1	-6	3	26
	1	-6	4	33
	1	-6	5	39
	1	-6	6	45
	1	-6	7	49
	1	-6	8	53
	1	-6	9	56
	1	-6	10	59
	1	-6	11	61
	1	-6	12	63
	2	-12	1	5
	2	-12	3	14
	2	-12	5	22
	2	-12	7	30
	2	-12	9	36
	2	-12	11	42
0	1	7	8	0
	0	-7	1	90
	1	-7	1	82
	1	0	0	0
	2	-7	1	74
	3	-7	1	67
	4	-7	1	61
	5	-7	1	55
	6	-7	1	50
	7	-7	1	45
	8	-7	1	41
	9	-7	1	38
	10	-7	1	35
	11	-7	1	33
	12	-7	1	31
1	0	7	8	90
	-7	-12	1	31

H	K U	L V	Φ W	φ2 φ1
	-7	-11	1	33
	-7	-10	1	35
	-7	-9	1	38
	-7	-8	1	41
	-7	-7	1	45
	-7	-6	1	50
	-7	-5	1	55
	-7	-4	1	61
	-7	-3	1	67
	-7	-2	1	74
	-7	-1	1	82
	-7	0	1	90
	0	-1	0	0
0	7	1	82	0
	0	-1	7	90
	1	-1	7	82
	1	0	0	0
	2	-1	7	74
	3	-1	7	67
	4	-1	7	61
	5	-1	7	55
	6	-1	7	50
	7	-1	7	45
	8	-1	7	41
	9	-1	7	38
	10	-1	7	35
	11	-1	7	33
	12	-1	7	31
7	0	1	82	90
	-1	-12	7	31
	-1	-11	7	33
	-1	-10	7	35
	-1	-9	7	38
	-1	-8	7	41
	-1	-7	7	45
	-1	-6	7	50
	-1	-5	7	55
	-1	-4	7	61
	-1	-3	7	67
	-1	-2	7	74
	-1	-1	7	82
	-1	0	7	90
	0	-1	0	0
1	7	0	90	8
	0	0	1	90
	7	-1	0	0

H	K U	L V	Φ W	φ2 φ1
	7	-1	1	8
	7	-1	2	16
	7	-1	3	23
	7	-1	4	29
	7	-1	5	35
	7	-1	6	40
	7	-1	7	45
	7	-1	8	49
	7	-1	9	52
	7	-1	10	55
	7	-1	11	57
	7	-1	12	59
7	1	0	90	82
	0	0	1	90
	1	-7	0	0
	1	-7	1	8
	1	-7	2	16
	1	-7	3	23
	1	-7	4	29
	1	-7	5	35
	1	-7	6	40
	1	-7	7	45
	1	-7	8	49
	1	-7	9	52
	1	-7	10	55
	1	-7	11	57
	1	-7	12	59
0	1	8	7	0
	0	-8	1	90
	1	-8	1	83
	1	0	0	0
	2	-8	1	76
	3	-8	1	70
	4	-8	1	64
	5	-8	1	58
	6	-8	1	53
	7	-8	1	49
	8	-8	1	45
	9	-8	1	42
	10	-8	1	39
	11	-8	1	36
	12	-8	1	34
1	0	8	7	90
	-8	-12	1	34
	-8	-11	1	36

H	K U	L V	Φ W	φ2 φ1
	-8	-10	1	39
	-8	-9	1	42
	-8	-8	1	45
	-8	-7	1	49
	-8	-6	1	53
	-8	-5	1	58
	-8	-4	1	64
	-8	-3	1	70
	-8	-2	1	76
	-8	-1	1	83
	-8	0	1	90
	0	-1	0	0
0	8	1	83	0
	0	-1	8	90
	1	-1	8	83
	1	0	0	0
	2	-1	8	76
	3	-1	8	70
	4	-1	8	64
	5	-1	8	58
	6	-1	8	53
	7	-1	8	49
	8	-1	8	45
	9	-1	8	42
	10	-1	8	39
	11	-1	8	36
	12	-1	8	34
8	0	1	83	90
	-1	-12	8	34
	-1	-11	8	36
	-1	-10	8	39
	-1	-9	8	42
	-1	-8	8	45
	-1	-7	8	49
	-1	-6	8	53
	-1	-5	8	58
	-1	-4	8	64
	-1	-3	8	70
	-1	-2	8	76
	-1	-1	8	83
	-1	0	8	90
	0	-1	0	0
1	8	0	90	7
	0	0	1	90
	8	-1	0	0
	8	-1	1	7

H	K U	L V	Φ W	φ2 φ1
	8	-1	2	14
	8	-1	3	20
	8	-1	4	26
	8	-1	5	32
	8	-1	6	37
	8	-1	7	41
	8	-1	8	45
	8	-1	9	48
	8	-1	10	51
	8	-1	11	54
	8	-1	12	56
8	1	0	90	83
	0	0	1	90
	1	-8	0	0
	1	-8	1	7
	1	-8	2	14
	1	-8	3	20
	1	-8	4	26
	1	-8	5	32
	1	-8	6	37
	1	-8	7	41
	1	-8	8	45
	1	-8	9	48
	1	-8	10	51
	1	-8	11	54
	1	-8	12	56
0	1	9	6	0
	0	-9	1	90
	1	-9	1	84
	1	0	0	0
	2	-9	1	78
	3	-9	1	72
	4	-9	1	66
	5	-9	1	61
	6	-9	1	56
	7	-9	1	52
	8	-9	1	49
	9	-9	1	45
	10	-9	1	42
	11	-9	1	39
	12	-9	1	37
1	0	9	6	90
	-9	-12	1	37
	-9	-11	1	39
	-9	-10	1	42

H	K U	L V	Φ W	φ2 φ1
	-9	-9	1	45
	-9	-8	1	49
	-9	-7	1	52
	-9	-6	1	56
	-9	-5	1	61
	-9	-4	1	66
	-9	-3	1	72
	-9	-2	1	78
	-9	-1	1	84
	-9	0	1	90
	0	-1	0	0
0	9	1	84	0
	0	-1	9	90
	1	-1	9	84
	1	0	0	0
	2	-1	9	78
	3	-1	9	72
	4	-1	9	66
	5	-1	9	61
	6	-1	9	56
	7	-1	9	52
	8	-1	9	49
	9	-1	9	45
	10	-1	9	42
	11	-1	9	39
	12	-1	9	37
9	0	1	84	90
	-1	-12	9	37
	-1	-11	9	39
	-1	-10	9	42
	-1	-9	9	45
	-1	-8	9	49
	-1	-7	9	52
	-1	-6	9	56
	-1	-5	9	61
	-1	-4	9	66
	-1	-3	9	72
	-1	-2	9	78
	-1	-1	9	84
	-1	0	9	90
	0	-1	0	0
1	9	0	90	6
	0	0	1	90
	9	-1	0	0
	9	-1	1	6
	9	-1	2	12

H	K U	L V	Φ W	φ2 φ1
	9	-1	3	18
	9	-1	4	24
	9	-1	5	29
	9	-1	6	34
	9	-1	7	38
	9	-1	8	41
	9	-1	9	45
	9	-1	10	48
	9	-1	11	51
	9	-1	12	53
9	1	0	90	84
	0	0	1	90
	1	-9	0	0
	1	-9	1	6
	1	-9	2	12
	1	-9	3	18
	1	-9	4	24
	1	-9	5	29
	1	-9	6	34
	1	-9	7	38
	1	-9	8	41
	1	-9	9	45
	1	-9	10	48
	1	-9	11	51
	1	-9	12	53
0	1	10	6	0
	0	-10	1	90
	1	-10	1	84
	1	0	0	0
	2	-10	1	79
	3	-10	1	73
	4	-10	1	68
	5	-10	1	64
	6	-10	1	59
	7	-10	1	55
	8	-10	1	51
	9	-10	1	48
	10	-10	1	45
	11	-10	1	42
	12	-10	1	40
1	0	10	6	90
	-10	-12	1	40
	-10	-11	1	42
	-10	-10	1	45
	-10	-9	1	48

TAB. V

FROM HKL-UVW INTO EULER ANGLES

H	K U	L V	Φ W	φ_2 φ_1
0	1	10		
1	0	10	6	90
	-10	-8	1	51
	-10	-7	1	55
	-10	-6	1	59
	-10	-5	1	64
	-10	-4	1	68
	-10	-3	1	73
	-10	-2	1	79
	-10	-1	1	84
	-10	0	1	90
	0	-1	0	0
0	10	1	84	0
	0	-1	10	90
	1	-1	10	84
	1	0	0	0
	2	-1	10	79
	3	-1	10	73
	4	-1	10	68
	5	-1	10	64
	6	-1	10	59
	7	-1	10	55
	8	-1	10	51
	9	-1	10	48
	10	-1	10	45
	11	-1	10	42
	12	-1	10	40
10	0	1	84	90
	-1	-12	10	40
	-1	-11	10	42
	-1	-10	10	45
	-1	-9	10	48
	-1	-8	10	51
	-1	-7	10	55
	-1	-6	10	59
	-1	-5	10	64
	-1	-4	10	68
	-1	-3	10	73
	-1	-2	10	79
	-1	-1	10	84
	-1	0	10	90
	0	-1	0	0
1	10	0	90	6
	0	0	1	90

H	K U	L V	Φ W	φ_2 φ_1
	10	-1	0	0
	10	-1	1	6
	10	-1	2	11
	10	-1	3	17
	10	-1	4	22
	10	-1	5	26
	10	-1	6	31
	10	-1	7	35
	10	-1	8	39
	10	-1	9	42
	10	-1	10	45
	10	-1	11	48
	10	-1	12	50
10	1	0	90	84
	0	0	1	90
	1	-10	0	0
	1	-10	1	6
	1	-10	2	11
	1	-10	3	17
	1	-10	4	22
	1	-10	5	26
	1	-10	6	31
	1	-10	7	35
	1	-10	8	39
	1	-10	9	42
	1	-10	10	45
	1	-10	11	48
	1	-10	12	50
0	1	11	5	0
	0	-11	1	90
	1	-11	1	85
	1	0	0	0
	2	-11	1	80
	3	-11	1	75
	4	-11	1	70
	5	-11	1	66
	6	-11	1	61
	7	-11	1	58
	8	-11	1	54
	9	-11	1	51
	10	-11	1	48
	11	-11	1	45
	12	-11	1	43
1	0	11	5	90
	-11	-12	1	43

H	K U	L V	Φ W	φ_2 φ_1
	-11	-11	1	45
	-11	-10	1	48
	-11	-9	1	51
	-11	-8	1	54
	-11	-7	1	58
	-11	-6	1	61
	-11	-5	1	66
	-11	-4	1	70
	-11	-3	1	75
	-11	-2	1	80
	-11	-1	1	85
	-11	0	1	90
	0	-1	0	0
0	11	1	85	0
	0	-1	11	90
	1	-1	11	85
	1	0	0	0
	2	-1	11	80
	3	-1	11	75
	4	-1	11	70
	5	-1	11	66
	6	-1	11	61
	7	-1	11	58
	8	-1	11	54
	9	-1	11	51
	10	-1	11	48
	11	-1	11	45
	12	-1	11	43
11	0	1	85	90
	-1	-12	11	43
	-1	-11	11	45
	-1	-10	11	48
	-1	-9	11	51
	-1	-8	11	54
	-1	-7	11	58
	-1	-6	11	61
	-1	-5	11	66
	-1	-4	11	70
	-1	-3	11	75
	-1	-2	11	80
	-1	-1	11	85
	-1	0	11	90
	0	-1	0	0
1	11	0	90	5
	0	0	1	90
	11	-1	0	0

H	K U	L V	Φ W	φ_2 φ_1
	11	-1	1	5
	11	-1	2	10
	11	-1	3	15
	11	-1	4	20
	11	-1	5	24
	11	-1	6	29
	11	-1	7	32
	11	-1	8	36
	11	-1	9	39
	11	-1	10	42
	11	-1	11	45
	11	-1	12	47
11	1	0	90	85
	0	0	1	90
	1	-11	0	0
	1	-11	1	5
	1	-11	2	10
	1	-11	3	15
	1	-11	4	20
	1	-11	5	24
	1	-11	6	29
	1	-11	7	32
	1	-11	8	36
	1	-11	9	39
	1	-11	10	42
	1	-11	11	45
	1	-11	12	47
0	1	12	5	0
	0	-12	1	90
	1	-12	1	85
	1	0	0	0
	2	-12	1	81
	3	-12	1	76
	4	-12	1	72
	5	-12	1	67
	6	-12	1	64
	7	-12	1	60
	8	-12	1	56
	9	-12	1	53
	10	-12	1	50
	11	-12	1	48
	12	-12	1	45
1	0	12	5	90
	-12	-12	1	45
	-12	-11	1	48

H	K U	L V	Φ W	φ_2 φ_1
	-12	-10	1	50
	-12	-9	1	53
	-12	-8	1	56
	-12	-7	1	60
	-12	-6	1	64
	-12	-5	1	67
	-12	-4	1	72
	-12	-3	1	76
	-12	-2	1	81
	-12	-1	1	85
	-12	0	1	90
	0	-1	0	0
0	12	1	85	0
	0	-1	12	90
	1	-1	12	85
	1	0	0	0
	2	-1	12	81
	3	-1	12	76
	4	-1	12	72
	5	-1	12	67
	6	-1	12	64
	7	-1	12	60
	8	-1	12	56
	9	-1	12	53
	10	-1	12	50
	11	-1	12	48
	12	-1	12	45
12	0	1	85	90
	-1	-12	12	45
	-1	-11	12	48
	-1	-10	12	50
	-1	-9	12	53
	-1	-8	12	56
	-1	-7	12	60
	-1	-6	12	64
	-1	-5	12	67
	-1	-4	12	72
	-1	-3	12	76
	-1	-2	12	81
	-1	-1	12	85
	-1	0	12	90
	0	-1	0	0
1	12	0	90	5
	0	0	1	90
	12	-1	0	0
	12	-1	1	5

H	K U	L V	Φ W	φ_2 φ_1
	12	-1	2	9
	12	-1	3	14
	12	-1	4	18
	12	-1	5	23
	12	-1	6	26
	12	-1	7	30
	12	-1	8	34
	12	-1	9	37
	12	-1	10	40
	12	-1	11	42
	12	-1	12	45
12	1	0	90	85
	0	0	1	90
	1	-12	0	0
	1	-12	1	5
	1	-12	2	9
	1	-12	3	14
	1	-12	4	18
	1	-12	5	23
	1	-12	6	26
	1	-12	7	30
	1	-12	8	34
	1	-12	9	37
	1	-12	10	40
	1	-12	11	42
	1	-12	12	45
0	2	3	34	0
	0	-3	2	90
	1	-12	8	86
	1	-9	6	85
	1	-6	4	82
	1	-3	2	74
	1	0	0	0
	2	-9	6	80
	2	-3	2	61
	3	-12	8	78
	3	-6	4	67
	3	-3	2	50
	4	-9	6	70
	4	-3	2	42
	5	-12	8	71
	5	-9	6	65
	5	-6	4	55
	5	-3	2	36
	6	-3	2	31
	7	-12	8	64

H	K U	L V	Φ W	φ_2 φ_1
	7	-9	6	57
	7	-6	4	46
	7	-3	2	27
	8	-9	6	54
	8	-3	2	24
	9	-12	8	58
	9	-6	4	39
	9	-3	2	22
	10	-9	6	47
	10	-3	2	20
	11	-12	8	53
	11	-9	6	45
	11	-6	4	33
	11	-3	2	18
	12	-3	2	17
2	0	3	34	90
	-12	-11	8	53
	-12	-9	8	58
	-12	-7	8	64
	-12	-5	8	71
	-12	-3	8	78
	-12	-1	8	86
	-9	-11	6	45
	-9	-10	6	47
	-9	-8	6	54
	-9	-7	6	57
	-9	-5	6	65
	-9	-4	6	70
	-9	-2	6	80
	-9	-1	6	85
	-6	-11	4	33
	-6	-9	4	39
	-6	-7	4	46
	-6	-5	4	55
	-6	-3	4	67
	-6	-1	4	82
	-3	-12	2	17
	-3	-11	2	18
	-3	-10	2	20
	-3	-9	2	22
	-3	-8	2	24
	-3	-7	2	27
	-3	-6	2	31
	-3	-5	2	36
	-3	-4	2	42
	-3	-3	2	50
	-3	-2	2	61
	-3	-1	2	74

H	K U	L V	Φ W	φ_2 φ_1
	-3	0	2	90
	0	-1	0	0
0	3	2	56	0
	0	-2	3	90
	1	-8	12	86
	1	-6	9	85
	1	-4	6	82
	1	-2	3	74
	1	0	0	0
	2	-6	9	80
	2	-2	3	61
	3	-8	12	78
	3	-4	6	67
	3	-2	3	50
	4	-6	9	70
	4	-2	3	42
	5	-8	12	71
	5	-6	9	65
	5	-4	6	55
	5	-2	3	36
	6	-2	3	31
	7	-8	12	64
	7	-6	9	57
	7	-4	6	46
	7	-2	3	27
	8	-6	9	54
	8	-2	3	24
	9	-8	12	58
	9	-4	6	39
	9	-2	3	22
	10	-6	9	47
	10	-2	3	20
	11	-8	12	53
	11	-6	9	45
	11	-4	6	33
	11	-2	3	18
	12	-2	3	17
3	0	2	56	90
	-8	-11	12	53
	-8	-9	12	58
	-8	-7	12	64
	-8	-5	12	71
	-8	-3	12	78
	-8	-1	12	86
	-6	-11	9	45
	-6	-10	9	47
	-6	-8	9	54

FROM HKL-UVW INTO EULER ANGLES

H	K U	L V	Φ W	φ2 φ1
0	2	3		
3	0	2	56	90
	-6	-7	9	57
	-6	-5	9	65
	-6	-4	9	70
	-6	-2	9	80
	-6	-1	9	85
	-4	-11	6	33
	-4	-9	6	39
	-4	-7	6	46
	-4	-5	6	55
	-4	-3	6	67
	-4	-1	6	82
	-2	-12	3	17
	-2	-11	3	18
	-2	-10	3	20
	-2	-9	3	22
	-2	-8	3	24
	-2	-7	3	27
	-2	-6	3	31
	-2	-5	3	36
	-2	-4	3	42
	-2	-3	3	50
	-2	-2	3	61
	-2	-1	3	74
	-2	0	3	90
	0	-1	0	0
2	3	0	90	34
	0	0	1	90
	3	-2	0	0
	3	-2	1	16
	3	-2	2	29
	3	-2	3	40
	3	-2	4	48
	3	-2	5	54
	3	-2	6	59
	3	-2	7	63
	3	-2	8	66
	3	-2	9	68
	3	-2	10	70
	3	-2	11	72
	3	-2	12	73
	6	-4	1	8
	6	-4	3	23
	6	-4	5	35
	6	-4	7	44

H	K U	L V	Φ W	φ2 φ1
	6	-4	9	51
	6	-4	11	57
	9	-6	1	5
	9	-6	2	10
	9	-6	4	20
	9	-6	5	25
	9	-6	7	33
	9	-6	8	36
	9	-6	10	43
	9	-6	11	45
	12	-8	1	4
	12	-8	3	12
	12	-8	5	19
	12	-8	7	26
	12	-8	9	32
	12	-8	11	37
3	2	0	90	56
	0	0	1	90
	2	-3	0	0
	2	-3	1	16
	2	-3	2	29
	2	-3	3	40
	2	-3	4	48
	2	-3	5	54
	2	-3	6	59
	2	-3	7	63
	2	-3	8	66
	2	-3	9	68
	2	-3	10	70
	2	-3	11	72
	2	-3	12	73
	4	-6	1	8
	4	-6	3	23
	4	-6	5	35
	4	-6	7	44
	4	-6	9	51
	4	-6	11	57
	6	-9	1	5
	6	-9	2	10
	6	-9	4	20
	6	-9	5	25
	6	-9	7	33
	6	-9	8	36
	6	-9	10	43
	6	-9	11	45
	8	-12	1	4
	8	-12	3	12
	8	-12	5	19

H	K U	L V	Φ W	φ2 φ1
	8	-12	7	26
	8	-12	9	32
	8	-12	11	37
0	2	5	22	0
	0	-5	2	90
	1	-10	4	85
	1	-5	2	79
	1	0	0	0
	2	-5	2	70
	3	-10	4	74
	3	-5	2	61
	4	-5	2	53
	5	-10	4	65
	5	-5	2	47
	6	-5	2	42
	7	-10	4	57
	7	-5	2	38
	8	-5	2	34
	9	-10	4	50
	9	-5	2	31
	10	-5	2	28
	11	-10	4	44
	11	-5	2	26
	12	-5	2	24
2	0	5	22	90
	-10	-11	4	44
	-10	-9	4	50
	-10	-7	4	57
	-10	-5	4	65
	-10	-3	4	74
	-10	-1	4	85
	-5	-12	2	24
	-5	-11	2	26
	-5	-10	2	28
	-5	-9	2	31
	-5	-8	2	34
	-5	-7	2	38
	-5	-6	2	42
	-5	-5	2	47
	-5	-4	2	53
	-5	-3	2	61
	-5	-2	2	70
	-5	-1	2	79
	-5	0	2	90
	0	-1	0	0

H	K U	L V	Φ W	φ2 φ1
0	5	2	68	0
	0	-2	5	90
	1	-4	10	85
	1	-2	5	79
	1	0	0	0
	2	-2	5	70
	3	-4	10	74
	3	-2	5	61
	4	-2	5	53
	5	-4	10	65
	5	-2	5	47
	6	-2	5	42
	7	-4	10	57
	7	-2	5	38
	8	-2	5	34
	9	-4	10	50
	9	-2	5	31
	10	-2	5	28
	11	-4	10	44
	11	-2	5	26
	12	-2	5	24
5	0	2	68	90
	-4	-11	10	44
	-4	-9	10	50
	-4	-7	10	57
	-4	-5	10	65
	-4	-3	10	74
	-4	-1	10	85
	-2	-12	5	24
	-2	-11	5	26
	-2	-10	5	28
	-2	-9	5	31
	-2	-8	5	34
	-2	-7	5	38
	-2	-6	5	42
	-2	-5	5	47
	-2	-4	5	53
	-2	-3	5	61
	-2	-2	5	70
	-2	-1	5	79
	-2	0	5	90
	0	-1	0	0
2	5	0	90	22
	0	0	1	90
	5	-2	0	0
	5	-2	1	11

H	K U	L V	Φ W	φ2 φ1
	5	-2	2	20
	5	-2	3	29
	5	-2	4	37
	5	-2	5	43
	5	-2	6	48
	5	-2	7	52
	5	-2	8	56
	5	-2	9	59
	5	-2	10	62
	5	-2	11	64
	5	-2	12	66
	10	-4	1	5
	10	-4	3	16
	10	-4	5	25
	10	-4	7	33
	10	-4	9	40
	10	-4	11	46
5	2	0	90	68
	0	0	1	90
	2	-5	0	0
	2	-5	1	11
	2	-5	2	20
	2	-5	3	29
	2	-5	4	37
	2	-5	5	43
	2	-5	6	48
	2	-5	7	52
	2	-5	8	56
	2	-5	9	59
	2	-5	10	62
	2	-5	11	64
	2	-5	12	66
	4	-10	1	5
	4	-10	3	16
	4	-10	5	25
	4	-10	7	33
	4	-10	9	40
	4	-10	11	46
0	2	7	16	0
	0	-7	2	90
	1	-7	2	82
	1	0	0	0
	2	-7	2	75
	3	-7	2	68
	4	-7	2	61
	5	-7	2	56

H	K U	L V	Φ W	φ2 φ1
	6	-7	2	51
	7	-7	2	46
	8	-7	2	42
	9	-7	2	39
	10	-7	2	36
	11	-7	2	33
	12	-7	2	31
2	0	7	16	90
	-7	-12	2	31
	-7	-11	2	33
	-7	-10	2	36
	-7	-9	2	39
	-7	-8	2	42
	-7	-7	2	46
	-7	-6	2	51
	-7	-5	2	56
	-7	-4	2	61
	-7	-3	2	68
	-7	-2	2	75
	-7	-1	2	82
	-7	0	2	90
	0	-1	0	0
0	7	2	74	0
	0	-2	7	90
	1	-2	7	82
	1	0	0	0
	2	-2	7	75
	3	-2	7	68
	4	-2	7	61
	5	-2	7	56
	6	-2	7	51
	7	-2	7	46
	8	-2	7	42
	9	-2	7	39
	10	-2	7	36
	11	-2	7	33
	12	-2	7	31
7	0	2	74	90
	-2	-12	7	31
	-2	-11	7	33
	-2	-10	7	36
	-2	-9	7	39
	-2	-8	7	42
	-2	-7	7	46
	-2	-6	7	51
	-2	-5	7	56

H	K U	L V	Φ W	φ2 φ1
	-2	-4	7	61
	-2	-3	7	68
	-2	-2	7	75
	-2	-1	7	82
	-2	0	7	90
	0	-1	0	0
2	7	0	90	16
	0	0	1	90
	7	-2	0	0
	7	-2	1	8
	7	-2	2	15
	7	-2	3	22
	7	-2	4	29
	7	-2	5	34
	7	-2	6	39
	7	-2	7	44
	7	-2	8	48
	7	-2	9	51
	7	-2	10	54
	7	-2	11	57
	7	-2	12	59
7	2	0	90	74
	0	0	1	90
	2	-7	0	0
	2	-7	1	8
	2	-7	2	15
	2	-7	3	22
	2	-7	4	29
	2	-7	5	34
	2	-7	6	39
	2	-7	7	44
	2	-7	8	48
	2	-7	9	51
	2	-7	10	54
	2	-7	11	57
	2	-7	12	59
0	2	9	13	0
	0	-9	2	90
	1	-9	2	84
	1	0	0	0
	2	-9	2	76
	3	-9	2	72
	4	-9	2	67
	5	-9	2	62
	6	-9	2	57

H	K U	L V	Φ W	φ2 φ1
	7	-9	2	53
	8	-9	2	49
	9	-9	2	46
	10	-9	2	43
	11	-9	2	40
	12	-9	2	38
2	0	9	13	90
	-9	-12	2	38
	-9	-11	2	40
	-9	-10	2	43
	-9	-9	2	46
	-9	-8	2	49
	-9	-7	2	53
	-9	-6	2	57
	-9	-5	2	62
	-9	-4	2	67
	-9	-3	2	72
	-9	-2	2	78
	-9	-1	2	84
	-9	0	2	90
	0	-1	0	0
0	9	2	77	0
	0	-2	9	90
	1	-2	9	84
	1	0	0	0
	2	-2	9	78
	3	-2	9	72
	4	-2	9	67
	5	-2	9	62
	6	-2	9	57
	7	-2	9	53
	8	-2	9	49
	9	-2	9	46
	10	-2	9	43
	11	-2	9	40
	12	-2	9	38
9	0	2	77	90
	-2	-12	9	38
	-2	-11	9	40
	-2	-10	9	43
	-2	-9	9	46
	-2	-8	9	49
	-2	-7	9	53
	-2	-6	9	57
	-2	-5	9	62
	-2	-4	9	67

TAB. V FROM HKL-UVW INTO EULER ANGLES V - 9

H	K / U	L / V	Φ / W	φ2 / φ1
0	2	9		
9	0	2	77	90
	-2	-3	9	72
	-2	-2	9	78
	-2	-1	9	84
	-2	0	9	90
	0	-1	0	0
2	9	0	90	13
	0	0	1	90
	9	-2	0	0
	9	-2	1	6
	9	-2	2	12
	9	-2	3	18
	9	-2	4	23
	9	-2	5	28
	9	-2	6	33
	9	-2	7	37
	9	-2	8	41
	9	-2	9	44
	9	-2	10	47
	9	-2	11	50
	9	-2	12	52
9	2	0	90	77
	0	0	1	90
	2	-9	0	0
	2	-9	1	6
	2	-9	2	12
	2	-9	3	18
	2	-9	4	23
	2	-9	5	28
	2	-9	6	33
	2	-9	7	37
	2	-9	8	41
	2	-9	9	44
	2	-9	10	47
	2	-9	11	50
	2	-9	12	52
0	2	11	10	0
	0	-11	2	90
	1	-11	2	85
	1	0	0	0
	2	-11	2	80
	3	-11	2	75

H	K / U	L / V	Φ / W	φ2 / φ1
	4	-11	2	70
	5	-11	2	66
	6	-11	2	62
	7	-11	2	58
	8	-11	2	54
	9	-11	2	51
	10	-11	2	48
	11	-11	2	45
	12	-11	2	43
2	0	11	10	90
	-11	-12	2	43
	-11	-11	2	45
	-11	-10	2	48
	-11	-9	2	51
	-11	-8	2	54
	-11	-7	2	58
	-11	-6	2	62
	-11	-5	2	66
	-11	-4	2	70
	-11	-3	2	75
	-11	-2	2	80
	-11	-1	2	85
	-11	0	2	90
	0	-1	0	0
0	11	2	80	0
	0	-2	11	90
	1	-2	11	85
	1	0	0	0
	2	-2	11	80
	3	-2	11	75
	4	-2	11	70
	5	-2	11	66
	6	-2	11	62
	7	-2	11	58
	8	-2	11	54
	9	-2	11	51
	10	-2	11	48
	11	-2	11	45
	12	-2	11	43
11	0	2	80	90
	-2	-12	11	43
	-2	-11	11	45
	-2	-10	11	48
	-2	-9	11	51
	-2	-8	11	54
	-2	-7	11	58

H	K / U	L / V	Φ / W	φ2 / φ1
	-2	-6	11	62
	-2	-5	11	66
	-2	-4	11	70
	-2	-3	11	75
	-2	-2	11	80
	-2	-1	11	85
	-2	0	11	90
	0	-1	0	0
2	11	0	90	10
	0	0	1	90
	11	-2	0	0
	11	-2	1	5
	11	-2	2	10
	11	-2	3	15
	11	-2	4	20
	11	-2	5	24
	11	-2	6	28
	11	-2	7	32
	11	-2	8	36
	11	-2	9	39
	11	-2	10	42
	11	-2	11	45
	11	-2	12	47
11	2	0	90	80
	0	0	1	90
	2	-11	0	0
	2	-11	1	5
	2	-11	2	10
	2	-11	3	15
	2	-11	4	20
	2	-11	5	24
	2	-11	6	28
	2	-11	7	32
	2	-11	8	36
	2	-11	9	39
	2	-11	10	42
	2	-11	11	45
	2	-11	12	47
0	3	4	37	0
	0	-4	3	90
	1	-12	9	86
	1	-8	6	84
	1	-4	3	79
	1	0	0	0
	2	-12	9	82

H	K / U	L / V	Φ / W	φ2 / φ1
	2	-4	3	68
	3	-8	6	73
	3	-4	3	59
	4	-12	9	75
	4	-4	3	51
	5	-12	9	72
	5	-8	6	63
	5	-4	3	45
	6	-4	3	40
	7	-12	9	65
	7	-8	6	55
	7	-4	3	36
	8	-12	9	62
	8	-4	3	32
	9	-8	6	48
	9	-4	3	29
	10	-12	9	56
	10	-4	3	27
	11	-12	9	54
	11	-8	6	42
	11	-4	3	24
	12	-4	3	23
3	0	4	37	90
	-12	-11	9	54
	-12	-10	9	56
	-12	-8	9	62
	-12	-7	9	65
	-12	-5	9	72
	-12	-4	9	75
	-12	-2	9	82
	-12	-1	9	86
	-8	-11	6	42
	-8	-9	6	48
	-8	-7	6	55
	-8	-5	6	63
	-8	-3	6	73
	-8	-1	6	84
	-4	-12	3	23
	-4	-11	3	24
	-4	-10	3	27
	-4	-9	3	29
	-4	-8	3	32
	-4	-7	3	36
	-4	-6	3	40
	-4	-5	3	45
	-4	-4	3	51
	-4	-3	3	59
	-4	-2	3	68

H	K / U	L / V	Φ / W	φ2 / φ1
	-4	-1	3	79
	-4	0	3	90
	0	-1	0	0
0	4	3	53	0
	0	-3	4	90
	1	-9	12	86
	1	-6	8	84
	1	-3	4	79
	1	0	0	0
	2	-9	12	82
	2	-3	4	68
	3	-6	8	73
	3	-3	4	59
	4	-9	12	75
	4	-3	4	51
	5	-9	12	72
	5	-6	8	63
	5	-3	4	45
	6	-3	4	40
	7	-9	12	65
	7	-6	8	55
	7	-3	4	36
	8	-9	12	62
	8	-3	4	32
	9	-6	8	48
	9	-3	4	29
	10	-9	12	56
	10	-3	4	27
	11	-9	12	54
	11	-6	8	42
	11	-3	4	24
	12	-3	4	23
4	0	3	53	90
	-9	-11	12	54
	-9	-10	12	56
	-9	-8	12	62
	-9	-7	12	65
	-9	-5	12	72
	-9	-4	12	75
	-9	-2	12	82
	-9	-1	12	86
	-6	-11	8	42
	-6	-9	8	48
	-6	-7	8	55
	-6	-5	8	63
	-6	-3	8	73
	-6	-1	8	84

H	K / U	L / V	Φ / W	φ2 / φ1
	-3	-12	4	23
	-3	-11	4	24
	-3	-10	4	27
	-3	-9	4	29
	-3	-8	4	32
	-3	-7	4	36
	-3	-6	4	40
	-3	-5	4	45
	-3	-4	4	51
	-3	-3	4	59
	-3	-2	4	68
	-3	-1	4	79
	-3	0	4	90
	0	-1	0	0
3	4	0	90	37
	0	0	1	90
	4	-3	0	0
	4	-3	1	11
	4	-3	2	22
	4	-3	3	31
	4	-3	4	39
	4	-3	5	45
	4	-3	6	50
	4	-3	7	54
	4	-3	8	58
	4	-3	9	61
	4	-3	10	63
	4	-3	11	66
	4	-3	12	67
	8	-6	1	6
	8	-6	3	17
	8	-6	5	27
	8	-6	7	35
	8	-6	9	42
	8	-6	11	48
	12	-9	1	4
	12	-9	2	8
	12	-9	4	15
	12	-9	5	18
	12	-9	7	25
	12	-9	8	28
	12	-9	10	34
	12	-9	11	36
4	3	0	90	53
	0	0	1	90
	3	-4	0	0
	3	-4	1	11

H	K / U	L / V	Φ / W	φ2 / φ1
	3	-4	2	22
	3	-4	3	31
	3	-4	4	39
	3	-4	5	45
	3	-4	6	50
	3	-4	7	54
	3	-4	8	58
	3	-4	9	61
	3	-4	10	63
	3	-4	11	66
	3	-4	12	67
	6	-8	1	6
	6	-8	3	17
	6	-8	5	27
	6	-8	7	35
	6	-8	9	42
	6	-8	11	48
	9	-12	1	4
	9	-12	2	8
	9	-12	4	15
	9	-12	5	18
	9	-12	7	25
	9	-12	8	28
	9	-12	10	34
	9	-12	11	36
0	3	5	31	0
	0	-5	3	90
	1	-10	6	85
	1	-5	3	80
	1	0	0	0
	2	-5	3	71
	3	-10	6	76
	3	-5	3	63
	4	-5	3	56
	5	-10	6	67
	5	-5	3	49
	6	-5	3	44
	7	-10	6	59
	7	-5	3	40
	8	-5	3	36
	9	-10	6	52
	9	-5	3	33
	10	-5	3	30
	11	-10	6	47
	11	-5	3	28
	12	-5	3	26

H	K / U	L / V	Φ / W	φ2 / φ1
3	0	5	31	90
	-10	-11	6	47
	-10	-9	6	52
	-10	-7	6	59
	-10	-5	6	67
	-10	-3	6	76
	-10	-1	6	85
	-5	-12	3	26
	-5	-11	3	28
	-5	-10	3	30
	-5	-9	3	33
	-5	-8	3	36
	-5	-7	3	40
	-5	-6	3	44
	-5	-5	3	49
	-5	-4	3	56
	-5	-3	3	63
	-5	-2	3	71
	-5	-1	3	80
	-5	0	3	90
	0	-1	0	0
0	5	3	59	0
	0	-3	5	90
	1	-6	10	85
	1	-3	5	80
	1	0	0	0
	2	-3	5	71
	3	-6	10	76
	3	-3	5	63
	4	-3	5	56
	5	-6	10	67
	5	-3	5	49
	6	-3	5	44
	7	-6	10	59
	7	-3	5	40
	8	-3	5	36
	9	-6	10	52
	9	-3	5	33
	10	-3	5	30
	11	-6	10	47
	11	-3	5	28
	12	-3	5	26
5	0	3	59	90
	-6	-11	10	47
	-6	-9	10	52
	-6	-7	10	59

B. V — FROM HKL-UVW INTO EULER ANGLES — V - 10

H	K / U	L / V	Φ / W	φ2 / φ1
	3	5		
	0	3	59	90
	-6	-5	10	67
	-6	-3	10	76
	-6	-1	10	85
	-3	-12	5	26
	-3	-11	5	28
	-3	-10	5	30
	-3	-9	5	33
	-3	-8	5	36
	-3	-7	5	40
	-3	-6	5	44
	-3	-5	5	49
	-3	-4	5	56
	-3	-3	5	63
	-3	-2	5	71
	-3	-1	5	80
	-3	0	5	90
	0	-1	0	0
	5	0	90	31
	0	0	1	90
	5	-3	0	0
	5	-3	1	10
	5	-3	2	19
	5	-3	3	27
	5	-3	4	34
	5	-3	5	41
	5	-3	6	46
	5	-3	7	50
	5	-3	8	54
	5	-3	9	57
	5	-3	10	60
	5	-3	11	62
	5	-3	12	64
	10	-6	1	5
	10	-6	3	14
	10	-6	5	23
	10	-6	7	31
	10	-6	9	38
	10	-6	11	43
5	3	0	90	59
	0	0	1	90
	3	-5	0	0
	3	-5	1	10
	3	-5	2	19

H	K / U	L / V	Φ / W	φ2 / φ1
	3	-5	3	27
	3	-5	4	34
	3	-5	5	41
	3	-5	6	46
	3	-5	7	50
	3	-5	8	54
	3	-5	9	57
	3	-5	10	60
	3	-5	11	62
	3	-5	12	64
	6	-10	1	5
	6	-10	3	14
	6	-10	5	23
	6	-10	7	31
	6	-10	9	38
	6	-10	11	43
0	3	7	23	0
	0	-7	3	90
	1	-7	3	83
	1	0	0	0
	2	-7	3	75
	3	-7	3	68
	4	-7	3	62
	5	-7	3	57
	6	-7	3	52
	7	-7	3	47
	8	-7	3	44
	9	-7	3	40
	10	-7	3	37
	11	-7	3	35
	12	-7	3	32
3	0	7	23	90
	-7	-12	3	32
	-7	-11	3	35
	-7	-10	3	37
	-7	-9	3	40
	-7	-8	3	44
	-7	-7	3	47
	-7	-6	3	52
	-7	-5	3	57
	-7	-4	3	62
	-7	-3	3	68
	-7	-2	3	75
	-7	-1	3	83
	-7	0	3	90
	0	-1	0	0

H	K / U	L / V	Φ / W	φ2 / φ1
0	7	3	67	0
	0	-3	7	90
	1	-3	7	83
	1	0	0	0
	2	-3	7	75
	3	-3	7	68
	4	-3	7	62
	5	-3	7	57
	6	-3	7	52
	7	-3	7	47
	8	-3	7	44
	9	-3	7	40
	10	-3	7	37
	11	-3	7	35
	12	-3	7	32
7	0	3	67	90
	-3	-12	7	32
	-3	-11	7	35
	-3	-10	7	37
	-3	-9	7	40
	-3	-8	7	44
	-3	-7	7	47
	-3	-6	7	52
	-3	-5	7	57
	-3	-4	7	62
	-3	-3	7	68
	-3	-2	7	75
	-3	-1	7	83
	-3	0	7	90
	0	-1	0	0
3	7	0	90	23
	0	0	1	90
	7	-3	0	0
	7	-3	1	7
	7	-3	2	15
	7	-3	3	22
	7	-3	4	28
	7	-3	5	33
	7	-3	6	38
	7	-3	7	43
	7	-3	8	46
	7	-3	9	50
	7	-3	10	53
	7	-3	11	55
	7	-3	12	58

H	K / U	L / V	Φ / W	φ2 / φ1
7	3	0	90	67
	0	0	1	90
	3	-7	0	0
	3	-7	1	7
	3	-7	2	15
	3	-7	3	22
	3	-7	4	28
	3	-7	5	33
	3	-7	6	38
	3	-7	7	43
	3	-7	8	46
	3	-7	9	50
	3	-7	10	53
	3	-7	11	55
	3	-7	12	58
0	3	8	21	0
	0	-8	3	90
	1	-8	3	83
	1	0	0	0
	2	-8	3	77
	3	-8	3	71
	4	-8	3	65
	5	-8	3	60
	6	-8	3	55
	7	-8	3	51
	8	-8	3	47
	9	-8	3	44
	10	-8	3	41
	11	-8	3	38
	12	-8	3	35
3	0	8	21	90
	-8	-12	3	35
	-8	-11	3	38
	-8	-10	3	41
	-8	-9	3	44
	-8	-8	3	47
	-8	-7	3	51
	-8	-6	3	55
	-8	-5	3	60
	-8	-4	3	65
	-8	-3	3	71
	-8	-2	3	77
	-8	-1	3	83
	-8	0	3	90
	0	-1	0	0

H	K / U	L / V	Φ / W	φ2 / φ1
0	8	3	69	0
	0	-3	8	90
	1	-3	8	83
	1	0	0	0
	2	-3	8	77
	3	-3	8	71
	4	-3	8	65
	5	-3	8	60
	6	-3	8	55
	7	-3	8	51
	8	-3	8	47
	9	-3	8	44
	10	-3	8	41
	11	-3	8	38
	12	-3	8	35
8	0	3	69	90
	-3	-12	8	35
	-3	-11	8	38
	-3	-10	8	41
	-3	-9	8	44
	-3	-8	8	47
	-3	-7	8	51
	-3	-6	8	55
	-3	-5	8	60
	-3	-4	8	65
	-3	-3	8	71
	-3	-2	8	77
	-3	-1	8	83
	-3	0	8	90
	0	-1	0	0
3	8	0	90	21
	0	0	1	90
	8	-3	0	0
	8	-3	1	7
	8	-3	2	13
	8	-3	3	19
	8	-3	4	25
	8	-3	5	30
	8	-3	6	35
	8	-3	7	39
	8	-3	8	43
	8	-3	9	46
	8	-3	10	49
	8	-3	11	52
	8	-3	12	55

H	K / U	L / V	Φ / W	φ2 / φ1
8	3	0	90	69
	0	0	1	90
	3	-8	0	0
	3	-8	1	7
	3	-8	2	13
	3	-8	3	19
	3	-8	4	25
	3	-8	5	30
	3	-8	6	35
	3	-8	7	39
	3	-8	8	43
	3	-8	9	46
	3	-8	10	49
	3	-8	11	52
	3	-8	12	55
0	3	10	17	0
	0	-10	3	90
	1	-10	3	85
	1	0	0	0
	2	-10	3	79
	3	-10	3	74
	4	-10	3	69
	5	-10	3	64
	6	-10	3	60
	7	-10	3	56
	8	-10	3	53
	9	-10	3	49
	10	-10	3	46
	11	-10	3	44
	12	-10	3	41
3	0	10	17	90
	-10	-12	3	41
	-10	-11	3	44
	-10	-10	3	46
	-10	-9	3	49
	-10	-8	3	53
	-10	-7	3	56
	-10	-6	3	60
	-10	-5	3	64
	-10	-4	3	69
	-10	-3	3	74
	-10	-2	3	79
	-10	-1	3	85
	-10	0	3	90
	0	-1	0	0

H	K / U	L / V	Φ / W	φ2 / φ1
0	10	3	73	0
	0	-3	10	90
	1	-3	10	85
	1	0	0	0
	2	-3	10	79
	3	-3	10	74
	4	-3	10	69
	5	-3	10	64
	6	-3	10	60
	7	-3	10	56
	8	-3	10	53
	9	-3	10	49
	10	-3	10	46
	11	-3	10	44
	12	-3	10	41
10	0	3	73	90
	-3	-12	10	41
	-3	-11	10	44
	-3	-10	10	46
	-3	-9	10	49
	-3	-8	10	53
	-3	-7	10	56
	-3	-6	10	60
	-3	-5	10	64
	-3	-4	10	69
	-3	-3	10	74
	-3	-2	10	79
	-3	-1	10	85
	-3	0	10	90
	0	-1	0	0
3	10	0	90	17
	0	0	1	90
	10	-3	0	0
	10	-3	1	5
	10	-3	2	11
	10	-3	3	16
	10	-3	4	21
	10	-3	5	26
	10	-3	6	30
	10	-3	7	34
	10	-3	8	37
	10	-3	9	41
	10	-3	10	44
	10	-3	11	46
	10	-3	12	49

H	K / U	L / V	Φ / W	φ2 / φ1
10	3	0	90	73
	0	0	1	90
	3	-10	0	0
	3	-10	1	5
	3	-10	2	11
	3	-10	3	16
	3	-10	4	21
	3	-10	5	26
	3	-10	6	30
	3	-10	7	34
	3	-10	8	37
	3	-10	9	41
	3	-10	10	44
	3	-10	11	46
	3	-10	12	49
0	3	11	15	0
	0	-11	3	90
	1	-11	3	85
	1	0	0	0
	2	-11	3	80
	3	-11	3	75
	4	-11	3	71
	5	-11	3	66
	6	-11	3	62
	7	-11	3	58
	8	-11	3	55
	9	-11	3	52
	10	-11	3	49
	11	-11	3	46
	12	-11	3	44
3	0	11	15	90
	-11	-12	3	44
	-11	-11	3	46
	-11	-10	3	49
	-11	-9	3	52
	-11	-8	3	55
	-11	-7	3	58
	-11	-6	3	62
	-11	-5	3	66
	-11	-4	3	71
	-11	-3	3	75
	-11	-2	3	80
	-11	-1	3	85
	-11	0	3	90
	0	-1	0	0

TAB. V — FROM HKL-UVW INTO EULER ANGLES — V - 11

H	K / U	L / V	Φ / W	φ2 / φ1
0	3	11		
0	11	3	75	0
	0	-3	11	90
	1	-3	11	85
	1	0	0	0
	2	-3	11	80
	3	-3	11	75
	4	-3	11	71
	5	-3	11	66
	6	-3	11	62
	7	-3	11	58
	8	-3	11	55
	9	-3	11	52
	10	-3	11	49
	11	-3	11	46
	12	-3	11	44
11	0	3	75	90
	-3	-12	11	44
	-3	-11	11	46
	-3	-10	11	49
	-3	-9	11	52
	-3	-8	11	55
	-3	-7	11	58
	-3	-6	11	62
	-3	-5	11	66
	-3	-4	11	71
	-3	-3	11	75
	-3	-2	11	80
	-3	-1	11	85
	-3	0	11	90
	0	-1	0	0
3	11	0	90	15
	0	0	1	90
	11	-3	0	0
	11	-3	1	5
	11	-3	2	10
	11	-3	3	15
	11	-3	4	19
	11	-3	5	24
	11	-3	6	28
	11	-3	7	32
	11	-3	8	35
	11	-3	9	38
	11	-3	10	41
	11	-3	11	44

H	K / U	L / V	Φ / W	φ2 / φ1
	11	-3	12	46
11	3	0	90	75
	0	0	1	90
	3	-11	0	0
	3	-11	1	5
	3	-11	2	10
	3	-11	3	15
	3	-11	4	19
	3	-11	5	24
	3	-11	6	28
	3	-11	7	32
	3	-11	8	35
	3	-11	9	38
	3	-11	10	41
	3	-11	11	44
	3	-11	12	46
0	4	5	39	0
	0	-5	4	90
	1	-10	8	86
	1	-5	4	81
	1	0	0	0
	2	-5	4	73
	3	-10	8	77
	3	-5	4	65
	4	-5	4	58
	5	-10	8	69
	5	-5	4	52
	6	-5	4	47
	7	-10	8	61
	7	-5	4	42
	8	-5	4	39
	9	-10	8	55
	9	-5	4	35
	10	-5	4	33
	11	-10	8	49
	11	-5	4	30
	12	-5	4	28
4	0	5	39	90
	-10	-11	8	49
	-10	-9	8	55
	-10	-7	8	61
	-10	-5	8	69
	-10	-3	8	77
	-10	-1	8	86
	-5	-12	4	28

H	K / U	L / V	Φ / W	φ2 / φ1
	-5	-11	4	30
	-5	-10	4	33
	-5	-9	4	35
	-5	-8	4	39
	-5	-7	4	42
	-5	-6	4	47
	-5	-5	4	52
	-5	-4	4	58
	-5	-3	4	65
	-5	-2	4	73
	-5	-1	4	81
	-5	0	4	90
	0	-1	0	0
0	5	4	51	0
	0	-4	5	90
	1	-8	10	86
	1	-4	5	81
	1	0	0	0
	2	-4	5	73
	3	-8	10	77
	3	-4	5	65
	4	-4	5	58
	5	-8	10	69
	5	-4	5	52
	6	-4	5	47
	7	-8	10	61
	7	-4	5	42
	8	-4	5	39
	9	-8	10	55
	9	-4	5	35
	10	-4	5	33
	11	-8	10	49
	11	-4	5	30
	12	-4	5	28
5	0	4	51	90
	-8	-11	10	49
	-8	-9	10	55
	-8	-7	10	61
	-8	-5	10	69
	-8	-3	10	77
	-8	-1	10	86
	-4	-12	5	28
	-4	-11	5	30
	-4	-10	5	33
	-4	-9	5	35
	-4	-8	5	39
	-4	-7	5	42

H	K / U	L / V	Φ / W	φ2 / φ1
	-4	-6	5	47
	-4	-5	5	52
	-4	-4	5	58
	-4	-3	5	65
	-4	-2	5	73
	-4	-1	5	81
	-4	0	5	90
	0	-1	0	0
4	5	0	90	39
	0	0	1	90
	5	-4	0	0
	5	-4	1	9
	5	-4	2	17
	5	-4	3	25
	5	-4	4	32
	5	-4	5	38
	5	-4	6	43
	5	-4	7	48
	5	-4	8	51
	5	-4	9	55
	5	-4	10	57
	5	-4	11	60
	5	-4	12	62
	10	-8	1	4
	10	-8	3	13
	10	-8	5	21
	10	-8	7	29
	10	-8	9	35
	10	-8	11	41
5	4	0	90	51
	0	0	1	90
	4	-5	0	0
	4	-5	1	9
	4	-5	2	17
	4	-5	3	25
	4	-5	4	32
	4	-5	5	38
	4	-5	6	43
	4	-5	7	48
	4	-5	8	51
	4	-5	9	55
	4	-5	10	57
	4	-5	11	60
	4	-5	12	62
	8	-10	1	4
	8	-10	3	13
	8	-10	5	21

H	K / U	L / V	Φ / W	φ2 / φ1
	8	-10	7	29
	8	-10	9	35
	8	-10	11	41
0	4	7	30	0
	0	-7	4	90
	1	-7	4	83
	1	0	0	0
	2	-7	4	76
	3	-7	4	70
	4	-7	4	64
	5	-7	4	58
	6	-7	4	53
	7	-7	4	49
	8	-7	4	45
	9	-7	4	42
	10	-7	4	39
	11	-7	4	36
	12	-7	4	34
4	0	7	30	90
	-7	-12	4	34
	-7	-11	4	36
	-7	-10	4	39
	-7	-9	4	42
	-7	-8	4	45
	-7	-7	4	49
	-7	-6	4	53
	-7	-5	4	58
	-7	-4	4	64
	-7	-3	4	70
	-7	-2	4	76
	-7	-1	4	83
	-7	0	4	90
	0	-1	0	0
0	7	4	60	0
	0	-4	7	90
	1	-4	7	83
	1	0	0	0
	2	-4	7	76
	3	-4	7	70
	4	-4	7	64
	5	-4	7	58
	6	-4	7	53
	7	-4	7	49
	8	-4	7	45
	9	-4	7	42

H	K / U	L / V	Φ / W	φ2 / φ1
	10	-4	7	39
	11	-4	7	36
	12	-4	7	34
7	0	4	60	90
	-4	-12	7	34
	-4	-11	7	36
	-4	-10	7	39
	-4	-9	7	42
	-4	-8	7	45
	-4	-7	7	49
	-4	-6	7	53
	-4	-5	7	58
	-4	-4	7	64
	-4	-3	7	70
	-4	-2	7	76
	-4	-1	7	83
	-4	0	7	90
	0	-1	0	0
4	7	0	90	30
	0	0	1	90
	7	-4	0	0
	7	-4	1	7
	7	-4	2	14
	7	-4	3	20
	7	-4	4	26
	7	-4	5	32
	7	-4	6	37
	7	-4	7	41
	7	-4	8	45
	7	-4	9	48
	7	-4	10	51
	7	-4	11	54
	7	-4	12	56
7	4	0	90	60
	0	0	1	90
	4	-7	0	0
	4	-7	1	7
	4	-7	2	14
	4	-7	3	20
	4	-7	4	26
	4	-7	5	32
	4	-7	6	37
	4	-7	7	41
	4	-7	8	45
	4	-7	9	48
	4	-7	10	51

H	K / U	L / V	Φ / W	φ2 / φ1
	4	-7	11	54
	4	-7	12	56
0	4	9	24	0
	0	-9	4	90
	1	-9	4	84
	1	0	0	0
	2	-9	4	79
	3	-9	4	73
	4	-9	4	68
	5	-9	4	63
	6	-9	4	59
	7	-9	4	55
	8	-9	4	51
	9	-9	4	48
	10	-9	4	45
	11	-9	4	42
	12	-9	4	39
4	0	9	24	90
	-9	-12	4	39
	-9	-11	4	42
	-9	-10	4	45
	-9	-9	4	48
	-9	-8	4	51
	-9	-7	4	55
	-9	-6	4	59
	-9	-5	4	63
	-9	-4	4	68
	-9	-3	4	73
	-9	-2	4	79
	-9	-1	4	84
	-9	0	4	90
	0	-1	0	0
0	9	4	66	0
	0	-4	9	90
	1	-4	9	84
	1	0	0	0
	2	-4	9	79
	3	-4	9	73
	4	-4	9	68
	5	-4	9	63
	6	-4	9	59
	7	-4	9	55
	8	-4	9	51
	9	-4	9	48
	10	-4	9	45

H	K / U	L / V	Φ / W	φ2 / φ1
	11	-4	9	42
	12	-4	9	39
9	0	4	66	90
	-4	-12	9	39
	-4	-11	9	42
	-4	-10	9	45
	-4	-9	9	48
	-4	-8	9	51
	-4	-7	9	55
	-4	-6	9	59
	-4	-5	9	63
	-4	-4	9	68
	-4	-3	9	73
	-4	-2	9	79
	-4	-1	9	84
	-4	0	9	90
	0	-1	0	0
4	9	0	90	24
	0	0	1	90
	9	-4	0	0
	9	-4	1	5
	9	-4	2	11
	9	-4	3	17
	9	-4	4	22
	9	-4	5	27
	9	-4	6	31
	9	-4	7	35
	9	-4	8	39
	9	-4	9	42
	9	-4	10	45
	9	-4	11	48
	9	-4	12	51
9	4	0	90	66
	0	0	1	90
	4	-9	0	0
	4	-9	1	6
	4	-9	2	11
	4	-9	3	17
	4	-9	4	22
	4	-9	5	27
	4	-9	6	31
	4	-9	7	35
	4	-9	8	39
	4	-9	9	42
	4	-9	10	45
	4	-9	11	48

H	K U	L V	Φ W	φ2 φ1
0	4	9		
9	4	0	90	66
	4	-9	12	51
0	4	11	20	0
	0	-11	4	90
	1	-11	4	85
	1	0	0	0
	2	-11	4	80
	3	-11	4	76
	4	-11	4	71
	5	-11	4	67
	6	-11	4	63
	7	-11	4	59
	8	-11	4	56
	9	-11	4	52
	10	-11	4	49
	11	-11	4	47
	12	-11	4	44
4	0	11	20	90
	-11	-12	4	44
	-11	-11	4	47
	-11	-10	4	49
	-11	-9	4	52
	-11	-8	4	56
	-11	-7	4	59
	-11	-6	4	63
	-11	-5	4	67
	-11	-4	4	71
	-11	-3	4	76
	-11	-2	4	80
	-11	-1	4	85
	-11	0	4	90
	0	-1	0	0
0	11	4	70	0
	0	-4	11	90
	1	-4	11	85
	1	0	0	0
	2	-4	11	80
	3	-4	11	76
	4	-4	11	71
	5	-4	11	67
	6	-4	11	63
	7	-4	11	59

H	K U	L V	Φ W	φ2 φ1
	8	-4	11	56
	9	-4	11	52
	10	-4	11	49
	11	-4	11	47
	12	-4	11	44
11	0	4	70	90
	-4	-12	11	44
	-4	-11	11	47
	-4	-10	11	49
	-4	-9	11	52
	-4	-8	11	56
	-4	-7	11	59
	-4	-6	11	63
	-4	-5	11	67
	-4	-4	11	71
	-4	-3	11	76
	-4	-2	11	80
	-4	-1	11	85
	-4	0	11	90
	0	-1	0	0
4	11	0	90	20
	0	0	1	90
	11	-4	0	0
	11	-4	1	5
	11	-4	2	10
	11	-4	3	14
	11	-4	4	19
	11	-4	5	23
	11	-4	6	27
	11	-4	7	31
	11	-4	8	34
	11	-4	9	38
	11	-4	10	41
	11	-4	11	43
	11	-4	12	46
11	4	0	90	70
	0	0	1	90
	4	-11	0	0
	4	-11	1	5
	4	-11	2	10
	4	-11	3	14
	4	-11	4	19
	4	-11	5	23
	4	-11	6	27
	4	-11	7	31
	4	-11	8	34

H	K U	L V	Φ W	φ2 φ1
	4	-11	9	38
	4	-11	10	41
	4	-11	11	43
	4	-11	12	46
0	5	6	40	0
	0	-6	5	90
	1	-12	10	86
	1	-6	5	83
	1	0	0	0
	2	-6	5	76
	3	-12	10	79
	3	-6	5	69
	4	-6	5	63
	5	-12	10	72
	5	-6	5	57
	6	-6	5	52
	7	-12	10	66
	7	-6	5	48
	8	-6	5	44
	9	-12	10	60
	9	-6	5	41
	10	-6	5	38
	11	-12	10	55
	11	-6	5	35
	12	-6	5	33
5	0	6	40	90
	-12	-11	10	55
	-12	-9	10	60
	-12	-7	10	66
	-12	-5	10	72
	-12	-3	10	79
	-12	-1	10	86
	-6	-12	5	33
	-6	-11	5	35
	-6	-10	5	38
	-6	-9	5	41
	-6	-8	5	44
	-6	-7	5	48
	-6	-6	5	52
	-6	-5	5	57
	-6	-4	5	63
	-6	-3	5	69
	-6	-2	5	76
	-6	-1	5	83
	-6	0	5	90
	0	-1	0	0

H	K U	L V	Φ W	φ2 φ1
0	6	5	50	0
	0	-5	6	90
	1	-10	12	86
	1	-5	6	83
	1	0	0	0
	2	-5	6	76
	3	-10	12	79
	3	-5	6	69
	4	-5	6	63
	5	-10	12	72
	5	-5	6	57
	6	-5	6	52
	7	-10	12	66
	7	-5	6	48
	8	-5	6	44
	9	-10	12	60
	9	-5	6	41
	10	-5	6	38
	11	-10	12	55
	11	-5	6	35
	12	-5	6	33
6	0	5	50	90
	-10	-11	12	55
	-10	-9	12	60
	-10	-7	12	66
	-10	-5	12	72
	-10	-3	12	79
	-10	-1	12	86
	-5	-12	6	33
	-5	-11	6	35
	-5	-10	6	38
	-5	-9	6	41
	-5	-8	6	44
	-5	-7	6	48
	-5	-6	6	52
	-5	-5	6	57
	-5	-4	6	63
	-5	-3	6	69
	-5	-2	6	76
	-5	-1	6	83
	-5	0	6	90
	0	-1	0	0
5	6	0	90	40
	0	0	1	90
	6	-5	0	0
	6	-5	1	7

H	K U	L V	Φ W	φ2 φ1
	6	-5	2	14
	6	-5	3	21
	6	-5	4	27
	6	-5	5	33
	6	-5	6	38
	6	-5	7	42
	6	-5	8	46
	6	-5	9	49
	6	-5	10	52
	6	-5	11	55
	6	-5	12	57
	12	-10	1	4
	12	-10	3	11
	12	-10	5	18
	12	-10	7	24
	12	-10	9	30
	12	-10	11	35
6	5	0	90	50
	0	0	1	90
	5	-6	0	0
	5	-6	1	7
	5	-6	2	14
	5	-6	3	21
	5	-6	4	27
	5	-6	5	33
	5	-6	6	38
	5	-6	7	42
	5	-6	8	46
	5	-6	9	49
	5	-6	10	52
	5	-6	11	55
	5	-6	12	57
	10	-12	1	4
	10	-12	3	11
	10	-12	5	16
	10	-12	7	24
	10	-12	9	30
	10	-12	11	35
0	5	7	36	0
	0	-7	5	90
	1	-7	5	83
	1	0	0	0
	2	-7	5	77
	3	-7	5	71
	4	-7	5	65
	5	-7	5	60

H	K U	L V	Φ W	φ2 φ1
	6	-7	5	55
	7	-7	5	51
	8	-7	5	47
	9	-7	5	44
	10	-7	5	41
	11	-7	5	38
	12	-7	5	36
5	0	7	36	90
	-7	-12	5	36
	-7	-11	5	38
	-7	-10	5	41
	-7	-9	5	44
	-7	-8	5	47
	-7	-7	5	51
	-7	-6	5	55
	-7	-5	5	60
	-7	-4	5	65
	-7	-3	5	71
	-7	-2	5	77
	-7	-1	5	83
	-7	0	5	90
	0	-1	0	0
0	7	5	54	0
	0	-5	7	90
	1	-5	7	83
	1	0	0	0
	2	-5	7	77
	3	-5	7	71
	4	-5	7	65
	5	-5	7	60
	6	-5	7	55
	7	-5	7	51
	8	-5	7	47
	9	-5	7	44
	10	-5	7	41
	11	-5	7	38
	12	-5	7	36
7	0	5	54	90
	-5	-12	7	36
	-5	-11	7	38
	-5	-10	7	41
	-5	-9	7	44
	-5	-8	7	47
	-5	-7	7	51
	-5	-6	7	55
	-5	-5	7	60

H	K U	L V	Φ W	φ2 φ1
	-5	-4	7	65
	-5	-3	7	71
	-5	-2	7	77
	-5	-1	7	83
	-5	0	7	90
	0	-1	0	0
5	7	0	90	36
	0	0	1	90
	7	-5	0	0
	7	-5	1	7
	7	-5	2	13
	7	-5	3	19
	7	-5	4	25
	7	-5	5	30
	7	-5	6	35
	7	-5	7	39
	7	-5	8	43
	7	-5	9	46
	7	-5	10	49
	7	-5	11	52
	7	-5	12	54
7	5	0	90	54
	0	0	1	90
	5	-7	0	0
	5	-7	1	7
	5	-7	2	13
	5	-7	3	19
	5	-7	4	25
	5	-7	5	30
	5	-7	6	35
	5	-7	7	39
	5	-7	8	43
	5	-7	9	46
	5	-7	10	49
	5	-7	11	52
	5	-7	12	54
0	5	8	32	0
	0	-8	5	90
	1	-8	5	84
	1	0	0	0
	2	-8	5	78
	3	-8	5	72
	4	-8	5	67
	5	-8	5	62
	6	-8	5	58

H	K U	L V	Φ W	φ2 φ1
	7	-8	5	53
	8	-8	5	50
	9	-8	5	46
	10	-8	5	43
	11	-8	5	41
	12	-8	5	38
5	0	8	32	90
	-8	-12	5	38
	-8	-11	5	41
	-8	-10	5	43
	-8	-9	5	46
	-8	-8	5	50
	-8	-7	5	53
	-8	-6	5	58
	-8	-5	5	62
	-8	-4	5	67
	-8	-3	5	72
	-8	-2	5	78
	-8	-1	5	84
	-8	0	5	90
	0	-1	0	0
0	8	5	58	0
	0	-5	8	90
	1	-5	8	84
	1	0	0	0
	2	-5	8	78
	3	-5	8	72
	4	-5	8	67
	5	-5	8	62
	6	-5	8	58
	7	-5	8	53
	8	-5	8	50
	9	-5	8	46
	10	-5	8	43
	11	-5	8	41
	12	-5	8	38
8	0	5	58	90
	-5	-12	8	38
	-5	-11	8	41
	-5	-10	8	43
	-5	-9	8	46
	-5	-8	8	50
	-5	-7	8	53
	-5	-6	8	58
	-5	-5	8	62
	-5	-4	8	67

TAB. V

FROM HKL-UVW INTO EULER ANGLES

H	K / U	L / V	Φ / W	φ_2 / φ_1
0	5	8		
8	0	5	58	90
	-5	-3	8	72
	-5	-2	8	78
	-5	-1	8	84
	-5	0	8	90
	0	-1	0	0
5	8	0	90	32
	0	0	1	90
	8	-5	0	0
	8	-5	1	6
	8	-5	2	12
	8	-5	3	18
	8	-5	4	23
	8	-5	5	28
	8	-5	6	32
	8	-5	7	37
	8	-5	8	40
	8	-5	9	44
	8	-5	10	47
	8	-5	11	49
	8	-5	12	52
8	5	0	90	58
	0	0	1	90
	5	-8	0	0
	5	-8	1	6
	5	-8	2	12
	5	-8	3	18
	5	-8	4	23
	5	-8	5	28
	5	-8	6	32
	5	-8	7	37
	5	-8	8	40
	5	-8	9	44
	5	-8	10	47
	5	-8	11	49
	5	-8	12	52
0	5	9	29	0
	0	-9	5	90
	1	-9	5	84
	1	0	0	0
	2	-9	5	79
	3	-9	5	74

H	K / U	L / V	Φ / W	φ_2 / φ_1
	4	-9	5	69
	5	-9	5	64
	6	-9	5	60
	7	-9	5	56
	8	-9	5	52
	9	-9	5	49
	10	-9	5	46
	11	-9	5	43
	12	-9	5	41
5	0	9	29	90
	-9	-12	5	41
	-9	-11	5	43
	-9	-10	5	46
	-9	-9	5	49
	-9	-8	5	52
	-9	-7	5	56
	-9	-6	5	60
	-9	-5	5	64
	-9	-4	5	69
	-9	-3	5	74
	-9	-2	5	79
	-9	-1	5	84
	-9	0	5	90
	0	-1	0	0
0	9	5	61	0
	0	-5	9	90
	1	-5	9	84
	1	0	0	0
	2	-5	9	79
	3	-5	9	74
	4	-5	9	69
	5	-5	9	64
	6	-5	9	60
	7	-5	9	56
	8	-5	9	52
	9	-5	9	49
	10	-5	9	46
	11	-5	9	43
	12	-5	9	41
9	0	5	61	90
	-5	-12	9	41
	-5	-11	9	43
	-5	-10	9	46
	-5	-9	9	49
	-5	-8	9	52
	-5	-7	9	56

H	K / U	L / V	Φ / W	φ_2 / φ_1
	-5	-6	9	60
	-5	-5	9	64
	-5	-4	9	69
	-5	-3	9	74
	-5	-2	9	79
	-5	-1	9	84
	-5	0	9	90
	0	-1	0	0
5	9	0	90	29
	0	0	1	90
	9	-5	0	0
	9	-5	1	6
	9	-5	2	11
	9	-5	3	16
	9	-5	4	21
	9	-5	5	26
	9	-5	6	30
	9	-5	7	34
	9	-5	8	38
	9	-5	9	41
	9	-5	10	44
	9	-5	11	47
	9	-5	12	49
9	5	0	90	61
	0	0	1	90
	5	-9	0	0
	5	-9	1	6
	5	-9	2	11
	5	-9	3	16
	5	-9	4	21
	5	-9	5	26
	5	-9	6	30
	5	-9	7	34
	5	-9	8	38
	5	-9	9	41
	5	-9	10	44
	5	-9	11	47
	5	-9	12	49
0	5	11	24	0
	0	-11	5	90
	1	-11	5	85
	1	0	0	0
	2	-11	5	81
	3	-11	5	76
	4	-11	5	72

H	K / U	L / V	Φ / W	φ_2 / φ_1
	5	-11	5	68
	6	-11	5	64
	7	-11	5	60
	8	-11	5	56
	9	-11	5	53
	10	-11	5	50
	11	-11	5	48
	12	-11	5	45
5	0	11	24	90
	-11	-12	5	45
	-11	-11	5	48
	-11	-10	5	50
	-11	-9	5	53
	-11	-8	5	56
	-11	-7	5	60
	-11	-6	5	64
	-11	-5	5	68
	-11	-4	5	72
	-11	-3	5	76
	-11	-2	5	81
	-11	-1	5	85
	-11	0	5	90
	0	-1	0	0
0	11	5	66	0
	0	-5	11	90
	1	-5	11	85
	1	0	0	0
	2	-5	11	81
	3	-5	11	76
	4	-5	11	72
	5	-5	11	68
	6	-5	11	64
	7	-5	11	60
	8	-5	11	56
	9	-5	11	53
	10	-5	11	50
	11	-5	11	48
	12	-5	11	45
11	0	5	66	90
	-5	-12	11	45
	-5	-11	11	48
	-5	-10	11	50
	-5	-9	11	53
	-5	-8	11	56
	-5	-7	11	60
	-5	-6	11	64

H	K / U	L / V	Φ / W	φ_2 / φ_1
	-5	-5	11	68
	-5	-4	11	72
	-5	-3	11	76
	-5	-2	11	81
	-5	-1	11	85
	-5	0	11	90
	0	-1	0	0
5	11	0	90	24
	0	0	1	90
	11	-5	0	0
	11	-5	1	5
	11	-5	2	9
	11	-5	3	14
	11	-5	4	18
	11	-5	5	22
	11	-5	6	26
	11	-5	7	30
	11	-5	8	34
	11	-5	9	37
	11	-5	10	40
	11	-5	11	42
	11	-5	12	45
11	5	0	90	66
	0	0	1	90
	5	-11	0	0
	5	-11	1	5
	5	-11	2	9
	5	-11	3	14
	5	-11	4	18
	5	-11	5	22
	5	-11	6	26
	5	-11	7	30
	5	-11	8	34
	5	-11	9	37
	5	-11	10	40
	5	-11	11	42
	5	-11	12	45
0	5	12	23	0
	0	-12	5	90
	1	-12	5	86
	1	0	0	0
	2	-12	5	81
	3	-12	5	77
	4	-12	5	73
	5	-12	5	69

H	K / U	L / V	W	φ_2 / φ_1
	6	-12	5	65
	7	-12	5	62
	8	-12	5	58
	9	-12	5	55
	10	-12	5	52
	11	-12	5	50
	12	-12	5	47
5	0	12	23	90
	-12	-12	5	47
	-12	-11	5	50
	-12	-10	5	52
	-12	-9	5	55
	-12	-8	5	58
	-12	-7	5	62
	-12	-6	5	65
	-12	-5	5	69
	-12	-4	5	73
	-12	-3	5	77
	-12	-2	5	81
	-12	-1	5	86
	-12	0	5	90
	0	-1	0	0
0	12	5	67	0
	0	-5	12	90
	1	-5	12	86
	1	0	0	0
	2	-5	12	81
	3	-5	12	77
	4	-5	12	73
	5	-5	12	69
	6	-5	12	65
	7	-5	12	62
	8	-5	12	58
	9	-5	12	55
	10	-5	12	52
	11	-5	12	50
	12	-5	12	47
12	0	5	67	90
	-5	-12	12	47
	-5	-11	12	50
	-5	-10	12	52
	-5	-9	12	55
	-5	-8	12	58
	-5	-7	12	62
	-5	-6	12	65
	-5	-5	12	69

H	K / U	L / V	Φ / W	φ_2 / φ_1
	-5	-4	12	73
	-5	-3	12	77
	-5	-2	12	81
	-5	-1	12	86
	-5	0	12	90
	0	-1	0	0
5	12	0	90	23
	0	0	1	90
	12	-5	0	0
	12	-5	1	4
	12	-5	2	9
	12	-5	3	13
	12	-5	4	17
	12	-5	5	21
	12	-5	6	25
	12	-5	7	28
	12	-5	8	32
	12	-5	9	35
	12	-5	10	38
	12	-5	11	40
	12	-5	12	43
12	5	0	90	67
	0	0	1	90
	5	-12	0	0
	5	-12	1	4
	5	-12	2	9
	5	-12	3	13
	5	-12	4	17
	5	-12	5	21
	5	-12	6	25
	5	-12	7	28
	5	-12	8	32
	5	-12	9	35
	5	-12	10	38
	5	-12	11	40
	5	-12	12	43
0	6	7	41	0
	0	-7	6	90
	1	-7	6	84
	1	0	0	0
	2	-7	6	78
	3	-7	6	72
	4	-7	6	67
	5	-7	6	62
	6	-7	6	57

H	K / U	L / V	Φ / W	φ_2 / φ_1
	7	-7	6	53
	8	-7	6	49
	9	-7	6	46
	10	-7	6	43
	11	-7	6	40
	12	-7	6	38
6	0	7	41	90
	-7	-12	6	38
	-7	-11	6	40
	-7	-10	6	43
	-7	-9	6	46
	-7	-8	6	49
	-7	-7	6	53
	-7	-6	6	57
	-7	-5	6	62
	-7	-4	6	67
	-7	-3	6	72
	-7	-2	6	78
	-7	-1	6	84
	-7	0	6	90
	0	-1	0	0
0	7	6	49	0
	0	-6	7	90
	1	-6	7	84
	1	0	0	0
	2	-6	7	78
	3	-6	7	72
	4	-6	7	67
	5	-6	7	62
	6	-6	7	57
	7	-6	7	53
	8	-6	7	49
	9	-6	7	46
	10	-6	7	43
	11	-6	7	40
	12	-6	7	38
7	0	6	49	90
	-6	-12	7	38
	-6	-11	7	40
	-6	-10	7	43
	-6	-9	7	46
	-6	-8	7	49
	-6	-7	7	53
	-6	-6	7	57
	-6	-5	7	62
	-6	-4	7	67

TAB. V FROM HKL-UVW INTO EULER ANGLES

H	K / U	L / V	Φ / W	φ_2 / φ_1
0	6	7		
7	0	6	49	90
	-6	-3	7	72
	-6	-2	7	78
	-6	-1	7	84
	-6	0	7	90
	0	-1	0	0
6	7	0	90	41
	0	0	1	90
	7	-6	0	0
	7	-6	1	6
	7	-6	2	12
	7	-6	3	18
	7	-6	4	23
	7	-6	5	28
	7	-6	6	33
	7	-6	7	37
	7	-6	8	41
	7	-6	9	44
	7	-6	10	47
	7	-6	11	50
	7	-6	12	52
7	6	0	90	49
	0	0	1	90
	6	-7	0	0
	6	-7	1	6
	6	-7	2	12
	6	-7	3	18
	6	-7	4	23
	6	-7	5	28
	6	-7	6	33
	6	-7	7	37
	6	-7	8	41
	6	-7	9	44
	6	-7	10	47
	6	-7	11	50
	6	-7	12	52
0	6	11	29	0
	0	-11	6	90
	1	-11	6	85
	1	0	0	0
	2	-11	6	81
	3	-11	6	77

H	K / U	L / V	Φ / W	φ_2 / φ_1
	4	-11	6	72
	5	-11	6	68
	6	-11	6	64
	7	-11	6	61
	8	-11	6	57
	9	-11	6	54
	10	-11	6	51
	11	-11	6	49
	12	-11	6	46
6	0	11	29	90
	-11	-12	6	46
	-11	-11	6	49
	-11	-10	6	51
	-11	-9	6	54
	-11	-8	6	57
	-11	-7	6	61
	-11	-6	6	64
	-11	-5	6	68
	-11	-4	6	72
	-11	-3	6	77
	-11	-2	6	81
	-11	-1	6	85
	-11	0	6	90
	0	-1	0	0
0	11	6	61	0
	0	-6	11	90
	1	-6	11	85
	1	0	0	0
	2	-6	11	81
	3	-6	11	77
	4	-6	11	72
	5	-6	11	68
	6	-6	11	64
	7	-6	11	61
	8	-6	11	57
	9	-6	11	54
	10	-6	11	51
	11	-6	11	49
	12	-6	11	46
11	0	6	61	90
	-6	-12	11	46
	-6	-11	11	49
	-6	-10	11	51
	-6	-9	11	54
	-6	-8	11	57
	-6	-7	11	61

H	K / U	L / V	Φ / W	φ_2 / φ_1
	-6	-6	11	64
	-6	-5	11	68
	-6	-4	11	72
	-6	-3	11	77
	-6	-2	11	81
	-6	-1	11	85
	-6	0	11	90
	0	-1	0	0
6	11	0	90	29
	0	0	1	90
	11	-6	0	0
	11	-6	1	5
	11	-6	2	9
	11	-6	3	13
	11	-6	4	18
	11	-6	5	22
	11	-6	6	26
	11	-6	7	29
	11	-6	8	33
	11	-6	9	36
	11	-6	10	39
	11	-6	11	41
	11	-6	12	44
11	6	0	90	61
	0	0	1	90
	6	-11	0	0
	6	-11	1	5
	6	-11	2	9
	6	-11	3	13
	6	-11	4	18
	6	-11	5	22
	6	-11	6	26
	6	-11	7	29
	6	-11	8	33
	6	-11	9	36
	6	-11	10	39
	6	-11	11	41
	6	-11	12	44
0	7	8	41	0
	0	-8	7	90
	1	-8	7	85
	1	0	0	0
	2	-8	7	79
	3	-8	7	74
	4	-8	7	69

H	K / U	L / V	Φ / W	φ_2 / φ_1
	5	-8	7	65
	6	-8	7	61
	7	-8	7	57
	8	-8	7	53
	9	-8	7	50
	10	-8	7	47
	11	-8	7	44
	12	-8	7	42
7	0	8	41	90
	-8	-12	7	42
	-8	-11	7	44
	-8	-10	7	47
	-8	-9	7	50
	-8	-8	7	53
	-8	-7	7	57
	-8	-6	7	61
	-8	-5	7	65
	-8	-4	7	69
	-8	-3	7	74
	-8	-2	7	79
	-8	-1	7	85
	-8	0	7	90
	0	-1	0	0
0	8	7	49	0
	0	-7	8	90
	1	-7	8	85
	1	0	0	0
	2	-7	8	79
	3	-7	8	74
	4	-7	8	69
	5	-7	8	65
	6	-7	8	61
	7	-7	8	57
	8	-7	8	53
	9	-7	8	50
	10	-7	8	47
	11	-7	8	44
	12	-7	8	42
8	0	7	49	90
	-7	-12	8	42
	-7	-11	8	44
	-7	-10	8	47
	-7	-9	8	50
	-7	-8	8	53
	-7	-7	8	57
	-7	-6	8	61

H	K / U	L / V	Φ / W	φ_2 / φ_1
	-7	-5	8	65
	-7	-4	8	69
	-7	-3	8	74
	-7	-2	8	79
	-7	-1	8	85
	-7	0	8	90
	0	-1	0	0
7	8	0	90	41
	0	0	1	90
	8	-7	0	0
	8	-7	1	5
	8	-7	2	11
	8	-7	3	16
	8	-7	4	21
	8	-7	5	25
	8	-7	6	29
	8	-7	7	33
	8	-7	8	37
	8	-7	9	40
	8	-7	10	43
	8	-7	11	46
	8	-7	12	48
8	7	0	90	49
	0	0	1	90
	7	-8	0	0
	7	-8	1	5
	7	-8	2	11
	7	-8	3	16
	7	-8	4	21
	7	-8	5	25
	7	-8	6	29
	7	-8	7	33
	7	-8	8	37
	7	-8	9	40
	7	-8	10	43
	7	-8	11	46
	7	-8	12	48
0	7	9	38	0
	0	-9	7	90
	1	-9	7	85
	1	0	0	0
	2	-9	7	80
	3	-9	7	75
	4	-9	7	71
	5	-9	7	66

H	K / U	L / V	Φ / W	φ_2 / φ_1
	6	-9	7	62
	7	-9	7	58
	8	-9	7	55
	9	-9	7	52
	10	-9	7	49
	11	-9	7	46
	12	-9	7	44
7	0	9	38	90
	-9	-12	7	44
	-9	-11	7	46
	-9	-10	7	49
	-9	-9	7	52
	-9	-8	7	55
	-9	-7	7	58
	-9	-6	7	62
	-9	-5	7	66
	-9	-4	7	71
	-9	-3	7	75
	-9	-2	7	80
	-9	-1	7	85
	-9	0	7	90
	0	-1	0	0
0	9	7	52	0
	0	-7	9	90
	1	-7	9	85
	1	0	0	0
	2	-7	9	80
	3	-7	9	75
	4	-7	9	71
	5	-7	9	66
	6	-7	9	62
	7	-7	9	58
	8	-7	9	55
	9	-7	9	52
	10	-7	9	49
	11	-7	9	46
	12	-7	9	44
9	0	7	52	90
	-7	-12	9	44
	-7	-11	9	46
	-7	-10	9	49
	-7	-9	9	52
	-7	-8	9	55
	-7	-7	9	58
	-7	-6	9	62
	-7	-5	9	66

H	K / U	L / V	Φ / W	φ_2 / φ_1
	-7	-4	9	71
	-7	-3	9	75
	-7	-2	9	80
	-7	-1	9	85
	-7	0	9	90
	0	-1	0	0
7	9	0	90	38
	0	0	1	90
	9	-7	0	0
	9	-7	1	5
	9	-7	2	10
	9	-7	3	15
	9	-7	4	19
	9	-7	5	24
	9	-7	6	28
	9	-7	7	32
	9	-7	8	35
	9	-7	9	38
	9	-7	10	41
	9	-7	11	44
	9	-7	12	46
9	7	0	90	52
	0	0	1	90
	7	-9	0	0
	7	-9	1	5
	7	-9	2	10
	7	-9	3	15
	7	-9	4	19
	7	-9	5	24
	7	-9	6	28
	7	-9	7	32
	7	-9	8	35
	7	-9	9	38
	7	-9	10	41
	7	-9	11	44
	7	-9	12	46
0	7	10	35	0
	0	-10	7	90
	1	-10	7	85
	1	0	0	0
	2	-10	7	81
	3	-10	7	76
	4	-10	7	72
	5	-10	7	68
	6	-10	7	64

H	K / U	L / V	Φ / W	φ_2 / φ_1
	7	-10	7	60
	8	-10	7	57
	9	-10	7	54
	10	-10	7	51
	11	-10	7	48
	12	-10	7	45
7	0	10	35	90
	-10	-12	7	45
	-10	-11	7	48
	-10	-10	7	51
	-10	-9	7	54
	-10	-8	7	57
	-10	-7	7	60
	-10	-6	7	64
	-10	-5	7	68
	-10	-4	7	72
	-10	-3	7	76
	-10	-2	7	81
	-10	-1	7	85
	-10	0	7	90
	0	-1	0	0
0	10	7	55	0
	0	-7	10	90
	1	-7	10	85
	1	0	0	0
	2	-7	10	81
	3	-7	10	76
	4	-7	10	72
	5	-7	10	68
	6	-7	10	64
	7	-7	10	60
	8	-7	10	57
	9	-7	10	54
	10	-7	10	51
	11	-7	10	48
	12	-7	10	45
10	0	7	55	90
	-7	-12	10	45
	-7	-11	10	48
	-7	-10	10	51
	-7	-9	10	54
	-7	-8	10	57
	-7	-7	10	60
	-7	-6	10	64
	-7	-5	10	68
	-7	-4	10	72

H	K U	L V	Φ W	φ2 φ1
0	7	10		
10	0	7	55	90
	-7	-3	10	76
	-7	-2	10	81
	-7	-1	10	85
	-7	0	10	90
	0	-1	0	0
7	10	0	90	35
	0	0	1	90
	10	-7	0	0
	10	-7	1	5
	10	-7	2	9
	10	-7	3	14
	10	-7	4	18
	10	-7	5	22
	10	-7	6	26
	10	-7	7	30
	10	-7	8	33
	10	-7	9	36
	10	-7	10	39
	10	-7	11	42
	10	-7	12	45
10	7	0	90	55
	0	0	1	90
	7	-10	0	0
	7	-10	1	5
	7	-10	2	9
	7	-10	3	14
	7	-10	4	18
	7	-10	5	22
	7	-10	6	26
	7	-10	7	30
	7	-10	8	33
	7	-10	9	36
	7	-10	10	39
	7	-10	11	42
	7	-10	12	45
0	7	11	32	0
	0	-11	7	90
	1	-11	7	86
	1	0	0	0
	2	-11	7	81
	3	-11	7	77

H	K U	L V	Φ W	φ2 φ1
	4	-11	7	73
	5	-11	7	69
	6	-11	7	65
	7	-11	7	62
	8	-11	7	58
	9	-11	7	55
	10	-11	7	53
	11	-11	7	50
	12	-11	7	47
7	0	11	32	90
	-11	-12	7	47
	-11	-11	7	50
	-11	-10	7	53
	-11	-9	7	55
	-11	-8	7	58
	-11	-7	7	62
	-11	-6	7	65
	-11	-5	7	69
	-11	-4	7	73
	-11	-3	7	77
	-11	-2	7	81
	-11	-1	7	86
	-11	0	7	90
	0	-1	0	0
0	11	7	58	0
	0	-7	11	90
	1	-7	11	86
	1	0	0	0
	2	-7	11	81
	3	-7	11	77
	4	-7	11	73
	5	-7	11	69
	6	-7	11	65
	7	-7	11	62
	8	-7	11	58
	9	-7	11	55
	10	-7	11	53
	11	-7	11	50
	12	-7	11	47
11	0	7	58	90
	-7	-12	11	47
	-7	-11	11	50
	-7	-10	11	53
	-7	-9	11	55
	-7	-8	11	58
	-7	-7	11	62

H	K U	L V	Φ W	φ2 φ1
	-7	-6	11	65
	-7	-5	11	69
	-7	-4	11	73
	-7	-3	11	77
	-7	-2	11	81
	-7	-1	11	86
	-7	0	11	90
	0	-1	0	0
7	11	0	90	32
	0	0	1	90
	11	-7	0	0
	11	-7	1	4
	11	-7	2	9
	11	-7	3	13
	11	-7	4	17
	11	-7	5	21
	11	-7	6	25
	11	-7	7	28
	11	-7	8	32
	11	-7	9	35
	11	-7	10	37
	11	-7	11	40
	11	-7	12	43
11	7	0	90	56
	0	0	1	90
	7	-11	0	0
	7	-11	1	4
	7	-11	2	9
	7	-11	3	13
	7	-11	4	17
	7	-11	5	21
	7	-11	6	25
	7	-11	7	28
	7	-11	8	32
	7	-11	9	35
	7	-11	10	37
	7	-11	11	40
	7	-11	12	43
0	7	12	30	0
	0	-12	7	90
	1	-12	7	86
	1	0	0	0
	2	-12	7	82
	3	-12	7	78
	4	-12	7	74

H	K U	L V	Φ W	φ2 φ1
	5	-12	7	70
	6	-12	7	67
	7	-12	7	63
	8	-12	7	60
	9	-12	7	57
	10	-12	7	54
	11	-12	7	52
	12	-12	7	49
7	0	12	30	90
	-12	-12	7	49
	-12	-11	7	52
	-12	-10	7	54
	-12	-9	7	57
	-12	-8	7	60
	-12	-7	7	63
	-12	-6	7	67
	-12	-5	7	70
	-12	-4	7	74
	-12	-3	7	78
	-12	-2	7	82
	-12	-1	7	86
	-12	0	7	90
	0	-1	0	0
0	12	7	60	0
	0	-7	12	90
	1	-7	12	86
	1	0	0	0
	2	-7	12	82
	3	-7	12	78
	4	-7	12	74
	5	-7	12	70
	6	-7	12	67
	7	-7	12	63
	8	-7	12	60
	9	-7	12	57
	10	-7	12	54
	11	-7	12	52
	12	-7	12	49
12	0	7	60	90
	-7	-12	12	49
	-7	-11	12	52
	-7	-10	12	54
	-7	-9	12	57
	-7	-8	12	60
	-7	-7	12	63
	-7	-6	12	67

H	K U	L V	Φ W	φ2 φ1
	-7	-5	12	70
	-7	-4	12	74
	-7	-3	12	78
	-7	-2	12	82
	-7	-1	12	86
	-7	0	12	90
	0	-1	0	0
7	12	0	90	30
	0	0	1	90
	12	-7	0	0
	12	-7	1	4
	12	-7	2	8
	12	-7	3	12
	12	-7	4	16
	12	-7	5	20
	12	-7	6	23
	12	-7	7	27
	12	-7	8	30
	12	-7	9	33
	12	-7	10	36
	12	-7	11	38
	12	-7	12	41
12	7	0	90	60
	0	0	1	90
	7	-12	0	0
	7	-12	1	4
	7	-12	2	8
	7	-12	3	12
	7	-12	4	16
	7	-12	5	20
	7	-12	6	23
	7	-12	7	27
	7	-12	8	30
	7	-12	9	33
	7	-12	10	36
	7	-12	11	38
	7	-12	12	41
0	8	9	42	0
	0	-9	8	90
	1	-9	8	85
	1	0	0	0
	2	-9	8	81
	3	-9	8	76
	4	-9	8	72
	5	-9	8	67

H	K U	L V	Φ W	φ2 φ1
	6	-9	8	64
	7	-9	8	60
	8	-9	8	56
	9	-9	8	53
	10	-9	8	50
	11	-9	8	48
	12	-9	8	45
8	0	9	42	90
	-9	-12	8	45
	-9	-11	8	48
	-9	-10	8	50
	-9	-9	8	53
	-9	-8	8	56
	-9	-7	8	60
	-9	-6	8	64
	-9	-5	8	67
	-9	-4	8	72
	-9	-3	8	76
	-9	-2	8	81
	-9	-1	8	85
	-9	0	8	90
	0	-1	0	0
0	9	8	48	0
	0	-8	9	90
	1	-8	9	85
	1	0	0	0
	2	-8	9	81
	3	-8	9	76
	4	-8	9	72
	5	-8	9	67
	6	-8	9	64
	7	-8	9	60
	8	-8	9	56
	9	-8	9	53
	10	-8	9	50
	11	-8	9	48
	12	-8	9	45
9	0	8	48	90
	-8	-12	9	45
	-8	-11	9	48
	-8	-10	9	50
	-8	-9	9	53
	-8	-8	9	56
	-8	-7	9	60
	-8	-6	9	64
	-8	-5	9	67

H	K U	L V	Φ W	φ2 φ1
	-8	-4	9	72
	-8	-3	9	76
	-8	-2	9	81
	-8	-1	9	85
	-8	0	9	90
	0	-1	0	0
8	9	0	90	42
	0	0	1	90
	9	-8	0	0
	9	-8	1	5
	9	-8	2	9
	9	-8	3	14
	9	-8	4	18
	9	-8	5	23
	9	-8	6	26
	9	-8	7	30
	9	-8	8	34
	9	-8	9	37
	9	-8	10	40
	9	-8	11	42
	9	-8	12	45
9	8	0	90	46
	0	0	1	90
	8	-9	0	0
	8	-9	1	5
	8	-9	2	9
	8	-9	3	14
	8	-9	4	18
	8	-9	5	23
	8	-9	6	26
	8	-9	7	30
	8	-9	8	34
	8	-9	9	37
	8	-9	10	40
	8	-9	11	42
	8	-9	12	45
0	8	11	36	0
	0	-11	8	90
	1	-11	8	86
	1	0	0	0
	2	-11	8	82
	3	-11	8	78
	4	-11	8	74
	5	-11	8	70
	6	-11	8	66

H	K U	L V	Φ W	φ2 φ1
	7	-11	8	53
	8	-11	8	50
	9	-11	8	57
	10	-11	8	54
	11	-11	8	51
	12	-11	8	49
8	0	11	36	90
	-11	-12	8	49
	-11	-11	8	51
	-11	-10	8	54
	-11	-9	8	57
	-11	-8	8	60
	-11	-7	8	63
	-11	-6	8	66
	-11	-5	8	70
	-11	-4	8	74
	-11	-3	8	78
	-11	-2	8	82
	-11	-1	8	86
	-11	0	8	90
	0	-1	0	0
0	11	8	54	0
	0	-8	11	90
	1	-8	11	86
	1	0	0	0
	2	-8	11	82
	3	-8	11	78
	4	-8	11	74
	5	-8	11	70
	6	-8	11	66
	7	-8	11	63
	8	-8	11	60
	9	-8	11	57
	10	-8	11	54
	11	-8	11	51
	12	-8	11	49
11	0	8	54	90
	-8	-12	11	49
	-8	-11	11	51
	-8	-10	11	54
	-8	-9	11	57
	-8	-8	11	60
	-8	-7	11	63
	-8	-6	11	66
	-8	-5	11	70
	-8	-4	11	74

FROM HKL-UVW INTO EULER ANGLES

H	K / U	L / V	Φ / W	φ2 / φ1
0	8	11		
1	0	8	54	90
	-8	-3	11	78
	-8	-2	11	82
	-8	-1	11	86
	-8	0	11	90
	0	-1	0	0
8	11	0	90	36
	0	0	1	90
	11	-8	0	0
	11	-8	1	4
	11	-8	2	8
	11	-8	3	12
	11	-8	4	16
	11	-8	5	20
	11	-8	6	24
	11	-8	7	27
	11	-8	8	30
	11	-8	9	33
	11	-8	10	36
	11	-8	11	39
	11	-8	12	41
1	8	0	90	54
	0	0	1	90
	8	-11	0	0
	8	-11	1	4
	8	-11	2	8
	8	-11	3	12
	8	-11	4	16
	8	-11	5	20
	8	-11	6	24
	8	-11	7	27
	8	-11	8	30
	8	-11	9	33
	8	-11	10	36
	8	-11	11	39
	8	-11	12	41
0	9	10	42	0
	0	-10	9	90
	1	-10	9	86
	1	0	0	0
	2	-10	9	82
	3	-10	9	77

H	K / U	L / V	Φ / W	φ2 / φ1
	4	-10	9	73
	5	-10	9	70
	6	-10	9	66
	7	-10	9	63
	8	-10	9	59
	9	-10	9	56
	10	-10	9	53
	11	-10	9	51
	12	-10	9	48
9	0	10	42	90
	-10	-12	9	48
	-10	-11	9	51
	-10	-10	9	53
	-10	-9	9	56
	-10	-8	9	59
	-10	-7	9	63
	-10	-6	9	66
	-10	-5	9	70
	-10	-4	9	73
	-10	-3	9	77
	-10	-2	9	82
	-10	-1	9	86
	-10	0	9	90
	0	-1	0	0
0	10	9	48	0
	0	-9	10	90
	1	-9	10	86
	1	0	0	0
	2	-9	10	82
	3	-9	10	77
	4	-9	10	73
	5	-9	10	70
	6	-9	10	66
	7	-9	10	63
	8	-9	10	59
	9	-9	10	56
	10	-9	10	53
	11	-9	10	51
	12	-9	10	48
10	0	9	48	90
	-9	-12	10	48
	-9	-11	10	51
	-9	-10	10	53
	-9	-9	10	56
	-9	-8	10	59
	-9	-7	10	63

H	K / U	L / V	Φ / W	φ2 / φ1
	-9	-6	10	66
	-9	-5	10	70
	-9	-4	10	73
	-9	-3	10	77
	-9	-2	10	82
	-9	-1	10	86
	-9	0	10	90
	0	-1	0	0
9	10	0	90	42
	0	0	1	90
	10	-9	0	0
	10	-9	1	4
	10	-9	2	8
	10	-9	3	13
	10	-9	4	17
	10	-9	5	20
	10	-9	6	24
	10	-9	7	27
	10	-9	8	31
	10	-9	9	34
	10	-9	10	37
	10	-9	11	39
	10	-9	12	42
10	9	0	90	48
	0	0	1	90
	9	-10	0	0
	9	-10	1	4
	9	-10	2	8
	9	-10	3	13
	9	-10	4	17
	9	-10	5	20
	9	-10	6	24
	9	-10	7	27
	9	-10	8	31
	9	-10	9	34
	9	-10	10	37
	9	-10	11	39
	9	-10	12	42
0	9	11	39	0
	0	-11	9	90
	1	-11	9	86
	1	0	0	0
	2	-11	9	82
	3	-11	9	78
	4	-11	9	74

H	K / U	L / V	Φ / W	φ2 / φ1
	5	-11	9	71
	6	-11	9	67
	7	-11	9	64
	8	-11	9	61
	9	-11	9	58
	10	-11	9	55
	11	-11	9	52
	12	-11	9	50
9	0	11	39	90
	-11	-12	9	50
	-11	-11	9	52
	-11	-10	9	55
	-11	-9	9	58
	-11	-8	9	61
	-11	-7	9	64
	-11	-6	9	67
	-11	-5	9	71
	-11	-4	9	74
	-11	-3	9	76
	-11	-2	9	82
	-11	-1	9	86
	-11	0	9	90
	0	-1	0	0
0	11	9	51	0
	0	-9	11	90
	1	-9	11	86
	1	0	0	0
	2	-9	11	82
	3	-9	11	78
	4	-9	11	74
	5	-9	11	71
	6	-9	11	67
	7	-9	11	64
	8	-9	11	61
	9	-9	11	58
	10	-9	11	55
	11	-9	11	52
	12	-9	11	50
11	0	9	51	90
	-9	-12	11	50
	-9	-11	11	52
	-9	-10	11	55
	-9	-9	11	58
	-9	-8	11	61
	-9	-7	11	64
	-9	-6	11	67

H	K / U	L / V	Φ / W	φ2 / φ1
	-9	-5	11	71
	-9	-4	11	74
	-9	-3	11	78
	-9	-2	11	82
	-9	-1	11	86
	-9	0	11	90
	0	-1	0	0
9	11	0	90	39
	0	0	1	90
	11	-9	0	0
	11	-9	1	4
	11	-9	2	8
	11	-9	3	12
	11	-9	4	16
	11	-9	5	19
	11	-9	6	23
	11	-9	7	26
	11	-9	8	29
	11	-9	9	32
	11	-9	10	35
	11	-9	11	38
	11	-9	12	40
11	9	0	90	51
	0	0	1	90
	9	-11	0	0
	9	-11	1	4
	9	-11	2	8
	9	-11	3	12
	9	-11	4	16
	9	-11	5	19
	9	-11	6	23
	9	-11	7	26
	9	-11	8	29
	9	-11	9	32
	9	-11	10	35
	9	-11	11	38
	9	-11	12	40
0	10	11	42	0
	0	-11	10	90
	1	-11	10	66
	1	0	0	0
	2	-11	10	82
	3	-11	10	79
	4	-11	10	75
	5	-11	10	71

H	K / U	L / V	Φ / W	φ2 / φ1
	6	-11	10	68
	7	-11	10	65
	8	-11	10	62
	9	-11	10	59
	10	-11	10	56
	11	-11	10	54
	12	-11	10	51
10	0	11	42	90
	-11	-12	10	51
	-11	-11	10	54
	-11	-10	10	56
	-11	-9	10	59
	-11	-8	10	62
	-11	-7	10	65
	-11	-6	10	68
	-11	-5	10	71
	-11	-4	10	75
	-11	-3	10	79
	-11	-2	10	82
	-11	-1	10	86
	-11	0	10	90
	0	-1	0	0
0	11	10	48	0
	0	-10	11	90
	1	-10	11	86
	1	0	0	0
	2	-10	11	82
	3	-10	11	79
	4	-10	11	75
	5	-10	11	71
	6	-10	11	68
	7	-10	11	65
	8	-10	11	62
	9	-10	11	59
	10	-10	11	56
	11	-10	11	54
	12	-10	11	51
11	0	10	48	90
	-10	-12	11	51
	-10	-11	11	54
	-10	-10	11	56
	-10	-9	11	59
	-10	-8	11	62
	-10	-7	11	65
	-10	-6	11	68
	-10	-5	11	71

H	K / U	L / V	Φ / W	φ2 / φ1
	-10	-4	11	75
	-10	-3	11	79
	-10	-2	11	82
	-10	-1	11	86
	-10	0	11	90
	0	-1	0	0
10	11	0	90	42
	0	0	1	90
	11	-10	0	0
	11	-10	1	4
	11	-10	2	8
	11	-10	3	11
	11	-10	4	15
	11	-10	5	19
	11	-10	6	22
	11	-10	7	25
	11	-10	8	28
	11	-10	9	31
	11	-10	10	34
	11	-10	11	36
	11	-10	12	39
11	10	0	90	48
	0	0	1	90
	10	-11	0	0
	10	-11	1	4
	10	-11	2	8
	10	-11	3	11
	10	-11	4	15
	10	-11	5	19
	10	-11	6	22
	10	-11	7	25
	10	-11	8	28
	10	-11	9	31
	10	-11	10	34
	10	-11	11	36
	10	-11	12	39
0	11	12	43	0
	0	-12	11	90
	1	-12	11	66
	1	0	0	0
	2	-12	11	83
	3	-12	11	80
	4	-12	11	76
	5	-12	11	73
	6	-12	11	70

H	K / U	L / V	Φ / W	φ2 / φ1
	7	-12	11	67
	8	-12	11	64
	9	-12	11	61
	10	-12	11	58
	11	-12	11	56
	12	-12	11	54
11	0	12	43	90
	-12	-12	11	54
	-12	-11	11	56
	-12	-10	11	58
	-12	-9	11	61
	-12	-8	11	64
	-12	-7	11	67
	-12	-6	11	70
	-12	-5	11	73
	-12	-4	11	76
	-12	-3	11	80
	-12	-2	11	83
	-12	-1	11	86
	-12	0	11	90
	0	-1	0	0
0	12	11	47	0
	0	-11	12	90
	1	-11	12	86
	1	0	0	0
	2	-11	12	83
	3	-11	12	80
	4	-11	12	76
	5	-11	12	73
	6	-11	12	70
	7	-11	12	67
	8	-11	12	64
	9	-11	12	61
	10	-11	12	58
	11	-11	12	56
	12	-11	12	54
12	0	11	47	90
	-11	-12	12	54
	-11	-11	12	56
	-11	-10	12	58
	-11	-9	12	61
	-11	-8	12	64
	-11	-7	12	67
	-11	-6	12	70
	-11	-5	12	73
	-11	-4	12	76

TAB. V — FROM HKL-UVW INTO EULER ANGLES

H	K / U	L / V	Φ / W	φ2 / φ1
0	11	12		
12	0	11	47	90
	-11	-3	12	80
	-11	-2	12	83
	-11	-1	12	86
	-11	0	12	90
	0	-1	0	0
11	12	0	90	43
	0	0	1	90
	12	-11	0	0
	12	-11	1	4
	12	-11	2	7
	12	-11	3	10
	12	-11	4	14
	12	-11	5	17
	12	-11	6	20
	12	-11	7	23
	12	-11	8	26
	12	-11	9	29
	12	-11	10	32
	12	-11	11	34
	12	-11	12	36
12	11	0	90	47
	0	0	1	90
	11	-12	0	0
	11	-12	1	4
	11	-12	2	7
	11	-12	3	10
	11	-12	4	14
	11	-12	5	17
	11	-12	6	20
	11	-12	7	23
	11	-12	8	26
	11	-12	9	29
	11	-12	10	32
	11	-12	11	34
	11	-12	12	36
1	1	1	55	45
	-5	-7	12	85
	-5	-6	11	87
	-4	-7	11	81
	-4	-5	9	86
	-3	-8	11	75

H	K / U	L / V	Φ / W	φ2 / φ1
	-3	-7	10	77
	-3	-5	8	82
	-3	-4	7	85
	-2	-9	11	70
	-2	-7	9	72
	-2	-5	7	76
	-2	-3	5	83
	-1	-11	12	64
	-1	-10	11	65
	-1	-9	10	65
	-1	-8	9	66
	-1	-7	8	67
	-1	-6	7	68
	-1	-5	6	69
	-1	-4	5	71
	-1	-3	4	74
	-1	-2	3	79
	-1	-1	2	90
	0	-1	1	60
	1	-12	11	56
	1	-11	10	55
	1	-10	9	55
	1	-9	8	54
	1	-8	7	53
	1	-7	6	52
	1	-6	5	51
	1	-5	4	49
	1	-4	3	46
	1	-3	2	41
	1	-2	1	30
	1	-1	0	0
	2	-11	9	50
	2	-9	7	48
	2	-7	5	44
	2	-5	3	37
	2	-3	1	19
	3	-11	8	45
	3	-10	7	43
	3	-8	5	38
	3	-7	4	35
	3	-5	2	23
	3	-4	1	14
	4	-11	7	39
	4	-9	5	34
	4	-7	3	25
	4	-5	1	11
	5	-12	7	35
	5	-11	6	33
	5	-9	4	26

H	K / U	L / V	Φ / W	φ2 / φ1
	5	-8	3	22
	5	-7	2	16
	5	-6	1	9
	6	-11	5	27
	6	-7	1	8
	7	-12	5	25
	7	-11	4	21
	7	-10	3	17
	7	-9	2	12
	7	-8	1	7
	8	-11	3	15
	8	-9	1	6
	9	-11	2	10
	9	-10	1	5
	10	-11	1	5
	11	-12	1	4
1	1	2	35	45
	-10	-12	11	86
	-9	-11	10	85
	-8	-10	9	85
	-7	-11	9	80
	-7	-9	8	84
	-6	-8	7	83
	-5	-11	8	73
	-5	-9	7	77
	-5	-7	6	82
	-4	-10	7	71
	-4	-6	5	81
	-3	-11	7	65
	-3	-7	5	72
	-3	-5	4	78
	-2	-12	7	60
	-2	-8	5	64
	-2	-4	3	75
	-1	-11	6	56
	-1	-9	5	57
	-1	-7	4	59
	-1	-5	3	61
	-1	-3	2	68
	-1	-1	1	90
	0	-2	1	51
	1	-11	5	46
	1	-9	4	44
	1	-7	3	43
	1	-5	2	39
	1	-3	1	31
	1	-1	0	0

H	K / U	L / V	Φ / W	φ2 / φ1
	2	-12	5	41
	2	-8	3	36
	2	-4	1	22
	3	-11	4	35
	3	-7	2	26
	3	-5	1	17
	4	-10	3	28
	4	-6	1	14
	5	-11	3	25
	5	-9	2	19
	5	-7	1	12
	6	-8	1	10
	7	-11	2	15
	7	-9	1	9
	8	-10	1	8
	9	-11	1	7
	10	-12	1	6
1	2	1	66	27
	-2	-5	12	86
	-1	-5	11	84
	-1	-4	9	85
	-1	-3	7	87
	-1	-2	5	90
	0	-1	2	76
	1	-6	11	73
	1	-5	9	72
	1	-4	7	71
	1	-3	5	68
	1	-2	3	61
	1	-1	1	39
	2	-7	12	69
	2	-5	8	65
	2	-3	4	54
	2	-1	0	0
	3	-7	11	64
	3	-5	7	57
	3	-4	5	51
	3	-2	1	17
	4	-7	10	59
	4	-5	6	49
	4	-3	2	24
	5	-8	11	56
	5	-7	9	52
	5	-6	7	47
	5	-4	3	28
	5	-3	1	11
	6	-7	8	46
	6	-5	4	30

H	K / U	L / V	Φ / W	φ2 / φ1
	7	-9	11	50
	7	-8	9	45
	7	-6	5	31
	7	-5	3	21
	7	-4	1	8
	8	-9	10	44
	8	-7	6	33
	8	-5	2	13
	9	-10	11	44
	9	-8	7	33
	9	-7	5	26
	9	-5	1	6
	10	-11	12	43
	10	-9	8	34
	10	-7	4	20
	11	-10	9	35
	11	-9	7	29
	11	-8	5	22
	11	-7	3	14
	11	-6	1	5
	12	-11	10	35
	12	-7	2	9
2	1	1	66	63
	-4	-3	11	86
	-3	-5	11	75
	-3	-4	10	76
	-3	-2	8	87
	-2	-7	11	66
	-2	-5	9	70
	-2	-3	7	77
	-2	-1	5	90
	-1	-10	12	57
	-1	-9	11	58
	-1	-8	10	59
	-1	-7	9	59
	-1	-6	8	61
	-1	-5	7	62
	-1	-4	6	65
	-1	-3	5	68
	-1	-2	4	73
	-1	-1	3	82
	0	-1	1	51
	1	-12	10	44
	1	-11	9	44
	1	-10	8	43
	1	-9	7	42
	1	-8	6	41
	1	-7	5	39

H	K / U	L / V	Φ / W	φ2 / φ1
	1	-6	4	37
	1	-5	3	34
	1	-4	2	29
	1	-3	1	19
	1	-2	0	0
	2	-11	7	36
	2	-9	5	31
	2	-7	3	25
	2	-5	1	12
	3	-11	5	26
	3	-10	4	23
	3	-8	2	14
	3	-7	1	8
	4	-11	3	16
	4	-9	1	6
	5	-12	2	10
	5	-11	1	5
1	1	3	25	45
	-10	-11	7	88
	-9	-12	7	83
	-7	-11	6	79
	-7	-8	5	87
	-6	-9	5	80
	-5	-7	4	81
	-4	-11	5	67
	-4	-5	3	84
	-3	-12	5	62
	-3	-9	4	66
	-3	-3	2	90
	-2	-7	3	63
	-1	-11	4	53
	-1	-8	3	55
	-1	-5	2	59
	-1	-2	1	73
	0	-3	1	48
	1	-10	3	42
	1	-7	2	40
	1	-4	1	34
	1	-1	0	0
	2	-11	3	37
	2	-5	1	25
	3	-9	2	29
	3	-6	1	20
	4	-7	1	17
	5	-11	2	23
	5	-8	1	14
	6	-9	1	12

H	K / U	L / V	Φ / W	φ2 / φ1
	7	-10	1	11
	8	-11	1	10
	9	-12	1	9
1	3	1	72	16
	-1	-3	10	90
	0	-1	3	84
	1	-4	11	79
	1	-3	6	77
	1	-2	5	73
	1	-1	2	59
	2	-3	7	69
	2	-1	1	25
	3	-5	12	71
	3	-4	9	66
	3	-2	3	42
	3	-1	0	0
	4	-5	11	65
	4	-3	5	45
	5	-4	7	51
	5	-3	4	36
	5	-2	1	11
	6	-5	9	52
	7	-6	11	53
	7	-5	8	46
	7	-4	5	34
	7	-3	2	15
	8	-5	7	39
	8	-3	1	7
	9	-7	12	49
	9	-5	6	32
	9	-4	3	18
	10	-7	11	45
	11	-7	10	40
	11	-6	7	31
	11	-5	4	19
	11	-4	1	5
	12	-7	9	35
	12	-5	3	14
3	1	1	72	72
	-3	-2	11	85
	-3	-1	10	90
	-2	-5	11	70
	-2	-3	9	77
	-2	-1	7	88
	-1	-9	12	57
	-1	-8	11	58
	-1	-7	10	59

H	K / U	L / V	Φ / W	φ2 / φ1
	-1	-6	9	60
	-1	-5	8	62
	-1	-4	7	65
	-1	-3	6	68
	-1	-2	5	73
	-1	-1	4	81
	0	-1	1	48
	1	-12	9	39
	1	-11	8	38
	1	-10	7	37
	1	-9	6	35
	1	-8	5	34
	1	-7	4	31
	1	-6	3	28
	1	-5	2	23
	1	-4	1	14
	1	-3	0	0
	2	-11	5	25
	2	-9	3	19
	2	-7	1	8
	3	-11	2	10
	3	-10	1	6
1	1	4	19	45
	-9	-11	5	85
	-8	-12	5	79
	-7	-9	4	83
	-5	-11	4	71
	-5	-7	3	81
	-4	-8	3	73
	-3	-5	2	77
	-2	-10	3	58
	-2	-2	1	90
	-1	-11	3	52
	-1	-7	2	55
	-1	-3	1	65
	0	-4	1	47
	1	-9	2	40
	1	-5	1	35
	1	-1	0	0
	2	-6	1	28
	3	-11	2	31
	3	-7	1	23
	4	-8	1	19
	5	-9	1	17
	6	-10	1	15
	7	-11	1	13
	8	-12	1	12

AB. V

FROM HKL-UVW INTO EULER ANGLES

V – 18

H	K / U	L / V	Φ / W	φ_2 / φ_1
1	1	4		
1	4	1	76	14
	0	-1	4	87
	1	-3	11	81
	1	-2	7	79
	1	-1	3	69
	2	-3	10	75
	2	-1	2	43
	3	-2	5	57
	3	-1	1	18
	4	-3	8	61
	4	-1	0	0
	5	-4	11	63
	5	-3	7	52
	5	-2	3	30
	7	-4	9	50
	7	-3	5	34
	7	-2	1	8
	8	-5	12	54
	8	-3	4	26
	9	-5	11	49
	9	-4	7	37
	10	-3	2	11
	11	-5	9	38
	11	-4	5	24
	11	-3	1	5
	12	-5	8	33
4	1	1	76	76
	-2	-3	11	78
	-2	-1	9	87
	-1	-8	12	59
	-1	-7	11	60
	-1	-6	10	62
	-1	-5	9	64
	-1	-4	8	66
	-1	-3	7	70
	-1	-2	6	75
	-1	-1	5	82
	0	-1	1	47
	1	-12	8	35
	1	-11	7	33
	1	-10	6	32
	1	-9	5	30
	1	-8	4	27
	1	-7	3	24
	1	-6	2	19

H	K / U	L / V	Φ / W	φ_2 / φ_1
	1	-5	1	11
	1	-4	0	0
	2	-11	3	15
	2	-9	1	6
1	1	5	16	45
	-9	-11	4	85
	-7	-8	3	86
	-5	-10	3	72
	-5	-5	2	90
	-4	-11	3	66
	-3	-7	2	69
	-2	-3	1	79
	-1	-9	2	52
	-1	-4	1	60
	0	-5	1	46
	1	-11	2	41
	1	-6	1	37
	1	-1	0	0
	2	-7	1	30
	3	-8	1	25
	4	-9	1	22
	5	-10	1	19
	6	-11	1	17
	7	-12	1	15
1	5	1	79	11
	0	-1	5	88
	1	-2	9	81
	1	-1	4	74
	2	-1	3	55
	3	-2	7	65
	3	-1	2	33
	4	-3	11	68
	4	-1	1	14
	5	-3	10	62
	5	-2	5	44
	5	-1	0	0
	7	-3	8	48
	7	-2	3	23
	8	-3	7	40
	9	-4	11	49
	9	-2	1	6
	10	-3	5	26
	11	-4	9	38
	11	-3	4	20

H	K / U	L / V	Φ / W	φ_2 / φ_1
5	1	1	79	79
	-2	-1	11	87
	-1	-7	12	61
	-1	-6	11	63
	-1	-5	10	65
	-1	-4	9	68
	-1	-3	8	71
	-1	-2	7	76
	-1	-1	6	83
	0	-1	1	46
	1	-12	7	31
	1	-11	6	29
	1	-10	5	27
	1	-9	4	24
	1	-8	3	21
	1	-7	2	16
	1	-6	1	10
	1	-5	0	0
	2	-11	1	5
1	1	6	13	45
	-8	-10	3	84
	-7	-11	3	76
	-5	-7	2	81
	-3	-9	2	64
	-3	-3	1	90
	-2	-4	1	72
	-1	-11	2	51
	-1	-5	1	57
	0	-6	1	46
	1	-7	1	38
	1	-1	0	0
	2	-8	1	32
	3	-9	1	27
	4	-10	1	24
	5	-11	1	21
	6	-12	1	19
1	6	1	81	9
	0	-1	6	88
	1	-2	11	83
	1	-1	5	77
	2	-1	4	62
	3	-2	9	70
	3	-1	3	44
	4	-1	2	26
	5	-2	7	53

H	K / U	L / V	Φ / W	φ_2 / φ_1
	5	-1	1	11
	6	-1	0	0
	7	-3	11	56
	7	-2	5	35
	8	-3	10	50
	9	-2	3	18
	10	-3	8	38
	11	-3	7	32
	11	-2	1	5
6	1	1	81	81
	-1	-6	12	65
	-1	-5	11	67
	-1	-4	10	70
	-1	-3	9	73
	-1	-2	8	77
	-1	-1	7	83
	0	-1	1	46
	1	-12	6	27
	1	-11	5	25
	1	-10	4	22
	1	-9	3	19
	1	-8	2	14
	1	-7	1	8
	1	-6	0	0
1	1	7	11	45
	-10	-11	3	87
	-7	-7	2	90
	-5	-9	2	74
	-3	-11	2	61
	-3	-4	1	82
	-2	-5	1	67
	-1	-6	1	55
	0	-7	1	46
	1	-8	1	38
	1	-1	0	0
	2	-9	1	33
	3	-10	1	29
	4	-11	1	25
	5	-12	1	23
1	7	1	82	8
	0	-1	7	89
	1	-1	6	79
	2	-1	5	67
	3	-2	11	74
	3	-1	4	52

H	K / U	L / V	Φ / W	φ_2 / φ_1
	4	-1	3	36
	5	-2	9	60
	5	-1	2	22
	6	-1	1	9
	7	-2	7	44
	7	-1	0	0
	9	-2	5	29
	10	-3	11	47
	11	-3	10	42
	11	-2	3	15
7	1	1	82	82
	-1	-5	12	68
	-1	-4	11	71
	-1	-3	10	74
	-1	-2	9	79
	-1	-1	8	84
	0	-1	1	46
	1	-12	5	23
	1	-11	4	20
	1	-10	3	17
	1	-9	2	13
	1	-8	1	7
	1	-7	0	0
1	1	8	10	45
	-7	-9	2	83
	-5	-11	2	70
	-4	-4	1	90
	-3	-5	1	76
	-2	-6	1	64
	-1	-7	1	54
	0	-8	1	45
	1	-9	1	39
	1	-1	0	0
	2	-10	1	34
	3	-11	1	30
	4	-12	1	27
1	8	1	83	7
	0	-1	8	69
	1	-1	7	81
	2	-1	6	71
	3	-1	5	58
	4	-1	4	45
	5	-2	11	65
	5	-1	3	31
	6	-1	2	18

H	K / U	L / V	Φ / W	φ_2 / φ_1
	7	-2	9	52
	7	-1	1	8
	8	-1	0	0
	9	-2	7	38
	11	-2	5	24
8	1	1	83	83
	-1	-4	12	72
	-1	-3	11	76
	-1	-2	10	80
	-1	-1	9	85
	0	-1	1	45
	1	-12	4	19
	1	-11	3	15
	1	-10	2	11
	1	-9	1	6
	1	-8	0	0
1	1	9	9	45
	-9	-9	2	90
	-7	-11	2	78
	-4	-5	1	84
	-3	-6	1	72
	-2	-7	1	61
	-1	-8	1	52
	0	-9	1	45
	1	-10	1	40
	1	-1	0	0
	2	-11	1	35
	3	-12	1	31
1	9	1	84	6
	0	-1	9	89
	1	-1	8	82
	2	-1	7	73
	3	-1	6	63
	4	-1	5	51
	5	-1	4	38
	6	-1	3	26
	7	-2	11	57
	7	-1	2	16
	8	-1	1	7
	9	-2	9	45
	9	-1	0	0
	11	-2	7	32
9	1	1	84	84
	-1	-3	12	77

H	K / U	L / V	Φ / W	φ_2 / φ_1
	-1	-2	11	80
	-1	-1	10	85
	0	-1	1	45
	1	-12	3	14
	1	-11	2	10
	1	-10	1	6
	1	-9	0	0
1	1	10	8	45
	-9	-11	2	84
	-5	-5	1	90
	-4	-6	1	79
	-3	-7	1	68
	-2	-8	1	59
	-1	-9	1	52
	0	-10	1	45
	1	-11	1	40
	1	-1	0	0
	2	-12	1	36
1	10	1	84	6
	0	-1	10	89
	1	-1	9	83
	2	-1	8	75
	3	-1	7	66
	4	-1	6	56
	5	-1	5	45
	6	-1	4	34
	7	-1	3	23
	8	-1	2	14
	9	-2	11	50
	9	-1	1	6
	10	-1	0	0
	11	-2	9	39
10	1	1	84	84
	-1	-2	12	81
	-1	-1	11	85
	0	-1	1	45
	1	-12	2	9
	1	-11	1	5
	1	-10	0	0
1	1	11	7	45
	-11	-11	2	90
	-5	-6	1	85
	-4	-7	1	75

H	K / U	L / V	Φ / W	φ_2 / φ_1
	-3	-8	1	66
	-2	-9	1	58
	-1	-10	1	51
	0	-11	1	45
	1	-12	1	40
	1	-1	0	0
1	11	1	85	5
	0	-1	11	90
	1	-1	10	84
	2	-1	9	77
	3	-1	8	69
	4	-1	7	60
	5	-1	6	50
	6	-1	5	40
	7	-1	4	30
	8	-1	3	20
	9	-1	2	13
	10	-1	1	6
	11	-2	11	45
	11	-1	0	0
11	1	1	85	85
	-1	-1	12	86
	0	-1	1	45
	1	-12	1	5
	1	-11	0	0
1	1	12	7	45
	-6	-6	1	90
	-5	-7	1	81
	-4	-8	1	72
	-3	-9	1	64
	-2	-10	1	56
	-1	-11	1	50
	0	-12	1	45
	1	-1	0	0
1	12	1	85	5
	0	-1	12	90
	1	-1	11	84
	2	-1	10	78
	3	-1	9	71
	4	-1	8	63
	5	-1	7	54
	6	-1	6	45
	7	-1	5	35
	8	-1	4	26

TAB. V FROM HKL-UVW INTO EULER ANGLES

H	K / U	L / V	Φ / W	φ2 / φ1
1	1	12		
1	12	1	85	5
	9	-1	3	18
	10	-1	2	11
	11	-1	1	5
	12	-1	0	0
12	1	1	85	85
	0	-1	1	45
	1	-12	0	0
1	2	2	48	27
	-4	-9	11	88
	-2	-11	12	79
	-2	-10	11	80
	-2	-9	10	81
	-2	-8	9	82
	-2	-7	8	83
	-2	-6	7	85
	-2	-5	6	87
	-2	-4	5	90
	0	-1	1	72
	2	-12	11	64
	2	-11	10	63
	2	-10	9	63
	2	-9	8	62
	2	-8	7	60
	2	-7	6	59
	2	-6	5	56
	2	-5	4	53
	2	-4	3	48
	2	-3	2	41
	2	-2	1	27
	2	-1	0	0
	4	-11	9	55
	4	-9	7	51
	4	-7	5	45
	4	-5	3	35
	4	-3	1	15
	6	-11	8	46
	6	-10	7	44
	6	-8	5	37
	6	-7	4	32
	6	-5	2	19
	6	-4	1	11
	8	-11	7	38

H	K / U	L / V	Φ / W	φ2 / φ1
	8	-9	5	31
	8	-7	3	21
	8	-5	1	8
	10	-12	7	33
	10	-11	6	30
	10	-9	4	22
	10	-8	3	18
	10	-7	2	13
	10	-6	1	7
	12	-11	5	23
	12	-7	1	6
2	1	2	48	63
	-8	-6	11	83
	-7	-10	12	70
	-7	-8	11	75
	-7	-6	10	81
	-7	-4	9	88
	-6	-10	11	67
	-5	-12	11	60
	-5	-8	9	68
	-5	-6	8	74
	-5	-4	7	82
	-4	-10	9	59
	-4	-6	7	69
	-4	-2	5	90
	-3	-10	8	55
	-3	-8	7	58
	-3	-4	5	72
	-3	-2	4	85
	-2	-10	7	49
	-2	-6	5	56
	-2	-2	3	77
	-1	-12	7	42
	-1	-10	6	43
	-1	-8	5	45
	-1	-6	4	47
	-1	-4	3	52
	-1	-2	2	63
	0	-2	1	37
	1	-12	5	31
	1	-10	4	30
	1	-8	3	28
	1	-6	2	25
	1	-4	1	18
	1	-2	0	0
	2	-10	3	22
	2	-6	1	12
	3	-10	2	15

H	K / U	L / V	Φ / W	φ2 / φ1
	3	-8	1	9
	4	-10	1	7
	5	-12	1	6
2	2	1	71	45
	-2	-3	10	86
	-1	-5	12	77
	-1	-4	10	79
	-1	-3	8	81
	-1	-2	6	84
	-1	-1	4	90
	0	-1	2	72
	1	-7	12	66
	1	-6	10	65
	1	-5	8	63
	1	-4	6	61
	1	-3	4	56
	1	-2	2	45
	1	-1	0	0
	2	-7	10	59
	2	-5	6	52
	2	-3	2	31
	3	-8	10	54
	3	-7	8	50
	3	-5	4	37
	3	-4	2	23
	4	-9	10	49
	4	-7	6	39
	4	-5	2	18
	5	-11	12	48
	5	-9	8	41
	5	-8	6	35
	5	-7	4	27
	5	-6	2	15
	6	-11	10	41
	6	-7	2	13
	7	-12	10	38
	7	-11	8	34
	7	-10	6	28
	7	-9	4	21
	7	-8	2	11
	8	-11	6	25
	8	-9	2	10
	9	-11	4	17
	9	-10	2	9
	10	-11	2	8
	11	-12	2	7

H	K / U	L / V	Φ / W	φ2 / φ1
1	2	3	37	27
	-5	-11	9	88
	-3	-9	7	83
	-3	-6	5	90
	-2	-11	6	77
	-2	-5	4	86
	-1	-10	7	73
	-1	-7	5	75
	-1	-4	3	80
	0	-3	2	68
	1	-11	7	64
	1	-8	5	62
	1	-5	3	58
	1	-2	1	43
	2	-7	4	54
	2	-1	0	0
	3	-12	7	56
	3	-9	5	51
	3	-3	1	23
	4	-11	6	50
	4	-5	2	30
	5	-7	3	33
	5	-4	1	15
	6	-9	4	35
	7	-11	5	37
	7	-8	3	27
	7	-5	1	11
	8	-7	2	18
	9	-12	5	32
	9	-6	1	9
	10	-11	4	26
	11	-10	3	19
	11	-7	1	7
	12	-9	2	13
2	1	3	37	63
	-12	-9	11	82
	-11	-8	10	82
	-10	-7	9	83
	-9	-12	10	68
	-9	-6	8	84
	-8	-11	9	67
	-8	-5	7	86
	-7	-10	8	67
	-7	-4	6	87
	-6	-9	7	65
	-6	-3	5	90

H	K / U	L / V	Φ / W	φ2 / φ1
	-5	-11	7	57
	-5	-8	6	64
	-4	-7	5	62
	-3	-9	5	51
	-3	-6	4	59
	-2	-11	5	43
	-2	-5	3	55
	-1	-10	4	38
	-1	-7	3	41
	-1	-4	2	47
	-1	-1	1	75
	0	-3	1	32
	1	-11	3	26
	1	-8	2	24
	1	-5	1	19
	1	-2	0	0
	2	-7	1	13
	3	-12	2	15
	3	-9	1	10
	4	-11	1	8
1	3	2	58	18
	-1	-7	11	84
	-1	-5	8	86
	-1	-3	5	90
	0	-2	3	80
	1	-7	10	75
	1	-5	7	73
	1	-3	4	68
	1	-1	1	43
	2	-8	11	71
	2	-4	5	62
	3	-7	9	65
	3	-5	6	58
	3	-1	0	0
	4	-6	7	56
	4	-2	1	15
	5	-9	11	60
	5	-7	8	54
	5	-3	2	23
	6	-8	9	52
	6	-4	3	27
	7	-9	10	51
	7	-5	4	30
	7	-3	1	9
	8	-10	11	50
	8	-6	5	32
	9	-11	12	50
	9	-7	6	33

H	K / U	L / V	Φ / W	φ2 / φ1
	9	-5	3	19
	10	-8	7	35
	10	-4	1	6
	11	-9	8	35
	11	-7	5	25
	11	-5	2	11
	12	-10	9	36
3	1	2	58	72
	-7	-3	12	87
	-6	-4	11	82
	-5	-9	12	64
	-5	-7	11	69
	-5	-3	9	83
	-4	-10	11	58
	-4	-6	9	67
	-4	-2	7	86
	-3	-11	10	51
	-3	-7	8	59
	-3	-5	7	65
	-3	-1	5	90
	-2	-12	9	45
	-2	-8	7	50
	-2	-4	5	62
	-1	-11	7	39
	-1	-9	6	41
	-1	-7	5	43
	-1	-5	4	47
	-1	-3	3	55
	-1	-1	2	75
	0	-2	1	32
	1	-11	4	24
	1	-9	3	22
	1	-7	2	19
	1	-5	1	13
	1	-3	0	0
	2	-12	3	16
	2	-8	1	8
	3	-11	1	6
2	3	1	74	34
	-1	-3	11	86
	-1	-2	8	88
	0	-1	3	60
	1	-4	10	74
	1	-3	7	71
	1	-2	4	65
	1	-1	1	37
	2	-5	11	69

H	K / U	L / V	Φ / W	φ2 / φ1
	2	-3	5	57
	3	-5	9	61
	3	-4	6	53
	3	-2	0	0
	4	-5	7	50
	4	-3	1	12
	5	-7	11	55
	5	-6	8	48
	5	-4	2	18
	6	-7	9	46
	6	-5	3	22
	7	-8	10	45
	7	-6	4	24
	7	-5	1	7
	8	-9	11	44
	8	-7	5	26
	9	-10	12	44
	9	-8	6	28
	9	-7	3	15
	10	-9	7	29
	10	-7	1	5
	11	-10	8	29
	11	-9	5	20
	11	-8	2	9
	12	-11	9	30
3	2	1	74	56
	-2	-3	12	84
	-1	-4	11	76
	-1	-3	9	78
	-1	-2	7	81
	-1	-1	5	87
	0	-1	2	68
	1	-7	11	61
	1	-6	9	59
	1	-5	7	57
	1	-4	5	53
	1	-3	3	46
	1	-2	1	25
	2	-9	12	55
	2	-7	8	50
	2	-5	4	38
	2	-3	0	0
	3	-10	11	49
	3	-8	7	41
	3	-7	5	35
	3	-5	1	10
	4	-11	10	42
	4	-9	6	33

H	K / U	L / V	Φ / W	φ2 / φ1
	4	-7	2	14
	5	-12	9	36
	5	-11	7	31
	5	-9	3	17
	5	-8	1	6
	6	-11	4	18
	7	-12	3	13
	7	-11	1	5
1	2	4	29	27
	-4	-12	7	83
	-4	-8	5	90
	-2	-11	6	76
	-2	-9	5	78
	-2	-7	4	81
	-2	-5	3	85
	0	-2	1	66
	2	-11	5	57
	2	-9	4	55
	2	-7	3	51
	2	-5	2	45
	2	-3	1	33
	2	-1	0	0
	4	-12	5	49
	4	-8	3	41
	4	-4	1	21
	6	-11	4	39
	6	-7	2	26
	6	-5	1	15
	8	-10	3	23
	8	-6	1	12
	10	-11	3	24
	10	-9	2	18
	10	-7	1	10
	12	-8	1	8
2	1	4	29	63
	-11	-10	8	76
	-11	-6	7	88
	-10	-8	7	79
	-9	-10	7	71
	-8	-12	7	63
	-8	-4	5	90
	-7	-10	6	65
	-7	-6	5	78
	-6	-8	5	66
	-5	-6	4	69
	-4	-12	5	49

AB. V — FROM HKL-UVW INTO EULER ANGLES — V - 20

H	K U	L V	Φ W	φ2 φ1
1	2	4		
2	1	4	29	63
	-4	-4	3	74
	-3	-10	4	47
	-3	-2	2	84
	-2	-8	3	44
	-1	-10	3	36
	-1	-6	2	40
	-1	-2	1	57
	0	-4	1	30
	1	-10	2	24
	1	-6	1	19
	1	-2	0	0
	2	-8	1	14
	3	-10	1	11
	4	-12	1	9
1	4	2	64	14
	0	-1	2	84
	2	-6	11	74
	2	-5	9	73
	2	-4	7	69
	2	-3	5	64
	2	-2	3	54
	2	-1	1	27
	4	-7	12	67
	4	-5	8	60
	4	-3	4	44
	4	-1	0	0
	6	-7	11	58
	6	-5	7	48
	6	-4	5	39
	6	-2	1	10
	8	-7	10	50
	8	-5	6	37
	8	-3	2	15
	10	-8	11	46
	10	-7	9	41
	10	-6	7	35
	10	-4	3	17
	10	-3	1	6
	12	-7	8	34
	12	-5	4	19
4	1	2	64	76
	-5	-4	12	79
	-5	-2	11	87

H	K U	L V	Φ W	φ2 φ1
	-4	-6	11	68
	-4	-2	9	84
	-3	-10	11	54
	-3	-8	10	58
	-3	-4	8	70
	-3	-2	7	81
	-2	-10	9	47
	-2	-6	7	56
	-2	-2	5	75
	-1	-12	8	38
	-1	-10	7	39
	-1	-8	6	42
	-1	-6	5	45
	-1	-4	4	51
	-1	-2	3	63
	0	-2	1	30
	1	-12	4	21
	1	-10	3	19
	1	-8	2	16
	1	-6	1	10
	1	-4	0	0
	2	-10	1	6
2	4	1	77	27
	-1	-2	10	90
	0	-1	4	84
	1	-3	10	78
	1	-2	6	74
	1	-1	2	57
	2	-3	8	69
	2	-1	0	0
	3	-4	10	66
	3	-2	2	30
	4	-5	12	65
	4	-3	4	40
	5	-4	6	44
	5	-3	2	19
	6	-5	8	47
	7	-6	10	49
	7	-5	6	36
	7	-4	2	14
	8	-7	12	50
	8	-5	4	24
	9	-7	10	43
	9	-5	2	11
	10	-7	8	34
	11	-8	10	37
	11	-7	6	25
	11	-6	2	9

H	K U	L V	Φ W	φ2 φ1
	12	-7	4	16
4	2	1	77	63
	-2	-1	10	90
	-1	-4	12	76
	-1	-3	10	78
	-1	-2	8	81
	-1	-1	6	86
	0	-1	2	66
	1	-8	12	58
	1	-7	10	57
	1	-6	8	55
	1	-5	6	51
	1	-4	4	46
	1	-3	2	33
	1	-2	0	0
	2	-9	10	49
	2	-7	6	41
	2	-5	2	21
	3	-11	10	43
	3	-10	8	39
	3	-8	4	26
	3	-7	2	15
	4	-11	6	28
	4	-9	2	12
	5	-12	4	18
	5	-11	2	10
1	2	5	24	27
	-3	-11	5	80
	-2	-9	4	77
	-1	-12	5	70
	-1	-7	3	73
	-1	-2	1	90
	0	-5	2	65
	1	-8	3	59
	1	-3	1	48
	2	-11	4	56
	2	-1	0	0
	3	-4	1	29
	4	-7	2	36
	5	-10	3	39
	5	-5	1	20
	7	-11	3	33
	7	-6	1	15
	8	-9	2	24
	9	-7	1	12
	11	-8	1	10

H	K U	L V	Φ W	φ2 φ1
	12	-11	2	17
2	1	5	24	63
	-12	-11	7	75
	-11	-8	6	81
	-9	-7	5	80
	-8	-9	5	70
	-7	-11	5	61
	-7	-6	4	77
	-5	-10	4	56
	-5	-5	3	73
	-4	-7	3	59
	-3	-4	2	65
	-2	-11	3	39
	-2	-1	1	90
	-1	-8	2	36
	-1	-3	1	48
	0	-5	1	29
	1	-12	2	24
	1	-7	1	20
	1	-2	0	0
	2	-9	1	15
	3	-11	1	12
1	5	2	69	11
	0	-2	5	86
	1	-5	12	81
	1	-3	7	78
	1	-1	2	61
	2	-4	9	74
	3	-5	11	72
	3	-1	1	19
	4	-2	3	37
	5	-3	5	44
	5	-1	0	0
	6	-4	7	48
	7	-5	9	51
	7	-3	4	30
	8	-6	11	53
	8	-2	1	7
	9	-5	8	41
	10	-4	5	27
	11	-7	12	47
	11	-5	7	33
	11	-3	2	11
5	1	2	69	79
	-4	-2	11	84
	-3	-7	11	62

H	K U	L V	Φ W	φ2 φ1
	-3	-5	10	68
	-3	-1	8	87
	-2	-12	11	46
	-2	-8	9	52
	-2	-4	7	65
	-1	-11	8	39
	-1	-9	7	41
	-1	-7	6	44
	-1	-5	5	49
	-1	-3	4	57
	-1	-1	3	76
	0	-2	1	29
	1	-11	3	16
	1	-9	2	13
	1	-7	1	9
	1	-5	0	0
	2	-12	1	5
2	5	1	79	22
	0	-1	5	86
	1	-2	8	78
	1	-1	3	67
	2	-3	11	75
	2	-1	1	25
	3	-2	4	49
	4	-3	7	56
	5	-4	10	59
	5	-3	5	41
	5	-2	0	0
	7	-5	11	53
	7	-4	6	37
	7	-3	1	8
	8	-5	9	45
	9	-5	7	35
	9	-4	2	12
	11	-6	8	33
	11	-5	3	14
	12	-7	11	39
	12	-5	1	4
5	2	1	79	68
	-2	-1	12	89
	-1	-3	11	76
	-1	-2	9	81
	-1	-1	7	86
	0	-1	2	65
	1	-8	11	55
	1	-7	9	53
	1	-6	7	50

H	K U	L V	Φ W	φ2 φ1
	1	-5	5	45
	1	-4	3	37
	1	-3	1	18
	2	-11	12	48
	2	-9	8	42
	2	-7	4	29
	2	-5	0	0
	3	-11	7	32
	3	-10	5	26
	3	-8	1	7
	4	-11	2	10
1	2	6	20	27
	-6	-12	5	90
	-2	-11	4	75
	-2	-8	3	78
	-2	-5	2	86
	0	-3	1	65
	2	-10	3	54
	2	-7	2	49
	2	-4	1	39
	2	-1	0	0
	4	-11	3	45
	4	-5	1	26
	6	-9	2	31
	6	-6	1	20
	8	-7	1	16
	10	-11	2	22
	10	-8	1	13
	12	-9	1	11
2	1	6	20	63
	-12	-6	5	90
	-11	-8	5	81
	-9	-12	5	65
	-9	-6	4	83
	-7	-10	4	63
	-7	-4	3	87
	-5	-8	3	60
	-4	-10	3	50
	-3	-6	2	55
	-2	-2	1	73
	-1	-10	2	34
	-1	-4	1	42
	0	-6	1	28
	1	-8	1	21
	1	-2	0	0
	2	-10	1	16

H	K U	L V	Φ W	φ2 φ1
	3	-12	1	13
1	6	2	72	9
	0	-1	3	87
	2	-4	11	77
	2	-3	8	74
	2	-2	5	66
	2	-1	2	45
	4	-3	7	59
	4	-1	1	14
	6	-5	12	62
	6	-4	9	55
	6	-2	3	27
	6	-1	0	0
	8	-5	11	53
	8	-3	5	32
	10	-4	7	35
	10	-3	4	22
	10	-2	1	6
	12	-5	9	37
6	1	2	72	81
	-3	-4	11	73
	-3	-2	10	82
	-2	-10	11	51
	-2	-6	9	59
	-2	-2	7	77
	-1	-12	9	39
	-1	-10	8	41
	-1	-8	7	44
	-1	-6	6	48
	-1	-4	5	54
	-1	-2	4	67
	0	-2	1	26
	1	-12	3	15
	1	-10	2	12
	1	-8	1	7
	1	-6	0	0
2	6	1	81	18
	0	-1	6	87
	1	-2	10	81
	1	-1	4	73
	2	-1	2	42
	3	-2	6	60
	3	-1	0	0
	4	-3	10	65
	5	-3	8	55
	5	-2	2	21

H	K U	L V	Φ W	φ2 φ1
	7	-4	10	52
	7	-3	4	28
	8	-3	2	13
	9	-5	12	50
	9	-4	6	32
	11	-5	8	34
	11	-4	2	10
	12	-5	6	25
6	2	1	81	72
	-1	-3	12	78
	-1	-2	10	81
	-1	-1	8	86
	0	-1	2	65
	1	-9	12	54
	1	-8	10	52
	1	-7	8	49
	1	-6	6	45
	1	-5	4	39
	1	-4	2	26
	1	-3	0	0
	2	-11	10	42
	2	-9	6	34
	2	-7	2	16
	3	-11	4	20
	3	-10	2	11
1	2	7	18	27
	-1	-10	3	70
	-1	-3	1	82
	0	-7	2	65
	1	-11	3	59
	1	-4	1	51
	2	-1	0	0
	3	-5	1	34
	4	-9	2	41
	5	-6	1	25
	7	-7	1	19
	8	-11	2	29
	9	-8	1	16
	11	-9	1	13
2	1	7	18	63
	-12	-11	5	75
	-11	-6	4	88
	-9	-10	4	69
	-8	-5	3	85
	-7	-7	3	72

TAB. V FROM HKL-UVW INTO EULER ANGLES

H	K U	L V	Φ W	φ2 φ1
1	2	7		
2	1	7	18	63
	-5	-11	3	52
	-5	-4	2	78
	-3	-8	2	49
	-2	-3	1	61
	-1	-12	2	33
	-1	-5	1	39
	0	-7	1	28
	1	-9	1	21
	1	-2	0	0
	2	-11	1	17
1	7	2	74	8
	0	-2	7	88
	1	-3	10	82
	1	-1	3	70
	3	-1	2	34
	4	-2	5	51
	5	-3	8	57
	5	-1	1	12
	6	-4	11	60
	7	-3	7	45
	7	-1	0	0
	8	-2	3	21
	10	-4	9	42
	11	-5	12	47
	11	-3	5	25
	12	-2	1	5
7	1	2	74	82
	-3	-1	11	87
	-2	-8	11	56
	-2	-4	9	69
	-1	-11	9	41
	-1	-9	8	43
	-1	-7	7	47
	-1	-5	6	52
	-1	-3	5	61
	-1	-1	4	78
	0	-2	1	28
	1	-11	2	11
	1	-9	1	7
	1	-7	0	0
2	7	1	82	16
	0	-1	7	88

H	K U	L V	Φ W	φ2 φ1
	1	-2	12	83
	1	-1	5	76
	2	-1	3	54
	3	-2	8	67
	3	-1	1	18
	5	-3	11	63
	5	-2	4	37
	7	-3	7	43
	7	-2	0	0
	8	-3	5	31
	9	-4	10	46
	10	-3	1	6
	11	-4	6	27
	12	-5	11	41
7	2	1	82	74
	-1	-2	11	82
	-1	-1	9	86
	0	-1	2	65
	1	-9	11	51
	1	-8	9	49
	1	-7	7	45
	1	-6	5	40
	1	-5	3	31
	1	-4	1	14
	2	-11	8	36
	2	-9	4	24
	2	-7	0	0
	3	-11	1	5
1	2	8	16	27
	-4	-10	3	85
	-2	-11	3	74
	-2	-7	2	80
	0	-4	1	64
	2	-9	2	52
	2	-5	1	43
	2	-1	0	0
	4	-6	1	31
	6	-11	2	36
	6	-7	1	24
	8	-8	1	19
	10	-9	1	16
	12	-10	1	14
2	1	8	16	63
	-11	-10	4	75
	-8	-8	3	72

H	K U	L V	Φ W	φ2 φ1
	-7	-10	3	62
	-5	-6	2	67
	-3	-10	2	44
	-3	-2	1	83
	-2	-4	1	54
	-1	-6	1	37
	0	-8	1	27
	1	-10	1	22
	1	-2	0	0
	2	-12	1	16
1	8	2	76	7
	0	-1	4	88
	2	-3	11	78
	2	-2	7	73
	2	-1	3	56
	4	-3	10	67
	4	-1	2	27
	6	-2	5	40
	6	-1	1	10
	8	-3	8	45
	8	-1	0	0
	10	-4	11	47
	10	-3	7	35
	10	-2	3	17
8	1	2	76	83
	-2	-6	11	63
	-2	-2	9	79
	-1	-12	10	41
	-1	-10	9	43
	-1	-8	8	47
	-1	-6	7	51
	-1	-4	6	58
	-1	-2	5	70
	0	-2	1	27
	1	-12	2	10
	1	-10	1	6
	1	-8	0	0
2	8	1	83	14
	0	-1	8	88
	1	-1	6	79
	2	-1	4	62
	3	-2	10	71
	3	-1	2	33
	4	-1	0	0
	5	-2	6	49
	7	-3	10	53

H	K U	L V	Φ W	φ2 φ1
	7	-2	2	15
	8	-3	6	44
	10	-3	4	21
	11	-4	10	41
	11	-3	2	10
8	2	1	83	76
	-1	-2	12	82
	-1	-1	10	86
	0	-1	2	64
	1	-10	12	51
	1	-9	10	48
	1	-8	8	45
	1	-7	6	41
	1	-6	4	34
	1	-5	2	22
	1	-4	0	0
	2	-11	6	26
	2	-9	2	12
1	2	9	14	27
	-5	-11	3	88
	-1	-4	1	78
	0	-9	2	64
	1	-5	1	53
	2	-1	0	0
	3	-6	1	38
	4	-11	2	44
	5	-7	1	29
	7	-8	1	23
	9	-9	1	19
	11	-10	1	16
2	1	9	14	63
	-10	-7	3	82
	-8	-11	3	63
	-7	-4	2	87
	-5	-8	2	59
	-3	-12	2	41
	-3	-3	1	72
	-2	-5	1	49
	-1	-7	1	36
	0	-9	1	27
	1	-11	1	22
	1	-2	0	0
1	9	2	78	6
	0	-2	9	89

H	K U	L V	Φ W	φ2 φ1
	1	-1	4	75
	3	-1	3	45
	4	-2	7	60
	5	-3	11	65
	5	-1	2	22
	7	-3	10	55
	7	-1	1	8
	8	-2	5	32
	9	-1	0	0
	11	-3	8	36
	12	-2	3	14
9	1	2	78	84
	-2	-4	11	72
	-1	-11	10	43
	-1	-9	9	46
	-1	-7	8	50
	-1	-5	7	56
	-1	-3	6	65
	-1	-1	5	80
	0	-2	1	27
	1	-11	1	5
	1	-9	0	0
2	9	1	84	13
	0	-1	9	89
	1	-1	7	80
	2	-1	5	67
	3	-2	12	74
	3	-1	3	44
	4	-1	1	14
	5	-2	8	57
	7	-2	4	29
	8	-3	11	53
	9	-2	0	0
	10	-3	7	34
	11	-3	5	24
9	2	1	84	77
	-1	-1	11	86
	0	-1	2	64
	1	-10	11	48
	1	-9	9	45
	1	-8	7	41
	1	-7	5	36
	1	-6	3	26
	1	-5	1	11
	2	-11	4	20
	2	-9	0	0

H	K U	L V	Φ W	φ2 φ1
1	2	10	13	27
	-2	-9	2	76
	-2	-4	1	90
	0	-5	1	64
	2	-11	2	54
	2	-6	1	46
	2	-1	0	0
	4	-7	1	34
	6	-8	1	27
	8	-9	1	22
	10	-10	1	19
	12	-11	1	16
2	1	10	13	63
	-11	-8	3	81
	-10	-10	3	72
	-7	-6	2	76
	-5	-10	2	54
	-4	-2	1	90
	-3	-4	1	64
	-2	-6	1	46
	-1	-8	1	34
	0	-10	1	27
	1	-12	1	22
	1	-2	0	0
1	10	2	79	6
	0	-1	5	89
	2	-2	9	77
	2	-1	4	63
	4	-1	3	37
	6	-2	7	49
	6	-1	2	19
	8	-3	11	54
	8	-1	1	7
	10	-3	10	45
	10	-2	5	27
	10	-1	0	0
10	1	2	79	84
	-2	-2	11	81
	-1	-12	11	43
	-1	-10	10	46
	-1	-8	9	49
	-1	-6	8	54
	-1	-4	7	61
	-1	-2	6	73

H	K U	L V	Φ W	φ2 φ1
	0	-2	1	27
	1	-12	1	5
	1	-10	0	0
2	10	1	84	11
	0	-1	10	89
	1	-1	8	82
	2	-1	6	70
	3	-1	4	52
	4	-1	2	26
	5	-2	10	62
	5	-1	0	0
	7	-2	6	40
	9	-2	2	12
	10	-3	10	44
	11	-3	8	35
10	2	1	84	79
	-1	-1	12	86
	0	-1	2	64
	1	-11	12	48
	1	-10	10	45
	1	-9	8	42
	1	-8	6	37
	1	-7	4	30
	1	-6	2	18
	1	-5	0	0
	2	-11	2	10
1	2	11	11	27
	-4	-9	2	87
	-1	-5	1	75
	0	-11	2	64
	1	-6	1	55
	2	-1	0	0
	3	-7	1	41
	5	-8	1	32
	7	-9	1	26
	9	-10	1	22
	11	-11	1	19
2	1	11	11	63
	-11	-11	3	72
	-7	-8	2	68
	-5	-12	2	50
	-4	-3	1	80
	-3	-5	1	56
	-2	-7	1	43

H	K U	L V	Φ W	φ2 φ1
	-1	-9	1	33
	0	-11	1	27
	1	-2	0	0
1	11	2	80	5
	0	-2	11	89
	1	-1	5	78
	3	-1	4	53
	4	-2	9	66
	5	-1	3	31
	7	-1	2	16
	8	-2	7	41
	9	-1	1	6
	11	-3	11	45
	11	-1	0	0
	12	-2	5	23
11	1	2	80	35
	-1	-11	11	46
	-1	-9	10	49
	-1	-7	9	53
	-1	-5	8	59
	-1	-3	7	68
	-1	-1	6	82
	0	-2	1	27
	1	-11	0	0
2	11	1	85	10
	0	-1	11	89
	1	-1	9	83
	2	-1	7	73
	3	-1	5	58
	4	-1	3	36
	5	-2	12	66
	5	-1	1	11
	7	-2	8	48
	9	-2	4	24
	11	-3	11	44
	11	-2	0	0
11	2	1	85	80
	0	-1	2	64
	1	-11	11	45
	1	-10	9	42
	1	-9	7	38
	1	-8	5	32
	1	-7	3	23
	1	-6	1	9
	2	-11	0	0

TAB. V

FROM HKL-UVW INTO EULER ANGLES

H	K / U	L / V	Φ / W	φ_2 / φ_1
1	2	12	11	27
	-2	-11	2	74
	-2	-5	1	85
	0	-6	1	64
	2	-7	1	48
	2	-1	0	0
	4	-8	1	37
	6	-9	1	30
	8	-10	1	25
	10	-11	1	21
	12	-12	1	19
2	1	12	11	63
	-9	-6	2	83
	-7	-10	2	62
	-4	-4	1	72
	-3	-6	1	54
	-2	-8	1	41
	-1	-10	1	33
	0	-12	1	27
	1	-2	0	0
1	12	2	81	5
	0	-1	6	89
	2	-2	11	79
	2	-1	5	68
	4	-1	4	45
	6	-2	9	56
	6	-1	3	27
	8	-1	2	14
	10	-2	7	35
	10	-1	1	6
	12	-1	0	0
12	1	2	81	85
	-1	-12	12	46
	-1	-10	11	48
	-1	-8	10	52
	-1	-6	9	57
	-1	-4	8	64
	-1	-2	7	75
	0	-2	1	27
	1	-12	0	0
2	12	1	85	9
	0	-1	12	89
	1	-1	10	83

H	K / U	L / V	Φ / W	φ_2 / φ_1
	2	-1	8	75
	3	-1	6	63
	4	-1	4	44
	5	-1	2	21
	6	-1	0	0
	7	-2	10	54
	9	-2	6	33
	11	-2	2	10
12	2	1	85	81
	0	-1	2	64
	1	-12	12	45
	1	-11	10	42
	1	-10	8	39
	1	-9	6	34
	1	-8	4	26
	1	-7	2	16
	1	-6	0	0
1	3	3	47	18
	-3	-11	12	88
	-3	-10	11	89
	-3	-9	10	90
	0	-1	1	77
	3	-12	11	66
	3	-11	10	65
	3	-10	9	64
	3	-9	8	63
	3	-8	7	61
	3	-7	6	59
	3	-6	5	55
	3	-5	4	51
	3	-4	3	45
	3	-3	2	36
	3	-2	1	22
	3	-1	0	0
	6	-11	9	54
	6	-9	7	48
	6	-7	5	41
	6	-5	3	30
	6	-3	1	12
	9	-11	8	43
	9	-10	7	40
	9	-8	5	32
	9	-7	4	27
	9	-5	2	15
	9	-4	1	8
	12	-11	7	33

H	K / U	L / V	Φ / W	φ_2 / φ_1
	12	-9	5	26
	12	-7	3	17
	12	-5	1	6
3	1	3	47	72
	-9	-6	11	79
	-9	-3	10	90
	-8	-9	11	68
	-8	-3	9	89
	-7	-12	11	59
	-7	-9	10	65
	-7	-6	9	74
	-7	-3	8	87
	-6	-3	7	84
	-5	-12	9	52
	-5	-9	8	58
	-5	-6	7	67
	-5	-3	6	81
	-4	-9	7	53
	-4	-3	5	77
	-3	-12	7	43
	-3	-6	5	55
	-3	-3	4	71
	-2	-9	5	41
	-2	-3	3	62
	-1	-12	5	32
	-1	-9	4	34
	-1	-6	3	38
	-1	-3	2	47
	0	-3	1	26
	1	-12	3	19
	1	-9	2	17
	1	-6	1	13
	1	-3	0	0
	2	-9	1	9
	3	-12	1	6
3	3	1	77	45
	-1	-3	12	83
	-1	-2	9	86
	-1	-1	6	90
	0	-1	3	77
	1	-5	12	71
	1	-4	9	69
	1	-3	6	65
	1	-2	3	55
	1	-1	0	0
	2	-5	9	62
	2	-3	3	41

H	K / U	L / V	Φ / W	φ_2 / φ_1
	3	-7	12	60
	3	-5	6	47
	3	-4	3	32
	4	-7	9	50
	4	-5	3	26
	5	-9	12	51
	5	-8	9	45
	5	-7	6	36
	5	-6	3	22
	6	-7	3	19
	7	-11	12	44
	7	-10	9	38
	7	-9	6	29
	7	-8	3	16
	8	-11	9	35
	8	-9	3	14
	9	-11	6	24
	9	-10	3	13
	10	-11	3	12
	11	-12	3	11
1	3	4	38	16
	-3	-11	9	88
	-2	-6	5	90
	-1	-9	7	80
	-1	-5	4	84
	0	-4	3	75
	1	-11	8	71
	1	-7	5	69
	1	-3	2	60
	2	-10	7	66
	2	-2	1	33
	3	-5	3	48
	3	-1	0	0
	4	-6	5	52
	5	-11	7	54
	5	-7	4	43
	5	-3	1	16
	7	-9	5	40
	7	-5	2	21
	6	-12	7	45
	8	-4	1	10
	9	-11	6	39
	9	-7	3	24
	11	-9	4	26
	11	-5	1	6
	12	-8	3	19

H	K / U	L / V	Φ / W	φ_2 / φ_1
3	1	4	38	72
	-12	-8	11	78
	-11	-7	10	79
	-10	-6	9	80
	-9	-5	8	82
	-8	-12	9	59
	-8	-4	7	84
	-7	-11	8	57
	-7	-3	6	86
	-6	-10	7	56
	-6	-2	5	90
	-5	-9	6	54
	-4	-8	5	52
	-3	-11	5	40
	-3	-7	4	49
	-2	-6	3	44
	-1	-9	3	30
	-1	-5	2	36
	-1	-1	1	69
	0	-4	1	23
	1	-11	2	17
	1	-7	1	13
	1	-3	0	0
	2	-10	1	9
1	4	3	54	14
	-1	-8	11	86
	-1	-5	7	88
	0	-3	4	82
	1	-7	9	77
	1	-4	5	73
	1	-1	1	46
	2	-5	6	67
	3	-9	11	69
	3	-6	7	63
	4	-7	8	61
	4	-1	0	0
	5	-8	9	59
	5	-2	1	13
	6	-9	10	57
	6	-3	2	21
	7	-10	11	56
	7	-4	3	26
	8	-11	12	55
	8	-5	4	29
	9	-6	5	31
	9	-3	1	7
	10	-7	6	33

H	K / U	L / V	Φ / W	φ_2 / φ_1
	11	-8	7	34
	11	-5	3	17
	12	-9	8	36
4	1	3	54	76
	-7	-8	12	68
	-7	-5	11	77
	-7	-2	10	89
	-6	-9	11	62
	-5	-7	9	63
	-5	-4	8	75
	-4	-11	9	49
	-4	-5	7	66
	-3	-12	6	42
	-3	-9	7	47
	-3	-3	5	71
	-2	-7	5	44
	-2	-1	3	83
	-1	-11	5	31
	-1	-8	4	33
	-1	-5	3	39
	-1	-2	2	56
	0	-3	1	23
	1	-10	2	14
	1	-7	1	10
	1	-4	0	0
	2	-11	1	6
3	4	1	79	37
	-1	-2	11	88
	0	-1	4	82
	1	-3	9	74
	1	-2	5	69
	1	-1	1	36
	2	-3	6	61
	3	-5	11	64
	3	-4	7	56
	4	-5	8	53
	4	-3	0	0
	5	-6	9	50
	5	-4	1	9
	6	-7	10	49
	6	-5	2	15
	7	-8	11	47
	7	-6	3	18
	8	-9	12	46
	8	-7	4	21
	9	-8	5	23
	9	-7	1	5

H	K / U	L / V	Φ / W	φ_2 / φ_1
	10	-9	6	25
	11	-10	7	26
	11	-9	3	12
	12	-11	8	27
4	3	1	79	53
	-1	-2	10	84
	-1	-1	7	86
	0	-1	3	75
	1	-5	11	68
	1	-4	8	65
	1	-3	5	60
	1	-2	2	43
	2	-5	7	54
	2	-3	1	16
	3	-8	12	56
	3	-7	9	51
	3	-5	3	28
	3	-4	0	0
	4	-9	11	49
	4	-7	5	33
	5	-9	7	35
	5	-8	4	23
	5	-7	1	7
	6	-11	9	37
	7	-12	8	31
	7	-11	5	21
	7	-10	2	9
	8	-11	1	4
1	3	5	32	16
	-2	-11	7	83
	-1	-8	5	80
	-1	-3	2	90
	0	-5	3	74
	1	-12	7	70
	1	-7	4	67
	1	-2	1	50
	2	-9	5	63
	3	-11	6	61
	3	-1	0	0
	4	-3	1	22
	5	-5	2	31
	6	-7	3	35
	7	-9	4	36
	7	-4	1	13
	8	-11	5	40
	9	-6	3	27

H	K / U	L / V	Φ / W	φ_2 / φ_1
	10	-5	1	10
	11	-12	5	33
	11	-7	2	16
3	1	5	32	72
	-11	-12	9	65
	-11	-7	8	78
	-10	-5	7	89
	-9	-8	7	70
	-8	-11	7	59
	-7	-9	6	61
	-7	-4	5	80
	-6	-7	5	63
	-5	-5	4	67
	-4	-3	3	74
	-3	-11	4	38
	-3	-1	2	90
	-2	-9	3	35
	-1	-12	3	27
	-1	-7	2	31
	-1	-2	1	50
	0	-5	1	22
	1	-8	1	13
	1	-3	0	0
	2	-11	1	10
1	5	3	60	11
	-1	-7	12	88
	0	-3	5	84
	1	-5	8	78
	1	-2	3	68
	2	-7	11	75
	2	-1	1	28
	3	-3	4	53
	4	-5	7	59
	5	-7	10	62
	5	-4	5	46
	5	-1	0	0
	7	-8	11	57
	7	-5	6	42
	7	-2	1	9
	8	-7	9	49
	9	-6	7	39
	9	-3	2	14
	11	-7	8	37
	11	-4	3	17
	12	-9	11	43
	12	-3	1	5

TAB. V

FROM HKL–UVW INTO EULER ANGLES

H	K / U	L / V	Φ / W	φ_2 / φ_1
1	3	5		
5	1	3	60	79
	-6	-3	11	82
	-5	-11	12	55
	-5	-8	11	62
	-5	-2	9	85
	-4	-7	9	60
	-4	-1	7	89
	-3	-9	8	48
	-3	-6	7	57
	-2	-11	7	38
	-2	-5	5	52
	-1	-10	5	31
	-1	-7	4	35
	-1	-4	3	43
	-1	-1	2	71
	0	-3	1	22
	1	-11	2	12
	1	-8	1	8
	1	-5	0	0
3	5	1	80	31
	0	-1	5	84
	1	-3	12	79
	1	-2	7	75
	1	-1	2	56
	2	-3	9	70
	3	-4	11	67
	3	-2	1	16
	4	-3	3	31
	5	-4	5	39
	5	-3	0	0
	6	-5	7	43
	7	-6	9	45
	7	-5	4	25
	8	-7	11	47
	8	-5	1	6
	9	-7	8	36
	10	-7	5	23
	11	-9	12	41
	11	-8	7	28
	11	-7	2	9
5	3	1	80	59
	-1	-2	11	84
	-1	-1	8	88
	0	-1	3	74

H	K / U	L / V	Φ / W	φ_2 / φ_1
	1	-5	10	65
	1	-4	7	61
	1	-3	4	53
	1	-2	1	24
	2	-7	11	58
	2	-5	5	44
	3	-8	9	47
	3	-7	6	39
	3	-5	0	0
	4	-9	7	36
	4	-7	1	7
	5	-12	11	41
	5	-11	8	34
	5	-9	2	11
	6	-11	3	14
	7	-12	1	4
1	3	6	28	18
	-3	-11	6	87
	-3	-9	5	90
	0	-2	1	74
	3	-11	5	59
	3	-9	4	56
	3	-7	3	52
	3	-5	2	44
	3	-3	1	29
	3	-1	0	0
	6	-12	5	49
	6	-8	3	38
	6	-4	1	17
	9	-11	4	36
	9	-7	2	22
	9	-5	1	12
	12	-10	3	24
	12	-6	1	9
3	1	6	28	72
	-12	-6	7	83
	-11	-9	7	71
	-10	-12	7	61
	-9	-3	5	90
	-8	-6	5	74
	-7	-9	5	59
	-7	-3	4	86
	-6	-12	5	49
	-5	-9	4	51
	-5	-3	3	79
	-4	-6	3	55

H	K / U	L / V	Φ / W	φ_2 / φ_1
	-3	-3	2	66
	-2	-12	3	31
	-1	-9	2	28
	-1	-3	1	40
	0	-6	1	21
	1	-9	1	14
	1	-3	0	0
	2	-12	1	10
1	6	3	64	9
	0	-1	2	86
	3	-6	11	72
	3	-5	9	69
	3	-4	7	65
	3	-3	5	58
	3	-2	3	45
	3	-1	1	20
	6	-7	12	62
	6	-5	8	53
	6	-3	4	35
	6	-1	0	0
	9	-7	11	51
	9	-5	7	39
	9	-4	5	30
	9	-2	1	7
	12	-7	10	41
	12	-5	6	28
	12	-3	2	10
6	1	3	64	81
	-5	-6	12	69
	-5	-3	11	80
	-4	-9	11	56
	-4	-3	9	77
	-3	-12	10	45
	-3	-6	8	59
	-3	-3	7	72
	-2	-9	7	42
	-2	-3	5	65
	-1	-12	6	30
	-1	-9	5	33
	-1	-6	4	38
	-1	-3	3	50
	0	-3	1	21
	1	-12	2	11
	1	-9	1	7
	1	-6	0	0

H	K / U	L / V	Φ / W	φ_2 / φ_1
3	6	1	62	27
	0	-1	6	86
	1	-2	9	79
	1	-1	3	66
	2	-3	12	76
	2	-1	0	0
	3	-2	3	40
	4	-3	6	51
	5	-4	9	55
	5	-3	3	28
	6	-5	12	58
	7	-5	9	47
	7	-4	3	21
	8	-5	6	33
	9	-5	3	16
	10	-7	12	45
	11	-7	9	35
	11	-6	3	14
	12	-7	6	24
6	3	1	82	63
	-1	-2	12	84
	-1	-1	9	87
	0	-1	3	74
	1	-6	12	64
	1	-5	9	62
	1	-4	6	56
	1	-3	3	44
	1	-2	0	0
	2	-7	9	52
	2	-5	3	29
	3	-10	12	50
	3	-8	6	36
	3	-7	3	22
	4	-11	9	38
	4	-9	3	17
	5	-12	6	25
	5	-11	3	14
1	3	7	24	18
	-2	-11	5	83
	-1	-9	4	79
	0	-7	3	73
	1	-12	5	69
	1	-5	2	62
	2	-3	1	40
	3	-8	3	54

H	K / U	L / V	Φ / W	φ_2 / φ_1
	3	-1	0	0
	5	-11	4	50
	5	-4	1	22
	7	-7	2	29
	8	-5	1	15
	9	-10	3	32
	11	-6	1	11
	12	-11	3	26
3	1	7	24	72
	-11	-9	6	71
	-9	-8	5	69
	-8	-11	5	57
	-7	-7	4	66
	-5	-6	3	61
	-4	-9	3	45
	-3	-5	2	52
	-2	-1	1	83
	-1	-11	2	26
	-1	-4	1	35
	0	-7	1	20
	1	-10	1	14
	1	-3	0	0
1	7	3	67	8
	0	-3	7	87
	1	-4	9	81
	1	-1	2	62
	2	-5	11	77
	4	-1	1	15
	5	-2	3	32
	6	-3	5	40
	7	-4	7	45
	7	-1	0	0
	8	-5	9	49
	9	-6	11	51
	9	-3	4	25
	11	-5	8	37
	11	-2	1	6
7	1	3	67	82
	-5	-1	12	89
	-4	-5	11	70
	-3	-12	11	46
	-3	-9	10	52
	-3	-3	8	74
	-2	-7	7	49
	-2	-1	5	83
	-1	-11	6	31

H	K / U	L / V	Φ / W	φ_2 / φ_1
	-1	-8	5	35
	-1	-5	4	42
	-1	-2	3	61
	0	-3	1	20
	1	-10	1	6
	1	-7	0	0
3	7	1	83	23
	0	-1	7	87
	1	-2	11	81
	1	-1	4	72
	2	-1	1	24
	3	-2	5	55
	4	-3	9	62
	5	-3	6	46
	7	-4	7	41
	7	-3	0	0
	8	-5	11	50
	9	-5	8	38
	9	-4	1	6
	11	-6	9	36
	11	-5	2	9
7	3	1	83	67
	-1	-1	10	87
	0	-1	3	73
	1	-6	11	62
	1	-5	6	56
	1	-4	5	51
	1	-3	2	33
	2	-7	7	44
	2	-5	1	11
	3	-11	12	47
	3	-10	9	41
	3	-8	3	20
	3	-7	0	0
	4	-11	5	23
	5	-12	1	4
1	3	8	22	18
	-4	-12	5	90
	-1	-5	2	83
	0	-8	3	73
	1	-11	4	68
	1	-3	1	55
	3	-1	0	0
	4	-4	1	28
	5	-7	2	38

H	K / U	L / V	Φ / W	φ_2 / φ_1
	6	-10	3	43
	7	-5	1	18
	9	-11	3	34
	10	-6	1	13
	11	-9	2	22
3	1	6	22	72
	-12	-4	5	90
	-11	-7	5	77
	-9	-5	4	80
	-7	-11	4	53
	-7	-3	3	86
	-5	-9	3	50
	-4	-12	3	39
	-3	-7	2	44
	-2	-2	1	65
	-1	-5	1	32
	0	-8	1	20
	1	-11	1	14
	1	-3	0	0
1	8	3	70	7
	0	-3	8	88
	1	-2	5	77
	2	-1	2	45
	3	-3	7	66
	4	-5	12	70
	5	-4	9	60
	5	-1	1	12
	7	-5	11	57
	7	-2	3	24
	8	-1	0	0
	9	-3	5	30
	11	-4	7	33
	12	-3	4	19
8	1	3	70	83
	-4	-1	11	88
	-3	-9	11	54
	-3	-6	10	62
	-2	-11	9	42
	-2	-5	7	58
	-1	-10	6	33
	-1	-7	5	38
	-1	-4	4	48
	-1	-1	3	75
	0	-3	1	20
	1	-11	1	6
	1	-8	0	0

H	K / U	L / V	Φ / W	φ_2 / φ_1
3	8	1	83	21
	0	-1	8	88
	1	-1	5	76
	2	-1	2	42
	3	-2	7	64
	4	-3	12	68
	5	-3	9	58
	5	-2	1	11
	7	-4	11	54
	7	-3	3	22
	8	-3	0	0
	9	-4	5	27
	11	-5	7	30
	12	-5	4	17
8	3	1	83	69
	-1	-1	11	87
	0	-1	3	73
	1	-6	10	59
	1	-5	7	54
	1	-4	4	45
	1	-3	1	18
	2	-9	11	51
	2	-7	5	35
	3	-11	9	39
	3	-10	6	30
	3	-8	0	0
	4	-11	1	5
1	3	9	19	18
	-3	-11	4	87
	0	-3	1	73
	3	-10	3	56
	3	-7	2	50
	3	-4	1	36
	3	-1	0	0
	6	-11	3	43
	6	-5	1	23
	9	-9	2	28
	9	-6	1	16
	12	-7	1	13
3	1	9	19	72
	-12	-9	5	73
	-11	-12	5	62
	-9	-9	4	65
	-8	-3	3	88

FROM HKL-UVW INTO EULER ANGLES

H	K / U	L / V	Φ / W	φ_2 / φ_1
1	3	9		
3	1	9	19	72
	-7	-6	3	69
	-5	-12	3	43
	-5	-3	2	78
	-3	-9	2	38
	-2	-3	1	54
	-1	-6	1	29
	0	-9	1	19
	1	-12	1	14
	1	-3	0	0
1	9	3	72	6
	0	-1	3	88
	3	-4	11	74
	3	-3	8	69
	3	-2	5	59
	3	-1	2	34
	6	-3	7	50
	6	-1	1	10
	9	-5	12	53
	9	-4	9	45
	9	-2	3	19
	9	-1	0	0
	12	-5	11	43
	12	-3	5	23
9	1	3	72	84
	-3	-6	11	64
	-3	-3	10	76
	-2	-9	9	47
	-2	-3	7	69
	-1	-12	7	32
	-1	-9	6	36
	-1	-6	5	42
	-1	-3	4	56
	0	-3	1	19
	1	-12	1	5
	1	-9	0	0
3	9	1	84	18
	0	-1	9	88
	1	-1	6	78
	2	-1	3	54
	3	-2	9	69
	3	-1	0	0
	5	-3	12	65

H	K / U	L / V	Φ / W	φ_2 / φ_1
	5	-2	3	29
	7	-3	6	38
	8	-3	3	19
	9	-4	9	43
	11	-5	12	45
	11	-4	3	14
	12	-5	9	35
9	3	1	84	72
	-1	-1	12	87
	0	-1	3	73
	1	-7	12	60
	1	-6	9	56
	1	-5	6	50
	1	-4	3	36
	1	-3	0	0
	2	-9	9	45
	2	-7	3	23
	3	-11	6	28
	3	-10	3	16
1	3	10	18	18
	-1	-3	1	90
	0	-10	3	72
	1	-7	2	65
	2	-4	1	46
	3	-11	3	58
	3	-1	0	0
	5	-5	1	28
	7	-9	2	35
	8	-6	1	19
	11	-7	1	15
3	1	10	18	72
	-11	-7	4	77
	-8	-6	3	72
	-7	-9	3	58
	-5	-5	2	65
	-3	-11	2	35
	-3	-1	1	90
	-2	-4	1	46
	-1	-7	1	28
	0	-10	1	19
	1	-3	0	0
1	10	3	73	6
	0	-3	10	88
	1	-1	3	71

H	K / U	L / V	Φ / W	φ_2 / φ_1
	4	-1	2	27
	5	-2	5	45
	6	-3	8	53
	7	-4	11	57
	7	-1	1	8
	9	-3	7	38
	10	-1	0	0
	11	-2	3	16
10	1	3	73	84
	-3	-3	11	77
	-2	-7	9	54
	-2	-1	7	84
	-1	-11	7	34
	-1	-8	6	39
	-1	-5	5	47
	-1	-2	4	66
	0	-3	1	19
	1	-10	0	0
3	10	1	85	17
	0	-1	10	88
	1	-1	7	80
	2	-1	4	61
	3	-2	11	73
	3	-1	1	18
	5	-2	5	43
	7	-3	9	50
	8	-3	6	35
	10	-3	0	0
	11	-4	7	31
10	3	1	85	73
	0	-1	3	72
	1	-7	11	58
	1	-6	8	53
	1	-5	5	45
	1	-4	2	26
	2	-9	7	37
	2	-7	1	8
	3	-11	3	15
	3	-10	0	0
1	3	11	16	18
	-3	-10	3	88
	-1	-7	2	80
	0	-11	3	72
	1	-4	1	59

H	K / U	L / V	Φ / W	φ_2 / φ_1
	3	-1	0	0
	4	-5	1	34
	5	-9	2	44
	7	-6	1	23
	10	-7	1	17
	11	-11	2	27
3	1	11	16	72
	-11	-11	4	64
	-8	-9	3	61
	-7	-12	3	50
	-5	-7	2	55
	-3	-2	1	75
	-2	-5	1	41
	-1	-8	1	26
	0	-11	1	19
	1	-3	0	0
1	11	3	75	5
	0	-3	11	89
	1	-2	7	81
	2	-1	3	56
	3	-3	10	73
	5	-1	2	22
	7	-2	5	36
	8	-1	1	7
	9	-3	8	42
	11	-4	11	45
	11	-1	0	0
	12	-3	7	31
11	1	3	75	85
	-2	-11	11	47
	-2	-5	9	63
	-1	-10	7	36
	-1	-7	6	42
	-1	-4	5	53
	-1	-1	4	76
	0	-3	1	19
	1	-11	0	0
3	11	1	85	15
	0	-1	11	89
	1	-1	6	81
	2	-1	5	66
	3	-1	2	32
	5	-2	7	53
	7	-3	12	56
	7	-2	1	6

H	K / U	L / V	Φ / W	φ_2 / φ_1
	8	-3	9	47
	10	-3	3	16
	11	-4	11	43
	11	-3	0	0
11	3	1	85	75
	0	-1	3	72
	1	-7	10	55
	1	-6	7	49
	1	-5	4	38
	1	-4	1	14
	2	-11	11	45
	2	-9	5	29
	3	-11	0	0
1	3	12	15	18
	-3	-11	3	87
	0	-4	1	72
	3	-9	2	54
	3	-5	1	42
	3	-1	0	0
	6	-6	1	27
	9	-11	2	33
	9	-7	1	20
	12	-8	1	16
3	1	12	15	72
	-10	-6	3	78
	-8	-12	3	53
	-7	-3	2	85
	-5	-9	2	48
	-3	-3	1	64
	-2	-6	1	38
	-1	-9	1	26
	0	-12	1	19
	1	-3	0	0
1	12	3	76	5
	0	-1	4	89
	3	-3	11	74
	3	-2	7	66
	3	-1	3	45
	6	-3	10	59
	6	-1	2	19
	9	-2	5	29
	9	-1	1	6
	12	-3	8	34
	12	-1	0	0

H	K / U	L / V	Φ / W	φ_2 / φ_1
12	1	3	76	85
	-2	-9	11	52
	-2	-3	9	73
	-1	-12	8	35
	-1	-9	7	39
	-1	-6	6	46
	-1	-3	5	61
	0	-3	1	19
	1	-12	0	0
3	12	1	85	14
	0	-1	12	89
	1	-1	9	82
	2	-1	6	70
	3	-1	3	44
	4	-1	0	0
	5	-2	9	59
	7	-2	3	22
	8	-3	12	55
	10	-3	6	30
	11	-3	3	15
12	3	1	85	76
	0	-1	3	72
	1	-8	12	56
	1	-7	9	52
	1	-6	6	45
	1	-5	3	31
	1	-4	0	0
	2	-11	9	39
	2	-9	3	18
1	4	4	46	14
	0	-1	1	80
	4	-12	11	66
	4	-11	10	65
	4	-10	9	63
	4	-9	8	61
	4	-8	7	59
	4	-7	6	56
	4	-6	5	53
	4	-5	4	48
	4	-4	3	41
	4	-3	2	31
	4	-2	1	18
	4	-1	0	0
	8	-11	9	50

H	K / U	L / V	Φ / W	φ_2 / φ_1
	8	-9	7	44
	8	-7	5	36
	8	-5	3	25
	8	-3	1	9
	12	-11	8	38
	12	-10	7	35
	12	-8	5	27
	12	-7	4	23
	12	-5	2	12
	12	-4	1	6
4	1	4	46	76
	-11	-4	12	86
	-10	-4	11	85
	-9	-8	11	70
	-9	-4	10	83
	-6	-12	11	58
	-8	-4	9	81
	-7	-12	10	54
	-7	-8	9	64
	-7	-4	8	79
	-6	-4	7	76
	-5	-12	8	47
	-5	-8	7	56
	-5	-4	6	72
	-4	-12	7	42
	-4	-4	5	67
	-3	-8	5	45
	-3	-4	4	61
	-2	-12	5	32
	-2	-4	3	51
	-1	-12	4	26
	-1	-8	3	29
	-1	-4	2	37
	0	-4	1	20
	1	-12	2	13
	1	-8	1	10
	1	-4	0	0
	2	-12	1	7
4	4	1	80	45
	-1	-2	12	87
	-1	-1	8	90
	0	-1	4	80
	1	-4	12	74
	1	-3	8	71
	1	-2	4	62
	1	-1	0	0
	2	-5	12	68

H	K / U	L / V	Φ / W	φ_2 / φ_1
	2	-3	4	49
	3	-5	8	55
	3	-4	4	39
	4	-7	12	57
	4	-5	4	33
	5	-8	12	53
	5	-7	8	44
	5	-6	4	28
	6	-7	4	24
	7	-10	12	45
	7	-9	8	36
	7	-8	4	21
	8	-11	12	42
	8	-9	4	19
	9	-11	6	30
	9	-10	4	17
	10	-11	4	15
	11	-12	4	14
1	4	5	40	14
	-1	-11	9	83
	-1	-6	5	86
	0	-5	4	79
	1	-9	7	74
	1	-4	3	68
	2	-3	2	50
	3	-7	5	60
	3	-2	1	25
	4	-11	8	62
	4	-1	0	0
	5	-10	7	57
	5	-5	3	38
	7	-8	5	42
	7	-3	1	12
	6	-7	4	34
	9	-11	7	44
	10	-5	2	16
	11	-9	5	31
	11	-4	1	8
4	1	5	40	76
	-12	-7	11	77
	-11	-6	10	79
	-10	-5	9	80
	-9	-4	8	82
	-8	-3	7	85
	-7	-12	8	52
	-7	-2	6	89

TAB. V FROM HKL-UVW INTO EULER ANGLES

H	K U	L V	Φ W	φ2 φ1
1	4	5		
4	1	5	40	76
	-6	-11	7	50
	-5	-10	6	48
	-4	-9	5	45
	-3	-8	4	42
	-2	-7	3	37
	-1	-11	3	24
	-1	-6	2	29
	-1	-1	1	65
	0	-5	1	18
	1	-9	1	10
	1	-4	0	0
1	5	4	52	11
	-1	-7	9	88
	0	-4	5	83
	1	-9	11	79
	1	-5	6	76
	1	-1	1	47
	2	-6	7	71
	3	-7	8	67
	4	-8	9	64
	5	-9	10	62
	5	-1	0	0
	6	-10	11	61
	6	-2	1	11
	7	-11	12	59
	7	-3	2	19
	8	-4	3	24
	9	-5	4	27
	10	-6	5	30
	11	-7	6	32
	11	-3	1	6
	12	-8	7	34
5	1	4	52	79
	-8	-4	11	80
	-7	-9	11	62
	-7	-5	10	74
	-5	-11	9	49
	-5	-7	8	60
	-5	-3	7	78
	-4	-8	7	52
	-3	-5	5	56
	-3	-1	4	86
	-2	-10	5	34

H	K U	L V	Φ W	φ2 φ1
	-2	-2	3	68
	-1	-11	4	26
	-1	-7	3	30
	-1	-3	2	43
	0	-4	1	18
	1	-9	1	8
	1	-5	0	0
4	5	1	81	39
	0	-1	5	83
	1	-3	11	77
	1	-2	6	72
	1	-1	1	36
	2	-3	7	64
	3	-4	8	59
	4	-5	9	56
	5	-6	10	53
	5	-4	0	0
	6	-7	11	51
	6	-5	1	7
	7	-8	12	49
	7	-6	2	12
	8	-7	3	16
	9	-8	4	19
	10	-9	5	21
	11	-10	6	22
	11	-9	1	4
	12	-11	7	24
5	4	1	81	51
	-1	-1	9	89
	0	-1	4	79
	1	-4	11	71
	1	-3	7	67
	1	-2	3	54
	2	-5	10	63
	2	-3	2	29
	3	-5	5	41
	3	-4	1	11
	4	-7	8	45
	4	-5	0	0
	5	-9	11	48
	5	-8	7	37
	5	-7	3	19
	7	-11	9	35
	7	-10	5	23
	7	-9	1	5
	8	-11	4	17

H	K U	L V	Φ W	φ2 φ1
1	4	6	34	14
	-2	-10	7	88
	0	-3	2	78
	2	-11	7	70
	2	-8	5	66
	2	-5	3	59
	2	-2	1	36
	4	-7	4	52
	4	-1	0	0
	6	-12	7	55
	6	-9	5	48
	6	-3	1	15
	8	-11	6	45
	8	-5	2	21
	10	-7	3	25
	10	-4	1	9
	12	-9	4	27
4	1	6	34	76
	-11	-10	9	66
	-11	-4	8	85
	-9	-12	8	56
	-9	-6	7	74
	-8	-10	7	58
	-7	-8	6	60
	-7	-2	5	88
	-6	-6	5	64
	-5	-4	4	69
	-4	-2	3	80
	-3	-12	4	33
	-2	-10	3	30
	-1	-8	2	25
	-1	-2	1	46
	0	-6	1	17
	1	-10	1	10
	1	-4	0	0
1	6	4	57	9
	0	-2	3	85
	2	-7	10	75
	2	-5	7	72
	2	-3	4	63
	2	-1	1	29
	4	-8	11	68
	4	-4	5	52
	6	-7	9	57
	6	-5	6	47

H	K U	L V	Φ W	φ2 φ1
	6	-1	0	0
	8	-6	7	43
	8	-2	1	8
	10	-9	11	49
	10	-7	8	41
	10	-3	2	13
	12	-8	9	39
	12	-4	3	16
6	1	4	57	81
	-7	-6	12	72
	-7	-2	11	86
	-6	-8	11	62
	-5	-6	9	65
	-5	-2	8	83
	-4	-12	9	44
	-4	-4	7	69
	-3	-10	7	42
	-3	-2	5	76
	-2	-8	5	38
	-1	-10	4	26
	-1	-6	3	32
	-1	-2	2	53
	0	-4	1	17
	1	-10	1	7
	1	-6	0	0
4	6	1	82	34
	0	-1	6	85
	1	-2	8	76
	1	-1	2	56
	2	-3	10	72
	3	-4	12	69
	3	-2	0	0
	4	-3	2	22
	5	-4	4	32
	6	-5	6	38
	7	-6	8	41
	7	-5	2	13
	8	-7	10	44
	9	-8	12	45
	9	-7	6	28
	10	-7	2	9
	11	-9	10	36
	11	-8	4	17
6	4	1	82	56
	-1	-1	10	88
	0	-1	4	78

H	K U	L V	Φ W	φ2 φ1
	1	-4	10	69
	1	-3	6	63
	1	-2	2	42
	2	-5	8	57
	2	-3	0	0
	3	-7	10	53
	3	-5	2	19
	4	-9	12	51
	4	-7	4	27
	5	-9	6	31
	5	-8	2	12
	6	-11	8	33
	7	-12	6	24
	7	-11	2	9
1	4	7	30	14
	-1	-12	7	82
	-1	-5	3	88
	0	-7	4	78
	1	-9	5	72
	1	-2	1	54
	2	-11	6	69
	4	-1	0	0
	5	-3	1	19
	6	-5	2	29
	7	-7	3	35
	8	-9	4	38
	9	-11	5	41
	9	-4	1	11
	11	-8	3	25
4	1	7	30	76
	-11	-12	8	60
	-11	-5	7	81
	-10	-9	7	65
	-8	-3	5	84
	-7	-7	5	63
	-6	-11	5	47
	-5	-8	4	50
	-4	-5	3	57
	-3	-2	2	73
	-1	-10	2	23
	-1	-3	1	36
	0	-7	1	16
	1	-11	1	10
	1	-4	0	0

H	K U	L V	Φ W	φ2 φ1
1	7	4	61	8
	0	-4	7	86
	1	-7	12	82
	1	-3	5	76
	2	-2	3	57
	3	-5	8	68
	3	-1	1	20
	5	-7	11	65
	5	-3	4	41
	7	-5	7	46
	7	-1	0	0
	8	-4	5	34
	9	-7	10	49
	10	-2	1	6
	11	-5	6	31
	12	-8	11	44
7	1	4	61	62
	-6	-2	11	85
	-5	-9	11	57
	-5	-1	9	88
	-4	-8	9	55
	-3	-11	8	41
	-3	-7	7	51
	-2	-6	5	45
	-1	-9	4	28
	-1	-5	3	36
	-1	-1	2	70
	0	-4	1	16
	1	-11	1	6
	1	-7	0	0
4	7	1	83	30
	0	-1	7	86
	1	-2	10	80
	1	-1	3	66
	3	-2	2	29
	4	-3	5	45
	5	-4	8	52
	5	-3	1	10
	6	-5	11	55
	7	-5	7	39
	7	-4	0	0
	8	-5	3	18
	10	-7	9	37
	11	-8	12	42
	11	-7	5	21
	12	-7	1	4

H	K U	L V	Φ W	φ2 φ1
7	4	1	83	60
	-1	-1	11	88
	0	-1	4	78
	1	-4	9	66
	1	-3	5	58
	1	-2	1	24
	2	-5	6	49
	3	-8	11	53
	3	-7	7	43
	4	-9	8	39
	4	-7	0	0
	5	-11	9	37
	5	-9	1	6
	6	-11	2	9
1	4	8	27	14
	0	-2	1	77
	4	-11	5	59
	4	-9	4	55
	4	-7	3	50
	4	-5	2	41
	4	-3	1	25
	4	-1	0	0
	8	-12	5	46
	8	-8	3	34
	8	-4	1	14
	12	-11	4	31
	12	-7	2	18
	12	-5	1	10
4	1	8	27	76
	-12	-8	7	72
	-11	-12	7	60
	-11	-4	6	85
	-9	-4	5	81
	-8	-8	5	62
	-7	-12	5	48
	-7	-4	4	76
	-5	-12	4	40
	-5	-4	3	68
	-4	-8	3	44
	-3	-4	2	54
	-1	-12	2	21
	-1	-4	1	31
	0	-8	1	16
	1	-12	1	10
	1	-4	0	0

H	K U	L V	Φ W	φ2 φ1
1	8	4	64	7
	0	-1	2	87
	4	-6	11	69
	4	-5	9	65
	4	-4	7	60
	4	-3	5	52
	4	-2	3	38
	4	-1	1	15
	8	-7	12	57
	8	-5	8	46
	8	-3	4	28
	8	-1	0	0
	12	-7	11	44
	12	-5	7	32
	12	-4	5	24
	12	-2	1	5
8	1	4	64	83
	-5	-8	12	61
	-5	-4	11	75
	-4	-12	11	67
	-4	-4	9	71
	-3	-8	8	50
	-3	-4	7	65
	-2	-12	7	34
	-2	-4	5	56
	-1	-12	5	25
	-1	-8	4	30
	-1	-4	3	41
	0	-4	1	16
	1	-12	1	5
	1	-8	0	0
4	8	1	84	27
	0	-1	8	87
	1	-2	12	82
	1	-1	4	72
	2	-1	0	0
	3	-2	4	48
	4	-3	8	59
	5	-4	12	63
	5	-3	4	35
	7	-5	12	55
	7	-4	4	27
	8	-5	8	41
	9	-5	4	21
	11	-7	12	43
	11	-6	4	18

TAB. V

FROM HKL-UVW INTO EULER ANGLES

H	K U	L V	Φ W	φ2 φ1
1	4	8		
4	8	1	84	27
	12	-7	8	30
8	4	1	84	63
	-1	-1	12	88
	0	-1	4	77
	1	-5	12	68
	1	-4	8	63
	1	-3	4	52
	1	-2	0	0
	2	-7	12	59
	2	-5	4	37
	3	-8	8	43
	3	-7	4	28
	4	-11	12	46
	4	-9	4	22
	5	-12	8	32
	5	-11	4	18
1	4	9	25	14
	-1	-11	5	82
	0	-9	4	77
	1	-7	3	70
	2	-5	2	57
	3	-12	5	64
	3	-3	1	33
	4	-1	0	0
	5	-8	3	47
	7	-4	1	17
	8	-11	4	43
	10	-7	2	23
	11	-5	1	11
4	1	9	25	76
	-11	-10	6	64
	-9	-9	5	61
	-7	-8	4	58
	-5	-7	3	52
	-4	-11	3	37
	-3	-6	2	43
	-2	-1	1	79
	-1	-5	1	28
	0	-9	1	15
	1	-4	0	0

H	K U	L V	Φ W	φ2 φ1
1	9	4	66	6
	0	-4	9	87
	1	-5	11	83
	1	-1	2	63
	5	-1	1	12
	6	-2	3	28
	7	-3	5	37
	8	-4	7	42
	9	-5	9	46
	9	-1	0	0
	10	-6	11	49
	11	-3	4	21
9	1	4	66	84
	-5	-3	12	80
	-4	-8	11	58
	-3	-5	8	62
	-3	-1	7	85
	-2	-10	7	38
	-2	-2	5	72
	-1	-11	5	27
	-1	-7	4	33
	-1	-3	3	49
	0	-4	1	15
	1	-9	0	0
4	9	1	84	24
	0	-1	9	87
	1	-1	5	75
	2	-1	1	24
	3	-2	6	59
	4	-3	11	66
	5	-3	7	51
	7	-4	8	45
	9	-5	9	41
	9	-4	0	0
	11	-6	10	39
	11	-5	1	5
9	4	1	84	66
	0	-1	4	77
	1	-5	11	66
	1	-4	7	60
	1	-3	3	44
	2	-7	10	54
	2	-5	2	20
	3	-8	5	31
	3	-7	1	8

H	K U	L V	Φ W	φ2 φ1
	4	-11	8	35
	4	-9	0	0
	5	-12	3	13
1	4	10	22	14
	-2	-12	5	86
	0	-5	2	77
	2	-8	3	64
	2	-3	1	45
	4	-11	4	58
	4	-1	0	0
	6	-4	1	21
	8	-7	2	29
	10	-10	3	33
	10	-5	1	14
4	1	10	22	76
	-11	-6	5	76
	-9	-4	4	81
	-7	-12	4	47
	-7	-2	3	88
	-5	-10	3	43
	-3	-8	2	37
	-2	-2	1	61
	-1	-6	1	25
	0	-10	1	15
	1	-4	0	0
1	10	4	68	6
	0	-2	5	88
	2	-5	12	79
	2	-3	7	73
	2	-1	2	46
	4	-4	9	66
	6	-5	11	61
	6	-1	1	10
	8	-2	3	22
	10	-3	5	28
	10	-1	0	0
	12	-4	7	31
10	1	4	68	84
	-4	-4	11	73
	-3	-10	10	48
	-3	-2	8	79
	-2	-8	7	44
	-1	-10	5	29
	-1	-6	4	36

H	K U	L V	Φ W	φ2 φ1
	-1	-2	3	60
	0	-4	1	15
	1	-10	0	0
4	10	1	85	22
	0	-1	10	88
	1	-1	6	78
	2	-1	2	42
	3	-2	8	66
	5	-3	10	60
	5	-2	0	0
	7	-4	12	56
	7	-3	2	15
	9	-4	4	22
	11	-5	6	27
	12	-5	2	9
10	4	1	85	68
	0	-1	4	77
	1	-5	10	63
	1	-4	6	56
	1	-3	2	32
	2	-7	8	48
	2	-5	0	0
	3	-10	10	44
	3	-8	2	13
	4	-11	4	19
1	4	11	21	14
	-1	-8	3	84
	0	-11	4	77
	1	-3	1	59
	4	-1	0	0
	5	-4	1	26
	6	-7	2	37
	7	-10	3	43
	9	-5	1	16
	11	-11	3	33
4	1	11	21	76
	-12	-7	5	75
	-11	-11	5	61
	-9	-8	4	64
	-7	-5	3	70
	-5	-2	2	83
	-3	-10	2	32
	-2	-3	1	50
	-1	-7	1	24

H	K U	L V	Φ W	φ2 φ1
	0	-11	1	15
	1	-4	0	0
1	11	4	70	5
	0	-4	11	88
	1	-3	8	82
	2	-2	5	68
	3	-1	2	35
	5	-3	7	55
	7	-5	12	60
	7	-1	1	9
	8	-4	9	49
	10	-2	3	17
	11	-5	11	46
	11	-1	0	0
11	1	4	70	85
	-3	-11	11	48
	-3	-7	10	58
	-2	-6	7	52
	-1	-9	5	31
	-1	-5	4	41
	-1	-1	3	74
	0	-4	1	15
	1	-11	0	0
4	11	1	85	20
	0	-1	11	88
	1	-1	7	80
	2	-1	3	54
	3	-2	10	71
	5	-2	2	20
	7	-3	5	33
	8	-3	1	7
	9	-4	8	39
	11	-5	11	43
	11	-4	0	0
	12	-5	7	28
11	4	1	85	70
	0	-1	4	77
	1	-5	9	61
	1	-4	5	51
	1	-3	1	18
	2	-7	6	40
	3	-11	11	44
	3	-10	7	34
	4	-11	0	0

H	K U	L V	Φ W	φ2 φ1
1	4	12	19	14
	0	-3	1	77
	4	-10	3	56
	4	-7	2	48
	4	-4	1	32
	4	-1	0	0
	8	-11	3	42
	8	-5	1	19
	12	-9	2	24
	12	-6	1	13
4	1	12	19	76
	-12	-12	5	60
	-11	-4	4	84
	-9	-12	4	52
	-8	-4	3	78
	-7	-8	3	57
	-5	-4	2	67
	-3	-12	2	29
	-2	-4	1	42
	-1	-8	1	22
	0	-12	1	15
	1	-4	0	0
1	12	4	72	5
	0	-1	3	88
	4	-4	11	70
	4	-3	8	63
	4	-2	5	52
	4	-1	2	27
	8	-3	7	42
	8	-1	1	7
	12	-5	12	46
	12	-4	9	38
	12	-2	3	15
	12	-1	0	0
12	1	4	72	85
	-3	-8	11	56
	-3	-4	10	70
	-2	-12	9	39
	-2	-4	7	63
	-1	-12	6	28
	-1	-8	5	34
	-1	-4	4	47
	0	-4	1	15
	1	-12	0	0

H	K U	L V	Φ W	φ2 φ1
4	12	1	85	18
	0	-1	12	88
	1	-1	8	81
	2	-1	4	61
	3	-2	12	74
	3	-1	0	0
	5	-2	4	37
	7	-3	8	47
	8	-3	4	25
	9	-4	12	51
	11	-4	4	19
	12	-5	12	43
12	4	1	85	72
	0	-1	4	77
	1	-6	12	63
	1	-5	8	58
	1	-4	4	44
	1	-3	0	0
	2	-9	12	53
	2	-7	4	29
	3	-11	8	35
	3	-10	4	21
1	5	5	46	11
	0	-1	1	82
	5	-12	11	65
	5	-11	10	63
	5	-10	9	61
	5	-9	8	59
	5	-8	7	57
	5	-7	6	53
	5	-6	5	49
	5	-5	4	44
	5	-4	3	36
	5	-3	2	27
	5	-2	1	15
	5	-1	0	0
	10	-11	9	46
	10	-9	7	40
	10	-7	5	32
	10	-5	3	21
	10	-3	1	8
5	1	5	46	79
	-11	-5	12	81
	-10	-5	11	79

H	K U	L V	Φ W	φ2 φ1
	-9	-10	11	62
	-9	-5	10	77
	-8	-5	9	75
	-7	-10	9	56
	-7	-5	8	73
	-6	-5	7	69
	-5	-10	7	48
	-5	-5	6	65
	-4	-5	5	60
	-3	-10	5	37
	-3	-5	4	52
	-2	-5	3	43
	-1	-10	3	24
	-1	-5	2	31
	0	-5	1	16
	1	-10	1	8
	1	-5	0	0
5	5	1	82	45
	-1	-1	10	90
	0	-1	5	82
	1	-3	10	74
	1	-2	5	67
	1	-1	0	0
	2	-3	5	55
	3	-5	10	61
	3	-4	5	46
	4	-5	5	38
	5	-7	10	50
	5	-6	5	33
	6	-7	5	29
	7	-9	10	42
	7	-8	5	25
	8	-9	5	23
	9	-11	10	36
	9	-10	5	21
	10	-11	5	19
	11	-12	5	17
1	5	6	40	11
	-1	-7	6	88
	0	-6	5	81
	1	-11	9	77
	1	-5	4	72
	2	-4	3	59
	3	-9	7	66
	3	-3	2	41
	4	-2	1	20

H	K U	L V	Φ W	φ_2 φ_1
1	5	6	40	11
	5	-7	5	51
	5	-1	0	0
	7	-11	8	54
	7	-5	3	31
	8	-10	7	48
	9	-3	1	9
	10	-8	5	34
	11	-7	4	27
5	1	6	40	79
	-12	-6	11	78
	-11	-5	10	80
	-10	-4	9	82
	-9	-3	8	85
	-8	-2	7	88
	-6	-12	7	46
	-5	-11	6	43
	-4	-10	5	41
	-3	-9	4	37
	-2	-8	3	32
	-1	-7	2	25
	-1	-1	1	63
	0	-6	1	15
	1	-11	1	8
	1	-5	0	0
1	6	5	51	9
	-1	-9	11	88
	0	-5	6	84
	1	-6	7	78
	1	-1	1	48
	2	-7	8	73
	3	-8	9	70
	4	-9	10	67
	5	-10	11	65
	6	-11	12	64
	6	-1	0	0
	7	-2	1	10
	8	-3	2	17
	9	-4	3	22
	10	-5	4	26
	11	-6	5	29
	12	-7	6	31
6	1	5	51	81
	-8	-7	11	69
	-7	-8	10	62

H	K U	L V	Φ W	φ_2 φ_1
	-7	-3	9	81
	-5	-10	8	49
	-5	-5	7	66
	-4	-11	7	42
	-4	-1	5	87
	-3	-7	5	45
	-3	-2	4	74
	-2	-3	3	56
	-1	-9	3	24
	-1	-4	2	34
	0	-5	1	15
	1	-11	1	7
	1	-6	0	0
5	6	1	83	40
	0	-1	6	84
	1	-2	7	74
	1	-1	1	36
	2	-3	8	67
	3	-4	9	62
	4	-5	10	58
	5	-6	11	55
	6	-7	12	53
	6	-5	0	0
	7	-6	1	6
	8	-7	2	11
	9	-8	3	14
	10	-9	4	17
	11	-10	5	19
	12	-11	6	20
6	5	1	83	50
	-1	-1	11	89
	0	-1	5	81
	1	-3	9	72
	1	-2	4	62
	2	-3	3	40
	3	-5	7	51
	3	-4	2	22
	4	-7	11	54
	4	-5	1	9
	5	-8	10	47
	5	-7	5	30
	5	-6	0	0
	7	-10	8	34
	7	-9	3	15
	8	-11	7	27
	9	-11	1	4

H	K U	L V	Φ W	φ_2 φ_1
1	5	7	36	11
	-1	-11	8	85
	0	-7	5	81
	1	-10	7	76
	1	-3	2	65
	3	-2	1	27
	4	-5	3	46
	5	-8	5	53
	5	-1	0	0
	6	-11	7	56
	7	-7	4	40
	8	-3	1	11
	10	-9	5	36
	11	-12	7	42
	11	-5	2	16
5	1	7	36	79
	-11	-8	9	70
	-9	-11	8	56
	-9	-4	7	80
	-8	-9	7	59
	-7	-7	6	62
	-6	-5	5	66
	-5	-3	4	74
	-4	-1	3	88
	-2	-11	3	26
	-1	-9	2	21
	-1	-2	1	44
	0	-7	1	14
	1	-12	1	8
	1	-5	0	0
1	7	5	55	8
	-1	-7	10	90
	0	-5	7	85
	1	-8	11	81
	1	-3	4	74
	2	-1	1	30
	3	-4	5	60
	4	-7	9	66
	5	-5	6	52
	7	-6	7	48
	7	-1	0	0
	8	-9	11	56
	9	-7	8	45
	9	-2	1	8
	11	-8	9	43

H	K U	L V	Φ W	φ_2 φ_1
	11	-3	2	12
7	1	5	55	82
	-7	-11	12	56
	-7	-6	11	70
	-7	-1	10	90
	-5	-10	9	50
	-5	-5	8	67
	-4	-7	7	53
	-3	-4	5	60
	-2	-11	5	30
	-2	-1	3	79
	-1	-8	3	25
	-1	-3	2	41
	0	-5	1	14
	1	-12	1	6
	1	-7	0	0
5	7	1	83	36
	0	-1	7	85
	1	-2	9	78
	1	-1	2	55
	2	-3	11	73
	4	-3	1	11
	5	-4	3	25
	6	-5	5	33
	7	-6	7	38
	7	-5	0	0
	8	-7	9	41
	9	-8	11	43
	9	-7	4	19
	11	-9	8	30
	11	-8	1	4
7	5	1	83	54
	-1	-1	12	89
	0	-1	5	81
	1	-3	8	69
	1	-2	3	54
	2	-5	11	65
	2	-3	1	16
	3	-5	4	35
	4	-7	7	41
	5	-9	10	45
	5	-8	5	28
	5	-7	0	0
	7	-12	11	39
	7	-11	6	25
	7	-10	1	5

H	K U	L V	Φ W	φ_2 φ_1
1	5	8	33	11
	-1	-11	7	85
	0	-8	5	80
	1	-5	3	71
	2	-2	1	38
	3	-7	4	60
	4	-12	7	64
	5	-9	5	54
	5	-1	0	0
	7	-11	6	51
	7	-3	1	14
	9	-5	2	21
	11	-7	3	25
	12	-4	1	8
5	1	8	33	79
	-11	-9	8	66
	-10	-6	7	73
	-9	-11	7	55
	-7	-5	5	69
	-6	-10	5	47
	-5	-7	4	52
	-4	-4	3	61
	-3	-1	2	84
	-1	-11	2	19
	-1	-3	1	34
	0	-8	1	13
	1	-5	0	0
1	8	5	58	7
	0	-5	8	86
	1	-7	11	82
	1	-2	3	71
	3	-1	1	21
	4	-3	4	47
	5	-5	7	56
	6	-7	10	60
	7	-4	5	38
	8	-1	0	0
	9	-8	11	53
	10	-5	6	34
	11	-7	9	42
	11	-2	1	6
8	1	5	58	83
	-7	-4	12	78
	-6	-7	11	64

H	K U	L V	Φ W	φ_2 φ_1
	-5	-5	9	68
	-4	-3	7	73
	-3	-11	7	38
	-3	-1	5	84
	-2	-9	5	34
	-1	-12	4	22
	-1	-7	3	27
	-1	-2	2	52
	0	-5	1	13
	1	-8	0	0
5	8	1	84	32
	0	-1	8	86
	1	-2	11	80
	1	-1	3	65
	3	-2	1	16
	4	-3	4	39
	5	-4	7	48
	6	-5	10	52
	7	-5	5	30
	8	-5	0	0
	9	-7	11	44
	10	-7	6	26
	11	-8	9	34
	11	-7	1	4
8	5	1	84	58
	0	-1	5	80
	1	-4	12	72
	1	-3	7	66
	1	-2	2	42
	2	-5	9	60
	3	-7	11	56
	3	-5	1	10
	4	-7	3	21
	5	-9	5	26
	5	-8	0	0
	6	-11	7	29
	7	-12	4	16
1	5	9	30	11
	-1	-7	4	87
	0	-9	5	80
	1	-11	6	76
	1	-2	1	56
	5	-1	0	0
	6	-3	1	17
	7	-5	2	27

H	K U	L V	Φ W	φ_2 φ_1
	8	-7	3	33
	9	-9	4	37
	10	-11	5	40
	11	-4	1	10
5	1	9	30	79
	-12	-3	7	88
	-11	-8	7	68
	-6	-5	5	72
	-7	-10	5	50
	-5	-11	4	40
	-5	-2	3	81
	-4	-7	3	45
	-3	-3	2	60
	-1	-4	1	29
	0	-9	1	13
	1	-5	0	0
1	9	5	61	6
	0	-5	9	87
	1	-4	7	80
	2	-3	5	68
	3	-7	12	75
	3	-2	3	47
	4	-1	1	16
	5	-5	8	59
	7	-3	4	32
	8	-7	11	55
	9	-1	0	0
	10	-5	7	37
	11	-4	5	27
9	1	5	61	84
	-6	-1	11	88
	-5	-10	11	53
	-4	-9	9	50
	-3	-8	7	46
	-2	-7	5	40
	-1	-11	4	23
	-1	-6	3	30
	-1	-1	2	69
	0	-5	1	13
	1	-9	0	0
5	9	1	84	29
	0	-1	9	87
	1	-1	4	71
	3	-2	3	40
	4	-3	7	55

H	K U	L V	Φ W	φ_2 φ_1
	5	-4	11	60
	5	-3	2	19
	7	-5	10	50
	7	-4	1	7
	8	-5	5	28
	9	-5	0	0
	11	-7	8	32
	12	-7	3	12
9	5	1	84	61
	0	-1	5	80
	1	-4	11	70
	1	-3	6	69
	1	-2	1	24
	2	-5	7	53
	3	-7	8	47
	4	-9	9	43
	5	-11	10	40
	5	-9	0	0
	6	-11	1	5
1	5	10	27	11
	0	-2	1	80
	5	-11	5	57
	5	-9	4	53
	5	-7	3	46
	5	-5	2	37
	5	-3	1	22
	5	-1	0	0
	10	-12	5	42
	10	-8	3	30
	10	-4	1	12
5	1	10	27	79
	-12	-10	7	64
	-11	-5	6	78
	-9	-5	5	74
	-8	-10	5	53
	-7	-5	4	68
	-5	-5	3	59
	-4	-10	3	36
	-3	-5	2	46
	-1	-5	1	25
	0	-10	1	13
	1	-5	0	0
1	10	5	64	6
	0	-1	2	87

TAB. V FROM HKL-UVW INTO EULER ANGLES

H	K / U	L / V	Φ / W	φ2 / φ1
1	5	10		
1	10	5	64	6
	5	-6	11	66
	5	-5	9	61
	5	-4	7	56
	5	-3	5	47
	5	-2	3	33
	5	-1	1	12
	10	-7	12	52
	10	-5	8	41
	10	-3	4	24
	10	-1	0	0
10	1	5	64	84
	-5	-10	12	55
	-5	-5	11	70
	-4	-5	9	66
	-3	-10	8	43
	-3	-5	7	59
	-2	-5	5	49
	-1	-10	4	24
	-1	-5	3	34
	0	-5	1	13
	1	-10	0	0
5	10	1	85	27
	0	-1	10	87
	1	-1	5	75
	2	-1	0	0
	3	-2	5	55
	4	-3	10	64
	5	-3	5	41
	7	-4	5	32
	8	-5	10	47
	9	-5	5	26
	11	-6	5	22
	12	-7	10	36
10	5	1	85	63
	0	-1	5	80
	1	-4	10	68
	1	-3	5	58
	1	-2	0	0
	2	-5	5	43
	3	-8	10	50
	3	-7	5	33
	4	-9	5	27

H	K / U	L / V	Φ / W	φ2 / φ1
	5	-12	10	38
	5	-11	5	23
1	5	11	25	11
	0	-11	5	80
	1	-9	4	74
	2	-7	3	65
	3	-5	2	50
	4	-3	1	28
	5	-12	5	59
	5	-1	0	0
	7	-8	3	40
	9	-4	1	14
	11	-11	4	36
5	1	11	25	79
	-11	-11	6	59
	-9	-10	5	56
	-7	-9	4	52
	-5	-8	3	46
	-3	-7	2	37
	-2	-1	1	76
	-1	-6	1	23
	0	-11	1	12
	1	-5	0	0
1	11	5	66	5
	0	-5	11	88
	1	-1	2	64
	6	-1	1	10
	7	-2	3	25
	8	-3	5	34
	9	-4	7	39
	10	-5	9	43
	11	-6	11	46
	11	-1	0	0
11	1	5	66	85
	-5	-5	12	71
	-4	-11	11	49
	-4	-1	9	86
	-3	-7	8	53
	-3	-2	7	77
	-2	-3	5	63
	-1	-9	4	26
	-1	-4	3	40
	0	-5	1	12
	1	-11	0	0

H	K / U	L / V	Φ / W	φ2 / φ1
5	11	1	85	24
	0	-1	11	88
	1	-1	6	78
	2	-1	1	24
	3	-2	7	63
	5	-3	8	54
	7	-4	9	48
	9	-5	10	44
	11	-6	11	41
	11	-5	0	0
11	5	1	85	66
	0	-1	5	80
	1	-4	9	66
	1	-3	4	52
	2	-5	3	29
	3	-8	7	39
	3	-7	2	15
	4	-11	11	43
	4	-9	1	6
	5	-12	5	21
	5	-11	0	0
1	5	12	23	11
	-1	-7	3	87
	0	-12	5	80
	1	-5	2	69
	3	-3	1	36
	4	-8	3	54
	5	-1	0	0
	7	-11	4	49
	8	-4	1	17
	11	-7	2	23
5	1	12	23	79
	-11	-5	5	78
	-9	-3	4	83
	-5	-11	3	38
	-3	-9	2	32
	-2	-2	1	58
	-1	-7	1	21
	0	-12	1	12
	1	-5	0	0
1	12	5	67	5
	0	-5	12	88
	1	-3	7	81

H	K / U	L / V	Φ / W	φ2 / φ1
	2	-1	2	46
	3	-4	9	71
	5	-5	11	66
	7	-1	1	9
	9	-2	3	20
	11	-3	5	26
	12	-1	0	0
12	1	5	67	85
	-4	-7	11	61
	-3	-4	8	67
	-2	-11	7	35
	-2	-1	5	81
	-1	-8	4	29
	-1	-3	3	48
	0	-5	1	12
	1	-12	0	0
5	12	1	86	23
	0	-1	12	88
	1	-1	7	79
	2	-1	2	42
	3	-2	9	69
	5	-3	11	62
	7	-3	1	8
	9	-4	3	17
	11	-5	5	23
	12	-5	0	0
12	5	1	86	67
	0	-1	5	80
	1	-4	8	63
	1	-3	3	44
	2	-7	11	57
	2	-5	1	11
	3	-8	4	25
	4	-11	7	31
	5	-12	0	0
1	6	6	45	9
	0	-1	1	83
	6	-12	11	63
	6	-11	10	61
	6	-10	9	59
	6	-9	8	57
	6	-8	7	54
	6	-7	6	50
	6	-6	5	45

H	K / U	L / V	Φ / W	φ2 / φ1
	6	-5	4	40
	6	-4	3	33
	6	-3	2	24
	6	-2	1	13
	6	-1	0	0
	12	-11	9	43
	12	-9	7	36
	12	-7	5	28
	12	-5	3	18
	12	-3	1	6
6	1	6	45	81
	-11	-6	12	76
	-10	-6	11	75
	-9	-12	11	56
	-9	-6	10	72
	-8	-6	9	70
	-7	-12	9	50
	-7	-6	8	67
	-6	-6	7	63
	-5	-12	7	42
	-5	-6	6	59
	-4	-6	5	53
	-3	-12	5	32
	-3	-6	4	46
	-2	-6	3	37
	-1	-12	3	20
	-1	-6	2	26
	0	-6	1	13
	1	-12	1	7
	1	-6	0	0
6	6	1	83	45
	-1	-1	12	90
	0	-1	6	83
	1	-3	12	77
	1	-2	6	71
	1	-1	0	0
	2	-3	6	60
	3	-5	12	65
	3	-4	6	51
	4	-5	6	44
	5	-7	12	55
	5	-6	6	38
	6	-7	6	33
	7	-9	12	47
	7	-8	6	30
	8	-9	6	27
	9	-11	12	41

H	K / U	L / V	Φ / W	φ2 / φ1
	9	-10	6	24
	10	-11	6	22
	11	-12	6	20
1	6	7	41	9
	-1	-8	7	88
	0	-7	6	83
	1	-6	5	75
	2	-5	4	65
	3	-11	9	71
	3	-4	3	52
	4	-3	2	34
	5	-9	7	59
	5	-2	1	16
	6	-1	0	0
	7	-7	5	43
	9	-5	3	25
	10	-11	8	46
	11	-10	7	41
	11	-3	1	8
6	1	7	41	81
	-12	-5	11	80
	-11	-4	10	82
	-10	-3	9	85
	-9	-2	8	88
	-5	-12	6	40
	-4	-11	5	37
	-3	-10	4	33
	-2	-9	3	28
	-1	-8	2	22
	-1	-1	1	62
	0	-7	1	12
	1	-6	0	0
1	7	6	50	8
	0	-6	7	85
	1	-7	8	79
	1	-1	1	49
	2	-8	9	75
	3	-9	10	72
	4	-10	11	70
	5	-11	12	68
	7	-1	0	0
	8	-2	1	9
	9	-3	2	16
	10	-4	3	21
	11	-5	4	24

H	K / U	L / V	Φ / W	φ2 / φ1
	12	-6	5	27
7	1	6	50	82
	-9	-3	11	83
	-8	-10	11	59
	-7	-11	10	53
	-7	-5	9	71
	-5	-7	7	56
	-5	-1	6	88
	-4	-2	5	78
	-3	-9	5	38
	-3	-3	4	64
	-2	-4	3	47
	-1	-11	3	20
	-1	-5	2	29
	0	-6	1	12
	1	-7	0	0
6	7	1	84	41
	0	-1	7	85
	1	-2	8	76
	1	-1	1	36
	2	-3	9	69
	3	-4	10	64
	4	-5	11	60
	5	-6	12	57
	7	-6	0	0
	8	-7	1	5
	9	-8	2	9
	10	-9	3	13
	11	-10	4	15
	12	-11	5	17
7	6	1	84	49
	0	-1	6	83
	1	-3	11	75
	1	-2	5	67
	2	-3	4	46
	3	-5	9	58
	3	-4	3	31
	4	-5	2	17
	5	-7	7	39
	5	-6	1	7
	6	-7	0	0
	7	-10	11	42
	7	-9	5	24
	8	-11	10	37
	9	-11	3	12

H	K / U	L / V	Φ / W	φ2 / φ1
1	6	8	37	9
	0	-4	3	82
	2	-11	8	74
	2	-7	5	69
	2	-3	2	59
	4	-10	7	64
	4	-2	1	21
	6	-5	3	36
	6	-1	0	0
	8	-8	5	42
	10	-11	7	45
	10	-7	4	31
	10	-3	1	9
6	1	8	37	81
	-11	-6	9	75
	-10	-12	9	56
	-9	-10	8	58
	-9	-2	7	88
	-8	-8	7	60
	-7	-6	6	64
	-6	-4	5	70
	-5	-2	4	80
	-2	-12	3	23
	-1	-10	2	19
	-1	-2	1	42
	0	-8	1	12
	1	-6	0	0
1	8	6	53	7
	0	-3	4	86
	2	-7	9	76
	2	-4	5	68
	2	-1	1	31
	4	-5	6	58
	6	-9	11	69
	6	-6	7	52
	8	-7	8	49
	8	-1	0	0
	10	-8	9	46
	10	-2	1	7
	12	-9	10	44
	12	-3	2	11
8	1	6	53	83
	-8	-2	11	86
	-7	-10	11	57

TAB. V

FROM HKL-UVW INTO EULER ANGLES

H	K U	L V	Φ W	φ2 φ1
1	6	8		
8	1	6	53	83
	-7	-4	10	76
	-5	-8	8	54
	-5	-2	7	81
	-4	-10	7	43
	-3	-6	5	48
	-2	-2	3	65
	-1	-10	3	21
	-1	-4	2	33
	0	-6	1	12
	1	-8	0	0
6	8	1	84	37
	0	-1	8	86
	1	-2	10	79
	1	-1	2	55
	2	-3	12	74
	4	-3	0	0
	5	-4	2	17
	6	-5	4	27
	7	-6	6	33
	8	-7	8	37
	9	-8	10	40
	9	-7	2	10
	10	-9	12	42
	11	-9	6	23
8	6	1	84	53
	0	-1	6	82
	1	-3	10	73
	1	-2	4	61
	2	-3	2	29
	3	-5	6	46
	3	-4	0	0
	4	-7	10	51
	5	-8	8	41
	5	-7	2	13
	7	-11	10	38
	7	-10	4	18
	8	-11	2	8
1	6	9	34	9
	0	-3	2	82
	3	-11	7	69
	3	-8	5	64

H	K U	L V	Φ W	φ2 φ1
	3	-5	3	55
	3	-2	1	29
	6	-7	4	45
	6	-1	0	0
	9	-12	7	49
	9	-9	5	41
	9	-3	1	11
	12	-11	6	38
	12	-5	2	16
6	1	9	34	81
	-11	-6	8	74
	-10	-3	7	84
	-9	-9	7	59
	-7	-12	6	45
	-7	-3	5	79
	-6	-9	5	49
	-5	-6	4	54
	-4	-3	3	67
	-1	-12	2	17
	-1	-3	1	33
	0	-9	1	11
	1	-6	0	0
1	9	6	56	6
	0	-2	3	86
	3	-7	10	73
	3	-5	7	67
	3	-3	4	55
	3	-1	1	21
	6	-8	11	63
	6	-4	5	43
	9	-7	9	48
	9	-5	6	37
	9	-1	0	0
	12	-6	7	34
	12	-2	1	6
9	1	6	56	84
	-7	-9	12	60
	-7	-3	11	80
	-6	-12	11	50
	-5	-9	9	52
	-5	-3	8	76
	-4	-6	7	57
	-3	-3	5	66
	-2	-12	5	27
	-1	-9	3	22
	-1	-3	2	40

H	K U	L V	Φ W	φ2 φ1
	0	-6	1	11
	1	-9	0	0
6	9	1	85	34
	0	-1	9	86
	1	-2	12	81
	1	-1	3	65
	3	-2	0	0
	4	-3	3	31
	5	-4	6	43
	6	-5	9	49
	7	-6	12	53
	7	-5	3	19
	9	-7	9	38
	10	-7	3	14
	11	-8	6	24
9	6	1	85	56
	0	-1	6	82
	1	-3	9	71
	1	-2	3	54
	2	-5	12	66
	2	-3	0	0
	3	-5	3	27
	4	-7	6	37
	5	-9	9	41
	5	-8	3	18
	6	-11	12	44
	7	-12	9	33
	7	-11	3	13
1	6	10	31	9
	0	-5	3	82
	2	-12	7	74
	2	-7	4	68
	2	-2	1	40
	4	-9	5	61
	6	-11	6	56
	6	-1	0	0
	8	-3	1	13
	10	-5	2	20
	12	-7	3	24
6	1	10	31	81
	-11	-4	7	81
	-10	-10	7	59
	-8	-2	5	86
	-7	-8	5	55

H	K U	L V	Φ W	φ2 φ1
	-5	-10	4	40
	-4	-6	3	46
	-3	-2	2	69
	-1	-4	1	27
	0	-10	1	11
	1	-6	0	0
1	10	6	59	6
	0	-3	5	87
	2	-5	8	75
	2	-2	3	58
	4	-7	11	70
	4	-1	1	16
	6	-3	4	37
	8	-5	7	44
	10	-7	10	48
	10	-4	5	29
	10	-1	0	0
10	1	6	59	84
	-7	-2	12	85
	-6	-6	11	67
	-5	-4	9	72
	-4	-2	7	79
	-3	-12	7	35
	-2	-10	5	31
	-1	-8	3	24
	-1	-2	2	51
	0	-6	1	11
	1	-10	0	0
6	10	1	85	31
	0	-1	10	87
	1	-1	4	71
	3	-2	2	29
	4	-3	6	50
	5	-4	10	58
	5	-3	0	0
	7	-5	8	43
	8	-5	2	12
	10	-7	10	39
	11	-7	4	17
10	6	1	85	59
	0	-1	6	82
	1	-3	8	69
	1	-2	2	42
	2	-5	10	62
	3	-7	12	58

H	K U	L V	Φ W	φ2 φ1
	3	-5	0	0
	4	-7	2	14
	5	-9	4	21
	6	-11	6	26
	7	-12	2	8
1	6	11	29	9
	-1	-9	5	87
	0	-11	6	82
	1	-2	1	58
	6	-1	0	0
	7	-3	1	16
	8	-5	2	25
	9	-7	3	32
	10	-9	4	36
	11	-11	5	39
6	1	11	29	81
	-12	-5	7	78
	-11	-11	7	58
	-8	-7	5	62
	-7	-2	4	84
	-5	-3	3	71
	-4	-9	3	37
	-3	-4	2	50
	-1	-5	1	23
	0	-11	1	11
	1	-6	0	0
1	11	6	61	5
	0	-6	11	88
	1	-5	9	82
	2	-4	7	74
	3	-3	5	60
	4	-2	3	39
	5	-7	12	68
	5	-1	1	13
	7	-5	8	51
	9	-3	4	26
	11	-7	11	47
	11	-1	0	0
11	1	6	61	85
	-5	-11	11	50
	-4	-10	9	47
	-3	-9	7	43
	-2	-8	5	36
	-1	-7	3	26

H	K U	L V	Φ W	φ2 φ1
	-1	-1	2	68
	0	-6	1	11
	1	-11	0	0
6	11	1	85	29
	0	-1	11	88
	1	-1	5	75
	3	-2	4	48
	4	-3	9	61
	5	-3	3	27
	7	-4	2	14
	8	-5	7	37
	9	-5	1	6
	11	-7	11	40
	11	-6	0	0
	12	-7	5	20
11	6	1	85	61
	0	-1	6	82
	1	-3	7	66
	1	-2	1	24
	2	-5	8	56
	3	-7	9	50
	4	-9	10	46
	5	-11	11	42
	6	-11	0	0
1	6	12	27	9
	0	-2	1	82
	6	-11	5	55
	6	-9	4	50
	6	-7	3	43
	6	-5	2	33
	6	-3	1	19
	6	-1	0	0
	12	-12	5	39
	12	-8	3	27
	12	-4	1	10
6	1	12	27	81
	-12	-12	7	57
	-11	-6	6	73
	-9	-6	5	68
	-8	-12	5	46
	-7	-6	4	62
	-5	-6	3	52
	-4	-12	3	31
	-3	-6	2	39

H	K U	L V	Φ W	φ2 φ1
	-1	-6	1	21
	0	-12	1	11
	1	-6	0	0
1	12	6	64	5
	0	-1	2	88
	6	-6	11	62
	6	-5	9	58
	6	-4	7	51
	6	-3	5	42
	6	-2	3	29
	6	-1	1	10
	12	-7	12	47
	12	-5	8	36
	12	-3	4	20
	12	-1	0	0
12	1	6	64	85
	-5	-12	12	49
	-5	-6	11	66
	-4	-6	9	61
	-3	-12	8	37
	-3	-6	7	54
	-2	-6	5	44
	-1	-12	4	21
	-1	-6	3	30
	0	-6	1	11
	1	-12	0	0
6	12	1	86	27
	0	-1	12	88
	1	-1	6	77
	2	-1	0	0
	3	-2	6	59
	4	-3	12	68
	5	-3	6	46
	7	-4	6	37
	8	-5	12	52
	9	-5	6	30
	11	-6	6	26
	12	-7	12	41
12	6	1	86	63
	0	-1	6	82
	1	-4	12	72
	1	-3	6	63
	1	-2	0	0
	2	-5	6	48
	3	-8	12	55

H	K U	L V	Φ W	φ2 φ1
	3	-7	6	38
	4	-9	6	31
	5	-12	12	43
	5	-11	6	26
1	7	7	45	8
	0	-1	1	84
	7	-12	11	61
	7	-11	10	59
	7	-10	9	57
	7	-9	8	54
	7	-8	7	51
	7	-7	6	47
	7	-6	5	42
	7	-5	4	36
	7	-4	3	29
	7	-3	2	21
	7	-2	1	11
	7	-1	0	0
7	1	7	45	82
	-11	-7	12	72
	-10	-7	11	70
	-9	-7	10	68
	-8	-7	9	65
	-7	-7	8	62
	-6	-7	7	58
	-5	-7	6	54
	-4	-7	5	48
	-3	-7	4	41
	-2	-7	3	32
	-1	-7	2	23
	0	-7	1	11
	1	-7	0	0
7	7	1	84	45
	0	-1	7	84
	1	-2	7	79
	1	-1	0	0
	2	-3	7	63
	3	-4	7	55
	4	-5	7	48
	5	-6	7	42
	6	-7	7	37
	7	-8	7	34
	8	-9	7	30
	9	-10	7	28
	10	-11	7	25

TAB. V

FROM HKL-UVW INTO EULER ANGLES

H	K / U	L / V	Φ / W	φ_2 / φ_1
1	7	7		
7	7	1	84	45
	11	-12	7	23
1	7	8	41	8
	-1	-9	8	89
	0	-8	7	84
	1	-7	6	78
	2	-6	5	69
	3	-5	4	59
	4	-4	3	45
	5	-11	9	64
	5	-3	2	29
	6	-2	1	14
	7	-9	7	52
	7	-1	0	0
	9	-7	5	37
	11	-5	3	21
7	1	8	41	82
	-12	-4	11	82
	-11	-3	10	85
	-10	-2	9	88
	-4	-12	5	34
	-3	-11	4	30
	-2	-10	3	25
	-1	-9	2	19
	-1	-1	1	61
	0	-8	1	11
	1	-7	0	0
1	8	7	49	7
	0	-7	8	85
	1	-8	9	81
	1	-1	1	50
	2	-9	10	77
	3	-10	11	74
	4	-11	12	71
	8	-1	0	0
	9	-2	1	8
	10	-3	2	14
	11	-4	3	19
	12	-5	4	23
8	1	7	49	83
	-9	-5	11	75

H	K / U	L / V	Φ / W	φ_2 / φ_1
	-7	-7	9	63
	-6	-1	7	88
	-5	-9	7	48
	-5	-2	6	80
	-4	-3	5	69
	-3	-11	5	32
	-3	-4	4	56
	-2	-5	3	40
	-1	-6	2	24
	0	-7	1	11
	1	-8	0	0
7	8	1	85	41
	0	-1	8	85
	1	-2	9	77
	1	-1	1	35
	2	-3	10	71
	3	-4	11	66
	4	-5	12	62
	8	-7	0	0
	9	-8	1	5
	10	-9	2	8
	11	-10	3	11
	12	-11	4	14
8	7	1	85	49
	0	-1	7	84
	1	-2	6	70
	2	-3	5	55
	3	-5	11	63
	3	-4	4	39
	4	-5	3	25
	5	-7	9	47
	5	-6	2	14
	6	-7	1	6
	7	-9	7	32
	7	-8	0	0
	9	-11	5	19
1	7	9	38	8
	0	-9	7	84
	1	-4	3	72
	3	-3	2	44
	4	-7	5	59
	5	-11	8	63
	5	-2	1	17
	7	-10	7	54
	7	-1	0	0

H	K / U	L / V	Φ / W	φ_2 / φ_1
	8	-5	3	29
	11	-8	5	34
	12	-3	1	7
7	1	9	38	82
	-11	-4	9	81
	-10	-11	9	57
	-9	-9	8	59
	-8	-7	7	63
	-7	-5	6	68
	-6	-3	5	75
	-5	-1	4	87
	-1	-11	2	17
	-1	-2	1	41
	0	-9	1	10
	1	-7	0	0
1	9	7	52	6
	0	-7	9	86
	1	-4	5	77
	2	-1	1	31
	3	-5	6	65
	4	-9	11	70
	5	-6	7	58
	7	-7	8	53
	9	-8	9	49
	9	-1	0	0
	11	-9	10	47
	11	-2	1	6
9	1	7	52	84
	-8	-5	11	74
	-7	-7	10	64
	-5	-11	8	44
	-5	-4	7	69
	-3	-8	5	40
	-3	-1	4	83
	-2	-3	3	54
	-1	-12	3	18
	-1	-5	2	27
	0	-7	1	10
	1	-9	0	0
7	9	1	85	38
	0	-1	9	86
	1	-2	11	80
	1	-1	2	55
	5	-4	1	9
	6	-5	3	21

H	K / U	L / V	Φ / W	φ_2 / φ_1
	7	-6	5	29
	8	-7	7	34
	9	-8	9	37
	9	-7	0	0
	10	-9	11	39
	11	-9	4	16
9	7	1	85	52
	0	-1	7	84
	1	-3	12	76
	1	-2	5	66
	2	-3	3	40
	3	-5	8	54
	3	-4	1	11
	5	-8	11	50
	5	-7	4	25
	7	-10	7	30
	7	-9	0	0
	8	-11	5	20
1	7	10	35	8
	-1	-7	5	90
	0	-10	7	83
	1	-3	2	68
	4	-2	1	22
	5	-5	3	43
	6	-8	5	51
	7	-11	7	55
	7	-1	0	0
	9	-7	4	35
	11	-3	1	9
7	1	10	35	82
	-11	-3	8	84
	-9	-7	7	65
	-7	-11	6	46
	-7	-1	5	90
	-6	-8	5	51
	-5	-5	4	59
	-4	-2	3	75
	-1	-3	1	31
	0	-10	1	10
	1	-7	0	0
1	10	7	55	6
	0	-7	10	87
	1	-5	7	80
	2	-3	4	65

H	K / U	L / V	Φ / W	φ_2 / φ_1
	3	-8	11	74
	3	-1	1	22
	5	-4	5	49
	7	-7	9	55
	8	-5	6	41
	10	-1	0	0
	11	-6	7	36
10	1	7	55	84
	-7	-7	11	65
	-5	-6	8	61
	-4	-9	7	45
	-3	-5	5	52
	-2	-1	3	78
	-1	-11	3	19
	-1	-4	2	32
	0	-7	1	10
	1	-10	0	0
7	10	1	85	35
	0	-1	10	87
	1	-1	3	65
	4	-3	2	22
	5	-4	5	38
	6	-5	8	46
	7	-6	11	50
	7	-5	1	7
	9	-7	7	32
	10	-7	0	0
	11	-8	3	12
10	7	1	85	55
	0	-1	7	83
	1	-3	11	75
	1	-2	4	61
	2	-3	1	16
	3	-5	5	41
	4	-7	9	48
	5	-8	6	33
	7	-11	7	28
	7	-10	0	0
1	7	11	33	8
	0	-11	7	83
	1	-8	5	77
	2	-5	3	64
	3	-2	1	30
	5	-7	4	51

H	K / U	L / V	Φ / W	φ_2 / φ_1
	7	-12	7	56
	7	-1	0	0
	8	-9	5	45
	10	-3	1	10
	11	-11	6	42
7	1	11	33	82
	-11	-11	8	58
	-10	-7	7	67
	-7	-6	5	62
	-5	-9	4	42
	-4	-5	3	52
	-3	-1	2	81
	-1	-4	1	26
	0	-11	1	10
	1	-7	0	0
1	11	7	58	5
	0	-7	11	87
	1	-2	3	72
	4	-1	1	16
	5	-3	4	42
	6	-5	7	52
	7	-7	10	57
	9	-4	5	32
	11	-8	11	48
	11	-1	0	0
11	1	7	58	85
	-7	-7	12	66
	-6	-11	11	51
	-5	-8	9	55
	-5	-1	8	87
	-4	-5	7	61
	-3	-2	5	74
	-1	-10	3	20
	-1	-3	2	39
	0	-7	1	10
	1	-11	0	0
7	11	1	86	32
	0	-1	11	87
	1	-1	4	71
	3	-2	1	16
	4	-3	5	45
	5	-4	9	55
	7	-5	6	35
	10	-7	7	30
	11	-8	11	39

H	K / U	L / V	Φ / W	φ_2 / φ_1
	11	-7	0	0
11	7	1	86	58
	0	-1	7	83
	1	-3	10	73
	1	-2	3	54
	3	-5	2	19
	4	-7	5	32
	5	-9	8	38
	5	-8	1	6
	6	-11	11	41
	7	-12	7	27
	7	-11	0	0
1	7	12	31	8
	0	-12	7	83
	1	-7	4	76
	2	-2	1	41
	3	-9	5	67
	5	-11	6	61
	7	-1	0	0
	9	-3	1	12
	11	-5	2	19
7	1	12	31	82
	-11	-7	7	69
	-8	-4	5	74
	-7	-11	5	45
	-5	-1	3	87
	-4	-8	3	39
	-3	-3	2	57
	-1	-5	1	22
	0	-12	1	9
	1	-7	0	0
1	12	7	60	5
	0	-7	12	88
	1	-3	5	78
	3	-2	3	48
	4	-5	8	65
	5	-1	1	13
	7	-7	11	59
	8	-3	4	29
	11	-5	7	35
	12	-1	0	0
12	1	7	60	85
	-6	-5	11	71

H	K / U	L / V	Φ / W	φ_2 / φ_1
	-5	-3	9	76
	-4	-1	7	85
	-2	-11	5	28
	-1	-9	3	21
	-1	-2	2	50
	0	-7	1	9
	1	-12	0	0
7	12	1	86	30
	0	-1	12	88
	1	-1	5	75
	3	-2	3	40
	4	-3	8	58
	5	-3	1	10
	7	-5	11	52
	8	-5	4	23
	11	-7	7	28
	12	-7	0	0
12	7	1	86	60
	0	-1	7	83
	1	-3	9	71
	1	-2	2	42
	2	-5	11	64
	4	-7	1	7
	5	-9	3	16
	6	-11	5	22
	7	-12	0	0
1	8	8	45	7
	0	-1	1	85
	8	-12	11	59
	8	-11	10	57
	8	-10	9	54
	8	-9	8	51
	8	-8	7	48
	8	-7	6	44
	8	-6	5	39
	8	-5	4	33
	8	-4	3	27
	8	-3	2	19
	8	-2	1	10
	8	-1	0	0
8	1	8	45	83
	-11	-6	12	69
	-10	-8	11	67
	-9	-8	10	64

TAB. V

FROM HKL-UVW INTO EULER ANGLES

H	K / U	L / V	Φ / W	φ_2 / φ_1
1	8	8		
8	1	8	45	83
	-8	-8	9	61
	-7	-8	8	58
	-6	-8	7	54
	-5	-8	6	49
	-4	-8	5	43
	-3	-8	4	37
	-2	-8	3	29
	-1	-8	2	20
	0	-8	1	10
	1	-8	0	0
8	8	1	85	45
	0	-1	8	85
	1	-2	8	75
	1	-1	0	0
	2	-3	8	66
	3	-4	8	58
	4	-5	8	52
	5	-6	8	46
	6	-7	8	41
	7	-8	8	37
	8	-9	8	34
	9	-10	8	31
	10	-11	8	28
	11	-12	8	26
1	8	9	42	7
	-1	-10	9	89
	0	-9	8	85
	1	-8	7	79
	2	-7	6	72
	3	-6	5	64
	4	-5	4	53
	5	-4	3	39
	6	-3	2	25
	7	-11	9	58
	7	-2	1	12
	8	-1	0	0
	9	-9	7	46
	11	-7	5	32
8	1	9	42	83
	-12	-3	11	85
	-11	-2	10	88

H	K / U	L / V	Φ / W	φ_2 / φ_1
	-3	-12	4	27
	-2	-11	3	23
	-1	-10	2	17
	-1	-1	1	60
	0	-9	1	10
	1	-8	0	0
1	9	8	49	6
	0	-8	9	86
	1	-9	10	82
	1	-1	1	50
	2	-10	11	78
	3	-11	12	75
	9	-1	0	0
	10	-2	1	7
	11	-3	2	13
	12	-4	3	18
9	1	8	49	84
	-9	-7	11	68
	-7	-9	9	56
	-7	-1	8	89
	-6	-2	7	82
	-5	-11	7	42
	-5	-3	6	73
	-4	-4	5	62
	-3	-5	4	49
	-2	-6	3	35
	-1	-7	2	21
	0	-8	1	10
	1	-9	0	0
8	9	1	85	42
	0	-1	9	86
	1	-2	10	78
	1	-1	1	35
	2	-3	11	72
	3	-4	12	68
	9	-8	0	0
	10	-9	1	4
	11	-10	2	8
	12	-11	3	10
9	8	1	85	48
	0	-1	8	85
	1	-2	7	73
	2	-3	6	59
	3	-4	5	45
	4	-5	4	32

H	K / U	L / V	Φ / W	φ_2 / φ_1
	5	-7	11	52
	5	-6	3	21
	6	-7	2	12
	7	-9	9	38
	7	-8	1	5
	8	-9	0	0
	9	-11	7	26
1	8	10	39	7
	0	-5	4	84
	2	-9	7	74
	2	-4	3	63
	4	-3	2	36
	6	-7	5	49
	6	-2	1	14
	8	-11	8	54
	8	-1	0	0
	10	-10	7	45
	10	-5	3	24
8	1	10	39	83
	-11	-12	10	57
	-11	-2	9	88
	-10	-10	9	59
	-9	-8	8	62
	-8	-6	7	66
	-7	-4	6	72
	-6	-2	5	81
	-1	-12	2	15
	-1	-2	1	41
	0	-10	1	9
	1	-8	0	0
1	10	8	51	6
	0	-4	5	86
	2	-9	11	78
	2	-5	6	72
	2	-1	1	31
	4	-6	7	63
	6	-7	8	57
	8	-8	9	53
	10	-9	10	50
	10	-1	0	0
	12	-10	11	47
	12	-2	1	6
10	1	8	51	84
	-8	-8	11	63

H	K / U	L / V	Φ / W	φ_2 / φ_1
	-7	-10	10	54
	-7	-2	9	84
	-5	-6	7	59
	-3	-10	5	34
	-3	-2	4	72
	-2	-4	3	45
	-1	-6	2	24
	0	-8	1	9
	1	-10	0	0
8	10	1	86	39
	0	-1	10	86
	1	-2	12	80
	1	-1	2	55
	5	-4	0	0
	6	-5	2	14
	7	-6	4	24
	8	-7	6	30
	9	-8	8	34
	10	-9	10	37
	11	-10	12	39
	11	-9	2	8
10	8	1	86	51
	0	-1	8	84
	1	-2	6	70
	2	-3	4	48
	3	-5	10	60
	3	-4	2	22
	4	-5	0	0
	5	-7	6	35
	7	-10	10	39
	7	-9	2	10
	8	-11	8	31
1	8	11	36	7
	0	-11	8	84
	1	-7	5	78
	2	-3	2	55
	3	-10	7	70
	5	-2	1	18
	7	-5	3	34
	8	-1	0	0
	9	-8	5	40
	11	-11	7	44
	12	-7	4	28

H	K / U	L / V	Φ / W	φ_2 / φ_1
8	1	11	36	83
	-11	-11	9	58
	-9	-5	7	72
	-7	-10	6	48
	-6	-7	5	54
	-5	-4	4	64
	-4	-1	3	84
	-1	-3	1	31
	0	-11	1	9
	1	-8	0	0
1	11	8	54	5
	0	-8	11	87
	1	-3	4	76
	3	-1	1	22
	4	-4	5	55
	5	-7	9	63
	7	-5	6	45
	10	-6	7	39
	11	-9	11	49
	11	-1	0	0
11	1	8	54	85
	-7	-11	11	53
	-7	-3	10	79
	-5	-9	8	49
	-5	-1	7	86
	-4	-12	7	37
	-3	-7	5	43
	-2	-2	3	64
	-1	-5	2	27
	0	-8	1	9
	1	-11	0	0
8	11	1	86	36
	0	-1	11	87
	1	-1	3	65
	4	-3	1	11
	5	-4	4	32
	6	-5	7	42
	7	-6	10	47
	9	-7	5	24
	11	-9	11	38
	11	-8	0	0
11	8	1	86	54
	0	-1	8	84
	1	-2	5	66

H	K / U	L / V	Φ / W	φ_2 / φ_1
	2	-3	2	29
	3	-5	7	50
	4	-7	12	56
	5	-8	9	44
	5	-7	1	7
	7	-11	11	40
	7	-10	3	14
	8	-11	0	0
1	8	12	34	7
	0	-3	2	84
	4	-11	7	67
	4	-8	5	61
	4	-5	3	50
	4	-2	1	23
	8	-7	4	39
	8	-1	0	0
	12	-12	7	43
	12	-9	5	35
	12	-3	1	8
8	1	12	34	83
	-11	-8	8	65
	-10	-4	7	78
	-9	-12	7	49
	-7	-4	5	71
	-6	-12	5	39
	-5	-8	4	44
	-4	-4	3	57
	-1	-4	1	25
	0	-12	1	9
	1	-8	0	0
1	12	8	56	5
	0	-2	3	87
	4	-7	10	69
	4	-5	7	62
	4	-3	4	49
	4	-1	1	16
	8	-8	11	57
	8	-4	5	36
	12	-7	9	41
	12	-5	6	30
	12	-1	0	0
12	1	8	56	85
	-7	-12	12	52
	-7	-4	11	76

H	K / U	L / V	Φ / W	φ_2 / φ_1
	-5	-12	9	43
	-5	-4	8	70
	-4	-8	7	48
	-3	-4	5	58
	-1	-12	3	17
	-1	-4	2	32
	0	-8	1	9
	1	-12	0	0
8	12	1	86	34
	0	-1	12	87
	1	-1	4	71
	3	-2	0	0
	4	-3	4	39
	5	-4	8	51
	6	-5	12	57
	7	-5	4	25
	9	-7	12	47
	10	-7	4	18
	11	-8	8	31
12	8	1	86	56
	0	-1	8	84
	1	-3	12	76
	1	-2	4	61
	2	-3	0	0
	3	-5	4	35
	4	-7	8	45
	5	-9	12	50
	5	-8	4	23
	7	-12	12	41
	7	-11	4	17
1	9	9	45	6
	0	-1	1	86
	9	-12	11	56
	9	-11	10	54
	9	-10	9	52
	9	-9	8	49
	9	-8	7	45
	9	-7	6	41
	9	-6	5	36
	9	-5	4	31
	9	-4	3	24
	9	-3	2	17
	9	-2	1	9
	9	-1	0	0

H	K / U	L / V	Φ / W	φ_2 / φ_1
9	1	9	45	84
	-11	-9	12	65
	-10	-9	11	63
	-9	-9	10	61
	-8	-9	9	58
	-7	-9	8	54
	-6	-9	7	50
	-5	-9	6	45
	-4	-9	5	40
	-3	-9	4	33
	-2	-9	3	26
	-1	-9	2	18
	0	-9	1	9
	1	-9	0	0
9	9	1	86	45
	0	-1	9	86
	1	-2	9	77
	1	-1	0	0
	2	-3	9	69
	3	-4	9	61
	4	-5	9	55
	5	-6	9	49
	6	-7	9	44
	7	-8	9	40
	8	-9	9	37
	9	-10	9	34
	10	-11	9	31
	11	-12	9	29
1	9	10	42	6
	-1	-11	10	89
	0	-10	9	85
	1	-9	8	81
	2	-8	7	75
	3	-7	6	67
	4	-6	5	58
	5	-5	4	47
	6	-4	3	35
	7	-3	2	22
	8	-2	1	10
	9	-11	9	59
	9	-1	0	0
	11	-9	7	41
9	1	10	42	84
	-12	-2	11	88

TAB. V — FROM HKL-UVW INTO EULER ANGLES

H	K / U	L / V	Φ / W	φ_2 / φ_1
1	9	10		
9	1	10	42	84
	-2	-12	3	21
	-1	-11	2	15
	-1	-1	1	59
	0	-10	1	9
	1	-9	0	0
1	10	9	48	6
	0	-9	10	86
	1	-10	11	82
	1	-1	1	51
	2	-11	12	79
	10	-1	0	0
	11	-2	1	7
	12	-3	2	12
0	1	9	48	84
	-9	-9	11	61
	-8	-1	9	89
	-7	-11	9	50
	-7	-2	8	83
	-6	-3	7	76
	-5	-4	6	67
	-4	-5	5	56
	-3	-6	4	43
	-2	-7	3	31
	-1	-8	2	19
	0	-9	1	9
	1	-10	0	0
9	10	1	86	42
	0	-1	10	86
	1	-2	11	79
	1	-1	1	35
	2	-3	12	74
	10	-9	0	0
	11	-10	1	4
	12	-11	2	7
0	9	1	86	48
	0	-1	9	85
	1	-2	8	75
	2	-3	7	63
	3	-4	6	50
	4	-5	5	38
	5	-6	4	27

H	K / U	L / V	Φ / W	φ_2 / φ_1
	6	-7	3	18
	7	-9	11	44
	7	-8	2	11
	8	-9	1	5
	9	-11	9	32
	9	-10	0	0
1	9	11	39	6
	0	-11	9	85
	1	-5	4	76
	3	-4	3	54
	4	-9	7	66
	5	-3	2	31
	7	-2	1	12
	8	-7	5	42
	9	-1	0	0
	11	-11	8	46
	12	-5	3	21
9	1	11	39	84
	-11	-11	10	58
	-10	-9	9	61
	-9	-7	8	65
	-8	-5	7	70
	-7	-3	6	77
	-6	-1	5	88
	-1	-2	1	40
	0	-11	1	8
	1	-9	0	0
1	11	9	51	5
	0	-9	11	87
	1	-5	6	79
	2	-1	1	32
	3	-6	7	69
	5	-7	8	61
	7	-8	9	56
	9	-9	10	53
	11	-10	11	50
	11	-1	0	0
11	1	9	51	85
	-8	-11	11	54
	-7	-4	9	74
	-5	-8	7	50
	-4	-1	5	84
	-3	-12	5	29
	-3	-3	4	62

H	K / U	L / V	Φ / W	φ_2 / φ_1
	-2	-5	3	39
	-1	-7	2	21
	0	-9	1	8
	1	-11	0	0
9	11	1	86	39
	0	-1	11	87
	1	-1	2	55
	6	-5	1	7
	7	-6	3	18
	8	-7	5	25
	9	-8	7	30
	10	-9	9	34
	11	-10	11	37
	11	-9	0	0
11	9	1	86	51
	0	-1	9	85
	1	-2	7	73
	2	-3	5	54
	3	-5	12	64
	3	-4	3	31
	4	-5	1	9
	5	-7	8	43
	7	-9	4	19
	8	-11	11	39
	9	-11	0	0
1	9	12	37	6
	0	-4	3	85
	3	-11	8	72
	3	-7	5	66
	3	-3	2	45
	6	-10	7	59
	6	-2	1	15
	9	-5	3	28
	9	-1	0	0
	12	-8	5	33
9	1	12	37	84
	-11	-9	9	63
	-9	-3	7	80
	-8	-12	7	46
	-7	-9	6	51
	-6	-6	5	57
	-5	-3	4	70
	-1	-3	1	30
	0	-12	1	8

H	K / U	L / V	Φ / W	φ_2 / φ_1
	1	-9	0	0
1	12	9	53	5
	0	-3	4	87
	3	-7	9	72
	3	-4	5	62
	3	-1	1	22
	6	-5	6	50
	9	-9	11	55
	9	-6	7	43
	12	-7	8	39
	12	-1	0	0
12	1	9	53	85
	-8	-3	11	80
	-7	-6	10	67
	-5	-12	8	41
	-5	-3	7	74
	-3	-9	5	36
	-2	-3	3	53
	-1	-6	2	23
	0	-9	1	8
	1	-12	0	0
9	12	1	86	37
	0	-1	12	87
	1	-1	3	65
	4	-3	0	0
	5	-4	3	25
	6	-5	6	38
	7	-6	9	44
	8	-7	12	49
	9	-7	3	15
	11	-9	9	32
12	9	1	86	53
	0	-1	9	85
	1	-2	6	70
	2	-3	3	40
	3	-5	9	57
	3	-4	0	0
	5	-8	12	52
	5	-7	3	19
	7	-10	6	26
	8	-11	3	12
1	10	10	45	6
	0	-1	1	86

H	K / U	L / V	Φ / W	φ_2 / φ_1
	10	-12	11	54
	10	-11	10	52
	10	-10	9	49
	10	-9	8	46
	10	-8	7	43
	10	-7	6	38
	10	-6	5	34
	10	-5	4	28
	10	-4	3	22
	10	-3	2	15
	10	-2	1	8
	10	-1	0	0
10	1	10	45	84
	-11	-10	12	62
	-10	-10	11	60
	-9	-10	10	57
	-8	-10	9	54
	-7	-10	8	51
	-6	-10	7	47
	-5	-10	6	42
	-4	-10	5	36
	-3	-10	4	30
	-2	-10	3	23
	-1	-10	2	16
	0	-10	1	8
	1	-10	0	0
10	10	1	86	45
	0	-1	10	86
	1	-2	10	78
	1	-1	0	0
	2	-3	10	71
	3	-4	10	64
	4	-5	10	58
	5	-6	10	52
	6	-7	10	47
	7	-8	10	43
	8	-9	10	40
	9	-10	10	37
	10	-11	10	34
	11	-12	10	32
1	10	11	42	6
	-1	-12	11	89
	0	-11	10	86
	1	-10	9	82
	2	-9	8	76

H	K / U	L / V	Φ / W	φ_2 / φ_1
	3	-8	7	70
	4	-7	6	62
	5	-6	5	53
	6	-5	4	43
	7	-4	3	31
	8	-3	2	20
	9	-2	1	9
	10	-1	0	0
	11	-11	9	48
10	1	11	42	84
	-1	-12	2	14
	-1	-1	1	59
	0	-11	1	8
	1	-10	0	0
1	11	10	48	5
	0	-10	11	87
	1	-11	12	83
	1	-1	1	51
	11	-1	0	0
	12	-2	1	6
11	1	10	48	85
	-9	-11	11	56
	-9	-1	10	89
	-8	-2	9	84
	-7	-3	8	78
	-6	-4	7	70
	-5	-5	6	61
	-4	-6	5	50
	-3	-7	4	39
	-2	-8	3	27
	-1	-9	2	17
	0	-10	1	8
	1	-11	0	0
10	11	1	86	42
	0	-1	11	87
	1	-2	12	80
	1	-1	1	35
	11	-10	0	0
	12	-11	1	4
11	10	1	86	48
	0	-1	10	86
	1	-2	9	77
	2	-3	8	66
	3	-4	7	55

H	K / U	L / V	Φ / W	φ_2 / φ_1
	4	-5	6	43
	5	-6	5	33
	6	-7	4	24
	7	-8	3	16
	8	-9	2	9
	9	-11	11	38
	9	-10	1	4
	10	-11	0	0
1	10	12	40	6
	0	-6	5	86
	2	-11	9	78
	2	-5	4	68
	4	-4	3	47
	6	-9	7	58
	6	-3	2	26
	8	-2	1	11
	10	-7	5	36
	10	-1	0	0
10	1	12	40	84
	-12	-12	11	58
	-11	-10	10	60
	-10	-8	9	64
	-9	-6	8	68
	-8	-4	7	74
	-7	-2	6	82
	-1	-2	1	39
	0	-12	1	7
	1	-10	0	0
1	12	10	50	5
	0	-5	6	87
	2	-6	7	75
	2	-1	1	32
	4	-7	8	66
	6	-8	9	60
	8	-9	10	56
	10	-10	11	53
	12	-11	12	50
	12	-1	0	0
12	1	10	50	85
	-9	-2	11	85
	-7	-6	9	65
	-5	-10	7	44
	-4	-2	5	76
	-3	-4	4	54

H	K / U	L / V	Φ / W	φ_2 / φ_1
	-2	-6	3	14
	-1	-8	2	18
	0	-10	1	7
	1	-12	0	0
10	12	1	86	40
	0	-1	12	87
	1	-1	2	55
	6	-5	0	0
	7	-6	2	12
	8	-7	4	21
	9	-8	6	27
	10	-9	8	31
	11	-10	10	34
	12	-11	12	36
12	10	1	86	50
	0	-1	10	86
	1	-2	8	75
	2	-3	6	59
	3	-4	4	39
	4	-5	2	17
	5	-7	10	49
	5	-6	0	0
	7	-9	6	28
	9	-11	2	8
1	11	11	45	5
	0	-1	1	86
	11	-12	11	52
	11	-11	10	50
	11	-10	9	47
	11	-9	8	44
	11	-8	7	40
	11	-7	6	36
	11	-6	5	32
	11	-5	4	26
	11	-4	3	21
	11	-3	2	14
	11	-2	1	7
	11	-1	0	0
11	1	11	45	85
	-11	-11	12	60
	-10	-11	11	57
	-9	-11	10	54
	-8	-11	9	51
	-7	-11	8	48

TAB. V

FROM HKL-UVW INTO EULER ANGLES

H	K / U	L / V	Φ / W	φ2 / φ1
1	11	11		
11	1	11	45	85
	-6	-11	7	43
	-5	-11	6	39
	-4	-11	5	34
	-3	-11	4	28
	-2	-11	3	21
	-1	-11	2	15
	0	-11	1	7
	1	-11	0	0
11	11	1	86	45
	0	-1	11	86
	1	-2	11	79
	1	-1	0	0
	2	-3	11	72
	3	-4	11	66
	4	-5	11	60
	5	-6	11	55
	6	-7	11	50
	7	-8	11	46
	8	-9	11	43
	9	-10	11	39
	10	-11	11	37
	11	-12	11	34
1	11	12	43	5
	0	-12	11	86
	1	-11	10	82
	2	-10	9	78
	3	-9	8	72
	4	-8	7	66
	5	-7	6	58
	6	-6	5	49
	7	-5	4	39
	8	-4	3	28
	9	-3	2	18
	10	-2	1	8
	11	-1	0	0
11	1	12	43	85
	-1	-1	1	58
	0	-12	1	7
	1	-11	0	0
1	12	11	48	5
	0	-11	12	87
	1	-1	1	51
	12	-1	0	0
12	1	11	48	85
	-10	-1	11	89
	-9	-2	10	85
	-8	-3	9	79
	-7	-4	8	73
	-6	-5	7	65
	-5	-6	6	56
	-4	-7	5	46
	-3	-8	4	35
	-2	-9	3	25
	-1	-10	2	15
	0	-11	1	7
	1	-12	0	0
11	12	1	86	43
	0	-1	12	87
	1	-1	1	35
	12	-11	0	0
12	11	1	86	47
	0	-1	11	86
	1	-2	10	78
	2	-3	9	68
	3	-4	8	58
	4	-5	7	48
	5	-6	6	38
	6	-7	5	29
	7	-8	4	21
	8	-9	3	14
	9	-10	2	8
	10	-11	1	4
	11	-12	0	0
1	12	12	45	5
	0	-1	1	87
	12	-12	11	50
	12	-11	10	48
	12	-10	9	45
	12	-9	8	42
	12	-8	7	38
	12	-7	6	34
	12	-6	5	30
	12	-5	4	25
	12	-4	3	19
	12	-3	2	13
	12	-2	1	7
	12	-1	0	0
12	1	12	45	85
	-11	-12	12	57
	-10	-12	11	54
	-9	-12	10	52
	-8	-12	9	48
	-7	-12	8	45
	-6	-12	7	41
	-5	-12	6	36
	-4	-12	5	31
	-3	-12	4	26
	-2	-12	3	20
	-1	-12	2	13
	0	-12	1	7
	1	-12	0	0
12	12	1	87	45
	0	-1	12	87
	1	-2	12	80
	1	-1	0	0
	2	-3	12	74
	3	-4	12	68
	4	-5	12	62
	5	-6	12	57
	6	-7	12	53
	7	-8	12	49
	8	-9	12	45
	9	-10	12	42
	10	-11	12	39
	11	-12	12	36
2	2	3	43	45
	-7	-11	12	81
	-7	-8	10	87
	-6	-9	10	82
	-5	-7	8	83
	-4	-11	10	71
	-4	-5	6	85
	-3	-12	10	66
	-3	-9	8	70
	-3	-3	4	90
	-2	-7	6	68
	-1	-11	8	59
	-1	-8	6	60
	-1	-5	4	64
	-1	-2	2	76
	0	-3	2	54
	1	-10	6	48
	1	-7	4	46
	1	-4	2	40
	1	-1	0	0
	2	-11	6	44
	2	-5	2	30
	3	-9	4	34
	3	-6	2	25
	4	-7	2	21
	5	-11	4	27
	5	-8	2	18
	6	-9	2	15
	7	-10	2	14
	8	-11	2	12
	9	-12	2	11
2	3	2	61	34
	-2	-6	11	82
	-1	-6	10	78
	-1	-4	7	80
	-1	-2	4	87
	0	-2	3	72
	1	-8	11	67
	1	-6	8	66
	1	-4	5	62
	1	-2	2	50
	2	-6	7	58
	2	-2	1	22
	3	-10	12	60
	3	-8	9	56
	3	-4	3	36
	3	-2	0	0
	4	-10	11	55
	4	-6	5	41
	5	-8	7	43
	5	-6	4	31
	5	-4	1	10
	6	-10	9	44
	7	-12	11	45
	7	-10	8	39
	7	-8	5	29
	7	-6	2	14
	8	-10	7	33
	8	-6	1	7
	9	-10	6	28
	9	-8	3	16
	11	-12	7	27
	11	-10	4	17
	11	-8	1	5
	12	-10	3	12
3	2	2	61	56
	-4	-5	11	81
	-4	-3	9	88
	-2	-9	12	65
	-2	-8	11	66
	-2	-7	10	68
	-2	-6	9	69
	-2	-5	8	72
	-2	-4	7	75
	-2	-3	6	79
	-2	-2	5	84
	0	-1	1	54
	2	-12	9	43
	2	-11	8	42
	2	-10	7	40
	2	-9	6	39
	2	-8	5	36
	2	-7	4	33
	2	-6	3	29
	2	-5	2	23
	2	-4	1	14
	2	-3	0	0
	4	-11	5	27
	4	-9	3	19
	4	-7	1	8
	6	-11	2	10
	6	-10	1	6
2	2	5	29	45
	-9	-11	8	85
	-7	-8	6	87
	-5	-10	6	74
	-5	-5	4	90
	-4	-11	6	68
	-3	-7	4	71
	-2	-3	2	80
	-1	-9	4	55
	-1	-4	2	62
	0	-5	2	49
	1	-11	4	44
	1	-6	2	39
	1	-1	0	0
	2	-7	2	33
	3	-8	2	28
	4	-9	2	24
	5	-10	2	21
	6	-11	2	19
	7	-12	2	17
2	5	2	70	22
	-1	-4	11	87
	0	-2	5	82
	1	-4	9	76
	1	-2	4	69
	2	-2	3	51
	3	-4	7	60
	3	-2	2	31
	4	-6	11	63
	4	-2	1	13
	5	-6	10	57
	5	-4	5	41
	5	-2	0	0
	7	-6	8	44
	7	-4	3	22
	8	-6	7	38
	9	-8	11	46
	9	-4	1	6
	10	-6	5	25
	11	-8	9	36
	11	-6	4	19
5	2	2	70	68
	-2	-7	12	66
	-2	-6	11	68
	-2	-5	10	70
	-2	-4	9	73
	-2	-3	8	77
	-2	-2	7	82
	-2	-1	6	88
	0	-1	1	49
	2	-12	7	32
	2	-11	6	30
	2	-10	5	28
	2	-9	4	25
	2	-8	3	21
	2	-7	2	16
	2	-6	1	10
	2	-5	0	0
	4	-11	1	5
2	2	7	22	45
	-10	-11	6	87
	-7	-7	4	90
	-5	-9	4	75
	-3	-11	4	62
	-3	-4	2	82
	-2	-5	2	68
	-1	-6	2	56
	0	-7	2	47
	1	-8	2	40
	1	-1	0	0
	2	-9	2	34
	3	-10	2	30
	4	-11	2	27
	5	-12	2	24
2	7	2	75	16
	0	-2	7	86
	1	-2	6	76
	2	-2	5	65
	3	-4	11	71
	3	-2	4	50
	4	-2	3	35
	5	-4	9	58
	5	-2	2	21
	6	-2	1	9
	7	-4	7	43
	7	-2	0	0
	9	-4	5	28
	10	-6	11	45
	11	-6	10	40
	11	-4	3	15
7	2	2	75	74
	-2	-5	12	71
	-2	-4	11	74
	-2	-3	10	77
	-2	-2	9	82
	-2	-1	8	87
	0	-1	1	47
	2	-12	5	23
	2	-11	4	20
	2	-10	3	17
	2	-9	2	13
	2	-8	1	7
	2	-7	0	0
2	2	9	17	45
	-9	-9	4	90
	-7	-11	4	78
	-4	-5	2	84
	-3	-6	2	72
	-2	-7	2	62
	-1	-8	2	53
	0	-9	2	46
	1	-10	2	41
	1	-1	0	0
	2	-11	2	36
	3	-12	2	32
2	9	2	78	13
	0	-2	9	87
	1	-2	8	80
	2	-2	7	72
	3	-2	6	61
	4	-2	5	50
	5	-2	4	38
	6	-2	3	26
	7	-4	11	56
	7	-2	2	16
	8	-2	1	7
	9	-4	9	44
	9	-2	0	0
	11	-4	7	32
9	2	2	78	77
	-2	-3	12	79
	-2	-2	11	82
	-2	-1	10	87
	0	-1	1	46
	2	-12	3	14
	2	-11	2	10
	2	-10	1	6
	2	-9	0	0
2	2	11	14	45
	-11	-11	4	90
	-5	-6	2	85
	-4	-7	2	75
	-3	-8	2	66
	-2	-9	2	58
	-1	-10	2	52
	0	-11	2	46

TAB. V

FROM HKL-UVW INTO EULER ANGLES

H	K / U	L / V	Φ / W	φ_2 / φ_1
2	2	11	14	45
	1	-12	2	41
	1	-1	0	0
2	11	2	80	10
	0	-2	11	88
	1	-2	10	82
	2	-2	9	76
	3	-2	8	68
	4	-2	7	59
	5	-2	6	49
	6	-2	5	39
	7	-2	4	29
	8	-2	3	20
	9	-2	2	12
	10	-2	1	6
	11	-4	11	44
	11	-2	0	0
11	2	2	80	80
	-2	-1	12	87
	0	-1	1	46
	2	-12	1	5
	2	-11	0	0
2	3	3	50	34
	-3	-10	12	79
	-3	-9	11	80
	-3	-8	10	82
	-3	-7	9	83
	-3	-6	8	85
	-3	-5	7	88
	0	-1	1	67
	3	-12	10	55
	3	-11	9	54
	3	-10	8	52
	3	-9	7	51
	3	-8	6	48
	3	-7	5	46
	3	-6	4	42
	3	-5	3	37
	3	-4	2	29
	3	-3	1	17
	3	-2	0	0
	6	-11	7	39
	6	-9	5	33
	6	-7	3	24

H	K / U	L / V	Φ / W	φ_2 / φ_1
	6	-5	1	10
	9	-11	5	26
	9	-10	4	22
	9	-8	2	12
	9	-7	1	7
	12	-11	3	14
	12	-9	1	5
3	2	3	50	56
	-7	-6	11	86
	-5	-9	11	72
	-5	-6	9	79
	-4	-9	10	68
	-4	-3	6	88
	-3	-12	11	60
	-3	-6	7	70
	-3	-3	5	83
	-2	-9	8	58
	-2	-3	4	75
	-1	-12	9	51
	-1	-9	7	53
	-1	-6	5	56
	-1	-3	3	64
	0	-3	2	46
	1	-12	7	41
	1	-9	5	39
	1	-6	3	35
	1	-3	1	23
	2	-9	4	31
	2	-3	0	0
	3	-12	5	29
	3	-6	1	11
	4	-9	2	15
	5	-12	3	17
	5	-9	1	7
	7	-12	1	5
3	3	2	65	45
	-3	-5	12	84
	-2	-4	9	82
	-1	-7	12	72
	-1	-5	9	74
	-1	-3	6	78
	-1	-1	3	90
	0	-2	3	67
	1	-9	12	62
	1	-7	9	60
	1	-5	6	57
	1	-3	3	50

H	K / U	L / V	Φ / W	φ_2 / φ_1
	1	-1	0	0
	2	-8	9	55
	2	-4	3	38
	3	-11	12	53
	3	-7	6	43
	3	-5	3	30
	4	-10	9	45
	4	-6	3	25
	5	-11	9	41
	5	-9	6	34
	5	-7	3	21
	6	-8	3	19
	7	-11	6	28
	7	-9	3	16
	8	-10	3	15
	9	-11	3	13
	10	-12	3	12
2	3	4	42	34
	-4	-12	11	79
	-3	-10	9	77
	-2	-8	7	75
	-1	-10	8	68
	-1	-6	5	72
	-1	-2	2	85
	0	-4	3	64
	1	-10	7	59
	1	-6	4	55
	1	-2	1	38
	2	-8	5	51
	3	-10	6	48
	3	-2	0	0
	4	-12	7	46
	4	-4	1	15
	5	-6	2	22
	6	-8	3	25
	7	-10	4	28
	7	-6	1	9
	8	-12	5	29
	9	-10	3	19
	10	-8	1	7
	11	-10	2	11
3	2	4	42	56
	-10	-9	12	84
	-10	-7	11	89
	-8	-10	11	77
	-8	-6	9	88

H	K / U	L / V	Φ / W	φ_2 / φ_1
	-6	-11	10	69
	-6	-7	8	78
	-6	-5	7	85
	-4	-12	9	60
	-4	-8	7	67
	-4	-4	5	82
	-2	-11	7	52
	-2	-9	6	55
	-2	-7	5	58
	-2	-5	4	63
	-2	-3	3	73
	0	-2	1	42
	2	-11	4	30
	2	-9	3	28
	2	-7	2	23
	2	-5	1	16
	2	-3	0	0
	4	-12	3	20
	4	-8	1	10
	6	-11	1	7
2	4	3	56	27
	-3	-6	10	90
	-2	-5	8	87
	-1	-7	10	79
	-1	-4	6	83
	0	-3	4	74
	1	-8	10	70
	1	-5	6	67
	1	-2	2	53
	2	-7	8	63
	2	-1	0	0
	3	-9	10	61
	3	-3	2	31
	4	-11	12	60
	4	-5	4	40
	5	-7	6	44
	5	-4	2	21
	6	-9	8	46
	7	-11	10	47
	7	-8	6	36
	7	-5	2	16
	8	-7	4	25
	9	-12	10	42
	9	-6	2	13
	10	-11	8	35
	11	-10	6	27
	11	-7	2	11
	12	-9	4	18

H	K / U	L / V	Φ / W	φ_2 / φ_1
4	2	3	56	63
	-7	-4	12	88
	-6	-3	10	90
	-5	-8	12	71
	-4	-7	10	70
	-3	-9	10	61
	-3	-6	8	67
	-2	-11	10	53
	-2	-5	6	64
	-1	-10	8	49
	-1	-7	6	51
	-1	-4	4	57
	-1	-1	2	79
	0	-3	2	42
	1	-11	6	35
	1	-8	4	32
	1	-5	2	26
	1	-2	0	0
	2	-7	2	19
	3	-12	4	22
	3	-9	2	14
	4	-11	2	12
3	4	2	68	37
	-2	-4	11	86
	-2	-3	9	89
	0	-1	2	74
	2	-7	11	64
	2	-6	9	62
	2	-5	7	59
	2	-4	5	53
	2	-3	3	44
	2	-2	1	21
	4	-9	12	56
	4	-7	8	49
	4	-5	4	35
	4	-3	0	0
	6	-10	11	48
	6	-8	7	38
	6	-7	5	31
	6	-5	1	8
	8	-11	10	40
	8	-9	6	29
	8	-7	2	11
	10	-12	9	33
	10	-11	7	27
	10	-9	3	14
	10	-8	1	5

H	K / U	L / V	Φ / W	φ_2 / φ_1
	12	-11	4	15
4	3	2	68	53
	-3	-4	12	84
	-2	-2	7	87
	-1	-6	11	70
	-1	-4	8	73
	-1	-2	5	79
	0	-2	3	64
	1	-8	10	57
	1	-6	7	54
	1	-4	4	49
	1	-2	1	26
	2	-10	11	52
	2	-6	5	42
	3	-10	9	45
	3	-8	6	38
	3	-4	0	0
	4	-10	7	36
	4	-6	1	9
	5	-12	8	34
	5	-8	2	13
	6	-10	3	16
	7	-12	4	17
	7	-10	1	5
2	3	5	36	34
	-7	-12	10	87
	-6	-11	9	86
	-5	-10	8	84
	-4	-9	7	82
	-3	-8	6	79
	-2	-7	5	75
	-1	-11	7	66
	-1	-6	4	70
	0	-5	3	62
	1	-9	5	56
	1	-4	2	48
	2	-3	1	27
	3	-7	3	39
	3	-2	0	0
	4	-11	5	42
	5	-10	4	35
	5	-5	1	14
	7	-8	2	18
	8	-7	1	9
	9	-11	3	21
	11	-9	1	7

H	K / U	L / V	Φ / W	φ_2 / φ_1
3	2	5	36	56
	-7	-12	9	68
	-6	-11	8	67
	-5	-10	7	65
	-4	-9	6	63
	-3	-8	5	60
	-2	-7	4	55
	-1	-11	5	45
	-1	-6	3	49
	-1	-1	1	81
	0	-5	2	39
	1	-9	3	33
	1	-4	1	24
	2	-3	0	0
	3	-7	1	13
	4	-11	2	17
	5	-10	1	9
2	5	3	61	22
	-1	-5	9	85
	0	-3	5	79
	1	-7	11	74
	1	-4	6	71
	1	-1	1	41
	2	-5	7	65
	3	-6	8	61
	4	-7	9	58
	5	-8	10	56
	5	-2	0	0
	6	-9	11	55
	6	-3	1	10
	7	-10	12	53
	7	-4	2	16
	8	-5	3	20
	9	-6	4	23
	10	-7	5	26
	11	-8	6	28
	11	-5	1	5
	12	-9	7	29
5	2	3	61	68
	-5	-4	11	82
	-4	-5	10	75
	-3	-9	11	60
	-3	-3	7	78
	-2	-7	8	58
	-2	-1	4	88
	-1	-11	9	46

H	K / U	L / V	Φ / W	φ_2 / φ_1
	-1	-8	7	49
	-1	-5	5	59
	-1	-2	3	67
	0	-3	2	39
	1	-10	5	31
	1	-7	3	27
	1	-4	1	16
	2	-11	4	29
	2	-5	0	0
	3	-9	1	7
3	5	2	71	31
	-1	-3	9	86
	0	-2	5	79
	1	-5	11	74
	1	-3	6	69
	1	-1	1	38
	2	-4	7	63
	3	-5	8	59
	4	-6	9	56
	5	-7	10	53
	5	-3	0	0
	6	-8	11	51
	6	-4	1	8
	7	-9	12	50
	7	-5	2	14
	8	-6	3	18
	9	-7	4	20
	10	-8	5	23
	11	-9	6	24
	11	-7	1	5
	12	-10	7	26
5	3	2	71	59
	-2	-4	11	78
	-1	-5	10	70
	-1	-3	7	74
	-1	-1	4	80
	0	-2	3	62
	1	-9	11	55
	1	-7	8	52
	1	-5	5	48
	1	-3	2	34
	2	-8	7	43
	2	-4	1	13
	3	-11	9	41
	3	-7	3	23
	3	-5	0	0
	4	-10	5	26

TAB. V

FROM HKL-UVW INTO EULER ANGLES

H	K U	L V	Φ W	φ2 φ1
2	3	5		
5	3	2	71	59
	5	-11	4	19
	5	-9	1	6
2	3	6	31	34
	-6	-10	7	88
	-3	-12	7	73
	-3	-10	6	75
	-3	-8	5	79
	-3	-6	4	84
	0	-2	1	60
	3	-12	5	47
	3	-10	4	44
	3	-8	3	40
	3	-6	2	34
	3	-4	1	22
	3	-2	0	0
	6	-10	3	29
	6	-6	1	13
	9	-10	2	17
	9	-8	1	9
	12	-10	1	7
3	2	6	31	56
	-10	-12	9	76
	-10	-9	8	83
	-8	-9	7	77
	-6	-12	7	64
	-6	-6	5	80
	-4	-9	5	62
	-4	-3	3	87
	-2	-12	5	48
	-2	-9	4	51
	-2	-6	3	56
	-2	-3	2	70
	0	-3	1	38
	2	-12	3	28
	2	-9	2	24
	2	-6	1	18
	2	-3	0	0
	4	-9	1	11
	6	-12	1	8
2	6	3	65	18
	0	-1	2	82

H	K U	L V	Φ W	φ2 φ1
	3	-7	12	69
	3	-6	10	67
	3	-5	8	63
	3	-4	6	58
	3	-3	4	49
	3	-2	2	32
	3	-1	0	0
	6	-7	10	54
	6	-5	6	42
	6	-3	2	18
	9	-8	10	45
	9	-7	8	39
	9	-5	4	24
	9	-4	2	13
	12	-9	10	38
	12	-7	6	26
	12	-5	2	10
6	2	3	65	72
	-5	-3	12	85
	-4	-3	10	82
	-3	-6	10	67
	-3	-3	8	78
	-2	-9	10	54
	-2	-3	6	72
	-1	-12	10	45
	-1	-9	8	47
	-1	-6	6	51
	-1	-3	4	60
	0	-3	2	38
	1	-12	6	30
	1	-9	4	27
	1	-6	2	20
	1	-3	0	0
	2	-9	2	14
	3	-12	2	10
3	6	2	73	27
	0	-1	3	82
	2	-5	12	72
	2	-4	9	69
	2	-3	6	63
	2	-2	3	49
	2	-1	0	0
	4	-5	9	58
	4	-3	3	32
	6	-7	12	56
	6	-5	6	39
	6	-4	3	24

H	K U	L V	Φ W	φ2 φ1
	8	-7	9	42
	8	-5	3	18
	10	-9	12	44
	10	-8	9	37
	10	-7	6	27
	10	-6	3	15
	12	-7	3	13
6	3	2	73	63
	-3	-2	12	88
	-2	-2	9	85
	-1	-6	12	69
	-1	-4	9	72
	-1	-2	6	78
	0	-2	3	60
	1	-10	12	53
	1	-8	9	51
	1	-6	6	47
	1	-4	3	38
	1	-2	0	0
	2	-10	9	44
	2	-6	3	27
	3	-10	6	31
	3	-8	3	20
	4	-10	3	16
	5	-12	3	14
2	3	7	27	34
	-4	-9	5	81
	-3	-5	3	88
	-1	-11	5	64
	-1	-4	2	72
	0	-7	3	59
	1	-10	4	54
	1	-3	1	41
	3	-2	0	0
	4	-5	1	20
	5	-8	2	27
	6	-11	3	31
	7	-7	1	13
	10	-9	1	9
	11	-12	2	15
3	2	7	27	56
	-11	-8	7	88
	-9	-11	7	75
	-8	-9	6	77
	-7	-7	5	80

H	K U	L V	Φ W	φ2 φ1
	-6	-5	4	85
	-2	-11	4	47
	-1	-9	3	43
	-1	-2	1	63
	0	-7	2	37
	1	-12	3	32
	1	-5	1	25
	2	-3	0	0
	3	-8	1	15
	5	-11	1	10
2	7	3	68	16
	-1	-4	10	89
	0	-3	7	84
	1	-5	11	79
	1	-2	4	71
	2	-1	1	26
	3	-3	5	56
	4	-5	9	62
	5	-4	6	48
	7	-5	7	43
	7	-2	0	0
	8	-7	11	51
	9	-6	8	40
	9	-3	1	7
	11	-7	9	38
	11	-4	2	10
7	2	3	68	74
	-3	-6	11	67
	-2	-11	12	52
	-2	-5	8	64
	-1	-10	9	46
	-1	-7	7	50
	-1	-4	5	57
	-1	-1	3	78
	0	-3	2	37
	1	-11	5	26
	1	-8	3	22
	1	-5	1	12
	2	-7	0	0
	3	-12	1	5
3	7	2	75	23
	-1	-3	12	89
	0	-2	7	84
	1	-3	9	77
	1	-1	2	58
	2	-4	11	73

H	K U	L V	Φ W	φ2 φ1
	4	-2	1	13
	5	-3	3	28
	6	-4	5	36
	7	-5	7	41
	7	-3	0	0
	8	-6	9	44
	9	-7	11	46
	9	-5	4	22
	11	-7	8	33
	11	-5	1	5
7	3	2	75	67
	-1	-5	11	70
	-1	-3	8	74
	-1	-1	5	84
	0	-2	3	59
	1	-9	10	50
	1	-7	7	47
	1	-5	4	40
	1	-3	1	18
	2	-12	11	44
	2	-8	5	32
	3	-11	6	29
	3	-7	0	0
	4	-10	1	5
2	3	8	24	34
	-2	-12	5	68
	-1	-10	4	64
	-1	-2	1	83
	0	-8	3	59
	1	-6	2	49
	2	-4	1	32
	3	-10	3	42
	3	-2	0	0
	5	-6	1	18
	7	-10	2	23
	8	-8	1	12
	11	-10	1	9
3	2	8	24	56
	-12	-10	7	84
	-10	-9	6	82
	-8	-8	5	80
	-6	-11	5	64
	-6	-7	4	76
	-4	-6	3	69
	-2	-9	3	49

H	K U	L V	Φ W	φ2 φ1
	-2	-5	2	58
	0	-4	1	36
	2	-11	2	25
	2	-7	1	19
	2	-3	0	0
	4	-10	1	13
2	8	3	70	14
	0	-3	8	85
	1	-4	10	80
	1	-1	2	60
	2	-5	12	76
	4	-1	0	0
	5	-2	2	22
	6	-3	4	33
	7	-4	6	39
	8	-5	8	43
	9	-6	10	46
	9	-3	2	13
	10	-7	12	48
	11	-5	6	28
8	2	3	70	76
	-3	-3	10	78
	-2	-7	10	59
	-2	-1	6	86
	-1	-11	10	46
	-1	-8	8	49
	-1	-5	6	54
	-1	-2	4	68
	0	-3	2	36
	1	-10	4	23
	1	-7	2	17
	1	-4	0	0
	2	-11	2	11
3	8	2	77	21
	0	-1	4	85
	2	-3	9	72
	2	-2	5	63
	2	-1	1	25
	4	-3	6	52
	6	-5	11	57
	6	-4	7	46
	8	-5	8	42
	8	-3	0	0
	10	-6	9	39
	10	-4	1	5
	12	-7	10	37

H	K U	L V	Φ W	φ2 φ1
	12	-5	2	9
8	3	2	77	69
	-2	-2	11	84
	-1	-4	10	72
	-1	-2	7	78
	0	-2	3	59
	1	-10	11	49
	1	-8	8	46
	1	-6	5	41
	1	-4	2	27
	2	-10	7	36
	2	-6	1	9
	3	-10	3	16
	3	-8	0	0
2	3	9	22	34
	-6	-11	5	85
	-3	-10	4	74
	-3	-7	3	80
	0	-3	1	58
	3	-11	3	43
	3	-8	2	38
	3	-5	1	27
	3	-2	0	0
	6	-7	1	17
	9	-12	2	21
	9	-9	1	12
	12	-11	1	9
3	2	9	22	56
	-9	-9	5	79
	-7	-12	5	66
	-6	-9	4	69
	-5	-6	3	75
	-4	-3	2	87
	-1	-12	3	41
	-1	-3	1	54
	0	-9	2	36
	1	-6	1	26
	2	-3	0	0
	3	-9	1	16
	5	-12	1	12
2	9	3	72	13
	0	-1	3	86
	3	-4	10	70
	3	-3	7	64

V -

H	K U	L V	Φ W	φ2 φ1
	3	-2	4	51
	3	-1	1	18
	6	-5	11	59
	6	-3	5	39
	9	-5	9	44
	9	-4	6	33
	9	-2	0	0
	12	-5	7	30
	12	-3	1	5
9	2	3	72	77
	-3	-3	11	79
	-2	-9	12	57
	-2	-3	8	73
	-1	-12	11	45
	-1	-9	9	48
	-1	-6	7	53
	-1	-3	5	63
	0	-3	2	36
	1	-12	5	24
	1	-9	3	19
	1	-6	1	10
	2	-9	0	0
3	9	2	78	18
	0	-2	9	86
	1	-3	12	81
	1	-1	3	68
	3	-1	0	0
	4	-2	3	35
	5	-3	6	47
	6	-4	9	59
	7	-5	12	56
	7	-3	3	22
	9	-5	9	42
	10	-4	3	16
	11	-5	6	27
9	3	2	78	72
	-1	-5	12	70
	-1	-3	9	73
	-1	-1	6	84
	0	-2	3	53
	1	-11	12	49
	1	-9	9	46
	1	-7	6	41
	1	-5	3	31
	1	-3	0	0
	2	-12	9	37

TAB. V

FROM HKL-UVW INTO EULER ANGLES

```
 H  K  L   Φ   φ2
    U  V   W   φ1
------------------
 2  3  9
========
 9  3  2  78  72
------------------
    2 -8   3  20
    3-11   3  15

 2  3 10  20  34
==================
   -7-12   5  87
   -5-10   4  83
   -3 -8   3  78
   -1 -6   2  67
    0-10   3  58
    1 -4   1  44
    3 -2   0   0
    4 -6   1  24
    5-10   2  31
    7 -8   1  16
   10-10   1  12
 3  2 10  20  56
------------------
   -6-11   4  64
   -4 -9   3  59
   -2-12   3  45
   -2 -7   2  51
   -2 -2   1  79
    0 -5   1  35
    2 -8   1  21
    2 -3   0   0
    4-11   1  15
 2 10  3  74  11
------------------
    0 -3  10  87
    1 -2   6  78
    2 -1   2  44
    3 -3   8  67
    5 -4  10  61
    5 -1   0   0
    7 -5  12  58
    7 -2   2  16
    9 -3   4  24
   11 -4   6  28
   12 -3   2  10
10  2  3  74  79
------------------
   -2 -5  10  67
   -1-10  10  47
   -1 -7   8  51
```

```
 H  K  L   Φ   φ2
    U  V   W   φ1
------------------
   -1 -4   6  59
   -1 -1   4  79
    0 -3   2  35
    1-11   4  21
    1 -8   2  19
    1 -5   0   0
 3 10  2  79  17
------------------
    0 -1   5  87
    2 -3  12  77
    2 -2   7  71
    2 -1   2  43
    4 -3   9  63
    6 -4  11  58
    6 -2   1   9
    8 -3   3  20
   10 -4   5  25
   10 -3   0   0
   12 -5   7  29
10  3  2  79  73
------------------
   -1 -4  11  72
   -1 -2   8  79
    0 -2   3  58
    1-10  10  46
    1 -8   7  42
    1 -6   4  34
    1 -4   1  14
    2-10   5  27
    3-10   0   0

 2  3 11  18  34
==================
   -1 -3   1  75
    0-11   3  58
    1 -8   2  51
    2 -5   1  36
    3 -2   0   0
    5 -7   1  22
    7-12   2  27
    8 -9   1  15
   11-11   1  12
 3  2 11  18  56
------------------
  -11-11   5  79
  -10 -7   4  89
   -7 -6   3  83
   -5 -9   3  64
   -4 -5   2  73
```

```
 H  K  L   Φ   φ2
    U  V   W   φ1
------------------
   -1 -4   1  49
    0-11   2  35
    1 -7   1  27
    2 -3   0   0
    3-10   1  18
 2 11  3  75  10
------------------
    0 -3  11  87
    1 -1   3  69
    4 -1   1  14
    5 -2   4  38
    6 -3   7  48
    7 -4  10  54
    9 -3   5  29
   11 -5  11  44
   11 -2   0   0
11  2  3  75  80
------------------
   -2 -7  12  62
   -2 -1   8  86
   -1-11  11  47
   -1 -8   9  50
   -1 -5   7  57
   -1 -2   5  71
    0 -3   2  35
    1-10   3  17
    1 -7   1   8
    2-11   0   0
 3 11  2  80  15
------------------
    0 -2  11  87
    1 -1   4  73
    3 -1   1  18
    4 -2   5  49
    5 -3   9  58
    7 -3   6  39
   10 -4   7  34
   11 -5  11  43
   11 -3   0   0
11  3  2  80  75
------------------
   -1 -3  10  75
   -1 -1   7  84
    0 -2   3  58
    1-11  11  46
    1 -9   8  42
    1 -7   5  36
    1 -5   2  22
    2-12   7  30
```

```
 H  K  L   Φ   φ2
    U  V   W   φ1
------------------
    2 -8   1   7
    3-11   0   0

 2  3 12  17  34
==================
   -3-10   3  74
   -3 -6   2  83
    0 -4   1  57
    3-10   2  41
    3 -6   1  31
    3 -2   0   0
    6 -8   1  20
    9-10   1  15
   12-12   1  12
 3  2 12  17  56
------------------
  -12-12   5  79
  -10 -9   4  82
   -8 -6   3  87
   -4-12   3  53
   -2 -9   2  47
   -2 -3   1  68
    0 -6   1  35
    2 -9   1  22
    2 -3   0   0
    4-12   1  16
 2 12  3  76   9
------------------
    0 -1   4  88
    3 -3  10  71
    3 -2   6  62
    3 -1   2  33
    6 -3   8  52
    6 -1   0   0
    9 -4  10  47
    9 -2   2  13
   12 -5  12  44
   12 -3   4  18
12  2  3  76  81
------------------
   -2 -3  10  76
   -1-12  12  47
   -1 -9  10  50
   -1 -6   8  55
   -1 -3   6  66
    0 -3   2  35
    1-12   4  19
    1 -9   2  13
    1 -6   0   0
```

```
 H  K  L   Φ   φ2
    U  V   W   φ1
------------------
 3 12  2  81  14
------------------
    0 -1   6  88
    2 -2   9  75
    2 -1   3  54
    4 -3  12  69
    4 -1   0   0
    6 -2   3  26
    8 -3   6  36
   10 -4   9  41
   10 -3   3  16
   12 -5  12  43
12  3  2  81  76
------------------
   -1 -4  12  73
   -1 -2   9  79
    0 -2   3  57
    1-12  12  46
    1-10   9  43
    1 -8   6  37
    1 -6   3  27
    1 -4   0   0
    2-10   3  17

 2  4  5  42  27
==================
   -3-11  10  82
   -2 -9   8  79
   -1-12  10  73
   -1 -7   6  76
   -1 -2   2  90
    0 -5   4  70
    1 -8   6  64
    1 -3   2  53
    2-11   8  61
    2 -1   0   0
    3 -4   2  34
    4 -7   4  42
    5-10   6  45
    5 -5   2  24
    7-11   6  39
    7 -6   2  19
    8 -9   4  28
    9 -7   2  15
   11 -8   2  13
   12-11   4  21
 4  2  5  42  63
------------------
  -11 -8  12  83
```

```
 H  K  L   Φ   φ2
    U  V   W   φ1
------------------
   -9 -7  10  82
   -8 -9  10  73
   -7-11  10  66
   -7 -6   8  79
   -5-10   8  61
   -5 -5   6  76
   -4 -7   6  64
   -3 -4   4  70
   -2-11   6  45
   -2 -1   2  90
   -1 -8   4  42
   -1 -3   2  53
    0 -5   2  34
    1-12   4  28
    1 -7   2  24
    1 -2   0   0
    2 -9   2  19
    3-11   2  15
 2  5  4  53  22
------------------
   -2 -8  11  85
   -1 -6   8  83
    0 -4   5  77
    1-10  12  73
    1 -6   7  70
    1 -2   2  56
    2 -8   9  67
    3-10  11  65
    3 -2   1  19
    4 -4   3  36
    5 -6   5  42
    5 -2   0   0
    6 -8   7  46
    7-10   9  48
    7 -6   4  30
    8-12  11  49
    8 -4   1   8
    9-10   8  40
   10 -8   5  27
   11-10   7  32
   11 -6   2  11
 5  2  4  53  68
------------------
   -6 -7  11  73
   -6 -5  10  79
   -4-12  11  55
   -4 -8   9  62
   -4 -4   7  76
   -2-11   8  46
```

```
 H  K  L   Φ   φ2
    U  V   W   φ1
------------------
   -2 -9   7  49
   -2 -7   6  52
   -2 -5   5  58
   -2 -3   4  68
   -2 -1   3  87
    0 -2   1  34
    2-11   3  19
    2 -9   2  15
    2 -7   1  10
    2 -5   0   0
    4-12   1   6
 4  5  2  73  39
------------------
   -1 -4  12  82
   -1 -2   7  86
    0 -2   5  77
    1 -4   8  69
    1 -2   3  57
    2 -6  11  65
    2 -2   1  20
    3 -4   4  41
    4 -6   7  47
    5 -8  10  50
    5 -6   5  34
    5 -4   0   0
    7-10  11  45
    7 -8   6  31
    7 -6   1   6
    8-10   9  37
    9-10   7  29
    9 -8   2  10
   11-12   8  28
   11-10   3  12
   12-10   1   4
 5  4  2  73  51
------------------
   -2 -3  11  85
   -2 -2   9  88
    0 -1   2  70
    2 -8  11  57
    2 -7   9  55
    2 -6   7  51
    2 -5   5  45
    2 -4   3  36
    2 -3   1  16
    4-11  12  49
    4 -9   8  41
    4 -7   4  28
    4 -5   0   0
```

```
 H  K  L   Φ   φ2
    U  V   W   φ1
------------------
    6-11   7  31
    6-10   5  24
    6 -8   1   6
    8-11   2   9

 2  4  7  33  27
==================
   -1-10   6  72
   -1 -3   2  83
    0 -7   4  67
    1-11   6  62
    1 -4   2  54
    2 -1   0   0
    3 -5   2  37
    4 -9   4  44
    5 -6   2  27
    7 -7   2  22
    8-11   4  32
    9 -8   2  18
   11 -9   2  15
 4  2  7  33  63
------------------
  -12-11  10  76
  -11 -6   8  88
   -9-10   8  72
   -8 -5   6  85
   -7 -7   6  74
   -5-11   6  56
   -5 -4   4  80
   -3 -8   4  52
   -2 -3   2  64
   -1-12   4  36
   -1 -5   2  43
    0 -7   2  31
    1 -9   2  24
    1 -2   0   0
    2-11   2  19
 2  7  4  61  16
------------------
   -1 -6  11  87
    0 -4   7  82
    1 -6  10  77
    1 -2   3  66
    3 -2   2  34
    4 -4   5  49
    5 -6   8  55
    5 -2   1  12
    6 -8  11  58
    7 -6   7  44
```

TAB. V

FROM HKL-UVW INTO EULER ANGLES

H	K / U	L / V	Φ / W	φ2 / φ1
2	4	7		
2	7	4	61	16
	7	-2	0	0
	8	-4	3	21
	10	-8	9	41
	11	-10	12	46
	11	-6	5	25
	12	-4	1	5
7	2	4	61	74
	-4	-8	11	62
	-4	-4	9	75
	-2	-11	9	46
	-2	-9	8	48
	-2	-7	7	52
	-2	-5	6	58
	-2	-3	5	68
	-2	-1	4	85
	0	-2	1	31
	2	-11	2	12
	2	-9	1	7
	2	-7	0	0
4	7	2	76	30
	-1	-2	9	89
	0	-2	7	82
	1	-4	12	77
	1	-2	5	70
	2	-2	3	49
	3	-4	8	61
	3	-2	1	16
	5	-6	11	57
	5	-4	4	33
	7	-6	7	39
	7	-4	0	0
	8	-6	5	27
	9	-8	10	41
	10	-6	1	5
	11	-8	6	25
	12	-10	11	36
7	4	2	76	60
	-2	-2	11	86
	0	-1	2	67
	2	-9	11	52
	2	-8	9	49
	2	-7	7	46

H	K / U	L / V	Φ / W	φ2 / φ1
	2	-6	5	40
	2	-5	3	30
	2	-4	1	13
	4	-11	8	36
	4	-9	4	23
	4	-7	0	0
	6	-11	1	5
2	4	9	26	27
	-5	-11	6	88
	-1	-4	2	79
	0	-9	4	66
	1	-5	2	55
	2	-1	0	0
	3	-6	2	40
	4	-11	4	47
	5	-7	2	31
	7	-8	2	25
	9	-9	2	20
	11	-10	2	17
4	2	9	26	63
	-10	-7	6	82
	-8	-11	6	65
	-7	-4	4	87
	-5	-8	4	61
	-3	-12	4	44
	-3	-3	2	73
	-2	-5	2	51
	-1	-7	2	38
	0	-9	2	29
	1	-11	2	24
	1	-2	0	0
2	9	4	67	13
	0	-4	9	85
	1	-2	4	72
	3	-2	3	44
	4	-4	7	58
	5	-6	11	63
	5	-2	2	22
	7	-6	10	53
	7	-2	1	9
	8	-4	5	32
	9	-2	0	0
	11	-6	8	36
	12	-4	3	15

H	K / U	L / V	Φ / W	φ2 / φ1
9	2	4	67	77
	-4	-4	11	76
	-2	-11	10	47
	-2	-9	9	50
	-2	-7	8	54
	-2	-5	7	60
	-2	-3	6	69
	-2	-1	5	84
	0	-2	1	29
	2	-11	1	6
	2	-9	0	0
4	9	2	79	24
	0	-2	9	85
	1	-2	7	76
	2	-2	5	63
	3	-4	12	70
	3	-2	3	41
	4	-2	1	13
	5	-4	8	53
	7	-4	4	27
	8	-6	11	49
	9	-4	0	0
	10	-6	7	32
	11	-6	5	22
9	4	2	79	66
	-2	-1	11	89
	0	-1	2	66
	2	-10	11	48
	2	-9	9	45
	2	-8	7	41
	2	-7	5	35
	2	-6	3	26
	2	-5	1	11
	4	-11	4	19
	4	-9	0	0
2	4	11	22	27
	-4	-9	4	88
	-1	-5	2	76
	0	-11	4	65
	1	-6	2	56
	2	-1	0	0
	3	-7	2	42
	5	-8	2	33
	7	-9	2	27

H	K / U	L / V	Φ / W	φ2 / φ1
	9	-10	2	23
	11	-11	2	20
4	2	11	22	63
	-11	-11	6	73
	-7	-8	4	69
	-5	-12	4	51
	-4	-3	2	80
	-3	-5	2	59
	-2	-7	2	45
	-1	-9	2	35
	0	-11	2	28
	1	-2	0	0
2	11	4	70	10
	0	-4	11	86
	1	-2	5	76
	3	-2	4	52
	4	-4	9	64
	5	-2	3	31
	7	-2	2	16
	8	-4	7	41
	9	-2	1	7
	11	-6	11	44
	11	-2	0	0
	12	-4	5	23
11	2	4	70	80
	-2	-11	11	48
	-2	-9	10	51
	-2	-7	9	56
	-2	-5	8	62
	-2	-3	7	71
	-2	-1	6	84
	0	-2	1	28
	2	-11	0	0
4	11	2	80	20
	0	-2	11	86
	1	-2	9	80
	2	-2	7	70
	3	-2	5	55
	4	-2	3	34
	5	-4	12	64
	5	-2	1	11
	7	-4	8	46
	9	-4	4	22
	11	-6	11	42
	11	-4	0	0

H	K / U	L / V	Φ / W	φ2 / φ1
11	4	2	80	70
	0	-1	2	65
	2	-11	11	45
	2	-10	9	42
	2	-9	7	38
	2	-8	5	32
	2	-7	3	23
	2	-6	1	9
	4	-11	0	0
2	5	5	47	22
	0	-1	1	75
	5	-12	10	56
	5	-11	9	55
	5	-10	8	53
	5	-9	7	50
	5	-8	6	47
	5	-7	5	43
	5	-6	4	38
	5	-5	3	32
	5	-4	2	24
	5	-3	1	13
	5	-2	0	0
	10	-11	7	36
	10	-9	5	28
	10	-7	3	19
	10	-5	1	7
5	2	5	47	68
	-10	-5	12	87
	-9	-5	11	85
	-8	-5	10	83
	-7	-10	11	66
	-7	-5	9	81
	-6	-5	8	78
	-5	-10	9	59
	-5	-5	7	74
	-4	-5	6	69
	-3	-10	7	49
	-3	-5	5	63
	-2	-5	4	54
	-1	-10	5	37
	-1	-5	3	44
	0	-5	2	30
	1	-10	3	23
	1	-5	1	15
	2	-5	0	0

H	K / U	L / V	Φ / W	φ2 / φ1
	3	-10	1	7
5	5	2	74	45
	-1	-3	10	82
	-1	-1	5	90
	0	-2	5	75
	1	-5	10	68
	1	-3	5	61
	1	-1	0	0
	2	-4	5	51
	3	-7	10	56
	3	-5	5	43
	4	-6	5	36
	5	-9	10	46
	5	-7	5	31
	6	-8	5	28
	7	-11	10	39
	7	-9	5	25
	8	-10	5	22
	9	-11	5	20
	10	-12	5	18
2	5	6	42	22
	-3	-12	11	84
	-2	-10	9	82
	-1	-8	7	79
	0	-6	5	73
	1	-10	8	69
	1	-4	3	62
	2	-2	1	30
	3	-6	4	50
	4	-10	7	55
	5	-8	5	45
	5	-2	0	0
	7	-10	6	41
	7	-4	1	11
	9	-12	7	39
	9	-6	2	16
	11	-8	3	19
	12	-6	1	6
5	2	6	42	68
	-10	-11	12	70
	-10	-8	11	77
	-8	-7	9	75
	-6	-9	8	63
	-6	-6	7	72
	-4	-11	7	50

H	K / U	L / V	Φ / W	φ2 / φ1
	-4	-5	5	67
	-2	-10	5	41
	-2	-7	4	46
	-2	-4	3	57
	-2	-1	2	86
	0	-3	1	28
	2	-11	2	15
	2	-8	1	10
	2	-5	0	0
2	6	5	52	18
	-1	-8	10	83
	-1	-3	4	90
	0	-5	6	78
	1	-7	8	73
	1	-2	2	58
	2	-9	10	70
	3	-11	12	68
	3	-1	0	0
	4	-3	2	28
	5	-5	4	39
	6	-7	6	44
	7	-9	8	47
	7	-4	2	18
	8	-11	10	49
	9	-8	6	35
	10	-5	2	13
	11	-12	10	42
	11	-7	4	22
6	2	5	52	72
	-7	-9	12	68
	-7	-4	10	83
	-6	-7	10	70
	-5	-5	8	73
	-4	-3	6	78
	-3	-11	8	47
	-3	-1	4	90
	-2	-9	6	44
	-1	-12	6	35
	-1	-7	4	39
	-1	-2	2	58
	0	-5	2	28
	1	-8	2	18
	1	-3	0	0
	2	-11	2	13
5	6	2	76	40
	0	-1	3	78

H	K / U	L / V	Φ / W	φ2 / φ1
	2	-5	10	65
	2	-4	7	60
	2	-3	4	50
	2	-2	1	20
	4	-7	11	56
	4	-5	5	39
	6	-8	9	64
	6	-7	6	34
	6	-5	0	0
	8	-9	7	31
	8	-7	1	6
	10	-12	11	36
	10	-11	8	29
	10	-9	2	9
	12	-11	3	11
6	5	2	76	30
	-2	-2	11	89
	-1	-2	8	84
	0	-2	5	73
	1	-6	12	67
	1	-4	7	63
	1	-2	2	43
	2	-6	9	58
	3	-8	11	55
	3	-4	1	12
	4	-6	3	23
	5	-8	5	29
	5	-6	0	0
	6	-10	7	32
	7	-12	9	34
	7	-10	4	19
	8	-10	1	5
2	5	7	38	22
	-4	-11	9	89
	-3	-10	8	86
	-2	-9	7	85
	-1	-8	6	78
	0	-7	5	72
	1	-6	4	64
	2	-5	3	59
	3	-11	7	59
	3	-4	2	38
	4	-3	1	19
	5	-9	5	46
	5	-2	0	0
	7	-7	3	28

TAB. V FROM HKL-UVW INTO EULER ANGLES V - 38

H	K / U	L / V	Φ / W	φ2 / φ1
2	5	7	38	22
	9	-5	1	9
	10	-11	5	32
	11	-10	4	25
5	2	7	38	68
	-12	-5	10	89
	-5	-12	7	51
	-4	-11	6	48
	-3	-10	5	45
	-2	-9	4	41
	-1	-8	3	35
	-1	-1	1	71
	0	-7	2	27
	1	-6	1	15
	2	-5	0	0
	3	-11	1	8
2	7	5	56	16
	-1	-4	6	89
	0	-5	7	81
	1	-6	8	75
	1	-1	1	44
	2	-7	9	71
	3	-8	10	67
	4	-9	11	65
	5	-10	12	63
	7	-2	0	0
	8	-3	1	8
	9	-4	2	14
	10	-5	3	18
	11	-6	4	22
	12	-7	5	24
7	2	5	56	74
	-7	-3	11	86
	-5	-10	11	58
	-5	-5	9	73
	-4	-11	10	52
	-3	-7	7	55
	-3	-2	5	80
	-2	-3	4	64
	-1	-9	5	36
	-1	-4	3	46
	0	-5	2	27
	1	-11	3	19
	1	-6	1	11
	2	-7	0	0

H	K / U	L / V	Φ / W	φ2 / φ1
5	7	2	77	36
	0	-2	7	81
	1	-3	8	73
	1	-1	1	36
	2	-4	9	67
	3	-5	10	62
	4	-6	11	59
	5	-7	12	57
	7	-5	0	0
	8	-6	1	6
	9	-7	2	10
	10	-8	3	14
	11	-9	4	16
	12	-10	5	18
7	5	2	77	54
	-1	-3	11	81
	-1	-1	6	88
	0	-2	5	72
	1	-5	9	63
	1	-3	4	54
	2	-4	3	35
	3	-7	7	44
	3	-5	2	19
	4	-10	11	47
	4	-6	1	8
	5	-11	10	41
	5	-9	5	27
	5	-7	0	0
	7	-11	3	13
2	5	8	34	22
	-3	-10	7	86
	-1	-6	4	80
	0	-8	5	72
	1	-10	6	67
	1	-2	1	47
	2	-12	7	63
	5	-2	0	0
	6	-4	1	14
	7	-6	2	22
	8	-8	3	27
	9	-10	4	31
	10	-12	5	33
	11	-6	1	8

H	K / U	L / V	Φ / W	φ2 / φ1
5	2	8	34	68
	-10	-11	9	68
	-10	-7	8	79
	-8	-8	7	70
	-6	-5	5	75
	-4	-10	5	49
	-4	-2	3	86
	-2	-11	4	37
	-2	-7	3	43
	-2	-3	2	60
	0	-4	1	26
	2	-9	1	11
	2	-5	0	0
2	8	5	59	14
	-1	-6	10	88
	0	-5	8	83
	1	-4	6	75
	2	-3	4	60
	3	-7	10	68
	3	-2	2	35
	4	-1	0	0
	5	-5	6	49
	7	-8	10	53
	7	-3	2	17
	8	-7	8	45
	10	-5	4	23
	11	-9	10	42
	11	-4	2	11
8	2	5	59	76
	-7	-2	12	89
	-5	-10	12	59
	-4	-9	10	56
	-3	-8	8	53
	-2	-7	6	48
	-1	-11	6	34
	-1	-6	4	40
	-1	-1	2	73
	0	-5	2	26
	1	-9	2	15
	1	-4	0	0
5	8	2	78	32
	0	-1	4	83
	2	-4	11	71
	2	-3	7	65
	2	-2	3	48

H	K / U	L / V	Φ / W	φ2 / φ1
	4	-5	10	59
	4	-3	2	22
	6	-5	5	33
	6	-4	1	8
	8	-7	8	38
	8	-5	0	0
	10	-9	11	40
	10	-8	7	29
	10	-7	3	14
8	5	2	78	58
	-1	-2	9	83
	0	-2	5	72
	1	-6	11	63
	1	-4	6	57
	1	-2	1	25
	2	-6	7	49
	3	-8	8	44
	4	-10	9	41
	5	-12	10	39
	5	-8	0	0
	6	-10	1	5
	7	-12	2	8
2	5	9	31	22
	-4	-11	7	88
	-1	-5	3	81
	0	-9	5	71
	1	-4	2	58
	3	-3	1	27
	4	-7	3	43
	5	-11	5	48
	5	-2	0	0
	7	-10	4	37
	8	-5	1	12
	11	-8	2	16
5	2	9	31	68
	-10	-11	8	67
	-9	-9	7	70
	-8	-7	6	73
	-7	-5	5	78
	-6	-3	4	86
	-1	-11	3	31
	-1	-2	1	53
	0	-9	2	25
	1	-7	1	16
	2	-5	0	0

H	K / U	L / V	Φ / W	φ2 / φ1
	3	-12	1	9
2	9	5	62	13
	0	-5	9	84
	1	-3	5	74
	2	-1	1	28
	3	-4	6	61
	4	-7	11	67
	5	-5	7	53
	7	-6	8	48
	9	-7	9	45
	9	-2	0	0
	11	-8	10	42
	11	-3	1	6
9	2	5	62	77
	-5	-5	11	73
	-4	-7	10	62
	-3	-4	7	68
	-2	-11	8	41
	-2	-1	4	83
	-1	-8	5	37
	-1	-3	3	52
	0	-5	2	25
	1	-12	3	16
	1	-7	1	9
	2	-9	0	0
5	9	2	79	29
	0	-2	9	84
	1	-3	11	78
	1	-1	2	56
	5	-3	1	10
	6	-4	3	23
	7	-5	5	31
	8	-6	7	36
	9	-7	9	39
	9	-5	0	0
	10	-8	11	42
	11	-7	4	17
9	5	2	79	61
	-1	-3	12	80
	-1	-1	7	87
	0	-2	5	71
	1	-5	8	59
	1	-3	3	45
	2	-8	11	55
	2	-4	1	13

H	K / U	L / V	Φ / W	φ2 / φ1
	3	-7	4	28
	4	-10	7	34
	5	-11	5	23
	5	-9	0	0
2	5	10	28	22
	0	-2	1	71
	5	-12	5	49
	5	-10	4	45
	5	-8	3	40
	5	-6	2	32
	5	-4	1	19
	5	-2	0	0
	10	-10	3	26
	10	-6	1	10
5	2	10	28	68
	-12	-5	7	89
	-10	-10	7	69
	-10	-5	6	86
	-8	-5	5	81
	-6	-10	5	56
	-6	-5	4	74
	-4	-5	3	63
	-2	-10	3	37
	-2	-5	2	47
	0	-5	1	24
	2	-10	1	12
	2	-5	0	0
2	10	5	64	11
	0	-1	2	85
	5	-7	12	65
	5	-6	10	61
	5	-5	8	57
	5	-4	6	50
	5	-3	4	39
	5	-2	2	23
	5	-1	0	0
	10	-7	10	45
	10	-5	6	32
	10	-3	2	12
10	2	5	64	79
	-5	-5	12	74
	-4	-5	10	70
	-3	-10	10	50
	-3	-5	8	64

H	K / U	L / V	Φ / W	φ2 / φ1
	-2	-5	6	56
	-1	-10	6	35
	-1	-5	4	43
	0	-5	2	24
	1	-10	2	13
	1	-5	0	0
5	10	2	80	27
	0	-1	5	85
	2	-3	10	73
	2	-2	5	62
	2	-1	0	0
	4	-3	5	46
	6	-5	10	53
	6	-4	5	35
	8	-5	5	28
	10	-7	10	40
	10	-6	5	24
	12	-7	5	20
10	5	2	80	63
	-1	-2	10	82
	0	-2	5	71
	1	-6	10	60
	1	-4	5	52
	1	-2	0	0
	2	-6	5	39
	3	-10	10	45
	3	-8	5	31
	4	-10	5	25
	5	-12	5	21
2	5	11	26	22
	-1	-4	2	83
	0	-11	5	70
	1	-7	3	63
	2	-3	1	37
	3	-10	4	54
	5	-2	0	0
	7	-5	1	15
	9	-8	2	22
	11	-11	3	26
	12	-7	1	9
5	2	11	26	68
	-11	-11	7	69
	-9	-5	5	63
	-7	-10	5	60

H	K / U	L / V	Φ / W	φ2 / φ1
	-6	-7	4	65
	-5	-4	3	75
	-1	-3	1	63
	0	-11	2	24
	1	-6	1	16
	2	-5	0	0
2	11	5	66	10
	0	-5	11	86
	1	-2	4	73
	3	-1	1	19
	4	-3	5	51
	5	-5	9	59
	7	-4	6	41
	10	-5	7	36
	11	-7	11	45
	11	-2	0	0
11	2	5	66	80
	-4	-3	10	78
	-3	-11	11	50
	-3	-1	7	87
	-2	-9	8	46
	-1	-12	7	33
	-1	-7	5	39
	-1	-2	3	61
	0	-5	2	24
	1	-8	1	8
	2	-11	0	0
5	11	2	81	24
	0	-2	11	86
	1	-1	3	66
	4	-2	1	13
	5	-3	4	35
	6	-4	7	45
	7	-5	10	50
	9	-5	5	26
	11	-7	11	41
	11	-5	0	0
11	5	2	81	66
	-1	-1	8	86
	0	-2	5	70
	1	-7	12	61
	1	-5	7	55
	1	-3	2	33
	2	-8	9	48
	3	-11	11	45

TAB. V

FROM HKL-UVW INTO EULER ANGLES

H	K U	L V	Φ W	φ2 φ1
2	5	11		
11	5	2	81	66
	3	-7	1	8
	4	-10	3	16
	5	-11	0	0
2	5	12	24	22
	0	-12	5	70
	1	-10	4	65
	2	-8	3	57
	3	-6	2	44
	4	-4	1	25
	5	-2	0	0
	7	-10	3	36
	9	-6	1	13
5	2	12	24	68
	-12	-12	7	69
	-10	-11	6	66
	-8	-10	5	63
	-6	-9	4	58
	-4	-8	3	51
	-2	-7	2	40
	-2	-1	1	86
	0	-6	1	24
	2	-11	1	13
	2	-5	0	0
2	12	5	68	9
	0	-5	12	86
	1	-1	2	62
	6	-1	0	0
	7	-2	2	17
	8	-3	4	27
	9	-4	6	34
	10	-5	8	39
	11	-6	10	42
	12	-7	12	45
12	2	5	68	81
	-4	-1	10	88
	-3	-7	10	59
	-3	-2	8	80
	-2	-3	6	68
	-1	-9	6	37
	-1	-4	4	49

H	K U	L V	Φ W	φ2 φ1
	0	-5	2	24
	1	-11	2	11
	1	-6	0	0
5	12	2	81	23
	0	-1	6	86
	2	-2	7	70
	2	-1	1	24
	4	-3	8	59
	6	-4	9	52
	8	-5	10	47
	10	-6	11	44
	12	-7	12	41
	12	-5	0	0
12	5	2	81	67
	-1	-2	11	83
	0	-2	5	70
	1	-6	9	57
	1	-4	4	45
	2	-6	3	26
	3	-10	7	34
	3	-8	2	13
	4	-10	1	5
	5	-12	0	0
2	6	7	42	18
	-2	-11	10	84
	-1	-9	8	81
	0	-7	6	76
	1	-12	10	72
	1	-5	4	67
	2	-3	2	46
	3	-8	6	59
	3	-1	0	0
	5	-11	8	55
	5	-4	2	26
	7	-7	4	34
	8	-5	2	18
	9	-10	6	37
	11	-6	2	14
	12	-11	6	31
6	2	7	42	72
	-11	-9	12	74
	-9	-8	10	72
	-8	-11	10	62
	-7	-7	8	70

H	K U	L V	Φ W	φ2 φ1
	-5	-6	6	65
	-4	-9	6	51
	-3	-5	4	58
	-2	-1	2	84
	-1	-11	4	31
	-1	-4	2	41
	0	-7	2	24
	1	-10	2	17
	1	-3	0	0
2	7	6	51	16
	-1	-10	12	83
	-1	-4	5	89
	0	-6	7	80
	1	-8	9	75
	1	-2	2	60
	2	-10	11	72
	4	-2	1	16
	5	-4	3	33
	6	-6	5	41
	7	-8	7	45
	7	-2	0	0
	8	-10	9	48
	9	-12	11	50
	9	-6	4	27
	11	-10	8	38
	11	-4	1	6
7	2	6	51	74
	-8	-5	11	80
	-6	-12	11	55
	-6	-9	10	62
	-6	-3	8	83
	-4	-7	7	58
	-2	-11	6	38
	-2	-8	5	42
	-2	-5	4	51
	-2	-2	3	71
	0	-3	1	24
	2	-10	1	7
	2	-7	0	0
6	7	2	78	41
	-1	-2	10	87
	0	-2	7	80
	1	-4	11	73
	1	-2	4	63
	2	-2	1	20
	3	-4	5	46

H	K U	L V	Φ W	φ2 φ1
	4	-6	9	53
	5	-6	6	39
	7	-8	7	34
	7	-6	0	0
	8	-10	11	42
	9	-10	6	32
	9	-8	1	5
	11	-12	9	30
	11	-10	2	8
7	6	2	78	49
	0	-1	3	76
	2	-6	11	63
	2	-5	8	58
	2	-4	5	50
	2	-3	2	30
	4	-7	7	42
	4	-5	1	9
	6	-11	12	45
	6	-10	9	39
	6	-8	3	17
	6	-7	0	0
	8	-11	5	21
	10	-12	1	4
2	6	9	35	18
	-3	-11	8	87
	0	-3	2	75
	3	-10	6	60
	3	-7	4	54
	3	-4	2	40
	3	-1	0	0
	6	-11	6	49
	6	-5	2	26
	9	-9	4	31
	9	-6	2	18
	12	-7	2	14
6	2	9	35	72
	-12	-9	10	75
	-11	-12	10	66
	-9	-9	8	68
	-8	-3	6	88
	-7	-6	6	72
	-5	-12	6	47
	-5	-3	4	80
	-3	-9	4	43
	-2	-3	2	58

H	K U	L V	Φ W	φ2 φ1
	-1	-6	2	33
	0	-9	2	22
	1	-12	2	17
	1	-3	0	0
2	9	6	57	13
	0	-2	3	83
	3	-8	11	70
	3	-6	8	66
	3	-4	5	58
	3	-2	2	35
	6	-6	7	49
	6	-2	1	11
	9	-10	12	53
	9	-8	9	46
	9	-4	3	20
	9	-2	0	0
	12	-10	11	43
	12	-6	5	25
9	2	6	57	77
	-6	-6	11	71
	-6	-3	10	82
	-4	-9	9	54
	-4	-3	7	76
	-2	-12	7	37
	-2	-9	6	41
	-2	-6	5	48
	-2	-3	4	62
	0	-3	1	22
	2	-12	1	6
	2	-9	0	0
6	9	2	80	34
	-1	-2	12	89
	0	-2	9	83
	1	-2	6	72
	2	-2	3	48
	3	-4	9	63
	3	-2	0	0
	5	-6	12	58
	5	-4	3	26
	7	-6	6	34
	8	-6	3	17
	9	-8	9	38
	11	-10	12	40
	11	-8	3	13
	12	-10	9	31

H	K U	L V	Φ W	φ2 φ1
9	6	2	80	56
	0	-1	3	75
	2	-7	12	60
	2	-6	9	56
	2	-5	6	49
	2	-4	3	35
	2	-3	0	0
	4	-9	9	43
	4	-7	3	21
	6	-11	6	26
	6	-10	3	15
2	6	11	30	18
	-3	-10	6	88
	-1	-7	4	81
	0	-11	6	74
	1	-4	2	61
	3	-1	0	0
	4	-5	2	37
	5	-9	4	47
	7	-6	2	25
	10	-7	2	19
	11	-11	4	30
6	2	11	30	72
	-11	-11	8	67
	-8	-9	6	63
	-7	-12	6	53
	-5	-7	4	58
	-3	-2	2	77
	-2	-5	2	44
	-1	-8	2	29
	0	-11	2	21
	1	-3	0	0
2	11	6	62	10
	0	-6	11	85
	1	-4	7	78
	2	-2	3	56
	3	-6	10	70
	5	-2	2	23
	7	-4	5	37
	8	-2	1	8
	9	-6	8	42
	11	-8	11	46
	11	-2	0	0
	12	-6	7	32

H	K U	L V	Φ W	φ2 φ1
11	2	6	62	80
	-4	-11	11	51
	-4	-5	9	68
	-2	-10	7	40
	-2	-7	6	46
	-2	-4	5	58
	-2	-1	4	82
	0	-3	1	21
	2	-11	0	0
6	11	2	81	29
	0	-2	11	85
	1	-2	8	77
	2	-2	5	62
	3	-2	2	29
	5	-4	7	48
	7	-6	12	53
	7	-4	1	7
	8	-6	9	43
	10	-6	3	15
	11	-8	11	40
	11	-6	0	0
11	6	2	81	61
	0	-1	3	74
	2	-7	10	55
	2	-6	7	49
	2	-5	4	37
	2	-4	1	13
	4	-11	11	44
	4	-9	5	27
	6	-11	0	0
2	7	7	46	16
	0	-1	1	79
	7	-12	10	54
	7	-11	9	52
	7	-10	8	50
	7	-9	7	47
	7	-8	6	43
	7	-7	5	39
	7	-6	4	34
	7	-5	3	27
	7	-4	2	20
	7	-3	1	10
	7	-2	0	0

H	K U	L V	Φ W	φ2 φ1
7	2	7	46	74
	-10	-7	12	77
	-9	-7	11	74
	-8	-7	10	72
	-7	-7	9	69
	-6	-7	8	65
	-5	-7	7	61
	-4	-7	6	56
	-3	-7	5	50
	-2	-7	4	42
	-1	-7	3	33
	0	-7	2	22
	1	-7	1	11
	2	-7	0	0
7	7	2	79	45
	-1	-1	7	90
	0	-2	7	79
	1	-3	7	68
	1	-1	0	0
	2	-4	7	59
	3	-5	7	52
	4	-6	7	45
	5	-7	7	40
	6	-8	7	36
	7	-9	7	32
	8	-10	7	29
	9	-11	7	27
	10	-12	7	25
2	7	8	42	16
	-2	-12	11	85
	-1	-10	9	82
	0	-8	7	78
	1	-6	5	71
	2	-4	3	56
	3	-10	6	65
	3	-2	1	23
	5	-6	4	45
	7	-10	7	40
	7	-2	0	0
	8	-8	5	37
	10	-4	1	8
	11	-10	6	34
7	2	8	42	74
	-10	-9	11	70

TAB. V FROM HKL-UVW INTO EULER ANGLES

H	K / U	L / V	Φ / W	φ2 / φ1
2	7	8		
7	2	8	42	74
	-8	-8	9	68
	-6	-11	8	53
	-6	-7	7	64
	-4	-6	5	58
	-2	-9	4	36
	-2	-5	3	46
	-2	-1	2	82
	0	-4	1	21
	2	-11	1	8
	2	-7	0	0
2	8	7	50	14
	-1	-5	6	88
	0	-7	8	81
	1	-9	10	76
	1	-2	2	61
	2	-11	12	74
	4	-1	0	0
	5	-3	2	25
	6	-5	4	37
	7	-7	6	43
	8	-9	8	47
	9	-11	10	49
	9	-4	2	15
	11	-8	6	32
8	2	7	50	76
	-8	-3	10	86
	-7	-7	10	69
	-6	-11	10	55
	-5	-8	8	58
	-4	-5	6	64
	-3	-2	4	77
	-1	-10	4	29
	-1	-3	2	45
	0	-7	2	21
	1	-11	2	14
	1	-4	0	0
7	8	2	79	41
	0	-1	4	81
	2	-4	9	66
	2	-3	5	56
	2	-2	1	20
	4	-5	6	44

H	K / U	L / V	Φ / W	φ2 / φ1
	6	-8	11	49
	6	-7	7	38
	8	-9	8	34
	8	-7	0	0
	10	-11	9	32
	10	-9	1	4
	12	-11	2	7
8	7	2	79	49
	-1	-2	11	86
	0	-2	7	78
	1	-4	10	70
	1	-2	3	55
	3	-4	2	22
	4	-6	5	35
	5	-8	8	41
	5	-6	1	7
	6	-10	11	44
	7	-10	7	30
	7	-8	0	0
	8	-10	3	13
2	7	9	39	16
	-3	-12	10	89
	-2	-11	9	86
	-1	-10	8	82
	0	-9	7	77
	1	-8	6	72
	2	-7	5	64
	3	-6	4	55
	4	-5	3	42
	5	-4	2	28
	6	-3	1	14
	7	-11	7	49
	7	-2	0	0
	9	-9	5	36
	11	-7	3	21
7	2	9	39	74
	-3	-12	5	37
	-2	-11	4	32
	-1	-10	3	27
	-1	-1	1	67
	0	-9	2	20
	1	-8	1	11
	2	-7	0	0

H	K / U	L / V	Φ / W	φ2 / φ1
2	9	7	53	13
	-1	-6	8	88
	0	-7	9	82
	1	-8	10	78
	1	-1	1	46
	2	-9	11	74
	3	-10	12	71
	9	-2	0	0
	10	-3	1	7
	11	-4	2	12
	12	-5	3	16
9	2	7	53	77
	-7	-7	11	69
	-5	-9	9	56
	-5	-2	7	84
	-4	-3	6	75
	-3	-11	7	41
	-3	-4	5	63
	-2	-5	4	48
	-1	-6	3	34
	0	-7	2	20
	1	-8	1	9
	2	-9	0	0
7	9	2	80	38
	0	-2	9	82
	1	-3	10	75
	1	-1	1	36
	2	-4	11	70
	3	-5	12	66
	9	-7	0	0
	10	-8	1	5
	11	-9	2	8
	12	-10	3	11
9	7	2	80	52
	-1	-1	8	89
	0	-2	7	77
	1	-3	6	64
	2	-4	5	49
	3	-7	11	57
	3	-5	4	35
	4	-6	3	23
	5	-9	9	42
	5	-7	2	13
	6	-8	1	6
	7	-11	7	29

H	K / U	L / V	Φ / W	φ2 / φ1
	7	-9	0	0
2	7	10	36	16
	-1	-4	3	88
	0	-10	7	77
	1	-6	4	69
	2	-2	1	34
	3	-8	5	59
	5	-10	6	53
	7	-12	7	50
	7	-2	0	0
	9	-4	1	10
	11	-6	2	16
7	2	10	36	74
	-10	-10	9	66
	-10	-5	8	81
	-8	-7	7	69
	-6	-4	5	75
	-4	-11	5	42
	-2	-8	3	36
	-2	-3	2	56
	0	-5	1	19
	2	-12	1	8
	2	-7	0	0
2	10	7	56	11
	0	-7	10	84
	1	-3	4	72
	3	-2	2	36
	4	-5	6	56
	5	-8	10	62
	5	-1	0	0
	7	-7	8	50
	8	-3	2	16
	10	-9	10	46
	11	-5	4	22
10	2	7	56	79
	-7	-7	12	69
	-6	-5	10	73
	-5	-3	8	79
	-4	-1	6	88
	-2	-11	6	35
	-1	-9	4	29
	-1	-2	2	54
	0	-7	2	19
	1	-12	2	11

H	K / U	L / V	Φ / W	φ2 / φ1
	1	-5	0	0
7	10	2	81	35
	0	-1	5	84
	2	-3	8	67
	2	-2	3	48
	4	-5	11	61
	4	-3	1	11
	6	-5	4	28
	8	-7	7	34
	10	-9	10	37
	10	-8	5	22
	10	-7	0	0
10	7	2	81	55
	-1	-2	12	85
	0	-2	7	77
	1	-4	9	67
	1	-2	2	42
	2	-6	11	61
	4	-6	1	8
	5	-8	3	18
	6	-10	5	24
	7	-12	7	27
	7	-10	0	0
2	7	11	33	16
	-1	-6	4	85
	0	-11	7	77
	1	-5	3	67
	3	-4	2	42
	4	-9	5	55
	5	-3	1	18
	7	-2	0	0
	8	-7	3	29
	11	-11	5	34
	12	-5	1	8
7	2	11	33	74
	-11	-11	9	65
	-10	-9	8	68
	-9	-7	7	71
	-8	-5	6	77
	-7	-3	5	84
	-1	-2	1	48
	0	-11	2	19
	1	-9	1	11
	2	-7	0	0

H	K / U	L / V	Φ / W	φ2 / φ1
2	11	7	58	10
	0	-7	11	84
	1	-4	6	77
	2	-1	1	29
	3	-5	7	65
	5	-6	8	58
	7	-7	9	53
	9	-8	10	49
	11	-9	11	46
	11	-2	0	0
11	2	7	58	80
	-5	-11	11	53
	-5	-4	9	74
	-3	-8	7	48
	-3	-1	5	86
	-2	-3	4	61
	-1	-12	5	27
	-1	-5	3	37
	0	-7	2	19
	1	-9	1	7
	2	-11	0	0
7	11	2	81	32
	0	-2	11	84
	1	-1	2	56
	6	-4	1	8
	7	-5	3	19
	8	-6	5	27
	9	-7	7	32
	10	-8	9	36
	11	-9	11	38
	11	-7	0	0
11	7	2	81	58
	-1	-1	9	88
	0	-2	7	77
	1	-5	12	69
	1	-3	5	59
	2	-4	3	34
	3	-7	8	47
	3	-5	1	10
	5	-11	11	43
	5	-9	4	21
	7	-11	0	0

H	K / U	L / V	Φ / W	φ2 / φ1
2	7	12	31	16
	-2	-8	5	88
	-1	-10	6	81
	0	-12	7	76
	1	-2	1	52
	7	-2	0	0
	8	-4	1	12
	9	-6	2	21
	10	-8	3	26
	11	-10	4	30
	12	-12	5	33
7	2	12	31	74
	-10	-7	7	74
	-6	-9	5	54
	-6	-3	4	81
	-4	-4	3	65
	-2	-11	3	30
	-2	-5	2	42
	0	-6	1	18
	2	-7	0	0
2	12	7	60	9
	0	-7	12	85
	1	-6	10	80
	2	-5	8	73
	3	-4	6	62
	4	-3	4	46
	5	-2	2	24
	6	-1	0	0
	7	-7	10	55
	9	-5	6	36
	11	-3	2	11
12	2	7	60	81
	-5	-12	12	51
	-4	-11	10	49
	-3	-10	8	45
	-2	-9	6	39
	-1	-8	4	31
	-1	-1	2	70
	0	-7	2	18
	1	-6	0	0
7	12	2	82	30
	0	-1	6	85
	2	-3	11	74

H	K / U	L / V	Φ / W	φ2 / φ1
	2	-2	5	62
	4	-3	4	39
	6	-5	9	50
	6	-4	3	23
	8	-5	2	18
	10	-7	7	30
	10	-6	1	5
	12	-7	0	0
12	7	2	82	60
	0	-2	7	76
	1	-4	8	64
	1	-2	1	24
	2	-6	9	56
	3	-8	10	50
	4	-10	11	46
	5	-12	12	49
	7	-12	0	0
2	8	9	42	14
	-1	-11	10	83
	0	-9	8	80
	1	-7	6	73
	2	-5	4	62
	3	-12	10	69
	3	-3	2	39
	4	-1	0	0
	5	-8	6	53
	7	-4	2	21
	8	-11	8	49
	10	-7	4	27
	11	-5	2	14
8	2	9	42	76
	-11	-10	12	68
	-9	-9	10	66
	-7	-8	8	63
	-5	-7	6	58
	-4	-11	6	42
	-3	-6	4	49
	-2	-1	2	81
	-1	-5	2	33
	0	-9	2	19
	1	-4	0	0
2	9	8	49	13
	-1	-6	7	88
	0	-6	9	82

TAB. V

FROM HKL-UVW INTO EULER ANGLES

H	K / U	L / V	Φ / W	φ_2 / φ_1
2	8	9		
2	9	8	49	13
	1	-10	11	78
	1	-2	2	62
	5	-2	1	14
	6	-4	3	31
	7	-6	5	39
	8	-8	7	44
	9	-10	9	47
	9	-2	0	0
	10	-12	11	50
	11	-6	4	24
9	2	8	49	77
	-10	-3	12	87
	-8	-8	11	67
	-6	-5	8	71
	-4	-10	7	46
	-4	-2	5	81
	-2	-11	5	33
	-2	-7	4	40
	-2	-3	3	58
	0	-4	1	19
	2	-9	0	0
8	9	2	81	42
	0	-2	9	82
	1	-2	5	68
	2	-2	1	20
	3	-4	6	51
	4	-6	11	58
	5	-6	7	43
	7	-8	8	38
	9	-10	9	34
	9	-8	0	0
	11	-12	10	32
	11	-10	1	4
9	8	2	81	48
	0	-1	4	80
	2	-5	11	66
	2	-4	7	59
	2	-3	3	40
	4	-7	10	52
	4	-5	2	18
	6	-8	5	27
	6	-7	1	6

H	K / U	L / V	Φ / W	φ_2 / φ_1
	8	-11	8	31
	8	-9	0	0
	10	-12	3	11
2	8	11	37	14
	-1	-8	6	84
	0	-11	8	79
	1	-3	2	63
	4	-1	0	0
	5	-4	2	30
	6	-7	4	42
	7	-10	6	47
	9	-5	2	19
	11	-11	6	37
8	2	11	37	76
	-12	-7	10	77
	-11	-11	10	64
	-9	-8	8	67
	-7	-5	6	73
	-5	-2	4	84
	-3	-10	4	37
	-2	-3	2	54
	-1	-7	2	27
	0	-11	2	17
	1	-4	0	0
2	11	8	54	10
	0	-8	11	84
	1	-6	8	78
	2	-4	5	66
	3	-2	2	37
	5	-6	7	55
	7	-10	12	60
	7	-2	1	10
	8	-8	9	50
	10	-4	3	19
	11	-10	11	47
	11	-2	0	0
11	2	8	54	80
	-6	-11	11	54
	-6	-7	10	65
	-4	-6	7	59
	-2	-9	5	36
	-2	-5	4	47
	-2	-1	3	80
	0	-4	1	17

H	K / U	L / V	Φ / W	φ_2 / φ_1
	2	-11	0	0
8	11	2	82	36
	0	-2	11	84
	1	-2	7	74
	2	-2	3	47
	3	-4	10	65
	5	-4	2	18
	7	-6	5	29
	8	-6	1	6
	9	-8	8	34
	11	-10	11	37
	11	-8	0	0
	12	-10	7	24
11	8	2	82	54
	0	-1	4	79
	2	-5	9	60
	2	-4	5	49
	2	-3	1	16
	4	-7	6	37
	6	-11	11	42
	6	-10	7	31
	8	-11	0	0
2	9	9	46	13
	0	-1	1	81
	9	-12	10	51
	9	-11	9	48
	9	-10	8	46
	9	-9	7	42
	9	-8	6	39
	9	-7	5	34
	9	-6	4	29
	9	-5	3	23
	9	-4	2	16
	9	-3	1	8
	9	-2	0	0
9	2	9	46	77
	-10	-9	12	68
	-9	-9	11	66
	-8	-9	10	63
	-7	-9	9	60
	-6	-9	8	56
	-5	-9	7	52
	-4	-9	6	47
	-3	-9	5	41

H	K / U	L / V	Φ / W	φ_2 / φ_1
	-2	-9	4	34
	-1	-9	3	26
	0	-9	2	18
	1	-9	1	9
	2	-9	0	0
9	9	2	81	45
	-1	-1	9	90
	0	-2	9	81
	1	-3	9	73
	1	-1	0	0
	2	-4	9	65
	3	-5	9	58
	4	-6	9	52
	5	-7	9	47
	6	-8	9	43
	7	-9	9	39
	8	-10	9	36
	9	-11	9	33
	10	-12	9	30
2	9	10	43	13
	-1	-12	11	84
	0	-10	9	81
	1	-8	7	75
	2	-6	5	66
	3	-4	3	49
	4	-2	1	19
	5	-10	8	59
	7	-6	4	36
	9	-2	0	0
	10	-10	7	41
	11	-8	5	31
9	2	10	43	77
	-10	-10	11	65
	-8	-9	9	62
	-6	-8	7	58
	-4	-7	5	51
	-2	-11	4	30
	-2	-6	3	39
	-2	-1	2	80
	0	-5	1	17
	2	-9	0	0
2	10	9	49	11
	-1	-7	8	88
	0	-9	10	82

H	K / U	L / V	Φ / W	φ_2 / φ_1
	1	-11	12	79
	1	-2	2	63
	5	-1	0	0
	6	-3	2	22
	7	-5	4	34
	8	-7	6	41
	9	-9	8	45
	10	-11	10	48
	11	-4	2	13
10	2	9	49	79
	-8	-5	10	76
	-7	-10	10	58
	-5	-11	8	47
	-5	-2	6	83
	-4	-7	6	53
	-3	-3	4	66
	-1	-4	2	36
	0	-9	2	17
	1	-5	0	0
9	10	2	82	42
	0	-1	5	82
	2	-4	11	69
	2	-3	6	60
	2	-2	1	20
	4	-5	7	48
	6	-7	8	41
	8	-9	9	37
	10	-11	10	34
	10	-9	0	0
	12	-11	1	4
10	9	2	82	48
	0	-2	9	81
	1	-2	4	62
	3	-4	3	31
	4	-6	7	45
	5	-8	11	50
	5	-6	2	15
	7	-10	10	40
	7	-8	1	5
	8	-10	5	22
	9	-10	0	0
2	9	11	40	13
	-1	-12	10	84
	0	-11	9	80

H	K / U	L / V	Φ / W	φ_2 / φ_1
	1	-10	8	76
	2	-9	7	70
	3	-8	6	63
	4	-7	5	55
	5	-6	4	45
	6	-5	3	34
	7	-4	2	22
	8	-3	1	10
	9	-2	0	0
	11	-11	7	40
9	2	11	40	77
	-1	-12	3	22
	-1	-1	1	64
	0	-11	2	16
	1	-10	1	9
	2	-9	0	0
2	11	9	51	10
	-1	-8	10	88
	0	-9	11	83
	1	-10	12	80
	1	-1	1	48
	11	-2	0	0
	12	-3	1	6
11	2	9	51	80
	-7	-11	11	56
	-7	-2	9	86
	-6	-3	8	80
	-5	-4	7	71
	-4	-5	6	61
	-3	-6	5	50
	-2	-7	4	38
	-1	-8	3	27
	0	-9	2	16
	1	-10	1	7
	2	-11	0	0
9	11	2	82	39
	0	-2	11	83
	1	-3	12	78
	1	-1	1	36
	11	-9	0	0
	12	-10	1	4
11	9	2	82	51
	-1	-1	10	89
	0	-2	9	80

H	K / U	L / V	Φ / W	φ_2 / φ_1
	1	-3	8	70
	2	-4	7	58
	3	-5	6	46
	4	-6	5	35
	5	-7	4	25
	6	-8	3	17
	7	-11	11	41
	7	-9	2	10
	8	-10	1	5
	9	-11	0	0
2	9	12	38	13
	0	-4	3	80
	3	-10	7	66
	3	-6	4	57
	3	-2	1	26
	6	-8	5	47
	9	-10	6	42
	9	-2	0	0
	12	-12	7	39
	12	-4	1	7
9	2	12	38	77
	-12	-12	11	63
	-10	-9	9	66
	-10	-3	8	87
	-8	-6	7	70
	-6	-3	5	79
	-4	-12	5	37
	-2	-9	3	31
	-2	-3	2	53
	0	-6	1	16
	2	-9	0	0
2	12	9	54	9
	0	-3	4	84
	3	-8	10	71
	3	-5	6	63
	3	-2	2	37
	6	-7	8	55
	6	-1	0	0
	9	-9	10	50
	9	-3	2	15
	12	-11	12	48
	12	-5	4	21
12	2	9	54	81
	-7	-12	12	54

H	K / U	L / V	Φ / W	φ_2 / φ_1
	-7	-3	10	82
	-6	-9	10	58
	-5	-6	8	63
	-4	-3	6	73
	-1	-12	4	23
	-1	-3	2	42
	0	-9	2	16
	1	-6	0	0
9	12	2	82	37
	0	-1	6	84
	2	-3	9	69
	2	-2	3	47
	4	-5	12	63
	4	-3	0	0
	6	-5	3	21
	8	-7	6	30
	10	-9	9	34
	10	-8	3	13
	12	-11	12	37
12	9	2	82	53
	0	-2	9	80
	1	-4	12	73
	1	-2	3	54
	3	-4	0	0
	4	-6	3	23
	5	-8	6	33
	6	-10	9	38
	7	-12	12	41
	7	-10	3	14
2	10	11	43	11
	0	-11	10	82
	1	-9	8	77
	2	-7	6	69
	3	-5	4	56
	4	-3	2	33
	5	-12	10	64
	5	-1	0	0
	7	-8	6	46
	9	-4	2	17
	11	-11	8	42
10	2	11	43	79
	-11	-11	12	64
	-9	-10	10	61
	-7	-9	8	58

TAB. V FROM HKL-UVW INTO EULER ANGLES

```
H  K  L   Φ   φ2
   U  V   W   φ1
------------------
 2 10 11
=======
10  2 11  43 79
------------------
  -5 -8  6  52
  -3 -7  4  43
  -2 -1  2  79
  -1 -6  2  27
   0-11  2  15
   1 -5  0   0
 2 11 10  48 10
------------------
  -1 -8  9  88
   0-10 11  83
   1 -2  2  63
   6 -2  1  12
   7 -4  3  28
   8 -6  5  37
   9 -8  7  42
  10-10  9  46
  11-12 11  49
  11 -2  0   0
 1  2 10  48 80
------------------
 -10 -5 12  79
  -8-11 11  58
  -6 -7  8  62
  -6 -2  7  85
  -4 -3  5  72
  -2 -9  4  32
  -2 -4  3  48
   0 -5  1  15
   2-11  0   0
 0 11  2  82 42
------------------
   0 -2 11  83
   1 -2  6  71
   2 -2  1  20
   3 -4  7  55
   5 -6  8  46
   7 -8  9  41
   9-10 10  37
  11-12 11  34
  11-10  0   0
 1 10  2  82 48
------------------
   0 -1  5  82
   2 -4  9  65
   2 -3  4  49
```

```
H  K  L   Φ   φ2
   U  V   W   φ1
------------------
   4 -5  3  25
   6 -8  7  35
   6 -7  2  12
   8-11 11  39
   8 -9  1   5
  10-12  5  18
  10-11  0   0
 2 11 11  45 10
================
   0 -1  1  83
  11-12 10  47
  11-11  9  45
  11-10  8  42
  11 -9  7  38
  11 -8  6  34
  11 -7  5  30
  11 -6  4  25
  11 -5  3  20
  11 -4  2  14
  11 -3  1   7
  11 -2  0   0
11  2 11  45 80
------------------
 -10-11 12  62
  -9-11 11  59
  -8-11 10  56
  -7-11  9  53
  -6-11  8  49
  -5-11  7  45
  -4-11  6  40
  -3-11  5  34
  -2-11  4  28
  -1-11  3  22
   0-11  2  15
   1-11  1   7
   2-11  0   0
11 11  2  83 45
------------------
  -1 -1 11  90
   0 -2 11  83
   1 -3 11  76
   1 -1  0   0
   2 -4 11  69
   3 -5 11  63
   4 -6 11  57
   5 -7 11  53
   6 -8 11  48
   7 -9 11  44
```

```
H  K  L   Φ   φ2
   U  V   W   φ1
------------------
   8-10 11  41
   9-11 11  38
  10-12 11  35

 2 11 12  43 10
================
   0-12 11  82
   1-10  9  78
   2 -8  7  72
   3 -6  5  61
   4 -4  3  43
   5 -2  1  16
   7-10  8  54
   9 -6  4  31
  11 -2  0   0
11  2 12  43 80
------------------
 -10-11 11  61
  -8-10  9  58
  -6 -9  7  53
  -4 -8  5  46
  -2 -7  3  34
  -2 -1  2  78
   0 -6  1  14
   2-11  0   0
 2 12 11  48  9
------------------
  -1 -9 10  88
   0-11 12  84
   1 -2  2  64
   6 -1  0   0
   7 -3  2  20
   8 -5  4  32
   9 -7  6  39
  10 -9  8  44
  11-11 10  47
12  2 11  48 81
------------------
  -8 -7 10  67
  -7 -2  8  86
  -5 -3  6  75
  -4 -9  6  45
  -3 -4  4  57
  -1 -5  2  29
   0-11  2  14
   1 -6  0   0
11 12  2  83 43
------------------
   0 -1  6  84
```

```
H  K  L   Φ   φ2
   U  V   W   φ1
------------------
   2 -3  7  64
   2 -2  1  20
   4 -5  8  52
   6 -7  9  45
   8 -9 10  40
  10-11 11  37
  12-11  0   0
12 11  2  83 47
------------------
   0 -2 11  82
   1 -2  5  67
   3 -4  4  39
   4 -6  9  52
   5 -6  3  21
   7 -8  2  11
   8-10  7  29
   9-10  1   4
  11-12  0   0

 3  3  4  47 45
================
  -7 -9 12  85
  -5-11 12  76
  -5 -7  9  83
  -4 -8  9  77
  -3 -5  6  80
  -2-10  9  65
  -2 -2  3  90
  -1-11  9  60
  -1 -7  6  63
  -1 -3  3  71
   0 -4  3  56
   1 -9  6  49
   1 -5  3  44
   1 -1  0   0
   2 -6  3  36
   3-11  6  40
   3 -7  3  30
   4 -8  3  26
   5 -9  3  23
   6-10  3  20
   7-11  3  18
   8-12  3  16
 3  4  3  59 37
------------------
  -3 -6 11  85
  -2 -3  6  88
  -1 -6  9  75
  -1 -3  5  80
```

```
H  K  L   Φ   φ2
   U  V   W   φ1
------------------
   0 -3  4  69
   1 -9 11  64
   1 -6  7  62
   1 -3  3  53
   2 -9 10  59
   2 -3  2  34
   3 -6  5  44
   3 -3  1  16
   4 -9  8  47
   4 -3  0   0
   5-12 11  49
   5 -9  7  41
   5 -6  3  25
   7-12  9  39
   7 -9  5  28
   7 -6  1   7
   8 -9  4  22
   9-12  7  30
  10 -9  2  10
  11-12  5  20
  11 -9  1   5
 4  3  3  59 53
------------------
  -3 -8 12  72
  -3 -7 11  73
  -3 -6 10  76
  -3 -5  9  78
  -3 -4  8  81
  -3 -3  7  86
   0 -1  1  56
   3-12  8  39
   3-11  7  38
   3-10  6  36
   3 -9  5  33
   3 -8  4  30
   3 -7  3  25
   3 -6  2  19
   3 -5  1  11
   3 -4  0   0
   6-11  3  16
   6 -9  1   6

 3  3  5  40 45
================
  -9-11 12  86
  -7 -8  9  87
  -5-10  9  76
  -5 -5  6  90
  -4-11  9  70
```

```
H  K  L   Φ   φ2
   U  V   W   φ1
------------------
  -3 -7  6  73
  -2 -3  3  81
  -1 -9  6  59
  -1 -4  3  65
   0 -5  3  53
   1-11  6  48
   1 -6  3  43
   1 -1  0   0
   2 -7  3  36
   3 -8  3  31
   4 -9  3  27
   5-10  3  24
   6-11  3  21
   7-12  3  19
 3  5  3  63 31
------------------
  -1 -6 11  80
  -1 -3  6  84
   0 -3  5  75
   1 -6  9  69
   1 -3  4  62
   2 -3  3  46
   3 -6  7  54
   3 -3  2  29
   4 -9 11  57
   4 -3  1  13
   5 -9 10  52
   5 -6  5  37
   5 -3  0   0
   7 -9  8  40
   7 -6  3  20
   8 -9  7  34
   9-12 11  42
   9 -6  1   6
  10 -9  5  23
  11-12  9  33
  11 -9  4  18
 5  3  3  63 59
------------------
  -3 -7 12  72
  -3 -6 11  74
  -3 -5 10  76
  -3 -4  9  79
  -3 -3  8  83
  -3 -2  7  89
   0 -1  1  53
   3-12  7  34
   3-11  6  32
   3-10  5  29
```

```
H  K  L   Φ   φ2
   U  V   W   φ1
------------------
   3 -9  4  26
   3 -8  3  22
   3 -7  2  17
   3 -6  1  10
   3 -5  0   0
   6-11  1   5

 3  3  7  31 45
================
 -10-11  9  88
  -7 -7  6  90
  -5 -9  6  76
  -3-11  6  64
  -3 -4  3  83
  -2 -5  3  70
  -1 -6  3  59
   0 -7  3  49
   1 -8  3  42
   1 -1  0   0
   2 -9  3  37
   3-10  3  32
   4-11  3  29
   5-12  3  26
 3  7  3  68 23
------------------
  -1 -3  8  88
   0 -3  7  81
   1 -3  6  72
   2 -3  5  61
   3 -6 11  67
   3 -3  4  48
   4 -3  3  34
   5 -6  9  54
   5 -3  2  20
   6 -3  1   9
   7 -6  7  41
   7 -3  0   0
   9 -6  5  27
  10 -9 11  43
  11 -9 10  36
  11 -6  3  14
 7  3  3  68 67
------------------
  -3 -5 12  75
  -3 -4 11  78
  -3 -3 10  82
  -3 -2  9  86
   0 -1  1  49
   3-12  5  24
```

```
         V   -  42
H  K  L   Φ   φ2
   U  V   W   φ1
------------------
   3-11  4  21
   3-10  3  17
   3 -9  2  13
   3 -8  1   7
   3 -7  0   0

 3  3  8  28 45
================
  -7 -9  6  84
  -5-11  6  72
  -4 -4  3  90
  -3 -5  3  78
  -2 -6  3  66
  -1 -7  3  56
   0 -8  3  49
   1 -9  3  42
   1 -1  0   0
   2-10  3  37
   3-11  3  33
   4-12  3  30
 3  8  3  71 21
------------------
  -1 -3  9  89
   0 -3  8  83
   1 -3  7  75
   2 -3  6  65
   3 -3  5  54
   4 -3  4  41
   5 -6 11  60
   5 -3  3  29
   6 -3  2  18
   7 -6  9  48
   7 -3  1   8
   8 -3  0   0
   9 -6  7  35
  11 -6  5  23
 8  3  3  71 69
------------------
  -3 -4 12  78
  -3 -3 11  81
  -3 -2 10  86
   0 -1  1  49
   3-12  4  19
   3-11  3  16
   3-10  2  12
   3 -9  1   6
   3 -8  0   0
```

TAB. V

FROM HKL-UVW INTO EULER ANGLES

H	K / U	L / V	Φ / W	φ_2 / φ_1
3	3	10	23	45
	-9	-11	6	85
	-5	-5	3	90
	-4	-6	3	80
	-3	-7	3	70
	-2	-8	3	61
	-1	-9	3	54
	0	-10	3	47
	1	-11	3	42
	1	-1	0	0
	2	-12	3	38
3	10	3	74	17
	0	-3	10	85
	1	-3	9	79
	2	-3	8	72
	3	-3	7	63
	4	-3	6	53
	5	-3	5	43
	6	-3	4	32
	7	-3	3	22
	8	-3	2	14
	9	-6	11	48
	9	-3	1	6
	10	-3	0	0
	11	-6	9	37
10	3	3	74	73
	-3	-2	12	85
	-3	-1	11	90
	0	-1	1	47
	3	-12	2	10
	3	-11	1	5
	3	-10	0	0
3	3	11	21	45
	-11	-11	6	90
	-5	-6	3	85
	-4	-7	3	76
	-3	-8	3	67
	-2	-9	3	59
	-1	-10	3	53
	0	-11	3	47
	1	-12	3	42
	1	-1	0	0

H	K / U	L / V	Φ / W	φ_2 / φ_1
3	11	3	75	15
	0	-3	11	86
	1	-3	10	80
	2	-3	9	74
	3	-3	8	66
	4	-3	7	57
	5	-3	6	48
	6	-3	5	38
	7	-3	4	29
	8	-3	3	20
	9	-3	2	12
	10	-3	1	6
	11	-6	11	43
	11	-3	0	0
11	3	3	75	75
	-3	-1	12	89
	0	-1	1	47
	3	-12	1	5
	3	-11	0	0
3	4	4	51	37
	-4	-9	12	82
	-4	-8	11	84
	-4	-7	10	86
	-4	-6	9	88
	0	-1	1	65
	4	-12	9	48
	4	-11	8	46
	4	-10	7	44
	4	-9	6	42
	4	-8	5	39
	4	-7	4	35
	4	-6	3	29
	4	-5	2	22
	4	-4	1	13
	4	-3	0	0
	8	-11	5	26
	8	-9	3	18
	8	-7	1	7
	12	-11	2	9
	12	-10	1	5
4	3	4	51	53
	-5	-8	11	76
	-5	-4	8	89
	-4	-4	7	85

H	K / U	L / V	Φ / W	φ_2 / φ_1
	-3	-8	9	68
	-3	-4	6	80
	-2	-12	11	59
	-2	-4	5	73
	-1	-12	10	55
	-1	-8	7	57
	-1	-4	4	63
	0	-4	3	50
	1	-12	8	45
	1	-8	5	42
	1	-4	2	34
	2	-12	7	40
	2	-4	1	16
	3	-8	3	25
	3	-4	0	0
	4	-12	5	28
	5	-12	4	22
	5	-8	1	8
	7	-12	2	11
	8	-12	1	5
4	4	3	62	45
	-4	-5	12	87
	-3	-3	8	90
	-2	-7	12	75
	-1	-8	12	70
	-1	-5	8	73
	-1	-2	4	81
	0	-3	4	65
	1	-10	12	60
	1	-7	8	58
	1	-4	4	52
	1	-1	0	0
	2	-11	12	56
	2	-5	4	42
	3	-9	8	47
	3	-6	4	35
	4	-7	4	30
	5	-11	8	39
	5	-8	4	26
	6	-9	4	23
	7	-10	4	21
	8	-11	4	19
	9	-12	4	17
3	4	5	45	37
	-5	-10	11	83
	-4	-7	8	85

H	K / U	L / V	Φ / W	φ_2 / φ_1
	-3	-4	5	90
	-2	-11	10	71
	-1	-8	7	68
	-1	-3	3	77
	0	-5	4	62
	1	-12	9	58
	1	-7	5	55
	1	-2	1	35
	2	-9	6	50
	3	-11	7	48
	4	-3	0	0
	5	-5	1	11
	6	-7	2	17
	7	-9	3	21
	8	-11	4	24
	9	-8	1	7
	11	-12	3	15
4	3	5	45	53
	-9	-8	12	87
	-7	-9	11	79
	-6	-7	9	81
	-5	-5	7	84
	-4	-3	5	90
	-3	-11	9	61
	-2	-9	7	59
	-1	-12	8	51
	-1	-7	5	55
	-1	-2	2	71
	0	-5	3	47
	1	-8	4	39
	1	-3	1	25
	2	-11	5	35
	3	-4	0	0
	4	-7	1	10
	5	-10	2	14
	7	-11	1	6
3	5	4	56	31
	-3	-7	11	86
	-2	-6	9	83
	-1	-9	12	75
	-1	-5	7	79
	0	-4	5	71
	1	-7	8	65
	1	-3	3	57
	2	-10	11	63
	2	-2	1	24
	3	-5	4	43

H	K / U	L / V	Φ / W	φ_2 / φ_1
	4	-8	7	48
	5	-11	10	51
	5	-7	5	38
	5	-3	0	0
	7	-9	6	34
	7	-5	1	8
	8	-12	9	40
	9	-11	7	32
	9	-7	2	12
	11	-9	3	15
	12	-8	1	5
5	3	4	56	59
	-4	-8	11	70
	-3	-11	12	62
	-3	-7	9	68
	-2	-6	7	64
	-1	-9	8	53
	-1	-5	5	58
	-1	-1	2	82
	0	-4	3	47
	1	-11	7	40
	1	-7	4	37
	1	-3	1	21
	2	-10	5	32
	3	-5	0	0
	4	-8	1	8
	5	-11	2	11
4	5	3	65	39
	-2	-5	11	83
	-1	-4	8	79
	0	-3	5	71
	1	-8	12	66
	1	-5	7	63
	1	-2	2	47
	2	-7	9	59
	3	-9	11	57
	3	-3	1	15
	4	-5	3	28
	5	-7	5	34
	5	-4	0	0
	6	-9	7	37
	7	-11	9	39
	7	-8	4	23
	8	-7	1	6
	9	-12	8	31
	10	-11	5	21
	11	-10	2	8

H	K / U	L / V	Φ / W	φ_2 / φ_1
5	4	3	65	51
	-2	-5	10	76
	-1	-7	11	68
	-1	-4	7	72
	-1	-1	3	87
	0	-3	4	62
	1	-8	9	55
	1	-5	5	51
	1	-2	1	27
	2	-7	6	45
	3	-12	11	47
	3	-9	7	41
	4	-11	8	39
	4	-5	0	0
	5	-7	1	7
	6	-9	2	12
	7	-11	3	14
	9	-12	1	4
3	4	6	40	37
	-6	-12	11	82
	-4	-9	8	80
	-2	-12	9	68
	-2	-9	7	71
	-2	-6	5	76
	-2	-3	3	88
	0	-3	2	60
	2	-12	7	51
	2	-9	5	48
	2	-6	3	42
	2	-3	1	25
	4	-9	4	36
	4	-3	0	0
	6	-12	5	33
	6	-6	1	11
	8	-9	2	15
	10	-12	3	17
	10	-9	1	7
4	3	6	40	53
	-9	-10	11	81
	-9	-8	10	86
	-6	-10	9	73
	-6	-6	7	84
	-3	-12	8	58
	-3	-10	7	60
	-3	-8	6	64

H	K / U	L / V	Φ / W	φ_2 / φ_1
	-3	-6	5	69
	-3	-4	4	77
	0	-2	1	44
	3	-12	4	29
	3	-10	3	26
	3	-8	2	21
	3	-6	1	13
	3	-4	0	0
	6	-10	1	8
3	6	4	59	27
	-2	-7	12	85
	-2	-5	9	88
	0	-2	3	76
	2	-9	12	67
	2	-7	9	65
	2	-5	6	60
	2	-3	3	48
	2	-1	0	0
	4	-8	9	56
	4	-4	3	33
	6	-11	12	54
	6	-7	6	39
	6	-5	3	25
	8	-10	9	42
	8	-6	3	20
	10	-11	9	37
	10	-9	6	28
	10	-7	3	16
	12	-8	3	14
6	3	4	59	63
	-5	-6	12	77
	-4	-4	9	80
	-3	-10	12	61
	-3	-2	6	86
	-2	-8	9	59
	-1	-10	9	51
	-1	-6	6	55
	-1	-2	3	69
	0	-4	3	44
	1	-10	6	37
	1	-6	3	31
	1	-2	0	0
	2	-8	3	23
	3	-10	3	19
	4	-12	3	16

H	K / U	L / V	Φ / W	φ_2 / φ_1
4	6	3	67	34
	0	-1	2	76
	3	-8	12	62
	3	-7	10	60
	3	-6	8	56
	3	-5	6	51
	3	-4	4	43
	3	-3	2	28
	3	-2	0	0
	6	-9	10	47
	6	-7	6	36
	6	-5	2	16
	9	-11	10	39
	9	-10	8	34
	9	-8	4	20
	9	-7	2	11
	12	-11	6	22
	12	-9	2	8
6	4	3	67	56
	-4	-3	12	89
	-3	-3	10	86
	-2	-3	8	81
	-1	-6	10	68
	-1	-3	6	73
	0	-3	4	60
	1	-9	10	55
	1	-6	6	50
	1	-3	2	35
	2	-9	8	43
	2	-3	0	0
	3	-12	10	43
	3	-6	2	18
	4	-9	4	24
	5	-12	6	27
	5	-9	2	12
	7	-12	2	9
3	4	7	36	37
	-5	-12	9	78
	-4	-11	8	76
	-3	-10	7	73
	-2	-9	6	70
	-1	-8	5	65
	0	-7	4	59
	1	-6	3	50
	2	-5	2	37

AB. V — FROM HKL-UVW INTO EULER ANGLES — V - 44

H	K / U	L / V	Φ / W	φ2 / φ1
3	4	7	36	37
	3	-11	5	44
	3	-4	1	20
	4	-3	0	0
	5	-9	3	29
	7	-7	1	10
	10	-11	2	13
	11	-10	1	7
4	3	7	36	53
	-5	-12	8	64
	-4	-11	7	62
	-3	-10	6	59
	-2	-9	5	55
	-1	-8	4	50
	-1	-1	1	83
	0	-7	3	43
	1	-6	2	33
	2	-5	1	18
	3	-11	3	26
	3	-4	0	0
	5	-9	1	10
3	7	4	62	23
	-1	-3	6	88
	0	-4	7	79
	1	-5	8	72
	1	-1	1	41
	2	-6	9	68
	3	-7	10	64
	4	-8	11	61
	5	-9	12	59
	7	-3	0	0
	8	-4	1	7
	9	-5	2	12
	10	-6	3	16
	11	-7	4	19
	12	-8	5	22
7	3	4	62	67
	-5	-3	11	86
	-3	-5	9	71
	-2	-10	11	56
	-2	-2	5	79
	-1	-11	10	49
	-1	-7	7	53
	-1	-3	4	62
	0	-4	3	43

H	K / U	L / V	Φ / W	φ2 / φ1
	1	-9	5	33
	1	-5	2	24
	2	-6	1	10
	3	-11	3	17
	3	-7	0	0
4	7	3	70	30
	-1	-2	6	89
	0	-3	7	79
	1	-4	8	72
	1	-1	1	38
	2	-5	9	66
	3	-6	10	62
	4	-7	11	59
	5	-8	12	57
	7	-4	0	0
	8	-5	1	6
	9	-6	2	11
	10	-7	3	15
	11	-8	4	18
	12	-9	5	20
7	4	3	70	60
	-3	-3	11	85
	-1	-5	9	68
	-1	-2	5	77
	0	-3	4	59
	1	-10	11	52
	1	-7	7	49
	1	-4	3	39
	2	-11	10	45
	2	-5	2	22
	3	-9	5	30
	3	-6	1	9
	4	-7	0	0
	5	-11	3	15
3	4	8	32	37
	-8	-12	9	87
	-4	-11	7	76
	-4	-9	6	79
	-4	-7	5	84
	0	-2	1	58
	4	-11	4	38
	4	-9	3	33
	4	-7	2	27
	4	-5	1	17
	4	-3	0	0

H	K / U	L / V	Φ / W	φ2 / φ1
	8	-12	3	23
	8	-8	1	10
	12	-11	1	7
4	3	8	32	53
	-11	-12	10	81
	-8	-8	7	83
	-7	-12	8	70
	-5	-12	7	63
	-5	-4	4	88
	-4	-8	5	67
	-3	-4	3	76
	-1	-12	5	46
	-1	-4	2	55
	0	-8	3	41
	1	-12	4	36
	1	-4	1	26
	3	-4	0	0
	4	-8	1	12
	5	-12	2	17
	7	-12	1	8
3	8	4	65	21
	0	-1	2	81
	4	-7	11	63
	4	-6	9	60
	4	-5	7	55
	4	-4	5	47
	4	-3	3	35
	4	-2	1	14
	8	-9	12	51
	8	-7	8	42
	8	-5	4	26
	8	-3	0	0
	12	-10	11	39
	12	-8	7	29
	12	-7	5	22
	12	-5	1	5
8	3	4	65	69
	-4	-4	11	79
	-3	-8	12	64
	-3	-4	9	75
	-2	-4	7	69
	-1	-12	11	48
	-1	-8	8	51
	-1	-4	5	58
	0	-4	3	41
	1	-12	7	34

H	K / U	L / V	Φ / W	φ2 / φ1
	1	-8	4	29
	1	-4	1	15
	2	-12	5	25
	3	-8	0	0
	4	-12	1	5
4	8	3	71	27
	-1	-4	12	86
	0	-3	8	81
	1	-5	12	76
	1	-2	4	67
	2	-1	0	0
	3	-3	4	46
	4	-5	8	55
	5	-7	12	59
	5	-4	4	34
	7	-8	12	52
	7	-5	4	26
	8	-7	8	39
	9	-6	4	21
	11	-10	12	41
	11	-7	4	18
	12	-9	8	30
8	4	3	71	63
	-2	-5	12	74
	-1	-7	12	65
	-1	-4	8	70
	-1	-1	4	84
	0	-3	4	58
	1	-11	12	51
	1	-8	6	48
	1	-5	4	41
	1	-2	0	0
	2	-7	4	31
	3	-12	6	35
	3	-9	4	24
	4	-11	4	20
3	4	9	29	37
	-5	-12	7	77
	-3	-9	5	74
	-2	-3	2	87
	-1	-6	3	66
	0	-9	4	57
	1	-12	5	52
	1	-3	1	38
	4	-3	0	0

H	K / U	L / V	Φ / W	φ2 / φ1
	5	-6	1	15
	6	-9	2	22
	7	-12	3	26
	9	-9	1	9
4	3	9	29	53
	-12	-11	9	85
	-9	-12	8	76
	-9	-9	7	83
	-6	-7	5	79
	-3	-11	5	56
	-3	-8	4	61
	-3	-5	3	70
	0	-3	1	41
	3	-10	2	23
	3	-7	1	16
	3	-4	0	0
	6	-11	1	9
3	9	4	67	18
	-1	-5	12	87
	0	-4	9	83
	1	-3	6	74
	2	-2	3	52
	3	-5	9	66
	3	-1	0	0
	5	-7	12	62
	5	-3	3	30
	7	-5	6	38
	8	-4	3	20
	9	-7	9	42
	11	-9	12	44
	11	-5	3	15
	12	-8	9	35
9	3	4	67	72
	-3	-7	12	66
	-2	-6	9	63
	-1	-9	9	50
	-1	-5	6	56
	-1	-1	3	79
	0	-4	3	41
	1	-11	6	31
	1	-7	3	25
	1	-3	0	0
	2	-10	3	18
4	9	3	73	24
	0	-1	3	83

H	K / U	L / V	Φ / W	φ2 / φ1
	3	-5	11	67
	3	-4	8	62
	3	-3	5	53
	3	-2	2	30
	6	-5	7	44
	6	-3	1	9
	9	-8	12	48
	9	-7	9	40
	9	-5	3	17
	9	-4	0	0
	12	-9	11	38
	12	-7	5	21
9	4	3	73	66
	-2	-3	10	80
	-1	-6	11	66
	-1	-3	7	72
	0	-3	4	57
	1	-9	9	47
	1	-6	5	42
	1	-3	1	18
	2	-9	6	35
	3	-12	7	31
	4	-9	0	0
	5	-12	1	5
3	4	10	27	37
	-6	-8	5	90
	-4	-7	4	84
	-2	-11	5	66
	-2	-6	3	73
	0	-5	2	56
	2	-9	3	44
	2	-4	1	29
	4	-3	0	0
	6	-7	1	14
	8	-11	2	19
	10	-10	1	9
4	3	10	27	53
	-11	-12	8	80
	-10	-10	7	83
	-9	-8	6	86
	-8	-6	5	90
	-1	-12	4	45
	-1	-2	1	66
	0	-10	3	40
	1	-8	2	33

H	K / U	L / V	Φ / W	φ2 / φ1
	2	-6	1	20
	3	-4	0	0
	5	-10	1	11
3	10	4	69	17
	0	-2	5	84
	2	-5	11	74
	2	-3	6	67
	2	-1	1	26
	4	-4	7	56
	6	-5	8	50
	8	-6	9	46
	10	-7	10	43
	10	-3	0	0
	12	-8	11	40
	12	-4	1	5
10	3	4	69	73
	-3	-2	9	84
	-2	-8	11	59
	-1	-10	10	49
	-1	-6	7	54
	-1	-2	4	69
	0	-4	3	40
	1	-10	5	28
	1	-6	2	20
	2	-8	1	7
	3	-10	0	0
4	10	3	74	22
	0	-3	10	84
	1	-4	12	79
	1	-1	2	58
	5	-2	0	0
	6	-3	2	17
	7	-4	4	27
	8	-5	6	34
	9	-6	8	36
	10	-7	10	41
	11	-8	12	43
	11	-5	2	10
10	4	3	74	68
	-2	-1	8	89
	-1	-5	10	68
	-1	-2	6	77
	0	-3	4	56
	1	-10	10	47
	1	-7	6	42

H	K / U	L / V	Φ / W	φ2 / φ1
	1	-4	2	27
	2	-11	6	37
	2	-5	0	0
	3	-9	2	12
3	4	11	24	37
	-1	-2	1	81
	0	-11	4	56
	1	-9	3	49
	2	-7	2	40
	3	-5	1	24
	4	-3	0	0
	5	-12	3	33
	7	-8	1	13
	11	-11	1	9
4	3	11	24	53
	-11	-11	7	83
	-9	-10	6	80
	-7	-9	5	76
	-5	-8	4	71
	-3	-7	3	62
	-1	-6	2	49
	0	-11	3	39
	1	-5	1	28
	3	-4	0	0
	4	-9	1	14
3	11	4	71	15
	0	-4	11	85
	1	-1	2	60
	6	-2	1	10
	7	-3	3	23
	8	-4	5	31
	9	-5	7	37
	10	-6	9	40
	11	-7	11	43
	11	-3	0	0
11	3	4	71	75
	-3	-5	12	72
	-3	-1	9	89
	-2	-2	7	79
	-1	-11	11	48
	-1	-7	6	53
	-1	-3	5	64
	0	-4	3	39
	1	-9	4	25

TAB. V FROM HKL-UVW INTO EULER ANGLES V - 45

H	K / U	L / V	Φ / W	φ_2 / φ_1
3	4	11		
11	3	4	71	75
	1	-5	1	12
	3	-11	0	0
4	11	3	76	20
	0	-3	11	85
	1	-2	6	75
	2	-1	1	25
	3	-3	7	62
	5	-4	8	54
	7	-5	9	48
	9	-6	10	44
	11	-7	11	42
	11	-4	0	0
11	4	3	76	70
	-1	-4	9	70
	-1	-1	5	83
	0	-3	4	56
	1	-11	11	47
	1	-8	7	43
	1	-5	3	32
	2	-7	2	16
	3	-12	5	23
	3	-9	1	6
	4	-11	0	0
3	4	12	23	37
	-4	-12	5	73
	-4	-9	4	78
	-4	-6	3	87
	0	-3	1	55
	4	-12	3	37
	4	-9	2	31
	4	-6	1	21
	4	-3	0	0
	8	-9	1	12
	12	-12	1	9
4	3	12	23	53
	-12	-12	7	82
	-9	-8	5	86
	-6	-12	5	65
	-3	-12	4	53
	-3	-8	3	59

H	K / U	L / V	Φ / W	φ_2 / φ_1
	-3	-4	2	75
	0	-4	1	39
	3	-12	2	25
	3	-8	1	18
	3	-4	0	0
	6	-12	1	11
3	12	4	72	14
	0	-1	3	86
	4	-5	12	68
	4	-4	9	63
	4	-3	6	54
	4	-2	3	36
	4	-1	0	0
	8	-5	9	47
	8	-3	3	20
	12	-7	12	43
	12	-5	6	26
	12	-4	3	14
12	3	4	72	76
	-3	-4	12	76
	-2	-4	9	70
	-1	-12	12	48
	-1	-8	9	52
	-1	-4	6	60
	0	-4	3	39
	1	-12	6	28
	1	-8	3	22
	1	-4	0	0
	2	-12	3	15
4	12	3	77	18
	0	-1	4	86
	3	-4	12	72
	3	-3	8	65
	3	-2	4	50
	3	-1	0	0
	6	-5	12	59
	6	-3	4	32
	9	-5	8	39
	9	-4	4	23
	12	-7	12	42
	12	-5	4	18
12	4	3	77	72
	-2	-3	12	80
	-1	-6	12	66
	-1	-3	8	73

H	K / U	L / V	Φ / W	φ_2 / φ_1
	0	-3	4	55
	1	-12	12	47
	1	-9	8	43
	1	-6	4	34
	1	-3	0	0
	2	-9	4	24
	3	-12	4	18
3	5	5	49	31
	-5	-9	12	89
	0	-1	1	69
	5	-12	9	49
	5	-11	8	47
	5	-10	7	44
	5	-9	6	42
	5	-8	5	38
	5	-7	4	34
	5	-6	3	28
	5	-5	2	21
	5	-4	1	12
	5	-3	0	0
	10	-11	5	25
	10	-9	3	17
	10	-7	1	6
5	3	5	49	59
	-8	-5	11	89
	-7	-5	10	87
	-6	-5	9	84
	-5	-10	11	67
	-5	-5	8	81
	-4	-5	7	76
	-3	-10	9	59
	-3	-5	6	71
	-2	-5	5	64
	-1	-10	7	49
	-1	-5	4	54
	0	-5	3	43
	1	-10	5	36
	1	-5	2	29
	2	-5	1	14
	3	-10	3	21
	3	-5	0	0
	5	-10	1	7
5	5	3	67	45
	-3	-3	10	90
	-1	-5	10	75

H	K / U	L / V	Φ / W	φ_2 / φ_1
	-1	-2	5	83
	0	-3	5	69
	1	-7	10	62
	1	-4	5	57
	1	-1	0	0
	2	-5	5	48
	3	-9	10	52
	3	-6	5	40
	4	-7	5	35
	5	-11	10	44
	5	-8	5	31
	6	-9	5	27
	7	-10	5	24
	8	-11	5	22
	9	-12	5	20
3	5	6	44	31
	-5	-9	10	89
	-2	-12	11	74
	-1	-9	8	72
	-1	-3	3	81
	0	-6	5	67
	1	-9	7	61
	1	-3	2	50
	2	-12	9	59
	3	-3	1	19
	4	-6	3	33
	5	-9	5	39
	5	-3	0	0
	6	-12	7	42
	7	-9	4	28
	8	-6	1	8
	10	-12	5	26
	11	-9	2	12
5	3	6	44	59
	-9	-7	11	85
	-6	-12	11	65
	-6	-8	9	74
	-6	-4	7	88
	-3	-11	8	56
	-3	-9	7	58
	-3	-7	6	63
	-3	-5	5	69
	-3	-3	4	80
	0	-2	1	40
	3	-11	3	21
	3	-9	2	17

H	K / U	L / V	Φ / W	φ_2 / φ_1
	3	-7	1	11
	3	-5	0	0
	6	-12	1	6
3	6	5	53	27
	-2	-9	12	82
	-1	-7	9	79
	-1	-2	3	90
	0	-5	6	73
	1	-8	9	68
	1	-3	3	59
	2	-11	12	66
	2	-1	0	0
	3	-4	3	40
	4	-7	6	48
	5	-10	9	51
	5	-5	3	29
	7	-11	9	45
	7	-6	3	23
	8	-9	6	34
	9	-7	3	19
	11	-8	3	16
	12	-11	6	26
6	3	5	53	63
	-7	-6	12	82
	-5	-10	12	66
	-5	-5	9	79
	-4	-7	9	68
	-3	-4	6	73
	-2	-11	9	51
	-2	-1	3	90
	-1	-8	6	48
	-1	-3	3	59
	0	-5	3	40
	1	-12	6	34
	1	-7	3	29
	1	-2	0	0
	2	-9	3	23
	3	-11	3	19
5	6	3	69	40
	0	-1	2	73
	3	-8	11	58
	3	-7	9	55
	3	-6	7	51
	3	-5	5	44
	3	-4	3	33
	3	-3	1	14

H	K / U	L / V	Φ / W	φ_2 / φ_1
	6	-11	12	48
	6	-9	6	40
	6	-7	4	25
	6	-5	0	0
	9	-11	7	28
	9	-10	5	22
	9	-8	1	5
	12	-11	2	8
6	5	3	69	50
	-3	-3	11	88
	-2	-3	9	84
	-1	-6	12	73
	-1	-3	7	77
	0	-3	5	67
	1	-6	8	59
	1	-3	3	47
	2	-9	11	55
	2	-3	1	17
	3	-6	4	33
	4	-9	7	38
	5	-12	10	41
	5	-9	5	28
	5	-6	0	0
	7	-12	6	25
	7	-9	1	5
	9	-12	2	8
3	5	7	40	31
	-5	-11	10	85
	-3	-8	7	82
	-1	-12	9	69
	-1	-5	4	75
	0	-7	5	65
	1	-9	6	60
	1	-2	1	40
	2	-11	7	56
	5	-3	0	0
	6	-5	1	11
	7	-7	2	18
	8	-9	3	22
	9	-11	4	25
	11	-8	1	7
5	3	7	40	59
	-10	-9	11	81
	-7	-7	8	79
	-6	-11	9	66

H	K / U	L / V	Φ / W	φ_2 / φ_1
	-5	-8	7	69
	-4	-5	5	74
	-3	-2	3	88
	-1	-10	5	44
	-1	-3	2	57
	0	-7	3	38
	1	-11	4	32
	1	-4	1	22
	3	-5	0	0
	4	-9	1	9
3	7	5	57	23
	-2	-7	11	86
	-1	-6	9	82
	0	-5	7	77
	1	-9	12	73
	1	-4	5	67
	2	-3	3	50
	3	-7	8	60
	3	-2	1	19
	5	-10	11	57
	5	-5	4	36
	7	-8	7	41
	7	-3	0	0
	8	-7	5	31
	9	-11	10	44
	10	-5	1	6
	11	-9	6	28
7	3	5	57	67
	-4	-9	11	63
	-3	-8	9	60
	-2	-7	7	56
	-1	-11	8	45
	-1	-6	5	49
	-1	-1	2	78
	0	-5	3	38
	1	-9	4	29
	1	-4	1	16
	3	-7	0	0
	4	-11	1	6
5	7	3	71	36
	-1	-4	11	83
	0	-3	7	77
	1	-5	10	71
	1	-2	3	58
	3	-3	2	27
	4	-5	5	41

H	K / U	L / V	Φ / W	φ_2 / φ_1
	5	-7	8	46
	5	-4	1	9
	6	-9	11	49
	7	-8	7	36
	7	-5	0	0
	8	-7	3	17
	10	-11	9	33
	11	-10	5	20
	12	-9	1	4
7	5	3	71	54
	-1	-4	9	74
	-1	-1	4	87
	0	-3	5	65
	1	-8	11	59
	1	-5	6	54
	1	-2	1	26
	2	-7	7	47
	3	-9	8	43
	4	-11	9	40
	5	-7	0	0
	6	-9	1	6
	7	-11	2	9
3	5	8	36	31
	-4	-12	9	80
	-3	-11	8	77
	-2	-10	7	74
	-1	-9	6	70
	0	-8	5	64
	1	-7	4	57
	2	-6	3	47
	3	-5	2	33
	4	-4	1	17
	5	-11	5	40
	5	-3	0	0
	7	-9	3	26
	9	-7	1	9
5	3	8	36	59
	-4	-12	7	55
	-3	-11	6	52
	-2	-10	5	48
	-1	-9	4	43
	-1	-1	1	79
	0	-8	3	37
	1	-7	2	28
	2	-6	1	15

TAB. V — FROM HKL-UVW INTO EULER ANGLES — V - 46

H	K / U	L / V	Φ / W	φ_2 / φ_1
3	5	8		
5	3	8	36	59
	3	-5	0	0
	5	-11	1	8
3	8	5	60	21
	-1	-4	7	87
	0	-5	8	79
	1	-6	9	74
	1	-1	1	42
	2	-7	10	70
	3	-8	11	66
	4	-9	12	64
	8	-3	0	0
	9	-4	1	7
	10	-5	2	12
	11	-6	3	16
	12	-7	4	19
8	3	5	60	69
	-5	-5	11	77
	-3	-7	9	62
	-3	-2	6	83
	-2	-3	5	70
	-1	-9	7	45
	-1	-4	4	54
	0	-5	3	37
	1	-11	5	29
	1	-6	2	21
	2	-7	1	9
	3	-8	0	0
5	8	3	72	32
	-1	-2	7	88
	0	-3	8	79
	1	-4	9	73
	1	-1	1	37
	2	-5	10	68
	3	-6	11	64
	4	-7	12	61
	8	-5	0	0
	9	-6	1	6
	10	-7	2	10
	11	-8	3	13
	12	-9	4	16

H	K / U	L / V	Φ / W	φ_2 / φ_1
8	5	3	72	58
	-1	-5	11	72
	-1	-2	6	80
	0	-3	5	64
	1	-7	9	56
	1	-4	4	47
	2	-5	3	31
	3	-9	7	39
	3	-6	2	17
	4	-7	1	7
	5	-11	5	24
	5	-8	0	0
3	5	9	33	31
	-7	-12	9	89
	-1	-12	7	68
	-1	-3	2	79
	0	-9	5	63
	1	-6	3	54
	2	-3	1	29
	3	-9	4	46
	5	-12	5	41
	5	-3	0	0
	7	-6	1	11
	9	-9	2	17
	11	-12	3	19
	12	-9	1	7
5	3	9	33	59
	-9	-9	8	78
	-9	-6	7	88
	-6	-11	7	64
	-6	-5	5	83
	-3	-10	5	53
	-3	-7	4	59
	-3	-4	3	71
	0	-3	1	36
	3	-11	2	19
	3	-8	1	12
	3	-5	0	0
3	9	5	62	18
	-1	-3	6	90
	0	-5	9	81
	1	-7	12	77
	1	-2	3	65
	3	-1	0	0

H	K / U	L / V	Φ / W	φ_2 / φ_1
	4	-3	3	36
	5	-5	6	47
	6	-7	9	52
	7	-9	12	55
	7	-4	3	23
	9	-8	9	43
	10	-5	3	17
	11	-7	6	28
9	3	5	62	72
	-5	-5	12	77
	-4	-3	9	81
	-3	-11	12	55
	-3	-1	6	90
	-2	-9	9	52
	-1	-12	9	43
	-1	-7	6	47
	-1	-2	3	65
	0	-5	3	36
	1	-8	3	23
	1	-3	0	0
	2	-11	3	17
5	9	3	74	29
	0	-1	3	81
	3	-5	10	64
	3	-4	7	58
	3	-3	4	46
	3	-2	1	16
	6	-7	11	53
	6	-5	5	34
	9	-8	9	39
	9	-7	6	29
	9	-5	0	0
	12	-9	7	26
	12	-7	1	4
9	5	3	74	61
	-2	-3	11	82
	-1	-3	8	76
	0	-3	5	63
	1	-9	12	56
	1	-6	7	52
	1	-3	2	34
	2	-9	9	47
	3	-12	11	44
	3	-6	1	9
	4	-9	3	18
	5	-12	5	22

H	K / U	L / V	Φ / W	φ_2 / φ_1
	5	-9	0	0
3	5	10	30	31
	-5	-11	7	84
	-5	-9	6	88
	0	-2	1	63
	5	-11	4	39
	5	-9	3	34
	5	-7	2	27
	5	-5	1	16
	5	-3	0	0
	10	-12	3	22
	10	-8	1	9
5	3	10	30	59
	-12	-10	9	82
	-8	-10	7	72
	-7	-5	5	86
	-5	-5	4	78
	-4	-10	5	57
	-3	-5	3	65
	-1	-5	2	46
	0	-10	3	35
	1	-5	1	22
	3	-5	0	0
	4	-10	1	11
3	10	5	64	17
	0	-1	2	83
	5	-7	11	61
	5	-6	9	57
	5	-5	7	51
	5	-4	5	43
	5	-3	3	30
	5	-2	1	12
	10	-9	12	48
	10	-7	8	37
	10	-5	4	22
	10	-3	0	0
10	3	5	64	73
	-4	-5	11	73
	-3	-10	12	57
	-3	-5	9	69
	-2	-5	7	61
	-1	-10	8	44
	-1	-5	5	51
	0	-5	3	35

H	K / U	L / V	Φ / W	φ_2 / φ_1
	1	-10	4	24
	1	-5	1	12
	3	-10	0	0
5	10	3	75	27
	0	-3	10	83
	1	-2	5	71
	2	-1	0	0
	3	-3	5	52
	4	-5	10	61
	5	-4	5	40
	7	-5	5	31
	8	-7	10	45
	9	-6	5	26
	11	-7	5	22
	12	-9	10	35
10	5	3	75	63
	-1	-4	10	73
	-1	-1	5	85
	0	-3	5	63
	1	-8	10	54
	1	-5	5	46
	1	-2	0	0
	2	-7	5	36
	3	-12	10	41
	3	-9	5	29
	4	-11	5	24
3	5	11	28	31
	-3	-7	4	83
	-1	-6	3	71
	0	-11	5	62
	1	-5	2	51
	3	-4	1	25
	4	-9	3	38
	5	-3	0	0
	8	-7	1	12
	11	-11	2	16
5	3	11	28	59
	-11	-11	8	78
	-10	-9	7	60
	-9	-7	6	84
	-8	-5	5	89
	-1	-2	1	61
	0	-11	3	34
	1	-9	2	27

H	K / U	L / V	Φ / W	φ_2 / φ_1
	2	-7	1	17
	3	-5	0	0
	5	-12	1	9
3	11	5	66	15
	0	-5	11	84
	1	-3	6	75
	2	-1	1	26
	3	-4	7	63
	5	-5	8	55
	7	-6	9	50
	9	-7	10	46
	11	-8	11	43
	11	-3	0	0
11	3	5	66	75
	-3	-4	9	73
	-2	-11	11	50
	-2	-1	5	85
	-1	-8	7	46
	-1	-3	4	59
	0	-5	3	34
	1	-12	5	25
	1	-7	2	17
	2	-9	1	7
	3	-11	0	0
5	11	3	76	24
	0	-3	11	84
	1	-1	2	57
	6	-3	1	9
	7	-4	3	21
	8	-5	5	29
	9	-6	7	34
	10	-7	9	38
	11	-8	11	40
	11	-5	0	0
11	5	3	76	66
	-2	-1	9	89
	-1	-5	12	71
	-1	-2	7	79
	0	-3	5	62
	1	-7	8	51
	1	-4	3	37
	2	-11	11	46
	2	-5	1	11
	3	-9	4	24
	5	-11	0	0

H	K / U	L / V	Φ / W	φ_2 / φ_1
3	5	12	26	31
	-5	-9	5	88
	-2	-6	3	79
	-1	-9	4	68
	0	-12	5	62
	1	-3	1	44
	5	-3	0	0
	6	-6	1	16
	7	-9	2	23
	8	-12	3	28
	11	-9	1	9
5	3	12	26	59
	-12	-8	7	88
	-6	-10	5	64
	-3	-11	4	49
	-3	-7	3	57
	-3	-3	2	77
	0	-4	1	34
	3	-9	1	14
	3	-5	0	0
3	12	5	68	14
	0	-5	12	85
	1	-4	9	79
	2	-3	6	68
	3	-2	3	44
	4	-1	0	0
	5	-5	9	58
	7	-3	3	23
	8	-7	12	54
	10	-5	6	31
	11	-4	3	16
12	3	5	68	76
	-3	-8	12	61
	-2	-7	9	57
	-1	-11	9	43
	-1	-6	6	49
	-1	-1	3	77
	0	-5	3	34
	1	-9	3	20
	1	-4	0	0
5	12	3	77	23
	0	-1	4	85
	3	-4	11	69

H	K / U	L / V	Φ / W	φ_2 / φ_1
	3	-3	7	61
	3	-2	3	41
	6	-5	10	54
	6	-3	2	17
	9	-5	5	27
	9	-4	1	6
	12	-7	8	31
	12	-5	0	0
12	5	3	77	67
	-1	-3	9	76
	0	-3	5	62
	1	-9	11	52
	1	-6	6	46
	1	-3	1	18
	2	-9	7	38
	3	-12	8	34
	5	-12	0	0
3	6	7	44	27
	-1	-10	9	75
	-1	-3	3	84
	0	-7	6	70
	1	-11	9	66
	1	-4	3	58
	2	-1	0	0
	3	-5	3	41
	4	-9	6	49
	5	-6	3	31
	7	-7	3	25
	8	-11	6	36
	9	-8	3	20
	11	-9	3	17
6	3	7	44	63
	-11	-6	12	89
	-9	-10	12	74
	-6	-5	9	86
	-7	-7	9	76
	-5	-11	9	60
	-5	-4	6	81
	-3	-6	6	56
	-2	-3	3	68
	-1	-12	6	40
	-1	-5	3	47
	0	-7	3	35
	1	-9	3	27
	1	-2	0	0

TAB. V

FROM HKL-UVW INTO EULER ANGLES

H	K / U	L / V	Φ / W	φ_2 / φ_1
3	6	7		
6	3	7	44	63
	2	-11	3	22
3	7	6	52	23
	-1	-9	11	79
	-1	-3	4	87
	0	-6	7	75
	1	-9	10	71
	1	-3	3	61
	3	-3	2	33
	4	-6	5	47
	5	-9	8	51
	5	-3	1	12
	6	-12	11	54
	7	-9	7	42
	7	-3	0	0
	8	-6	3	21
	10	-12	9	39
	11	-9	5	25
	12	-6	1	5
7	3	6	52	67
	-6	-8	11	70
	-6	-4	9	83
	-3	-11	9	52
	-3	-9	8	55
	-3	-7	7	59
	-3	-5	6	66
	-3	-3	5	76
	0	-2	1	35
	3	-11	2	13
	3	-9	1	8
	3	-7	0	0
6	7	3	72	41
	-2	-3	11	88
	-1	-3	9	83
	0	-3	7	75
	1	-6	12	70
	1	-3	5	63
	2	-3	3	42
	3	-6	8	54
	3	-3	1	14
	5	-9	11	50
	5	-6	4	29
	7	-9	7	33

H	K / U	L / V	Φ / W	φ_2 / φ_1
	7	-6	0	0
	8	-9	5	24
	9	-12	10	36
	10	-9	1	4
	11	-12	6	21
7	6	3	72	49
	0	-1	2	70
	3	-9	11	53
	3	-8	9	50
	3	-7	7	45
	3	-6	5	39
	3	-5	3	29
	3	-4	1	12
	6	-11	8	34
	6	-9	4	21
	6	-7	0	0
	9	-11	1	4
3	6	8	40	27
	-4	-10	9	86
	-2	-11	9	77
	-2	-7	6	82
	0	-4	3	69
	2	-9	6	58
	2	-5	3	49
	2	-1	0	0
	4	-6	3	37
	6	-11	6	42
	6	-7	3	29
	8	-8	3	24
	10	-9	3	20
	12	-10	3	17
6	3	8	40	63
	-11	-10	12	78
	-8	-8	9	76
	-7	-10	9	67
	-5	-6	6	71
	-3	-10	6	51
	-3	-2	3	85
	-2	-4	3	60
	-1	-6	3	44
	0	-8	3	33
	1	-10	3	26
	1	-2	0	0
	2	-12	3	22

H	K / U	L / V	Φ / W	φ_2 / φ_1
3	8	6	55	21
	-2	-6	9	89
	0	-3	4	78
	2	-9	11	69
	2	-6	7	65
	2	-3	3	51
	4	-9	10	61
	4	-3	2	27
	6	-6	5	38
	6	-3	1	10
	8	-9	8	43
	8	-3	0	0
	10	-12	11	45
	10	-9	7	34
	10	-6	3	18
8	3	6	55	69
	-6	-6	11	75
	-3	-12	10	50
	-3	-10	9	53
	-3	-8	8	57
	-3	-6	7	62
	-3	-4	6	70
	-3	-2	5	82
	0	-2	1	33
	3	-12	2	11
	3	-10	1	7
	3	-8	0	0
6	8	3	73	37
	-2	-3	12	89
	-1	-3	10	85
	0	-3	8	78
	1	-3	6	67
	2	-3	4	51
	3	-6	10	60
	3	-3	2	26
	4	-3	0	0
	5	-6	6	39
	7	-9	10	44
	7	-6	2	13
	8	-9	8	35
	10	-9	4	17
	11	-12	10	33
	11	-9	2	8
8	6	3	73	53
	0	-1	2	69

H	K / U	L / V	Φ / W	φ_2 / φ_1
	3	-10	12	52
	3	-9	10	49
	3	-8	8	46
	3	-7	6	40
	3	-6	4	32
	3	-5	2	20
	3	-4	0	0
	6	-11	6	27
	6	-9	2	11
3	6	10	34	27
	-2	-9	6	78
	-2	-4	3	90
	0	-5	3	67
	2	-11	6	58
	2	-6	3	50
	2	-1	0	0
	4	-7	3	39
	6	-8	3	31
	8	-9	3	26
	10	-10	3	22
	12	-11	3	19
6	3	10	34	63
	-11	-8	9	82
	-10	-10	9	75
	-7	-6	6	78
	-5	-10	6	58
	-4	-2	3	90
	-3	-4	3	67
	-2	-6	3	50
	-1	-8	3	39
	0	-10	3	31
	1	-12	3	26
	1	-2	0	0
3	10	6	60	17
	0	-3	5	81
	2	-6	9	71
	2	-3	4	59
	4	-3	3	36
	6	-6	7	47
	6	-3	2	19
	8	-9	11	51
	8	-3	1	8
	10	-9	10	43
	10	-6	5	27
	10	-3	0	0

H	K / U	L / V	Φ / W	φ_2 / φ_1
10	3	6	60	73
	-6	-2	11	89
	-3	-12	11	50
	-3	-10	10	53
	-3	-8	9	57
	-3	-6	8	62
	-3	-4	7	70
	-3	-2	6	81
	0	-2	1	31
	3	-12	1	5
	3	-10	0	0
6	10	3	76	31
	-1	-3	12	87
	0	-3	10	81
	1	-3	8	74
	2	-3	6	62
	3	-3	4	45
	4	-3	2	23
	5	-6	10	54
	5	-3	0	0
	7	-6	6	34
	9	-6	2	11
	10	-9	10	38
	11	-9	8	30
10	6	3	76	59
	0	-1	2	67
	3	-11	12	48
	3	-10	10	46
	3	-9	8	42
	3	-8	6	36
	3	-7	4	29
	3	-6	2	17
	3	-5	0	0
	6	-11	2	9
3	6	11	31	27
	-4	-9	6	88
	-1	-5	3	77
	0	-11	6	67
	1	-6	3	58
	2	-1	0	0
	3	-7	3	45
	5	-8	3	36
	7	-9	3	29
	9	-10	3	25

H	K / U	L / V	Φ / W	φ_2 / φ_1
	11	-11	3	21
6	3	11	31	63
	-11	-11	9	74
	-7	-8	6	71
	-5	-12	6	54
	-4	-3	3	81
	-3	-5	3	61
	-2	-7	3	47
	-1	-9	3	37
	0	-11	3	30
	1	-2	0	0
3	11	6	62	15
	0	-6	11	83
	1	-3	5	73
	3	-3	4	51
	4	-6	9	62
	5	-3	3	31
	7	-3	2	17
	8	-6	7	40
	9	-3	1	7
	11	-9	11	44
	11	-3	0	0
	12	-6	5	23
11	3	6	62	75
	-3	-11	11	52
	-3	-9	10	55
	-3	-7	9	60
	-3	-5	8	66
	-3	-3	7	75
	-3	-1	6	89
	0	-2	1	30
	3	-11	0	0
6	11	3	77	29
	0	-3	11	83
	1	-3	9	76
	2	-3	7	66
	3	-3	5	52
	4	-3	3	32
	5	-6	12	60
	5	-3	1	10
	7	-6	8	42
	9	-6	4	21
	11	-9	11	39
	11	-6	0	0

H	K / U	L / V	Φ / W	φ_2 / φ_1
11	6	3	77	61
	0	-1	2	67
	3	-11	11	46
	3	-10	9	42
	3	-9	7	36
	3	-8	5	31
	3	-7	3	22
	3	-6	1	9
	6	-11	0	0
3	7	7	47	23
	0	-1	1	74
	7	-12	9	48
	7	-11	8	45
	7	-10	7	43
	7	-9	6	39
	7	-8	5	35
	7	-7	4	31
	7	-6	3	25
	7	-5	2	18
	7	-4	1	10
	7	-3	0	0
7	3	7	47	67
	-9	-7	12	80
	-8	-7	11	78
	-7	-7	10	75
	-6	-7	9	72
	-5	-7	8	68
	-4	-7	7	63
	-3	-7	6	57
	-2	-7	5	50
	-1	-7	4	42
	0	-7	3	32
	1	-7	2	22
	2	-7	1	11
	3	-7	0	0
7	7	3	73	45
	-1	-2	7	84
	0	-3	7	74
	1	-4	7	64
	1	-1	0	0
	2	-5	7	56
	3	-6	7	49
	4	-7	7	43
	5	-8	7	39

H	K / U	L / V	Φ / W	φ_2 / φ_1
	6	-9	7	35
	7	-10	7	31
	8	-11	7	29
	9	-12	7	26
3	7	8	44	23
	-1	-11	10	77
	-1	-3	3	87
	0	-8	7	73
	1	-5	4	64
	2	-2	1	29
	3	-7	5	59
	4	-12	9	57
	5	-9	6	47
	7	-11	7	43
	7	-3	0	0
	9	-5	1	8
	11	-7	2	13
7	3	8	44	67
	-10	-6	11	84
	-9	-11	12	69
	-7	-5	8	81
	-6	-10	9	62
	-5	-7	7	66
	-4	-4	5	74
	-1	-11	5	37
	-1	-3	2	51
	0	-8	3	31
	1	-5	1	16
	3	-7	0	0
	4	-12	1	7
3	8	7	51	21
	-1	-4	5	86
	0	-7	8	77
	1	-10	11	79
	1	-3	3	63
	3	-2	1	20
	4	-5	4	49
	5	-8	7	50
	6	-11	10	54
	7	-7	5	36
	8	-3	0	0
	10	-9	6	32
	11	-12	9	39
	11	-5	1	6

```
H  K  L   Φ  φ2
   U  V   W  φ1
---------------
3  7  8
=======
8  3  7  51 69
---------------
  -9 -4 12  88
  -7 -7 11  74
  -6 -5  9  78
  -5 -3  7  83
  -2-11  7  43
  -1 -9  5  39
  -1 -2  2  60
   0 -7  3  31
   1-12  4  24
   1 -5  1  14
   3 -8  0   0
7  8  3  74 41
---------------
   0 -3  8  77
   1 -5 11  71
   1 -2  3  56
   3 -3  1  14
   4 -5  4  33
   5 -7  7  41
   6 -9 10  45
   7 -8  5  26
   8 -7  0   0
   9-12 11  38
  10-11  6  23
  11-10  1   4
8  7  3  74 49
---------------
  -1 -4 12  79
  -1 -1  5  89
   0 -3  7  73
   1 -5  9  65
   1 -2  2  44
   2 -7 11  60
   4 -5  1   9
   5 -7  3  20
   6 -9  5  26
   7-11  7  29
   7 -8  0   0
   9-12  4  16

3  7  9  40 23
===============
  -5-12 11  90
  -3 -9  8  86
  -1 -6  5  79
```

```
H  K  L   Φ  φ2
   U  V   W  φ1
---------------
   0 -9  7  72
   1-12  9  68
   1 -3  2  56
   4 -3  1  18
   5 -6  3  34
   6 -9  5  41
   7-12  7  44
   7 -3  0   0
   9 -9  4  28
  11 -6  1   7
7  3  9  40 67
---------------
  -9-12 11  66
  -9 -9 10  73
  -6 -7  7  69
  -3-11  6  46
  -3 -8  5  51
  -3 -5  4  61
  -3 -2  3  82
   0 -3  1  29
   3-10  1   8
   3 -7  0   0
3  9  7  54 18
---------------
  -1 -9 12  83
   0 -7  9  79
   1 -5  6  71
   2 -3  3  53
   3 -8  9  64
   3 -1  0   0
   5-11 12  61
   5 -4  3  32
   7 -7  6  40
   8 -5  3  22
   9-10  9  44
  11 -6  3  17
  12-11  9  37
9  3  7  54 72
---------------
  -7 -7 12  73
  -5 -6  9  70
  -4 -9  9  57
  -3 -5  6  63
  -2 -1  3  85
  -1-11  6  36
  -1 -4  3  47
   0 -7  3  29
   1-10  3  21
   1 -3  0   0
```

```
H  K  L   Φ  φ2
   U  V   W  φ1
---------------
7  9  3  75 38
---------------
   0 -1  3  79
   3 -6 11  62
   3 -5  8  57
   3 -4  5  47
   3 -3  2  26
   6 -7  7  39
   6 -5  1   8
   9-11 12  42
   9-10  9  35
   9 -8  3  14
   9 -7  0   0
  12-11  5  18
9  7  3  75 52
---------------
  -1 -3 10  80
   0 -3  7  72
   1 -6 11  65
   1 -3  4  54
   2 -3  1  16
   3 -6  5  38
   4 -9  9  44
   5 -9  6  31
   7-12  7  28
   7 -9  0   0
   9-12  1   4

3  7 10  37 23
===============
  -2-12  9  79
  -1-11  8  76
   0-10  7  71
   1 -9  6  66
   2 -8  5  59
   3 -7  4  50
   4 -6  3  39
   5 -5  2  27
   6 -4  1  13
   7 -3  0   0
   9-11  5  33
  11 -9  3  20
7  3 10  37 67
---------------
  -2-12  5  39
  -1-11  4  34
  -1 -1  1  72
   0-10  3  28
   1 -9  2  21
```

```
H  K  L   Φ  φ2
   U  V   W  φ1
---------------
   2 -8  1  11
   3 -7  0   0
3 10  7  56 17
---------------
  -1 -6  9  86
   0 -7 10  81
   1 -8 11  76
   1 -1  1  44
   2 -9 12  73
  10 -3  0   0
  11 -4  1   6
  12 -5  2  11
10 3  7  56 73
---------------
  -5 -9 11  62
  -5 -2  8  87
  -4 -3  7  78
  -3-11  9  48
  -3 -4  6  68
  -2 -5  5  55
  -1 -6  4  41
   0 -7  3  28
   1 -8  2  17
   2 -9  1   7
   3-10  0   0
7 10  3  76 35
---------------
  -1 -2  9  88
   0 -3 10  81
   1 -4 11  75
   1 -1  1  36
   2 -5 12  70
  10 -7  0   0
  11 -8  1   4
  12 -9  2   8
10 7  3  76 55
---------------
  -1 -2  8  83
   0 -3  7  71
   1 -4  6  58
   2 -5  5  44
   3 -9 11  51
   3 -6  4  32
   4 -7  3  21
   5-11  9  38
   5 -8  2  12
   6 -9  1   5
   7-10  0   0
```

```
H  K  L   Φ  φ2
   U  V   W  φ1
---------------

3  7 11  35 23
===============
  -5-12  9  90
  -2 -7  5  84
  -1 -9  6  76
   0-11  7  71
   1 -2  1  46
   7 -3  0   0
   8 -5  1  11
   9 -7  2  18
  10 -9  3  22
  11-11  4  26
7  3 11  35 67
---------------
 -11-11 10  72
  -8 -7  7  75
  -5 -3  4  84
  -4 -9  5  53
  -3 -4  3  65
  -1 -5  2  40
   0-11  3  28
   1 -6  1  17
   3 -7  0   0
3 11  7  58 15
---------------
   0 -7 11  82
   1 -6  9  76
   2 -5  7  68
   3 -4  5  56
   4 -3  3  37
   5 -9 12  63
   5 -2  1  12
   7 -7  8  48
   9 -5  4  25
  11-10 11  44
  11 -3  0   0
11 3  7  58 75
---------------
  -4-11 11  53
  -3-10  9  50
  -2 -9  7  45
  -1 -8  5  38
  -1 -1  2  73
   0 -7  3  28
   1 -6  1  11
   3-11  0   0
```

```
H  K  L   Φ  φ2
   U  V   W  φ1
---------------
7 11  3  77 32
---------------
   0 -3 11  82
   1 -2  5  70
   3 -3  4  45
   4 -5  9  57
   5 -4  3  26
   7 -5  2  13
   8 -7  7  34
   9 -6  1   5
  11-10 11  38
  11 -7  0   0
  12 -9  5  19
11 7  3  77 58
---------------
  -1 -1  6  87
   0 -3  7  71
   1 -5  8  60
   1 -2  1  25
   2 -7  9  53
   3 -9 10  48
   4-11 11  45
   7-11  0   0

3  7 12  32 23
===============
  -1 -3  2  86
   0-12  7  70
   1 -9  5  64
   2 -6  3  53
   3 -3  1  25
   5 -9  4  43
   7 -3  0   0
   8-12  5  38
  10 -6  1   9
7  3 12  32 67
---------------
 -12 -8  9  81
  -9-11  8  66
  -9 -7  7  78
  -6 -6  5  71
  -3 -9  4  46
  -3 -5  3  59
   0 -4  1  27
   3-11  1   9
   3 -7  0   0
3 12  7  60 14
---------------
  -1 -5  9  89
```

```
H  K  L   Φ  φ2
   U  V   W  φ1
---------------
   0 -7 12  83
   1 -2  3  67
   4 -1  0   0
   5 -3  3  32
   6 -5  6  44
   7 -7  9  51
   8 -9 12  54
   9 -4  3  20
  11 -8  9  39
12 3  7  60 76
---------------
  -5 -8 12  65
  -4 -5  9  69
  -3 -2  6  80
  -1-10  6  36
  -1 -3  3  52
   0 -7  3  27
   1-11  3  18
   1 -4  0   0
7 12  3  78 30
---------------
   0 -1  4  83
   3 -4  9  63
   3 -3  5  51
   3 -2  1  16
   6 -5  6  39
   9 -8 11  44
   9 -7  7  32
  12 -9  8  29
  12 -7  0   0
12 7  3  78 60
---------------
  -1 -3 11  79
   0 -3  7  70
   1 -6 10  61
   1 -3  3  45
   3 -6  2  17
   4 -9  5  28
   5-12  8  32
   5 -9  1   6
   7-12  0   0

3  8  8  47 21
===============
   0 -1  1  76
   8-12  9  46
   6-11  8  44
   8-10  7  41
   8 -9  6  38
```

```
H  K  L   Φ  φ2
   U  V   W  φ1
---------------
   8 -8  5  34
   8 -7  4  29
   8 -6  3  23
   8 -5  2  17
   8 -4  1   9
   8 -3  0   0
8  3  8  47 69
---------------
  -9 -8 12  75
  -8 -8 11  73
  -7 -8 10  70
  -6 -8  9  66
  -5 -8  8  62
  -4 -8  7  58
  -3 -8  6  52
  -2 -8  5  45
  -1 -8  4  38
   0 -8  3  29
   1 -8  2  19
   2 -8  1   9
   3 -8  0   0
8  8  3  75 45
---------------
  -1 -2  8  85
   0 -3  8  76
   1 -4  8  67
   1 -1  0   0
   2 -5  8  59
   3 -6  8  52
   4 -7  8  47
   5 -8  8  42
   6 -9  8  38
   7-10  8  35
   8-11  8  32
   9-12  8  29

3  8  9  44 21
===============
  -1-12 11  78
  -1 -3  3  88
   0 -9  8  75
   1 -6  5  67
   2 -3  2  45
   3 -9  7  60
   5-12  9  56
   5 -3  1  14
   7 -6  3  27
   8 -3  0   0
   9 -9  5  32
```

TAB. V FROM HKL-UVW INTO EULER ANGLES

H	K / U	L / V	Φ / W	φ2 / φ1
3	8	9	44	21
	11	-12	7	35
	12	-9	4	22
8	3	9	44	69
	-9	-9	11	72
	-9	-6	10	80
	-6	-11	9	58
	-6	-5	7	76
	-3	-10	6	46
	-3	-7	5	53
	-3	-4	4	65
	0	-3	1	27
	3	-11	1	7
	3	-8	0	0
3	9	8	50	18
	-1	-5	6	85
	0	-8	9	78
	1	-11	12	74
	1	-3	3	64
	3	-1	0	0
	4	-4	3	38
	5	-7	6	48
	6	-10	9	53
	7	-5	3	26
	9	-11	9	44
	10	-6	3	19
	11	-9	6	31
9	3	8	50	72
	-9	-5	12	83
	-7	-11	12	62
	-7	-3	9	87
	-5	-9	9	59
	-4	-12	9	49
	-3	-7	6	54
	-2	-2	3	72
	-1	-5	3	42
	0	-8	3	27
	1	-11	3	20
	1	-3	0	0
8	9	3	76	42
	0	-1	3	78
	3	-6	10	59
	3	-5	7	52
	3	-4	4	40

H	K / U	L / V	Φ / W	φ2 / φ1
	3	-3	1	14
	6	-9	11	47
	6	-7	5	29
	9	-11	9	33
	9	-10	6	25
	9	-8	0	0
	12	-11	1	4
9	8	3	76	48
	-1	-3	11	82
	0	-3	8	75
	1	-3	5	61
	2	-3	2	30
	3	-6	7	48
	4	-9	12	53
	5	-9	9	43
	5	-6	1	8
	7	-12	11	40
	7	-9	3	15
	8	-9	0	0
	9	-12	5	19
3	8	10	41	21
	-4	-11	10	90
	-2	-8	7	85
	0	-5	4	74
	2	-12	9	66
	2	-7	5	61
	2	-2	1	31
	4	-9	6	53
	6	-11	7	49
	8	-3	0	0
	10	-5	1	8
	12	-7	2	13
8	3	10	41	69
	-10	-10	11	71
	-7	-8	8	68
	-5	-10	7	55
	-4	-6	5	61
	-3	-2	3	80
	-1	-4	2	42
	0	-10	3	26
	1	-6	1	14
	3	-8	0	0
3	10	8	53	17
	-2	-9	12	87

H	K / U	L / V	Φ / W	φ2 / φ1
	0	-4	5	80
	2	-7	8	69
	2	-3	3	54
	4	-10	11	64
	4	-2	1	16
	6	-5	4	35
	8	-8	7	42
	10	-11	10	45
	10	-7	5	29
	10	-3	0	0
10	3	8	53	73
	-7	-6	11	75
	-6	-4	9	79
	-5	-2	7	87
	-2	-12	7	39
	-1	-10	5	34
	-1	-2	2	57
	0	-8	3	26
	1	-6	1	12
	3	-10	0	0
8	10	3	77	39
	0	-3	10	80
	1	-2	4	64
	3	-3	2	26
	4	-5	6	45
	5	-7	10	51
	5	-4	0	0
	7	-8	8	38
	8	-7	2	11
	10	-11	10	35
	11	-10	4	15
10	8	3	77	51
	-1	-1	6	89
	0	-3	8	74
	1	-5	10	66
	1	-2	2	43
	2	-7	12	61
	4	-5	0	0
	5	-7	2	13
	6	-9	4	21
	7	-11	6	25
	9	-12	2	8
3	8	11	38	21
	-1	-12	9	77

H	K / U	L / V	Φ / W	φ2 / φ1
	0	-11	6	74
	1	-10	7	69
	2	-9	6	63
	3	-8	5	55
	4	-7	4	46
	5	-6	3	36
	6	-5	2	24
	7	-4	1	12
	8	-3	0	0
	11	-11	5	30
8	3	11	38	69
	-1	-12	4	31
	-1	-1	1	70
	0	-11	3	25
	1	-10	2	19
	2	-9	1	10
	3	-8	0	0
3	11	8	55	15
	-1	-7	10	86
	0	-8	11	81
	1	-9	12	77
	1	-1	1	45
	11	-3	0	0
	12	-4	1	6
11	3	8	55	75
	-6	-2	9	88
	-5	-11	11	55
	-5	-3	8	81
	-4	-4	7	72
	-3	-5	6	61
	-2	-6	5	49
	-1	-7	4	37
	0	-8	3	25
	1	-9	2	15
	2	-10	1	7
	3	-11	0	0
8	11	3	78	36
	-1	-2	10	88
	0	-3	11	81
	1	-4	12	76
	1	-1	1	36
	11	-8	0	0
	12	-9	1	4

H	K / U	L / V	Φ / W	φ2 / φ1
11	8	3	78	54
	-1	-2	9	84
	0	-3	8	74
	1	-4	7	62
	2	-5	6	50
	3	-6	5	38
	4	-7	4	27
	5	-11	11	44
	5	-8	3	18
	6	-9	2	11
	7	-10	1	5
	8	-11	0	0
3	8	12	35	21
	-4	-12	9	68
	0	-3	2	73
	4	-12	7	57
	4	-9	5	51
	4	-6	3	41
	4	-3	1	20
	8	-9	4	33
	8	-3	0	0
	12	-12	5	29
	12	-6	1	7
8	3	12	35	69
	-12	-12	11	70
	-9	-8	8	73
	-9	-4	7	87
	-6	-12	7	53
	-6	-4	5	79
	-3	-12	5	40
	-3	-8	4	47
	-3	-4	3	63
	0	-4	1	25
	3	-12	1	8
	3	-8	0	0
3	12	8	57	14
	0	-2	3	82
	4	-9	12	67
	4	-7	9	63
	4	-5	6	55
	4	-3	3	38
	4	-1	0	0
	8	-8	9	48
	8	-4	3	22

H	K / U	L / V	Φ / W	φ2 / φ1
	12	-11	12	45
	12	-7	6	28
	12	-5	3	16
12	3	8	57	76
	-7	-4	12	81
	-5	-12	12	54
	-5	-4	9	76
	-4	-8	9	58
	-3	-4	6	66
	-1	-12	6	32
	-1	-4	3	44
	0	-8	3	25
	1	-12	3	17
	1	-4	0	0
8	12	3	78	34
	0	-1	4	82
	3	-5	12	67
	3	-4	8	60
	3	-3	4	44
	3	-2	0	0
	6	-7	12	54
	6	-5	4	28
	9	-8	8	34
	9	-7	4	20
	12	-11	12	37
	12	-9	4	15
12	8	3	78	56
	-1	-3	12	81
	0	-3	8	73
	1	-6	12	66
	1	-3	4	53
	2	-3	0	0
	3	-6	4	32
	4	-9	8	40
	5	-12	12	44
	5	-9	4	22
	7	-12	4	16
3	9	10	43	18
	-1	-3	3	90
	0	-10	9	76
	1	-7	6	70
	2	-4	3	54
	3	-11	9	64
	3	-1	0	0

H	K / U	L / V	Φ / W	φ2 / φ1
	5	-5	3	35
	7	-9	6	43
	8	-6	3	25
	11	-7	3	19
9	3	10	43	72
	-11	-7	12	80
	-8	-6	9	76
	-7	-9	9	64
	-5	-5	6	70
	-3	-11	6	43
	-3	-1	3	90
	-2	-4	3	54
	-1	-7	3	35
	0	-10	3	25
	1	-3	0	0
3	10	9	49	17
	-1	-6	7	85
	0	-9	10	79
	1	-3	3	65
	4	-3	2	29
	5	-6	5	45
	6	-9	8	52
	7	-12	11	55
	7	-3	1	10
	9	-9	7	40
	10	-3	0	0
	11	-6	3	18
10	3	9	49	73
	-9	-3	11	89
	-6	-7	9	67
	-3	-11	7	44
	-3	-8	6	49
	-3	-5	5	59
	-3	-2	4	79
	0	-3	1	25
	3	-10	0	0
9	10	3	77	42
	0	-3	10	79
	1	-3	7	69
	2	-3	4	50
	3	-6	11	61
	3	-3	1	14
	5	-6	5	34
	7	-9	9	39
	8	-9	6	27

H	K / U	L / V	Φ / W	φ2 / φ1
	10	-9	0	0
	11	-12	7	24
10	9	3	77	48
	0	-1	3	76
	3	-7	11	57
	3	-6	8	52
	3	-5	5	42
	3	-4	2	22
	6	-9	7	34
	6	-7	1	6
	9	-11	3	12
	9	-10	0	0
3	9	11	41	18
	-3	-10	9	89
	-1	-7	6	82
	0	-11	9	76
	1	-4	3	64
	3	-1	0	0
	4	-5	3	41
	5	-9	6	50
	7	-6	3	28
	10	-7	3	21
	11	-11	6	33
9	3	11	41	72
	-11	-11	12	69
	-8	-9	9	66
	-7	-12	9	56
	-5	-7	6	61
	-3	-2	3	78
	-2	-5	3	48
	-1	-8	3	32
	0	-11	3	24
	1	-3	0	0
3	11	9	52	15
	0	-9	11	80
	1	-6	7	74
	2	-3	3	55
	3	-9	10	68
	5	-3	2	24
	7	-6	5	37
	8	-3	1	9
	9	-9	8	43
	11	-12	11	46
	11	-3	0	0

AB. V — FROM HKL-UVW INTO EULER ANGLES — V - 50

H	K / U	L / V	Φ / W	φ_2 / φ_1
3	9	11		
3	11	9	52	15
	12	-9	7	33
11	3	9	52	75
	-6	-11	11	57
	-6	-5	9	74
	-3	-10	7	45
	-3	-7	6	52
	-3	-4	5	64
	-3	-1	4	88
	0	-3	1	24
	3	-11	0	0
9	11	3	78	39
	0	-3	11	80
	1	-3	8	72
	2	-3	5	56
	3	-3	2	26
	5	-6	7	43
	7	-9	12	48
	7	-6	1	6
	8	-9	9	38
	10	-9	3	13
	11	-12	11	35
	11	-9	0	0
11	9	3	78	51
	0	-1	3	76
	3	-7	10	54
	3	-6	7	48
	3	-5	4	35
	3	-4	1	12
	6	-11	11	42
	6	-9	5	25
	9	-11	0	0
3	10	10	46	17
	0	-1	1	78
	10	-12	9	44
	10	-11	8	41
	10	-10	7	38
	10	-9	6	34
	10	-8	5	30
	10	-7	4	26
	10	-6	3	20

H	K / U	L / V	Φ / W	φ_2 / φ_1
	10	-5	2	14
	10	-4	1	7
	10	-3	0	0
10	3	10	46	73
	-9	-10	12	67
	-8	-10	11	64
	-7	-10	10	61
	-6	-10	9	58
	-5	-10	8	54
	-4	-10	7	49
	-3	-10	6	44
	-2	-10	5	38
	-1	-10	4	31
	0	-10	3	23
	1	-10	2	16
	2	-10	1	8
	3	-10	0	0
10	10	3	78	45
	-1	-2	10	86
	0	-3	10	78
	1	-4	10	71
	1	-1	0	0
	2	-5	10	64
	3	-6	10	58
	4	-7	10	53
	5	-8	10	48
	6	-9	10	44
	7	-10	10	40
	8	-11	10	37
	9	-12	10	35
3	10	11	44	17
	0	-11	10	78
	1	-8	7	72
	2	-5	4	60
	3	-2	1	23
	5	-7	5	47
	7	-12	9	52
	8	-9	6	40
	10	-3	0	0
	11	-11	7	37
10	3	11	44	73
	-10	-7	11	77
	-7	-6	8	72
	-5	-9	7	55

H	K / U	L / V	Φ / W	φ_2 / φ_1
	-4	-5	5	63
	-3	-1	3	89
	-1	-4	2	39
	0	-11	3	22
	1	-7	1	12
	3	-10	0	0
3	11	10	49	15
	-1	-7	8	85
	0	-10	11	80
	1	-3	3	66
	4	-2	1	17
	5	-5	4	41
	6	-8	7	50
	7	-11	10	54
	9	-7	5	32
	11	-3	0	0
11	3	10	49	75
	-9	-7	12	75
	-7	-11	11	59
	-6	-8	9	63
	-5	-5	7	69
	-4	-2	5	82
	-1	-3	2	45
	0	-10	3	22
	1	-7	1	11
	3	-11	0	0
10	11	3	79	42
	0	-3	11	80
	1	-2	4	63
	3	-3	1	14
	4	-5	5	39
	5	-7	9	48
	7	-8	6	30
	10	-11	7	26
	11	-10	0	0
11	10	3	79	48
	-1	-1	7	89
	0	-3	10	78
	1	-2	3	55
	4	-5	2	18
	5	-7	5	31
	6	-9	8	37
	7	-11	11	41
	7	-8	1	5
	9	-12	7	26

H	K / U	L / V	Φ / W	φ_2 / φ_1
	10	-11	0	0
3	10	12	41	17
	-2	-9	8	87
	0	-6	5	77
	2	-9	7	67
	2	-3	2	48
	4	-12	9	62
	6	-3	1	13
	8	-6	3	26
	10	-9	5	32
	10	-3	0	0
	12	-12	7	36
10	3	12	41	73
	-12	-4	11	89
	-9	-10	10	65
	-6	-8	7	61
	-3	-10	5	41
	-3	-6	4	51
	-3	-2	3	77
	0	-4	1	22
	3	-10	0	0
3	12	10	51	14
	0	-5	6	81
	2	-8	9	71
	2	-3	3	55
	4	-11	12	67
	4	-1	0	0
	6	-4	3	30
	8	-7	6	39
	10	-10	9	44
	10	-5	3	19
12	3	10	51	76
	-9	-4	12	84
	-7	-12	12	57
	-7	-2	9	89
	-5	-10	9	54
	-3	-8	6	48
	-2	-2	3	69
	-1	-6	3	35
	0	-10	3	22
	1	-4	0	0
10	12	3	79	40
	0	-1	4	81

H	K / U	L / V	Φ / W	φ_2 / φ_1
	3	-5	10	62
	3	-4	6	51
	3	-3	2	26
	6	-7	8	42
	6	-5	0	0
	9	-10	10	37
	9	-8	2	10
	12	-11	4	14
12	10	3	79	50
	0	-3	10	77
	1	-3	6	64
	2	-3	2	30
	3	-6	8	51
	5	-9	10	45
	5	-6	0	0
	7	-12	12	42
	7	-9	2	10
	9	-12	4	15
3	11	11	46	15
	0	-1	1	79
	11	-12	9	42
	11	-11	8	39
	11	-10	7	36
	11	-9	6	33
	11	-8	5	29
	11	-7	4	24
	11	-6	3	19
	11	-5	2	13
	11	-4	1	7
	11	-3	0	0
11	3	11	46	75
	-9	-11	12	64
	-8	-11	11	61
	-7	-11	10	58
	-6	-11	9	54
	-5	-11	8	50
	-4	-11	7	45
	-3	-11	6	40
	-2	-11	5	35
	-1	-11	4	28
	0	-11	3	21
	1	-11	2	14
	2	-11	1	7
	3	-11	0	0

H	K / U	L / V	Φ / W	φ_2 / φ_1
11	11	3	79	45
	-1	-2	11	86
	0	-3	11	79
	1	-4	11	72
	1	-1	0	0
	2	-5	11	66
	3	-6	11	60
	4	-7	11	55
	5	-8	11	51
	6	-9	11	47
	7	-10	11	43
	8	-11	11	40
	9	-12	11	37
3	11	12	44	15
	0	-12	11	79
	1	-9	8	74
	2	-6	5	64
	3	-3	2	38
	5	-9	7	55
	7	-3	1	11
	8	-12	9	50
	10	-6	3	21
	11	-3	0	0
11	3	12	44	75
	-9	-11	11	63
	-9	-7	10	73
	-6	-6	7	67
	-3	-9	5	43
	-3	-5	4	55
	-3	-1	3	88
	0	-4	1	21
	3	-11	0	0
3	12	11	48	14
	-1	-8	9	85
	0	-11	12	81
	1	-3	3	67
	4	-1	0	0
	5	-4	3	35
	6	-7	6	47
	7	-10	9	53
	9	-5	3	22
	11	-11	9	42

H	K / U	L / V	Φ / W	φ_2 / φ_1
12	3	11	48	76
	-9	-8	12	71
	-7	-5	9	75
	-5	-2	6	85
	-3	-10	6	42
	-2	-3	3	59
	-1	-7	3	32
	0	-11	3	21
	1	-4	0	0
11	12	3	80	43
	0	-1	4	81
	3	-5	9	59
	3	-4	5	46
	3	-3	1	13
	6	-7	6	34
	9	-11	11	38
	9	-10	7	28
	12	-11	0	0
12	11	3	80	47
	0	-3	11	79
	1	-3	7	68
	2	-3	3	41
	3	-6	10	58
	5	-6	2	15
	7	-9	5	24
	8	-9	1	5
	9	-12	8	29
	11	-12	0	0
4	4	5	49	45
	-7	-8	12	87
	-5	-10	12	78
	-5	-5	8	90
	-4	-11	12	73
	-3	-7	8	75
	-2	-3	4	82
	-1	-9	8	62
	-1	-4	4	68
	0	-5	4	56
	1	-11	8	52
	1	-6	4	47
	1	-1	0	0
	2	-7	4	40
	3	-8	4	34
	4	-9	4	30

H	K / U	L / V	Φ / W	φ_2 / φ_1
	5	-10	4	27
	6	-11	4	24
	7	-12	4	22
4	5	4	58	39
	-3	-4	8	89
	-2	-4	7	84
	-1	-8	11	72
	-1	-4	6	76
	0	-4	5	67
	1	-8	9	61
	1	-4	4	55
	2	-4	3	41
	3	-8	7	48
	3	-4	2	26
	4	-12	11	51
	4	-4	1	12
	5	-12	10	46
	5	-8	5	34
	5	-4	0	0
	7	-12	8	36
	7	-8	3	19
	8	-12	7	31
	9	-8	1	6
	10	-12	5	21
	11	-12	4	16
5	4	4	58	51
	-4	-7	12	78
	-4	-6	11	80
	-4	-5	10	83
	-4	-4	9	87
	0	-1	1	56
	4	-12	7	35
	4	-11	6	33
	4	-10	5	30
	4	-9	4	26
	4	-8	3	22
	4	-7	2	16
	4	-6	1	9
	4	-5	0	0
	8	-11	1	5
4	4	7	39	45
	-10	-11	12	88
	-7	-7	8	90
	-5	-9	8	77
	-3	-11	8	66

TAB. V FROM HKL-UVW INTO EULER ANGLES

H	K / U	L / V	Φ / W	φ_2 / φ_1
4	4	7	39	45
	-3	-4	4	84
	-2	-5	4	72
	-1	-6	4	61
	0	-7	4	52
	1	-8	4	45
	1	-1	0	0
	2	-9	4	39
	3	-10	4	35
	4	-11	4	31
	5	-12	4	28
4	7	4	64	30
	-2	-4	9	89
	-1	-4	8	83
	0	-4	7	76
	1	-4	6	67
	2	-4	5	56
	3	-8	11	62
	3	-4	4	44
	4	-4	3	32
	5	-8	9	50
	5	-4	2	19
	6	-4	1	9
	7	-8	7	38
	7	-4	0	0
	9	-8	5	25
	10	-12	11	40
	11	-12	10	36
	11	-8	3	14
7	4	4	64	60
	-4	-5	12	80
	-4	-4	11	83
	-4	-3	10	87
	0	-1	1	52
	4	-12	5	24
	4	-11	4	21
	4	-10	3	17
	4	-9	2	13
	4	-8	1	7
	4	-7	0	0
4	4	9	32	45
	-9	-9	8	90
	-7	-11	8	79
	-4	-5	4	85

H	K / U	L / V	Φ / W	φ_2 / φ_1
	-3	-6	4	74
	-2	-7	4	65
	-1	-8	4	57
	0	-9	4	50
	1	-10	4	44
	1	-1	0	0
	2	-11	4	39
	3	-12	4	35
4	9	4	68	24
	-1	-4	10	86
	0	-4	9	81
	1	-4	8	74
	2	-4	7	65
	3	-4	6	56
	4	-4	5	46
	5	-4	4	35
	6	-4	3	24
	7	-8	11	51
	7	-4	2	15
	8	-4	1	7
	9	-8	9	40
	9	-4	0	0
	11	-8	7	30
9	4	4	68	66
	-4	-3	12	85
	-4	-2	11	89
	0	-1	1	50
	4	-12	3	14
	4	-11	2	10
	4	-10	1	6
	4	-9	0	0
4	4	11	27	45
	-11	-11	8	90
	-5	-6	4	85
	-4	-7	4	76
	-3	-8	4	68
	-2	-9	4	60
	-1	-10	4	54
	0	-11	4	48
	1	-12	4	44
	1	-1	0	0
4	11	4	71	20
	-1	-4	12	88
	0	-4	11	83

H	K / U	L / V	Φ / W	φ_2 / φ_1
	1	-4	10	78
	2	-4	9	71
	3	-4	8	64
	4	-4	7	55
	5	-4	6	46
	6	-4	5	37
	7	-4	4	28
	8	-4	3	20
	9	-4	2	12
	10	-4	1	6
	11	-8	11	42
	11	-4	0	0
11	4	4	71	70
	0	-1	1	48
	4	-12	1	5
	4	-11	0	0
4	5	5	52	39
	-5	-8	12	86
	-5	-7	11	88
	0	-1	1	64
	5	-12	8	42
	5	-11	7	39
	5	-10	6	37
	5	-9	5	34
	5	-8	4	30
	5	-7	3	25
	5	-6	2	18
	5	-5	1	10
	5	-4	0	0
	10	-11	3	15
	10	-9	1	5
5	4	5	52	51
	-6	-5	10	89
	-5	-5	9	86
	-4	-5	8	82
	-3	-10	11	67
	-3	-5	7	77
	-2	-5	6	71
	-1	-10	9	58
	-1	-5	5	63
	0	-5	4	52
	1	-10	7	46
	1	-5	3	40
	2	-5	2	26
	3	-10	5	33

H	K / U	L / V	Φ / W	φ_2 / φ_1
	3	-5	1	12
	4	-5	0	0
	5	-10	3	19
	7	-10	1	6
5	5	4	61	45
	-3	-5	10	83
	-2	-2	5	90
	-1	-7	10	70
	-1	-3	5	76
	0	-4	5	64
	1	-9	10	58
	1	-5	5	54
	1	-1	0	0
	2	-6	5	45
	3	-11	10	49
	3	-7	5	39
	4	-8	5	34
	5	-9	5	30
	6	-10	5	27
	7	-11	5	24
	8	-12	5	22
4	5	6	47	39
	-5	-8	10	85
	-4	-10	11	78
	-3	-6	7	82
	-1	-10	9	66
	-1	-4	4	73
	0	-6	5	61
	1	-8	6	55
	1	-2	1	34
	2	-10	7	51
	3	-12	8	48
	5	-4	0	0
	6	-6	1	9
	7	-8	2	15
	8	-10	3	18
	9	-12	4	21
	11	-10	1	5
5	4	6	47	51
	-6	-9	11	78
	-4	-7	8	75
	-2	-11	9	59
	-2	-8	7	62
	-2	-5	5	69
	-2	-2	3	86

H	K / U	L / V	Φ / W	φ_2 / φ_1
	0	-3	2	49
	2	-10	5	37
	2	-7	3	31
	2	-4	1	17
	4	-11	4	26
	4	-5	0	0
	6	-9	1	7
4	6	5	55	34
	-3	-8	12	82
	-2	-7	10	80
	-1	-6	8	76
	0	-5	6	69
	1	-9	10	64
	1	-4	4	58
	2	-3	2	36
	3	-7	6	49
	3	-2	0	0
	4	-11	10	52
	5	-10	8	45
	5	-5	2	19
	7	-8	4	25
	8	-7	2	13
	9	-11	6	28
	11	-9	2	10
6	4	5	55	56
	-4	-9	12	70
	-3	-8	10	68
	-2	-7	8	64
	-1	-11	10	55
	-1	-6	6	59
	-1	-1	2	83
	0	-5	4	49
	1	-9	6	42
	1	-4	2	32
	2	-3	0	0
	3	-7	2	18
	4	-11	4	23
	5	-10	2	12
5	6	4	63	40
	-2	-5	10	82
	-2	-3	7	87
	0	-2	3	69
	2	-9	11	59
	2	-7	8	56
	2	-5	5	50
	2	-3	2	33

H	K / U	L / V	Φ / W	φ_2 / φ_1
	4	-8	7	44
	4	-4	1	11
	6	-11	9	41
	6	-7	3	20
	6	-5	0	0
	8	-10	5	24
	10	-11	4	17
	10	-9	1	5
6	5	4	63	50
	-4	-4	11	88
	-1	-6	9	69
	-1	-2	4	79
	0	-4	5	61
	1	-10	11	56
	1	-6	6	52
	1	-2	1	27
	2	-8	7	47
	3	-10	8	43
	4	-12	9	41
	5	-6	0	0
	6	-8	1	6
	7	-10	2	10
	8	-12	3	13
4	5	7	42	39
	-2	-11	9	68
	-1	-9	7	65
	-1	-2	2	81
	0	-7	5	59
	1	-12	8	55
	1	-5	3	49
	2	-3	1	23
	3	-8	4	39
	5	-11	5	35
	5	-4	0	0
	7	-7	1	9
	9	-10	2	13
	12	-11	1	5
5	4	7	42	51
	-9	-8	11	88
	-8	-11	12	79
	-7	-7	9	85
	-5	-6	7	81
	-4	-9	8	69
	-3	-5	5	75
	-1	-11	7	52

H	K / U	L / V	Φ / W	φ_2 / φ_1
	-1	-4	3	61
	0	-7	4	47
	1	-10	5	41
	1	-3	1	27
	4	-5	0	0
	5	-8	1	9
	6	-11	2	14
4	7	5	58	30
	-1	-8	12	78
	-1	-3	5	84
	0	-5	7	73
	1	-7	9	68
	1	-2	2	52
	2	-9	11	64
	4	-3	1	13
	5	-5	3	27
	6	-7	5	34
	7	-9	7	38
	7	-4	0	0
	8	-11	9	40
	9	-8	4	22
	11	-12	8	31
	11	-7	1	5
7	4	5	58	60
	-5	-5	11	82
	-4	-3	8	86
	-2	-9	10	60
	-1	-12	11	53
	-1	-7	7	56
	-1	-2	3	71
	0	-5	4	47
	1	-8	5	38
	1	-3	1	21
	2	-11	6	34
	4	-7	0	0
	5	-10	1	6
5	7	4	65	36
	-1	-5	10	79
	0	-4	7	73
	1	-7	11	68
	1	-3	4	60
	2	-2	1	22
	3	-5	5	46
	4	-8	9	51
	5	-7	6	39
	7	-9	7	35

H	K / U	L / V	Φ / W	φ_2 / φ_1
	7	-5	0	0
	8	-12	11	42
	9	-11	8	33
	9	-7	1	6
	11	-9	2	9
7	5	4	65	54
	-2	-6	11	73
	-1	-5	8	68
	-1	-1	3	86
	0	-4	5	59
	1	-11	12	54
	1	-7	7	51
	1	-3	2	36
	2	-10	9	47
	3	-5	1	11
	4	-8	3	21
	5	-11	5	25
	5	-7	0	0
	8	-12	1	4
4	5	8	39	39
	-7	-12	11	89
	-5	-12	10	77
	-4	-8	7	81
	-3	-4	4	89
	-1	-12	8	62
	-1	-4	3	70
	0	-8	5	58
	1	-12	7	54
	1	-4	2	44
	3	-4	1	18
	4	-8	3	31
	5	-12	5	35
	5	-4	0	0
	7	-12	4	26
	8	-8	1	8
	11	-12	2	11
5	4	8	39	51
	-8	-12	11	76
	-8	-8	9	85
	-4	-11	8	65
	-4	-9	7	68
	-4	-7	6	73
	-4	-5	5	80
	0	-2	1	46
	4	-11	3	23

TAB. V FROM HKL-UVW INTO EULER ANGLES

H	K / U	L / V	Φ / W	φ2 / φ1
4	5	8		
5	4	8	39	51
	4	-9	2	19
	4	-7	1	11
	4	-5	0	0
	8	-12	1	6
4	8	5	61	27
	-1	-7	12	81
	-1	-2	4	90
	0	-5	8	76
	1	-8	12	72
	1	-3	4	64
	2	-1	0	0
	3	-4	4	46
	4	-7	8	54
	5	-10	12	57
	5	-5	4	34
	7	-11	12	51
	7	-6	4	27
	8	-9	8	39
	9	-7	4	22
	11	-8	4	19
	12	-11	8	30
8	4	5	61	63
	-5	-5	12	81
	-4	-7	12	72
	-3	-4	8	76
	-2	-11	12	57
	-2	-1	4	90
	-1	-8	8	54
	-1	-3	4	64
	0	-5	4	46
	1	-12	8	39
	1	-7	4	34
	1	-2	0	0
	2	-9	4	27
	3	-11	4	22
5	8	4	67	32
	0	-1	2	76
	4	-8	11	57
	4	-7	9	54
	4	-6	7	49
	4	-5	5	42
	4	-4	3	31

H	K / U	L / V	Φ / W	φ2 / φ1
	4	-3	1	12
	8	-11	12	46
	8	-9	8	37
	8	-7	4	22
	8	-5	0	0
	12	-11	7	25
	12	-10	5	19
	12	-8	1	4
8	5	4	67	58
	-3	-4	11	81
	-2	-4	9	77
	-1	-8	12	64
	-1	-4	7	69
	0	-4	5	58
	1	-8	8	50
	1	-4	3	40
	2	-12	11	47
	2	-4	1	14
	3	-8	4	27
	4	-12	7	32
	5	-12	5	23
	5	-8	0	0
	7	-12	1	4
4	5	9	35	39
	-3	-12	8	70
	-2	-11	7	66
	-1	-10	6	62
	0	-9	5	57
	1	-8	4	50
	2	-7	3	41
	3	-6	2	30
	4	-5	1	15
	5	-4	0	0
	7	-11	3	23
	9	-9	1	8
5	4	9	35	51
	-3	-12	7	58
	-2	-11	6	55
	-1	-10	5	50
	-1	-1	1	85
	0	-9	4	44
	1	-8	3	37
	2	-7	2	27
	3	-6	1	15
	4	-5	0	0

H	K / U	L / V	Φ / W	φ2 / φ1
	7	-11	1	8
4	9	5	63	24
	-1	-4	8	85
	0	-5	9	79
	1	-6	10	73
	1	-1	1	40
	2	-7	11	69
	3	-8	12	66
	9	-4	0	0
	10	-5	1	6
	11	-6	2	10
	12	-7	3	14
9	4	5	63	66
	-3	-7	11	67
	-3	-2	7	86
	-2	-3	6	74
	-1	-9	9	52
	-1	-4	5	60
	0	-5	4	44
	1	-11	7	37
	1	-6	3	30
	2	-7	2	17
	3	-8	1	7
	4	-9	0	0
5	9	4	69	29
	-1	-3	8	86
	0	-4	9	79
	1	-5	10	73
	1	-1	1	38
	2	-6	11	68
	3	-7	12	65
	9	-5	0	0
	10	-6	1	5
	11	-7	2	9
	12	-8	3	13
9	5	4	69	61
	-2	-2	7	84
	-1	-7	11	64
	-1	-3	6	72
	0	-4	5	57
	1	-9	9	49
	1	-5	4	41
	2	-6	3	27
	3	-11	7	34
	3	-7	2	16

H	K / U	L / V	Φ / W	φ2 / φ1
	4	-8	1	7
	5	-9	0	0
4	5	10	33	39
	-5	-12	8	76
	-5	-10	7	80
	-5	-8	6	84
	0	-2	1	56
	5	-12	4	33
	5	-10	3	29
	5	-8	2	23
	5	-6	1	14
	5	-4	0	0
	10	-10	1	8
5	4	10	33	51
	-10	-10	9	85
	-6	-10	7	73
	-6	-5	5	89
	-4	-5	4	79
	-2	-10	5	55
	-2	-5	3	64
	0	-5	2	44
	2	-10	3	32
	2	-5	1	20
	4	-5	0	0
	6	-10	1	9
4	10	5	65	22
	0	-1	2	80
	5	-8	12	60
	5	-7	10	57
	5	-6	8	52
	5	-5	6	46
	5	-4	4	36
	5	-3	2	21
	5	-2	0	0
	10	-9	10	41
	10	-7	6	29
	10	-5	2	11
10	4	5	65	68
	-4	-5	12	77
	-3	-5	10	72
	-2	-5	8	66
	-1	-10	10	51
	-1	-5	6	57
	0	-5	4	44

H	K / U	L / V	Φ / W	φ2 / φ1
	1	-10	6	34
	1	-5	2	24
	2	-5	0	0
	3	-10	2	12
5	10	4	70	27
	0	-2	5	80
	2	-5	10	69
	2	-3	5	59
	2	-1	0	0
	4	-4	5	45
	6	-7	10	51
	6	-5	5	35
	8	-6	5	28
	10	-9	10	39
	10	-7	5	24
	12	-8	5	20
10	5	4	70	63
	-3	-2	10	88
	-1	-6	10	65
	-1	-2	5	76
	0	-4	5	56
	1	-10	10	49
	1	-6	5	42
	1	-2	0	0
	2	-8	5	33
	3	-10	5	27
	4	-12	5	23
4	5	11	30	39
	-7	-12	8	83
	-5	-7	5	87
	-2	-5	3	75
	-1	-8	4	62
	0	-11	5	55
	1	-3	1	37
	5	-4	0	0
	6	-7	1	12
	7	-10	2	19
	11	-11	1	7
5	4	11	30	51
	-11	-11	9	85
	-9	-8	7	87
	-3	-10	5	59
	-2	-3	2	75
	-1	-7	3	51

H	K / U	L / V	Φ / W	φ2 / φ1
	0	-11	4	43
	1	-4	1	28
	4	-5	0	0
	5	-9	1	11
4	11	5	67	20
	0	-5	11	82
	1	-4	8	75
	2	-3	5	62
	3	-2	2	32
	5	-5	7	50
	7	-8	12	54
	7	-3	1	8
	8	-7	9	45
	10	-5	3	16
	11	-9	11	42
	11	-4	0	0
11	4	5	67	70
	-2	-7	10	62
	-1	-11	11	50
	-1	-6	7	55
	-1	-1	3	80
	0	-5	4	43
	1	-9	5	32
	1	-4	1	15
	4	-11	0	0
5	11	4	72	24
	0	-4	11	82
	1	-3	7	74
	2	-2	3	50
	3	-5	10	66
	5	-3	2	20
	7	-5	5	32
	8	-4	1	7
	9	-7	8	37
	11	-9	11	40
	11	-5	0	0
	12	-8	7	27
11	5	4	72	66
	-1	-5	9	66
	-1	-1	4	83
	0	-4	5	55
	1	-11	11	48
	1	-7	6	43
	1	-3	1	19
	2	-10	7	37

H	K / U	L / V	Φ / W	φ2 / φ1
	5	-11	0	0
4	5	12	28	39
	-9	-12	8	88
	-6	-12	7	79
	-5	-6	5	84
	-1	-4	2	68
	0	-12	5	55
	1	-8	3	48
	2	-4	1	28
	3	-12	4	41
	5	-4	0	0
	7	-8	1	11
	9	-12	2	16
	12	-12	1	7
5	4	12	28	51
	-8	-11	7	76
	-4	-10	5	63
	-4	-7	4	71
	-4	-4	3	84
	0	-3	1	42
	4	-11	2	21
	4	-8	1	14
	4	-5	0	0
4	12	5	68	18
	-1	-3	8	90
	0	-5	12	83
	1	-2	4	70
	3	-1	0	0
	4	-3	4	42
	5	-5	8	54
	6	-7	12	59
	7	-4	4	29
	9	-8	12	49
	10	-5	4	21
	11	-7	8	34
12	4	5	68	72
	-4	-3	12	83
	-3	-1	8	90
	-2	-9	12	59
	-1	-12	12	49
	-1	-7	8	54
	-1	-2	4	70
	0	-5	4	42
	1	-8	4	29

H	K / U	L / V	Φ / W	φ2 / φ1
	1	-3	0	0
	2	-11	4	21
5	12	4	73	23
	0	-1	3	83
	4	-5	10	62
	4	-4	7	54
	4	-3	4	41
	4	-2	1	13
	8	-7	11	49
	8	-5	5	29
	12	-8	9	34
	12	-7	6	25
	12	-5	0	0
12	5	4	73	67
	-2	-4	11	76
	-1	-4	8	68
	0	-4	5	55
	1	-12	12	48
	1	-8	7	43
	1	-4	2	27
	2	-12	9	38
	3	-8	1	7
	4	-12	3	14
	5	-12	0	0
4	6	7	46	34
	-4	-9	10	83
	-3	-5	6	88
	-1	-11	10	69
	-1	-4	4	76
	0	-7	6	65
	1	-10	8	60
	1	-3	2	48
	3	-2	0	0
	4	-5	2	25
	5	-8	4	33
	6	-11	6	37
	7	-7	2	16
	10	-9	2	12
	11	-12	4	19
6	4	7	46	56
	-8	-9	12	80
	-7	-7	10	82
	-6	-5	8	86
	-2	-11	8	54

TAB. V

FROM HKL-UVW INTO EULER ANGLES

H	K / U	L / V	Φ / W	φ2 / φ1
4	6	7		
6	4	7	46	56
	-1	-9	6	50
	-1	-2	2	68
	0	-7	4	44
	1	-12	6	38
	1	-5	2	31
	2	-3	0	0
	3	-8	2	19
	5	-11	2	13
4	7	6	53	30
	-1	-8	10	76
	-1	-2	3	88
	0	-6	7	71
	1	-10	11	67
	1	-4	4	60
	2	-2	1	25
	3	-6	5	48
	4	-10	9	53
	5	-8	6	42
	7	-10	7	38
	7	-4	0	0
	9	-12	8	36
	9	-6	1	7
	11	-8	2	10
7	4	6	53	60
	-6	-6	11	81
	-4	-11	12	63
	-4	-5	8	77
	-2	-10	9	56
	-2	-7	7	60
	-2	-4	5	68
	0	-3	2	44
	2	-11	5	31
	2	-8	3	25
	2	-5	1	13
	4	-7	0	0
	6	-12	1	5
6	7	4	67	41
	-1	-6	12	76
	-1	-2	5	84
	0	-4	7	71
	1	-6	9	65
	1	-2	2	47

H	K / U	L / V	Φ / W	φ2 / φ1
	2	-8	11	61
	4	-4	1	11
	5	-6	3	23
	6	-8	5	29
	7	-10	7	33
	7	-6	0	0
	8	-12	9	35
	9	-10	4	18
	11	-10	1	4
7	6	4	67	49
	-2	-5	11	78
	-2	-3	8	84
	0	-2	3	65
	2	-9	10	53
	2	-7	7	49
	2	-5	4	41
	2	-3	1	17
	4	-12	11	46
	4	-8	5	32
	6	-11	6	28
	6	-7	0	0
	8	-10	1	5
4	6	9	39	34
	-6	-11	10	86
	-3	-10	8	77
	-3	-7	6	82
	0	-3	2	63
	3	-11	6	48
	3	-8	4	43
	3	-5	2	31
	3	-2	0	0
	6	-7	2	20
	9	-12	4	24
	9	-9	2	14
	12	-11	2	11
6	4	9	39	56
	-9	-9	10	81
	-7	-12	10	69
	-6	-9	8	72
	-5	-6	6	77
	-4	-3	4	88
	-1	-12	6	45
	-1	-3	2	59
	0	-9	4	41
	1	-6	2	30

H	K / U	L / V	Φ / W	φ2 / φ1
	2	-3	0	0
	3	-9	2	19
	5	-12	2	14
4	9	6	59	24
	0	-2	3	77
	3	-8	10	63
	3	-6	7	58
	3	-4	4	47
	3	-2	1	18
	6	-10	11	53
	6	-6	5	36
	9	-10	9	41
	9	-8	6	31
	9	-4	0	0
	12	-10	7	29
	12	-6	1	5
9	4	6	59	66
	-6	-3	11	89
	-4	-9	12	65
	-4	-3	8	83
	-2	-12	11	52
	-2	-9	9	55
	-2	-6	7	60
	-2	-3	5	72
	0	-3	2	41
	2	-12	5	26
	2	-9	3	21
	2	-6	1	11
	4	-9	0	0
6	9	4	70	34
	-1	-2	6	88
	0	-4	9	77
	1	-6	12	72
	1	-2	3	59
	3	-2	0	0
	4	-4	3	30
	5	-6	6	41
	6	-8	9	45
	7	-10	12	48
	7	-6	3	19
	9	-10	9	36
	10	-8	3	14
	11	-10	6	24
9	6	4	70	56
	-2	-5	12	77

H	K / U	L / V	Φ / W	φ2 / φ1
	-2	-3	9	82
	0	-2	3	63
	2	-11	12	51
	2	-9	9	48
	2	-7	6	43
	2	-5	3	31
	2	-3	0	0
	4	-12	9	38
	4	-8	3	20
	6	-11	3	14
4	6	11	33	34
	-1	-3	2	77
	0	-11	6	61
	1	-8	4	54
	2	-5	2	39
	3	-2	0	0
	5	-7	2	24
	7	-12	4	30
	8	-9	2	17
	11	-11	2	13
6	4	11	33	56
	-11	-11	10	81
	-10	-7	8	89
	-7	-6	6	84
	-5	-9	6	67
	-4	-5	4	75
	-1	-4	2	53
	0	-11	4	39
	1	-7	2	30
	2	-3	0	0
	3	-10	2	20
4	11	6	63	20
	-1	-4	8	87
	0	-6	11	81
	1	-2	3	64
	4	-2	1	14
	5	-4	4	37
	6	-6	7	46
	7	-8	10	50
	9	-6	5	28
	11	-10	11	42
	11	-4	0	0
11	4	6	63	70
	-4	-7	12	69

H	K / U	L / V	Φ / W	φ2 / φ1
	-2	-11	11	52
	-2	-8	9	56
	-2	-5	7	63
	-2	-2	5	78
	0	-3	2	39
	2	-10	3	18
	2	-7	1	9
	4	-11	0	0
6	11	4	72	29
	-1	-2	7	89
	0	-4	11	81
	1	-2	4	66
	3	-2	1	16
	4	-4	5	44
	5	-6	9	52
	7	-6	6	35
	10	-8	7	30
	11	-10	11	39
	11	-6	0	0
11	6	4	72	61
	-2	-3	10	81
	0	-2	3	61
	2	-11	11	47
	2	-9	8	43
	2	-7	5	36
	2	-5	2	21
	4	-12	7	31
	4	-8	1	7
	6	-11	0	0
4	7	7	49	30
	0	-1	1	69
	7	-12	8	41
	7	-11	7	39
	7	-10	6	36
	7	-9	5	32
	7	-8	4	28
	7	-7	3	23
	7	-6	2	16
	7	-5	1	9
	7	-4	0	0
7	4	7	49	60
	-8	-7	12	82
	-7	-7	11	80
	-6	-7	10	77

H	K / U	L / V	Φ / W	φ2 / φ1
	-5	-7	9	73
	-4	-7	8	69
	-3	-7	7	64
	-2	-7	6	57
	-1	-7	5	50
	0	-7	4	41
	1	-7	3	31
	2	-7	2	21
	3	-7	1	10
	4	-7	0	0
7	7	4	68	45
	-2	-2	7	90
	-1	-3	7	79
	0	-4	7	69
	1	-5	7	61
	1	-1	0	0
	2	-6	7	53
	3	-7	7	47
	4	-8	7	42
	5	-9	7	37
	6	-10	7	34
	7	-11	7	31
	8	-12	7	28
4	7	8	45	30
	-4	-8	9	88
	-1	-12	11	72
	-1	-4	4	79
	0	-8	7	68
	1	-12	10	64
	1	-4	3	56
	3	-4	2	32
	4	-8	5	43
	5	-12	8	48
	5	-4	1	13
	7	-12	7	39
	7	-4	0	0
	8	-8	3	21
	11	-12	5	24
	12	-8	1	6
7	4	8	45	60
	-8	-8	11	79
	-4	-11	9	59
	-4	-9	8	63
	-4	-7	7	67
	-4	-5	6	74

H	K / U	L / V	Φ / W	φ2 / φ1
	-4	-3	5	85
	0	-2	1	39
	4	-11	2	14
	4	-9	1	8
	4	-7	0	0
4	8	7	52	27
	-1	-10	12	77
	-1	-3	4	85
	0	-7	8	73
	1	-11	12	69
	1	-4	4	62
	2	-1	0	0
	3	-5	4	46
	4	-9	8	53
	5	-6	4	35
	7	-7	4	26
	8	-11	8	40
	9	-8	4	24
	11	-9	4	20
8	4	7	52	63
	-8	-5	12	87
	-7	-7	12	78
	-5	-11	12	63
	-5	-4	8	82
	-3	-8	8	60
	-2	-3	4	71
	-1	-12	8	45
	-1	-5	4	52
	0	-7	4	39
	1	-9	4	31
	1	-2	0	0
	2	-11	4	25
7	8	4	69	41
	0	-1	2	73
	4	-9	11	53
	4	-8	9	49
	4	-7	7	44
	4	-6	5	38
	4	-5	3	27
	4	-4	1	11
	8	-11	8	33
	8	-9	4	20
	8	-7	0	0
	12	-11	1	4

H	K / U	L / V	Φ / W	φ2 / φ1
8	7	4	69	49
	-2	-4	11	82
	-1	-4	9	76
	0	-4	7	68
	1	-8	12	62
	1	-4	5	56
	2	-4	3	37
	3	-8	8	47
	3	-4	1	12
	5	-12	11	44
	5	-8	4	25
	7	-12	7	29
	7	-8	0	0
	8	-12	5	20
	10	-12	1	4
4	7	9	42	30
	-1	-11	9	71
	-1	-2	2	88
	0	-9	7	67
	1	-7	5	60
	2	-5	3	47
	3	-12	8	54
	3	-3	1	20
	5	-8	4	36
	7	-4	0	0
	8	-11	5	31
	10	-7	1	7
7	4	9	42	60
	-9	-9	11	79
	-7	-8	9	76
	-5	-7	7	71
	-4	-11	8	58
	-3	-6	5	64
	-1	-5	3	49
	0	-9	4	37
	1	-4	1	21
	4	-7	0	0
	5	-11	1	7
4	9	7	55	24
	-3	-8	12	88
	-1	-5	7	88
	0	-7	9	76
	1	-9	11	71
	1	-2	2	55

TAB. V FROM HKL-UVW INTO EULER ANGLES

H	K / U	L / V	Φ / W	φ_2 / φ_1
4	7	9		
4	9	7	55	24
	5	-3	1	12
	6	-5	3	26
	7	-7	5	34
	8	-9	7	38
	9	-11	9	41
	9	-4	0	0
	11	-8	4	20
9	4	7	55	66
	-5	-8	11	69
	-4	-5	8	73
	-3	-2	5	84
	-1	-10	7	45
	-1	-3	3	58
	0	-7	4	37
	1	-11	5	30
	1	-4	1	17
	4	-9	0	0
7	9	4	71	38
	0	-4	9	76
	1	-3	5	64
	2	-2	1	21
	3	-5	6	49
	4	-8	11	55
	5	-7	7	42
	7	-9	8	37
	9	-11	9	35
	9	-7	0	0
	11	-9	1	4
9	7	4	71	52
	-1	-5	11	74
	-1	-1	4	88
	0	-4	7	67
	1	-7	10	60
	1	-3	3	47
	3	-5	2	20
	4	-8	5	31
	5	-11	8	36
	5	-7	1	7
	7	-9	0	0
	8	-12	3	12

H	K / U	L / V	Φ / W	φ_2 / φ_1
4	7	10	39	30
	-5	-10	9	88
	-2	-6	5	81
	-1	-8	6	72
	0	-10	7	66
	1	-12	8	62
	1	-2	1	41
	7	-4	0	0
	8	-6	1	9
	9	-8	2	15
	10	-10	3	19
	11	-12	4	22
7	4	10	39	60
	-10	-10	11	78
	-8	-11	10	71
	-6	-7	7	74
	-4	-3	4	84
	-2	-9	5	49
	-2	-4	3	63
	0	-5	2	36
	2	-11	3	24
	2	-6	1	14
	4	-7	0	0
4	10	7	57	22
	-1	-8	12	82
	0	-7	10	78
	1	-6	8	72
	2	-5	6	63
	3	-4	4	48
	4	-3	2	26
	5	-9	10	56
	5	-2	0	0
	7	-7	6	38
	9	-5	2	13
	10	-11	10	42
	11	-10	8	34
10	4	7	57	68
	-4	-11	12	59
	-3	-10	10	56
	-2	-9	8	51
	-1	-8	6	45
	-1	-1	2	77
	0	-7	4	36
	1	-6	2	22

H	K / U	L / V	Φ / W	φ_2 / φ_1
	2	-5	0	0
	3	-11	2	12
7	10	4	72	35
	-2	-3	11	90
	0	-2	5	78
	2	-5	9	65
	2	-3	4	51
	4	-4	3	30
	6	-7	7	40
	6	-5	2	15
	8	-10	11	43
	8	-6	1	6
	10	-11	10	36
	10	-9	5	22
	10	-7	0	0
10	7	4	72	55
	-1	-2	6	80
	0	-4	7	66
	1	-6	8	57
	1	-2	1	25
	2	-8	9	51
	3	-10	10	47
	4	-12	11	44
	7	-10	0	0
	8	-12	1	4
4	7	11	36	30
	-1	-12	8	69
	0	-11	7	65
	1	-10	6	60
	2	-9	5	54
	3	-8	4	46
	4	-7	3	36
	5	-6	2	25
	6	-5	1	12
	7	-4	0	0
	11	-11	3	19
7	4	11	36	60
	-1	-12	5	40
	-1	-1	1	78
	0	-11	4	35
	1	-10	3	29
	2	-9	2	21
	3	-8	1	11
	4	-7	0	0

H	K / U	L / V	Φ / W	φ_2 / φ_1
4	11	7	59	20
	-1	-6	10	85
	0	-7	11	79
	1	-8	12	75
	1	-1	1	42
	11	-4	0	0
	12	-5	1	5
11	4	7	59	70
	-5	-2	9	89
	-4	-3	8	81
	-3	-11	11	54
	-3	-4	7	71
	-2	-5	6	60
	-1	-6	5	48
	0	-7	4	35
	1	-8	3	24
	2	-9	2	14
	3	-10	1	6
	4	-11	0	0
7	11	4	73	32
	-1	-3	10	86
	0	-4	11	79
	1	-5	12	74
	1	-1	1	37
	11	-7	0	0
	12	-8	1	4
11	7	4	73	58
	-2	-2	9	86
	-1	-3	8	77
	0	-4	7	65
	1	-5	6	53
	2	-6	5	40
	3	-11	11	47
	3	-7	4	29
	4	-8	3	19
	5	-9	2	12
	6	-10	1	5
	7	-11	0	0
4	7	12	34	30
	-3	-12	8	77
	-2	-4	3	87
	-1	-8	5	71
	0	-12	7	65

H	K / U	L / V	Φ / W	φ_2 / φ_1
	1	-4	2	51
	4	-4	1	18
	5	-8	3	33
	6	-12	5	39
	7	-4	0	0
	9	-12	4	28
	11	-8	1	8
7	4	12	34	60
	-12	-12	11	77
	-12	-9	10	84
	-8	-7	7	80
	-4	-11	6	55
	-4	-8	5	61
	-4	-5	4	72
	0	-3	1	35
	4	-10	1	10
	4	-7	0	0
4	12	7	61	18
	0	-7	12	81
	1	-5	8	75
	2	-3	4	58
	3	-8	12	69
	3	-1	0	0
	5	-4	4	37
	7	-7	8	46
	8	-5	4	26
	9	-10	12	50
	11	-6	4	20
	12	-11	12	43
12	4	7	61	72
	-5	-6	12	73
	-4	-9	12	62
	-3	-5	8	67
	-2	-1	4	86
	-1	-11	8	42
	-1	-4	4	53
	0	-7	4	35
	1	-10	4	25
	1	-3	0	0
7	12	4	74	30
	0	-1	3	81
	4	-6	11	60
	4	-5	8	54
	4	-4	5	44
	4	-3	2	23

H	K / U	L / V	Φ / W	φ_2 / φ_1
	8	-7	7	35
	8	-5	1	6
	12	-11	12	38
	12	-10	9	31
	12	-8	3	12
	12	-7	0	0
12	7	4	74	60
	-1	-4	10	74
	0	-4	7	65
	1	-8	11	57
	1	-4	4	46
	2	-4	1	13
	3	-8	5	32
	4	-12	9	37
	5	-12	6	26
	7	-12	0	0
4	8	9	45	27
	-5	-11	12	88
	-1	-4	4	81
	0	-9	8	70
	1	-5	4	61
	2	-1	0	0
	3	-6	4	47
	4	-11	8	53
	5	-7	4	37
	7	-8	4	30
	9	-9	4	25
	11	-10	4	22
8	4	9	45	63
	-10	-7	12	84
	-8	-11	12	70
	-7	-4	8	88
	-5	-8	8	67
	-3	-12	8	50
	-3	-3	4	77
	-2	-5	4	58
	-1	-7	4	44
	0	-9	4	35
	1	-11	4	29
	1	-2	0	0
4	9	8	51	24
	-1	-4	5	84
	0	-8	9	74
	1	-4	4	64

H	K / U	L / V	Φ / W	φ_2 / φ_1
	3	-4	3	42
	4	-8	7	53
	5	-12	11	56
	5	-4	2	23
	7	-12	10	49
	7	-4	1	9
	8	-8	5	31
	9	-4	0	0
	11	-12	8	35
	12	-8	3	15
9	4	8	51	66
	-8	-4	11	88
	-4	-11	10	57
	-4	-9	9	60
	-4	-7	8	65
	-4	-5	7	72
	-4	-3	6	82
	0	-2	1	35
	4	-11	1	6
	4	-9	0	0
8	9	4	72	42
	-1	-4	11	81
	0	-4	9	74
	1	-4	7	65
	2	-4	5	52
	3	-8	12	59
	3	-4	3	33
	4	-4	1	11
	5	-8	8	43
	7	-8	4	22
	8	-12	11	40
	9	-8	0	0
	10	-12	7	26
	11	-12	5	18
9	8	4	72	48
	0	-1	2	70
	4	-10	11	49
	4	-9	9	45
	4	-8	7	40
	4	-7	5	34
	4	-6	3	24
	4	-5	1	9
	8	-11	4	17
	6	-9	0	0

H	K / U	L / V	Φ / W	φ_2 / φ_1
4	8	11	39	27
	-4	-9	8	88
	-1	-5	4	78
	0	-11	8	69
	1	-6	4	61
	2	-1	0	0
	3	-7	4	47
	5	-8	4	38
	7	-9	4	32
	9	-10	4	27
	11	-11	4	23
8	4	11	39	63
	-11	-11	12	75
	-7	-8	8	72
	-5	-12	8	56
	-4	-3	4	82
	-3	-5	4	64
	-2	-7	4	50
	-1	-9	4	40
	0	-11	4	33
	1	-2	0	0
4	11	8	56	20
	-1	-4	6	87
	0	-8	11	78
	1	-4	5	69
	3	-4	4	49
	4	-8	9	59
	5	-4	3	31
	7	-4	2	17
	8	-8	7	40
	9	-4	1	7
	11	-12	11	43
	11	-4	0	0
	12	-6	5	23
11	4	8	56	70
	-4	-11	11	56
	-4	-9	10	60
	-4	-7	9	64
	-4	-5	8	71
	-4	-3	7	80
	0	-2	1	33
	4	-11	0	0

TAB. V FROM HKL-UVW INTO EULER ANGLES

H	K / U	L / V	Φ / W	φ2 / φ1
4	8	11		
8	11	4	74	36
	0	-4	11	78
	1	-4	9	71
	2	-4	7	61
	3	-4	5	47
	4	-4	3	29
	5	-8	12	55
	5	-4	1	9
	7	-8	8	39
	9	-8	4	19
	11	-12	11	36
	11	-8	0	0
11	8	4	74	54
	0	-1	2	69
	4	-11	11	46
	4	-10	9	42
	4	-9	7	37
	4	-8	5	31
	4	-7	3	21
	4	-6	1	8
	8	-11	0	0
4	9	9	48	24
	0	-1	1	73
	9	-12	8	40
	9	-11	7	37
	9	-10	6	33
	9	-9	5	30
	9	-8	4	25
	9	-7	3	20
	9	-6	2	14
	9	-5	1	8
	9	-4	0	0
9	4	9	48	66
	-8	-9	12	73
	-7	-9	11	70
	-6	-9	10	67
	-5	-9	9	63
	-4	-9	8	59
	-3	-9	7	54
	-2	-9	6	48
	-1	-9	5	41
	0	-9	4	33

H	K / U	L / V	Φ / W	φ2 / φ1
	1	-9	3	25
	2	-9	2	17
	3	-9	1	8
	4	-9	0	0
9	9	4	73	45
	-2	-2	9	90
	-1	-3	9	81
	0	-4	9	73
	1	-5	9	66
	1	-1	0	0
	2	-6	9	59
	3	-7	9	53
	4	-8	9	48
	5	-9	9	44
	6	-10	9	40
	7	-11	9	37
	8	-12	9	34
4	9	10	45	24
	-1	-4	4	83
	0	-10	9	72
	1	-6	5	65
	2	-2	1	28
	3	-8	6	55
	5	-10	7	49
	7	-12	8	45
	9	-4	0	0
	11	-6	1	7
9	4	10	45	66
	-10	-5	11	88
	-8	-7	10	78
	-6	-4	7	83
	-4	-11	8	54
	-2	-8	5	48
	-2	-3	3	66
	0	-5	2	32
	2	-12	3	20
	2	-7	1	11
	4	-9	0	0
4	10	9	50	22
	-1	-5	6	83
	0	-9	10	76
	1	-4	4	65
	3	-3	2	34
	4	-7	6	51

H	K / U	L / V	Φ / W	φ2 / φ1
	5	-11	10	56
	5	-2	0	0
	7	-10	8	46
	8	-5	2	16
	11	-8	4	22
10	4	9	50	68
	-8	-7	12	77
	-7	-5	10	81
	-6	-3	8	87
	-1	-11	6	38
	-1	-2	2	60
	0	-9	4	32
	1	-7	2	21
	2	-5	0	0
	3	-12	2	12
9	10	4	73	42
	-2	-3	12	88
	0	-2	5	76
	2	-5	8	60
	2	-3	3	42
	4	-8	11	54
	4	-4	1	10
	6	-7	4	25
	8	-10	7	30
	10	-11	5	19
	10	-9	0	0
10	9	4	73	48
	-1	-2	7	84
	0	-4	9	72
	1	-6	11	66
	1	-2	2	44
	5	-6	1	8
	6	-8	3	17
	7	-10	5	23
	8	-12	7	27
	9	-10	0	0
4	9	11	42	24
	0	-11	9	72
	1	-9	7	66
	2	-7	5	58
	3	-5	3	43
	4	-3	1	17
	5	-12	8	52
	7	-8	4	32

H	K / U	L / V	Φ / W	φ2 / φ1
	9	-4	0	0
	11	-11	5	27
9	4	11	42	66
	-9	-10	11	72
	-7	-9	9	68
	-5	-8	7	63
	-3	-7	5	55
	-2	-1	2	88
	-1	-6	3	42
	0	-11	4	31
	1	-5	1	17
	4	-9	0	0
4	11	9	52	20
	-1	-7	9	83
	0	-9	11	78
	1	-2	2	57
	6	-3	1	11
	7	-5	3	25
	8	-7	5	32
	9	-9	7	37
	10	-11	9	41
	11	-4	0	0
11	4	9	52	70
	-8	-5	12	83
	-5	-11	11	58
	-5	-2	7	89
	-4	-7	6	63
	-3	-3	5	74
	-1	-4	3	48
	0	-9	4	31
	1	-5	1	14
	4	-11	0	0
9	11	4	74	39
	0	-4	11	78
	1	-3	6	67
	2	-2	1	20
	3	-5	7	53
	5	-7	8	45
	7	-9	9	40
	9	-11	10	37
	11	-9	0	0
11	9	4	74	51
	-1	-1	5	88
	0	-4	9	72

H	K / U	L / V	Φ / W	φ2 / φ1
	1	-3	4	55
	3	-5	3	28
	4	-8	7	40
	5	-11	11	44
	5	-7	2	14
	7	-9	1	5
	8	-12	5	20
	9	-11	0	0
4	9	12	39	24
	-3	-12	10	82
	-3	-8	7	87
	0	-4	3	71
	3	-12	8	59
	3	-8	5	53
	3	-4	2	36
	6	-12	7	47
	6	-4	1	13
	9	-8	3	22
	9	-4	0	0
	12	-12	5	26
9	4	12	39	66
	-12	-6	11	88
	-8	-9	9	71
	-4	-12	7	50
	-4	-9	6	55
	-4	-6	5	64
	-4	-3	4	80
	0	-3	1	30
	4	-12	1	7
	4	-9	0	0
4	12	9	55	18
	0	-3	4	79
	3	-10	12	68
	3	-7	8	63
	3	-4	4	50
	3	-1	0	0
	6	-11	12	58
	6	-5	4	34
	9	-9	8	41
	9	-6	4	25
	12	-7	4	20
12	4	9	55	72
	-8	-3	12	89
	-7	-6	12	77

H	K / U	L / V	Φ / W	φ2 / φ1
	-5	-12	12	56
	-5	-3	8	83
	-3	-9	8	52
	-2	-3	4	66
	-1	-6	4	42
	0	-9	4	30
	1	-12	4	23
	1	-3	0	0
9	12	4	75	37
	0	-1	3	79
	4	-7	12	59
	4	-6	9	54
	4	-5	6	45
	4	-4	3	29
	4	-3	0	0
	8	-9	9	38
	8	-7	3	16
	12	-11	6	21
	12	-10	3	11
12	9	4	75	53
	-1	-4	12	78
	0	-4	9	71
	1	-4	6	59
	2	-4	3	35
	3	-8	9	49
	3	-4	0	0
	5	-12	12	45
	5	-8	3	18
	7	-12	6	24
	8	-12	3	12
4	10	11	44	22
	-1	-4	4	84
	0	-11	10	74
	1	-7	6	68
	2	-3	2	44
	3	-10	8	60
	5	-2	0	0
	7	-5	2	19
	9	-8	4	27
	11	-11	6	31
	12	-7	2	12
10	4	11	44	68
	-9	-5	10	85
	-7	-10	10	65

H	K / U	L / V	Φ / W	φ2 / φ1
	-6	-7	8	70
	-5	-4	6	78
	-1	-3	2	50
	0	-11	4	29
	1	-8	2	20
	2	-5	0	0
4	11	10	49	20
	-1	-6	7	83
	0	-10	11	77
	1	-4	4	66
	3	-2	1	21
	4	-6	5	49
	5	-10	9	56
	7	-8	6	40
	10	-10	7	36
	11	-4	0	0
11	4	10	49	70
	-8	-3	10	90
	-6	-11	11	60
	-4	-9	8	56
	-2	-12	7	41
	-2	-7	5	48
	-2	-2	3	73
	0	-5	2	29
	2	-8	1	9
	4	-11	0	0
10	11	4	75	42
	-1	-2	8	86
	0	-4	11	77
	1	-2	3	56
	4	-4	1	10
	5	-6	4	28
	6	-8	7	36
	7	-10	10	41
	9	-10	5	21
	11	-10	0	0
11	10	4	75	48
	0	-2	5	74
	2	-7	12	62
	2	-5	7	55
	2	-3	2	30
	4	-8	9	47
	6	-11	11	43
	6	-7	1	6
	8	-10	3	14

H	K / U	L / V	Φ / W	φ2 / φ1
	10	-11	0	0
4	11	11	47	20
	0	-1	1	76
	11	-12	8	37
	11	-11	7	34
	11	-10	6	31
	11	-9	5	27
	11	-8	4	23
	11	-7	3	18
	11	-6	2	12
	11	-5	1	6
	11	-4	0	0
11	4	11	47	70
	-8	-11	12	65
	-7	-11	11	62
	-6	-11	10	59
	-5	-11	9	55
	-4	-11	8	51
	-3	-11	7	46
	-2	-11	6	40
	-1	-11	5	34
	0	-11	4	28
	1	-11	3	21
	2	-11	2	14
	3	-11	1	7
	4	-11	0	0
11	11	4	76	45
	-2	-2	11	90
	-1	-3	11	83
	0	-4	11	76
	1	-5	11	70
	1	-1	0	0
	2	-6	11	64
	3	-7	11	58
	4	-8	11	53
	5	-9	11	49
	6	-10	11	45
	7	-11	11	42
	8	-12	11	39
4	11	12	44	20
	-1	-4	4	86
	0	-12	11	75
	1	-8	7	70

AB. V FROM HKL-UVW INTO EULER ANGLES V - 56

H	K / U	L / V	Φ / W	φ_2 / φ_1
4	11	12	44	20
	2	-4	3	53
	3	-12	10	64
	5	-4	2	25
	7	-8	5	38
	8	-4	1	9
	9	-12	8	42
	11	-4	0	0
	12	-12	7	33
1	4	12	44	70
	-8	-11	11	64
	-8	-5	9	81
	-4	-10	7	51
	-4	-7	6	59
	-4	-4	5	72
	0	-3	1	27
	4	-11	0	0
4	12	11	49	18
	-3	-10	12	89
	-1	-7	8	83
	0	-11	12	78
	1	-4	4	67
	3	-1	0	0
	4	-5	4	45
	5	-9	8	54
	7	-6	4	32
	10	-7	4	24
	11	-11	8	37
12	4	11	49	72
	-8	-9	12	69
	-7	-12	12	60
	-5	-7	8	64
	-3	-2	4	80
	-2	-5	4	52
	-1	-8	4	36
	0	-11	4	27
	1	-3	0	0
11	12	4	76	43
	0	-1	3	78
	4	-7	10	53
	4	-6	7	46
	4	-5	4	33
	4	-4	1	10
	8	-11	11	40

H	K / U	L / V	Φ / W	φ_2 / φ_1
	8	-9	5	23
	12	-11	0	0
12	11	4	76	47
	0	-4	11	75
	1	-4	8	66
	2	-4	5	50
	3	-4	2	22
	5	-8	7	38
	7	-12	12	42
	7	-8	1	6
	8	-12	9	33
	10	-12	3	11
	11	-12	0	0
5	5	6	50	45
	-5	-7	10	84
	-3	-9	10	72
	-3	-3	5	90
	-2	-4	5	78
	-1	-11	10	62
	-1	-5	5	67
	0	-6	5	57
	1	-7	5	49
	1	-1	0	0
	2	-8	5	43
	3	-9	5	38
	4	-10	5	34
	5	-11	5	30
	6	-12	5	27
5	6	5	57	40
	-4	-5	10	89
	-3	-5	9	85
	-2	-5	8	80
	-1	-5	7	74
	0	-5	6	66
	1	-10	11	61
	1	-5	5	56
	2	-5	4	45
	3	-10	9	51
	3	-5	3	33
	4	-5	2	21
	5	-10	7	39
	5	-5	1	10
	6	-5	0	0
	7	-10	5	27
	9	-10	3	15

H	K / U	L / V	Φ / W	φ_2 / φ_1
	11	-10	1	5
6	5	5	57	50
	-5	-6	12	84
	-5	-5	11	87
	0	-1	1	57
	5	-12	6	30
	5	-11	5	27
	5	-10	4	24
	5	-9	3	19
	5	-8	2	14
	5	-7	1	8
	5	-6	0	0
5	5	7	45	45
	-7	-7	10	90
	-5	-9	10	79
	-3	-11	10	68
	-3	-4	5	84
	-2	-5	5	73
	-1	-6	5	63
	0	-7	5	55
	1	-8	5	48
	1	-1	0	0
	2	-9	5	42
	3	-10	5	37
	4	-11	5	34
	5	-12	5	30
5	7	5	60	36
	-3	-5	10	88
	-2	-5	9	83
	-1	-5	8	77
	0	-5	7	70
	1	-5	6	62
	2	-5	5	52
	3	-10	11	57
	3	-5	4	41
	4	-5	3	29
	5	-10	9	46
	5	-5	2	16
	6	-5	1	8
	7	-10	7	35
	7	-5	0	0
	9	-10	5	24
	11	-10	3	13

H	K / U	L / V	Φ / W	φ_2 / φ_1
7	5	5	60	54
	-5	-5	12	85
	-5	-4	11	88
	0	-1	1	55
	5	-12	5	25
	5	-11	4	21
	5	-10	3	17
	5	-9	2	13
	5	-8	1	7
	5	-7	0	0
5	5	8	41	45
	-7	-9	10	85
	-5	-11	10	74
	-4	-4	5	90
	-3	-5	5	79
	-2	-6	5	69
	-1	-7	5	61
	0	-8	5	53
	1	-9	5	47
	1	-1	0	0
	2	-10	5	42
	3	-11	5	37
	4	-12	5	34
5	8	5	62	32
	-3	-5	11	90
	-2	-5	10	85
	-1	-5	9	80
	0	-5	8	74
	1	-5	7	66
	2	-5	6	57
	3	-5	5	47
	4	-5	4	37
	5	-10	11	53
	5	-5	3	26
	6	-5	2	16
	7	-10	9	42
	7	-5	1	8
	8	-5	0	0
	9	-10	7	31
	11	-10	5	21
8	5	5	62	58
	-5	-4	12	87
	0	-1	1	53
	5	-12	4	19

H	K / U	L / V	Φ / W	φ_2 / φ_1
	5	-11	3	16
	5	-10	2	11
	5	-9	1	6
	5	-8	0	0
5	5	9	38	45
	-9	-9	10	90
	-7	-11	10	80
	-4	-5	5	85
	-3	-6	5	75
	-2	-7	5	66
	-1	-8	5	59
	0	-9	5	52
	1	-10	5	46
	1	-1	0	0
	2	-11	5	41
	3	-12	5	37
5	9	5	64	29
	-2	-5	11	87
	-1	-5	10	82
	0	-5	9	76
	1	-5	8	70
	2	-5	7	62
	3	-5	6	53
	4	-5	5	43
	5	-5	4	33
	6	-5	3	23
	7	-10	11	48
	7	-5	2	15
	8	-5	1	7
	9	-10	9	38
	9	-5	0	0
	11	-10	7	28
9	5	5	64	61
	-5	-3	12	89
	0	-1	1	52
	5	-12	3	14
	5	-11	2	10
	5	-10	1	6
	5	-9	0	0
5	5	11	33	45
	-11	-11	10	90
	-5	-6	5	86
	-4	-7	5	77

H	K / U	L / V	Φ / W	φ_2 / φ_1
	-3	-8	5	69
	-2	-9	5	62
	-1	-10	5	55
	0	-11	5	50
	1	-12	5	45
	1	-1	0	0
5	11	5	68	24
	-1	-5	12	85
	0	-5	11	80
	1	-5	10	75
	2	-5	9	68
	3	-5	8	61
	4	-5	7	53
	5	-5	6	44
	6	-5	5	36
	7	-5	4	27
	8	-5	3	19
	9	-5	2	12
	10	-5	1	6
	11	-10	11	40
	11	-5	0	0
11	5	5	68	66
	0	-1	1	50
	5	-12	1	5
	5	-11	0	0
5	5	12	31	45
	-6	-6	5	90
	-5	-7	5	82
	-4	-8	5	74
	-3	-9	5	67
	-2	-10	5	60
	-1	-11	5	54
	0	-12	5	49
	1	-1	0	0
5	12	5	69	23
	0	-5	12	81
	1	-5	11	76
	2	-5	10	71
	3	-5	9	64
	4	-5	8	57
	5	-5	7	49
	6	-5	6	41
	7	-5	5	33
	8	-5	4	25

H	K / U	L / V	Φ / W	φ_2 / φ_1
	9	-5	3	17
	10	-5	2	11
	11	-5	1	5
	12	-5	0	0
12	5	5	69	67
	0	-1	1	49
	5	-12	0	0
5	6	6	52	40
	0	-1	1	63
	6	-12	7	36
	6	-11	6	33
	6	-10	5	30
	6	-9	4	26
	6	-8	3	21
	6	-7	2	16
	6	-6	1	8
	6	-5	0	0
	12	-11	1	4
6	5	6	52	50
	-7	-6	12	90
	-6	-6	11	87
	-5	-6	10	84
	-4	-6	9	80
	-3	-6	8	75
	-2	-6	7	69
	-1	-12	11	58
	-1	-6	6	62
	0	-6	5	54
	1	-12	9	49
	1	-6	4	44
	2	-6	3	33
	3	-12	7	36
	3	-6	2	21
	4	-6	1	10
	5	-12	5	27
	5	-6	0	0
	7	-12	3	15
	9	-12	1	5
6	6	5	59	45
	-5	-5	12	90
	-3	-7	12	79
	-2	-3	6	84
	-1	-9	12	68
	-1	-4	6	73

H	K / U	L / V	Φ / W	φ_2 / φ_1
	0	-5	6	63
	1	-11	12	59
	1	-6	6	55
	1	-1	0	0
	2	-7	6	48
	3	-8	6	42
	4	-9	6	37
	5	-10	6	33
	6	-11	6	30
	7	-12	6	27
5	6	7	48	40
	-3	-8	9	77
	-2	-3	4	86
	-1	-12	11	65
	-1	-5	5	70
	0	-7	6	61
	1	-9	7	55
	1	-2	1	33
	2	-11	8	51
	6	-5	0	0
	7	-7	1	8
	8	-9	2	13
	9	-11	3	16
6	5	7	48	50
	-7	-7	11	87
	-5	-8	10	78
	-4	-5	7	82
	-1	-10	8	57
	-1	-3	3	68
	0	-7	5	51
	1	-11	7	46
	1	-4	2	36
	3	-5	1	13
	4	-9	3	23
	5	-6	0	0
	8	-11	1	6
5	7	6	55	36
	-2	-6	11	77
	-1	-7	9	73
	0	-6	7	68
	1	-11	12	64
	1	-5	5	59
	2	-4	3	43
	3	-9	8	52
	3	-3	1	16

TAB. V FROM HKL-UVW INTO EULER ANGLES V - 57

H	K / U	L / V	Φ / W	φ_2 / φ_1
5	6	7		
5	7	6	55	36
	5	-7	4	31
	7	-11	7	35
	7	-5	0	0
	8	-10	5	26
	10	-8	1	5
7	5	6	55	54
	-3	-9	11	67
	-2	-8	9	64
	-1	-7	7	59
	-1	-1	2	85
	0	-6	5	51
	1	-11	8	46
	1	-5	3	38
	2	-4	1	15
	3	-9	4	28
	5	-7	0	0
	7	-11	1	5
6	7	5	62	41
	-1	-7	11	73
	-1	-2	4	83
	0	-5	7	68
	1	-8	10	62
	1	-3	3	52
	3	-4	2	25
	4	-7	5	37
	5	-10	8	41
	5	-5	1	9
	7	-11	7	33
	7	-6	0	0
	8	-9	3	16
	12	-11	1	4
7	6	5	62	49
	-3	-4	9	84
	-1	-8	11	67
	-1	-3	5	74
	0	-5	6	61
	1	-7	7	53
	1	-2	1	28
	2	-9	8	48
	3	-11	9	45
	6	-7	0	0
	7	-9	1	6

H	K / U	L / V	Φ / W	φ_2 / φ_1
	8	-11	2	10
5	6	8	44	40
	-6	-11	12	82
	-4	-6	7	86
	-2	-9	8	70
	-2	-5	5	77
	0	-4	3	59
	2	-11	7	49
	2	-7	4	44
	2	-3	1	22
	4	-10	5	37
	6	-5	0	0
	8	-8	1	7
	10	-11	2	11
6	5	8	44	50
	-8	-8	11	86
	-7	-6	9	89
	-2	-12	9	58
	-1	-10	7	55
	-1	-2	2	73
	0	-8	5	49
	1	-6	3	39
	2	-4	1	18
	3	-10	4	31
	5	-6	0	0
	7	-10	1	7
5	8	6	58	32
	-2	-7	11	81
	-2	-4	7	87
	0	-3	4	71
	2	-8	9	61
	2	-5	5	54
	2	-2	1	23
	4	-7	6	45
	6	-12	11	49
	6	-9	7	40
	8	-11	8	37
	8	-5	0	0
	10	-7	1	6
	12	-9	2	9
8	5	6	58	58
	-5	-4	10	86
	-2	-10	11	60
	-1	-8	8	57

H	K / U	L / V	Φ / W	φ_2 / φ_1
	-1	-2	3	72
	0	-6	5	49
	1	-10	7	43
	1	-4	2	31
	3	-6	1	10
	4	-10	3	19
	5	-8	0	0
6	8	5	63	37
	-3	-4	10	90
	-1	-3	6	82
	0	-5	8	71
	1	-7	10	66
	1	-2	2	48
	2	-9	12	62
	4	-3	0	0
	5	-5	2	18
	6	-7	4	26
	7	-9	6	31
	8	-11	8	35
	9	-8	2	11
	11	-12	6	23
8	6	5	63	53
	-4	-3	10	90
	-1	-7	10	66
	-1	-2	4	77
	0	-5	6	59
	1	-8	8	52
	1	-3	2	37
	2	-11	10	48
	3	-4	0	0
	4	-7	2	16
	5	-10	4	22
	7	-11	2	10
5	6	9	41	40
	-3	-11	9	71
	-3	-8	7	75
	-3	-5	5	83
	0	-3	2	58
	3	-10	5	41
	3	-7	3	34
	3	-4	1	17
	6	-11	4	28
	6	-5	0	0
	9	-9	1	7

H	K / U	L / V	Φ / W	φ_2 / φ_1
6	5	9	41	50
	-9	-9	11	86
	-7	-6	8	89
	-5	-12	10	68
	-3	-9	7	65
	-2	-3	3	77
	-1	-6	4	57
	0	-9	5	48
	1	-12	6	43
	1	-3	1	27
	5	-6	0	0
	6	-9	1	8
	7	-12	2	13
5	9	6	60	29
	0	-2	3	74
	3	-9	11	61
	3	-7	8	57
	3	-5	5	49
	3	-3	2	30
	6	-8	7	42
	6	-4	1	9
	9	-11	9	38
	9	-7	3	17
	9	-5	0	0
	12	-10	5	21
9	5	6	60	61
	-5	-3	10	89
	-4	-6	11	75
	-3	-3	7	82
	-1	-9	9	55
	-1	-3	4	65
	0	-6	5	48
	1	-9	6	40
	1	-3	1	20
	2	-12	7	35
	5	-9	0	0
	6	-12	1	5
6	9	5	65	34
	-1	-6	12	79
	0	-5	9	74
	1	-4	6	65
	2	-3	3	45
	3	-7	9	57
	3	-2	0	0
	5	-10	12	54

H	K / U	L / V	Φ / W	φ_2 / φ_1
	5	-5	3	25
	7	-8	6	33
	8	-7	3	17
	9	-11	9	36
	11	-9	3	13
9	6	5	65	56
	-2	-7	12	70
	-1	-6	9	66
	-1	-1	3	85
	0	-5	6	58
	1	-9	9	51
	1	-4	3	40
	2	-3	0	0
	3	-7	3	24
	4	-11	6	30
	5	-10	3	17
5	6	10	38	40
	-6	-10	9	83
	-4	-5	5	89
	-2	-10	7	67
	-2	-5	4	76
	0	-5	3	57
	2	-10	5	46
	2	-5	2	34
	4	-5	1	15
	6	-10	3	24
	6	-5	0	0
	10	-10	1	7
6	5	10	38	50
	-10	-10	11	86
	-5	-12	9	68
	-5	-10	8	71
	-5	-8	7	75
	-5	-6	6	82
	0	-2	1	47
	5	-12	3	21
	5	-10	2	17
	5	-8	1	10
	5	-6	0	0
5	10	6	62	27
	-2	-5	10	88
	0	-3	5	77
	2	-7	10	67
	2	-4	5	58

H	K / U	L / V	Φ / W	φ_2 / φ_1
	2	-1	0	0
	4	-5	5	44
	6	-9	10	50
	6	-6	5	35
	8	-7	5	29
	10	-11	10	39
	10	-8	5	24
	12	-9	5	21
10	5	6	62	63
	-3	-6	10	70
	-2	-2	5	81
	-1	-10	10	53
	-1	-4	5	61
	0	-6	5	47
	1	-8	5	37
	1	-2	0	0
	2	-10	5	30
	3	-12	5	25
6	10	5	67	31
	0	-1	2	77
	5	-9	12	56
	5	-8	10	52
	5	-7	8	48
	5	-6	6	42
	5	-5	4	32
	5	-4	2	19
	5	-3	0	0
	10	-11	10	37
	10	-9	6	26
	10	-7	2	10
10	6	5	67	59
	-3	-5	12	78
	-2	-5	10	73
	-1	-5	8	67
	0	-5	6	57
	1	-10	10	50
	1	-5	4	42
	2	-5	2	22
	3	-10	6	33
	3	-5	0	0
	5	-10	2	11
5	6	11	35	40
	-1	-12	7	60
	0	-11	6	56

H	K / U	L / V	Φ / W	φ_2 / φ_1
	1	-10	5	50
	2	-9	4	43
	3	-8	3	35
	4	-7	2	25
	5	-6	1	13
	6	-5	0	0
	11	-11	1	6
6	5	11	35	50
	-1	-12	6	50
	-1	-1	1	86
	0	-11	5	46
	1	-10	4	40
	2	-9	3	32
	3	-8	2	23
	4	-7	1	12
	5	-6	0	0
5	11	6	64	24
	-1	-5	10	84
	0	-6	11	79
	1	-7	12	74
	1	-1	1	40
	11	-5	0	0
	12	-6	1	5
11	5	6	64	66
	-4	-2	9	89
	-3	-3	8	81
	-2	-4	7	70
	-1	-11	11	52
	-1	-5	6	58
	0	-6	5	46
	1	-7	4	33
	2	-8	3	22
	3	-9	2	13
	4	-10	1	6
	5	-11	0	0
6	11	5	68	29
	-1	-4	10	84
	0	-5	11	79
	1	-6	12	74
	1	-1	1	38
	11	-6	0	0
	12	-7	1	4
11	6	5	68	61
	-3	-2	9	88

H	K / U	L / V	Φ / W	φ_2 / φ_1
	-2	-3	8	79
	-1	-4	7	68
	0	-5	6	56
	1	-11	11	49
	1	-6	5	43
	2	-7	4	31
	3	-8	3	21
	4	-9	2	12
	5	-10	1	6
	6	-11	0	0
5	6	12	33	40
	-6	-11	8	81
	-6	-9	7	55
	0	-2	1	55
	6	-11	3	25
	6	-9	2	19
	6	-7	1	11
	6	-5	0	0
	12	-12	1	6
6	5	12	33	50
	-12	-12	11	86
	-8	-12	9	76
	-7	-6	6	39
	-5	-6	5	31
	-4	-12	7	63
	-3	-6	4	70
	-1	-6	3	34
	0	-12	5	45
	1	-6	2	35
	3	-6	1	16
	4	-12	3	25
	5	-6	0	0
	8	-12	1	7
5	12	6	65	23
	0	-1	2	80
	6	-8	11	55
	6	-7	9	50
	6	-6	7	44
	6	-5	5	36
	6	-4	3	28
	6	-3	1	9
	12	-11	12	41
	12	-9	8	31
	12	-7	4	18
	12	-5	0	0

TAB. V

FROM HKL-UVW INTO EULER ANGLES

H	K / U	L / V	Φ / W	φ2 / φ1
5	6	12		
12	5	6	65	67
	-3	-6	11	70
	-2	-6	9	64
	-1	-12	12	51
	-1	-6	7	56
	0	-6	5	45
	1	-12	8	38
	1	-6	3	29
	2	-6	1	10
	3	-12	4	20
	5	-12	0	0
6	12	5	70	27
	-1	-2	6	90
	0	-5	12	80
	1	-3	6	71
	2	-1	0	0
	3	-4	6	55
	4	-7	12	62
	5	-5	6	44
	7	-6	6	36
	8	-9	12	49
	9	-7	6	30
	11	-8	6	26
	12	-11	12	39
12	6	5	70	63
	-3	-4	12	80
	-2	-1	6	90
	-1	-8	12	62
	-1	-3	6	71
	0	-5	6	55
	1	-12	12	49
	1	-7	6	44
	1	-2	0	0
	2	-9	6	36
	3	-11	6	30
5	7	7	51	36
	0	-1	1	66
	7	-12	7	35
	7	-11	6	33
	7	-10	5	29
	7	-9	4	25
	7	-8	3	20

H	K / U	L / V	Φ / W	φ2 / φ1
	7	-7	2	15
	7	-6	1	8
	7	-5	0	0
7	5	7	51	54
	-7	-7	12	84
	-6	-7	11	81
	-5	-7	10	78
	-4	-7	9	74
	-3	-7	8	69
	-2	-7	7	63
	-1	-7	6	57
	0	-7	5	49
	1	-7	4	39
	2	-7	3	29
	3	-7	2	19
	4	-7	1	9
	5	-7	0	0
7	7	5	63	45
	-2	-3	7	85
	-1	-4	7	75
	0	-5	7	66
	1	-6	7	58
	1	-1	0	0
	2	-7	7	51
	3	-8	7	45
	4	-9	7	40
	5	-10	7	36
	6	-11	7	33
	7	-12	7	30
5	7	8	47	36
	-5	-9	11	86
	-3	-7	8	82
	-1	-5	5	73
	0	-8	7	64
	1	-11	9	60
	1	-3	2	47
	4	-4	1	14
	5	-7	3	27
	6	-10	5	33
	7	-5	0	0
	9	-11	4	22
	11	-9	1	6
7	5	8	47	54
	-5	-9	10	72

H	K / U	L / V	Φ / W	φ2 / φ1
	-4	-12	11	64
	-3	-7	7	68
	-2	-2	3	84
	-1	-5	4	57
	0	-8	5	46
	1	-11	6	41
	1	-3	1	24
	5	-7	0	0
	6	-10	1	7
5	8	7	53	32
	-1	-9	11	74
	-1	-2	3	87
	0	-7	8	70
	1	-5	5	61
	2	-3	2	37
	3	-8	7	52
	5	-11	9	48
	5	-4	1	11
	7	-7	3	21
	8	-5	0	0
	9	-10	5	26
	12	-11	4	17
8	5	7	53	58
	-5	-6	10	79
	-4	-9	11	68
	-3	-5	7	73
	-1	-11	9	52
	-1	-4	4	60
	0	-7	5	46
	1	-10	6	40
	1	-3	1	22
	5	-8	0	0
	6	-11	1	6
7	8	5	65	41
	-1	-6	11	75
	0	-5	8	70
	1	-4	5	58
	2	-3	2	32
	3	-7	7	48
	4	-11	12	52
	5	-10	9	44
	5	-5	1	9
	7	-8	3	17
	8	-7	0	0
	9	-11	5	22

H	K / U	L / V	Φ / W	φ2 / φ1
8	7	5	65	49
	-1	-6	10	71
	-1	-1	3	88
	0	-5	7	64
	1	-9	11	59
	1	-4	4	50
	2	-3	1	17
	3	-7	5	37
	4	-11	9	42
	5	-10	6	32
	7	-8	0	0
	9	-11	1	4
5	7	9	44	36
	-5	-8	9	87
	-3	-12	11	74
	-2	-5	5	80
	-1	-7	6	69
	0	-9	7	63
	1	-11	8	58
	1	-2	1	36
	7	-5	0	0
	8	-7	1	8
	9	-9	2	13
	10	-11	3	17
7	5	9	44	54
	-7	-10	11	76
	-5	-11	10	67
	-4	-7	7	72
	-3	-3	4	83
	-1	-4	3	58
	0	-9	5	45
	1	-5	2	32
	3	-6	1	12
	4	-11	3	21
	5	-7	0	0
5	9	7	56	29
	-1	-8	11	77
	0	-7	9	73
	1	-6	7	66
	2	-5	5	55
	3	-11	12	61
	3	-4	3	38
	4	-3	1	14
	5	-9	8	48

H	K / U	L / V	Φ / W	φ2 / φ1
	7	-7	4	27
	9	-5	0	0
	10	-11	7	31
	11	-10	5	23
9	5	7	56	61
	-3	-10	11	61
	-2	-9	9	58
	-1	-8	7	52
	-1	-1	2	81
	0	-7	5	45
	1	-6	3	32
	2	-5	1	13
	3	-11	4	24
	5	-9	0	0
7	9	5	66	38
	-1	-2	5	85
	0	-5	9	73
	1	-3	4	59
	3	-4	3	34
	4	-7	7	46
	5	-10	11	50
	5	-5	2	17
	7	-11	10	42
	7	-6	1	7
	8	-9	5	25
	9	-7	0	0
	12	-11	3	11
9	7	5	66	52
	-3	-4	11	84
	-1	-3	6	75
	0	-5	7	63
	1	-7	8	55
	1	-2	1	26
	2	-9	9	50
	3	-11	10	46
	7	-9	0	0
	8	-11	1	5
5	7	10	41	36
	-4	-10	9	80
	-3	-5	5	87
	-1	-5	4	71
	0	-10	7	62
	1	-5	3	51
	3	-5	2	30

H	K / U	L / V	Φ / W	φ2 / φ1
	4	-10	5	40
	5	-5	1	12
	7	-5	0	0
	8	-10	3	20
	12	-10	1	6
7	5	10	41	54
	-10	-8	11	88
	-5	-11	9	66
	-5	-9	8	70
	-5	-7	7	75
	-5	-5	6	83
	0	-2	1	43
	5	-11	2	15
	5	-9	1	9
	5	-7	0	0
5	10	7	58	27
	-1	-3	5	86
	0	-7	10	75
	1	-4	5	66
	2	-1	0	0
	3	-5	5	50
	4	-9	10	57
	5	-6	5	40
	7	-7	5	32
	8	-11	10	44
	9	-8	5	27
	11	-9	5	23
10	5	7	58	63
	-5	-4	10	84
	-3	-8	10	64
	-2	-3	5	73
	-1	-12	10	49
	-1	-5	5	56
	0	-7	5	43
	1	-9	5	35
	1	-2	0	0
	2	-11	5	29
7	10	5	68	35
	0	-1	2	75
	5	-9	11	52
	5	-8	9	48
	5	-7	7	43
	5	-6	5	36
	5	-5	3	25
	5	-4	1	10

H	K / U	L / V	Φ / W	φ2 / φ1
	10	-11	8	31
	10	-9	4	18
	10	-7	0	0
10	7	5	68	55
	-2	-5	11	76
	-1	-5	9	70
	0	-5	7	62
	1	-10	12	56
	1	-5	5	49
	2	-5	3	32
	3	-10	8	41
	3	-5	1	11
	5	-10	4	21
	7	-10	0	0
5	7	11	38	36
	-5	-9	8	85
	-1	-4	3	73
	0	-11	7	61
	1	-7	4	53
	2	-3	1	26
	3	-10	5	45
	7	-5	0	0
	9	-8	1	8
	11	-11	2	12
7	5	11	38	54
	-11	-11	12	83
	-7	-10	9	74
	-6	-7	7	79
	-5	-4	5	88
	-1	-3	2	60
	0	-11	5	42
	1	-8	3	34
	2	-5	1	17
	5	-7	0	0
	7	-12	1	7
5	11	7	60	24
	-1	-4	7	85
	0	-7	11	77
	1	-3	4	65
	3	-2	1	16
	4	-5	5	45
	5	-8	9	53
	7	-7	6	37
	10	-9	7	32

H	K / U	L / V	Φ / W	φ2 / φ1
	11	-12	11	40
	11	-5	0	0
11	5	7	60	66
	-5	-3	10	87
	-2	-11	11	54
	-1	-9	8	50
	-1	-2	3	68
	0	-7	5	42
	1	-12	7	36
	1	-5	2	25
	3	-8	1	8
	5	-11	0	0
7	11	5	69	32
	-1	-3	8	85
	0	-5	11	77
	1	-2	3	59
	4	-3	1	12
	5	-5	4	32
	6	-7	7	40
	7	-9	10	45
	9	-8	5	24
	11	-12	11	37
	11	-7	0	0
11	7	5	69	58
	-1	-7	12	67
	-1	-2	5	78
	0	-5	7	61
	1	-8	9	55
	1	-3	2	35
	2	-11	11	49
	4	-7	1	8
	5	-10	3	16
	7	-11	0	0
5	7	12	36	36
	0	-12	7	60
	1	-11	6	55
	2	-10	5	49
	3	-9	4	42
	4	-8	3	33
	5	-7	2	23
	6	-6	1	12
	7	-5	0	0

TAB. V FROM HKL-UVW INTO EULER ANGLES

H	K U	L V	Φ W	φ2 φ1
5	7	12		
7	5	12	36	54
	-1	-1	1	82
	0	-12	5	41
	1	-11	4	36
	2	-10	3	29
	3	-9	2	21
	4	-8	1	11
	5	-7	0	0
5	12	7	62	23
	-2	-5	10	90
	-1	-6	11	84
	0	-7	12	79
	1	-1	1	41
	12	-5	0	0
12	5	7	62	67
	-4	-3	9	83
	-3	-4	8	74
	-2	-5	7	64
	-1	-6	6	53
	0	-7	5	41
	1	-8	4	30
	2	-9	3	21
	3	-10	2	12
	4	-11	1	6
	5	-12	0	0
7	12	5	70	30
	-1	-4	11	84
	0	-5	12	79
	1	-1	1	38
	12	-7	0	0
12	7	5	70	60
	-3	-2	10	89
	-2	-3	9	81
	-1	-4	8	71
	0	-5	7	60
	1	-6	6	48
	2	-7	5	37
	3	-8	4	27
	4	-9	3	18
	5	-10	2	11
	6	-11	1	5
	7	-12	0	0

H	K U	L V	Φ W	φ2 φ1
5	8	8	50	32
	0	-1	1	68
	8	-12	7	35
	8	-11	6	32
	8	-10	5	28
	8	-9	4	24
	8	-8	3	20
	8	-7	2	14
	8	-6	1	7
	8	-5	0	0
8	5	8	50	58
	-7	-8	12	79
	-6	-8	11	76
	-5	-8	10	73
	-4	-8	9	68
	-3	-8	8	64
	-2	-8	7	58
	-1	-8	6	52
	0	-8	5	44
	1	-8	4	36
	2	-8	3	27
	3	-8	2	17
	4	-8	1	8
	5	-8	0	0
8	8	5	66	45
	-2	-3	8	85
	-1	-4	8	76
	0	-5	8	68
	1	-6	8	60
	1	-1	0	0
	2	-7	8	54
	3	-8	8	48
	4	-9	8	44
	5	-10	8	40
	6	-11	8	36
	7	-12	8	33
5	8	9	46	32
	-4	-11	12	82
	-3	-6	7	86
	-1	-5	5	75
	0	-9	8	67
	1	-4	3	54
	3	-3	1	18

H	K U	L V	Φ W	φ2 φ1
	4	-7	4	36
	5	-11	7	44
	7	-10	5	32
	8	-5	0	0
	11	-8	1	6
8	5	9	46	58
	-8	-7	11	84
	-7	-5	9	88
	-1	-11	7	48
	-1	-2	2	67
	0	-9	5	42
	1	-7	3	33
	2	-5	1	15
	3	-12	4	25
	5	-8	0	0
5	9	8	52	29
	-1	-3	4	83
	0	-8	9	71
	1	-5	5	62
	2	-2	1	25
	3	-7	6	52
	4	-12	11	56
	5	-9	7	45
	7	-11	8	41
	9	-5	0	0
	11	-7	1	6
9	5	8	52	61
	-7	-5	11	86
	-5	-7	10	74
	-4	-4	7	80
	-1	-11	8	48
	-1	-3	3	61
	0	-8	5	42
	1	-5	2	28
	3	-7	1	9
	4	-12	3	17
	5	-9	0	0
8	9	5	67	42
	-3	-4	12	88
	-1	-3	7	81
	0	-5	9	71
	1	-7	11	66
	1	-2	2	46
	5	-5	1	9
	6	-7	3	20

H	K U	L V	Φ W	φ2 φ1
	7	-9	5	26
	8	-11	7	30
	9	-8	0	0
	11	-12	4	15
9	8	5	67	48
	-1	-2	5	81
	0	-5	8	67
	1	-8	11	61
	1	-3	3	46
	3	-4	1	12
	4	-7	4	29
	5	-10	7	35
	7	-11	5	23
	8	-9	0	0
5	8	10	43	32
	-6	-10	11	89
	-2	-10	9	75
	-2	-5	5	83
	0	-5	4	66
	2	-10	7	56
	2	-5	3	45
	4	-5	2	26
	6	-10	5	35
	6	-5	1	11
	8	-5	0	0
	10	-10	3	18
8	5	10	43	58
	-5	-12	10	63
	-5	-10	9	66
	-5	-8	8	70
	-5	-6	7	77
	-5	-4	6	85
	0	-2	1	41
	5	-12	2	13
	5	-10	1	7
	5	-8	0	0
5	10	8	54	27
	-2	-7	10	84
	0	-4	5	74
	2	-9	10	65
	2	-5	5	57
	2	-1	0	0
	4	-6	5	44
	6	-11	10	50

H	K U	L V	Φ W	φ2 φ1
	6	-7	5	36
	8	-8	5	30
	10	-9	5	25
	12	-10	5	22
10	5	8	54	63
	-5	-6	10	76
	-3	-10	10	58
	-3	-2	5	86
	-2	-4	5	66
	-1	-6	5	51
	0	-8	5	41
	1	-10	5	33
	1	-2	0	0
	2	-12	5	28
8	10	5	69	39
	0	-1	2	74
	5	-10	12	52
	5	-9	10	48
	5	-8	8	44
	5	-7	6	38
	5	-6	4	29
	5	-5	2	17
	5	-4	0	0
	10	-11	6	24
	10	-9	2	9
10	8	5	69	51
	-2	-5	12	78
	-1	-5	10	73
	0	-5	8	66
	1	-5	6	55
	2	-5	4	40
	3	-10	10	48
	3	-5	2	20
	4	-5	0	0
	5	-10	6	31
	7	-10	2	10
5	8	11	41	32
	-5	-12	11	83
	-3	-5	5	89
	-2	-7	6	78
	-1	-9	7	70
	0	-11	8	65
	1	-2	1	39
	8	-5	0	0

H	K U	L V	Φ W	φ2 φ1
	9	-7	1	8
	10	-9	2	13
	11	-11	3	17
8	5	11	41	58
	-8	-7	9	83
	-4	-9	7	63
	-3	-4	4	74
	-1	-5	3	51
	0	-11	5	39
	1	-6	2	29
	3	-7	1	12
	5	-8	0	0
5	11	8	56	24
	0	-8	11	76
	1	-7	9	71
	2	-6	7	63
	3	-5	5	51
	4	-4	3	34
	5	-11	12	58
	5	-3	1	12
	7	-9	8	44
	9	-7	4	23
	11	-5	0	0
11	5	8	56	66
	-3	-11	11	56
	-2	-10	9	53
	-1	-9	7	47
	-1	-1	2	78
	0	-8	5	39
	1	-7	3	28
	2	-6	1	11
	5	-11	0	0
8	11	5	70	36
	-1	-2	6	87
	0	-5	11	76
	1	-3	5	64
	3	-4	4	42
	4	-7	9	53
	5	-5	3	25
	7	-6	2	13
	8	-9	7	32
	9	-7	1	5
	11	-8	0	0
	12	-11	5	18

H	K U	L V	Φ W	φ2 φ1
11	8	5	70	54
	-1	-3	7	76
	0	-5	8	65
	1	-7	9	57
	1	-2	1	26
	2	-9	10	52
	3	-11	11	48
	8	-11	0	0
5	8	12	38	32
	-4	-11	9	80
	-4	-6	7	86
	0	-3	2	64
	4	-10	5	43
	4	-7	3	34
	4	-4	1	16
	8	-11	4	27
	8	-5	0	0
	12	-9	1	6
8	5	12	38	58
	-9	-12	11	73
	-7	-8	8	77
	-5	-4	5	85
	-3	-12	7	53
	-2	-4	3	64
	-1	-8	4	46
	0	-12	5	38
	1	-4	1	22
	5	-8	0	0
	6	-12	1	7
5	12	8	58	23
	0	-2	3	78
	4	-9	11	61
	4	-7	8	56
	4	-5	5	46
	4	-3	2	26
	8	-8	7	38
	8	-4	1	7
	12	-11	9	35
	12	-7	3	14
	12	-5	0	0
12	5	8	58	67
	-5	-4	10	81
	-4	-8	11	66

H	K U	L V	Φ W	φ2 φ1
	-3	-4	7	73
	-1	-12	9	45
	-1	-4	4	55
	0	-8	5	38
	1	-12	6	32
	1	-4	1	16
	5	-12	0	0
8	12	5	71	34
	0	-5	12	78
	1	-4	8	70
	2	-3	4	52
	3	-7	12	63
	3	-2	0	0
	5	-5	4	31
	7	-8	8	40
	8	-7	4	22
	9	-11	12	43
	11	-9	4	17
12	8	5	71	56
	-1	-6	12	71
	-1	-1	4	86
	0	-5	8	64
	1	-9	12	58
	1	-4	4	47
	2	-3	0	0
	3	-7	4	29
	4	-11	8	37
	5	-10	4	21
5	9	9	49	29
	0	-1	1	70
	9	-12	7	34
	9	-11	6	31
	9	-10	5	28
	9	-9	4	23
	9	-8	3	19
	9	-7	2	13
	9	-6	1	7
	9	-5	0	0
9	5	9	49	61
	-7	-9	12	76
	-6	-9	11	71
	-5	-9	10	68
	-4	-9	9	64
	-3	-9	8	59

TAB. V FROM HKL-UVW INTO EULER ANGLES

```
 H  K  L  Φ  φ2
    U  V  W  φ1
---------------
 5  9  9
===============
 9  5  9 49 61
---------------
   -2 -9  7 53
   -1 -9  6 47
    0 -9  5 40
    1 -9  4 32
    2 -9  3 24
    3 -9  2 16
    4 -9  1  8
    5 -9  0  0
 9  9  5 69 45
---------------
   -2 -3  9 86
   -1 -4  9 78
    0 -5  9 70
    1 -6  9 63
    1 -1  0  0
    2 -7  9 57
    3 -8  9 51
    4 -9  9 46
    5-10  9 42
    6-11  9 39
    7-12  9 36

 5  9 10 46 29
===============
   -4-10 11 85
   -1 -5  5 77
    0-10  9 69
    1 -5  4 59
    3 -5  3 40
    4-10  7 49
    5 -5  2 22
    7 -5  1  9
    8-10  5 30
    9 -5  0  0
   12-10  3 15
 9  5 10 46 61
---------------
   -5-11 10 63
   -5 -9  9 67
   -5 -7  8 72
   -5 -5  7 79
   -5 -3  6 89
    0 -2  1 39
    5-11  1  7
    5 -9  0  0
```

```
 H  K  L  Φ  φ2
    U  V  W  φ1
---------------
 5 10  9 51 27
---------------
   -1 -4  5 82
    0 -9 10 73
    1 -5  5 64
    2 -1  0  0
    3 -6  5 50
    4-11 10 56
    5 -7  5 40
    7 -8  5 33
    9 -9  5 28
   11-10  5 24
10  5  9 51 63
---------------
   -7 -4 10 88
   -5 -8 10 69
   -3-12 10 54
   -3 -3  5 78
   -2 -5  5 61
   -1 -7  5 48
    0 -9  5 39
    1-11  5 32
    1 -2  0  0
 9 10  5 70 42
---------------
    0 -1  2 73
    5-10 11 48
    5 -9  9 45
    5 -8  7 39
    5 -7  5 32
    5 -6  3 22
    5 -5  1  9
   10-11  4 16
   10 -9  0  0
10  9  5 70 48
---------------
   -1 -5 11 75
    0 -5  9 69
    1 -5  7 60
    2 -5  5 47
    3-10 12 54
    3 -5  3 29
    4 -5  1  9
    5-10  8 38
    7-10  4 19
    9-10  0  0
```

```
 H  K  L  Φ  φ2
    U  V  W  φ1
---------------
 5  9 11 43 29
===============
   -2 -5  5 85
   -1 -8  7 74
    0-11  9 68
    1 -3  2 51
    5 -4  1 13
    6 -7  3 27
    7-10  5 34
    9 -5  0  0
   11-11  4 21
 9  5 11 43 61
---------------
   -9 -8 11 81
   -7 -5  8 85
   -3-10  7 55
   -2 -3  3 69
   -1 -7  4 46
    0-11  5 37
    1 -4  1 20
    5 -9  0  0
 5 11  9 53 24
---------------
    0 -9 11 75
    1 -7  8 69
    2 -5  5 58
    3 -3  2 32
    5 -8  7 48
    7 -4  1  9
    8-11  9 43
   10 -7  3 17
   11 -5  0  0
11  5  9 53 66
---------------
   -5 -7 10 71
   -4-11 11 59
   -3 -6  7 64
   -2 -1  3 89
   -1 -5  4 50
    0 -9  5 37
    1 -4  1 17
    5-11  0  0
 9 11  5 71 39
---------------
    0 -5 11 75
    1 -4  7 66
    2 -3  3 43
    3 -7 10 57
```

```
 H  K  L  Φ  φ2
    U  V  W  φ1
---------------
    5 -5  2 17
    7 -8  5 27
    8 -7  1  6
    9-11  8 31
   11 -9  0  0
11  9  5 71 51
---------------
   -1 -1  4 88
    0 -5  9 68
    1 -4  5 55
    2 -3  1 16
    3 -7  6 41
    4-11 11 47
    5-10  7 34
    9-11  0  0

 5  9 12 41 29
===============
   -3 -9  8 82
    0 -4  3 67
    3-11  7 53
    3 -7  4 46
    3 -3  1 21
    6-10  5 37
    9 -5  0  0
   12 -8  1  6
 9  5 12 41 61
---------------
  -11 -9 12 82
   -8-12 11 69
   -7 -9  9 72
   -6 -6  7 78
   -5 -3  5 89
   -1 -3  2 55
    0-12  5 36
    1 -9  3 29
    2 -6  1 14
    5 -9  0  0
 5 12  9 55 23
---------------
    0 -3  4 77
    3 -8  9 62
    3 -5  5 52
    3 -2  1 19
    6 -7  6 42
    9-12 11 46
    9 -9  7 36
   12-11  8 32
   12 -5  0  0
```

```
 H  K  L  Φ  φ2
    U  V  W  φ1
---------------
12  5  9 55 67
---------------
   -7 -3 11 90
   -5 -6 10 73
   -4 -3  7 82
   -1-12  8 42
   -1 -3  3 57
    0 -9  5 36
    1 -6  2 22
    3 -9  1  7
    5-12  0  0
 9 12  5 72 37
---------------
   -1 -3  9 84
    0 -5 12 77
    1 -2  3 58
    4 -3  0  0
    5 -5  3 24
    6 -7  6 35
    7 -9  9 41
    8-11 12 44
    9 -8  3 15
   11-12  9 31
12  9  5 72 53
---------------
   -1 -2  6 81
    0 -5  9 67
    1 -8 12 61
    1 -3  3 47
    3 -4  0  0
    4 -7  3 22
    5-10  6 30
    7-11  3 14

 5 10 11 45 27
===============
   -4 -9 10 88
   -1 -5  5 79
    0-11 10 71
    1 -6  5 63
    2 -1  0  0
    3 -7  5 50
    5 -8  5 41
    7 -9  5 34
    9-10  5 29
   11-11  5 25
10  5 11 45 63
---------------
   -7 -8 10 74
```

```
 H  K  L  Φ  φ2
    U  V  W  φ1
---------------
   -5-12 10 59
   -4 -3  5 83
   -3 -5  5 66
   -2 -7  5 53
   -1 -9  5 43
    0-11  5 35
    1 -2  0  0
 5 11 10 50 24
---------------
   -1 -5  6 82
    0-10 11 74
    1 -5  5 65
    3 -5  4 47
    4-10  9 56
    5 -5  3 30
    7 -5  2 17
    8-10  7 39
    9 -5  1  7
   11 -5  0  0
   12-10  5 23
11  5 10 50 66
---------------
   -5-11 11 61
   -5 -9 10 65
   -5 -7  9 70
   -5 -5  8 77
   -5 -3  7 86
    0 -2  1 35
    5-11  0  0
10 11  5 71 42
---------------
    0 -5 11 74
    1 -5  9 67
    2 -5  7 57
    3 -5  5 43
    4 -5  3 27
    5-10 12 51
    5 -5  1  8
    7-10  8 35
    9-10  4 17
   11-10  0  0
11 10  5 71 48
---------------
    0 -1  2 71
    5-11 11 45
    5-10  9 41
    5 -9  7 36
    5 -8  5 30
    5 -7  3 20
```

```
 H  K  L  Φ  φ2
    U  V  W  φ1
---------------
    5 -6  1  8
   10-11  0  0

 5 10 12 43 27
===============
   -2-11 10 78
   -2 -5  5 87
    0 -6  5 70
    2 -7  5 56
    2 -1  0  0
    4 -8  5 46
    6 -9  5 38
    8-10  5 32
   10-11  5 28
   12-12  5 24
10  5 12 43 63
---------------
   -9 -6 10 85
   -7-10 10 68
   -4 -4  5 76
   -3 -6  5 61
   -2 -8  5 50
   -1-10  5 41
    0-12  5 34
    1 -2  0  0
 5 12 10 52 23
---------------
   -2 -5  7 90
    0 -5  6 76
    2-10 11 68
    2 -5  5 59
    4 -5  4 42
    6-10  9 50
    6 -5  3 27
    8 -5  2 15
   10-10  7 34
   10 -5  1  6
   12 -5  0  0
12  5 10 52 67
---------------
   -5-12 12 59
   -5-10 11 62
   -5 -8 10 67
   -5 -6  9 72
   -5 -4  8 80
    0 -2  1 34
    5-12  0  0
```

```
 H  K  L  Φ  φ2
    U  V  W  φ1
---------------
10 12  5 72 40
---------------
    0 -5 12 76
    1 -5 10 69
    2 -5  8 61
    3 -5  6 49
    4 -5  4 34
    5 -5  2 17
    6 -5  0  0
    7-10 10 42
    9-10  6 25
   11-10  2  8
12 10  5 72 50
---------------
    0 -1  2 70
    5-12 12 45
    5-11 10 42
    5-10  8 38
    5 -9  6 32
    5 -8  4 24
    5 -7  2 14
    5 -6  0  0

 5 11 11 48 24
===============
    0 -1  1 73
   11-12  7 32
   11-11  6 29
   11-10  5 26
   11 -9  4 21
   11 -8  3 17
   11 -7  2 12
   11 -6  1  6
   11 -5  0  0
11  5 11 48 66
---------------
   -7-11 12 66
   -6-11 11 63
   -5-11 10 60
   -4-11  9 56
   -3-11  8 51
   -2-11  7 46
   -1-11  6 40
    0-11  5 34
    1-11  4 27
    2-11  3 21
    3-11  2 14
    4-11  1  7
    5-11  0  0
```

TAB. V FROM HKL-UVW INTO EULER ANGLES

H	K / U	L / V	Φ / W	φ2 / φ1
5	11	11		
11	11	5	72	45
	-2	-3	11	86
	-1	-4	11	80
	0	-5	11	73
	1	-6	11	67
	1	-1	0	0
	2	-7	11	61
	3	-8	11	56
	4	-9	11	51
	5	-10	11	47
	6	-11	11	44
	7	-12	11	41
5	11	12	45	24
	-1	-5	5	81
	0	-12	11	72
	1	-7	6	66
	2	-2	1	28
	3	-9	7	57
	5	-11	8	51
	11	-5	0	0
11	5	12	45	66
	-8	-4	9	89
	-7	-11	11	65
	-4	-8	7	60
	-3	-3	4	75
	-1	-5	3	46
	0	-12	5	33
	1	-7	2	23
	3	-9	1	8
	5	-11	0	0
5	12	11	50	23
	-1	-6	7	81
	0	-11	12	75
	1	-5	5	67
	3	-4	3	42
	4	-9	8	56
	5	-3	1	13
	8	-7	4	27
	11	-11	7	33
	12	-5	0	0

H	K / U	L / V	Φ / W	φ2 / φ1
12	5	11	50	67
	-8	-5	11	84
	-7	-3	9	90
	-1	-2	2	61
	0	-11	5	33
	1	-9	3	24
	2	-7	1	10
	5	-12	0	0
11	12	5	73	43
	-1	-2	7	85
	0	-5	12	75
	1	-3	5	62
	3	-4	3	33
	4	-7	8	47
	5	-5	1	8
	7	-11	11	42
	8	-9	4	19
	12	-11	0	0
12	11	5	73	47
	-1	-3	9	81
	0	-5	11	72
	1	-2	2	44
	6	-7	1	6
	7	-9	3	15
	8	-11	5	21
	11	-12	0	0
5	12	12	47	23
	0	-1	1	74
	12	-12	7	31
	12	-11	6	28
	12	-10	5	25
	12	-9	4	21
	12	-8	3	16
	12	-7	2	11
	12	-6	1	6
	12	-5	0	0
12	5	12	47	67
	-7	-12	12	63
	-6	-12	11	60
	-5	-12	10	56
	-4	-12	9	52
	-3	-12	8	48
	-2	-12	7	43

H	K / U	L / V	Φ / W	φ2 / φ1
	-1	-12	6	37
	0	-12	5	32
	1	-12	4	25
	2	-12	3	19
	3	-12	2	13
	4	-12	1	6
	5	-12	0	0
12	12	5	74	45
	-2	-3	12	87
	-1	-4	12	80
	0	-5	12	74
	1	-6	12	68
	1	-1	0	0
	2	-7	12	63
	3	-8	12	58
	4	-9	12	54
	5	-10	12	50
	6	-11	12	46
	7	-12	12	43
6	6	7	50	45
	-7	-7	12	90
	-5	-9	12	80
	-3	-11	12	70
	-3	-4	6	85
	-2	-5	6	75
	-1	-6	6	66
	0	-7	6	58
	1	-8	6	51
	1	-1	0	0
	2	-9	6	45
	3	-10	6	40
	4	-11	6	36
	5	-12	6	33
6	7	6	57	41
	-5	-6	12	90
	-4	-6	11	86
	-3	-6	10	82
	-2	-6	9	77
	-1	-6	8	72
	0	-6	7	65
	1	-6	6	57
	2	-6	5	48
	3	-12	11	52
	3	-6	4	38
	4	-6	3	27

H	K / U	L / V	Φ / W	φ2 / φ1
	5	-12	9	43
	5	-6	2	17
	6	-6	1	8
	7	-12	7	32
	7	-6	0	0
	9	-12	5	22
	11	-12	3	12
7	6	6	57	49
	0	-1	1	58
	6	-12	5	25
	6	-11	4	21
	6	-10	3	17
	6	-9	2	13
	6	-8	1	7
	6	-7	0	0
6	6	11	38	45
	-11	-11	12	90
	-5	-6	6	86
	-4	-7	6	78
	-3	-8	6	70
	-2	-9	6	63
	-1	-10	6	57
	0	-11	6	52
	1	-12	6	47
	1	-1	0	0
6	11	6	64	29
	-1	-6	12	81
	0	-6	11	77
	1	-6	10	71
	2	-6	9	65
	3	-6	8	58
	4	-6	7	51
	5	-6	6	42
	6	-6	5	34
	7	-6	4	26
	8	-6	3	19
	9	-6	2	12
	10	-6	1	5
	11	-12	11	38
	11	-6	0	0
11	6	6	64	61
	0	-1	1	52
	6	-12	1	5
	6	-11	0	0

H	K / U	L / V	Φ / W	φ2 / φ1
6	7	7	53	41
	0	-1	1	63
	7	-12	6	30
	7	-11	5	27
	7	-10	4	23
	7	-9	3	19
	7	-8	2	13
	7	-7	1	7
	7	-6	0	0
7	6	7	53	49
	-6	-7	12	85
	-5	-7	11	82
	-4	-7	10	78
	-3	-7	9	73
	-2	-7	8	68
	-1	-7	7	62
	0	-7	6	55
	1	-7	5	46
	2	-7	4	37
	3	-7	3	27
	4	-7	2	18
	5	-7	1	8
	6	-7	0	0
7	7	6	59	45
	-3	-3	7	90
	-2	-4	7	80
	-1	-5	7	71
	0	-6	7	63
	1	-7	7	55
	1	-1	0	0
	2	-8	7	49
	3	-9	7	44
	4	-10	7	40
	5	-11	7	36
	6	-12	7	33
6	7	8	49	41
	-5	-6	9	89
	-3	-10	11	74
	-2	-4	5	81
	-1	-6	6	68
	0	-8	7	61
	1	-10	8	56
	1	-2	1	33

H	K / U	L / V	Φ / W	φ2 / φ1
	2	-12	9	52
	7	-6	0	0
	8	-8	1	7
	9	-10	2	11
	10	-12	3	14
7	6	8	49	49
	-4	-10	11	71
	-2	-11	10	62
	-2	-7	7	67
	-2	-3	4	80
	0	-4	3	53
	2	-9	5	39
	2	-5	2	27
	4	-6	1	10
	6	-11	3	18
	6	-7	0	0
6	8	7	55	37
	-2	-9	12	75
	-1	-8	10	72
	0	-7	8	67
	1	-6	6	59
	2	-5	4	47
	3	-11	10	54
	3	-4	2	27
	4	-3	0	0
	5	-9	6	38
	7	-7	2	14
	10	-11	4	18
	11	-10	2	9
8	6	7	55	53
	-3	-10	12	67
	-2	-9	10	64
	-1	-8	8	59
	-1	-1	2	85
	0	-7	6	53
	1	-6	4	42
	2	-5	2	25
	3	-11	6	35
	3	-4	0	0
	5	-9	2	13
7	8	6	61	41
	-2	-5	9	80
	0	-3	4	67
	2	-10	11	57
	2	-7	7	53

H	K / U	L / V	Φ / W	φ2 / φ1
	2	-4	3	40
	4	-11	10	48
	4	-5	2	20
	6	-9	5	29
	6	-6	1	8
	8	-7	0	0
	10	-11	3	13
8	7	6	61	49
	-3	-6	11	79
	-2	-2	5	88
	-1	-4	6	71
	0	-6	7	61
	1	-8	8	54
	1	-2	1	28
	2	-10	9	49
	3	-12	10	46
	7	-8	0	0
	8	-10	1	5
	9	-12	2	9
6	7	9	46	41
	-6	-9	11	85
	-5	-6	8	89
	-1	-12	10	63
	-1	-3	3	74
	0	-9	7	59
	1	-6	4	50
	2	-3	1	22
	3	-9	5	41
	5	-12	6	36
	7	-6	0	0
	9	-9	1	6
	11	-12	2	10
7	6	9	46	49
	-6	-11	12	75
	-3	-10	9	66
	-3	-7	7	71
	-3	-4	5	81
	0	-3	2	51
	3	-11	5	34
	3	-8	3	28
	3	-5	1	14
	6	-7	0	0
	9	-12	1	5

H	K / U	L / V	Φ / W	φ2 / φ1
6	9	7	57	34
	-3	-5	9	89
	-1	-4	6	79
	0	-7	9	70
	1	-10	12	66
	1	-3	3	55
	3	-2	0	0
	4	-5	3	30
	5	-8	6	40
	6	-11	9	44
	7	-7	3	20
	10	-9	3	15
	11	-12	6	24
9	6	7	57	36
	-6	-5	12	87
	-2	-11	12	61
	-1	-9	9	57
	-1	-2	3	73
	0	-7	6	51
	1	-12	9	45
	1	-5	3	37
	2	-3	0	0
	3	-8	3	23
	5	-11	3	17
7	9	6	62	38
	-3	-5	11	87
	0	-2	3	70
	3	-9	10	55
	3	-7	7	50
	3	-5	4	40
	3	-3	1	19
	6	-12	11	46
	6	-8	5	30
	9	-11	6	26
	9	-7	0	0
	12	-10	1	4
9	7	6	62	52
	-3	-3	8	87
	-1	-9	12	64
	-1	-3	5	73
	0	-6	7	59
	1	-9	9	53
	1	-3	2	37
	2	-12	11	40
	4	-6	1	9

H	K / U	L / V	Φ / W	φ2 / φ1
6	7	9		
9	7	6	62	52
	5	-9	3	18
	6	-12	5	23
	7	-9	0	0
6	7	10	43	41
	-1	-12	9	62
	-1	-2	2	80
	0	-10	7	58
	1	-8	5	51
	2	-6	3	39
	3	-4	1	17
	5	-10	4	30
	7	-6	0	0
	10	-10	1	6
7	6	10	43	49
	-8	-9	11	84
	-6	-8	9	81
	-4	-7	7	75
	-2	-11	8	59
	-2	-6	5	66
	0	-5	3	49
	2	-9	4	36
	2	-4	1	19
	6	-7	0	0
	8	-11	1	6
6	10	7	59	31
	-1	-5	8	80
	0	-7	10	73
	1	-9	12	69
	1	-2	2	51
	5	-3	0	0
	6	-5	2	17
	7	-7	4	26
	8	-9	6	31
	9	-11	8	35
	11	-8	2	10
0	6	7	59	59
	-4	-5	10	79
	-3	-2	6	89
	-1	-10	10	55
	-1	-3	4	66

H	K / U	L / V	Φ / W	φ2 / φ1
	0	-7	6	49
	1	-11	8	43
	1	-4	2	31
	3	-5	0	0
	4	-9	2	13
7	10	6	64	35
	-2	-4	9	86
	0	-3	5	73
	2	-8	11	63
	2	-5	6	56
	2	-2	1	22
	4	-7	7	47
	6	-9	8	41
	8	-11	9	38
	10	-7	0	0
	12	-9	1	4
10	7	6	64	55
	-1	-8	11	64
	-1	-2	4	77
	0	-6	7	58
	1	-10	10	52
	1	-4	3	41
	3	-6	2	19
	4	-10	5	28
	5	-8	1	7
	7	-10	0	0
6	7	11	40	41
	-3	-10	8	71
	-2	-3	3	85
	-1	-7	5	64
	0	-11	7	57
	1	-4	2	43
	4	-5	1	14
	5	-9	3	26
	7	-6	0	0
	11	-11	1	6
7	6	11	40	49
	-8	-9	10	84
	-7	-12	11	75
	-5	-7	7	79
	-2	-5	4	68
	-1	-8	5	55
	0	-11	6	48
	1	-3	1	28

H	K / U	L / V	Φ / W	φ2 / φ1
	6	-7	0	0
	7	-10	1	7
6	11	7	61	29
	-1	-2	4	89
	0	-7	11	75
	1	-5	7	68
	2	-3	3	47
	3	-8	10	61
	5	-4	2	20
	7	-7	5	31
	8	-5	1	7
	9	-10	8	36
	11	-6	0	0
	12	-11	7	27
11	6	7	61	61
	-3	-5	9	74
	-1	-11	11	54
	-1	-4	5	62
	0	-7	6	48
	1	-10	7	41
	1	-3	1	20
	6	-11	0	0
7	11	6	65	32
	0	-6	11	75
	1	-5	8	68
	2	-4	5	55
	3	-3	2	28
	5	-7	7	44
	7	-11	12	48
	7	-5	1	7
	8	-10	9	39
	10	-8	3	15
	11	-7	0	0
11	7	6	65	58
	-1	-7	10	64
	-1	-1	3	85
	0	-6	7	57
	1	-11	11	51
	1	-5	4	43
	2	-4	1	14
	3	-9	5	31
	7	-11	0	0

H	K / U	L / V	Φ / W	φ2 / φ1
6	7	12	38	41
	-8	-12	11	85
	-5	-6	6	89
	-4	-12	9	72
	-3	-6	5	79
	-1	-6	4	64
	0	-12	7	56
	1	-6	3	47
	3	-6	2	28
	4	-12	5	37
	5	-6	1	12
	7	-6	0	0
	8	-12	3	20
	12	-12	1	6
7	6	12	38	49
	-6	-11	9	73
	-6	-9	8	77
	-6	-7	7	83
	0	-2	1	47
	6	-11	2	15
	6	-9	1	9
	6	-7	0	0
6	12	7	62	27
	-1	-3	6	86
	0	-7	12	77
	1	-4	6	68
	2	-1	0	0
	3	-5	6	54
	4	-9	12	61
	5	-6	6	43
	7	-7	6	36
	8	-11	12	48
	9	-8	6	30
	11	-9	6	26
12	6	7	62	63
	-5	-4	12	84
	-3	-8	12	67
	-2	-3	6	75
	-1	-12	12	53
	-1	-5	6	59
	0	-7	6	47
	1	-9	6	39
	1	-2	0	0
	2	-11	6	32

H	K / U	L / V	Φ / W	φ2 / φ1
7	12	6	67	30
	0	-1	2	77
	6	-9	11	51
	6	-8	9	47
	6	-7	7	41
	6	-6	5	34
	6	-5	3	23
	6	-4	1	9
	12	-11	8	29
	12	-9	4	16
	12	-7	0	0
12	7	6	67	60
	-2	-6	11	71
	-1	-6	9	64
	0	-6	7	56
	1	-12	12	50
	1	-6	5	44
	2	-6	3	28
	3	-12	8	36
	3	-6	1	9
	5	-12	4	19
	7	-12	0	0
6	8	9	48	37
	-3	-9	10	77
	-2	-3	4	88
	-1	-6	6	71
	0	-9	8	63
	1	-12	10	59
	1	-3	2	46
	4	-3	0	0
	5	-6	2	19
	6	-9	4	28
	7	-12	6	32
	9	-9	2	12
8	6	9	48	53
	-6	-7	10	82
	-3	-11	10	63
	-3	-8	8	67
	-3	-5	6	75
	0	-3	2	48
	3	-10	4	29
	3	-7	2	20
	3	-4	0	0
	6	-11	2	12

H	K / U	L / V	Φ / W	φ2 / φ1
6	9	8	54	34
	-1	-10	12	72
	-1	-2	3	86
	0	-8	9	68
	1	-6	6	61
	2	-4	3	44
	3	-10	9	54
	3	-2	0	0
	5	-6	3	26
	7	-10	6	33
	8	-8	3	19
	11	-10	3	14
9	6	8	54	56
	-6	-7	12	81
	-4	-6	9	76
	-2	-9	9	60
	-2	-5	6	68
	0	-4	3	48
	2	-11	6	36
	2	-7	3	28
	2	-3	0	0
	4	-10	3	19
8	9	6	64	42
	-3	-4	10	88
	0	-2	3	68
	3	-10	11	54
	3	-8	8	50
	3	-6	5	42
	3	-4	2	25
	6	-10	7	35
	6	-6	1	8
	9	-10	3	14
	9	-8	0	0
9	8	6	64	48
	-2	-6	11	76
	-2	-3	7	83
	0	-3	4	63
	2	-9	9	51
	2	-6	5	44
	2	-3	1	17
	4	-9	6	36
	6	-12	7	31
	8	-9	0	0
	10	-12	1	4

H	K / U	L / V	Φ / W	φ2 / φ1
6	8	11	42	37
	-1	-2	2	82
	0	-11	8	61
	1	-9	6	55
	2	-7	4	46
	3	-5	2	29
	4	-3	0	0
	5	-12	6	39
	7	-8	2	16
	11	-11	2	11
8	6	11	42	53
	-9	-10	12	82
	-7	-9	10	79
	-5	-8	8	74
	-3	-7	6	67
	-1	-6	4	55
	0	-11	6	45
	1	-5	2	33
	3	-4	0	0
	4	-9	2	17
6	11	8	57	29
	-2	-4	7	89
	-1	-6	9	79
	0	-8	11	74
	1	-2	2	52
	6	-4	1	9
	7	-6	3	22
	8	-8	5	29
	9	-10	7	33
	10	-12	9	36
	11	-6	0	0
11	6	8	57	61
	-6	-5	12	84
	-2	-11	11	56
	-2	-7	8	61
	-2	-3	5	74
	0	-4	3	45
	2	-9	4	28
	2	-5	1	13
	6	-11	0	0
8	11	6	66	36
	-1	-2	5	86
	0	-6	11	74

H	K / U	L / V	Φ / W	φ2 / φ1
	1	-4	6	64
	2	-2	1	21
	3	-6	7	52
	5	-8	8	48
	7	-10	9	40
	9	-12	10	37
	11	-8	0	0
11	8	6	66	56
	-2	-4	9	78
	0	-3	4	61
	2	-11	11	50
	2	-8	7	45
	2	-5	3	32
	4	-7	2	15
	6	-12	5	22
	6	-9	1	6
	6	-11	0	0
6	9	10	47	34
	-5	-10	12	85
	-3	-8	9	81
	-1	-6	6	73
	0	-10	9	66
	1	-4	3	53
	3	-2	0	0
	4	-6	3	32
	5	-10	6	40
	7	-8	3	22
	10	-10	3	16
9	6	10	47	56
	-6	-11	12	70
	-4	-9	9	67
	-2	-12	9	54
	-2	-7	6	60
	-2	-2	3	82
	0	-5	3	44
	2	-8	3	28
	2	-3	0	0
	4	-11	3	20
6	10	9	52	31
	-1	-3	4	82
	0	-9	10	70
	1	-6	6	63
	2	-3	2	38
	3	-9	8	55

TAB. V — FROM HKL-UVW INTO EULER ANGLES — V - 63

H	K / U	L / V	Φ / W	φ2 / φ1
6	9	10		
6	10	9	52	31
	5	-12	10	50
	5	-3	0	0
	7	-6	2	16
	9	-9	4	22
	11	-12	6	26
	12	-9	2	10
10	6	9	52	59
	-6	-5	10	85
	-3	-10	10	61
	-3	-7	8	66
	-3	-4	6	76
	0	-3	2	44
	3	-11	4	25
	3	-8	2	17
	3	-5	0	0
9	10	6	66	42
	-2	-3	8	87
	0	-3	5	70
	2	-9	12	60
	2	-6	7	54
	2	-3	2	32
	4	-9	9	48
	6	-12	11	44
	6	-6	1	7
	8	-9	3	15
	10	-12	5	19
	10	-9	0	0
10	9	6	66	48
	-3	-4	11	85
	0	-2	3	66
	3	-10	10	49
	3	-8	7	44
	3	-6	4	34
	3	-4	1	12
	6	-10	5	26
	9	-10	0	0
6	9	11	45	34
	-1	-3	3	79
	0	-11	9	65
	1	-8	6	58

H	K / U	L / V	Φ / W	φ2 / φ1
	2	-5	3	44
	3	-2	0	0
	5	-7	3	28
	7	-12	6	34
	8	-9	3	20
	11	-11	3	16
9	6	11	45	56
	-10	-7	12	89
	-7	-6	9	85
	-5	-9	9	70
	-4	-5	6	77
	-1	-4	3	57
	0	-11	6	43
	1	-7	3	34
	2	-3	0	0
	3	-10	3	23
6	11	9	54	29
	-1	-6	8	79
	0	-9	11	72
	1	-3	3	58
	4	-3	1	14
	5	-6	4	34
	6	-9	7	42
	7	-12	10	46
	9	-9	5	27
	11	-6	0	0
11	6	9	54	61
	-6	-7	12	78
	-3	-11	11	59
	-3	-8	9	63
	-3	-5	7	71
	-3	-2	5	87
	0	-3	2	43
	3	-10	3	20
	3	-7	1	9
	6	-11	0	0
9	11	6	67	39
	-1	-3	7	82
	0	-6	11	72
	1	-3	4	58
	3	-3	1	14
	4	-6	5	38
	5	-9	9	46
	7	-9	6	30
	10	-12	7	26

H	K / U	L / V	Φ / W	φ2 / φ1
	11	-9	0	0
11	9	6	67	51
	-3	-3	10	88
	0	-2	3	65
	3	-11	11	49
	3	-9	8	44
	3	-7	5	37
	3	-5	2	21
	6	-12	7	30
	6	-8	1	6
	9	-11	0	0
6	10	11	47	31
	-3	-7	8	85
	-1	-6	6	75
	0	-11	10	68
	1	-5	4	58
	3	-4	2	31
	4	-9	6	46
	5	-3	0	0
	8	-7	2	15
	11	-11	4	20
10	6	11	47	59
	-9	-7	12	85
	-8	-5	10	89
	-1	-2	2	66
	0	-11	6	41
	1	-9	4	34
	2	-7	2	21
	3	-5	0	0
	5	-12	2	12
6	11	10	51	29
	-1	-4	5	81
	0	-10	11	71
	1	-6	6	64
	2	-2	1	25
	3	-8	7	54
	5	-10	8	48
	7	-12	9	44
	11	-6	0	0
11	6	10	51	61
	-6	-4	9	87
	-4	-11	11	61
	-2	-8	7	56

H	K / U	L / V	Φ / W	φ2 / φ1
	-2	-3	4	72
	0	-5	3	41
	2	-12	5	29
	2	-7	2	20
	4	-9	1	7
	6	-11	0	0
10	11	6	68	42
	-1	-4	9	79
	0	-6	11	71
	1	-2	2	46
	6	-6	1	7
	7	-8	3	17
	8	-10	5	23
	9	-12	7	27
	11	-10	0	0
11	10	6	68	48
	-2	-5	12	80
	-2	-2	7	89
	0	-3	5	68
	2	-7	8	53
	2	-4	3	37
	4	-11	11	48
	4	-5	1	10
	6	-9	4	22
	10	-11	0	0
6	11	11	49	29
	0	-1	1	70
	11	-12	6	27
	11	-11	5	24
	11	-10	4	20
	11	-9	3	16
	11	-8	2	11
	11	-7	1	6
	11	-6	0	0
11	6	11	49	61
	-6	-11	12	67
	-5	-11	11	64
	-4	-11	10	60
	-3	-11	9	56
	-2	-11	8	51
	-1	-11	7	45
	0	-11	6	40
	1	-11	5	33
	2	-11	4	27

H	K / U	L / V	Φ / W	φ2 / φ1
	3	-11	3	20
	4	-11	2	13
	5	-11	1	6
	6	-11	0	0
11	11	6	69	45
	-3	-3	11	90
	-2	-4	11	83
	-1	-5	11	77
	0	-6	11	70
	1	-7	11	64
	1	-1	0	0
	2	-8	11	59
	3	-9	11	54
	4	-10	11	50
	5	-11	11	46
	6	-12	11	43
6	11	12	46	29
	-3	-6	7	89
	-1	-6	6	76
	0	-12	11	69
	1	-6	5	62
	3	-6	4	45
	4	-12	9	53
	5	-6	3	30
	7	-6	2	17
	8	-12	7	37
	9	-6	1	7
	11	-6	0	0
	12	-12	5	23
11	6	12	46	61
	-6	-11	11	66
	-6	-9	10	70
	-6	-7	9	75
	-6	-5	8	82
	0	-2	1	38
	6	-11	0	0
6	12	11	51	27
	-4	-9	12	88
	-1	-5	6	80
	0	-11	12	72
	1	-6	6	65
	2	-1	0	0
	3	-7	6	53
	5	-8	6	44

H	K / U	L / V	Φ / W	φ2 / φ1
	7	-9	6	37
	9	-10	6	32
	11	-11	6	28
12	6	11	51	63
	-7	-8	12	75
	-5	-12	12	61
	-4	-3	6	83
	-3	-5	6	68
	-2	-7	6	55
	-1	-9	6	46
	0	-11	6	38
	1	-2	0	0
11	12	6	70	43
	0	-1	2	72
	6	-11	11	45
	6	-10	9	41
	6	-9	7	35
	6	-8	5	28
	6	-7	3	19
	6	-6	1	7
	12	-11	0	0
12	11	6	70	47
	0	-6	11	69
	1	-6	9	62
	2	-6	7	52
	3	-6	5	40
	4	-6	3	24
	5	-12	12	46
	5	-6	1	8
	7	-12	8	32
	9	-12	4	16
	11	-12	0	0
7	7	8	51	45
	-4	-4	7	90
	-3	-5	7	81
	-2	-6	7	73
	-1	-7	7	65
	0	-8	7	58
	1	-9	7	52
	1	-1	0	0
	2	-10	7	47
	3	-11	7	42
	4	-12	7	39

H	K / U	L / V	Φ / W	φ2 / φ1
7	8	7	57	41
	-4	-7	12	84
	-3	-7	11	80
	-2	-7	10	75
	-1	-7	9	70
	0	-7	8	64
	1	-7	7	57
	2	-7	6	50
	3	-7	5	41
	4	-7	4	32
	5	-7	3	23
	6	-7	2	15
	7	-7	1	7
	8	-7	0	0
8	7	7	57	49
	0	-1	1	58
	7	-12	4	19
	7	-11	3	16
	7	-10	2	11
	7	-9	1	6
	7	-6	0	0
7	7	9	48	45
	-4	-5	7	86
	-3	-6	7	77
	-2	-7	7	70
	-1	-8	7	62
	0	-9	7	56
	1	-10	7	51
	1	-1	0	0
	2	-11	7	46
	3	-12	7	42
7	9	7	58	38
	-3	-7	12	82
	-2	-7	11	78
	-1	-7	10	73
	0	-7	9	68
	1	-7	8	62
	2	-7	7	54
	3	-7	6	47
	4	-7	5	38
	5	-7	4	30
	6	-7	3	21
	7	-7	2	13
	8	-7	1	6

H	K / U	L / V	Φ / W	φ2 / φ1
	9	-7	0	0
9	7	7	58	52
	0	-1	1	56
	7	-12	3	14
	7	-11	2	10
	7	-10	1	5
	7	-9	0	0
7	7	10	45	45
	-5	-5	7	90
	-4	-6	7	82
	-3	-7	7	74
	-2	-8	7	67
	-1	-9	7	60
	0	-10	7	55
	1	-11	7	50
	1	-1	0	0
	2	-12	7	45
7	10	7	60	33
	-2	-7	12	80
	-1	-7	11	76
	0	-7	10	71
	1	-7	9	63
	2	-7	8	58
	3	-7	7	51
	4	-7	6	43
	5	-7	5	35
	6	-7	4	27
	7	-7	3	20
	8	-7	2	12
	9	-7	1	6
	10	-7	0	0
10	7	7	60	53
	0	-1	1	53
	7	-12	2	9
	7	-11	1	5
	7	-10	0	0
7	7	11	42	45
	-5	-6	7	86
	-4	-7	7	79
	-3	-8	7	71
	-2	-9	7	65
	-1	-10	7	59

TAB. V

FROM HKL-UVW INTO EULER ANGLES

H	K / U	L / V	Φ / W	φ_2 / φ_1
7	7	11	42	45
	0	-11	7	53
	1	-12	7	49
	1	-1	0	0
7	11	7	62	32
	-1	-7	12	78
	0	-7	11	73
	1	-7	10	68
	2	-7	9	62
	3	-7	8	55
	4	-7	7	48
	5	-7	6	40
	6	-7	5	33
	7	-7	4	25
	8	-7	3	18
	9	-7	2	11
	10	-7	1	5
	11	-7	0	0
11	7	7	62	58
	0	-1	1	53
	7	-12	1	5
	7	-11	0	0
7	7	12	40	45
	-6	-6	7	90
	-5	-7	7	83
	-4	-8	7	76
	-3	-9	7	69
	-2	-10	7	63
	-1	-11	7	57
	0	-12	7	52
	1	-1	0	0
7	12	7	63	30
	0	-7	12	75
	1	-7	11	70
	2	-7	10	65
	3	-7	9	59
	4	-7	8	52
	5	-7	7	45
	6	-7	6	38
	7	-7	5	30
	8	-7	4	23
	9	-7	3	17
	10	-7	2	10

H	K / U	L / V	Φ / W	φ_2 / φ_1
	11	-7	1	5
	12	-7	0	0
12	7	7	63	60
	0	-1	1	52
	7	-12	0	0
7	8	8	53	41
	0	-1	1	62
	8	-12	5	24
	8	-11	4	21
	8	-10	3	17
	8	-9	2	12
	8	-8	1	6
	8	-7	0	0
8	7	8	53	49
	-5	-8	12	80
	-4	-8	11	76
	-3	-8	10	72
	-2	-8	9	67
	-1	-8	8	62
	0	-8	7	56
	1	-8	6	48
	2	-8	5	40
	3	-8	4	32
	4	-8	3	23
	5	-8	2	15
	6	-8	1	7
	7	-8	0	0
8	8	7	58	45
	-3	-4	8	86
	-2	-5	8	77
	-1	-6	8	69
	0	-7	8	62
	1	-8	8	56
	1	-1	0	0
	2	-9	8	50
	3	-10	8	46
	4	-11	8	42
	5	-12	8	38
7	8	9	50	41
	-5	-8	11	84
	-2	-5	6	77
	-1	-7	7	67

H	K / U	L / V	Φ / W	φ_2 / φ_1
	0	-9	8	61
	1	-11	9	56
	1	-2	1	32
	8	-7	0	0
	9	-9	1	6
	10	-11	2	10
8	7	9	50	49
	-4	-7	9	77
	-3	-3	5	88
	-1	-4	4	66
	0	-9	7	54
	1	-5	3	42
	3	-6	2	22
	4	-11	5	31
	5	-7	1	9
	7	-8	0	0
7	9	8	55	38
	-1	-9	11	71
	0	-8	9	66
	1	-7	7	59
	2	-6	5	49
	3	-5	3	34
	4	-4	1	12
	5	-11	8	42
	7	-9	4	24
	9	-7	0	0
9	7	8	55	52
	-2	-10	11	64
	-1	-9	9	59
	-1	-1	2	86
	0	-8	7	54
	1	-7	5	45
	2	-6	3	32
	3	-5	1	12
	5	-11	4	23
	7	-9	0	0
8	9	7	60	42
	-4	-5	11	89
	-1	-3	5	78
	0	-7	9	66
	1	-4	4	54
	3	-5	3	32
	4	-9	7	42
	5	-6	2	17
	7	-7	1	7

H	K / U	L / V	Φ / W	φ_2 / φ_1
	8	-11	5	24
	9	-8	0	0
9	8	7	60	48
	-2	-3	6	83
	-1	-5	7	69
	0	-7	8	61
	1	-9	[illegible]	55
	1	-2	1	28
	2	-11	10	50
	8	-9	0	0
	9	-11	1	5
7	8	10	47	41
	-4	-9	10	78
	-2	-12	11	67
	-2	-7	7	72
	0	-5	4	59
	2	-8	5	45
	2	-3	1	22
	4	-11	6	39
	8	-7	0	0
	10	-10	1	6
8	7	10	47	49
	-5	-10	11	74
	-3	-8	8	70
	-2	-2	3	87
	-1	-6	5	61
	0	-10	7	52
	1	-4	2	37
	4	-6	1	11
	5	-10	3	21
	7	-8	0	0
7	10	8	57	35
	-4	-6	11	89
	-2	-5	8	83
	0	-4	5	69
	2	-11	12	61
	2	-7	7	56
	2	-3	2	35
	4	-10	9	50
	6	-5	1	9
	8	-8	3	16
	10	-11	5	22
	10	-7	0	0

H	K / U	L / V	Φ / W	φ_2 / φ_1
10	7	8	57	55
	-1	-10	10	57
	-1	-2	3	73
	0	-8	7	52
	1	-6	4	41
	2	-4	1	15
	3	-10	5	31
	7	-10	0	0
8	10	7	61	39
	-1	-2	4	84
	0	-7	10	69
	1	-5	6	60
	2	-3	2	34
	3	-8	8	51
	5	-11	10	47
	5	-4	0	0
	7	-7	2	13
	9	-10	4	19
	12	-11	2	8
10	8	7	61	51
	-3	-5	10	80
	-1	-4	6	70
	0	-7	8	59
	1	-10	10	53
	1	-3	2	38
	4	-5	0	0
	5	-8	2	14
	6	-11	4	20
7	8	11	44	41
	-7	-9	11	88
	-5	-8	9	83
	-3	-7	7	77
	-1	-6	5	66
	0	-11	8	58
	1	-5	3	47
	3	-4	1	16
	4	-9	4	33
	8	-7	0	0
	11	-11	1	5
8	7	11	44	49
	-1	-2	2	74
	0	-11	7	51
	1	-9	5	44

H	K / U	L / V	Φ / W	φ_2 / φ_1
	2	-7	3	33
	3	-5	1	14
	5	-12	4	25
	7	-8	0	0
7	11	8	58	32
	-1	-3	5	83
	0	-8	11	72
	1	-5	6	63
	2	-2	1	23
	3	-7	7	53
	5	-9	8	46
	7	-11	9	42
	11	-7	0	0
11	7	8	58	58
	-4	-4	9	83
	-1	-11	11	56
	-1	-3	4	67
	0	-8	7	51
	1	-5	3	37
	3	-7	2	17
	4	-12	5	26
	5	-9	1	7
	7	-11	0	0
8	11	7	63	36
	-2	-3	7	89
	-1	-5	9	78
	0	-7	11	72
	1	-2	2	49
	6	-5	1	8
	7	-7	3	19
	6	-9	5	26
	9	-11	7	30
	11	-8	0	0
11	8	7	63	54
	-4	-5	12	83
	-1	-3	5	72
	0	-7	8	58
	1	-11	11	53
	1	-4	3	41
	3	-5	1	11
	4	-9	4	25
	8	-11	0	0

H	K / U	L / V	Φ / W	φ_2 / φ_1
7	8	12	42	41
	-8	-11	12	86
	-4	-10	9	75
	-4	-7	7	81
	0	-3	2	57
	4	-11	5	36
	4	-8	3	29
	4	-5	1	13
	8	-7	0	0
	12	-12	1	5
8	7	12	42	49
	-6	-12	11	73
	-5	-8	8	77
	-4	-4	5	87
	-1	-4	3	63
	0	-12	7	49
	1	-8	4	42
	2	-4	1	19
	3	-12	5	34
	7	-8	0	0
	9	-12	1	6
7	12	8	60	30
	0	-2	3	74
	4	-9	10	55
	4	-7	7	49
	4	-5	4	38
	4	-3	1	13
	8	-12	11	44
	8	-8	5	28
	12	-11	6	24
	12	-7	0	0
12	7	8	60	60
	-5	-4	11	86
	-3	-4	8	78
	-1	-12	12	55
	-1	-4	5	63
	0	-6	7	49
	1	-12	9	44
	1	-4	2	30
	4	-8	1	7
	5	-12	3	15
	7	-12	0	0

H	K / U	L / V	Φ / W	φ_2 / φ_1
8	12	7	64	34
	-3	-5	12	89
	-1	-4	8	81
	0	-7	12	74
	1	-3	4	61
	3	-2	0	0
	4	-5	4	36
	5	-8	8	46
	6	-11	12	50
	7	-7	4	25
	10	-9	4	18
	11	-12	8	29
12	8	7	64	56
	-1	-9	12	63
	-1	-2	4	76
	0	-7	8	57
	1	-12	12	52
	1	-5	4	43
	2	-3	0	0
	3	-8	4	28
	5	-11	4	20
7	9	9	52	38
	0	-1	1	64
	9	-12	5	24
	9	-11	4	20
	9	-10	3	16
	9	-9	2	11
	9	-8	1	6
	9	-7	0	0
9	7	9	52	52
	-5	-9	12	75
	-4	-9	11	72
	-3	-9	10	68
	-2	-9	9	63
	-1	-9	8	58
	0	-9	7	51
	1	-9	6	45
	2	-9	5	37
	3	-9	4	30
	4	-9	3	22
	5	-9	2	14
	6	-9	1	7
	7	-9	0	0

TAB. V FROM HKL-UVW INTO EULER ANGLES V - 65

H / U	K / V	L / W	Φ	φ_2 / φ_1
7	9	9		
9	9	7	61	45
-3	-4	9		86
-2	-5	9		78
-1	-6	9		71
0	-7	9		64
1	-8	9		58
1	-1	0		0
2	-9	9		53
3	-10	9		48
4	-11	9		44
5	-12	9		41
7	9	10	49	38
-3	-11	12		75
-2	-4	5		82
-1	-7	7		69
0	-10	9		63
1	-3	2		45
5	-5	1		11
6	-8	3		22
7	-11	5		28
9	-7	0		0
9	7	10	49	52
-6	-8	11		80
-5	-5	8		85
-1	-3	3		66
0	-10	7		50
1	-7	4		41
2	-4	1		17
3	-11	5		32
7	-9	0		0
7	10	9	54	35
-1	-2	3		85
0	-9	10		67
1	-7	7		61
2	-5	4		48
3	-12	11		56
3	-3	1		17
5	-8	5		36
8	-11	6		30
10	-7	0		0

H / U	K / V	L / W	Φ	φ_2 / φ_1
10	7	9	54	55
-5	-7	11		78
-3	-6	8		72
-1	-5	5		60
0	-9	7		50
1	-4	2		33
4	-7	1		9
5	-11	3		17
7	-10	0		0
9	10	7	63	42
-3	-5	11		85
-1	-4	7		76
0	-7	10		67
1	-3	3		51
4	-5	2		20
5	-8	5		32
6	-11	8		37
7	-7	1		7
10	-9	0		0
11	-12	3		12
10	9	7	63	48
-1	-2	4		80
0	-7	9		63
1	-5	5		52
2	-3	1		18
3	-8	6		40
5	-11	7		34
9	-10	0		0
7	9	11	46	38
-3	-5	6		85
-2	-7	7		74
-1	-9	8		67
0	-11	9		62
1	-2	1		35
9	-7	0		0
10	-9	1		6
11	-11	2		10
9	7	11	46	52
-8	-7	11		88
-4	-9	9		70
-3	-4	5		79
-1	-5	4		59
0	-11	7		48

H / U	K / V	L / W	Φ	φ_2 / φ_1
1	-6	3		38
3	-7	2		21
5	-8	1		8
7	-9	0		0
7	11	9	55	32
0	-9	11		70
1	-8	9		65
2	-7	7		57
3	-6	5		47
4	-5	3		31
5	-4	1		11
7	-11	8		39
9	-9	4		21
11	-7	0		0
11	7	9	55	58
-2	-11	11		58
-1	-10	9		54
-1	-1	2		83
0	-9	7		48
1	-8	5		40
2	-7	3		28
3	-6	1		10
7	-11	0		0
9	11	7	64	39
-1	-3	6		80
0	-7	11		70
1	-4	5		59
3	-5	4		39
4	-9	9		49
5	-6	3		24
7	-7	2		13
8	-11	7		31
9	-8	1		5
11	-9	0		0
11	9	7	64	51
-2	-3	7		82
-1	-5	8		70
0	-7	9		62
1	-9	10		56
1	-2	1		27
2	-11	11		51
9	-11	0		0

H / U	K / V	L / W	Φ	φ_2 / φ_1
7	9	12	44	38
-6	-10	11		85
-3	-11	10		73
-3	-7	7		79
0	-4	3		61
3	-9	5		43
3	-5	2		28
6	-6	1		10
9	-11	3		17
9	-7	0		0
9	7	12	44	52
-5	-9	9		73
-3	-3	4		85
-2	-6	5		64
-1	-9	6		53
0	-12	7		47
1	-3	1		26
7	-9	0		0
8	-12	1		6
7	12	9	57	30
0	-3	4		72
3	-10	11		60
3	-7	7		54
3	-4	3		38
6	-11	10		48
6	-5	2		17
9	-9	5		26
9	-6	1		6
12	-7	0		0
12	7	9	57	60
-5	-3	9		90
-3	-9	11		64
-2	-3	5		75
-1	-6	6		57
0	-9	7		47
1	-12	8		41
1	-3	1		21
7	-12	0		0
9	12	7	65	37
0	-7	12		72
1	-6	9		66
2	-5	6		55
3	-4	3		35

H / U	K / V	L / W	Φ	φ_2 / φ_1
4	-3	0		0
5	-9	9		47
7	-7	3		19
10	-11	6		24
11	-10	3		13
12	9	7	65	53
-1	-8	12		66
-1	-1	3		87
0	-7	9		61
1	-6	6		51
2	-5	3		32
3	-11	9		43
3	-4	0		0
5	-9	3		18
7	10	10	51	35
0	-1	1		66
10	-12	5		23
10	-11	4		20
10	-10	3		16
10	-9	2		11
10	-8	1		6
10	-7	0		0
10	7	10	51	55
-5	-10	12		71
-4	-10	11		67
-3	-10	10		63
-2	-10	9		59
-1	-10	8		54
0	-10	7		48
1	-10	6		42
2	-10	5		35
3	-10	4		28
4	-10	3		20
5	-10	2		13
6	-10	1		6
7	-10	0		0
10	10	7	64	45
-3	-4	10		86
-2	-5	10		79
-1	-6	10		72
0	-7	10		66
1	-8	10		60
1	-1	0		0
2	-9	10		55

H / U	K / V	L / W	Φ	φ_2 / φ_1
3	-10	10		51
4	-11	10		46
5	-12	10		43
7	10	11	48	35
-3	-10	11		78
-2	-3	4		89
-1	-7	7		71
0	-11	10		65
1	-4	3		52
4	-5	2		24
5	-9	5		36
7	-6	1		8
10	-7	0		0
11	-11	3		15
10	7	11	48	55
-5	-7	9		77
-2	-5	5		66
-1	-8	6		53
0	-11	7		46
1	-3	1		24
7	-10	0		0
7	11	10	53	32
-1	-3	4		81
0	-10	11		69
1	-7	7		62
2	-4	3		45
3	-11	10		56
5	-5	2		20
7	-9	5		30
8	-6	1		7
11	-7	0		0
11	7	10	53	58
-5	-5	9		82
-3	-11	11		61
-2	-4	5		70
-1	-7	6		55
0	-10	7		46
1	-3	1		22
7	-11	0		0
10	11	7	65	42
0	-7	11		69
1	-6	8		62
2	-5	5		49

H / U	K / V	L / W	Φ	φ_2 / φ_1
3	-4	2		24
5	-9	7		38
7	-7	1		6
10	-11	3		13
11	-10	0		0
11	10	7	65	48
-1	-1	3		89
0	-7	10		65
1	-6	7		57
2	-5	4		41
3	-11	11		50
3	-4	1		13
5	-9	5		29
10	-11	0		0
7	10	12	45	35
-6	-9	11		89
-4	-8	9		84
-2	-7	7		76
0	-6	5		64
2	-11	8		55
2	-5	3		43
4	-4	1		14
6	-9	4		29
10	-7	0		0
10	7	12	45	55
-1	-2	2		69
0	-12	7		45
1	-10	5		39
2	-8	3		29
3	-6	1		12
7	-10	0		0
7	12	10	54	30
-2	-8	11		80
0	-5	6		71
2	-7	7		59
2	-2	1		24
4	-9	8		51
6	-11	9		46
12	-7	0		0
12	7	10	54	60
-4	-6	9		74
-3	-2	5		88
-1	-4	4		59

H / U	K / V	L / W	Φ	φ_2 / φ_1
0	-10	7		45
1	-6	3		33
3	-8	2		16
5	-10	1		6
7	-12	0		0
10	12	7	66	40
-2	-3	8		87
-1	-5	10		77
0	-7	12		71
1	-2	2		47
6	-5	0		0
7	-7	2		13
8	-9	4		20
9	-11	6		25
12	10	7	66	50
-1	-3	6		76
0	-7	10		64
1	-4	4		50
3	-5	2		21
4	-9	6		35
5	-6	0		0
8	-11	2		9
7	11	11	50	32
0	-1	1		68
11	-12	5		23
11	-11	4		19
11	-10	3		15
11	-9	2		11
11	-8	1		6
11	-7	0		0
11	7	11	50	58
-5	-11	12		67
-4	-11	11		64
-3	-11	10		60
-2	-11	9		55
-1	-11	8		50
0	-11	7		48
1	-11	6		39
2	-11	5		32
3	-11	4		26
4	-11	3		19
5	-11	2		12
6	-11	1		6
7	-11	0		0

TAB. V — FROM HKL-UVW INTO EULER ANGLES — V - 66

H	K / U	L / V	Φ / W	φ_2 / φ_1
7	11	11		
11	11	7	66	45
	-3	-4	11	87
	-2	-5	11	80
	-1	-6	11	74
	0	-7	11	68
	1	-8	11	62
	1	-1	0	0
	2	-9	11	57
	3	-10	11	53
	4	-11	11	49
	5	-12	11	45
7	11	12	47	32
	-3	-9	10	80
	-1	-7	7	73
	0	-12	11	67
	1	-5	4	57
	3	-3	1	18
	4	-8	5	42
	7	-11	6	35
	11	-7	0	0
11	7	12	47	58
	-5	-11	11	66
	-3	-9	8	61
	-2	-2	3	81
	-1	-7	5	52
	0	-12	7	43
	1	-5	2	30
	4	-8	1	9
	7	-11	0	0
7	12	11	52	30
	-1	-4	5	80
	0	-11	12	70
	1	-7	7	64
	2	-3	2	38
	3	-10	9	56
	7	-5	1	8
	9	-8	3	18
	11	-11	5	23
	12	-7	0	0
12	7	11	52	60
	-6	-7	11	78

H	K / U	L / V	Φ / W	φ_2 / φ_1
	-5	-4	8	85
	-1	-3	3	61
	0	-11	7	43
	1	-8	4	35
	2	-5	1	13
	7	-12	0	0
11	12	7	67	43
	-1	-2	5	84
	0	-7	12	70
	1	-5	7	62
	2	-3	2	32
	3	-8	9	52
	5	-11	11	47
	7	-7	1	6
	9	-10	3	14
	12	-11	0	0
12	11	7	67	47
	-1	-4	8	75
	0	-7	11	67
	1	-3	3	49
	4	-5	1	10
	5	-8	4	25
	6	-11	7	32
	11	-12	0	0
7	12	12	49	30
	0	-1	1	69
	12	-12	5	22
	12	-11	4	18
	12	-10	3	14
	12	-9	2	10
	12	-8	1	5
	12	-7	0	0
12	7	12	49	60
	-5	-12	12	64
	-4	-12	11	60
	-3	-12	10	56
	-2	-12	9	52
	-1	-12	8	47
	0	-12	7	42
	1	-12	6	36
	2	-12	5	30
	3	-12	4	24
	4	-12	3	18
	5	-12	2	12

H	K / U	L / V	Φ / W	φ_2 / φ_1
	6	-12	1	6
	7	-12	0	0
12	12	7	68	45
	-3	-4	12	87
	-2	-5	12	81
	-1	-6	12	75
	0	-7	12	69
	1	-8	12	64
	1	-1	0	0
	2	-9	12	59
	3	-10	12	55
	4	-11	12	51
	5	-12	12	47
8	8	9	51	45
	-4	-5	8	86
	-3	-6	8	78
	-2	-7	8	71
	-1	-8	8	64
	0	-9	8	58
	1	-10	8	53
	1	-1	0	0
	2	-11	8	48
	3	-12	8	44
8	9	8	56	42
	-3	-8	12	78
	-2	-8	11	74
	-1	-8	10	69
	0	-8	9	64
	1	-8	8	58
	2	-8	7	51
	3	-8	6	44
	4	-8	5	36
	5	-8	4	28
	6	-8	3	20
	7	-8	2	13
	8	-8	1	6
	9	-8	0	0
9	8	8	56	48
	0	-1	1	58
	8	-12	3	14
	8	-11	2	10
	8	-10	1	5
	8	-9	0	0

H	K / U	L / V	Φ / W	φ_2 / φ_1
8	8	11	46	45
	-5	-6	8	86
	-4	-7	8	79
	-3	-8	8	72
	-2	-9	8	66
	-1	-10	8	60
	0	-11	8	55
	1	-12	8	51
	1	-1	0	0
8	11	8	60	36
	-1	-8	12	74
	0	-8	11	70
	1	-8	10	65
	2	-8	9	59
	3	-8	8	52
	4	-8	7	46
	5	-8	6	39
	6	-8	5	31
	7	-8	4	24
	8	-8	3	17
	9	-8	2	11
	10	-8	1	5
	11	-8	0	0
11	8	8	60	54
	0	-1	1	55
	8	-12	1	5
	8	-11	0	0
8	9	9	53	42
	0	-1	1	62
	9	-12	4	19
	9	-11	3	15
	9	-10	2	11
	9	-9	1	6
	9	-8	0	0
9	8	9	53	48
	-4	-9	12	75
	-3	-9	11	71
	-2	-9	10	67
	-1	-9	9	62
	0	-9	8	56
	1	-9	7	50
	2	-9	6	43

H	K / U	L / V	Φ / W	φ_2 / φ_1
	3	-9	5	36
	4	-9	4	28
	5	-9	3	20
	6	-9	2	13
	7	-9	1	6
	8	-9	0	0
9	9	8	58	45
	-4	-4	9	90
	-3	-5	9	82
	-2	-6	9	75
	-1	-7	9	68
	0	-8	9	62
	1	-9	9	56
	1	-1	0	0
	2	-10	9	51
	3	-11	9	47
	4	-12	9	43
8	9	10	50	42
	-3	-4	6	87
	-2	-6	7	75
	-1	-8	8	66
	0	-10	9	60
	1	-12	10	56
	1	-2	1	32
	9	-8	0	0
	10	-10	1	5
	11	-12	2	9
9	8	10	50	48
	-6	-7	11	85
	-2	-9	9	65
	-2	-4	5	76
	0	-5	4	54
	2	-11	7	44
	2	-6	3	34
	4	-7	2	18
	6	-8	1	7
	8	-9	0	0
8	10	9	55	39
	-1	-10	12	70
	0	-9	10	65
	1	-8	8	59
	2	-7	6	51
	3	-6	4	39
	4	-5	2	21

H	K / U	L / V	Φ / W	φ_2 / φ_1
	5	-4	0	0
	7	-11	6	31
	9	-9	2	11
10	8	9	55	51
	-2	-11	12	63
	-1	-10	10	60
	-1	-1	2	86
	0	-9	8	54
	1	-8	6	47
	2	-7	4	36
	3	-6	2	20
	4	-5	0	0
	7	-11	2	11
9	10	8	59	42
	-2	-7	11	76
	-2	-3	6	86
	0	-4	5	65
	2	-9	9	54
	2	-5	4	44
	4	-6	3	27
	6	-11	7	35
	6	-7	2	14
	8	-8	1	6
	10	-9	0	0
10	9	8	59	48
	-2	-4	7	79
	-1	-6	8	68
	0	-8	9	60
	1	-10	10	55
	1	-2	1	28
	2	-12	11	51
	9	-10	0	0
	10	-12	1	4
8	9	11	48	42
	-5	-9	11	81
	-4	-5	7	88
	-1	-4	4	71
	0	-11	9	59
	1	-7	5	51
	2	-3	1	21
	3	-10	6	42
	9	-8	0	0
	11	-11	1	5

H	K / U	L / V	Φ / W	φ_2 / φ_1
9	8	11	48	48
	-6	-7	10	85
	-1	-3	3	69
	0	-11	8	53
	1	-8	5	46
	2	-5	2	28
	5	-7	1	9
	7	-12	3	17
	8	-9	0	0
8	11	9	57	36
	-3	-6	10	85
	-1	-5	7	76
	0	-9	11	68
	1	-4	4	57
	3	-3	1	16
	4	-7	5	39
	5	-11	9	46
	7	-10	6	32
	11	-8	0	0
11	8	9	57	54
	-1	-11	11	58
	-1	-2	3	74
	0	-9	8	53
	1	-7	5	44
	2	-5	2	25
	3	-12	7	36
	5	-8	1	7
	8	-11	0	0
9	11	8	61	39
	-1	-5	8	75
	0	-8	11	66
	1	-3	3	52
	4	-4	1	12
	5	-7	4	29
	6	-10	7	36
	9	-11	5	22
	11	-9	0	0
11	9	6	61	51
	-3	-7	12	76
	-2	-2	5	87
	-1	-5	7	68
	0	-8	9	59
	1	-11	11	54
	1	-3	2	36

H	K / U	L / V	Φ / W	φ_2 / φ_1
	5	-7	1	8
	6	-10	3	17
	9	-11	0	0
8	9	12	45	42
	-3	-12	11	70
	-3	-8	8	75
	-3	-4	5	87
	0	-4	3	58
	3	-12	7	44
	3	-8	4	37
	3	-4	1	16
	6	-12	5	30
	9	-8	0	0
	12	-12	1	5
9	8	12	45	48
	-8	-9	12	85
	-4	-12	11	68
	-4	-9	9	72
	-4	-6	7	80
	0	-3	2	52
	4	-12	5	31
	4	-9	3	24
	4	-6	1	11
	8	-9	0	0
8	12	9	56	34
	-3	-7	12	84
	0	-3	4	71
	3	-11	12	59
	3	-8	8	54
	3	-5	4	42
	3	-2	0	0
	6	-7	4	28
	9	-12	8	34
	9	-9	4	21
	12	-11	4	16
12	8	9	56	56
	-5	-6	12	81
	-4	-3	8	88
	-1	-12	12	56
	-1	-3	4	68
	0	-9	8	52
	1	-6	4	40
	2	-3	0	0
	3	-9	4	27

TAB. V FROM HKL-UVW INTO EULER ANGLES

```
 H  K  L   Φ  φ2
    U  V   W  φ1
----------------
 8  9 12
========
12  8  9  58 56
----------------
    5-12   4  20
 9 12  8  62 37
----------------
    0 -2   3  71
    4-11  12  54
    4 -9   9  50
    4 -7   6  43
    4 -5   3  29
    4 -3   0   0
    8-12   9  37
    8 -8   3  17
   12-11   3  12
12  9  8  62 53
----------------
   -5 -4  12  89
   -3 -4   9  82
   -1 -4   6  69
    0 -8   9  58
    1-12  12  53
    1 -4   3  42
    3 -4   0   0
    4 -8   3  21
    5-12   6  28
    7-12   3  14
 8 10 11  49 39
================
   -5 -7  10  88
   -2 -5   6  79
   -1 -8   8  68
    0-11  10  62
    1 -3   2  45
    5 -4   0   0
    6 -7   2  16
    7-10   4  24
   11-11   2  10
10  8 11  49 51
----------------
   -3-10  10  66
   -2 -3   4  78
   -1 -7   6  59
    0-11   8  51
    1 -4   2  35
    4 -5   0   0
    5 -9   2  15
```

```
 H  K  L   Φ  φ2
    U  V   W  φ1
----------------
 8 11 10  54 36
----------------
   -1 -2   3  84
    0-10  11  67
    1 -8   8  61
    2 -6   5  50
    3 -4   2  27
    5-10   7  41
    7 -6   1   8
   10-10   3  15
   11 -8   0   0
11  8 10  54 54
----------------
   -4 -7  10  75
   -2-11  11  61
   -2 -6   7  67
    0 -5   4  51
    2 -9   5  36
    2 -4   1  16
    8-11   0   0
10 11  8  62 42
----------------
   -1 -2   4  82
    0 -8  11  67
    1 -6   7  59
    2 -4   3  39
    3-10  10  52
    5 -6   2  16
    7-10   5  25
    8 -8   1   6
   11-10   0   0
11 10  8  62 48
----------------
   -2 -5   9  77
    0 -4   5  62
    2-11  11  53
    2 -7   6  46
    2 -3   1  18
    4-10   7  38
   10-11   0   0
 8 11 11  51 36
================
    0 -1   1  65
   11-12   4  18
   11-11   3  14
   11-10   2  10
   11 -9   1   5
   11 -8   0   0
```

```
 H  K  L   Φ  φ2
    U  V   W  φ1
----------------
11  8 11  51 54
----------------
   -4-11  12  67
   -3-11  11  63
   -2-11  10  59
   -1-11   9  54
    0-11   8  49
    1-11   7  44
    2-11   6  37
    3-11   5  31
    4-11   4  25
    5-11   3  18
    6-11   2  12
    7-11   1   6
    8-11   0   0
11 11  8  63 45
----------------
   -4 -4  11  90
   -3 -5  11  83
   -2 -6  11  77
   -1 -7  11  71
    0 -8  11  65
    1 -9  11  60
    1 -1   0   0
    2-10  11  56
    3-11  11  51
    4-12  11  48
 8 11 12  49 36
================
   -2 -4   5  84
   -1 -8   8  70
    0-12  11  64
    1 -4   3  52
    4 -4   1  13
    5 -8   4  31
    6-12   7  38
    9-12   5  25
   11 -8   0   0
11  8 12  49 54
----------------
   -8 -7  12  87
   -4-11  11  66
   -4 -8   9  71
   -4 -5   7  80
    0 -3   2  48
    4-10   3  21
    4 -7   1   9
    8-11   0   0
```

```
 H  K  L   Φ  φ2
    U  V   W  φ1
----------------
 8 12 11  53 34
----------------
   -1 -3   4  81
    0-11  12  68
    1 -8   8  62
    2 -5   4  49
    3 -2   0   0
    5 -7   4  32
    7-12   8  39
    8 -9   4  23
   11-11   4  18
12  8 11  53 56
----------------
   -7 -6  12  86
   -5 -9  12  73
   -4 -5   8  79
   -1 -4   4  61
    0-11   8  48
    1 -7   4  38
    2 -3   0   0
    3-10   4  27
11 12  8  64 43
----------------
    0 -2   3  68
    4-11  11  50
    4 -9   8  45
    4 -7   5  36
    4 -5   2  19
    8-12   7  29
    8 -8   1   6
   12-11   0   0
12 11  8  64 47
----------------
   -3 -4  10  85
   -1 -4   7  74
    0 -8  11  64
    1 -4   4  51
    3 -4   1  13
    4 -8   5  33
    5-12   9  39
    7-12   6  26
   11-12   0   0

 9  9 10  52 45
================
   -5 -5   9  90
   -4 -6   9  83
   -3 -7   9  76
   -2 -8   9  70
```

```
 H  K  L   Φ  φ2
    U  V   W  φ1
----------------
   -1 -9   9  64
    0-10   9  58
    1-11   9  53
    1 -1   0   0
    2-12   9  49
 9 10  9  56 42
----------------
   -2 -9  12  73
   -1 -9  11  68
    0 -9  10  63
    1 -9   9  58
    2 -9   8  52
    3 -9   7  46
    4 -9   6  39
    5 -9   5  32
    6 -9   4  25
    7 -9   3  18
    8 -9   2  11
    9 -9   1   5
   10 -9   0   0
10  9  9  56 48
----------------
    0 -1   1  58
    9-12   2   9
    9-11   1   5
    9-10   0   0

 9  9 11  49 45
================
   -5 -6   9  87
   -4 -7   9  80
   -3 -8   9  73
   -2 -9   9  67
   -1-10   9  62
    0-11   9  57
    1-12   9  52
    1 -1   0   0
 9 11  9  58 39
----------------
   -1 -9  12  71
    0 -9  11  66
    1 -9  10  61
    2 -9   9  56
    3 -9   8  50
    4 -9   7  43
    5 -9   6  37
    6 -9   5  30
    7 -9   4  23
    8 -9   3  17
```

```
 H  K  L   Φ  φ2
    U  V   W  φ1
----------------
    9 -9   2  11
   10 -9   1   5
   11 -9   0   0
11  9  9  58 51
----------------
    0 -1   1  57
    9-12   1   5
    9-11   0   0

 9 10 10  53 42
================
    0 -1   1  62
   10-12   3  14
   10-11   2  10
   10-10   1   5
   10 -9   0   0
10  9 10  53 48
----------------
   -3-10  12  70
   -2-10  11  66
   -1-10  10  62
    0-10   9  56
    1-10   8  51
    2-10   7  45
    3-10   6  38
    4-10   5  32
    5-10   4  25
    6-10   3  18
    7-10   2  12
    8-10   1   6
    9-10   0   0
10 10  9  58 45
----------------
   -4 -5  10  87
   -3 -6  10  80
   -2 -7  10  73
   -1 -8  10  67
    0 -9  10  62
    1-10  10  57
    1 -1   0   0
    2-11  10  52
    3-12  10  48

 9 10 11  51 42
================
   -3 -5   7  83
   -2 -7   8  73
   -1 -9   9  66
    0-11  10  60
```

```
 H  K  L   Φ  φ2
    U  V   W  φ1
----------------
    1 -2   1  32
   10 -9   0   0
   11-11   1   5
10  9 11  51 48
----------------
   -4 -9  11  74
   -3 -4   6  63
   -1 -5   5  65
    0-11   9  55
    1 -6   4  45
    3 -7   3  28
    5 -8   2  16
    7 -9   1   6
    9-10   0   0
 9 11 10  55 39
----------------
    0-10  11  65
    1 -9   9  60
    2 -8   7  52
    3 -7   5  42
    4 -6   3  28
    5 -5   1  10
    9-11   4  19
   11 -9   0   0
11  9 10  55 51
----------------
   -1-11  11  60
   -1 -1   2  87
    0-10   9  55
    1 -9   7  48
    2 -8   5  39
    3 -7   3  27
    4 -6   1  10
    9-11   0   0
10 11  9  59 42
----------------
   -1 -4   6  74
    0 -9  11  65
    1 -5   5  55
    3 -6   4  37
    4-11   9  45
    5 -7   3  23
    7 -8   2  12
    9 -9   1   5
   11-10   0   0
11 10  9  59 48
----------------
   -3 -3   7  89
   -2 -5   8  76
```

```
 H  K  L   Φ  φ2
    U  V   W  φ1
----------------
   -1 -7   9  67
    0 -9  10  60
    1-11  11  56
    1 -2   1  29
   10-11   0   0

 9 10 12  48 42
================
   -2 -9   9  69
   -2 -3   4  84
    0 -6   5  59
    2 -9   6  47
    2 -3   1  21
    4-12   7  40
   10 -9   0   0
   12-12   1   5
10  9 12  48 48
----------------
   -6 -8  11  83
   -3-10  10  68
   -3 -6   7  75
    0 -4   3  54
    3-10   5  35
    3 -6   2  23
    6 -8   1   8
    9-10   0   0
 9 12 10  56 37
----------------
   -4 -7  12  86
   -2 -6   9  80
    0 -5   6  67
    2 -9   9  57
    2 -4   3  42
    4 -3   0   0
    6 -7   3  22
    8-11   6  29
   10-10   3  14
12  9 10  56 53
----------------
   -1-12  12  58
   -1 -2   3  74
    0-10   9  54
    1 -8   6  46
    2 -6   3  31
    3 -4   0   0
    5-10   3  18
10 12  9  60 40
----------------
   -3 -5  10  86
```

TAB. V

FROM HKL-UVW INTO EULER ANGLES

H	K U	L V	Φ W	φ_2 φ_1
9	10	12		
10	12	9	60	40
	0	-3	4	67
	3	-10	10	53
	3	-7	6	46
	3	-4	2	25
	6	-11	8	38
	6	-5	0	0
	9	-9	2	10
12	10	9	60	50
	-2	-3	6	82
	-1	-6	8	67
	0	-9	10	59
	1	-12	12	55
	1	-3	2	38
	5	-6	0	0
	6	-9	2	12
	7	-12	4	19
9	11	11	52	39
	0	-1	1	63
	11	-12	3	13
	11	-11	2	9
	11	-10	1	5
	11	-9	0	0
11	9	11	52	51
	-3	-11	12	66
	-2	-11	11	62
	-1	-11	10	58
	0	-11	9	53
	1	-11	8	48
	2	-11	7	42
	3	-11	6	36
	4	-11	5	30
	5	-11	4	23
	6	-11	3	17
	7	-11	2	11
	8	-11	1	5
	9	-11	0	0
11	11	9	60	45
	-4	-5	11	87
	-3	-6	11	81
	-2	-7	11	74

H	K U	L V	Φ W	φ_2 φ_1
	-1	-8	11	69
	0	-9	11	63
	1	-10	11	59
	1	-1	0	0
	2	-11	11	54
	3	-12	11	50
9	11	12	50	39
	-5	-9	12	83
	-2	-6	7	76
	-1	-9	9	67
	0	-12	11	62
	1	-3	2	44
	6	-6	1	9
	7	-9	3	19
	8	-12	5	25
	11	-9	0	0
11	9	12	50	51
	-3	-11	11	65
	-3	-7	8	71
	-3	-3	5	86
	0	-4	3	52
	3	-9	4	31
	3	-5	1	13
	9	-11	0	0
9	12	11	54	37
	-1	-2	3	84
	0	-11	12	66
	1	-9	9	61
	2	-7	6	52
	3	-5	3	35
	4	-3	0	0
	5	-12	9	45
	7	-8	3	20
	11	-11	3	14
12	9	11	54	53
	-5	-8	12	77
	-3	-7	9	71
	-1	-6	6	61
	0	-11	9	52
	1	-5	3	39
	3	-4	0	0
	4	-9	3	21

H	K U	L V	Φ W	φ_2 φ_1
11	12	9	61	43
	-3	-4	9	87
	0	-3	4	66
	3	-11	11	53
	3	-8	7	46
	3	-5	3	32
	6	-7	2	14
	9	-12	5	21
	9	-9	1	5
	12	-11	0	0
12	11	9	61	47
	-1	-3	5	75
	0	-9	11	62
	1	-6	6	53
	2	-3	1	18
	3	-9	7	43
	5	-12	8	37
	11	-12	0	0
10	10	11	52	45
	-5	-6	10	87
	-4	-7	10	80
	-3	-8	10	74
	-2	-9	10	69
	-1	-10	10	63
	0	-11	10	58
	1	-12	10	54
	1	-1	0	0
10	11	10	56	42
	-1	-10	12	68
	0	-10	11	63
	1	-10	10	58
	2	-10	9	53
	3	-10	8	47
	4	-10	7	41
	5	-10	6	35
	6	-10	5	28
	7	-10	4	22
	8	-10	3	16
	9	-10	2	10
	10	-10	1	5
	11	-10	0	0
11	10	10	56	48
	0	-1	1	58

H	K U	L V	Φ W	φ_2 φ_1
	10	-12	1	4
	10	-11	0	0
10	11	11	54	42
	0	-1	1	62
	11	-12	2	9
	11	-11	1	5
	11	-10	0	0
11	10	11	54	48
	-2	-11	12	66
	-1	-11	11	61
	0	-11	10	57
	1	-11	9	52
	2	-11	8	46
	3	-11	7	41
	4	-11	6	35
	5	-11	5	28
	6	-11	4	22
	7	-11	3	16
	8	-11	2	10
	9	-11	1	5
	10	-11	0	0
11	11	10	57	45
	-5	-5	11	90
	-4	-6	11	84
	-3	-7	11	78
	-2	-8	11	72
	-1	-9	11	67
	0	-10	11	62
	1	-11	11	57
	1	-1	0	0
	2	-12	11	53
10	11	12	51	42
	-3	-6	8	80
	-2	-8	9	71
	-1	-10	10	65
	0	-12	11	60
	1	-2	1	32
	11	-10	0	0
	12	-12	1	4
11	10	12	51	48
	-4	-4	7	88
	-2	-11	11	64

H	K U	L V	Φ W	φ_2 φ_1
	-2	-5	6	73
	0	-6	5	55
	2	-7	4	38
	4	-8	3	24
	6	-9	2	14
	8	-10	1	6
	10	-11	0	0
10	12	11	55	40
	0	-11	12	64
	1	-10	10	60
	2	-9	8	53
	3	-8	6	45
	4	-7	4	33
	5	-6	2	18
	6	-5	0	0
	11	-11	2	9
12	10	11	55	50
	-1	-12	12	60
	-1	-1	2	87
	0	-11	10	55
	1	-10	8	50
	2	-9	6	42
	3	-8	4	31
	4	-7	2	17
	5	-6	0	0
11	12	10	58	43
	-2	-4	7	81
	0	-5	6	64
	2	-11	11	55
	2	-6	5	47
	4	-7	4	31
	6	-8	3	20
	8	-9	2	11
	10	-10	1	5
	12	-11	0	0
12	11	10	56	47
	-3	-4	8	84
	-2	-6	9	74
	-1	-8	10	66
	0	-10	11	60
	1	-12	12	56
	1	-2	1	29
	11	-12	0	0

H	K U	L V	Φ W	φ_2 φ_1
11	11	12	52	45
	-6	-6	11	90
	-5	-7	11	84
	-4	-8	11	78
	-3	-9	11	73
	-2	-10	11	68
	-1	-11	11	63
	0	-12	11	59
	1	-1	0	0
11	12	11	56	43
	0	-11	12	63
	1	-11	11	58
	2	-11	10	54
	3	-11	9	48
	4	-11	8	43
	5	-11	7	37
	6	-11	6	31
	7	-11	5	26
	8	-11	4	20
	9	-11	3	14
	10	-11	2	9
	11	-11	1	4
	12	-11	0	0
12	11	11	56	47
	0	-1	1	59
	11	-12	0	0
11	12	12	54	43
	0	-1	1	61
	12	-12	1	4
	12	-11	0	0
12	11	12	54	47
	-1	-12	12	61
	0	-12	11	57
	1	-12	10	53
	2	-12	9	48
	3	-12	8	42
	4	-12	7	37
	5	-12	6	31
	6	-12	5	26
	7	-12	4	20
	8	-12	3	15
	9	-12	2	9

H	K U	L V	Φ W	φ_2 φ_1
	10	-12	1	5
	11	-12	0	0
12	12	11	57	45
	-5	-6	12	87
	-4	-7	12	82
	-3	-8	12	76
	-2	-9	12	71
	-1	-10	12	66
	0	-11	12	61
	1	-12	12	57
	1	-1	0	0

H	K U	L V	Φ W	φ_2 φ_1

TABLE VI

From Euler Angles into Miller Indices

The relationships between Euler angles and Miller indices are given by Eq. (37). The present table allows to numerically find the Miller indices (HKL) [UVW] from a given set of Euler angles $\varphi_1\ \phi\ \varphi_2$ for $|H, K, L, U, V, W| \leqslant 12$. Since for cubic crystal and orthgonal sample symmetry Euler angles need to be considered only in the range $H^\circ = 0 \leqslant \varphi_1\ \phi\ \varphi_2 \leqslant \pi/2$, one obtains the following limits on the indices

$$0 \leqslant H, K, L, \qquad V \leqslant 0, \qquad 0 \leqslant W, \qquad (HV - KU) \leqslant 0. \tag{a}$$

The table is made up of blocks containing all orientations in 10° intervals of $\varphi_1\ \phi\ \varphi_2$ with the lowest set $\varphi_1, \phi, \varphi_2$ underlined on top of each block. For example, the block 30 50 70 contains all orientations with $30° \leqslant \varphi_1 \leqslant 39°$; $50° \leqslant \phi \leqslant 59°$ and $70° \leqslant \varphi_2 \leqslant 79°$. Within each block the orientations are ordered according to increasing φ_1. The (HKL) [UVW] values are written on the right of the angles $\varphi_1\ \phi\ \varphi_2$.

Since only $|H, K, L, U, V, W| \leqslant 12$ are considered here, the description of an orientation given by an arbitrary set $\varphi_2\ \phi\ \varphi_2$ of Euler angles by such a set of Miller indices is, in general, only an approximation. Thus one has to pick from the listed sets of Euler angles that one which comes closest to the given set. This may require that one has to check in different blocks.

If one has obtained a set of indices (HKL) [UVW], one can find the symmetrical equivalent sets of indices by permutation of the indices HKL and of the indices UVW or by permutation and additional change of signs of UVW. These permutation must be carried out for UVW in the same manner as for HKL, otherwise no equivalent orientations would be obtained (even if the orthogonality is retained). Furthermore, the above conditions (a) must be fulfilled after permutation (otherwise this orientation would not lie in the range H°). The Euler angles corresponding to these (symmetrically equivalent) Miller indices can then be obtained from Table V.

TAB. VI FROM EULER ANGLES INTO HKL-UVW VI - 2

φ_1	Φ	φ_2	H	K	L	U	V	W
0	0	0						
0	0	0	0	0	1	1	0	0
0	5	0	0	1	11	1	0	0
0	5	0	0	1	12	1	0	0
0	6	0	0	1	9	1	0	0
0	6	0	0	1	10	1	0	0
0	7	0	0	1	8	1	0	0
0	8	0	0	1	7	1	0	0
0	9	0	0	1	6	1	0	0
2	0	2	0	0	1	12	-1	0
3	0	3	0	0	1	9	-1	0
3	0	3	0	0	1	10	-1	0
3	0	3	0	0	1	11	-1	0
4	0	4	0	0	1	7	-1	0
4	0	4	0	0	1	8	-1	0
5	0	5	0	0	1	11	-2	0
5	0	5	0	0	1	6	-1	0
6	0	6	0	0	1	9	-2	0
6	0	6	0	0	1	5	-1	0
7	0	7	0	0	1	4	-1	0
8	0	8	0	0	1	10	-3	0
8	0	8	0	0	1	7	-2	0
8	0	8	0	0	1	11	-3	0
9	0	9	0	0	1	3	-1	0
0	0	40						
0	7	45	1	1	11	1	-1	0
0	7	45	1	1	12	1	-1	0
0	8	45	1	1	10	1	-1	0
0	9	45	1	1	9	1	-1	0
0	0	90						
0	5	90	1	0	11	0	-1	0
0	5	90	1	0	12	0	-1	0
0	6	90	1	0	9	0	-1	0
0	6	90	1	0	10	0	-1	0
0	7	90	1	0	8	0	-1	0
0	8	90	1	0	7	0	-1	0
0	9	90	1	0	6	0	-1	0
0	10	0						
0	10	0	0	2	11	1	0	0
0	11	0	0	1	5	1	0	0
0	13	0	0	2	9	1	0	0
0	14	0	0	1	4	1	0	0

φ_1	Φ	φ_2	H	K	L	U	V	W
0	10	0						
0	15	0	0	3	11	1	0	0
0	16	0	0	2	7	1	0	0
0	17	0	0	3	10	1	0	0
0	18	0	0	1	3	1	0	0
0	10	10						
0	15	18	1	3	12	3	-1	0
0	16	18	1	3	11	3	-1	0
0	18	18	1	3	10	3	-1	0
0	19	14	1	4	12	4	-1	0
0	10	20						
0	11	27	1	2	11	2	-1	0
0	11	27	1	2	12	2	-1	0
0	13	27	1	2	10	2	-1	0
0	14	27	1	2	9	2	-1	0
0	16	27	1	2	8	2	-1	0
0	18	27	1	2	7	2	-1	0
0	10	30						
0	17	34	2	3	12	3	-2	0
0	18	34	2	3	11	3	-2	0
0	10	40						
0	10	45	1	1	8	1	-1	0
0	11	45	1	1	7	1	-1	0
0	13	45	1	1	6	1	-1	0
0	14	45	2	2	11	1	-1	0
0	16	45	1	1	5	1	-1	0
0	17	45	2	2	9	1	-1	0
0	19	45	1	1	4	1	-1	0
0	10	50						
0	17	56	3	2	12	2	-3	0
0	18	56	3	2	11	2	-3	0
0	10	60						
0	11	63	2	1	11	1	-2	0
0	11	63	2	1	12	1	-2	0
0	13	63	2	1	10	1	-2	0
0	14	63	2	1	9	1	-2	0
0	16	63	2	1	8	1	-2	0

φ_1	Φ	φ_2	H	K	L	U	V	W
0	10	60						
0	18	63	2	1	7	1	-2	0
0	10	70						
0	15	72	3	1	12	1	-3	0
0	16	72	3	1	11	1	-3	0
0	18	72	3	1	10	1	-3	0
0	19	76	4	1	12	1	-4	0
0	10	90						
0	10	90	2	0	11	0	-1	0
0	11	90	1	0	5	0	-1	0
0	13	90	2	0	9	0	-1	0
0	14	90	1	0	4	0	-1	0
0	15	90	3	0	11	0	-1	0
0	16	90	2	0	7	0	-1	0
0	17	90	3	0	10	0	-1	0
0	18	90	1	0	3	0	-1	0
0	20	0						
0	20	0	0	4	11	1	0	0
0	21	0	0	3	8	1	0	0
0	22	0	0	2	5	1	0	0
0	23	0	0	3	7	1	0	0
0	23	0	0	5	12	1	0	0
0	24	0	0	4	9	1	0	0
0	24	0	0	5	11	1	0	0
0	27	0	0	1	2	1	0	0
0	27	9	1	6	12	6	-1	0
0	29	0	0	5	9	1	0	0
0	29	0	0	6	11	1	0	0
0	29	9	1	6	11	6	-1	0
0	20	10						
0	21	14	1	4	11	4	-1	0
0	22	14	1	4	10	4	-1	0
0	22	18	1	3	8	3	-1	0
0	23	11	1	5	12	5	-1	0
0	24	18	1	3	7	3	-1	0
0	25	11	1	5	11	5	-1	0
0	25	14	1	4	9	4	-1	0
0	27	11	1	5	10	5	-1	0
0	27	14	1	4	8	4	-1	0
0	28	18	1	3	6	3	-1	0
9	28	18	1	3	6	12	-6	1

φ_1	Φ	φ_2	H	K	L	U	V	W
0	20	20						
0	20	27	1	2	6	2	-1	0
0	22	27	2	4	11	2	-1	0
0	24	22	2	5	12	5	-2	0
0	24	27	1	2	5	2	-1	0
0	26	22	2	5	11	5	-2	0
0	26	27	2	4	9	2	-1	0
0	28	22	2	5	10	5	-2	0
8	29	27	1	2	4	12	-8	1
9	26	22	2	5	11	12	-7	1
0	20	30						
0	20	34	2	3	10	3	-2	0
0	22	34	2	3	9	3	-2	0
0	23	37	3	4	12	4	-3	0
0	24	34	2	3	8	3	-2	0
0	24	37	3	4	11	4	-3	0
0	26	31	3	5	12	5	-3	0
0	27	34	2	3	7	3	-2	0
0	27	37	3	4	10	4	-3	0
0	28	31	3	5	11	5	-3	0
0	28	39	4	5	12	5	-4	0
0	29	37	3	4	9	4	-3	0
7	28	39	4	5	12	12	-12	1
9	22	34	2	3	9	12	-11	1
9	23	37	3	4	12	12	-12	1
9	24	34	2	3	8	11	-10	1
9	24	37	3	4	11	11	-11	1
9	26	31	3	5	12	11	-9	1
9	27	34	2	3	7	10	-9	1
9	27	37	3	4	10	10	-10	1
9	29	37	3	4	9	9	-9	1
0	20	40						
0	21	45	3	3	11	1	-1	0
0	22	45	2	2	7	1	-1	0
0	23	45	3	3	10	1	-1	0
0	25	45	1	1	3	1	-1	0
0	27	45	4	4	11	1	-1	0
0	28	45	3	3	6	1	-1	0
0	29	45	2	2	5	1	-1	0
9	25	45	1	1	3	9	-12	1
0	20	50						
0	20	56	3	2	10	2	-3	0

TAB. VI FROM EULER ANGLES INTO HKL-UVW

φ_1	Φ	φ_2	H	K	L	U	V	W
0	20	50						
0	22	56	3	2	9	2	-3	0
0	23	53	4	3	12	3	-4	0
0	24	53	4	3	11	3	-4	0
0	24	56	3	2	8	2	-3	0
0	26	59	5	3	12	3	-5	0
0	27	53	4	3	10	3	-4	0
0	27	56	3	2	7	2	-3	0
0	28	51	5	4	12	4	-5	0
0	28	59	5	3	11	3	-5	0
0	29	53	4	3	9	3	-4	0
9	28	59	5	3	11	5	-12	1
9	29	53	4	3	9	6	-11	1
0	20	60						
0	20	63	2	1	6	1	-2	0
0	22	63	4	2	11	1	-2	0
0	24	63	2	1	5	1	-2	0
0	24	68	5	2	12	2	-5	0
0	26	63	4	2	9	1	-2	0
0	26	68	5	2	11	2	-5	0
0	28	68	5	2	10	2	-5	0
9	29	63	2	1	4	4	-12	1
0	20	70						
0	21	76	4	1	11	1	-4	0
0	22	72	3	1	8	1	-3	0
0	22	76	4	1	10	1	-4	0
0	23	79	5	1	12	1	-5	0
0	24	72	3	1	7	1	-3	0
0	25	76	4	1	9	1	-4	0
0	25	79	5	1	11	1	-5	0
0	27	76	4	1	8	1	-4	0
0	27	79	5	1	10	1	-5	0
0	28	72	3	1	6	1	-3	0
0	20	80						
0	27	81	6	1	12	1	-6	0
0	29	81	6	1	11	1	-6	0
0	20	90						
0	20	90	4	0	11	0	-1	0
0	21	90	3	0	8	0	-1	0
0	22	90	2	0	5	0	-1	0
0	23	90	3	0	7	0	-1	0

φ_1	Φ	φ_2	H	K	L	U	V	W
0	20	90						
0	23	90	5	0	12	0	-1	0
0	24	90	4	0	9	0	-1	0
0	24	90	5	0	11	0	-1	0
0	27	90	1	0	2	0	-1	0
0	29	90	5	0	9	0	-1	0
0	29	90	6	0	11	0	-1	0
0	30	0						
0	30	0	0	4	7	1	0	0
0	30	0	0	7	12	1	0	0
0	31	0	0	3	5	1	0	0
0	31	8	1	7	12	7	-1	0
0	31	9	1	6	10	6	-1	0
0	32	0	0	5	8	1	0	0
0	32	0	0	7	11	1	0	0
0	33	8	1	7	11	7	-1	0
0	34	0	0	2	3	1	0	0
0	34	7	1	8	12	8	-1	0
0	34	9	1	6	9	6	-1	0
0	35	0	0	7	10	1	0	0
0	35	8	1	7	10	7	-1	0
0	36	0	0	5	7	1	0	0
0	36	0	0	8	11	1	0	0
0	36	7	1	8	11	8	-1	0
0	37	0	0	3	4	1	0	0
0	37	6	1	9	12	9	-1	0
0	37	9	1	6	8	6	-1	0
0	38	0	0	7	9	1	0	0
0	38	8	1	7	9	7	-1	0
0	39	0	0	4	5	1	0	0
0	39	0	0	9	11	1	0	0
0	39	6	1	9	11	9	-1	0
0	39	7	1	8	10	8	-1	0
7	38	8	1	7	9	12	-3	1
8	34	7	1	8	12	12	-3	1
9	35	8	1	7	10	11	-3	1
9	37	9	1	6	8	10	-3	1
0	30	10						
0	30	11	1	5	9	5	-1	0
0	30	14	1	4	7	4	-1	0
0	30	18	2	6	11	3	-1	0
0	31	16	2	7	12	7	-2	0
0	32	18	1	3	5	3	-1	0
0	33	11	1	5	8	5	-1	0
0	33	16	2	7	11	7	-2	0

φ_1	Φ	φ_2	H	K	L	U	V	W
0	30	10						
0	34	14	1	4	6	4	-1	0
0	35	18	2	6	9	3	-1	0
0	36	11	1	5	7	5	-1	0
0	36	16	2	7	10	7	-2	0
0	37	14	2	8	11	4	-1	0
0	38	13	2	9	12	9	-2	0
0	38	18	1	3	4	3	-1	0
0	39	16	2	7	9	7	-2	0
7	38	13	2	9	12	12	-4	1
8	33	11	1	5	8	12	-4	1
8	33	16	2	7	11	12	-5	1
8	38	18	1	3	4	11	-5	1
9	34	14	1	4	6	10	-4	1
0	30	20						
0	31	22	2	5	9	5	-2	0
0	31	27	3	6	11	2	-1	0
0	32	23	3	7	12	7	-3	0
0	33	27	2	4	7	2	-1	0
0	34	22	2	5	8	5	-2	0
0	34	27	3	6	10	2	-1	0
0	35	21	3	8	12	8	-3	0
0	35	23	3	7	11	7	-3	0
0	37	23	3	7	10	7	-3	0
0	37	27	1	2	3	2	-1	0
0	38	21	3	8	11	8	-3	0
0	38	22	2	5	7	5	-2	0
0	39	24	4	9	12	9	-4	0
0	39	27	4	8	11	2	-1	0
7	35	21	3	8	12	12	-6	1
8	34	22	2	5	8	11	-6	1
9	32	23	3	7	12	10	-6	1
9	37	27	1	2	3	9	-6	1
9	38	22	2	5	7	9	-5	1
0	30	30						
0	30	31	3	5	10	5	-3	0
0	30	39	4	5	11	5	-4	0
0	31	34	2	3	6	3	-2	0
0	32	37	3	4	8	4	-3	0
0	33	31	3	5	9	5	-3	0
0	33	34	4	6	11	3	-2	0
0	33	39	4	5	10	5	-4	0
0	34	30	4	7	12	7	-4	0
0	35	39	4	5	9	5	-4	0
0	36	30	4	7	11	7	-4	0

φ_1	Φ	φ_2	H	K	L	U	V	W
0	30	30						
0	36	31	3	5	6	5	-3	0
0	36	34	2	3	5	3	-2	0
0	36	36	5	7	12	7	-5	0
0	36	37	3	4	7	4	-3	0
0	36	32	5	8	12	8	-5	0
0	36	36	5	7	11	7	-5	0
0	39	30	4	7	10	7	-4	0
0	39	39	4	5	8	5	-4	0
6	38	32	5	8	12	12	-9	1
7	30	39	4	5	11	11	-11	1
7	31	34	2	3	6	12	-10	1
7	32	37	3	4	8	12	-11	1
7	33	31	3	5	9	12	-9	1
7	36	34	2	3	5	11	-9	1
7	36	37	3	4	7	11	-10	1
8	33	39	4	5	10	10	-10	1
8	34	30	4	7	12	11	-8	1
8	35	39	4	5	9	9	-9	1
8	36	36	5	7	11	9	-8	1
8	39	39	4	5	8	8	-8	1
9	30	31	3	5	10	10	-8	1
9	31	34	2	3	6	9	-8	1
9	36	31	3	5	8	9	-7	1
9	36	34	2	3	5	8	-7	1
0	30	40						
0	31	45	3	3	7	1	-1	0
0	31	45	5	5	12	1	-1	0
0	32	45	4	4	9	1	-1	0
0	33	40	5	6	12	6	-5	0
0	33	45	5	5	11	1	-1	0
0	35	40	5	6	11	6	-5	0
0	35	45	1	1	2	1	-1	0
0	38	40	5	6	10	6	-5	0
0	38	41	6	7	12	7	-6	0
0	38	45	5	5	9	1	-1	0
0	38	45	6	6	11	1	-1	0
0	38	49	7	6	12	6	-7	0
0	39	45	4	4	7	1	-1	0
6	33	40	5	6	12	12	-12	1
6	35	40	5	6	11	11	-11	1
6	35	45	1	1	2	10	-12	1
6	38	41	6	7	12	12	-12	1
7	35	45	1	1	2	9	-11	1
7	38	40	5	6	10	10	-10	1
8	35	45	1	1	2	8	-10	1
9	35	45	1	1	2	7	-9	1

φ_1	Φ	φ_2	H	K	L	U	V	W
0	30	40						
9	38	49	7	6	12	6	-9	1
0	30	50						
0	30	51	5	4	11	4	-5	0
0	30	59	5	3	10	3	-5	0
0	31	56	3	2	6	2	-3	0
0	32	53	4	3	8	3	-4	0
0	33	50	6	5	12	5	-6	0
0	33	51	5	4	10	4	-5	0
0	33	56	6	4	11	2	-3	0
0	33	59	5	3	9	3	-5	0
0	35	50	6	5	11	5	-6	0
0	35	51	5	4	9	4	-5	0
0	36	53	4	3	7	3	-4	0
0	36	54	7	5	12	5	-7	0
0	36	56	3	2	5	2	-3	0
0	36	59	5	3	8	3	-5	0
0	38	50	6	5	10	5	-6	0
0	38	54	7	5	11	5	-7	0
0	38	58	8	5	12	5	-8	0
0	39	51	5	4	8	4	-5	0
7	33	50	6	5	12	8	-12	1
7	38	54	7	5	11	7	-12	1
7	38	58	8	5	12	6	-12	1
8	31	56	3	2	6	6	-12	1
8	32	53	4	3	8	7	-12	1
8	35	51	5	4	9	7	-11	1
8	36	59	5	3	8	5	-11	1
9	33	51	5	4	10	6	-10	1
9	36	56	3	2	5	5	-10	1
0	30	60						
0	31	63	6	3	11	1	-2	0
0	31	68	5	2	9	2	-5	0
0	32	67	7	3	12	3	-7	0
0	33	63	4	2	7	1	-2	0
0	34	60	7	4	12	4	-7	0
0	34	63	6	3	10	1	-2	0
0	34	68	5	2	8	2	-5	0
0	35	67	7	3	11	3	-7	0
0	35	69	8	3	12	3	-8	0
0	36	60	7	4	11	4	-7	0
0	37	63	2	1	3	1	-2	0
0	37	67	7	3	10	3	-7	0
0	38	68	5	2	7	2	-5	0
0	38	69	8	3	11	3	-8	0

φ_1	Φ	φ_2	H	K	L	U	V	W
0	30	60						
0	39	60	7	4	10	4	-7	0
0	39	63	8	4	11	1	-2	0
0	39	66	9	4	12	4	-9	0
7	39	66	9	4	12	4	-12	1
8	35	69	8	3	12	3	-12	1
8	37	63	2	1	3	4	-11	1
8	38	68	5	2	7	3	-11	1
9	31	68	5	2	9	3	-12	1
9	32	67	7	3	12	3	-11	1
0	30	70						
0	30	72	6	2	11	1	-3	0
0	30	76	4	1	7	1	-4	0
0	30	79	5	1	9	1	-5	0
0	31	74	7	2	12	2	-7	0
0	32	72	3	1	5	1	-3	0
0	33	74	7	2	11	2	-7	0
0	33	79	5	1	8	1	-5	0
0	34	76	4	1	6	1	-4	0
0	35	72	6	2	9	1	-3	0
0	36	74	7	2	10	2	-7	0
0	36	79	5	1	7	1	-5	0
0	37	76	8	2	11	1	-4	0
0	38	72	3	1	4	1	-3	0
0	38	77	9	2	12	2	-9	0
0	39	74	7	2	9	2	-7	0
8	36	74	7	2	10	2	-12	1
8	36	79	5	1	7	1	-12	1
9	38	72	3	1	4	2	-10	1
0	30	80						
0	31	81	6	1	10	1	-6	0
0	31	82	7	1	12	1	-7	0
0	33	82	7	1	11	1	-7	0
0	34	81	6	1	9	1	-6	0
0	34	83	8	1	12	1	-8	0
0	35	82	7	1	10	1	-7	0
0	36	83	8	1	11	1	-6	0
0	37	81	6	1	8	1	-6	0
0	37	84	9	1	12	1	-9	0
0	38	82	7	1	9	1	-7	0
0	39	83	8	1	10	1	-8	0
0	39	84	9	1	11	1	-9	0
8	37	84	9	1	12	0	-12	1
8	39	84	9	1	11	0	-11	1
9	31	82	7	1	12	0	-12	1

φ_1	Φ	φ_2	H	K	L	U	V	W
0	30	80						
9	34	83	8	1	12	0	-12	1
9	36	83	8	1	11	0	-11	1
9	39	83	8	1	10	0	-10	1
0	30	90						
0	30	90	4	0	7	0	-1	0
0	30	90	7	0	12	0	-1	0
0	31	90	3	0	5	0	-1	0
0	32	90	5	0	8	0	-1	0
0	32	90	7	0	11	0	-1	0
0	34	90	2	0	3	0	-1	0
0	35	90	7	0	10	0	-1	0
0	36	90	5	0	7	0	-1	0
0	36	90	8	0	11	0	-1	0
0	37	90	3	0	4	0	-1	0
0	38	90	7	0	9	0	-1	0
0	39	90	4	0	5	0	-1	0
0	39	90	9	0	11	0	-1	0
0	40	0						
0	40	0	0	5	6	1	0	0
0	40	6	1	10	12	10	-1	0
0	41	0	0	6	7	1	0	0
0	41	0	0	7	8	1	0	0
0	41	8	1	7	8	7	-1	0
0	41	9	1	6	7	6	-1	0
0	42	0	0	8	9	1	0	0
0	42	0	0	9	10	1	0	0
0	42	0	0	10	11	1	0	0
0	42	6	1	9	10	9	-1	0
0	42	6	1	10	11	10	-1	0
0	42	7	1	8	9	8	-1	0
0	43	0	0	11	12	1	0	0
0	43	5	1	11	12	11	-1	0
0	45	0	0	1	1	1	0	0
0	45	5	1	11	11	11	-1	0
0	45	5	1	12	12	12	-1	0
0	45	6	1	9	9	9	-1	0
0	45	6	1	10	10	10	-1	0
0	45	7	1	8	8	8	-1	0
0	45	8	1	7	7	7	-1	0
0	45	9	1	6	6	6	-1	0
0	47	0	0	12	11	1	0	0
0	48	0	0	9	8	1	0	0
0	48	0	0	10	9	1	0	0
0	48	0	0	11	10	1	0	0

φ_1	Φ	φ_2	H	K	L	U	V	W
0	40	0						
0	48	5	1	11	10	11	-1	0
0	48	5	1	12	11	12	-1	0
0	48	6	1	10	9	10	-1	0
0	48	9	2	12	11	6	-1	0
0	49	0	0	7	6	1	0	0
0	49	0	0	8	7	1	0	0
0	49	6	1	9	8	9	-1	0
0	49	7	1	8	7	8	-1	0
6	45	9	1	6	6	12	-3	1
6	48	5	1	11	10	12	-2	1
7	45	0	0	1	1	11	-1	1
7	45	0	0	1	1	12	-1	1
7	45	5	1	11	11	11	-2	1
7	45	5	1	12	12	12	-2	1
7	48	6	1	10	9	11	-2	1
7	49	6	1	9	8	10	-2	1
8	41	9	1	6	7	11	-3	1
8	43	5	1	11	12	10	-2	1
8	45	0	0	1	1	10	-1	1
8	45	6	1	10	10	10	-2	1
8	49	7	1	8	7	9	-2	1
9	42	6	1	10	11	9	-2	1
9	45	0	0	1	1	9	-1	1
9	45	6	1	9	9	9	-2	1
0	40	10						
0	40	11	1	5	6	5	-1	0
0	40	13	2	9	11	9	-2	0
0	40	14	1	4	5	4	-1	0
0	41	17	3	10	12	10	-3	0
0	41	18	3	9	11	3	-1	0
0	42	14	2	8	9	4	-1	0
0	42	16	2	7	8	7	-2	0
0	42	18	2	6	7	3	-1	0
0	43	10	2	11	12	11	-2	0
0	43	11	2	10	11	5	-1	0
0	43	13	2	9	10	9	-2	0
0	43	18	3	9	10	3	-1	0
0	44	15	3	11	12	11	-3	0
0	44	17	3	10	11	10	-3	0
0	45	10	2	11	11	11	-2	0
0	46	11	1	5	5	5	-1	0
0	46	13	2	9	9	9	-2	0
0	46	14	1	4	4	4	-1	0
0	46	15	3	11	11	11	-3	0
0	46	16	2	7	7	7	-2	0
0	46	17	3	10	10	10	-3	0

φ_1	Φ	φ_2	H	K	L	U	V	W
0	40	10						
0	47	18	1	3	3	3	-1	0
0	48	10	2	11	10	11	-2	0
0	48	14	3	12	11	4	-1	0
0	49	11	2	10	9	5	-1	0
0	49	13	2	9	8	9	-2	0
0	49	15	3	11	10	11	-3	0
0	49	17	3	10	9	10	-3	0
0	49	18	4	12	11	3	-1	0
6	46	14	1	4	4	12	-4	1
6	47	18	1	3	3	12	-5	1
7	45	10	2	11	11	11	-3	1
7	46	15	3	11	11	11	-4	1
7	46	17	3	10	10	10	-4	1
8	40	14	1	4	5	11	-4	1
8	42	16	2	7	8	10	-4	1
8	46	11	1	5	5	10	-3	1
8	46	13	2	9	9	9	-3	1
8	47	18	1	3	3	9	-4	1
9	40	11	1	5	6	9	-3	1
9	46	14	1	4	4	8	-3	1
0	40	20						
0	40	23	3	7	9	7	-3	0
0	40	27	3	6	8	2	-1	0
0	41	21	3	8	10	8	-3	0
0	41	29	5	9	12	9	-5	0
0	42	22	2	5	6	5	-2	0
0	42	24	4	9	11	9	-4	0
0	42	27	2	4	5	2	-1	0
0	43	27	5	10	12	2	-1	0
0	43	29	5	9	11	9	-5	0
0	44	20	4	11	12	11	-4	0
0	44	21	3	8	9	8	-3	0
0	44	22	4	10	11	5	-2	0
0	44	23	3	7	8	7	-3	0
0	44	27	3	6	7	2	-1	0
0	45	24	4	9	10	9	-4	0
0	45	24	5	11	12	11	-5	0
0	45	27	4	8	9	2	-1	0
0	45	27	5	10	11	2	-1	0
0	46	29	5	9	10	9	-5	0
0	46	29	6	11	12	11	-6	0
0	47	20	4	11	11	11	-4	0
0	47	21	3	8	8	8	-3	0
0	47	22	2	5	5	5	-2	0
0	47	23	3	7	7	7	-3	0
0	47	23	5	12	12	12	-5	0

φ_1	Φ	φ_2	H	K	L	U	V	W
0	40	20						
0	48	24	4	9	9	9	-4	0
0	48	24	5	11	11	11	-5	0
0	48	27	1	2	2	2	-1	0
0	49	20	4	11	10	11	-4	0
0	49	29	5	9	9	9	-5	0
0	49	29	6	11	11	11	-6	0
6	41	29	5	9	12	12	-8	1
6	42	22	2	5	6	12	-6	1
6	47	20	4	11	11	11	-5	1
6	47	23	5	12	12	12	-6	1
6	48	24	5	11	11	11	-6	1
6	48	27	1	2	2	12	-7	1
6	49	29	6	11	11	11	-7	1
7	40	23	3	7	9	11	-6	1
7	45	24	4	9	10	11	-6	1
7	46	29	6	11	12	9	-6	1
7	47	22	2	5	5	10	-5	1
7	48	27	1	2	2	10	-6	1
7	49	29	5	9	9	9	-6	1
8	41	21	3	8	10	10	-5	1
8	44	23	3	7	8	9	-5	1
8	48	24	4	9	9	9	-5	1
8	48	27	1	2	2	8	-5	1
9	44	20	4	11	12	8	-4	1
9	46	29	5	9	10	7	-5	1
9	47	21	3	8	8	8	-4	1
0	40	30						
0	40	31	3	5	7	5	-3	0
0	40	37	3	4	6	4	-3	0
0	41	32	5	8	11	8	-5	0
0	41	36	5	7	10	7	-5	0
0	42	30	4	7	9	7	-4	0
0	42	34	2	3	4	3	-2	0
0	42	37	6	6	11	4	-3	0
0	42	39	4	5	7	5	-4	0
0	43	32	5	8	10	6	-5	0
0	44	31	3	5	6	5	-3	0
0	44	36	5	7	9	7	-5	0
0	44	38	7	9	12	9	-7	0
0	45	30	4	7	8	7	-4	0
0	45	34	6	9	11	3	-2	0
0	45	35	7	10	12	10	-7	0
0	45	37	3	4	5	4	-3	0
0	46	32	5	8	9	8	-5	0
0	46	34	4	6	7	3	-2	0
0	46	38	7	9	11	9	-7	0

φ_1	Φ	φ_2	H	K	L	U	V	W
0	40	30						
0	47	31	6	10	11	5	-3	0
0	47	32	7	11	12	11	-7	0
0	47	34	6	9	10	3	-2	0
0	47	36	5	7	8	7	-5	0
0	47	39	4	5	6	5	-4	0
0	48	35	7	10	11	10	-7	0
0	48	37	6	8	9	4	-3	0
0	49	30	4	7	7	7	-4	0
0	49	30	7	12	12	12	-7	0
0	49	31	3	5	5	5	-3	0
0	49	36	8	11	12	11	-8	0
0	49	38	7	9	10	9	-7	0
0	49	39	8	10	11	5	-4	0
5	42	39	4	5	7	12	-11	1
5	47	39	4	5	6	11	-10	1
5	49	30	7	12	12	12	-6	1
6	41	36	5	7	10	12	-10	1
6	45	30	4	7	8	12	-8	1
6	46	32	5	8	9	11	-8	1
6	46	38	7	9	11	10	-9	1
6	49	31	3	5	5	10	-7	1
7	40	31	3	5	7	11	-8	1
7	40	37	3	4	6	10	-9	1
7	42	30	4	7	9	10	-7	1
7	45	37	3	4	5	9	-8	1
8	44	31	3	5	6	8	-6	1
8	44	36	5	7	9	8	-7	1
8	48	35	7	10	11	7	-6	1
9	42	34	2	3	4	7	-6	1
9	42	39	4	5	7	7	-7	1
9	47	39	4	5	6	6	-6	1
9	49	30	4	7	7	7	-5	1
0	40	40						
0	40	41	6	7	11	7	-6	0
0	40	45	3	3	5	1	-1	0
0	40	45	7	7	12	1	-1	0
0	40	49	7	6	11	6	-7	0
0	41	40	5	6	9	6	-5	0
0	41	45	5	5	8	1	-1	0
0	42	41	7	8	12	8	-7	0
0	42	45	7	7	11	1	-1	0
0	42	49	8	7	12	7	-8	0
0	43	41	6	7	10	7	-6	0
0	43	45	2	2	3	1	-1	0
0	43	49	7	6	10	6	-7	0
0	44	40	5	6	8	6	-5	0

φ_1	Φ	φ_2	H	K	L	U	V	W
0	40	40						
0	44	41	7	8	11	8	-7	0
0	44	49	8	7	11	7	-8	0
0	45	42	8	9	12	9	-8	0
0	45	45	5	5	7	1	-1	0
0	45	45	7	7	10	1	-1	0
0	45	48	9	8	12	8	-9	0
0	46	41	6	7	9	7	-6	0
0	46	45	8	8	11	1	-1	0
0	46	49	7	6	9	6	-7	0
0	47	41	7	8	10	8	-7	0
0	47	45	3	3	4	1	-1	0
0	47	49	8	7	10	7	-8	0
0	48	40	5	6	7	6	-5	0
0	48	42	8	9	11	9	-8	0
0	48	42	9	10	12	10	-9	0
0	48	45	7	7	9	1	-1	0
0	48	48	9	8	11	8	-9	0
0	48	48	10	9	12	9	-10	0
0	49	41	6	7	8	7	-6	0
0	49	45	4	4	5	1	-1	0
0	49	45	9	9	11	1	-1	0
0	49	49	7	6	8	6	-7	0
5	42	41	7	8	12	12	-12	1
5	44	41	7	8	11	11	-11	1
5	45	42	8	9	12	12	-12	1
5	46	49	7	6	9	9	-12	1
5	48	42	8	9	11	11	-11	1
5	48	42	9	10	12	12	-12	1
6	40	41	6	7	11	11	-11	1
6	42	49	8	7	12	9	-12	1
6	43	41	6	7	10	10	-10	1
6	43	49	7	6	10	8	-11	1
6	46	41	6	7	9	9	-9	1
6	47	41	7	8	10	10	-10	1
7	40	49	7	6	11	7	-10	1
7	41	40	5	6	9	9	-9	1
7	44	40	5	6	8	8	-8	1
7	49	41	6	7	8	8	-8	1
8	48	40	5	6	7	7	-7	1
8	48	48	10	9	12	6	-8	1
0	40	50						
0	40	53	4	3	6	3	-4	0
0	40	59	5	3	7	3	-5	0
0	41	50	6	5	9	5	-6	0
0	41	54	7	5	10	5	-7	0
0	41	58	8	5	11	5	-8	0

φ_1 Φ φ_2	H	K	L	U	V	W
0 40 50						
0 42 51	5	4	7	4	-5	0
0 42 53	8	6	11	3	-4	0
0 42 56	3	2	4	2	-3	0
0 43 58	8	5	10	5	-8	0
0 44 50	6	5	8	5	-6	0
0 44 52	9	7	12	7	-9	0
0 44 54	7	5	9	5	-7	0
0 44 59	5	3	6	3	-5	0
0 45 53	4	3	5	3	-4	0
0 45 55	10	7	12	7	-10	0
0 45 56	9	6	11	2	-3	0
0 46 52	9	7	11	7	-9	0
0 46 56	6	4	7	2	-3	0
0 46 58	8	5	9	5	-8	0
0 47 51	5	4	6	4	-5	0
0 47 54	7	5	8	5	-7	0
0 47 56	9	6	10	2	-3	0
0 47 58	11	7	12	7	-11	0
0 47 59	10	6	11	3	-5	0
0 48 50	6	5	7	5	-6	0
0 48 53	8	6	9	3	-4	0
0 48 55	10	7	11	7	-10	0
0 49 51	10	8	11	4	-5	0
0 49 52	9	7	10	7	-9	0
0 49 54	11	8	12	8	-11	0
0 49 59	5	3	5	3	-5	0
6 44 52	9	7	12	8	-12	1
6 44 59	5	3	6	6	-12	1
6 45 53	4	3	5	7	-11	1
6 48 50	6	5	7	8	-11	1
7 42 56	3	2	4	6	-11	1
7 43 58	8	5	10	5	-10	1
7 47 51	5	4	6	6	-9	1
7 47 54	7	5	8	6	-10	1
7 49 59	5	3	5	5	-10	1
8 40 53	4	3	6	6	-10	1
8 41 50	6	5	9	6	-9	1
8 46 52	9	7	11	5	-8	1
9 40 59	5	3	7	4	-9	1
9 41 54	7	5	10	5	-9	1
9 42 51	5	4	7	5	-8	1
9 47 58	11	7	12	4	-8	1
9 49 54	11	8	12	4	-7	1
0 40 60						
0 40 63	6	3	8	1	-2	0
0 40 67	7	3	9	3	-7	0

φ_1 Φ φ_2	H	K	L	U	V	W
0 40 60						
0 41 61	9	5	12	5	-9	0
0 41 69	8	3	10	3	-8	0
0 42 60	7	4	9	4	-7	0
0 42 63	4	2	5	1	-2	0
0 42 66	9	4	11	4	-9	0
0 42 68	5	2	6	2	-5	0
0 43 61	9	5	11	5	-9	0
0 43 63	10	5	12	1	-2	0
0 44 63	6	3	7	1	-2	0
0 44 67	7	3	8	3	-7	0
0 44 68	10	4	11	2	-5	0
0 44 69	8	3	9	3	-8	0
0 45 60	7	4	8	4	-7	0
0 45 63	8	4	9	1	-2	0
0 45 63	10	5	11	1	-2	0
0 45 66	9	4	10	4	-9	0
0 45 66	11	5	12	5	-11	0
0 46 61	9	5	10	5	-9	0
0 46 61	11	6	12	6	-11	0
0 47 67	7	3	7	3	-7	0
0 47 67	12	5	12	5	-12	0
0 47 68	5	2	5	2	-5	0
0 47 69	8	3	8	3	-8	0
0 48 63	2	1	2	1	-2	0
0 48 66	9	4	9	4	-9	0
0 48 66	11	5	11	5	-11	0
0 49 60	7	4	7	4	-7	0
0 49 60	12	7	12	7	-12	0
0 49 61	9	5	9	5	-9	0
0 49 61	11	6	11	6	-11	0
6 47 67	12	5	12	4	-12	1
6 48 63	2	1	2	5	-12	1
6 49 60	12	7	12	6	-12	1
6 49 61	11	6	11	5	-11	1
7 42 60	7	4	9	5	-11	1
7 44 67	7	3	8	4	-12	1
7 44 69	8	3	9	3	-11	1
7 46 61	9	5	10	5	-11	1
7 47 68	5	2	5	3	-10	1
7 48 63	2	1	2	4	-10	1
7 48 66	11	5	11	4	-11	1
8 40 67	7	3	9	3	-10	1
8 45 60	7	4	8	4	-9	1
8 45 66	11	5	12	3	-9	1
8 48 66	9	4	9	3	-9	1
8 49 61	9	5	9	4	-9	1
9 47 69	8	3	8	2	-8	1
9 48 63	2	1	2	3	-8	1

φ_1 Φ φ_2	H	K	L	U	V	W
0 40 70						
0 40 76	4	1	5	1	-4	0
0 40 77	9	2	11	2	-9	0
0 40 79	5	1	6	1	-5	0
0 41 72	9	3	11	1	-3	0
0 41 73	10	3	12	3	-10	0
0 42 72	6	2	7	1	-3	0
0 42 74	7	2	8	2	-7	0
0 42 76	8	2	9	1	-4	0
0 43 72	9	3	10	1	-3	0
0 43 77	9	2	10	2	-9	0
0 43 79	10	2	11	1	-5	0
0 44 70	11	4	12	4	-11	0
0 44 73	10	3	11	3	-10	0
0 44 75	11	3	12	3	-11	0
0 46 73	10	3	10	3	-10	0
0 46 74	7	2	7	2	-7	0
0 46 75	11	3	11	3	-11	0
0 46 76	4	1	4	1	-4	0
0 46 77	9	2	9	2	-9	0
0 46 79	5	1	5	1	-5	0
0 47 70	11	4	11	4	-11	0
0 47 72	3	1	3	1	-3	0
0 48 76	12	3	11	1	-4	0
0 49 70	11	4	10	4	-11	0
0 49 72	12	4	11	1	-3	0
0 49 73	10	3	9	3	-10	0
0 49 75	11	3	10	3	-11	0
0 49 77	9	2	8	2	-9	0
0 49 79	10	2	9	1	-5	0
6 47 72	3	1	3	3	-12	1
7 46 75	11	3	11	2	-11	1
7 46 76	4	1	4	2	-12	1
7 47 70	11	4	11	3	-11	1
8 40 79	5	1	6	1	-11	1
8 42 74	7	2	8	2	-11	1
8 46 73	10	3	10	2	-10	1
8 46 79	5	1	5	1	-10	1
9 40 77	9	2	11	1	-10	1
9 46 77	9	2	9	1	-9	1
9 47 72	3	1	3	2	-9	1
9 49 70	11	4	10	2	-8	1
0 40 80						
0 40 84	10	1	12	1	-10	0
0 41 81	6	1	7	1	-6	0
0 41 82	7	1	8	1	-7	0

φ_1 Φ φ_2	H	K	L	U	V	W
0 40 80						
0 42 83	8	1	9	1	-8	0
0 42 84	9	1	10	1	-9	0
0 42 84	10	1	11	1	-10	0
0 43 80	11	2	12	2	-11	0
0 43 85	11	1	12	1	-11	0
0 45 80	11	2	11	2	-11	0
0 45 81	6	1	6	1	-6	0
0 45 82	7	1	7	1	-7	0
0 45 83	8	1	8	1	-8	0
0 45 84	9	1	9	1	-9	0
0 45 84	10	1	10	1	-10	0
0 45 85	11	1	11	1	-11	0
0 45 85	12	1	12	1	-12	0
0 48 80	11	2	10	2	-11	0
0 48 81	12	2	11	1	-6	0
0 48 84	10	1	9	1	-10	0
0 48 85	11	1	10	1	-11	0
0 48 85	12	1	11	1	-12	0
0 49 83	8	1	7	1	-8	0
0 49 84	9	1	8	1	-9	0
7 40 84	10	1	12	0	-12	1
7 43 85	11	1	12	0	-12	1
7 45 80	11	2	11	1	-11	1
7 45 81	6	1	6	1	-12	1
7 45 85	11	1	11	0	-11	1
7 45 85	12	1	12	0	-12	1
7 48 85	12	1	11	0	-11	1
8 42 84	10	1	11	0	-11	1
8 45 84	10	1	10	0	-10	1
8 48 85	11	1	10	0	-10	1
9 42 84	9	1	10	0	-10	1
9 45 84	9	1	9	0	-9	1
9 48 84	10	1	9	0	-9	1
0 40 90						
0 40 90	5	0	6	0	-1	0
0 41 90	6	0	7	0	-1	0
0 41 90	7	0	8	0	-1	0
0 42 90	8	0	9	0	-1	0
0 42 90	9	0	10	0	-1	0
0 42 90	10	0	11	0	-1	0
0 43 90	11	0	12	0	-1	0
0 45 90	1	0	1	0	-1	0
0 47 90	12	0	11	0	-1	0
0 48 90	9	0	8	0	-1	0
0 48 90	10	0	9	0	-1	0
0 48 90	11	0	10	0	-1	0

TAB. VI FROM EULER ANGLES INTO HKL-UVW

φ_1	Φ	φ_2	H	K	L	U	V	W
0 40 90								
0	49	90	7	0	6	0	-1	0
0	49	90	8	0	7	0	-1	0
7	45	90	1	0	1	-1	-12	1
7	45	90	1	0	1	-1	-11	1
8	45	90	1	0	1	-1	-10	1
9	45	90	1	0	1	-1	-9	1
0 50 0								
0	50	0	0	6	5	1	0	0
0	50	5	1	12	10	12	-1	0
0	50	8	1	7	6	7	-1	0
0	51	0	0	5	4	1	0	0
0	51	0	0	11	9	1	0	0
0	51	5	1	11	9	11	-1	0
0	51	6	1	10	8	10	-1	0
0	51	9	1	6	5	6	-1	0
0	52	0	0	9	7	1	0	0
0	52	6	1	9	7	9	-1	0
0	53	0	0	4	3	1	0	0
0	53	5	1	12	9	12	-1	0
0	53	7	1	8	6	8	-1	0
0	54	0	0	7	5	1	0	0
0	54	0	0	11	8	1	0	0
0	54	5	1	11	8	11	-1	0
0	54	9	2	12	9	6	-1	0
0	55	0	0	10	7	1	0	0
0	55	6	1	10	7	10	-1	0
0	55	8	1	7	5	7	-1	0
0	56	0	0	3	2	1	0	0
0	56	5	1	12	8	12	-1	0
0	56	6	1	9	6	9	-1	0
0	57	9	1	6	4	6	-1	0
0	58	0	0	8	5	1	0	0
0	58	0	0	11	7	1	0	0
0	58	5	1	11	7	11	-1	0
0	58	7	1	8	5	8	-1	0
0	59	0	0	5	3	1	0	0
0	59	6	1	10	6	10	-1	0
6	51	6	1	10	8	12	-2	1
6	52	6	1	9	7	11	-2	1
6	56	6	1	9	6	12	-2	1
6	58	7	1	8	5	11	-2	1
7	53	7	1	8	6	10	-2	1
8	55	8	1	7	5	9	-2	1
8	57	9	1	6	4	8	-2	1
9	50	8	1	7	6	8	-2	1
0 50 10								
0	50	14	2	8	7	4	-1	0
0	50	18	3	9	8	3	-1	0
0	51	10	2	11	9	11	-2	0
0	51	14	3	12	10	4	-1	0
0	51	16	2	7	6	7	-2	0
0	52	11	1	5	4	5	-1	0
0	52	15	3	11	9	11	-3	0
0	52	18	2	6	5	3	-1	0
0	53	13	2	9	7	9	-2	0
0	53	17	3	10	8	10	-3	0
0	54	10	2	11	8	11	-2	0
0	54	14	1	4	3	4	-1	0
0	54	18	3	9	7	3	-1	0
0	55	15	3	11	8	11	-3	0
0	55	18	4	12	9	3	-1	0
0	56	11	2	10	7	5	-1	0
0	56	16	2	7	5	7	-2	0
0	56	17	3	10	7	10	-3	0
0	57	13	2	9	6	9	-2	0
0	57	14	3	12	8	4	-1	0
0	58	10	2	11	7	11	-2	0
0	58	15	3	11	7	11	-3	0
0	58	18	1	3	2	3	-1	0
0	59	14	2	8	5	4	-1	0
6	51	10	2	11	9	12	-3	1
6	51	16	2	7	6	11	-4	1
6	52	11	1	5	4	11	-3	1
6	55	15	3	11	8	12	-4	1
6	56	17	3	10	7	11	-4	1
6	58	18	1	3	2	10	-4	1
7	53	13	2	9	7	10	-3	1
7	54	14	1	4	3	9	-3	1
8	56	16	2	7	5	8	-3	1
9	52	15	3	11	9	8	-3	1
9	58	18	1	3	2	7	-3	1
0 50 20								
0	50	22	4	10	9	5	-2	0
0	50	23	5	12	11	12	-5	0
0	50	24	5	11	10	11	-5	0
0	51	21	3	8	7	8	-3	0
0	51	24	4	9	8	9	-4	0
0	51	27	5	10	9	2	-1	0
0	51	27	6	12	11	2	-1	0
0	51	29	6	11	10	11	-6	0
0	52	20	4	11	9	11	-4	0
0	52	23	3	7	6	7	-3	0
0	52	23	5	12	10	12	-5	0
0	52	27	4	8	7	2	-1	0
0	52	29	5	9	8	9	-5	0
0	53	22	2	5	4	5	-2	0
0	53	24	5	11	9	11	-5	0
0	53	27	3	6	5	2	-1	0
0	54	27	5	10	8	2	-1	0
0	54	29	6	11	9	11	-6	0
0	55	21	3	8	6	8	-3	0
0	55	23	5	12	9	12	-5	0
0	55	24	4	9	7	9	-4	0
0	56	20	4	11	8	11	-4	0
0	56	24	5	11	8	11	-5	0
0	56	27	2	4	3	2	-1	0
0	56	29	5	9	7	9	-5	0
0	57	22	4	10	7	5	-2	0
0	57	23	3	7	5	7	-3	0
0	57	29	6	11	8	11	-6	0
0	58	23	5	12	8	12	-5	0
0	58	27	5	10	7	2	-1	0
0	59	20	4	11	7	11	-4	0
0	59	24	4	9	6	9	-4	0
0	59	27	3	6	4	2	-1	0
5	52	23	3	7	6	12	-6	1
5	59	20	4	11	7	12	-5	1
5	59	24	4	9	6	12	-6	1
6	51	21	3	8	7	11	-5	1
6	52	23	5	12	10	10	-5	1
6	52	29	5	9	8	11	-7	1
6	57	23	3	7	5	10	-5	1
7	50	24	5	11	10	9	-5	1
7	56	20	4	11	8	9	-4	1
7	58	23	5	12	8	8	-4	1
8	53	22	2	5	4	8	-4	1
9	51	24	4	9	8	7	-4	1
9	53	24	5	11	9	7	-4	1
9	57	29	6	11	8	6	-4	1
0 50 30								
0	50	32	5	8	8	8	-5	0
0	50	32	7	11	11	11	-7	0
0	50	34	2	3	3	3	-2	0
0	50	39	9	11	12	11	-9	0
0	51	35	7	10	10	10	-7	0
0	51	36	5	7	7	7	-5	0
0	51	36	8	11	11	11	-8	0
0	51	37	3	4	4	4	-3	0
0	52	30	7	12	11	12	-7	0
0	52	31	6	10	9	5	-3	0
0	52	38	7	9	9	9	-7	0
0	52	39	4	5	5	5	-4	0
0	52	39	9	11	11	11	-9	0
0	53	30	4	7	6	7	-4	0
0	53	32	5	8	7	8	-5	0
0	53	32	7	11	10	11	-7	0
0	53	34	8	12	11	3	-2	0
0	54	30	7	12	10	12	-7	0
0	54	34	6	9	8	3	-2	0
0	54	35	7	10	9	10	-7	0
0	54	36	8	11	10	11	-8	0
0	54	37	9	12	11	4	-3	0
0	55	32	7	11	9	11	-7	0
0	55	34	4	6	5	3	-2	0
0	55	36	5	7	6	7	-5	0
0	55	37	6	8	7	4	-3	0
0	55	38	7	9	8	9	-7	0
0	55	39	8	10	9	5	-4	0
0	55	39	9	11	10	11	-9	0
0	56	31	3	5	4	5	-3	0
0	56	37	9	12	10	4	-3	0
0	57	30	7	12	9	12	-7	0
0	57	34	6	9	7	3	-2	0
0	57	35	7	10	8	10	-7	0
0	57	36	8	11	9	11	-8	0
0	58	30	4	7	5	7	-4	0
0	58	32	5	8	6	8	-5	0
0	58	32	7	11	8	11	-7	0
0	58	34	8	12	9	3	-2	0
0	58	38	7	9	7	9	-7	0
0	58	39	4	5	4	5	-4	0
0	58	39	9	11	9	11	-9	0
0	59	31	6	10	7	5	-3	0
0	59	37	3	4	3	4	-3	0
5	50	34	2	3	3	12	-9	1
5	51	36	8	11	11	11	-9	1
5	51	37	3	4	4	12	-10	1
5	52	39	4	5	5	10	-9	1
5	52	39	9	11	11	11	-10	1
5	55	36	5	7	6	10	-8	1
5	56	31	3	5	4	12	-8	1
5	58	30	4	7	5	11	-7	1
5	58	39	9	11	9	10	-9	1
5	59	37	3	4	3	11	-9	1
6	50	32	7	11	11	11	-8	1

φ_1	Φ	φ_2	H	K	L	U	V	W
0	50	30						
6	51	35	7	10	10	10	-8	1
6	52	38	7	9	9	9	-8	1
6	57	30	7	12	9	9	-6	1
6	58	32	5	8	6	10	-7	1
6	58	38	7	9	7	8	-7	1
6	58	39	4	5	4	9	-8	1
7	50	32	5	8	8	8	-6	1
7	50	34	2	3	3	9	-7	1
7	51	37	3	4	4	8	-7	1
7	53	30	4	7	6	9	-6	1
7	53	32	7	11	10	8	-6	1
7	59	37	3	4	3	7	-6	1
8	51	36	5	7	7	7	-6	1
8	52	30	7	12	11	7	-5	1
8	54	36	8	11	10	7	-6	1
8	56	31	3	5	4	7	-5	1
9	50	39	9	11	12	6	-6	1
9	51	37	3	4	4	12	-11	2
9	52	39	9	11	11	11	-11	2
9	55	37	6	8	7	11	-10	2
9	57	35	7	10	8	6	-5	1
9	58	32	5	8	6	12	-9	2
0	50	40						
0	50	41	7	8	9	8	-7	0
0	50	42	8	9	10	9	-8	0
0	50	45	5	5	6	1	-1	0
0	50	45	6	6	7	1	-1	0
0	50	48	9	8	10	8	-9	0
0	50	49	8	7	9	7	-8	0
0	51	42	9	10	11	10	-9	0
0	51	42	10	11	12	11	-10	0
0	51	45	7	7	8	1	-1	0
0	51	45	8	8	9	1	-1	0
0	51	48	10	9	11	9	-10	0
0	51	48	11	10	12	10	-11	0
0	52	40	5	6	6	6	-5	0
0	52	45	9	9	10	1	-1	0
0	52	45	10	10	11	1	-1	0
0	52	45	11	11	12	1	-1	0
0	53	41	6	7	7	7	-6	0
0	53	41	7	8	8	8	-7	0
0	53	42	8	9	9	9	-8	0
0	53	42	9	10	10	10	-9	0
0	53	48	9	8	9	8	-9	0
0	53	48	10	9	10	9	-10	0
0	53	49	7	6	7	6	-7	0

φ_1	Φ	φ_2	H	K	L	U	V	W
0	50	40						
0	53	49	8	7	8	7	-8	0
0	54	42	10	11	11	11	-10	0
0	54	43	11	12	12	12	-11	0
0	54	47	12	11	12	11	-12	0
0	54	48	11	10	11	10	-11	0
0	55	40	10	12	11	6	-5	0
0	55	45	1	1	1	1	-1	0
0	56	42	8	9	8	9	-8	0
0	56	42	9	10	9	10	-9	0
0	56	42	10	11	10	11	-10	0
0	56	43	11	12	11	12	-11	0
0	56	47	12	11	11	11	-12	0
0	56	48	9	8	8	8	-9	0
0	56	48	10	9	9	9	-10	0
0	56	48	11	10	10	10	-11	0
0	57	40	5	6	5	6	-5	0
0	57	41	6	7	6	7	-6	0
0	57	41	7	8	7	8	-7	0
0	57	45	11	11	10	1	-1	0
0	57	45	12	12	11	1	-1	0
0	57	49	7	6	6	6	-7	0
0	57	49	8	7	7	7	-8	0
0	58	43	11	12	10	12	-11	0
0	58	45	8	8	7	1	-1	0
0	58	45	9	9	8	1	-1	0
0	58	45	10	10	9	1	-1	0
0	58	47	12	11	10	11	-12	0
0	59	42	9	10	8	10	-9	0
0	59	42	10	11	9	11	-10	0
0	59	45	6	6	5	1	-1	0
0	59	45	7	7	6	1	-1	0
0	59	48	10	9	8	9	-10	0
0	59	48	11	10	9	10	-11	0
4	51	42	10	11	12	12	-12	1
4	52	40	5	6	6	12	-11	1
4	54	43	11	12	12	12	-12	1
4	55	45	1	1	1	11	-12	1
4	56	43	11	12	11	11	-11	1
4	56	48	11	10	10	10	-12	1
4	59	48	10	9	8	10	-12	1
5	50	42	8	9	10	10	-10	1
5	51	42	9	10	11	11	-11	1
5	53	42	9	10	10	10	-10	1
5	54	42	10	11	11	11	-11	1
5	54	47	12	11	12	10	-12	1
5	54	48	11	10	11	9	-11	1
5	55	45	1	1	1	10	-11	1
5	55	45	1	1	1	9	-10	1

φ_1	Φ	φ_2	H	K	L	U	V	W
0	50	40						
5	56	42	9	10	9	9	-9	1
5	56	42	10	11	10	10	-10	1
5	56	48	9	6	8	8	-10	1
5	56	48	10	9	9	9	-11	1
5	57	40	5	6	5	11	-10	1
5	58	43	11	12	10	10	-10	1
5	59	42	10	11	9	9	-9	1
6	50	41	7	8	9	9	-9	1
6	51	48	10	9	11	7	-9	1
6	51	48	11	10	12	8	-10	1
6	53	41	7	8	8	8	-8	1
6	53	42	8	9	9	9	-9	1
6	53	48	9	6	9	7	-9	1
6	53	48	10	9	10	8	-10	1
6	55	45	1	1	1	8	-9	1
6	56	42	8	9	8	8	-8	1
6	57	49	8	7	7	7	-9	1
6	59	42	9	10	8	8	-8	1
7	50	48	9	8	10	6	-8	1
7	53	41	6	7	7	7	-7	1
7	53	49	8	7	8	6	-8	1
7	55	45	1	1	1	7	-8	1
7	57	41	7	8	7	7	-7	1
7	57	49	7	6	6	6	-8	1
8	52	40	5	6	6	6	-6	1
8	53	49	7	6	7	5	-7	1
8	55	45	1	1	1	6	-7	1
8	57	41	6	7	6	6	-6	1
9	50	42	8	9	10	11	-12	2
9	50	49	8	7	9	5	-7	1
9	54	42	10	11	11	11	-12	2
9	54	47	12	11	12	9	-12	2
9	55	40	10	12	11	11	-11	2
9	55	45	1	1	1	5	-6	1
9	56	43	11	12	11	10	-11	2
9	56	48	10	9	9	9	-12	2
0	50	50						
0	50	51	11	9	12	9	-11	0
0	50	56	3	2	3	2	-3	0
0	50	58	8	5	8	5	-8	0
0	50	58	11	7	11	7	-11	0
0	51	53	4	3	4	3	-4	0
0	51	54	7	5	7	5	-7	0
0	51	54	11	8	11	8	-11	0
0	51	55	10	7	10	7	-10	0
0	52	50	6	5	6	5	-6	0

φ_1	Φ	φ_2	H	K	L	U	V	W
0	50	50						
0	52	51	5	4	5	4	-5	0
0	52	51	11	9	11	9	-11	0
0	52	52	9	7	9	7	-9	0
0	52	59	10	6	9	3	-5	0
0	53	56	12	8	11	2	-3	0
0	53	58	8	5	7	5	-8	0
0	53	58	11	7	10	7	-11	0
0	54	53	12	9	11	3	-4	0
0	54	54	11	6	10	8	-11	0
0	54	55	10	7	9	7	-10	0
0	54	56	9	6	8	2	-3	0
0	55	50	12	10	11	5	-6	0
0	55	51	10	8	9	4	-5	0
0	55	51	11	9	10	9	-11	0
0	55	52	9	7	8	7	-9	0
0	55	53	8	6	7	3	-4	0
0	55	54	7	5	6	5	-7	0
0	55	56	6	4	5	2	-3	0
0	55	58	11	7	9	7	-11	0
0	56	53	12	9	10	3	-4	0
0	56	59	5	3	4	3	-5	0
0	57	50	6	5	5	5	-6	0
0	57	54	11	8	9	8	-11	0
0	57	55	10	7	8	7	-10	0
0	57	56	9	6	7	2	-3	0
0	58	51	5	4	4	4	-5	0
0	58	51	11	9	9	9	-11	0
0	58	52	9	7	7	7	-9	0
0	58	56	12	8	9	2	-3	0
0	58	58	8	5	6	5	-8	0
0	58	58	11	7	8	7	-11	0
0	59	53	4	3	3	3	-4	0
0	59	59	10	6	7	3	-5	0
5	50	56	3	2	3	7	-12	1
5	51	53	4	3	4	8	-12	1
5	52	50	6	5	6	9	-12	1
5	52	51	11	9	11	8	-11	1
5	55	54	7	5	6	7	-11	1
5	58	51	5	4	4	8	-11	1
5	58	51	11	9	9	9	-12	1
5	58	52	9	7	7	7	-10	1
6	50	58	11	7	11	6	-11	1
6	51	54	11	8	11	7	-11	1
6	51	55	10	7	10	6	-10	1
6	52	51	5	4	5	7	-10	1
6	53	58	8	5	7	6	-11	1
6	59	53	4	3	3	6	-9	1
7	50	56	3	2	3	5	-9	1

TAB. VI

FROM EULER ANGLES INTO HKL-UVW

φ_1 Φ φ_2	H	K	L	U	V	W
0 50 50						
7 52 52	9	7	9	6	-9	1
7 57 54	11	8	9	5	-8	1
7 58 58	11	7	8	5	-9	1
8 50 58	8	5	8	4	-8	1
8 51 53	4	3	4	5	-8	1
8 56 59	5	3	4	4	-8	1
8 57 50	6	5	5	5	-7	1
9 51 54	7	5	7	4	-7	1
9 54 55	10	7	9	4	-7	1
9 58 51	5	4	4	4	-6	1
0 50 60						
0 50 66	11	5	10	5	-11	0
0 50 67	12	5	11	5	-12	0
0 50 68	10	4	9	2	-5	0
0 51 61	11	6	10	6	-11	0
0 51 63	10	5	9	1	-2	0
0 51 63	12	6	11	1	-2	0
0 51 66	9	4	8	4	-9	0
0 51 69	8	3	7	3	-8	0
0 52 60	12	7	11	7	-12	0
0 52 61	9	5	8	5	-9	0
0 52 63	8	4	7	1	-2	0
0 52 67	7	3	6	3	-7	0
0 52 67	12	5	10	5	-12	0
0 53 60	7	4	6	4	-7	0
0 53 63	6	3	5	1	-2	0
0 53 66	11	5	9	5	-11	0
0 53 68	5	2	4	2	-5	0
0 54 60	12	7	10	7	-12	0
0 54 61	11	6	9	6	-11	0
0 54 63	10	5	8	1	-2	0
0 55 66	9	4	7	4	-9	0
0 55 67	12	5	9	5	-12	0
0 55 69	8	3	6	3	-8	0
0 56 61	9	5	7	5	-9	0
0 56 63	4	2	3	1	-2	0
0 56 66	11	5	8	5	-11	0
0 57 60	12	7	9	7	-12	0
0 57 61	11	6	8	6	-11	0
0 57 67	7	3	5	3	-7	0
0 57 68	10	4	7	2	-5	0
0 58 60	7	4	5	4	-7	0
0 58 63	10	5	7	1	-2	0
0 58 67	12	5	8	5	-12	0
0 59 63	6	3	4	1	-2	0
0 59 66	9	4	6	4	-9	0

φ_1 Φ φ_2	H	K	L	U	V	W
0 50 60						
5 53 60	7	4	6	6	-12	1
6 51 66	9	4	8	4	-11	1
6 53 68	5	2	4	4	-12	1
6 54 60	12	7	10	5	-10	1
6 57 67	7	3	5	4	-11	1
6 58 60	7	4	5	5	-10	1
7 51 61	11	6	10	4	-9	1
7 55 67	12	5	9	3	-9	1
7 55 69	8	3	6	3	-10	1
8 52 67	7	3	6	3	-9	1
9 52 61	9	5	8	3	-7	1
9 54 61	11	6	9	3	-7	1
0 50 70						
0 50 72	9	3	8	1	-3	0
0 50 76	8	2	7	1	-4	0
0 51 74	7	2	6	2	-7	0
0 51 76	12	3	10	1	-4	0
0 52 70	11	4	9	4	-11	0
0 52 72	6	2	5	1	-3	0
0 52 75	11	3	9	3	-11	0
0 52 79	5	1	4	1	-5	0
0 53 73	10	3	8	3	-10	0
0 53 77	9	2	7	2	-9	0
0 54 72	9	3	7	1	-3	0
0 54 76	4	1	3	1	-4	0
0 55 72	12	4	9	1	-3	0
0 55 75	11	3	8	3	-11	0
0 56 70	11	4	8	4	-11	0
0 56 73	10	3	7	3	-10	0
0 56 74	7	2	5	2	-7	0
0 56 79	10	2	7	1	-5	0
0 57 76	12	3	8	1	-4	0
0 57 77	9	2	6	2	-9	0
0 58 72	3	1	2	1	-3	0
0 58 75	11	3	7	3	-11	0
0 59 70	11	4	7	4	-11	0
0 59 76	8	2	5	1	-4	0
6 54 76	4	1	3	2	-11	1
6 57 77	9	2	6	2	-12	1
6 58 72	3	1	2	3	-11	1
6 59 70	11	4	7	3	-10	1
7 51 74	7	2	6	2	-10	1
7 55 75	11	3	8	2	-10	1
7 56 73	10	3	7	2	-9	1
8 52 79	5	1	4	1	-9	1
8 58 72	3	1	2	2	-8	1

φ_1 Φ φ_2	H	K	L	U	V	W
0 50 70						
9 53 77	9	2	7	1	-8	1
0 50 80						
0 50 82	7	1	6	1	-7	0
0 50 85	12	1	10	1	-12	0
0 51 80	11	2	9	2	-11	0
0 51 81	6	1	5	1	-6	0
0 51 84	10	1	8	1	-10	0
0 51 85	11	1	9	1	-11	0
0 52 84	9	1	7	1	-9	0
0 53 83	8	1	6	1	-8	0
0 53 85	12	1	9	1	-12	0
0 54 80	11	2	8	2	-11	0
0 54 81	12	2	9	1	-6	0
0 54 85	11	1	8	1	-11	0
0 55 82	7	1	5	1	-7	0
0 55 84	10	1	7	1	-10	0
0 56 84	9	1	6	1	-9	0
0 56 85	12	1	8	1	-12	0
0 57 81	6	1	4	1	-6	0
0 58 80	11	2	7	2	-11	0
0 58 83	8	1	5	1	-8	0
0 58 85	11	1	7	1	-11	0
0 59 84	10	1	6	1	-10	0
6 55 82	7	1	5	1	-12	1
7 50 85	12	1	10	0	-10	1
7 51 80	11	2	9	1	-10	1
7 51 81	6	1	5	1	-11	1
7 57 81	6	1	4	1	-10	1
7 58 80	11	2	7	1	-9	1
8 51 85	11	1	9	0	-9	1
8 53 85	12	1	9	0	-9	1
9 51 84	10	1	8	0	-8	1
9 54 85	11	1	8	0	-8	1
9 56 85	12	1	8	0	-8	1
0 50 90						
0 50 90	6	0	5	0	-1	0
0 51 90	5	0	4	0	-1	0
0 51 90	11	0	9	0	-1	0
0 52 90	9	0	7	0	-1	0
0 53 90	4	0	3	0	-1	0
0 54 90	7	0	5	0	-1	0
0 54 90	11	0	8	0	-1	0
0 55 90	10	0	7	0	-1	0
0 56 90	3	0	2	0	-1	0

φ_1 Φ φ_2	H	K	L	U	V	W
0 50 90						
0 58 90	8	0	5	0	-1	0
0 58 90	11	0	7	0	-1	0
0 59 90	5	0	3	0	-1	0
0 60 0						
0 60 0	0	7	4	1	0	0
0 60 0	0	12	7	1	0	0
0 60 5	1	12	7	12	-1	0
0 60 9	2	12	7	6	-1	0
0 61 0	0	9	5	1	0	0
0 61 0	0	11	6	1	0	0
0 61 5	1	11	6	11	-1	0
0 61 6	1	9	5	9	-1	0
0 61 8	1	7	4	7	-1	0
0 63 0	0	2	1	1	0	0
0 64 5	1	12	6	12	-1	0
0 64 6	1	10	5	10	-1	0
0 64 7	1	8	4	8	-1	0
0 64 9	1	6	3	6	-1	0
0 66 0	0	9	4	1	0	0
0 66 0	0	11	5	1	0	0
0 66 5	1	11	5	11	-1	0
0 66 6	1	9	4	9	-1	0
0 67 0	0	7	3	1	0	0
0 67 0	0	12	5	1	0	0
0 67 5	1	12	5	12	-1	0
0 67 8	1	7	3	7	-1	0
0 68 0	0	5	2	1	0	0
0 68 6	1	10	4	10	-1	0
0 68 9	2	12	5	6	-1	0
0 69 0	0	8	3	1	0	0
5 64 7	1	8	4	12	-2	1
6 61 8	1	7	4	10	-2	1
6 67 8	1	7	3	11	-2	1
7 64 9	1	6	3	9	-2	1
9 67 5	1	12	5	7	-1	1
0 60 10						
0 60 11	1	5	3	5	-1	0
0 60 14	3	12	7	4	-1	0
0 60 17	3	10	6	10	-3	0
0 61 16	2	7	4	7	-2	0
0 61 18	4	12	7	3	-1	0
0 62 10	2	11	6	11	-2	0
0 62 13	2	9	5	9	-2	0
0 62 15	3	11	6	11	-3	0

TAB. VI — FROM EULER ANGLES INTO HKL-UVW — VI - 10

φ_1	Φ	φ_2	H	K	L	U	V	W
0	60	10						
0	62	18	3	9	5	3	-1	0
0	64	11	2	10	5	5	-1	0
0	64	14	1	4	2	4	-1	0
0	64	17	3	10	5	10	-3	0
0	65	18	2	6	3	3	-1	0
0	66	10	2	11	5	11	-2	0
0	66	15	3	11	5	11	-3	0
0	67	13	2	9	4	9	-2	0
0	67	18	3	9	4	3	-1	0
0	68	14	3	12	5	4	-1	0
0	68	16	2	7	3	7	-2	0
0	68	18	4	12	5	3	-1	0
0	69	11	1	5	2	5	-1	0
0	69	17	3	10	4	10	-3	0
5	60	11	1	5	3	12	-3	1
5	61	16	2	7	4	12	-4	1
5	69	17	3	10	4	12	-4	1
6	62	13	2	9	5	11	-3	1
6	64	14	1	4	2	10	-3	1
7	62	15	3	11	6	9	-3	1
7	68	16	2	7	3	9	-3	1
7	69	11	1	5	2	8	-2	1
8	60	17	3	10	6	8	-3	1
8	62	10	2	11	6	8	-2	1
9	60	11	1	5	3	7	-2	1
9	67	13	2	9	4	7	-2	1
0	60	20						
0	60	21	3	8	5	8	-3	0
0	60	24	5	11	7	11	-5	0
0	60	29	5	9	6	9	-5	0
0	61	22	2	5	3	5	-2	0
0	61	27	4	8	5	2	-1	0
0	61	29	6	11	7	11	-6	0
0	62	23	3	7	4	7	-3	0
0	62	23	5	12	7	12	-5	0
0	62	27	5	10	6	2	-1	0
0	62	27	6	12	7	2	-1	0
0	63	20	4	11	6	11	-4	0
0	63	24	4	9	5	9	-4	0
0	64	24	5	11	6	11	-5	0
0	64	29	5	9	5	9	-5	0
0	64	29	6	11	6	11	-6	0
0	65	21	3	8	4	8	-3	0
0	65	22	4	10	5	5	-2	0
0	65	23	5	12	6	12	-5	0
0	66	27	1	2	1	2	-1	0

φ_1	Φ	φ_2	H	K	L	U	V	W
0	60	20						
0	67	20	4	11	5	11	-4	0
0	68	23	3	7	3	7	-3	0
0	68	24	4	9	4	9	-4	0
0	68	24	5	11	5	11	-5	0
0	68	29	6	11	5	11	-6	0
0	69	23	5	12	5	12	-5	0
0	69	29	5	9	4	9	-5	0
4	68	29	6	11	5	12	-7	1
5	61	22	2	5	3	11	-5	1
5	64	24	5	11	6	12	-6	1
5	64	29	6	11	6	10	-6	1
5	65	21	3	8	4	12	-5	1
5	66	27	1	2	1	11	-6	1
5	69	23	5	12	5	11	-5	1
5	69	29	5	9	4	10	-6	1
6	63	24	4	9	5	10	-5	1
6	66	27	1	2	1	9	-5	1
6	68	24	5	11	5	10	-5	1
7	60	21	3	8	5	9	-4	1
7	61	29	6	11	7	8	-5	1
7	62	23	3	7	4	8	-4	1
7	64	29	5	9	5	8	-5	1
7	68	24	4	9	4	8	-4	1
8	66	27	1	2	1	7	-4	1
8	67	20	4	11	5	7	-3	1
9	60	29	5	9	6	6	-4	1
9	65	23	5	12	6	6	-3	1
9	66	27	1	2	1	12	-7	2
9	68	23	3	7	3	6	-3	1
9	69	29	5	9	4	11	-7	2
0	60	30						
0	60	30	7	12	8	12	-7	0
0	60	35	7	10	7	10	-7	0
0	60	36	5	7	5	7	-5	0
0	60	36	8	11	8	11	-8	0
0	61	34	2	3	2	3	-2	0
0	61	39	8	10	7	5	-4	0
0	61	39	9	11	8	11	-9	0
0	62	32	5	8	5	8	-5	0
0	62	32	7	11	7	11	-7	0
0	62	37	9	12	8	4	-3	0
0	62	38	7	9	6	9	-7	0
0	63	30	7	12	7	12	-7	0
0	63	31	3	5	3	5	-3	0
0	63	36	8	11	7	11	-8	0
0	63	37	6	8	5	4	-3	0

φ_1	Φ	φ_2	H	K	L	U	V	W
0	60	30						
0	64	30	4	7	4	7	-4	0
0	64	34	8	12	7	3	-2	0
0	64	35	7	10	6	10	-7	0
0	64	39	9	11	7	11	-9	0
0	65	32	7	11	6	11	-7	0
0	65	34	6	9	5	3	-2	0
0	65	36	5	7	4	7	-5	0
0	65	37	9	12	7	4	-3	0
0	65	39	4	5	3	5	-4	0
0	66	36	8	11	6	11	-8	0
0	66	38	7	9	5	9	-7	0
0	67	30	7	12	6	12	-7	0
0	67	31	6	10	5	5	-3	0
0	67	32	5	8	4	8	-5	0
0	67	34	4	6	3	3	-2	0
0	67	39	9	11	6	11	-9	0
0	68	35	7	10	5	10	-7	0
0	68	37	3	4	2	4	-3	0
0	69	32	7	11	5	11	-7	0
0	69	39	8	10	5	5	-4	0
4	62	38	7	9	6	12	-10	1
4	64	35	7	10	6	12	-9	1
4	67	32	5	8	4	12	-8	1
5	60	36	8	11	8	10	-8	1
5	61	34	2	3	2	11	-8	1
5	62	32	7	11	7	10	-7	1
5	63	30	7	12	7	11	-7	1
5	64	39	9	11	7	9	-8	1
5	68	37	3	4	2	10	-8	1
6	60	35	7	10	7	9	-7	1
6	63	31	3	5	3	9	-6	1
6	65	36	5	7	4	9	-7	1
6	65	39	4	5	3	8	-7	1
7	61	34	2	3	2	8	-6	1
7	65	32	7	11	6	7	-5	1
7	66	38	7	9	5	7	-6	1
8	60	36	5	7	5	6	-5	1
8	61	39	8	10	7	12	-11	2
8	62	32	5	8	5	7	-5	1
8	63	36	8	11	7	6	-5	1
8	65	39	4	5	3	11	-10	2
8	67	34	4	6	3	12	-9	2
8	68	37	3	4	2	6	-5	1
9	64	30	4	7	4	6	-4	1
9	65	36	5	7	4	11	-9	2
9	67	30	7	12	6	6	-4	1
9	69	39	8	10	5	10	-9	2

φ_1	Φ	φ_2	H	K	L	U	V	W
0	60	40						
0	60	40	10	12	9	6	-5	0
0	60	42	8	9	7	9	-8	0
0	60	45	11	11	9	1	-1	0
0	60	48	9	8	7	8	-9	0
0	61	41	7	8	6	8	-7	0
0	61	43	11	12	9	12	-11	0
0	61	45	5	5	4	1	-1	0
0	61	45	9	9	7	1	-1	0
0	61	47	12	11	9	11	-12	0
0	61	49	8	7	6	7	-8	0
0	62	41	6	7	5	7	-6	0
0	62	42	10	11	8	11	-10	0
0	62	45	4	4	3	1	-1	0
0	62	48	11	10	8	10	-11	0
0	62	49	7	6	5	6	-7	0
0	63	40	5	6	4	6	-5	0
0	63	42	9	10	7	10	-9	0
0	63	45	7	7	5	1	-1	0
0	63	45	11	11	8	1	-1	0
0	63	48	10	9	7	9	-10	0
0	64	42	8	9	6	9	-8	0
0	64	43	11	12	8	12	-11	0
0	64	45	10	10	7	1	-1	0
0	64	47	12	11	8	11	-12	0
0	64	48	9	8	6	8	-9	0
0	65	41	7	8	5	8	-7	0
0	65	42	10	11	7	11	-10	0
0	65	45	3	3	2	1	-1	0
0	65	48	11	10	7	10	-11	0
0	65	49	8	7	5	7	-8	0
0	66	40	10	12	7	6	-5	0
0	66	42	9	10	6	10	-9	0
0	66	45	8	8	5	1	-1	0
0	66	45	11	11	7	1	-1	0
0	66	48	10	9	6	9	-10	0
0	67	41	6	7	4	7	-6	0
0	67	42	8	9	5	9	-8	0
0	67	43	11	12	7	12	-11	0
0	67	45	5	5	3	1	-1	0
0	67	47	12	11	7	11	-12	0
0	67	48	9	8	5	8	-9	0
0	67	49	7	6	4	6	-7	0
0	68	42	10	11	6	11	-10	0
0	68	45	7	7	4	1	-1	0
0	68	45	12	12	7	1	-1	0
0	68	48	11	10	6	10	-11	0
0	69	40	5	6	3	6	-5	0

φ_1	Φ	φ_2	H	K	L	U	V	W
0	60	40						
0	69	41	7	8	4	8	-7	0
0	69	45	9	9	5	1	-1	0
0	69	45	11	11	6	1	-1	0
0	69	49	8	7	4	7	-8	0
4	62	41	6	7	5	12	-11	1
4	64	48	9	8	6	10	-12	1
4	65	49	8	7	5	9	-11	1
4	67	41	6	7	4	11	-10	1
4	69	41	7	8	4	12	-11	1
4	69	49	8	7	4	10	-12	1
5	60	48	9	8	7	9	-11	1
5	61	43	11	12	9	9	-9	1
5	61	49	8	7	6	8	-10	1
5	63	40	5	6	4	10	-9	1
5	67	49	7	6	4	8	-10	1
5	69	40	5	6	3	9	-8	1
6	62	42	10	11	8	8	-8	1
6	62	49	7	6	5	7	-9	1
6	64	43	11	12	8	8	-8	1
6	65	42	10	11	7	7	-7	1
6	67	43	11	12	7	7	-7	1
7	60	42	8	9	7	7	-7	1
7	63	42	9	10	7	7	-7	1
7	66	42	9	10	6	6	-6	1
7	68	42	10	11	6	6	-6	1
8	61	41	7	8	6	6	-6	1
8	64	42	8	9	6	6	-6	1
8	69	40	5	6	3	12	-11	2
9	61	49	8	7	6	9	-12	2
9	62	41	6	7	5	5	-5	1
9	65	41	7	8	5	5	-5	1
9	67	42	8	9	5	5	-5	1
0	60	50						
0	60	50	12	10	9	5	-6	0
0	60	54	7	5	5	5	-7	0
0	60	54	11	8	8	8	-11	0
0	60	55	10	7	7	7	-10	0
0	61	51	10	8	7	4	-5	0
0	61	51	11	9	8	9	-11	0
0	61	56	3	2	2	2	-3	0
0	62	52	9	7	6	7	-9	0
0	62	53	12	9	8	3	-4	0
0	62	58	8	5	5	5	-8	0
0	62	58	11	7	7	7	-11	0
0	63	50	6	5	4	5	-6	0
0	63	53	8	6	5	3	-4	0

φ_1	Φ	φ_2	H	K	L	U	V	W
0	60	50						
0	63	54	11	8	7	8	-11	0
0	63	59	5	3	3	3	-5	0
0	64	51	11	9	7	9	-11	0
0	64	55	10	7	6	7	-10	0
0	64	56	12	8	7	2	-3	0
0	65	51	5	4	3	4	-5	0
0	65	53	12	9	7	3	-4	0
0	65	54	7	5	4	5	-7	0
0	65	56	9	6	5	2	-3	0
0	65	58	11	7	6	7	-11	0
0	66	50	12	10	7	5	-6	0
0	66	52	9	7	5	7	-9	0
0	66	54	11	8	6	8	-11	0
0	67	51	11	9	6	9	-11	0
0	67	56	6	4	3	2	-3	0
0	67	58	8	5	4	5	-8	0
0	67	59	10	6	5	3	-5	0
0	68	53	4	3	2	3	-4	0
0	68	55	10	7	5	7	-10	0
0	69	50	6	5	3	5	-6	0
0	69	51	10	8	5	4	-5	0
0	69	58	11	7	5	7	-11	0
4	65	51	5	4	3	9	-12	1
4	65	54	7	5	4	8	-12	1
4	67	58	8	5	4	7	-12	1
5	60	54	11	8	8	8	-12	1
5	60	55	10	7	7	7	-11	1
5	62	58	11	7	7	7	-12	1
5	63	59	5	3	3	6	-11	1
5	66	52	9	7	5	8	-11	1
5	68	53	4	3	2	7	-10	1
5	69	50	6	5	3	7	-9	1
6	61	56	3	2	2	6	-10	1
6	62	58	8	5	5	5	-9	1
6	63	50	6	5	4	6	-8	1
6	66	54	11	8	6	6	-9	1
6	67	51	11	9	6	6	-8	1
7	60	54	7	5	5	5	-8	1
7	64	55	10	7	6	5	-8	1
8	61	51	11	9	8	5	-7	1
8	61	56	3	2	2	4	-7	1
8	69	50	6	5	3	9	-12	2
8	69	58	11	7	5	4	-7	1
9	60	55	10	7	7	7	-12	2
9	62	52	9	7	6	4	-6	1
9	66	50	12	10	7	8	-11	2
9	67	56	6	4	3	7	-12	2
9	68	53	4	3	2	4	-6	1

φ_1	Φ	φ_2	H	K	L	U	V	W
0	60	60						
0	60	60	12	7	8	7	-12	0
0	60	61	9	5	6	5	-9	0
0	60	66	11	5	7	5	-11	0
0	60	69	8	3	5	3	-8	0
0	61	61	11	6	7	6	-11	0
0	61	63	8	4	5	1	-2	0
0	61	68	5	2	3	2	-5	0
0	62	63	10	5	6	1	-2	0
0	62	63	12	6	7	1	-2	0
0	62	67	7	3	4	3	-7	0
0	62	67	12	5	7	5	-12	0
0	63	60	12	7	7	7	-12	0
0	63	66	9	4	5	4	-9	0
0	64	60	7	4	4	4	-7	0
0	64	61	9	5	5	5	-9	0
0	64	61	11	6	6	6	-11	0
0	64	66	11	5	6	5	-11	0
0	65	67	12	5	6	5	-12	0
0	65	68	10	4	5	2	-5	0
0	65	69	8	3	4	3	-8	0
0	66	63	2	1	1	1	-2	0
0	67	60	12	7	6	7	-12	0
0	68	61	11	6	5	6	-11	0
0	68	66	9	4	4	4	-9	0
0	68	66	11	5	5	5	-11	0
0	68	67	7	3	3	3	-7	0
0	69	61	9	5	4	5	-9	0
0	69	67	12	5	5	5	-12	0
5	60	61	9	5	6	6	-12	1
5	64	61	11	6	6	6	-12	1
5	65	69	8	3	4	4	-12	1
5	66	63	2	1	1	5	-11	1
5	68	66	11	5	5	5	-12	1
6	62	67	12	5	7	4	-11	1
6	64	61	9	5	5	5	-10	1
6	64	66	11	5	6	4	-10	1
6	66	63	2	1	1	4	-9	1
6	68	61	11	6	5	5	-10	1
6	68	66	9	4	4	4	-10	1
7	60	60	12	7	8	4	-6	1
7	61	68	5	2	3	3	-9	1
7	63	66	9	4	5	3	-8	1
7	64	60	7	4	4	4	-8	1
7	68	67	7	3	3	3	-8	1
7	69	61	9	5	4	4	-8	1
8	60	66	11	5	7	3	-8	1
8	66	63	2	1	1	3	-7	1

φ_1	Φ	φ_2	H	K	L	U	V	W
0	60	60						
9	60	69	8	3	5	2	-7	1
9	67	60	12	7	6	3	-6	1
0	60	70						
0	60	73	10	3	6	3	-10	0
0	60	76	12	3	7	1	-4	0
0	60	79	5	1	3	1	-5	0
0	61	72	12	4	7	1	-3	0
0	61	74	7	2	4	2	-7	0
0	62	72	9	3	5	1	-3	0
0	62	75	11	3	6	3	-11	0
0	62	77	9	2	5	2	-9	0
0	63	70	11	4	6	4	-11	0
0	64	73	10	3	5	3	-10	0
0	64	76	4	1	2	1	-4	0
0	64	79	10	2	5	1	-5	0
0	65	72	6	2	3	1	-3	0
0	66	75	11	3	5	3	-11	0
0	67	70	11	4	5	4	-11	0
0	67	72	9	3	4	1	-3	0
0	67	77	9	2	4	2	-9	0
0	68	72	12	4	5	1	-3	0
0	68	74	7	2	3	2	-7	0
0	68	76	12	3	5	1	-4	0
0	69	73	10	3	4	3	-10	0
0	69	79	5	1	2	1	-5	0
5	60	73	10	3	6	3	-12	1
5	68	74	7	2	3	3	-12	1
5	69	79	5	1	2	2	-12	1
6	64	76	4	1	2	2	-10	1
6	67	77	9	2	4	2	-11	1
7	61	74	7	2	4	2	-9	1
7	66	75	11	3	5	2	-9	1
7	69	73	10	3	4	2	-8	1
8	60	79	5	1	3	1	-8	1
9	62	77	9	2	5	1	-7	1
9	63	70	11	4	6	2	-7	1
9	69	79	5	1	2	1	-7	1
0	60	80						
0	60	81	12	2	7	1	-6	0
0	60	85	12	1	7	1	-12	0
0	61	82	7	1	4	1	-7	0
0	61	84	9	1	5	1	-9	0
0	61	85	11	1	6	1	-11	0
0	62	80	11	2	6	2	-11	0

φ_1	Φ	φ_2	H	K	L	U	V	W
0	60	80						
0	64	81	6	1	3	1	-6	0
0	64	83	8	1	4	1	-8	0
0	64	84	10	1	5	1	-10	0
0	64	85	12	1	6	1	-12	0
0	66	80	11	2	5	2	-11	0
0	66	84	9	1	4	1	-9	0
0	66	85	11	1	5	1	-11	0
0	67	82	7	1	3	1	-7	0
0	67	85	12	1	5	1	-12	0
0	68	81	12	2	5	1	-6	0
0	68	84	10	1	4	1	-10	0
5	64	83	8	1	4	1	-12	1
6	61	82	7	1	4	1	-11	1
6	67	82	7	1	3	1	-10	1
7	64	81	6	1	3	1	-9	1
8	66	80	11	2	5	1	-8	1
9	60	85	12	1	7	0	-7	1
0	60	90						
0	60	90	7	0	4	0	-1	0
0	60	90	12	0	7	0	-1	0
0	61	90	9	0	5	0	-1	0
0	61	90	11	0	6	0	-1	0
0	63	90	2	0	1	0	-1	0
0	66	90	9	0	4	0	-1	0
0	66	90	11	0	5	0	-1	0
0	67	90	7	0	3	0	-1	0
0	67	90	12	0	5	0	-1	0
0	68	90	5	0	2	0	-1	0
0	69	90	8	0	3	0	-1	0
0	70	0						
0	70	0	0	11	4	1	0	0
0	70	5	1	11	4	11	-1	0
0	70	7	1	8	3	8	-1	0
0	72	0	0	3	1	1	0	0
0	72	5	1	12	4	12	-1	0
0	72	6	1	9	3	9	-1	0
0	72	9	1	6	2	6	-1	0
0	73	0	0	10	3	1	0	0
0	73	6	1	10	3	10	-1	0
0	74	0	0	7	2	1	0	0
0	74	8	1	7	2	7	-1	0
0	75	0	0	11	3	1	0	0
0	75	5	1	11	3	11	-1	0
0	76	0	0	4	1	1	0	0

φ_1	Φ	φ_2	H	K	L	U	V	W
0	70	0						
0	76	5	1	12	3	12	-1	0
0	76	7	1	8	2	8	-1	0
0	76	9	2	12	3	6	-1	0
0	77	0	0	9	2	1	0	0
0	78	6	1	9	2	9	-1	0
0	79	0	0	5	1	1	0	0
0	79	6	1	10	2	10	-1	0
5	74	8	1	7	2	12	-2	1
6	72	9	1	6	2	10	-2	1
6	76	5	1	12	3	9	-1	1
7	72	5	1	12	4	8	-1	1
7	75	5	1	11	3	8	-1	1
7	79	6	1	10	2	8	-1	1
8	73	6	1	10	3	7	-1	1
8	78	6	1	9	2	7	-1	1
9	70	5	1	11	4	7	-1	1
0	70	10						
0	70	10	2	11	4	11	-2	0
0	70	14	2	8	3	4	-1	0
0	71	15	3	11	4	11	-3	0
0	72	13	2	9	3	9	-2	0
0	72	14	3	12	4	4	-1	0
0	72	18	1	3	1	3	-1	0
0	74	11	2	10	3	5	-1	0
0	74	17	3	10	3	10	-3	0
0	75	10	2	11	3	11	-2	0
0	75	15	3	11	3	11	-3	0
0	75	16	2	7	2	7	-2	0
0	76	14	1	4	1	4	-1	0
0	77	18	4	12	3	3	-1	0
0	78	13	2	9	2	9	-2	0
0	78	18	3	9	2	3	-1	0
0	79	11	1	5	1	5	-1	0
0	79	17	3	10	2	10	-3	0
5	72	13	2	9	3	12	-3	1
5	72	18	1	3	1	11	-4	1
5	76	14	1	4	1	11	-3	1
6	74	17	3	10	3	9	-3	1
6	75	15	3	11	3	10	-3	1
6	79	11	1	5	1	9	-2	1
7	70	10	2	11	4	9	-2	1
7	72	18	1	3	1	8	-3	1
7	78	13	2	9	2	8	-2	1
8	76	14	1	4	1	7	-2	1
9	75	16	2	7	2	6	-2	1
9	79	17	3	10	2	6	-2	1

φ_1	Φ	φ_2	H	K	L	U	V	W
0	70	20						
0	70	22	2	5	2	5	-2	0
0	70	27	5	10	4	2	-1	0
0	70	27	6	12	5	2	-1	0
0	71	20	4	11	4	11	-4	0
0	71	21	3	8	3	8	-3	0
0	71	27	4	8	3	2	-1	0
0	72	24	5	11	4	11	-5	0
0	72	29	6	11	4	11	-6	0
0	73	23	5	12	4	12	-5	0
0	73	24	4	9	3	9	-4	0
0	73	27	3	6	2	2	-1	0
0	74	22	4	10	3	5	-2	0
0	74	29	5	9	3	9	-5	0
0	75	23	3	7	2	7	-3	0
0	75	27	5	10	3	2	-1	0
0	76	20	4	11	3	11	-4	0
0	76	24	5	11	3	11	-5	0
0	77	21	3	8	2	6	-3	0
0	77	23	5	12	3	12	-5	0
0	77	27	2	4	1	2	-1	0
0	77	29	6	11	3	11	-6	0
0	79	22	2	5	1	5	-2	0
0	79	24	4	9	2	9	-4	0
0	79	29	5	9	2	9	-5	0
4	74	29	5	9	3	12	-7	1
4	79	22	2	5	1	12	-5	1
5	75	23	3	7	2	11	-5	1
5	77	21	3	8	2	10	-4	1
6	70	22	2	5	2	9	-4	1
6	71	20	4	11	4	10	-4	1
6	77	23	5	12	3	9	-4	1
7	72	24	5	11	4	8	-4	1
8	71	21	3	8	3	7	-3	1
8	79	22	2	5	1	7	-3	1
9	73	24	4	9	3	6	-3	1
9	76	24	5	11	3	6	-3	1
9	77	21	3	8	2	12	-5	2
9	77	27	2	4	1	11	-6	2
0	70	30						
0	70	30	4	7	3	7	-4	0
0	70	30	7	12	5	12	-7	0
0	70	34	6	9	4	3	-2	0
0	70	36	8	11	5	11	-8	0
0	71	31	3	5	2	5	-3	0
0	71	34	8	12	5	3	-2	0

φ_1	Φ	φ_2	H	K	L	U	V	W
0	70	30						
0	71	36	5	7	3	7	-5	0
0	71	38	7	9	4	9	-7	0
0	71	39	9	11	5	11	-9	0
0	72	32	5	6	3	8	-5	0
0	72	35	7	10	4	10	-7	0
0	72	37	9	12	5	4	-3	0
0	73	32	7	11	4	11	-7	0
0	73	37	6	8	3	4	-3	0
0	73	39	4	5	2	5	-4	0
0	74	30	7	12	4	12	-7	0
0	74	34	2	3	1	3	-2	0
0	74	36	8	11	4	11	-8	0
0	74	39	9	11	4	11	-9	0
0	75	37	9	12	4	4	-3	0
0	75	38	7	9	3	9	-7	0
0	76	30	4	7	2	7	-4	0
0	76	31	6	10	3	5	-3	0
0	76	35	7	10	3	10	-7	0
0	77	32	7	11	3	11	-7	0
0	77	36	5	7	2	7	-5	0
0	77	39	8	10	3	5	-4	0
0	78	30	7	12	3	12	-7	0
0	78	32	5	8	2	8	-5	0
0	78	34	8	12	3	3	-2	0
0	78	36	8	11	3	11	-8	0
0	78	39	9	11	3	11	-9	0
0	79	37	3	4	1	4	-3	0
4	71	36	5	7	3	12	-9	1
4	71	38	7	9	4	11	-9	1
4	73	32	7	11	4	12	-8	1
4	73	39	4	5	2	12	-10	1
4	76	35	7	10	3	11	-8	1
4	78	36	8	11	3	12	-9	1
5	70	36	8	11	5	9	-7	1
5	71	31	3	5	2	11	-7	1
5	74	34	2	3	1	10	-7	1
5	76	30	4	7	2	10	-6	1
5	77	32	7	11	3	9	-6	1
5	79	37	3	4	1	9	-7	1
6	70	30	4	7	3	8	-5	1
6	71	39	9	11	5	8	-7	1
6	72	32	5	8	3	9	-6	1
6	72	35	7	10	4	8	-6	1
6	73	39	4	5	2	7	-6	1
6	74	30	7	12	4	8	-5	1
6	77	36	5	7	2	8	-6	1
6	78	39	9	11	3	7	-6	1
7	74	34	2	3	1	7	-5	1

φ_1	Φ	φ_2	H	K	L	U	V	W
0	70	30						
8	71	31	3	5	2	6	-4	1
8	73	37	6	8	3	11	-9	2
8	75	38	7	9	3	6	-5	1
8	76	35	7	10	3	12	-9	2
8	78	32	5	8	2	6	-4	1
9	71	36	5	7	3	5	-4	1
9	74	34	2	3	1	11	-8	2
9	74	36	8	11	4	5	-4	1
9	79	37	3	4	1	5	-4	1
0	70	40						
0	70	42	9	10	5	10	-9	0
0	70	43	11	12	6	12	-11	0
0	70	47	12	11	6	11	-12	0
0	70	48	10	9	5	9	-10	0
0	71	42	10	11	5	11	-10	0
0	71	45	2	2	1	1	-1	0
0	71	48	11	10	5	10	-11	0
0	72	40	10	12	5	6	-5	0
0	72	41	6	7	3	7	-6	0
0	72	42	8	9	4	9	-8	0
0	72	45	11	11	5	1	-1	0
0	72	48	9	8	4	8	-9	0
0	72	49	7	6	3	6	-7	0
0	73	42	9	10	4	10	-9	0
0	73	43	11	12	5	12	-11	0
0	73	45	7	7	3	1	-1	0
0	73	45	9	9	4	1	-1	0
0	73	47	12	11	5	11	-12	0
0	73	48	10	9	4	9	-10	0
0	74	41	7	8	3	8	-7	0
0	74	45	5	5	2	1	-1	0
0	74	45	12	12	5	1	-1	0
0	74	49	8	7	3	7	-8	0
0	75	42	10	11	4	11	-10	0
0	75	45	8	8	3	1	-1	0
0	75	48	11	10	4	10	-11	0
0	76	40	5	6	2	6	-5	0
0	76	42	8	9	3	9	-8	0
0	76	43	11	12	4	12	-11	0
0	76	45	11	11	4	1	-1	0
0	76	47	12	11	4	11	-12	0
0	76	48	9	8	3	8	-9	0
0	77	42	9	10	3	10	-9	0
0	77	45	3	3	1	1	-1	0
0	77	48	10	9	3	9	-10	0
0	78	41	6	7	2	7	-6	0

φ_1	Φ	φ_2	H	K	L	U	V	W
0	70	40						
0	78	45	10	10	3	1	-1	0
0	78	49	7	6	2	6	-7	0
0	79	40	10	12	3	6	-5	0
0	79	41	7	8	2	8	-7	0
0	79	42	10	11	3	11	-10	0
0	79	45	7	7	2	1	-1	0
0	79	45	11	11	3	1	-1	0
0	79	48	11	10	3	10	-11	0
0	79	49	8	7	2	7	-6	0
4	72	41	6	7	3	10	-9	1
4	72	49	7	6	3	9	-11	1
4	74	41	7	6	3	11	-10	1
4	76	42	8	9	3	12	-11	1
4	78	49	7	6	2	10	-12	1
4	79	41	7	8	2	10	-9	1
5	78	41	6	7	2	9	-6	1
5	79	48	11	10	3	7	-8	1
6	73	47	12	11	5	6	-7	1
6	75	48	11	10	4	6	-7	1
6	76	40	5	6	2	8	-7	1
6	76	47	12	11	4	7	-8	1
6	77	48	10	9	3	6	-7	1
7	70	43	11	12	6	6	-6	1
7	71	45	2	2	1	11	-12	2
7	79	41	7	8	2	12	-11	2
7	79	49	8	7	2	5	-6	1
8	70	47	12	11	6	5	-6	1
8	71	42	10	11	5	5	-5	1
8	71	45	2	2	1	10	-11	2
8	71	48	11	10	5	5	-6	1
8	72	40	10	12	5	11	-10	2
8	73	43	11	12	5	5	-5	1
8	73	48	10	9	4	5	-6	1
8	76	48	9	8	3	5	-6	1
8	78	41	6	7	2	11	-10	2
9	70	42	9	10	5	5	-5	1
9	70	48	10	9	5	4	-5	1
9	71	45	2	2	1	9	-10	2
9	72	48	9	8	4	4	-5	1
9	74	49	8	7	3	4	-5	1
9	76	40	5	6	2	10	-9	2
9	78	49	7	6	2	4	-5	1
0	70	50						
0	70	54	11	8	5	8	-11	0
0	70	56	9	6	4	2	-3	0
0	71	51	11	9	5	9	-11	0

φ_1	Φ	φ_2	H	K	L	U	V	W
0	70	50						
0	71	52	9	7	4	7	-9	0
0	71	54	7	5	3	5	-7	0
0	71	56	12	8	5	2	-3	0
0	71	59	5	3	2	3	-5	0
0	72	50	12	10	5	5	-6	0
0	72	53	12	9	5	3	-4	0
0	72	55	10	7	4	7	-10	0
0	72	58	8	5	3	5	-8	0
0	73	51	5	4	2	4	-5	0
0	73	53	8	6	3	3	-4	0
0	73	58	11	7	4	7	-11	0
0	74	51	11	9	4	9	-11	0
0	74	54	11	8	4	8	-11	0
0	74	56	3	2	1	2	-3	0
0	75	52	9	7	3	7	-9	0
0	75	53	12	9	4	3	-4	0
0	76	50	6	5	2	5	-6	0
0	76	55	10	7	3	7	-10	0
0	76	59	10	6	3	3	-5	0
0	77	51	10	8	3	4	-5	0
0	77	54	7	5	2	5	-7	0
0	77	58	11	7	3	7	-11	0
0	78	51	11	9	3	9	-11	0
0	78	54	11	8	3	8	-11	0
0	78	56	12	8	3	2	-3	0
0	78	58	8	5	2	5	-8	0
0	79	50	12	10	3	5	-6	0
0	79	53	4	3	1	3	-4	0
4	72	55	10	7	4	8	-12	1
4	75	52	9	7	3	9	-12	1
4	79	53	4	3	1	8	-11	1
5	73	58	11	7	4	6	-10	1
5	74	51	11	9	4	7	-9	1
5	74	56	3	2	1	7	-11	1
5	76	50	6	5	2	8	-10	1
5	76	55	10	7	3	6	-9	1
5	78	54	11	8	3	7	-10	1
5	78	58	8	5	2	6	-10	1
6	71	54	7	5	3	6	-9	1
6	71	59	5	3	2	5	-9	1
6	73	51	5	4	2	6	-8	1
6	74	56	3	2	1	5	-8	1
7	71	52	9	7	4	5	-7	1
7	72	58	8	5	3	4	-7	1
7	79	53	4	3	1	5	-7	1
8	74	54	11	8	4	4	-6	1
8	77	51	10	8	3	9	-12	2
6	77	54	7	5	2	4	-6	1

φ_1	Φ	φ_2	H	K	L	U	V	W
0	70	50						
6	78	58	8	5	2	7	-12	2
9	71	54	7	5	3	7	-11	2
9	73	51	5	4	2	8	-11	2
9	76	59	10	6	3	6	-11	2
9	79	53	4	3	1	7	-10	2
0	70	60						
0	70	60	7	4	3	4	-7	0
0	70	60	12	7	5	7	-12	0
0	70	63	10	5	4	1	-2	0
0	70	63	12	6	5	1	-2	0
0	70	68	5	2	2	2	-5	0
0	71	63	8	4	3	1	-2	0
0	71	69	8	3	3	3	-8	0
0	72	61	11	6	4	6	-11	0
0	72	66	11	5	4	5	-11	0
0	73	63	6	3	2	1	-2	0
0	73	66	9	4	3	4	-9	0
0	73	67	12	5	4	5	-12	0
0	74	60	12	7	4	7	-12	0
0	74	61	9	5	3	5	-9	0
0	74	68	10	4	3	2	-5	0
0	75	63	10	5	3	1	-2	0
0	75	67	7	3	2	3	-7	0
0	76	60	7	4	2	4	-7	0
0	76	66	11	5	3	5	-11	0
0	77	61	11	6	3	6	-11	0
0	77	63	4	2	1	1	-2	0
0	77	67	12	5	3	5	-12	0
0	77	69	8	3	2	3	-8	0
0	78	60	12	7	3	7	-12	0
0	79	61	9	5	2	5	-9	0
0	79	66	9	4	2	4	-9	0
0	79	68	5	2	1	2	-5	0
5	70	60	12	7	5	6	-11	1
5	70	68	5	2	2	4	-11	1
5	73	66	9	4	3	5	-12	1
5	75	67	7	3	2	4	-10	1
5	76	60	7	4	2	6	-11	1
6	71	69	8	3	3	3	-9	1
6	78	60	12	7	3	5	-9	1
7	72	61	11	6	4	4	-8	1
7	73	67	12	5	4	3	-8	1
7	79	68	5	2	1	3	-8	1
9	70	60	7	4	3	3	-6	1
9	74	61	9	5	3	3	-6	1
9	77	61	11	6	3	3	-6	1

φ_1	Φ	φ_2	H	K	L	U	V	W
0	70	60						
9	77	69	8	3	2	2	-6	1
0	70	70						
0	70	76	8	2	3	1	-4	0
0	71	70	11	4	4	4	-11	0
0	71	75	11	3	4	3	-11	0
0	72	72	3	1	1	1	-3	0
0	72	76	12	3	4	1	-4	0
0	72	77	9	2	3	2	-9	0
0	74	73	10	3	3	3	-10	0
0	74	79	10	2	3	1	-5	0
0	75	74	7	2	2	2	-7	0
0	75	75	11	3	3	3	-11	0
0	76	70	11	4	3	4	-11	0
0	76	76	4	1	1	1	-4	0
0	77	72	12	4	3	1	-3	0
0	78	72	9	3	2	1	-3	0
0	78	77	9	2	2	2	-9	0
0	79	73	10	3	2	3	-10	0
0	79	79	5	1	1	1	-5	0
5	71	70	11	4	4	4	-12	1
5	74	73	10	3	3	3	-11	1
5	75	75	11	3	3	3	-12	1
5	79	79	5	1	1	2	-11	1
6	72	72	3	1	1	3	-10	1
6	76	70	11	4	3	3	-9	1
6	76	76	4	1	1	2	-9	1
6	78	77	9	2	2	2	-10	1
7	75	74	7	2	2	2	-8	1
8	72	72	3	1	1	2	-7	1
0	70	80						
0	70	80	11	2	4	2	-11	0
0	70	83	8	1	3	1	-8	0
0	70	85	11	1	4	1	-11	0
0	72	81	6	1	2	1	-6	0
0	72	84	9	1	3	1	-9	0
0	72	85	12	1	4	1	-12	0
0	73	84	10	1	3	1	-10	0
0	74	82	7	1	2	1	-7	0
0	75	80	11	2	3	2	-11	0
0	75	85	11	1	3	1	-11	0
0	76	81	12	2	3	1	-6	0
0	76	83	8	1	2	1	-8	0
0	76	85	12	1	3	1	-12	0
0	78	84	9	1	2	1	-9	0

φ_1	Φ	φ_2	H	K	L	U	V	W
0	70	80						
0	79	84	10	1	2	1	-10	0
5	72	84	9	1	3	1	-12	1
5	78	84	9	1	2	1	-11	1
5	79	84	10	1	2	1	-12	1
6	70	83	8	1	3	1	-11	1
6	76	83	8	1	2	1	-10	1
7	72	81	6	1	2	1	-8	1
7	74	82	7	1	2	1	-9	1
8	75	80	11	2	3	1	-7	1
0	70	90						
0	70	90	11	0	4	0	-1	0
0	72	90	3	0	1	0	-1	0
0	73	90	10	0	3	0	-1	0
0	74	90	7	0	2	0	-1	0
0	75	90	11	0	3	0	-1	0
0	76	90	4	0	1	0	-1	0
0	77	90	9	0	2	0	-1	0
0	79	90	5	0	1	0	-1	0
0	80	0						
0	80	0	0	11	2	1	0	0
0	80	5	1	11	2	11	-1	0
0	81	0	0	6	1	1	0	0
0	81	5	1	12	2	12	-1	0
0	81	9	1	6	1	6	-1	0
0	82	0	0	7	1	1	0	0
0	82	8	1	7	1	7	-1	0
0	83	0	0	8	1	1	0	0
0	83	7	1	8	1	8	-1	0
0	84	0	0	9	1	1	0	0
0	84	0	0	10	1	1	0	0
0	84	6	1	9	1	9	-1	0
0	84	6	1	10	1	10	-1	0
0	85	0	0	11	1	1	0	0
0	85	0	0	12	1	1	0	0
0	85	5	1	11	1	11	-1	0
0	85	5	1	12	1	12	-1	0
0	85	9	2	12	1	6	-1	0
5	81	9	1	6	1	11	-2	1
5	85	5	1	12	1	11	-1	1
6	80	5	1	11	2	9	-1	1
6	81	5	1	12	2	10	-1	1
6	84	6	1	10	1	9	-1	1
6	85	5	1	11	1	10	-1	1
7	84	6	1	9	1	8	-1	1

φ_1	Φ	φ_2	H	K	L	U	V	W
0	80	0						
8	83	7	1	8	1	7	-1	1
9	82	8	1	7	1	6	-1	1
0	80	10						
0	80	10	2	11	2	11	-2	0
0	80	15	3	11	2	11	-3	0
0	81	14	3	12	2	4	-1	0
0	81	18	2	6	1	3	-1	0
0	82	16	2	7	1	7	-2	0
0	83	14	2	8	1	4	-1	0
0	84	11	2	10	1	5	-1	0
0	84	13	2	9	1	9	-2	0
0	84	18	3	9	1	3	-1	0
0	85	10	2	11	1	11	-2	0
0	85	14	3	12	1	4	-1	0
0	85	15	3	11	1	11	-3	0
0	85	17	3	10	1	10	-3	0
0	85	18	4	12	1	3	-1	0
6	80	10	2	11	2	10	-2	1
6	82	16	2	7	1	10	-3	1
8	85	15	3	11	1	7	-2	1
0	80	20						
0	80	20	4	11	2	11	-4	0
0	80	27	5	10	2	2	-1	0
0	81	23	5	12	2	12	-5	0
0	81	24	5	11	2	11	-5	0
0	81	29	6	11	2	11	-6	0
0	82	27	3	6	1	2	-1	0
0	83	21	3	8	1	8	-3	0
0	83	23	3	7	1	7	-3	0
0	84	24	4	9	1	9	-4	0
0	84	27	4	8	1	2	-1	0
0	84	29	5	9	1	9	-5	0
0	85	20	4	11	1	11	-4	0
0	85	22	4	10	1	5	-2	0
0	85	24	5	11	1	11	-5	0
0	85	27	5	10	1	2	-1	0
0	85	29	6	11	1	11	-6	0
0	86	23	5	12	1	12	-5	0
0	86	27	6	12	1	2	-1	0
5	84	24	4	9	1	11	-5	1
6	83	23	3	7	1	9	-4	1
6	85	29	6	11	1	9	-5	1
7	81	29	6	11	2	7	-4	1
7	84	29	5	9	1	7	-4	1

φ_1	Φ	φ_2	H	K	L	U	V	W
0	80	20						
7	85	20	4	11	1	8	-3	1
8	86	23	5	12	1	7	-3	1
9	83	23	3	7	1	11	-5	2
9	85	22	4	10	1	12	-5	2
0	80	30						
0	80	31	3	5	1	5	-3	0
0	80	34	6	9	2	3	-2	0
0	80	38	7	9	2	9	-7	0
0	81	32	7	11	2	11	-7	0
0	81	35	7	10	2	10	-7	0
0	81	39	4	5	1	5	-4	0
0	82	30	7	12	2	12	-7	0
0	82	34	4	6	1	3	-2	0
0	82	36	8	11	2	11	-8	0
0	82	37	9	12	2	4	-3	0
0	82	39	9	11	2	11	-9	0
0	83	30	4	7	1	7	-4	0
0	83	36	5	7	1	7	-5	0
0	84	32	5	8	1	8	-5	0
0	84	37	6	8	1	4	-3	0
0	85	31	6	10	1	5	-3	0
0	85	34	6	9	1	3	-2	0
0	85	35	7	10	1	10	-7	0
0	85	38	7	9	1	9	-7	0
0	86	30	7	12	1	12	-7	0
0	86	32	7	11	1	11	-7	0
0	86	34	8	12	1	3	-2	0
0	86	36	8	11	1	11	-8	0
0	86	37	9	12	1	4	-3	0
0	86	39	8	10	1	5	-4	0
0	86	39	9	11	1	11	-9	0
4	81	39	4	5	1	11	-9	1
4	82	39	9	11	2	12	-10	1
4	83	30	4	7	1	12	-7	1
4	83	36	5	7	1	11	-8	1
4	84	32	5	8	1	11	-7	1
5	80	36	7	9	2	10	-8	1
5	82	30	7	12	2	10	-6	1
6	80	31	3	5	1	6	-5	1
6	82	36	8	11	2	8	-6	1
7	81	39	4	5	1	6	-5	1
7	85	35	7	10	1	7	-5	1
7	86	39	9	11	1	6	-5	1
8	80	38	7	9	2	11	-9	2
8	81	32	7	11	2	6	-4	1
8	86	39	8	10	1	11	-9	2

TAB. VI FROM EULER ANGLES INTO HKL-UVW

φ_1	Φ	φ_2	H	K	L	U	V	W
0	80	30						
9	80	31	3	5	1	11	-7	2
9	82	34	4	6	1	10	-7	2
9	85	38	7	9	1	5	-4	1
0	80	40						
0	80	43	11	12	3	12	-11	0
0	80	45	4	4	1	1	-1	0
0	80	47	12	11	3	11	-12	0
0	81	42	8	9	2	9	-8	0
0	81	45	9	9	2	1	-1	0
0	81	46	9	8	2	8	-9	0
0	82	42	9	10	2	10	-9	0
0	82	42	10	11	2	11	-10	0
0	82	45	5	5	1	1	-1	0
0	82	48	10	9	2	9	-10	0
0	82	48	11	10	2	10	-11	0
0	83	40	5	6	1	6	-5	0
0	83	43	11	12	2	12	-11	0
0	83	45	6	6	1	1	-1	0
0	83	45	11	11	2	1	-1	0
0	83	47	12	11	2	11	-12	0
0	84	41	6	7	1	7	-6	0
0	84	45	7	7	1	1	-1	0
0	84	49	7	6	1	6	-7	0
0	85	41	7	8	1	8	-7	0
0	85	42	8	9	1	9	-8	0
0	85	45	8	8	1	1	-1	0
0	85	48	9	8	1	8	-9	0
0	85	49	8	7	1	7	-8	0
0	86	40	10	12	1	6	-5	0
0	86	42	9	10	1	10	-9	0
0	86	42	10	11	1	11	-10	0
0	86	43	11	12	1	12	-11	0
0	86	45	9	9	1	1	-1	0
0	86	45	10	10	1	1	-1	0
0	86	45	11	11	1	1	-1	0
0	86	47	12	11	1	11	-12	0
0	86	48	10	9	1	9	-10	0
0	86	48	11	10	1	10	-11	0
0	87	45	12	12	1	1	-1	0
4	81	42	8	9	2	11	-10	1
4	82	42	9	10	2	12	-11	1
4	83	47	12	11	2	9	-10	1
4	85	42	8	9	1	10	-9	1
4	86	42	9	10	1	11	-10	1
4	86	42	10	11	1	12	-11	1
4	86	47	12	11	1	10	-11	1

φ_1	Φ	φ_2	H	K	L	U	V	W
0	80	40						
4	86	48	11	10	1	9	-10	1
5	80	47	12	11	3	8	-9	1
5	82	48	10	9	2	7	-8	1
5	82	48	11	10	2	8	-9	1
5	84	41	6	7	1	8	-7	1
5	85	41	7	8	1	9	-8	1
5	85	48	9	8	1	7	-8	1
5	86	48	10	9	1	8	-9	1
6	81	48	9	8	2	6	-7	1
6	83	40	5	6	1	7	-6	1
6	85	49	8	7	1	6	-7	1
7	84	49	7	6	1	5	-6	1
7	86	42	9	10	1	12	-11	2
8	85	41	7	6	1	10	-9	2
8	85	42	8	9	1	11	-10	2
8	86	47	12	11	1	9	-10	2
9	84	41	6	7	1	9	-8	2
9	86	48	11	10	1	8	-9	2
0	80	50						
0	80	52	9	7	2	7	-9	0
0	80	56	9	6	2	2	-3	0
0	80	59	5	3	1	3	-5	0
0	81	51	5	4	1	4	-5	0
0	81	55	10	7	2	7	-10	0
0	81	58	11	7	2	7	-11	0
0	82	51	11	9	2	9	-11	0
0	82	53	12	9	2	3	-4	0
0	82	54	11	8	2	8	-11	0
0	82	56	6	4	1	2	-3	0
0	83	50	6	5	1	5	-6	0
0	83	54	7	5	1	5	-7	0
0	84	53	8	6	1	3	-4	0
0	84	58	8	5	1	5	-8	0
0	85	52	9	7	1	7	-9	0
0	85	55	10	7	1	7	-10	0
0	85	56	9	6	1	2	-3	0
0	85	59	10	6	1	3	-5	0
0	86	50	12	10	1	5	-6	0
0	86	51	10	8	1	4	-5	0
0	86	51	11	9	1	9	-11	0
0	86	53	12	9	1	3	-4	0
0	86	54	11	8	1	8	-11	0
0	86	56	12	8	1	2	-3	0
0	86	58	11	7	1	7	-11	0
4	80	59	5	3	1	7	-12	1
4	83	50	6	5	1	9	-11	1

φ_1	Φ	φ_2	H	K	L	U	V	W
0	80	50						
5	81	51	5	4	1	7	-9	1
5	82	51	11	9	2	8	-10	1
5	83	54	7	5	1	7	-10	1
6	80	52	9	7	2	6	-8	1
6	86	58	11	7	1	5	-8	1
7	80	59	5	3	1	4	-7	1
7	86	54	11	6	1	5	-7	1
8	81	55	10	7	2	4	-6	1
8	84	53	8	6	1	8	-11	2
8	85	59	10	6	1	7	-12	2
8	86	50	12	10	1	9	-11	2
9	82	56	6	4	1	7	-11	2
9	83	50	6	5	1	4	-5	1
9	86	51	11	9	1	4	-5	1
0	80	60						
0	80	63	10	5	2	1	-2	0
0	81	61	11	6	2	6	-11	0
0	81	66	11	5	2	5	-11	0
0	81	67	12	5	2	5	-12	0
0	82	60	12	7	2	7	-12	0
0	82	63	6	3	1	1	-2	0
0	83	60	7	4	1	4	-7	0
0	83	67	7	3	1	3	-7	0
0	83	69	8	3	1	3	-8	0
0	84	61	9	5	1	5	-9	0
0	84	63	8	4	1	1	-2	0
0	84	66	9	4	1	4	-9	0
0	85	61	11	6	1	6	-11	0
0	85	63	10	5	1	1	-2	0
0	85	66	11	5	1	5	-11	0
0	85	68	10	4	1	2	-5	0
0	86	60	12	7	1	7	-12	0
0	86	63	12	6	1	1	-2	0
0	86	67	12	5	1	5	-12	0
4	83	67	7	3	1	5	-12	1
5	81	67	12	5	2	4	-10	1
5	83	69	8	3	1	4	-11	1
5	84	61	9	5	1	6	-11	1
6	83	60	7	4	1	5	-9	1
6	85	66	11	5	1	4	-9	1
7	86	60	12	7	1	4	-7	1
8	81	66	11	5	2	3	-7	1
8	84	66	9	4	1	3	-7	1
9	83	60	7	4	1	6	-11	2

φ_1	Φ	φ_2	H	K	L	U	V	W
0	80	70						
0	80	70	11	4	2	4	-11	0
0	80	75	11	3	2	3	-11	0
0	81	72	6	2	1	1	-3	0
0	81	76	12	3	2	1	-4	0
0	82	74	7	2	1	2	-7	0
0	83	76	8	2	1	1	-4	0
0	84	72	9	3	1	1	-3	0
0	84	77	9	2	1	2	-9	0
0	84	79	10	2	1	1	-5	0
0	85	70	11	4	1	4	-11	0
0	85	72	12	4	1	1	-3	0
0	85	73	10	3	1	3	-10	0
0	85	75	11	3	1	3	-11	0
0	85	76	12	3	1	1	-4	0
5	82	74	7	2	1	3	-11	1
7	80	75	11	3	2	2	-8	1
8	85	73	10	3	1	2	-7	1
9	80	70	11	4	2	2	-6	1
0	80	80						
0	80	80	11	2	2	2	-11	0
0	80	85	11	1	2	1	-11	0
0	81	81	6	1	1	1	-6	0
0	81	85	12	1	2	1	-12	0
0	82	82	7	1	1	1	-7	0
0	83	83	8	1	1	1	-8	0
0	84	84	9	1	1	1	-9	0
0	84	84	10	1	1	1	-10	0
0	85	80	11	2	1	2	-11	0
0	85	81	12	2	1	1	-6	0
0	85	85	11	1	1	1	-11	0
0	85	85	12	1	1	1	-12	0
5	80	80	11	2	2	2	-12	1
5	84	84	10	1	1	1	-11	1
5	85	85	11	1	1	1	-12	1
6	83	83	8	1	1	1	-9	1
6	84	84	9	1	1	1	-10	1
7	82	82	7	1	1	1	-8	1
8	81	81	6	1	1	1	-7	1
9	84	84	10	1	1	1	-12	2
9	85	80	11	2	1	1	-6	1
0	80	90						
0	80	90	11	0	2	0	-1	0
0	81	90	6	0	1	0	-1	0

φ_1	Φ	φ_2	H	K	L	U	V	W
0	80	90						
0	82	90	7	0	1	0	-1	0
0	83	90	8	0	1	0	-1	0
0	84	90	9	0	1	0	-1	0
0	84	90	10	0	1	0	-1	0
0	85	90	11	0	1	0	-1	0
0	85	90	12	0	1	0	-1	0
0	90	0						
0	90	0	0	1	0	1	0	0
0	90	5	1	11	0	11	-1	0
0	90	5	1	12	0	12	-1	0
0	90	6	1	9	0	9	-1	0
0	90	6	1	10	0	10	-1	0
0	90	7	1	8	0	8	-1	0
0	90	8	1	7	0	7	-1	0
0	90	9	1	6	0	6	-1	0
5	90	0	0	1	0	11	0	1
5	90	0	0	1	0	12	0	1
5	90	5	1	11	0	11	-1	1
5	90	5	1	12	0	12	-1	1
5	90	9	1	6	0	12	-2	1
6	90	0	0	1	0	9	0	1
6	90	0	0	1	0	10	0	1
6	90	6	1	9	0	9	-1	1
6	90	6	1	10	0	10	-1	1
7	90	0	0	1	0	8	0	1
7	90	7	1	8	0	8	-1	1
8	90	0	0	1	0	7	0	1
8	90	8	1	7	0	7	-1	1
9	90	0	0	1	0	6	0	1
9	90	5	1	12	0	12	-1	2
9	90	9	1	6	0	6	-1	1
0	90	10						
0	90	10	2	11	0	11	-2	0
0	90	11	1	5	0	5	-1	0
0	90	13	2	9	0	9	-2	0
0	90	14	1	4	0	4	-1	0
0	90	15	3	11	0	11	-3	0
0	90	16	2	7	0	7	-2	0
0	90	17	3	10	0	10	-3	0
0	90	18	1	3	0	3	-1	0
5	90	10	2	11	0	11	-2	1
5	90	14	1	4	0	12	-3	1
5	90	15	3	11	0	11	-3	1
5	90	17	3	10	0	10	-3	1

φ_1	Φ	φ_2	H	K	L	U	V	W
0	90	10						
5	90	18	1	3	0	12	-4	1
6	90	11	1	5	0	10	-2	1
6	90	13	2	9	0	9	-2	1
6	90	18	1	3	0	9	-3	1
7	90	14	1	4	0	8	-2	1
8	90	16	2	7	0	7	-2	1
9	90	14	1	4	0	12	-3	2
9	90	18	1	3	0	6	-2	1
0	90	20						
0	90	20	4	11	0	11	-4	0
0	90	21	3	8	0	8	-3	0
0	90	22	2	5	0	5	-2	0
0	90	23	3	7	0	7	-3	0
0	90	23	5	12	0	12	-5	0
0	90	24	4	9	0	9	-4	0
0	90	24	5	11	0	11	-5	0
0	90	27	1	2	0	2	-1	0
0	90	29	5	9	0	9	-5	0
0	90	29	6	11	0	11	-6	0
4	90	23	5	12	0	12	-5	1
4	90	27	1	2	0	12	-6	1
5	90	20	4	11	0	11	-4	1
5	90	22	2	5	0	10	-4	1
5	90	24	5	11	0	11	-5	1
5	90	27	1	2	0	10	-5	1
5	90	29	6	11	0	11	-6	1
6	90	24	4	9	0	9	-4	1
6	90	27	1	2	0	8	-4	1
6	90	29	5	9	0	9	-5	1
7	90	21	3	8	0	8	-3	1
7	90	23	3	7	0	7	-3	1
8	90	27	1	2	0	6	-3	1
9	90	23	5	12	0	12	-5	2
9	90	24	5	11	0	11	-5	2
9	90	29	6	11	0	11	-6	2
0	90	30						
0	90	30	4	7	0	7	-4	0
0	90	30	7	12	0	12	-7	0
0	90	31	3	5	0	5	-3	0
0	90	32	5	6	0	6	-5	0
0	90	32	7	11	0	11	-7	0
0	90	34	2	3	0	3	-2	0
0	90	35	7	10	0	10	-7	0
0	90	36	5	7	0	7	-5	0

φ_1	Φ	φ_2	H	K	L	U	V	W
0	90	30						
0	90	36	8	11	0	11	-8	0
0	90	37	3	4	0	4	-3	0
0	90	38	7	9	0	9	-7	0
0	90	39	4	5	0	5	-4	0
0	90	39	9	11	0	11	-9	0
4	90	30	7	12	0	12	-7	1
4	90	32	7	11	0	11	-7	1
4	90	34	2	3	0	12	-6	1
4	90	36	8	11	0	11	-8	1
4	90	37	3	4	0	12	-9	1
4	90	39	4	5	0	10	-8	1
4	90	39	9	11	0	11	-9	1
5	90	31	3	5	0	10	-6	1
5	90	34	2	3	0	9	-6	1
5	90	35	7	10	0	10	-7	1
5	90	38	7	9	0	9	-7	1
6	90	32	5	8	0	8	-5	1
6	90	37	3	4	0	8	-6	1
7	90	30	4	7	0	7	-4	1
7	90	36	5	7	0	7	-5	1
8	90	30	7	12	0	12	-7	2
8	90	34	2	3	0	6	-4	1
8	90	36	8	11	0	11	-8	2
8	90	37	3	4	0	12	-9	2
8	90	39	9	11	0	11	-9	2
9	90	32	7	11	0	11	-7	2
9	90	35	7	10	0	10	-7	2
9	90	39	4	5	0	5	-4	1
0	90	40						
0	90	40	5	6	0	6	-5	0
0	90	41	6	7	0	7	-6	0
0	90	41	7	8	0	8	-7	0
0	90	42	8	9	0	9	-8	0
0	90	42	9	10	0	10	-9	0
0	90	42	10	11	0	11	-10	0
0	90	43	11	12	0	12	-11	0
0	90	45	1	1	0	1	-1	0
0	90	47	12	11	0	11	-12	0
0	90	48	9	8	0	8	-9	0
0	90	48	10	9	0	9	-10	0
0	90	48	11	10	0	10	-11	0
0	90	49	7	6	0	6	-7	0
0	90	49	8	7	0	7	-8	0
3	90	45	1	1	0	12	-12	1
4	90	40	5	6	0	12	-10	1
4	90	42	9	10	0	10	-9	1

φ_1	Φ	φ_2	H	K	L	U	V	W
0	90	40						
4	90	42	10	11	0	11	-10	1
4	90	43	11	12	0	12	-11	1
4	90	45	1	1	0	9	-9	1
4	90	45	1	1	0	11	-11	1
4	90	45	1	1	0	10	-10	1
4	90	47	12	11	0	11	-12	1
4	90	48	10	9	0	9	-10	1
4	90	48	11	10	0	10	-11	1
5	90	41	7	8	0	8	-7	1
5	90	42	8	9	0	9	-8	1
5	90	45	1	1	0	8	-8	1
5	90	48	9	8	0	8	-9	1
5	90	49	8	7	0	7	-8	1
6	90	41	6	7	0	7	-6	1
6	90	45	1	1	0	7	-7	1
6	90	49	7	6	0	6	-7	1
7	90	40	5	6	0	6	-5	1
7	90	43	11	12	0	12	-11	2
7	90	45	1	1	0	11	-11	2
7	90	45	1	1	0	6	-6	1
7	90	47	12	11	0	11	-12	2
8	90	42	9	10	0	10	-9	2
8	90	42	10	11	0	11	-10	2
8	90	45	1	1	0	5	-5	1
8	90	48	10	9	0	9	-10	2
8	90	48	11	10	0	10	-11	2
9	90	42	8	9	0	9	-8	2
9	90	45	1	1	0	9	-9	2
9	90	48	9	8	0	8	-9	2
0	90	50						
0	90	50	6	5	0	5	-6	0
0	90	51	5	4	0	4	-5	0
0	90	51	11	9	0	9	-11	0
0	90	52	9	7	0	7	-9	0
0	90	53	4	3	0	3	-4	0
0	90	54	7	5	0	5	-7	0
0	90	54	11	6	0	6	-11	0
0	90	55	10	7	0	7	-10	0
0	90	56	3	2	0	2	-3	0
0	90	58	8	5	0	5	-8	0
0	90	58	11	7	0	7	-11	0
0	90	59	5	3	0	3	-5	0
4	90	50	6	5	0	10	-12	1
4	90	51	5	4	0	6	-10	1
4	90	51	11	9	0	9	-11	1
4	90	53	4	3	0	9	-12	1

φ_1 Φ φ_2	H	K	L	U	V	W
0 90 50						
4 90 54	11	8	0	8	-11	1
4 90 56	3	2	0	8	-12	1
4 90 58	11	7	0	7	-11	1
5 90 52	9	7	0	7	-9	1
5 90 55	10	7	0	7	-10	1
5 90 56	3	2	0	6	-9	1
5 90 59	5	3	0	6	-10	1
6 90 53	4	3	0	6	-8	1
6 90 58	8	5	0	5	-8	1
7 90 50	6	5	0	5	-6	1
7 90 54	7	5	0	5	-7	1
8 90 51	11	9	0	9	-11	2
8 90 53	4	3	0	9	-12	2
8 90 54	11	8	0	8	-11	2
8 90 56	3	2	0	4	-6	1
9 90 51	5	4	0	4	-5	1
9 90 55	10	7	0	7	-10	2
9 90 58	11	7	0	7	-11	2
0 90 60						
0 90 60	7	4	0	4	-7	0
0 90 60	12	7	0	7	-12	0
0 90 61	9	5	0	5	-9	0
0 90 61	11	6	0	6	-11	0
0 90 63	2	1	0	1	-2	0
0 90 66	9	4	0	4	-9	0
0 90 66	11	5	0	5	-11	0
0 90 67	7	3	0	3	-7	0
0 90 67	12	5	0	5	-12	0
0 90 68	5	2	0	2	-5	0
0 90 69	8	3	0	3	-8	0
4 90 60	12	7	0	7	-12	1
4 90 63	2	1	0	6	-12	1
4 90 67	12	5	0	5	-12	1
5 90 61	11	6	0	6	-11	1
5 90 63	2	1	0	5	-10	1
5 90 66	11	5	0	5	-11	1
5 90 68	5	2	0	4	-10	1
6 90 61	9	5	0	5	-9	1
6 90 63	2	1	0	4	-8	1
6 90 66	9	4	0	4	-9	1
7 90 60	7	4	0	4	-7	1
7 90 67	7	3	0	3	-7	1
7 90 69	8	3	0	3	-8	1
8 90 60	12	7	0	7	-12	2
8 90 63	2	1	0	3	-6	1
9 90 61	11	6	0	6	-11	2

φ_1 Φ φ_2	H	K	L	U	V	W
0 90 60						
9 90 66	11	5	0	5	-11	2
9 90 67	12	5	0	5	-12	2
0 90 70						
0 90 70	11	4	0	4	-11	0
0 90 72	3	1	0	1	-3	0
0 90 73	10	3	0	3	-10	0
0 90 74	7	2	0	2	-7	0
0 90 75	11	3	0	3	-11	0
0 90 76	4	1	0	1	-4	0
0 90 77	9	2	0	2	-9	0
0 90 79	5	1	0	1	-5	0
5 90 70	11	4	0	4	-11	1
5 90 72	3	1	0	4	-12	1
5 90 73	10	3	0	3	-10	1
5 90 75	11	3	0	3	-11	1
5 90 76	4	1	0	3	-12	1
6 90 72	3	1	0	3	-9	1
6 90 77	9	2	0	2	-9	1
6 90 79	5	1	0	2	-10	1
7 90 76	4	1	0	2	-8	1
8 90 74	7	2	0	2	-7	1
9 90 72	3	1	0	2	-6	1
9 90 76	4	1	0	3	-12	2
0 90 80						
0 90 80	11	2	0	2	-11	0
0 90 81	6	1	0	1	-6	0
0 90 82	7	1	0	1	-7	0
0 90 83	8	1	0	1	-8	0
0 90 84	9	1	0	1	-9	0
0 90 84	10	1	0	1	-10	0
0 90 85	11	1	0	1	-11	0
0 90 85	12	1	0	1	-12	0
5 90 80	11	2	0	2	-11	1
5 90 81	6	1	0	2	-12	1
5 90 85	11	1	0	1	-11	1
5 90 85	12	1	0	1	-12	1
6 90 84	9	1	0	1	-9	1
6 90 84	10	1	0	1	-10	1
7 90 83	8	1	0	1	-8	1
8 90 82	7	1	0	1	-7	1
9 90 81	6	1	0	1	-6	1
9 90 85	12	1	0	1	-12	2

φ_1 Φ φ_2	H	K	L	U	V	W
0 90 90						
0 90 90	1	0	0	0	-1	0
5 90 90	1	0	0	0	-12	1
5 90 90	1	0	0	0	-11	1
6 90 90	1	0	0	0	-9	1
6 90 90	1	0	0	0	-10	1
7 90 90	1	0	0	0	-8	1
8 90 90	1	0	0	0	-7	1
9 90 90	1	0	0	0	-6	1
10 0 10						
10 0 10	0	0	1	11	-4	0
10 0 10	0	0	1	8	-3	0
11 0 11	0	0	1	12	-5	0
11 0 11	0	0	1	5	-2	0
12 0 12	0	0	1	11	-5	0
12 0 12	0	0	1	9	-4	0
12 0 12	0	0	1	7	-3	0
13 0 13	0	0	1	2	-1	0
14 0 14	0	0	1	11	-6	0
15 0 15	0	0	1	9	-5	0
15 0 15	0	0	1	12	-7	0
15 0 15	0	0	1	5	-3	0
15 0 15	0	0	1	7	-4	0
16 0 16	0	0	1	11	-7	0
16 0 16	0	0	1	8	-5	0
17 0 17	0	0	1	10	-7	0
17 0 17	0	0	1	3	-2	0
18 0 18	0	0	1	11	-8	0
18 0 18	0	0	1	4	-3	0
18 0 18	0	0	1	7	-5	0
19 0 19	0	0	1	9	-7	0
19 0 19	0	0	1	5	-4	0
10 10 0						
15 18 0	0	1	3	12	-3	1
16 18 0	0	1	3	11	-3	1
18 18 0	0	1	3	10	-3	1
19 14 0	0	1	4	12	-4	1
10 10 10						
13 19 14	1	4	12	12	-6	1
13 19 18	1	3	9	12	-7	1
15 18 18	1	3	10	11	-7	1
16 15 18	1	3	12	12	-8	1

φ_1 Φ φ_2	H	K	L	U	V	W
10 10 10						
16 19 16	1	3	9	9	-6	1
17 16 18	1	3	11	10	-7	1
19 18 18	1	3	10	8	-6	1
19 19 14	1	4	12	8	-5	1
10 10 20						
13 16 27	1	2	7	11	-9	1
14 16 27	1	2	6	12	-10	1
16 13 27	1	2	10	12	-11	1
16 14 27	1	2	9	11	-10	1
16 16 27	1	2	8	10	-9	1
16 18 27	1	2	7	9	-8	1
19 11 27	1	2	11	11	-11	1
19 11 27	1	2	12	12	-12	1
19 13 27	1	2	10	10	-10	1
19 14 27	1	2	9	9	-9	1
19 16 27	1	2	8	8	-8	1
19 18 27	1	2	7	7	-7	1
10 10 30						
12 17 34	2	3	12	12	-12	1
12 18 34	2	3	11	11	-11	1
15 17 34	2	3	12	9	-10	1
15 18 34	2	3	11	8	-9	1
10 10 40						
12 19 45	1	1	4	8	-12	1
13 19 45	1	1	4	7	-11	1
15 16 45	1	1	5	7	-12	1
15 19 45	1	1	4	6	-10	1
17 16 45	1	1	5	6	-11	1
17 19 45	1	1	4	5	-9	1
19 13 45	1	1	6	6	-12	1
19 16 45	1	1	5	5	-10	1
19 19 45	1	1	4	4	-8	1
10 10 50						
16 17 56	3	2	12	4	-12	1
18 18 56	3	2	11	3	-10	1
10 10 60						
17 18 63	2	1	7	2	-11	1
18 16 63	2	1	8	2	-12	1

φ1	Φ	φ2	H	K	L	U	V	W
10	**10**	**70**						
14	19	72	3	1	9	1	-12	1
15	19	76	4	1	12	0	-12	1
19	15	72	3	1	12	0	-12	1
19	16	72	3	1	11	0	-11	1
19	18	72	3	1	10	0	-10	1
19	19	72	3	1	9	0	-9	1
10	**10**	**90**						
15	18	90	1	0	3	-3	-12	1
16	18	90	1	0	3	-3	-11	1
18	18	90	1	0	3	-3	-10	1
19	14	90	1	0	4	-4	-12	1
10	**20**	**0**						
10	27	9	1	6	12	12	-4	1
11	27	0	0	1	2	11	-2	1
11	27	0	0	1	2	12	-2	1
13	27	0	0	1	2	10	-2	1
14	27	0	0	1	2	9	-2	1
16	27	0	0	1	2	8	-2	1
16	29	9	1	6	11	7	-3	1
18	27	0	0	1	2	7	-2	1
19	27	9	1	6	12	6	-3	1
10	**20**	**10**						
10	27	14	1	4	8	12	-5	1
11	24	18	1	3	7	11	-6	1
12	27	11	1	5	10	10	-4	1
12	28	18	1	3	6	9	-5	1
13	22	18	1	3	8	10	-6	1
14	22	14	1	4	10	10	-5	1
14	27	14	1	4	8	8	-4	1
15	24	18	1	3	7	8	-5	1
16	21	14	1	4	11	9	-5	1
17	23	11	1	5	12	8	-4	1
17	25	14	1	4	9	7	-4	1
17	28	18	1	3	6	6	-4	1
18	22	18	1	3	8	7	-5	1
18	27	14	1	4	8	12	-7	2
10	**20**	**20**						
10	24	27	1	2	5	11	-8	1
10	28	22	2	5	10	10	-6	1

φ1	Φ	φ2	H	K	L	U	V	W
10	**20**	**20**						
10	29	27	1	2	4	10	-7	1
11	20	27	1	2	6	12	-9	1
12	24	27	1	2	5	9	-7	1
12	29	27	1	2	4	8	-6	1
13	20	27	1	2	6	10	-8	1
13	24	22	2	5	12	9	-6	1
15	24	27	1	2	5	7	-6	1
15	26	22	2	5	11	7	-5	1
15	29	27	1	2	4	6	-5	1
16	20	27	1	2	6	8	-7	1
17	24	27	1	2	5	12	-11	2
17	26	27	2	4	9	11	-10	2
18	29	27	1	2	4	10	-9	2
19	28	22	2	5	10	5	-4	1
10	**20**	**30**						
11	28	39	4	5	12	7	-8	1
12	20	34	2	3	10	10	-10	1
12	22	34	2	3	9	9	-9	1
12	23	37	3	4	12	8	-9	1
12	24	34	2	3	8	8	-8	1
12	28	31	3	5	11	8	-7	1
13	24	37	3	4	11	7	-8	1
13	27	34	2	3	7	7	-7	1
14	27	37	3	4	10	6	-7	1
15	27	34	2	3	7	11	-12	2
15	29	37	3	4	9	5	-6	1
16	20	34	2	3	10	7	-8	1
16	26	31	3	5	12	6	-6	1
16	28	31	3	5	11	11	-11	2
16	28	39	4	5	12	9	-12	2
17	22	34	2	3	9	6	-7	1
18	24	34	2	3	8	5	-6	1
19	27	37	3	4	10	6	-11	2
10	**20**	**40**						
10	25	45	1	1	3	8	-11	1
11	25	45	1	1	3	7	-10	1
12	25	45	1	1	3	6	-9	1
14	25	45	1	1	3	5	-8	1
17	25	45	1	1	3	4	-7	1
17	29	45	2	2	5	7	-12	2
19	29	45	2	2	5	6	-11	2

φ1	Φ	φ2	H	K	L	U	V	W
10	**20**	**50**						
11	23	53	4	3	12	6	-12	1
11	27	53	4	3	10	5	-10	1
12	22	56	3	2	9	5	-12	1
14	24	53	4	3	11	4	-9	1
14	26	59	5	3	12	3	-9	1
14	28	51	5	4	12	4	-8	1
15	20	56	3	2	10	4	-11	1
15	27	56	3	2	7	3	-8	1
16	22	56	3	2	9	3	-9	1
16	29	53	4	3	9	3	-7	1
17	28	59	5	3	11	2	-7	1
18	23	53	4	3	12	3	-8	1
19	24	56	3	2	8	2	-7	1
10	**20**	**60**						
12	24	63	2	1	5	3	-11	1
12	28	68	5	2	10	2	-10	1
13	20	63	2	1	6	3	-12	1
13	24	68	5	2	12	2	-11	1
14	29	63	2	1	4	2	-8	1
15	24	63	2	1	5	2	-9	1
16	26	68	5	2	11	1	-8	1
19	29	63	2	1	4	1	-6	1
10	**20**	**70**						
10	27	76	4	1	8	1	-12	1
10	28	72	3	1	6	2	-12	1
12	23	79	5	1	12	0	-12	1
12	25	79	5	1	11	0	-11	1
13	27	79	5	1	10	0	-10	1
14	22	72	3	1	8	1	-11	1
14	24	72	3	1	7	1	-10	1
14	28	72	3	1	6	1	-9	1
15	21	76	4	1	11	0	-11	1
15	22	76	4	1	10	0	-10	1
15	25	76	4	1	9	0	-9	1
16	27	76	4	1	8	0	-8	1
10	**20**	**80**						
11	27	81	6	1	12	0	-12	1
11	29	81	6	1	11	0	-11	1

φ1	Φ	φ2	H	K	L	U	V	W
10	**20**	**90**						
11	27	90	1	0	2	-2	-11	1
11	27	90	1	0	2	-2	-12	1
13	27	90	1	0	2	-2	-10	1
14	27	90	1	0	2	-2	-9	1
16	27	90	1	0	2	-2	-8	1
18	27	90	1	0	2	-2	-7	1
10	**30**	**0**						
10	33	8	1	7	11	10	-3	1
11	34	9	1	6	9	9	-3	1
12	31	8	1	7	12	9	-3	1
12	39	6	1	9	11	7	-2	1
13	31	9	1	6	10	8	-3	1
14	39	7	1	8	10	6	-2	1
15	37	6	1	9	12	6	-2	1
16	34	9	1	6	9	12	-5	2
17	34	0	0	2	3	12	-3	2
17	36	8	1	7	9	5	-2	1
18	34	0	0	2	3	11	-3	2
18	36	7	1	8	11	5	-2	1
19	31	8	1	7	12	11	-5	2
10	**30**	**10**						
10	30	11	1	5	9	11	-4	1
10	32	18	1	3	5	10	-5	1
10	36	16	2	7	10	9	-4	1
10	38	18	1	3	4	8	-4	1
11	30	14	1	4	7	9	-4	1
11	36	11	1	5	7	8	-3	1
12	31	16	2	7	12	8	-4	1
14	33	11	1	5	8	7	-3	1
14	35	18	2	6	9	12	-7	2
14	39	16	2	7	9	6	-3	1
15	34	14	1	4	6	6	-3	1
16	36	11	1	5	7	11	-5	2
16	36	16	2	7	10	11	-6	2
16	38	18	1	3	4	5	-3	1
17	30	11	1	5	9	6	-3	1
18	35	18	2	6	9	9	-6	2
19	30	18	2	6	11	10	-7	2
19	37	14	2	8	11	9	-5	2
19	38	18	1	3	4	12	-8	3

φ_1	Φ	φ_2	H	K	L	U	V	W
10	30	20						
11	35	23	3	7	11	8	-5	1
12	31	22	2	5	9	8	-5	1
12	38	21	3	8	11	7	-4	1
13	37	23	3	7	10	6	-4	1
13	37	27	1	2	3	12	-9	2
13	39	24	4	9	12	6	-4	1
15	33	27	2	4	7	11	-9	2
15	37	27	1	2	3	5	-4	1
16	31	22	2	5	9	11	-8	2
18	33	27	2	4	7	9	-8	2
18	35	23	3	7	11	9	-7	2
18	37	27	1	2	3	8	-7	2
19	34	27	3	6	10	12	-11	3
19	37	27	1	2	3	11	-10	3
19	38	22	2	5	7	4	-3	1
10	30	30						
10	32	37	3	4	8	8	-8	1
10	36	37	3	4	7	7	-7	1
11	33	31	3	5	9	7	-6	1
11	39	34	4	6	9	12	-11	2
11	39	39	4	5	8	11	-12	2
12	30	39	4	5	11	6	-7	1
12	36	30	4	7	11	6	-5	1
12	36	36	5	7	12	6	-6	1
12	38	36	5	7	11	11	-11	2
13	31	34	2	3	6	6	-6	1
13	33	34	4	6	11	11	-11	2
13	36	37	3	4	7	10	-11	2
14	33	39	4	5	10	5	-6	1
14	36	34	2	3	5	5	-5	1
14	39	34	4	6	9	9	-9	2
15	35	39	4	5	9	4	-5	1
15	39	30	4	7	10	9	-8	2
16	30	31	3	5	10	5	-5	1
16	38	32	5	8	12	4	-4	1
17	31	34	2	3	6	9	-10	2
17	32	37	3	4	8	4	-5	1
17	33	31	3	5	9	9	-9	2
17	33	34	4	6	11	8	-9	2
17	36	31	3	5	8	4	-4	1
18	34	30	4	7	12	4	-4	1
18	36	34	2	3	5	7	-8	2
18	39	39	4	5	8	3	-4	1
19	30	39	4	5	11	7	-10	2
19	33	31	3	5	9	11	-12	3

φ_1	Φ	φ_2	H	K	L	U	V	W
10	30	30						
19	36	30	4	7	11	11	-11	3
19	39	30	4	7	10	10	-10	3
10	30	40						
10	35	45	1	1	2	6	-6	1
11	33	40	5	6	12	6	-7	1
12	35	45	1	1	2	5	-7	1
12	38	41	6	7	12	5	-6	1
13	35	40	5	6	11	5	-6	1
14	35	45	1	1	2	4	-6	1
15	35	45	1	1	2	7	-11	2
15	38	40	5	6	10	4	-5	1
15	38	49	7	6	12	6	-11	2
17	35	45	1	1	2	3	-5	1
19	33	40	5	6	12	6	-9	2
19	35	45	1	1	2	5	-9	2
10	30	50						
10	36	53	4	3	7	5	-9	1
10	38	50	6	5	10	5	-8	1
11	30	51	5	4	11	5	-9	1
11	30	59	5	3	10	4	-10	1
11	31	56	3	2	6	4	-9	1
11	36	54	7	5	12	4	-8	1
11	39	51	5	4	8	4	-7	1
12	32	53	4	3	8	4	-8	1
12	33	59	5	3	9	3	-8	1
12	35	50	6	5	11	4	-7	1
13	36	56	3	2	5	3	-7	1
14	39	56	6	4	9	5	-12	2
15	35	51	5	4	9	3	-6	1
15	36	59	5	3	8	2	-6	1
17	32	53	4	3	8	5	-12	2
17	36	56	3	2	5	4	-11	2
17	38	54	7	5	11	2	-5	1
18	31	56	3	2	6	2	-6	1
18	36	53	4	3	7	2	-5	1
19	33	59	5	3	9	3	-11	2
19	39	51	5	4	8	4	-9	2
19	39	56	6	4	9	3	-9	2
10	30	60						
10	34	60	7	4	12	4	-10	1
10	37	63	2	1	3	3	-9	1
10	38	69	8	3	11	2	-9	1

φ_1	Φ	φ_2	H	K	L	U	V	W
10	30	60						
11	34	68	5	2	8	2	-9	1
11	36	60	7	4	11	3	-8	1
11	37	67	7	3	10	2	-8	1
13	37	63	2	1	3	2	-7	1
14	39	60	7	4	10	2	-6	1
15	37	63	2	1	3	3	-12	2
15	38	68	5	2	7	1	-6	1
16	31	68	5	2	9	1	-7	1
17	35	67	7	3	11	1	-6	1
19	33	63	4	2	7	2	-11	2
19	37	63	2	1	3	1	-5	1
19	38	69	8	3	11	1	-10	2
10	30	70						
10	30	76	4	1	7	1	-11	1
10	32	72	3	1	5	2	-11	1
10	34	76	4	1	6	1	-10	1
11	33	74	7	2	11	1	-9	1
11	39	74	7	2	9	1	-8	1
13	30	79	5	1	9	0	-9	1
13	32	72	3	1	5	1	-8	1
13	33	79	5	1	8	0	-8	1
13	38	72	3	1	4	1	-7	1
14	36	79	5	1	7	0	-7	1
16	30	76	4	1	7	0	-7	1
16	38	77	9	2	12	0	-6	1
17	34	76	4	1	6	0	-6	1
17	35	72	6	2	9	1	-12	2
17	37	76	8	2	11	0	-11	2
17	38	72	3	1	4	1	-11	2
18	31	74	7	2	12	0	-6	1
19	33	74	7	2	11	0	-11	2
19	33	79	5	1	8	-1	-11	2
19	36	74	7	2	10	0	-5	1
10	30	80						
10	33	82	7	1	11	0	-11	1
10	35	82	7	1	10	0	-10	1
10	38	82	7	1	9	0	-9	1
11	31	81	6	1	10	0	-10	1
11	34	81	6	1	9	0	-9	1
12	37	81	6	1	8	0	-8	1
15	39	83	8	1	10	-1	-12	2
17	34	81	6	1	9	-1	-12	2
17	38	82	7	1	9	-1	-11	2
19	37	81	6	1	8	-1	-10	2

φ_1	Φ	φ_2	H	K	L	U	V	W
10	30	90						
17	34	90	2	0	3	-3	-12	2
18	34	90	2	0	3	-3	-11	2
10	40	0						
10	42	6	1	9	10	8	-2	1
10	45	0	0	1	1	8	-1	1
10	45	7	1	8	8	6	-2	1
11	40	6	1	10	12	8	-2	1
11	45	0	0	1	1	7	-1	1
11	45	8	1	7	7	7	-2	1
12	42	7	1	8	9	7	-2	1
12	46	6	1	10	9	12	-3	2
13	45	0	0	1	1	6	-1	1
13	45	5	1	12	12	12	-3	2
13	45	9	1	6	6	6	-2	1
13	49	6	1	9	8	11	-3	2
14	41	8	1	7	6	6	-2	1
14	45	0	0	1	1	11	-2	2
14	45	5	1	11	11	11	-3	2
14	49	7	1	8	7	10	-3	2
15	45	6	1	10	10	10	-3	2
16	41	9	1	6	7	5	-2	1
16	45	0	0	1	1	5	-1	1
17	45	0	0	1	1	9	-2	2
17	45	6	1	9	9	9	-3	2
18	43	5	1	11	12	9	-3	2
18	45	9	1	6	6	12	-5	3
18	49	6	1	9	8	12	-4	3
19	45	0	0	1	1	4	-1	1
19	45	5	1	12	12	12	-4	3
19	45	7	1	8	8	8	-3	2
19	49	7	1	6	7	11	-4	3
10	40	10						
10	40	13	2	9	11	8	-3	1
10	46	16	2	7	7	7	-3	1
10	49	17	3	10	9	7	-3	1
11	44	15	3	11	12	7	-3	1
12	40	14	1	4	5	7	-3	1
12	46	14	1	4	4	12	-5	2
12	47	18	1	3	3	6	-3	1
12	48	10	2	11	10	6	-2	1
13	41	17	3	10	12	6	-3	1
13	46	15	3	11	11	11	-5	2
13	49	11	2	10	9	11	-4	2

φ_1	Φ	φ_2	H	K	L	U	V	W
10	40	10						
14	42	14	2	8	9	11	-5	2
14	42	18	2	6	7	11	-6	2
14	45	10	2	11	11	11	-4	2
14	46	17	3	10	10	10	-5	2
14	49	13	2	9	8	5	-2	1
15	46	11	1	5	5	5	-2	1
15	47	18	1	3	3	9	-5	2
16	40	14	1	4	5	10	-5	2
16	43	10	2	11	12	5	-2	1
16	46	13	2	9	9	9	-4	2
17	43	11	2	10	11	9	-4	2
17	47	18	1	3	3	12	-7	3
17	49	15	3	11	10	4	-2	1
18	42	18	2	6	7	8	-5	2
18	46	14	1	4	4	4	-2	1
18	49	17	3	10	9	11	-6	3
19	43	13	2	9	10	4	-2	1
19	43	18	3	9	10	11	-7	3
19	46	15	3	11	11	11	-6	3
10	40	20						
10	47	23	3	7	7	7	-4	1
11	42	22	2	5	6	7	-4	1
11	47	23	5	12	12	12	-7	2
11	48	27	1	2	2	6	-4	1
11	49	29	6	11	11	11	-8	2
12	44	22	4	10	11	12	-7	2
12	47	20	4	11	11	11	-6	2
12	48	24	5	11	11	11	-7	2
13	41	21	3	8	10	12	-7	2
13	43	29	5	9	11	5	-4	1
13	44	23	3	7	8	11	-7	2
13	47	22	2	5	5	5	-3	1
13	48	27	1	2	2	10	-7	2
13	49	29	5	9	9	9	-7	2
14	44	21	3	8	9	5	-3	1
14	48	24	4	9	9	9	-6	2
15	42	27	2	4	5	9	-7	2
15	46	29	5	9	10	12	-10	3
15	48	27	1	2	2	4	-3	1
16	42	22	2	5	6	9	-6	2
16	47	23	5	12	12	12	-8	3
16	49	29	6	11	11	11	-9	3
17	40	27	3	6	8	12	-10	3
17	42	24	4	9	11	4	-3	1
17	44	27	3	6	7	11	-9	3
17	46	29	6	11	12	7	-6	2

φ_1	Φ	φ_2	H	K	L	U	V	W
10	40	20						
17	47	21	3	8	8	8	-5	2
17	48	24	5	11	11	11	-8	3
18	40	23	3	7	9	4	-3	1
18	47	20	4	11	11	11	-7	3
18	47	23	3	7	7	7	-5	2
18	48	27	1	2	2	10	-8	3
19	42	22	2	5	6	11	-8	3
19	42	27	2	4	5	7	-6	2
19	44	22	4	10	11	7	-5	2
19	47	22	2	5	5	10	-7	3
19	49	29	5	9	9	9	-8	3
10	40	30						
10	44	38	7	9	12	6	-6	1
10	46	38	7	9	11	11	-11	2
10	49	30	7	12	12	12	-9	2
10	49	39	8	10	11	11	-11	2
11	40	31	3	5	7	6	-5	1
11	40	37	3	4	6	6	-6	1
11	42	34	2	3	4	11	-10	2
11	42	37	6	8	11	11	-11	2
11	43	32	5	8	10	6	-5	1
11	45	37	3	4	5	5	-5	1
11	49	38	7	9	10	5	-5	1
12	41	36	5	7	10	5	-5	1
12	44	31	3	5	6	11	-9	2
12	46	34	4	6	7	10	-9	2
12	48	37	6	8	9	9	-9	2
12	49	31	3	5	5	5	-4	1
13	41	32	5	8	11	10	-9	2
13	42	39	4	5	7	9	-10	2
13	44	36	5	7	9	9	-9	2
13	45	30	4	7	8	5	-4	1
13	49	36	8	11	12	4	-4	1
14	45	35	7	10	12	4	-4	1
14	47	36	5	7	8	4	-4	1
14	49	30	7	12	12	12	-10	3
15	40	37	3	4	6	8	-9	2
15	42	34	2	3	4	4	-4	1
15	45	37	3	4	5	11	-12	3
15	47	31	6	10	11	8	-7	2
15	47	39	4	5	6	7	-8	2
15	48	35	7	10	11	11	-11	3
16	42	37	6	8	11	7	-8	2
16	45	34	6	9	11	11	-11	3
16	46	34	4	6	7	7	-7	2
16	47	34	6	9	10	10	-10	3

φ_1	Φ	φ_2	H	K	L	U	V	W
10	40	30						
16	49	30	4	7	7	7	-6	2
16	49	39	8	10	11	6	-7	2
17	40	37	3	4	6	10	-12	3
17	41	32	5	8	11	11	-11	3
17	44	36	5	7	9	10	-11	3
17	44	38	7	9	12	9	-11	3
17	45	37	3	4	5	6	-7	2
17	49	31	3	5	5	10	-9	3
18	40	31	3	5	7	7	-7	2
18	43	32	5	8	10	10	-10	3
18	46	32	5	8	9	3	-3	1
18	47	32	7	11	12	3	-3	1
18	47	39	4	5	6	8	-10	3
18	49	30	7	12	12	12	-11	4
19	42	34	2	3	4	9	-10	3
19	46	34	4	6	7	11	-12	4
19	48	37	6	6	9	5	-6	2
10	40	40						
10	46	41	6	7	9	11	-12	2
10	49	49	7	6	8	4	-6	1
11	43	45	2	2	3	9	-12	2
11	44	40	5	6	8	10	-11	2
11	45	48	9	8	12	4	-6	1
11	47	49	8	7	10	4	-6	1
11	49	41	6	7	8	9	-10	2
12	43	45	2	2	3	8	-11	2
13	42	41	7	8	12	4	-5	1
13	48	40	5	6	7	8	-9	2
14	40	41	6	7	11	4	-5	1
14	43	45	2	2	3	7	-10	2
14	46	49	7	6	9	3	-5	1
14	49	41	6	7	8	10	-12	3
15	43	45	2	2	3	6	-9	2
16	44	41	7	8	11	3	-4	1
16	45	42	8	9	12	3	-4	1
16	47	45	3	3	4	8	-12	3
16	48	40	5	6	7	9	-11	3
17	41	40	5	6	9	3	-4	1
17	43	41	6	7	10	3	-4	1
17	48	48	9	8	11	7	-12	3
18	43	45	2	2	3	5	-8	2
18	47	45	3	3	4	7	-11	3
18	49	49	7	6	8	6	-11	3
19	40	45	3	3	5	7	-12	3
19	42	49	8	7	12	2	-4	1
19	43	49	7	6	10	2	-4	1

φ_1	Φ	φ_2	H	K	L	U	V	W
10	40	50						
10	42	56	3	2	4	4	-8	1
10	45	53	4	3	5	4	-7	1
11	44	59	5	3	6	3	-7	1
12	41	56	8	5	11	3	-7	1
12	44	54	7	5	9	3	-6	1
12	45	55	10	7	12	3	-6	1
12	47	59	10	6	11	5	-12	2
12	48	53	8	6	9	6	-11	2
13	40	53	4	3	6	3	-6	1
13	41	50	6	5	9	7	-12	2
13	43	58	8	5	10	5	-12	2
13	46	56	6	4	7	5	-11	2
13	48	50	6	5	7	3	-5	1
14	42	51	5	4	7	6	-11	2
14	45	53	4	3	5	5	-10	2
14	49	59	5	3	5	2	-5	1
15	41	54	7	5	10	5	-11	2
15	46	58	8	5	9	2	-5	1
15	49	51	10	8	11	5	-9	2
16	42	56	3	2	4	2	-5	1
17	42	53	8	6	11	4	-9	2
17	44	59	5	3	6	3	-9	2
17	47	51	5	4	6	2	-4	1
17	49	52	9	7	10	2	-4	1
18	44	50	6	5	8	2	-4	1
19	46	56	6	4	7	3	-8	2
10	40	60						
10	42	68	5	2	6	2	-8	1
10	49	60	7	4	7	3	-7	1
11	45	66	9	4	10	2	-7	1
11	47	67	7	3	7	2	-7	1
12	46	63	2	1	2	2	-6	1
12	49	60	12	7	12	5	-12	2
13	47	67	12	5	12	3	-12	2
13	49	61	11	6	11	4	-11	2
14	41	61	9	5	12	2	-6	1
14	41	69	8	3	10	1	-6	1
14	45	60	7	4	8	4	-11	2
14	48	66	11	5	11	3	-11	2
15	42	63	4	2	5	3	-11	2
15	42	68	5	2	6	2	-11	2
15	47	68	5	2	5	1	-5	1
15	48	63	2	1	2	3	-10	2
16	44	67	7	3	6	1	-5	1
16	49	61	9	5	9	3	-9	2

TAB. VI FROM EULER ANGLES INTO HKL-UVW

φ_1	Φ	φ_2	H	K	L	U	V	W
10	40	60						
17	42	66	9	4	11	1	-5	1
17	48	66	9	4	9	2	-9	2
18	48	63	2	1	2	1	-4	1
18	49	60	12	7	12	4	-12	3
19	42	63	4	2	5	2	-9	2
19	47	67	12	5	12	2	-12	3
19	47	69	8	3	8	1	-8	2
10	40	70						
10	40	76	4	1	5	1	-9	1
10	46	76	4	1	4	1	-8	1
11	46	74	7	2	7	1	-7	1
11	49	75	11	3	10	1	-7	1
12	44	73	10	3	11	1	-7	1
13	46	76	4	1	4	1	-12	2
13	47	72	3	1	3	1	-6	1
14	46	75	11	3	11	1	-11	2
14	47	70	11	4	11	2	-11	2
15	40	79	5	1	6	0	-6	1
15	43	79	10	2	11	0	-11	2
16	40	77	9	2	11	0	-11	2
16	46	73	10	3	10	1	-10	2
16	46	79	5	1	5	0	-5	1
17	42	72	6	2	7	1	-10	2
17	43	77	9	2	10	0	-5	1
17	47	72	3	1	3	1	-9	2
17	49	79	10	2	9	0	-9	2
18	40	76	4	1	5	0	-5	1
18	46	77	9	2	9	0	-9	2
19	42	76	8	2	9	0	-9	2
19	47	72	3	1	3	1	-12	3
19	49	77	9	2	8	0	-4	1
10	40	80						
10	42	83	8	1	9	0	-9	1
10	45	83	8	1	8	0	-8	1
10	49	84	9	1	8	0	-8	1
11	41	82	7	1	8	0	-8	1
11	45	82	7	1	7	0	-7	1
11	49	83	8	1	7	0	-7	1
12	41	81	6	1	7	0	-7	1
13	45	81	6	1	6	0	-6	1
13	45	85	12	1	12	-1	-12	2
14	42	84	10	1	11	-1	-12	2
14	43	80	11	2	12	0	-6	1
14	48	81	12	2	11	0	-11	2

φ_1	Φ	φ_2	H	K	L	U	V	W
10	40	80						
15	42	84	9	1	10	-1	-11	2
15	45	80	11	2	11	0	-11	2
15	45	85	11	1	11	-1	-11	2
15	48	80	11	2	10	0	-5	1
15	48	85	12	1	11	-1	-10	2
16	45	84	10	1	10	-1	-10	2
17	42	83	8	1	9	-1	-10	2
17	48	85	11	1	10	-1	-9	2
18	45	84	9	1	9	-1	-9	2
19	41	82	7	1	8	-1	-9	2
19	48	84	10	1	9	-1	-6	2
10	40	90						
10	45	90	1	0	1	-1	-8	1
11	45	90	1	0	1	-1	-7	1
13	45	90	1	0	1	-1	-6	1
14	45	90	1	0	1	-2	-11	2
16	45	90	1	0	1	-1	-5	1
17	45	90	1	0	1	-2	-9	2
19	45	90	1	0	1	-1	-4	1
10	50	0						
10	51	9	1	6	5	7	-2	1
11	53	7	1	8	6	12	-3	2
12	55	8	1	7	5	11	-3	2
13	57	9	1	6	4	10	-3	2
15	54	9	2	12	9	9	-3	2
16	50	8	1	7	6	9	-3	2
16	56	5	1	12	8	4	-1	1
16	57	9	1	6	4	12	-4	3
16	58	5	1	11	7	4	-1	1
16	59	6	1	10	6	4	-1	1
17	51	9	1	6	5	8	-3	2
17	56	0	0	3	2	12	-2	3
18	56	0	0	3	2	11	-2	3
10	50	10						
10	54	10	2	11	8	7	-2	1
11	52	11	1	5	4	6	-2	1
11	56	17	3	10	7	12	-5	2
11	57	13	2	9	6	6	-2	1
11	58	18	1	3	2	11	-5	2
11	59	14	2	8	5	11	-4	2
12	53	13	2	9	7	11	-4	2
12	58	15	3	11	7	5	-2	1

φ_1	Φ	φ_2	H	K	L	U	V	W
10	50	10						
13	52	18	2	6	5	10	-5	2
13	54	14	1	4	3	5	-2	1
14	56	16	2	7	5	9	-4	2
15	50	14	2	8	7	9	-4	2
15	58	18	1	3	2	4	-2	1
16	51	16	2	7	6	4	-2	1
16	53	13	2	9	7	12	-5	3
16	53	17	3	10	8	4	-2	1
16	56	11	2	10	7	6	-3	2
16	57	14	3	12	8	12	-5	3
17	54	14	1	4	3	11	-5	3
17	54	18	3	9	7	11	-6	3
17	59	14	2	8	5	7	-3	2
18	52	18	2	6	5	7	-4	2
18	56	16	2	7	5	10	-5	3
19	50	18	3	9	8	10	-6	3
19	51	14	3	12	10	10	-5	3
19	52	11	1	5	4	7	-3	2
19	54	10	2	11	8	10	-4	3
19	58	18	1	3	2	9	-5	3
10	50	20						
10	55	21	3	8	6	6	-3	1
11	52	20	4	11	9	6	-3	1
11	53	22	2	5	4	11	-6	2
11	56	27	2	4	3	11	-7	2
12	52	23	3	7	6	5	-3	1
12	55	24	4	9	7	5	-3	1
12	56	24	5	11	8	5	-3	1
13	50	23	5	12	11	5	-3	1
13	56	27	2	4	3	9	-6	2
13	57	22	4	10	7	9	-5	2
14	54	29	6	11	9	4	-3	1
14	56	29	5	9	7	4	-3	1
14	58	23	5	12	8	12	-7	3
14	59	27	3	6	4	12	-8	3
15	51	24	4	9	8	12	-8	3
15	52	23	5	12	10	8	-5	2
16	50	22	4	10	9	8	-5	2
16	53	27	3	6	5	11	-8	3
16	56	27	2	4	3	7	-5	2
16	59	27	3	6	4	10	-7	3
17	50	24	5	11	10	7	-5	2
17	53	24	5	11	9	10	-7	3
17	56	20	4	11	8	7	-4	2
18	55	21	3	8	6	10	-6	3
18	56	27	2	4	3	12	-9	4

φ_1	Φ	φ_2	H	K	L	U	V	W
10	50	20						
18	59	24	4	9	6	3	-2	1
19	53	22	2	5	4	3	-2	1
19	53	27	3	6	5	9	-7	3
19	55	23	5	12	9	3	-2	1
19	57	23	3	7	5	3	-2	1
10	50	30						
10	50	34	2	3	3	6	-5	1
10	51	36	8	11	11	11	-10	2
10	52	31	6	10	9	12	-9	2
10	52	39	4	5	5	5	-5	1
10	53	30	4	7	6	11	-8	2
10	55	34	4	6	5	11	-9	2
10	55	39	9	11	10	5	-5	1
10	59	31	6	10	7	11	-8	2
10	59	37	3	4	3	10	-9	2
11	50	32	7	11	11	11	-9	2
11	51	35	7	10	10	10	-9	2
11	52	38	7	9	9	9	-9	2
11	53	32	5	8	7	5	-4	1
11	55	32	7	11	9	5	-4	1
11	55	39	8	10	9	9	-9	2
11	58	39	9	11	9	9	-9	2
12	50	34	2	3	3	9	-8	2
12	55	36	7	9	8	4	-4	1
12	56	31	3	5	4	9	-7	2
12	58	39	4	5	4	4	-4	1
13	51	37	3	4	4	4	-4	1
13	52	39	9	11	11	11	-12	3
13	58	30	4	7	5	4	-3	1
13	58	36	7	9	7	7	-7	2
14	50	32	5	8	8	8	-7	2
14	50	34	2	3	3	12	-11	3
14	51	36	8	11	11	11	-11	3
14	54	34	6	9	8	11	-10	3
14	54	37	9	12	11	11	-11	3
14	55	37	6	8	7	7	-7	2
14	56	37	9	12	10	10	-10	3
15	50	32	7	11	11	11	-10	3
15	51	36	5	7	7	7	-7	2
15	52	39	4	5	5	10	-11	3
15	54	36	8	11	10	10	-10	3
15	56	31	3	5	4	11	-9	3
15	57	34	6	9	7	10	-9	3
16	51	35	7	10	10	10	-10	3
16	52	31	6	10	9	7	-6	2
16	52	38	7	9	9	9	-10	3

φ_1 Φ φ_2	H	K	L	U	V	W
10 50 30						
16 55 36	5	7	6	3	-3	1
16 57 36	8	11	9	3	-3	1
16 58 34	8	12	9	12	-11	4
16 58 39	4	5	4	11	-12	4
16 59 37	3	4	3	3	-3	1
17 50 34	2	3	3	3	-3	1
17 53 32	5	8	7	12	-11	4
17 54 35	7	10	9	3	-3	1
17 57 30	7	12	9	6	-5	2
17 58 39	9	11	9	8	-9	3
17 59 31	6	10	7	6	-5	2
18 51 36	8	11	11	11	-12	4
18 51 37	3	4	4	8	-9	3
18 52 30	7	12	11	9	-8	3
18 52 39	4	5	5	5	-6	2
18 53 34	8	12	11	11	-11	4
18 55 37	6	8	7	10	-11	4
18 57 35	7	10	8	8	-8	3
19 50 32	7	11	11	11	-11	4
19 50 39	9	11	12	7	-9	3
19 54 34	6	9	8	8	-8	3
19 55 34	4	6	5	5	-5	2
19 55 39	9	11	10	9	-11	4
19 58 39	4	5	4	7	-8	3
10 50 40						
10 50 41	7	8	9	10	-11	2
10 53 42	9	10	10	10	-11	2
10 54 48	11	10	11	8	-11	2
10 55 45	1	1	1	9	-11	2
10 56 42	10	11	10	9	-10	2
10 56 48	9	8	8	8	-11	2
10 57 40	5	6	5	5	-5	1
11 53 42	8	9	9	9	-10	2
11 55 45	1	1	1	4	-5	1
11 56 42	9	10	9	8	-9	2
11 57 49	8	7	7	7	-10	2
11 58 43	11	12	10	8	-9	2
12 53 41	7	8	8	8	-9	2
12 53 48	10	9	10	7	-10	2
12 55 45	1	1	1	7	-9	2
12 57 41	6	7	6	11	-12	3
12 59 42	10	11	9	7	-8	2
13 53 41	6	7	7	7	-8	2
13 53 48	9	8	9	6	-9	2
13 56 42	8	9	8	7	-8	2
13 57 49	7	6	6	6	-9	2

φ_1 Φ φ_2	H	K	L	U	V	W
10 50 40						
14 51 48	11	10	12	6	-9	2
14 53 42	9	10	10	10	-12	3
14 55 45	1	1	1	3	-4	1
14 56 43	11	12	11	9	-11	3
14 56 48	9	8	8	8	-12	3
14 59 42	9	10	8	6	-7	2
15 53 42	8	9	9	9	-11	3
15 53 49	8	7	8	5	-8	2
15 54 47	12	11	12	8	-12	3
15 55 45	1	1	1	8	-11	3
15 57 40	5	6	5	9	-10	3
15 57 41	7	8	7	6	-7	2
16 51 48	10	9	11	5	-8	2
16 52 40	5	6	6	6	-7	2
16 54 48	11	10	11	7	-11	3
16 55 45	1	1	1	5	-7	2
16 56 42	10	11	10	8	-10	3
16 57 49	8	7	7	7	-11	3
17 53 41	7	8	8	8	-10	3
17 55 45	1	1	1	7	-10	3
17 57 41	6	7	6	5	-6	2
17 57 49	7	6	6	6	-10	3
18 50 48	9	8	10	4	-7	2
18 53 48	10	9	10	6	-10	3
18 53 49	7	6	7	4	-7	2
18 55 40	10	12	11	5	-6	2
18 56 42	9	10	9	7	-9	3
19 53 41	6	7	7	7	-9	3
19 53 42	8	9	9	9	-12	4
19 55 45	1	1	1	2	-3	1
19 57 49	8	7	7	7	-12	4
10 50 50						
10 52 50	6	5	6	4	-6	1
10 55 51	11	9	10	4	-6	1
10 55 58	11	7	9	3	-6	1
10 58 52	9	7	7	7	-11	2
10 58 58	8	5	6	3	-6	1
11 50 56	3	2	3	3	-6	1
11 51 53	4	3	4	7	-12	2
11 52 51	11	9	11	7	-11	2
11 55 51	10	8	9	7	-11	2
11 56 59	5	3	4	5	-11	2
11 59 53	4	3	3	3	-5	1
12 50 58	11	7	11	5	-11	2
12 51 54	11	8	11	6	-11	2
12 52 51	5	4	5	3	-5	1

φ_1 Φ φ_2	H	K	L	U	V	W
10 50 50						
12 55 52	9	7	8	3	-5	1
12 55 56	6	4	5	5	-10	2
13 50 51	11	9	12	3	-5	1
13 51 55	10	7	10	5	-10	2
13 55 53	8	6	7	5	-9	2
13 59 59	10	6	7	4	-9	2
14 52 52	9	7	9	5	-9	2
14 57 50	6	5	5	5	-8	2
14 58 52	9	7	7	7	-12	3
15 50 56	3	2	3	4	-9	2
15 52 50	6	5	6	7	-12	3
15 55 54	7	5	6	2	-4	1
15 57 55	10	7	8	2	-4	1
16 51 53	4	3	4	2	-4	1
16 54 54	11	8	10	2	-4	1
16 58 51	5	4	4	4	-7	2
16 59 53	4	3	3	6	-11	3
17 50 56	3	2	3	5	-12	3
17 50 58	8	5	8	3	-8	2
17 52 51	11	9	11	6	-11	3
17 52 59	10	6	9	3	-8	2
17 54 55	10	7	9	5	-11	3
17 55 50	12	10	11	4	-7	2
17 57 56	9	6	7	5	-11	3
17 58 58	11	7	8	3	-7	2
18 51 54	11	8	11	5	-11	3
18 55 56	6	4	5	3	-7	2
18 56 53	12	9	10	5	-10	3
19 50 58	11	7	11	4	-11	3
19 51 54	7	5	7	3	-7	2
19 52 51	5	4	5	5	-10	3
19 54 56	9	6	8	4	-10	3
19 57 50	6	5	5	5	-9	3
19 58 58	8	5	6	4	-10	3
19 59 53	4	3	3	3	-6	2
10 50 60						
10 50 67	12	5	11	2	-7	1
10 53 68	5	2	4	2	-7	1
11 55 69	8	3	6	3	-12	2
11 56 66	11	5	8	2	-6	1
11 59 66	9	4	6	2	-6	1
12 50 68	10	4	9	3	-12	2
12 56 63	4	2	3	4	-11	2
12 57 68	10	4	7	3	-11	2
13 52 60	12	7	11	2	-5	1
13 52 67	7	3	6	3	-11	2

φ_1 Φ φ_2	H	K	L	U	V	W
10 50 60						
13 53 60	7	4	6	2	-5	1
13 56 61	9	5	7	2	-5	1
13 57 61	11	6	8	2	-5	1
14 51 69	8	3	7	1	-5	1
14 56 63	4	2	3	3	-9	2
15 53 68	5	2	4	2	-9	2
16 54 60	12	7	10	3	-8	2
16 57 67	7	3	5	1	-4	1
16 58 67	12	5	8	1	-4	1
16 59 63	6	3	4	4	-12	3
17 52 61	9	5	8	4	-12	3
17 53 66	11	5	9	1	-4	1
17 55 66	9	4	7	1	-4	1
19 53 63	6	3	5	3	-11	3
19 53 68	5	2	4	2	-11	3
19 56 63	4	2	3	2	-7	2
19 59 63	6	3	4	3	-10	3
10 50 70						
10 54 76	4	1	3	1	-7	1
11 56 74	7	2	5	1	-6	1
11 56 79	10	2	7	1	-12	2
11 58 75	11	3	7	1	-6	1
12 53 73	10	3	8	1	-6	1
13 52 72	6	2	5	2	-11	2
13 58 72	3	1	2	1	-5	1
14 50 76	8	2	7	1	-11	2
14 52 70	11	4	9	1	-5	1
14 54 76	4	1	3	1	-10	2
14 59 70	11	4	7	2	-9	2
15 55 75	11	3	8	1	-9	2
15 59 76	8	2	5	1	-9	2
16 56 72	3	1	2	2	-12	3
17 56 73	10	3	7	1	-8	2
17 57 76	12	3	8	1	-12	3
18 52 72	6	2	5	1	-8	2
18 52 79	5	1	4	0	-4	1
19 56 74	7	2	5	1	-11	3
19 56 79	10	2	7	0	-7	2
19 58 72	3	1	2	1	-7	2
10 50 80						
10 52 84	9	1	7	0	-7	1
10 55 84	10	1	7	0	-7	1
10 58 85	11	1	7	0	-7	1
11 56 84	9	1	6	0	-6	1

TAB. VI

FROM EULER ANGLES INTO HKL-UVW

φ_1 Φ φ_2	H	K	L	U	V	W
10 50 80						
11 59 84	10	1	6	0	-6	1
12 50 82	7	1	6	0	-6	1
12 53 83	8	1	6	0	-6	1
13 58 83	8	1	5	0	-5	1
14 55 82	7	1	5	0	-5	1
15 51 81	6	1	5	0	-5	1
16 51 80	11	2	9	0	-9	2
16 54 81	12	2	9	0	-9	2
17 54 80	11	2	8	0	-4	1
17 56 85	12	1	8	-1	-12	3
17 57 81	6	1	4	0	-4	1
18 50 85	12	1	10	-1	-8	2
18 52 84	9	1	7	-1	-12	3
19 55 84	10	1	7	-1	-11	3
19 58 80	11	2	7	0	-7	2
10 50 90						
17 56 90	3	0	2	-2	-12	3
18 56 90	3	0	2	-2	-11	3
10 60 0						
10 64 5	1	12	6	6	-1	1
10 64 9	1	6	3	12	-3	2
10 66 5	1	11	5	6	-1	1
10 68 6	1	10	4	6	-1	1
11 60 9	2	12	7	11	-3	2
11 63 0	0	2	1	11	-1	2
11 63 0	0	2	1	12	-1	2
12 64 6	1	10	5	5	-1	1
12 66 6	1	9	4	5	-1	1
13 60 5	1	12	7	5	-1	1
13 61 5	1	11	6	5	-1	1
13 63 0	0	2	1	10	-1	2
14 63 0	0	2	1	9	-1	2
15 64 7	1	8	4	4	-1	1
15 67 8	1	7	3	4	-1	1
16 61 6	1	9	5	4	-1	1
16 63 0	0	2	1	8	-1	2
17 68 9	2	12	5	7	-2	2
18 63 0	0	2	1	7	-1	2
10 60 10						
10 64 14	1	4	2	6	-2	1
10 65 18	2	6	3	12	-5	2
10 68 16	2	7	3	11	-4	2

φ_1 Φ φ_2	H	K	L	U	V	W
10 60 10						
11 69 11	1	5	2	11	-3	2
12 61 16	2	7	4	5	-2	1
12 64 11	2	10	5	10	-3	2
12 64 17	3	10	5	5	-2	1
13 65 18	2	6	3	9	-4	2
14 60 11	1	5	3	9	-3	2
15 64 14	1	4	2	8	-3	2
15 67 13	2	9	4	12	-4	3
15 67 18	3	9	4	11	-5	3
16 66 14	3	12	5	11	-4	3
17 60 11	1	5	3	11	-4	3
17 62 15	3	11	6	7	-3	2
17 62 18	3	9	5	10	-5	3
17 64 14	1	4	2	10	-4	3
19 60 17	3	10	6	6	-3	2
19 64 14	1	4	2	12	-5	4
19 66 10	2	11	5	3	-1	1
10 60 20						
10 61 22	2	5	3	6	-3	1
10 63 24	4	9	5	11	-6	2
11 65 22	4	10	5	10	-5	2
11 69 23	5	12	5	10	-5	2
12 60 21	3	8	5	10	-5	2
12 62 23	3	7	4	9	-5	2
12 64 29	6	11	6	9	-6	2
12 68 24	5	11	5	9	-5	2
13 66 27	1	2	1	8	-5	2
13 69 29	5	9	4	12	-6	3
14 63 20	4	11	6	4	-2	1
14 63 24	4	9	5	12	-7	3
14 65 21	3	8	4	4	-2	1
14 66 27	1	2	1	11	-7	3
14 68 23	3	7	3	11	-6	3
15 64 29	5	9	5	7	-5	2
15 68 24	4	9	4	7	-4	2
16 60 21	3	8	5	11	-6	3
16 61 22	2	5	3	7	-4	2
16 62 23	3	7	4	10	-6	3
16 67 20	4	11	5	10	-5	3
17 60 29	5	9	6	9	-7	3
17 66 27	1	2	1	3	-2	1
17 69 23	5	12	5	9	-5	3
18 65 23	5	12	6	12	-7	4
19 60 21	3	8	5	12	-7	4
19 61 27	4	8	5	11	-8	4
19 62 23	3	7	4	11	-7	4

φ_1 Φ φ_2	H	K	L	U	V	W
10 60 20						
19 64 29	6	11	6	8	-6	3
19 68 24	5	11	5	6	-5	3
10 60 30						
10 61 34	2	3	2	5	-4	1
10 63 30	7	12	7	10	-7	2
10 67 31	6	10	5	10	-7	2
10 68 35	7	10	5	5	-4	1
11 60 36	8	11	8	9	-8	2
11 62 32	7	11	7	9	-7	2
11 63 37	6	8	5	9	-6	2
11 66 38	7	9	5	12	-11	3
11 67 34	4	6	3	9	-7	2
11 68 37	3	4	2	6	-7	2
12 60 35	7	10	7	8	-7	2
12 61 34	2	3	2	12	-10	3
12 61 39	9	11	8	4	-4	1
12 62 37	9	12	8	12	-11	3
12 69 32	7	11	5	4	-3	1
13 60 30	7	12	8	4	-3	1
13 60 36	5	7	5	11	-10	3
13 61 39	8	10	7	7	-7	2
13 63 31	3	5	3	4	-3	1
13 64 39	9	11	7	7	-7	2
13 65 34	6	9	5	11	-9	3
13 65 37	9	12	7	11	-10	3
14 61 34	2	3	2	7	-6	2
14 64 30	4	7	4	11	-8	3
14 67 39	9	11	6	3	-3	1
15 62 38	7	9	6	3	-3	1
15 65 32	7	11	6	10	-8	3
15 65 39	4	5	3	3	-3	1
15 68 37	3	4	2	12	-11	4
16 61 34	2	3	2	9	-8	3
16 62 32	5	8	5	6	-5	2
16 67 30	7	12	6	12	-9	4
16 67 34	4	6	3	6	-5	2
17 60 36	8	11	8	8	-8	3
17 61 34	2	3	2	11	-10	4
17 62 37	9	12	8	8	-8	3
17 63 30	7	12	7	9	-7	3
17 65 34	6	9	5	8	-7	3
17 66 38	7	9	5	5	-5	2
17 69 39	8	10	5	5	-5	2
18 60 36	5	7	5	5	-5	2
18 62 32	7	11	7	8	-7	3
18 63 31	3	5	3	11	-9	4

φ_1 Φ φ_2	H	K	L	U	V	W
10 60 30						
18 63 37	6	8	5	5	-5	2
18 64 34	8	12	7	10	-9	4
18 68 35	7	10	5	10	-9	4
19 61 39	8	10	7	9	-10	4
19 63 36	8	11	7	7	-7	3
19 64 30	4	7	4	5	-4	2
19 65 37	9	12	7	7	-7	3
19 67 31	6	10	5	5	-4	2
19 67 32	5	8	4	12	-10	5
10 60 40						
10 60 40	10	12	9	9	-9	2
10 62 49	7	6	5	8	-11	2
10 67 47	12	11	7	4	-5	1
10 68 46	11	10	6	4	-5	1
11 63 40	5	6	4	4	-4	1
11 67 41	6	7	4	4	-4	1
11 69 41	7	8	4	4	-4	1
12 63 42	9	10	7	11	-12	3
12 65 45	3	3	2	10	-12	3
12 66 48	10	9	6	3	-4	1
12 67 48	9	8	5	3	-4	1
12 69 49	8	7	4	3	-4	1
13 61 41	7	8	6	10	-11	3
13 64 47	12	11	8	3	-4	1
13 65 42	10	11	7	10	-11	3
13 65 45	3	3	2	9	-11	3
13 65 48	11	10	7	3	-4	1
13 66 40	10	12	7	7	-7	2
14 61 43	11	12	9	6	-7	2
14 64 42	8	9	6	9	-10	3
14 67 43	11	12	7	9	-10	3
14 69 40	5	6	3	3	-3	1
15 65 45	3	3	2	8	-10	3
15 66 42	9	10	6	8	-9	3
15 67 42	8	9	5	11	-12	4
16 62 41	6	7	5	8	-9	3
16 62 42	10	11	8	5	-6	2
16 65 45	3	3	2	7	-9	3
17 60 42	8	9	7	5	-6	2
17 62 45	4	4	3	9	-12	4
17 63 40	5	6	4	10	-11	4
17 64 48	9	8	6	2	-3	1
17 65 41	7	6	5	7	-8	3
17 65 49	8	7	5	2	-3	1
17 67 49	7	6	4	2	-3	1
17 68 42	10	11	6	7	-8	3

φ_1 Φ φ_2	H	K	L	U	V	W
10 60 40						
18 61 47	12	11	9	2	-3	1
18 62 48	11	10	8	2	-3	1
18 63 48	10	9	7	2	-3	1
18 67 41	6	7	4	9	-10	4
19 62 45	4	4	3	8	-11	4
19 64 43	11	12	8	4	-5	2
19 65 45	3	3	2	6	-8	3
19 66 42	9	10	6	10	-12	5
10 60 50						
10 61 56	3	2	2	6	-11	2
10 63 50	6	5	4	7	-10	2
10 63 53	8	6	5	7	-11	2
10 63 59	5	3	3	3	-6	1
10 69 51	10	8	5	7	-10	2
11 62 58	8	5	5	5	-10	2
11 63 54	11	8	7	3	-5	1
11 65 54	7	5	4	3	-5	1
11 67 59	10	6	5	5	-10	2
11 68 55	10	7	5	3	-5	1
12 60 50	12	10	9	6	-9	2
12 65 51	5	4	3	6	-9	2
12 67 56	6	4	3	5	-9	2
13 60 54	7	5	5	5	-9	2
13 63 50	6	5	4	8	-12	3
13 68 53	4	3	2	5	-8	2
14 61 51	10	8	7	5	-8	2
14 61 56	3	2	2	2	-4	1
14 62 53	12	9	8	7	-12	3
14 65 51	5	4	3	7	-11	3
14 65 58	11	7	6	2	-4	1
14 67 58	8	5	4	2	-4	1
15 66 54	11	8	6	4	-7	2
16 62 58	8	5	5	5	-11	3
16 63 53	8	6	5	4	-7	2
16 68 53	4	3	2	6	-10	3
16 69 58	11	7	5	5	-10	3
17 60 54	7	5	5	5	-10	3
17 61 51	11	9	8	6	-10	3
17 63 59	5	3	3	3	-7	2
17 65 56	9	6	5	5	-10	3
17 68 53	4	3	2	7	-12	4
17 69 50	6	5	3	2	-3	1
18 62 52	9	7	6	5	-9	3
18 65 53	12	9	7	5	-9	3
18 67 56	6	4	3	3	-6	2
19 60 50	12	10	9	7	-12	4

φ_1 Φ φ_2	H	K	L	U	V	W
10 60 50						
19 61 56	3	2	2	4	-9	3
19 62 58	6	5	5	5	-12	4
19 64 55	10	7	6	3	-6	2
10 60 60						
10 62 67	7	3	4	2	-6	1
10 64 61	9	5	5	5	-11	2
10 65 67	12	5	6	2	-6	1
10 66 63	2	1	1	5	-12	2
10 68 66	9	4	4	4	-11	2
12 62 67	12	5	7	3	-10	2
12 65 68	10	4	5	3	-10	2
12 66 63	2	1	1	2	-5	1
12 68 61	11	6	5	4	-9	2
13 64 60	7	4	4	4	-9	2
13 64 66	11	5	6	3	-9	2
13 68 67	7	3	3	3	-9	2
14 64 61	9	5	5	5	-12	3
14 66 63	2	1	1	3	-8	2
14 68 66	9	4	4	4	-12	3
15 60 60	12	7	6	5	-12	3
15 65 69	8	3	4	1	-4	1
16 61 68	5	2	3	1	-4	1
16 66 63	2	1	1	4	-11	3
16 69 61	9	5	4	3	-7	2
17 62 67	7	3	4	3	-11	3
17 63 66	9	4	5	2	-7	2
17 64 60	7	4	4	4	-10	3
17 68 67	7	3	3	3	-10	3
19 66 63	2	1	1	1	-3	1
19 67 60	12	7	6	5	-12	4
10 60 70						
10 64 76	4	1	2	1	-6	1
10 65 72	6	2	3	3	-12	2
12 60 79	5	1	3	1	-11	2
12 61 74	7	2	4	2	-11	2
12 64 73	10	3	5	1	-5	1
12 68 74	7	2	3	1	-5	1
13 64 79	10	2	5	1	-10	2
13 69 79	5	1	2	1	-9	2
14 65 72	6	2	3	2	-9	2
15 67 70	11	4	5	1	-4	1
16 62 77	9	2	5	1	-12	3
16 64 76	4	1	2	1	-8	2
16 69 79	5	1	2	1	-11	3

φ_1 Φ φ_2	H	K	L	U	V	W
10 60 70						
17 62 72	9	3	5	2	-11	3
17 66 75	11	3	5	1	-7	2
18 60 76	12	3	7	1	-11	3
18 63 70	11	4	6	2	-10	3
18 67 72	9	3	4	2	-10	3
10 60 80						
11 61 85	11	1	6	0	-6	1
11 64 81	6	1	3	1	-12	2
11 64 85	12	1	6	0	-6	1
11 68 81	12	2	5	1	-11	2
12 66 85	11	1	5	0	-5	1
12 67 85	12	1	5	0	-5	1
13 61 84	9	1	5	0	-5	1
13 64 84	10	1	5	0	-5	1
15 66 84	9	1	4	0	-4	1
15 68 84	10	1	4	0	-4	1
16 61 82	7	1	4	0	-4	1
16 64 83	8	1	4	0	-4	1
18 60 81	12	2	7	0	-7	2
10 60 90						
11 63 90	2	0	1	-1	-11	2
11 63 90	2	0	1	-1	-12	2
13 63 90	2	0	1	-1	-10	2
14 63 90	2	0	1	-1	-9	2
16 63 90	2	0	1	-1	-8	2
18 63 90	2	0	1	-1	-7	2
10 70 0						
10 72 6	1	9	3	6	-1	1
10 76 7	1	8	2	6	-1	1
12 70 7	1	8	3	5	-1	1
12 74 8	1	7	2	5	-1	1
13 76 9	2	12	3	9	-2	2
14 72 9	1	6	2	4	-1	1
14 78 6	1	9	2	12	-2	3
15 72 0	0	3	1	12	-1	3
15 72 5	1	12	4	12	-2	3
16 72 0	0	3	1	11	-1	3
16 73 6	1	10	3	11	-2	3
17 70 5	1	11	4	10	-2	3
17 76 7	1	8	2	10	-2	3
18 72 0	0	3	1	10	-1	3
18 76 9	2	12	3	12	-3	4

φ_1 Φ φ_2	H	K	L	U	V	W
10 70 0						
19 70 7	1	6	3	12	-3	4
19 72 6	1	9	3	9	-2	3
19 76 0	0	4	1	12	-1	4
19 76 5	1	12	3	6	-1	2
10 70 10						
10 71 15	3	11	4	6	-2	1
10 74 11	2	10	3	12	-3	2
11 76 14	1	4	1	10	-3	2
12 75 15	3	11	3	9	-3	2
13 70 14	2	8	3	9	-3	2
14 72 18	1	3	1	12	-5	3
14 74 17	3	10	3	8	-3	2
14 75 10	2	11	3	4	-1	1
14 79 11	1	5	1	4	-1	1
15 72 18	1	3	1	7	-3	2
15 75 16	2	7	2	11	-4	3
16 70 10	2	11	4	7	-2	2
16 74 11	2	10	3	7	-2	2
16 78 13	2	9	2	7	-2	2
16 78 18	3	9	2	10	-4	3
18 72 13	2	9	3	3	-1	1
18 72 18	1	3	1	9	-4	3
18 76 14	1	4	1	3	-1	1
18 77 18	4	12	3	12	-5	4
19 72 18	1	3	1	11	-5	4
10 70 20						
10 74 22	4	10	3	11	-5	2
10 77 29	6	11	3	5	-3	1
10 79 29	5	9	2	5	-3	1
11 77 27	2	4	1	9	-5	2
12 71 20	4	11	4	9	-4	2
12 79 22	2	5	1	9	-4	2
13 70 22	2	5	2	4	-2	1
13 73 27	3	6	2	12	-7	3
13 75 23	3	7	2	4	-2	1
13 79 24	4	9	2	4	-2	1
14 77 27	2	4	1	7	-4	2
14 79 22	2	5	1	11	-5	3
16 72 29	6	11	4	3	-2	1
16 77 27	2	4	1	12	-7	4
17 73 24	4	9	3	9	-5	3
17 74 22	4	10	3	6	-3	2
17 77 23	5	12	3	6	-3	2
17 79 29	5	9	2	11	-7	4

φ_1	Φ	φ_2	H	K	L	U	V	W
10	70	20						
18	71	21	3	8	3	6	-3	2
18	71	27	4	8	3	11	-7	4
18	73	27	3	6	2	8	-5	3
19	70	22	2	5	2	11	-6	4
19	77	27	2	4	1	5	-3	2
10	70	30						
10	72	32	5	8	3	10	-7	2
10	73	39	4	5	2	9	-8	2
10	77	36	5	7	2	9	-7	2
11	70	30	4	7	3	9	-6	2
11	75	37	9	12	4	12	-10	3
11	76	31	6	10	3	9	-6	2
11	77	39	8	10	3	8	-7	2
12	73	39	4	5	2	11	-10	3
12	74	30	7	12	4	12	-8	3
12	74	34	2	3	1	4	-3	1
12	79	37	3	4	1	11	-9	3
13	70	36	8	11	5	7	-6	2
13	72	32	5	8	3	11	-8	3
13	73	37	6	8	3	7	-6	2
13	77	32	7	11	3	7	-5	2
13	78	39	9	11	3	10	-9	3
14	70	34	6	9	4	10	-8	3
14	71	31	3	5	2	7	-5	2
14	75	38	7	9	3	9	-8	3
14	77	36	5	7	2	10	-8	3
14	78	32	5	8	2	10	-7	3
15	70	30	4	7	3	10	-7	3
15	72	35	7	10	4	6	-5	2
15	72	37	9	12	5	9	-8	3
15	74	34	2	3	1	9	-7	3
15	77	39	8	10	3	11	-10	4
15	78	34	8	12	3	12	-9	4
15	79	37	3	4	1	6	-5	2
16	72	32	5	8	3	12	-9	4
16	75	37	9	12	4	8	-7	3
16	76	30	4	7	2	3	-2	1
16	77	36	5	7	2	11	-9	4
16	78	30	7	12	3	3	-2	1
17	71	34	8	12	5	11	-9	4
17	71	36	5	7	3	8	-7	3
17	71	39	9	11	5	5	-5	2
17	73	37	6	8	3	10	-9	4
18	70	30	4	7	3	11	-8	4
18	70	36	8	11	5	12	-11	5
18	71	31	3	5	2	8	-6	3

φ_1	Φ	φ_2	H	K	L	U	V	W
10	70	30						
18	75	38	7	9	3	12	-11	5
18	77	36	5	7	2	12	-10	5
18	79	37	3	4	1	7	-6	3
19	70	34	6	9	4	7	-6	3
19	74	36	8	11	4	9	-8	4
19	77	32	7	11	3	12	-9	5
10	70	40						
10	71	45	2	2	1	8	-9	2
10	73	42	9	10	4	4	-4	1
10	75	42	10	11	4	4	-4	1
10	76	43	11	12	4	4	-4	1
10	79	40	10	12	3	9	-8	2
11	71	45	2	2	1	7	-8	2
11	72	42	8	9	4	4	-4	1
11	76	40	5	6	2	12	-11	3
11	76	47	12	11	4	10	-12	3
11	77	45	3	3	1	11	-12	3
12	72	49	7	6	3	3	-4	1
12	77	45	3	3	1	10	-11	3
12	77	48	10	9	3	9	-11	3
13	71	45	2	2	1	6	-7	2
13	77	45	3	3	1	9	-10	3
13	79	49	8	7	2	8	-10	3
14	72	41	6	7	3	3	-3	1
14	74	41	7	8	3	3	-3	1
14	75	48	11	10	4	8	-10	3
14	76	42	8	9	3	3	-3	1
14	77	42	9	10	3	3	-3	1
14	77	45	3	3	1	8	-9	3
14	79	40	10	12	3	12	-11	4
14	79	42	10	11	3	3	-3	1
15	71	45	2	2	1	5	-6	2
15	73	47	12	11	5	7	-9	3
15	76	48	9	8	3	7	-9	3
16	70	42	9	10	5	10	-11	4
16	70	47	12	11	6	9	-12	4
16	74	49	8	7	3	9	-12	4
16	77	45	3	3	1	7	-8	3
17	71	42	10	11	5	9	-10	4
17	71	45	2	2	1	9	-11	4
17	72	40	10	12	5	5	-5	2
17	72	48	9	8	4	8	-11	4
17	73	48	10	9	4	6	-8	3
17	78	49	7	6	2	6	-8	3
18	71	45	2	2	1	4	-5	2
18	72	42	8	9	4	11	-12	5

φ_1	Φ	φ_2	H	K	L	U	V	W
10	70	40						
18	74	45	5	5	2	10	-12	5
18	79	48	11	10	3	4	-5	2
19	70	43	11	12	6	6	-7	3
19	70	48	10	9	5	7	-10	4
19	73	42	9	10	4	10	-11	5
19	73	43	11	12	5	8	-9	4
19	76	48	9	8	3	9	-12	5
19	77	45	3	3	1	6	-7	3
10	70	50						
10	74	56	3	2	1	3	-5	1
10	79	50	12	10	3	7	-9	2
11	73	53	8	6	3	6	-9	2
11	78	54	11	8	3	6	-9	2
12	71	52	9	7	4	8	-12	3
12	73	58	11	7	4	5	-9	2
12	75	53	12	9	4	8	-12	3
12	76	50	6	5	2	3	-4	1
12	76	55	10	7	3	5	-8	2
12	78	51	11	9	3	3	-4	1
13	71	59	5	3	2	2	-4	1
13	74	56	3	2	1	7	-12	3
13	77	51	10	8	3	5	-7	2
13	77	54	7	5	2	7	-11	3
14	70	56	9	6	4	6	-11	3
14	72	50	12	10	5	5	-7	2
14	72	53	12	9	5	7	-11	3
14	74	51	11	9	4	5	-7	2
14	74	56	3	2	1	4	-7	2
15	79	50	12	10	3	9	-12	4
16	71	51	11	9	5	2	-3	1
16	75	52	9	7	3	2	-3	1
16	78	56	12	8	3	7	-12	4
16	79	53	4	3	1	2	-3	1
17	72	58	8	5	3	3	-6	2
18	74	56	3	2	1	6	-11	4
18	75	53	12	9	4	5	-8	3
18	78	54	11	8	3	5	-8	3
19	71	59	5	3	2	5	-11	4
19	73	58	11	7	4	4	-8	3
19	76	50	6	5	2	7	-10	4
19	77	54	7	5	2	3	-5	2
10	70	60						
10	70	68	5	2	2	2	-6	1
10	77	63	4	2	1	5	-11	2

φ_1	Φ	φ_2	H	K	L	U	V	W
10	70	60						
10	79	68	5	2	1	4	-11	2
11	70	60	12	7	5	5	-10	2
11	76	66	11	5	3	2	-5	1
11	79	66	9	4	2	2	-5	1
12	71	69	8	3	3	3	-10	2
12	74	68	10	4	3	3	-9	2
12	77	63	4	2	1	4	-9	2
13	74	60	12	7	4	2	-4	1
13	76	60	7	4	2	2	-4	1
13	79	61	9	5	2	2	-4	1
14	73	63	6	3	2	5	-12	3
14	73	67	12	5	4	4	-12	3
15	70	60	7	4	3	5	-11	3
15	77	63	4	2	1	3	-7	2
16	70	68	5	2	2	2	-7	2
16	71	69	8	3	3	3	-11	3
16	73	63	6	3	2	4	-10	3
16	77	69	8	3	2	3	-10	3
17	76	60	12	7	3	3	-6	2
18	70	60	12	7	5	4	-9	3
18	73	66	9	4	3	1	-3	1
18	74	61	9	5	3	4	-9	3
18	75	67	7	3	2	1	-3	1
18	77	67	12	5	3	1	-3	1
18	79	68	5	2	1	1	-3	1
19	71	69	8	3	3	3	-12	4
19	72	66	11	5	4	1	-3	1
19	79	66	9	4	2	4	-11	4
10	70	70						
10	72	72	3	1	1	3	-11	2
10	72	77	9	2	3	1	-6	1
10	74	73	10	3	3	3	-12	2
10	78	77	9	2	2	2	-11	2
10	79	79	5	1	1	1	-6	1
11	70	76	8	2	3	2	-11	2
11	76	76	4	1	1	1	-5	1
12	71	75	11	3	4	1	-5	1
13	75	74	7	2	2	2	-9	2
14	72	72	3	1	1	1	-4	1
14	78	77	9	2	2	2	-12	3
14	79	73	10	3	2	1	-4	1
15	72	76	12	3	4	2	-12	3
15	74	79	10	2	3	1	-8	2
15	76	76	4	1	1	2	-11	3
15	78	72	9	3	2	3	-11	3
16	76	70	11	4	3	2	-7	2

φ_1 Φ φ_2	H	K	L	U	V	W
10 70 70						
16 79 79	5	1	1	1	-7	2
17 70 76	8	2	3	1	-7	2
17 75 74	7	2	2	2	-10	3
18 77 72	12	4	3	3	-12	4
19 72 72	3	1	1	2	-9	3
19 72 77	9	2	3	1	-9	3
19 76 76	4	1	1	1	-6	2
10 70 80						
10 76 83	8	1	2	1	-12	2
11 74 82	7	1	2	1	-11	2
13 76 81	12	2	3	1	-9	2
15 70 85	11	1	4	0	-4	1
15 72 81	6	1	2	1	-12	3
15 72 85	12	1	4	0	-4	1
17 75 80	11	2	3	1	-10	3
19 73 84	10	1	3	0	-3	1
19 75 85	11	1	3	0	-3	1
19 76 81	12	2	3	1	-12	4
19 76 85	12	1	3	0	-3	1
10 70 90						
15 72 90	3	0	1	-1	-12	3
16 72 90	3	0	1	-1	-11	3
18 72 90	3	0	1	-1	-10	3
19 76 90	4	0	1	-1	-12	4
10 80 0						
10 85 9	2	12	1	11	-2	2
11 81 9	1	6	1	5	-1	1
11 85 5	1	12	1	10	-1	2
13 85 5	1	11	1	9	-1	2
14 81 5	1	12	2	8	-1	2
14 84 6	1	10	1	8	-1	2
15 82 8	1	7	1	11	-2	3
16 80 5	1	11	2	7	-1	2
16 84 6	1	9	1	7	-1	2
18 81 9	1	6	1	9	-2	3
18 83 7	1	8	1	6	-1	2
18 85 5	1	12	1	9	-1	3
10 80 10						
1C 81 18	2	6	1	11	-4	2
10 83 14	2	8	1	11	-3	2

φ_1 Φ φ_2	H	K	L	U	V	W
10 80 10						
11 85 10	2	11	1	5	-1	1
12 80 10	2	11	2	9	-2	2
12 84 11	2	10	1	9	-2	2
13 81 18	2	6	1	8	-3	2
14 84 13	2	9	1	4	-1	1
14 84 18	3	9	1	11	-4	3
15 83 14	2	8	1	7	-2	2
15 85 14	3	12	1	11	-3	3
16 81 14	3	12	2	10	-3	3
16 85 15	3	11	1	10	-3	3
18 80 15	3	11	2	3	-1	1
18 82 16	2	7	1	3	-1	1
18 85 17	3	10	1	3	-1	1
19 84 18	3	9	1	8	-3	3
19 85 18	4	12	1	11	-4	4
10 80 20						
11 80 20	4	11	2	5	-2	1
11 83 21	3	8	1	5	-2	1
12 84 29	5	9	1	12	-7	3
13 81 24	5	11	2	4	-2	1
14 82 27	3	6	1	11	-6	3
14 85 29	6	11	1	7	-4	2
15 81 29	6	11	2	10	-6	3
15 85 22	4	10	1	7	-3	2
16 82 27	3	6	1	9	-5	3
17 83 21	3	8	1	12	-5	4
17 86 23	5	12	1	9	-4	3
18 84 27	4	8	1	11	-6	4
19 84 29	5	9	1	5	-3	2
10 80 30						
10 83 30	4	7	1	5	-3	1
10 84 37	6	8	1	9	-7	2
10 86 30	7	12	1	5	-3	1
11 80 38	7	9	2	12	-10	3
11 81 35	7	10	2	4	-3	1
11 83 36	5	7	1	4	-3	1
11 86 36	8	11	1	4	-3	1
12 81 39	4	5	1	7	-6	2
12 82 30	7	12	2	8	-5	2
12 85 31	6	10	1	8	-5	2
12 85 35	7	10	1	11	-8	3
13 80 34	6	9	2	11	-8	3
13 82 34	4	6	1	7	-5	2
13 82 37	9	12	2	10	-8	3

φ_1 Φ φ_2	H	K	L	U	V	W
10 80 30						
14 85 34	6	9	1	10	-7	3
14 86 39	8	10	1	6	-5	2
15 86 37	9	12	1	9	-7	3
16 80 31	3	5	1	3	-2	1
16 81 39	4	5	1	8	-7	3
16 84 32	5	8	1	3	-2	1
16 85 38	7	9	1	11	-9	4
16 86 32	7	11	1	3	-2	1
17 80 34	6	9	2	8	-6	3
17 82 34	4	6	1	11	-8	4
17 84 37	6	6	1	5	-4	2
17 85 31	6	10	1	11	-7	4
18 82 36	8	11	2	5	-4	2
18 63 30	4	7	1	6	-5	3
18 86 34	8	12	1	10	-7	4
18 86 39	9	11	1	7	-6	3
19 81 32	7	11	2	7	-5	3
19 81 39	4	5	1	9	-8	4
19 83 36	5	7	1	9	-7	4
19 85 34	6	9	1	7	-5	3
10 80 40						
10 85 42	8	9	1	12	-11	3
11 81 48	9	8	2	10	-12	3
11 83 40	5	6	1	8	-7	2
11 83 47	12	11	2	7	-8	2
11 85 41	7	8	1	11	-10	3
11 86 48	10	9	1	7	-8	2
12 82 48	11	10	2	6	-7	2
12 84 49	7	6	1	9	-11	3
12 85 48	9	8	1	6	-7	2
12 86 40	10	12	1	7	-6	2
13 80 43	11	12	3	3	-3	1
13 84 41	6	7	1	10	-9	3
14 80 45	4	4	1	11	-12	4
14 83 40	5	6	1	9	-8	3
14 85 41	7	8	1	12	-11	4
14 85 49	8	7	1	5	-6	2
14 86 47	12	11	1	8	-9	3
15 80 45	4	4	1	10	-11	4
15 80 47	12	11	3	5	-6	2
15 82 48	10	9	2	5	-6	2
15 84 41	6	7	1	11	-10	4
16 86 48	11	10	1	7	-8	3
17 80 45	4	4	1	9	-10	4
17 82 45	5	5	1	11	-12	5
17 83 40	5	6	1	10	-9	4

φ_1 Φ φ_2	H	K	L	U	V	W
10 80 40						
17 84 41	6	7	1	12	-11	5
17 84 49	7	6	1	4	-5	2
18 81 48	9	8	2	4	-5	2
18 82 48	11	10	2	10	-12	5
18 86 46	10	9	1	6	-7	3
19 80 45	4	4	1	8	-9	4
19 82 45	5	5	1	10	-11	5
19 83 40	5	6	1	11	-10	5
19 85 49	8	7	1	9	-11	5
10 80 50						
10 81 58	11	7	2	3	-5	1
10 82 51	11	9	2	7	-9	2
1C 84 58	8	5	1	3	-5	1
10 86 51	10	8	1	7	-9	2
11 80 59	5	3	1	5	-9	2
11 81 51	5	4	1	3	-4	1
11 85 52	9	7	1	3	-4	1
12 82 56	6	4	1	5	-8	2
12 86 53	12	9	1	8	-11	3
13 60 52	9	7	2	5	-7	2
13 84 53	8	6	1	5	-7	2
13 85 56	9	6	1	7	-11	3
14 80 59	5	3	1	6	-11	3
14 82 53	12	9	2	7	-10	3
14 85 59	10	6	1	4	-7	2
14 66 54	11	8	1	7	-10	3
15 80 56	9	6	2	6	-10	3
15 83 50	6	5	1	7	-9	3
16 82 54	11	8	2	2	-3	1
16 83 54	7	5	1	2	-3	1
16 84 58	8	5	1	7	-12	4
16 85 55	10	7	1	2	-3	1
17 81 51	5	4	1	8	-11	4
17 82 51	11	9	2	6	-8	3
17 86 50	12	10	1	4	-5	2
17 86 56	12	8	1	7	-11	4
18 81 55	10	7	2	5	-8	3
18 84 53	8	6	1	7	-10	4
18 85 56	9	6	1	5	-8	3
19 81 51	5	4	1	5	-7	3
19 82 56	6	4	1	3	-5	2
19 66 51	11	9	1	7	-9	4
19 86 53	12	9	1	5	-7	3
19 86 58	11	7	1	3	-5	2

φ_1	Φ	φ_2	H	K	L	U	V	W
10	80	60						
11	83	67	7	3	1	2	-5	1
11	86	67	12	5	1	2	-5	1
13	81	61	11	6	2	2	-4	1
13	81	67	12	5	2	3	-8	2
13	84	66	9	4	1	5	-12	3
13	85	68	10	4	1	3	-8	2
14	82	63	6	3	1	5	-11	3
15	85	66	11	5	1	3	-7	2
16	81	66	11	5	2	4	-10	3
16	86	60	12	7	1	5	-9	3
17	82	63	6	3	1	4	-9	3
18	83	69	8	3	1	1	-3	1
18	84	63	8	4	1	5	-11	4
19	85	68	10	4	1	4	-11	4
10	80	70						
10	84	79	10	2	1	2	-11	2
11	81	72	6	2	1	3	-10	2
11	84	77	9	2	1	1	-5	1
12	83	76	8	2	1	2	-9	2
14	82	74	7	2	1	1	-4	1
14	85	75	11	3	1	1	-4	1
15	85	73	10	3	1	3	-11	3
16	81	72	6	2	1	2	-7	2
16	84	72	9	3	1	3	-10	3
17	81	76	12	3	2	2	-10	3
18	84	79	10	2	1	1	-6	2
18	85	70	11	4	1	1	-3	1
18	85	76	12	3	1	2	-9	3
10	80	80						
10	84	84	9	1	1	1	-11	2
11	83	83	8	1	1	1	-10	2
13	82	82	7	1	1	1	-9	2
14	81	81	6	1	1	1	-8	2
14	84	84	9	1	1	1	-12	3
15	83	83	8	1	1	1	-11	3
16	85	81	12	2	1	1	-7	2
17	82	82	7	1	1	1	-10	3
19	81	81	6	1	1	1	-9	3
19	83	83	8	1	1	1	-12	4
10	90	0						
10	90	0	0	1	0	11	0	2

φ_1	Φ	φ_2	H	K	L	U	V	W
10	90	0						
10	90	5	1	11	0	11	-1	2
11	90	0	0	1	0	5	0	1
11	90	6	1	10	0	10	-1	2
12	90	6	1	9	0	9	-1	2
13	90	0	0	1	0	9	0	2
14	90	0	0	1	0	4	0	1
14	90	5	1	12	0	12	-1	3
14	90	7	1	8	0	8	-1	2
14	90	9	1	6	0	12	-2	3
15	90	0	0	1	0	11	0	3
15	90	5	1	11	0	11	-1	3
16	90	0	0	1	0	7	0	2
16	90	8	1	7	0	7	-1	2
17	90	0	0	1	0	10	0	3
17	90	6	1	10	0	10	-1	3
18	90	0	0	1	0	3	0	1
18	90	5	1	12	0	12	-1	4
18	90	6	1	9	0	9	-1	3
18	90	9	1	6	0	6	-1	2
10	90	10						
10	90	10	2	11	0	11	-2	2
10	90	15	3	11	0	11	-3	2
11	90	11	1	5	0	5	-1	1
11	90	17	3	10	0	10	-3	2
12	90	13	2	9	0	9	-2	2
12	90	18	1	3	0	9	-3	2
13	90	18	1	3	0	12	-4	3
14	90	14	1	4	0	4	-1	1
15	90	10	2	11	0	11	-2	3
15	90	15	3	11	0	11	-3	3
15	90	16	2	7	0	7	-2	2
16	90	11	1	5	0	10	-2	3
16	90	17	3	10	0	10	-3	3
18	90	13	2	9	0	9	-2	3
18	90	14	1	4	0	12	-3	4
18	90	18	1	3	0	3	-1	1
19	90	15	3	11	0	11	-3	4
10	90	20						
10	90	20	4	11	0	11	-4	2
10	90	27	1	2	0	10	-5	2
11	90	22	2	5	0	5	-2	1
11	90	24	4	9	0	9	-4	2
11	90	29	5	9	0	9	-5	2
13	90	21	3	8	0	8	-3	2

φ_1	Φ	φ_2	H	K	L	U	V	W
10	90	20						
13	90	23	5	12	0	12	-5	3
13	90	27	1	2	0	4	-2	1
13	90	29	6	11	0	11	-6	3
14	90	20	4	11	0	11	-4	3
14	90	24	5	11	0	11	-5	3
15	90	23	3	7	0	7	-3	2
15	90	27	1	2	0	10	-5	3
16	90	22	2	5	0	10	-4	3
16	90	29	5	9	0	9	-5	3
17	90	23	5	12	0	12	-5	4
17	90	24	4	9	0	9	-4	3
17	90	27	1	2	0	6	-3	2
18	90	24	5	11	0	11	-5	4
18	90	29	6	11	0	11	-6	4
19	90	20	4	11	0	11	-4	4
19	90	21	3	8	0	8	-3	3
19	90	27	1	2	0	8	-4	3
10	90	30						
10	90	31	3	5	0	5	-3	1
10	90	34	2	3	0	9	-6	2
10	90	38	7	9	0	9	-7	2
11	90	37	3	4	0	4	-3	1
12	90	30	7	12	0	12	-7	3
12	90	32	5	8	0	8	-5	2
12	90	34	2	3	0	12	-8	3
12	90	36	8	11	0	11	-8	3
12	90	39	9	11	0	11	-9	3
13	90	32	7	11	0	11	-7	3
13	90	36	5	7	0	7	-5	2
13	90	39	4	5	0	10	-8	3
14	90	30	4	7	0	7	-4	2
14	90	31	3	5	0	10	-6	3
14	90	35	7	10	0	10	-7	3
15	90	37	3	4	0	12	-9	4
15	90	38	7	9	0	9	-7	3
16	90	30	7	12	0	12	-7	4
16	90	34	2	3	0	3	-2	1
16	90	36	8	11	0	11	-8	4
16	90	39	9	11	0	11	-9	4
17	90	32	7	11	0	11	-7	4
17	90	37	3	4	0	8	-6	3
17	90	39	4	5	0	5	-4	2
18	90	32	5	8	0	8	-5	3
18	90	35	7	10	0	10	-7	4
18	90	37	3	4	0	12	-9	5
19	90	31	3	5	0	5	-3	2

φ_1	Φ	φ_2	H	K	L	U	V	W
10	90	30						
19	90	34	2	3	0	12	-8	5
19	90	36	5	7	0	7	-5	3
19	90	38	7	9	0	9	-7	4
19	90	39	9	11	0	11	-9	5
10	90	40						
10	90	43	11	12	0	12	-11	3
10	90	45	1	1	0	4	-4	1
10	90	47	12	11	0	11	-12	3
11	90	40	5	6	0	12	-10	3
11	90	41	7	8	0	8	-7	2
11	90	42	10	11	0	11	-10	3
11	90	45	1	1	0	11	-11	3
11	90	45	1	1	0	7	-7	2
11	90	48	11	10	0	10	-11	3
11	90	49	8	7	0	7	-8	2
12	90	41	6	7	0	7	-6	2
12	90	45	1	1	0	10	-10	3
12	90	49	7	6	0	6	-7	2
13	90	42	9	10	0	10	-9	3
13	90	45	1	1	0	3	-3	1
13	90	48	10	9	0	9	-10	3
14	90	40	5	6	0	6	-5	2
14	90	42	8	9	0	9	-8	3
14	90	43	11	12	0	12	-11	4
14	90	45	1	1	0	11	-11	4
14	90	47	12	11	0	11	-12	4
14	90	48	9	8	0	8	-9	3
15	90	42	10	11	0	11	-10	4
15	90	45	1	1	0	8	-8	3
15	90	48	11	10	0	10	-11	4
16	90	41	7	8	0	8	-7	3
16	90	45	1	1	0	12	-12	5
16	90	45	1	1	0	5	-5	2
16	90	49	8	7	0	7	-8	3
17	90	42	9	10	0	10	-9	4
17	90	43	11	12	0	12	-11	5
17	90	45	1	1	0	9	-9	4
17	90	45	1	1	0	7	-7	3
17	90	47	12	11	0	11	-12	5
17	90	48	10	9	0	9	-10	4
18	90	40	5	6	0	12	-10	5
18	90	41	6	7	0	7	-6	3
18	90	42	8	9	0	9	-8	4
18	90	45	1	1	0	11	-11	5
18	90	48	9	8	0	8	-9	4
18	90	49	7	6	0	6	-7	3

φ_1	Φ	φ_2	H	K	L	U	V	W
10	90	40						
19	90	42	10	11	0	11	-10	5
19	90	45	1	1	0	2	-2	1
19	90	48	11	10	0	10	-11	5
10	90	50						
10	90	52	9	7	0	7	-9	2
10	90	56	3	2	0	6	-9	2
10	90	59	5	3	0	3	-5	1
11	90	50	6	5	0	10	-12	3
11	90	53	4	3	0	3	-4	1
12	90	51	11	9	0	9	-11	3
12	90	54	11	8	0	8	-11	3
12	90	56	3	2	0	8	-12	3
12	90	58	8	5	0	5	-8	2
13	90	51	5	4	0	8	-10	3
13	90	54	7	5	0	5	-7	2
13	90	58	11	7	0	7	-11	3
14	90	50	6	5	0	5	-6	2
14	90	55	10	7	0	7	-10	3
14	90	59	5	3	0	6	-10	3
15	90	52	9	7	0	7	-9	3
15	90	53	4	3	0	9	-12	4
16	90	51	11	9	0	9	-11	4
16	90	54	11	8	0	8	-11	4
16	90	56	3	2	0	2	-3	1
17	90	51	5	4	0	4	-5	2
17	90	53	4	3	0	6	-8	3
17	90	58	11	7	0	7	-11	4
18	90	50	6	5	0	10	-12	5
18	90	53	4	3	0	9	-12	5
18	90	55	10	7	0	7	-10	4
18	90	58	8	5	0	5	-8	3
19	90	51	11	9	0	9	-11	5
19	90	52	9	7	0	7	-9	4
19	90	54	7	5	0	5	-7	3
19	90	56	3	2	0	8	-12	5
19	90	59	5	3	0	3	-5	2
10	90	60						
10	90	63	2	1	0	5	-10	2
11	90	61	9	5	0	5	-9	2
11	90	66	9	4	0	4	-9	2
11	90	68	5	2	0	2	-5	1
12	90	60	12	7	0	7	-12	3
13	90	61	11	6	0	6	-11	3
13	90	63	2	1	0	2	-4	1

φ_1	Φ	φ_2	H	K	L	U	V	W
10	90	60						
13	90	67	12	5	0	5	-12	3
13	90	69	8	3	0	3	-8	2
14	90	60	7	4	0	4	-7	2
14	90	66	11	5	0	5	-11	3
15	90	63	2	1	0	5	-10	3
15	90	67	7	3	0	3	-7	2
16	90	60	12	7	0	7	-12	4
16	90	61	9	5	0	5	-9	3
16	90	68	5	2	0	4	-10	3
17	90	63	2	1	0	3	-6	2
17	90	66	9	4	0	4	-9	3
17	90	67	12	5	0	5	-12	4
18	90	61	11	6	0	6	-11	4
18	90	66	11	5	0	5	-11	4
19	90	63	2	1	0	4	-8	3
19	90	69	8	3	0	3	-8	3
10	90	70						
10	90	70	11	4	0	4	-11	2
10	90	75	11	3	0	3	-11	2
11	90	73	10	3	0	3	-10	2
11	90	79	5	1	0	1	-5	1
12	90	72	3	1	0	3	-9	2
12	90	77	9	2	0	2	-9	2
13	90	72	3	1	0	4	-12	3
14	90	70	11	4	0	4	-11	3
14	90	76	4	1	0	1	-4	1
15	90	74	7	2	0	2	-7	2
15	90	75	11	3	0	3	-11	3
16	90	73	10	3	0	3	-10	3
16	90	79	5	1	0	2	-10	3
18	90	72	3	1	0	1	-3	1
18	90	76	4	1	0	3	-12	4
18	90	77	9	2	0	2	-9	3
19	90	70	11	4	0	4	-11	4
19	90	75	11	3	0	3	-11	4
10	90	80						
10	90	80	11	2	0	2	-11	2
10	90	85	11	1	0	1	-11	2
11	90	84	10	1	0	1	-10	2
12	90	84	9	1	0	1	-9	2
14	90	81	6	1	0	2	-12	3
14	90	83	8	1	0	1	-8	2
14	90	85	12	1	0	1	-12	3
15	90	80	11	2	0	2	-11	3

φ_1	Φ	φ_2	H	K	L	U	V	W
10	90	80						
15	90	85	11	1	0	1	-11	3
16	90	82	7	1	0	1	-7	2
17	90	84	10	1	0	1	-10	3
18	90	81	6	1	0	1	-6	2
18	90	84	9	1	0	1	-9	3
18	90	85	12	1	0	1	-12	4
10	90	90						
10	90	90	1	0	0	0	-11	2
11	90	90	1	0	0	0	-5	1
13	90	90	1	0	0	0	-9	2
14	90	90	1	0	0	0	-4	1
15	90	90	1	0	0	0	-11	3
16	90	90	1	0	0	0	-7	2
17	90	90	1	0	0	0	-10	3
18	90	90	1	0	0	0	-3	1
20	0	0						
27	9	0	0	1	6	12	-6	1
29	9	0	0	1	6	11	-6	1
20	0	20						
20	0	20	0	0	1	11	-9	0
20	0	20	0	0	1	6	-5	0
20	0	20	0	0	1	7	-6	0
21	0	21	0	0	1	9	-8	0
21	0	21	0	0	1	8	-7	0
21	0	21	0	0	1	12	-11	0
21	0	21	0	0	1	10	-9	0
21	0	21	0	0	1	11	-10	0
23	0	23	0	0	1	1	-1	0
24	0	24	0	0	1	7	-8	0
24	0	24	0	0	1	8	-9	0
24	0	24	0	0	1	10	-11	0
24	0	24	0	0	1	9	-10	0
24	0	24	0	0	1	11	-12	0
25	0	25	0	0	1	5	-6	0
25	0	25	0	0	1	6	-7	0
25	0	25	0	0	1	9	-11	0
26	0	26	0	0	1	4	-5	0
26	0	26	0	0	1	7	-9	0
27	0	27	0	0	1	5	-7	0
27	0	27	0	0	1	8	-11	0
27	0	27	0	0	1	3	-4	0
28	0	28	0	0	1	7	-10	0

φ_1	Φ	φ_2	H	K	L	U	V	W
20	0	20						
28	0	28	0	0	1	2	-3	0
29	0	29	0	0	1	7	-11	0
29	0	29	0	0	1	5	-6	0
20	0	90						
27	9	90	1	0	6	-6	-12	1
29	9	90	1	0	6	-6	-11	1
20	10	0						
21	14	0	0	1	4	11	-4	1
22	14	0	0	1	4	10	-4	1
22	18	0	0	1	3	8	-3	1
23	11	0	0	1	5	12	-5	1
24	18	0	0	1	3	7	-3	1
25	11	0	0	1	5	11	-5	1
25	14	0	0	1	4	9	-4	1
27	11	0	0	1	5	10	-5	1
27	14	0	0	1	4	8	-4	1
28	18	0	0	1	3	6	-3	1
20	10	10						
20	15	18	1	3	12	9	-7	1
23	16	18	1	3	11	7	-6	1
23	19	18	1	3	9	6	-5	1
24	19	14	1	4	12	12	-9	2
27	15	18	1	3	12	6	-6	1
27	16	18	1	3	11	11	-11	2
28	18	18	1	3	10	5	-5	1
26	19	18	1	3	9	9	-9	2
20	10	20						
21	11	27	1	2	12	10	-11	1
22	11	27	1	2	11	9	-10	1
22	13	27	1	2	10	8	-9	1
23	14	27	1	2	9	7	-8	1
24	16	27	1	2	8	6	-7	1
25	11	27	1	2	12	8	-10	1
25	18	27	1	2	7	5	-6	1
26	11	27	1	2	11	7	-9	1
27	13	27	1	2	10	6	-8	1
29	14	27	1	2	9	5	-7	1
29	18	27	1	2	7	8	-11	2

φ_1	Φ	φ_2	H	K	L	U	V	W
20	**10**	**30**						
20	17	34	2	3	12	6	-8	1
22	18	34	2	3	11	5	-7	1
27	18	34	2	3	11	7	-12	2
20	**10**	**40**						
21	13	45	1	1	6	5	-11	1
22	16	45	1	1	5	4	-9	1
23	11	45	1	1	7	5	-12	1
23	19	45	1	1	4	3	-7	1
24	13	45	1	1	6	4	-10	1
25	11	45	1	1	7	4	-11	1
25	16	45	1	1	5	3	-8	1
27	10	45	1	1	8	4	-12	1
27	13	45	1	1	6	3	-9	1
28	19	45	1	1	4	2	-6	1
29	11	45	1	1	7	3	-10	1
20	**10**	**50**						
22	17	56	3	2	12	2	-9	1
27	18	56	3	2	11	1	-7	1
20	**10**	**60**						
21	18	63	2	1	7	1	-9	1
22	13	63	2	1	10	1	-12	1
22	14	63	2	1	9	1	-11	1
22	16	63	2	1	8	1	-10	1
27	11	63	2	1	11	0	-11	1
27	11	63	2	1	12	0	-12	1
27	13	63	2	1	10	0	-10	1
27	14	63	2	1	9	0	-9	1
27	16	63	2	1	8	0	-8	1
28	18	63	2	1	7	0	-7	1
20	**10**	**70**						
22	19	76	4	1	12	-1	-8	1
26	15	72	3	1	12	-1	-9	1
26	16	72	3	1	11	-1	-8	1
28	18	72	3	1	10	-1	-7	1
29	19	72	3	1	9	-1	-6	1
29	19	76	4	1	12	-3	-12	2

φ_1	Φ	φ_2	H	K	L	U	V	W
20	**10**	**90**						
21	14	90	1	0	4	-4	-11	1
22	14	90	1	0	4	-4	-10	1
22	18	90	1	0	3	-3	-6	1
23	11	90	1	0	5	-5	-12	1
24	18	90	1	0	3	-3	-7	1
25	11	90	1	0	5	-5	-11	1
25	14	90	1	0	4	-4	-9	1
27	11	90	1	0	5	-5	-10	1
27	14	90	1	0	4	-4	-8	1
28	18	90	1	0	3	-3	-6	1
20	**20**	**0**						
20	27	0	0	1	2	6	-2	1
22	27	0	0	1	2	11	-4	2
24	22	0	0	2	5	12	-5	2
24	27	0	0	1	2	5	-2	1
25	29	9	1	6	11	8	-5	2
26	22	0	0	2	5	11	-5	2
26	27	0	0	1	2	9	-4	2
27	27	9	1	6	12	12	-6	3
28	22	0	0	2	5	10	-5	2
20	**20**	**10**						
21	22	14	1	4	10	6	-4	1
22	22	18	1	3	8	11	-9	2
22	24	18	1	3	7	5	-4	1
22	27	11	1	5	10	5	-3	1
22	28	18	1	3	6	9	-7	2
23	23	11	1	5	12	11	-7	2
23	25	14	1	4	9	10	-7	2
24	28	18	1	3	6	12	-10	3
25	27	14	1	4	8	4	-3	1
26	21	14	1	4	11	5	-4	1
26	24	18	1	3	7	12	-11	3
28	22	18	1	3	8	4	-4	1
28	25	11	1	5	11	4	-3	1
29	22	14	1	4	10	8	-7	2
29	24	18	1	3	7	7	-7	2
29	28	18	1	3	6	3	-3	1
20	**20**	**20**						
20	20	27	1	2	6	6	-6	1
20	22	27	2	4	11	11	-11	2
20	24	27	1	2	5	5	-5	1

φ_1	Φ	φ_2	H	K	L	U	V	W
20	**20**	**20**						
20	26	27	2	4	9	9	-9	2
21	29	27	1	2	4	4	-4	1
22	20	27	1	2	6	10	-11	2
22	26	22	2	5	11	9	-8	2
23	22	27	2	4	11	9	-10	2
24	24	27	1	2	5	8	-9	2
24	29	27	1	2	4	10	-11	3
25	24	22	2	5	12	4	-4	1
25	26	27	2	4	9	7	-8	2
26	20	27	1	2	6	4	-5	1
26	26	22	2	5	11	11	-11	3
26	28	22	2	5	10	10	-10	3
26	29	27	1	2	4	6	-7	2
27	22	27	2	4	11	7	-9	2
28	29	27	1	2	4	8	-10	3
29	24	27	1	2	5	3	-4	1
20	**20**	**30**						
20	27	34	2	3	7	4	-5	1
21	22	34	2	3	9	9	-12	2
21	23	37	3	4	12	4	-6	1
22	29	37	3	4	9	6	-9	2
23	24	34	2	3	8	7	-10	2
23	26	31	3	5	12	7	-9	2
24	20	34	2	3	10	4	-6	1
24	24	37	3	4	11	3	-5	1
25	28	31	3	5	11	3	-4	1
26	29	37	3	4	9	7	-12	3
27	22	34	2	3	9	3	-5	1
27	27	34	2	3	7	5	-8	2
28	26	31	3	5	12	8	-12	3
28	26	39	4	5	12	2	-4	1
29	27	37	3	4	10	2	-4	1
20	**20**	**40**						
20	25	45	1	1	3	3	-6	1
21	29	45	2	2	5	5	-10	2
23	25	45	1	1	3	5	-11	2
24	22	45	2	2	7	5	-12	2
24	29	45	2	2	5	4	-9	2
25	25	45	1	1	3	2	-5	1
27	22	45	2	2	7	4	-11	2
28	29	45	2	2	5	3	-8	2
29	25	45	1	1	3	3	-9	2

φ_1	Φ	φ_2	H	K	L	U	V	W
20	**20**	**50**						
20	27	53	4	3	10	2	-6	1
21	20	56	3	2	10	2	-8	1
21	28	51	5	4	12	4	-11	2
23	29	53	4	3	9	3	-10	2
25	23	53	4	3	12	3	-12	2
25	24	56	3	2	8	2	-11	2
25	27	56	3	2	7	1	-5	1
26	22	56	3	2	9	1	-6	1
27	26	59	5	3	11	1	-9	2
28	24	53	4	3	11	1	-5	1
20	**20**	**60**						
20	24	63	2	1	5	1	-7	1
21	20	63	2	1	6	1	-8	1
24	24	63	2	1	5	1	-12	2
24	24	68	5	2	12	0	-6	1
24	26	63	4	2	9	1	-11	2
24	26	68	5	2	11	0	-11	2
24	28	68	5	2	10	0	-5	1
24	29	63	2	1	4	1	-10	2
28	20	63	2	1	6	0	-6	1
28	22	63	4	2	11	0	-11	2
29	24	63	2	1	5	0	-5	1
29	26	63	4	2	9	0	-9	2
20	**20**	**70**						
20	22	72	3	1	8	0	-8	1
20	24	72	3	1	7	0	-7	1
21	23	79	5	1	12	-1	-7	1
21	27	76	4	1	6	-1	-12	2
21	28	72	3	1	6	0	-6	1
23	25	79	5	1	11	-1	-6	1
24	21	76	4	1	11	-1	-7	1
25	22	76	4	1	10	-1	-6	1
25	27	79	5	1	10	-1	-5	1
26	24	72	3	1	7	-1	-11	2
28	25	76	4	1	9	-1	-5	1
28	28	72	3	1	6	-1	-9	2
20	**20**	**80**						
21	27	81	6	1	12	-1	-6	1
23	29	81	6	1	11	-1	-5	1

20 20 90

φ_1	Φ	φ_2	H	K	L	U	V	W
20	27	90	1	0	2	-2	-6	1
22	27	90	1	0	2	-4	-11	2
24	22	90	2	0	5	-5	-12	2
24	27	90	1	0	2	-2	-5	1
26	22	90	2	0	5	-5	-11	2
26	27	90	1	0	2	-4	-9	2
28	22	90	2	0	5	-5	-10	2

20 30 0

φ_1	Φ	φ_2	H	K	L	U	V	W
20	31	9	1	6	10	10	-5	2
20	34	0	0	2	3	10	-3	2
21	37	9	1	6	8	4	-2	1
21	39	6	1	9	11	12	-5	3
22	34	0	0	2	3	9	-3	2
22	35	8	1	7	10	4	-2	1
23	34	7	1	8	12	4	-2	1
23	37	0	0	3	4	12	-4	3
24	31	9	1	6	10	12	-7	3
24	34	0	0	2	3	8	-3	2
24	37	0	0	3	4	11	-4	3
24	39	7	1	8	10	10	-5	3
26	31	0	0	3	5	12	-5	3
27	34	0	0	2	3	7	-3	2
27	37	0	0	3	4	10	-4	3
28	31	0	0	3	5	11	-5	3
28	36	7	1	8	11	12	-7	4
28	37	6	1	9	12	9	-5	3
28	39	0	0	4	5	12	-5	4
29	34	9	1	6	9	3	-2	1
29	37	0	0	3	4	9	-4	3
29	38	8	1	7	9	8	-5	3

20 30 10

φ_1	Φ	φ_2	H	K	L	U	V	W
21	31	16	2	7	12	9	-6	2
21	33	11	1	5	8	9	-5	2
21	38	18	1	3	4	7	-5	2
21	39	16	2	7	9	11	-7	3
22	32	18	1	3	5	4	-3	1
24	38	18	1	3	4	9	-7	3
25	30	14	1	4	7	11	-8	3
25	30	18	2	6	11	7	-6	2
25	33	11	1	5	8	11	-7	3
25	34	14	1	4	6	10	-7	3
26	31	16	2	7	12	10	-8	3
26	35	18	2	6	9	6	-5	2

20 30 10

φ_1	Φ	φ_2	H	K	L	U	V	W
26	38	13	2	9	12	3	-2	1
26	38	18	1	3	4	11	-9	4
27	30	11	1	5	9	7	-5	2
27	32	18	1	3	5	9	-8	3
27	34	14	1	4	6	12	-9	4
28	39	16	2	7	9	5	-4	2
29	30	14	1	4	7	6	-5	2
29	33	16	2	7	11	8	-7	3

20 30 20

φ_1	Φ	φ_2	H	K	L	U	V	W
20	35	21	3	8	12	4	-3	1
20	37	23	3	7	10	11	-9	3
21	31	27	3	6	11	11	-11	3
22	33	27	2	4	7	7	-7	2
22	34	22	2	5	8	7	-6	2
22	34	27	3	6	10	10	-10	3
22	35	23	3	7	11	10	-9	3
22	39	24	4	9	12	9	-8	3
23	37	27	1	2	3	3	-3	1
23	39	27	4	8	11	11	-11	4
24	38	21	3	8	11	6	-5	2
25	31	27	3	6	11	9	-10	3
25	32	23	3	7	12	3	-3	1
25	38	22	2	5	7	11	-10	4
26	34	27	3	6	10	8	-9	3
26	35	23	3	7	11	11	-11	4
26	37	27	1	2	3	10	-11	4
26	39	24	4	9	12	12	-12	5
27	31	22	2	5	9	3	-3	1
27	33	27	2	4	7	5	-6	2
27	34	22	2	5	8	8	-8	3
27	37	23	3	7	10	5	-5	2
27	37	27	1	2	3	7	-8	3
27	39	27	4	6	11	9	-10	4
28	38	22	2	5	7	7	-7	3
29	31	27	3	6	11	7	-9	3
29	35	21	3	8	12	12	-12	5

20 30 30

φ_1	Φ	φ_2	H	K	L	U	V	W
20	36	37	3	4	7	3	-4	1
20	39	34	4	6	9	6	-7	2
21	36	34	2	3	5	9	-11	3
22	30	31	3	5	10	10	-12	3
22	31	34	2	3	6	3	-4	1
22	39	30	4	7	10	11	-12	4
23	32	37	3	4	6	8	-12	3

20 30 30

φ_1	Φ	φ_2	H	K	L	U	V	W
23	33	39	4	5	10	5	-8	2
23	35	39	4	5	9	7	-11	3
23	36	36	5	7	12	5	-7	2
24	33	34	4	6	11	5	-7	2
24	39	34	4	6	9	9	-12	4
25	36	30	4	7	11	5	-6	2
26	36	31	3	5	8	7	-9	3
26	38	36	5	7	11	2	-3	1
26	39	39	4	5	8	7	-12	4
27	30	31	3	5	10	5	-7	2
27	32	37	3	4	8	4	-7	2
27	36	34	2	3	5	2	-3	1
27	38	32	5	8	12	8	-11	4
28	34	30	4	7	12	9	-12	4
29	31	34	2	3	6	6	-10	3
29	33	31	3	5	9	2	-3	1
29	33	39	4	5	10	5	-10	3
29	36	37	3	4	7	5	-9	3

20 30 40

φ_1	Φ	φ_2	H	K	L	U	V	W
20	38	41	6	7	12	6	-12	3
22	35	45	1	1	2	2	-4	1
24	38	40	5	6	10	6	-10	3
25	33	40	5	6	12	6	-11	3
25	35	40	5	6	11	4	-7	2
25	35	45	1	1	2	5	-11	3
26	31	45	3	3	7	5	-12	3
26	35	45	1	1	2	3	-7	2
26	35	45	1	1	2	4	-10	3
28	38	41	6	7	12	3	-6	2
28	39	45	4	4	7	5	-12	4
29	31	45	3	3	7	4	-11	3

20 30 50

φ_1	Φ	φ_2	H	K	L	U	V	W
20	33	51	5	4	10	2	-5	1
20	33	56	6	4	11	3	-10	2
21	36	54	7	5	12	3	-9	2
21	36	50	6	5	10	5	-12	3
22	30	59	5	3	10	1	-5	1
22	38	58	8	5	12	1	-4	1
23	35	50	6	5	11	3	-8	2
23	39	51	5	4	8	4	-11	3
24	31	56	3	2	6	2	-9	2
24	36	56	3	2	5	1	-4	1
25	33	50	6	5	12	4	-12	3
26	32	53	4	3	8	1	-4	1

20 30 50

φ_1	Φ	φ_2	H	K	L	U	V	W
26	36	53	4	3	7	3	-11	3
27	35	51	5	4	9	2	-7	2
28	30	51	5	4	11	1	-4	1
28	31	56	3	2	6	2	-12	3
26	36	59	5	3	8	1	-7	2
29	36	54	7	5	12	2	-10	3

20 30 60

φ_1	Φ	φ_2	H	K	L	U	V	W
21	36	60	7	4	11	2	-9	2
21	37	67	7	3	10	1	-9	2
24	33	63	4	2	7	1	-9	2
24	37	63	2	1	3	1	-6	2
24	39	60	7	4	10	2	-11	3
25	31	68	5	2	9	0	-9	2
25	35	69	8	3	12	0	-4	1
25	38	69	8	3	11	0	-11	3
26	34	63	6	3	10	1	-12	3
26	34	68	5	2	8	0	-4	1
26	37	63	2	1	3	1	-11	3
27	32	67	7	3	12	0	-4	1
27	36	68	5	2	7	0	-7	2
28	35	67	7	3	11	0	-11	3
28	37	67	7	3	10	0	-10	3
29	36	60	7	4	11	1	-10	3

20 30 70

φ_1	Φ	φ_2	H	K	L	U	V	W
20	39	74	7	2	9	0	-9	2
21	30	72	6	2	11	0	-11	2
21	36	79	5	1	7	-1	-9	2
22	32	72	3	1	5	0	-5	1
22	35	72	6	2	9	0	-9	2
23	30	76	4	1	7	-1	-10	2
23	38	72	3	1	4	0	-4	1
25	34	76	4	1	6	-1	-8	2
26	36	79	5	1	7	-2	-11	3
27	32	72	3	1	5	-1	-12	3
27	37	76	8	2	11	-1	-7	2
27	39	74	7	2	9	-1	-10	3
29	30	72	6	2	11	-1	-8	2
29	30	79	5	1	9	-1	-4	1

20 30 80

φ_1	Φ	φ_2	H	K	L	U	V	W
22	31	82	7	1	12	-1	-5	1
23	37	81	6	1	8	-2	-12	3
25	34	83	6	1	12	-1	-4	1

φ_1	Φ	φ_2	H	K	L	U	V	W
20	30	80						
26	33	82	7	1	11	-1	-4	1
27	31	81	6	1	10	-1	-4	1
20	30	90						
20	34	90	2	0	3	-3	-10	2
22	34	90	2	0	3	-3	-9	2
23	37	90	3	0	4	-4	-12	3
24	34	90	2	0	3	-3	-8	2
24	37	90	3	0	4	-4	-11	3
26	31	90	3	0	5	-5	-12	3
27	34	90	2	0	3	-3	-7	2
27	37	90	3	0	4	-4	-10	3
28	31	90	3	0	5	-5	-11	3
28	39	90	4	0	5	-5	-12	4
29	37	90	3	0	4	-4	-9	3
20	40	0						
20	42	6	1	10	11	8	-3	2
20	48	9	2	12	11	7	-3	2
21	41	8	1	7	8	11	-5	3
21	45	0	0	1	1	11	-3	3
21	45	5	1	11	11	11	-4	3
21	45	8	1	7	7	7	-3	2
22	42	6	1	9	10	7	-3	2
22	45	0	0	1	1	7	-2	2
22	45	6	1	10	10	10	-4	3
23	45	0	0	1	1	10	-3	3
23	49	7	1	8	7	12	-5	4
24	45	6	1	9	9	9	-4	3
24	45	9	1	6	6	6	-3	2
25	41	9	1	6	7	9	-5	3
25	42	7	1	8	9	6	-3	2
25	45	0	0	1	1	3	-1	1
25	45	5	1	12	12	12	-5	4
26	40	6	1	10	12	6	-3	2
26	45	5	1	11	11	11	-5	4
27	45	0	0	1	1	11	-4	4
27	45	7	1	8	8	8	-4	3
28	43	5	1	11	12	8	-4	3
28	45	0	0	1	1	8	-3	3
28	45	6	1	10	10	10	-5	4
28	45	9	1	6	6	12	-7	5
29	41	8	1	7	8	5	-3	2
29	45	0	0	1	1	5	-2	2
29	45	8	1	7	7	7	-4	3

φ_1	Φ	φ_2	H	K	L	U	V	W
20	40	10						
20	40	11	1	5	6	4	-2	1
20	45	10	2	11	11	11	-5	3
20	46	16	2	7	7	7	-4	2
20	46	17	3	10	10	10	-6	3
21	41	18	3	9	11	10	-7	3
21	42	14	2	8	9	7	-4	2
21	44	15	3	11	12	10	-6	3
21	46	11	1	5	5	10	-5	3
22	40	13	2	9	11	7	-4	2
22	47	18	1	3	3	3	-2	1
22	48	14	3	12	11	9	-5	3
22	49	11	2	10	9	6	-3	2
23	42	16	2	7	8	3	-2	1
23	44	17	3	10	11	3	-2	1
23	46	13	2	9	9	9	-5	3
23	46	14	1	4	4	12	-7	4
24	46	15	3	11	11	11	-7	4
24	49	13	2	9	8	11	-6	4
24	49	18	4	12	11	10	-7	4
25	40	14	1	4	5	3	-2	1
25	43	18	3	9	10	8	-6	3
25	45	10	2	11	11	11	-6	4
25	46	14	1	4	4	8	-5	3
26	41	17	3	10	12	6	-6	3
26	42	18	2	6	7	5	-4	2
26	46	17	3	10	10	10	-7	4
26	47	18	1	3	3	12	-9	5
27	40	11	1	5	6	11	-7	4
27	42	14	2	8	9	10	-7	4
27	46	11	1	5	5	5	-3	2
27	46	14	1	4	4	12	-8	5
27	46	16	2	7	7	7	-5	3
27	47	18	1	3	3	9	-7	4
28	41	18	3	9	11	7	-6	3
28	48	10	2	11	10	7	-4	3
29	46	13	2	9	9	9	-6	4
29	46	15	3	11	11	11	-8	5
29	49	17	3	10	9	4	-3	2
20	40	20						
20	40	27	3	6	8	10	-9	3
20	44	27	3	6	7	9	-8	3
20	48	24	4	9	9	9	-7	3
20	49	29	6	11	11	11	-10	4
21	41	29	5	9	12	3	-3	1
21	42	27	2	4	5	12	-11	4

φ_1	Φ	φ_2	H	K	L	U	V	W
20	40	20						
21	43	29	5	9	11	11	-11	4
21	47	23	5	12	12	12	-9	4
21	48	24	5	11	11	11	-9	4
21	48	27	1	2	2	8	-7	3
21	49	20	4	11	10	3	-2	1
22	44	21	3	8	9	12	-9	4
22	45	27	4	6	9	11	-10	4
22	46	29	5	9	10	5	-5	2
22	48	27	1	2	2	10	-9	4
23	46	29	6	11	12	12	-12	5
23	47	20	4	11	11	11	-8	4
23	47	21	3	8	8	8	-6	3
23	48	27	1	2	2	12	-11	5
23	49	29	5	9	9	9	-9	4
24	40	27	3	6	8	8	-8	3
24	42	27	2	4	5	5	-5	2
24	43	27	5	10	12	12	-12	5
24	47	22	2	5	5	5	-4	2
24	49	29	6	11	11	11	-11	5
25	44	20	4	11	12	5	-4	2
25	44	27	3	6	7	7	-7	3
25	45	27	4	8	9	9	-9	4
25	45	27	5	10	11	11	-11	5
25	47	23	3	7	7	7	-6	3
25	47	23	5	12	12	12	-10	5
25	48	24	4	9	9	9	-6	4
26	48	24	5	11	11	11	-10	5
27	42	24	4	9	11	11	-11	5
27	43	29	5	9	11	6	-7	3
27	44	21	3	8	9	7	-6	3
27	44	22	4	10	11	9	-6	4
27	47	20	4	11	11	11	-9	5
27	48	27	1	2	2	2	-2	1
27	49	29	6	11	11	11	-12	6
28	40	23	3	7	9	9	-9	4
28	42	27	2	4	5	8	-9	4
28	43	27	5	10	12	10	-11	5
28	45	24	4	9	10	2	-2	1
28	45	24	5	11	12	2	-2	1
28	47	22	2	5	5	10	-9	5
28	47	23	5	12	12	12	-11	6
28	49	29	5	9	9	9	-10	5
29	40	27	3	6	8	6	-7	3
29	44	23	3	7	8	2	-2	1
29	45	27	5	10	11	9	-10	5
29	47	21	3	8	8	6	-7	4
29	48	24	5	11	11	11	-11	6

φ_1	Φ	φ_2	H	K	L	U	V	W
20	40	30						
20	41	36	5	7	10	6	-10	3
20	42	30	4	7	9	3	-3	1
20	45	34	6	9	11	8	-9	3
20	47	31	6	10	11	11	-11	4
21	45	30	4	7	8	8	-8	3
21	45	37	3	4	5	7	-9	3
21	47	39	4	5	6	9	-12	4
21	49	31	3	5	5	5	-5	2
22	40	31	3	5	7	8	-9	3
22	42	34	2	3	4	5	-6	2
22	47	34	6	9	10	7	-8	3
22	47	36	5	7	8	9	-11	4
22	49	30	7	12	12	12	-12	5
22	49	38	7	9	10	6	-8	3
23	42	39	4	5	7	2	-3	1
23	49	30	4	7	7	7	-7	3
24	45	30	4	7	8	11	-12	5
24	45	37	3	4	5	8	-11	4
24	48	35	7	10	11	4	-5	2
24	49	39	8	10	11	7	-10	4
25	40	31	3	5	7	9	-11	4
25	40	37	3	4	6	2	-3	1
25	42	34	2	3	4	6	-8	3
25	46	34	4	6	7	4	-5	2
25	49	31	3	5	5	10	-11	5
25	49	36	8	11	12	9	-12	5
26	43	32	5	6	10	4	-5	2
26	44	31	3	5	6	10	-12	5
27	47	36	5	7	8	5	-7	3
28	42	34	2	3	4	7	-10	4
28	44	31	3	5	6	7	-9	4
28	44	38	7	9	12	3	-5	2
28	45	34	6	9	11	5	-7	3
28	46	37	6	8	9	6	-9	4
28	49	30	4	7	7	7	-8	4
28	49	31	3	5	5	5	-6	3
28	49	38	7	9	10	7	-11	5
29	42	34	2	3	4	8	-12	5
29	42	37	6	8	11	3	-5	2
29	45	35	7	10	12	6	-9	4
20	40	40						
20	47	45	3	3	4	6	-10	3
21	40	45	3	3	5	6	-11	3
21	43	45	2	2	3	4	-7	2
21	47	49	8	7	10	5	-10	3

20 40 40

φ_1	Φ	φ_2	H	K	L	U	V	W
21	48	42	8	9	11	2	-3	1
21	48	42	9	10	12	2	-3	1
22	44	40	5	6	8	2	-3	1
22	46	41	6	7	9	2	-3	1
22	47	41	7	8	10	2	-3	1
22	49	45	4	4	5	7	-12	4
23	47	45	3	3	4	5	-9	3
23	48	48	10	9	12	3	-6	2
24	40	45	3	3	5	5	-10	3
24	45	48	9	8	12	4	-9	3
24	49	45	4	4	5	6	-11	4
25	43	45	2	2	3	3	-6	2
25	44	49	8	7	11	5	-12	4
26	40	41	6	7	11	5	-9	3
26	47	45	3	3	4	4	-8	3
27	40	45	3	3	5	4	-9	3
27	43	45	2	2	3	5	-11	4
27	49	45	4	4	5	5	-10	4
27	49	49	7	6	8	2	-5	2
28	40	49	7	6	11	1	-3	1
28	41	40	5	6	9	6	-11	4
28	46	49	7	6	9	3	-8	3
28	48	48	9	8	11	2	-5	2
29	42	41	7	8	12	4	-8	3

20 40 50

φ_1	Φ	φ_2	H	K	L	U	V	W
20	42	56	3	2	4	4	-12	3
20	47	56	9	6	10	4	-11	3
20	48	53	8	6	9	3	-7	2
21	40	53	4	3	6	3	-8	2
21	44	54	7	5	9	4	-11	3
21	44	59	5	3	6	3	-11	3
21	46	52	9	7	11	3	-7	2
21	47	59	10	6	11	2	-7	2
21	49	54	11	8	12	4	-10	3
21	49	59	5	3	5	3	-10	3
22	40	59	5	3	7	1	-4	1
23	42	56	3	2	4	2	-7	2
23	45	56	9	6	11	3	-10	3
23	48	50	6	5	7	4	-9	3
24	47	54	7	5	8	1	-3	1
24	48	55	10	7	11	1	-3	1
25	45	53	4	3	5	1	-3	1
25	46	58	8	5	9	3	-12	4
26	40	53	4	3	6	3	-10	3
26	44	52	9	7	12	1	-3	1
26	47	51	5	4	6	4	-11	4

20 40 50

φ_1	Φ	φ_2	H	K	L	U	V	W
27	41	50	6	5	9	1	-3	1
27	42	51	5	4	7	1	-3	1
28	42	56	3	2	4	2	-9	3
28	47	56	9	6	10	2	-8	3
29	40	53	4	3	6	3	-12	4
29	41	58	8	5	11	1	-6	2
29	45	55	10	7	12	2	-8	3
29	48	53	8	6	9	3	-10	4
29	49	59	5	3	5	1	-5	2

20 40 60

φ_1	Φ	φ_2	H	K	L	U	V	W
20	43	61	9	5	11	1	-4	1
20	44	68	10	4	11	1	-6	2
20	45	66	9	4	10	2	-12	3
20	49	61	11	6	11	3	-11	3
21	42	60	7	4	9	1	-4	1
21	48	66	11	5	11	2	-11	3
21	49	60	7	4	7	2	-7	2
22	40	63	6	3	8	2	-12	3
22	44	63	6	3	7	2	-11	3
22	47	67	7	3	7	1	-7	2
22	48	63	2	1	2	2	-10	3
23	45	66	11	5	12	1	-7	2
23	47	68	5	2	5	1	-10	3
24	42	63	4	2	5	1	-7	2
24	49	60	12	7	12	3	-12	4
24	49	61	9	5	9	2	-9	3
25	47	67	12	5	12	1	-12	4
25	48	63	2	1	2	1	-6	2
25	48	66	9	4	9	1	-9	3
26	40	63	6	3	8	1	-10	3
26	41	69	8	3	10	0	-10	3
27	44	63	6	3	7	1	-9	3
27	44	69	8	3	9	0	-3	1
27	48	66	11	5	11	1	-11	4
27	49	61	11	6	11	2	-11	4
28	42	63	4	2	5	1	-12	4
28	42	68	5	2	6	0	-3	1
28	48	63	2	1	2	1	-8	3
29	40	67	7	3	9	0	-3	1
29	41	61	9	5	12	1	-9	3
29	44	68	10	4	11	0	-11	4
29	45	63	8	4	9	1	-11	4
29	47	69	8	3	8	0	-6	3

20 40 70

φ_1	Φ	φ_2	H	K	L	U	V	W
20	46	76	4	1	4	0	-4	1
21	42	74	7	2	8	0	-4	1
21	44	75	11	3	12	0	-4	1
21	46	75	11	3	11	0	-11	3
21	47	70	11	4	11	1	-11	3
21	48	76	12	3	11	0	-11	3
22	40	77	9	2	11	-1	-12	3
22	41	73	10	3	12	0	-4	1
22	44	73	10	3	11	0	-11	3
22	46	74	7	2	7	0	-7	2
22	49	75	11	3	10	0	-10	3
23	46	73	10	3	10	0	-10	3
24	40	76	4	1	5	-1	-11	3
24	41	72	9	3	11	0	-11	3
24	42	72	6	2	7	0	-7	2
24	46	79	5	1	5	-1	-10	3
25	40	79	5	1	6	-1	-7	2
25	43	72	9	3	10	0	-10	3
25	49	73	10	3	9	0	-3	1
26	46	76	4	1	4	-1	-12	4
26	46	77	9	2	9	-1	-9	3
26	47	72	3	1	3	0	-3	1
27	43	79	10	2	11	-1	-6	2
27	44	70	11	4	12	0	-3	1
27	49	72	12	4	11	0	-11	4
28	46	75	11	3	11	-1	-11	4
28	47	70	11	4	11	0	-11	4
29	40	76	4	1	5	-1	-6	2
29	46	76	4	1	4	-1	-8	3
29	49	70	11	4	10	0	-5	2

20 40 80

φ_1	Φ	φ_2	H	K	L	U	V	W
20	45	81	6	1	6	-1	-12	3
20	45	83	8	1	8	-1	-8	2
20	45	85	12	1	12	-2	-12	3
21	42	84	9	1	10	-2	-12	3
21	45	85	11	1	11	-2	-11	3
21	49	84	9	1	8	-1	-7	2
22	41	81	6	1	7	-1	-6	2
22	45	80	11	2	11	-1	-11	3
23	42	83	8	1	9	-2	-11	3
23	45	82	7	1	7	-1	-7	2
23	45	84	10	1	10	-2	-10	3
24	49	83	8	1	7	-1	-6	2
25	41	82	7	1	8	-2	-10	3
25	48	85	12	1	11	-2	-9	3

20 40 80

φ_1	Φ	φ_2	H	K	L	U	V	W
26	45	81	6	1	6	-1	-6	2
26	45	84	9	1	9	-2	-9	3
26	45	85	12	1	12	-3	-12	4
27	42	83	8	1	9	-3	-12	4
27	48	85	11	1	10	-2	-6	3
28	41	81	6	1	7	-2	-9	3
28	45	80	11	2	11	-2	-11	4
28	45	85	11	1	11	-3	-11	4
29	45	83	8	1	8	-2	-8	3
29	48	81	12	2	11	-1	-5	2

20 40 90

φ_1	Φ	φ_2	H	K	L	U	V	W
21	45	90	1	0	1	-3	-11	3
22	45	90	1	0	1	-2	-7	2
23	45	90	1	0	1	-3	-10	3
25	45	90	1	0	1	-1	-3	1
27	45	90	1	0	1	-4	-11	4
28	45	90	1	0	1	-3	-6	3
29	45	90	1	0	1	-2	-5	2

20 50 0

φ_1	Φ	φ_2	H	K	L	U	V	W
20	56	0	0	3	2	10	-2	3
21	50	8	1	7	6	10	-4	3
21	54	9	2	12	9	12	-5	4
21	56	6	1	9	6	3	-1	1
21	58	7	1	8	5	3	-1	1
22	51	9	1	6	5	9	-4	3
22	53	5	1	12	9	3	-1	1
22	54	5	1	11	6	3	-1	1
22	55	6	1	10	7	3	-1	1
22	56	0	0	3	2	9	-2	3
23	53	0	0	4	3	12	-3	4
24	50	8	1	7	6	11	-5	4
24	53	0	0	4	3	11	-3	4
24	56	0	0	3	2	8	-2	3
26	51	9	1	6	5	10	-5	4
26	59	0	0	5	3	12	-3	5
27	50	8	1	7	6	12	-6	5
27	53	0	0	4	3	10	-3	4
27	56	0	0	3	2	7	-2	3
28	51	0	0	5	4	12	-4	5
28	59	0	0	5	3	11	-3	5
29	51	9	1	6	5	11	-6	5
29	53	0	0	4	3	9	-3	4
29	57	9	1	6	4	2	-1	1
29	59	6	1	10	6	10	-4	5

φ_1	Φ	φ_2	H	K	L	U	V	W
20	**50**	**10**						
20	55	18	4	12	9	12	-7	4
20	57	13	2	9	6	9	-4	3
21	54	14	1	4	3	6	-3	2
22	52	18	2	6	5	11	-7	4
22	54	18	3	9	7	8	-5	3
22	56	11	2	10	7	11	-5	4
22	56	16	2	7	5	11	-6	4
22	57	14	3	12	8	8	-4	3
23	58	18	1	3	2	5	-3	2
23	59	14	2	8	5	10	-5	4
24	52	11	1	5	4	8	-4	3
24	52	15	3	11	9	5	-3	2
24	56	16	2	7	5	12	-7	5
25	50	14	2	8	7	5	-3	2
25	55	18	4	12	9	9	-6	4
25	57	13	2	9	6	12	-6	5
25	58	15	3	11	7	9	-5	4
26	50	18	3	9	8	7	-5	3
26	54	14	1	4	3	7	-4	3
27	51	16	2	7	6	9	-6	4
27	52	11	1	5	4	9	-5	4
27	58	18	1	3	2	6	-4	3
28	52	18	2	6	5	4	-3	2
28	57	14	3	12	8	12	-7	6
29	53	17	3	10	8	10	-7	5
29	54	14	1	4	3	8	-5	4
29	58	10	2	11	7	2	-1	1
20	**50**	**20**						
20	51	21	3	8	7	3	-2	1
20	52	27	4	8	7	11	-9	4
20	55	24	4	9	7	11	-8	4
20	59	27	3	6	4	8	-6	3
21	52	23	3	7	6	8	-6	3
21	56	27	2	4	3	5	-4	2
22	50	22	4	10	9	11	-8	4
22	54	27	5	10	8	12	-10	5
22	57	29	6	11	8	7	-6	3
23	50	24	5	11	10	12	-10	5
23	51	24	4	9	8	5	-4	2
23	53	27	3	6	5	7	-6	3
23	56	20	4	11	8	12	-8	5
23	56	24	5	11	8	9	-7	4
23	56	29	5	9	7	11	-10	5
23	58	27	5	10	7	11	-9	5
24	51	27	5	10	9	11	-10	5

φ_1	Φ	φ_2	H	K	L	U	V	W
20	**50**	**20**						
25	51	29	6	11	10	2	-2	1
25	52	20	4	11	9	7	-5	3
25	52	23	3	7	6	11	-9	5
25	52	29	5	9	8	2	-2	1
25	54	27	5	10	8	10	-9	5
25	56	27	2	4	3	8	-7	4
25	59	27	3	6	4	6	-5	3
26	53	27	3	6	5	12	-11	6
26	55	24	4	9	7	6	-5	3
26	57	22	4	10	7	4	-3	2
26	58	23	5	12	8	4	-3	2
27	50	23	5	12	11	8	-7	4
27	52	23	5	12	10	6	-5	3
27	53	22	2	5	4	10	-8	5
27	54	29	6	11	9	9	-9	5
27	55	21	3	8	6	4	-3	2
27	56	27	2	4	3	11	-10	6
27	56	29	5	9	7	7	-7	4
27	58	27	5	10	7	9	-8	5
28	51	27	5	10	9	9	-9	5
28	51	27	6	12	11	11	-11	6
28	52	27	4	8	7	7	-7	4
28	57	23	3	7	5	11	-9	6
28	59	27	3	6	4	10	-9	6
29	53	27	3	6	5	5	-5	3
29	57	29	6	11	8	8	-8	5
29	59	24	4	9	6	12	-10	7
20	**50**	**30**						
20	50	32	5	8	8	8	-8	3
20	51	35	7	10	10	10	-11	4
20	51	36	5	7	7	7	-8	3
20	52	38	7	9	9	9	-11	4
20	53	32	7	11	10	5	-5	2
20	54	37	9	12	11	7	-8	3
20	57	34	6	9	7	7	-7	3
20	59	37	3	4	3	11	-12	5
21	53	32	5	8	7	7	-7	3
21	55	32	7	11	9	9	-9	4
21	55	39	8	10	9	4	-5	2
21	58	34	8	12	9	9	-9	4
21	58	38	7	9	7	6	-7	3
21	58	39	4	5	4	10	-12	5
22	50	34	2	3	3	9	-10	4
22	51	37	3	4	4	4	-5	2
22	52	31	6	10	9	9	-9	4
22	56	37	9	12	10	6	-7	3

φ_1	Φ	φ_2	H	K	L	U	V	W
20	**50**	**30**						
22	57	35	7	10	8	10	-11	5
22	58	30	4	7	5	9	-8	4
22	59	37	3	4	3	8	-9	4
23	50	32	7	11	11	11	-12	5
23	51	35	7	10	10	10	-12	5
23	52	30	7	12	11	11	-11	5
23	53	34	8	12	11	8	-9	4
23	58	32	5	8	6	2	-2	1
23	58	32	7	11	8	2	-2	1
23	58	39	9	11	9	7	-9	4
24	50	32	5	8	8	8	-9	4
24	50	34	2	3	3	6	-7	3
24	52	38	7	9	9	9	-12	5
24	54	30	7	12	10	2	-2	1
24	55	38	7	9	8	7	-9	4
24	56	31	3	5	4	2	-2	1
24	57	34	6	9	7	11	-12	6
25	50	39	9	11	12	8	-12	5
25	51	36	5	7	7	7	-9	4
25	52	39	4	5	5	5	-7	3
25	53	30	4	7	6	2	-2	1
25	55	34	4	6	5	7	-8	4
25	59	37	3	4	3	5	-6	3
26	50	34	2	3	3	9	-11	5
26	51	37	3	4	4	8	-11	5
26	52	31	6	10	9	11	-12	6
26	53	32	5	8	7	9	-10	5
26	54	34	6	9	8	5	-6	3
26	55	36	5	7	6	8	-10	5
26	57	30	7	12	9	9	-9	5
26	58	39	4	5	4	3	-4	2
26	59	31	6	10	7	7	-7	4
27	54	36	8	11	10	3	-4	2
27	55	37	6	8	7	3	-4	2
27	58	30	4	7	5	5	-5	3
28	50	32	5	8	8	8	-10	5
28	55	34	4	6	5	9	-11	6
28	55	39	9	11	10	4	-6	3
28	58	34	8	12	9	6	-7	4
28	59	37	3	4	3	7	-9	5
29	50	34	2	3	3	3	-4	2
29	51	36	5	7	7	7	-10	5
29	51	37	3	4	4	4	-6	3
29	56	37	9	12	10	8	-11	6
20	**50**	**40**						
20	53	48	9	8	9	5	-9	3

φ_1	Φ	φ_2	H	K	L	U	V	W
20	**50**	**40**						
20	54	47	12	11	12	7	-12	4
20	56	42	8	9	8	6	-8	3
20	56	43	11	12	11	8	-11	4
20	56	43	11	12	10	6	-8	3
21	52	40	5	6	6	6	-8	3
21	53	41	7	8	8	8	-11	4
21	55	45	1	1	1	7	-11	4
21	57	40	5	6	5	4	-5	2
21	57	49	7	6	6	6	-11	4
22	50	49	8	7	9	3	-6	2
22	54	46	11	10	11	6	-11	4
22	55	45	1	1	1	5	-8	3
22	56	42	10	11	10	7	-10	4
22	57	41	6	7	6	9	-12	5
23	53	41	6	7	7	7	-10	4
23	53	49	8	7	8	4	-8	3
23	55	45	1	1	1	3	-5	2
23	57	41	7	8	7	5	-7	3
23	59	42	10	11	9	5	-7	3
24	51	48	11	10	12	4	-8	3
24	53	41	7	8	8	8	-12	5
25	53	48	10	9	10	5	-10	4
25	55	45	1	1	1	4	-7	3
25	55	45	1	1	1	7	-12	5
25	56	42	9	10	9	6	-9	4
25	57	49	7	6	6	6	-12	5
26	52	40	5	6	6	6	-9	4
26	54	47	12	11	12	6	-12	5
26	55	45	1	1	1	5	-9	4
26	56	43	11	12	11	7	-11	5
27	50	45	5	5	6	6	-12	5
27	53	41	6	7	7	7	-11	5
27	53	49	7	6	7	3	-7	3
27	55	45	1	1	1	6	-11	5
27	57	40	5	6	5	7	-10	5
27	57	41	6	7	6	4	-6	3
27	59	42	9	10	8	4	-6	3
27	59	45	6	6	5	7	-12	6
28	51	48	10	9	11	3	-7	3
28	53	48	9	8	9	4	-9	4
28	54	48	11	10	11	5	-11	5
28	56	42	8	9	8	5	-8	4
28	56	42	10	11	10	6	-10	5
28	59	48	10	9	8	1	-2	1
29	58	47	12	11	10	1	-2	1
29	59	48	11	10	9	1	-2	1

φ_1	Φ	φ_2	H	K	L	U	V	W
20	50	50						
20	51	55	10	7	10	4	-10	3
20	55	51	10	8	9	3	-6	2
20	58	56	12	8	9	5	-12	4
21	52	50	6	5	6	3	-6	2
21	54	53	12	9	11	4	-9	3
21	56	59	5	3	4	1	-3	1
22	51	53	4	3	4	5	-12	4
22	52	52	9	7	9	4	-9	3
22	53	58	8	5	7	1	-3	1
22	53	58	11	7	10	1	-3	1
22	58	51	5	4	4	4	-8	3
23	50	56	3	2	3	1	-3	1
23	52	51	11	9	11	5	-11	4
23	55	52	9	7	8	5	-11	4
23	55	56	6	4	5	4	-11	4
23	57	56	9	6	7	3	-8	3
24	57	50	6	5	5	5	-10	4
25	51	53	4	3	4	3	-8	3
25	51	54	11	8	11	4	-11	4
25	52	59	10	6	9	3	-11	4
25	55	53	8	6	7	2	-5	2
25	57	54	11	8	9	2	-5	2
25	59	53	4	3	3	3	-7	3
26	50	58	11	7	11	3	-11	4
26	52	51	5	4	5	2	-5	2
26	58	51	5	4	4	4	-9	4
26	58	56	11	7	8	4	-12	5
27	50	58	8	5	8	2	-8	3
27	52	50	6	5	6	5	-12	5
27	53	56	12	8	11	3	-10	4
27	55	51	11	9	10	3	-7	3
27	57	50	6	5	5	5	-11	5
27	58	56	12	8	9	3	-9	4
28	51	53	4	3	4	4	-12	5
28	51	55	10	7	10	3	-10	4
28	54	56	9	6	8	2	-7	3
28	55	54	7	5	6	3	-9	4
28	55	58	11	7	9	2	-7	3
29	50	56	3	2	3	3	-12	5
29	51	54	7	5	7	2	-7	3
20	50	60						
20	51	61	11	6	10	2	-7	2
20	54	61	11	6	9	3	-10	3
21	50	68	10	4	9	1	-7	2
21	57	60	12	7	9	1	-3	1

φ_1	Φ	φ_2	H	K	L	U	V	W
20	50	60						
21	58	60	7	4	5	1	-3	1
21	59	66	9	4	6	2	-9	3
22	55	67	12	5	9	1	-6	2
22	56	63	4	2	3	3	-12	4
22	57	68	10	4	7	1	-6	2
23	53	63	6	3	5	2	-9	3
23	59	63	6	3	4	2	-8	3
24	50	67	12	5	11	1	-9	3
24	51	69	8	3	7	1	-12	4
24	56	61	9	5	7	3	-11	4
25	52	63	8	4	7	2	-11	4
25	53	60	7	4	6	2	-8	3
26	56	63	4	2	3	1	-5	2
26	59	66	9	4	6	2	-12	5
28	52	61	9	5	8	1	-5	2
28	54	63	10	5	8	2	-12	5
28	56	66	11	5	8	1	-7	3
28	57	61	11	6	6	2	-9	4
29	51	61	11	6	10	2	-12	5
29	53	63	6	3	5	1	-7	3
29	57	67	7	3	5	1	-9	4
29	58	63	10	5	7	2	-11	5
20	50	70						
20	50	72	9	3	8	1	-11	3
20	53	77	9	2	7	0	-7	2
21	50	76	6	2	7	0	-7	2
21	54	72	9	3	7	1	-10	3
22	51	76	12	3	10	0	-10	3
22	57	77	9	2	6	0	-3	1
22	58	72	3	1	2	1	-9	3
23	54	76	4	1	3	0	-3	1
23	55	72	12	4	9	1	-12	4
24	51	74	7	2	6	0	-3	1
24	52	75	11	3	9	0	-3	1
24	58	72	3	1	2	1	-11	4
24	59	70	11	4	7	1	-8	3
25	55	75	11	3	8	0	-8	3
25	57	76	12	3	8	0	-8	3
26	52	79	5	1	4	-1	-11	4
26	53	73	10	3	8	0	-8	3
26	59	76	8	2	5	0	-5	2
27	50	72	9	3	8	0	-6	3
27	56	74	7	2	5	0	-5	2
28	52	72	6	2	5	0	-5	2
28	56	73	10	3	7	0	-7	3
28	58	75	11	3	7	0	-7	3

φ_1	Φ	φ_2	H	K	L	U	V	W
20	50	70						
29	50	76	8	2	7	-1	-10	4
29	54	72	9	3	7	0	-7	3
29	56	79	10	2	7	-1	-9	4
20	50	80						
20	50	82	7	1	6	-1	-11	3
20	58	85	11	1	7	-1	-10	3
21	51	85	11	1	9	-1	-7	2
21	53	83	8	1	6	-1	-10	3
22	56	84	9	1	6	-1	-9	3
22	58	83	8	1	5	-1	-12	4
23	53	85	12	1	9	-1	-6	2
23	54	81	12	2	9	-1	-12	4
24	51	81	6	1	5	-1	-9	3
24	51	84	10	1	8	-1	-6	2
24	59	84	10	1	6	-1	-8	3
25	55	82	7	1	5	-1	-8	3
26	57	81	6	1	4	-1	-10	4
27	51	80	11	2	9	-1	-6	3
27	52	84	9	1	7	-1	-5	2
27	54	85	11	1	8	-1	-5	2
27	56	84	9	1	6	-2	-12	5
27	58	80	11	2	7	-1	-12	5
27	58	83	8	1	5	-1	-7	3
29	50	82	7	1	6	-1	-5	2
29	51	85	11	1	9	-3	-12	5
20	50	90						
20	56	90	3	0	2	-2	-10	3
22	56	90	3	0	2	-2	-9	3
23	53	90	4	0	3	-3	-12	4
24	53	90	4	0	3	-3	-11	4
24	56	90	3	0	2	-2	-6	3
26	59	90	5	0	3	-3	-12	5
27	53	90	4	0	3	-3	-10	4
27	56	90	3	0	2	-2	-7	3
28	51	90	5	0	4	-4	-12	5
28	59	90	5	0	3	-3	-11	5
29	53	90	4	0	3	-3	-9	4
20	60	0						
20	61	8	1	7	4	3	-1	1
20	63	0	0	2	1	6	-1	2
20	64	5	1	12	6	12	-3	4
20	64	9	1	6	3	3	-1	1

φ_1	Φ	φ_2	H	K	L	U	V	W
20	60	0						
20	67	5	1	12	5	9	-2	3
21	66	6	1	9	4	11	-3	4
22	63	0	0	2	1	11	-2	4
24	60	9	2	12	7	5	-2	2
24	63	0	0	2	1	5	-1	2
24	64	6	1	10	5	10	-3	4
24	64	7	1	8	4	12	-4	5
24	68	0	0	5	2	12	-2	5
25	67	8	1	7	3	9	-3	4
26	61	5	1	11	6	9	-3	4
26	63	0	0	2	1	9	-2	4
26	67	5	1	12	5	11	-3	5
26	68	0	0	5	2	11	-2	5
27	61	6	1	9	5	11	-4	5
27	68	9	2	12	5	6	-3	4
28	64	7	1	8	4	8	-3	4
28	66	6	1	9	4	6	-2	3
28	66	0	0	5	2	10	-2	5
28	68	6	1	10	4	10	-3	5
29	60	5	1	12	7	8	-3	4
29	64	5	1	12	6	6	-2	3
20	60	10						
20	60	14	3	12	7	9	-4	3
20	61	18	4	12	7	11	-6	4
20	67	18	3	9	4	8	-4	3
21	61	16	2	7	4	8	-4	3
21	68	18	4	12	5	10	-5	4
22	64	17	3	10	5	10	-5	4
22	67	13	2	9	4	5	-2	2
23	62	10	2	11	6	5	-2	2
23	62	15	3	11	6	12	-6	5
23	62	18	3	9	5	7	-4	3
23	64	11	2	10	5	5	-2	2
23	68	14	3	12	5	7	-3	3
24	65	18	2	6	3	9	-5	4
25	61	16	2	7	4	11	-6	5
26	61	18	4	12	7	8	-5	4
26	65	18	2	6	3	12	-7	6
26	66	15	3	11	5	2	-1	1
26	68	16	2	7	3	2	-1	1
26	69	17	3	10	4	2	-1	1
27	60	17	3	10	6	10	-6	5
27	64	14	1	4	2	2	-1	1
27	69	11	1	5	2	10	-4	5
28	62	13	2	9	5	2	-1	1
28	62	18	3	9	5	11	-7	6

TAB. VI FROM EULER ANGLES INTO HKL-UVW

φ_1	Φ	φ_2	H	K	L	U	V	W
20	60	10						
29	68	16	4	12	5	7	-4	4
20	60	20						
20	61	22	2	5	3	8	-5	3
20	61	29	6	11	7	5	-4	2
20	66	27	1	2	1	10	-7	4
20	68	23	3	7	3	5	-3	2
21	60	29	5	9	6	12	-10	5
21	62	27	5	10	6	12	-9	5
21	65	22	4	10	5	5	-3	2
21	66	27	1	2	1	7	-5	3
22	61	27	4	8	5	9	-7	4
22	62	23	3	7	4	12	-8	5
22	65	21	3	8	4	12	-7	5
22	66	27	1	2	1	11	-8	5
23	61	22	2	5	3	9	-6	4
23	64	29	5	9	5	6	-5	3
24	62	27	5	10	6	10	-8	5
24	66	27	1	2	1	4	-3	2
24	68	24	4	9	4	6	-4	3
25	65	23	5	12	6	6	-4	3
25	69	23	5	12	5	8	-5	4
26	61	22	2	5	3	10	-7	5
26	62	27	6	12	7	11	-9	6
26	64	29	6	11	6	7	-6	4
26	65	21	3	8	4	8	-5	4
26	66	27	1	2	1	9	-7	5
27	61	27	4	8	5	7	-6	4
27	61	29	6	11	7	12	-11	7
27	68	23	3	7	3	9	-6	5
27	68	24	5	11	5	7	-5	4
28	61	22	2	5	3	11	-8	6
28	63	20	4	11	6	9	-6	5
28	64	29	5	9	5	11	-10	7
28	66	27	1	2	1	5	-4	3
29	61	22	2	5	3	12	-9	7
29	62	27	5	10	6	8	-7	5
29	65	21	3	8	4	12	-8	7
29	65	22	4	10	5	10	-7	6
29	66	27	1	2	1	11	-9	7
20	60	30						
20	60	35	7	10	7	7	-7	3
20	63	31	3	5	3	7	-6	3
20	67	34	4	6	3	9	-8	4
21	62	32	5	8	5	11	-10	5

φ_1	Φ	φ_2	H	K	L	U	V	W
20	60	30						
21	65	39	4	5	3	10	-11	5
21	66	36	8	11	6	2	-2	1
21	68	37	3	4	2	2	-2	1
22	61	34	2	3	2	2	-2	1
22	61	39	9	11	8	9	-11	5
22	64	35	7	10	6	2	-2	1
22	65	36	5	7	4	2	-2	1
22	67	32	5	8	4	8	-7	4
22	67	34	4	6	3	12	-11	6
23	63	30	7	12	7	8	-7	4
23	63	31	3	5	3	10	-9	5
23	63	37	6	8	5	11	-12	6
23	65	39	4	5	3	7	-8	4
23	67	30	7	12	6	6	-5	3
24	60	30	7	12	8	12	-11	6
24	60	36	5	7	5	9	-10	5
24	60	36	8	11	8	7	-8	4
24	64	39	9	11	7	5	-6	3
24	65	37	9	12	7	10	-11	6
24	69	32	7	11	5	9	-8	5
24	69	39	8	10	5	10	-11	6
25	62	32	7	11	7	7	-7	4
25	64	30	4	7	4	9	-8	5
25	64	34	8	12	7	7	-7	4
25	65	34	6	9	5	5	-5	3
25	66	38	7	9	5	8	-9	5
25	67	32	5	8	4	12	-11	7
25	68	35	7	10	5	5	-5	3
26	62	32	5	8	5	5	-5	3
26	62	38	7	9	6	9	-11	6
26	63	36	8	11	7	8	-9	5
26	63	37	6	8	5	6	-7	4
26	67	31	6	10	5	10	-9	6
26	67	39	9	11	6	10	-12	7
27	60	35	7	10	7	6	-7	4
27	61	34	2	3	2	11	-12	7
27	68	37	3	4	2	10	-11	7
28	60	30	7	12	8	8	-8	5
28	61	34	2	3	2	9	-10	6
28	65	32	7	11	6	3	-3	2
28	65	39	4	5	3	4	-5	3
28	67	34	4	6	3	3	-3	2
29	60	36	5	7	5	4	-5	3
29	61	34	2	3	2	7	-8	5
29	61	39	9	11	8	5	-7	4
29	62	37	9	12	8	4	-5	3
29	63	31	3	5	3	3	-3	2
29	64	34	8	12	7	11	-12	8

φ_1	Φ	φ_2	H	K	L	U	V	W
20	60	30						
29	67	30	7	12	6	12	-11	8
29	68	37	3	4	2	8	-9	6
29	69	39	8	10	5	5	-6	4
20	60	40						
20	61	41	7	8	6	4	-5	2
20	63	40	5	6	4	6	-7	3
20	63	42	9	10	7	4	-5	2
20	66	40	10	12	7	8	-9	4
20	67	42	8	9	5	6	-7	3
20	67	45	5	5	3	9	-12	5
20	69	41	7	8	4	8	-9	4
20	69	49	8	7	4	8	-12	5
21	61	43	11	12	9	9	-12	5
21	62	45	4	4	3	7	-10	4
21	65	45	3	3	2	5	-7	3
22	61	45	5	5	4	8	-12	5
22	65	41	7	8	5	9	-11	5
22	67	45	5	5	3	8	-11	5
22	68	48	11	10	6	6	-9	4
22	69	40	5	6	3	9	-10	5
23	62	45	4	4	3	6	-9	4
23	67	41	6	7	4	5	-6	3
23	67	48	9	8	5	7	-11	5
23	68	42	10	11	6	8	-10	5
24	60	42	8	9	7	8	-11	5
24	61	45	5	5	4	7	-11	5
24	63	40	5	6	4	8	-10	5
24	65	42	10	11	7	3	-4	2
24	67	45	5	5	3	7	-10	5
25	60	40	10	12	9	3	-4	2
25	62	41	6	7	5	3	-4	2
25	62	42	10	11	8	7	-10	5
25	64	42	8	9	6	3	-4	2
25	65	45	3	3	2	4	-6	3
25	66	40	10	12	7	9	-11	6
25	67	47	12	11	7	5	-8	4
25	69	40	5	6	3	6	-7	4
25	69	49	8	7	4	5	-8	4
26	62	45	4	4	3	5	-8	4
26	64	47	12	11	8	7	-12	6
26	66	48	10	9	6	6	-10	5
26	67	42	8	9	5	7	-9	5
27	61	45	5	5	4	6	-10	5
27	67	45	5	5	3	6	-9	5
27	68	42	10	11	6	9	-12	7
27	69	41	7	8	4	4	-5	3

φ_1	Φ	φ_2	H	K	L	U	V	W
20	60	40						
28	60	48	9	8	7	1	-2	1
28	61	49	8	7	6	1	-2	1
28	62	49	7	6	5	1	-2	1
28	65	45	3	3	2	7	-11	6
28	67	49	7	6	4	6	-11	6
28	68	45	7	7	4	8	-12	7
28	69	40	5	6	3	9	-11	7
29	61	41	7	8	6	6	-9	5
29	64	43	11	12	8	8	-12	7
29	65	48	11	10	7	5	-9	5
29	67	41	6	7	4	6	-8	5
29	67	48	9	8	5	4	-7	4
29	69	49	8	7	4	7	-12	7
20	60	50						
20	61	51	10	8	7	6	-11	4
20	64	56	12	8	7	5	-11	4
20	69	51	10	8	5	3	-5	2
21	60	54	7	5	5	5	-11	4
21	62	53	12	9	8	4	-8	3
21	65	54	7	5	4	4	-8	3
21	66	50	12	10	7	3	-5	2
21	67	51	11	9	6	3	-5	2
21	68	55	10	7	5	5	-10	4
22	63	53	8	6	5	5	-10	4
22	63	59	5	3	3	3	-8	3
22	66	54	11	8	6	6	-12	5
22	67	59	10	6	5	2	-5	2
23	61	56	3	2	2	2	-5	2
23	62	52	9	7	6	6	-12	5
23	67	56	8	5	4	5	-12	5
24	65	56	9	6	5	3	-7	3
24	67	56	6	4	3	4	-9	4
25	60	54	7	5	5	5	-12	5
25	63	54	11	8	7	4	-9	4
25	65	54	7	5	4	5	-11	5
25	69	50	6	5	3	7	-12	6
26	63	59	5	3	3	3	-9	4
26	66	52	9	7	5	1	-2	1
26	68	53	4	3	2	1	-2	1
27	61	56	3	2	2	4	-11	5
27	63	50	6	5	4	1	-2	1
27	64	51	11	9	7	1	-2	1
27	65	51	5	4	3	1	-2	1
27	67	56	6	4	3	5	-12	6
27	67	58	8	5	4	3	-8	4
28	62	53	12	9	8	5	-12	6

φ_1	Φ	φ_2	H	K	L	U	V	W
20	60	50						
28	64	55	10	7	6	4	-10	5
28	64	56	12	8	7	3	-8	4
28	69	50	6	5	3	5	-9	5
29	61	56	3	2	2	2	-6	3
29	63	59	5	3	3	3	-10	5
20	60	60						
20	60	61	9	5	6	1	-3	1
20	61	61	11	6	7	1	-3	1
20	65	67	12	5	6	3	-12	4
21	60	69	8	3	5	1	-6	2
21	62	67	12	5	7	2	-9	3
21	64	60	7	4	4	4	-11	4
21	68	61	11	6	5	3	-8	3
21	68	67	7	3	3	3	-11	4
22	61	63	8	4	5	3	-11	4
22	64	66	11	5	6	2	-8	3
23	61	68	5	2	3	2	-11	4
23	66	63	2	1	1	3	-10	4
24	62	67	7	3	4	1	-5	2
24	64	60	7	4	4	4	-12	5
24	65	68	10	4	5	1	-5	2
24	68	67	7	3	3	3	-12	5
25	60	66	11	5	7	1	-5	2
25	62	63	10	5	6	3	-12	5
25	65	69	8	3	4	2	-12	5
25	66	63	2	1	1	2	-7	3
26	66	63	2	1	1	3	-11	5
27	61	63	8	4	5	2	-9	4
27	61	68	5	2	3	1	-7	3
27	69	61	9	5	4	2	-6	3
28	67	60	12	7	6	2	-6	3
29	60	69	8	3	5	1	-11	5
29	65	69	8	3	4	1	-8	4
29	66	63	2	1	1	1	-4	2
20	60	70						
20	65	72	6	2	3	1	-6	2
20	68	76	12	3	5	1	-9	3
20	69	73	10	3	4	1	-6	2
21	64	76	4	1	2	1	-12	4
21	68	72	12	4	5	2	-11	4
22	60	79	5	1	3	0	-3	1
22	68	74	7	2	3	1	-8	3
23	62	72	9	3	5	1	-8	3
24	64	73	10	3	5	1	-10	4

φ_1	Φ	φ_2	H	K	L	U	V	W
20	60	70						
24	64	79	10	2	5	0	-5	2
25	61	72	12	4	7	1	-10	4
25	62	77	9	2	5	0	-5	2
25	66	75	11	3	5	1	-12	5
25	67	72	9	3	4	1	-7	3
27	60	76	12	3	7	0	-7	3
27	65	72	6	2	3	1	-9	4
28	69	73	10	3	4	1	-10	5
29	67	77	9	2	4	0	-2	1
29	68	72	12	4	5	1	-8	4
29	69	79	5	1	2	0	-2	1
20	60	80						
20	67	82	7	1	3	0	-3	1
21	60	85	12	1	7	-1	-9	3
21	62	80	11	2	6	0	-3	1
21	64	81	6	1	3	0	-3	1
21	64	85	12	1	6	-1	-12	4
23	61	84	9	1	5	-1	-11	4
24	64	84	10	1	5	-1	-10	4
24	66	80	11	2	5	0	-5	2
24	68	81	12	2	5	0	-5	2
25	64	83	8	1	4	-1	-12	5
26	61	85	11	1	6	-1	-7	3
28	60	85	12	1	7	-2	-11	5
28	61	82	7	1	4	-1	-9	4
29	67	85	12	1	5	-1	-8	4
29	68	84	10	1	4	-1	-10	5
20	60	90						
20	63	90	2	0	1	-1	-6	2
22	63	90	2	0	1	-2	-11	4
24	63	90	2	0	1	-1	-5	2
24	68	90	5	0	2	-2	-12	5
26	63	90	2	0	1	-2	-9	4
26	68	90	5	0	2	-2	-11	5
28	68	90	5	0	2	-2	-10	5
20	70	0						
21	74	8	1	7	2	8	-2	3
21	76	0	0	4	1	11	-1	4
22	72	0	0	3	1	8	-1	3
22	72	9	1	6	2	10	-3	4
22	75	5	1	11	3	5	-1	2
22	76	0	0	4	1	10	-1	4

φ_1	Φ	φ_2	H	K	L	U	V	W
20	70	0						
22	78	6	1	9	2	5	-1	2
23	72	6	1	9	3	12	-3	5
23	79	0	0	5	1	12	-1	5
24	70	7	1	8	3	7	-2	3
24	72	0	0	3	1	7	-1	3
25	74	8	1	7	2	11	-3	5
25	76	0	0	4	1	9	-1	4
25	79	0	0	5	1	11	-1	5
27	72	5	1	12	4	4	-1	2
27	72	9	1	6	2	6	-2	3
27	76	0	0	4	1	8	-1	4
27	76	7	1	8	2	4	-1	2
27	79	0	0	5	1	10	-1	5
27	79	6	1	10	2	10	-2	5
28	72	0	0	3	1	6	-1	3
29	76	5	1	12	3	9	-2	5
20	70	10						
20	72	14	3	12	4	8	-3	3
20	75	15	3	11	3	8	-3	3
20	79	11	1	5	1	11	-3	4
20	79	17	3	10	2	8	-3	3
21	75	16	2	7	2	5	-2	2
22	70	14	2	8	3	5	-2	2
22	74	17	3	10	3	7	-3	3
22	78	18	3	9	2	7	-3	3
23	70	10	2	11	4	12	-4	5
23	71	15	3	11	4	7	-3	3
23	77	18	4	12	3	9	-4	4
23	79	11	1	5	1	7	-2	3
24	74	11	2	10	3	9	-3	4
24	76	14	1	4	1	11	-4	5
25	72	18	1	3	1	2	-1	1
25	79	17	3	10	2	10	-4	5
26	72	14	3	12	4	12	-5	6
26	76	14	1	4	1	8	-3	4
26	78	13	2	9	2	6	-2	3
26	79	11	1	5	1	10	-3	5
27	78	18	3	9	2	11	-5	6
28	70	14	2	8	3	11	-5	6
28	74	11	2	10	3	11	-4	6
28	75	16	2	7	2	9	-4	5
29	75	10	2	11	3	9	-3	5
29	75	15	3	11	3	7	-3	4
29	79	17	3	10	2	12	-5	7

φ_1	Φ	φ_2	H	K	L	U	V	W
20	70	20						
20	70	27	5	10	4	12	-8	5
20	71	20	4	11	4	8	-4	3
20	72	24	5	11	4	5	-3	2
21	71	27	4	8	3	9	-6	4
21	73	24	4	9	3	12	-7	5
21	76	24	5	11	3	7	-4	3
21	77	29	6	11	3	9	-6	4
22	70	22	2	5	2	7	-4	3
22	75	23	3	7	2	9	-5	4
22	75	27	5	10	3	11	-7	5
22	79	24	4	9	2	11	-6	5
23	71	21	3	8	3	11	-6	5
23	79	29	5	9	2	6	-4	3
24	70	27	5	10	4	10	-7	5
24	73	27	3	6	2	6	-4	3
24	77	27	2	4	1	8	-5	4
25	70	22	2	5	2	10	-6	5
25	73	23	5	12	4	12	-7	6
25	76	20	4	11	3	2	-1	1
25	77	21	3	8	2	2	-1	1
25	77	27	2	4	1	11	-7	6
25	79	22	2	5	1	2	-1	1
26	70	27	6	12	5	11	-8	6
26	71	27	4	8	3	7	-5	4
26	74	29	5	9	3	12	-9	7
26	75	27	5	10	3	9	-6	5
27	72	24	5	11	4	12	-8	7
27	73	27	3	6	2	10	-7	6
27	74	22	4	10	3	7	-4	4
27	77	23	5	12	3	9	-5	5
27	79	24	4	9	2	7	-4	4
28	70	27	5	10	4	8	-6	5
28	71	20	4	11	4	7	-4	4
28	75	23	3	7	2	5	-3	3
29	71	21	3	8	3	5	-3	3
29	73	23	5	12	4	8	-5	5
29	74	29	5	9	3	9	-7	6
29	76	24	5	11	3	8	-5	5
20	70	30						
20	70	30	4	7	3	12	-9	5
20	71	31	3	5	2	9	-7	4
20	71	36	5	7	3	11	-10	5
20	73	39	4	5	2	2	-2	1
20	74	34	2	3	1	11	-9	5
20	74	39	9	11	4	2	-2	1

TAB. VI

FROM EULER ANGLES INTO HKL-UVW

φ_1	Φ	φ_2	H	K	L	U	V	W
20	70	30						
20	78	34	8	12	3	9	-7	4
21	71	38	7	9	4	2	-2	1
21	75	37	9	12	4	12	-11	6
21	79	37	3	4	1	8	-7	4
22	71	34	8	12	5	8	-7	4
22	72	35	7	10	4	10	-9	5
22	74	34	2	3	1	6	-5	3
22	78	32	5	8	2	4	-3	2
23	71	31	3	5	2	10	-8	5
23	74	30	7	12	4	4	-3	2
23	76	31	6	10	3	4	-3	2
23	79	37	3	4	1	9	-8	5
24	70	34	6	9	4	11	-10	6
24	71	31	3	5	2	11	-9	6
24	72	37	9	12	5	5	-5	3
24	74	34	2	3	1	7	-6	4
25	70	36	8	11	5	5	-5	3
25	76	30	4	7	2	11	-8	6
25	79	37	3	4	1	10	-9	6
26	71	31	3	5	2	12	-10	7
26	73	37	6	8	3	3	-3	2
26	74	34	2	3	1	8	-7	5
26	75	38	7	9	3	3	-3	2
26	77	32	7	11	3	5	-4	3
26	77	39	8	10	3	3	-3	2
26	78	39	9	11	3	3	-3	2
26	79	37	3	4	1	11	-10	7
27	71	36	5	7	3	3	-3	2
27	71	39	9	11	5	7	-8	5
27	76	30	4	7	2	8	-6	5
27	79	37	3	4	1	12	-11	8
28	73	39	4	5	2	11	-12	8
28	74	34	2	3	1	9	-8	6
28	78	34	8	12	3	6	-5	4
29	73	39	4	5	2	9	-10	7
29	74	34	2	3	1	10	-9	7
29	74	34	2	3	1	11	-10	8
29	74	36	8	11	4	4	-4	3
29	75	37	9	12	4	4	-4	3
29	78	30	7	12	3	12	-9	8
29	78	32	5	8	2	10	-8	7
20	70	40						
20	71	48	11	10	5	5	-7	3
20	74	45	5	5	2	9	-11	5
20	74	49	8	7	3	5	-7	3
20	76	40	5	6	2	2	-2	1
20	78	41	6	7	2	2	-2	1
20	79	41	7	8	2	2	-2	1
21	71	45	2	2	1	7	-9	4
21	72	41	6	7	3	11	-12	6
21	72	49	7	6	3	6	-9	4
21	73	47	12	11	5	8	-11	5
21	75	42	10	11	4	9	-10	5
21	78	49	7	6	2	8	-11	5
22	70	42	9	10	5	5	-6	3
22	72	42	8	9	4	7	-8	4
22	74	45	5	5	2	8	-10	5
22	76	47	12	11	4	3	-4	2
22	77	45	3	3	1	5	-6	3
22	79	49	8	7	2	3	-4	2
23	71	45	2	2	1	3	-4	2
23	73	48	10	9	4	7	-10	5
23	74	41	7	8	3	10	-11	6
23	76	43	11	12	4	8	-9	5
24	70	47	12	11	6	4	-6	3
24	72	41	6	7	3	8	-9	5
24	72	48	9	8	4	4	-6	3
24	77	42	9	10	3	11	-12	7
24	77	45	3	3	1	9	-11	6
25	71	45	2	2	1	8	-11	6
25	72	40	10	12	5	9	-10	6
25	73	42	9	10	4	6	-7	4
25	74	45	5	5	2	7	-9	5
25	76	42	8	9	3	9	-10	6
25	79	45	7	7	2	10	-12	7
26	72	42	8	9	4	10	-12	7
26	73	45	7	7	3	9	-12	7
26	74	41	7	8	3	7	-6	5
26	74	49	8	7	3	6	-9	5
26	77	45	3	3	1	4	-5	3
26	79	40	10	12	3	3	-3	2
26	79	42	10	11	3	10	-11	7
26	79	48	11	10	3	9	-12	7
27	71	42	10	11	5	4	-5	3
27	71	45	2	2	1	5	-7	4
27	73	48	10	9	4	8	-12	7
27	77	42	9	10	3	8	-9	6
27	79	45	7	7	2	9	-11	7
28	70	43	11	12	6	6	-8	5
28	71	45	2	2	1	7	-10	6
28	74	45	5	5	2	6	-6	5
28	75	42	10	11	4	5	-6	4
29	70	48	10	9	5	3	-5	3
29	72	41	6	7	3	5	-6	4
29	72	49	7	6	3	3	-5	3
29	73	45	7	7	3	8	-11	7
29	74	49	8	7	3	7	-11	7
29	75	45	8	8	3	9	-12	8
29	76	40	5	6	2	10	-11	8
29	76	42	6	9	3	6	-7	5
29	77	45	3	3	1	7	-9	6
29	79	45	7	7	2	6	-10	7
20	70	50						
20	70	56	9	6	4	4	-8	3
20	71	52	9	7	4	3	-5	2
20	73	53	8	6	3	3	-5	2
20	74	51	11	9	4	8	-12	5
21	71	56	12	6	5	5	-10	4
21	74	54	11	8	4	4	-7	3
21	76	55	10	7	3	4	-7	3
21	77	51	10	8	3	6	-9	4
21	79	53	4	3	1	7	-11	5
22	72	53	12	9	5	4	-7	3
22	78	56	12	8	3	5	-9	4
23	71	59	5	3	2	3	-7	3
23	76	50	6	5	2	4	-6	3
24	72	50	12	10	5	5	-8	4
24	72	58	8	5	3	5	-11	5
24	73	51	5	4	2	6	-10	5
24	75	53	12	9	4	7	-12	6
25	72	55	10	7	4	1	-2	1
25	74	56	3	2	1	1	-2	1
25	77	51	10	8	3	7	-11	6
25	77	58	11	7	3	1	-2	1
25	78	51	11	9	3	6	-9	5
25	78	58	8	5	2	1	-2	1
26	70	54	11	8	5	1	-2	1
26	71	54	7	5	3	1	-2	1
26	71	59	5	3	2	4	-10	5
27	73	53	8	6	3	6	-11	6
27	77	54	7	5	2	5	-9	5
27	78	54	11	8	3	4	-7	4
28	73	51	5	4	2	4	-7	4
28	74	51	11	9	4	3	-5	3
28	75	52	9	7	3	7	-12	7
28	79	53	4	3	1	3	-5	3
29	71	56	12	8	5	3	-7	4
29	73	58	11	7	4	3	-7	4
29	76	50	6	5	2	5	-8	5
29	76	59	10	6	3	3	-7	4
20	70	60						
20	71	63	8	4	3	4	-11	4
20	73	63	6	3	2	3	-8	3
21	70	68	5	2	2	2	-6	3
21	72	61	11	6	4	2	-5	2
21	77	63	4	2	1	2	-5	2
22	70	60	7	4	3	2	-5	2
22	74	61	9	5	3	5	-12	5
22	77	61	11	6	3	3	-7	3
23	70	63	10	5	4	4	-12	5
23	76	60	7	4	2	4	-9	4
23	79	61	9	5	2	5	-11	5
24	71	63	8	4	3	3	-9	4
24	75	63	10	5	3	4	-11	5
24	76	66	11	5	3	3	-9	4
25	70	68	5	2	2	2	-9	4
26	74	60	12	7	4	5	-12	6
26	77	63	4	2	1	3	-8	4
26	79	66	9	4	2	2	-6	3
26	79	68	5	2	1	3	-10	5
27	70	60	12	7	5	3	-8	4
27	70	63	10	5	4	3	-10	5
27	73	63	6	3	2	2	-6	3
27	73	67	12	5	4	1	-4	2
27	74	68	10	4	3	1	-4	2
27	77	69	8	3	2	1	-4	2
28	70	68	5	2	2	2	-10	5
28	77	63	4	2	1	4	-11	6
28	78	60	12	7	3	4	-9	5
28	79	61	9	5	2	3	-7	4
29	75	63	10	5	3	3	-9	5
29	75	67	7	3	2	3	-11	6
29	79	68	5	2	1	2	-7	4
20	70	70						
20	75	74	7	2	2	2	-11	4
20	78	72	9	3	2	2	-8	3
21	74	79	10	2	3	1	-11	4
21	79	79	5	1	1	1	-8	3
22	72	76	12	3	4	1	-8	3
23	72	72	3	1	1	1	-5	2
23	75	74	7	2	2	2	-12	5
23	76	70	11	4	3	3	-12	5
24	72	77	9	2	3	1	-12	5
24	76	76	4	1	1	1	-7	3
24	77	72	12	4	3	2	-9	4
24	79	79	5	1	1	1	-9	4

φ1 Φ φ2	H	K	L	U	V	W
20 70 70						
25 71 75	11	3	4	1	-9	4
25 72 72	3	1	1	2	-11	5
27 76 76	4	1	1	1	-8	4
27 79 73	10	3	2	2	-10	5
27 79 79	5	1	1	1	-10	5
28 72 72	3	1	1	1	-6	3
28 72 76	12	3	4	1	-12	6
29 79 79	5	1	1	1	-11	6
20 70 80						
20 70 83	8	1	3	0	-3	1
27 76 83	8	1	2	0	-2	1
27 78 84	9	1	2	0	-2	1
27 79 84	10	1	2	0	-2	1
28 70 80	11	2	4	0	-2	1
28 72 81	6	1	2	0	-2	1
28 72 85	12	1	4	-1	-12	6
28 74 82	7	1	2	0	-2	1
20 70 90						
21 76 90	4	0	1	-1	-11	4
22 72 90	3	0	1	-1	-8	3
22 76 90	4	0	1	-1	-10	4
23 79 90	5	0	1	-1	-12	5
24 72 90	3	0	1	-1	-7	3
25 76 90	4	0	1	-1	-9	4
25 79 90	5	0	1	-1	-11	5
27 76 90	4	0	1	-1	-8	4
27 79 90	5	0	1	-1	-10	5
28 72 90	3	0	1	-1	-6	3
20 80 0						
20 85 5	1	11	1	8	-1	3
21 85 9	2	12	1	5	-1	2
22 82 8	1	7	1	5	-1	2
23 80 5	1	11	2	12	-2	5
23 84 6	1	10	1	7	-1	3
24 83 7	1	8	1	11	-2	5
26 81 9	1	6	1	4	-1	2
26 84 6	1	9	1	6	-1	3
26 85 5	1	12	1	8	-1	4
27 81 0	0	6	1	12	-1	6
27 81 5	1	12	2	6	-1	3
29 81 0	0	6	1	11	-1	6
29 82 8	1	7	1	9	-2	5

φ1 Φ φ2	H	K	L	U	V	W
20 80 10						
20 80 10	2	11	2	8	-2	3
21 81 18	2	6	1	5	-2	2
21 83 14	2	6	1	10	-3	4
22 85 14	3	12	1	7	-2	3
24 84 13	2	9	1	11	-3	5
24 85 10	2	11	1	9	-2	4
25 81 18	2	6	1	12	-5	6
25 85 18	4	12	1	8	-3	4
26 81 14	3	12	2	6	-2	3
26 84 11	2	10	1	4	-1	2
27 82 16	2	7	1	11	-4	6
28 81 18	2	6	1	7	-3	4
29 80 10	2	11	2	7	-2	4
29 84 13	2	9	1	7	-2	4
29 84 18	3	9	1	5	-2	3
20 80 20						
20 80 27	5	10	2	12	-7	5
20 85 20	4	11	1	5	-2	2
20 85 29	6	11	1	12	-7	5
21 82 27	3	6	1	7	-4	3
21 84 27	4	8	1	9	-5	4
22 80 20	4	11	2	9	-4	4
22 83 21	3	8	1	7	-3	3
22 85 22	4	10	1	9	-4	4
22 85 27	5	10	1	11	-6	5
23 86 23	5	12	1	11	-5	5
24 80 27	5	10	2	10	-6	5
24 81 23	5	12	2	2	-1	1
24 82 27	3	6	1	12	-7	6
24 83 23	3	7	1	2	-1	1
24 84 24	4	9	1	2	-1	1
24 85 24	5	11	1	2	-1	1
26 81 24	5	11	2	9	-5	5
26 85 27	5	10	1	9	-5	5
26 86 27	6	12	1	11	-6	6
27 83 21	3	8	1	9	-4	5
27 84 27	4	8	1	7	-4	4
27 85 22	4	10	1	11	-5	6
27 85 29	6	11	1	5	-3	3
28 80 27	5	10	2	8	-5	5
28 82 27	3	6	1	5	-3	3
28 84 29	5	9	1	8	-5	5
28 85 20	4	11	1	12	-5	7
29 81 29	6	11	2	3	-2	2

φ1 Φ φ2	H	K	L	U	V	W
20 80 30						
21 81 39	4	5	1	10	-9	5
21 82 37	9	12	2	6	-5	3
21 83 30	4	7	1	11	-7	5
21 85 38	7	9	1	6	-5	3
22 81 35	7	10	2	10	-8	5
22 81 39	4	5	1	11	-10	6
22 82 34	4	6	1	4	-3	2
22 85 35	7	10	1	4	-3	2
23 80 31	3	5	1	10	-7	5
23 82 30	7	12	2	6	-4	3
23 84 37	6	8	1	11	-9	6
23 86 30	7	12	1	8	-5	4
24 81 39	4	5	1	12	-11	7
24 82 36	8	11	2	12	-10	7
24 85 34	6	9	1	11	-8	6
24 86 36	8	11	1	9	-7	5
24 86 39	8	10	1	7	-6	4
25 80 31	3	5	1	7	-5	4
25 86 34	6	12	1	7	-5	4
25 86 37	9	12	1	5	-4	3
25 86 39	9	11	1	8	-7	5
26 80 34	6	9	2	5	-4	3
26 84 32	5	8	1	10	-7	6
27 81 32	7	11	2	8	-6	5
27 84 37	6	8	1	6	-5	4
28 80 31	3	5	1	11	-8	7
28 81 35	7	10	2	6	-5	4
28 82 34	4	6	1	9	-7	6
28 86 30	7	12	1	11	-7	7
29 82 36	6	11	2	7	-6	5
29 83 30	4	7	1	3	-2	2
29 85 31	6	10	1	3	-2	2
29 85 38	7	9	1	7	-6	5
20 80 40						
20 81 42	8	9	2	2	-2	1
20 82 42	9	10	2	2	-2	1
20 82 42	10	11	2	2	-2	1
20 83 40	5	6	1	12	-11	6
20 83 43	11	12	2	2	-2	1
20 83 45	6	6	1	11	-12	6
21 80 45	4	4	1	7	-8	4
21 82 45	5	5	1	9	-10	5
21 83 47	12	11	2	5	-6	3
21 85 48	9	8	1	5	-6	3
21 86 40	10	12	1	8	-7	4

φ1 Φ φ2	H	K	L	U	V	W
20 80 40						
21 86 47	12	11	1	7	-8	4
22 82 48	10	9	2	8	-10	5
22 83 45	6	6	1	10	-11	6
23 82 45	5	5	1	8	-9	5
23 84 45	7	7	1	11	-12	7
24 80 45	4	4	1	6	-7	4
24 80 47	12	11	3	7	-9	5
24 83 45	6	6	1	9	-10	6
24 84 49	7	6	1	7	-9	5
24 86 48	11	10	1	6	-7	4
25 82 45	5	5	1	7	-8	5
25 82 48	11	10	2	4	-5	3
25 84 45	7	7	1	10	-11	7
25 85 49	8	7	1	4	-5	3
26 85 45	8	8	1	11	-12	8
26 85 46	9	8	1	9	-11	7
27 81 48	9	8	2	6	-8	5
27 83 45	6	6	1	8	-9	6
27 86 40	10	12	1	9	-8	6
27 86 48	10	9	1	5	-6	4
28 80 43	11	12	3	9	-10	7
28 80 45	4	4	1	5	-6	4
28 84 45	7	7	1	9	-10	7
28 85 45	8	8	1	10	-11	8
29 80 47	12	11	3	9	-12	8
29 82 45	5	5	1	6	-7	5
29 83 47	12	11	2	8	-10	7
29 86 45	9	9	1	11	-12	9
29 86 47	12	11	1	6	-7	5
20 80 50						
20 85 52	9	7	1	6	-11	5
21 80 56	9	6	2	4	-7	3
21 81 58	11	7	2	5	-9	4
21 84 56	6	5	1	4	-7	3
21 85 59	10	6	1	5	-9	4
22 83 50	6	5	1	3	-4	2
22 86 51	10	8	1	3	-4	2
23 80 52	9	7	2	4	-6	3
23 81 51	5	4	1	7	-10	5
23 82 53	12	9	2	4	-6	3
23 86 56	12	8	1	5	-8	4
24 80 59	5	3	1	1	-2	1
24 81 55	10	7	2	6	-10	5
24 82 56	6	4	1	7	-12	6
25 82 51	11	9	2	5	-7	4
25 83 54	7	5	1	7	-11	6

φ_1	Φ	φ_2	H	K	L	U	V	W
20	80	50						
25	85	52	9	7	1	5	-7	4
26	80	56	9	6	2	6	-11	6
26	84	58	8	5	1	5	-9	5
26	85	59	10	6	1	6	-11	6
26	86	53	12	9	1	7	-10	6
27	81	55	10	7	2	7	-12	7
27	82	56	6	4	1	4	-7	4
27	83	50	6	5	1	8	-11	7
27	85	56	9	6	1	3	-5	3
27	86	58	11	7	1	7	-12	7
28	83	54	7	5	1	5	-8	5
28	85	55	10	7	1	7	-11	7
28	86	50	12	10	1	7	-9	6
29	80	52	9	7	2	7	-11	7
29	81	51	5	4	1	2	-3	2
29	84	53	8	6	1	2	-3	2
29	84	58	8	5	1	6	-11	7
29	86	54	11	8	1	2	-3	2
20	80	60						
20	83	67	7	3	1	3	-8	3
20	84	66	9	4	1	2	-5	2
21	80	63	10	5	2	5	-12	5
21	85	66	11	5	1	5	-12	5
22	82	63	6	3	1	3	-7	3
22	84	63	8	4	1	4	-9	4
22	86	60	12	7	1	6	-11	5
23	83	67	7	3	1	4	-11	5
23	85	63	10	5	1	5	-11	5
24	82	60	12	7	2	1	-2	1
24	83	60	7	4	1	1	-2	1
24	84	61	9	5	1	1	-2	1
24	85	61	11	6	1	1	-2	1
25	80	63	10	5	2	4	-10	5
25	82	63	6	3	1	5	-12	6
25	86	67	12	5	1	3	-8	4
26	81	67	12	5	2	2	-6	3
26	86	63	12	6	1	5	-11	6
27	81	61	11	6	2	4	-9	5
27	85	63	10	5	1	4	-9	5
28	84	63	8	4	1	3	-7	4
29	82	63	6	3	1	2	-5	3
29	85	66	11	5	1	2	-5	3
20	80	70						
20	81	72	6	2	1	3	-11	4

φ_1	Φ	φ_2	H	K	L	U	V	W
20	80	70						
20	84	77	9	2	1	2	-11	4
21	85	72	12	4	1	3	-10	4
22	80	75	11	3	2	1	-5	2
22	83	76	8	2	1	1	-5	2
23	80	70	11	4	2	2	-7	3
23	84	72	9	3	1	2	-7	3
24	82	74	7	2	1	2	-9	4
26	81	72	6	2	1	1	-4	2
26	84	77	9	2	1	1	-6	3
26	85	73	10	3	1	1	-4	2
27	81	76	12	3	2	1	-6	3
28	83	76	8	2	1	2	-11	6
28	84	72	9	3	1	3	-11	6
29	85	72	12	4	1	2	-7	4
29	85	75	11	3	1	2	-9	5
20	80	80						
20	82	82	7	1	1	1	-11	4
22	81	81	6	1	1	1	-10	4
23	82	82	7	1	1	1	-12	5
23	85	80	11	2	1	1	-7	3
25	81	81	6	1	1	1	-11	5
26	85	81	12	2	1	1	-8	4
27	60	85	11	1	2	0	-2	1
27	81	81	6	1	1	1	-12	6
27	81	85	12	1	2	0	-2	1
20	80	90						
27	81	90	6	0	1	-1	-12	6
29	81	90	6	0	1	-1	-11	6
20	90	0						
20	90	0	0	1	0	11	0	4
20	90	5	1	11	0	11	-1	4
20	90	7	1	8	0	8	-1	3
21	90	0	0	1	0	8	0	3
22	90	0	0	1	0	5	0	2
22	90	6	1	10	0	10	-1	4
22	90	9	1	6	0	12	-2	5
23	90	0	0	1	0	12	0	5
23	90	0	0	1	0	7	0	3
23	90	5	1	12	0	12	-1	5
23	90	8	1	7	0	7	-1	3
24	90	0	0	1	0	11	0	5
24	90	0	0	1	0	9	0	4

φ_1	Φ	φ_2	H	K	L	U	V	W
20	90	0						
24	90	5	1	11	0	11	-1	5
24	90	6	1	9	0	9	-1	4
26	90	5	1	12	0	12	-1	6
26	90	6	1	10	0	10	-1	5
26	90	7	1	8	0	8	-1	4
26	90	9	1	6	0	6	-1	3
27	90	0	0	1	0	2	0	1
29	90	0	0	1	0	11	0	6
29	90	0	0	1	0	9	0	5
29	90	5	1	11	0	11	-1	6
29	90	6	1	9	0	9	-1	5
29	90	8	1	7	0	7	-1	4
20	90	10						
20	90	10	2	11	0	11	-2	4
20	90	14	1	4	0	8	-2	3
21	90	11	1	5	0	5	-1	2
21	90	17	3	10	0	10	-3	4
22	90	14	1	4	0	12	-3	5
22	90	16	2	7	0	7	-2	3
22	90	18	1	3	0	12	-4	5
23	90	13	2	9	0	9	-2	4
23	90	18	1	3	0	9	-3	4
24	90	10	2	11	0	11	-2	5
24	90	15	3	11	0	11	-3	5
25	90	18	1	3	0	6	-2	3
26	90	11	1	5	0	10	-2	5
26	90	14	1	4	0	4	-1	2
26	90	17	3	10	0	10	-3	5
28	90	10	2	11	0	11	-2	6
28	90	13	2	9	0	9	-2	5
28	90	15	3	11	0	11	-3	6
28	90	18	1	3	0	9	-3	5
29	90	16	2	7	0	7	-2	4
29	90	18	1	3	0	12	-4	7
20	90	20						
20	90	22	2	5	0	5	-2	2
20	90	27	1	2	0	10	-5	4
20	90	27	1	2	0	12	-6	5
21	90	23	5	12	0	12	-5	5
21	90	29	5	9	0	9	-5	4
22	90	23	3	7	0	7	-3	3
22	90	24	4	9	0	9	-4	4
22	90	24	5	11	0	11	-5	5
22	90	29	6	11	0	11	-6	5

φ_1	Φ	φ_2	H	K	L	U	V	W
20	90	20						
23	90	20	4	11	0	11	-4	5
24	90	27	1	2	0	2	-1	1
25	90	21	3	8	0	8	-3	4
25	90	22	2	5	0	10	-4	5
25	90	23	5	12	0	12	-5	6
26	90	24	5	11	0	11	-5	6
26	90	29	5	9	0	9	-5	5
26	90	29	6	11	0	11	-6	6
27	90	20	4	11	0	11	-4	6
27	90	24	4	9	0	9	-4	5
28	90	23	3	7	0	7	-3	4
28	90	23	5	12	0	12	-5	7
28	90	27	1	2	0	10	-5	6
28	90	27	1	2	0	12	-6	7
29	90	22	2	5	0	5	-2	3
29	90	27	1	2	0	8	-4	5
29	90	29	6	11	0	11	-6	7
20	90	30						
20	90	30	4	7	0	7	-4	3
20	90	30	7	12	0	12	-7	5
20	90	34	2	3	0	9	-6	4
20	90	36	8	11	0	11	-8	5
21	90	32	7	11	0	11	-7	5
21	90	39	4	5	0	10	-8	5
22	90	35	7	10	0	10	-7	5
22	90	37	3	4	0	4	-3	2
23	90	30	7	12	0	12	-7	6
23	90	31	3	5	0	10	-6	5
23	90	32	5	8	0	8	-5	4
23	90	34	2	3	0	6	-4	3
23	90	39	9	11	0	11	-9	6
24	90	36	8	11	0	11	-8	6
24	90	38	7	9	0	9	-7	5
25	90	32	7	11	0	11	-7	6
25	90	34	2	3	0	9	-6	5
25	90	36	5	7	0	7	-5	4
25	90	37	3	4	0	12	-9	7
25	90	39	4	5	0	5	-4	3
26	90	30	4	7	0	7	-4	4
26	90	34	2	3	0	12	-8	7
26	90	35	7	10	0	10	-7	6
26	90	39	9	11	0	11	-9	7
27	90	30	7	12	0	12	-7	7
27	90	31	3	5	0	5	-3	3
27	90	36	8	11	0	11	-8	7
27	90	37	3	4	0	8	-6	5

φ_1	Φ	φ_2	H	K	L	U	V	W
20	90	30						
28	90	32	5	8	0	8	-5	5
28	90	32	7	11	0	11	-7	7
28	90	37	3	4	0	12	-9	8
28	90	38	7	9	0	9	-7	6
29	90	34	2	3	0	3	-2	2
29	90	39	4	5	0	10	-8	7
29	90	39	9	11	0	11	-9	8
20	90	40						
20	90	42	9	10	0	10	-9	5
20	90	43	11	12	0	12	-11	6
20	90	47	12	11	0	11	-12	6
20	90	48	10	9	0	9	-10	5
21	90	40	5	6	0	6	-5	3
21	90	41	7	8	0	8	-7	4
21	90	45	1	1	0	11	-11	6
21	90	45	1	1	0	9	-9	5
21	90	49	8	7	0	7	-8	4
22	90	42	10	11	0	11	-10	6
22	90	45	1	1	0	12	-12	7
22	90	45	1	1	0	7	-7	4
22	90	48	11	10	0	10	-11	6
23	90	41	6	7	0	7	-6	4
23	90	42	8	9	0	9	-8	5
23	90	43	11	12	0	12	-11	7
23	90	45	1	1	0	5	-5	3
23	90	47	12	11	0	11	-12	7
23	90	48	9	8	0	8	-9	5
23	90	49	7	6	0	6	-7	4
24	90	40	5	6	0	12	-10	7
24	90	42	9	10	0	10	-9	6
24	90	45	1	1	0	11	-11	7
24	90	45	1	1	0	8	-8	5
24	90	48	10	9	0	9	-10	6
25	90	41	7	8	0	8	-7	5
25	90	42	10	11	0	11	-10	7
25	90	45	1	1	0	3	-3	2
25	90	48	11	10	0	10	-11	7
25	90	49	8	7	0	7	-8	5
26	90	42	8	9	0	9	-8	6
26	90	43	11	12	0	12	-11	8
26	90	45	1	1	0	10	-10	7
26	90	47	12	11	0	11	-12	8
26	90	48	9	8	0	8	-9	6
27	90	40	5	6	0	6	-5	4
27	90	42	9	10	0	10	-9	7
27	90	45	1	1	0	7	-7	5
20	90	40						
27	90	45	1	1	0	11	-11	6
27	90	48	10	9	0	9	-10	7
28	90	41	6	7	0	7	-6	5
28	90	42	10	11	0	11	-10	8
28	90	45	1	1	0	4	-4	3
28	90	48	11	10	0	10	-11	8
28	90	49	7	6	0	6	-7	5
29	90	41	7	8	0	8	-7	6
29	90	43	11	12	0	12	-11	9
29	90	45	1	1	0	5	-5	4
29	90	45	1	1	0	9	-9	7
29	90	47	12	11	0	11	-12	9
29	90	49	8	7	0	7	-8	6
20	90	50						
20	90	54	11	8	0	8	-11	5
20	90	56	3	2	0	6	-9	4
21	90	50	6	5	0	5	-6	3
21	90	51	5	4	0	8	-10	5
21	90	58	11	7	0	7	-11	5
22	90	53	4	3	0	3	-4	2
22	90	55	10	7	0	7	-10	5
23	90	51	11	9	0	9	-11	6
23	90	56	3	2	0	4	-6	3
23	90	58	8	5	0	5	-8	4
23	90	59	5	3	0	6	-10	5
24	90	50	6	5	0	10	-12	7
24	90	52	9	7	0	7	-9	5
24	90	54	11	8	0	8	-11	6
25	90	51	5	4	0	4	-5	3
25	90	53	4	3	0	9	-12	7
25	90	54	7	5	0	5	-7	4
25	90	56	3	2	0	6	-9	5
25	90	58	11	7	0	7	-11	6
26	90	51	11	9	0	9	-11	7
26	90	55	10	7	0	7	-10	6
26	90	56	3	2	0	8	-12	7
27	90	50	6	5	0	5	-6	4
27	90	53	4	3	0	6	-8	5
27	90	54	11	8	0	8	-11	7
27	90	59	5	3	0	3	-5	3
28	90	52	9	7	0	7	-9	6
28	90	53	4	3	0	9	-12	8
28	90	58	8	5	0	5	-6	5
28	90	58	11	7	0	7	-11	7
29	90	51	5	4	0	6	-10	7
29	90	51	11	9	0	9	-11	8
20	90	50						
29	90	56	3	2	0	2	-3	2
20	90	60						
20	90	60	7	4	0	4	-7	3
20	90	60	12	7	0	7	-12	5
20	90	63	2	1	0	6	-12	5
20	90	63	2	1	0	5	-10	4
20	90	68	5	2	0	2	-5	2
21	90	61	9	5	0	5	-9	4
21	90	67	12	5	0	5	-12	5
22	90	61	11	6	0	6	-11	5
22	90	66	9	4	0	4	-9	4
22	90	66	11	5	0	5	-11	5
22	90	67	7	3	0	3	-7	3
23	90	60	12	7	0	7	-12	6
24	90	63	2	1	0	1	-2	1
25	90	67	12	5	0	5	-12	6
25	90	68	5	2	0	4	-10	5
25	90	69	8	3	0	3	-8	4
26	90	60	7	4	0	4	-7	4
26	90	61	9	5	0	5	-9	5
26	90	61	11	6	0	6	-11	6
26	90	66	11	5	0	5	-11	6
27	90	60	12	7	0	7	-12	7
27	90	66	9	4	0	4	-9	5
28	90	63	2	1	0	5	-10	6
28	90	63	2	1	0	6	-12	7
28	90	67	7	3	0	3	-7	4
28	90	67	12	5	0	5	-12	7
29	90	61	11	6	0	6	-11	7
29	90	63	2	1	0	4	-8	5
29	90	68	5	2	0	2	-5	3
20	90	70						
20	90	76	4	1	0	2	-8	3
21	90	73	10	3	0	3	-10	4
21	90	79	5	1	0	1	-5	2
22	90	72	3	1	0	4	-12	5
22	90	74	7	2	0	2	-7	3
22	90	76	4	1	0	3	-12	5
23	90	70	11	4	0	4	-11	5
23	90	72	3	1	0	3	-9	4
23	90	77	9	2	0	2	-9	4
24	90	75	11	3	0	3	-11	5
25	90	72	3	1	0	2	-6	3
26	90	73	10	3	0	3	-10	5
20	90	70						
26	90	76	4	1	0	1	-4	2
26	90	79	5	1	0	2	-10	5
27	90	70	11	4	0	4	-11	6
28	90	72	3	1	0	3	-9	5
28	90	75	11	3	0	3	-11	6
28	90	77	9	2	0	2	-9	5
29	90	72	3	1	0	4	-12	7
29	90	74	7	2	0	2	-7	4
20	90	80						
20	90	80	11	2	0	2	-11	4
20	90	83	8	1	0	1	-8	3
20	90	85	11	1	0	1	-11	4
22	90	81	6	1	0	2	-12	5
22	90	84	10	1	0	1	-10	4
23	90	82	7	1	0	1	-7	3
23	90	85	12	1	0	1	-12	5
24	90	80	11	2	0	2	-11	5
24	90	84	9	1	0	1	-9	4
24	90	85	11	1	0	1	-11	5
26	90	81	6	1	0	1	-6	3
26	90	83	8	1	0	1	-8	4
26	90	84	10	1	0	1	-10	5
26	90	85	12	1	0	1	-12	6
26	90	80	11	2	0	2	-11	6
29	90	82	7	1	0	1	-7	4
29	90	84	9	1	0	1	-9	5
29	90	85	11	1	0	1	-11	6
20	90	90						
20	90	90	1	0	0	0	-11	4
21	90	90	1	0	0	0	-8	3
22	90	90	1	0	0	0	-5	2
23	90	90	1	0	0	0	-12	5
23	90	90	1	0	0	0	-7	3
24	90	90	1	0	0	0	-11	5
24	90	90	1	0	0	0	-9	4
27	90	90	1	0	0	0	-2	1
29	90	90	1	0	0	0	-11	6
29	90	90	1	0	0	0	-9	5
30	0	0						
31	8	0	0	1	7	12	-7	1
31	9	0	0	1	6	10	-6	1
33	8	0	0	1	7	11	-7	1

TAB. VI

FROM EULER ANGLES INTO HKL-UVW

φ1	Φ	φ2	H	K	L	U	V	W
30	0	0						
34	7	0	0	1	8	12	-8	1
34	9	0	0	1	6	9	-6	1
35	8	0	0	1	7	10	-7	1
36	7	0	0	1	8	11	-8	1
37	6	0	0	1	9	12	-9	1
37	9	0	0	1	6	8	-6	1
38	8	0	0	1	7	9	-7	1
39	6	0	0	1	9	11	-9	1
39	7	0	0	1	8	10	-8	1
30	0	30						
30	0	30	0	0	1	7	-12	0
30	0	30	0	0	1	3	-5	0
30	0	30	0	0	1	4	-7	0
30	0	30	0	0	1	5	-9	0
31	0	31	0	0	1	6	-11	0
32	0	32	0	0	1	1	-2	0
33	0	33	0	0	1	3	-7	0
33	0	33	0	0	1	4	-9	0
33	0	33	0	0	1	5	-11	0
34	0	34	0	0	1	5	-12	0
34	0	34	0	0	1	2	-5	0
35	0	35	0	0	1	3	-8	0
35	0	35	0	0	1	4	-11	0
36	0	36	0	0	1	1	-3	0
37	0	37	0	0	1	3	-11	0
37	0	37	0	0	1	2	-7	0
37	0	37	0	0	1	3	-10	0
38	0	38	0	0	1	1	-4	0
39	0	39	0	0	1	2	-9	0
39	0	39	0	0	1	1	-5	0
30	0	40						
31	9	45	1	1	9	3	-12	1
35	9	45	1	1	9	2	-11	1
36	8	45	1	1	10	2	-12	1
30	0	90						
31	8	90	1	0	7	-7	-12	1
31	9	90	1	0	6	-6	-10	1
33	8	90	1	0	7	-7	-11	1
34	7	90	1	0	8	-8	-12	1
34	9	90	1	0	6	-6	-9	1
35	8	90	1	0	7	-7	-10	1
36	7	90	1	0	8	-8	-11	1

φ1	Φ	φ2	H	K	L	U	V	W
30	0	90						
37	6	90	1	0	9	-9	-12	1
37	9	90	1	0	6	-6	-8	1
38	8	90	1	0	7	-7	-9	1
39	6	90	1	0	9	-9	-11	1
39	7	90	1	0	8	-8	-10	1
30	10	0						
30	11	0	0	1	5	9	-5	1
30	14	0	0	1	4	7	-4	1
30	18	0	0	1	3	11	-6	2
31	16	0	0	2	7	12	-7	2
32	18	0	0	1	3	5	-3	1
33	11	0	0	1	5	8	-5	1
33	16	0	0	2	7	11	-7	2
34	14	0	0	1	4	6	-4	1
35	18	0	0	1	3	9	-6	2
36	11	0	0	1	5	7	-5	1
36	16	0	0	2	7	10	-7	2
37	14	0	0	1	4	11	-8	2
38	13	0	0	2	9	12	-9	2
38	18	0	0	1	3	4	-3	1
39	16	0	0	2	7	9	-7	2
30	10	10						
32	19	14	1	4	12	4	-4	1
33	15	18	1	3	12	9	-11	2
34	16	18	1	3	11	4	-5	1
35	18	18	1	3	10	7	-9	2
36	19	18	1	3	9	3	-4	1
30	10	20						
30	11	27	1	2	12	6	-9	1
31	16	27	1	2	8	4	-6	1
32	11	27	1	2	11	5	-8	1
34	13	27	1	2	10	4	-7	1
34	18	27	1	2	7	3	-5	1
36	16	27	1	2	8	6	-11	2
37	11	27	1	2	12	4	-8	1
38	14	27	1	2	9	3	-6	1
30	10	30						
31	17	34	2	3	12	3	-6	1
36	18	34	2	3	11	2	-5	1

φ1	Φ	φ2	H	K	L	U	V	W
30	10	40						
30	10	45	1	1	8	3	-11	1
30	16	45	1	1	5	2	-7	1
31	19	45	1	1	4	3	-11	2
32	13	45	1	1	6	2	-8	1
32	17	45	2	2	9	3	-12	2
33	11	45	1	1	7	2	-9	1
34	10	45	1	1	8	2	-10	1
35	19	45	1	1	4	1	-5	1
36	17	45	2	2	9	2	-11	2
37	16	45	1	1	5	1	-6	1
38	11	45	1	1	7	1	-8	1
38	13	45	1	1	6	1	-7	1
39	10	45	1	1	8	1	-9	1
30	10	50						
35	17	56	3	2	12	0	-6	1
35	18	56	3	2	11	0	-11	2
30	10	60						
33	11	63	2	1	11	-1	-9	1
33	11	63	2	1	12	-1	-10	1
33	18	63	2	1	7	-1	-12	2
34	13	63	2	1	10	-1	-6	1
36	14	63	2	1	9	-1	-7	1
37	16	63	2	1	8	-1	-6	1
39	18	63	2	1	7	-1	-5	1
30	10	70						
35	18	72	3	1	10	-3	-11	2
38	15	72	3	1	12	-2	-6	1
38	19	72	3	1	9	-3	-9	2
30	10	90						
30	11	90	1	0	5	-5	-9	1
30	14	90	1	0	4	-4	-7	1
30	18	90	1	0	3	-6	-11	2
31	16	90	2	0	7	-7	-12	2
32	18	90	1	0	3	-3	-5	1
33	11	90	1	0	5	-5	-8	1
33	16	90	2	0	7	-7	-11	2
34	14	90	1	0	4	-4	-6	1
35	18	90	1	0	3	-6	-9	2
36	11	90	1	0	5	-5	-7	1

φ1	Φ	φ2	H	K	L	U	V	W
30	10	90						
36	16	90	2	0	7	-7	-10	2
37	14	90	1	0	4	-8	-11	2
38	13	90	2	0	9	-9	-12	2
38	18	90	1	0	3	-3	-4	1
39	16	90	2	0	7	-7	-9	2
30	20	0						
31	22	0	0	2	5	9	-5	2
31	27	0	0	1	2	11	-6	3
32	23	0	0	3	7	12	-7	3
32	29	9	1	6	11	9	-7	3
33	27	0	0	1	2	7	-4	2
33	27	9	1	6	12	6	-5	2
34	22	0	0	2	5	8	-5	2
34	27	0	0	1	2	10	-6	3
35	21	0	0	3	8	12	-8	3
35	23	0	0	3	7	11	-7	3
36	29	9	1	6	11	10	-9	4
37	23	0	0	3	7	10	-7	3
37	27	0	0	1	2	3	-2	1
36	21	0	0	3	8	11	-8	3
38	22	0	0	2	5	7	-5	2
39	24	0	0	4	9	12	-9	4
39	27	0	0	1	2	11	-8	4
39	27	9	1	6	12	12	-12	5
39	29	9	1	6	11	11	-11	5
30	20	10						
30	27	11	1	5	10	10	-8	3
31	27	14	1	4	8	12	-11	4
32	24	18	1	3	7	9	-10	3
33	21	14	1	4	11	11	-11	3
33	22	14	1	4	10	10	-10	3
33	25	14	1	4	9	3	-3	1
34	22	18	1	3	8	9	-11	3
34	27	14	1	4	6	8	-8	3
36	23	11	1	5	12	3	-3	1
36	25	11	1	5	11	11	-11	4
36	28	18	1	3	6	9	-11	4
37	21	14	1	4	11	6	-7	2
37	27	11	1	5	10	5	-5	2
38	22	18	1	3	8	5	-7	2
36	28	18	1	3	6	6	-8	3

TAB. VI

FROM EULER ANGLES INTO HKL-UVW

φ_1	Φ	φ_2	H	K	L	U	V	W
30	20	20						
31	20	27	1	2	6	6	-9	2
31	26	27	2	4	9	5	-7	2
32	28	22	2	5	10	5	-6	2
33	22	27	2	4	11	5	-8	2
33	24	27	1	2	5	7	-11	3
33	29	27	1	2	4	2	-3	1
36	24	22	2	5	12	7	-10	3
36	24	27	1	2	5	4	-7	2
37	26	22	2	5	11	2	-3	1
39	20	27	1	2	6	2	-4	1
39	24	27	1	2	5	5	-10	3
39	29	27	1	2	4	6	-11	4
30	20	30						
31	20	34	2	3	10	5	-10	2
31	23	37	3	4	12	4	-9	2
31	27	34	2	3	7	6	-11	3
32	24	34	2	3	8	2	-4	1
33	24	37	3	4	11	5	-12	3
37	23	37	3	4	12	4	-12	3
38	22	34	2	3	9	3	-8	2
38	28	31	3	5	11	4	-9	3
38	29	37	3	4	9	1	-3	1
30	20	40						
30	22	45	2	2	7	3	-10	2
30	28	45	3	3	8	4	-12	3
33	28	45	3	3	8	3	-11	3
33	29	45	2	2	5	2	-7	2
34	22	45	2	2	7	2	-9	2
34	25	45	1	1	3	1	-4	1
37	25	45	1	1	3	2	-11	3
37	28	45	3	3	8	2	-10	3
38	23	45	3	3	10	2	-12	3
39	29	45	2	2	5	1	-6	2
30	20	50						
32	27	56	3	2	7	1	-12	3
33	27	53	4	3	10	1	-8	2
34	26	59	5	3	12	0	-4	1
34	28	59	5	3	11	0	-11	3
35	20	56	3	2	10	0	-5	1
36	22	56	3	2	9	0	-9	2
36	24	56	3	2	8	0	-4	1

φ_1	Φ	φ_2	H	K	L	U	V	W
30	20	50						
37	27	56	3	2	7	0	-7	2
39	23	53	4	3	12	0	-4	1
39	24	53	4	3	11	0	-11	3
30	20	60						
30	29	63	2	1	4	0	-4	1
34	20	63	2	1	6	-1	-10	2
35	22	63	4	2	11	-1	-9	2
36	24	63	2	1	5	-1	-8	2
36	29	63	2	1	4	-1	-10	3
37	28	68	5	2	10	-2	-10	3
38	26	63	4	2	9	-1	-7	2
39	24	63	2	1	5	-2	-11	3
30	20	70						
31	27	76	4	1	8	-1	-4	1
31	28	72	3	1	6	-2	-12	3
32	21	76	4	1	11	-3	-10	2
32	22	72	3	1	8	-1	-5	1
32	23	79	5	1	12	-3	-9	2
35	24	72	3	1	7	-1	-4	1
36	27	79	5	1	10	-4	-10	3
37	25	76	4	1	9	-4	-11	3
38	23	79	5	1	12	-5	-11	3
39	22	72	3	1	8	-4	-12	3
30	20	80						
31	27	81	6	1	12	-4	-12	3
37	29	81	6	1	11	-4	-9	3
39	27	81	6	1	12	-3	-6	2
30	20	90						
31	22	90	2	0	5	-5	-9	2
31	27	90	1	0	2	-6	-11	3
32	23	90	3	0	7	-7	-12	3
33	27	90	1	0	2	-4	-7	2
34	22	90	2	0	5	-5	-6	2
34	27	90	1	0	2	-6	-10	3
35	21	90	3	0	8	-8	-12	3
35	23	90	3	0	7	-7	-11	3
37	23	90	3	0	7	-7	-10	3
37	27	90	1	0	2	-2	-3	1
38	21	90	3	0	6	-8	-11	3
38	22	90	2	0	5	-5	-7	2

φ_1	Φ	φ_2	H	K	L	U	V	W
30	20	90						
39	24	90	4	0	9	-9	-12	4
39	27	90	1	0	2	-6	-11	4
30	30	0						
30	31	0	0	3	5	10	-5	3
30	39	0	0	4	5	11	-5	4
31	34	0	0	2	3	6	-3	2
31	39	6	1	9	11	5	-3	2
32	37	0	0	3	4	8	-4	3
33	31	0	0	3	5	9	-5	3
33	34	0	0	2	3	11	-6	4
33	37	6	1	9	12	12	-8	5
33	39	0	0	4	5	10	-5	4
34	30	0	0	4	7	12	-7	4
34	36	7	1	8	11	7	-5	3
35	34	7	1	8	12	12	-9	5
35	35	8	1	7	10	9	-7	4
35	39	0	0	4	5	9	-5	4
36	30	0	0	4	7	11	-7	4
36	31	0	0	3	5	8	-5	3
36	34	0	0	2	3	5	-3	2
36	36	0	0	5	7	12	-7	5
36	37	0	0	3	4	7	-4	3
36	37	9	1	6	8	6	-5	3
38	32	0	0	5	8	12	-8	5
38	34	9	1	6	9	12	-11	6
38	36	0	0	5	7	11	-7	5
39	30	0	0	4	7	10	-7	4
39	34	7	1	8	12	8	-7	4
39	39	0	0	4	5	8	-5	4
30	30	10						
30	30	18	2	6	11	11	-11	4
30	31	16	2	7	12	11	-10	4
31	32	18	1	3	5	5	-5	2
31	35	18	2	6	9	9	-9	4
33	30	11	1	5	9	8	-7	3
33	31	16	2	7	12	12	-12	5
33	32	18	1	3	5	11	-12	5
33	38	18	1	3	4	2	-2	1
34	33	16	2	7	11	11	-11	5
34	36	16	2	7	10	2	-2	1
35	30	14	1	4	7	7	-7	3
35	32	18	1	3	5	6	-7	3
36	34	14	1	4	6	2	-2	1
36	36	11	1	5	7	10	-9	5

φ_1	Φ	φ_2	H	K	L	U	V	W
30	30	10						
36	39	16	2	7	9	9	-9	5
37	30	11	1	5	9	9	-9	4
37	30	18	2	6	11	4	-5	2
37	37	14	2	8	11	11	-11	6
38	30	14	1	4	7	8	-9	4
38	32	18	1	3	5	7	-9	4
38	33	11	1	5	6	2	-2	1
39	38	13	2	9	12	12	-12	7
39	38	18	1	3	4	9	-11	6
30	30	20						
30	37	27	1	2	3	4	-5	2
30	38	21	3	8	11	11	-11	5
31	34	22	2	5	6	9	-10	4
31	34	27	3	6	10	6	-8	3
32	33	27	2	4	7	8	-11	4
32	37	27	1	2	3	9	-12	5
32	38	22	2	5	7	10	-11	5
32	39	27	4	8	11	7	-9	4
33	34	22	2	5	6	10	-12	5
33	35	21	3	8	12	8	-9	4
33	37	23	3	7	10	9	-11	5
33	37	27	1	2	3	5	-7	3
35	37	27	1	2	3	6	-9	4
36	31	27	3	6	11	5	-8	3
36	38	21	3	8	11	5	-6	3
36	39	24	4	9	12	3	-4	2
37	31	22	2	5	9	7	-10	4
37	33	27	2	4	7	3	-5	2
37	37	27	1	2	3	7	-11	5
38	32	23	3	7	12	8	-12	5
38	38	22	2	5	7	3	-4	2
38	39	27	4	8	11	5	-8	4
39	34	27	3	6	10	4	-7	3
39	37	23	3	7	10	4	-6	3
30	30	30						
30	33	34	4	6	11	7	-12	4
30	35	39	4	5	9	3	-6	2
31	39	34	4	6	9	3	-5	2
31	39	39	4	5	8	4	-8	3
33	32	37	3	4	6	4	-9	3
33	33	39	4	5	10	5	-12	4
33	34	30	4	7	12	5	-8	3
33	36	31	3	5	8	3	-5	2
33	36	36	5	7	12	4	-8	3

φ_1	Φ	φ_2	H	K	L	U	V	W
30	30	30						
34	30	31	3	5	10	5	-9	3
34	31	34	2	3	6	3	-6	2
34	38	32	5	8	12	4	-7	3
35	36	34	2	3	5	5	-10	4
35	39	39	4	5	8	5	-12	5
36	36	30	4	7	11	4	-7	3
37	30	39	4	5	11	1	-3	1
37	36	37	3	4	7	2	-5	2
38	32	37	3	4	8	4	-11	4
39	30	31	3	5	10	5	-11	4
39	33	34	4	6	11	2	-5	2
39	34	30	4	7	12	6	-12	5
39	36	34	2	3	5	3	-7	3
30	30	40						
31	35	45	1	1	2	1	-3	1
31	39	45	4	4	7	4	-11	4
32	31	45	3	3	7	3	-10	3
34	38	40	5	6	10	2	-5	2
35	32	45	4	4	9	3	-12	4
35	35	40	5	6	11	3	-8	3
35	35	45	1	1	2	3	-11	4
35	39	45	4	4	7	3	-10	4
36	35	45	1	1	2	2	-8	3
37	31	45	3	3	7	2	-9	3
37	38	41	6	7	12	4	-12	5
37	38	45	5	5	9	3	-12	5
39	32	45	4	4	9	2	-11	4
39	35	45	1	1	2	1	-5	2
39	39	45	4	4	7	2	-9	4
30	30	50						
30	33	56	6	4	11	1	-7	2
30	39	56	6	4	9	1	-6	2
32	33	51	5	4	10	2	-10	3
32	35	50	6	5	11	2	-9	3
33	36	53	4	3	7	1	-6	2
33	36	56	3	2	5	1	-9	3
34	38	54	7	5	11	1	-8	3
35	30	59	5	3	10	0	-10	3
35	33	50	6	5	12	1	-6	2
36	32	53	4	3	8	1	-12	4
36	33	59	5	3	9	0	-3	1
36	36	54	7	5	12	1	-11	4
37	35	51	5	4	9	1	-8	3
37	36	59	5	3	8	0	-8	3

φ_1	Φ	φ_2	H	K	L	U	V	W
30	30	50						
38	31	56	3	2	6	0	-3	1
38	38	58	8	5	12	0	-12	5
39	33	56	6	4	11	0	-11	4
39	36	56	3	2	5	0	-5	2
30	30	60						
30	31	63	6	3	11	0	-11	3
30	39	66	9	4	12	0	-3	1
31	31	68	5	2	9	-1	-11	3
31	33	63	4	2	7	0	-7	2
31	34	63	6	3	10	0	-10	3
31	38	69	8	3	11	-1	-12	4
32	37	63	2	1	3	0	-3	1
33	39	63	8	4	11	0	-11	4
34	37	67	7	3	10	-1	-11	4
35	34	60	7	4	12	0	-3	1
35	36	60	7	4	11	0	-11	4
35	38	68	5	2	7	-1	-8	3
36	33	63	4	2	7	-1	-12	4
36	39	60	7	4	10	0	-5	2
37	31	63	6	3	11	-1	-9	3
37	34	68	5	2	8	-2	-11	4
38	37	63	2	1	3	-1	-10	4
39	34	63	6	3	10	-1	-8	3
39	37	67	7	3	10	-2	-12	5
30	30	70						
30	31	74	7	2	12	-2	-11	3
30	34	76	4	1	6	-2	-10	3
30	38	72	3	1	4	-1	-9	3
31	32	72	3	1	5	-1	-7	2
31	38	77	9	2	12	-2	-9	3
32	39	74	7	2	9	-2	-11	4
33	34	76	4	1	6	-3	-12	4
33	35	72	6	2	9	-1	-6	2
34	33	79	5	1	8	-1	-3	1
35	32	72	3	1	5	-2	-9	3
36	30	76	4	1	7	-1	-3	1
36	36	74	7	2	10	-2	-8	3
36	38	72	3	1	4	-1	-5	2
37	37	76	8	2	11	-3	-10	4
37	38	77	9	2	12	-4	-12	5
37	39	74	7	2	9	-3	-12	5
38	32	72	3	1	5	-3	-11	4

φ_1	Φ	φ_2	H	K	L	U	V	W
30	30	80						
30	37	84	9	1	12	-1	-3	1
31	35	82	7	1	10	-1	-3	1
31	36	83	8	1	11	-1	-3	1
33	34	81	6	1	9	-1	-3	1
39	31	82	7	1	12	-4	-8	3
39	34	83	8	1	12	-6	-12	5
30	30	90						
30	31	90	3	0	5	-5	-10	3
30	39	90	4	0	5	-5	-11	4
31	34	90	2	0	3	-3	-6	2
32	37	90	3	0	4	-4	-8	3
33	31	90	3	0	5	-5	-9	3
33	34	90	2	0	3	-6	-11	4
33	39	90	4	0	5	-5	-10	4
34	30	90	4	0	7	-7	-12	4
35	39	90	4	0	5	-5	-9	4
36	30	90	4	0	7	-7	-11	4
36	31	90	3	0	5	-5	-8	3
36	34	90	2	0	3	-3	-5	2
36	36	90	5	0	7	-7	-12	5
36	37	90	3	0	4	-4	-7	3
38	32	90	5	0	8	-8	-12	5
38	36	90	5	0	7	-7	-11	5
39	30	90	4	0	7	-7	-10	4
39	39	90	4	0	5	-5	-8	4
30	40	0						
30	45	5	1	12	12	12	-6	5
31	42	6	1	10	11	7	-4	3
31	45	0	0	1	1	12	-5	5
31	45	0	0	1	1	7	-3	3
31	45	6	1	9	9	9	-5	4
32	42	7	1	8	9	11	-7	5
32	45	0	0	1	1	9	-4	4
32	45	5	1	11	11	11	-6	5
32	48	9	2	12	11	8	-5	4
33	40	0	0	5	6	12	-6	5
33	45	0	0	1	1	11	-5	5
33	45	7	1	8	8	8	-5	4
33	45	9	1	6	6	6	-4	3
34	41	9	1	6	7	4	-3	2
34	45	5	1	12	12	12	-7	6
34	45	6	1	10	10	10	-6	5
35	40	0	0	5	6	11	-6	5

φ_1	Φ	φ_2	H	K	L	U	V	W
30	40	0						
35	42	6	1	9	10	6	-4	3
35	45	0	0	1	1	2	-1	1
36	40	6	1	10	12	10	-7	5
36	45	5	1	11	11	11	-7	6
36	45	6	1	9	9	9	-6	5
36	45	8	1	7	7	7	-5	4
36	45	9	1	6	6	12	-9	7
37	41	8	1	7	6	9	-7	5
38	40	0	0	5	6	10	-6	5
38	41	0	0	6	7	12	-7	6
38	45	0	0	1	1	11	-6	6
38	45	0	0	1	1	9	-5	5
38	45	5	1	12	12	12	-8	7
38	45	6	1	10	10	10	-7	6
38	49	0	0	7	6	12	-6	7
39	42	7	1	8	9	5	-4	3
39	43	5	1	11	12	7	-5	4
39	45	0	0	1	1	7	-4	4
39	45	7	1	6	8	8	-6	5
39	48	9	2	12	11	9	-7	6
30	40	10						
30	45	10	2	11	11	11	-7	5
30	46	17	3	10	10	10	-8	5
31	40	11	1	5	6	7	-5	3
31	40	14	1	4	5	11	-9	5
31	42	18	2	6	7	12	-11	6
31	43	10	2	11	12	9	-6	4
31	43	13	2	9	10	11	-8	5
31	46	14	1	4	4	4	-3	2
31	49	13	2	9	8	6	-4	3
32	41	17	3	10	12	10	-9	5
32	46	11	1	5	5	10	-7	5
32	47	18	1	3	3	9	-8	5
32	49	15	3	11	10	9	-7	5
32	49	18	4	12	11	7	-6	4
33	41	18	3	9	11	11	-11	6
33	43	11	2	10	11	4	-3	2
33	46	15	3	11	11	11	-9	6
33	47	18	1	3	3	12	-11	7
34	40	11	1	5	6	10	-8	5
34	40	13	2	9	11	6	-5	3
34	40	14	1	4	5	8	-7	4
34	42	16	2	7	8	11	-10	6
34	42	18	2	6	7	7	-7	4
34	45	10	2	11	11	11	-8	6
34	46	13	2	9	9	9	-7	5

FROM EULER ANGLES INTO HKL-UVW

φ1	Φ	φ2	H	K	L	U	V	W
30	**40**	**10**						
34	46	16	2	7	7	7	-6	4
34	46	17	3	10	10	10	-9	6
34	49	11	2	10	9	7	-5	4
35	43	18	3	9	10	5	-5	3
35	46	14	1	4	4	12	-10	7
35	48	14	3	12	11	5	-4	3
36	41	17	3	10	12	12	-12	7
36	43	13	2	9	10	7	-6	4
36	46	11	1	5	5	5	-4	3
36	46	14	1	4	4	8	-7	5
36	46	15	3	11	11	11	-10	7
36	47	18	1	3	3	3	-3	2
37	42	16	2	7	8	8	-8	5
37	42	18	2	6	7	9	-10	6
37	44	17	3	10	11	11	-11	7
37	48	10	2	11	10	8	-6	5
37	49	18	4	12	11	11	-11	8
38	40	14	1	4	5	5	-5	3
38	44	15	3	11	12	3	-3	2
38	45	10	2	11	11	11	-9	7
38	46	14	1	4	4	12	-11	8
38	46	17	3	10	10	10	-10	7
39	42	14	2	8	9	3	-3	2
39	46	13	2	9	9	9	-8	6
39	46	15	3	11	11	11	-11	8
39	46	16	2	7	7	7	-7	5
39	49	13	2	9	8	7	-6	5
30	**40**	**20**						
30	46	29	5	9	10	8	-10	5
30	46	29	6	11	12	5	-6	3
30	48	24	4	9	9	9	-9	5
30	48	27	1	2	2	10	-11	6
31	41	21	3	8	10	2	-2	1
31	44	22	4	10	11	11	-11	6
31	44	27	3	6	7	5	-6	3
31	47	20	4	11	11	11	-10	6
31	47	23	3	7	7	7	-7	4
31	47	23	5	12	12	12	-12	7
31	48	27	1	2	2	8	-9	5
31	49	29	5	9	9	9	-11	6
32	42	24	4	9	11	7	-8	4
32	43	27	5	10	12	8	-10	5
32	44	21	3	8	9	9	-9	5
32	47	22	2	5	5	5	-5	3
32	48	24	5	11	11	11	-12	7
32	48	27	1	2	2	6	-7	4

φ1	Φ	φ2	H	K	L	U	V	W
30	**40**	**20**						
33	44	20	4	11	12	12	-12	7
33	48	24	4	9	9	9	-10	6
33	48	27	1	2	2	10	-12	7
34	40	23	3	7	9	5	-6	3
34	42	27	2	4	5	3	-4	2
34	43	29	5	9	11	7	-10	5
34	45	27	5	10	11	7	-9	5
34	47	20	4	11	11	11	-11	7
34	47	21	3	8	8	8	-8	5
34	49	29	5	9	9	9	-12	7
35	44	21	3	8	9	11	-12	7
35	47	23	3	7	7	7	-8	5
35	48	27	1	2	2	4	-5	3
36	44	27	3	6	7	8	-11	6
36	47	22	2	5	5	10	-11	7
36	49	20	4	11	10	10	-10	7
37	40	27	3	6	8	4	-6	3
37	41	29	5	9	12	6	-10	5
37	45	27	4	8	9	5	-7	4
37	46	29	6	11	12	8	-12	7
37	47	20	4	11	11	11	-12	8
37	48	24	4	9	9	9	-11	7
37	48	27	1	2	2	6	-8	5
38	43	27	5	10	12	6	-9	5
38	44	20	4	11	12	7	-8	5
38	47	21	3	8	8	8	-9	6
38	47	22	2	5	5	5	-6	4
38	48	27	1	2	2	8	-11	7
39	42	22	2	5	6	9	-12	7
39	42	27	2	4	5	7	-11	6
39	47	23	3	7	7	7	-9	6
30	**40**	**30**						
30	41	36	5	7	10	3	-5	2
31	42	30	4	7	9	8	-11	5
31	47	31	6	10	11	3	-4	2
31	49	36	8	11	12	5	-8	4
32	45	30	4	7	8	3	-4	2
32	46	32	5	8	9	7	-10	5
32	47	34	6	9	10	4	-6	3
32	48	37	6	8	9	7	-12	6
32	49	30	4	7	7	7	-9	5
33	40	37	3	4	6	6	-12	5
33	44	31	3	5	6	4	-6	3
33	46	34	4	6	7	5	-6	4
33	47	36	5	7	8	6	-10	5
34	45	34	6	9	11	7	-12	6

φ1	Φ	φ2	H	K	L	U	V	W
30	**40**	**30**						
34	47	39	4	5	6	1	-2	1
34	49	31	3	5	5	5	-7	4
35	42	39	4	5	7	5	-11	5
35	43	32	5	8	10	6	-10	5
35	45	37	3	4	5	1	-2	1
35	46	38	7	9	11	1	-2	1
35	47	32	7	11	12	7	-11	6
36	40	37	3	4	6	4	-9	4
36	42	30	4	7	9	5	-8	4
36	44	36	5	7	9	1	-2	1
36	48	35	7	10	11	5	-9	5
36	49	30	4	7	7	7	-10	6
37	46	34	4	6	7	6	-11	6
38	42	34	2	3	4	1	-2	1
38	46	32	5	8	9	4	-7	4
38	49	31	3	5	5	5	-8	5
38	49	36	8	11	12	6	-12	7
39	41	32	5	8	11	1	-2	1
39	42	37	6	8	11	5	-12	6
39	42	39	4	5	7	3	-8	4
39	44	31	3	5	6	5	-9	5
39	45	30	4	7	8	7	-12	7
39	49	30	4	7	7	7	-11	7
30	**40**	**40**						
30	43	41	6	7	10	5	-10	4
30	43	45	2	2	3	2	-5	2
30	45	42	8	9	12	6	-12	5
30	45	45	5	5	7	5	-12	5
30	47	45	3	3	4	3	-7	3
30	49	45	4	4	5	4	-9	4
31	40	45	3	3	5	3	-8	3
31	45	48	9	8	12	4	-12	5
33	44	41	7	8	11	4	-9	4
33	44	49	8	7	11	2	-7	3
33	48	40	5	6	7	1	-2	1
33	49	41	6	7	8	1	-2	1
34	41	40	5	6	9	3	-7	3
34	41	45	5	5	8	4	-12	5
34	42	49	8	7	12	3	-12	5
34	43	45	2	2	3	3	-9	4
34	45	45	5	5	7	4	-11	5
34	46	49	7	6	9	3	-11	5
34	49	45	4	4	5	3	-8	4
35	48	48	10	9	12	3	-10	5
36	40	45	3	3	5	2	-7	3
36	42	41	7	8	12	4	-11	5

φ1	Φ	φ2	H	K	L	U	V	W
30	**40**	**40**						
36	43	49	7	6	10	2	-9	4
36	46	41	6	7	9	5	-12	6
36	47	45	3	3	4	2	-6	3
37	41	45	5	5	8	3	-11	5
37	44	40	5	6	8	4	-10	5
37	45	42	8	9	12	3	-8	4
37	45	45	5	5	7	3	-10	5
37	47	49	8	7	10	1	-4	2
39	43	41	6	7	10	2	-6	3
39	47	41	7	8	10	4	-11	6
39	49	49	7	6	8	2	-9	5
30	**40**	**50**						
30	42	56	3	2	4	2	-11	4
30	47	58	11	7	12	1	-5	2
31	44	50	6	5	8	3	-10	4
31	46	56	6	4	7	1	-5	2
31	47	51	5	4	6	2	-7	3
32	40	59	5	3	7	1	-11	4
32	44	54	7	5	9	1	-5	2
32	49	52	9	7	10	3	-11	5
33	42	53	8	6	11	1	-5	2
33	46	58	8	5	9	1	-7	3
34	45	56	9	6	11	1	-7	3
34	47	59	10	6	11	1	-9	4
35	45	53	4	3	5	2	-11	5
35	49	51	10	8	11	1	-4	2
36	46	50	6	5	7	1	-4	2
36	49	59	5	3	5	1	-10	5
37	47	51	5	4	6	2	-10	5
38	40	59	5	3	7	0	-7	3
38	46	52	9	7	11	1	-6	3
38	46	56	6	4	7	1	-12	6
39	41	58	8	5	11	0	-11	5
39	44	50	6	5	8	1	-6	3
39	45	53	4	3	5	1	-8	4
39	45	55	10	7	12	1	-10	5
30	**40**	**60**						
30	47	68	5	2	5	0	-5	2
30	48	63	2	1	2	1	-10	4
30	49	60	12	7	12	2	-12	5
31	42	66	9	4	11	0	-11	4
31	44	67	7	3	8	0	-8	3
31	46	63	2	1	2	1	-12	5
31	49	60	7	4	7	1	-7	3

φ_1	Φ	φ_2	H	K	L	U	V	W
30	40	60						
32	45	66	9	4	10	0	-5	2
32	47	67	7	3	7	0	-7	3
32	47	67	12	5	12	0	-12	5
32	49	61	9	5	9	1	-9	4
33	40	63	6	3	8	0	-8	3
33	45	66	11	5	12	0	-12	5
33	48	66	9	4	9	0	-9	4
33	49	61	11	6	11	1	-11	5
34	42	63	4	2	5	0	-5	2
34	43	63	10	5	12	0	-12	5
34	48	66	11	5	11	0	-11	5
35	44	63	6	3	7	0	-7	3
35	45	63	8	4	9	0	-9	4
35	45	63	10	5	11	0	-11	5
36	41	61	9	5	12	0	-12	5
36	49	60	12	7	12	1	-12	6
37	42	60	7	4	9	0	-9	4
37	43	61	9	5	11	0	-11	5
37	44	67	7	3	8	-1	-11	5
37	47	67	12	5	12	-1	-12	6
37	47	68	5	2	5	-1	-10	5
37	48	63	2	1	2	0	-2	1
38	46	61	11	6	12	0	-2	1
38	47	69	8	3	8	-1	-8	4
39	45	60	7	4	8	0	-2	1
39	46	61	9	5	10	0	-2	1
30	40	70						
30	43	77	9	2	10	-2	-11	4
31	42	72	6	2	7	-1	-11	4
31	46	73	10	3	10	-1	-10	4
31	46	79	5	1	5	-1	-5	2
32	40	79	5	1	6	-2	-8	3
32	41	72	9	3	11	-1	-8	3
32	46	76	4	1	4	-2	-12	5
32	47	72	3	1	3	-1	-12	5
32	48	76	12	3	11	-1	-7	3
33	42	76	8	2	9	-1	-5	2
33	46	74	7	2	7	-1	-7	3
33	49	77	9	2	8	-2	-11	5
34	46	77	9	2	9	-2	-9	4
34	47	70	11	4	11	-1	-11	5
34	47	72	3	1	3	-1	-9	4
35	43	72	9	3	10	-1	-7	3
35	46	75	11	3	11	-2	-11	5
36	42	74	7	2	8	-2	-9	4
36	49	72	12	4	11	-1	-8	4

φ_1	Φ	φ_2	H	K	L	U	V	W
30	40	70						
36	49	79	10	2	9	-1	-4	2
37	40	76	4	1	5	-2	-7	3
37	40	79	5	1	6	-3	-9	4
37	46	76	4	1	4	-1	-4	2
37	46	79	5	1	5	-3	-10	5
38	46	73	10	3	10	-2	-10	5
38	47	72	3	1	3	-1	-6	3
39	43	77	9	2	10	-2	-6	3
39	44	73	10	3	11	-1	-4	2
30	40	80						
30	41	82	7	1	8	-3	-11	4
30	45	84	10	1	10	-3	-10	4
31	45	85	12	1	12	-4	-12	5
31	48	84	10	1	9	-2	-7	3
32	45	81	6	1	6	-3	-12	5
32	45	82	7	1	7	-2	-7	3
32	48	80	11	2	10	-2	-9	4
32	49	83	8	1	7	-3	-11	5
33	41	81	6	1	7	-3	-10	4
33	45	84	9	1	9	-3	-9	4
34	41	82	7	1	8	-4	-12	5
34	43	80	11	2	12	-2	-7	3
34	45	80	11	2	11	-3	-11	5
34	45	85	11	1	11	-4	-11	5
35	48	85	12	1	11	-3	-8	4
35	49	84	9	1	8	-2	-6	3
36	45	84	10	1	10	-4	-10	5
36	45	85	12	1	12	-5	-12	6
37	41	81	6	1	7	-4	-11	5
37	45	81	6	1	6	-2	-6	3
37	45	83	8	1	8	-3	-8	4
39	40	84	10	1	12	-1	-2	1
39	45	85	11	1	11	-5	-11	6
39	48	85	11	1	10	-3	-7	4
30	40	90						
31	45	90	1	0	1	-3	-7	3
31	45	90	1	0	1	-5	-12	5
32	45	90	1	0	1	-4	-9	4
33	40	90	5	0	6	-6	-12	5
33	45	90	1	0	1	-5	-11	5
35	40	90	5	0	6	-6	-11	5
35	45	90	1	0	1	-1	-2	1
38	40	90	5	0	6	-6	-10	5
38	41	90	6	0	7	-7	-12	6

φ_1	Φ	φ_2	H	K	L	U	V	W
30	40	90						
38	45	90	1	0	1	-5	-9	5
38	45	90	1	0	1	-6	-11	6
38	49	90	7	0	6	-6	-12	7
39	45	90	1	0	1	-4	-7	4
30	50	0						
30	51	0	0	5	4	11	-4	5
30	55	8	1	7	5	2	-1	1
30	56	5	1	12	8	12	-5	6
30	59	0	0	5	3	10	-3	5
31	51	6	1	10	8	2	-1	1
31	51	9	1	6	5	12	-7	6
31	52	6	1	9	7	2	-1	1
31	53	7	1	8	6	2	-1	1
31	56	0	0	3	2	6	-2	3
32	50	5	1	12	10	2	-1	1
32	51	5	1	11	9	2	-1	1
32	53	0	0	4	3	8	-3	4
32	58	5	1	11	7	9	-4	5
33	50	0	0	6	5	12	-5	6
33	51	0	0	5	4	10	-4	5
33	56	0	0	3	2	11	-4	6
33	59	0	0	5	3	9	-3	5
34	56	6	1	9	6	12	-6	7
34	58	7	1	8	5	10	-5	6
35	50	0	0	6	5	11	-5	6
35	51	0	0	5	4	9	-4	5
36	53	0	0	4	3	7	-3	4
36	54	0	0	7	5	12	-5	7
36	55	6	1	10	7	11	-6	7
36	56	0	0	3	2	5	-2	3
36	59	0	0	5	3	8	-3	5
37	54	9	2	12	9	3	-2	2
37	56	6	1	9	6	9	-5	6
37	59	6	1	10	6	6	-3	4
38	50	0	0	6	5	10	-5	6
38	54	0	0	7	5	11	-5	7
38	58	0	0	8	5	12	-5	8
38	58	7	1	8	5	7	-4	5
39	51	0	0	5	4	8	-4	5
39	53	5	1	12	9	12	-7	8
39	54	5	1	11	8	10	-6	7
39	57	9	1	6	4	12	-8	9
30	50	10						
30	51	14	3	12	10	6	-4	3

φ_1	Φ	φ_2	H	K	L	U	V	W
30	50	10						
30	52	11	1	5	4	10	-6	5
30	58	18	1	3	2	7	-5	4
31	50	18	3	9	6	11	-9	6
31	54	14	1	4	3	9	-6	5
32	50	14	2	8	7	11	-8	6
32	52	11	1	5	4	11	-7	6
32	54	18	3	9	7	5	-4	3
32	58	18	1	3	2	8	-6	5
33	51	16	2	7	6	5	-4	3
33	52	15	3	11	9	12	-9	7
33	54	14	1	4	3	10	-7	6
33	58	18	1	3	2	9	-7	6
34	52	11	1	5	4	12	-8	7
34	54	14	1	4	3	11	-8	7
34	55	18	4	12	9	6	-5	4
35	52	18	2	6	5	9	-8	6
35	53	17	3	10	8	6	-5	4
35	57	13	2	9	6	3	-2	2
35	58	18	1	3	2	10	-8	7
35	58	18	1	3	2	11	-9	8
35	59	14	2	8	5	3	-2	2
36	54	14	1	4	3	12	-9	8
36	56	11	2	10	7	3	-2	2
36	58	18	1	3	2	12	-10	9
37	52	15	3	11	9	7	-6	5
37	54	10	2	11	8	3	-2	2
37	54	18	3	9	7	12	-11	9
37	58	15	3	11	7	4	-3	3
38	50	18	3	9	8	4	-4	3
38	51	16	2	7	6	11	-10	8
38	57	14	3	12	8	4	-3	3
39	51	14	3	12	10	8	-7	6
39	52	18	2	6	5	5	-5	4
30	50	20						
30	50	24	5	11	10	5	-5	3
30	53	22	2	5	4	7	-6	4
30	54	27	5	10	8	6	-8	5
31	51	24	4	9	8	8	-8	5
31	56	20	4	11	8	5	-4	3
31	56	27	2	4	3	3	-3	2
31	56	29	5	9	7	10	-11	7
31	57	23	3	7	5	8	-7	5
31	59	24	4	9	6	9	-8	6
32	51	21	3	6	7	10	-9	6
32	51	27	6	12	11	9	-10	6
32	52	20	4	11	9	8	-7	5

φ1	Φ	φ2	H	K	L	U	V	W
30	50	20						
32	53	22	2	5	4	11	-10	7
32	53	24	5	11	9	3	-3	2
32	55	23	5	12	9	12	-11	8
32	58	27	5	10	7	7	-7	5
33	50	23	5	12	11	11	-11	7
33	51	27	5	10	9	7	-8	5
33	52	23	3	7	6	3	-3	2
33	57	29	6	11	8	9	-10	7
33	59	27	3	6	4	4	-4	3
34	50	22	4	10	9	3	-3	2
34	52	23	5	12	10	10	-10	7
34	53	27	3	6	5	8	-9	6
34	54	29	6	11	9	5	-6	4
34	55	21	3	8	6	10	-9	7
34	55	24	4	9	7	7	-7	5
34	56	24	5	11	8	4	-4	3
34	57	22	4	10	7	11	-10	8
35	51	24	4	9	8	11	-12	8
35	52	27	4	8	7	5	-6	4
35	56	27	2	4	3	10	-11	8
35	58	23	5	12	8	12	-11	9
36	51	21	3	8	7	7	-7	5
36	53	22	2	5	4	4	-4	3
36	54	27	5	10	8	6	-7	5
36	55	23	5	12	9	9	-9	7
36	56	27	2	4	3	7	-8	6
36	57	23	3	7	5	5	-5	4
36	57	29	6	11	8	10	-12	9
36	59	24	4	9	6	6	-6	5
37	51	27	6	12	11	7	-9	6
37	52	20	4	11	9	9	-9	7
37	59	27	3	6	4	10	-11	9
38	55	21	3	8	6	6	-6	5
38	55	24	4	9	7	8	-9	7
38	56	29	5	9	7	3	-4	3
38	57	22	4	10	7	7	-7	6
38	58	23	5	12	8	8	-8	7
39	50	24	5	11	10	8	-10	7
39	51	21	3	8	7	11	-12	9
39	52	23	3	7	6	10	-12	9
39	59	27	3	6	4	6	-7	6
30	50	30						
30	52	39	4	5	5	5	-8	4
30	53	32	7	11	10	7	-9	5
30	54	35	7	10	9	8	-11	6
30	57	34	6	9	7	4	-5	3

φ1	Φ	φ2	H	K	L	U	V	W
30	50	30						
30	58	38	7	9	7	5	-7	4
30	58	39	9	11	9	6	-9	5
30	59	37	3	4	3	9	-12	7
31	55	32	7	11	9	4	-5	3
31	55	36	5	7	6	5	-7	4
31	55	39	8	10	9	7	-11	6
31	58	30	4	7	5	11	-12	8
31	58	39	4	5	4	8	-12	7
31	59	31	6	10	7	8	-9	6
32	50	32	5	8	8	8	-11	6
32	53	34	8	12	11	5	-7	4
32	56	31	3	5	4	9	-11	7
32	57	36	8	11	9	7	-10	6
33	50	34	2	3	3	6	-9	5
33	51	36	5	7	7	7	-11	6
33	54	34	6	9	8	7	-10	6
34	52	39	4	5	5	5	-9	5
34	55	38	7	9	8	3	-5	3
34	56	31	3	5	4	7	-9	6
34	58	30	4	7	5	6	-7	5
34	58	34	8	12	9	9	-12	8
34	58	39	4	5	4	5	-8	5
34	59	37	3	4	3	2	-3	2
35	50	32	5	8	8	8	-12	7
35	51	36	5	7	7	7	-12	7
35	51	37	3	4	4	4	-7	4
35	54	37	9	12	11	3	-5	3
35	55	36	5	7	6	7	-11	7
35	57	35	7	10	8	2	-3	2
35	59	31	6	10	7	9	-11	8
36	53	30	4	7	6	9	-12	8
36	54	35	7	10	9	5	-8	5
36	55	34	4	6	5	2	-3	2
36	58	39	4	5	4	7	-12	8
37	50	34	2	3	3	3	-5	3
37	52	39	4	5	5	5	-10	6
37	53	32	5	8	7	2	-3	2
37	58	32	5	8	6	8	-11	8
37	58	39	9	11	9	5	-9	6
38	52	30	7	12	11	2	-3	2
38	52	31	6	10	9	2	-3	2
38	53	30	4	7	6	7	-10	7
38	55	37	6	8	7	5	-9	6
38	56	31	3	5	4	5	-7	5
38	57	30	7	12	9	3	-4	3
38	58	30	4	7	5	7	-9	7
38	58	38	7	9	7	4	-7	5
39	50	34	2	3	3	6	-11	7

φ1	Φ	φ2	H	K	L	U	V	W
30	50	30						
39	51	37	3	4	4	4	-8	5
39	52	39	4	5	5	5	-11	7
39	53	34	8	12	11	7	-12	8
39	55	32	7	11	9	7	-11	8
39	55	39	8	10	9	3	-6	4
39	57	36	8	11	9	4	-7	5
39	59	37	3	4	3	7	-12	9
30	50	40						
30	50	45	5	5	6	5	-11	5
30	52	40	5	6	6	6	-10	5
30	53	41	6	7	7	7	-12	6
30	55	45	1	1	1	1	-2	1
30	59	45	6	6	5	6	-11	6
31	50	49	8	7	9	4	-11	5
31	54	47	12	11	12	5	-12	6
31	56	43	11	12	11	6	-11	6
31	58	43	11	12	10	4	-7	4
32	50	41	7	8	9	1	-2	1
32	50	42	8	9	10	1	-2	1
32	51	42	9	10	11	1	-2	1
32	51	42	10	11	12	1	-2	1
32	53	48	10	9	10	4	-10	5
32	53	49	8	7	8	3	-8	4
32	56	42	9	10	9	5	-9	5
32	57	41	6	7	6	7	-12	7
32	57	41	7	8	7	4	-7	4
33	50	45	6	6	7	5	-12	6
33	52	40	5	6	6	6	-11	6
33	55	40	10	12	11	4	-7	4
33	55	45	1	1	1	5	-11	6
33	57	40	5	6	5	3	-5	3
33	59	45	6	6	5	5	-10	6
33	59	45	7	7	6	6	-12	7
34	50	45	5	5	6	4	-10	5
34	50	48	9	8	10	2	-6	3
34	55	45	1	1	1	4	-9	5
35	54	48	11	10	11	4	-11	6
35	55	45	1	1	1	3	-7	4
35	55	45	1	1	1	5	-12	7
35	56	42	10	11	10	5	-10	6
35	59	42	9	10	8	6	-11	7
36	50	45	6	6	7	4	-11	6
36	52	40	5	6	6	6	-12	7
36	53	48	9	8	9	3	-9	5
36	56	42	8	9	8	4	-8	5
36	59	45	7	7	6	5	-11	7

φ1	Φ	φ2	H	K	L	U	V	W
30	50	40						
37	53	49	7	6	7	2	-7	4
37	54	47	12	11	12	4	-12	7
37	55	45	1	1	1	2	-5	3
37	56	43	11	12	11	5	-11	7
37	59	42	10	11	9	3	-6	4
37	59	45	6	6	5	4	-9	6
38	50	45	5	5	6	3	-9	5
38	51	48	11	10	12	2	-7	4
38	53	48	10	9	10	3	-10	6
38	55	45	1	1	1	3	-8	5
38	57	41	6	7	6	3	-6	4
38	58	45	8	8	7	5	-12	8
39	51	45	7	7	8	4	-12	7
39	55	45	1	1	1	4	-11	7
39	56	42	9	10	9	4	-9	6
39	57	40	5	6	5	5	-10	7
30	50	50						
30	52	51	11	9	11	4	-11	5
30	52	52	9	7	9	3	-9	4
30	57	50	6	5	5	5	-12	6
30	58	51	5	4	4	4	-10	5
30	59	53	4	3	3	3	-8	4
31	50	51	11	9	12	3	-9	4
31	50	56	3	2	3	2	-9	4
31	51	54	11	8	11	3	-11	5
31	55	50	12	10	11	3	-8	4
31	56	53	12	9	10	2	-6	3
31	57	55	10	7	8	3	-10	5
31	58	58	8	5	6	1	-4	2
31	59	59	10	6	7	1	-4	2
32	50	58	11	7	11	2	-11	5
32	55	52	9	7	8	2	-6	3
32	55	56	6	4	5	1	-4	2
32	56	59	5	3	4	2	-10	5
33	52	50	6	5	6	2	-6	3
33	52	51	5	4	5	3	-10	5
33	54	55	10	7	9	1	-4	2
33	58	51	5	4	4	4	-11	6
33	59	53	4	3	3	3	-9	5
34	51	53	4	3	4	1	-4	2
35	50	56	3	2	3	1	-6	3
35	51	55	10	7	10	2	-10	5
35	55	53	8	6	7	3	-11	6
35	58	51	5	4	4	4	-12	7
36	50	58	8	5	6	1	-8	4
36	52	51	11	9	11	3	-11	6

φ_1	Φ	φ_2	H	K	L	U	V	W
30	50	50						
36	54	54	11	8	10	2	-9	5
36	54	56	9	6	8	2	-11	6
36	55	51	10	8	9	2	-7	4
36	57	54	11	8	9	3	-12	7
36	59	53	4	3	3	3	-10	6
37	51	54	11	8	11	2	-11	6
37	52	52	9	7	9	2	-9	5
37	56	59	5	3	4	1	-7	4
37	57	56	9	6	7	1	-5	3
38	52	50	6	5	6	3	-12	7
38	53	56	12	8	11	1	-7	4
38	55	54	7	5	6	1	-5	3
38	59	53	4	3	3	3	-11	7
39	50	56	3	2	3	1	-9	5
39	50	58	11	7	11	1	-11	6
39	51	54	7	5	7	1	-7	4
39	54	53	12	9	11	1	-5	3
39	55	51	11	9	10	2	-8	5
39	59	53	4	3	3	3	-12	8
30	50	60						
30	55	66	9	4	7	1	-11	5
31	51	69	8	3	7	0	-7	3
31	52	63	8	4	7	1	-9	4
31	53	60	7	4	6	2	-11	5
31	59	63	6	3	4	1	-6	3
32	50	68	10	4	9	0	-9	4
32	51	63	10	5	9	1	-11	5
32	56	61	9	5	7	1	-6	3
32	56	63	4	2	3	1	-8	4
32	58	67	12	5	8	1	-12	6
33	50	67	12	5	11	0	-11	5
33	54	60	12	7	10	1	-6	3
33	54	63	10	5	8	1	-10	5
33	55	69	8	3	6	0	-2	1
34	52	67	12	5	10	0	-2	1
34	53	63	6	3	5	1	-12	6
34	53	68	5	2	4	0	-2	1
34	58	60	7	4	5	2	-11	6
35	50	66	11	5	10	0	-2	1
35	51	66	9	4	8	0	-2	1
35	52	60	12	7	11	1	-8	4
35	52	67	7	3	6	0	-2	1
35	56	63	4	2	3	1	-11	6
35	58	63	10	5	7	1	-9	5
36	55	67	12	5	9	0	-9	5
36	57	68	10	4	7	0	-7	4

φ_1	Φ	φ_2	H	K	L	U	V	W
30	50	60						
37	53	66	11	5	9	0	-9	5
37	55	66	9	4	7	0	-7	4
37	59	63	6	3	4	1	-10	6
38	50	68	10	4	9	-1	-11	6
38	51	63	12	6	11	0	-11	6
38	57	67	7	3	5	0	-5	3
38	58	60	7	4	5	1	-8	5
38	58	67	12	5	8	0	-8	5
39	51	63	10	5	9	0	-9	5
39	51	69	8	3	7	-1	-9	5
39	52	63	8	4	7	0	-7	4
39	56	66	11	5	8	0	-8	5
30	50	70						
30	52	79	5	1	4	-1	-7	3
30	55	72	12	4	9	0	-9	4
31	52	70	11	4	9	0	-9	4
31	54	76	4	1	3	-1	-11	5
32	57	76	12	3	8	-1	-12	6
32	58	72	3	1	2	0	-2	1
33	54	76	4	1	3	-1	-8	4
33	56	70	11	4	8	0	-2	1
34	52	79	5	1	4	-2	-10	5
34	53	73	10	3	8	-1	-10	5
34	53	77	9	2	7	-1	-6	3
34	59	76	8	2	5	-1	-11	6
35	51	76	12	3	10	-1	-6	3
35	52	72	6	2	5	-1	-12	6
35	56	79	10	2	7	-2	-11	6
35	59	70	11	4	7	0	-7	4
36	54	72	9	3	7	-1	-11	6
36	56	74	7	2	5	-1	-9	5
37	55	75	11	3	8	-1	-7	4
37	57	77	9	2	6	-2	-12	7
38	51	74	7	2	6	-2	-11	6
38	58	75	11	3	7	-1	-8	5
39	52	72	6	2	5	-1	-7	4
39	53	73	10	3	8	-2	-12	7
39	54	76	4	1	3	-1	-5	3
30	50	80						
30	55	82	7	1	5	-2	-11	5
31	59	84	10	1	6	-2	-10	5
32	55	84	10	1	7	-1	-4	2
32	56	85	12	1	8	-1	-4	2
32	57	81	6	1	4	-1	-6	3

φ_1	Φ	φ_2	H	K	L	U	V	W
30	50	80						
33	53	83	8	1	6	-1	-4	2
34	50	85	12	1	10	-2	-6	3
34	51	81	6	1	5	-1	-4	2
34	51	84	10	1	8	-3	-10	5
34	58	83	8	1	5	-2	-9	5
35	59	84	10	1	6	-3	-12	7
36	53	85	12	1	9	-3	-9	5
36	54	80	11	2	8	-2	-9	5
37	54	85	11	1	8	-4	-12	7
38	50	82	7	1	6	-3	-9	5
38	51	60	11	2	9	-2	-7	4
38	58	83	8	1	5	-3	-11	7
39	51	85	11	1	9	-2	-5	3
30	50	90						
30	51	90	5	0	4	-4	-11	5
30	59	90	5	0	3	-3	-10	5
31	56	90	3	0	2	-2	-6	3
32	53	90	4	0	3	-3	-8	4
33	50	90	6	0	5	-5	-12	6
33	51	90	5	0	4	-4	-10	5
33	56	90	3	0	2	-4	-11	6
33	59	90	5	0	3	-3	-9	5
35	50	90	6	0	5	-5	-11	6
35	51	90	5	0	4	-4	-9	5
36	53	90	4	0	3	-3	-7	4
36	54	90	7	0	5	-5	-12	7
36	56	90	3	0	2	-2	-5	3
36	59	90	5	0	3	-3	-8	5
38	50	90	6	0	5	-5	-10	6
38	54	90	7	0	5	-5	-11	7
38	58	90	8	0	5	-5	-12	8
39	51	90	5	0	4	-4	-8	5
30	60	0						
30	64	9	1	6	3	9	-4	5
31	61	8	1	7	4	11	-5	6
31	63	0	0	2	1	11	-3	6
31	68	0	0	5	2	9	-2	5
31	68	6	1	10	4	12	-4	7
32	61	6	1	9	5	7	-3	4
32	64	7	1	8	4	12	-5	7
32	67	0	0	7	3	12	-3	7
32	67	8	1	7	3	5	-2	3
33	63	0	0	2	1	7	-2	4
33	64	6	1	10	5	5	-2	3

φ_1	Φ	φ_2	H	K	L	U	V	W
30	60	0						
34	60	0	0	7	4	12	-4	7
34	61	6	1	7	4	8	-4	5
34	63	0	0	2	1	10	-3	6
34	66	5	1	11	5	8	-3	5
34	68	0	0	5	2	6	-2	5
34	68	9	2	12	5	9	-4	6
35	60	5	1	12	7	11	-5	7
35	64	9	1	6	3	6	-3	4
35	67	0	0	7	3	11	-3	7
35	69	0	0	8	3	12	-3	8
36	60	0	0	7	4	11	-4	7
36	60	9	2	12	7	9	-5	6
36	64	5	1	12	6	12	-5	8
37	61	6	1	9	5	10	-5	7
37	63	0	0	2	1	3	-1	2
37	66	6	1	9	4	7	-3	5
37	67	0	0	7	3	10	-3	7
37	67	8	1	7	3	11	-5	8
38	64	7	1	8	4	4	-2	3
38	68	0	0	5	2	7	-2	5
38	69	0	0	8	3	11	-3	8
39	60	0	0	7	4	10	-4	7
39	61	5	1	11	6	4	-2	3
39	63	0	0	2	1	11	-4	8
39	64	9	1	6	3	9	-5	7
39	66	0	0	9	4	12	-4	9
39	66	5	1	11	5	9	-4	7
39	68	9	2	12	5	10	-5	8
30	60	10						
30	64	17	3	10	5	5	-3	3
30	67	16	3	9	4	5	-3	3
30	69	11	1	5	2	7	-3	4
31	62	15	3	11	6	5	-3	3
31	68	14	3	12	5	10	-5	6
32	60	14	3	12	7	5	-3	3
32	62	10	2	11	6	12	-6	7
32	65	18	2	6	3	3	-2	2
32	67	13	2	9	4	8	-4	5
33	69	11	1	5	2	11	-5	7
34	61	16	2	7	4	3	-2	2
34	64	14	1	4	2	12	-7	8
34	68	18	4	12	5	11	-7	6
35	64	14	1	4	2	10	-6	7
35	67	18	3	9	4	12	-8	9
36	60	17	3	10	6	4	-3	3
36	62	18	3	9	5	4	-3	3

φ_1	Φ	φ_2	H	K	L	U	V	W
30	60	10						
36	66	10	2	11	5	10	-5	7
36	67	13	2	9	4	11	-6	8
37	61	18	4	12	7	5	-4	4
37	62	10	2	11	6	7	-4	5
37	64	14	1	4	2	8	-5	6
37	64	17	3	10	5	10	-7	8
37	69	11	1	5	2	4	-2	3
38	65	18	2	6	3	12	-9	10
38	67	18	3	9	4	7	-5	6
38	68	16	2	7	3	11	-7	9
39	60	11	1	5	3	9	-6	7
39	60	14	3	12	7	11	-8	9
39	64	11	2	10	5	5	-3	4
39	64	14	1	4	2	6	-4	5
39	65	18	2	6	3	9	-7	8
30	60	20						
30	60	29	5	9	6	3	-3	2
30	61	27	4	8	5	12	-11	8
30	62	27	6	12	7	9	-8	6
30	66	27	1	2	1	6	-5	4
30	68	24	4	9	4	11	-8	7
31	61	29	6	11	7	7	-7	5
31	65	23	5	12	6	12	-9	8
31	66	27	1	2	1	7	-6	5
32	60	24	5	11	7	10	-9	7
32	67	20	4	11	5	3	-2	2
33	64	29	5	9	5	5	-5	4
33	66	27	1	2	1	8	-7	6
33	66	27	1	2	1	9	-8	7
33	69	23	5	12	5	7	-5	5
34	61	27	4	8	5	5	-5	4
34	64	29	6	11	6	6	-6	5
34	66	27	1	2	1	10	-9	8
34	68	23	3	7	3	4	-3	3
35	62	27	5	10	6	6	-6	5
35	65	21	3	8	4	4	-3	3
35	66	27	1	2	1	11	-10	9
35	66	27	1	2	1	12	-11	10
35	68	24	4	9	4	5	-4	4
36	61	29	6	11	7	9	-10	8
36	62	27	6	12	7	7	-7	6
36	65	22	4	10	5	5	-4	4
36	65	23	5	12	6	6	-5	5
36	68	24	5	11	5	6	-5	5
37	60	24	5	11	7	7	-7	6
37	63	20	4	11	6	5	-4	4
30	60	20						
38	60	29	5	9	6	9	-11	9
38	64	29	5	9	5	9	-10	9
38	64	29	6	11	6	11	-12	11
38	68	23	3	7	3	11	-9	10
38	68	29	6	11	5	1	-1	1
38	69	29	5	9	4	1	-1	1
39	61	27	4	8	5	8	-9	8
39	62	27	5	10	6	10	-11	10
39	65	21	3	8	4	12	-10	11
39	66	27	1	2	1	1	-1	1
30	60	30						
30	62	38	7	9	6	6	-8	5
30	63	30	7	12	7	7	-7	5
30	63	36	8	11	7	9	-11	7
30	67	39	9	11	6	7	-9	6
31	60	36	8	11	8	6	-8	5
31	61	34	2	3	2	5	-6	4
31	62	32	5	8	5	9	-10	7
31	63	37	6	8	5	7	-9	6
31	64	39	9	11	7	8	-11	7
31	65	39	4	5	3	9	-12	8
31	67	32	5	8	4	4	-4	3
31	68	35	7	10	5	10	-11	8
31	68	37	3	4	2	6	-7	5
32	64	30	4	7	4	4	-4	3
32	67	31	6	10	5	5	-5	4
32	69	32	7	11	5	5	-5	4
33	61	34	2	3	2	8	-10	7
33	62	32	7	11	7	6	-7	5
33	63	31	3	5	3	11	-12	9
33	65	34	6	9	5	7	-8	6
33	65	36	5	7	4	9	-11	8
33	68	37	3	4	2	10	-12	9
34	61	39	8	10	7	2	-3	2
34	63	31	3	5	3	8	-9	7
34	65	39	4	5	3	5	-7	5
34	66	38	7	9	5	3	-4	3
34	67	30	7	12	6	6	-6	5
34	67	34	4	6	3	9	-10	8
35	60	35	7	10	7	5	-7	5
35	60	36	5	7	5	7	-10	7
35	63	37	6	8	5	6	-11	8
35	65	36	5	7	4	7	-9	7
35	65	37	9	12	7	3	-4	3
35	68	37	3	4	2	4	-5	4
36	61	34	2	3	2	3	-4	3
30	60	30						
36	61	39	9	11	8	6	-10	7
36	64	30	4	7	4	11	-12	10
36	64	34	8	12	7	4	-5	4
36	65	34	6	9	5	9	-11	9
36	67	34	4	6	3	6	-7	6
36	68	35	7	10	5	5	-6	5
37	62	32	5	8	5	4	-5	4
37	62	37	9	12	8	8	-12	9
37	63	31	3	5	3	5	-6	5
37	65	39	4	5	3	6	-9	7
37	66	36	8	11	6	9	-12	10
37	67	31	6	10	5	10	-11	10
37	67	32	5	8	4	8	-9	6
37	69	32	7	11	5	11	-12	11
38	60	30	7	12	8	4	-5	4
38	63	30	7	12	7	6	-7	6
38	64	30	4	7	4	7	-8	7
38	64	35	7	10	6	6	-11	9
38	67	39	9	11	6	4	-6	5
38	68	37	3	4	2	6	-8	7
38	69	39	8	10	5	5	-7	6
39	60	36	8	11	8	5	-8	6
39	61	34	2	3	2	7	-10	6
39	64	39	9	11	7	3	-5	4
39	65	32	7	11	6	8	-10	9
39	65	36	5	7	4	5	-7	6
39	65	39	4	5	3	7	-11	9
39	67	34	4	6	3	9	-11	10
30	60	40						
30	61	45	5	5	4	5	-9	5
30	62	45	4	4	3	4	-7	4
30	63	45	7	7	5	7	-12	7
30	65	45	3	3	2	3	-5	3
30	67	42	8	9	5	8	-11	7
31	64	48	9	8	6	6	-12	7
31	67	45	5	5	3	5	-8	5
31	68	45	7	7	4	7	-11	7
32	60	42	6	9	7	3	-5	3
32	61	43	11	12	9	3	-5	3
32	63	42	9	10	7	5	-8	5
32	65	41	7	8	5	2	-3	2
32	65	49	8	7	5	5	-10	6
32	66	42	9	10	6	2	-3	2
32	67	43	11	12	7	2	-3	2
32	67	47	12	11	7	6	-11	7
32	67	49	7	6	4	4	-8	5
30	60	40						
33	62	41	6	7	5	7	-11	7
33	63	40	5	6	4	2	-3	2
33	63	45	7	7	5	6	-11	7
33	64	47	12	11	8	4	-8	5
33	66	45	8	8	5	7	-12	8
33	67	41	6	7	4	7	-10	7
33	69	40	5	6	3	3	-4	3
33	69	41	7	8	4	8	-11	8
34	61	45	5	5	4	4	-8	5
34	63	48	10	9	7	5	-11	7
34	65	45	3	3	2	5	-9	6
34	66	48	10	9	6	3	-6	4
34	68	45	7	7	4	6	-10	7
35	62	45	4	4	3	3	-6	4
35	64	42	8	9	6	6	-10	7
35	67	41	6	7	4	6	-12	9
35	67	45	5	5	3	4	-7	5
35	67	48	9	8	5	5	-10	7
36	63	45	7	7	5	5	-10	7
36	64	43	11	12	8	4	-7	5
36	64	48	9	8	6	4	-9	6
36	66	45	8	8	5	6	-11	8
36	69	45	9	9	5	7	-12	9
37	61	47	12	11	9	5	-12	8
37	62	41	6	7	5	4	-7	5
37	63	42	9	10	7	6	-11	8
37	65	49	8	7	5	3	-7	5
37	68	45	7	7	4	5	-9	7
37	68	48	11	10	6	2	-4	3
37	69	49	8	7	4	2	-4	3
38	60	40	10	12	9	6	-11	8
38	62	48	11	10	8	4	-10	7
38	65	42	10	11	7	5	-9	7
38	65	45	3	3	2	2	-4	3
38	69	41	7	8	4	4	-6	5
39	61	45	5	5	4	3	-7	5
39	62	42	10	11	8	2	-4	3
39	62	45	4	4	3	5	-11	8
39	64	47	12	11	8	5	-12	9
39	69	45	9	9	5	6	-11	9
30	60	50						
30	65	56	9	6	5	4	-11	6
30	67	51	11	9	6	6	-12	7
31	65	58	11	7	6	3	-9	5
31	69	51	10	8	5	5	-10	6
32	63	59	5	3	3	3	-11	6

φ_1	Φ	φ_2	H	K	L	U	V	W
30	**60**	**50**						
32	65	53	12	9	7	2	-5	3
32	66	54	11	8	6	2	-5	3
32	67	58	8	5	4	4	-12	7
32	68	55	10	7	5	2	-5	3
33	61	56	3	2	2	2	-7	4
33	67	59	10	6	5	3	-10	6
33	69	50	6	5	3	3	-6	4
34	63	59	5	3	3	3	-12	7
34	68	53	4	3	2	5	-12	8
35	66	50	12	10	7	4	-9	6
35	67	56	6	4	3	1	-3	2
35	69	58	11	7	5	1	-3	2
36	61	56	3	2	2	2	-8	5
36	65	54	7	5	4	1	-3	2
36	68	53	4	3	2	4	-10	7
37	62	52	9	7	6	1	-3	2
37	63	53	8	6	5	1	-3	2
37	67	51	11	9	6	3	-7	5
38	60	50	12	10	9	1	-3	2
38	61	51	10	8	7	1	-3	2
38	61	51	11	9	8	1	-3	2
38	68	53	4	3	2	3	-8	6
38	69	50	6	5	3	4	-9	7
39	61	56	3	2	2	2	-9	6
39	65	51	5	4	3	4	-11	8
30	**60**	**60**						
30	60	60	12	7	8	1	-4	2
30	62	67	12	5	7	1	-8	4
30	63	66	9	4	5	1	-6	3
31	61	68	5	2	3	1	-10	5
31	66	63	2	1	1	2	-9	5
31	68	61	11	6	5	2	-7	4
32	62	63	12	6	7	2	-11	6
33	62	67	7	3	4	1	-9	5
33	64	66	11	5	6	1	-7	4
34	61	63	8	4	5	1	-7	4
34	65	66	10	4	5	1	-10	6
34	65	69	8	3	4	1	-12	7
34	66	63	2	1	1	1	-5	3
34	69	61	9	5	4	3	-11	7
35	60	61	9	5	6	2	-12	7
36	60	66	11	5	7	1	-12	7
36	66	63	2	1	1	2	-11	7
36	67	60	12	7	6	3	-12	8
37	60	69	8	3	5	0	-5	3
37	62	63	10	5	6	1	-8	5

φ_1	Φ	φ_2	H	K	L	U	V	W
30	**60**	**60**						
37	63	66	9	4	5	1	-11	7
37	66	63	2	1	1	1	-6	4
38	65	67	12	5	6	1	-12	8
39	61	68	5	2	3	0	-3	2
39	62	63	12	6	7	1	-9	6
39	66	63	2	1	1	1	-7	5
30	**60**	**70**						
30	62	75	11	3	6	0	-2	1
30	65	72	6	2	3	1	-12	6
31	60	73	10	3	6	0	-2	1
31	60	79	5	1	3	-1	-10	5
31	61	74	7	2	4	0	-2	1
31	67	72	9	3	4	1	-11	6
32	67	70	11	4	5	1	-9	5
34	66	75	11	3	5	0	-5	3
34	68	76	12	3	5	0	-5	3
35	61	72	12	4	7	0	-7	4
35	64	73	10	3	5	0	-5	3
35	64	79	10	2	5	-1	-10	6
36	60	76	12	3	7	-1	-10	6
36	62	72	9	3	5	0	-5	3
37	62	77	9	2	5	-1	-8	5
37	68	74	7	2	3	0	-3	2
38	64	76	4	1	2	-1	-12	8
38	65	72	6	2	3	0	-3	2
39	63	70	11	4	6	0	-3	2
39	64	76	4	1	2	-1	-10	7
39	69	79	5	1	2	-1	-11	8
30	**60**	**80**						
30	61	84	9	1	5	-1	-6	3
30	64	81	6	1	3	-1	-12	6
30	64	83	8	1	4	-1	-8	4
30	64	85	12	1	6	-1	-6	3
31	60	81	12	2	7	-1	-8	4
31	67	82	7	1	3	-1	-11	6
33	64	81	6	1	3	-1	-9	5
33	66	80	11	2	5	-1	-12	7
33	66	84	9	1	4	-1	-7	4
34	64	83	8	1	4	-2	-12	7
34	64	84	10	1	5	-1	-5	3
35	67	82	7	1	3	-1	-8	5
35	67	85	12	1	5	-2	-11	7
36	61	85	11	1	6	-2	-8	5
36	68	84	10	1	4	-1	-6	4

φ_1	Φ	φ_2	H	K	L	U	V	W
30	**60**	**80**						
37	64	85	12	1	6	-3	-12	8
37	68	81	12	2	5	-1	-9	6
38	64	81	6	1	3	-1	-6	4
38	66	84	9	1	4	-2	-10	7
39	60	81	12	2	7	-2	-9	6
39	66	80	11	2	5	-1	-7	5
30	**60**	**90**						
31	63	90	2	0	1	-3	-11	6
31	68	90	5	0	2	-2	-9	5
32	67	90	7	0	3	-3	-12	7
33	63	90	2	0	1	-2	-7	4
34	60	90	7	0	4	-4	-12	7
34	63	90	2	0	1	-3	-10	6
34	68	90	5	0	2	-2	-8	5
35	67	90	7	0	3	-3	-11	7
35	69	90	8	0	3	-3	-12	8
36	60	90	7	0	4	-4	-11	7
37	63	90	2	0	1	-1	-3	2
37	67	90	7	0	3	-3	-10	7
38	68	90	5	0	2	-2	-7	5
38	69	90	8	0	3	-3	-11	8
39	60	90	7	0	4	-4	-10	7
39	63	90	2	0	1	-4	-11	8
39	66	90	9	0	4	-4	-12	9
30	**70**	**0**						
30	70	7	1	6	3	9	-3	5
30	72	0	0	3	1	11	-2	6
30	76	0	0	4	1	7	-1	4
30	79	0	0	5	1	9	-1	5
31	74	0	0	7	2	12	-2	7
31	75	5	1	11	3	12	-3	7
32	72	0	0	3	1	5	-1	3
32	72	9	1	6	2	8	-3	5
32	78	6	1	9	2	8	-2	5
33	70	7	1	8	3	11	-4	7
33	74	0	0	7	2	11	-2	7
33	76	9	2	12	3	3	-1	2
33	79	0	0	5	1	8	-1	5
34	72	6	1	9	3	3	-1	2
34	74	8	1	7	2	3	-1	2
34	76	0	0	4	1	6	-1	4
34	76	5	1	12	3	12	-3	8
35	70	5	1	11	4	3	-1	2
35	72	0	0	3	1	9	-2	6

φ_1	Φ	φ_2	H	K	L	U	V	W
30	**70**	**0**						
35	72	9	1	6	2	10	-4	7
35	76	7	1	8	2	10	-3	7
36	74	0	0	7	2	10	-2	7
36	75	5	1	11	3	7	-2	5
36	78	6	1	9	2	11	-3	8
36	79	0	0	5	1	7	-1	5
37	72	9	1	6	2	12	-5	9
37	76	0	0	4	1	11	-2	8
37	79	6	1	10	2	4	-1	3
38	72	0	0	3	1	4	-1	3
38	72	5	1	12	4	12	-4	9
38	73	6	1	10	3	9	-3	7
38	77	0	0	9	2	12	-2	9
39	74	0	0	7	2	9	-2	7
30	**70**	**10**						
30	72	13	2	9	3	12	-5	7
30	76	14	1	4	1	5	-2	3
31	70	10	2	11	4	5	-2	3
31	71	15	3	11	4	8	-4	5
31	72	18	1	3	1	11	-6	7
32	72	18	1	3	1	9	-5	6
32	74	17	3	10	3	6	-3	4
32	77	18	4	12	3	6	-3	4
32	78	13	2	9	2	11	-4	7
33	70	14	2	8	3	6	-3	4
33	72	13	2	9	3	9	-4	6
33	76	14	1	4	1	12	-5	8
33	79	11	1	5	1	3	-1	2
34	72	18	1	3	1	7	-4	5
34	76	14	1	4	1	7	-3	5
35	72	18	1	3	1	12	-7	9
35	75	16	2	7	2	4	-2	3
35	78	18	3	9	2	4	-2	3
36	72	14	3	12	4	4	-2	3
36	72	18	1	3	1	5	-3	4
37	71	15	3	11	4	9	-5	7
37	74	17	3	10	3	11	-6	9
37	76	14	1	4	1	9	-4	7
38	75	10	2	11	3	5	-2	4
38	75	15	3	11	3	6	-3	5
38	76	14	1	4	1	11	-5	9
38	78	13	2	9	2	5	-2	4
38	79	11	1	5	1	11	-4	9
39	70	14	2	8	3	7	-4	6
39	72	13	2	9	3	6	-3	5
39	72	18	1	3	1	8	-5	7

φ_1	Φ	φ_2	H	K	L	U	V	W
30 70 10								
39	77	18	4	12	3	9	-5	8
30 70 20								
30	70	27	6	12	5	9	-7	6
30	71	27	4	8	3	12	-9	8
30	72	29	6	11	4	10	-8	7
30	73	24	4	9	3	3	-2	2
30	77	27	2	4	1	3	-2	2
31	70	22	2	5	2	3	-2	2
31	75	27	5	10	3	7	-5	5
31	77	23	5	12	3	12	-7	8
31	79	29	5	9	2	7	-5	5
32	72	24	5	11	4	7	-5	5
32	73	27	3	6	2	4	-3	3
32	77	29	6	11	3	4	-3	3
32	79	24	4	9	2	10	-6	7
33	75	23	3	7	2	11	-7	8
33	79	22	2	5	1	11	-6	8
34	71	27	4	8	3	5	-4	4
34	73	23	5	12	4	12	-8	9
34	74	22	4	10	3	8	-5	6
34	74	29	5	9	3	6	-5	5
34	76	24	5	11	3	9	-6	7
34	77	27	2	4	1	10	-7	8
35	70	27	5	10	4	6	-5	5
35	71	21	3	8	3	9	-6	7
35	72	29	6	11	4	7	-6	6
35	75	27	5	10	3	12	-9	10
35	79	22	2	5	1	9	-5	7
36	70	22	2	5	2	11	-8	9
36	70	27	6	12	5	7	-6	6
36	75	23	3	7	2	6	-4	5
36	77	27	2	4	1	7	-5	6
36	79	29	5	9	2	8	-6	7
37	71	20	4	11	4	6	-4	5
37	72	24	5	11	4	9	-7	8
37	73	27	3	6	2	10	-8	9
37	77	21	3	8	2	12	-7	10
37	77	27	2	4	1	11	-8	10
37	79	22	2	5	1	7	-4	6
38	70	22	2	5	2	8	-6	7
38	73	24	4	9	3	12	-9	11
38	74	22	4	10	3	9	-6	8
38	76	24	5	11	3	10	-7	9
39	70	27	5	10	4	10	-9	10
39	70	27	6	12	5	12	-11	12
39	71	27	4	8	3	8	-7	8

φ_1	Φ	φ_2	H	K	L	U	V	W
30 70 20								
39	72	29	6	11	4	11	-10	11
39	73	27	3	6	2	6	-5	6
39	74	29	5	9	3	9	-8	9
39	77	21	3	8	2	10	-6	9
39	77	29	6	11	3	11	-9	11
39	79	22	2	5	1	12	-7	11
39	79	29	5	9	2	9	-7	9
30 70 30								
30	70	34	6	9	4	4	-4	3
30	72	35	7	10	4	4	-4	3
30	74	34	2	3	1	12	-11	9
30	76	31	6	10	3	11	-9	8
31	71	34	8	12	5	5	-5	4
31	71	39	9	11	5	9	-11	8
31	72	37	9	12	5	11	-12	9
31	73	39	4	5	2	7	-6	6
31	74	30	7	12	4	12	-10	9
32	70	36	8	11	5	8	-9	7
32	78	30	7	12	3	9	-7	7
33	71	36	5	7	3	10	-11	9
33	73	37	6	8	3	11	-12	10
33	76	30	4	7	2	5	-4	4
33	78	32	5	8	2	6	-5	5
34	73	39	4	5	2	5	-6	5
34	76	31	6	10	3	7	-6	6
34	77	32	7	11	3	8	-7	7
34	78	34	8	12	3	9	-8	8
35	71	38	7	9	4	9	-11	9
35	72	37	9	12	5	6	-7	6
35	73	37	6	8	3	8	-9	8
35	74	30	7	12	4	8	-7	7
35	75	38	7	9	3	9	-10	9
35	77	39	8	10	3	10	-11	10
35	78	39	9	11	3	11	-12	11
36	70	34	6	9	4	9	-10	9
36	71	36	5	7	3	7	-8	7
36	72	35	7	10	4	10	-11	10
36	74	36	8	11	4	11	-12	11
36	76	30	4	7	2	12	-10	11
36	76	35	7	10	3	1	-1	1
36	77	36	5	7	2	1	-1	1
36	78	36	8	11	3	1	-1	1
36	79	37	3	4	1	1	-1	1
37	71	38	7	9	4	7	-9	8
37	72	32	5	8	3	1	-1	1
37	73	32	7	11	4	1	-1	1

φ_1	Φ	φ_2	H	K	L	U	V	W
30 70 30								
37	73	39	4	5	2	8	-10	9
37	74	34	2	3	1	1	-1	1
37	74	39	9	11	4	9	-11	10
37	78	34	8	12	3	12	-11	12
38	70	30	4	7	3	1	-1	1
38	70	30	7	12	5	1	-1	1
38	71	31	3	5	2	1	-1	1
38	74	30	7	12	4	12	-11	12
38	75	37	9	12	4	8	-9	9
38	76	31	6	10	3	10	-9	10
38	77	32	7	11	3	11	-10	11
38	77	39	8	10	3	7	-8	8
38	78	32	5	8	2	8	-7	8
38	78	39	9	11	3	8	-9	9
39	73	37	6	8	3	5	-6	6
39	74	36	8	11	4	7	-8	8
39	75	38	7	9	3	6	-7	7
39	76	30	4	7	2	7	-6	7
39	78	30	7	12	3	6	-5	6
30 70 40								
30	71	48	11	10	5	5	-8	5
30	73	42	9	10	4	8	-10	7
30	75	48	11	10	4	2	-3	2
30	76	48	9	8	3	2	-3	2
30	78	41	6	7	2	11	-12	9
30	78	49	7	6	2	2	-3	2
30	79	42	10	11	3	7	-8	6
30	79	49	8	7	2	7	-10	7
31	71	45	2	2	1	2	-3	2
31	73	45	7	7	3	7	-10	7
31	74	45	5	5	2	5	-7	5
31	76	40	5	6	2	8	-9	7
31	79	48	11	10	3	5	-7	5
32	70	42	9	10	5	5	-7	5
32	70	47	12	11	6	7	-12	8
32	75	45	8	8	3	8	-11	8
32	77	45	3	3	1	3	-4	3
32	78	41	6	7	2	9	-10	8
32	79	41	7	8	2	10	-11	9
32	79	45	7	7	2	7	-9	7
33	72	41	6	7	3	7	-9	7
33	72	42	8	9	4	3	-4	3
33	73	43	11	12	5	3	-4	3
33	74	41	7	8	3	4	-5	4
33	76	42	8	9	3	9	-11	9
33	76	43	11	12	4	4	-5	4

φ_1	Φ	φ_2	H	K	L	U	V	W
30 70 40								
33	76	47	12	11	4	8	-12	9
34	71	45	2	2	1	7	-11	8
34	72	40	10	12	5	4	-5	4
34	72	48	9	8	4	4	-7	5
34	72	49	7	6	3	6	-11	8
34	73	45	9	9	4	8	-12	9
34	76	40	5	6	2	6	-7	6
34	77	42	9	10	3	5	-6	5
34	77	48	10	9	3	6	-9	7
34	78	41	6	7	2	7	-8	7
34	79	41	7	8	2	8	-9	6
35	70	43	11	12	6	6	-9	7
35	71	42	10	11	5	7	-10	8
35	71	45	2	2	1	5	-8	6
35	73	45	7	7	3	6	-9	7
35	75	45	8	8	3	7	-10	8
35	77	45	3	3	1	8	-11	9
35	78	45	10	10	3	9	-12	10
35	79	49	8	7	2	4	-6	5
36	71	48	11	10	5	5	-9	7
36	72	41	6	7	3	9	-12	10
36	74	45	5	5	2	4	-6	5
36	75	42	10	11	4	6	-8	7
36	76	40	5	6	2	10	-12	11
36	77	45	3	3	1	5	-7	6
36	79	45	7	7	2	6	-8	7
37	71	45	2	2	1	3	-5	4
37	73	45	9	9	4	7	-11	9
37	78	45	10	10	3	8	-11	10
37	79	40	10	12	3	9	-10	10
37	79	45	11	11	3	9	-12	11
37	79	48	11	10	3	6	-9	8
38	70	48	10	9	5	5	-10	8
38	71	45	2	2	1	7	-12	10
38	74	41	7	8	3	9	-12	11
38	75	45	8	8	3	6	-9	8
38	76	47	12	11	4	5	-8	7
38	77	45	3	3	1	7	-10	9
38	79	41	7	8	2	6	-7	7
39	70	42	9	10	5	5	-8	7
39	71	45	2	2	1	4	-7	6
39	72	49	7	6	3	3	-6	5
39	73	45	7	7	3	5	-8	7
39	74	45	5	5	2	7	-11	10
39	76	40	5	6	2	4	-5	5
39	76	45	11	11	4	8	-12	11
39	77	42	9	10	3	7	-9	9
39	78	41	6	7	2	5	-6	6

φ_1	Φ	φ_2	H	K	L	U	V	W
30 70 40								
39	78	49	7	6	2	6	-10	9
39	79	42	10	11	3	4	-5	5
30 70 50								
30	72	53	12	9	5	5	-10	6
30	79	50	12	10	3	2	-3	2
31	70	56	9	6	4	2	-5	3
31	71	52	9	7	4	4	-8	5
31	72	58	8	5	3	2	-5	3
31	73	51	5	4	2	6	-11	7
31	74	54	11	8	4	4	-8	5
31	74	56	3	2	1	5	-11	7
31	75	52	9	7	3	5	-9	6
31	79	53	4	3	1	7	-12	8
32	72	50	12	10	5	5	-9	6
32	73	53	8	6	3	3	-6	4
32	76	50	6	5	2	6	-10	7
32	76	55	10	7	3	3	-6	4
32	78	56	12	8	3	3	-6	4
33	74	56	3	2	1	4	-9	6
33	79	53	4	3	1	4	-7	5
34	71	51	11	9	5	5	-10	7
34	71	59	5	3	2	1	-3	2
34	76	50	6	5	2	7	-12	9
35	74	56	3	2	1	3	-7	5
35	75	53	12	9	4	2	-4	3
35	77	54	7	5	2	2	-4	3
35	78	51	11	9	3	3	-5	4
35	79	53	4	3	1	5	-9	7
36	71	52	9	7	4	5	-11	8
36	73	51	5	4	2	2	-4	3
36	74	56	3	2	1	5	-12	9
36	76	59	10	6	3	3	-8	6
37	71	56	12	8	5	4	-11	8
37	74	54	11	8	4	4	-9	7
37	79	53	4	3	1	6	-11	9
38	70	56	9	6	4	4	-12	9
36	72	50	12	10	5	5	-10	8
38	74	56	3	2	1	2	-5	4
38	75	52	9	7	3	3	-6	5
38	76	55	10	7	3	5	-11	9
38	78	54	11	8	3	3	-6	5
39	72	58	8	5	3	3	-9	7
39	78	58	8	5	2	5	-12	10

φ_1	Φ	φ_2	H	K	L	U	V	W
30 70 60								
30	70	60	7	4	3	3	-9	5
30	70	63	12	6	5	3	-11	6
30	70	68	5	2	2	2	-11	6
30	76	60	7	4	2	2	-5	3
31	71	63	8	4	3	2	-7	4
31	72	61	11	6	4	4	-12	7
31	73	63	6	3	2	3	-10	6
31	73	66	9	4	3	3	-12	7
31	77	61	11	6	3	3	-8	5
32	70	68	5	2	2	2	-12	7
32	74	60	12	7	4	3	-8	5
32	75	67	7	3	2	2	-8	5
32	78	60	12	7	3	5	-12	8
32	79	68	5	2	1	3	-11	7
33	70	63	10	5	4	2	-8	5
33	77	63	4	2	1	1	-3	2
34	74	61	9	5	3	1	-3	2
34	77	67	12	5	3	3	-12	8
34	79	61	9	5	2	4	-10	7
35	71	63	8	4	3	3	-12	8
35	73	66	9	4	3	2	-9	6
35	79	66	9	4	2	2	-7	5
36	70	63	12	6	5	2	-9	6
36	72	61	11	6	4	2	-7	5
36	75	63	10	5	3	2	-7	5
36	76	60	7	4	2	4	-11	8
36	77	69	8	3	2	2	-10	7
37	70	60	12	7	5	2	-7	5
37	72	66	11	5	4	2	-10	7
37	74	60	12	7	4	4	-12	9
37	74	68	10	4	3	2	-11	8
37	76	66	11	5	3	1	-4	3
37	79	68	5	2	1	1	-4	3
38	73	63	6	3	2	1	-4	3
38	73	67	12	5	4	2	-12	9
38	77	61	11	6	3	3	-9	7
38	77	67	12	5	3	2	-9	7
39	70	60	7	4	3	1	-4	3
39	77	63	4	2	1	3	-10	8
30 70 70								
30	76	76	4	1	1	1	-9	5
31	72	72	3	1	1	1	-7	4
31	78	72	9	3	2	1	-5	3
31	79	79	5	1	1	1	-12	7
32	76	70	11	4	3	1	-5	3

φ_1	Φ	φ_2	H	K	L	U	V	W
30 70 70								
32	76	76	4	1	1	1	-10	6
33	76	76	4	1	1	1	-11	7
34	72	72	3	1	1	1	-8	5
34	77	72	12	4	3	1	-6	4
34	79	73	10	3	2	1	-6	4
35	72	72	3	1	1	1	-9	6
35	74	79	10	2	3	0	-3	2
35	76	76	4	1	1	1	-12	8
36	70	76	8	2	3	0	-3	2
36	72	77	9	2	3	0	-3	2
37	72	72	3	1	1	1	-10	7
37	78	72	9	3	2	2	-12	9
38	72	72	3	1	1	1	-11	8
39	71	75	11	3	4	0	-4	3
39	72	72	3	1	1	1	-12	9
39	72	76	12	3	4	0	-4	3
30 70 80								
31	70	85	11	1	4	-1	-9	5
32	72	84	9	1	3	-1	-12	7
33	70	83	8	1	3	-1	-10	6
34	72	85	12	1	4	-1	-8	5
34	73	84	10	1	3	-1	-11	7
35	75	80	11	2	3	0	-3	2
35	76	81	12	2	3	0	-3	2
35	76	85	12	1	3	-1	-12	8
36	72	84	9	1	3	-1	-9	6
36	75	85	11	1	3	-1	-10	7
38	70	83	8	1	3	-1	-7	5
39	72	81	6	1	2	-1	-12	9
39	72	85	12	1	4	-2	-12	9
39	73	84	10	1	3	-1	-8	6
39	76	85	12	1	3	-1	-9	7
30 70 90								
30	72	90	3	0	1	-2	-11	6
30	76	90	4	0	1	-1	-7	4
30	79	90	5	0	1	-1	-9	5
31	74	90	7	0	2	-2	-12	7
32	72	90	3	0	1	-1	-5	3
33	74	90	7	0	2	-2	-11	7
33	79	90	5	0	1	-1	-8	5
34	76	90	4	0	1	-1	-6	4
35	72	90	3	0	1	-2	-9	6
36	74	90	7	0	2	-2	-10	7
36	79	90	5	0	1	-1	-7	5

φ_1	Φ	φ_2	H	K	L	U	V	W
30 70 90								
37	76	90	4	0	1	-2	-11	8
38	72	90	3	0	1	-1	-4	3
38	77	90	9	0	2	-2	-12	9
39	74	90	7	0	2	-2	-9	7
30 80 0								
30	85	5	1	11	1	7	-1	4
31	80	5	1	11	2	5	-1	3
31	81	0	0	6	1	10	-1	6
31	82	0	0	7	1	12	-1	7
31	83	7	1	8	1	5	-1	3
32	81	9	1	6	1	11	-3	7
32	84	6	1	9	1	11	-2	7
33	82	0	0	7	1	11	-1	7
33	85	9	2	12	1	9	-2	6
34	81	0	0	6	1	9	-1	6
34	83	0	0	8	1	12	-1	8
34	84	6	1	10	1	6	-1	4
35	81	5	1	12	2	10	-2	7
35	81	9	1	6	1	7	-2	5
35	82	0	0	7	1	10	-1	7
35	85	5	1	12	1	7	-1	5
36	82	8	1	7	1	4	-1	3
36	83	0	0	8	1	11	-1	8
37	81	0	0	6	1	8	-1	6
37	84	0	0	9	1	12	-1	9
38	81	9	1	6	1	10	-3	8
38	82	0	0	7	1	9	-1	7
38	83	7	1	8	1	9	-2	7
38	84	6	1	9	1	5	-1	4
39	83	0	0	8	1	10	-1	8
39	84	0	0	9	1	11	-1	9
39	84	6	1	10	1	11	-2	9
30 80 10								
30	85	14	3	12	1	10	-3	6
31	82	16	2	7	1	8	-3	5
31	85	17	3	10	1	11	-4	7
32	81	18	2	6	1	9	-4	6
32	85	15	3	11	1	3	-1	2
33	83	14	2	8	1	3	-1	2
34	80	15	3	11	2	10	-4	7
34	81	18	2	6	1	11	-5	8
34	84	13	2	9	1	10	-3	7
35	84	11	2	10	1	11	-3	8
35	84	18	3	9	1	12	-5	9

φ_1	Φ	φ_2	H	K	L	U	V	W
30 80 10								
35	85	17	3	10	1	8	-3	6
36	81	14	3	12	2	8	-3	6
36	85	10	2	11	1	4	-1	3
37	82	16	2	7	1	5	-2	4
37	85	18	4	12	1	5	-2	4
38	84	18	3	9	1	7	-3	6
39	80	10	2	11	2	6	-2	5
39	80	15	3	11	2	7	-3	6
30 80 20								
30	83	21	3	8	1	11	-5	7
30	84	27	4	8	1	12	-7	8
30	86	27	6	12	1	9	-5	6
32	84	29	5	9	1	11	-7	8
32	85	27	5	10	1	7	-4	5
33	82	27	3	6	1	8	-5	6
33	85	20	4	11	1	7	-3	5
34	80	20	4	11	2	4	-2	3
35	80	27	5	10	2	6	-4	5
35	81	24	5	11	2	5	-3	4
35	82	27	3	6	1	11	-7	9
35	84	27	4	8	1	5	-3	4
36	83	23	3	7	1	11	-6	9
36	85	27	5	10	1	12	-7	10
37	85	29	6	11	1	8	-5	7
37	86	27	6	12	1	7	-4	6
38	83	23	3	7	1	9	-5	8
39	84	24	4	9	1	11	-6	10
39	85	20	4	11	1	9	-4	8
30 80 30								
30	82	30	7	12	2	10	-7	7
30	82	37	9	12	2	8	-7	6
30	83	36	5	7	1	11	-9	8
30	84	32	5	8	1	7	-5	5
30	86	32	7	11	1	10	-7	7
30	86	39	8	10	1	8	-7	6
30	86	39	9	11	1	9	-8	7
31	80	31	3	5	1	4	-3	3
31	80	34	6	9	2	12	-10	9
31	85	34	6	9	1	4	-3	3
31	86	34	8	12	1	11	-8	8
32	81	32	7	11	2	9	-7	7
32	82	34	4	6	1	5	-4	4
32	85	35	7	10	1	9	-7	7
32	86	36	8	11	1	5	-4	4
32	86	37	9	12	1	11	-9	9
33	83	36	5	7	1	6	-5	5
33	84	37	6	8	1	7	-6	6
34	80	34	6	9	2	7	-6	6
34	81	35	7	10	2	8	-7	7
34	82	36	8	11	2	9	-8	8
34	82	37	9	12	2	10	-9	9
34	84	32	5	8	1	11	-8	9
34	85	38	7	9	1	8	-7	7
34	86	39	8	10	1	9	-8	8
34	86	39	9	11	1	10	-9	9
35	86	32	7	11	1	7	-5	6
36	80	31	3	5	1	9	-7	8
36	80	38	7	9	2	1	-1	1
36	81	32	7	11	2	10	-8	9
36	81	39	4	5	1	1	-1	1
36	82	34	4	6	1	11	-9	10
36	82	39	9	11	2	1	-1	1
37	81	35	7	10	2	10	-9	10
37	82	36	8	11	2	11	-10	11
37	82	37	9	12	2	12	-11	12
37	83	30	4	7	1	10	-7	9
37	84	37	6	8	1	8	-7	8
37	85	38	7	9	1	9	-8	9
37	86	39	8	10	1	10	-9	10
37	86	39	9	11	1	11	-10	11
38	80	34	6	9	2	9	-8	9
38	81	32	7	11	2	11	-9	11
38	82	34	4	6	1	6	-5	6
38	83	36	5	7	1	7	-6	7
38	85	34	6	9	1	9	-7	9
38	85	35	7	10	1	5	-4	5
38	86	36	8	11	1	11	-9	11
38	86	37	9	12	1	6	-5	6
39	80	31	3	5	1	5	-4	5
39	82	30	7	12	2	4	-3	4
39	83	30	4	7	1	7	-5	7
39	84	32	5	8	1	4	-3	4
39	85	31	6	10	1	10	-7	10
39	85	38	7	9	1	10	-9	11
39	86	32	7	11	1	11	-8	11
39	86	34	8	12	1	4	-3	4
39	86	39	8	10	1	11	-10	12
30 80 40								
30	80	45	4	4	1	9	-11	6
30	81	45	9	9	2	10	-12	9
30	83	45	6	6	1	7	-8	6
30	84	45	7	7	1	8	-9	7
31	81	48	9	8	2	8	-11	8
31	82	48	10	9	2	3	-4	3
31	84	49	7	6	1	3	-4	3
31	85	45	8	8	1	9	-10	8
31	86	40	10	12	1	10	-9	8
31	86	45	9	9	1	10	-11	9
32	81	42	8	9	2	11	-12	10
32	85	48	9	8	1	4	-5	4
32	85	49	8	7	1	7	-9	7
32	86	45	10	10	1	11	-12	10
32	86	48	10	9	1	9	-11	9
33	80	45	4	4	1	4	-5	4
33	61	45	9	9	2	9	-11	9
33	82	45	5	5	1	5	-6	5
33	83	45	6	6	1	6	-7	6
33	86	48	11	10	1	5	-6	5
34	80	43	11	12	3	6	-7	6
34	81	42	8	9	2	9	-10	9
34	82	42	9	10	2	10	-11	10
34	82	42	10	11	2	11	-12	11
34	84	45	7	7	1	7	-8	7
34	85	45	8	8	1	8	-9	8
34	86	40	10	12	1	11	-10	10
34	86	45	9	9	1	9	-10	9
34	86	45	10	10	1	10	-11	10
34	86	45	11	11	1	11	-12	11
35	82	48	11	10	2	6	-6	7
35	83	45	11	11	2	10	-12	11
35	85	41	7	8	1	1	-1	1
35	85	42	8	9	1	1	-1	1
35	86	42	9	10	1	1	-1	1
35	86	42	10	11	1	1	-1	1
35	86	43	11	12	1	1	-1	1
36	80	45	4	4	1	7	-9	8
36	81	45	9	9	2	8	-10	9
36	82	45	5	5	1	9	-11	10
36	83	40	5	6	1	1	-1	1
36	84	41	6	7	1	1	-1	1
36	86	40	10	12	1	12	-11	12
36	87	45	12	12	1	11	-12	12
37	82	42	9	10	2	8	-9	9
37	82	42	10	11	2	9	-10	10
37	83	43	11	12	2	10	-11	11
37	84	45	7	7	1	6	-7	7
37	84	49	7	6	1	8	-11	10
37	85	45	8	8	1	7	-8	8
37	86	45	9	9	1	8	-9	9
37	86	45	10	10	1	9	-10	10
37	86	45	11	11	1	10	-11	11
38	80	43	11	12	3	9	-11	11
38	81	42	8	9	2	7	-8	8
38	82	45	5	5	1	4	-5	5
38	83	45	6	6	1	5	-6	6
38	83	45	11	11	2	9	-11	11
38	85	48	9	8	1	7	-9	9
38	86	47	12	11	1	5	-6	6
38	86	48	10	9	1	4	-5	5
38	86	48	11	10	1	9	-11	11
39	80	45	4	4	1	3	-4	4
39	81	45	9	9	2	7	-9	9
39	82	48	11	10	2	8	-11	11
39	83	47	12	11	2	3	-4	4
39	84	49	7	6	1	5	-7	7
39	85	49	8	7	1	3	-4	4
39	86	45	11	11	1	9	-10	11
39	87	45	12	12	1	10	-11	12
30 80 50								
30	83	50	6	5	1	5	-7	5
30	85	52	9	7	1	7	-10	7
31	82	54	11	8	2	6	-10	7
31	82	56	6	4	1	5	-9	6
31	86	51	10	8	1	8	-11	8
31	86	51	11	9	1	3	-4	3
32	86	58	11	7	1	4	-7	5
33	82	53	12	9	2	5	-8	6
33	82	56	6	4	1	6	-11	8
33	85	55	10	7	1	5	-8	6
33	85	56	9	6	1	7	-12	9
34	80	59	5	3	1	5	-11	8
34	81	58	11	7	2	2	-4	3
34	83	50	6	5	1	7	-10	8
35	80	52	9	7	2	3	-5	4
35	80	56	9	6	2	2	-4	3
35	81	51	5	4	1	7	-11	9
35	82	51	11	9	2	4	-6	5
35	83	54	7	5	1	3	-5	4
35	86	51	10	8	1	5	-7	6
35	86	56	12	8	1	3	-5	4
36	60	59	5	3	1	4	-9	7
37	81	51	5	4	1	5	-8	7
37	82	54	11	8	2	4	-7	6
37	85	56	9	6	1	4	-7	6

φ_1	Φ	φ_2	H	K	L	U	V	W
30 80 50								
38	82	53	12	9	2	6	-10	9
38	84	53	8	6	1	7	-11	10
38	86	58	11	7	1	5	-9	8
39	80	59	5	3	1	3	-7	6
39	83	54	7	5	1	7	-12	11
39	86	50	12	10	1	3	-4	4
39	86	51	10	8	1	7	-10	10
39	86	51	11	9	1	8	-11	11
30 80 60								
30	83	69	8	3	1	3	-10	6
31	80	63	10	5	2	3	-8	5
31	84	66	9	4	1	3	-8	5
31	86	63	12	6	1	4	-9	6
31	86	67	12	5	1	4	-11	7
32	84	63	8	4	1	5	-12	8
32	85	68	10	4	1	1	-3	2
33	81	66	11	5	2	1	-3	2
33	83	67	7	3	1	1	-3	2
33	85	63	10	5	1	3	-7	5
34	81	67	12	5	2	3	-10	7
35	83	69	8	3	1	2	-7	5
35	84	66	9	4	1	4	-11	8
36	82	63	6	3	1	3	-8	6
37	81	61	11	6	2	2	-5	4
37	83	60	7	4	1	5	-11	9
37	84	63	8	4	1	2	-5	4
38	82	63	6	3	1	4	-11	9
38	85	63	10	5	1	5	-12	10
38	86	63	12	6	1	3	-7	6
39	80	63	10	5	2	2	-6	5
39	83	60	7	4	1	4	-9	8
39	83	69	8	3	1	3	-11	9
39	85	66	11	5	1	3	-8	7
30 80 70								
30	80	75	11	3	2	2	-12	7
30	84	79	10	2	1	1	-7	4
31	82	74	7	2	1	1	-5	3
31	85	76	12	3	1	1	-5	3
32	80	70	11	4	2	2	-8	5
34	81	72	6	2	1	2	-9	6
34	83	76	8	2	1	1	-6	4
34	85	70	11	4	1	3	-10	7
35	85	72	12	4	1	3	-11	8
36	80	75	11	3	2	1	-7	5
36	82	74	7	2	1	2	-11	8
36	84	72	9	3	1	1	-4	3
36	84	77	9	2	1	1	-7	5
37	81	76	12	3	2	1	-8	6
37	84	79	10	2	1	1	-8	6
37	85	73	10	3	1	2	-9	7
38	80	70	11	4	2	2	-9	7
38	85	75	11	3	1	1	-5	4
39	81	72	6	2	1	1	-5	4
39	85	76	12	3	1	2	-11	9
30 80 80								
32	85	80	11	2	1	1	-8	5
34	85	81	12	2	1	1	-9	6
38	85	80	11	2	1	1	-9	7
39	85	81	12	2	1	1	-10	8
30 80 90								
31	81	90	6	0	1	-1	-10	6
31	82	90	7	0	1	-1	-12	7
33	82	90	7	0	1	-1	-11	7
34	81	90	6	0	1	-1	-9	6
34	83	90	8	0	1	-1	-12	8
35	82	90	7	0	1	-1	-10	7
36	83	90	8	0	1	-1	-11	8
37	81	90	6	0	1	-1	-8	6
37	84	90	9	0	1	-1	-12	9
38	82	90	7	0	1	-1	-9	7
39	83	90	8	0	1	-1	-10	8
39	84	90	9	0	1	-1	-11	9
30 90 0								
30	90	0	0	1	0	12	0	7
30	90	0	0	1	0	7	0	4
30	90	5	1	12	0	12	-1	7
30	90	9	1	6	0	12	-2	7
31	90	0	0	1	0	5	0	3
31	90	6	1	10	0	10	-1	6
32	90	0	0	1	0	11	0	7
32	90	0	0	1	0	8	0	5
32	90	5	1	11	0	11	-1	7
32	90	7	1	8	0	8	-1	5
33	90	9	1	6	0	6	-1	4
34	90	0	0	1	0	3	0	2
34	90	5	1	12	0	12	-1	8
34	90	6	1	9	0	9	-1	6
35	90	0	0	1	0	10	0	7
35	90	6	1	10	0	10	-1	7
35	90	8	1	7	0	7	-1	5
36	90	0	0	1	0	7	0	5
36	90	0	0	1	0	11	0	8
36	90	5	1	11	0	11	-1	8
36	90	9	1	6	0	12	-2	9
37	90	0	0	1	0	4	0	3
37	90	5	1	12	0	12	-1	9
37	90	7	1	8	0	8	-1	6
38	90	0	0	1	0	9	0	7
38	90	6	1	9	0	9	-1	7
39	90	0	0	1	0	5	0	4
39	90	0	0	1	0	11	0	9
39	90	5	1	11	0	11	-1	9
39	90	6	1	10	0	10	-1	8
39	90	9	1	6	0	6	-1	5
30 90 10								
30	90	11	1	5	0	5	-1	3
30	90	14	1	4	0	12	-3	7
30	90	17	3	10	0	10	-3	6
31	90	14	1	4	0	8	-2	5
32	90	10	2	11	0	11	-2	7
32	90	15	3	11	0	11	-3	7
32	90	18	1	3	0	3	-1	2
33	90	13	2	9	0	9	-2	6
33	90	14	1	4	0	12	-3	8
34	90	11	1	5	0	10	-2	7
34	90	16	2	7	0	7	-2	5
34	90	17	3	10	0	10	-3	7
35	90	15	3	11	0	11	-3	8
35	90	18	1	3	0	12	-4	9
36	90	10	2	11	0	11	-2	8
36	90	14	1	4	0	4	-1	3
36	90	18	1	3	0	9	-3	7
37	90	13	2	9	0	9	-2	7
37	90	17	3	10	0	10	-3	8
38	90	11	1	5	0	5	-1	4
38	90	15	3	11	0	11	-3	9
38	90	18	1	3	0	6	-2	5
39	90	10	2	11	0	11	-2	9
39	90	14	1	4	0	12	-3	10
39	90	16	2	7	0	7	-2	6
30 90 20								
30	90	21	3	6	0	6	-3	5
30	90	24	5	11	0	11	-5	7
30	90	29	5	9	0	9	-5	6
31	90	20	4	11	0	11	-4	7
31	90	24	4	9	0	9	-4	6
31	90	27	1	2	0	6	-3	4
32	90	23	5	12	0	12	-5	8
32	90	27	1	2	0	10	-5	7
33	90	22	2	5	0	10	-4	7
33	90	23	3	7	0	7	-3	5
33	90	29	6	11	0	11	-6	8
34	90	20	4	11	0	11	-4	8
34	90	24	5	11	0	11	-5	8
34	90	27	1	2	0	4	-2	3
34	90	29	5	9	0	9	-5	7
35	90	21	3	8	0	8	-3	6
35	90	23	5	12	0	12	-5	9
35	90	24	4	9	0	9	-4	7
36	90	27	1	2	0	10	-5	8
36	90	29	6	11	0	11	-6	9
37	90	22	2	5	0	5	-2	4
37	90	24	5	11	0	11	-5	9
37	90	27	1	2	0	6	-3	5
38	90	20	4	11	0	11	-4	9
38	90	23	3	7	0	7	-3	6
38	90	23	5	12	0	12	-5	10
38	90	27	1	2	0	8	-4	7
36	90	29	5	9	0	9	-5	8
39	90	21	3	8	0	8	-3	7
39	90	24	4	9	0	9	-4	8
39	90	27	1	2	0	12	-6	11
39	90	27	1	2	0	10	-5	9
39	90	29	6	11	0	11	-6	10
30 90 30								
30	90	30	7	12	0	12	-7	8
30	90	35	7	10	0	10	-7	7
30	90	36	5	7	0	7	-5	5
30	90	36	8	11	0	11	-8	8
31	90	31	3	5	0	10	-6	7
31	90	37	3	4	0	4	-3	3
32	90	30	4	7	0	7	-4	5
32	90	32	5	8	0	8	-5	6
32	90	32	7	11	0	11	-7	8
32	90	34	2	3	0	12	-8	9
32	90	38	7	9	0	9	-7	7

φ_1	Φ	φ_2	H	K	L	U	V	W
30	90	30						
32	90	39	4	5	0	5	-4	4
32	90	39	9	11	0	11	-9	9
33	90	30	7	12	0	12	-7	9
33	90	34	2	3	0	9	-6	7
33	90	35	7	10	0	10	-7	8
33	90	36	8	11	0	11	-8	9
34	90	31	3	5	0	5	-3	4
34	90	37	3	4	0	12	-9	10
35	90	32	7	11	0	11	-7	9
35	90	34	2	3	0	6	-4	5
35	90	36	5	7	0	7	-5	6
35	90	37	3	4	0	8	-6	7
35	90	38	7	9	0	9	-7	8
35	90	39	4	5	0	10	-8	9
35	90	39	9	11	0	11	-9	10
36	90	30	7	12	0	12	-7	10
36	90	34	2	3	0	9	-6	8
36	90	35	7	10	0	10	-7	9
36	90	36	8	11	0	11	-8	10
36	90	37	3	4	0	12	-9	11
37	90	30	4	7	0	7	-4	6
37	90	32	5	8	0	8	-5	7
37	90	32	7	11	0	11	-7	10
37	90	34	2	3	0	12	-8	11
38	90	30	7	12	0	12	-7	11
38	90	31	3	5	0	10	-6	9
38	90	38	7	9	0	9	-7	9
38	90	39	4	5	0	5	-4	5
38	90	39	9	11	0	11	-9	11
39	90	35	7	10	0	10	-7	10
39	90	36	5	7	0	7	-5	7
39	90	36	8	11	0	11	-8	11
39	90	37	3	4	0	4	-3	4
30	90	40						
30	90	40	5	6	0	12	-10	9
30	90	42	8	9	0	9	-8	7
30	90	45	1	1	0	11	-11	9
30	90	48	9	8	0	8	-9	7
31	90	42	9	10	0	10	-9	8
31	90	42	10	11	0	11	-10	9
31	90	45	1	1	0	6	-6	5
31	90	45	1	1	0	7	-7	6
31	90	48	10	9	0	9	-10	8
31	90	48	11	10	0	10	-11	9
32	90	43	11	12	0	12	-11	10
32	90	45	1	1	0	9	-9	8

φ_1	Φ	φ_2	H	K	L	U	V	W
30	90	40						
32	90	45	1	1	0	8	-8	7
32	90	45	1	1	0	10	-10	9
32	90	47	12	11	0	11	-12	10
33	90	40	5	6	0	6	-5	5
33	90	41	6	7	0	7	-6	6
33	90	41	7	8	0	8	-7	7
33	90	45	1	1	0	11	-11	10
33	90	45	1	1	0	12	-12	11
33	90	49	7	6	0	6	-7	6
33	90	49	8	7	0	7	-8	7
34	90	42	8	9	0	9	-8	8
34	90	42	9	10	0	10	-9	9
34	90	42	10	11	0	11	-10	10
34	90	43	11	12	0	12	-11	11
34	90	47	12	11	0	11	-12	11
34	90	48	9	8	0	8	-9	8
34	90	48	10	9	0	9	-10	9
34	90	48	11	10	0	10	-11	10
35	90	40	5	6	0	12	-10	11
35	90	45	1	1	0	1	-1	1
36	90	42	10	11	0	11	-10	11
36	90	43	11	12	0	12	-11	12
36	90	47	12	11	0	11	-12	12
36	90	48	11	10	0	10	-11	11
37	90	41	6	7	0	7	-6	7
37	90	41	7	8	0	8	-7	8
37	90	42	8	9	0	9	-8	9
37	90	42	9	10	0	10	-9	10
37	90	48	9	8	0	8	-9	9
37	90	48	10	9	0	9	-10	10
37	90	49	7	6	0	6	-7	7
37	90	49	8	7	0	7	-8	8
38	90	40	5	6	0	6	-5	6
38	90	45	1	1	0	9	-9	10
38	90	45	1	1	0	11	-11	12
38	90	45	1	1	0	10	-10	11
39	90	42	9	10	0	10	-9	11
39	90	42	10	11	0	11	-10	12
39	90	45	1	1	0	7	-7	8
39	90	45	1	1	0	8	-8	9
39	90	48	10	9	0	9	-10	11
39	90	48	11	10	0	10	-11	12
30	90	50						
30	90	50	6	5	0	10	-12	9
30	90	54	7	5	0	5	-7	5
30	90	54	11	8	0	8	-11	8

φ_1	Φ	φ_2	H	K	L	U	V	W
30	90	50						
30	90	55	10	7	0	7	-10	7
31	90	53	4	3	0	3	-4	3
31	90	59	5	3	0	6	-10	7
32	90	51	5	4	0	4	-5	4
32	90	51	11	9	0	9	-11	9
32	90	52	9	7	0	7	-9	7
32	90	56	3	2	0	8	-12	9
32	90	58	8	5	0	5	-8	6
32	90	58	11	7	0	7	-11	8
33	90	50	6	5	0	5	-6	5
33	90	54	11	8	0	8	-11	9
33	90	55	10	7	0	7	-10	8
33	90	56	3	2	0	6	-9	7
34	90	53	4	3	0	9	-12	10
34	90	59	5	3	0	3	-5	4
35	90	50	6	5	0	10	-12	11
35	90	51	5	4	0	8	-10	9
35	90	51	11	9	0	9	-11	10
35	90	52	9	7	0	7	-9	8
35	90	53	4	3	0	6	-8	7
35	90	54	7	5	0	5	-7	6
35	90	56	3	2	0	4	-6	5
35	90	58	11	7	0	7	-11	9
36	90	53	4	3	0	9	-12	11
36	90	54	11	8	0	8	-11	10
36	90	55	10	7	0	7	-10	9
36	90	56	3	2	0	6	-9	8
37	90	56	3	2	0	8	-12	11
37	90	58	8	5	0	5	-8	7
37	90	58	11	7	0	7	-11	10
38	90	50	6	5	0	5	-6	6
38	90	51	5	4	0	4	-5	5
38	90	51	11	9	0	9	-11	11
38	90	52	9	7	0	7	-9	9
38	90	59	5	3	0	6	-10	9
39	90	53	4	3	0	3	-4	4
39	90	54	7	5	0	5	-7	7
39	90	54	11	8	0	8	-11	11
39	90	55	10	7	0	7	-10	10
30	90	60						
30	90	60	12	7	0	7	-12	8
30	90	61	9	5	0	5	-9	6
30	90	66	11	5	0	5	-11	7
30	90	69	8	3	0	3	-8	5
31	90	63	2	1	0	3	-6	4
31	90	66	9	4	0	4	-9	6

φ_1	Φ	φ_2	H	K	L	U	V	W
30	90	60						
32	90	60	7	4	0	4	-7	5
32	90	63	2	1	0	5	-10	7
32	90	67	12	5	0	5	-12	8
33	90	60	12	7	0	7	-12	9
33	90	61	11	6	0	6	-11	8
33	90	67	7	3	0	3	-7	5
33	90	68	5	2	0	4	-10	7
34	90	61	9	5	0	5	-9	7
34	90	63	2	1	0	2	-4	3
34	90	66	11	5	0	5	-11	8
35	90	66	9	4	0	4	-9	7
35	90	67	12	5	0	5	-12	9
35	90	69	8	3	0	3	-8	6
36	90	60	12	7	0	7	-12	10
36	90	61	11	6	0	6	-11	9
36	90	63	2	1	0	5	-10	8
37	90	60	7	4	0	4	-7	6
37	90	63	2	1	0	3	-6	5
37	90	66	11	5	0	5	-11	9
37	90	68	5	2	0	2	-5	4
38	90	60	12	7	0	7	-12	11
38	90	61	9	5	0	5	-9	8
38	90	63	2	1	0	4	-8	7
38	90	67	7	3	0	3	-7	6
38	90	67	12	5	0	5	-12	10
39	90	61	11	6	0	6	-11	10
39	90	63	2	1	0	6	-12	11
39	90	63	2	1	0	5	-10	9
39	90	66	9	4	0	4	-9	8
39	90	69	8	3	0	3	-8	7
30	90	70						
30	90	73	10	3	0	3	-10	6
30	90	76	4	1	0	3	-12	7
30	90	79	5	1	0	1	-5	3
31	90	70	11	4	0	4	-11	7
31	90	76	4	1	0	2	-8	5
32	90	72	3	1	0	1	-3	2
32	90	75	11	3	0	3	-11	7
33	90	76	4	1	0	3	-12	8
33	90	77	9	2	0	2	-9	6
34	90	70	11	4	0	4	-11	8
34	90	73	10	3	0	3	-10	7
34	90	74	7	2	0	2	-7	5
34	90	79	5	1	0	2	-10	7
35	90	72	3	1	0	4	-12	9
35	90	75	11	3	0	3	-11	8

φ_1	Φ	φ_2	H	K	L	U	V	W
30	90	70						
36	90	72	3	1	0	3	-9	7
36	90	76	4	1	0	1	-4	3
37	90	73	10	3	0	3	-10	8
37	90	77	9	2	0	2	-9	7
38	90	70	11	4	0	4	-11	9
38	90	72	3	1	0	2	-6	5
38	90	75	11	3	0	3	-11	9
38	90	79	5	1	0	1	-5	4
39	90	74	7	2	0	2	-7	6
39	90	76	4	1	0	3	-12	10
30	90	80						
30	90	81	6	1	0	2	-12	7
30	90	85	12	1	0	1	-12	7
31	90	84	10	1	0	1	-10	6
32	90	80	11	2	0	2	-11	7
32	90	83	8	1	0	1	-8	5
32	90	85	11	1	0	1	-11	7
33	90	81	6	1	0	1	-6	4
34	90	84	9	1	0	1	-9	6
34	90	85	12	1	0	1	-12	8
35	90	82	7	1	0	1	-7	5
35	90	84	10	1	0	1	-10	7
36	90	80	11	2	0	2	-11	8
36	90	81	6	1	0	2	-12	9
36	90	85	11	1	0	1	-11	8
37	90	83	8	1	0	1	-8	6
37	90	85	12	1	0	1	-12	9
38	90	84	9	1	0	1	-9	7
39	90	80	11	2	0	2	-11	9
39	90	81	6	1	0	1	-6	5
39	90	84	10	1	0	1	-10	8
39	90	85	11	1	0	1	-11	9
30	90	90						
30	90	90	1	0	0	0	-12	7
30	90	90	1	0	0	0	-7	4
31	90	90	1	0	0	0	-5	3
32	90	90	1	0	0	0	-11	7
32	90	90	1	0	0	0	-8	5
34	90	90	1	0	0	0	-3	2
35	90	90	1	0	0	0	-10	7
36	90	90	1	0	0	0	-7	5
36	90	90	1	0	0	0	-11	8
37	90	90	1	0	0	0	-4	3
38	90	90	1	0	0	0	-9	7

φ_1	Φ	φ_2	H	K	L	U	V	W
30	90	90						
39	90	90	1	0	0	0	-5	4
39	90	90	1	0	0	0	-11	9
40	0	0						
40	6	0	0	1	10	12	-10	1
41	8	0	0	1	7	8	-7	1
41	9	0	0	1	6	7	-6	1
42	6	0	0	1	9	10	-9	1
42	6	0	0	1	10	11	-10	1
42	7	0	0	1	8	9	-8	1
43	5	0	0	1	11	12	-11	1
45	5	0	0	1	11	11	-11	1
45	5	0	0	1	12	12	-12	1
45	6	0	0	1	9	9	-9	1
45	6	0	0	1	10	10	-10	1
45	7	0	0	1	8	8	-8	1
45	8	0	0	1	7	7	-7	1
45	9	0	0	1	6	6	-6	1
48	5	0	0	1	11	10	-11	1
48	5	0	0	1	12	11	-12	1
48	6	0	0	1	10	9	-10	1
48	9	0	0	1	6	11	-12	2
49	6	0	0	1	9	8	-9	1
49	7	0	0	1	8	7	-8	1
40	0	40						
40	0	40	0	0	1	1	-6	0
40	0	40	0	0	1	2	-11	0
40	7	45	1	1	11	1	-12	1
40	8	45	1	1	10	1	-11	1
40	9	45	1	1	9	1	-10	1
41	0	41	0	0	1	1	-8	0
41	0	41	0	0	1	1	-7	0
42	0	42	0	0	1	1	-10	0
42	0	42	0	0	1	1	-11	0
42	0	42	0	0	1	1	-9	0
43	0	43	0	0	1	1	-12	0
45	0	45	0	0	1	0	-1	0
45	7	45	1	1	11	0	-11	1
45	7	45	1	1	12	0	-12	1
45	8	45	1	1	10	0	-10	1
45	9	45	1	1	9	0	-9	1
47	0	47	0	0	1	-1	-12	0
48	0	48	0	0	1	-1	-10	0
48	0	48	0	0	1	-1	-11	0
48	0	48	0	0	1	-1	-9	0

φ_1	Φ	φ_2	H	K	L	U	V	W
40	0	40						
49	0	49	0	0	1	-1	-6	0
49	0	49	0	0	1	-1	-7	0
40	0	90						
40	6	90	1	0	10	-10	-12	1
41	8	90	1	0	7	-7	-8	1
41	9	90	1	0	6	-6	-7	1
42	6	90	1	0	9	-9	-10	1
42	6	90	1	0	10	-10	-11	1
42	7	90	1	0	8	-8	-9	1
43	5	90	1	0	11	-11	-12	1
45	5	90	1	0	11	-11	-11	1
45	5	90	1	0	12	-12	-12	1
45	6	90	1	0	9	-9	-9	1
45	6	90	1	0	10	-10	-10	1
45	7	90	1	0	8	-8	-8	1
45	8	90	1	0	7	-7	-7	1
45	9	90	1	0	6	-6	-6	1
48	5	90	1	0	11	-11	-10	1
48	5	90	1	0	12	-12	-11	1
48	6	90	1	0	10	-10	-9	1
48	9	90	1	0	6	-12	-11	2
49	6	90	1	0	9	-9	-8	1
49	7	90	1	0	8	-8	-7	1
40	10	0						
40	11	0	0	1	5	6	-5	1
40	13	0	0	2	9	11	-9	2
40	14	0	0	1	4	5	-4	1
41	17	0	0	3	10	12	-10	3
41	18	0	0	1	3	11	-9	3
42	14	0	0	1	4	9	-8	2
42	16	0	0	2	7	8	-7	2
42	18	0	0	1	3	7	-6	2
43	10	0	0	2	11	12	-11	2
43	11	0	0	1	5	11	-10	2
43	13	0	0	2	9	10	-9	2
43	18	0	0	1	3	10	-9	3
44	15	0	0	3	11	12	-11	3
44	17	0	0	3	10	11	-10	3
45	10	0	0	2	11	11	-11	2
46	11	0	0	1	5	5	-5	1
46	13	0	0	2	9	9	-9	2
46	14	0	0	1	4	4	-4	1
46	15	0	0	3	11	11	-11	3
46	16	0	0	2	7	7	-7	2

φ_1	Φ	φ_2	H	K	L	U	V	W
40	10	0						
46	17	0	0	3	10	10	-10	3
47	18	0	0	1	3	3	-3	1
48	10	0	0	2	11	10	-11	2
48	14	0	0	1	4	11	-12	3
49	11	0	0	1	5	9	-10	2
49	13	0	0	2	9	8	-9	2
49	15	0	0	3	11	10	-11	3
49	17	0	0	3	10	9	-10	3
49	18	0	0	1	3	11	-12	4
40	10	10						
42	15	18	1	3	12	3	-5	1
42	19	14	1	4	12	8	-11	3
44	16	18	1	3	11	5	-9	2
45	19	18	1	3	9	6	-11	3
46	18	18	1	3	10	2	-4	1
48	19	14	1	4	12	4	-7	2
40	10	20						
41	11	27	1	2	11	3	-7	1
41	18	27	1	2	7	4	-9	2
43	16	27	1	2	8	2	-5	1
44	14	27	1	2	9	4	-11	2
46	13	27	1	2	10	2	-6	1
48	11	27	1	2	12	2	-7	1
40	10	30						
41	17	34	2	3	12	3	-10	2
40	10	40						
40	19	45	1	1	4	1	-9	2
41	14	45	2	2	11	1	-12	2
41	16	45	1	1	5	1	-11	2
41	17	45	2	2	9	1	-10	2
45	10	45	1	1	8	0	-8	1
46	11	45	1	1	7	0	-7	1
46	13	45	1	1	6	0	-6	1
46	14	45	2	2	11	0	-11	2
46	16	45	1	1	5	0	-5	1
46	17	45	2	2	9	0	-9	2
47	19	45	1	1	4	0	-4	1

φ_1	Φ	φ_2	H	K	L	U	V	W
40	10	50						
47	17	56	3	2	12	-2	-9	2
49	18	56	3	2	11	-1	-4	1
40	10	60						
41	11	63	2	1	12	-2	-8	1
41	14	63	2	1	9	-3	-12	2
43	11	63	2	1	11	-2	-7	1
44	16	63	2	1	8	-3	-10	2
49	14	63	2	1	9	-2	-5	1
49	18	63	2	1	7	-3	-8	2
40	10	70						
41	16	72	3	1	11	-2	-5	1
42	19	76	4	1	12	-2	-4	1
43	19	72	3	1	9	-5	-12	3
48	15	72	3	1	12	-5	-9	2
40	10	90						
40	11	90	1	0	5	-5	-6	1
40	13	90	2	0	9	-9	-11	2
40	14	90	1	0	4	-4	-5	1
41	17	90	3	0	10	-10	-12	3
41	18	90	1	0	3	-9	-11	3
42	14	90	1	0	4	-8	-9	2
42	16	90	2	0	7	-7	-8	2
42	18	90	1	0	3	-6	-7	2
43	10	90	2	0	11	-11	-12	2
43	11	90	1	0	5	-10	-11	2
43	13	90	2	0	9	-9	-10	2
43	18	90	1	0	3	-9	-10	3
44	15	90	3	0	11	-11	-12	3
44	17	90	3	0	10	-10	-11	3
45	10	90	2	0	11	-11	-11	2
46	11	90	1	0	5	-5	-5	1
46	13	90	2	0	9	-9	-9	2
46	14	90	1	0	4	-4	-4	1
46	15	90	3	0	11	-11	-11	3
46	16	90	2	0	7	-7	-7	2
46	17	90	3	0	10	-10	-10	3
47	18	90	1	0	3	-3	-3	1
48	10	90	2	0	11	-11	-10	2
48	14	90	1	0	4	-12	-11	3
49	11	90	1	0	5	-10	-9	2
49	13	90	2	0	9	-9	-8	2

φ_1	Φ	φ_2	H	K	L	U	V	W
40	10	90						
49	15	90	3	0	11	-11	-10	3
49	17	90	3	0	10	-10	-9	3
49	18	90	1	0	3	-12	-11	4
40	20	0						
40	23	0	0	3	7	9	-7	3
40	27	0	0	1	2	8	-6	3
41	21	0	0	3	8	10	-8	3
41	29	0	0	5	9	12	-9	5
42	22	0	0	2	5	6	-5	2
42	24	0	0	4	9	11	-9	4
42	27	0	0	1	2	5	-4	2
43	27	0	0	1	2	12	-10	5
43	27	9	1	6	12	6	-7	3
43	29	0	0	5	9	11	-9	5
44	20	0	0	4	11	12	-11	4
44	21	0	0	3	8	9	-8	3
44	22	0	0	2	5	11	-10	4
44	23	0	0	3	7	8	-7	3
44	27	0	0	1	2	7	-6	3
45	24	0	0	4	9	10	-9	4
45	24	0	0	5	11	12	-11	5
45	27	0	0	1	2	11	-10	5
45	27	0	0	1	2	9	-8	4
46	29	0	0	5	9	10	-9	5
46	29	0	0	6	11	12	-11	6
47	20	0	0	4	11	11	-11	4
47	21	0	0	3	8	8	-8	3
47	22	0	0	2	5	5	-5	2
47	23	0	0	3	7	7	-7	3
47	23	0	0	5	12	12	-12	5
48	24	0	0	4	9	9	-9	4
48	24	0	0	5	11	11	-11	5
48	27	0	0	1	2	2	-2	1
49	20	0	0	4	11	10	-11	4
49	29	0	0	5	9	9	-9	5
49	29	0	0	6	11	11	-11	6
40	20	10						
40	24	18	1	3	7	2	-3	1
40	25	11	1	5	11	7	-8	3
41	27	14	1	4	8	4	-5	2
42	27	11	1	5	10	10	-12	5
43	21	14	1	4	11	7	-10	3
43	22	18	1	3	8	6	-10	3
43	25	14	1	4	9	8	-11	4

φ_1	Φ	φ_2	H	K	L	U	V	W
40	20	10						
44	28	18	1	3	6	3	-5	2
45	22	14	1	4	10	2	-3	1
46	27	11	1	5	10	5	-7	3
46	27	14	1	4	8	8	-12	5
47	25	14	1	4	9	5	-8	3
49	23	11	1	5	12	7	-11	4
49	28	18	1	3	6	6	-12	5
40	20	20						
40	26	27	2	4	9	3	-6	2
40	28	22	2	5	10	5	-8	3
41	29	27	1	2	4	4	-8	3
42	22	27	2	4	11	3	-7	2
44	24	22	2	5	12	3	-6	2
45	20	27	1	2	6	4	-11	3
45	28	22	2	5	10	5	-10	4
46	29	27	1	2	4	2	-5	2
47	26	27	2	4	9	4	-11	4
48	24	27	1	2	5	1	-3	1
49	20	27	1	2	6	2	-7	2
49	28	22	2	5	10	5	-12	5
40	20	30						
40	24	37	3	4	11	2	-7	2
41	27	34	2	3	7	1	-3	1
41	28	39	4	5	12	3	-12	4
42	24	34	2	3	8	3	-10	3
43	22	34	2	3	9	3	-11	3
44	20	34	2	3	10	1	-4	1
44	26	31	3	5	12	1	-3	1
44	27	37	3	4	10	2	-9	3
48	28	39	4	5	12	1	-8	3
49	24	34	2	3	8	1	-6	2
49	24	37	3	4	11	1	-9	3
40	20	40						
40	22	45	2	2	7	1	-8	2
40	25	45	1	1	3	1	-7	2
42	21	45	3	3	11	1	-12	3
42	23	45	3	3	10	1	-11	3
42	25	45	1	1	3	1	-10	3
42	28	45	3	3	8	1	-9	3
44	27	45	4	4	11	1	-12	4
44	29	45	2	2	5	1	-11	4
47	21	45	3	3	11	0	-11	3

φ_1	Φ	φ_2	H	K	L	U	V	W
40	20	40						
47	22	45	2	2	7	0	-7	2
47	23	45	3	3	10	0	-10	3
48	25	45	1	1	3	0	-3	1
48	27	45	4	4	11	0	-11	4
49	28	45	3	3	8	0	-8	3
49	29	45	2	2	5	0	-5	2
40	20	50						
40	27	53	4	3	10	0	-10	3
41	22	56	3	2	9	-1	-12	3
41	29	53	4	3	9	0	-3	1
42	28	51	5	4	12	0	-3	1
43	27	56	3	2	7	-1	-9	3
45	20	56	3	2	10	-2	-12	3
45	27	53	4	3	10	-1	-12	4
47	27	56	3	2	7	-2	-11	4
49	24	53	4	3	11	-1	-6	2
49	24	56	3	2	8	-2	-9	3
49	26	59	5	3	12	-3	-11	4
40	20	60						
40	24	68	5	2	12	-2	-7	2
40	29	63	2	1	4	-1	-6	2
42	20	63	2	1	6	-1	-4	1
43	26	68	5	2	11	-1	-3	1
44	26	63	4	2	9	-3	-12	4
44	29	63	2	1	4	-2	-8	3
45	22	63	4	2	11	-2	-7	2
47	28	68	5	2	10	-2	-5	2
47	29	63	2	1	4	-3	-10	4
49	29	63	2	1	4	-4	-12	5
40	20	70						
40	27	76	4	1	6	-5	-12	4
40	28	72	3	1	6	-1	-3	1
43	22	76	4	1	10	-5	-10	3
43	25	76	4	1	9	-3	-6	2
44	22	72	3	1	8	-3	-7	2
44	27	76	4	1	8	-4	-8	3
45	24	72	3	1	7	-4	-9	3
46	25	79	5	1	11	-5	-8	3
46	27	79	5	1	10	-3	-5	2
47	22	76	4	1	10	-7	-12	4
48	27	76	4	1	8	-7	-12	5

φ_1	Φ	φ_2	H	K	L	U	V	W
40 20 80								
46	27	81	6	1	12	-8	-12	5
40 20 90								
40	23	90	3	0	7	-7	-9	3
40	27	90	1	0	2	-6	-8	3
41	21	90	3	0	8	-8	-10	3
41	29	90	5	0	9	-9	-12	5
42	22	90	2	0	5	-5	-6	2
42	24	90	4	0	9	-9	-11	4
42	27	90	1	0	2	-4	-5	2
43	27	90	1	0	2	-10	-12	5
43	29	90	5	0	9	-9	-11	5
44	20	90	4	0	11	-11	-12	4
44	21	90	3	0	8	-8	-9	3
44	22	90	2	0	5	-10	-11	4
44	23	90	3	0	7	-7	-8	3
44	27	90	1	0	2	-6	-7	3
45	24	90	4	0	9	-9	-10	4
45	24	90	5	0	11	-11	-12	5
45	27	90	1	0	2	-10	-11	5
45	27	90	1	0	2	-8	-9	4
46	29	90	5	0	9	-9	-10	5
46	29	90	6	0	11	-11	-12	6
47	20	90	4	0	11	-11	-11	4
47	21	90	3	0	8	-8	-8	3
47	22	90	2	0	5	-5	-5	2
47	23	90	3	0	7	-7	-7	3
47	23	90	5	0	12	-12	-12	5
48	24	90	4	0	9	-9	-9	4
48	24	90	5	0	11	-11	-11	5
48	27	90	1	0	2	-2	-2	1
49	20	90	4	0	11	-11	-10	4
49	29	90	5	0	9	-9	-9	5
49	29	90	6	0	11	-11	-11	6
40 30 0								
40	31	0	0	3	5	7	-5	3
40	31	9	1	6	10	2	-2	1
40	36	7	1	8	11	9	-8	5
40	37	0	0	3	4	6	-4	3
41	31	8	1	7	12	2	-2	1
41	32	0	0	5	8	11	-8	5
41	34	9	1	6	9	9	-9	5
41	36	0	0	5	7	10	-7	5
42	30	0	0	4	7	9	-7	4

φ_1	Φ	φ_2	H	K	L	U	V	W
40 30 0								
42	33	8	1	7	11	11	-11	6
42	34	0	0	2	3	4	-3	2
42	37	0	0	3	4	11	-8	6
42	37	9	1	6	8	8	-8	5
42	39	0	0	4	5	7	-5	4
43	32	0	0	5	8	10	-8	5
43	34	7	1	8	12	12	-12	7
43	35	8	1	7	10	5	-5	3
44	31	0	0	3	5	6	-5	3
44	36	0	0	5	7	9	-7	5
44	36	7	1	8	11	11	-11	7
44	38	0	0	7	9	12	-9	7
44	38	8	1	7	9	3	-3	2
45	30	0	0	4	7	8	-7	4
45	33	8	1	7	11	8	-9	5
45	34	0	0	2	3	11	-9	6
45	34	9	1	6	9	6	-7	4
45	35	0	0	7	10	12	-10	7
45	37	0	0	3	4	5	-4	3
45	37	6	1	9	12	3	-3	2
45	37	9	1	6	8	10	-11	7
45	39	7	1	8	10	10	-10	7
46	32	0	0	5	8	9	-8	5
46	34	0	0	2	3	7	-6	4
46	38	0	0	7	9	11	-9	7
46	39	6	1	9	11	11	-11	8
47	31	0	0	3	5	11	-10	6
47	32	0	0	7	11	12	-11	7
47	34	0	0	2	3	10	-9	6
47	36	0	0	5	7	8	-7	5
47	39	0	0	4	5	6	-5	4
48	35	0	0	7	10	11	-10	7
48	37	0	0	3	4	9	-8	6
49	30	0	0	4	7	7	-7	4
49	30	0	0	7	12	12	-12	7
49	31	0	0	3	5	5	-5	3
49	34	9	1	6	9	9	-12	7
49	36	0	0	8	11	12	-11	8
49	38	0	0	7	9	10	-9	7
49	39	0	0	4	5	11	-10	8
49	39	7	1	8	10	6	-7	5
40 30 10								
40	30	11	1	5	9	10	-11	5
40	32	18	1	3	5	8	-11	5
40	35	18	2	6	9	3	-4	2
40	36	11	1	5	7	7	-7	4

φ_1	Φ	φ_2	H	K	L	U	V	W
40 30 10								
40	38	18	1	3	4	7	-9	5
41	30	14	1	4	7	9	-11	5
42	33	16	2	7	11	3	-4	2
42	36	11	1	5	7	11	-12	7
42	37	14	2	8	11	6	-7	4
42	38	13	2	9	12	9	-10	6
42	39	16	2	7	9	4	-5	3
43	38	18	1	3	4	5	-7	4
45	34	14	1	4	6	8	-11	6
45	38	18	1	3	4	8	-12	7
46	36	11	1	5	7	4	-5	3
47	30	18	2	6	11	5	-9	4
47	37	14	2	8	11	7	-10	6
47	38	13	2	9	12	6	-8	5
48	34	14	1	4	6	6	-9	5
48	38	18	1	3	4	3	-5	3
49	35	18	2	6	9	6	-11	6
49	39	16	2	7	9	7	-11	7
40 30 20								
41	35	21	3	6	12	4	-6	3
43	31	22	2	5	9	4	-7	3
43	32	23	3	7	12	5	-9	4
43	37	27	1	2	3	1	-2	1
44	33	27	2	4	7	4	-9	4
45	31	27	3	6	11	3	-7	3
46	35	23	3	7	11	1	-2	1
46	38	21	3	8	11	4	-7	4
46	38	22	2	5	7	5	-9	5
47	34	22	2	5	8	1	-2	1
47	39	24	4	9	12	6	-12	7
47	39	27	4	8	11	3	-7	4
48	31	22	2	5	9	5	-11	5
40 30 30								
40	31	34	2	3	6	3	-8	3
40	36	31	3	5	8	5	-11	5
41	33	31	3	5	9	5	-12	5
41	35	39	4	5	9	2	-7	3
41	39	30	4	7	10	1	-2	1
42	36	34	2	3	5	4	-11	5
42	36	36	5	7	12	3	-9	4
43	38	32	5	8	12	4	-10	5
43	39	34	4	6	9	3	-8	4
44	31	34	2	3	6	3	-10	4
44	36	37	3	4	7	3	-11	5

φ_1	Φ	φ_2	H	K	L	U	V	W
40 30 30								
44	39	39	4	5	6	1	-4	2
45	38	36	5	7	11	3	-10	5
46	33	31	3	5	9	3	-9	4
46	36	30	4	7	11	3	-8	4
47	31	34	2	3	6	3	-12	5
47	36	31	3	5	6	2	-6	3
48	36	34	2	3	5	1	-4	2
48	39	34	4	6	9	3	-11	6
49	36	36	5	7	12	2	-10	5
40 30 40								
41	35	45	1	1	2	2	-12	5
41	38	45	5	5	9	2	-11	5
42	31	45	3	3	7	1	-8	3
43	35	40	5	6	11	2	-9	4
43	35	45	1	1	2	1	-7	3
44	32	45	4	4	9	1	-10	4
44	35	45	1	1	2	1	-9	4
45	33	45	5	5	11	1	-12	5
45	39	45	4	4	7	1	-8	4
46	35	45	1	1	2	1	-11	5
46	38	40	5	6	10	2	-10	5
46	38	45	5	5	9	1	-10	5
47	38	41	6	7	12	1	-6	3
47	38	45	6	6	11	1	-12	6
47	38	49	7	6	12	0	-2	1
49	31	45	3	3	7	0	-7	3
49	31	45	5	5	12	0	-12	5
40 30 50								
40	35	50	6	5	11	1	-10	4
41	32	53	4	3	8	0	-8	3
41	36	54	7	5	12	0	-12	5
41	39	56	6	4	9	0	-9	4
42	38	54	7	5	11	0	-11	5
43	30	51	5	4	11	0	-11	4
43	36	53	4	3	7	0	-7	3
43	36	59	5	3	8	-1	-9	4
44	33	51	5	4	10	0	-5	2
44	35	51	5	4	9	0	-9	4
45	33	50	6	5	12	0	-12	5
45	36	56	3	2	5	-1	-11	5
45	39	56	6	4	9	-1	-12	6
46	30	59	5	3	10	-1	-5	2
46	32	53	4	3	8	-1	-12	5
46	35	50	6	5	11	0	-11	5

φ_1	Φ	φ_2	H	K	L	U	V	W
40	30	50						
46	38	58	8	5	12	-1	-8	4
46	39	51	5	4	8	0	-2	1
47	38	50	6	5	10	0	-2	1
48	31	56	3	2	6	-2	-12	5
48	36	59	5	3	8	-2	-10	5
49	36	56	3	2	5	-1	-6	3
40	30	60						
40	35	67	7	3	11	-1	-5	2
40	35	69	8	3	12	-3	-12	5
40	36	60	7	4	11	-1	-12	5
40	39	63	8	4	11	-1	-9	4
41	37	63	2	1	3	-1	-7	3
41	38	68	5	2	7	-2	-9	4
43	33	63	4	2	7	-1	-5	2
43	34	68	5	2	8	-2	-7	3
43	37	63	2	1	3	-2	-11	5
45	38	68	5	2	7	-3	-10	5
46	32	67	7	3	12	-3	-9	4
47	31	63	6	3	11	-2	-7	3
47	35	69	8	3	12	-3	-8	4
47	37	63	2	1	3	-1	-4	2
48	38	68	5	2	7	-4	-11	6
49	34	68	5	2	8	-4	-10	5
49	39	60	7	4	10	-2	-9	5
40	30	70						
40	30	79	5	1	9	-5	-11	4
40	38	72	3	1	4	-3	-11	5
42	31	74	7	2	12	-2	-5	2
42	36	74	7	2	10	-4	-11	5
43	35	72	6	2	9	-3	-9	4
44	30	72	6	2	11	-2	-5	2
44	36	79	5	1	7	-1	-2	1
44	38	72	3	1	4	-2	-6	3
45	30	79	5	1	9	-4	-7	3
46	34	76	4	1	6	-1	-2	1
47	30	76	4	1	7	-6	-11	5
47	33	79	5	1	8	-6	-10	5
47	35	72	6	2	9	-5	-12	6
48	33	74	7	2	11	-1	-2	1
49	38	72	3	1	4	-3	-7	4
40	30	80						
40	31	81	6	1	10	-5	-10	4

φ_1	Φ	φ_2	H	K	L	U	V	W
40	30	80						
40	39	84	9	1	11	-1	-2	1
41	38	82	7	1	9	-1	-2	1
41	39	83	8	1	10	-1	-2	1
42	33	82	7	1	11	-5	-9	4
42	37	81	6	1	8	-1	-2	1
45	31	82	7	1	12	-7	-11	5
45	34	81	6	1	9	-7	-12	6
46	35	82	7	1	10	-7	-11	6
46	37	84	9	1	12	-8	-12	7
48	31	81	6	1	10	-4	-6	3
48	36	83	8	1	11	-7	-10	6
49	34	81	6	1	9	-6	-9	5
49	34	83	8	1	12	-9	-12	7
40	30	90						
40	31	90	3	0	5	-5	-7	3
40	37	90	3	0	4	-4	-6	3
41	32	90	5	0	8	-8	-11	5
41	36	90	5	0	7	-7	-10	5
42	30	90	4	0	7	-7	-9	4
42	34	90	2	0	3	-3	-4	2
42	37	90	3	0	4	-8	-11	6
42	39	90	4	0	5	-5	-7	4
43	32	90	5	0	8	-8	-10	5
44	31	90	3	0	5	-5	-6	3
44	36	90	5	0	7	-7	-9	5
44	38	90	7	0	9	-9	-12	7
45	30	90	4	0	7	-7	-8	4
45	34	90	2	0	3	-9	-11	6
45	35	90	7	0	10	-10	-12	7
45	37	90	3	0	4	-4	-5	3
46	32	90	5	0	8	-8	-9	5
46	34	90	2	0	3	-6	-7	4
46	38	90	7	0	9	-9	-11	7
47	31	90	3	0	5	-10	-11	6
47	32	90	7	0	11	-11	-12	7
47	34	90	2	0	3	-9	-10	6
47	36	90	5	0	7	-7	-8	5
47	39	90	4	0	5	-5	-6	4
48	35	90	7	0	10	-10	-11	7
48	37	90	3	0	4	-8	-9	6
49	30	90	4	0	7	-7	-7	4
49	30	90	7	0	12	-12	-12	7
49	31	90	3	0	5	-5	-5	3
49	36	90	8	0	11	-11	-12	8
49	38	90	7	0	9	-9	-10	7
49	39	90	4	0	5	-10	-11	8

φ_1	Φ	φ_2	H	K	L	U	V	W
40	40	0						
40	41	0	0	6	7	11	-7	6
40	45	0	0	1	1	5	-3	3
40	45	0	0	1	1	12	-7	7
40	45	5	1	11	11	11	-8	7
40	45	9	1	6	6	6	-5	4
40	49	0	0	7	6	11	-6	7
41	40	0	0	5	6	9	-6	5
41	41	9	1	6	7	11	-10	7
41	42	6	1	9	10	11	-9	7
41	45	0	0	1	1	8	-5	5
41	45	6	1	9	9	9	-7	6
42	41	0	0	7	8	12	-8	7
42	45	0	0	1	1	11	-7	7
42	45	5	1	12	12	12	-9	8
42	45	8	1	7	7	7	-6	5
42	49	0	0	8	7	12	-7	8
43	41	0	0	6	7	10	-7	6
43	41	9	1	6	7	7	-7	5
43	42	6	1	10	11	6	-5	4
43	45	0	0	1	1	3	-2	2
43	45	6	1	10	10	10	-8	7
43	45	9	1	6	6	12	-11	9
43	49	0	0	7	6	10	-6	7
44	40	0	0	5	6	6	-6	5
44	41	0	0	7	8	11	-8	7
44	45	5	1	11	11	11	-9	8
44	45	7	1	8	8	8	-7	6
44	48	9	2	12	11	10	-9	8
44	49	0	0	8	7	11	-7	8
45	41	8	1	7	8	4	-4	3
45	42	0	0	8	9	12	-9	8
45	45	0	0	1	1	7	-5	5
45	45	0	0	1	1	10	-7	7
45	45	5	1	12	12	12	-10	9
45	45	6	1	9	9	9	-8	7
45	45	9	1	6	6	6	-6	5
45	48	0	0	9	8	12	-8	9
46	41	0	0	6	7	9	-7	6
46	41	9	1	6	7	10	-11	8
46	42	7	1	8	9	9	-9	7
46	45	0	0	1	1	11	-8	8
46	45	6	1	10	10	10	-9	8
46	49	0	0	7	6	9	-6	7
47	40	6	1	10	12	4	-4	3
47	41	0	0	7	8	10	-8	7
47	42	6	1	9	10	5	-5	4
47	45	0	0	1	1	4	-3	3

φ_1	Φ	φ_2	H	K	L	U	V	W
40	40	0						
47	45	5	1	11	11	11	-10	9
47	45	8	1	7	7	7	-7	6
47	48	9	2	12	11	11	-11	10
47	49	0	0	8	7	10	-7	8
48	40	0	0	5	6	7	-6	5
48	42	0	0	8	9	11	-9	8
48	42	0	0	9	10	12	-10	9
48	42	6	1	10	11	11	-11	9
48	45	0	0	1	1	9	-7	7
48	45	5	1	12	12	12	-11	10
48	45	7	1	8	8	8	-8	7
48	48	0	0	9	8	11	-8	9
48	48	0	0	10	9	12	-9	10
49	41	0	0	6	7	8	-7	6
49	43	5	1	11	12	6	-6	5
49	45	0	0	1	1	5	-4	4
49	45	0	0	1	1	11	-9	9
49	45	6	1	9	9	9	-9	8
49	45	6	1	10	10	10	-10	9
49	49	0	0	7	6	8	-6	7
40	40	10						
40	40	13	2	9	11	11	-11	7
40	44	17	3	10	11	8	-9	6
40	46	11	1	5	5	10	-9	7
40	47	18	1	3	3	9	-10	7
40	49	17	3	10	9	9	-9	7
41	40	11	1	5	6	3	-3	2
41	41	18	3	9	11	4	-5	3
41	43	13	2	9	10	10	-10	7
41	46	14	1	4	4	4	-4	3
41	46	17	3	10	10	10	-11	8
41	47	18	1	3	3	6	-7	5
41	49	11	2	10	9	6	-7	6
41	49	15	3	11	10	5	-5	4
42	40	14	1	4	5	7	-8	5
42	43	11	2	10	11	11	-11	8
42	45	10	2	11	11	11	-10	8
42	46	13	2	9	9	9	-9	7
42	46	15	3	11	11	11	-12	9
42	48	10	2	11	10	9	-8	7
42	48	14	3	12	11	11	-11	9
43	42	16	2	7	8	5	-6	4
43	43	10	2	11	12	4	-4	3
43	43	18	3	9	10	7	-9	6
43	46	16	2	7	7	7	-6	6
43	47	18	1	3	3	9	-11	8

φ_1	Φ	φ_2	H	K	L	U	V	W
40	40	10						
44	40	14	1	4	5	9	-11	7
44	46	11	1	5	5	5	-5	4
44	46	14	1	4	4	8	-9	7
44	46	17	3	10	10	10	-12	9
44	49	13	2	9	8	8	-8	7
45	40	13	2	9	11	5	-6	4
45	45	10	2	11	11	11	-11	9
45	47	18	1	3	3	3	-4	3
45	49	11	2	10	9	9	-9	8
45	49	17	3	10	9	5	-6	5
45	49	18	4	12	11	4	-5	4
46	42	18	2	6	7	2	-3	2
46	43	11	2	10	11	7	-8	6
46	46	11	1	5	5	10	-11	9
46	46	13	2	9	9	9	-10	8
46	48	10	2	11	10	10	-10	9
47	44	17	3	10	11	5	-7	5
47	45	10	2	11	11	11	-12	10
47	46	16	2	7	7	7	-9	7
47	48	14	3	12	11	6	-7	6
47	49	13	2	9	8	9	-10	9
48	40	11	1	5	6	8	-10	7
48	41	17	3	10	12	2	-3	2
48	42	16	2	7	8	7	-10	7
48	46	13	2	9	9	9	-11	9
48	46	14	1	4	4	4	-5	4
48	47	18	1	3	3	6	-9	7
48	49	11	2	10	9	10	-11	10
49	42	14	2	8	9	8	-11	8
49	43	13	2	9	10	3	-4	3
49	46	11	1	5	5	5	-6	5
49	48	10	2	11	10	11	-12	11
40	40	20						
40	46	29	5	9	10	3	-5	3
40	48	24	4	9	9	9	-12	8
40	49	20	4	11	10	7	-8	6
41	40	23	3	7	9	6	-9	5
41	42	22	2	5	6	7	-10	6
41	44	27	3	6	7	3	-5	3
41	45	27	5	10	11	5	-8	5
41	47	21	3	8	8	8	-10	7
41	48	27	1	2	2	2	-3	2
42	40	27	3	6	8	6	-11	6
42	42	27	2	4	5	4	-7	4
42	44	20	4	11	12	9	-12	8
43	42	24	4	9	11	3	-5	3

φ_1	Φ	φ_2	H	K	L	U	V	W
40	40	20						
43	44	23	3	7	8	7	-11	7
43	47	22	2	5	5	5	-7	5
43	47	23	3	7	7	7	-10	7
44	40	23	3	7	9	7	-12	7
44	44	22	4	10	11	2	-3	2
44	47	21	3	8	8	8	-11	8
44	48	27	1	2	2	6	-10	7
45	42	22	2	5	6	5	-8	5
45	42	27	2	4	5	5	-10	6
45	44	21	3	8	9	2	-3	2
45	45	24	4	9	10	7	-12	8
45	46	29	6	11	12	3	-6	4
45	47	23	3	7	7	7	-11	8
45	48	27	1	2	2	4	-7	5
46	41	29	5	9	12	3	-7	4
46	43	27	5	10	12	4	-8	5
46	47	21	3	8	8	8	-12	9
46	48	27	1	2	2	6	-11	8
47	44	23	3	7	8	5	-9	6
47	45	27	4	8	9	3	-6	4
47	47	22	2	5	5	5	-8	6
48	47	23	3	7	7	7	-12	9
48	48	27	1	2	2	2	-4	3
49	40	27	3	6	8	2	-5	3
49	41	21	3	8	10	6	-11	7
49	44	27	3	6	7	4	-9	6
49	45	24	4	9	10	5	-10	7
49	46	29	5	9	10	4	-10	7
49	49	20	4	11	10	4	-6	5
40	40	30						
40	40	31	3	5	7	1	-2	1
40	41	36	5	7	10	4	-10	5
40	47	34	6	9	10	5	-10	6
41	49	30	4	7	7	7	-12	8
42	40	37	3	4	6	2	-6	3
42	44	31	3	5	6	6	-12	7
42	47	32	7	11	12	4	-8	5
42	49	31	3	5	5	5	-9	6
43	44	38	7	9	12	3	-9	5
43	45	30	4	7	8	4	-8	5
43	45	35	7	10	12	2	-5	3
44	45	34	6	9	11	2	-5	3
44	46	32	5	8	9	5	-11	7
44	49	31	3	5	5	5	-10	7
45	43	32	5	8	10	2	-5	3
45	49	38	7	9	10	1	-3	2

φ_1	Φ	φ_2	H	K	L	U	V	W
40	40	30						
45	49	39	8	10	11	1	-3	2
46	42	34	2	3	4	4	-12	7
46	42	37	6	8	11	2	-7	4
46	47	31	6	10	11	4	-9	6
46	48	37	6	8	9	1	-3	2
47	42	30	4	7	9	2	-5	3
47	47	36	5	7	8	1	-3	2
47	49	31	3	5	5	5	-11	8
48	40	37	3	4	6	2	-9	5
48	42	34	2	3	4	3	-10	6
48	45	30	4	7	8	5	-12	8
48	45	37	3	4	5	3	-11	7
48	46	34	4	6	7	1	-3	2
48	47	39	4	5	6	3	-12	8
49	42	39	4	5	7	1	-5	3
49	49	31	3	5	5	5	-12	9
40	40	40						
40	43	45	2	2	3	1	-4	2
40	47	45	3	3	4	3	-11	6
40	48	42	9	10	12	4	-12	7
40	49	45	4	4	5	2	-7	4
41	41	40	5	6	9	3	-10	5
41	46	41	6	7	9	3	-9	5
42	41	45	5	5	8	2	-10	5
42	42	49	8	7	12	1	-8	4
42	45	45	5	5	7	2	-9	5
42	48	42	8	9	11	3	-10	6
42	48	45	7	7	9	3	-12	7
43	40	41	6	7	11	1	-4	2
43	40	45	3	3	5	1	-6	3
44	43	45	2	2	3	2	-11	6
44	44	40	5	6	8	2	-7	4
44	44	49	8	7	11	1	-9	5
44	45	42	8	9	12	3	-12	7
44	47	45	3	3	4	1	-5	3
45	45	45	7	7	10	2	-12	7
45	47	41	7	8	10	2	-8	5
46	43	45	2	2	3	1	-7	4
46	48	45	7	7	9	2	-11	7
46	48	48	9	8	11	1	-8	5
47	41	45	5	5	8	1	-9	5
47	44	41	7	8	11	1	-5	3
47	48	42	9	10	12	2	-9	6
47	49	45	4	4	5	1	-6	4
48	40	45	3	3	5	1	-11	6
48	40	49	7	6	11	0	-11	6

φ_1	Φ	φ_2	H	K	L	U	V	W
40	40	40						
46	43	45	2	2	3	1	-10	6
48	45	45	5	5	7	1	-8	5
49	42	45	7	7	11	1	-12	7
49	42	49	8	7	12	0	-12	7
49	43	49	7	6	10	0	-5	3
49	44	40	5	6	8	2	-11	7
49	47	45	3	3	4	1	-9	6
40	40	50						
40	44	59	5	3	6	0	-2	1
41	42	51	5	4	7	1	-10	5
41	43	58	8	5	10	0	-2	1
41	47	54	7	5	8	1	-11	6
41	47	59	10	6	11	0	-11	6
41	49	52	9	7	10	1	-7	4
42	42	56	3	2	4	0	-2	1
42	46	58	8	5	9	0	-9	5
43	41	50	6	5	9	1	-12	6
43	41	54	7	5	10	0	-2	1
43	45	56	9	6	11	0	-11	6
43	47	58	11	7	12	0	-12	7
43	49	59	5	3	5	0	-5	3
44	40	53	4	3	6	0	-2	1
44	40	59	5	3	7	-1	-10	5
44	46	56	6	4	7	0	-7	4
44	47	56	9	6	10	0	-5	3
45	42	53	8	6	11	0	-11	6
45	44	54	7	5	9	0	-9	5
45	45	55	10	7	12	0	-12	7
46	47	54	7	5	8	0	-8	5
46	48	50	6	5	7	1	-11	7
46	48	55	10	7	11	0	-11	7
47	42	51	5	4	7	0	-7	4
47	44	52	9	7	12	0	-12	7
47	45	53	4	3	5	0	-5	3
48	41	50	6	5	9	0	-9	5
48	46	52	9	7	11	0	-11	7
48	46	58	8	5	9	-1	-11	7
48	48	53	8	6	9	0	-3	2
48	49	54	11	8	12	0	-3	2
49	44	50	6	5	8	0	-8	5
49	47	51	5	4	6	0	-3	2
49	49	59	5	3	5	-1	-10	7
40	40	60						
40	44	63	6	3	7	-1	-12	6

φ_1	Φ	φ_2	H	K	L	U	V	W
40	40	60						
40	48	66	11	5	11	-1	-11	6
40	49	61	9	5	9	0	-9	5
40	49	61	11	6	11	0	-11	6
41	42	68	5	2	6	-2	-10	5
41	43	63	10	5	12	-1	-10	5
41	48	66	9	4	9	-1	-9	5
41	49	60	7	4	7	0	-7	4
42	41	69	8	3	10	-1	-4	2
42	42	63	4	2	5	-1	-8	4
42	42	66	9	4	11	-1	-6	3
42	47	67	7	3	7	-1	-7	4
42	48	63	2	1	2	-1	-12	7
42	49	60	12	7	12	0	-12	7
43	45	63	10	5	11	-1	-9	5
43	47	67	12	5	12	-2	-12	7
43	48	63	2	1	2	-1	-10	6
44	40	63	6	3	8	-1	-6	3
44	45	63	8	4	9	-1	-7	4
44	47	68	5	2	5	-1	-5	3
45	42	63	4	2	5	-2	-11	6
45	47	69	8	3	8	-2	-8	5
45	48	63	2	1	2	-1	-8	5
45	49	61	11	6	11	-1	-11	7
46	40	67	7	3	9	-3	-11	6
46	42	68	5	2	6	-2	-7	4
46	43	61	9	5	11	-1	-7	4
46	44	69	8	3	9	-3	-10	6
46	45	66	11	5	12	-1	-5	3
46	48	66	11	5	11	-2	-11	7
47	44	63	6	3	7	-1	-5	3
47	48	63	2	1	2	-1	-6	4
47	49	60	12	7	12	-1	-12	8
47	49	61	9	5	9	-1	-9	6
48	45	66	9	4	10	-2	-8	5
48	47	67	12	5	12	-3	-12	8
48	48	66	9	4	9	-2	-9	6
49	42	60	7	4	9	-1	-5	3
49	47	68	5	2	5	-3	-10	7
49	48	63	2	1	2	-2	-10	7
40	40	70						
40	46	75	11	3	11	-3	-11	6
40	47	70	11	4	11	-2	-11	6
40	49	77	9	2	8	-2	-7	4
41	40	79	5	1	6	-4	-10	5
41	41	73	10	3	12	-3	-10	5
41	42	72	6	2	7	-1	-4	2

φ_1	Φ	φ_2	H	K	L	U	V	W
40	40	70						
41	46	77	9	2	9	-3	-9	5
41	47	72	3	1	3	-2	-9	5
41	49	70	11	4	10	-2	-12	7
42	40	76	4	1	5	-3	-8	4
42	42	76	8	2	9	-4	-11	6
42	46	74	7	2	7	-2	-7	4
42	46	76	4	1	4	-4	-12	7
42	48	76	12	3	11	-3	-10	6
43	40	79	5	1	6	-5	-11	6
43	43	72	9	3	10	-3	-11	6
43	43	79	10	2	11	-3	-7	4
43	44	75	11	3	12	-3	-9	5
43	46	79	5	1	5	-2	-5	3
43	47	72	3	1	3	-3	-12	7
44	46	73	10	3	10	-3	-10	6
44	49	73	10	3	9	-3	-11	7
45	40	76	4	1	5	-4	-9	5
45	46	75	11	3	11	-4	-11	7
45	46	76	4	1	4	-3	-8	5
45	49	75	11	3	10	-1	-3	2
46	40	79	5	1	6	-6	-12	7
46	42	74	7	2	8	-2	-5	3
46	47	70	11	4	11	-3	-11	7
46	49	77	9	2	8	-4	-10	7
47	46	76	4	1	4	-5	-12	8
47	46	77	9	2	9	-4	-9	6
47	47	72	3	1	3	-1	-3	2
47	49	79	10	2	9	-5	-11	8
48	40	76	4	1	5	-5	-10	6
48	41	72	9	3	11	-2	-5	3
48	46	79	5	1	5	-5	-10	7
48	49	70	11	4	10	-2	-7	5
49	42	76	8	2	9	-3	-6	4
49	46	73	10	3	10	-4	-10	7
49	49	73	10	3	9	-3	-8	6
40	40	80						
40	41	81	6	1	7	-5	-12	6
40	45	80	11	2	11	-4	-11	6
40	45	84	9	1	9	-4	-9	5
40	49	83	8	1	7	-2	-5	3
41	45	82	7	1	7	-3	-7	4
41	45	85	12	1	12	-6	-12	7
42	45	81	6	1	6	-5	-12	7
42	45	84	10	1	10	-5	-10	6
42	49	84	9	1	8	-5	-11	7
43	45	83	8	1	8	-4	-6	5

φ_1	Φ	φ_2	H	K	L	U	V	W
40	40	80						
43	45	85	11	1	11	-6	-11	7
43	48	84	10	1	9	-3	-6	4
45	45	80	11	2	11	-5	-11	7
45	45	84	9	1	9	-5	-9	6
45	45	85	12	1	12	-7	-12	8
45	48	81	12	2	11	-4	-9	6
46	43	80	11	2	12	-4	-8	5
46	45	81	6	1	6	-3	-6	4
46	48	85	12	1	11	-4	-7	5
47	45	84	10	1	10	-6	-10	7
48	45	82	7	1	7	-4	-7	5
46	45	85	11	1	11	-7	-11	8
48	45	85	12	1	12	-8	-12	9
48	48	80	11	2	10	-2	-4	3
48	49	83	8	1	7	-5	-9	7
49	45	80	11	2	11	-6	-11	8
49	45	83	8	1	8	-5	-8	6
49	49	84	9	1	8	-3	-5	4
40	40	90						
40	41	90	6	0	7	-7	-11	6
40	45	90	1	0	1	-7	-12	7
40	45	90	1	0	1	-3	-5	3
40	49	90	7	0	6	-6	-11	7
41	40	90	5	0	6	-6	-9	5
41	45	90	1	0	1	-5	-8	5
42	41	90	7	0	8	-8	-12	7
42	45	90	1	0	1	-7	-11	7
42	49	90	8	0	7	-7	-12	8
43	41	90	6	0	7	-7	-10	6
43	45	90	1	0	1	-2	-3	2
43	49	90	7	0	6	-6	-10	7
44	40	90	5	0	6	-6	-8	5
44	41	90	7	0	8	-8	-11	7
44	49	90	8	0	7	-7	-11	8
45	42	90	8	0	9	-9	-12	8
45	45	90	1	0	1	-5	-7	5
45	45	90	1	0	1	-7	-10	7
45	48	90	9	0	8	-8	-12	9
46	41	90	6	0	7	-7	-9	6
46	45	90	1	0	1	-8	-11	8
46	49	90	7	0	6	-6	-9	7
47	41	90	7	0	8	-8	-10	7
47	45	90	1	0	1	-3	-4	3
47	49	90	8	0	7	-7	-10	8
48	40	90	5	0	6	-6	-7	5
48	42	90	8	0	9	-9	-11	8

φ_1	Φ	φ_2	H	K	L	U	V	W
40	40	90						
48	42	90	9	0	10	-10	-12	9
48	45	90	1	0	1	-7	-9	7
48	48	90	9	0	8	-8	-11	9
46	48	90	10	0	9	-9	-12	10
49	41	90	6	0	7	-7	-8	6
49	45	90	1	0	1	-4	-5	4
49	45	90	1	0	1	-9	-11	9
49	49	90	7	0	6	-6	-8	7
40	50	0						
40	53	0	0	4	3	6	-3	4
40	59	0	0	5	3	7	-3	5
41	50	0	0	6	5	9	-5	6
41	54	0	0	7	5	10	-5	7
41	56	5	1	12	6	12	-7	9
41	57	9	1	6	4	10	-7	8
41	58	0	0	8	5	11	-5	8
42	51	0	0	5	4	7	-4	5
42	53	0	0	4	3	11	-6	8
42	56	0	0	3	2	4	-2	3
42	58	5	1	11	7	5	-3	4
42	58	7	1	8	5	11	-7	9
43	53	5	1	12	9	9	-6	7
43	56	6	1	9	6	6	-4	5
43	57	9	1	6	4	8	-6	7
43	58	0	0	8	5	10	-5	8
44	50	0	0	6	5	8	-5	6
44	52	0	0	9	7	12	-7	9
44	53	7	1	8	6	12	-9	10
44	54	0	0	7	5	9	-5	7
44	59	0	0	5	3	6	-3	5
44	59	6	1	10	6	8	-5	7
45	53	0	0	4	3	5	-3	4
45	54	5	1	11	8	7	-5	6
45	55	0	0	10	7	12	-7	10
45	55	8	1	7	5	9	-7	8
45	56	0	0	3	2	11	-6	9
46	52	0	0	9	7	11	-7	9
46	53	7	1	8	6	10	-6	9
46	56	0	0	3	2	7	-4	6
46	58	0	0	8	5	9	-5	8
47	51	0	0	5	4	6	-4	5
47	51	6	1	10	8	12	-10	11
47	52	6	1	9	7	11	-9	10
47	54	0	0	7	5	8	-5	7
47	56	0	0	3	2	10	-6	9
47	57	9	1	6	4	6	-5	6

φ_1	Φ	φ_2	H	K	L	U	V	W
40	50	0						
47	58	0	0	11	7	12	-7	11
47	58	7	1	8	5	4	-3	4
47	59	0	0	5	3	11	-6	10
48	50	0	0	6	5	7	-5	6
48	51	9	1	6	5	1	-1	1
48	53	0	0	4	3	9	-6	8
48	54	9	2	12	9	12	-11	12
48	55	0	0	10	7	11	-7	10
48	55	8	1	7	5	7	-6	7
48	56	6	1	9	6	9	-7	9
48	58	5	1	11	7	11	-8	11
48	59	6	1	10	6	10	-7	10
49	50	8	1	7	6	1	-1	1
49	51	0	0	5	4	11	-8	10
49	52	0	0	9	7	10	-7	9
49	52	6	1	9	7	9	-8	9
49	53	7	1	8	6	8	-7	8
49	54	0	0	11	8	12	-8	11
49	54	5	1	11	8	11	-9	11
49	55	6	1	10	7	5	-4	5
49	56	5	1	12	8	4	-3	4
49	57	9	1	6	4	10	-9	11
49	59	0	0	5	3	5	-3	5
40	50	10						
40	54	18	3	9	7	7	-7	6
41	51	16	2	7	6	6	-6	5
41	55	18	4	12	9	9	-9	8
42	52	18	2	6	5	11	-12	10
42	53	17	3	10	8	8	-8	7
42	59	14	2	8	5	11	-9	10
43	50	14	2	8	7	7	-7	6
43	52	15	3	11	9	9	-9	8
43	57	13	2	9	6	12	-10	11
43	58	18	1	3	2	1	-1	1
44	50	18	3	9	8	9	-11	9
44	51	14	3	12	10	10	-10	9
44	52	18	2	6	5	6	-7	6
44	54	18	3	9	7	9	-10	9
44	56	16	2	7	5	1	-1	1
44	56	17	3	10	7	1	-1	1
44	58	15	3	11	7	11	-10	11
45	51	16	2	7	6	7	-8	7
45	53	17	3	10	8	10	-11	10
45	55	15	3	11	8	1	-1	1
45	57	14	3	12	8	12	-11	12
45	59	14	2	8	5	8	-7	8

φ_1	Φ	φ_2	H	K	L	U	V	W
40	50	10						
46	52	15	3	11	9	11	-12	11
46	53	13	2	9	7	1	-1	1
46	54	14	1	4	3	1	-1	1
46	56	11	2	10	7	10	-9	10
46	57	13	2	9	6	9	-8	9
46	58	10	2	11	7	11	-9	11
47	50	14	2	8	7	8	-9	8
47	52	11	1	5	4	1	-1	1
47	52	18	2	6	5	7	-9	8
47	54	10	2	11	8	11	-10	11
48	50	18	3	9	8	5	-7	6
48	51	10	2	11	9	1	-1	1
48	51	16	2	7	6	8	-10	9
48	57	14	3	12	8	8	-8	9
48	58	15	3	11	7	7	-7	8
49	50	14	2	8	7	9	-11	10
49	52	18	2	6	5	8	-11	10
49	57	13	2	9	6	6	-6	7
49	58	10	2	11	7	9	-6	10
49	59	14	2	8	5	5	-5	6
40	50	20						
40	51	27	5	10	9	5	-7	5
40	52	27	4	8	7	8	-11	8
40	53	22	2	5	4	9	-10	8
40	53	27	3	6	5	3	-4	3
40	56	20	4	11	8	8	-8	7
40	56	27	2	4	3	4	-5	4
40	58	27	5	10	7	5	-6	5
41	52	20	4	11	9	10	-11	9
41	52	29	5	9	8	7	-11	8
41	55	24	4	9	7	9	-11	9
41	57	23	3	7	5	7	-8	7
41	59	24	4	9	6	9	-10	9
42	50	23	5	12	11	3	-4	3
42	51	24	4	9	8	3	-4	3
42	52	23	3	7	6	7	-9	7
42	52	23	5	12	10	4	-5	4
42	53	22	2	5	4	5	-6	5
42	54	29	6	11	9	6	-9	7
42	55	23	5	12	9	6	-7	6
42	56	27	2	4	3	9	-12	10
42	57	22	4	10	7	10	-11	10
42	59	20	4	11	7	1	-1	1
42	59	27	3	6	4	8	-10	9
43	51	21	3	8	7	4	-5	4
43	53	24	5	11	9	8	-11	9

φ_1	Φ	φ_2	H	K	L	U	V	W
40	50	20						
43	55	21	3	6	6	6	-9	8
43	56	20	4	11	8	11	-12	11
44	51	27	6	12	11	5	-8	6
44	51	29	6	11	10	7	-12	9
44	54	27	5	10	8	4	-6	5
44	56	24	5	11	8	7	-9	8
44	56	27	2	4	3	5	-7	6
44	57	23	3	7	5	9	-11	10
44	58	27	5	10	7	8	-11	10
45	52	29	5	9	8	5	-9	7
45	53	27	3	6	5	7	-11	9
45	55	21	3	8	6	10	-12	11
46	50	22	4	10	9	7	-10	8
46	52	27	4	8	7	3	-5	4
46	53	22	2	5	4	6	-8	7
46	54	29	6	11	9	7	-12	10
46	55	23	5	12	9	9	-12	11
46	56	27	2	4	3	6	-9	8
46	58	23	5	12	8	4	-5	5
47	50	24	5	11	10	3	-5	4
47	52	23	3	7	6	4	-6	5
47	56	27	2	4	3	7	-11	10
47	59	24	4	9	6	3	-4	4
48	51	29	6	11	10	5	-10	8
48	53	22	2	5	4	7	-10	9
48	53	24	5	11	9	5	-8	7
48	53	27	3	6	5	4	-7	6
48	56	29	5	9	7	5	-9	8
48	57	22	4	10	7	3	-4	4
48	59	27	3	6	4	2	-3	3
49	51	24	4	9	8	7	-12	10
49	53	22	2	5	4	8	-12	11
49	56	20	4	11	8	3	-4	4
40	50	30						
40	56	31	3	5	4	8	-12	9
40	57	34	6	9	7	5	-8	6
40	58	30	4	7	5	8	-11	9
40	58	32	5	8	6	6	-9	7
41	54	36	8	11	10	5	-10	7
41	58	39	4	5	4	2	-4	3
41	59	37	3	4	3	5	-9	7
42	50	34	2	3	3	3	-6	4
42	51	37	3	4	4	4	-9	6
42	52	39	4	5	5	5	-12	8
42	53	30	4	7	6	5	-8	6
42	55	38	7	9	8	5	-11	8

φ_1	Φ	φ_2	H	K	L	U	V	W
40	50	30						
42	55	39	9	11	10	3	-7	5
42	56	37	9	12	10	2	-4	3
42	58	32	7	11	8	7	-11	9
42	58	34	8	12	9	3	-5	4
43	55	36	5	7	6	2	-4	3
43	56	31	3	5	4	3	-5	4
43	58	39	9	11	9	4	-9	7
44	50	39	9	11	12	1	-3	2
44	51	37	3	4	4	4	-10	7
44	54	34	6	9	8	2	-4	3
44	57	34	6	9	7	6	-11	9
44	59	37	3	4	3	3	-6	5
45	53	32	7	11	10	2	-4	3
45	54	37	9	12	11	5	-12	9
45	55	34	4	6	5	5	-10	8
45	58	32	5	8	6	4	-7	6
46	50	34	2	3	3	3	-7	5
46	51	37	3	4	4	4	-11	8
46	54	30	7	12	10	6	-11	9
46	57	36	8	11	9	5	-11	9
46	58	32	7	11	8	5	-9	8
46	58	39	4	5	4	5	-12	10
47	55	32	7	11	9	3	-6	5
47	55	37	6	8	7	2	-5	4
47	58	38	7	9	7	3	-7	6
47	59	37	3	4	3	4	-9	8
48	50	34	2	3	3	3	-8	6
48	51	37	3	4	4	4	-12	9
48	53	30	4	7	6	3	-6	5
48	53	32	5	8	7	5	-11	9
48	54	35	7	10	9	2	-5	4
48	56	31	3	5	4	4	-8	7
48	57	30	7	12	9	6	-11	10
48	58	39	4	5	4	3	-8	7
49	53	34	8	12	11	2	-5	4
49	55	34	4	6	5	3	-7	6
49	55	38	7	9	8	2	-6	5
49	58	32	5	8	6	6	-12	11
49	59	37	3	4	3	5	-12	11
40	50	40						
40	50	45	6	6	7	3	-10	6
40	53	49	8	7	8	2	-6	5
40	59	45	7	7	6	4	-10	7
41	54	48	11	10	11	3	-11	7
41	55	45	1	1	1	1	-3	2
41	56	42	10	11	10	4	-10	7

φ_1	Φ	φ_2	H	K	L	U	V	W
40	50	40						
41	57	41	7	8	7	3	-7	5
42	50	49	8	7	9	1	-5	3
42	51	45	7	7	8	3	-11	7
42	54	47	12	11	12	3	-12	8
42	58	45	8	8	7	4	-11	8
42	59	45	6	6	5	3	-8	6
43	50	45	5	5	6	2	-8	5
43	53	48	9	8	9	2	-9	6
43	55	45	1	1	1	3	-10	7
43	56	43	11	12	11	4	-11	8
43	57	41	6	7	6	5	-12	9
43	58	45	9	9	8	4	-12	9
44	50	48	9	8	10	2	-11	7
44	51	45	8	8	9	3	-12	8
44	55	45	1	1	1	2	-7	5
44	56	42	8	9	8	3	-8	6
44	59	42	9	10	8	2	-5	4
44	59	45	7	7	6	3	-9	7
45	50	45	6	6	7	2	-9	6
45	51	48	10	9	11	1	-6	4
45	53	48	10	9	10	2	-10	7
45	55	40	10	12	11	3	-8	6
45	55	45	1	1	1	3	-11	8
45	57	40	5	6	5	2	-5	4
45	59	42	10	11	9	4	-11	9
46	53	49	7	6	7	1	-7	5
46	54	48	11	10	11	2	-11	8
46	55	45	1	1	1	1	-4	3
46	56	42	9	10	9	3	-9	7
46	58	45	8	8	7	3	-10	8
47	51	45	7	7	8	2	-10	7
47	56	42	10	11	10	3	-10	8
47	58	43	11	12	10	2	-6	5
47	58	45	9	9	8	3	-11	9
48	51	45	8	8	9	2	-11	8
48	53	49	8	7	8	1	-8	6
48	54	47	12	11	12	2	-12	9
48	55	45	1	1	1	2	-9	7
48	56	43	11	12	11	3	-11	9
48	57	41	6	7	6	2	-6	5
48	58	45	10	10	9	3	-12	10
48	59	45	6	6	5	2	-7	6
49	50	45	5	5	6	1	-7	5
49	52	45	9	9	10	2	-12	9
49	55	45	1	1	1	1	-5	4
49	59	45	7	7	6	2	-8	7

φ_1	Φ	φ_2	H	K	L	U	V	W
40	50	50						
40	51	53	4	3	4	2	-12	7
40	52	51	5	4	5	1	-5	3
40	53	58	8	5	7	1	-10	6
40	55	58	11	7	9	1	-8	5
40	56	59	5	3	4	1	-11	7
40	58	56	12	8	9	1	-6	4
41	50	56	3	2	3	1	-12	7
41	57	55	10	7	8	1	-6	4
42	51	53	4	3	4	1	-8	5
42	51	55	10	7	10	1	-10	6
42	52	51	11	9	11	2	-11	7
42	55	50	12	10	11	2	-9	6
42	55	53	8	6	7	1	-6	4
42	55	56	6	4	5	1	-9	6
43	59	59	10	6	7	1	-11	8
44	50	58	8	5	8	0	-8	5
44	51	54	11	8	11	1	-11	7
44	52	50	6	5	6	1	-6	4
44	52	59	10	6	9	0	-3	2
45	50	58	11	7	11	0	-11	7
45	51	53	4	3	4	1	-12	8
45	52	52	9	7	9	1	-9	6
45	55	52	9	7	8	1	-7	5
45	57	56	9	6	7	1	-12	9
46	50	56	3	2	3	0	-3	2
46	52	51	5	4	5	1	-10	7
46	53	58	8	5	7	0	-7	5
46	53	58	11	7	10	0	-10	7
46	55	54	7	5	6	1	-11	8
46	56	53	12	9	10	1	-8	6
47	55	51	10	8	9	1	-8	6
47	56	59	5	3	4	0	-4	3
48	51	55	10	7	10	0	-10	7
48	52	51	11	9	11	1	-11	8
48	53	56	12	8	11	0	-11	8
48	54	56	9	6	6	0	-4	3
48	55	51	11	9	10	1	-9	7
48	55	58	11	7	9	0	-9	7
49	51	54	7	5	7	0	-7	5
49	51	54	11	8	11	0	-11	8
49	52	50	6	5	6	1	-12	9
49	55	56	6	4	5	0	-5	4
49	58	58	8	5	6	0	-6	5
49	59	59	10	6	7	0	-7	6

φ_1	Φ	φ_2	H	K	L	U	V	W
40	50	60						
40	53	63	6	3	5	0	-5	3
41	51	61	11	6	10	0	-5	3
41	54	63	10	5	8	0	-8	5
41	57	60	12	7	9	1	-12	8
42	52	61	9	5	8	0	-8	5
42	55	67	12	5	9	-1	-12	8
42	56	63	4	2	3	0	-3	2
43	51	69	8	3	7	-2	-11	7
43	52	60	12	7	11	0	-11	7
43	54	61	11	6	9	0	-3	2
43	58	63	10	5	7	0	-7	5
44	53	60	7	4	6	0	-3	2
44	59	63	6	3	4	0	-4	3
45	52	63	8	4	7	-1	-12	8
45	54	60	12	7	10	0	-10	7
45	55	66	9	4	7	-1	-10	7
45	56	61	9	5	7	0	-7	5
45	57	61	11	6	8	0	-4	3
45	57	67	7	3	5	-1	-11	8
45	58	67	12	5	8	-1	-12	9
46	51	63	12	6	11	-1	-9	6
47	57	60	12	7	9	0	-9	7
47	58	60	7	4	5	0	-5	4
48	51	63	10	5	9	-1	-7	5
48	52	61	9	5	8	-1	-11	8
48	53	63	6	3	5	-1	-8	6
49	53	68	5	2	4	-2	-9	7
49	57	67	7	3	5	-1	-6	5
49	58	63	10	5	7	-1	-12	10
40	50	70						
41	53	77	9	2	7	-3	-11	7
41	56	73	10	3	7	-1	-6	4
41	57	77	9	2	6	-2	-9	6
41	58	72	3	1	2	-1	-9	6
42	50	72	9	3	8	-1	-5	3
42	51	74	7	2	6	-2	-8	5
42	54	76	4	1	3	-3	-12	8
42	55	72	12	4	9	-1	-6	4
43	52	79	5	1	4	-1	-3	2
44	52	72	6	2	5	-2	-9	6
44	54	76	4	1	3	-2	-7	5
44	57	76	12	3	8	-1	-4	3
45	50	76	8	2	7	-1	-3	2
45	52	75	11	3	9	-3	-10	7
45	58	72	3	1	2	-2	-12	9

φ_1	Φ	φ_2	H	K	L	U	V	W
40	50	70						
45	58	75	11	3	7	-2	-9	7
46	56	74	7	2	5	-1	-4	3
47	52	72	6	2	5	-3	-11	8
47	54	72	9	3	7	-1	-4	3
47	54	76	4	1	3	-3	-9	7
47	58	72	3	1	2	-1	-5	4
48	51	76	12	3	10	-3	-8	6
48	52	70	11	4	9	-1	-4	3
48	53	77	9	2	7	-2	-5	4
48	56	73	10	3	7	-3	-11	9
48	57	77	9	2	6	-2	-6	5
48	59	70	11	4	7	-1	-6	5
48	59	76	8	2	5	-2	-7	6
49	50	72	9	3	8	-4	-12	9
49	52	79	5	1	4	-5	-11	9
49	54	76	4	1	3	-4	-11	9
49	55	75	11	3	8	-2	-6	5
40	50	80						
40	52	84	9	1	7	-3	-8	5
40	56	84	9	1	6	-1	-3	2
41	53	65	12	1	9	-5	-12	8
42	51	61	6	1	5	-4	-11	7
42	54	81	12	2	9	-1	-3	2
42	57	81	6	1	4	-3	-10	7
43	54	65	11	1	8	-3	-7	5
43	56	85	12	1	8	-5	-12	9
44	50	85	12	1	10	-5	-10	7
44	57	61	6	1	4	-4	-12	9
45	51	61	6	1	5	-3	-7	5
45	55	64	10	1	7	-4	-9	7
47	50	62	7	1	6	-2	-4	3
48	53	63	8	1	6	-3	-6	5
48	56	65	12	1	8	-4	-8	7
48	56	60	11	2	7	-3	-8	7
49	51	61	6	1	5	-5	-10	8
49	54	65	11	1	8	-5	-9	8
40	50	90						
40	53	90	4	0	3	-3	-6	4
40	59	90	5	0	3	-3	-7	5
41	50	90	6	0	5	-5	-9	6
41	54	90	7	0	5	-5	-10	7
41	58	90	8	0	5	-5	-11	8
42	51	90	5	0	4	-4	-7	5
42	53	90	4	0	3	-6	-11	6

φ_1	Φ	φ_2	H	K	L	U	V	W
40	50	90						
42	56	90	3	0	2	-2	-4	3
43	58	90	8	0	5	-5	-10	8
44	50	90	6	0	5	-5	-8	6
44	52	90	9	0	7	-7	-12	9
44	54	90	7	0	5	-5	-9	7
44	59	90	5	0	3	-3	-6	5
45	53	90	4	0	3	-3	-5	4
45	55	90	10	0	7	-7	-12	10
45	56	90	3	0	2	-6	-11	9
46	52	90	9	0	7	-7	-11	9
46	56	90	3	0	2	-4	-7	6
46	58	90	8	0	5	-5	-9	8
47	51	90	5	0	4	-4	-6	5
47	54	90	7	0	5	-5	-8	7
47	56	90	3	0	2	-6	-10	9
47	58	90	11	0	7	-7	-12	11
47	59	90	5	0	3	-6	-11	10
48	50	90	6	0	5	-5	-7	6
48	53	90	4	0	3	-6	-9	8
48	55	90	10	0	7	-7	-11	10
49	51	90	5	0	4	-8	-11	10
49	52	90	9	0	7	-7	-10	9
49	54	90	11	0	8	-8	-12	11
49	59	90	5	0	3	-3	-5	5
40	60	0						
40	63	0	0	2	1	8	-3	6
40	67	0	0	7	3	9	-3	7
40	67	8	1	7	3	6	-3	5
41	61	0	0	9	5	12	-5	9
41	61	8	1	7	4	5	-3	4
41	64	6	1	10	5	10	-5	8
41	64	9	1	6	3	12	-7	10
41	69	0	0	8	3	10	-3	8
42	60	0	0	7	4	9	-4	7
42	63	0	0	2	1	5	-2	4
42	64	5	1	12	6	6	-3	5
42	66	0	0	9	4	11	-4	9
42	66	6	1	9	4	8	-4	7
42	68	0	0	5	2	6	-2	5
42	68	9	2	12	5	11	-6	10
43	61	0	0	9	5	11	-5	9
43	63	0	0	2	1	12	-5	10
43	66	5	1	11	5	10	-5	9
44	61	8	1	7	4	12	-8	11
44	63	0	0	2	1	7	-3	6
44	64	7	1	8	4	12	-7	11
44	67	0	0	7	3	8	-3	7
44	68	0	0	5	2	11	-4	10
44	69	0	0	8	3	9	-3	8
45	60	0	0	7	4	8	-4	7
45	63	0	0	2	1	11	-5	10
45	63	0	0	2	1	9	-4	8
45	64	9	1	6	3	3	-2	3
45	66	0	0	9	4	10	-4	9
45	66	0	0	11	5	12	-5	11
45	67	8	1	7	3	7	-4	7
45	68	9	2	12	5	12	-7	12
46	60	9	2	12	7	4	-3	4
46	61	0	0	9	5	10	-5	9
46	61	0	0	11	6	12	-6	11
46	61	8	1	7	4	7	-5	7
46	64	7	1	8	4	8	-5	8
46	66	5	1	11	5	11	-6	11
46	66	6	1	9	4	9	-5	9
46	67	5	1	12	5	2	-1	2
46	68	6	1	10	4	2	-1	2
47	61	5	1	11	6	11	-7	11
47	61	6	1	9	5	3	-2	3
47	64	5	1	12	6	12	-7	12
47	64	6	1	10	5	5	-3	5
47	67	0	0	7	3	7	-3	7
47	67	0	0	12	5	12	-5	12
47	68	0	0	5	2	5	-2	5
47	69	0	0	8	3	8	-3	8
48	60	5	1	12	7	3	-2	3
48	63	0	0	2	1	2	-1	2
48	66	0	0	9	4	9	-4	9
48	66	0	0	11	5	11	-5	11
49	60	0	0	7	4	7	-4	7
49	60	0	0	12	7	12	-7	12
49	61	0	0	9	5	9	-5	9
49	61	0	0	11	6	11	-6	11
49	61	8	1	7	4	9	-7	10
49	66	6	1	9	4	10	-6	11
49	67	8	1	7	3	8	-5	9
40	60	10						
40	62	15	3	11	6	8	-6	7
40	68	16	2	7	3	9	-6	8
40	69	17	3	10	4	12	-8	11
41	61	16	2	7	4	10	-6	9
41	64	14	1	4	2	10	-7	9
41	66	10	2	11	5	7	-4	6
41	69	11	1	5	2	9	-5	8
42	60	11	1	5	3	7	-5	6
42	62	10	2	11	6	9	-6	8
42	62	13	2	9	5	11	-8	10
42	65	18	2	6	3	6	-5	6
42	67	18	3	9	4	9	-7	9
42	68	18	4	12	5	4	-3	4
43	60	11	1	5	3	12	-9	11
43	60	17	3	10	6	10	-9	10
43	61	18	4	12	7	12	-11	12
43	62	18	3	9	5	9	-8	9
43	64	17	3	10	5	5	-4	5
43	66	15	3	11	5	11	-8	11
43	68	16	2	7	3	7	-5	7
43	69	17	3	10	4	10	-7	10
44	60	14	3	12	7	6	-5	6
44	61	16	2	7	4	7	-6	7
44	62	15	3	11	6	11	-9	11
44	64	14	1	4	2	4	-3	4
44	67	13	2	9	4	3	-2	3
44	67	18	3	9	4	11	-9	12
44	68	14	3	12	5	3	-2	3
44	69	11	1	5	2	5	-3	5
45	62	13	2	9	5	9	-7	9
45	64	11	2	10	5	10	-7	10
45	65	18	2	6	3	9	-8	10
45	66	10	2	11	5	11	-7	11
46	60	11	1	5	3	5	-4	5
46	61	16	2	7	4	11	-10	12
46	61	18	4	12	7	7	-7	8
46	62	10	2	11	6	11	-8	11
46	64	14	1	4	2	10	-8	11
46	66	15	3	11	5	9	-7	10
46	69	17	3	10	4	8	-6	9
47	60	17	3	10	6	6	-6	7
47	62	18	3	9	5	5	-5	6
47	69	11	1	5	2	11	-7	12
48	62	13	2	9	5	7	-6	8
48	64	14	1	4	2	6	-5	7
48	64	17	3	10	5	10	-9	12
48	68	16	2	7	3	5	-4	6
48	69	11	1	5	2	6	-4	7
49	60	11	1	5	3	8	-7	9
49	61	16	2	7	4	4	-4	5
49	65	18	2	6	3	3	-3	4
49	68	18	4	12	5	9	-8	12
40	60	20						
40	60	24	5	11	7	11	-12	11
40	63	24	4	9	5	1	-1	1
40	64	24	5	11	6	1	-1	1
40	68	24	4	9	4	9	-8	9
40	68	24	5	11	5	11	-10	11
41	61	22	2	5	3	1	-1	1
41	62	23	3	7	4	1	-1	1
41	62	23	5	12	7	1	-1	1
41	65	22	4	10	5	10	-9	10
41	65	23	5	12	6	12	-11	12
41	68	23	3	7	3	7	-6	7
41	69	23	5	12	5	6	-5	6
42	60	21	3	8	5	1	-1	1
42	60	29	5	9	6	6	-8	7
42	63	20	4	11	6	11	-10	11
42	64	29	6	11	6	5	-6	6
42	65	21	3	8	4	8	-7	8
42	67	20	4	11	5	11	-9	11
43	62	27	6	12	7	5	-6	6
43	64	29	5	9	5	4	-5	5
43	66	27	1	2	1	10	-11	12
43	68	23	3	7	3	10	-9	11
44	62	27	5	10	6	4	-5	5
44	65	23	5	12	6	6	-6	7
44	66	27	1	2	1	8	-9	10
44	66	27	1	2	1	9	-10	11
44	68	24	5	11	5	5	-5	6
45	60	24	5	11	7	4	-5	5
45	66	27	1	2	1	7	-8	9
45	67	20	4	11	5	8	-7	9
46	61	27	4	8	5	3	-4	4
46	63	20	4	11	6	6	-6	7
46	65	22	4	10	5	5	-5	6
46	66	27	1	2	1	6	-7	8
46	68	24	4	9	4	4	-4	5
47	61	29	6	11	7	2	-3	3
47	65	21	3	8	4	4	-4	5
47	66	27	1	2	1	5	-6	7
48	62	27	6	12	7	8	-11	12
48	64	29	5	9	5	7	-10	11
48	68	23	3	7	3	3	-3	4
49	60	29	5	9	6	3	-5	5
49	66	27	1	2	1	4	-5	6
49	69	23	5	12	5	5	-5	7

40 60 30

φ_1	Φ	φ_2	H	K	L	U	V	W
40	62	32	7	11	7	5	-7	6
40	62	38	7	9	6	3	-5	4
40	63	31	3	5	3	7	-9	8
40	64	30	4	7	4	10	-12	11
40	66	36	8	11	6	7	-10	9
40	68	37	3	4	2	8	-11	10
40	69	32	7	11	5	6	-7	7
41	60	36	5	7	5	3	-5	4
41	61	34	2	3	2	4	-6	5
41	64	35	7	10	6	6	-9	8
41	67	30	7	12	6	6	-7	7
42	62	32	5	8	5	7	-10	9
42	63	31	3	5	3	9	-12	11
42	65	36	5	7	4	8	-12	11
42	66	38	7	9	5	7	-11	10
42	67	31	6	10	5	5	-6	6
42	67	32	5	8	4	4	-5	5
43	60	35	7	10	7	4	-7	6
43	61	34	2	3	2	5	-8	7
43	62	37	9	12	8	4	-7	6
43	67	34	4	6	3	3	-4	4
43	68	35	7	10	5	5	-7	7
44	60	30	7	12	8	8	-12	11
44	61	34	2	3	2	6	-10	9
44	64	30	4	7	4	3	-4	4
44	65	32	7	11	6	5	-7	7
44	68	37	3	4	2	2	-3	3
44	69	39	8	10	5	5	-8	8
45	61	34	2	3	2	7	-12	11
45	63	30	7	12	7	5	-7	7
45	65	34	6	9	5	2	-3	3
45	66	36	8	11	6	5	-8	8
45	69	32	7	11	5	7	-9	10
46	60	36	5	7	5	5	-10	9
46	60	36	8	11	8	4	-8	7
46	62	38	7	9	6	6	-12	11
46	63	31	3	5	3	2	-3	3
46	64	34	8	12	7	5	-8	8
46	65	36	5	7	4	3	-5	5
46	66	38	7	9	5	4	-7	7
46	67	32	5	8	4	8	-11	12
46	67	39	9	11	6	5	-9	9
47	61	39	8	10	7	5	-11	10
47	62	32	5	8	5	3	-5	5
47	64	35	7	10	6	4	-7	7
47	65	37	9	12	7	5	-9	9
47	65	39	4	5	3	1	-2	2

40 60 30

φ_1	Φ	φ_2	H	K	L	U	V	W
47	67	30	7	12	6	6	-8	9
47	67	34	4	6	3	6	-9	10
48	62	32	7	11	7	4	-7	7
48	63	37	6	8	5	1	-2	2
48	65	32	7	11	6	7	-11	12
48	67	31	6	10	5	5	-7	8
48	68	35	7	10	5	5	-8	9
48	68	37	3	4	2	6	-10	11
48	69	39	8	10	5	5	-9	10
49	60	30	7	12	8	4	-7	7
49	63	36	8	11	7	1	-2	2
49	64	39	9	11	7	4	-9	9
49	67	32	5	8	4	4	-6	7
49	68	37	3	4	2	4	-7	8

40 60 40

φ_1	Φ	φ_2	H	K	L	U	V	W
40	61	41	7	8	6	2	-4	3
40	63	45	7	7	5	4	-9	7
40	63	48	10	9	7	3	-8	6
40	66	45	8	8	5	5	-10	8
40	67	45	5	5	3	3	-6	5
40	69	40	5	6	3	6	-9	6
41	61	45	9	9	7	5	-12	9
41	62	41	6	7	5	5	-10	8
41	63	40	5	6	4	6	-11	9
41	65	45	3	3	2	5	-11	9
41	65	48	11	10	7	2	-5	4
41	67	49	7	6	4	2	-5	4
42	60	42	8	9	7	4	-9	7
42	62	45	4	4	3	2	-5	4
42	64	42	8	9	6	3	-6	5
42	65	49	8	7	5	4	-11	9
42	68	45	7	7	4	4	-8	7
42	69	45	9	9	5	5	-10	9
43	61	47	12	11	9	3	-9	7
43	64	45	10	10	7	5	-12	10
43	65	45	3	3	2	3	-7	6
43	69	45	11	11	6	6	-12	11
44	61	45	9	9	7	4	-11	9
44	63	40	5	6	4	4	-8	7
44	64	48	9	8	6	2	-6	5
44	65	41	7	8	5	5	-10	9
44	66	42	9	10	6	6	-12	11
44	66	45	8	8	5	4	-9	8
44	66	48	10	9	6	3	-8	7
44	67	45	5	5	3	5	-11	10
44	69	40	5	6	3	3	-5	5

40 60 40

φ_1	Φ	φ_2	H	K	L	U	V	W
44	69	41	7	8	4	4	-7	7
44	69	49	8	7	4	5	-12	11
45	61	45	5	5	4	2	-6	5
45	62	49	7	6	5	3	-11	9
45	63	45	7	7	5	3	-8	7
45	64	43	11	12	8	4	-9	8
45	65	45	3	3	2	4	-10	9
45	66	45	11	11	7	5	-12	11
46	60	40	10	12	9	3	-7	6
46	61	43	11	12	9	3	-8	7
46	61	49	8	7	6	3	-12	10
46	62	48	11	10	8	2	-7	6
46	64	45	10	10	7	4	-11	10
46	67	42	8	9	5	1	-2	2
46	67	49	7	6	4	4	-12	11
46	68	42	10	11	6	1	-2	2
46	69	45	9	9	5	4	-9	9
46	69	45	11	11	6	5	-11	11
47	62	45	4	4	3	3	-9	8
47	66	40	10	12	7	1	-2	2
47	67	41	6	7	4	1	-2	2
47	67	43	11	12	7	5	-11	11
47	68	45	7	7	4	3	-7	7
47	68	45	12	12	7	5	-12	12
47	69	49	8	7	4	3	-8	8
48	61	41	7	8	6	4	-11	10
48	61	45	9	9	7	3	-10	9
48	62	49	7	6	5	2	-9	8
48	63	45	11	11	8	4	-12	11
48	65	41	7	8	5	3	-7	7
48	66	42	9	10	6	4	-9	9
48	66	45	8	8	5	3	-8	8
48	67	45	5	5	3	2	-5	5
48	67	48	9	8	5	1	-3	3
48	68	48	11	10	6	4	-11	11
48	69	40	5	6	3	6	-11	12
49	61	45	5	5	4	3	-11	10
49	61	49	8	7	6	2	-10	9
49	65	42	10	11	7	2	-5	5
49	66	45	11	11	7	4	-11	11
49	66	48	10	9	6	3	-10	10
49	67	47	12	11	7	1	-3	3
49	67	49	7	6	4	2	-7	7

40 60 50

φ_1	Φ	φ_2	H	K	L	U	V	W
40	61	56	3	2	2	2	-10	7
40	65	56	9	6	5	1	-4	3

40 60 50

φ_1	Φ	φ_2	H	K	L	U	V	W
40	67	58	8	5	4	1	-4	3
40	69	51	10	8	5	2	-5	4
41	63	50	6	5	4	4	-12	9
41	63	54	11	8	7	1	-4	3
41	64	55	10	7	6	1	-4	3
41	65	51	5	4	3	3	-9	7
41	68	55	10	7	5	3	-10	8
41	69	50	6	5	3	5	-12	10
42	61	56	3	2	2	2	-11	8
42	62	53	12	9	8	1	-4	3
42	67	59	10	6	5	1	-5	4
42	68	53	4	3	2	2	-6	5
43	61	56	3	2	2	2	-12	9
43	63	50	6	5	4	3	-10	8
43	64	56	12	8	7	1	-5	4
43	65	53	12	9	7	3	-11	9
43	65	58	11	7	6	1	-5	4
43	67	56	6	4	3	3	-12	10
44	67	51	11	9	6	3	-9	8
45	65	51	5	4	3	2	-7	6
45	66	54	11	8	6	2	-8	7
45	67	56	6	4	3	2	-9	8
45	68	53	4	3	2	3	-10	9
46	66	52	9	7	5	3	-11	10
47	63	50	6	5	4	2	-8	7
47	65	51	5	4	3	3	-12	11
47	65	54	7	5	4	2	-10	9
47	67	58	8	5	4	2	-12	11
48	63	53	8	6	5	2	-11	10
48	69	51	10	8	5	3	-10	10
49	62	52	9	7	6	2	-12	11
49	67	51	11	9	6	3	-11	11
49	68	53	4	3	2	1	-4	4
49	68	55	10	7	5	1	-5	5
49	69	58	11	7	5	2	-11	11

40 60 60

φ_1	Φ	φ_2	H	K	L	U	V	W
40	60	61	9	5	6	1	-9	6
41	61	61	11	6	7	1	-10	7
41	62	67	12	5	7	0	-7	5
41	65	69	8	3	4	0	-4	3
41	66	63	2	1	1	1	-8	6
41	69	61	9	5	4	1	-5	4
42	60	66	11	5	7	0	-7	5
42	66	63	2	1	1	1	-9	7
43	62	67	7	3	4	0	-4	3
43	66	63	2	1	1	1	-10	8

φ_1	Φ	φ_2	H	K	L	U	V	W
40	60	60						
43	68	61	11	6	5	1	-6	5
44	60	60	12	7	8	1	-12	9
44	63	66	9	4	5	0	-5	4
44	65	68	10	4	5	0	-5	4
44	66	63	2	1	1	1	-12	10
44	66	63	2	1	1	1	-11	9
44	67	60	12	7	6	1	-6	5
45	60	69	8	3	5	-1	-9	7
45	65	67	12	5	6	0	-6	5
46	61	63	8	4	5	0	-5	4
46	61	68	5	2	3	-1	-11	9
46	64	66	11	5	6	0	-6	5
47	62	63	10	5	6	0	-6	5
47	62	63	12	6	7	0	-7	6
48	60	61	9	5	6	0	-6	5
48	61	61	11	6	7	0	-7	6
48	65	69	8	3	4	-1	-12	11
49	60	60	12	7	8	0	-8	7
49	61	68	5	2	3	-1	-8	7
49	62	67	7	3	4	-1	-11	10
49	68	61	11	6	5	1	-11	11
49	68	67	7	3	3	0	-1	1
49	69	61	9	5	4	1	-9	9
49	69	67	12	5	5	0	-1	1
40	60	70						
40	69	73	10	3	4	0	-4	3
41	62	77	9	2	5	-2	-11	8
41	67	72	9	3	4	0	-4	3
41	69	79	5	1	2	-1	-9	7
42	61	72	12	4	7	-1	-11	8
42	64	76	4	1	2	-1	-8	6
42	68	72	12	4	5	0	-5	4
43	60	79	5	1	3	-1	-4	3
43	62	72	9	3	5	-1	-12	9
43	64	79	10	2	5	-1	-5	4
43	67	70	11	4	5	0	-5	4
43	68	76	12	3	5	-1	-11	9
44	64	73	10	3	5	-1	-10	8
44	69	79	5	1	2	-1	-7	6
45	64	76	4	1	2	-1	-6	5
45	65	72	6	2	3	-1	-12	10
46	61	74	7	2	4	-2	-11	9
46	66	75	11	3	5	-1	-8	7
46	68	74	7	2	3	-1	-10	9
46	69	79	5	1	2	-2	-12	11
47	62	72	9	3	5	-1	-7	6

φ_1	Φ	φ_2	H	K	L	U	V	W
40	60	70						
47	64	76	4	1	2	-2	-10	9
47	65	72	6	2	3	-1	-9	8
47	67	77	9	2	4	-2	-11	10
48	60	79	5	1	3	-3	-9	8
48	61	74	7	2	4	-2	-9	8
49	68	72	12	4	5	-1	-12	12
49	68	76	12	3	5	-1	-6	6
49	69	73	10	3	4	-1	-10	10
49	69	79	5	1	2	-1	-5	5
40	60	80						
40	61	84	9	1	5	-2	-7	5
40	62	80	11	2	6	-2	-10	7
40	66	85	11	1	5	-1	-4	3
41	61	82	7	1	4	-3	-11	8
41	64	83	8	1	4	-1	-4	3
42	64	81	6	1	3	-2	-9	7
42	67	82	7	1	3	-1	-5	4
43	61	85	11	1	6	-3	-9	7
43	64	84	10	1	5	-3	-10	8
44	64	85	12	1	6	-2	-6	5
44	68	84	10	1	4	-2	-8	7
45	60	81	12	2	7	-3	-10	8
45	61	82	7	1	4	-2	-6	5
45	64	81	6	1	3	-3	-12	10
46	61	84	9	1	5	-3	-8	7
46	62	80	11	2	6	-2	-7	6
46	66	80	11	2	5	-2	-9	8
46	67	82	7	1	3	-3	-12	11
47	61	85	11	1	6	-4	-10	9
47	64	83	8	1	4	-4	-12	11
48	67	85	12	1	5	-1	-3	3
48	68	84	10	1	4	-3	-10	10
49	60	81	12	2	7	-4	-11	10
49	64	84	10	1	5	-2	-5	5
49	64	85	12	1	6	-5	-12	12
49	66	84	9	1	4	-1	-3	3
49	66	85	11	1	5	-4	-11	11
49	67	82	7	1	3	-2	-7	7
49	68	81	12	2	5	-1	-4	4
40	60	90						
40	63	90	2	0	1	-3	-8	6
40	67	90	7	0	3	-3	-9	7
41	61	90	9	0	5	-5	-12	9
41	69	90	8	0	3	-3	-10	8

φ_1	Φ	φ_2	H	K	L	U	V	W
40	60	90						
42	60	90	7	0	4	-4	-9	7
42	63	90	2	0	1	-2	-5	4
42	66	90	9	0	4	-4	-11	9
42	68	90	5	0	2	-2	-6	5
43	61	90	9	0	5	-5	-11	9
43	63	90	2	0	1	-5	-12	10
44	63	90	2	0	1	-3	-7	6
44	67	90	7	0	3	-3	-8	7
44	68	90	5	0	2	-4	-11	10
44	69	90	8	0	3	-3	-9	8
45	60	90	7	0	4	-4	-8	7
45	63	90	2	0	1	-5	-11	10
45	63	90	2	0	1	-4	-9	8
45	66	90	9	0	4	-4	-10	9
45	66	90	11	0	5	-5	-12	11
46	61	90	9	0	5	-5	-10	9
46	61	90	11	0	6	-6	-12	11
47	67	90	7	0	3	-3	-7	7
47	67	90	12	0	5	-5	-12	12
47	68	90	5	0	2	-2	-5	5
47	69	90	8	0	3	-3	-8	8
48	63	90	2	0	1	-1	-2	2
48	66	90	9	0	4	-4	-9	9
48	66	90	11	0	5	-5	-11	11
49	60	90	7	0	4	-4	-7	7
49	60	90	12	0	7	-7	-12	12
49	61	90	9	0	5	-5	-9	9
49	61	90	11	0	6	-6	-11	11
40	70	0						
40	76	0	0	4	1	5	-1	4
40	76	7	1	8	2	6	-2	5
40	77	0	0	9	2	11	-2	9
40	79	0	0	5	1	6	-1	5
41	72	0	0	3	1	11	-3	9
41	73	0	0	10	3	12	-3	10
42	72	0	0	3	1	7	-2	6
42	72	5	1	12	4	8	-3	7
42	74	0	0	7	2	8	-2	7
42	74	8	1	7	2	10	-4	9
42	75	5	1	11	3	9	-3	8
42	76	0	0	4	1	9	-2	8
43	72	0	0	3	1	10	-3	9
43	72	6	1	9	3	12	-5	11
43	77	0	0	9	2	10	-2	9
43	79	0	0	5	1	11	-2	10
44	70	0	0	11	4	12	-4	11

φ_1	Φ	φ_2	H	K	L	U	V	W
40	70	0						
44	73	0	0	10	3	11	-3	10
44	75	0	0	11	3	12	-3	11
44	76	9	2	12	3	12	-5	12
45	70	7	1	8	3	2	-1	2
45	72	6	1	9	3	9	-4	9
45	72	9	1	6	2	2	-1	2
45	73	6	1	10	3	5	-2	5
45	74	8	1	7	2	7	-3	7
45	75	5	1	11	3	11	-4	11
45	76	5	1	12	3	3	-1	3
45	76	7	1	8	2	6	-3	8
45	78	6	1	9	2	3	-1	3
45	79	6	1	10	2	10	-3	10
46	70	5	1	11	4	11	-5	11
46	72	5	1	12	4	12	-5	12
46	73	0	0	10	3	10	-3	10
46	74	0	0	7	2	7	-2	7
46	75	0	0	11	3	11	-3	11
46	76	0	0	4	1	4	-1	4
46	77	0	0	9	2	9	-2	9
46	79	0	0	5	1	5	-1	5
47	70	0	0	11	4	11	-4	11
47	72	0	0	3	1	3	-1	3
47	74	8	1	7	2	11	-5	12
47	76	7	1	8	2	10	-4	11
47	76	9	2	12	3	9	-4	10
48	76	0	0	4	1	11	-3	12
49	70	0	0	11	4	10	-4	11
49	70	5	1	11	4	6	-4	9
49	72	0	0	3	1	11	-4	12
49	73	0	0	10	3	9	-3	10
49	75	0	0	11	3	10	-3	11
49	77	0	0	9	2	8	-2	9
49	79	0	0	5	1	9	-2	10
49	79	6	1	10	2	6	-2	7
40	70	10						
40	71	15	3	11	4	10	-6	9
40	72	18	1	3	1	11	-7	10
40	75	16	2	7	2	11	-6	10
40	79	11	1	5	1	8	-3	7
41	70	10	2	11	4	8	-4	7
42	72	18	1	3	1	3	-2	3
42	77	18	4	12	3	12	-7	12
42	78	18	3	9	2	9	-5	9
43	70	14	2	8	3	8	-5	8
43	71	15	3	11	4	11	-7	11

φ_1	Φ	φ_2	H	K	L	U	V	W
40	70	10						
43	72	14	3	12	4	12	-7	12
43	74	17	3	10	3	5	-3	5
43	75	15	3	11	3	11	-6	11
43	75	16	2	7	2	7	-4	7
43	76	14	1	4	1	2	-1	2
43	79	17	3	10	2	2	-1	2
44	70	10	2	11	4	11	-6	11
44	72	13	2	9	3	9	-5	9
44	74	11	2	10	3	2	-1	2
44	75	10	2	11	3	11	-5	11
44	78	13	2	9	2	9	-4	9
44	79	11	1	5	1	5	-2	5
45	72	18	1	3	1	10	-7	11
45	75	16	2	7	2	10	-6	11
46	70	14	2	8	3	9	-6	10
46	72	18	1	3	1	7	-5	8
47	72	14	3	12	4	8	-5	9
47	78	18	3	9	2	5	-3	6
48	70	14	2	8	3	10	-7	12
48	72	18	1	3	1	4	-3	5
48	74	17	3	10	3	9	-6	11
48	75	10	2	11	3	6	-3	7
48	75	15	3	11	3	5	-3	6
48	79	11	1	5	1	7	-3	8
49	72	18	1	3	1	9	-7	12
49	76	14	1	4	1	9	-5	11
49	79	11	1	5	1	9	-4	11
40	70	20						
40	72	24	5	11	4	11	-9	11
40	73	24	4	9	3	9	-7	9
40	75	27	5	10	3	5	-4	5
40	76	24	5	11	3	11	-8	11
40	77	27	2	4	1	4	-3	4
41	70	22	2	5	2	5	-4	5
41	71	21	3	8	3	4	-3	4
41	71	27	4	8	3	11	-10	12
41	73	23	5	12	4	4	-3	4
41	74	22	4	10	3	10	-7	10
41	75	23	3	7	2	7	-5	7
41	77	23	5	12	3	3	-2	3
41	79	22	2	5	1	5	-3	5
41	79	24	4	9	2	3	-2	3
42	71	20	4	11	4	11	-8	11
42	73	27	3	6	2	8	-7	9
42	76	20	4	11	3	11	-7	11
42	77	21	3	8	2	8	-5	8
42	77	29	6	11	3	7	-6	8
42	79	29	5	9	2	10	-8	11
43	74	22	4	10	3	11	-8	12
43	77	27	2	4	1	9	-7	10
44	70	22	2	5	2	7	-6	8
44	70	27	6	12	5	5	-5	6
44	72	29	6	11	4	4	-4	5
44	73	24	4	9	3	6	-5	7
44	73	27	3	6	2	10	-9	12
44	75	23	3	7	2	8	-6	9
44	76	20	4	11	3	9	-6	10
44	77	27	2	4	1	5	-4	6
45	70	27	5	10	4	4	-4	5
45	75	27	5	10	3	8	-7	10
45	79	22	2	5	1	8	-5	9
46	70	22	2	5	2	9	-8	11
46	71	20	4	11	4	5	-4	6
46	71	27	4	8	3	3	-3	4
46	74	29	5	9	3	3	-3	4
46	75	23	3	7	2	9	-7	11
46	77	21	3	8	2	6	-4	7
47	77	27	2	4	1	6	-5	8
48	71	21	3	8	3	7	-6	9
48	73	24	4	9	3	9	-8	12
48	76	20	4	11	3	7	-5	9
49	70	27	6	12	5	6	-9	12
49	73	23	5	12	4	8	-7	11
49	73	27	3	6	2	2	-2	3
49	77	27	2	4	1	7	-6	10
49	79	22	2	5	1	3	-2	4
49	79	24	4	9	2	8	-6	11
40	70	30						
40	71	34	8	12	5	7	-8	8
40	72	35	7	10	4	6	-7	7
40	74	39	9	11	4	7	-9	9
40	78	32	5	8	2	10	-9	11
41	70	34	6	9	4	5	-6	6
41	71	36	5	7	3	4	-5	5
41	72	37	9	12	5	7	-9	9
41	73	39	4	5	2	3	-4	4
41	76	30	4	7	2	9	-8	10
42	70	36	8	11	5	3	-4	4
42	71	38	7	9	4	5	-7	7
42	75	38	7	9	3	9	-11	12
43	71	34	8	12	5	9	-11	12
43	71	39	9	11	5	2	-3	3
43	72	35	7	10	4	8	-10	11
43	78	39	9	11	3	5	-6	7
44	72	37	9	12	5	8	-11	12
44	73	37	6	8	3	7	-9	10
44	74	30	7	12	4	4	-4	5
44	74	34	2	3	1	9	-10	12
44	74	34	2	3	1	8	-9	11
44	78	30	7	12	3	9	-8	11
44	78	34	8	12	3	3	-3	4
45	70	34	6	9	4	6	-8	9
45	73	39	4	5	2	7	-10	11
45	74	34	2	3	1	7	-8	10
45	74	39	9	11	4	5	-7	8
45	75	37	9	12	4	4	-5	6
45	76	31	6	10	3	3	-3	4
45	77	32	7	11	3	3	-3	4
45	77	39	8	10	3	4	-5	6
46	71	36	5	7	3	5	-7	8
46	74	34	2	3	1	6	-7	9
46	79	37	3	4	1	8	-9	12
47	73	39	4	5	2	4	-6	7
47	74	36	8	11	4	3	-4	5
47	75	38	7	9	3	3	-4	5
47	79	37	3	4	1	7	-8	11
48	70	34	6	9	4	7	-10	12
48	74	34	2	3	1	5	-6	8
48	78	32	5	8	2	2	-2	3
48	78	39	9	11	3	7	-9	12
49	71	36	5	7	3	6	-9	11
49	71	38	7	9	4	3	-5	6
49	76	30	4	7	2	2	-2	3
49	79	37	3	4	1	6	-7	10
40	70	40						
40	70	47	12	11	6	3	-6	5
40	72	42	8	9	4	8	-12	11
40	73	45	9	9	4	6	-10	9
40	76	42	8	9	3	3	-4	4
40	76	43	11	12	4	8	-11	11
40	76	48	9	8	3	7	-12	11
40	78	45	10	10	3	7	-10	10
40	79	45	7	7	2	5	-7	7
40	79	45	11	11	3	8	-11	11
41	70	43	11	12	6	6	-10	9
41	71	45	2	2	1	6	-11	10
41	71	45	2	2	1	5	-9	8
41	71	48	11	10	5	5	-10	9
41	72	45	11	11	5	7	-12	11
41	74	41	7	8	3	5	-7	7
41	75	42	10	11	4	7	-10	10
41	77	45	3	3	1	2	-3	3
41	79	48	11	10	3	7	-11	11
41	79	49	8	7	2	5	-8	8
42	72	40	10	12	5	7	-10	10
42	72	41	6	7	3	2	-3	3
42	73	42	9	10	4	2	-3	3
42	73	43	11	12	5	7	-11	11
42	75	45	8	8	3	5	-8	8
42	76	45	11	11	4	7	-11	11
42	76	47	12	11	4	7	-12	12
42	77	48	10	9	3	3	-5	5
42	78	41	6	7	2	8	-10	11
42	78	49	7	6	2	4	-7	7
42	79	40	10	12	3	6	-7	8
43	71	42	10	11	5	3	-5	5
43	72	42	8	9	4	5	-8	8
43	73	45	7	7	3	4	-7	7
43	74	45	5	5	2	3	-5	5
43	74	45	12	12	5	7	-12	12
43	75	48	11	10	4	6	-11	11
43	76	48	9	8	3	5	-9	9
43	79	45	11	11	3	7	-10	11
44	72	45	11	11	5	6	-11	11
44	73	45	9	9	4	5	-9	9
44	73	47	12	11	5	1	-2	2
44	73	48	10	9	4	1	-2	2
44	74	49	8	7	3	1	-2	2
44	76	40	5	6	2	6	-8	9
44	77	45	3	3	1	7	-11	12
44	78	45	10	10	3	6	-9	10
44	79	49	8	7	2	6	-10	11
45	70	42	9	10	5	5	-9	9
45	70	43	11	12	6	6	-11	11
45	71	45	2	2	1	1	-2	2
45	71	48	11	10	5	5	-11	11
45	72	48	9	8	4	4	-9	9
45	72	49	7	6	3	3	-7	7
45	74	41	7	8	3	6	-9	10
45	76	45	11	11	4	6	-10	11
45	77	45	3	3	1	5	-8	9
45	78	49	7	6	2	6	-11	12
45	79	45	7	7	2	4	-6	7
46	70	47	12	11	6	5	-12	12
46	74	45	5	5	2	5	-9	10
46	74	45	12	12	5	6	-11	12

φ_1	Φ	φ_2	H	K	L	U	V	W
40	70	40						
46	76	43	11	12	4	4	-6	7
46	78	41	6	7	2	3	-4	5
47	70	48	10	9	5	2	-5	5
47	72	45	11	11	5	5	-10	11
47	73	43	11	12	5	4	-7	8
47	75	45	8	8	3	4	-7	8
47	75	48	11	10	4	4	-8	9
47	76	42	8	9	3	6	-9	11
47	79	45	11	11	3	6	-9	11
48	70	42	9	10	5	5	-10	11
48	71	45	2	2	1	5	-11	12
48	73	45	9	9	4	4	-8	9
48	76	48	9	8	3	3	-6	7
48	78	45	10	10	3	5	-8	10
48	79	42	10	11	3	5	-7	9
49	71	45	2	2	1	4	-9	10
49	72	40	10	12	5	3	-5	6
49	72	48	9	8	4	4	-10	11
49	73	45	7	7	3	3	-6	7
49	76	45	11	11	4	5	-9	11
49	79	41	7	8	2	6	-8	11
40	70	50						
40	71	54	7	5	3	4	-11	9
40	73	53	8	6	3	3	-7	6
40	73	58	11	7	4	2	-6	5
40	74	51	11	9	4	4	-8	7
40	78	56	12	8	3	4	-9	8
41	71	51	11	9	5	3	-7	6
41	71	59	5	3	2	3	-11	9
41	73	51	5	4	2	4	-9	8
41	74	56	3	2	1	3	-8	7
41	77	54	7	5	2	5	-11	10
41	78	58	8	5	2	4	-10	9
42	72	50	12	10	5	5	-11	10
42	74	54	11	8	4	4	-10	9
42	74	56	3	2	1	4	-11	10
42	76	59	10	6	3	3	-9	8
42	78	51	11	9	3	6	-11	11
42	79	50	12	10	3	7	-12	12
43	70	56	9	6	4	2	-7	6
43	71	54	7	5	3	3	-9	8
43	71	59	5	3	2	2	-8	7
43	77	51	10	8	3	1	-2	2
43	79	53	4	3	1	1	-2	2
44	72	55	10	7	4	4	-12	11
44	74	51	11	9	4	5	-11	11
44	75	52	9	7	3	4	-9	9
44	76	55	10	7	3	2	-5	5
44	77	54	7	5	2	3	-7	7
44	78	54	11	8	3	5	-11	11
44	78	56	12	8	3	5	-12	12
44	78	58	8	5	2	3	-8	8
45	72	50	12	10	5	5	-12	12
45	73	51	5	4	2	2	-5	5
45	75	53	12	9	4	5	-12	12
45	77	58	11	7	3	4	-11	11
45	79	50	12	10	3	5	-9	10
46	73	53	8	6	3	3	-8	8
46	74	54	11	8	4	4	-11	11
46	74	56	3	2	1	1	-3	3
46	76	59	10	6	3	3	-10	10
47	71	51	11	9	5	4	-11	11
47	71	52	9	7	4	1	-3	3
47	71	54	7	5	3	2	-7	7
47	71	56	12	8	5	1	-4	4
47	72	53	12	9	5	1	-3	3
47	72	55	10	7	4	3	-10	10
47	72	58	8	5	3	1	-4	4
47	73	58	11	7	4	3	-11	11
47	77	54	7	5	2	4	-10	11
48	70	54	11	8	5	3	-11	11
48	70	56	9	6	4	2	-9	9
48	71	59	5	3	2	1	-5	5
48	76	59	10	6	3	3	-11	12
48	77	58	11	7	3	3	-9	10
48	78	51	11	9	3	3	-6	7
49	73	51	5	4	2	4	-11	12
49	73	53	8	6	3	3	-9	10
49	74	56	3	2	1	3	-10	11
49	75	53	12	9	4	3	-8	9
49	78	58	8	5	2	2	-6	7
49	79	53	4	3	1	4	-9	11
40	70	60						
40	75	67	7	3	2	1	-5	4
40	76	60	7	4	2	2	-6	5
41	71	63	8	4	3	1	-5	4
41	75	63	10	5	3	3	-12	10
41	77	63	4	2	1	2	-7	6
41	77	69	8	3	2	1	-6	5
41	79	66	9	4	2	2	-8	7
42	70	63	10	5	4	1	-6	5
42	73	66	9	4	3	1	-6	5
42	74	68	10	4	3	1	-7	6
42	77	61	11	6	3	3	-10	9
42	79	68	5	2	1	2	-9	8
43	72	61	11	6	4	2	-9	8
43	72	66	11	5	4	1	-7	6
43	73	67	12	5	4	1	-8	7
43	77	63	4	2	1	3	-11	10
44	70	63	12	6	5	1	-7	6
44	73	63	6	3	2	2	-10	9
44	74	61	9	5	3	3	-12	11
44	75	67	7	3	2	2	-12	11
45	70	60	7	4	3	2	-11	10
45	78	60	12	7	3	1	-3	3
45	79	61	9	5	2	1	-3	3
45	79	66	9	4	2	2	-9	9
45	79	68	5	2	1	1	-5	5
46	74	60	12	7	4	1	-4	4
46	75	63	10	5	3	1	-5	5
46	76	60	7	4	2	2	-7	7
46	76	66	11	5	3	2	-11	11
46	77	61	11	6	3	3	-11	11
46	77	63	4	2	1	1	-4	4
46	77	67	12	5	3	1	-6	6
46	77	69	8	3	2	1	-8	8
47	72	61	11	6	4	2	-11	11
47	73	63	6	3	2	1	-6	6
47	73	66	9	4	3	1	-9	9
47	74	61	9	5	3	2	-9	9
47	74	68	10	4	3	1	-10	10
47	75	67	7	3	2	1	-7	7
48	70	60	12	7	5	1	-6	6
48	71	63	8	4	3	1	-8	8
48	72	66	11	5	4	1	-11	11
48	73	67	12	5	4	1	-12	12
48	79	66	9	4	2	2	-10	11
48	79	68	5	2	1	2	-11	12
49	70	60	7	4	3	1	-7	7
49	70	63	10	5	4	1	-10	10
49	70	63	12	6	5	1	-12	12
49	70	68	5	2	2	0	-1	1
49	71	69	8	3	3	0	-1	1
49	76	60	7	4	2	2	-8	9
49	77	63	4	2	1	2	-9	10
49	77	69	8	3	2	1	-10	11
40	70	70						
41	78	72	9	3	2	1	-7	6
42	79	73	10	3	2	1	-8	7
43	76	70	11	4	3	1	-8	7
43	77	72	12	4	3	1	-9	8
45	72	77	9	2	3	-1	-12	11
46	70	76	8	2	3	-1	-11	10
46	78	72	9	3	2	1	-9	9
46	78	77	9	2	2	0	-1	1
46	79	73	10	3	2	1	-10	10
46	79	79	5	1	1	0	-1	1
47	74	73	10	3	3	0	-1	1
47	74	79	10	2	3	-1	-10	10
47	75	74	7	2	2	0	-1	1
47	75	75	11	3	3	0	-1	1
47	76	70	11	4	3	1	-11	11
47	76	76	4	1	1	0	-1	1
47	77	72	12	4	3	1	-12	12
48	71	70	11	4	4	0	-1	1
48	71	75	11	3	4	-1	-11	11
48	72	72	3	1	1	0	-1	1
48	72	76	12	3	4	-1	-12	12
48	72	77	9	2	3	-1	-9	9
49	70	76	8	2	3	-1	-8	8
49	78	72	9	3	2	1	-11	12
40	70	80						
41	70	85	11	1	4	-1	-5	4
41	72	61	6	1	2	-1	-10	8
41	74	62	7	1	2	-1	-11	9
41	76	83	8	1	2	-1	-12	10
42	70	83	8	1	3	-2	-11	9
42	72	84	9	1	3	-1	-6	5
42	75	85	11	1	3	-1	-7	6
43	74	82	7	1	2	-1	-9	8
43	76	83	8	1	2	-1	-10	9
43	78	84	9	1	2	-1	-11	10
43	79	64	10	1	2	-1	-12	11
44	72	81	6	1	2	-1	-8	7
46	76	65	12	1	3	-1	-6	6
46	78	64	9	1	2	-1	-9	9
46	79	84	10	1	2	-1	-10	10
47	72	84	9	1	3	-2	-9	9
47	72	65	12	1	4	-1	-4	4
47	73	84	10	1	3	-1	-5	5
47	74	62	7	1	2	-1	-7	7
47	75	60	11	2	3	-1	-11	11
47	75	85	11	1	3	-2	-11	11
47	76	81	12	2	3	-1	-12	12

φ_1	Φ	φ_2	H	K	L	U	V	W
40	70	80						
47	76	83	8	1	2	-1	-8	8
48	70	80	11	2	4	-2	-11	11
48	70	83	8	1	3	-1	-4	4
48	70	85	11	1	4	-3	-11	11
48	72	81	6	1	2	-1	-6	6
49	79	84	10	1	2	-1	-8	9
40	70	90						
40	76	90	4	0	1	-1	-5	4
40	77	90	9	0	2	-2	-11	9
40	79	90	5	0	1	-1	-6	5
41	72	90	3	0	1	-3	-11	9
41	73	90	10	0	3	-3	-12	10
42	72	90	3	0	1	-2	-7	6
42	74	90	7	0	2	-2	-8	7
42	76	90	4	0	1	-2	-9	8
43	72	90	3	0	1	-3	-10	9
43	77	90	9	0	2	-2	-10	9
43	79	90	5	0	1	-2	-11	10
44	70	90	11	0	4	-4	-12	11
44	73	90	10	0	3	-3	-11	10
44	75	90	11	0	3	-3	-12	11
46	73	90	10	0	3	-3	-10	10
46	74	90	7	0	2	-2	-7	7
46	75	90	11	0	3	-3	-11	11
46	76	90	4	0	1	-1	-4	4
46	77	90	9	0	2	-2	-9	9
46	79	90	5	0	1	-1	-5	5
47	70	90	11	0	4	-4	-11	11
47	72	90	3	0	1	-1	-3	3
48	76	90	4	0	1	-3	-11	12
49	70	90	11	0	4	-4	-10	11
49	72	90	3	0	1	-4	-11	12
49	73	90	10	0	3	-3	-9	10
49	75	90	11	0	3	-3	-10	11
49	77	90	9	0	2	-2	-8	9
49	79	90	5	0	1	-2	-9	10
40	80	0						
40	84	0	0	10	1	12	-1	10
40	85	5	1	11	1	6	-1	5
41	80	5	1	11	2	8	-2	7
41	81	0	0	6	1	7	-1	6
41	82	0	0	7	1	8	-1	7
42	82	8	1	7	1	11	-3	10
42	83	0	0	8	1	9	-1	8

φ_1	Φ	φ_2	H	K	L	U	V	W
40	80	0						
42	84	0	0	9	1	10	-1	9
42	84	0	0	10	1	11	-1	10
43	80	0	0	11	2	12	-2	11
43	85	0	0	11	1	12	-1	11
44	81	9	1	6	1	3	-1	3
44	82	8	1	7	1	7	-2	7
44	85	9	2	12	1	4	-1	4
45	80	0	0	11	2	11	-2	11
45	80	5	1	11	2	11	-3	11
45	81	0	0	6	1	6	-1	6
45	81	5	1	12	2	4	-1	4
45	82	0	0	7	1	7	-1	7
45	83	0	0	8	1	8	-1	8
45	83	7	1	8	1	4	-1	4
45	84	0	0	9	1	9	-1	9
45	84	0	0	10	1	10	-1	10
45	84	6	1	9	1	9	-2	9
45	84	6	1	10	1	5	-1	5
45	85	0	0	11	1	11	-1	11
45	85	0	0	12	1	12	-1	12
45	85	5	1	11	1	11	-2	11
45	85	5	1	12	1	6	-1	6
47	82	8	1	7	1	10	-3	11
48	80	0	0	11	2	10	-2	11
48	81	0	0	6	1	11	-2	12
48	84	0	0	10	1	9	-1	10
48	85	0	0	11	1	10	-1	11
48	85	0	0	12	1	11	-1	12
49	83	0	0	8	1	7	-1	8
49	84	0	0	9	1	8	-1	9
40	80	10						
40	84	11	2	10	1	7	-2	6
41	81	14	3	12	2	10	-4	9
41	82	16	2	7	1	12	-5	11
41	83	14	2	8	1	11	-4	10
42	81	18	2	6	1	2	-1	2
43	80	15	3	11	2	11	-5	11
43	81	14	3	12	2	12	-5	12
43	82	16	2	7	1	7	-3	7
43	84	18	3	9	1	9	-4	9
43	85	15	3	11	1	11	-4	11
43	85	17	3	10	1	5	-2	5
43	85	18	4	12	1	12	-5	12
44	80	10	2	11	2	11	-4	11
44	83	14	2	8	1	8	-3	8
44	84	11	2	10	1	10	-3	10

φ_1	Φ	φ_2	H	K	L	U	V	W
40	80	10						
44	84	13	2	9	1	3	-1	3
44	85	10	2	11	1	11	-3	11
44	85	14	3	12	1	3	-1	3
45	84	18	3	9	1	11	-5	12
46	82	16	2	7	1	9	-4	10
47	85	15	3	11	1	8	-3	9
47	85	18	4	12	1	7	-3	8
48	85	10	2	11	1	7	-2	8
49	80	10	2	11	2	5	-2	6
49	80	15	3	11	2	4	-2	5
49	83	14	2	8	1	5	-2	6
40	80	20						
40	80	27	5	10	2	10	-7	10
40	81	29	6	11	2	11	-8	11
40	82	27	3	6	1	3	-2	3
40	84	29	5	9	1	3	-2	3
40	85	29	6	11	1	11	-7	11
41	81	23	5	12	2	12	-7	12
41	81	24	5	11	2	11	-7	11
41	83	23	3	7	1	7	-4	7
41	84	24	4	9	1	9	-5	9
41	84	27	4	8	1	8	-5	8
41	85	24	5	11	1	11	-6	11
41	85	27	5	10	1	5	-3	5
41	86	27	6	12	1	12	-7	12
42	80	20	4	11	2	11	-6	11
42	83	21	3	8	1	2	-1	2
42	85	22	4	10	1	2	-1	2
42	86	23	5	12	1	2	-1	2
43	81	29	6	11	2	8	-6	9
43	84	27	4	8	1	11	-7	12
43	85	20	4	11	1	11	-5	11
44	81	23	5	12	2	10	-6	11
44	65	24	5	11	1	9	-5	10
45	81	24	5	11	2	6	-4	7
45	82	27	3	6	1	10	-7	12
45	84	24	4	9	1	7	-4	8
46	80	20	4	11	2	7	-4	8
46	80	27	5	10	2	4	-3	5
46	83	23	3	7	1	5	-3	6
46	86	27	6	12	1	5	-3	6
47	81	23	5	12	2	8	-5	10
47	82	27	3	6	1	7	-5	9
47	85	27	5	10	1	6	-5	10
48	81	29	6	11	2	5	-4	7
48	84	27	4	8	1	3	-2	4

φ_1	Φ	φ_2	H	K	L	U	V	W
40	80	20						
48	85	24	5	11	1	7	-4	9
48	85	29	6	11	1	3	-2	4
40	80	30						
40	80	34	6	9	2	11	-10	12
40	84	37	6	8	1	9	-8	10
40	86	30	7	12	1	3	-2	3
41	80	31	3	5	1	11	-9	12
41	82	34	4	6	1	7	-6	8
41	83	36	5	7	1	8	-7	9
42	83	30	4	7	1	11	-8	12
42	84	37	6	8	1	10	-9	12
42	86	36	8	11	1	6	-5	7
43	80	31	3	5	1	6	-5	7
43	85	31	6	10	1	7	-5	8
43	85	34	6	9	1	5	-4	6
44	82	34	4	6	1	8	-7	10
44	84	32	5	8	1	9	-7	11
44	86	37	9	12	1	7	-6	9
45	80	31	3	5	1	7	-6	9
45	82	34	4	6	1	9	-8	12
45	83	30	4	7	1	4	-3	5
45	86	32	7	11	1	4	-3	5
47	82	36	8	11	2	2	-2	3
47	82	37	9	12	2	2	-2	3
47	86	34	8	12	1	9	-7	12
47	86	36	8	11	1	7	-6	10
48	80	34	6	9	2	2	-2	3
48	81	35	7	10	2	2	-2	3
48	84	32	5	8	1	5	-4	7
49	81	39	4	5	1	7	-8	12
49	85	34	6	9	1	6	-5	9
49	86	37	9	12	1	8	-7	12
40	80	40						
40	81	48	9	8	2	2	-3	3
40	82	48	10	9	2	7	-10	10
40	83	43	11	12	2	8	-9	10
40	86	45	9	9	1	7	-8	9
40	86	45	10	10	1	8	-9	10
41	80	47	12	11	3	2	-3	3
41	82	42	9	10	2	6	-7	8
41	82	42	10	11	2	7	-8	9
41	83	45	6	6	1	9	-11	12
41	83	45	11	11	2	8	-10	11
41	85	45	8	8	1	6	-7	8

TAB. VI

FROM EULER ANGLES INTO HKL-UVW

φ_1	Φ	φ_2	H	K	L	U	V	W
40	80	40						
42	80	45	4	4	1	8	-11	12
42	82	45	5	5	1	7	-9	10
42	84	45	7	7	1	5	-6	7
42	84	49	7	6	1	7	-10	11
42	87	45	12	12	1	9	-10	12
43	81	42	8	9	2	5	-6	7
43	81	45	9	9	2	6	-8	9
43	86	45	10	10	1	7	-8	10
43	86	45	11	11	1	8	-9	11
43	86	48	11	10	1	4	-5	6
44	80	45	4	4	1	5	-7	8
44	83	45	6	6	1	4	-5	6
44	83	45	11	11	2	7	-9	11
44	86	45	9	9	1	6	-7	9
44	86	48	10	9	1	7	-9	11
45	80	45	4	4	1	7	-10	12
45	82	48	10	9	2	4	-6	7
45	83	43	11	12	2	6	-7	9
45	85	48	9	8	1	3	-4	5
45	87	45	12	12	1	8	-9	12
46	80	43	11	12	3	3	-4	5
46	82	42	10	11	2	5	-6	8
46	82	45	5	5	1	3	-4	5
46	85	45	8	8	1	5	-6	8
46	86	45	11	11	1	7	-8	11
47	81	45	9	9	2	5	-7	9
47	83	45	6	6	1	7	-9	12
47	85	49	8	7	1	5	-7	9
47	86	45	10	10	1	6	-7	10
48	82	42	9	10	2	4	-5	7
48	83	45	11	11	2	6	-8	11
48	84	45	7	7	1	4	-5	7
48	84	49	7	6	1	2	-3	4
48	86	47	12	11	1	4	-5	7
49	80	45	4	4	1	2	-3	4
49	82	48	11	10	2	2	-3	4
49	86	45	9	9	1	5	-6	9
49	87	45	12	12	1	7	-8	12
40	80	50						
40	83	50	6	5	1	2	-3	3
40	85	52	9	7	1	2	-3	3
40	86	53	12	9	1	2	-3	3
40	86	54	11	8	1	7	-11	11
41	80	59	5	3	1	5	-12	11
41	81	51	5	4	1	3	-5	5
41	82	51	11	9	2	7	-11	11

φ_1	Φ	φ_2	H	K	L	U	V	W
40	80	50						
41	82	53	12	9	2	7	-12	12
41	83	54	7	5	1	4	-7	7
41	84	53	8	6	1	5	-8	8
41	85	55	10	7	1	3	-5	5
41	85	56	9	6	1	5	-9	9
41	86	56	12	8	1	7	-12	12
41	86	58	11	7	1	6	-11	11
42	80	52	9	7	2	5	-9	9
42	81	55	10	7	2	1	-2	2
42	82	54	11	6	2	6	-11	11
42	82	56	6	4	1	1	-2	2
42	84	58	8	5	1	1	-2	2
42	85	59	10	6	1	1	-2	2
43	80	56	9	6	2	4	-9	9
43	81	58	11	7	2	5	-11	11
44	80	59	5	3	1	2	-5	5
44	85	56	9	6	1	6	-11	12
44	86	54	11	8	1	5	-8	9
45	81	51	5	4	1	4	-7	8
45	83	54	7	5	1	5	-9	10
45	86	56	12	8	1	4	-7	8
46	82	51	11	9	2	3	-5	6
46	84	53	8	6	1	3	-5	6
47	80	59	5	3	1	3	-8	9
47	81	58	11	7	2	3	-7	8
47	83	50	6	5	1	5	-8	10
48	81	51	5	4	1	5	-9	11
48	85	55	10	7	1	4	-7	9
49	80	52	9	7	2	2	-4	5
49	80	56	9	6	2	2	-5	6
49	86	50	12	10	1	5	-7	10
40	80	60						
40	84	61	9	5	1	5	-11	10
41	83	67	7	3	1	3	-10	9
42	85	61	11	6	1	5	-11	11
42	86	60	12	7	1	1	-2	2
43	82	60	12	7	2	5	-12	12
43	83	60	7	4	1	3	-7	7
43	84	61	9	5	1	4	-9	9
43	84	63	8	4	1	3	-8	8
43	85	63	10	5	1	2	-5	5
43	85	66	11	5	1	4	-11	11
43	86	63	12	6	1	5	-12	12
44	81	61	11	6	2	4	-11	11
44	82	63	6	3	1	1	-3	3
44	83	67	7	3	1	2	-7	7

φ_1	Φ	φ_2	H	K	L	U	V	W
40	80	60						
44	84	66	9	4	1	1	-3	3
44	85	68	10	4	1	3	-10	10
44	86	67	12	5	1	1	-3	3
45	80	63	10	5	2	3	-10	10
45	81	66	11	5	2	3	-11	11
45	81	67	12	5	2	1	-4	4
45	83	69	8	3	1	1	-4	4
46	82	60	12	7	2	4	-10	11
46	84	63	8	4	1	4	-11	12
46	85	61	11	6	1	4	-9	10
47	83	67	7	3	1	3	-11	12
47	84	61	9	5	1	3	-7	8
48	81	66	11	5	2	2	-8	9
48	85	68	10	4	1	2	-7	8
48	86	63	12	6	1	2	-5	6
49	81	61	11	6	2	2	-6	7
49	83	60	7	4	1	2	-5	6
40	80	70						
40	82	74	7	2	1	1	-6	5
40	85	70	11	4	1	2	-7	6
41	83	76	8	2	1	1	-7	6
41	84	77	9	2	1	1	-8	7
42	80	70	11	4	2	2	-10	9
42	80	75	11	3	2	1	-9	8
42	81	72	6	2	1	2	-11	10
42	84	79	10	2	1	1	-9	8
43	81	76	12	3	2	1	-10	9
44	85	70	11	4	1	3	-11	11
44	85	72	12	4	1	1	-4	4
45	80	70	11	4	2	2	-11	11
45	81	72	6	2	1	1	-6	6
45	82	74	7	2	1	1	-7	7
45	83	76	8	2	1	1	-8	8
45	84	72	9	3	1	2	-9	9
45	84	77	9	2	1	1	-9	9
45	84	79	10	2	1	1	-10	10
45	85	73	10	3	1	1	-5	5
45	85	75	11	3	1	2	-11	11
45	85	76	12	3	1	1	-6	6
46	80	75	11	3	2	1	-11	11
46	81	76	12	3	2	1	-12	12
48	83	76	8	2	1	1	-9	10
48	84	77	9	2	1	1	-10	11
48	84	79	10	2	1	1	-11	12
49	81	72	6	2	1	1	-7	8
49	82	74	7	2	1	1	-8	9

φ_1	Φ	φ_2	H	K	L	U	V	W
40	80	70						
49	85	75	11	3	1	1	-6	7
40	80	80						
42	85	80	11	2	1	1	-10	9
42	85	81	12	2	1	1	-11	10
45	83	83	8	1	1	0	-1	1
45	84	84	9	1	1	0	-1	1
45	84	84	10	1	1	0	-1	1
45	85	80	11	2	1	1	-11	11
45	85	81	12	2	1	1	-12	12
45	85	85	11	1	1	0	-1	1
45	85	85	12	1	1	0	-1	1
46	80	80	11	2	2	0	-1	1
46	80	85	11	1	2	-1	-11	11
46	81	81	6	1	1	0	-1	1
46	81	85	12	1	2	-1	-12	12
46	82	82	7	1	1	0	-1	1
46	81	85	12	1	2	-1	-10	11
49	80	85	11	1	2	-1	-9	10
40	80	90						
40	84	90	10	0	1	-1	-12	10
41	81	90	6	0	1	-1	-7	6
41	82	90	7	0	1	-1	-8	7
42	83	90	8	0	1	-1	-9	8
42	84	90	9	0	1	-1	-10	9
42	84	90	10	0	1	-1	-11	10
43	80	90	11	0	2	-2	-12	11
43	85	90	11	0	1	-1	-12	11
45	80	90	11	0	2	-2	-11	11
45	81	90	6	0	1	-1	-6	6
45	82	90	7	0	1	-1	-7	7
45	83	90	8	0	1	-1	-8	8
45	84	90	9	0	1	-1	-9	9
45	84	90	10	0	1	-1	-10	10
45	85	90	11	0	1	-1	-11	11
45	85	90	12	0	1	-1	-12	12
46	80	90	11	0	2	-2	-10	11
48	81	90	6	0	1	-2	-11	12
48	84	90	10	0	1	-1	-9	10
48	85	90	11	0	1	-1	-10	11
48	85	90	12	0	1	-1	-11	12
49	83	90	8	0	1	-1	-7	8
49	84	90	9	0	1	-1	-8	9

φ1	Φ	φ2	H	K	L	U	V	W
40	90	0						
40	90	0	0	1	0	6	0	5
40	90	5	1	12	0	12	-1	10
40	90	8	1	7	0	7	-1	6
41	90	0	0	1	0	8	0	7
41	90	0	0	1	0	7	0	6
41	90	6	1	9	0	9	-1	8
41	90	7	1	8	0	8	-1	7
42	90	0	0	1	0	11	0	10
42	90	0	0	1	0	10	0	9
42	90	0	0	1	0	9	0	8
42	90	5	1	11	0	11	-1	10
42	90	5	1	12	0	12	-1	11
42	90	6	1	10	0	10	-1	9
42	90	9	1	6	0	12	-2	11
43	90	0	0	1	0	12	0	11
45	90	0	0	1	0	1	0	1
45	90	5	1	11	0	11	-1	11
45	90	5	1	12	0	12	-1	12
45	90	6	1	9	0	9	-1	9
45	90	6	1	10	0	10	-1	10
45	90	7	1	8	0	8	-1	8
45	90	8	1	7	0	7	-1	7
45	90	9	1	6	0	6	-1	6
47	90	0	0	1	0	11	0	12
47	90	5	1	11	0	11	-1	12
48	90	0	0	1	0	9	0	10
48	90	0	0	1	0	10	0	11
48	90	0	0	1	0	8	0	9
48	90	6	1	9	0	9	-1	10
48	90	6	1	10	0	10	-1	11
48	90	7	1	8	0	8	-1	9
49	90	0	0	1	0	7	0	8
49	90	0	0	1	0	6	0	7
49	90	8	1	7	0	7	-1	8
49	90	9	1	6	0	6	-1	7
40	90	10						
40	90	14	1	4	0	8	-2	7
40	90	18	1	3	0	9	-3	6
41	90	11	1	5	0	10	-2	9
41	90	13	2	9	0	9	-2	8
41	90	15	3	11	0	11	-3	10
41	90	17	3	10	0	10	-3	9
41	90	16	1	3	0	12	-4	11
42	90	10	2	11	0	11	-2	10
42	90	14	1	4	0	12	-3	11

φ1	Φ	φ2	H	K	L	U	V	W
40	90	10						
43	90	18	1	3	0	3	-1	3
44	90	11	1	5	0	5	-1	5
44	90	13	2	9	0	9	-2	9
44	90	14	1	4	0	4	-1	4
44	90	15	3	11	0	11	-3	11
44	90	16	2	7	0	7	-2	7
44	90	17	3	10	0	10	-3	10
45	90	10	2	11	0	11	-2	11
46	90	15	3	11	0	11	-3	12
46	90	17	3	10	0	10	-3	11
47	90	10	2	11	0	11	-2	12
47	90	11	1	5	0	10	-2	11
47	90	13	2	9	0	9	-2	10
47	90	18	1	3	0	9	-3	10
48	90	14	1	4	0	8	-2	9
48	90	16	2	7	0	7	-2	8
48	90	18	1	3	0	6	-2	7
49	90	17	3	10	0	10	-3	12
49	90	18	1	3	0	9	-3	11
40	90	20						
40	90	22	2	5	0	10	-4	9
40	90	23	5	12	0	12	-5	11
40	90	24	5	11	0	11	-5	10
41	90	20	4	11	0	11	-4	10
41	90	29	5	9	0	9	-5	9
41	90	29	6	11	0	11	-6	11
42	90	24	4	9	0	9	-4	9
42	90	24	5	11	0	11	-5	11
42	90	27	1	2	0	2	-1	2
43	90	20	4	11	0	11	-4	11
43	90	21	3	8	0	8	-3	8
43	90	22	2	5	0	5	-2	5
43	90	23	3	7	0	7	-3	7
43	90	23	5	12	0	12	-5	12
44	90	29	5	9	0	9	-5	10
44	90	29	6	11	0	11	-6	12
45	90	24	4	9	0	9	-4	10
45	90	24	5	11	0	11	-5	12
45	90	27	1	2	0	8	-4	9
45	90	27	1	2	0	10	-5	11
46	90	20	4	11	0	11	-4	12
46	90	21	3	8	0	8	-3	9
46	90	22	2	5	0	10	-4	11
46	90	23	3	7	0	7	-3	8
46	90	27	1	2	0	6	-3	7
47	90	27	1	2	0	10	-5	12

φ1	Φ	φ2	H	K	L	U	V	W
40	90	20						
47	90	29	5	9	0	9	-5	11
48	90	22	2	5	0	5	-2	6
48	90	24	4	9	0	9	-4	11
48	90	27	1	2	0	4	-2	5
49	90	21	3	8	0	8	-3	10
49	90	29	5	9	0	9	-5	12
40	90	30						
40	90	32	5	8	0	6	-5	8
40	90	32	7	11	0	11	-7	11
40	90	34	2	3	0	3	-2	3
40	90	39	9	11	0	11	-9	12
41	90	30	4	7	0	7	-4	7
41	90	30	7	12	0	12	-7	12
41	90	31	3	5	0	5	-3	5
41	90	36	8	11	0	11	-8	12
41	90	38	7	9	0	9	-7	10
41	90	39	4	5	0	10	-6	11
42	90	35	7	10	0	10	-7	11
42	90	37	3	4	0	8	-6	9
43	90	31	3	5	0	10	-6	11
43	90	32	7	11	0	11	-7	12
43	90	34	2	3	0	9	-6	10
43	90	36	5	7	0	7	-5	8
43	90	39	4	5	0	5	-4	6
44	90	32	5	8	0	6	-5	9
44	90	34	2	3	0	6	-4	7
44	90	38	7	9	0	9	-7	11
45	90	30	4	7	0	7	-4	8
45	90	34	2	3	0	9	-6	11
45	90	35	7	10	0	10	-7	12
45	90	37	3	4	0	4	-3	5
46	90	31	3	5	0	5	-3	6
46	90	36	5	7	0	7	-5	9
46	90	38	7	9	0	9	-7	12
47	90	32	5	8	0	8	-5	10
48	90	30	4	7	0	7	-4	9
48	90	34	2	3	0	3	-2	4
48	90	37	3	4	0	8	-6	11
48	90	39	4	5	0	5	-4	7
49	90	32	5	8	0	8	-5	11
49	90	36	5	7	0	7	-5	10
40	90	40						
40	90	41	7	8	0	8	-7	9
40	90	42	8	9	0	9	-8	10

φ1	Φ	φ2	H	K	L	U	V	W
40	90	40						
40	90	45	1	1	0	5	-5	6
40	90	45	1	1	0	6	-6	7
40	90	46	9	8	0	8	-9	10
40	90	49	8	7	0	7	-8	9
41	90	41	6	7	0	7	-6	8
41	90	45	1	1	0	9	-9	11
41	90	45	1	1	0	4	-4	5
41	90	49	7	6	0	6	-7	8
42	90	40	5	6	0	6	-5	7
42	90	42	8	9	0	9	-8	11
42	90	42	9	10	0	10	-9	12
42	90	45	1	1	0	7	-7	9
42	90	48	9	8	0	8	-9	11
42	90	46	10	9	0	9	-10	12
43	90	41	7	6	0	6	-7	10
43	90	45	1	1	0	3	-3	4
43	90	49	8	7	0	7	-8	10
44	90	41	6	7	0	7	-6	9
44	90	45	1	1	0	8	-8	11
44	90	49	7	6	0	6	-7	9
45	90	42	8	9	0	9	-8	12
45	90	45	1	1	0	5	-5	7
45	90	45	1	1	0	7	-7	10
45	90	48	9	8	0	8	-9	12
46	90	40	5	6	0	6	-5	8
46	90	41	7	8	0	8	-7	11
46	90	49	8	7	0	7	-8	11
47	90	41	6	7	0	7	-6	10
47	90	45	1	1	0	2	-2	3
47	90	49	7	6	0	6	-7	10
48	90	41	7	8	0	8	-7	12
48	90	45	1	1	0	7	-7	11
48	90	49	8	7	0	7	-8	12
49	90	40	5	6	0	6	-5	9
49	90	45	1	1	0	5	-5	[illegible]
40	90	50						
40	90	51	11	9	0	9	-11	12
40	90	56	3	2	0	2	-3	3
40	90	56	6	5	0	5	-6	8
40	90	58	11	7	0	7	-11	11
41	90	51	5	4	0	8	-10	11
41	90	52	9	7	0	7	-9	10
41	90	54	11	[illegible]	0	8	-11	12
41	90	59	5	3	0	3	-5	5
42	90	50	6	5	0	5	-6	7
42	90	53	4	3	0	6	-8	9

φ_1	Φ	φ_2	H	K	L	U	V	W
40	90	50						
42	90	55	10	7	0	7	-10	11
43	90	51	5	4	0	4	-5	6
43	90	54	7	5	0	5	-7	8
43	90	56	3	2	0	6	-9	10
43	90	58	11	7	0	7	-11	12
43	90	59	5	3	0	6	-10	11
44	90	52	9	7	0	7	-9	11
44	90	56	3	2	0	4	-6	7
44	90	58	8	5	0	5	-8	9
45	90	53	4	3	0	3	-4	5
45	90	55	10	7	0	7	-10	12
45	90	56	3	2	0	6	-9	11
46	90	50	6	5	0	5	-6	8
46	90	52	9	7	0	7	-9	12
46	90	54	7	5	0	5	-7	9
46	90	59	5	3	0	3	-5	6
47	90	58	8	5	0	5	-8	10
48	90	51	5	4	0	4	-5	7
48	90	53	4	3	0	6	-8	11
48	90	56	3	2	0	2	-3	4
49	90	50	6	5	0	5	-6	9
49	90	54	7	5	0	5	-7	10
49	90	58	8	5	0	5	-8	11
40	90	60						
40	90	66	11	5	0	5	-11	10
40	90	67	12	5	0	5	-12	11
40	90	68	5	2	0	4	-10	9
41	90	60	7	4	0	4	-7	7
41	90	60	12	7	0	7	-12	12
41	90	61	9	5	0	5	-9	9
41	90	61	11	6	0	6	-11	11
42	90	63	2	1	0	1	-2	2
42	90	66	9	4	0	4	-9	9
42	90	66	11	5	0	5	-11	11
43	90	67	7	3	0	3	-7	7
43	90	67	12	5	0	5	-12	12
43	90	68	5	2	0	2	-5	5
43	90	69	8	3	0	3	-8	8
44	90	61	9	5	0	5	-9	10
44	90	61	11	6	0	6	-11	12
45	90	60	7	4	0	4	-7	8
45	90	63	2	1	0	4	-8	9
45	90	63	2	1	0	5	-10	11
45	90	66	9	4	0	4	-9	10
45	90	66	11	5	0	5	-11	12
46	90	63	2	1	0	3	-6	7

φ_1	Φ	φ_2	H	K	L	U	V	W
40	90	60						
46	90	67	7	3	0	3	-7	8
46	90	68	5	2	0	4	-10	11
46	90	69	8	3	0	3	-8	9
47	90	61	9	5	0	5	-9	11
47	90	63	2	1	0	5	-10	12
48	90	60	7	4	0	4	-7	9
48	90	63	2	1	0	2	-4	5
48	90	66	9	4	0	4	-9	11
48	90	68	5	2	0	2	-5	6
49	90	61	9	5	0	5	-9	12
49	90	69	8	3	0	3	-6	10
40	90	70						
40	90	72	3	1	0	3	-9	8
40	90	76	4	1	0	2	-8	7
41	90	70	11	4	0	4	-11	10
41	90	72	3	1	0	4	-12	11
41	90	73	10	3	0	3	-10	9
41	90	75	11	3	0	3	-11	10
41	90	77	9	2	0	2	-9	8
41	90	79	5	1	0	2	-10	9
42	90	76	4	1	0	3	-12	11
43	90	70	11	4	0	4	-11	11
43	90	72	3	1	0	1	-3	3
44	90	73	10	3	0	3	-10	10
44	90	74	7	2	0	2	-7	7
44	90	75	11	3	0	3	-11	11
44	90	76	4	1	0	1	-4	4
44	90	77	9	2	0	2	-9	9
44	90	79	5	1	0	1	-5	5
46	90	70	11	4	0	4	-11	12
46	90	73	10	3	0	3	-10	11
46	90	75	11	3	0	3	-11	12
47	90	72	3	1	0	3	-9	10
47	90	77	9	2	0	2	-9	10
47	90	79	5	1	0	2	-10	11
48	90	72	3	1	0	2	-6	7
48	90	74	7	2	0	2	-7	8
48	90	76	4	1	0	2	-6	9
49	90	72	3	1	0	3	-9	11
49	90	73	10	3	0	3	-10	12
40	90	80						
40	90	82	7	1	0	1	-7	6
40	90	85	12	1	0	1	-12	10
41	90	83	8	1	0	1	-8	7

φ_1	Φ	φ_2	H	K	L	U	V	W
40	90	80						
41	90	84	9	1	0	1	-9	8
42	90	80	11	2	0	2	-11	10
42	90	81	6	1	0	2	-12	11
42	90	84	10	1	0	1	-10	9
42	90	85	11	1	0	1	-11	10
42	90	85	12	1	0	1	-12	11
45	90	80	11	2	0	2	-11	11
45	90	81	6	1	0	1	-6	6
45	90	82	7	1	0	1	-7	7
45	90	83	8	1	0	1	-8	8
45	90	84	9	1	0	1	-9	9
45	90	84	10	1	0	1	-10	10
45	90	85	11	1	0	1	-11	11
45	90	85	12	1	0	1	-12	12
47	90	80	11	2	0	2	-11	12
47	90	85	11	1	0	1	-11	12
48	90	83	8	1	0	1	-8	9
48	90	84	9	1	0	1	-9	10
48	90	84	10	1	0	1	-10	11
49	90	81	6	1	0	1	-6	7
49	90	82	7	1	0	1	-7	8
40	90	90						
40	90	90	1	0	0	0	-6	5
41	90	90	1	0	0	0	-8	7
41	90	90	1	0	0	0	-7	6
42	90	90	1	0	0	0	-9	8
42	90	90	1	0	0	0	-11	10
42	90	90	1	0	0	0	-10	9
43	90	90	1	0	0	0	-12	11
45	90	90	1	0	0	0	-1	1
47	90	90	1	0	0	0	-11	12
48	90	90	1	0	0	0	-8	9
48	90	90	1	0	0	0	-10	11
48	90	90	1	0	0	0	-9	10
49	90	90	1	0	0	0	-7	8
49	90	90	1	0	0	0	-6	7
50	0	0						
50	5	0	0	1	12	10	-12	1
50	8	0	0	1	7	6	-7	1
51	5	0	0	1	11	9	-11	1
51	6	0	0	1	10	8	-10	1
51	9	0	0	1	6	5	-6	1
52	6	0	0	1	9	7	-9	1
53	5	0	0	1	12	9	-12	1

φ_1	Φ	φ_2	H	K	L	U	V	W
50	0	0						
53	7	0	0	1	6	6	-8	1
54	5	0	0	1	11	6	-11	1
54	9	0	0	1	6	9	-12	2
55	6	0	0	1	10	7	-10	1
55	8	0	0	1	7	5	-7	1
56	5	0	0	1	12	8	-12	1
56	6	0	0	1	9	6	-9	1
57	9	0	0	1	6	4	-6	1
56	5	0	0	1	11	7	-11	1
58	7	0	0	1	8	5	-8	1
59	6	0	0	1	10	6	-10	1
50	0	40						
50	7	45	1	1	12	-1	-11	1
51	7	45	1	1	11	-1	-10	1
52	8	45	1	1	10	-1	-9	1
52	9	45	1	1	9	-1	-8	1
56	7	45	1	1	12	-2	-10	1
58	7	45	1	1	11	-2	-9	1
59	8	45	1	1	10	-2	-8	1
50	0	50						
50	0	50	0	0	1	-2	-11	0
50	0	50	0	0	1	-1	-6	0
51	0	51	0	0	1	-2	-9	0
51	0	51	0	0	1	-1	-5	0
52	0	52	0	0	1	-1	-4	0
53	0	53	0	0	1	-3	-10	0
53	0	53	0	0	1	-2	-7	0
53	0	53	0	0	1	-3	-11	0
54	0	54	0	0	1	-1	-3	0
55	0	55	0	0	1	-4	-11	0
55	0	55	0	0	1	-3	-8	0
56	0	56	0	0	1	-5	-12	0
56	0	56	0	0	1	-2	-5	0
57	0	57	0	0	1	-5	-11	0
57	0	57	0	0	1	-3	-7	0
57	0	57	0	0	1	-4	-9	0
58	0	58	0	0	1	-1	-2	0
59	0	59	0	0	1	-6	-11	0
50	0	90						
50	5	90	1	0	12	-12	-10	1
50	8	90	1	0	7	-7	-6	1
51	5	90	1	0	11	-11	-9	1

φ1	Φ	φ2	H	K	L	U	V	W
50	0	90						
51	6	90	1	0	10	-10	-8	1
51	9	90	1	0	6	-6	-5	1
52	6	90	1	0	9	-9	-7	1
53	5	90	1	0	12	-12	-9	1
53	7	90	1	0	8	-8	-6	1
54	5	90	1	0	11	-11	-8	1
54	9	90	1	0	6	-12	-9	2
55	6	90	1	0	10	-10	-7	1
55	8	90	1	0	7	-7	-5	1
56	5	90	1	0	12	-12	-8	1
56	6	90	1	0	9	-9	-6	1
57	9	90	1	0	6	-6	-4	1
58	5	90	1	0	11	-11	-7	1
58	7	90	1	0	8	-8	-5	1
59	6	90	1	0	10	-10	-6	1
50	10	0						
50	14	0	0	1	4	7	-8	2
50	18	0	0	1	3	8	-9	3
51	10	0	0	2	11	9	-11	2
51	14	0	0	1	4	10	-12	3
51	16	0	0	2	7	6	-7	2
52	11	0	0	1	5	4	-5	1
52	15	0	0	3	11	9	-11	3
52	18	0	0	1	3	5	-6	2
53	13	0	0	2	9	7	-9	2
53	17	0	0	3	10	8	-10	3
54	10	0	0	2	11	8	-11	2
54	14	0	0	1	4	3	-4	1
54	18	0	0	1	3	7	-9	3
55	15	0	0	3	11	8	-11	3
55	18	0	0	1	3	9	-12	4
56	11	0	0	1	5	7	-10	2
56	16	0	0	2	7	5	-7	2
56	17	0	0	3	10	7	-10	3
57	13	0	0	2	9	6	-9	2
57	14	0	0	1	4	8	-12	3
58	10	0	0	2	11	7	-11	2
58	15	0	0	3	11	7	-11	3
58	18	0	0	1	3	2	-3	1
59	14	0	0	1	4	5	-8	2
50	10	10						
50	19	18	1	3	9	3	-7	2
54	15	18	1	3	12	3	-9	2
56	19	14	1	4	12	4	-10	3

φ1	Φ	φ2	H	K	L	U	V	W
50	10	10						
56	19	18	1	3	9	3	-10	3
58	18	18	1	3	10	3	-11	3
59	16	18	1	3	11	1	-4	1
50	10	20						
51	18	27	1	2	7	1	-4	1
52	16	27	1	2	8	2	-9	2
53	14	27	1	2	9	1	-5	1
54	13	27	1	2	10	2	-11	2
55	11	27	1	2	11	1	-6	1
59	18	27	1	2	7	1	-11	3
50	10	30						
51	18	34	2	3	11	1	-8	2
57	17	34	2	3	12	0	-4	1
58	18	34	2	3	11	0	-11	3
50	10	40						
51	13	45	1	1	6	-1	-11	2
52	14	45	2	2	11	-1	-10	2
52	16	45	1	1	5	-1	-9	2
52	19	45	1	1	4	-1	-11	3
53	17	45	2	2	9	-1	-6	2
54	10	45	1	1	8	-1	-7	1
55	11	45	1	1	7	-1	-6	1
55	19	45	1	1	4	-1	-7	2
57	13	45	1	1	6	-1	-5	1
58	14	45	2	2	11	-2	-9	2
58	19	45	1	1	4	-2	-10	3
50	10	50						
53	17	56	3	2	12	-4	-12	3
50	10	60						
50	11	63	2	1	11	-5	-12	2
52	18	63	2	1	7	-5	-11	3
54	11	63	2	1	12	-3	-6	1
54	13	63	2	1	10	-5	-10	2
54	16	63	2	1	8	-2	-4	1
58	11	63	2	1	11	-3	-5	1
59	14	63	2	1	9	-5	-8	2

φ1	Φ	φ2	H	K	L	U	V	W
50	10	70						
50	16	72	3	1	11	-7	-12	3
52	19	76	4	1	12	-9	-12	4
53	15	72	3	1	12	-8	-12	3
54	19	72	3	1	9	-2	-3	1
55	16	72	3	1	11	-5	-7	2
57	19	76	4	1	12	-7	-8	3
58	18	72	3	1	10	-7	-9	3
50	10	90						
50	14	90	1	0	4	-8	-7	2
50	18	90	1	0	3	-9	-8	3
51	10	90	2	0	11	-11	-9	2
51	14	90	1	0	4	-12	-10	3
51	16	90	2	0	7	-7	-6	2
52	11	90	1	0	5	-5	-4	1
52	15	90	3	0	11	-11	-9	3
52	18	90	1	0	3	-6	-5	2
53	13	90	2	0	9	-9	-7	2
53	17	90	3	0	10	-10	-8	3
54	10	90	2	0	11	-11	-8	2
54	14	90	1	0	4	-4	-3	1
54	18	90	1	0	3	-9	-7	3
55	15	90	3	0	11	-11	-8	3
55	18	90	1	0	3	-12	-9	4
56	11	90	1	0	5	-10	-7	2
56	16	90	2	0	7	-7	-5	2
56	17	90	3	0	10	-10	-7	3
57	13	90	2	0	9	-9	-6	2
57	14	90	1	0	4	-12	-6	3
58	10	90	2	0	11	-11	-7	2
58	15	90	3	0	11	-11	-7	3
58	18	90	1	0	3	-3	-2	1
59	14	90	1	0	4	-8	-5	2
50	20	0						
50	22	0	0	2	5	9	-10	4
50	23	0	0	5	12	11	-12	5
50	24	0	0	5	11	10	-11	5
50	27	9	1	6	12	6	-9	4
51	21	0	0	3	8	7	-8	3
51	24	0	0	4	9	8	-9	4
51	27	0	0	1	2	9	-10	5
51	27	0	0	1	2	11	-12	6
51	29	0	0	6	11	10	-11	6
52	20	0	0	4	11	9	-11	4

φ1	Φ	φ2	H	K	L	U	V	W
50	20	0						
52	23	0	0	3	7	6	-7	3
52	23	0	0	5	12	10	-12	5
52	27	0	0	1	2	7	-8	4
52	29	0	0	5	9	8	-9	5
53	22	0	0	2	5	4	-5	2
53	24	0	0	5	11	9	-11	5
53	27	0	0	1	2	5	-6	3
54	27	0	0	1	2	8	-10	5
54	29	0	0	6	11	9	-11	6
55	21	0	0	3	8	6	-8	3
55	23	0	0	5	12	9	-12	5
55	24	0	0	4	9	7	-9	4
55	27	9	1	6	12	6	-11	5
56	20	0	0	4	11	8	-11	4
56	24	0	0	5	11	8	-11	5
56	27	0	0	1	2	3	-4	2
56	29	0	0	5	9	7	-9	5
57	22	0	0	2	5	7	-10	4
57	23	0	0	3	7	5	-7	3
57	29	0	0	6	11	8	-11	6
58	23	0	0	5	12	8	-12	5
58	27	0	0	1	2	7	-10	5
58	29	9	1	6	11	1	-2	1
59	20	0	0	4	11	7	-11	4
59	24	0	0	4	9	6	-9	4
59	27	0	0	1	2	4	-6	3
50	20	10						
50	24	18	1	3	7	5	-11	4
50	25	11	1	5	11	3	-5	2
50	27	14	1	4	6	4	-7	3
52	28	18	1	3	6	3	-7	3
53	27	11	1	5	10	5	-9	4
54	23	11	1	5	12	4	-6	3
54	24	18	1	3	7	3	-8	3
55	22	18	1	3	8	1	-3	1
55	27	14	1	4	8	4	-9	4
56	26	18	1	3	6	3	-9	4
57	25	14	1	4	9	2	-5	2
57	27	11	1	5	10	5	-11	5
58	22	14	1	4	10	4	-11	4
59	21	14	1	4	11	1	-3	1
59	25	11	1	5	11	5	-12	5
59	27	14	1	4	8	4	-11	5
59	26	18	1	3	6	3	-11	5

φ_1	Φ	φ_2	H	K	L	U	V	W
50	20	20						
51	29	27	1	2	4	2	-7	3
54	20	27	1	2	6	2	-10	3
54	26	22	2	5	11	3	-10	4
55	26	27	2	4	9	1	-5	2
55	29	27	1	2	4	2	-9	4
56	22	27	2	4	11	1	-6	2
56	24	27	1	2	5	2	-11	4
57	24	22	2	5	12	2	-8	3
57	29	27	1	2	4	2	-11	5
59	24	27	1	2	5	1	-8	3
50	20	30						
51	28	31	3	5	11	1	-5	2
52	29	37	3	4	9	1	-12	5
54	27	34	2	3	7	1	-10	4
55	23	37	3	4	12	0	-3	1
55	28	39	4	5	12	0	-12	5
56	24	37	3	4	11	0	-11	4
56	27	37	3	4	10	0	-5	2
57	29	37	3	4	9	0	-9	4
58	20	34	2	3	10	0	-10	3
58	22	34	2	3	9	0	-3	1
59	24	34	2	3	8	0	-8	3
59	27	34	2	3	7	0	-7	3
50	20	40						
53	21	45	3	3	11	-1	-10	3
53	25	45	1	1	3	-1	-11	4
54	23	45	3	3	10	-1	-9	3
54	27	45	4	4	11	-1	-10	4
55	25	45	1	1	3	-1	-8	3
55	29	45	2	2	5	-1	-9	4
56	22	45	2	2	7	-1	-6	2
56	28	45	3	3	8	-1	-7	3
59	21	45	3	3	11	-2	-9	3
50	20	50						
51	20	56	3	2	10	-2	-7	2
54	22	56	3	2	9	-1	-3	1
56	29	53	4	3	9	-3	-11	5
57	26	59	5	3	12	-3	-7	3
58	24	56	3	2	8	-2	-5	2
59	20	56	3	2	10	-4	-9	3
59	23	53	4	3	12	-3	-8	3

φ_1	Φ	φ_2	H	K	L	U	V	W
50	20	60						
50	20	63	2	1	6	-4	-10	3
51	22	63	4	2	11	-5	-12	4
51	24	68	5	2	12	-4	-6	3
51	26	63	4	2	9	-2	-5	2
55	20	63	2	1	6	-3	-6	2
56	24	63	2	1	5	-5	-10	4
57	29	63	2	1	4	-1	-2	1
58	24	68	5	2	12	-6	-9	4
59	22	63	4	2	11	-3	-5	2
59	24	63	2	1	5	-4	-7	3
50	20	70						
50	21	76	4	1	11	-2	-3	1
50	22	72	3	1	8	-5	-9	3
51	28	72	3	1	6	-5	-9	4
52	24	72	3	1	7	-3	-5	2
52	25	76	4	1	9	-5	-7	3
52	25	79	5	1	11	-7	-9	4
53	22	72	3	1	8	-7	-11	4
53	27	79	5	1	10	-8	-10	5
54	27	76	4	1	8	-3	-4	2
55	28	72	3	1	6	-4	-6	3
56	25	79	5	1	11	-9	-10	5
57	24	72	3	1	7	-8	-11	5
58	23	79	5	1	12	-2	-2	1
58	25	76	4	1	9	-7	-8	4
59	25	79	5	1	11	-11	-11	6
59	27	79	5	1	10	-5	-5	3
59	28	72	3	1	6	-7	-9	5
50	20	80						
50	29	81	6	1	11	-3	-4	2
52	27	81	6	1	12	-5	-6	3
57	27	81	6	1	12	-12	-12	7
58	29	81	6	1	11	-11	-11	7
50	20	90						
50	22	90	2	0	5	-10	-9	4
50	23	90	5	0	12	-12	-11	5
50	24	90	5	0	11	-11	-10	5
51	21	90	3	0	8	-8	-7	3
51	24	90	4	0	9	-9	-8	4
51	27	90	1	0	2	-12	-11	6
51	27	90	1	0	2	-10	-9	5

φ_1	Φ	φ_2	H	K	L	U	V	W
50	20	90						
51	29	90	6	0	11	-11	-10	6
52	20	90	4	0	11	-11	-9	4
52	23	90	3	0	7	-7	-6	3
52	23	90	5	0	12	-12	-10	5
52	27	90	1	0	2	-8	-7	4
52	29	90	5	0	9	-9	-8	5
53	22	90	2	0	5	-5	-4	2
53	24	90	5	0	11	-11	-9	5
53	27	90	1	0	2	-6	-5	3
54	27	90	1	0	2	-10	-8	5
54	29	90	6	0	11	-11	-9	6
55	21	90	3	0	8	-8	-6	3
55	23	90	5	0	12	-12	-9	5
55	24	90	4	0	9	-9	-7	4
56	20	90	4	0	11	-11	-8	4
56	24	90	5	0	11	-11	-8	5
56	27	90	1	0	2	-4	-3	2
56	29	90	5	0	9	-9	-7	5
57	22	90	2	0	5	-10	-7	4
57	23	90	3	0	7	-7	-5	3
57	29	90	6	0	11	-11	-8	6
58	23	90	5	0	12	-12	-8	5
58	27	90	1	0	2	-10	-7	5
59	20	90	4	0	11	-11	-7	4
59	24	90	4	0	9	-9	-6	4
59	27	90	1	0	2	-6	-4	3
50	30	0						
50	32	0	0	5	8	8	-8	5
50	32	0	0	7	11	11	-11	7
50	34	0	0	2	3	3	-3	2
50	34	7	1	8	12	4	-5	3
50	39	0	0	9	11	12	-11	9
51	33	8	1	7	11	5	-7	4
51	35	0	0	7	10	10	-10	7
51	35	8	1	7	10	6	-8	5
51	36	0	0	5	7	7	-7	5
51	36	0	0	8	11	11	-11	8
51	37	0	0	3	4	4	-4	3
52	30	0	0	7	12	11	-12	7
52	31	0	0	3	5	9	-10	6
52	38	0	0	7	9	9	-9	7
52	39	0	0	4	5	5	-5	4
52	39	0	0	9	11	11	-11	9
53	30	0	0	4	7	6	-7	4
53	32	0	0	5	8	7	-8	5
53	32	0	0	7	11	10	-11	7

φ_1	Φ	φ_2	H	K	L	U	V	W
50	30	0						
53	34	0	0	2	3	11	-12	8
53	37	9	1	6	8	2	-3	2
54	30	0	0	7	12	10	-12	7
54	34	0	0	2	3	8	-9	6
54	35	0	0	7	10	9	-10	7
54	36	0	0	8	11	10	-11	8
54	37	0	0	3	4	11	-12	9
54	38	8	1	7	9	7	-10	7
54	39	6	1	9	11	3	-4	3
54	39	7	1	8	10	8	-11	8
55	32	0	0	7	11	9	-11	7
55	34	0	0	2	3	5	-6	4
55	34	9	1	6	9	3	-5	3
55	35	8	1	7	10	7	-11	7
55	36	0	0	5	7	6	-7	5
55	36	7	1	8	11	2	-3	2
55	37	0	0	3	4	7	-8	6
55	38	0	0	7	9	8	-9	7
55	39	0	0	4	5	9	-10	8
55	39	0	0	9	11	10	-11	9
56	31	0	0	3	5	4	-5	3
56	31	9	1	6	10	6	-11	6
56	33	8	1	7	11	7	-12	7
56	37	0	0	3	4	10	-12	9
57	30	0	0	7	12	9	-12	7
57	34	0	0	2	3	7	-9	6
57	35	0	0	7	10	8	-10	7
57	36	0	0	8	11	9	-11	8
58	30	0	0	4	7	5	-7	4
58	32	0	0	5	8	6	-8	5
58	32	0	0	7	11	8	-11	7
58	34	0	0	2	3	9	-12	8
58	38	0	0	7	9	7	-9	7
58	39	0	0	4	5	4	-5	4
58	39	0	0	9	11	9	-11	9
59	31	0	0	3	5	7	-10	6
59	37	0	0	3	4	3	-4	3
59	37	6	1	9	12	6	-10	7
59	38	8	1	7	9	4	-7	5
50	30	10						
50	32	18	1	3	5	1	-2	1
50	36	16	2	7	10	7	-12	7
51	33	11	1	5	8	7	-11	6
52	31	16	2	7	12	1	-2	1
52	34	14	1	4	6	4	-7	4
53	36	11	1	5	7	5	-8	5

φ_1 Φ φ_2	H K L	U V W
50 30 10		
53 36 16	2 7 10	5-10 6
54 30 14	1 4 7	1 -2 1
54 33 11	1 5 8	5 -9 5
54 35 18	2 6 9	3 -7 4
54 38 18	1 3 4	5-11 7
55 33 16	2 7 11	4 -9 5
55 34 14	1 4 6	6-12 7
55 39 16	2 7 9	3 -6 4
56 30 11	1 5 9	1 -2 1
56 36 11	1 5 7	6-11 7
57 38 13	2 9 12	3 -6 4
59 34 14	1 4 6	2 -5 3
59 36 16	2 7 10	3 -8 5
50 30 20		
50 37 23	3 7 10	3 -7 4
50 37 27	1 2 3	4-11 6
51 35 21	3 8 12	4 -9 5
53 32 23	3 7 12	2 -6 3
53 38 22	2 5 7	2 -5 3
53 39 24	4 9 12	3 -8 5
54 33 27	2 4 7	1 -4 2
54 37 27	1 2 3	2 -7 4
55 38 21	3 8 11	3 -8 5
56 37 27	1 2 3	3-12 7
57 35 21	3 8 12	4-12 7
58 31 22	2 5 9	1 -4 2
58 31 27	3 6 11	1 -6 3
58 34 27	3 6 10	2-11 6
58 37 27	1 2 3	1 -5 3
59 37 23	3 7 10	2 -8 5
59 38 22	2 5 7	3-11 7
59 39 24	4 9 12	3-12 8
50 30 30		
50 35 39	4 5 9	1 -8 4
50 36 37	3 4 7	1 -6 3
51 34 30	4 7 12	1 -4 2
53 38 36	5 7 11	1 -7 4
54 33 31	3 5 9	1 -6 3
54 33 34	4 6 11	1 -8 4
54 36 30	4 7 11	2 -9 5
54 39 39	4 5 8	1-12 7
55 30 39	4 5 11	0-11 5
55 36 36	5 7 12	1-11 6
56 33 39	4 5 10	0 -2 1

φ_1 Φ φ_2	H K L	U V W
50 30 30		
56 36 34	2 3 5	1 -9 5
57 35 39	4 5 9	0 -9 5
57 36 31	3 5 8	1 -7 4
58 32 37	3 4 8	0 -2 1
58 39 39	4 5 8	0 -8 5
59 36 37	3 4 7	0 -7 4
50 30 40		
50 32 45	4 4 9	0 -9 4
50 33 45	5 5 11	0-11 5
50 35 40	5 6 11	1-10 5
51 35 45	1 1 2	0 -2 1
52 38 45	5 5 9	0 -9 5
52 38 45	6 6 11	0-11 6
52 39 45	4 4 7	0 -7 4
54 31 45	5 5 12	-1-11 5
55 33 40	5 6 12	0 -2 1
55 33 45	5 5 11	-1-10 5
56 35 40	5 6 11	0-11 6
56 35 45	1 1 2	-1-11 6
56 38 41	6 7 12	0-12 7
57 32 45	4 4 9	-1 -8 4
57 35 45	1 1 2	-1 -9 5
57 38 40	5 6 10	0 -5 3
57 38 45	6 6 11	-1-10 6
59 31 45	3 3 7	-1 -6 3
59 35 45	1 1 2	-1 -7 4
59 38 45	5 5 9	-1 -8 5
50 30 50		
50 35 50	6 5 11	-1-12 6
50 35 51	5 4 9	-1-10 5
50 36 53	4 3 7	-1 -8 4
51 30 51	5 4 11	-1 -7 3
51 31 56	3 2 6	-2 -9 4
52 36 59	5 3 8	-3-11 6
53 33 56	6 4 11	-1 -4 2
53 33 59	5 3 9	-3-10 5
53 38 58	8 5 12	-3-12 7
54 33 50	6 5 12	-1 -6 3
55 32 53	4 3 8	-1 -4 2
55 33 51	5 4 10	-2-10 5
55 35 51	5 4 9	-2-11 6
55 36 53	4 3 7	-2 -9 5
55 36 56	3 2 5	-2 -7 4
55 36 59	5 3 8	-4-12 7

φ_1 Φ φ_2	H K L	U V W
50 30 50		
57 30 59	5 3 10	-4-10 5
58 35 51	5 4 9	-3-12 7
59 30 51	5 4 11	-3-10 5
59 33 59	5 3 9	-3 -7 4
59 36 53	4 3 7	-3-10 6
59 39 56	6 4 9	-1 -3 2
50 30 60		
50 34 63	6 3 10	-2 -6 3
50 39 63	8 4 11	-2 -7 4
50 39 66	9 4 12	-4-12 7
51 37 63	2 1 3	-3 -9 5
51 38 68	5 2 7	-5-12 7
52 33 63	4 2 7	-3 -8 4
53 31 68	5 2 9	-1 -2 1
53 35 67	7 3 11	-4 -9 5
53 35 69	8 3 12	-6-12 7
54 31 63	6 3 11	-5-12 6
55 34 60	7 4 12	-4-11 6
55 37 63	2 1 3	-2 -5 3
55 39 66	9 4 12	-4 -9 6
56 33 63	4 2 7	-5-11 6
56 39 63	8 4 11	-5-12 8
58 34 63	6 3 10	-5-10 6
59 32 67	7 3 12	-3 -5 3
59 37 63	2 1 3	-3 -6 4
50 30 70		
50 30 76	4 1 7	-5 -8 4
50 30 79	5 1 9	-7-10 5
52 33 79	5 1 8	-5 -7 4
52 38 72	3 1 4	-4 -8 5
53 30 72	6 2 11	-7-12 6
53 38 77	9 2 12	-2 -3 2
54 31 74	7 2 12	-6 -9 5
54 37 76	8 2 11	-2 -3 2
54 38 72	3 1 4	-5 -9 6
55 33 79	5 1 8	-9-11 7
56 34 76	4 1 6	-9-12 8
56 36 74	7 2 10	-2 -3 2
56 38 72	3 1 4	-6-10 7
57 30 76	4 1 7	-4 -5 3
57 38 72	3 1 4	-7-11 8
58 30 72	6 2 11	-5 -7 4
58 34 76	4 1 6	-8-10 7
58 35 72	6 2 9	-2 -3 2

φ_1 Φ φ_2	H K L	U V W
50 30 70		
59 36 79	5 1 7	-8 -9 7
59 38 72	3 1 4	-8-12 9
50 30 80		
51 37 84	9 1 12	-7 -9 6
52 33 82	7 1 11	-4 -5 3
54 34 81	6 1 9	-5 -6 4
54 36 83	8 1 11	-6 -7 5
55 31 81	6 1 10	-7 -8 5
56 37 81	6 1 8	-10-12 9
57 31 82	7 1 12	-3 -3 2
57 34 83	8 1 12	-4 -4 3
57 37 84	9 1 12	-6 -6 5
57 38 82	7 1 9	-10-11 9
57 39 83	8 1 10	-11-12 10
58 33 82	7 1 11	-11-11 8
58 36 83	8 1 11	-11-11 9
58 37 81	6 1 8	-9-10 8
58 39 84	9 1 11	-11-11 10
59 31 81	6 1 10	-10-10 7
59 34 81	6 1 9	-9 -9 7
59 35 82	7 1 10	-5 -5 4
59 38 82	7 1 9	-9 -9 8
59 39 83	8 1 10	-10-10 9
50 30 90		
50 32 90	5 0 8	-8 -8 5
50 32 90	7 0 11	-11-11 7
50 34 90	2 0 3	-3 -3 2
50 39 90	9 0 11	-11-12 9
51 35 90	7 0 10	-10-10 7
51 36 90	5 0 7	-7 -7 5
51 36 90	8 0 11	-11-11 8
51 37 90	3 0 4	-4 -4 3
52 30 90	7 0 12	-12-11 7
52 31 90	3 0 5	-10 -9 6
52 38 90	7 0 9	-9 -9 7
52 39 90	4 0 5	-5 -5 4
52 39 90	9 0 11	-11-11 9
53 30 90	4 0 7	-7 -6 4
53 32 90	5 0 8	-8 -7 5
53 32 90	7 0 11	-11-10 7
53 34 90	2 0 3	-12-11 8
54 30 90	7 0 12	-12-10 7
54 34 90	2 0 3	-9 -8 6
54 35 90	7 0 10	-10 -9 7

φ_1	Φ	φ_2	H	K	L	U	V	W
50	30	90						
54	36	90	8	0	11	-11	-10	8
54	37	90	3	0	4	-12	-11	9
55	32	90	7	0	11	-11	-9	7
55	34	90	2	0	3	-6	-5	4
55	36	90	5	0	7	-7	-6	5
55	37	90	3	0	4	-8	-7	6
55	38	90	7	0	9	-9	-8	7
55	39	90	4	0	5	-10	-9	8
55	39	90	9	0	11	-11	-10	9
56	31	90	3	0	5	-5	-4	3
56	37	90	3	0	4	-12	-10	9
57	30	90	7	0	12	-12	-9	7
57	34	90	2	0	3	-9	-7	6
57	35	90	7	0	10	-10	-8	7
57	36	90	8	0	11	-11	-9	8
58	30	90	4	0	7	-7	-5	4
58	32	90	5	0	8	-8	-6	5
58	32	90	7	0	11	-11	-8	7
58	34	90	2	0	3	-12	-9	8
58	38	90	7	0	9	-9	-7	7
58	39	90	4	0	5	-5	-4	4
58	39	90	9	0	11	-11	-9	9
59	31	90	3	0	5	-10	-7	6
59	37	90	3	0	4	-4	-3	3
50	40	0						
50	41	0	0	7	8	9	-8	7
50	42	0	0	8	9	10	-9	8
50	45	0	0	1	1	7	-6	6
50	45	0	0	1	1	6	-5	5
50	45	5	1	11	11	11	-11	10
50	45	5	1	12	12	12	-12	11
50	45	9	1	6	6	6	-7	6
50	48	0	0	9	8	10	-8	9
50	49	0	0	8	7	9	-7	8
50	49	6	1	9	8	1	-1	1
50	49	7	1	8	7	1	-1	1
51	42	0	0	9	10	11	-10	9
51	42	0	0	10	11	12	-11	10
51	45	0	0	1	1	8	-7	7
51	45	0	0	1	1	9	-8	8
51	45	7	1	8	8	8	-9	8
51	45	8	1	7	7	7	-8	7
51	48	0	0	10	9	11	-9	10
51	48	0	0	11	10	12	-10	11
51	48	5	1	11	10	1	-1	1
51	48	5	1	12	11	1	-1	1
51	48	6	1	10	9	1	-1	1
52	40	0	0	5	6	6	-6	5
52	41	8	1	7	8	7	-9	7
52	41	9	1	6	7	3	-4	3
52	45	0	0	1	1	11	-10	10
52	45	0	0	1	1	12	-11	11
52	45	0	0	1	1	10	-9	9
52	45	5	1	11	11	11	-12	11
52	45	6	1	9	9	9	-10	9
52	45	6	1	10	10	10	-11	10
53	41	0	0	6	7	7	-7	6
53	41	0	0	7	8	8	-8	7
53	42	0	0	8	9	9	-9	8
53	42	0	0	9	10	10	-10	9
53	42	6	1	9	10	9	-11	9
53	42	6	1	10	11	5	-6	5
53	42	7	1	8	9	4	-5	4
53	48	0	0	9	8	9	-8	9
53	48	0	0	10	9	10	-9	10
53	49	0	0	7	6	7	-6	7
53	49	0	0	8	7	8	-7	8
54	42	0	0	10	11	11	-11	10
54	43	0	0	11	12	12	-12	11
54	45	6	1	9	9	9	-11	10
54	45	6	1	10	10	10	-12	11
54	45	7	1	8	6	8	-10	9
54	45	8	1	7	7	7	-9	8
54	45	9	1	6	6	6	-8	7
54	47	0	0	12	11	12	-11	12
54	48	0	0	11	10	11	-10	11
55	40	0	0	5	6	11	-12	10
55	45	0	0	1	1	1	-1	1
56	42	0	0	8	9	8	-9	8
56	42	0	0	9	10	9	-10	9
56	42	0	0	10	11	10	-11	10
56	43	0	0	11	12	11	-12	11
56	45	6	1	9	9	9	-12	11
56	47	0	0	12	11	11	-11	12
56	48	0	0	9	8	8	-8	9
56	48	0	0	10	9	9	-9	10
56	48	0	0	11	10	10	-10	11
57	40	0	0	5	6	5	-6	5
57	41	0	0	6	7	6	-7	6
57	41	0	0	7	8	7	-8	7
57	45	0	0	1	1	10	-11	11
57	45	0	0	1	1	11	-12	12
57	45	7	1	8	8	8	-11	10
57	45	8	1	7	7	7	-10	9
57	45	9	1	6	6	6	-9	8
57	49	0	0	7	6	6	-6	7
57	49	0	0	8	7	7	-7	6
58	40	6	1	10	12	6	-9	7
58	42	6	1	9	10	4	-6	5
58	42	7	1	8	9	7	-11	9
58	43	0	0	11	12	10	-12	11
58	43	5	1	11	12	5	-7	6
58	45	0	0	1	1	8	-9	9
58	45	0	0	1	1	7	-8	8
58	45	0	0	1	1	9	-10	10
58	47	0	0	12	11	10	-11	12
59	41	8	1	7	8	3	-5	4
59	41	9	1	6	7	5	-9	7
59	42	0	0	9	10	8	-10	9
59	42	0	0	10	11	9	-11	10
59	45	0	0	1	1	5	-6	6
59	45	0	0	1	1	6	-7	7
59	45	7	1	8	8	8	-12	11
59	45	8	1	7	7	7	-11	10
59	45	9	1	6	6	6	-10	9
59	48	0	0	10	9	8	-9	10
59	48	0	0	11	10	9	-10	11
50	40	10						
50	40	14	1	4	5	2	-3	2
50	41	18	3	9	11	5	-9	6
50	44	15	3	11	12	8	-12	9
50	46	14	1	4	4	8	-11	9
50	46	16	2	7	7	7	-10	8
50	49	13	2	9	8	10	-12	11
50	49	15	3	11	10	6	-8	7
51	40	11	1	5	6	5	-7	5
51	46	13	2	9	9	9	-12	10
51	47	18	1	3	3	3	-5	4
52	44	17	3	10	11	7	-12	9
52	46	16	2	7	7	7	-11	9
52	49	17	3	10	9	6	-9	8
53	42	14	2	8	9	5	-8	6
53	46	11	1	5	5	5	-7	6
53	46	14	1	4	4	4	-6	5
53	48	14	3	12	11	7	-10	9
54	40	11	1	5	6	7	-11	8
54	43	10	2	11	12	7	-10	8
54	46	16	2	7	7	7	-12	10
54	47	18	1	3	3	6	-11	9
54	49	15	3	11	10	7	-11	10
54	49	18	4	12	11	5	-9	8
55	40	13	2	9	11	4	-7	5
55	42	18	2	6	7	5	-11	6
55	44	15	3	11	12	5	-9	7
55	47	18	1	3	3	3	-6	5
55	49	17	3	10	9	7	-12	11
56	42	16	2	7	8	2	-4	3
56	43	11	2	10	11	3	-5	4
56	46	14	1	4	4	4	-7	6
57	40	14	1	4	5	5	-10	7
57	46	11	1	5	5	5	-8	7
59	40	11	1	5	6	2	-4	3
59	42	16	2	6	7	3	-8	6
59	43	13	2	9	10	5	-10	8
59	46	11	1	5	5	5	-9	8
59	46	14	1	4	4	4	-8	7
59	47	18	1	3	3	3	-7	6
50	40	20						
50	42	22	2	5	6	3	-6	4
50	45	27	5	10	11	3	-7	5
50	47	22	2	5	5	5	-9	7
51	43	29	5	9	11	1	-3	2
51	45	24	5	11	12	5	-11	8
51	48	27	1	2	2	4	-9	7
52	42	24	4	9	11	5	-12	8
53	41	21	3	8	10	4	-9	6
53	41	29	5	9	12	3	-11	7
53	44	20	4	11	12	2	-4	3
53	44	23	3	7	8	3	-7	5
53	45	27	4	8	9	4	-11	8
53	46	29	6	11	12	4	-12	9
53	47	22	2	5	5	5	-10	8
53	48	27	1	2	2	2	-5	4
55	42	22	2	5	6	4	-10	7
55	45	24	4	9	10	3	-8	6
55	47	22	2	5	5	5	-11	9
55	48	27	1	2	2	4	-11	9
56	40	23	3	7	9	1	-3	2
56	43	27	5	10	12	2	-7	5
56	44	21	3	6	9	5	-12	9
56	47	22	2	5	5	5	-12	10
56	48	27	1	2	2	2	-6	5
56	49	20	4	11	10	5	-10	9
57	44	23	3	7	8	4	-12	9
57	45	24	5	11	12	3	-9	7
58	40	27	3	6	8	2	-9	6

φ_1 Φ φ_2	H	K	L	U	V	W
50 40 20						
58 42 24	4	9	11	2	-7	5
58 44 27	3	6	7	1	-4	3
59 46 29	5	9	10	1	-5	4
59 48 27	1	2	2	2	-7	6
50 40 30						
50 44 31	3	5	6	1	-3	2
50 45 37	3	4	5	2	-9	6
51 40 37	3	4	6	2	-12	7
51 41 36	5	7	10	1	-5	3
51 42 34	2	3	4	2	-8	5
51 47 39	4	5	6	2	-10	7
52 48 35	7	10	11	1	-4	3
52 49 36	8	11	12	1	-4	3
53 47 34	6	9	10	1	-4	3
54 42 30	4	7	9	3	-12	8
54 46 32	5	8	9	1	-4	3
55 42 34	2	3	4	1	-6	4
55 42 37	6	8	11	1	-9	6
55 42 39	4	5	7	1	-12	8
55 45 35	7	10	12	2	-11	8
55 47 39	4	5	6	1	-8	6
56 40 31	3	5	7	2	-11	7
56 43 32	5	8	10	2	-10	7
56 45 30	4	7	8	1	-4	3
57 47 32	7	11	12	1	-5	4
58 44 36	5	7	9	1	-11	8
58 45 34	6	9	11	1	-8	6
58 45 37	3	4	5	1	-12	9
58 47 31	6	10	11	1	-5	4
59 42 34	2	3	4	1	-10	7
59 42 39	4	5	7	0	-7	5
59 44 31	3	5	6	2	-12	9
59 48 37	6	8	9	1	-12	10
50 40 40						
50 45 45	7	7	10	1	-11	7
50 46 41	6	7	9	1	-6	4
51 43 41	6	7	10	1	-8	5
51 44 49	8	7	11	0	-11	7
51 46 45	8	8	11	1	-12	8
51 46 49	7	6	9	0	-3	2
51 48 40	5	6	7	2	-11	8
51 48 42	8	9	11	1	-7	5
51 48 45	7	7	9	1	-10	7
52 40 45	7	7	12	0	-12	7

φ_1 Φ φ_2	H	K	L	U	V	W
50 40 40						
52 45 48	9	8	12	0	-3	2
52 47 49	8	7	10	0	-10	7
52 49 41	6	7	8	2	-12	9
52 49 45	4	4	5	1	-11	6
52 49 45	9	9	11	1	-12	9
53 40 45	3	3	5	0	-5	3
53 41 45	5	5	8	0	-8	5
53 42 45	7	7	11	0	-11	7
53 48 48	9	8	11	0	-11	6
53 49 49	7	6	8	0	-4	3
54 43 45	2	2	3	0	-3	2
54 48 48	10	9	12	0	-4	3
55 40 49	7	6	11	-1	-8	5
55 45 45	5	5	7	0	-7	5
55 45 45	7	7	10	0	-10	7
55 46 45	8	8	11	0	-11	6
55 48 40	5	6	7	1	-9	7
56 47 45	3	3	4	0	-4	3
56 48 45	7	7	9	0	-9	7
56 49 41	6	7	8	1	-10	8
56 49 45	4	4	5	0	-5	4
57 40 41	6	7	11	0	-11	7
57 40 45	7	7	12	-1	-11	7
57 42 41	7	8	12	0	-3	2
57 49 45	9	9	11	0	-11	9
58 41 40	5	6	9	0	-3	2
58 43 41	6	7	10	0	-10	7
58 44 41	7	8	11	0	-11	8
58 45 42	8	9	12	0	-4	3
59 40 45	3	3	5	-1	-9	6
59 42 45	7	7	11	-1	-10	7
59 43 45	2	2	3	-1	-11	8
59 43 49	7	6	10	-2	-11	8
59 44 40	5	6	6	0	-4	3
59 46 41	6	7	9	0	-9	7
59 47 41	7	8	10	0	-5	4
59 48 42	8	9	11	0	-11	9
59 48 42	9	10	12	0	-6	5
50 40 50						
50 46 56	6	4	7	-1	-9	6
50 49 52	9	7	10	0	-10	7
51 41 58	8	5	11	-1	-5	3
51 45 53	4	3	5	-1	-12	8
51 48 50	6	5	7	0	-7	5
51 49 51	10	6	11	0	-11	6
52 42 51	5	4	7	-1	-11	7

φ_1 Φ φ_2	H	K	L	U	V	W
50 40 50						
52 42 56	3	2	4	-2	-11	7
52 47 58	11	7	12	-1	-7	5
53 44 52	9	7	12	-1	-9	6
53 48 55	10	7	11	-1	-8	6
54 46 56	6	4	7	-2	-11	8
54 47 56	9	6	10	-2	-12	9
54 49 59	5	3	5	-1	-5	4
55 42 53	8	6	11	-1	-6	4
55 42 56	3	2	4	-2	-9	6
55 44 50	6	5	8	-1	-10	7
55 45 53	4	3	5	-1	-7	5
56 44 59	5	3	6	-3	-11	8
57 40 59	5	3	7	-1	-3	2
57 41 50	6	5	9	-1	-6	4
57 45 56	9	6	11	-1	-4	3
57 47 54	7	5	8	-1	-5	4
57 48 50	6	5	7	-1	-10	8
58 40 53	4	3	6	-3	-12	8
58 42 56	3	2	4	-2	-7	5
58 44 50	6	5	8	-2	-12	9
58 44 54	7	5	9	-1	-4	3
58 44 59	5	3	6	-3	-9	7
59 45 53	4	3	5	-2	-9	7
59 46 52	9	7	11	-1	-5	4
59 47 51	5	4	6	-2	-11	9
59 49 51	10	8	11	-1	-7	6
59 49 59	5	3	5	-3	-10	9
50 40 60						
50 42 68	5	2	6	-4	-11	7
50 43 63	10	5	12	-2	-8	5
50 44 68	10	4	11	-1	-3	2
50 45 63	8	4	9	-3	-12	8
50 47 67	7	3	7	-2	-7	5
50 49 60	7	4	7	-1	-7	5
51 40 63	6	3	8	-3	-10	6
51 40 67	7	3	9	-3	-8	5
51 44 67	7	3	8	-1	-3	2
51 48 66	11	5	11	-3	-11	6
51 49 61	11	6	11	-2	-11	6
52 47 67	12	5	12	-4	-12	9
52 47 69	8	3	8	-3	-8	6
52 48 63	2	1	2	-1	-4	3
52 49 60	12	7	12	-2	-12	9
53 42 63	4	2	5	-1	-3	2
53 44 69	8	3	9	-3	-7	5
53 45 63	10	5	11	-2	-7	5

φ_1 Φ φ_2	H	K	L	U	V	W
50 40 60						
53 49 61	9	5	9	-2	-9	7
54 45 66	9	4	10	-4	-11	8
54 47 68	5	2	5	-2	-5	4
54 48 66	9	4	9	-3	-9	7
55 41 61	9	5	12	-1	-3	2
55 41 69	8	3	10	-5	-10	7
55 42 66	9	4	11	-3	-7	5
55 43 61	9	5	11	-3	-10	7
55 48 63	2	1	2	-3	-10	8
56 44 63	6	3	7	-3	-8	6
56 47 67	12	5	12	-5	-12	10
56 48 66	11	5	11	-4	-11	9
56 49 60	12	7	12	-3	-12	10
56 49 61	11	6	11	-3	-11	9
57 42 68	5	2	6	-2	-4	3
57 47 67	7	3	7	-3	-7	6
57 49 60	7	4	7	-2	-7	6
58 42 60	7	4	9	-4	-11	8
58 44 69	8	3	9	-6	-11	9
58 45 63	8	4	9	-2	-5	4
58 47 69	8	3	8	-4	-8	7
58 48 63	2	1	2	-3	-8	7
59 45 60	7	4	8	-4	-11	9
59 45 63	10	5	11	-5	-12	10
59 47 68	5	2	5	-5	-10	9
59 48 63	2	1	2	-4	-10	9
59 48 66	9	4	9	-4	-9	8
59 49 61	9	5	9	-3	-9	8
50 40 70						
50 40 76	4	1	5	-6	-11	7
50 46 74	7	2	7	-3	-7	5
50 46 75	11	3	11	-5	-11	8
51 41 73	10	3	12	-3	-6	4
51 42 72	6	2	7	-4	-9	6
51 43 77	9	2	10	-4	-7	5
51 44 70	11	4	12	-4	-10	7
51 46 76	4	1	4	-2	-4	3
51 47 70	11	4	11	-4	-11	6
52 40 76	4	1	5	-7	-12	8
52 43 79	10	2	11	-5	-8	6
52 46 77	9	2	9	-5	-9	7
52 46 79	5	1	5	-3	-5	4
52 47 72	3	1	3	-5	-12	9
52 49 72	12	4	11	-2	-5	4
53 42 74	7	2	8	-6	-11	6
53 47 72	3	1	3	-4	-9	7

φ_1	Φ	φ_2	H	K	L	U	V	W
50	40	70						
53	49	79	10	2	9	-4	-7	6
54	43	72	9	3	10	-2	-4	3
54	46	73	10	3	10	-5	-10	8
54	46	75	11	3	11	-6	-11	9
54	46	76	4	1	4	-7	-12	10
55	44	73	10	3	11	-5	-9	7
55	44	75	11	3	12	-3	-5	4
55	47	70	11	4	11	-5	-11	9
56	41	72	9	3	11	-7	-12	9
56	46	74	7	2	7	-4	-7	6
56	46	76	4	1	4	-5	-8	7
56	46	77	9	2	9	-6	-9	8
56	46	79	5	1	5	-7	-10	9
56	49	70	11	4	10	-4	-9	8
58	42	72	6	2	7	-3	-5	4
58	42	76	8	2	9	-5	-7	6
58	43	77	9	2	10	-6	-8	7
58	43	79	10	2	11	-7	-9	8
58	46	73	10	3	10	-6	-10	9
58	46	75	11	3	11	-7	-11	10
58	46	76	4	1	4	-8	-12	11
58	47	72	3	1	3	-5	-9	8
58	49	77	9	2	8	-2	-3	3
58	49	79	10	2	9	-7	-10	10
59	44	70	11	4	12	-4	-7	6
59	47	70	11	4	11	-6	-11	10
59	47	72	3	1	3	-7	-12	11
59	48	76	12	3	11	-2	-3	3
59	49	73	10	3	9	-3	-5	5
59	49	75	11	3	10	-7	-11	11
50	40	80						
50	45	81	6	1	6	-7	-12	9
50	45	84	9	1	9	-6	-9	7
50	48	84	10	1	9	-7	-11	9
50	48	85	11	1	10	-4	-6	5
51	45	84	10	1	10	-7	-10	8
51	45	85	11	1	11	-8	-11	9
52	45	85	12	1	12	-9	-12	10
53	43	80	11	2	12	-6	-9	7
53	45	80	11	2	11	-7	-11	9
53	45	81	6	1	6	-4	-6	5
54	45	82	7	1	7	-5	-7	6
54	45	83	8	1	8	-6	-8	7
54	45	84	9	1	9	-7	-9	8
54	45	84	10	1	10	-8	-10	9
54	45	85	11	1	11	-9	-11	10

φ_1	Φ	φ_2	H	K	L	U	V	W
50	40	60						
54	45	65	12	1	12	-10	-12	11
56	45	80	11	2	11	-8	-11	10
56	45	81	6	1	6	-9	-12	11
56	48	84	10	1	9	-4	-5	5
56	48	85	11	1	10	-9	-11	11
56	48	65	12	1	11	-5	-6	6
56	49	83	8	1	7	-3	-4	4
56	49	84	9	1	8	-7	-9	9
57	45	84	10	1	10	-9	-10	10
57	45	85	11	1	11	-10	-11	11
57	45	85	12	1	12	-11	-12	12
57	48	61	12	2	11	-3	-4	4
58	40	84	10	1	12	-12	-12	11
58	43	80	11	2	12	-8	-10	9
58	43	85	11	1	12	-1	-1	1
58	45	82	7	1	7	-6	-7	7
58	45	63	8	1	8	-7	-8	8
58	45	84	9	1	9	-8	-9	9
58	48	80	11	2	10	-8	-11	11
59	42	84	9	1	10	-1	-1	1
59	42	84	10	1	11	-1	-1	1
59	45	80	11	2	11	-9	-11	11
59	45	81	6	1	6	-5	-6	6
50	40	90						
50	41	90	7	0	8	-8	-9	7
50	42	90	8	0	9	-9	-10	8
50	45	90	1	0	1	-5	-6	5
50	45	90	1	0	1	-6	-7	6
50	48	90	9	0	8	-8	-10	9
50	49	90	8	0	7	-7	-9	8
51	42	90	9	0	10	-10	-11	9
51	42	90	10	0	11	-11	-12	10
51	45	90	1	0	1	-8	-9	8
51	45	90	1	0	1	-7	-8	7
51	48	90	10	0	9	-9	-11	10
51	48	90	11	0	10	-10	-12	11
52	40	90	5	0	6	-6	-6	5
52	45	90	1	0	1	-10	-11	10
52	45	90	1	0	1	-9	-10	9
52	45	90	1	0	1	-11	-12	11
53	41	90	6	0	7	-7	-7	6
53	41	90	7	0	8	-8	-8	7
53	42	90	8	0	9	-9	-9	8
53	42	90	9	0	10	-10	-10	9
53	48	90	9	0	8	-8	-9	9
53	48	90	10	0	9	-9	-10	10

φ_1	Φ	φ_2	H	K	L	U	V	W
50	40	90						
53	49	90	7	0	6	-6	-7	7
53	49	90	8	0	7	-7	-8	8
54	42	90	10	0	11	-11	-11	10
54	43	90	11	0	12	-12	-12	11
54	47	90	12	0	11	-11	-12	12
54	48	90	11	0	10	-10	-11	11
55	40	90	5	0	6	-12	-11	10
55	45	90	1	0	1	-1	-1	1
56	42	90	6	0	9	-9	-8	6
56	42	90	9	0	10	-10	-9	9
56	42	90	10	0	11	-11	-10	10
56	43	90	11	0	12	-12	-11	11
56	47	90	12	0	11	-11	-11	12
56	48	90	9	0	8	-8	-8	9
56	46	90	10	0	9	-9	-9	10
56	48	90	11	0	10	-10	-10	11
57	40	90	5	0	6	-6	-5	5
57	41	90	6	0	7	-7	-6	6
57	41	90	7	0	8	-8	-7	7
57	45	90	1	0	1	-12	-11	12
57	45	90	1	0	1	-11	-10	11
57	49	90	7	0	6	-6	-6	7
57	49	90	8	0	7	-7	-7	8
58	43	90	11	0	12	-12	-10	11
58	45	90	1	0	1	-8	-7	8
58	45	90	1	0	1	-9	-8	9
58	45	90	1	0	1	-10	-9	10
58	47	90	12	0	11	-11	-10	12
59	42	90	9	0	10	-10	-8	9
59	42	90	10	0	11	-11	-9	10
59	45	90	1	0	1	-6	-5	6
59	45	90	1	0	1	-7	-6	7
59	48	90	10	0	9	-9	-8	10
59	48	90	11	0	10	-10	-9	11
50	50	0						
50	50	5	1	12	10	12	-11	12
50	51	0	0	11	9	12	-9	11
50	51	5	1	11	9	11	-10	11
50	51	6	1	10	8	10	-9	10
50	53	5	1	12	9	6	-5	6
50	54	9	2	12	9	9	-9	10
50	56	0	0	3	2	3	-2	3
50	58	0	0	8	5	8	-5	8
50	58	0	0	11	7	11	-7	11
51	53	0	0	4	3	4	-3	4
51	54	0	0	7	5	7	-5	7

φ_1	Φ	φ_2	H	K	L	U	V	W
50	50	0						
51	54	0	0	11	8	11	-8	11
51	55	0	0	10	7	10	-7	10
52	50	0	0	6	5	6	-5	6
52	51	0	0	5	4	5	-4	5
52	51	0	0	11	9	11	-9	11
52	52	0	0	9	7	9	-7	9
52	53	7	1	8	6	6	-6	7
52	55	8	1	7	5	5	-5	6
52	57	9	1	6	4	4	-4	5
52	58	5	1	11	7	6	-5	7
52	59	0	0	5	3	9	-6	10
53	50	5	1	12	10	10	-10	11
53	51	5	1	11	9	9	-9	10
53	51	6	1	10	8	8	-8	9
53	52	6	1	9	7	7	-7	8
53	56	0	0	3	2	11	-8	12
53	58	0	0	8	5	7	-5	8
53	58	0	0	11	7	10	-7	11
53	58	7	1	8	5	9	-8	11
54	53	0	0	4	3	11	-9	12
54	54	0	0	11	8	10	-8	11
54	55	0	0	10	7	9	-7	10
54	56	0	0	3	2	8	-6	9
55	50	0	0	6	5	11	-10	12
55	51	0	0	5	4	9	-8	10
55	51	0	0	11	9	10	-9	11
55	52	0	0	9	7	8	-7	9
55	53	0	0	4	3	7	-6	8
55	53	5	1	12	9	9	-9	11
55	54	0	0	7	5	6	-5	7
55	54	5	1	11	8	4	-4	5
55	54	9	2	12	9	6	-7	8
55	55	6	1	10	7	7	-7	9
55	56	0	0	3	2	5	-4	6
55	56	6	1	9	6	3	-3	4
55	58	0	0	11	7	9	-7	11
56	50	5	1	12	10	8	-9	10
56	51	5	1	11	9	7	-8	9
56	53	0	0	4	3	10	-9	12
56	55	8	1	7	5	8	-9	11
56	56	7	1	8	5	5	-5	7
56	59	0	0	5	3	4	-3	5
57	50	0	0	6	5	5	-5	6
57	51	6	1	10	8	6	-7	8
57	54	0	0	11	8	9	-8	11
57	55	0	0	10	7	8	-7	10
57	56	0	0	3	2	7	-6	9
57	58	5	1	12	8	8	-8	11

φ_1	Φ	φ_2	H	K	L	U	V	W
50	50	0						
57	57	9	1	6	4	6	-7	9
57	58	5	1	11	7	7	-7	10
58	51	0	0	5	4	4	-4	5
58	51	0	0	11	9	9	-9	11
58	52	0	0	9	7	7	-7	9
58	52	6	1	9	7	5	-6	7
58	53	7	1	8	6	4	-5	6
58	56	0	0	3	2	9	-8	12
58	58	0	0	8	5	6	-5	8
58	58	0	0	11	7	8	-7	11
58	59	6	1	10	6	2	-2	3
59	53	0	0	4	3	3	-3	4
59	59	0	0	5	3	7	-6	10
50	50	10						
50	51	16	2	7	6	9	-12	11
50	54	10	2	11	8	8	-8	9
50	55	18	4	12	9	3	-4	4
50	56	11	2	10	7	7	-7	8
50	58	18	1	3	2	8	-10	11
50	58	18	1	3	2	9	-11	12
51	58	18	1	3	2	7	-9	10
52	58	18	1	3	2	6	-8	9
53	50	18	3	9	8	6	-10	9
53	54	18	3	9	7	2	-3	3
53	57	13	2	9	6	9	-10	12
53	58	10	2	11	7	7	-7	9
53	59	14	2	8	5	7	-8	10
54	53	17	3	10	8	2	-3	3
54	58	18	1	3	2	5	-7	8
55	51	14	3	12	10	2	-3	3
55	52	15	3	11	9	2	-3	3
55	54	10	2	11	8	5	-6	7
55	54	14	1	4	3	8	-11	12
55	57	14	3	12	8	4	-5	6
56	54	14	1	4	3	7	-10	11
56	56	11	2	10	7	4	-5	6
56	58	15	3	11	7	3	-4	5
56	58	18	1	3	2	4	-6	7
57	54	14	1	4	3	6	-9	10
58	52	18	2	6	5	1	-2	2
58	55	18	4	12	9	6	-11	12
58	57	13	2	9	6	3	-4	5
58	58	10	2	11	7	5	-6	8
58	58	18	1	3	2	3	-5	6
59	52	11	1	5	4	7	-11	12
59	54	14	1	4	3	5	-8	9

φ_1	Φ	φ_2	H	K	L	U	V	W
50	50	20						
50	51	21	3	8	7	5	-8	7
50	51	27	5	10	9	3	-6	5
50	52	23	5	12	10	6	-10	9
50	54	27	5	10	8	6	-11	10
50	57	23	3	7	5	2	-3	3
50	58	27	5	10	7	3	-5	5
51	50	22	4	10	9	4	-7	6
51	52	23	3	7	6	5	-9	8
51	53	27	3	6	5	5	-10	9
51	55	21	3	8	6	2	-3	3
51	56	24	5	11	8	3	-5	5
52	52	29	5	9	8	3	-7	6
52	55	23	5	12	9	3	-5	5
52	57	29	6	11	8	1	-2	2
53	51	24	4	9	8	4	-8	7
53	51	27	6	12	11	3	-7	6
53	52	27	4	8	7	4	-9	8
53	56	27	2	4	3	1	-2	2
53	59	24	4	9	6	6	-10	11
54	51	21	3	8	7	6	-11	10
54	51	29	6	11	10	3	-8	7
54	52	23	3	7	6	6	-12	11
54	59	27	3	6	4	6	-11	12
55	55	24	4	9	7	1	-2	2
55	56	29	5	9	7	2	-5	5
56	50	22	4	10	9	5	-11	10
56	50	23	5	12	11	4	-9	8
56	50	24	5	11	10	4	-10	9
56	51	24	4	9	8	5	-12	11
56	51	27	5	10	9	4	-11	10
56	52	29	5	9	8	4	-12	11
56	53	22	2	5	4	1	-2	2
56	57	22	4	10	7	5	-9	10
56	58	23	5	12	8	4	-7	8
56	59	27	3	6	4	4	-8	9
57	52	20	4	11	9	1	-2	2
57	54	27	5	10	8	2	-5	5
57	57	23	3	7	5	5	-10	11
57	58	27	5	10	7	4	-9	10
58	53	24	5	11	9	2	-5	5
58	54	29	6	11	9	1	-3	3
58	56	24	5	11	8	5	-11	12
58	59	24	4	9	6	3	-6	7
59	52	23	5	12	10	2	-5	5
59	53	27	3	6	5	1	-3	3
59	56	20	4	11	8	4	-8	9

φ_1	Φ	φ_2	H	K	L	U	V	W
50	50	30						
50	52	31	6	10	9	5	-12	10
50	54	36	8	11	10	2	-6	5
50	57	35	7	10	8	4	-10	9
50	58	39	9	11	9	3	-9	8
51	50	34	2	3	3	3	-9	7
51	54	30	7	12	10	4	-9	8
51	55	39	8	10	9	2	-7	6
51	56	31	3	5	4	5	-11	10
51	58	39	4	5	4	4	-12	11
51	59	31	6	10	7	1	-2	2
52	50	34	2	3	3	3	-10	8
52	53	32	5	8	7	3	-8	7
52	54	37	9	12	11	2	-7	6
52	55	34	4	6	5	4	-11	10
52	55	36	5	7	6	3	-9	8
52	55	39	9	11	10	2	-8	7
52	58	30	4	7	5	1	-2	2
53	53	30	4	7	6	4	-10	9
53	58	32	7	11	8	3	-7	7
53	59	37	3	4	3	1	-3	3
54	50	34	2	3	3	3	-11	9
54	54	34	6	9	8	3	-10	9
54	55	37	6	8	7	3	-11	10
54	57	30	7	12	9	3	-7	7
54	58	32	5	8	6	2	-5	5
54	58	34	6	12	9	3	-8	8
54	58	38	7	9	7	2	-7	7
55	50	34	2	3	3	3	-12	10
55	52	31	6	10	9	3	-9	8
55	57	34	6	9	7	1	-3	3
55	58	39	4	5	4	1	-4	4
56	52	30	7	12	11	3	-10	9
56	53	32	7	11	10	3	-11	10
56	54	35	7	10	9	3	-12	11
56	57	35	7	10	8	2	-7	7
56	58	39	9	11	9	2	-9	9
57	55	32	7	11	9	2	-7	7
57	56	31	3	5	4	1	-3	3
57	56	37	9	12	10	2	-9	9
57	57	36	8	11	9	1	-4	4
58	55	34	4	6	5	1	-4	4
59	54	30	7	12	10	2	-7	7
59	55	36	5	7	6	1	-5	5
59	55	37	6	8	7	1	-6	6
59	55	38	7	9	8	1	-7	7
59	55	39	8	10	9	1	-8	8
59	55	34	8	12	9	3	-11	12

φ_1	Φ	φ_2	H	K	L	U	V	W
50	50	30						
59	59	37	3	4	3	2	-9	10
50	50	40						
50	53	48	9	8	9	1	-9	7
50	55	45	1	1	1	2	-11	9
50	57	41	7	8	7	2	-7	6
50	58	45	8	8	7	2	-9	8
51	50	45	6	6	7	1	-8	6
51	53	48	10	9	10	1	-10	8
51	55	45	1	1	1	1	-6	5
51	56	42	8	9	8	2	-8	7
51	57	40	5	6	5	3	-10	9
51	58	45	9	9	8	2	-10	9
51	59	46	10	9	8	2	-12	11
52	51	45	7	7	8	1	-9	7
52	54	46	11	10	11	1	-11	9
52	55	45	1	1	1	1	-7	6
52	56	42	9	10	9	2	-9	8
52	57	41	6	7	6	3	-12	11
52	58	45	10	10	9	2	-11	10
53	51	45	8	8	9	1	-10	8
53	52	45	9	9	10	1	-11	9
53	54	47	12	11	12	1	-12	10
53	55	40	10	12	11	2	-9	8
53	55	45	1	1	1	1	-8	7
53	56	42	10	11	10	2	-10	9
53	57	45	11	11	10	2	-12	11
54	50	46	9	8	10	0	-5	4
54	50	49	8	7	9	0	-9	7
54	52	45	10	10	11	1	-12	10
54	55	45	1	1	1	1	-9	8
54	56	43	11	12	11	2	-11	10
54	59	42	9	10	8	2	-9	9
55	51	48	10	9	11	0	-11	9
55	51	48	11	10	12	0	-6	5
55	53	49	7	6	7	0	-7	6
55	55	45	1	1	1	1	-11	10
55	55	45	1	1	1	1	-10	9
55	58	43	11	12	10	2	-11	11
55	59	42	10	11	9	1	-5	5
55	59	45	6	6	5	1	-6	6
55	59	45	7	7	6	1	-7	7
55	59	46	10	9	8	1	-10	10
56	50	41	7	8	9	1	-11	9
56	50	42	8	9	10	1	-12	10
56	53	48	9	8	9	0	-9	8
56	53	46	10	9	10	0	-10	9

φ_1	Φ	φ_2	H	K	L	U	V	W
50	50	40						
56	53	49	8	7	8	0	-8	7
56	55	45	1	1	1	1	-12	11
56	57	40	5	6	5	1	-5	5
56	58	45	8	8	7	1	-8	8
56	58	45	9	9	8	1	-9	9
56	58	47	12	11	10	1	-12	12
56	59	48	11	10	9	1	-11	11
57	50	45	5	5	6	0	-6	5
57	54	47	12	11	12	0	-12	11
57	54	48	11	10	11	0	-11	10
57	57	41	6	7	6	1	-6	6
57	57	41	7	8	7	1	-7	7
57	57	45	11	11	10	1	-11	11
57	57	45	12	12	11	1	-12	12
57	58	45	10	10	9	1	-10	10
58	50	45	6	6	7	0	-7	6
58	51	45	7	7	8	0	-8	7
58	51	45	8	8	9	0	-9	8
58	52	45	9	9	10	0	-10	9
58	52	45	10	10	11	0	-11	10
58	56	42	8	9	8	1	-8	8
58	56	42	9	10	9	1	-9	9
58	56	42	10	11	10	1	-10	10
58	56	43	11	12	11	1	-11	11
58	56	48	9	8	8	0	-1	1
58	56	48	10	9	9	0	-1	1
58	56	48	11	10	10	0	-1	1
58	57	49	7	6	6	0	-1	1
58	57	49	8	7	7	0	-1	1
59	52	45	11	11	12	0	-12	11
59	56	47	12	11	11	0	-1	1
59	59	45	6	6	5	1	-11	12
50	50	50						
50	50	58	11	7	11	-1	-11	8
50	51	53	4	3	4	0	-4	3
50	54	55	10	7	9	0	-9	7
50	55	50	12	10	11	1	-10	8
51	50	56	3	2	3	-1	-12	9
51	52	52	9	7	9	0	-9	7
51	54	54	11	8	10	0	-5	4
51	55	54	7	5	6	0	-6	5
51	57	56	9	6	7	0	-7	6
51	58	58	11	7	8	0	-8	7
52	50	51	11	9	12	0	-4	3
52	50	58	8	5	6	-1	-8	6
52	52	51	5	4	5	0	-5	4

φ_1	Φ	φ_2	H	K	L	U	V	W
50	50	50						
52	54	53	12	9	11	0	-11	9
52	57	55	10	7	8	0	-8	7
52	58	56	12	8	9	0	-9	8
53	50	56	3	2	3	-1	-9	7
53	52	51	11	9	11	0	-11	9
53	55	53	8	6	7	0	-7	6
53	56	59	5	3	4	-1	-9	8
53	57	54	11	8	9	0	-9	8
54	51	54	11	8	11	-1	-11	9
54	51	55	10	7	10	-1	-10	8
54	52	50	6	5	6	0	-6	5
54	55	51	10	8	9	0	-9	8
54	55	52	9	7	8	0	-8	7
54	55	58	11	7	9	-1	-10	9
54	56	53	12	9	10	0	-10	9
55	50	58	11	7	11	-2	-11	9
55	51	53	4	3	4	-1	-12	10
55	53	58	11	7	10	-1	-7	6
55	55	50	12	10	11	0	-11	10
55	55	51	11	9	10	0	-10	9
55	55	56	6	4	5	-1	-11	10
55	59	59	10	6	7	-1	-10	10
56	50	56	3	2	3	-1	-6	5
56	58	51	5	4	4	0	-1	1
56	58	52	9	7	7	0	-1	1
56	58	56	12	6	9	-1	-12	12
56	58	58	11	7	8	-1	-11	11
56	59	53	4	3	3	0	-1	1
57	51	53	4	3	4	-1	-8	7
57	51	54	7	5	7	-1	-7	6
57	57	50	6	5	5	0	-1	1
57	57	55	10	7	8	-1	-10	10
57	57	56	9	6	7	-1	-9	9
57	58	51	11	9	9	0	-1	1
57	58	58	8	5	6	-1	-8	8
58	50	56	3	2	3	-2	-9	8
58	50	58	8	5	8	-2	-8	7
58	52	50	6	5	6	-1	-12	11
58	52	51	5	4	5	-1	-10	9
58	52	51	11	9	11	-1	-11	10
58	52	52	9	7	9	-1	-9	8
58	55	58	11	7	9	-2	-11	11
58	56	53	12	9	10	-1	-12	12
58	56	59	5	3	4	-1	-5	5
58	57	54	11	8	9	-1	-11	11
59	51	53	4	3	4	-2	-12	11
59	51	54	11	8	11	-2	-11	10
59	51	55	10	7	10	-2	-10	9

φ_1	Φ	φ_2	H	K	L	U	V	W
50	50	50						
59	55	52	9	7	8	-1	-9	9
59	55	53	8	6	7	-1	-8	8
59	55	54	7	5	6	-1	-7	7
59	55	56	6	4	5	-1	-6	6
50	50	60						
50	53	66	11	5	9	-1	-5	4
50	55	69	8	3	6	-3	-12	10
51	53	63	6	3	5	-2	-11	9
51	54	63	10	5	8	-1	-6	5
51	56	63	4	2	3	-1	-7	6
51	57	68	10	4	7	-2	-9	8
51	59	63	6	3	4	-1	-10	9
52	52	63	8	4	7	-1	-5	4
52	52	67	7	3	6	-3	-11	9
52	53	68	5	2	4	-2	-7	6
52	56	61	9	5	7	-1	-8	7
52	59	66	9	4	6	-2	-12	11
53	55	69	8	3	6	-3	-10	9
53	56	63	4	2	3	-2	-11	10
53	56	66	11	5	8	-2	-10	9
53	58	60	7	4	5	-1	-12	11
54	51	63	10	5	9	-3	-12	10
55	51	63	12	6	11	-2	-7	6
55	52	67	7	3	6	-3	-9	8
55	53	68	5	2	4	-4	-12	11
55	58	67	12	5	8	-1	-4	4
55	59	63	6	3	4	-1	-6	6
55	59	66	9	4	6	-2	-9	9
56	51	61	11	6	10	-2	-8	7
56	53	60	7	4	6	-2	-10	9
56	56	66	11	5	8	-3	-11	11
56	57	61	11	6	8	-2	-11	11
56	57	67	7	3	5	-2	-7	7
56	57	68	10	4	7	-3	-10	10
56	58	60	7	4	5	-1	-7	7
56	58	63	10	5	7	-1	-5	5
57	51	66	9	4	8	-4	-11	10
57	55	67	12	5	9	-1	-3	3
57	55	69	8	3	6	-3	-8	8
57	56	63	4	2	3	-1	-4	4
57	57	60	12	7	9	-1	-6	6
58	53	68	5	2	4	-2	-5	5
58	54	63	10	5	8	-3	-10	10
58	55	66	9	4	7	-1	-3	3
58	56	61	9	5	7	-2	-9	9
59	52	67	7	3	6	-3	-7	7

φ_1	Φ	φ_2	H	K	L	U	V	W
50	50	60						
59	52	67	12	5	10	-5	-12	12
59	53	66	11	5	9	-4	-11	11
59	54	60	12	7	10	-1	-4	4
59	54	61	11	6	9	-3	-11	11
59	57	66	10	4	7	-4	-11	12
59	59	63	6	3	4	-2	-8	9
50	50	70						
50	56	72	3	1	2	-2	-8	7
50	58	75	11	3	7	-3	-10	9
51	51	74	7	2	6	-2	-5	4
51	58	72	3	1	2	-3	-11	10
52	52	75	11	3	9	-3	-7	6
52	52	79	5	1	4	-4	-8	7
52	55	72	12	4	9	-3	-9	8
52	56	74	7	2	5	-4	-11	10
53	58	75	11	3	7	-4	-11	11
53	59	76	8	2	5	-3	-8	8
54	50	72	9	3	8	-3	-7	6
54	51	76	12	3	10	-5	-10	9
54	56	79	10	2	7	-1	-2	2
54	57	76	12	3	8	-5	-12	12
54	57	77	9	2	6	-4	-9	9
54	59	70	11	4	7	-3	-11	11
55	50	76	8	2	7	-6	-11	10
55	51	74	7	2	6	-6	-12	11
55	55	75	11	3	8	-5	-11	11
55	56	73	10	3	7	-2	-5	5
55	56	74	7	2	5	-3	-7	7
55	58	72	3	1	2	-1	-3	3
56	52	79	5	1	4	-3	-5	5
56	53	77	9	2	7	-5	-9	9
56	54	76	4	1	3	-1	-2	2
56	55	72	12	4	9	-5	-12	12
56	56	70	11	4	8	-4	-11	11
56	59	76	8	2	5	-4	-9	10
57	51	76	12	3	10	-7	-12	12
57	52	75	11	3	9	-6	-11	11
57	53	73	10	3	8	-1	-2	2
57	54	72	9	3	7	-4	-9	9
58	50	76	8	2	7	-5	-8	8
58	51	74	7	2	6	-4	-7	7
58	52	70	11	4	9	-5	-11	11
58	56	74	7	2	5	-5	-10	11
58	57	76	12	3	8	-4	-8	9
58	58	72	3	1	2	-4	-10	11
59	50	72	9	3	8	-5	-9	9

TAB. VI FROM EULER ANGLES INTO HKL-UVW

φ_1 Φ φ_2	H	K	L	U	V	W
50 50 70						
59 58 72	3	1	2	-3	-7	8
59 59 76	8	2	5	-5	-10	12
50 50 80						
50 51 80	11	2	9	-3	-6	5
50 55 82	7	1	5	-5	-10	9
50 56 84	9	1	6	-6	-12	11
51 58 85	11	1	7	-6	-11	11
51 59 84	10	1	6	-1	-2	2
52 55 84	10	1	7	-3	-5	5
52 56 84	9	1	6	-5	-9	9
52 56 85	12	1	8	-7	-12	12
52 58 83	8	1	5	-1	-2	2
53 50 82	7	1	6	-7	-11	10
53 53 85	12	1	9	-2	-3	3
53 54 85	11	1	8	-7	-11	11
53 55 82	7	1	5	-4	-7	7
53 57 81	6	1	4	-1	-2	2
53 58 80	11	2	7	-5	-11	11
54 50 85	12	1	10	-3	-4	4
54 51 84	10	1	8	-7	-10	10
54 51 85	11	1	9	-8	-11	11
54 52 84	9	1	7	-2	-3	3
54 53 83	8	1	6	-5	-8	8
54 54 80	11	2	8	-6	-11	11
54 54 81	12	2	9	-7	-12	12
55 58 85	11	1	7	-5	-8	9
56 50 82	7	1	6	-5	-7	7
56 51 80	11	2	9	-7	-11	11
56 51 81	6	1	5	-2	-3	3
56 55 82	7	1	5	-7	-11	12
57 53 83	8	1	6	-7	-10	11
57 56 84	9	1	6	-4	-6	7
58 54 81	12	2	9	-6	-9	10
58 56 85	12	1	8	-3	-4	5
59 50 82	7	1	6	-8	-10	11
59 51 84	10	1	8	-5	-6	7
59 54 80	11	2	8	-4	-6	7
50 50 90						
50 51 90	11	0	9	-9	-12	11
50 56 90	3	0	2	-2	-3	3
50 58 90	8	0	5	-5	-8	8
50 58 90	11	0	7	-7	-11	11
51 53 90	4	0	3	-3	-4	4
51 54 90	7	0	5	-5	-7	7

φ_1 Φ φ_2	H	K	L	U	V	W
50 50 90						
51 54 90	11	0	8	-8	-11	11
51 55 90	10	0	7	-7	-10	10
52 50 90	6	0	5	-5	-6	6
52 51 90	5	0	4	-4	-5	5
52 51 90	11	0	9	-9	-11	11
52 52 90	9	0	7	-7	-9	9
52 59 90	5	0	3	-6	-9	10
53 56 90	3	0	2	-8	-11	12
53 58 90	8	0	5	-5	-7	8
53 58 90	11	0	7	-7	-10	11
54 53 90	4	0	3	-9	-11	12
54 54 90	11	0	8	-8	-10	11
54 55 90	10	0	7	-7	-9	10
54 56 90	3	0	2	-6	-8	9
55 50 90	6	0	5	-10	-11	12
55 51 90	5	0	4	-8	-9	10
55 51 90	11	0	9	-9	-10	11
55 52 90	9	0	7	-7	-6	9
55 53 90	4	0	3	-6	-7	8
55 54 90	7	0	5	-5	-6	7
55 56 90	3	0	2	-4	-5	6
55 56 90	11	0	7	-7	-9	11
56 53 90	4	0	3	-9	-10	12
56 59 90	5	0	3	-3	-4	5
57 50 90	6	0	5	-5	-5	6
57 54 90	11	0	8	-8	-9	11
57 55 90	10	0	7	-7	-8	10
57 56 90	3	0	2	-6	-7	9
58 51 90	5	0	4	-4	-4	5
58 51 90	11	0	9	-9	-9	11
58 52 90	9	0	7	-7	-7	9
58 56 90	3	0	2	-8	-9	12
58 58 90	8	0	5	-5	-6	8
58 58 90	11	0	7	-7	-6	11
59 53 90	4	0	3	-3	-3	4
59 59 90	5	0	3	-6	-7	10
50 60 0						
50 66 0	0	11	5	10	-5	11
50 67 0	0	12	5	11	-5	12
50 68 0	0	5	2	9	-4	10
51 61 0	0	11	6	10	-6	11
51 61 5	1	11	6	7	-5	8
51 63 0	0	2	1	11	-6	12
51 63 0	0	2	1	9	-5	10
51 64 5	1	12	6	6	-4	7
51 64 9	1	6	3	9	-7	11

φ_1 Φ φ_2	H	K	L	U	V	W
50 60 0						
51 66 0	0	9	4	8	-4	9
51 67 8	1	7	3	9	-6	11
51 69 0	0	8	3	7	-3	8
52 60 0	0	12	7	11	-7	12
52 61 0	0	9	5	8	-5	9
52 63 0	0	2	1	7	-4	8
52 64 6	1	10	5	10	-7	12
52 64 7	1	8	4	4	-3	5
52 67 0	0	7	3	6	-3	7
52 67 0	0	12	5	10	-5	12
53 60 0	0	7	4	6	-4	7
53 63 0	0	2	1	5	-3	6
53 64 9	1	6	3	6	-5	8
53 66 0	0	11	5	9	-5	11
53 68 0	0	5	2	4	-2	5
54 60 0	0	12	7	10	-7	12
54 61 0	0	11	6	9	-6	11
54 63 0	0	2	1	8	-5	10
55 60 9	2	12	7	7	-7	10
55 61 6	1	9	5	8	-7	11
55 66 0	0	9	4	7	-4	9
55 67 0	0	12	5	9	-5	12
55 69 0	0	8	3	6	-3	8
56 61 0	0	9	5	7	-5	9
56 63 0	0	2	1	3	-2	4
56 64 6	1	10	5	5	-4	7
56 66 0	0	11	5	8	-5	11
57 60 0	0	12	7	9	-7	12
57 61 0	0	11	6	8	-6	11
57 61 8	1	7	4	2	-2	3
57 64 7	1	8	4	8	-7	12
57 67 0	0	7	3	5	-3	7
57 68 0	0	5	2	7	-4	10
58 60 0	0	7	4	5	-4	7
58 63 0	0	2	1	7	-5	10
58 64 5	1	12	6	6	-5	9
58 64 9	1	6	3	3	-3	5
58 67 0	0	12	5	8	-5	12
59 60 5	1	12	7	7	-7	11
59 61 6	1	9	5	5	-5	8
59 63 0	0	2	1	4	-3	6
59 66 0	0	9	4	6	-4	9
50 60 10						
50 61 18	4	12	7	9	-10	12
50 64 11	2	10	5	5	-4	6
50 64 14	1	4	2	8	-7	10

φ_1 Φ φ_2	H	K	L	U	V	W
50 60 10						
50 66 15	3	11	5	7	-6	9
50 69 17	3	10	4	6	-5	8
51 60 14	3	12	7	7	-7	9
51 60 17	3	10	6	6	-9	11
51 62 15	3	11	6	3	-3	4
51 64 17	3	10	5	5	-5	7
51 66 10	2	11	5	4	-3	5
51 68 16	2	7	3	8	-7	11
51 69 11	1	5	2	7	-5	9
52 62 18	3	9	5	6	-7	9
52 67 18	3	9	4	2	-2	3
53 60 11	1	5	3	3	-3	4
53 62 13	2	9	5	5	-5	7
53 67 13	2	9	4	7	-6	10
53 69 11	1	5	2	8	-6	11
54 60 14	3	12	7	8	-9	12
54 64 14	1	4	2	2	-2	3
54 65 18	2	6	3	6	-7	10
54 66 14	3	12	5	8	-7	12
54 68 16	4	12	5	5	-5	8
55 61 16	2	7	4	5	-6	8
55 62 18	3	9	5	7	-9	12
55 66 15	3	11	5	5	-5	8
56 62 10	2	11	6	2	-2	3
56 68 16	2	7	3	3	-3	5
56 69 17	3	10	4	4	-4	7
57 60 11	1	5	3	7	-8	11
57 64 11	2	10	5	5	-5	8
57 64 17	3	10	5	5	-6	9
58 61 16	2	7	4	6	-8	11
58 61 18	4	12	7	2	-3	4
58 64 14	1	4	2	6	-7	11
58 65 18	2	6	3	3	-4	6
58 67 13	2	9	4	4	-4	7
58 68 14	3	12	5	5	-5	9
59 60 11	1	5	3	4	-5	7
59 60 17	3	10	6	2	-3	4
59 66 10	2	11	5	5	-5	9
59 66 18	4	12	5	6	-7	12
50 60 20						
50 62 27	5	10	6	6	-9	10
50 63 20	4	11	6	7	-8	10
50 65 23	5	12	6	6	-7	9
50 66 27	1	2	1	7	-9	11
50 67 20	4	11	5	5	-5	7
51 61 27	4	8	5	7	-11	12

φ_1	Φ	φ_2	H	K	L	U	V	W
50	60	20						
51	64	29	6	11	6	4	-6	7
51	65	21	3	8	4	8	-9	12
51	66	27	1	2	1	3	-4	5
51	68	24	4	9	4	7	-8	11
52	65	22	4	10	5	5	-6	8
52	66	27	1	2	1	5	-7	9
53	60	24	5	11	7	5	-8	9
53	61	22	2	5	3	7	-10	12
53	64	29	5	9	5	3	-5	6
53	68	24	5	11	5	4	-5	7
54	61	27	4	8	5	4	-7	8
54	62	27	6	12	7	3	-5	6
54	66	27	1	2	1	2	-3	4
54	67	20	4	11	5	7	-8	12
54	68	23	3	7	3	5	-6	9
55	61	22	2	5	3	6	-9	11
55	65	21	3	8	4	4	-5	7
55	65	23	5	12	6	6	-8	11
56	61	22	2	5	3	5	-8	10
56	66	27	1	2	1	5	-8	11
56	68	24	4	9	4	3	-4	6
57	60	29	5	9	6	3	-7	8
57	61	27	4	8	5	5	-10	12
57	65	22	4	10	5	5	-7	10
57	66	27	1	2	1	3	-5	7
57	69	23	5	12	5	4	-5	8
58	61	22	2	5	3	4	-7	9
58	62	27	5	10	6	2	-4	5
58	64	29	6	11	6	3	-6	8
59	62	23	3	7	4	5	-9	12
59	66	27	1	2	1	4	-7	10
50	60	30						
50	61	34	2	3	2	1	-2	2
50	62	37	9	12	8	4	-9	9
50	62	38	7	9	6	3	-7	7
50	64	30	4	7	4	5	-8	9
50	64	34	8	12	7	6	-11	12
50	66	38	7	9	5	5	-10	11
51	60	35	7	10	7	3	-7	7
51	61	39	8	10	7	3	-8	8
51	65	36	5	7	4	4	-8	9
51	67	30	7	12	6	6	-9	11
51	67	34	4	6	3	3	-5	6
52	60	36	5	7	5	2	-5	5
52	60	36	8	11	8	3	-8	8
52	61	39	9	11	8	1	-3	3

φ_1	Φ	φ_2	H	K	L	U	V	W
50	60	30						
52	63	30	7	12	7	4	-7	8
52	63	31	3	5	3	5	-9	10
52	66	36	8	11	6	3	-6	7
52	67	31	6	10	5	5	-8	10
52	68	35	7	10	5	5	-9	11
52	69	39	8	10	5	5	-10	12
53	62	32	5	8	5	5	-10	11
53	68	37	3	4	2	2	-4	5
54	62	37	9	12	8	4	-11	12
54	63	31	3	5	3	3	-6	7
54	65	34	6	9	5	5	-10	12
54	67	32	5	8	4	4	-7	9
55	60	30	7	12	8	4	-9	10
55	61	34	2	3	2	4	-10	11
55	62	32	7	11	7	3	-7	8
55	62	38	7	9	6	3	-9	10
55	65	32	7	11	6	2	-4	5
55	65	37	9	12	7	2	-5	6
56	61	34	2	3	2	3	-8	9
56	64	30	4	7	4	2	-4	5
56	64	35	7	10	6	2	-5	6
56	67	31	6	10	5	5	-9	12
56	67	34	4	6	3	3	-6	6
56	68	37	3	4	2	4	-9	12
57	60	36	5	7	5	3	-10	11
57	62	32	5	8	5	2	-5	6
57	63	31	3	5	3	4	-9	11
57	65	34	6	9	5	3	-7	9
57	65	39	4	5	3	3	-9	11
57	67	32	5	8	4	4	-8	11
58	60	35	7	10	7	2	-7	8
58	61	34	2	3	2	2	-6	7
58	67	39	9	11	6	1	-3	4
59	60	36	8	11	8	2	-8	9
59	63	30	7	12	7	3	-7	9
59	64	39	9	11	7	1	-4	5
59	65	39	4	5	3	2	-7	9
59	66	38	7	9	5	1	-3	4
59	68	37	3	4	2	2	-5	7
50	60	40						
50	60	45	11	11	9	3	-12	11
50	60	48	9	8	7	2	-11	10
50	63	40	5	6	4	2	-5	5
50	64	42	8	9	6	3	-8	8
50	64	43	11	12	8	4	-11	11
50	65	45	3	3	2	1	-3	3

φ_1	Φ	φ_2	H	K	L	U	V	W
50	60	40						
50	65	48	11	10	7	3	-11	11
50	65	49	8	7	5	1	-4	4
50	69	45	11	11	6	4	-10	11
51	63	42	9	10	7	1	-3	3
51	63	45	7	7	5	2	-7	7
51	63	45	11	11	8	3	-11	11
51	64	45	10	10	7	3	-10	10
51	64	47	12	11	8	1	-4	4
51	64	48	9	8	6	2	-9	9
51	68	45	12	12	7	4	-11	12
51	69	40	5	6	3	3	-6	7
51	69	45	9	9	5	3	-6	9
52	62	41	6	7	5	1	-3	3
52	62	42	10	11	8	3	-10	10
52	62	45	4	4	3	1	-4	4
52	63	48	10	9	7	1	-5	5
52	65	41	7	8	5	4	-11	12
52	67	43	11	12	7	3	-8	9
52	67	45	5	5	3	3	-9	10
53	60	40	10	12	9	3	-10	10
53	61	41	7	6	6	2	-7	7
53	61	43	11	12	9	3	-11	11
53	61	45	9	9	7	2	-9	9
53	61	47	12	11	9	1	-6	6
53	62	48	11	10	8	2	-11	11
53	62	49	7	6	5	1	-7	7
53	65	45	3	3	2	3	-11	12
53	66	45	11	11	7	3	-10	11
53	67	49	7	6	4	2	-9	10
53	68	45	7	7	4	2	-6	7
53	68	48	11	10	6	2	-7	8
53	69	41	7	8	4	4	-9	11
54	60	42	8	9	7	1	-4	4
54	60	45	11	11	9	2	-11	11
54	61	45	5	5	4	1	-5	5
54	61	49	8	7	6	1	-8	8
54	64	42	8	9	6	3	-10	11
54	66	42	9	10	6	2	-6	7
54	66	45	8	8	5	2	-7	8
54	69	45	11	11	6	3	-9	11
55	60	48	9	8	7	1	-9	9
55	64	45	10	10	7	2	-9	10
55	65	45	3	3	2	2	-8	9
55	68	45	12	12	7	3	-10	12
55	69	40	5	6	3	3	-7	9
56	62	45	4	4	3	2	-11	12
56	63	45	11	11	8	2	-10	11
56	69	49	8	7	4	1	-4	5

φ_1	Φ	φ_2	H	K	L	U	V	W
50	60	40						
57	61	41	7	6	6	2	-10	11
57	65	45	3	3	2	1	-5	6
57	65	46	11	10	7	1	-6	7
57	66	45	11	11	7	2	-9	11
57	67	45	5	5	3	1	-4	5
57	69	45	9	9	5	2	-7	9
58	61	45	5	5	4	1	-9	10
58	61	45	9	9	7	1	-8	9
58	62	45	4	4	3	1	-7	8
58	63	45	7	7	5	1	-6	7
58	65	41	7	8	5	1	-4	5
58	69	40	5	6	3	3	-8	11
59	60	45	11	11	9	1	-10	11
59	62	42	10	11	8	1	-6	7
59	63	40	5	6	4	2	-9	11
59	68	45	12	12	7	2	-9	12
59	69	45	11	11	6	2	-8	11
50	60	50						
50	66	50	12	10	7	1	-4	4
50	66	52	9	7	5	2	-9	9
50	66	54	11	8	6	2	-11	11
50	67	56	6	4	3	1	-6	6
50	67	58	8	5	4	1	-8	8
50	67	59	10	6	5	1	-10	10
51	64	51	11	9	7	2	-11	11
51	65	51	5	4	3	1	-5	5
51	65	53	12	9	7	1	-6	6
51	65	54	7	5	4	1	-7	7
51	65	56	9	6	5	1	-9	9
51	65	56	11	7	6	1	-11	11
52	63	50	6	5	4	1	-6	6
52	63	53	8	6	5	1	-8	8
52	64	55	10	7	6	1	-10	10
52	64	56	12	8	7	1	-12	12
52	68	53	4	3	2	2	-10	11
53	61	51	10	8	7	1	-10	10
53	62	52	9	7	6	1	-9	9
53	62	53	12	9	8	1	-12	12
53	62	58	8	5	5	0	-1	1
53	62	58	11	7	7	0	-1	1
53	63	54	11	8	7	1	-11	11
53	63	59	5	3	3	0	-1	1
53	67	56	6	4	3	1	-9	10
53	69	58	11	7	5	1	-8	9
54	61	51	11	9	8	1	-11	11
54	61	56	3	2	2	0	-1	1

φ_1	Φ	φ_2	H	K	L	U	V	W
50	60	50						
54	65	54	7	5	4	1	-11	12
54	68	53	4	3	2	1	-6	7
55	60	50	12	10	9	1	-12	12
55	60	54	7	5	5	0	-1	1
55	60	54	11	8	8	0	-1	1
55	60	55	10	7	7	0	-1	1
55	65	51	5	4	3	1	-8	9
55	66	52	9	7	5	1	-7	8
55	69	50	6	5	3	2	-9	11
55	69	51	10	8	5	1	-5	6
56	63	50	6	5	4	1	-10	11
56	64	51	11	9	7	1	-9	10
56	68	55	10	7	5	1	-10	12
57	64	56	12	8	7	0	-7	8
57	65	58	11	7	6	0	-6	7
57	67	59	10	6	5	0	-5	6
58	62	53	12	9	8	0	-8	9
58	63	54	11	8	7	0	-7	8
58	64	55	10	7	6	0	-6	7
58	65	56	9	6	5	0	-5	6
58	67	58	8	5	4	0	-4	5
59	60	50	12	10	9	0	-9	10
59	61	51	10	8	7	0	-7	8
59	61	51	11	9	8	0	-8	9
59	62	52	9	7	6	0	-6	7
59	63	53	8	6	5	0	-5	6
59	65	54	7	5	4	0	-4	5
59	69	50	6	5	3	1	-6	8
50	60	60						
50	60	66	11	5	7	-1	-9	8
50	67	60	12	7	6	1	-12	12
50	68	66	9	4	4	0	-1	1
50	68	66	11	5	5	0	-1	1
51	65	67	12	5	6	-1	-12	12
51	65	68	10	4	5	-1	-10	10
51	65	69	8	3	4	-1	-8	8
51	66	63	2	1	1	0	-1	1
52	63	60	12	7	7	0	-1	1
52	63	66	9	4	5	-1	-9	9
52	64	60	7	4	4	0	-1	1
52	64	61	9	5	5	0	-1	1
52	64	61	11	6	6	0	-1	1
52	64	66	11	5	6	-1	-11	11
53	61	68	5	2	3	-1	-5	5
53	62	63	10	5	6	-1	-10	10
53	62	63	12	6	7	-1	-12	12

φ_1	Φ	φ_2	H	K	L	U	V	W
50	60	60						
53	62	67	7	3	4	-1	-7	7
53	62	67	12	5	7	-1	-6	6
54	60	66	11	5	7	-2	-11	11
54	60	69	8	3	5	-1	-4	4
54	61	61	11	6	7	-1	-11	11
54	61	63	6	4	5	-1	-6	6
55	60	60	12	7	6	-1	-12	12
55	60	61	9	5	6	-1	-9	9
56	62	67	7	3	4	-2	-10	11
56	65	67	12	5	6	-1	-6	7
56	67	60	12	7	6	0	-6	7
56	68	61	11	6	5	0	-5	6
57	61	63	8	4	5	-2	-11	12
57	65	68	10	4	5	-1	-5	6
57	66	63	2	1	1	-1	-10	12
57	69	61	9	5	4	0	-4	5
58	61	68	5	2	3	-2	-7	6
58	64	66	11	5	6	-1	-5	6
58	65	69	8	3	4	-1	-4	5
58	66	63	2	1	1	-1	-9	11
59	62	63	12	6	7	-1	-5	6
59	66	63	2	1	1	-1	-7	9
59	66	63	2	1	1	-1	-8	10
50	60	70						
50	60	73	10	3	6	-3	-12	11
50	64	79	10	2	5	-3	-10	10
50	66	75	11	3	5	-2	-11	11
50	67	70	11	4	5	-1	-11	11
50	67	72	9	3	4	-1	-9	9
50	67	77	9	2	4	-2	-9	9
50	66	74	7	2	3	-1	-7	7
51	64	73	10	3	5	-1	-5	5
51	64	76	4	1	2	-1	-4	4
51	65	72	6	2	3	-1	-6	6
52	60	76	12	3	7	-1	-3	3
52	60	79	5	1	3	-2	-5	5
52	61	74	7	2	4	-2	-7	7
52	62	72	9	3	5	-2	-9	9
52	62	75	11	3	6	-3	-11	11
52	62	77	9	2	5	-1	-3	3
52	63	70	11	4	6	-2	-11	11
52	68	74	7	2	3	-2	-11	12
52	69	79	5	1	2	-2	-8	9
53	60	73	10	3	6	-3	-10	10
53	61	72	12	4	7	-1	-4	4
54	64	76	4	1	2	-3	-10	11

φ_1	Φ	φ_2	H	K	L	U	V	W
50	60	70						
54	65	72	6	2	3	-2	-9	10
54	67	77	9	2	4	-2	-7	8
54	68	72	12	4	5	-1	-7	8
54	69	73	10	3	4	-1	-6	7
55	60	79	5	1	3	-5	-11	12
55	62	72	9	3	5	-3	-11	12
55	62	75	11	3	6	-3	-9	10
55	67	70	11	4	5	-1	-6	7
56	63	70	11	4	6	-2	-8	9
56	64	76	4	1	2	-2	-6	7
56	64	79	10	2	5	-2	-5	6
56	67	72	9	3	4	-1	-5	6
57	60	73	10	3	6	-3	-8	9
57	60	79	5	1	3	-3	-6	7
57	64	73	10	3	5	-3	-10	12
57	68	74	7	2	3	-1	-4	5
57	68	76	12	3	5	-2	-7	9
57	69	79	5	1	2	-1	-3	4
58	61	74	7	2	4	-2	-5	6
58	64	76	4	1	2	-3	-6	10
59	66	75	11	3	5	-1	-3	4
59	68	72	12	4	5	-2	-9	12
59	69	73	10	3	4	-2	-8	11
50	60	80						
50	60	85	12	1	7	-1	-2	2
50	61	84	9	1	5	-4	-9	9
50	61	85	11	1	6	-5	-11	11
50	64	81	6	1	3	-1	-3	3
50	64	83	8	1	4	-3	-8	8
50	66	80	11	2	5	-3	-11	11
51	60	81	12	2	7	-5	-12	12
51	61	82	7	1	4	-3	-7	7
51	62	80	11	2	6	-4	-11	11
52	67	82	7	1	3	-3	-9	10
53	61	84	9	1	5	-5	-10	11
53	66	85	11	1	5	-3	-7	8
54	64	85	12	1	6	-3	-6	7
55	61	82	7	1	4	-4	-8	9
55	64	84	10	1	5	-5	-10	12
56	64	81	6	1	3	-4	-9	11
56	64	83	8	1	4	-2	-4	5
57	61	82	7	1	4	-5	-9	11
58	62	80	11	2	6	-2	-4	5
58	66	84	9	1	4	-4	-8	11
59	64	81	6	1	3	-3	-6	8
59	64	84	10	1	5	-3	-5	7

φ_1	Φ	φ_2	H	K	L	U	V	W
50	60	80						
59	68	81	12	2	5	-3	-7	10
50	60	90						
50	66	90	11	0	5	-5	-10	11
50	67	90	12	0	5	-5	-11	12
50	68	90	5	0	2	-4	-9	10
51	61	90	11	0	6	-6	-10	11
51	63	90	2	0	1	-6	-11	12
51	63	90	2	0	1	-5	-9	10
51	66	90	9	0	4	-4	-8	9
51	69	90	8	0	3	-3	-7	8
52	60	90	12	0	7	-7	-11	12
52	61	90	9	0	5	-5	-8	9
52	63	90	2	0	1	-4	-7	8
52	67	90	7	0	3	-3	-6	7
52	67	90	12	0	5	-5	-10	12
53	60	90	7	0	4	-4	-6	7
53	63	90	2	0	1	-3	-5	6
53	66	90	11	0	5	-5	-9	11
53	68	90	5	0	2	-2	-4	5
54	60	90	12	0	7	-7	-10	12
54	61	90	11	0	6	-6	-9	11
54	63	90	2	0	1	-5	-8	10
55	66	90	9	0	4	-4	-7	9
55	67	90	12	0	5	-5	-9	12
55	69	90	8	0	3	-3	-6	6
56	61	90	9	0	5	-5	-7	9
56	63	90	2	0	1	-2	-3	4
56	66	90	11	0	5	-5	-8	11
57	60	90	12	0	7	-7	-9	12
57	61	90	11	0	6	-6	-8	11
57	67	90	7	0	3	-3	-5	7
57	68	90	5	0	2	-4	-7	[illegible]
58	60	90	7	0	4	-4	-5	[illegible]
58	63	90	2	0	1	-5	-7	10
58	67	90	12	0	5	-5	-8	[illegible]
59	63	90	2	0	1	-3	-4	[illegible]
59	66	90	9	0	4	-4	-6	9
50	70	0						
50	72	0	0	3	1	8	-3	9
50	72	6	1	9	3	6	-3	7
50	76	0	0	4	1	7	-2	8
51	74	0	0	7	2	6	-2	7
51	74	8	1	7	2	4	-2	5
51	76	0	0	4	1	10	-3	12

φ_1	Φ	φ_2	H	K	L	U	V	W
50	70	0						
52	70	0	0	11	4	9	-4	11
52	72	0	0	3	1	5	-2	6
52	72	5	1	12	4	4	-2	5
52	75	0	0	11	3	9	-3	11
52	76	9	2	12	3	6	-3	8
52	79	0	0	5	1	4	-1	5
53	72	6	1	9	3	9	-5	12
53	72	9	1	6	2	8	-5	11
53	73	0	0	10	3	8	-3	10
53	73	6	1	10	3	6	-3	8
53	77	0	0	9	2	7	-2	9
54	72	0	0	3	1	7	-3	9
54	76	0	0	4	1	3	-1	4
54	79	6	1	10	2	8	-3	11
55	70	5	1	11	4	5	-3	7
55	72	0	0	3	1	9	-4	12
55	72	9	1	6	2	6	-4	9
55	75	0	0	11	3	8	-3	11
55	78	6	1	9	2	7	-3	10
56	70	0	0	11	4	8	-4	11
56	73	0	0	10	3	7	-3	10
56	74	0	0	7	2	5	-2	7
56	75	5	1	11	3	2	-1	3
56	76	7	1	8	2	2	-1	3
56	79	0	0	5	1	7	-2	10
57	70	7	1	8	3	7	-5	11
57	73	6	1	10	3	7	-4	11
57	74	8	1	7	2	5	-3	8
57	76	0	0	4	1	8	-3	12
57	77	0	0	9	2	6	-2	9
58	72	0	0	3	1	2	-1	3
58	75	0	0	11	3	7	-3	11
59	70	0	0	11	4	7	-4	11
59	72	6	1	9	3	3	-2	5
59	72	9	1	6	2	4	-3	7
59	76	0	0	4	1	5	-2	8
59	76	5	1	12	3	6	-3	10
50	70	10						
50	75	16	2	7	2	3	-2	4
50	76	14	1	4	1	7	-4	9
50	77	18	4	12	3	3	-2	4
50	78	13	2	9	2	4	-2	5
51	72	13	2	9	3	3	-2	4
51	72	18	1	3	1	5	-4	7
52	70	10	2	11	4	3	-2	4
52	72	18	1	3	1	6	-5	9

φ_1	Φ	φ_2	H	K	L	U	V	W
50	70	10						
52	76	14	1	4	1	5	-3	7
53	72	18	1	3	1	7	-6	11
53	74	17	3	10	3	4	-3	6
53	76	16	3	9	2	6	-4	9
54	72	14	3	12	4	4	-3	6
54	75	10	2	11	3	7	-4	10
54	76	14	1	4	1	8	-5	12
55	79	11	1	5	1	2	-1	3
56	78	13	2	9	2	7	-4	11
56	78	18	3	9	2	7	-5	12
57	75	15	3	11	3	4	-3	7
57	76	14	1	4	1	3	-2	5
58	74	11	2	10	3	7	-5	12
58	75	16	2	7	2	5	-4	9
58	79	17	3	10	2	6	-4	11
59	72	13	2	9	3	6	-5	11
59	72	18	1	3	1	1	-1	2
59	77	18	4	12	3	6	-5	12
50	70	20						
50	72	24	5	11	4	2	-2	3
50	77	27	2	4	1	8	-7	12
51	70	22	2	5	2	2	-2	3
51	70	27	5	10	4	6	-7	10
52	71	27	4	8	3	7	-8	12
52	72	29	6	11	4	5	-6	9
52	75	27	5	10	3	3	-3	5
52	77	21	3	8	2	4	-3	6
52	77	29	6	11	3	3	-3	5
53	73	24	4	9	3	3	-3	5
53	74	29	5	9	3	6	-7	11
53	79	22	2	5	1	7	-5	11
53	79	24	4	9	2	5	-4	6
54	71	21	3	8	3	3	-3	5
54	73	23	5	12	4	4	-4	7
54	76	20	4	11	3	5	-4	8
54	77	23	5	12	3	6	-5	10
55	70	27	6	12	5	3	-4	6
55	71	20	4	11	4	4	-4	7
55	71	27	4	8	3	4	-5	6
56	73	27	3	6	2	6	-7	12
56	79	22	2	5	1	4	-3	7
56	79	29	5	9	2	1	-1	2
57	70	22	2	5	2	5	-6	10
57	76	24	5	11	3	1	-1	2
57	77	21	3	8	2	6	-5	11
57	77	27	2	4	1	1	-1	2

φ_1	Φ	φ_2	H	K	L	U	V	W
50	70	20						
58	73	27	3	6	2	4	-5	9
58	74	22	4	10	3	1	-1	2
58	74	29	5	9	3	3	-4	7
58	75	23	3	7	2	1	-1	2
59	70	27	5	10	4	2	-3	5
59	71	27	4	8	3	5	-7	12
59	79	22	2	5	1	5	-4	10
50	70	30						
50	71	31	3	5	2	7	-9	12
50	73	39	4	5	2	5	-6	10
50	74	34	2	3	1	4	-5	7
50	79	37	3	4	1	5	-6	9
51	71	31	3	5	2	6	-6	11
51	72	35	7	10	4	2	-3	4
51	73	37	6	6	3	2	-3	4
51	77	39	8	10	3	5	-7	10
51	78	30	7	12	3	3	-3	5
52	71	34	8	12	5	2	-3	4
53	70	36	8	11	5	4	-7	9
53	71	31	3	5	2	5	-7	10
53	74	34	2	3	1	3	-4	6
53	74	39	9	11	4	3	-5	7
53	79	37	3	4	1	4	-5	8
54	74	30	7	12	4	4	-5	8
54	75	37	9	12	4	4	-6	9
54	76	31	6	10	3	5	-6	10
54	78	34	8	12	3	6	-7	12
55	71	38	7	9	4	4	-6	11
55	74	36	8	11	4	5	-8	12
56	71	31	3	5	2	4	-6	9
56	78	39	9	11	3	2	-3	5
56	79	37	3	4	1	3	-4	7
57	70	30	4	7	3	5	-8	12
57	71	39	9	11	5	3	-7	10
57	73	39	4	5	2	1	-2	3
57	74	34	2	3	1	2	-3	5
57	75	38	7	9	3	3	-5	8
57	76	30	4	7	2	5	-6	11
57	77	32	7	11	3	4	-5	9
57	77	36	5	7	2	5	-7	12
58	71	36	5	7	3	1	-2	3
58	72	37	9	12	5	1	-2	3
59	70	30	4	7	3	4	-7	11
59	70	34	6	9	4	1	-2	3
59	71	31	3	5	2	3	-5	8
59	75	37	9	12	4	4	-7	12

φ_1	Φ	φ_2	H	K	L	U	V	W
50	70	30						
59	77	36	5	7	2	4	-6	11
59	78	32	5	6	2	4	-5	10
50	70	40						
50	71	45	2	2	1	3	-7	6
50	72	41	6	7	3	5	-9	11
50	72	49	7	6	3	3	-8	9
50	74	45	12	12	5	5	-10	12
50	76	40	5	6	2	2	-3	4
50	76	47	12	11	4	2	-4	5
50	77	42	9	10	3	2	-3	4
50	77	45	3	3	1	4	-7	9
50	78	49	7	6	2	2	-4	5
51	71	42	10	11	5	5	-10	12
51	72	45	11	11	5	4	-9	11
51	74	45	5	5	2	2	-4	5
51	77	45	3	3	1	5	-9	12
51	79	40	10	12	3	3	-4	6
51	79	45	11	11	3	5	-6	11
52	70	47	12	11	6	2	-6	7
52	72	42	8	9	4	2	-4	5
52	75	45	6	8	3	3	-6	8
52	76	42	8	9	3	3	-5	7
52	77	46	10	9	3	3	-6	8
52	79	45	7	7	2	3	-5	7
53	72	49	7	6	3	3	-9	11
53	73	45	9	9	4	3	-7	9
53	76	43	11	12	4	4	-7	10
53	76	45	11	11	4	4	-8	11
53	76	48	9	8	3	4	-9	12
53	78	41	6	7	2	4	-6	9
53	78	45	10	10	3	4	-7	10
54	70	46	10	9	5	3	-10	12
54	71	45	2	2	1	3	-6	10
54	72	41	6	7	3	3	-6	8
54	73	42	9	10	4	4	-8	11
54	74	45	12	12	5	4	-9	12
55	75	48	11	10	4	2	-5	7
55	77	45	3	3	1	1	-2	3
55	79	45	11	11	3	4	-7	11
55	79	48	11	10	3	1	-2	3
55	79	49	8	7	2	1	-2	3
56	71	45	2	2	1	1	-3	4
56	72	45	11	11	5	3	-8	11
56	73	45	7	7	3	2	-5	7
56	74	41	7	8	3	1	-2	3
56	74	45	5	5	2	3	-7	10

φ_1	Φ	φ_2	H	K	L	U	V	W
50	70	40						
56	75	42	10	11	4	1	-2	3
56	76	40	5	6	2	4	-7	11
56	79	41	7	8	2	2	-3	5
57	71	42	10	11	5	2	-5	7
57	77	46	10	9	3	3	-7	11
58	74	45	12	12	5	3	-8	12
58	76	45	11	11	4	3	-7	11
58	78	45	10	10	3	3	-6	10
58	78	49	7	6	2	2	-5	8
59	71	45	2	2	1	2	-7	10
59	72	42	8	9	4	3	-8	12
59	73	45	9	9	4	2	-6	9
59	75	45	8	8	3	2	-5	8
59	76	42	8	9	3	3	-6	10
59	79	45	7	7	2	2	-4	7
50	70	50						
50	74	56	3	2	1	2	-7	8
50	78	54	11	8	3	2	-5	6
51	70	56	9	6	4	2	-11	12
51	72	55	10	7	4	2	-8	9
51	73	51	5	4	2	2	-6	7
51	76	55	10	7	3	3	-9	11
51	79	50	12	10	3	3	-6	8
51	79	53	4	3	1	3	-7	9
52	70	54	11	8	5	2	-9	10
52	71	59	5	3	2	1	-7	8
52	73	53	8	6	3	3	-10	12
53	73	58	11	7	4	1	-5	6
53	74	56	3	2	1	1	-4	5
53	77	58	11	7	3	2	-7	9
53	78	56	12	8	3	1	-3	4
54	71	54	7	5	3	1	-5	6
54	75	52	9	7	3	1	-3	4
54	77	54	7	5	2	1	-3	4
54	78	51	11	9	3	3	-7	10
54	79	53	4	3	1	2	-5	7
55	71	51	11	9	5	1	-4	5
55	71	59	5	3	2	1	-9	11
55	73	51	5	4	2	2	-7	9
55	74	51	11	9	4	1	-3	4
55	74	56	3	2	1	2	-9	12
55	76	50	6	5	2	3	-8	11
56	72	58	8	5	3	1	-7	9
56	79	53	4	3	1	3	-8	12
56	71	56	12	8	5	1	-9	12
58	76	50	6	5	2	2	-6	9

φ_1	Φ	φ_2	H	K	L	U	V	W
50	70	50						
58	76	55	10	7	3	1	-4	6
59	71	54	7	5	3	1	-6	11
59	74	56	3	2	1	1	-6	9
59	75	53	12	9	4	1	-4	6
50	70	60						
50	75	67	7	3	2	1	-9	10
50	79	68	5	2	1	1	-6	7
51	71	63	8	4	3	1	-11	12
51	73	63	6	3	2	1	-8	9
51	76	66	11	5	3	1	-7	8
51	77	63	4	2	1	1	-5	6
52	70	60	7	4	3	1	-10	11
52	74	61	9	5	3	1	-6	7
52	76	60	7	4	2	2	-9	11
52	77	67	12	5	3	1	-9	11
53	73	63	6	3	2	1	-10	12
53	79	68	5	2	1	1	-7	9
54	75	63	10	5	3	1	-8	10
55	70	63	12	6	5	0	-5	6
55	72	66	11	5	4	0	-4	5
55	73	67	12	5	4	0	-4	5
55	77	63	4	2	1	1	-6	8
55	79	61	9	5	2	2	-6	11
55	79	68	5	2	1	1	-8	11
56	70	63	10	5	4	0	-4	5
56	74	61	9	5	3	1	-9	12
56	74	68	10	4	3	0	-3	4
57	73	66	9	4	3	0	-3	4
57	74	60	12	7	4	1	-8	11
57	77	63	4	2	1	1	-7	10
58	71	63	8	4	3	0	-3	4
58	77	63	4	2	1	1	-8	12
59	70	60	7	4	3	0	-3	4
59	75	67	7	3	2	0	-2	3
59	77	69	8	3	2	0	-2	3
59	79	61	9	5	2	1	-5	8
50	70	70						
51	74	79	10	2	3	-1	-7	8
52	72	76	12	3	4	-1	-6	9
53	71	75	11	3	4	-1	-7	8
53	72	77	9	2	3	-1	-6	7
54	70	76	8	2	3	-1	-5	6
55	77	72	12	4	3	0	-3	4
56	76	70	11	4	3	0	-3	4

φ_1	Φ	φ_2	H	K	L	U	V	W
50	70	70						
57	72	72	3	1	1	-1	-9	12
57	72	77	9	2	3	-2	-9	12
58	72	72	3	1	1	-1	-8	11
58	78	72	9	3	2	0	-2	3
58	79	73	10	3	2	0	-2	3
59	70	76	8	2	3	-2	-7	10
59	72	72	3	1	1	-1	-7	10
59	74	79	10	2	3	-1	-4	6
59	76	76	4	1	1	-1	-8	12
50	70	80						
50	75	80	11	2	3	-1	-8	9
50	76	81	12	2	3	-1	-9	10
50	78	84	9	1	2	-1	-7	8
51	70	80	11	2	4	-2	-9	10
51	72	81	6	1	2	-2	-10	11
51	76	83	8	1	2	-1	-6	7
52	70	85	11	1	4	-2	-6	7
52	74	82	7	1	2	-1	-5	6
52	76	85	12	1	3	-2	-9	11
53	75	85	11	1	3	-1	-4	5
54	70	83	8	1	3	-3	-9	11
54	72	81	6	1	2	-1	-4	5
54	73	84	10	1	3	-2	-7	9
54	79	84	10	1	2	-1	-6	8
55	76	81	12	2	3	-1	-6	8
56	70	80	11	2	4	-2	-7	9
56	72	84	9	1	3	-1	-3	4
56	72	85	12	1	4	-3	-8	11
56	74	82	7	1	2	-2	-8	11
56	78	84	9	1	2	-1	-5	7
57	75	80	11	2	3	-1	-5	7
58	70	83	8	1	3	-2	-5	7
58	70	85	11	1	4	-3	-7	10
58	76	83	6	1	2	-1	-4	6
59	72	81	6	1	2	-2	-6	9
50	70	90						
50	72	90	3	0	1	-3	-6	9
50	76	90	4	0	1	-2	-7	8
51	74	90	7	0	2	-2	-6	7
51	76	90	4	0	1	-3	-10	12
52	70	90	11	0	4	-4	-9	11
52	72	90	3	0	1	-2	-5	6
52	75	90	11	0	3	-3	-9	11
52	79	90	5	0	1	-1	-4	5

φ_1	Φ	φ_2	H	K	L	U	V	W
50	70	90						
53	73	90	10	0	3	-3	-8	10
53	77	90	9	0	2	-2	-7	9
54	72	90	3	0	1	-3	-7	9
54	76	90	4	0	1	-1	-3	4
55	72	90	3	0	1	-4	-9	12
55	75	90	11	0	3	-3	-8	11
56	70	90	11	0	4	-4	-8	11
56	73	90	10	0	3	-3	-7	10
56	74	90	7	0	2	-2	-5	7
56	79	90	5	0	1	-2	-7	10
57	76	90	4	0	1	-3	-8	12
57	77	90	9	0	2	-2	-6	9
58	72	90	3	0	1	-1	-2	3
58	75	90	11	0	3	-3	-7	11
59	70	90	11	0	4	-4	-7	11
59	76	90	4	0	1	-2	-5	8
50	80	0						
50	81	9	1	6	1	6	-3	10
50	82	0	0	7	1	6	-1	7
50	84	6	1	10	1	9	-2	11
50	85	0	0	12	1	10	-1	12
50	85	5	1	11	1	5	-1	6
51	80	0	0	11	2	9	-2	11
51	81	0	0	6	1	5	-1	6
51	84	0	0	10	1	8	-1	10
51	84	6	1	9	1	4	-1	5
51	85	0	0	11	1	9	-1	11
52	82	6	1	7	1	3	-1	4
52	83	7	1	8	1	7	-2	9
52	84	0	0	9	1	7	-1	9
53	80	5	1	11	2	3	-1	4
53	81	9	1	6	1	5	-2	7
53	83	0	0	8	1	6	-1	8
53	85	0	0	12	1	9	-1	12
54	80	0	0	11	2	8	-2	11
54	81	0	0	6	1	9	-2	12
54	85	0	0	11	1	8	-1	11
54	85	5	1	12	1	5	-1	7
54	85	9	2	12	1	7	-2	10
55	82	0	0	7	1	5	-1	7
55	84	0	0	10	1	7	-1	10
56	81	5	1	12	2	6	-2	9
56	81	9	1	8	1	7	-3	11
56	84	0	0	9	1	6	-1	9
56	84	6	1	10	1	4	-1	6
56	85	0	0	12	1	8	-1	12

φ_1 Φ φ_2	H	K	L	U	V	W
50 80 0						
57 81 0	0	6	1	4	-1	6
57 84 6	1	9	1	7	-2	11
58 80 0	0	11	2	7	-2	11
58 83 0	0	8	1	5	-1	8
58 83 7	1	8	1	3	-1	5
58 85 0	0	11	1	7	-1	11
59 84 0	0	10	1	6	-1	10
50 80 10						
50 81 18	2	6	1	9	-5	12
50 85 17	3	10	1	7	-3	9
51 85 18	4	12	1	9	-4	12
52 81 18	2	6	1	7	-4	10
52 84 11	2	10	1	3	-1	4
53 83 14	2	8	1	7	-3	10
53 84 13	2	9	1	8	-3	11
53 85 15	3	11	1	5	-2	7
54 81 14	3	12	2	2	-1	3
54 82 16	2	7	1	2	-1	3
54 84 18	3	9	1	2	-1	3
55 81 18	2	6	1	5	-3	8
55 85 14	3	12	1	8	-3	12
57 84 13	2	9	1	5	-2	8
58 80 15	3	11	2	5	-3	9
58 85 10	2	11	1	3	-1	5
58 85 15	3	11	1	7	-3	12
59 80 10	2	11	2	4	-2	7
59 85 14	3	12	1	5	-2	9
50 80 20						
50 81 24	5	11	2	7	-5	10
50 83 23	3	7	1	8	-5	11
50 84 29	5	9	1	7	-5	10
51 82 27	3	6	1	4	-3	6
51 84 24	4	9	1	5	-3	7
52 81 23	5	12	2	6	-4	9
52 86 27	6	12	1	8	-5	12
53 81 29	6	11	2	7	-6	12
54 83 21	3	8	1	7	-4	11
54 85 20	4	11	1	2	-1	3
54 85 24	5	11	1	5	-3	8
55 80 20	4	11	2	3	-2	5
55 82 27	3	6	1	5	-4	9
55 83 23	3	7	1	3	-2	5
55 84 27	4	8	1	7	-5	12
55 84 29	5	9	1	4	-3	7

φ_1 Φ φ_2	H	K	L	U	V	W
50 80 20						
55 85 27	5	10	1	3	-2	5
56 85 22	4	10	1	7	-4	12
58 83 21	3	8	1	5	-3	9
59 81 23	5	12	2	4	-3	8
59 84 24	4	9	1	3	-2	6
59 84 27	4	8	1	4	-3	8
59 86 27	6	12	1	3	-2	6
50 80 30						
50 82 30	7	12	2	6	-5	9
50 85 31	6	10	1	4	-3	6
50 85 35	7	10	1	7	-6	11
51 81 39	4	5	1	6	-7	11
52 83 30	4	7	1	5	-4	8
52 84 32	5	8	1	6	-5	10
52 86 30	7	12	1	7	-5	11
53 81 39	4	5	1	5	-6	10
53 85 34	6	9	1	7	-6	12
55 83 30	4	7	1	6	-5	11
55 83 36	5	7	1	1	-1	2
55 84 37	6	8	1	1	-1	2
55 85 38	7	9	1	1	-1	2
55 86 32	7	11	1	5	-4	9
55 86 39	8	10	1	1	-1	2
55 86 39	9	11	1	1	-1	2
56 81 32	7	11	2	1	-1	2
56 81 39	4	5	1	4	-5	9
56 82 34	4	6	1	1	-1	2
57 86 34	8	12	1	6	-5	12
58 80 34	6	9	2	5	-6	12
58 85 31	6	10	1	5	-4	10
58 86 30	7	12	1	4	-3	8
59 81 39	4	5	1	3	-4	8
50 80 40						
50 82 45	5	5	1	5	-7	10
50 82 48	10	9	2	5	-8	11
50 86 45	11	11	1	6	-7	11
50 86 48	10	9	1	3	-4	6
51 81 42	8	9	2	3	-4	6
51 83 45	6	6	1	3	-4	6
52 81 45	9	9	2	4	-6	9
52 81 48	9	8	2	4	-7	10
52 83 43	11	12	2	4	-5	8
52 83 47	12	11	2	4	-6	9
52 85 45	8	8	1	4	-5	8

φ_1 Φ φ_2	H	K	L	U	V	W
50 80 40						
52 85 48	9	8	1	5	-7	11
52 86 45	10	10	1	5	-6	10
53 80 45	4	4	1	5	-8	12
53 83 40	5	6	1	6	-7	12
53 83 45	11	11	2	5	-7	11
53 87 45	12	12	1	6	-7	12
55 80 45	4	4	1	3	-5	8
55 82 42	10	11	2	3	-4	7
55 82 45	5	5	1	2	-3	5
55 83 40	5	6	1	5	-6	11
55 83 45	6	6	1	5	-7	12
55 84 45	7	7	1	3	-4	7
55 85 49	8	7	1	2	-3	5
55 86 40	10	12	1	1	-1	2
55 86 45	9	9	1	4	-5	9
55 86 45	11	11	1	5	-6	11
55 86 48	11	10	1	3	-4	7
57 80 45	4	4	1	4	-7	12
57 83 45	11	11	2	4	-6	11
57 84 41	6	7	1	5	-6	12
57 87 45	12	12	1	5	-6	12
58 80 47	12	11	3	3	-6	10
58 81 42	8	9	2	4	-6	11
58 81 45	9	9	2	3	-5	9
58 83 40	5	6	1	4	-5	10
58 84 49	7	6	1	3	-5	9
58 85 45	8	8	1	3	-4	8
58 86 45	10	10	1	4	-5	10
58 86 47	12	11	1	3	-4	8
59 80 43	11	12	3	3	-5	9
59 81 48	9	8	2	2	-4	7
59 85 48	9	8	1	2	-3	6
50 80 50						
50 86 54	11	8	1	3	-5	7
50 86 56	12	8	1	5	-9	12
51 82 56	6	4	1	4	-9	12
51 83 50	6	5	1	3	-5	7
52 86 53	12	9	1	5	-8	12
53 80 59	5	3	1	1	-3	4
53 82 56	6	4	1	3	-7	10
54 81 51	5	4	1	1	-2	3
54 82 53	12	9	2	1	-2	3
54 83 50	6	5	1	4	-7	11
54 85 52	9	7	1	3	-5	8
54 85 56	9	6	1	1	-2	3
54 86 51	11	9	1	2	-3	5

φ_1 Φ φ_2	H	K	L	U	V	W
50 80 50						
56 80 56	9	6	2	2	-6	9
56 84 58	8	5	1	3	-7	11
56 86 54	11	8	1	4	-7	12
57 80 52	9	7	2	3	-7	11
57 86 53	12	9	1	3	-5	9
58 82 51	11	9	2	2	-4	7
58 85 59	10	6	1	3	-7	12
59 86 50	12	10	1	2	-3	6
50 80 60						
50 82 60	12	7	2	3	-8	10
50 82 63	6	3	1	3	-10	12
50 85 61	11	6	1	3	-7	9
50 85 63	10	5	1	3	-8	10
51 83 67	7	3	1	1	-4	5
51 83 69	8	3	1	2	-9	11
52 80 63	10	5	2	1	-4	5
52 82 63	6	3	1	2	-7	9
52 84 63	8	4	1	1	-3	4
52 85 66	11	5	1	1	-3	4
53 83 60	7	4	1	3	-8	11
53 84 61	9	5	1	2	-5	7
54 83 69	8	3	1	1	-5	7
54 84 66	9	4	1	2	-7	10
55 81 61	11	6	2	2	-7	10
55 81 66	11	5	2	1	-5	7
55 86 63	12	6	1	3	-8	12
56 82 60	12	7	2	2	-6	9
56 82 63	6	3	1	1	-4	6
56 85 61	11	6	1	2	-5	8
56 85 68	10	4	1	1	-4	6
57 81 67	12	5	2	1	-6	9
57 86 67	12	5	1	2	-7	11
58 83 67	7	3	1	1	-5	8
58 85 63	10	5	1	1	-3	5
59 83 69	8	3	1	1	-6	10
59 84 63	8	4	1	2	-7	12
50 80 70						
50 84 72	9	3	1	1	-5	6
51 82 74	7	2	1	1	-9	11
51 83 76	8	2	1	1	-10	12
51 85 70	11	4	1	1	-4	5
52 81 72	6	2	1	1	-8	10
52 85 76	12	3	1	1	-7	9
53 85 72	12	4	1	2	-9	12

φ_1 Φ φ_2	H	K	L	U	V	W
50 80 70						
53 85 73	10	3	1	1	-6	8
54 81 72	6	2	1	1	-9	12
55 85 75	11	3	1	1	-7	10
56 84 72	9	3	1	1	-6	9
56 85 76	12	3	1	1	-8	12
57 81 76	12	3	2	0	-2	3
58 80 75	11	3	2	0	-2	3
58 85 72	12	4	1	1	-5	8
58 85 73	10	3	1	1	-7	11
50 80 80						
52 81 85	12	1	2	-1	-8	10
53 80 85	11	1	2	-1	-7	9
57 81 85	12	1	2	-1	-6	9
59 80 85	11	1	2	-1	-5	6
50 80 90						
50 82 90	7	0	1	-1	-6	7
50 85 90	12	0	1	-1	-10	12
51 80 90	11	0	2	-2	-9	11
51 81 90	6	0	1	-1	-5	6
51 84 90	10	0	1	-1	-8	10
51 85 90	11	0	1	-1	-9	11
52 84 90	9	0	1	-1	-7	9
53 83 90	8	0	1	-1	-6	8
53 85 90	12	0	1	-1	-9	12
54 80 90	11	0	2	-2	-8	11
54 81 90	6	0	1	-2	-9	12
54 85 90	11	0	1	-1	-8	11
55 82 90	7	0	1	-1	-5	7
55 84 90	10	0	1	-1	-7	10
56 84 90	9	0	1	-1	-6	9
56 85 90	12	0	1	-1	-8	12
57 81 90	6	0	1	-1	-4	6
58 80 90	11	0	2	-2	-7	11
58 83 90	8	0	1	-1	-5	8
58 85 90	11	0	1	-1	-7	11
59 84 90	10	0	1	-1	-6	10
50 90 0						
50 90 0	0	1	0	5	0	6
50 90 6	1	10	0	10	-1	12
51 90 0	0	1	0	9	0	11
51 90 0	0	1	0	4	0	5
51 90 6	1	9	0	9	-1	11

φ_1 Φ φ_2	H	K	L	U	V	W
50 90 0						
51 90 7	1	8	0	8	-1	10
52 90 0	0	1	0	7	0	9
52 90 8	1	7	0	7	-1	9
53 90 0	0	1	0	3	0	4
53 90 6	1	9	0	9	-1	12
53 90 9	1	6	0	6	-1	8
54 90 0	0	1	0	8	0	11
54 90 0	0	1	0	5	0	7
54 90 7	1	8	0	8	-1	11
55 90 0	0	1	0	7	0	10
55 90 8	1	7	0	7	-1	10
56 90 0	0	1	0	2	0	3
56 90 7	1	8	0	8	-1	12
56 90 9	1	6	0	6	-1	9
57 90 8	1	7	0	7	-1	11
58 90 0	0	1	0	7	0	11
58 90 0	0	1	0	5	0	8
59 90 0	0	1	0	3	0	5
59 90 8	1	7	0	7	-1	12
59 90 9	1	6	0	6	-1	10
50 90 10						
50 90 11	1	5	0	5	-1	6
50 90 13	2	9	0	9	-2	11
50 90 14	1	4	0	4	-1	5
51 90 16	2	7	0	7	-2	9
52 90 13	2	9	0	9	-2	12
52 90 18	1	3	0	3	-1	4
53 90 14	1	4	0	8	-2	11
54 90 11	1	5	0	5	-1	7
54 90 16	2	7	0	7	-2	10
55 90 18	1	3	0	6	-2	9
56 90 14	1	4	0	4	-1	6
57 90 11	1	5	0	5	-1	8
57 90 16	2	7	0	7	-2	11
58 90 18	1	3	0	3	-1	5
59 90 16	2	7	0	7	-2	12
50 90 20						
50 90 23	3	7	0	7	-3	9
50 90 27	1	2	0	6	-3	8
51 90 24	4	9	0	9	-4	12
51 90 27	1	2	0	6	-4	11
52 90 21	3	8	0	8	-3	11
52 90 22	2	5	0	5	-2	7
53 90 23	3	7	0	7	-3	10

φ_1 Φ φ_2	H	K	L	U	V	W
50 90 20						
53 90 27	1	2	0	2	-1	3
55 90 21	3	8	0	8	-3	12
55 90 23	3	7	0	7	-3	11
56 90 22	2	5	0	5	-2	8
56 90 27	1	2	0	6	-3	10
57 90 27	1	2	0	4	-2	7
58 90 23	3	7	0	7	-3	12
59 90 22	2	5	0	5	-2	9
59 90 27	1	2	0	6	-3	11
50 90 30						
50 90 31	3	5	0	5	-3	7
50 90 37	3	4	0	4	-3	6
51 90 30	4	7	0	7	-4	10
51 90 39	4	5	0	5	-4	8
52 90 32	5	8	0	8	-5	12
52 90 36	5	7	0	7	-5	11
54 90 30	4	7	0	7	-4	11
54 90 31	3	5	0	5	-3	8
54 90 34	2	3	0	3	-2	5
54 90 36	5	7	0	7	-5	12
54 90 37	3	4	0	4	-3	7
55 90 39	4	5	0	5	-4	9
56 90 30	4	7	0	7	-4	12
57 90 31	3	5	0	5	-3	9
57 90 34	2	3	0	6	-4	11
57 90 39	4	5	0	5	-4	10
58 90 37	3	4	0	4	-3	8
59 90 34	2	3	0	3	-2	6
50 90 40						
50 90 41	6	7	0	7	-6	11
50 90 45	1	1	0	3	-3	5
50 90 45	1	1	0	7	-7	12
50 90 49	7	6	0	6	-7	11
51 90 45	1	1	0	4	-4	7
52 90 40	5	6	0	6	-5	10
52 90 41	6	7	0	7	-6	12
52 90 45	1	1	0	6	-6	11
52 90 45	1	1	0	5	-5	9
52 90 49	7	6	0	6	-7	12
55 90 40	5	6	0	6	-5	11
55 90 45	1	1	0	1	-1	2
57 90 40	5	6	0	6	-5	12
57 90 45	1	1	0	5	-5	11
58 90 45	1	1	0	4	-4	9

φ_1 Φ φ_2	H	K	L	U	V	W
50 90 40						
59 90 45	1	1	0	3	-3	7
59 90 45	1	1	0	5	-5	12
50 90 50						
50 90 53	4	3	0	3	-4	6
50 90 59	5	3	0	3	-5	7
51 90 51	5	4	0	4	-5	8
52 90 50	6	5	0	5	-6	10
52 90 54	7	5	0	5	-7	11
52 90 56	8	5	0	5	-8	12
54 90 53	4	3	0	3	-4	7
54 90 54	7	5	0	5	-7	12
54 90 56	3	2	0	2	-3	5
54 90 59	5	3	0	3	-5	8
55 90 50	6	5	0	5	-6	11
55 90 51	5	4	0	4	-5	9
57 90 50	6	5	0	5	-6	12
57 90 51	5	4	0	4	-5	10
57 90 56	3	2	0	4	-6	11
57 90 59	5	3	0	3	-5	9
58 90 53	4	3	0	3	-4	8
59 90 56	3	2	0	2	-3	6
50 90 60						
50 90 63	2	1	0	3	-6	8
50 90 67	7	3	0	3	-7	9
51 90 60	7	4	0	4	-7	10
51 90 63	2	1	0	4	-8	11
51 90 66	9	4	0	4	-9	12
52 90 66	5	2	0	2	-5	7
52 90 69	8	3	0	3	-8	11
53 90 63	2	1	0	1	-2	3
53 90 67	7	3	0	3	-7	10
54 90 60	7	4	0	4	-7	11
55 90 67	7	3	0	3	-7	11
55 90 69	8	3	0	3	-8	12
56 90 60	7	4	0	4	-7	12
56 90 63	2	1	0	3	-6	10
56 90 66	5	2	0	2	-5	8
57 90 63	2	1	0	2	-4	7
58 90 67	7	3	0	3	-7	12
59 90 63	2	1	0	3	-6	11
59 90 66	5	2	0	2	-5	9

φ1	Φ	φ2	H	K	L	U	V	W
50	90	70						
50	90	76	4	1	0	1	-4	5
50	90	77	9	2	0	2	-9	11
50	90	79	5	1	0	1	-5	6
51	90	74	7	2	0	2	-7	9
52	90	72	3	1	0	1	-3	4
52	90	77	9	2	0	2	-9	12
53	90	76	4	1	0	2	-8	11
54	90	74	7	2	0	2	-7	10
54	90	79	5	1	0	1	-5	7
55	90	72	3	1	0	2	-6	9
56	90	76	4	1	0	1	-4	6
57	90	74	7	2	0	2	-7	11
57	90	79	5	1	0	1	-5	8
58	90	72	3	1	0	1	-3	5
59	90	74	7	2	0	2	-7	12
50	90	80						
50	90	84	10	1	0	1	-10	12
51	90	83	8	1	0	1	-8	10
51	90	84	9	1	0	1	-9	11
52	90	82	7	1	0	1	-7	9
53	90	81	6	1	0	1	-6	8
53	90	84	9	1	0	1	-9	12
54	90	83	8	1	0	1	-8	11
55	90	82	7	1	0	1	-7	10
56	90	81	6	1	0	1	-6	9
56	90	83	8	1	0	1	-8	12
57	90	82	7	1	0	1	-7	11
59	90	81	6	1	0	1	-6	10
59	90	82	7	1	0	1	-7	12
50	90	90						
50	90	90	1	0	0	0	-5	6
51	90	90	1	0	0	0	-9	11
51	90	90	1	0	0	0	-4	5
52	90	90	1	0	0	0	-7	9
53	90	90	1	0	0	0	-3	4
54	90	90	1	0	0	0	-8	11
54	90	90	1	0	0	0	-5	7
55	90	90	1	0	0	0	-7	10
56	90	90	1	0	0	0	-2	3
58	90	90	1	0	0	0	-7	11
58	90	90	1	0	0	0	-5	8
59	90	90	1	0	0	0	-3	5

φ1	Φ	φ2	H	K	L	U	V	W
60	0	0						
60	5	0	0	1	12	7	-12	1
60	9	0	0	1	6	7	-12	2
61	5	0	0	1	11	6	-11	1
61	6	0	0	1	9	5	-9	1
61	8	0	0	1	7	4	-7	1
64	5	0	0	1	12	6	-12	1
64	6	0	0	1	10	5	-10	1
64	7	0	0	1	8	4	-8	1
64	9	0	0	1	6	3	-6	1
66	5	0	0	1	11	5	-11	1
66	6	0	0	1	9	4	-9	1
67	5	0	0	1	12	5	-12	1
67	8	0	0	1	7	3	-7	1
68	6	0	0	1	10	4	-10	1
68	9	0	0	1	6	5	-12	2
60	0	40						
61	9	45	1	1	9	-2	-7	1
64	7	45	1	1	12	-3	-9	1
66	7	45	1	1	11	-3	-8	1
68	8	45	1	1	10	-3	-7	1
60	0	60						
60	0	60	0	0	1	-7	-12	0
60	0	60	0	0	1	-4	-7	0
60	0	60	0	0	1	-3	-5	0
60	0	60	0	0	1	-5	-9	0
61	0	61	0	0	1	-7	-11	0
61	0	61	0	0	1	-5	-8	0
62	0	62	0	0	1	-7	-10	0
62	0	62	0	0	1	-2	-3	0
63	0	63	0	0	1	-8	-11	0
63	0	63	0	0	1	-3	-4	0
63	0	63	0	0	1	-5	-7	0
64	0	64	0	0	1	-7	-9	0
64	0	64	0	0	1	-4	-5	0
65	0	65	0	0	1	-9	-11	0
65	0	65	0	0	1	-5	-6	0
65	0	65	0	0	1	-6	-7	0
66	0	66	0	0	1	-11	-12	0
66	0	66	0	0	1	-7	-8	0
66	0	66	0	0	1	-8	-9	0
66	0	66	0	0	1	-10	-11	0
66	0	66	0	0	1	-9	-10	0
67	0	67	0	0	1	-1	-1	0

φ1	Φ	φ2	H	K	L	U	V	W
60	0	60						
69	0	69	0	0	1	-8	-7	0
69	0	69	0	0	1	-12	-11	0
69	0	69	0	0	1	-9	-8	0
69	0	69	0	0	1	-11	-10	0
69	0	69	0	0	1	-10	-9	0
60	0	90						
60	5	90	1	0	12	-12	-7	1
60	9	90	1	0	6	-12	-7	2
61	5	90	1	0	11	-11	-6	1
61	6	90	1	0	9	-9	-5	1
61	8	90	1	0	7	-7	-4	1
64	5	90	1	0	12	-12	-6	1
64	6	90	1	0	10	-10	-5	1
64	7	90	1	0	8	-8	-4	1
64	9	90	1	0	6	-6	-3	1
66	5	90	1	0	11	-11	-5	1
66	6	90	1	0	9	-9	-4	1
67	5	90	1	0	12	-12	-5	1
67	8	90	1	0	7	-7	-3	1
68	6	90	1	0	10	-10	-4	1
68	9	90	1	0	6	-12	-5	2
60	10	0						
60	11	0	0	1	5	3	-5	1
60	14	0	0	1	4	7	-12	3
60	17	0	0	3	10	6	-10	3
61	16	0	0	2	7	4	-7	2
61	18	0	0	1	3	7	-12	4
62	10	0	0	2	11	6	-11	2
62	13	0	0	2	9	5	-9	2
62	15	0	0	3	11	6	-11	3
62	18	0	0	1	3	5	-9	3
64	11	0	0	1	5	5	-10	2
64	14	0	0	1	4	2	-4	1
64	17	0	0	3	10	5	-10	3
65	18	0	0	1	3	3	-6	2
66	10	0	0	2	11	5	-11	2
66	15	0	0	3	11	5	-11	3
67	13	0	0	2	9	4	-9	2
67	18	0	0	1	3	4	-9	3
68	14	0	0	1	4	5	-12	3
68	16	0	0	2	7	3	-7	2
68	18	0	0	1	3	5	-12	4
69	11	0	0	1	5	2	-5	1
69	17	0	0	3	10	4	-10	3

φ1	Φ	φ2	H	K	L	U	V	W
60	10	10						
65	16	18	1	3	10	1	-7	2
60	10	20						
64	11	27	1	2	11	0	-11	2
64	11	27	1	2	12	0	-6	1
64	13	27	1	2	10	0	-5	1
64	14	27	1	2	9	0	-9	2
64	16	27	1	2	8	0	-4	1
65	16	27	1	2	7	0	-7	2
60	10	40						
60	16	45	1	1	5	-1	-4	1
61	11	45	1	1	7	-3	-11	2
62	17	45	2	2	9	-2	-7	2
64	10	45	1	1	8	-2	-6	1
64	13	45	1	1	6	-3	-9	2
65	19	45	1	1	4	-1	-3	1
66	14	45	2	2	11	-3	-8	2
66	16	45	1	1	5	-4	-11	3
67	11	45	1	1	7	-2	-5	1
69	16	45	1	1	5	-3	-7	2
60	10	50						
64	18	56	3	2	11	-5	-9	3
68	17	56	3	2	12	-2	-3	1
60	10	60						
61	18	63	2	1	7	-2	-3	1
62	11	63	2	1	12	-7	-10	2
62	16	63	2	1	8	-7	-10	3
63	14	63	2	1	9	-8	-11	3
64	13	63	2	1	10	-3	-4	1
67	16	63	2	1	8	-5	-6	2
68	11	63	2	1	11	-7	-8	2
69	18	63	2	1	7	-9	-10	4
60	10	70						
60	19	76	4	1	12	-12	-12	5
61	16	72	3	1	11	-6	-9	3
62	19	72	3	1	9	-11	-12	5
64	15	72	3	1	12	-3	-3	1
64	16	72	3	1	11	-11	-11	4

φ_1	Φ	φ_2	H	K	L	U	V	W
60	10	70						
65	18	72	3	1	10	-5	-5	2
65	19	72	3	1	9	-9	-9	4
67	19	76	4	1	12	-5	-4	2
69	19	72	3	1	9	-7	-6	3
60	10	90						
60	11	90	1	0	5	-5	-3	1
60	14	90	1	0	4	-12	-7	3
60	17	90	3	0	10	-10	-6	3
61	16	90	2	0	7	-7	-4	2
61	18	90	1	0	3	-12	-7	4
62	10	90	2	0	11	-11	-6	2
62	13	90	2	0	9	-9	-5	2
62	15	90	3	0	11	-11	-6	3
62	18	90	1	0	3	-9	-5	3
64	11	90	1	0	5	-10	-5	2
64	14	90	1	0	4	-4	-2	1
64	17	90	3	0	10	-10	-5	3
65	18	90	1	0	3	-6	-3	2
66	10	90	2	0	11	-11	-5	2
66	15	90	3	0	11	-11	-5	3
67	13	90	2	0	9	-9	-4	2
67	18	90	1	0	3	-9	-4	3
68	14	90	1	0	4	-12	-5	3
68	16	90	2	0	7	-7	-3	2
68	18	90	1	0	3	-12	-5	4
69	11	90	1	0	5	-5	-2	1
69	17	90	3	0	10	-10	-4	3
60	20	0						
60	21	0	0	3	8	5	-8	3
60	24	0	0	5	11	7	-11	5
60	29	0	0	5	9	6	-9	5
61	22	0	0	2	5	3	-5	2
61	27	0	0	1	2	5	-8	4
61	29	0	0	6	11	7	-11	6
62	23	0	0	3	7	4	-7	3
62	23	0	0	5	12	7	-12	5
62	27	0	0	1	2	7	-12	6
62	27	0	0	1	2	6	-10	5
63	20	0	0	4	11	6	-11	4
63	24	0	0	4	9	5	-9	4
64	24	0	0	5	11	6	-11	5
64	29	0	0	5	9	5	-9	5
64	29	0	0	6	11	6	-11	6
65	21	0	0	3	8	4	-8	3

φ_1	Φ	φ_2	H	K	L	U	V	W
60	20	0						
65	22	0	0	2	5	5	-10	4
65	23	0	0	5	12	6	-12	5
66	27	0	0	1	2	1	-2	1
67	20	0	0	4	11	5	-11	4
68	23	0	0	3	7	3	-7	3
68	24	0	0	4	9	4	-9	4
68	24	0	0	5	11	5	-11	5
68	29	0	0	6	11	5	-11	6
69	23	0	0	5	12	5	-12	5
69	29	0	0	5	9	4	-9	5
60	20	10						
62	24	18	1	3	7	1	-5	2
64	22	14	1	4	10	2	-6	3
64	25	14	1	4	9	3	-12	5
65	25	11	1	5	11	2	-7	3
68	22	18	1	3	8	1	-11	4
69	23	11	1	5	12	1	-5	2
69	24	18	1	3	7	1	-12	5
60	20	20						
63	26	22	2	5	11	1	-7	3
65	20	27	1	2	6	0	-3	1
65	22	27	2	4	11	0	-11	4
65	24	22	2	5	12	1	-10	4
65	24	27	1	2	5	0	-5	2
66	26	27	2	4	9	0	-9	4
66	29	27	1	2	4	0	-2	1
60	20	30						
62	26	31	3	5	12	0	-12	5
62	28	31	3	5	11	0	-11	5
64	24	34	2	3	6	-1	-10	4
64	27	34	2	3	7	-1	-11	5
66	27	37	3	4	10	-2	-11	5
66	29	37	3	4	9	-1	-6	3
67	20	34	2	3	10	-1	-6	2
68	24	34	2	3	8	-2	-12	5
68	26	31	3	5	12	-1	-9	4
60	20	40						
60	27	45	4	4	11	-2	-9	4
61	23	45	3	3	10	-2	-6	3
62	22	45	2	2	7	-3	-11	4

φ_1	Φ	φ_2	H	K	L	U	V	W
60	20	40						
62	25	45	1	1	3	-3	-12	5
62	29	45	2	2	5	-1	-4	2
63	25	45	1	1	3	-2	-7	3
66	25	45	1	1	3	-3	-9	4
66	28	45	3	3	8	-2	-6	3
67	21	45	3	3	11	-3	-6	3
67	25	45	1	1	3	-4	-11	5
68	22	45	2	2	7	-2	-5	2
68	27	45	4	4	11	-3	-8	4
68	29	45	2	2	5	-4	-11	6
60	20	50						
61	26	59	5	3	11	-1	-2	1
61	29	53	4	3	9	-3	-8	4
62	24	53	4	3	11	-3	-7	3
63	27	56	3	2	7	-1	-2	1
63	28	51	5	4	12	-4	-10	5
64	20	56	3	2	10	-6	-11	4
64	24	56	3	2	8	-6	-11	5
64	26	59	5	3	12	-6	-10	5
65	23	53	4	3	12	-6	-12	5
66	22	56	3	2	9	-7	-12	5
56	27	53	4	3	10	-1	-2	1
69	22	56	3	2	9	-6	-9	4
69	24	56	3	2	8	-4	-6	3
60	20	60						
60	20	63	2	1	6	-5	-8	3
60	26	68	5	2	11	-7	-10	5
61	24	63	2	1	5	-7	-11	5
61	26	63	4	2	9	-5	-8	4
63	20	63	2	1	6	-7	-10	4
63	24	68	5	2	12	-8	-10	5
63	28	68	5	2	10	-4	-5	3
63	29	63	2	1	4	-8	-12	7
65	20	63	2	1	6	-9	-12	5
65	24	63	2	1	5	-3	-4	2
65	26	63	4	2	9	-8	-11	6
65	26	68	5	2	11	-6	-7	4
65	29	63	2	1	4	-7	-10	6
66	24	68	5	2	12	-10	-11	6
66	29	63	2	1	4	-6	-8	5
69	22	63	4	2	11	-7	-8	4
69	24	68	5	2	12	-12	-12	7
69	26	68	5	2	11	-11	-11	7
69	28	68	5	2	10	-10	-10	7

φ_1	Φ	φ_2	H	K	L	U	V	W
60	20	60						
69	29	63	2	1	4	-5	-6	4
60	20	70						
60	27	76	4	1	8	-11	-12	7
61	21	76	4	1	11	-11	-11	5
61	22	76	4	1	10	-2	-2	1
61	24	72	3	1	7	-5	-6	3
61	25	76	4	1	9	-9	-9	5
61	26	72	3	1	6	-10	-12	7
62	27	76	4	1	8	-8	-8	5
64	21	76	4	1	11	-9	-8	4
64	25	76	4	1	9	-11	-10	6
64	27	79	5	1	10	-12	-10	7
65	22	72	3	1	8	-2	-2	1
66	24	72	3	1	7	-7	-7	4
66	26	72	3	1	6	-3	-3	2
68	27	76	4	1	8	-5	-4	3
68	27	79	5	1	10	-7	-5	4
69	24	72	3	1	7	-9	-8	5
60	20	80						
62	27	81	6	1	12	-7	-6	4
62	29	81	6	1	11	-8	-7	5
68	27	81	6	1	12	-9	-6	5
60	20	90						
60	21	90	3	0	8	-8	-5	3
60	24	90	5	0	11	-11	-7	5
60	29	90	5	0	9	-9	-6	5
61	22	90	2	0	5	-5	-3	2
61	27	90	1	0	2	-6	-5	4
61	29	90	6	0	11	-11	-7	6
62	23	90	3	0	7	-7	-4	3
62	23	90	5	0	12	-12	-7	5
62	27	90	1	0	2	-12	-7	6
62	27	90	1	0	2	-10	-6	5
63	20	90	4	0	11	-11	-6	4
63	24	90	4	0	9	-9	-5	4
64	24	90	5	0	11	-11	-6	5
64	29	90	5	0	9	-9	-5	5
64	29	90	6	0	11	-11	-6	6
65	21	90	3	0	6	-6	-4	3
65	22	90	2	0	5	-10	-5	4
65	23	90	5	0	12	-12	-6	5
66	27	90	1	0	2	-2	-1	1

TAB. VI FROM EULER ANGLES INTO HKL-UVW

φ_1	Φ	φ_2	H	K	L	U	V	W
60	20	90						
67	20	90	4	0	11	-11	-5	4
68	23	90	3	0	7	-7	-3	3
68	24	90	4	0	9	-9	-4	4
68	24	90	5	0	11	-11	-5	5
68	29	90	6	0	11	-11	-5	6
69	23	90	5	0	12	-12	-5	5
69	29	90	5	0	9	-9	-4	5
60	30	0						
60	30	0	0	7	12	8	-12	7
60	35	0	0	7	10	7	-10	7
60	36	0	0	5	7	5	-7	5
60	36	0	0	8	11	8	-11	8
61	31	8	1	7	12	5	-11	6
61	31	9	1	6	10	4	-9	5
61	34	0	0	2	3	2	-3	2
61	34	7	1	8	12	4	-8	5
61	39	0	0	4	5	7	-10	8
61	39	0	0	9	11	8	-11	9
62	32	0	0	5	8	5	-8	5
62	32	0	0	7	11	7	-11	7
62	37	0	0	3	4	8	-12	9
62	38	0	0	7	9	6	-9	7
63	30	0	0	7	12	7	-12	7
63	31	0	0	3	5	3	-5	3
63	36	0	0	8	11	7	-11	8
63	37	0	0	3	4	5	-8	6
63	38	6	1	7	9	5	-11	8
63	39	7	1	8	10	2	-4	3
64	30	0	0	4	7	4	-7	4
64	33	8	1	7	11	2	-5	3
64	34	0	0	2	3	7	-12	8
64	34	9	1	6	9	3	-8	5
64	35	0	0	7	10	6	-10	7
64	37	9	1	6	8	4	-10	7
64	39	0	0	9	11	7	-11	9
65	32	0	0	7	11	6	-11	7
65	34	0	0	2	3	5	-9	6
65	36	0	0	5	7	4	-7	5
65	37	0	0	3	4	7	-12	9
65	39	0	0	4	5	3	-5	4
66	36	0	0	8	11	6	-11	8
66	37	6	1	9	12	3	-7	5
66	38	0	0	7	9	5	-9	7
66	39	6	1	9	11	4	-9	7
67	30	0	0	7	12	6	-12	7
67	31	0	0	3	5	5	-10	6

φ_1	Φ	φ_2	H	K	L	U	V	W
60	30	0						
67	31	6	1	7	12	3	-9	5
67	32	0	0	5	8	4	-8	5
67	34	0	0	2	3	3	-6	4
67	34	7	1	8	12	4	-11	7
67	39	0	0	9	11	6	-11	9
68	31	9	1	6	10	2	-7	4
68	35	0	0	7	10	5	-10	7
68	35	8	1	7	10	1	-3	2
68	37	0	0	3	4	2	-4	3
69	32	0	0	7	11	5	-11	7
69	34	9	1	6	9	3	-11	7
69	37	9	1	6	6	2	-7	5
69	39	0	0	4	5	5	-10	8
60	30	10						
60	33	11	1	5	8	3	-7	4
60	35	18	2	6	9	3	-10	6
60	38	18	1	3	4	1	-3	2
61	30	18	2	6	11	1	-4	2
61	32	18	1	3	5	3	-11	6
63	32	18	1	3	5	2	-9	5
63	37	14	2	8	11	1	-3	2
64	33	11	1	5	8	4	-12	7
64	39	16	2	7	9	2	-7	5
65	36	11	1	5	7	1	-3	2
66	34	14	1	4	6	2	-8	5
66	38	13	2	9	12	3	-10	7
66	38	16	1	3	4	2	-10	7
67	32	18	1	3	5	1	-7	4
67	33	16	2	7	11	1	-5	3
69	30	14	1	4	7	2	-11	6
69	36	16	2	7	10	1	-6	4
69	38	18	1	3	4	1	-7	5
60	30	20						
61	39	27	4	8	11	1	-6	4
62	33	27	2	4	7	1	-11	6
62	37	27	1	2	3	1	-6	5
63	34	22	2	5	8	2	-12	7
63	38	21	3	8	11	2	-9	6
64	32	23	3	7	12	1	-9	5
64	37	27	1	2	3	1	-11	7
64	38	22	2	5	7	1	-6	4
66	37	23	3	7	10	1	-9	6
67	31	27	3	6	11	0	-11	6
67	33	27	2	4	7	0	-7	4

φ_1	Φ	φ_2	H	K	L	U	V	W
60	30	20						
67	34	22	2	5	8	1	-10	6
67	34	27	3	6	10	0	-5	3
68	37	27	1	2	3	0	-3	2
69	38	21	3	6	11	1	-10	7
69	39	27	4	6	11	0	-11	8
60	30	30						
60	31	34	2	3	6	0	-2	1
60	36	30	4	7	11	1	-10	6
60	36	36	5	7	12	0	-12	7
61	33	34	4	6	11	0	-11	6
61	38	36	5	7	11	0	-11	7
62	30	39	4	5	11	-1	-8	4
62	35	39	4	5	9	-1	-10	6
62	36	34	2	3	5	0	-5	3
62	39	30	4	7	10	1	-12	8
62	39	39	4	5	6	-1	-12	8
63	30	31	3	5	10	0	-2	1
63	33	31	3	5	9	0	-9	5
63	39	34	4	6	9	0	-3	2
64	36	31	3	5	8	0	-8	5
64	36	32	5	8	12	0	-3	2
65	34	30	4	7	12	0	-12	7
65	36	30	4	7	11	0	-11	7
65	36	37	3	4	7	-1	-8	5
66	35	39	4	5	9	-2	-11	7
66	36	34	2	3	5	-1	-11	7
66	39	30	4	7	10	0	-10	7
68	33	31	3	5	9	-1	-12	7
69	36	30	4	7	11	-1	-12	8
60	30	40						
60	31	45	5	5	12	-2	-10	5
60	35	40	5	6	11	-1	-12	7
60	35	45	1	1	2	-2	-12	7
61	39	45	4	4	7	-1	-6	4
62	33	45	5	5	11	-2	-9	5
63	38	45	6	6	11	-2	-9	6
64	31	45	3	3	7	-3	-11	6
64	35	45	1	1	2	-2	-8	5
64	36	41	6	7	12	-1	-6	4
65	32	45	4	4	9	-2	-7	4
65	35	45	1	1	2	-3	-11	7
66	38	45	5	5	9	-2	-7	5
66	39	45	4	4	7	-3	-11	8
67	31	45	5	5	12	-3	-9	5

φ_1	Φ	φ_2	H	K	L	U	V	W
60	30	40						
67	38	40	5	6	10	-2	-10	7
68	35	45	1	1	2	-1	-3	2
69	33	45	5	5	11	-3	-8	5
60	30	50						
60	36	56	3	2	5	-3	-8	5
62	31	56	3	2	6	-4	-9	5
62	36	53	4	3	7	-4	-11	7
63	32	53	4	3	8	-5	-12	7
63	33	50	6	5	12	-4	-12	7
63	36	56	3	2	5	-4	-9	6
64	31	56	3	2	6	-6	-12	7
64	33	51	5	4	10	-2	-5	3
64	33	59	5	3	9	-6	-11	7
64	36	53	4	3	7	-5	-12	6
64	36	58	8	5	12	-2	-4	3
65	30	59	5	3	10	-3	-5	3
65	36	56	3	2	5	-5	-10	7
65	39	51	5	4	8	-4	-11	8
67	32	53	4	3	6	-4	-8	5
67	33	56	6	4	11	-5	-9	6
67	36	56	3	2	5	-6	-11	6
68	36	56	3	2	5	-7	-12	9
68	38	50	6	5	10	-5	-12	9
68	39	51	5	4	8	-4	-9	7
69	39	56	6	4	9	-7	-12	10
60	30	60						
60	34	68	5	2	8	-2	-3	2
61	31	63	6	3	11	-3	-5	3
61	34	60	7	4	12	-4	-8	5
62	37	63	2	1	3	-4	-7	5
63	35	69	6	3	12	-3	-4	3
63	39	60	7	4	10	-2	-4	3
64	33	63	4	2	7	-2	-3	2
64	37	63	2	1	3	-5	-8	6
64	39	63	8	4	11	-3	-5	4
64	39	66	9	4	12	-4	-6	5
65	35	67	7	3	11	-3	-4	3
65	37	63	2	1	3	-6	-9	7
66	32	67	7	3	12	-9	-11	8
67	31	68	5	2	9	-10	-11	8
67	34	63	6	3	10	-3	-4	3
67	37	63	2	1	3	-7	-10	8
67	37	63	2	1	3	-8	-11	9
68	34	68	5	2	8	-10	-11	9

φ_1	Φ	φ_2	H	K	L	U	V	W
60	30	60						
68	37	63	2	1	3	-9	-12	10
60	30	70						
60	30	76	4	1	7	-11	-12	8
60	30	79	5	1	9	-3	-3	2
60	34	76	4	1	6	-7	-8	6
61	32	72	3	1	5	-7	-9	6
61	33	79	5	1	8	-4	-4	3
62	36	79	5	1	7	-7	-7	6
63	30	72	6	2	11	-8	-9	6
63	30	76	4	1	7	-7	-7	5
63	32	72	3	1	5	-6	-7	5
63	38	77	9	2	12	-12	-12	11
64	34	76	4	1	6	-6	-6	5
64	37	76	8	2	11	-11	-11	10
65	30	76	4	1	7	-10	-9	7
65	31	74	7	2	12	-4	-4	3
65	32	72	3	1	5	-11	-12	9
65	33	74	7	2	11	-11	-11	9
66	33	79	5	1	8	-11	-9	8
66	34	76	4	1	6	-11	-10	9
66	35	72	6	2	9	-11	-12	10
66	36	74	7	2	10	-10	-10	9
66	36	79	5	1	7	-6	-5	5
66	38	77	9	2	12	-10	-9	9
67	30	72	6	2	11	-11	-11	8
67	32	72	3	1	5	-5	-5	4
67	37	76	8	2	11	-9	-8	8
67	39	74	7	2	9	-1	-1	1
68	30	79	5	1	9	-11	-8	7
68	33	74	7	2	11	-10	-9	8
68	35	72	6	2	9	-9	-9	8
69	33	79	5	1	8	-7	-5	5
69	34	76	4	1	6	-5	-4	4
69	36	74	7	2	10	-8	-7	7
69	38	72	3	1	4	-1	-1	1
60	30	80						
60	37	81	6	1	8	-8	-8	7
61	39	84	9	1	11	-10	-9	9
62	33	82	7	1	11	-7	-6	5
62	39	83	8	1	10	-9	-8	8
63	37	84	9	1	12	-11	-9	9
63	38	82	7	1	9	-8	-7	7
64	36	83	8	1	11	-5	-4	4
64	37	81	6	1	8	-7	-6	6

φ_1	Φ	φ_2	H	K	L	U	V	W
60	30	80						
65	34	83	8	1	12	-11	-8	8
65	35	82	7	1	10	-9	-7	7
65	39	84	9	1	11	-9	-7	8
66	39	83	6	1	10	-8	-6	7
67	33	82	7	1	11	-10	-7	7
67	34	81	6	1	9	-4	-3	3
68	38	82	7	1	9	-7	-5	6
69	31	81	6	1	10	-3	-2	2
69	31	82	7	1	12	-11	-7	7
60	30	90						
60	30	90	7	0	12	-12	-8	7
60	35	90	7	0	10	-10	-7	7
60	36	90	5	0	7	-7	-5	5
60	36	90	8	0	11	-11	-8	8
61	34	90	2	0	3	-3	-2	2
61	39	90	4	0	5	-10	-7	8
61	39	90	9	0	11	-11	-8	9
62	32	90	5	0	8	-8	-5	5
62	32	90	7	0	11	-11	-7	7
62	37	90	3	0	4	-12	-8	9
62	38	90	7	0	9	-9	-6	7
63	30	90	7	0	12	-12	-7	7
63	31	90	3	0	5	-5	-3	3
63	36	90	8	0	11	-11	-7	8
63	37	90	3	0	4	-8	-5	6
64	30	90	4	0	7	-7	-4	4
64	34	90	2	0	3	-12	-7	8
64	35	90	7	0	10	-10	-6	7
64	39	90	9	0	11	-11	-7	9
65	32	90	7	0	11	-11	-6	7
65	34	90	2	0	3	-9	-5	6
65	36	90	5	0	7	-7	-4	5
65	37	90	3	0	4	-12	-7	9
65	39	90	4	0	5	-5	-3	4
66	36	90	8	0	11	-11	-6	8
66	38	90	7	0	9	-9	-5	7
67	30	90	7	0	12	-12	-6	7
67	31	90	3	0	5	-10	-5	6
67	32	90	5	0	8	-8	-4	5
67	34	90	2	0	3	-6	-3	4
67	39	90	9	0	11	-11	-6	9
68	35	90	7	0	10	-10	-5	7
68	37	90	3	0	4	-4	-2	3
69	32	90	7	0	11	-11	-5	7
69	39	90	4	0	5	-10	-5	8

φ_1	Φ	φ_2	H	K	L	U	V	W
60	40	0						
60	40	0	0	5	6	9	-12	10
60	42	0	0	8	9	7	-9	8
60	45	0	0	1	1	9	-11	11
60	48	0	0	9	8	7	-8	9
61	41	0	0	7	8	6	-8	7
61	43	0	0	11	12	9	-12	11
61	45	0	0	1	1	4	-5	5
61	45	0	0	1	1	7	-9	9
61	45	8	1	7	7	7	-12	11
61	45	9	1	6	6	6	-11	10
61	47	0	0	12	11	9	-11	12
61	49	0	0	6	7	6	-7	8
62	41	0	0	6	7	5	-7	6
62	42	0	0	10	11	8	-11	10
62	42	6	1	10	11	4	-7	6
62	45	0	0	1	1	3	-4	4
62	48	0	0	11	10	8	-10	11
62	49	0	0	7	6	5	-6	7
63	40	0	0	5	6	4	-6	5
63	42	0	0	9	10	7	-10	9
63	45	0	0	1	1	5	-7	7
63	45	0	0	1	1	8	-11	11
63	45	9	1	6	6	6	-12	11
63	48	0	0	10	9	7	-9	10
64	41	8	1	7	8	5	-11	9
64	42	0	0	8	9	6	-9	8
64	42	7	1	8	9	3	-6	5
64	43	0	0	11	12	8	-12	11
64	45	0	0	1	1	7	-10	10
64	47	0	0	12	11	8	-11	12
64	48	0	0	9	8	6	-8	9
64	48	9	2	12	11	1	-2	2
65	41	0	0	7	8	5	-8	7
65	41	9	1	6	7	2	-5	4
65	42	0	0	10	11	7	-11	10
65	45	0	0	1	1	2	-3	3
65	48	0	0	11	10	7	-10	11
65	49	0	0	8	7	5	-7	8
66	40	0	0	5	6	7	-12	10
66	42	0	0	9	10	6	-10	9
66	43	5	1	11	12	4	-8	7
66	45	0	0	1	1	5	-6	6
66	45	0	0	1	1	7	-11	11
66	48	0	0	10	9	6	-9	10
67	41	0	0	6	7	4	-7	6
67	42	0	0	8	9	5	-9	8
67	42	6	1	9	10	3	-7	6

φ_1	Φ	φ_2	H	K	L	U	V	W
60	40	0						
67	43	0	0	11	12	7	-12	11
67	45	0	0	1	1	3	-5	5
67	47	0	0	12	11	7	-11	12
67	48	0	0	9	8	5	-8	9
67	49	0	0	7	6	4	-6	7
68	40	6	1	10	12	2	-5	4
68	42	0	0	10	11	6	-11	10
68	45	0	0	1	1	4	-7	7
68	45	0	0	1	1	7	-12	12
68	48	0	0	11	10	6	-10	11
69	40	0	0	5	6	3	-6	5
69	41	0	0	7	8	4	-8	7
69	41	6	1	7	6	2	-6	5
69	45	0	0	1	1	5	-9	9
69	45	0	0	1	1	6	-11	11
69	49	0	0	8	7	4	-7	8
60	40	10						
60	40	14	1	4	5	3	-7	5
60	44	17	3	10	11	2	-5	4
61	43	10	2	11	12	3	-6	5
61	46	11	1	5	5	5	-10	9
61	46	14	1	4	4	4	-9	8
61	47	16	1	3	3	3	-8	7
62	40	14	1	4	5	4	-11	8
62	41	17	3	10	12	4	-12	9
62	42	14	2	8	9	2	-5	4
62	49	13	2	9	8	1	-2	2
63	40	13	2	9	11	3	-8	6
63	46	11	1	5	5	5	-11	10
63	46	14	1	4	4	4	-10	9
63	47	16	1	3	3	3	-9	8
63	48	10	2	11	10	1	-2	2
63	49	11	2	10	9	1	-2	2
64	41	18	3	9	11	1	-4	3
64	43	11	2	10	11	5	-12	10
64	43	18	3	9	10	3	-11	9
64	44	15	3	11	12	2	-6	5
64	47	18	1	3	3	3	-10	9
65	42	16	2	7	8	3	-10	8
65	46	11	1	5	5	5	-12	11
65	46	14	1	4	4	4	-11	10
65	47	18	1	3	3	3	-11	10
65	49	17	3	10	9	1	-3	3
66	40	11	1	5	6	3	-9	7
66	43	13	2	9	10	2	-6	5
66	46	14	1	4	4	4	-12	11

φ_1	Φ	φ_2	H	K	L	U	V	W
60	40	10						
66	47	18	1	3	3	3	-12	11
66	49	15	3	11	10	1	-3	3
67	41	17	3	10	12	2	-9	7
67	42	18	2	6	7	1	-5	4
67	48	14	3	12	11	1	-3	3
67	49	18	4	12	11	1	-4	4
68	40	14	1	4	5	1	-4	3
69	42	14	2	8	9	3	-12	10
69	43	11	2	10	11	2	-7	6
60	40	20						
60	44	21	3	8	9	3	-9	7
60	44	22	4	10	11	3	-10	8
60	48	27	1	2	2	2	-8	7
61	41	21	3	8	10	2	-7	5
61	42	27	2	4	5	2	-11	8
61	45	27	4	8	9	1	-5	4
62	42	22	2	5	6	1	-4	3
62	46	29	6	11	12	1	-6	5
62	48	27	1	2	2	2	-9	8
63	45	27	5	10	11	1	-6	5
63	48	27	1	2	2	2	-10	9
63	48	27	1	2	2	2	-11	10
64	42	27	2	4	5	1	-8	6
64	44	20	4	11	12	3	-12	10
64	44	23	3	7	8	1	-5	4
64	48	27	1	2	2	2	-12	11
65	45	24	4	9	10	1	-6	5
66	41	21	3	8	10	2	-12	9
66	42	24	4	9	11	1	-9	7
66	44	27	3	6	7	1	-11	9
66	45	24	5	11	12	1	-7	6
66	49	20	4	11	10	1	-4	4
67	41	29	5	9	12	0	-4	3
67	44	21	3	8	9	1	-6	5
68	40	23	3	7	9	1	-12	9
68	43	29	5	9	11	0	-11	9
68	44	22	4	10	11	1	-7	6
69	40	27	3	6	8	0	-4	3
69	42	22	2	5	6	1	-10	8
69	46	29	5	9	10	0	-10	9
69	46	29	6	11	12	0	-12	11
60	40	30						
60	40	31	3	5	7	1	-9	6
60	40	37	3	4	6	0	-3	2

φ_1	Φ	φ_2	H	K	L	U	V	W
60	40	30						
60	42	30	4	7	9	1	-7	5
60	46	34	4	6	7	1	-10	8
60	47	36	5	7	8	1	-11	9
61	42	37	6	8	11	0	-11	8
61	44	31	3	5	6	1	-9	7
61	44	38	7	9	12	0	-4	3
61	47	39	4	5	6	0	-6	5
62	41	36	5	7	10	0	-10	7
62	45	37	3	4	5	0	-5	4
62	46	38	7	9	11	0	-11	9
62	49	39	8	10	11	0	-11	10
63	44	36	5	7	9	0	-9	7
63	48	37	6	8	9	0	-9	8
63	49	38	7	9	10	0	-10	9
64	42	34	2	3	4	0	-4	3
64	45	30	4	7	8	1	-12	10
64	45	35	7	10	12	0	-6	5
64	47	36	5	7	8	0	-8	7
64	49	36	8	11	12	0	-12	11
65	40	31	3	5	7	0	-7	5
65	41	32	5	8	11	0	-11	8
65	42	39	4	5	7	-1	-9	7
65	45	34	6	9	11	0	-11	9
65	46	34	4	6	7	0	-7	6
65	48	35	7	10	11	0	-11	10
66	43	32	5	8	10	0	-5	4
66	47	34	6	9	10	0	-10	9
66	47	39	4	5	6	-1	-10	9
67	42	30	4	7	9	0	-9	7
67	44	31	3	5	6	0	-6	5
67	46	32	5	8	9	0	-9	8
67	46	38	7	9	11	-1	-9	8
67	47	32	7	11	12	0	-12	11
68	40	37	3	4	6	-2	-12	9
68	42	34	2	3	4	-1	-10	8
68	42	39	4	5	7	-2	-11	9
68	45	30	4	7	8	0	-8	7
68	45	37	3	4	5	-1	-8	7
68	47	31	6	10	11	0	-11	10
68	49	39	6	10	11	-1	-8	6
69	40	31	3	5	7	-1	-12	9
69	44	36	5	7	9	-1	-7	6
69	46	34	4	6	7	-1	-11	10
69	49	30	4	7	7	0	-1	1
69	49	30	7	12	12	0	-1	1
69	49	31	3	5	5	0	-1	1
69	49	38	7	9	10	-1	-7	7

φ_1	Φ	φ_2	H	K	L	U	V	W
60	40	40						
60	43	45	2	2	3	-1	-6	6
60	45	45	7	7	10	-1	-9	7
60	46	45	8	8	11	-1	-10	8
60	47	45	3	3	4	-1	-11	9
61	41	45	5	5	6	-1	-7	5
61	47	49	6	7	10	-1	-6	5
61	48	40	5	6	7	0	-7	6
61	49	41	6	7	8	0	-8	7
62	43	41	6	7	10	-1	-12	9
62	48	45	7	7	9	-1	-8	7
62	49	45	4	4	5	-1	-9	6
62	49	45	9	9	11	-1	-10	9
62	49	49	7	6	8	-2	-11	10
63	40	45	7	7	12	-2	-10	7
63	42	49	8	7	12	-1	-4	3
63	45	45	5	5	7	-1	-6	5
63	46	41	6	7	9	-1	-12	10
63	47	45	3	3	4	-1	-7	6
64	40	41	6	7	11	-1	-7	5
64	43	45	2	2	3	-1	-5	4
65	40	45	3	3	5	-1	-4	3
65	42	45	7	7	11	-2	-9	7
65	47	45	3	3	4	-2	-10	9
65	48	40	5	6	7	-1	-12	11
66	43	45	2	2	3	-3	-12	10
66	43	49	7	6	10	-2	-6	5
66	44	41	7	8	11	-1	-6	5
66	46	45	8	8	11	-2	-9	8
66	46	49	7	6	9	-3	-10	9
67	45	45	7	7	10	-2	-8	7
67	47	41	7	8	10	-2	-12	11
67	49	45	9	9	11	-2	-9	9
67	49	49	7	6	8	-2	-7	7
68	40	49	7	6	11	-2	-5	4
68	43	45	2	2	3	-2	-7	6
68	45	45	5	5	7	-3	-11	10
68	45	48	9	8	12	-4	-12	11
68	48	48	10	9	12	-3	-10	10
68	49	41	6	7	8	-1	-6	6
68	49	45	4	4	5	-1	-4	4
69	40	45	7	7	12	-3	-9	7
69	41	45	5	5	8	-2	-6	5
69	48	42	9	10	12	-2	-9	9
69	48	48	9	8	11	-1	-3	3

φ_1	Φ	φ_2	H	K	L	U	V	W
60	40	50						
60	40	53	4	3	6	-3	-10	7
60	42	56	3	2	4	-4	-12	9
60	47	56	9	6	10	-2	-7	6
61	42	51	5	4	7	-1	-4	3
61	45	53	4	3	5	-3	-11	9
61	47	56	11	7	12	-3	-9	8
63	41	58	6	5	11	-4	-9	7
63	42	56	3	2	4	-2	-5	4
63	43	58	8	5	10	-5	-12	10
63	44	59	5	3	6	-3	-7	6
63	46	53	8	6	9	-3	-11	10
64	40	53	4	3	6	-3	-6	6
64	44	52	9	7	12	-2	-6	5
64	47	54	7	5	8	-4	-12	11
64	49	59	5	3	5	-2	-5	5
65	41	50	6	5	9	-3	-9	7
65	44	59	5	3	6	-6	-12	11
66	40	59	5	3	7	-6	-11	9
66	41	54	7	5	10	-5	-11	9
66	43	58	8	5	10	-5	-10	9
66	47	58	11	7	12	-5	-11	11
66	47	59	10	6	11	-1	-2	2
66	48	55	10	7	11	-2	-5	5
66	49	51	10	8	11	-3	-10	10
66	49	52	9	7	10	-1	-3	3
66	49	54	11	8	12	-4	-11	11
67	42	53	8	6	11	-3	-7	6
67	42	56	3	2	4	-4	-8	7
67	44	54	7	5	9	-5	-11	10
67	46	58	8	5	9	-1	-2	2
67	47	56	9	6	10	-4	-9	9
67	48	53	8	6	9	-3	-6	8
67	49	59	5	3	5	-5	-10	11
68	41	50	6	5	9	-5	-12	10
68	46	56	6	4	7	-1	-2	2
68	47	54	7	5	8	-3	-7	7
68	48	50	6	5	7	-1	-3	3
69	40	53	4	3	6	-3	-6	5
69	40	59	5	3	7	-5	-6	7
69	42	51	5	4	7	-4	-9	8
69	42	56	3	2	4	-6	-11	10
69	44	59	5	3	6	-3	-5	5
69	45	55	10	7	12	-1	-2	2
69	47	51	5	4	6	-2	-5	5

φ_1	Φ	φ_2	H	K	L	U	V	W
60	40	60						
60	40	63	6	3	8	-2	-4	3
60	44	63	6	3	7	-5	-11	9
60	45	66	11	5	12	-4	-8	7
60	47	67	12	5	12	-6	-12	11
60	48	63	2	1	2	-5	-12	11
60	48	66	11	5	11	-5	-11	10
60	49	60	12	7	12	-4	-12	11
60	49	61	11	6	11	-4	-11	10
61	40	67	7	3	9	-3	-5	4
61	41	69	8	3	10	-4	-6	5
61	42	63	4	2	5	-5	-10	8
61	43	63	10	5	12	-3	-6	5
62	44	67	7	3	8	-6	-10	9
62	47	69	8	3	8	-5	-8	8
63	42	66	9	4	11	-5	-8	7
63	42	68	5	2	6	-6	-9	8
63	46	61	9	5	10	-5	-11	10
63	47	67	7	3	7	-4	-7	7
63	47	67	12	5	12	-7	-12	12
63	47	68	5	2	5	-3	-5	5
63	48	63	2	1	2	-1	-2	2
63	48	66	9	4	9	-5	-9	9
63	46	66	11	5	11	-6	-11	11
64	42	60	7	4	9	-3	-6	5
64	42	63	4	2	5	-4	-7	6
64	49	60	7	4	7	-3	-7	7
64	49	60	12	7	12	-5	-12	12
64	49	61	9	5	9	-4	-9	9
64	49	61	11	6	11	-5	-11	11
65	44	68	10	4	11	-7	-10	10
65	44	69	8	3	9	-3	-4	4
65	45	66	11	5	12	-7	-11	11
66	40	67	7	3	9	-9	-12	11
66	42	63	4	2	5	-7	-11	10
66	44	67	7	3	8	-5	-7	7
66	45	63	10	5	11	-3	-5	5
66	45	66	9	4	10	-2	-3	3
66	46	61	11	6	12	-6	-11	11
66	47	68	5	2	5	-7	-10	11
66	47	69	8	3	8	-6	-8	9
66	48	66	11	5	11	-7	-11	12
67	40	63	6	3	8	-7	-10	9
67	42	66	5	2	6	-4	-5	5
67	45	60	7	4	8	-4	-7	7
67	45	63	8	4	9	-5	-8	8
67	46	61	9	5	10	-5	-9	9
67	48	63	2	1	2	-6	-10	11

φ_1	Φ	φ_2	H	K	L	U	V	W
60	40	60						
67	48	66	9	4	9	-6	-9	10
67	49	61	11	6	11	-6	-11	12
68	41	69	8	3	10	-7	-8	8
68	42	66	9	4	11	-7	-9	9
68	43	63	10	5	12	-7	-10	10
68	44	63	6	3	7	-2	-3	3
68	47	67	7	3	7	-5	-7	8
68	48	63	2	1	2	-5	-8	9
68	49	61	9	5	9	-5	-9	10
69	40	67	7	3	9	-6	-7	7
69	41	61	9	5	12	-8	-12	11
69	43	61	9	5	11	-2	-3	3
69	44	67	7	3	8	-9	-11	12
69	47	68	5	2	5	-4	-5	6
69	48	63	2	1	2	-4	-6	7
69	49	60	7	4	7	-4	-7	8
60	40	70						
60	46	77	9	2	9	-7	-9	9
60	46	79	5	1	5	-4	-5	5
60	49	70	11	4	10	-6	-11	11
60	49	72	12	4	11	-7	-12	12
61	41	72	9	3	11	-5	-7	6
61	41	73	10	3	12	-6	-8	7
61	43	79	10	2	11	-9	-10	10
61	46	73	10	3	10	-7	-10	10
61	46	74	7	2	7	-5	-7	7
61	46	75	11	3	11	-8	-11	11
61	46	76	4	1	4	-3	-4	4
62	42	72	6	2	7	-8	-11	10
62	43	77	9	2	10	-8	-9	9
62	46	79	5	1	5	-9	-10	11
62	47	70	11	4	11	-7	-11	11
62	47	72	3	1	3	-2	-3	3
63	40	79	5	1	6	-1	-1	1
63	42	76	8	2	9	-7	-8	8
63	44	73	10	3	11	-4	-5	5
63	44	75	11	3	12	-9	-11	11
63	46	77	9	2	9	-8	-9	10
63	49	75	11	3	10	-6	-8	9
64	40	77	9	2	11	-1	-1	1
64	42	74	7	2	8	-6	-7	7
64	43	72	9	3	10	-7	-9	9
64	43	79	10	2	11	-11	-11	12
64	44	70	11	4	12	-8	-11	11
64	46	73	10	3	10	-6	-10	11
64	46	75	11	3	11	-9	-11	12

φ_1	Φ	φ_2	H	K	L	U	V	W
60	40	70						
64	46	76	4	1	4	-7	-8	9
64	49	72	12	4	11	-5	-7	8
65	40	76	4	1	5	-1	-1	1
65	41	73	10	3	12	-9	-10	10
65	42	72	6	2	7	-5	-6	6
65	43	77	9	2	10	-10	-10	11
65	46	74	7	2	7	-6	-7	8
65	46	79	5	1	5	-5	-5	6
65	47	70	11	4	11	-8	-11	12
65	47	72	3	1	3	-7	-9	10
66	41	72	9	3	11	-8	-9	9
66	42	76	8	2	9	-9	-9	10
66	46	77	9	2	9	-9	-9	11
66	49	79	10	2	9	-3	-3	4
67	44	75	11	3	12	-6	-6	7
67	46	73	10	3	10	-9	-10	12
67	46	76	4	1	4	-4	-4	5
67	47	72	3	1	3	-5	-6	7
67	49	73	10	3	9	-6	-7	9
67	49	77	9	2	8	-8	-8	11
68	42	74	7	2	8	-8	-8	9
68	42	76	8	2	9	-11	-10	12
68	46	77	9	2	9	-10	-9	12
68	47	72	3	1	3	-6	-9	11
69	41	72	9	3	11	-11	-11	12
69	46	74	7	2	7	-7	-7	9
69	46	79	5	1	5	-6	-5	7
69	49	72	12	4	11	-8	-9	12
69	49	75	11	3	10	-5	-5	7
60	40	80						
60	40	84	10	1	12	-11	-10	10
60	42	83	8	1	9	-1	-1	1
60	45	84	10	1	10	-10	-10	11
60	45	85	11	1	11	-11	-11	12
61	41	82	7	1	8	-1	-1	1
61	43	80	11	2	12	-10	-11	11
61	45	83	8	1	6	-6	-8	9
61	45	84	9	1	9	-9	-9	10
61	48	84	10	1	9	-9	-9	11
61	48	85	11	1	10	-5	-5	6
62	41	81	6	1	7	-1	-1	1
62	45	80	11	2	11	-10	-11	12
62	45	82	7	1	7	-7	-7	8
62	45	84	10	1	10	-11	-10	12
62	48	80	11	2	10	-6	-7	8
62	49	84	9	1	8	-4	-4	5

φ_1	Φ	φ_2	H	K	L	U	V	W
60	40	80						
63	45	81	6	1	6	-6	-6	7
63	45	84	9	1	9	-10	-9	11
63	49	83	8	1	7	-7	-7	9
64	40	84	10	1	12	-10	-8	9
64	45	83	8	1	8	-9	-8	10
65	45	82	7	1	7	-6	-7	9
65	45	84	9	1	9	-11	-9	12
65	48	85	12	1	11	-6	-5	7
67	45	81	6	1	6	-7	-6	8
67	45	83	8	1	8	-10	-6	11
67	48	81	12	2	11	-8	-7	10
67	48	84	10	1	9	-5	-4	6
68	40	84	10	1	12	-9	-6	8
68	45	82	7	1	7	-9	-7	10
68	49	84	9	1	8	-9	-7	11
69	45	83	8	1	8	-11	-8	12
69	49	83	8	1	7	-4	-3	5
60	40	90						
60	40	90	5	0	6	-12	-9	10
60	42	90	8	0	9	-9	-7	8
60	45	90	1	0	1	-11	-9	11
60	48	90	9	0	8	-8	-7	9
61	41	90	7	0	6	-6	-6	7
61	43	90	11	0	12	-12	-9	11
61	45	90	1	0	1	-5	-4	5
61	45	90	1	0	1	-9	-7	9
61	47	90	12	0	11	-11	-9	12
61	49	90	8	0	7	-7	-6	8
62	41	90	6	0	7	-7	-5	6
62	42	90	10	0	11	-11	-6	10
62	45	90	1	0	1	-4	-3	4
62	48	90	11	0	10	-10	-8	11
62	49	90	7	0	6	-6	-5	7
63	40	90	5	0	6	-6	-4	5
63	42	90	9	0	10	-10	-7	9
63	45	90	1	0	1	-7	-5	7
63	45	90	1	0	1	-11	-8	11
63	48	90	10	0	9	-9	-7	10
64	42	90	8	0	9	-9	-6	8
64	43	90	11	0	12	-12	-8	11
64	45	90	1	0	1	-10	-7	10
64	47	90	12	0	11	-11	-8	12
64	48	90	9	0	8	-8	-6	9
65	41	90	7	0	8	-8	-5	7
65	42	90	10	0	11	-11	-7	10
65	45	90	1	0	1	-3	-2	3

TAB. VI FROM EULER ANGLES INTO HKL-UVW

φ1	Φ	φ2	H	K	L	U	V	W
60	40	90						
65	48	90	11	0	10	-10	-7	11
65	49	90	8	0	7	-7	-5	8
66	40	90	5	0	6	-12	-7	10
66	42	90	9	0	10	-10	-6	9
66	45	90	1	0	1	-8	-5	8
66	45	90	1	0	1	-11	-7	11
66	48	90	10	0	9	-9	-6	10
67	41	90	6	0	7	-7	-4	6
67	42	90	8	0	9	-9	-5	8
67	43	90	11	0	12	-12	-7	11
67	45	90	1	0	1	-5	-3	5
67	47	90	12	0	11	-11	-7	12
67	48	90	9	0	8	-8	-5	9
67	49	90	7	0	6	-6	-4	7
68	42	90	10	0	11	-11	-6	10
68	45	90	1	0	1	-7	-4	7
68	45	90	1	0	1	-12	-7	12
68	48	90	11	0	10	-10	-6	11
69	40	90	5	0	6	-6	-3	5
69	41	90	7	0	8	-8	-4	7
69	45	90	1	0	1	-9	-5	9
69	45	90	1	0	1	-11	-6	11
69	49	90	8	0	7	-7	-4	8
60	50	0						
60	50	0	0	6	5	9	-10	12
60	50	5	1	12	10	6	-8	9
60	54	0	0	7	5	5	-5	7
60	54	0	0	11	8	8	-8	11
60	55	0	0	10	7	7	-7	10
60	58	7	1	8	5	6	-7	10
61	51	0	0	5	4	7	-8	10
61	51	0	0	11	9	8	-9	11
61	51	5	1	11	9	5	-7	8
61	56	0	0	3	2	2	-2	3
62	52	0	0	9	7	6	-7	9
62	53	0	0	4	3	8	-9	12
62	53	5	1	12	9	3	-4	5
62	56	5	1	12	8	4	-5	7
62	58	0	0	8	5	5	-5	8
62	58	0	0	11	7	7	-7	11
63	50	0	0	6	5	4	-5	6
63	51	6	1	10	8	4	-6	7
63	53	0	0	4	3	5	-6	8
63	53	7	1	8	6	6	-9	11
63	54	0	0	11	8	7	-8	11
63	54	5	1	11	8	5	-7	9

φ1	Φ	φ2	H	K	L	U	V	W
60	50	0						
63	54	9	2	12	9	3	-5	6
63	56	6	1	9	6	6	-8	11
63	57	9	1	6	4	2	-3	4
63	59	0	0	5	3	3	-3	5
64	51	0	0	11	9	7	-9	11
64	51	9	1	6	5	6	-11	12
64	55	0	0	10	7	6	-7	10
64	56	0	0	3	2	7	-8	12
65	51	0	0	5	4	3	-4	5
65	51	9	1	6	5	5	-10	11
65	52	6	1	9	7	3	-5	6
65	53	0	0	4	3	7	-9	12
65	54	0	0	7	5	4	-5	7
65	55	6	1	10	7	2	-3	4
65	56	0	0	3	2	5	-6	9
65	58	0	0	11	7	6	-7	11
66	50	0	0	6	5	7	-10	12
66	50	5	1	12	10	4	-7	6
66	52	0	0	9	7	5	-7	9
66	54	0	0	11	8	6	-8	11
66	55	6	1	7	5	4	-7	9
67	51	0	0	11	9	6	-9	11
67	51	9	1	6	5	4	-9	10
67	56	0	0	3	2	3	-4	6
67	56	6	1	9	6	3	-5	7
67	58	0	0	8	5	4	-5	8
67	59	0	0	5	3	5	-6	10
68	50	6	1	7	6	5	-11	12
68	53	0	0	4	3	2	-3	4
68	53	7	1	6	6	2	-4	5
68	55	0	0	10	7	5	-7	10
68	57	9	1	6	4	4	-8	11
69	50	0	0	6	5	3	-5	6
69	51	0	0	5	4	5	-8	10
69	51	5	1	11	9	3	-6	7
69	56	5	1	12	8	4	-7	10
69	58	0	0	11	7	5	-7	11
60	50	10						
60	51	16	2	7	6	1	-2	2
60	54	10	2	11	8	7	-10	12
60	58	18	1	3	2	5	-9	11
60	59	14	2	8	5	2	-3	4
61	50	14	2	6	7	1	-2	2
61	52	11	1	5	4	6	-10	11
61	54	14	1	4	3	4	-7	8
61	54	16	3	9	7	5	-11	12

φ1	Φ	φ2	H	K	L	U	V	W
60	50	10						
62	52	11	1	5	4	5	-9	10
62	56	11	2	10	7	5	-8	10
63	54	14	1	4	3	3	-6	7
63	55	18	4	12	9	3	-7	8
63	56	16	2	7	5	5	-10	12
63	57	14	3	12	8	4	-7	9
63	58	15	3	11	7	5	-9	12
64	50	18	3	9	8	1	-3	3
64	52	11	1	5	4	4	-6	9
64	53	17	3	10	8	4	-10	11
64	54	18	3	9	7	3	-8	9
65	56	16	2	7	5	4	-9	11
65	58	10	2	11	7	3	-5	7
65	58	18	1	3	2	3	-7	9
66	54	10	2	11	8	2	-4	5
66	57	13	2	9	6	3	-6	8
67	51	14	3	12	10	4	-11	12
67	52	11	1	5	4	3	-7	8
67	54	14	1	4	3	2	-5	6
67	56	16	2	7	5	3	-8	10
67	57	14	3	12	8	4	-9	12
68	52	15	3	11	9	3	-9	10
68	52	18	2	6	5	3	-11	12
68	55	18	4	12	9	3	-10	12
68	58	15	3	11	7	2	-5	7
68	58	18	1	3	2	1	-3	4
68	59	14	2	8	5	3	-7	10
69	53	17	3	10	8	2	-7	8
69	54	14	1	4	3	3	-9	11
60	50	20						
60	56	27	2	4	3	4	-11	12
60	57	23	3	7	5	3	-7	8
60	59	27	3	6	4	2	-5	6
61	52	23	3	7	6	1	-3	3
61	55	21	3	8	6	4	-9	10
61	56	29	5	9	7	3	-11	12
61	58	23	5	12	8	4	-9	11
62	52	27	4	8	7	1	-4	4
62	52	29	5	9	8	1	-5	5
62	55	23	5	12	9	3	-8	9
63	51	21	3	8	7	1	-3	3
63	56	24	5	11	8	2	-6	7
63	56	27	2	4	3	2	-7	8
63	57	22	4	10	7	2	-5	6
63	59	24	4	9	6	3	-6	10
64	51	24	4	9	8	1	-4	4

φ1	Φ	φ2	H	K	L	U	V	W
60	50	20						
64	51	27	5	10	9	1	-5	5
64	51	29	6	11	10	1	-6	6
65	50	22	4	10	9	1	-4	4
65	50	24	5	11	10	1	-5	5
65	51	27	6	12	11	1	-6	6
65	53	22	2	5	4	3	-10	11
65	54	27	5	10	8	2	-9	10
65	55	21	3	8	6	2	-6	7
65	59	27	3	6	4	2	-7	9
66	53	27	3	6	5	2	-11	12
66	56	29	5	9	7	1	-6	7
66	58	27	5	10	7	1	-4	5
67	50	23	5	12	11	1	-5	5
67	53	22	2	5	4	2	-6	9
67	56	27	2	4	3	1	-5	6
67	57	23	3	7	5	1	-4	5
67	59	27	3	6	4	2	-9	12
68	52	23	5	12	10	2	-10	11
68	53	27	3	6	5	1	-8	9
69	52	27	4	8	7	1	-11	12
69	53	24	5	11	9	1	-7	8
69	55	21	3	8	6	2	-9	11
69	56	20	4	11	6	1	-4	5
60	50	30						
60	53	30	4	7	6	1	-4	4
60	55	39	9	11	10	1	-9	9
60	57	30	7	12	9	3	-10	11
61	53	32	5	8	7	1	-5	5
61	54	34	6	9	8	1	-6	6
61	54	35	7	10	9	1	-7	7
61	54	36	8	11	10	1	-8	8
61	54	37	9	12	11	1	-9	9
61	57	35	7	10	8	2	-11	12
61	58	32	5	8	6	2	-8	9
61	58	39	4	5	4	1	-8	9
61	58	39	9	11	9	1	-9	10
62	50	39	9	11	12	0	-12	11
62	53	32	7	11	10	1	-7	7
62	53	34	8	12	11	1	-8	8
62	58	36	7	9	7	1	-7	8
62	59	37	3	4	3	1	-6	7
63	52	31	6	10	9	1	-6	6
63	52	39	9	11	11	0	-1	1
63	56	31	3	5	4	2	-10	11
63	58	32	7	11	8	1	-5	6
64	52	30	7	12	11	1	-7	7

φ_1	Φ	φ_2	H	K	L	U	V	W
60	50	30						
64	52	38	7	9	9	0	-1	1
64	52	39	4	5	5	0	-1	1
64	55	34	4	6	5	1	-9	10
64	55	36	5	7	6	1	-11	12
64	58	30	4	7	5	2	-9	11
64	59	37	3	4	3	1	-9	11
65	51	36	8	11	11	0	-1	1
65	51	37	3	4	4	0	-1	1
65	55	32	7	11	9	1	-8	9
65	55	39	8	10	9	0	-9	10
65	55	39	9	11	10	0	-10	11
65	56	31	3	5	4	1	-7	8
66	51	35	7	10	10	0	-1	1
66	51	36	5	7	7	0	-1	1
66	54	37	9	12	11	0	-11	12
66	55	38	7	9	8	0	-8	9
66	57	34	6	9	7	1	-10	12
66	58	39	9	11	9	0	-9	11
67	50	34	2	3	3	0	-1	1
67	50	39	9	11	12	-1	-9	9
67	53	30	4	7	6	1	-10	11
67	54	35	7	10	9	0	-9	10
67	54	36	8	11	10	0	-10	11
67	55	37	6	8	7	0	-7	8
67	56	37	9	12	10	0	-5	6
67	58	39	4	5	4	0	-4	5
68	50	32	5	8	8	0	-1	1
68	50	32	7	11	11	0	-1	1
68	53	34	8	12	11	0	-11	12
68	54	34	6	9	8	0	-8	9
68	55	36	5	7	6	0	-6	7
68	57	36	8	11	9	0	-9	11
68	58	30	4	7	5	1	-7	9
68	58	36	7	9	7	0	-7	9
69	53	32	7	11	10	0	-10	11
69	55	34	4	6	5	0	-5	6
69	57	35	7	10	8	0	-4	5
69	59	31	6	10	7	1	-9	12
69	59	37	3	4	3	0	-3	4
60	50	40						
60	50	42	8	9	10	0	-10	9
60	51	42	9	10	11	0	-11	10
60	51	42	10	11	12	0	-12	11
60	55	40	10	12	11	1	-10	10
60	55	45	1	1	1	0	-1	1
60	58	47	12	11	10	0	-10	11

φ_1	Φ	φ_2	H	K	L	U	V	W
60	50	40						
60	59	48	10	9	8	0	-8	9
60	59	48	11	10	9	0	-9	10
61	50	41	7	8	9	0	-9	8
61	54	43	11	12	12	0	-1	1
61	54	47	12	11	12	-1	-12	12
61	54	48	11	10	11	-1	-11	11
61	57	40	5	6	5	1	-10	11
61	57	45	12	12	11	0	-11	12
62	50	45	5	5	6	-1	-11	10
62	53	41	7	8	8	0	-1	1
62	53	42	8	9	9	0	-1	1
62	53	42	9	10	10	0	-1	1
62	53	48	9	8	9	-1	-9	9
62	53	48	10	9	10	-1	-10	10
62	53	49	7	6	7	-1	-7	7
62	53	49	8	7	8	-1	-8	8
62	54	42	10	11	11	0	-1	1
62	57	45	11	11	10	0	-10	11
62	58	45	8	8	7	0	-7	8
62	58	45	9	9	8	0	-8	9
62	58	45	10	10	9	0	-9	10
63	52	40	5	6	6	0	-1	1
63	52	45	10	10	11	-1	-10	10
63	52	45	11	11	12	-1	-11	11
63	53	41	6	7	7	0	-1	1
63	56	42	9	10	9	0	-9	10
63	56	42	10	11	10	0	-10	11
63	56	43	11	12	11	0	-11	12
63	59	45	6	6	5	0	-5	6
63	59	45	7	7	6	0	-6	7
64	51	45	8	8	9	-1	-8	8
64	51	48	11	10	12	-2	-11	11
64	52	45	9	9	10	-1	-9	9
64	55	40	10	12	11	0	-11	12
64	55	45	1	1	1	-1	-11	12
64	56	42	8	9	8	0	-8	9
64	57	41	7	8	7	0	-7	8
64	58	43	11	12	10	0	-5	6
65	50	48	9	8	10	-2	-9	9
65	51	42	10	11	12	-1	-10	10
65	51	45	7	7	8	-1	-7	7
65	51	48	10	9	11	-1	-5	5
65	55	45	1	1	1	-1	-10	11
65	55	45	1	1	1	-1	-9	10
65	57	41	6	7	6	0	-6	7
65	59	42	9	10	8	0	-4	5
65	59	42	10	11	9	0	-9	11
66	50	42	8	9	10	-1	-6	6

φ_1	Φ	φ_2	H	K	L	U	V	W
60	50	40						
66	50	45	6	6	7	-1	-6	6
66	50	49	8	7	9	-1	-4	4
66	51	42	9	10	11	-1	-9	9
66	53	48	10	9	10	-2	-10	11
66	54	48	11	10	11	-2	-11	12
66	55	45	1	1	1	-1	-8	9
66	57	40	5	6	5	0	-5	6
66	57	45	12	12	11	-1	-10	12
66	58	47	12	11	10	-1	-8	10
67	50	41	7	8	9	-1	-7	7
67	50	45	5	5	6	-1	-5	5
67	53	48	9	8	9	-2	-9	10
67	53	49	8	7	8	-2	-8	9
67	55	45	1	1	1	-1	-7	8
67	57	45	11	11	10	-1	-9	11
67	58	45	10	10	9	-1	-8	10
67	59	48	11	10	9	-1	-7	9
68	52	45	11	11	12	-2	-10	11
68	53	49	7	6	7	-2	-7	8
68	55	45	1	1	1	-1	-6	7
68	56	42	9	10	9	-1	-9	11
68	56	42	10	11	10	-1	-10	12
68	58	45	9	9	8	-1	-7	9
68	59	45	6	6	5	-1	-9	12
68	59	48	10	9	8	-1	-6	8
69	52	45	10	10	11	-2	-9	10
69	55	45	1	1	1	-1	-5	6
69	56	42	8	9	8	-1	-8	10
69	58	45	8	8	7	-1	-6	8
60	50	50						
60	50	56	3	2	3	-3	-12	11
60	50	58	11	7	11	-3	-11	10
60	53	58	8	5	7	-1	-4	4
60	54	55	10	7	9	-1	-5	5
60	54	56	9	6	8	-2	-9	9
60	55	50	12	10	11	-1	-12	12
60	55	51	10	8	9	-1	-10	10
60	55	51	11	9	10	-1	-11	11
60	58	58	8	5	6	-2	-10	11
61	52	59	10	6	9	-3	-10	10
61	53	56	12	8	11	-1	-4	4
61	53	58	11	7	10	-3	-11	11
61	54	53	12	9	11	-1	-6	6
61	54	54	11	8	10	-2	-11	11
61	57	56	9	6	7	-2	-11	12
62	52	50	6	5	6	-1	-6	6

φ_1	Φ	φ_2	H	K	L	U	V	W
60	50	50						
62	52	51	11	9	11	-2	-11	11
62	56	59	5	3	4	-3	-11	12
63	51	53	4	3	4	-1	-4	4
63	51	54	7	5	7	-2	-7	7
63	51	54	11	8	11	-3	-11	11
63	51	55	10	7	10	-3	-10	10
63	52	51	5	4	5	-1	-5	5
63	52	52	9	7	9	-2	-9	9
63	55	51	10	8	9	-2	-11	12
64	50	56	3	2	3	-1	-3	3
64	50	58	8	5	8	-3	-8	8
64	50	58	11	7	11	-4	-11	11
64	55	52	9	7	8	-2	-10	11
64	55	53	8	6	7	-2	-9	10
64	55	54	7	5	6	-2	-8	9
64	55	56	6	4	5	-2	-7	8
64	56	59	5	3	4	-2	-6	7
65	50	51	11	9	12	-3	-11	11
66	52	51	11	9	11	-3	-11	12
66	52	59	10	6	9	-3	-7	8
66	59	59	10	6	7	-1	-3	4
67	50	58	11	7	11	-5	-11	12
67	51	54	11	8	11	-4	-11	12
67	51	55	10	7	10	-4	-10	11
67	52	51	5	4	5	-3	-10	11
67	54	54	11	8	10	-2	-6	7
67	55	53	8	6	7	-3	-10	12
67	55	54	7	5	6	-3	-9	11
67	58	58	11	7	8	-1	-3	4
68	50	56	3	2	3	-4	-9	10
68	50	58	8	5	6	-4	-8	9
68	51	53	4	3	4	-3	-8	9
68	52	52	9	7	9	-3	-9	10
68	53	58	8	5	7	-4	-9	11
68	54	56	9	6	8	-2	-5	6
68	55	56	6	4	5	-3	-8	10
68	56	59	5	3	4	-3	-7	9
68	58	56	12	8	9	-1	-3	4
69	51	54	7	5	7	-3	-7	8
69	52	50	6	5	6	-2	-6	7
60	50	60						
60	50	68	10	4	9	-1	-2	2
60	51	66	9	4	8	-4	-9	9
60	51	69	8	3	7	-1	-2	2
60	52	63	8	4	7	-3	-8	8
60	53	60	7	4	6	-2	-7	7

φ_1	Φ	φ_2	H	K	L	U	V	W
60	50	60						
60	57	67	7	3	5	-3	-8	9
60	58	60	7	4	5	-2	-9	10
60	59	66	9	4	6	-2	-6	7
61	50	66	11	5	10	-5	-11	11
61	50	67	12	5	11	-1	-2	2
61	51	61	11	6	10	-4	-11	11
61	51	63	10	5	9	-2	-5	5
61	51	63	12	6	11	-5	-12	12
61	52	60	12	7	11	-1	-3	3
61	52	61	9	5	8	-1	-3	3
61	56	61	9	5	7	-3	-10	11
61	56	63	4	2	3	-3	-9	10
61	57	61	11	6	8	-2	-7	8
61	59	63	6	3	4	-3	-10	12
62	52	67	12	5	10	-5	-10	11
62	53	68	5	2	4	-4	-8	9
62	55	69	8	3	6	-3	-6	7
63	52	63	8	4	7	-5	-11	12
63	53	60	7	4	6	-4	-11	12
63	54	61	11	6	9	-3	-8	9
63	57	67	7	3	5	-4	-9	11
64	53	66	11	5	9	-3	-6	7
64	56	63	4	2	3	-2	-5	6
64	57	60	12	7	9	-3	-9	11
64	58	63	10	5	7	-3	-8	10
65	50	66	11	5	10	-5	-9	10
65	59	66	9	4	6	-4	-9	12
66	52	67	7	3	6	-3	-5	6
66	53	63	6	3	5	-5	-10	12
66	54	63	10	5	8	-2	-4	5
66	58	67	12	5	8	-4	-8	11
67	52	67	12	5	10	-5	-8	10
67	56	63	4	2	3	-3	-6	8
68	51	63	12	6	11	-3	-5	6
68	53	60	7	4	6	-2	-4	5
68	53	63	6	3	5	-4	-7	9
68	53	68	5	2	4	-2	-3	4
69	51	63	10	5	9	-5	-8	10
69	55	66	9	4	7	-5	-8	11
69	59	63	6	3	4	-1	-2	3
60	50	70						
60	52	79	5	1	4	-5	-7	8
60	56	70	11	4	8	-4	-9	10
60	59	70	11	4	7	-2	-5	6
61	55	75	11	3	8	-3	-5	6
62	50	72	9	3	8	-7	-11	12

φ_1	Φ	φ_2	H	K	L	U	V	W
60	50	70						
62	51	74	7	2	6	-6	-9	10
62	52	79	5	1	4	-7	-9	11
62	54	76	4	1	3	-6	-9	11
62	56	73	10	3	7	-5	-9	11
62	57	77	9	2	6	-2	-3	4
62	58	72	3	1	2	-2	-4	5
63	52	70	11	4	9	-4	-7	8
63	53	77	9	2	7	-3	-4	5
63	54	72	9	3	7	-3	-5	6
63	54	76	4	1	3	-5	-7	9
64	50	76	8	2	7	-4	-5	6
64	52	75	11	3	9	-3	-4	5
64	56	70	11	4	8	-4	-7	9
64	56	74	7	2	5	-2	-3	4
64	58	72	3	1	2	-5	-9	12
65	58	72	3	1	2	-3	-5	7
66	54	76	4	1	3	-4	-5	7
66	55	72	12	4	9	-2	-3	4
66	57	76	12	3	8	-3	-4	6
67	58	72	3	1	2	-4	-6	9
68	52	72	6	2	5	-7	-9	12
68	52	79	5	1	4	-2	-2	3
68	54	76	4	1	3	-7	-8	12
68	56	73	10	3	7	-3	-4	6
69	50	76	8	2	7	-7	-7	10
69	51	76	12	3	10	-2	-2	3
69	53	77	9	2	7	-7	-7	11
69	56	79	10	2	7	-7	-7	12
69	58	72	3	1	2	-5	-7	11
60	50	80						
60	55	82	7	1	5	-3	-4	5
60	56	84	9	1	6	-7	-9	12
61	51	80	11	2	9	-4	-5	6
61	55	64	10	1	7	-5	-6	8
61	58	80	11	2	7	-2	-3	4
61	58	85	11	1	7	-4	-5	7
62	51	81	6	1	5	-7	-8	10
62	51	85	11	1	9	-3	-3	4
62	57	81	6	1	4	-6	-6	11
63	51	84	10	1	8	-8	-8	11
63	54	81	12	2	9	-5	-6	8
64	50	82	7	1	6	-3	-3	4
64	52	84	9	1	7	-7	-7	10
64	54	85	11	1	8	-2	-2	3
64	58	83	8	1	5	-6	-7	11
65	50	85	12	1	10	-7	-6	9

φ_1	Φ	φ_2	H	K	L	U	V	W
60	50	80						
65	53	83	8	1	6	-2	-2	3
65	54	80	11	2	8	-6	-7	10
65	55	84	10	1	7	-7	-7	11
65	57	81	6	1	4	-5	-6	9
66	51	81	6	1	5	-5	-5	7
66	56	84	9	1	6	-3	-3	5
66	58	85	11	1	7	-7	-7	12
67	53	85	12	1	9	-7	-6	10
67	55	82	7	1	5	-5	-5	8
67	59	84	10	1	6	-6	-6	11
68	56	83	8	1	5	-5	-5	9
69	51	81	6	1	5	-8	-7	11
69	52	84	9	1	7	-5	-4	7
69	57	81	6	1	4	-4	-4	7
60	50	90						
60	50	90	6	0	5	-10	-9	12
60	54	90	7	0	5	-5	-5	7
60	54	90	11	0	8	-8	-8	11
60	55	90	10	0	7	-7	-7	10
61	51	90	5	0	4	-8	-7	10
61	51	90	11	0	9	-9	-8	11
61	56	90	3	0	2	-2	-2	3
62	52	90	9	0	7	-7	-6	9
62	53	90	4	0	3	-9	-8	12
62	58	90	8	0	5	-5	-5	8
62	58	90	11	0	7	-7	-7	11
63	50	90	6	0	5	-5	-4	6
63	53	90	4	0	3	-6	-5	8
63	54	90	11	0	8	-8	-7	11
63	59	90	5	0	3	-3	-3	5
64	51	90	11	0	9	-9	-7	11
64	55	90	10	0	7	-7	-6	10
64	56	90	3	0	2	-8	-7	12
65	51	90	5	0	4	-4	-3	5
65	53	90	4	0	3	-9	-7	12
65	54	90	7	0	5	-5	-4	7
65	56	90	3	0	2	-6	-5	9
65	58	90	11	0	7	-7	-6	11
66	50	90	6	0	5	-10	-7	12
66	52	90	9	0	7	-7	-5	9
66	54	90	11	0	8	-8	-6	11
67	51	90	11	0	9	-9	-6	11
67	56	90	3	0	2	-4	-3	6
67	58	90	8	0	5	-5	-4	8
67	59	90	5	0	3	-6	-5	10
68	53	90	4	0	3	-3	-2	4

φ_1	Φ	φ_2	H	K	L	U	V	W
60	50	90						
68	55	90	10	0	7	-7	-5	10
69	50	90	6	0	5	-5	-3	6
69	51	90	5	0	4	-8	-5	10
69	58	90	11	0	7	-7	-5	11
60	60	0						
60	60	0	0	12	7	6	-7	12
60	61	0	0	9	5	6	-5	9
60	61	5	1	11	6	3	-3	5
60	64	7	1	8	4	4	-4	7
60	66	0	0	11	5	7	-5	11
60	69	0	0	8	3	5	-3	6
61	61	0	0	11	6	7	-6	11
61	63	0	0	2	1	5	-4	8
61	64	6	1	10	5	5	-5	9
61	68	0	0	5	2	3	-2	5
61	68	6	1	10	4	6	-5	11
62	60	9	2	12	7	3	-4	6
62	63	0	0	2	1	7	-6	12
62	63	0	0	2	1	6	-5	10
62	64	5	1	12	6	6	-6	11
62	64	9	1	6	3	6	-7	12
62	67	0	0	7	3	4	-3	7
62	67	0	0	12	5	7	-5	12
62	67	8	1	7	3	1	-1	2
62	68	9	2	12	5	1	-1	2
63	60	0	0	12	7	7	-7	12
63	66	0	0	9	4	5	-4	9
63	66	6	1	9	4	1	-1	2
64	60	0	0	7	4	4	-4	7
64	61	0	0	9	5	5	-5	9
64	61	0	0	11	6	6	-6	11
64	66	0	0	11	5	6	-5	11
64	66	5	1	11	5	1	-1	2
65	60	5	1	12	7	4	-5	8
65	61	6	1	7	4	5	-7	11
65	64	7	1	8	4	4	-5	9
65	64	9	1	6	3	3	-4	7
65	67	0	0	12	5	6	-5	12
65	68	0	0	5	2	5	-4	10
65	69	0	0	6	3	4	-3	8
66	63	0	0	2	1	1	-1	2
66	64	6	1	10	5	5	-6	11
66	67	5	1	12	5	5	-5	11
66	68	6	1	10	4	4	-4	9
67	60	0	0	12	7	6	-7	12
68	61	0	0	11	6	5	-6	11

φ_1	Φ	φ_2	H	K	L	U	V	W
60	60	0						
68	61	5	1	11	6	5	-7	12
68	61	6	1	9	5	2	-3	5
68	61	8	1	7	4	3	-5	8
68	66	0	0	9	4	4	-4	9
68	66	0	0	11	5	5	-5	11
68	67	0	0	7	3	3	-3	7
69	61	0	0	9	5	4	-5	9
69	64	7	1	8	4	4	-6	11
69	64	9	1	6	3	3	-5	9
69	67	0	0	12	5	5	-5	12
60	60	10						
60	64	14	1	4	2	4	-5	8
61	62	13	2	9	5	3	-4	6
61	64	11	2	10	5	5	-6	10
61	64	17	3	10	5	5	-7	11
61	69	11	1	5	2	1	-1	2
62	60	11	1	5	3	5	-7	10
62	62	15	3	11	6	4	-6	9
62	67	18	3	9	4	5	-7	12
62	68	16	2	7	3	4	-5	9
63	65	18	2	6	3	3	-5	8
63	66	15	3	11	5	3	-4	7
63	67	13	2	9	4	5	-6	11
64	64	14	1	4	2	2	-3	5
65	64	11	2	10	5	5	-7	12
66	61	16	2	7	4	1	-2	3
66	67	18	3	9	4	3	-5	9
67	60	14	3	12	7	1	-2	3
67	62	13	2	9	5	4	-7	11
67	64	14	1	4	2	4	-7	12
67	69	17	3	10	4	2	-3	6
68	60	11	1	5	3	1	-2	3
68	68	14	3	12	5	2	-3	6
69	61	18	4	12	7	3	-8	12
69	64	14	1	4	2	2	-4	7
69	65	18	2	6	3	3	-7	12
60	60	20						
60	65	21	3	8	4	4	-6	9
60	65	22	4	10	5	5	-8	12
61	60	29	5	9	6	3	-9	11
61	61	22	2	5	3	3	-6	8
61	61	29	6	11	7	3	-8	10
61	62	23	3	7	4	4	-8	11
61	62	27	6	12	7	4	-9	12

φ_1	Φ	φ_2	H	K	L	U	V	W
60	60	20						
61	66	27	1	2	1	1	-2	3
61	68	23	3	7	3	2	-3	5
61	68	24	5	11	5	3	-5	8
62	64	29	5	9	5	2	-5	7
62	67	20	4	11	5	2	-3	5
63	65	21	3	8	4	4	-7	11
64	60	21	3	8	5	4	-9	12
64	62	23	3	7	4	3	-7	10
64	63	20	4	11	6	1	-2	3
64	66	27	1	2	1	3	-7	11
64	69	23	5	12	5	3	-5	9
65	60	24	5	11	7	1	-3	4
65	61	22	2	5	3	2	-5	7
65	64	29	6	11	6	2	-6	9
65	66	27	1	2	1	2	-5	8
65	68	24	4	9	4	2	-4	7
65	69	29	5	9	4	3	-7	12
66	60	21	3	8	5	3	-8	11
66	63	24	4	9	5	3	-8	12
67	62	27	5	10	6	2	-7	10
67	68	23	3	7	3	3	-6	11
68	61	29	6	11	7	1	-5	7
68	62	23	3	7	4	2	-6	9
68	62	27	6	12	7	1	-4	6
68	68	24	5	11	5	2	-5	9
68	69	29	5	9	4	2	-6	11
69	63	24	4	9	5	2	-7	11
69	66	27	1	2	1	2	-7	12
60	60	30						
60	61	34	2	3	2	3	-10	12
60	61	39	8	10	7	1	-5	6
60	65	36	5	7	4	1	-3	4
60	67	34	4	6	3	3	-7	10
61	64	34	8	12	7	1	-3	4
62	60	36	5	7	5	1	-5	6
62	61	34	2	3	2	1	-4	5
62	62	32	7	11	7	2	-7	9
62	63	31	3	5	3	1	-3	4
62	63	37	6	8	5	2	-9	12
62	64	30	4	7	4	3	-8	11
62	67	34	4	6	3	3	-8	12
62	68	37	3	4	2	2	-6	9
63	64	35	7	10	6	2	-8	11
63	65	39	4	5	3	1	-5	7
64	66	36	8	11	6	1	-4	6
64	68	37	3	4	2	2	-7	11

φ_1	Φ	φ_2	H	K	L	U	V	W
60	60	30						
65	60	35	7	10	7	1	-7	9
65	60	36	8	11	8	1	-8	10
65	63	30	7	12	7	2	-7	10
65	65	34	6	9	5	1	-4	6
66	61	34	2	3	2	1	-6	8
66	62	32	5	8	5	1	-5	7
66	63	37	6	8	5	1	-7	10
66	65	37	9	12	7	1	-6	9
66	65	39	4	5	3	1	-8	12
67	61	34	2	3	2	1	-8	11
67	64	30	4	7	4	1	-4	6
68	61	39	9	11	8	0	-8	11
68	62	32	7	11	7	1	-7	10
68	65	32	7	11	6	1	-5	8
68	65	36	5	7	4	1	-7	11
69	61	39	8	10	7	0	-7	10
69	63	31	3	5	3	1	-6	9
60	60	40						
60	62	45	4	4	3	1	-10	12
60	63	45	11	11	8	1	-9	11
60	64	45	10	10	7	1	-8	10
60	65	45	3	3	2	1	-7	9
60	66	42	9	10	6	2	-9	12
60	66	45	8	8	5	1	-6	8
61	60	48	9	8	7	0	-7	8
61	61	49	8	7	6	0	-6	7
61	62	49	7	6	5	0	-5	6
61	67	41	6	7	4	2	-8	11
61	67	48	9	8	5	1	-8	11
61	68	45	7	7	4	1	-5	7
62	61	47	12	11	9	0	-9	11
62	62	41	6	7	5	1	-8	10
62	62	48	11	10	8	0	-4	5
62	65	42	10	11	7	1	-6	8
62	65	45	3	3	2	1	-9	12
62	66	45	11	11	7	1	-8	11
62	67	43	11	12	7	1	-5	7
62	67	45	5	5	3	1	-7	10
62	69	49	8	7	4	1	-8	12
63	60	45	11	11	9	0	-9	11
63	63	48	10	9	7	0	-7	9
63	64	48	9	8	6	0	-3	4
63	69	45	9	9	5	1	-6	9
64	61	45	5	5	4	0	-4	5
64	61	45	9	9	7	0	-7	9
64	64	47	12	11	8	0	-8	11

φ_1	Φ	φ_2	H	K	L	U	V	W
60	60	40						
64	65	49	8	7	5	0	-5	7
64	68	45	12	12	7	1	-8	12
64	69	45	11	11	6	1	-7	11
65	62	45	4	4	3	0	-3	4
65	63	45	11	11	8	0	-8	11
65	65	46	11	10	7	0	-7	10
65	67	41	6	7	4	1	-6	9
65	67	49	7	6	4	0	-2	3
66	60	42	8	9	7	0	-7	9
66	61	43	11	12	9	0	-3	4
66	63	45	7	7	5	0	-5	7
66	64	45	10	10	7	0	-7	10
66	66	48	10	9	6	0	-2	3
66	67	42	8	9	5	1	-7	11
67	60	40	10	12	9	0	-3	4
67	61	41	7	8	6	0	-3	4
67	62	42	10	11	8	0	-8	11
67	62	49	7	6	5	-1	-6	11
67	63	42	9	10	7	0	-7	10
67	65	45	3	3	2	0	-2	3
67	67	47	12	11	7	0	-7	11
67	67	48	9	6	5	0	-5	6
68	62	41	6	7	5	0	-5	7
68	64	42	8	9	6	0	-2	3
68	64	43	11	12	8	0	-2	3
68	66	45	8	8	5	0	-5	8
68	66	45	11	11	7	0	-7	11
68	68	48	11	10	6	0	-3	5
68	69	49	8	7	4	0	-4	7
69	60	45	11	11	9	-1	-8	11
69	60	48	9	8	7	-1	-5	7
69	63	40	5	6	4	0	-2	3
69	65	42	10	11	7	0	-7	11
69	67	45	5	5	3	0	-3	5
69	68	45	7	7	4	0	-4	7
69	68	45	12	12	7	0	-7	12
60	60	50						
60	67	56	6	4	3	0	-3	4
61	63	50	6	5	4	0	-4	5
61	65	53	12	9	7	0	-7	9
61	66	54	11	8	6	0	-3	4
61	69	58	11	7	5	0	-5	7
62	64	51	11	9	7	0	-7	9
62	65	51	5	4	3	0	-3	4
62	68	55	10	7	5	0	-5	7
63	64	56	12	8	7	-1	-9	12

φ_1	Φ	φ_2	H	K	L	U	V	W
60	60	50						
63	66	52	9	7	5	0	-5	7
64	62	52	9	7	6	-1	-9	12
64	64	55	10	7	6	-1	-8	11
64	65	58	11	7	6	-1	-7	10
64	66	50	12	10	7	0	-7	10
64	68	53	4	3	2	0	-2	3
65	61	56	3	2	2	-2	-9	12
65	67	51	11	9	6	0	-2	3
66	61	56	3	2	2	-2	-8	11
66	65	53	12	9	7	-1	-8	12
66	65	56	9	6	5	-1	-6	9
66	69	51	10	8	5	0	-5	8
67	60	50	12	10	9	-1	-6	8
67	67	59	10	6	5	-1	-5	8
67	69	50	6	5	3	0	-3	5
67	69	58	11	7	5	-1	-7	12
68	61	56	3	2	2	-2	-7	10
68	65	51	5	4	3	-1	-7	11
68	65	54	7	5	4	-1	-5	8
68	67	56	6	4	3	-1	-6	10
69	61	56	3	2	2	-2	-6	9
69	62	53	12	9	8	-1	-4	6
69	63	50	6	5	4	-1	-6	9
69	67	58	8	5	4	-1	-4	7
60	60	60						
60	61	68	5	2	3	-3	-9	11
60	63	66	9	4	5	-1	-4	5
61	62	63	10	5	6	-1	-4	5
61	66	63	2	1	1	-1	-6	8
62	60	69	8	3	5	-3	-7	9
62	61	61	11	6	7	-1	-4	5
62	62	67	7	3	4	-1	-3	4
62	66	63	2	1	1	-1	-5	7
63	60	60	12	7	8	-1	-4	5
64	61	63	8	4	5	-1	-3	4
64	62	67	12	5	7	-2	-5	7
64	65	67	12	5	6	-2	-6	9
64	65	69	8	3	4	-3	-8	12
64	67	60	12	7	6	-1	-6	9
64	69	61	9	5	4	-1	-7	11
65	60	61	9	5	6	-1	-3	4
65	66	63	2	1	1	-1	-4	6
66	65	68	10	4	5	-2	-5	8
66	66	63	2	1	1	-2	-7	11
67	61	68	5	2	3	-1	-2	3
67	62	63	12	6	7	-3	-8	12

φ_1	Φ	φ_2	H	K	L	U	V	W
60	60	60						
67	63	66	9	4	5	-3	-7	11
68	60	66	11	5	7	-1	-2	3
68	66	63	2	1	1	-1	-3	5
68	68	61	11	6	5	-1	-4	7
69	65	69	8	3	4	-2	-4	7
60	60	70						
60	60	79	5	1	3	-4	-7	9
60	62	75	11	3	6	-3	-7	9
60	65	72	6	2	3	-1	-3	4
60	67	77	9	2	4	-2	-5	7
61	64	73	10	3	5	-2	-5	7
61	68	76	12	3	5	-3	-8	12
62	60	73	10	3	6	-3	-6	8
62	60	79	5	1	3	-5	-8	11
62	61	72	12	4	7	-4	-9	12
62	61	74	7	2	4	-4	-6	11
62	62	77	9	2	5	-4	-7	10
62	67	70	11	4	5	-2	-7	10
62	69	79	5	1	2	-3	-7	11
63	63	70	11	4	6	-2	-5	7
63	64	76	4	1	2	-1	-2	3
63	67	72	9	3	4	-2	-6	9
64	64	79	10	2	5	-3	-5	8
64	68	74	7	2	3	-2	-5	8
65	60	76	12	3	7	-5	-8	12
65	62	72	9	3	5	-1	-2	3
65	69	79	5	1	2	-2	-4	7
66	62	75	11	3	6	-3	-5	8
66	67	72	9	3	4	-3	-7	12
67	61	72	12	4	7	-3	-5	8
67	65	72	6	2	3	-3	-6	10
67	68	74	7	2	3	-3	-6	11
68	61	74	7	2	4	-2	-3	5
68	62	77	9	2	5	-3	-4	7
68	64	76	4	1	2	-4	-6	11
68	69	79	5	1	2	-3	-5	10
69	60	76	12	3	7	-4	-5	9
69	63	70	11	4	6	-4	-7	12
69	64	73	10	3	5	-3	-5	9
69	67	77	9	2	4	-2	-3	6
69	69	73	10	3	4	-1	-2	4
60	60	80						
60	68	84	10	1	4	-1	-2	3
61	64	83	8	1	4	-5	-6	12

φ_1	Φ	φ_2	H	K	L	U	V	W
60	60	80						
61	64	85	12	1	6	-4	-6	9
61	66	80	11	2	5	-1	-2	3
61	67	82	7	1	3	-1	-2	3
61	67	85	12	1	5	-4	-7	11
62	66	84	9	1	4	-3	-5	8
63	66	85	11	1	5	-2	-3	5
65	64	81	6	1	3	-2	-3	5
65	64	83	8	1	4	-3	-4	7
66	64	84	10	1	5	-4	-5	9
66	64	85	12	1	6	-5	-6	11
67	67	85	12	1	5	-3	-4	6
68	61	85	11	1	6	-1	-1	2
68	62	80	11	2	6	-4	-5	9
68	68	81	12	2	5	-2	-3	6
69	61	84	9	1	5	-1	-1	2
69	64	81	6	1	3	-5	-6	12
60	60	90						
60	60	90	12	0	7	-7	-8	12
60	61	90	9	0	5	-5	-6	9
60	66	90	11	0	5	-5	-7	11
60	69	90	8	0	3	-3	-5	8
61	61	90	11	0	6	-6	-7	11
61	63	90	2	0	1	-4	-5	8
61	68	90	5	0	2	-2	-3	5
62	63	90	2	0	1	-6	-7	12
62	63	90	2	0	1	-5	-6	10
62	67	90	7	0	3	-3	-4	7
62	67	90	12	0	5	-5	-7	12
63	60	90	12	0	7	-7	-7	12
63	66	90	9	0	4	-4	-5	9
64	60	90	7	0	4	-4	-4	7
64	61	90	9	0	5	-5	-5	9
64	61	90	11	0	6	-6	-6	11
64	66	90	11	0	5	-5	-6	11
65	67	90	12	0	5	-5	-6	12
65	68	90	5	0	2	-4	-5	10
65	69	90	8	0	3	-3	-4	8
66	63	90	2	0	1	-1	-1	2
67	60	90	12	0	7	-7	-6	12
68	61	90	11	0	6	-6	-5	11
68	66	90	9	0	4	-4	-4	9
68	66	90	11	0	5	-5	-5	11
68	67	90	7	0	3	-3	-3	7
69	61	90	9	0	5	-5	-4	9
69	67	90	12	0	5	-5	-5	12

φ_1	Φ	φ_2	H	K	L	U	V	W
60	70	0						
60	70	5	1	11	4	7	-5	12
60	70	7	1	8	3	5	-4	9
60	73	0	0	10	3	6	-3	10
60	74	8	1	7	2	6	-4	11
60	76	0	0	4	1	7	-3	12
60	76	6	1	9	2	4	-2	7
60	79	0	0	5	1	3	-1	5
61	72	0	0	3	1	7	-4	12
61	74	0	0	7	2	4	-2	7
62	72	0	0	3	1	5	-3	9
62	72	9	1	6	2	6	-5	12
62	75	0	0	11	3	6	-3	11
62	76	9	2	12	3	3	-2	6
62	77	0	0	9	2	5	-2	9
63	70	0	0	11	4	6	-4	11
63	72	5	1	12	4	4	-3	8
63	79	6	1	10	2	2	-1	4
64	73	0	0	10	3	5	-3	10
64	76	0	0	4	1	2	-1	4
64	79	0	0	5	1	5	-2	10
65	72	0	0	3	1	3	-2	6
65	78	6	1	9	2	5	-3	11
66	70	7	1	8	3	3	-3	7
66	72	9	1	6	2	2	-2	5
66	75	0	0	11	3	5	-3	11
66	76	5	1	12	3	3	-2	7
67	70	0	0	11	4	5	-4	11
67	72	0	0	3	1	4	-3	9
67	76	7	1	8	2	4	-3	10
67	77	0	0	9	2	4	-2	9
68	70	5	1	11	4	2	-2	5
68	72	0	0	3	1	5	-4	12
68	74	0	0	7	2	3	-2	7
68	76	0	0	4	1	5	-3	12
69	72	6	1	9	3	3	-3	8
69	73	0	0	10	3	4	-3	10
69	79	0	0	5	1	2	-1	5
60	70	10						
60	70	14	2	8	3	1	-1	2
60	71	15	3	11	4	1	-1	2
61	74	11	2	10	3	5	-4	10
61	76	14	1	4	1	4	-3	8
61	78	13	2	9	2	3	-2	6
62	79	11	1	5	1	5	-3	10
63	72	14	3	12	4	4	-4	9

φ1	Φ	φ2	H	K	L	U	V	W
60	70	10						
63	74	17	3	10	3	3	-3	7
63	76	14	1	4	1	5	-4	11
63	79	17	3	10	2	4	-3	9
64	70	10	2	11	4	4	-4	9
64	72	13	2	9	3	3	-3	7
65	72	18	1	3	1	4	-5	11
65	75	16	2	7	2	2	-2	5
65	77	18	4	12	3	3	-3	8
65	79	11	1	5	1	3	-2	7
66	72	18	1	3	1	3	-4	9
66	75	15	3	11	3	3	-3	8
67	74	11	2	10	3	3	-3	8
68	72	14	3	12	4	4	-5	12
68	78	18	3	9	2	1	-1	3
68	79	11	1	5	1	4	-3	11
69	72	18	1	3	1	2	-3	7
69	75	10	2	11	3	1	-1	3
69	76	14	1	4	1	1	-1	3
60	70	20						
60	70	22	2	5	2	3	-4	7
60	71	21	3	8	3	5	-6	11
60	77	29	6	11	3	5	-6	12
61	75	27	5	10	3	4	-5	10
61	77	23	5	12	3	3	-3	7
62	70	27	6	12	5	4	-7	12
62	73	23	5	12	4	4	-5	10
62	73	24	4	9	3	3	-4	8
62	76	20	4	11	3	3	-3	7
63	70	22	2	5	2	4	-6	11
63	77	21	3	8	2	2	-2	5
63	79	24	4	9	2	2	-2	5
64	71	20	4	11	4	3	-4	8
64	74	29	5	9	3	3	-5	10
65	71	21	3	8	3	2	-3	6
65	77	27	2	4	1	4	-5	12
66	72	24	5	11	4	3	-5	10
66	72	29	6	11	4	1	-2	4
66	77	27	2	4	1	3	-4	10
66	77	29	6	11	3	2	-3	7
67	71	27	4	8	3	1	-2	4
67	73	24	4	9	3	3	-5	11
69	70	22	2	5	2	1	-2	4
69	70	27	5	10	4	2	-5	10
69	73	27	3	6	2	2	-4	9
69	77	23	5	12	3	3	-4	11
69	77	27	2	4	1	2	-3	8

φ1	Φ	φ2	H	K	L	U	V	W
60	70	30						
60	73	37	6	8	3	3	-6	10
60	74	30	7	12	4	4	-6	11
60	78	34	8	12	3	3	-4	8
61	72	32	5	8	3	4	-7	12
61	74	34	2	3	1	3	-5	9
61	74	36	8	11	4	2	-4	7
61	76	30	4	7	2	3	-4	8
61	79	37	3	4	1	2	-3	6
62	70	30	4	7	3	3	-6	10
62	75	38	7	9	3	3	-6	11
62	76	31	6	10	3	2	-3	6
62	77	36	5	7	2	3	-5	10
63	71	31	3	5	2	2	-4	7
63	71	34	8	12	5	3	-7	12
63	78	30	7	12	3	3	-4	9
64	70	36	8	11	5	1	-3	5
64	71	38	7	9	4	1	-3	5
64	72	32	5	8	3	3	-6	11
64	77	39	8	10	3	1	-2	4
64	79	37	3	4	1	3	-5	11
65	72	35	7	10	4	2	-5	9
65	73	39	4	5	2	2	-6	11
65	74	34	2	3	1	1	-2	4
65	78	32	5	8	2	2	-3	7
66	70	30	4	7	3	2	-5	9
66	71	39	9	11	5	1	-4	7
67	73	37	6	8	3	1	-3	6
67	74	39	9	11	4	1	-3	6
67	77	36	5	7	2	2	-4	9
67	78	34	8	12	3	3	-5	12
68	72	32	5	8	3	2	-5	10
69	71	31	3	5	2	1	-3	6
69	73	39	4	5	2	1	-4	8
69	74	34	2	3	1	2	-5	11
69	79	37	3	4	1	1	-2	5
60	70	40						
60	70	48	10	9	5	1	-5	7
60	73	42	9	10	4	2	-5	8
60	74	49	6	7	3	2	-7	11
60	76	40	5	6	2	2	-4	7
60	77	45	3	3	1	3	-7	12
60	79	45	11	11	3	3	-6	11
61	71	45	2	2	1	1	-4	6
61	72	40	10	12	5	2	-5	8
61	72	45	11	11	5	2	-7	11

φ1	Φ	φ2	H	K	L	U	V	W
60	70	40						
61	74	45	5	5	2	1	-3	5
61	76	48	9	8	3	1	-3	5
61	77	42	9	10	3	3	-6	11
62	70	47	12	11	6	1	-6	9
62	73	43	11	12	5	1	-3	5
62	75	48	11	10	4	2	-7	12
62	77	45	3	3	1	2	-5	9
62	79	40	10	12	3	3	-5	10
63	71	45	2	2	1	1	-5	8
63	72	41	6	7	3	1	-3	5
63	74	45	12	12	5	2	-7	12
63	78	41	6	7	2	1	-2	4
63	78	49	7	6	2	2	-6	11
63	79	42	10	11	3	1	-2	4
64	73	45	7	7	3	1	-4	7
64	76	45	11	11	4	2	-6	11
64	78	45	10	10	3	2	-5	10
65	71	45	2	2	1	1	-6	10
65	72	42	8	9	4	1	-4	7
65	74	49	8	7	3	1	-5	9
65	76	40	5	6	2	2	-5	10
65	77	45	3	3	1	1	-3	6
66	71	45	2	2	1	1	-7	12
66	73	45	9	9	4	1	-5	9
66	73	48	10	9	4	1	-6	11
66	76	47	12	11	4	1	-4	8
66	79	41	7	6	2	2	-4	9
66	79	45	11	11	3	2	-5	11
67	71	42	10	11	5	1	-5	9
67	72	45	11	11	5	1	-6	11
67	75	45	8	8	3	1	-4	8
68	74	45	5	5	2	1	-5	10
68	74	45	12	12	5	1	-6	12
68	79	45	7	7	2	1	-3	7
69	70	47	12	11	6	0	-6	11
69	70	48	10	9	5	0	-5	9
69	72	40	10	12	5	1	-5	10
69	77	42	9	10	3	1	-3	7
69	77	45	3	3	1	1	-4	9
60	70	50						
60	71	52	9	7	4	1	-7	10
60	77	58	11	7	3	1	-5	8
60	79	53	4	3	1	1	-3	5
61	72	53	12	9	5	1	-8	12
61	77	51	10	8	3	2	-7	12
62	71	59	5	3	2	0	-2	3

φ1	Φ	φ2	H	K	L	U	V	W
60	70	50						
62	78	54	11	8	3	1	-4	7
63	76	50	6	5	2	1	-4	7
63	77	54	7	5	2	1	-5	9
63	78	58	8	5	2	1	-6	11
64	71	56	12	8	5	0	-5	8
64	72	58	8	5	3	0	-3	5
64	79	50	12	10	3	1	-3	6
65	70	54	11	8	5	0	-5	8
65	71	54	7	5	3	0	-3	5
65	73	58	11	7	4	0	-4	7
65	75	52	9	7	3	1	-6	11
65	79	53	4	3	1	1	-4	8
66	72	55	10	7	4	0	-4	7
66	77	51	10	8	3	1	-5	10
66	78	56	12	8	3	1	-6	12
67	71	52	9	7	4	0	-4	7
67	72	53	12	9	5	0	-5	9
67	76	50	6	5	2	1	-6	12
67	76	59	10	6	3	0	-1	2
68	71	51	11	9	5	0	-5	9
68	74	56	3	2	1	0	-1	2
68	79	53	4	3	1	1	-5	11
69	73	53	8	6	3	0	-1	2
69	74	54	11	8	4	0	-1	2
60	70	60						
60	70	60	12	7	5	0	-5	7
60	73	63	6	3	2	0	-2	3
61	72	61	11	6	4	0	-2	3
61	78	60	12	7	3	1	-6	10
62	70	63	12	6	5	-1	-8	12
62	76	66	11	5	3	0	-3	5
62	77	67	12	5	3	0	-3	5
63	74	61	9	5	3	0	-3	5
63	75	63	10	5	3	0	-3	5
65	70	63	10	5	4	-1	-6	10
65	71	63	8	4	3	-1	-7	12
65	74	60	12	7	4	0	-4	7
65	79	68	5	2	1	0	-1	2
66	70	66	5	2	2	-2	-7	12
66	72	66	11	5	4	-1	-5	9
66	73	66	9	4	3	-1	-6	11
66	79	66	9	4	2	0	-1	2
67	76	60	7	4	2	0	-1	2
67	77	61	11	6	3	0	-1	2
68	70	60	7	4	3	-1	-5	9
68	70	68	5	2	2	-2	-6	11

φ_1	Φ	φ_2	H	K	L	U	V	W
60	70	60						
68	73	67	12	5	4	-1	-4	8
68	74	68	10	4	3	-1	-5	10
69	73	63	6	3	2	-1	-6	12
60	70	70						
60	72	72	3	1	1	-1	-6	9
60	72	76	12	3	4	-1	-4	6
60	76	76	4	1	1	-1	-7	11
61	79	79	5	1	1	-1	-7	12
62	72	72	3	1	1	-1	-5	8
62	76	76	4	1	1	-1	-6	10
63	72	77	9	2	3	-1	-3	5
63	79	79	5	1	1	-1	-6	11
64	71	75	11	3	4	-1	-3	5
64	76	76	4	1	1	-1	-5	9
65	72	72	3	1	1	-1	-4	7
65	79	79	5	1	1	-1	-5	10
66	76	76	4	1	1	-1	-4	8
66	77	72	12	4	3	-1	-6	12
67	74	79	10	2	3	-2	-5	10
68	70	76	8	2	3	-1	-2	4
68	72	72	3	1	1	-1	-3	6
68	79	79	5	1	1	-1	-4	9
60	70	80						
61	74	82	7	1	2	-1	-3	5
61	76	85	12	1	3	-1	-3	5
61	79	84	10	1	2	-1	-4	7
62	70	80	11	2	4	-2	-5	8
62	70	83	8	1	3	-3	-6	10
62	75	80	11	2	3	-2	-7	12
63	72	85	12	1	4	-2	-4	7
63	75	85	11	1	3	-2	-5	9
63	76	83	8	1	2	-2	-6	11
64	72	84	9	1	3	-3	-6	11
65	78	84	9	1	2	-1	-3	6
66	73	84	10	1	3	-1	-2	4
66	76	81	12	2	3	-1	-3	6
67	72	81	6	1	2	-1	-2	4
69	72	84	9	1	3	-2	-3	7
69	74	82	7	1	2	-2	-4	9
60	70	90						
60	73	90	10	0	3	-3	-6	10
60	76	90	4	0	1	-3	-7	12

φ_1	Φ	φ_2	H	K	L	U	V	W
60	70	90						
60	79	90	5	0	1	-1	-3	5
61	72	90	3	0	1	-4	-7	12
61	74	90	7	0	2	-2	-4	7
62	72	90	3	0	1	-3	-5	9
62	75	90	11	0	3	-3	-6	11
62	77	90	9	0	2	-2	-5	9
63	70	90	11	0	4	-4	-6	11
64	73	90	10	0	3	-3	-5	10
64	76	90	4	0	1	-1	-2	4
64	79	90	5	0	1	-2	-5	10
65	72	90	3	0	1	-2	-3	6
66	75	90	11	0	3	-3	-5	11
67	70	90	11	0	4	-4	-5	11
67	72	90	3	0	1	-3	-4	9
67	77	90	9	0	2	-2	-4	9
68	72	90	3	0	1	-4	-5	12
68	74	90	7	0	2	-2	-3	7
68	76	90	4	0	1	-3	-5	12
69	73	90	10	0	3	-3	-4	10
69	79	90	5	0	1	-1	-2	5
60	80	0						
60	81	0	0	6	1	7	-2	12
60	82	8	1	7	1	5	-2	9
60	85	0	0	12	1	7	-1	12
60	85	5	1	11	1	4	-1	7
61	82	0	0	7	1	4	-1	7
61	84	0	0	9	1	5	-1	9
61	85	0	0	11	1	6	-1	11
62	80	0	0	11	2	6	-2	11
62	81	9	1	6	1	2	-1	4
63	84	6	1	9	1	3	-1	6
63	85	5	1	12	1	4	-1	8
63	85	9	2	12	1	3	-1	6
64	81	0	0	6	1	3	-1	6
64	83	0	0	8	1	4	-1	8
64	84	0	0	10	1	5	-1	10
64	85	0	0	12	1	6	-1	12
65	83	7	1	8	1	5	-2	11
66	80	0	0	11	2	5	-2	11
66	80	5	1	11	2	4	-2	9
66	84	0	0	9	1	4	-1	9
66	84	6	1	10	1	3	-1	7
66	85	0	0	11	1	5	-1	11
67	82	0	0	7	1	3	-1	7
67	82	8	1	7	1	2	-1	5
67	85	0	0	12	1	5	-1	12

φ_1	Φ	φ_2	H	K	L	U	V	W
60	80	0						
68	81	0	0	6	1	5	-2	12
68	81	5	1	12	2	2	-1	5
66	84	0	0	10	1	4	-1	10
69	85	5	1	11	1	3	-1	8
60	80	10						
60	81	18	2	6	1	3	-2	6
61	85	18	4	12	1	2	-1	4
62	83	14	2	8	1	2	-1	4
62	84	11	2	10	1	5	-2	10
63	82	16	2	7	1	5	-3	11
65	81	18	2	6	1	4	-3	10
65	84	18	3	9	1	5	-3	12
66	85	10	2	11	1	5	-2	12
66	85	15	3	11	1	2	-1	5
67	82	16	2	7	1	3	-2	8
67	84	13	2	9	1	2	-1	5
68	80	10	2	11	2	3	-2	8
69	81	14	3	12	2	4	-3	12
69	84	18	3	9	1	3	-2	9
60	80	20						
60	84	29	5	9	1	5	-4	11
60	85	22	4	10	1	5	-3	10
61	85	29	6	11	1	4	-3	9
62	80	27	5	10	2	2	-2	5
62	81	29	6	11	2	2	-2	5
62	83	23	3	7	1	4	-3	9
62	86	23	5	12	1	5	-3	11
63	84	27	4	8	1	5	-4	12
64	80	20	4	11	2	5	-4	12
64	83	21	3	8	1	3	-2	7
64	85	27	5	10	1	4	-3	10
66	81	24	5	11	2	1	-1	3
66	82	27	3	6	1	1	-1	3
66	84	24	4	9	1	4	-3	11
68	83	21	3	8	1	4	-3	12
68	86	27	6	12	1	4	-3	12
69	86	23	5	12	1	3	-2	9
60	80	30						
61	81	35	7	10	2	4	-5	11
62	82	30	7	12	2	2	-2	5
63	80	34	6	9	2	3	-4	9
63	82	37	9	12	2	4	-5	12

φ_1	Φ	φ_2	H	K	L	U	V	W
60	80	30						
64	81	39	4	5	1	2	-3	7
65	82	36	8	11	2	3	-4	10
65	84	32	5	8	1	1	-1	3
65	85	34	6	9	1	1	-1	3
65	85	35	7	10	1	1	-1	3
65	86	36	8	11	1	1	-1	3
65	86	37	9	12	1	1	-1	3
66	80	38	7	9	2	3	-5	12
66	83	30	4	7	1	1	-1	3
67	80	31	3	5	1	3	-4	11
67	81	35	7	10	2	2	-3	8
69	82	34	4	6	1	3	-4	12
69	82	37	9	12	2	2	-3	9
60	80	40						
60	82	42	9	10	2	2	-3	6
60	83	45	6	6	1	2	-3	6
60	84	41	6	7	1	4	-5	11
60	86	45	11	11	1	4	-5	11
61	82	45	5	5	1	3	-5	10
61	86	45	9	9	1	3	-4	9
62	80	45	4	4	1	1	-2	4
62	82	48	10	9	2	1	-2	4
62	83	40	5	6	1	3	-4	9
62	85	41	7	8	1	4	-5	12
62	87	45	12	12	1	4	-5	12
63	83	45	11	11	2	3	-5	11
63	84	45	7	7	1	2	-3	7
63	85	49	8	7	1	3	-5	11
63	86	48	10	9	1	2	-3	7
64	83	43	11	12	2	2	-3	7
64	84	41	6	7	1	3	-4	10
64	86	45	10	10	1	3	-4	10
65	81	45	9	9	2	2	-4	9
65	82	48	11	10	2	2	-4	9
65	83	45	6	6	1	3	-5	12
66	81	48	9	8	2	2	-5	11
66	85	41	7	8	1	3	-4	11
66	85	45	8	8	1	2	-3	8
66	86	45	11	11	1	3	-4	11
66	86	48	11	10	1	2	-3	8
67	82	45	5	5	1	1	-2	5
67	83	40	5	6	1	2	-3	8
67	83	47	12	11	2	1	-2	5
67	84	49	7	6	1	1	-2	5
68	80	45	4	4	1	2	-5	12
68	80	47	12	11	3	1	-3	7

φ_1 Φ φ_2	H	K	L	U	V	W
60 80 40						
68 81 42	8	9	2	1	-2	5
68 85 42	6	9	1	3	-4	12
68 86 47	12	11	1	2	-3	9
68 87 45	12	12	1	3	-4	12
69 82 42	9	10	2	2	-4	11
69 83 45	11	11	2	2	-4	11
69 84 41	6	7	1	2	-3	9
69 86 45	9	9	1	2	-3	9
60 80 50						
60 80 56	9	6	2	2	-7	12
60 82 54	11	8	2	2	-5	9
60 84 58	8	5	1	2	-5	9
60 86 51	10	8	1	3	-5	10
61 81 55	10	7	2	2	-6	11
61 84 53	8	6	1	1	-2	4
61 85 55	10	7	1	1	-2	4
61 86 56	12	8	1	1	-2	4
62 83 50	6	5	1	1	-2	4
62 85 59	10	6	1	2	-5	10
63 81 51	5	4	1	2	-5	10
63 82 56	6	4	1	1	-3	6
64 80 52	9	7	2	1	-3	6
64 86 51	11	9	1	3	-5	12
65 80 59	5	3	1	1	-5	10
65 83 54	7	5	1	2	-5	11
66 84 58	8	5	1	1	-3	7
66 85 52	9	7	1	1	-2	5
66 85 56	9	6	1	2	-5	12
66 86 54	11	8	1	1	-2	5
67 81 51	5	4	1	1	-3	7
67 81 55	10	7	2	1	-4	9
69 81 58	11	7	2	1	-5	12
69 82 56	6	4	1	1	-4	10
69 83 54	7	5	1	1	-3	8
69 85 59	10	6	1	1	-3	6
60 80 60						
60 80 63	10	5	2	1	-6	10
60 84 66	9	4	1	1	-4	7
61 81 66	11	5	2	1	-7	12
62 82 63	6	3	1	1	-5	9
62 83 67	7	3	1	1	-6	11
63 84 61	9	5	1	1	-3	6
63 84 63	8	4	1	1	-4	8
63 85 68	10	4	1	1	-5	10

φ_1 Φ φ_2	H	K	L	U	V	W
60 80 60						
63 86 63	12	6	1	1	-3	6
63 86 67	12	5	1	1	-4	8
64 82 60	12	7	2	1	-4	8
64 82 63	6	3	1	1	-6	12
64 86 60	12	7	1	2	-5	11
66 83 60	7	4	1	1	-4	9
66 85 61	11	6	1	1	-3	7
68 84 63	8	4	1	1	-5	12
68 85 63	10	5	1	1	-4	10
60 80 70						
60 84 72	9	3	1	1	-7	12
61 85 70	11	4	1	1	-5	9
63 85 72	12	4	1	1	-6	12
64 83 76	8	2	1	0	-1	2
64 84 77	9	2	1	0	-1	2
64 84 79	10	2	1	0	-1	2
65 80 70	11	4	2	0	-1	2
65 81 72	6	2	1	0	-1	2
65 82 74	7	2	1	0	-1	2
60 80 80						
64 81 85	12	1	2	-1	-4	8
64 85 60	11	2	1	0	-1	2
64 85 81	12	2	1	0	-1	2
65 81 81	6	1	1	-1	-6	12
67 81 81	6	1	1	-1	-5	11
68 80 85	11	1	2	-1	-3	7
68 82 82	7	1	1	-1	-5	12
60 80 90						
60 81 90	6	0	1	-2	-7	12
60 65 90	12	0	1	-1	-7	12
61 82 90	7	0	1	-1	-4	7
61 84 90	9	0	1	-1	-5	9
61 85 90	11	0	1	-1	-6	11
62 80 90	11	0	2	-2	-6	11
64 81 90	6	0	1	-1	-3	6
64 83 90	8	0	1	-1	-4	8
64 84 90	10	0	1	-1	-5	10
64 85 90	12	0	1	-1	-6	12
66 80 90	11	0	2	-2	-5	11
66 84 90	9	0	1	-1	-4	9
66 85 90	11	0	1	-1	-5	11
67 82 90	7	0	1	-1	-3	7

φ_1 Φ φ_2	H	K	L	U	V	W
60 80 90						
67 85 90	12	0	1	-1	-5	12
68 81 90	6	0	1	-2	-5	12
68 84 90	10	0	1	-1	-4	10
60 90 0						
60 90 0	0	1	0	7	0	12
60 90 0	0	1	0	4	0	7
61 90 0	0	1	0	6	0	11
61 90 0	0	1	0	5	0	9
61 90 9	1	6	0	6	-1	11
63 90 0	0	1	0	1	0	2
63 90 9	1	6	0	6	-1	12
66 90 0	0	1	0	4	0	9
66 90 0	0	1	0	5	0	11
67 90 0	0	1	0	5	0	12
67 90 0	0	1	0	3	0	7
68 90 0	0	1	0	2	0	5
69 90 0	0	1	0	3	0	8
60 90 10						
60 90 11	1	5	0	5	-1	9
60 90 14	1	4	0	4	-1	7
60 90 18	1	3	0	6	-2	11
62 90 18	1	3	0	3	-1	6
63 90 11	1	5	0	5	-1	10
63 90 14	1	4	0	4	-1	8
65 90 11	1	5	0	5	-1	11
65 90 14	1	4	0	4	-1	9
66 90 18	1	3	0	3	-1	7
67 90 11	1	5	0	5	-1	12
68 90 14	1	4	0	4	-1	10
68 90 18	1	3	0	3	-1	8
69 90 14	1	4	0	4	-1	11
60 90 20						
62 90 22	2	5	0	5	-2	10
64 90 22	2	5	0	5	-2	11
64 90 27	1	2	0	4	-2	9
66 90 22	2	5	0	5	-2	12
66 90 27	1	2	0	2	-1	5
66 90 27	1	2	0	4	-2	11
60 90 30						
60 90 31	3	5	0	5	-3	10

φ_1 Φ φ_2	H	K	L	U	V	W
60 90 30						
60 90 39	4	5	0	5	-4	11
61 90 37	3	4	0	4	-3	9
62 90 31	3	5	0	5	-3	11
62 90 39	4	5	0	5	-4	12
63 90 34	2	3	0	3	-2	7
63 90 37	3	4	0	4	-3	10
64 90 31	3	5	0	5	-3	12
66 90 34	2	3	0	3	-2	8
66 90 37	3	4	0	4	-3	11
67 90 37	3	4	0	4	-3	12
68 90 34	2	3	0	3	-2	9
60 90 40						
61 90 45	1	1	0	2	-2	5
62 90 45	1	1	0	3	-3	8
63 90 45	1	1	0	4	-4	11
65 90 45	1	1	0	1	-1	3
67 90 45	1	1	0	3	-3	10
68 90 45	1	1	0	2	-2	7
69 90 45	1	1	0	3	-3	11
60 90 50						
60 90 51	5	4	0	4	-5	11
60 90 59	5	3	0	3	-5	10
61 90 53	4	3	0	3	-4	9
62 90 51	5	4	0	4	-5	12
62 90 59	5	3	0	3	-5	11
63 90 53	4	3	0	3	-4	10
63 90 56	3	2	0	2	-3	7
64 90 59	5	3	0	3	-5	12
66 90 53	4	3	0	3	-4	11
66 90 56	3	2	0	2	-3	8
67 90 53	4	3	0	3	-4	12
68 90 56	3	2	0	2	-3	9
60 90 60						
62 90 68	5	2	0	2	-5	10
64 90 63	2	1	0	2	-4	9
64 90 68	5	2	0	2	-5	11
66 90 63	2	1	0	1	-2	5
66 90 68	5	2	0	2	-5	12
68 90 63	2	1	0	2	-4	11

TAB. VI

FROM EULER ANGLES INTO HKL–UVW

φ_1	Φ	φ_2	H	K	L	U	V	W
60	90	70						
60	90	72	3	1	0	2	-6	11
60	90	76	4	1	0	1	-4	7
60	90	79	5	1	0	1	-5	9
62	90	72	3	1	0	1	-3	6
63	90	76	4	1	0	1	-4	8
63	90	79	5	1	0	1	-5	10
65	90	76	4	1	0	1	-4	9
65	90	79	5	1	0	1	-5	11
66	90	72	3	1	0	1	-3	7
67	90	79	5	1	0	1	-5	12
68	90	72	3	1	0	1	-3	8
68	90	76	4	1	0	1	-4	10
69	90	76	4	1	0	1	-4	11
60	90	80						
61	90	81	6	1	0	1	-6	11
63	90	81	6	1	0	1	-6	12
60	90	90						
60	90	90	1	0	0	0	-4	7
60	90	90	1	0	0	0	-7	12
61	90	90	1	0	0	0	-5	9
61	90	90	1	0	0	0	-6	11
63	90	90	1	0	0	0	-1	2
66	90	90	1	0	0	0	-5	11
66	90	90	1	0	0	0	-4	9
67	90	90	1	0	0	0	-5	12
67	90	90	1	0	0	0	-3	7
68	90	90	1	0	0	0	-2	5
69	90	90	1	0	0	0	-3	8
70	0	0						
70	5	0	0	1	11	4	-11	1
70	7	0	0	1	8	3	-8	1
72	5	0	0	1	12	4	-12	1
72	6	0	0	1	9	3	-9	1
72	9	0	0	1	6	2	-6	1
73	6	0	0	1	10	3	-10	1
74	8	0	0	1	7	2	-7	1
75	5	0	0	1	11	3	-11	1
76	5	0	0	1	12	3	-12	1
76	7	0	0	1	8	2	-8	1
76	9	0	0	1	6	3	-12	2
76	6	0	0	1	9	2	-9	1

φ_1	Φ	φ_2	H	K	L	U	V	W
70	0	0						
79	6	0	0	1	10	2	-10	1
70	0	40						
72	7	45	1	1	12	-4	-8	1
72	9	45	1	1	9	-3	-6	1
75	7	45	1	1	11	-4	-7	1
78	9	45	1	1	9	-7	-11	2
79	8	45	1	1	10	-4	-6	1
70	0	70						
70	0	70	0	0	1	-11	-9	0
70	0	70	0	0	1	-6	-5	0
70	0	70	0	0	1	-7	-6	0
71	0	71	0	0	1	-5	-4	0
71	0	71	0	0	1	-9	-7	0
72	0	72	0	0	1	-7	-5	0
72	0	72	0	0	1	-11	-6	0
72	0	72	0	0	1	-4	-3	0
73	0	73	0	0	1	-10	-7	0
73	0	73	0	0	1	-3	-2	0
74	0	74	0	0	1	-11	-7	0
74	0	74	0	0	1	-8	-5	0
75	0	75	0	0	1	-12	-7	0
75	0	75	0	0	1	-5	-3	0
75	0	75	0	0	1	-7	-4	0
75	0	75	0	0	1	-9	-5	0
76	0	76	0	0	1	-11	-6	0
77	0	77	0	0	1	-2	-1	0
78	0	78	0	0	1	-7	-3	0
78	0	78	0	0	1	-9	-4	0
78	0	78	0	0	1	-11	-5	0
79	0	79	0	0	1	-12	-5	0
79	0	79	0	0	1	-5	-2	0
70	0	90						
70	5	90	1	0	11	-11	-4	1
70	7	90	1	0	8	-8	-3	1
72	5	90	1	0	12	-12	-4	1
72	6	90	1	0	9	-9	-3	1
72	9	90	1	0	6	-6	-2	1
73	6	90	1	0	10	-10	-3	1
74	8	90	1	0	7	-7	-2	1
75	5	90	1	0	11	-11	-3	1
76	5	90	1	0	12	-12	-3	1
76	7	90	1	0	8	-8	-2	1

φ_1	Φ	φ_2	H	K	L	U	V	W
70	0	90						
76	9	90	1	0	6	-12	-3	2
78	6	90	1	0	9	-9	-2	1
79	6	90	1	0	10	-10	-2	1
70	10	0						
70	10	0	0	2	11	4	-11	2
70	14	0	0	1	4	3	-8	2
71	15	0	0	3	11	4	-11	3
72	13	0	0	2	9	3	-9	2
72	14	0	0	1	4	4	-12	3
72	18	0	0	1	3	1	-3	1
74	11	0	0	1	5	3	-10	2
74	17	0	0	3	10	3	-10	3
75	10	0	0	2	11	3	-11	2
75	15	0	0	3	11	3	-11	3
75	16	0	0	2	7	2	-7	2
76	14	0	0	1	4	1	-4	1
77	18	0	0	1	3	3	-12	4
78	13	0	0	2	9	2	-9	2
78	18	0	0	1	3	2	-9	3
79	11	0	0	1	5	1	-5	1
79	17	0	0	3	10	2	-10	3
70	10	10						
72	15	18	1	3	12	0	-4	1
72	16	18	1	3	11	0	-11	3
72	18	18	1	3	10	0	-10	3
73	19	18	1	3	9	0	-3	1
77	19	14	1	4	12	0	-3	1
70	10	20						
70	18	27	1	2	7	-1	-10	3
74	11	27	1	2	12	-2	-11	2
74	16	27	1	2	8	-2	-11	3
75	11	27	1	2	11	-1	-5	1
76	13	27	1	2	10	-2	-9	2
78	14	27	1	2	9	-1	-4	1
70	10	30						
74	17	34	2	3	12	-3	-10	3
75	18	34	2	3	11	-1	-3	1

φ_1	Φ	φ_2	H	K	L	U	V	W
70	10	40						
70	10	45	1	1	8	-5	-11	2
71	19	45	1	1	4	-5	-11	4
72	13	45	1	1	6	-2	-4	1
72	16	45	1	1	5	-5	-10	3
72	17	45	2	2	9	-3	-6	2
73	19	45	1	1	4	-4	-8	3
74	11	45	1	1	7	-5	-9	2
75	14	45	2	2	11	-4	-7	2
76	10	45	1	1	6	-3	-5	1
77	19	45	1	1	4	-3	-5	2
78	13	45	1	1	6	-7	-11	3
78	17	45	2	2	9	-7	-11	4
79	16	45	1	1	5	-2	-3	1
79	19	45	1	1	4	-8	-12	5
70	10	50						
73	18	56	3	2	11	-4	-5	2
79	17	56	3	2	12	-12	-12	5
79	18	56	3	2	11	-11	-11	5
70	10	60						
72	11	63	2	1	11	-11	-11	3
72	11	63	2	1	12	-4	-4	1
72	13	63	2	1	10	-10	-10	3
72	14	63	2	1	9	-3	-3	1
72	16	63	2	1	8	-8	-8	3
72	18	63	2	1	7	-7	-7	3
75	16	63	2	1	8	-11	-10	4
75	18	63	2	1	7	-12	-11	5
76	13	63	2	1	10	-7	-6	2
78	18	63	2	1	7	-5	-4	2
70	10	70						
72	18	72	3	1	10	-8	-6	3
73	19	72	3	1	9	-12	-9	5
75	16	72	3	1	11	-3	-2	1
77	18	72	3	1	10	-11	-7	4
78	15	72	3	1	12	-10	-6	3
78	19	72	3	1	9	-5	-3	2
78	19	76	4	1	12	-8	-4	3
70	10	90						
70	10	90	2	0	11	-11	-4	2

φ1 Φ φ2	H	K	L	U	V	W
70 10 90						
70 14 90	1	0	4	-8	-3	2
71 15 90	3	0	11	-11	-4	3
72 13 90	2	0	9	-9	-3	2
72 14 90	1	0	4	-12	-4	3
72 18 90	1	0	3	-3	-1	1
74 11 90	1	0	5	-10	-3	2
74 17 90	3	0	10	-10	-3	3
75 10 90	2	0	11	-11	-3	2
75 15 90	3	0	11	-11	-3	3
75 16 90	2	0	7	-7	-2	2
76 14 90	1	0	4	-4	-1	1
77 18 90	1	0	3	-12	-3	4
78 13 90	2	0	9	-9	-2	2
78 18 90	1	0	3	-9	-2	3
79 11 90	1	0	5	-5	-1	1
79 17 90	3	0	10	-10	-2	3
70 20 0						
70 22 0	0	2	5	2	-5	2
70 27 0	0	1	2	5	-12	6
70 27 0	0	1	2	4	-10	5
71 20 0	0	4	11	4	-11	4
71 21 0	0	3	8	3	-8	3
71 27 0	0	1	2	3	-8	4
72 24 0	0	5	11	4	-11	5
72 29 0	0	6	11	4	-11	6
73 23 0	0	5	12	4	-12	5
73 24 0	0	4	9	3	-9	4
73 27 0	0	1	2	2	-6	3
74 22 0	0	2	5	3	-10	4
74 29 0	0	5	9	3	-9	5
75 23 0	0	3	7	2	-7	3
75 27 0	0	1	2	3	-10	5
76 20 0	0	4	11	3	-11	4
76 24 0	0	5	11	3	-11	5
77 21 0	0	3	8	2	-8	3
77 23 0	0	5	12	3	-12	5
77 27 0	0	1	2	1	-4	2
77 29 0	0	6	11	3	-11	6
79 22 0	0	2	5	1	-5	2
79 24 0	0	4	9	2	-9	4
79 29 0	0	5	9	2	-9	5
70 20 10						
70 25 14	1	4	9	1	-7	3
73 22 18	1	3	8	0	-8	3

φ1 Φ φ2	H	K	L	U	V	W
70 20 10						
73 24 18	1	3	7	0	-7	3
74 25 11	1	5	11	1	-9	4
74 28 18	1	3	6	0	-2	1
77 21 14	1	4	11	0	-11	4
77 22 14	1	4	10	0	-5	2
77 25 14	1	4	9	0	-9	4
77 27 14	1	4	6	0	-2	1
79 24 18	1	3	7	-1	-9	4
70 20 20						
70 24 22	2	5	12	0	-12	5
70 24 27	1	2	5	-1	-12	5
70 26 22	2	5	11	0	-11	5
71 28 22	2	5	10	0	-2	1
73 24 27	1	2	5	-1	-7	3
75 20 27	1	2	6	-2	-11	4
76 22 27	2	4	11	-1	-5	2
76 29 27	1	2	4	-2	-11	6
77 24 27	1	2	5	-2	-9	4
78 20 27	1	2	6	-2	-6	3
78 29 27	1	2	4	-2	-9	5
79 26 27	2	4	9	-1	-4	2
70 20 30						
71 28 31	3	5	11	-1	-6	3
72 27 34	2	3	7	-1	-4	2
73 23 37	3	4	12	-4	-12	5
73 27 37	3	4	10	-2	-6	3
74 22 34	2	3	9	-3	-10	4
74 29 37	3	4	9	-3	-9	5
77 29 37	3	4	9	-5	-12	7
78 20 34	2	3	10	-3	-8	3
78 23 37	3	4	12	-4	-9	4
79 26 31	3	5	12	-2	-6	3
79 28 39	4	5	12	-6	-12	7
70 20 40						
70 23 45	3	3	10	-3	-7	3
71 29 45	2	2	5	-3	-7	4
72 28 45	3	3	8	-5	-11	6
73 25 45	1	1	3	-1	-2	1
74 29 45	2	2	5	-5	-10	6
75 22 45	2	2	7	-5	-9	4
76 21 45	3	3	11	-4	-7	3
76 27 45	4	4	11	-4	-7	4

φ1 Φ φ2	H	K	L	U	V	W
70 20 40						
78 28 45	3	3	8	-3	-5	3
79 25 45	1	1	3	-7	-11	6
70 20 50						
70 29 53	4	3	9	-3	-5	3
71 24 53	4	3	11	-5	-8	4
71 28 51	5	4	12	-4	-7	4
75 22 56	3	2	9	-5	-6	3
75 23 53	4	3	12	-3	-4	2
75 27 56	3	2	7	-9	-11	7
76 24 53	4	3	11	-7	-9	5
76 24 56	3	2	8	-6	-7	4
76 28 51	5	4	12	-8	-11	7
76 29 53	4	3	9	-9	-12	8
77 26 59	5	3	12	-3	-3	2
77 27 56	3	2	7	-8	-9	6
78 28 59	5	3	11	-11	-11	8
79 20 56	3	2	10	-2	-2	1
79 22 56	3	2	9	-9	-9	5
79 29 53	4	3	9	-6	-7	5
70 20 60						
70 24 63	2	1	5	-8	-9	5
71 29 63	2	1	4	-9	-10	7
73 20 63	2	1	6	-2	-2	1
73 22 63	4	2	11	-11	-11	6
73 24 63	2	1	5	-5	-5	3
73 26 63	4	2	9	-3	-3	2
74 28 68	5	2	10	-6	-5	4
74 29 63	2	1	4	-4	-4	3
75 24 63	2	1	5	-12	-11	7
75 26 68	5	2	11	-5	-4	3
76 29 63	2	1	4	-11	-10	6
77 24 63	2	1	5	-7	-6	4
78 29 63	2	1	4	-7	-6	5
79 29 63	2	1	4	-10	-8	7
70 20 70						
70 21 76	4	1	11	-7	-5	3
71 24 72	3	1	7	-11	-9	6
71 28 72	3	1	6	-11	-9	7
72 27 76	4	1	6	-12	-6	7
74 27 79	5	1	10	-9	-5	5
74 26 72	3	1	6	-8	-6	5
75 21 76	4	1	11	-12	-7	5

φ1 Φ φ2	H	K	L	U	V	W
70 20 70						
76 22 76	4	1	10	-11	-6	5
76 25 79	5	1	11	-2	-1	1
76 27 76	4	1	8	-7	-4	4
77 22 72	3	1	8	-11	-7	5
78 23 79	5	1	12	-11	-5	5
78 27 79	5	1	10	-11	-5	6
79 25 76	4	1	9	-2	-1	1
79 28 72	3	1	6	-5	-3	3
70 20 80						
71 29 81	6	1	11	-5	-3	3
73 27 81	6	1	12	-11	-6	6
78 29 81	6	1	11	-12	-5	7
70 20 90						
70 22 90	2	0	5	-5	-2	2
70 27 90	1	0	2	-12	-5	6
70 27 90	1	0	2	-10	-4	5
71 20 90	4	0	11	-11	-4	4
71 21 90	3	0	8	-8	-3	3
71 27 90	1	0	2	-8	-3	4
72 24 90	5	0	11	-11	-4	5
72 29 90	6	0	11	-11	-4	6
73 23 90	5	0	12	-12	-4	5
73 24 90	4	0	9	-9	-3	4
73 27 90	1	0	2	-6	-2	3
74 22 90	2	0	5	-10	-3	4
74 29 90	5	0	9	-9	-3	5
75 23 90	3	0	7	-7	-2	3
75 27 90	1	0	2	-10	-3	5
76 20 90	4	0	11	-11	-3	4
76 24 90	5	0	11	-11	-3	5
77 21 90	3	0	8	-8	-2	3
77 23 90	5	0	12	-12	-3	5
77 27 90	1	0	2	-4	-1	2
77 29 90	6	0	11	-11	-3	6
79 22 90	2	0	5	-5	-1	2
79 24 90	4	0	9	-9	-2	4
79 29 90	5	0	9	-9	-2	5
70 30 0						
70 30 0	0	4	7	3	-7	4
70 30 0	0	7	12	5	-12	7
70 34 0	0	2	3	4	-9	6
70 36 0	0	6	11	5	-11	8

φ_1	Φ	φ_2	H	K	L	U	V	W
70 30 0								
70	36	7	1	8	11	3	-10	7
71	31	0	0	3	5	2	-5	3
71	34	0	0	2	3	5	-12	8
71	36	0	0	5	7	3	-7	5
71	38	0	0	7	9	4	-9	7
71	39	0	0	9	11	5	-11	9
72	32	0	0	5	8	3	-8	5
72	35	0	0	7	10	4	-10	7
72	37	0	0	3	4	5	-12	9
72	37	6	1	9	12	3	-11	8
72	38	8	1	7	9	1	-4	3
73	32	0	0	7	11	4	-11	7
73	37	0	0	3	4	3	-8	6
73	39	0	0	4	5	2	-5	4
74	30	0	0	7	12	4	-12	7
74	31	9	1	6	10	2	-12	7
74	34	0	0	2	3	1	-3	2
74	36	0	0	8	11	4	-11	8
74	37	9	1	6	8	2	-11	8
74	39	0	0	9	11	4	-11	9
74	39	7	1	8	10	2	-9	7
75	37	0	0	3	4	4	-12	9
75	38	0	0	7	9	3	-9	7
76	30	0	0	4	7	2	-7	4
76	31	0	0	3	5	3	-10	6
76	31	8	1	7	12	1	-7	4
76	35	0	0	7	10	3	-10	7
76	39	6	1	9	11	1	-5	4
77	32	0	0	7	11	3	-11	7
77	33	8	1	7	11	1	-8	5
77	36	0	0	5	7	2	-7	5
77	39	0	0	4	5	3	-10	8
78	30	0	0	7	12	3	-12	7
78	32	0	0	5	8	2	-8	5
78	34	0	0	2	3	3	-12	8
78	36	0	0	8	11	3	-11	8
78	36	7	1	8	11	1	-7	5
78	39	0	0	9	11	3	-11	9
79	37	0	0	3	4	1	-4	3
70 30 10								
70	32	18	1	3	5	1	-12	7
70	34	14	1	4	6	2	-11	7
71	33	11	1	5	8	1	-5	3
71	38	18	1	3	4	1	-11	8
72	30	14	1	4	7	1	-9	5
72	39	16	2	7	9	1	-8	6

φ_1	Φ	φ_2	H	K	L	U	V	W
70 30 10								
74	30	18	2	6	11	0	-11	6
74	32	18	1	3	5	0	-5	3
75	35	18	2	6	9	0	-3	2
75	38	18	1	3	4	0	-4	3
76	30	11	1	5	9	1	-11	6
76	31	16	2	7	12	0	-12	7
76	36	11	1	5	7	1	-10	7
77	33	16	2	7	11	0	-11	7
77	36	16	2	7	10	0	-10	7
77	39	16	2	7	9	0	-9	7
78	30	14	1	4	7	0	-7	4
78	34	14	1	4	6	0	-3	2
79	37	14	2	8	11	0	-11	8
70 30 20								
70	32	23	3	7	12	0	-12	7
71	31	22	2	5	9	0	-9	5
71	35	23	3	7	11	0	-11	7
71	37	23	3	7	10	0	-10	7
71	39	24	4	9	12	0	-4	3
72	33	27	2	4	7	-1	-10	6
72	34	22	2	5	8	0	-6	5
72	38	22	2	5	7	0	-7	5
73	35	21	3	8	12	0	-3	2
73	37	27	1	2	3	-1	-10	7
74	38	21	3	8	11	0	-11	8
75	37	27	1	2	3	-1	-7	5
76	35	23	3	7	11	-1	-9	6
76	37	23	3	7	10	-1	-11	8
77	31	27	3	6	11	-1	-5	3
77	37	27	1	2	3	-2	-11	8
77	38	21	3	8	11	-1	-12	9
78	34	27	3	6	10	-2	-9	6
78	38	22	2	5	7	-1	-8	6
78	39	27	4	8	11	-1	-5	4
79	37	23	3	7	10	-2	-12	9
70 30 30								
70	35	39	4	5	9	-3	-12	8
70	36	31	3	5	6	-1	-9	6
70	36	34	2	3	5	-1	-6	4
70	36	37	3	4	7	-2	-9	6
71	34	30	4	7	12	-1	-8	5
72	39	30	4	7	10	-1	-8	6
73	31	34	2	3	6	-3	-12	7
73	36	37	3	4	7	-3	-10	7

φ_1	Φ	φ_2	H	K	L	U	V	W
70 30 30								
73	38	36	5	7	11	-1	-4	3
74	36	31	3	5	8	-2	-10	7
75	30	39	4	5	11	-2	-5	3
75	36	34	2	3	5	-2	-7	5
76	32	37	3	4	8	-4	-11	7
76	33	39	4	5	10	-5	-12	8
76	36	37	3	4	7	-4	-11	8
77	33	34	4	6	11	-1	-3	2
77	34	30	4	7	12	-3	-12	8
77	36	31	3	5	8	-3	-11	8
77	39	34	4	6	9	-3	-10	8
77	39	39	4	5	8	-5	-12	10
78	36	37	3	4	7	-5	-12	9
79	31	34	2	3	6	-3	-8	5
79	32	37	3	4	8	-4	-9	6
79	36	34	2	3	5	-3	-8	6
70 30 40								
70	31	45	3	3	7	-2	-5	3
70	38	45	6	6	11	-3	-6	6
71	35	45	1	1	2	-4	-10	7
72	35	45	1	1	2	-3	-7	5
72	38	41	6	7	12	-4	-12	9
72	39	45	4	4	7	-2	-5	4
73	35	45	1	1	2	-5	-11	8
73	38	49	7	6	12	-6	-11	9
74	31	45	5	5	12	-4	-8	5
74	32	45	4	4	9	-3	-6	4
75	35	45	1	1	2	-2	-4	3
75	38	45	5	5	9	-3	-6	5
76	31	45	3	3	7	-5	-9	6
76	38	40	5	6	10	-2	-5	4
77	33	45	5	5	11	-4	-7	5
77	35	45	1	1	2	-5	-9	7
77	38	49	7	6	12	-6	-9	8
77	39	45	4	4	7	-5	-9	8
78	35	45	1	1	2	-3	-5	4
78	38	45	6	6	11	-4	-7	6
79	32	45	4	4	9	-7	-11	8
79	38	41	6	7	12	-3	-6	5
70 30 50								
70	31	56	3	2	6	-2	-3	2
70	32	53	4	3	8	-7	-12	8
70	33	50	6	5	12	-3	-6	4
71	33	59	5	3	9	-3	-4	3

φ_1	Φ	φ_2	H	K	L	U	V	W
70 30 50								
71	38	50	6	5	10	-5	-10	8
72	30	59	5	3	10	-6	-10	7
72	39	56	6	4	9	-6	-9	8
73	33	51	5	4	10	-6	-10	7
73	38	58	8	5	12	-9	-12	11
73	39	51	5	4	8	-4	-7	6
74	38	54	7	5	11	-7	-10	9
75	30	51	5	4	11	-2	-3	2
75	33	56	6	4	11	-4	-5	4
75	36	50	6	5	10	-5	-8	7
76	31	56	3	2	6	-10	-12	9
76	32	53	4	3	8	-3	-4	3
76	33	50	6	5	12	-6	-12	9
76	39	51	5	4	8	-8	-12	11
77	31	56	3	2	6	-8	-9	7
77	38	58	8	5	12	-7	-8	8
77	39	56	6	4	9	-5	-6	6
78	30	59	5	3	10	-5	-5	4
78	33	59	5	3	9	-9	-9	8
79	33	51	5	4	10	-4	-5	4
79	36	59	5	3	8	-1	-1	1
79	38	54	7	5	11	-6	-7	7
70 30 60								
70	31	68	5	2	9	-9	-9	7
70	34	68	5	2	8	-8	-8	7
70	35	69	8	3	12	-12	-12	11
70	38	69	8	3	11	-1	-1	1
71	31	63	6	3	11	-7	-8	6
71	32	67	7	3	12	-6	-6	5
71	38	68	5	2	7	-1	-1	1
71	39	60	7	4	10	-8	-11	10
71	39	66	9	4	12	-8	-9	9
72	33	63	4	2	7	-9	-10	8
72	34	60	7	4	12	-4	-5	4
72	35	67	7	3	11	-11	-11	10
72	37	67	7	3	10	-1	-1	1
72	39	63	8	4	11	-7	-8	8
73	31	68	5	2	9	-8	-7	6
73	35	69	8	3	12	-9	-8	8
74	31	63	6	3	11	-11	-11	9
74	33	63	4	2	7	-7	-7	6
74	39	60	7	4	10	-6	-7	7
75	34	63	6	3	10	-10	-10	9
75	34	66	5	2	8	-6	-5	5
75	35	67	7	3	11	-8	-7	7
75	37	63	2	1	3	-1	-1	1

φ_1	Φ	φ_2	H	K	L	U	V	W
70	30	60						
75	39	63	8	4	11	-11	-11	12
76	33	63	4	2	7	-12	-11	10
77	34	60	7	4	12	-12	-12	11
78	31	68	5	2	9	-7	-5	5
78	32	67	7	3	12	-9	-7	7
78	34	63	6	3	10	-7	-6	6
78	36	60	7	4	11	-1	-1	1
78	39	60	7	4	10	-10	-10	11
79	34	68	5	2	8	-10	-7	8
79	35	69	8	3	12	-6	-4	5
70	30	70						
70	32	72	3	1	5	-9	-8	7
70	36	79	5	1	7	-11	-8	9
70	38	77	9	2	12	-8	-6	7
71	33	74	7	2	11	-9	-7	7
72	30	79	5	1	9	-8	-5	5
72	35	72	6	2	9	-7	-6	6
73	30	76	4	1	7	-3	-2	2
73	33	79	5	1	8	-10	-6	7
73	37	76	8	2	11	-7	-5	6
74	31	74	7	2	12	-10	-7	7
74	32	72	3	1	5	-4	-3	3
74	34	76	4	1	6	-9	-6	7
75	35	72	6	2	9	-12	-9	10
75	36	74	7	2	10	-6	-4	5
77	30	72	6	2	11	-3	-2	2
77	33	74	7	2	11	-8	-5	6
77	37	76	8	2	11	-12	-7	10
78	32	72	3	1	5	-11	-7	8
78	38	72	3	1	4	-12	-8	11
79	38	72	3	1	4	-11	-7	10
79	38	77	9	2	12	-6	-3	5
70	30	80						
70	37	81	6	1	8	-6	-4	5
70	37	84	9	1	12	-5	-3	4
70	39	84	9	1	11	-8	-5	7
71	34	83	6	1	12	-7	-4	5
72	36	83	8	1	11	-9	-5	7
72	39	83	8	1	10	-7	-4	6
74	31	82	7	1	12	-8	-4	5
74	34	81	6	1	9	-11	-6	8
75	35	82	7	1	10	-4	-2	3
75	37	81	6	1	8	-11	-6	9
75	38	82	7	1	9	-6	-3	5

φ_1	Φ	φ_2	H	K	L	U	V	W
70	30	80						
77	39	84	9	1	11	-7	-3	6
78	34	83	8	1	12	-10	-4	7
79	34	81	6	1	9	-7	-3	5
70	30	90						
70	30	90	4	0	7	-7	-3	4
70	30	90	7	0	12	-12	-5	7
70	34	90	2	0	3	-9	-4	6
70	36	90	8	0	11	-11	-5	8
71	31	90	3	0	5	-5	-2	3
71	34	90	2	0	3	-12	-5	6
71	36	90	5	0	7	-7	-3	5
71	38	90	7	0	9	-9	-4	7
71	39	90	9	0	11	-11	-5	9
72	32	90	5	0	8	-8	-3	5
72	35	90	7	0	10	-10	-4	7
72	37	90	3	0	4	-12	-5	9
73	32	90	7	0	11	-11	-4	7
73	37	90	3	0	4	-8	-3	6
73	39	90	4	0	5	-5	-2	4
74	30	90	7	0	12	-12	-4	7
74	34	90	2	0	3	-3	-1	2
74	36	90	8	0	11	-11	-4	6
74	39	90	9	0	11	-11	-4	9
75	37	90	3	0	4	-12	-4	9
75	38	90	7	0	9	-9	-3	7
76	30	90	4	0	7	-7	-2	4
76	31	90	3	0	5	-10	-3	6
76	35	90	7	0	10	-10	-3	7
77	32	90	7	0	11	-11	-3	7
77	36	90	5	0	7	-7	-2	5
77	39	90	4	0	5	-10	-3	8
78	30	90	7	0	12	-12	-3	7
78	32	90	5	0	8	-8	-2	5
78	34	90	2	0	3	-12	-3	8
78	36	90	8	0	11	-11	-3	6
78	39	90	9	0	11	-11	-3	9
79	37	90	3	0	4	-4	-1	3
70	40	0						
70	42	0	0	9	10	5	-10	9
70	42	6	1	10	11	3	-8	7
70	43	0	0	11	12	6	-12	11
70	47	0	0	12	11	6	-11	12
70	48	0	0	10	9	5	-9	10
71	41	9	1	6	7	3	-11	9

φ_1	Φ	φ_2	H	K	L	U	V	W
70	40	0						
71	42	0	0	10	11	5	-11	10
71	45	0	0	1	1	1	-2	2
71	48	0	0	11	10	5	-10	11
71	49	7	1	8	7	4	-11	12
72	40	0	0	5	6	5	-12	10
72	41	0	0	6	7	3	-7	6
72	42	0	0	8	9	4	-9	8
72	42	7	1	8	9	2	-7	6
72	43	5	1	11	12	3	-9	8
72	45	0	0	1	1	5	-11	11
72	48	0	0	9	8	4	-8	9
72	49	0	0	7	6	3	-6	7
73	42	0	0	9	10	4	-10	9
73	43	0	0	11	12	5	-12	11
73	45	0	0	1	1	3	-7	7
73	45	0	0	1	1	4	-9	9
73	47	0	0	12	11	5	-11	12
73	48	0	0	10	9	4	-9	10
74	41	0	0	7	8	3	-8	7
74	45	0	0	1	1	2	-5	5
74	45	0	0	1	1	5	-12	12
74	49	0	0	8	7	3	-7	8
74	49	7	1	8	7	3	-10	11
75	41	9	1	6	7	1	-6	5
75	42	0	0	10	11	4	-11	10
75	42	6	1	9	10	2	-8	7
75	45	0	0	1	1	3	-8	8
75	48	0	0	11	10	4	-10	11
75	49	6	1	9	8	3	-11	12
76	40	0	0	5	6	2	-6	5
76	42	0	0	8	9	3	-9	8
76	42	6	1	10	11	2	-9	8
76	43	0	0	11	12	4	-12	11
76	45	0	0	1	1	4	-11	11
76	47	0	0	12	11	4	-11	12
76	48	0	0	9	8	3	-8	9
77	42	0	0	9	10	3	-10	9
77	45	0	0	1	1	1	-3	3
77	48	0	0	10	9	3	-9	10
77	49	7	1	8	7	2	-9	10
78	40	6	1	10	12	2	-11	9
78	41	0	0	6	7	2	-7	6
78	41	8	1	7	8	1	-7	6
78	43	5	1	11	12	2	-10	9
78	45	0	0	1	1	3	-10	10
78	49	0	0	7	6	2	-6	7
78	49	6	1	9	8	2	-10	11
79	40	0	0	5	6	3	-12	10

φ_1	Φ	φ_2	H	K	L	U	V	W
70	40	0						
79	41	0	0	7	8	2	-8	7
79	42	0	0	10	11	3	-11	10
79	42	7	1	8	9	1	-8	7
79	45	0	0	1	1	2	-7	7
79	45	0	0	1	1	3	-11	11
79	48	0	0	11	10	3	-10	11
79	48	6	1	10	9	2	-11	12
79	49	0	0	8	7	2	-7	8
70	40	10						
70	40	13	2	9	11	2	-9	7
70	43	16	3	9	10	1	-7	6
71	42	16	2	7	8	1	-6	5
72	40	11	1	5	6	1	-5	4
72	42	18	2	6	7	1	-12	10
72	43	10	2	11	12	2	-8	7
72	44	17	3	10	11	1	-8	7
73	42	14	2	8	9	1	-7	6
74	40	14	1	4	5	1	-9	7
74	44	15	3	11	12	1	-9	8
75	43	13	2	9	10	1	-8	7
76	40	13	2	9	11	1	-10	8
76	41	18	3	9	11	0	-11	9
76	42	16	2	6	7	0	-7	6
76	43	18	3	9	10	0	-10	9
77	40	11	1	5	6	1	-11	9
77	41	17	3	10	12	0	-6	5
77	43	11	2	10	11	1	-9	8
77	47	18	1	3	3	0	-1	1
78	42	16	2	7	8	0	-8	7
78	43	10	2	11	12	1	-10	9
78	44	17	3	10	11	0	-11	10
78	46	17	3	10	10	0	-1	1
78	49	13	2	9	8	1	-10	11
78	49	18	4	12	11	0	-11	12
79	40	14	1	4	5	0	-5	4
79	44	15	3	11	12	0	-12	11
79	46	15	3	11	11	0	-1	1
79	46	16	2	7	7	0	-1	1
79	49	11	2	10	9	1	-11	12
79	49	17	3	10	9	0	-9	10
70	40	20						
70	42	27	2	4	5	0	-5	4
70	43	27	5	10	12	0	-6	5
70	44	20	4	11	12	1	-8	7

φ_1	Φ	φ_2	H	K	L	U	V	W
70	40	20						
70	44	27	3	6	7	0	-7	6
70	45	27	4	8	9	0	-9	8
70	49	29	5	9	9	0	-1	1
70	49	29	6	11	11	0	-1	1
71	45	27	5	10	11	0	-11	10
72	40	23	3	7	9	0	-9	7
72	42	24	4	9	11	0	-11	9
72	45	24	4	9	10	0	-10	9
72	45	24	5	11	12	0	-12	11
72	48	27	1	2	2	0	-1	1
73	42	22	2	5	6	0	-6	5
73	42	27	2	4	5	-1	-12	10
73	44	23	3	7	8	0	-8	7
73	48	24	4	9	9	0	-1	1
73	48	24	5	11	11	0	-1	1
74	41	21	3	8	10	0	-5	4
74	43	29	5	9	11	-1	-8	7
74	44	22	4	10	11	0	-11	10
74	47	23	3	7	7	0	-1	1
74	47	23	5	12	12	0	-1	1
75	44	20	4	11	12	0	-12	11
75	44	21	3	8	9	0	-9	8
75	44	27	3	6	7	-1	-10	9
75	47	22	2	5	5	0	-1	1
76	42	27	2	4	5	-1	-7	6
76	46	29	6	11	12	-1	-6	6
76	47	20	4	11	11	0	-1	1
76	47	21	3	8	8	0	-1	1
77	40	27	3	6	8	-2	-11	9
77	44	23	3	7	8	-1	-11	10
77	46	29	5	9	10	-1	-5	5
77	49	20	4	11	10	0	-10	11
78	43	27	5	10	12	-2	-11	10
78	44	21	3	8	9	-1	-12	11
79	40	23	3	7	9	-1	-6	5
79	42	22	2	5	6	-1	-8	7
79	42	27	2	4	5	-2	-9	8
79	45	27	5	10	11	-1	-5	5
79	48	27	1	2	2	-2	-11	12
70	40	30						
70	41	32	5	8	11	-1	-9	7
70	49	36	8	11	12	-1	-8	8
71	40	37	3	4	6	-2	-9	7
71	41	36	5	7	10	-1	-5	4
71	42	30	4	7	9	-1	-11	9
71	45	37	3	4	5	-2	-11	10

φ_1	Φ	φ_2	H	K	L	U	V	W
70	40	30						
71	48	35	7	10	11	-1	-7	7
71	48	37	6	8	9	-1	-6	6
72	42	34	2	3	4	-1	-6	5
72	44	31	3	5	6	-1	-9	8
72	45	30	4	7	8	-1	-12	11
73	44	38	7	9	12	-3	-11	10
73	47	32	7	11	12	-1	-7	7
73	47	34	6	9	10	-1	-6	6
73	47	36	5	7	8	-1	-5	5
73	47	39	4	5	6	-1	-4	4
74	44	31	3	5	6	-2	-12	11
74	44	36	5	7	9	-3	-12	11
74	46	38	7	9	11	-2	-7	7
75	40	31	3	5	7	-1	-5	4
75	42	34	2	3	4	-2	-8	7
75	43	32	5	8	10	-2	-10	9
75	46	32	5	8	9	-1	-5	5
75	47	31	6	10	11	-1	-6	6
75	49	38	7	9	10	-3	-11	12
76	40	37	3	4	6	-2	-6	5
76	45	35	7	10	12	-2	-7	7
76	46	34	4	6	7	-1	-4	4
77	45	37	3	4	5	-1	-3	3
77	48	37	6	8	9	-3	-9	10
78	41	32	5	8	11	-2	-7	6
78	47	39	4	5	6	-4	-10	11
78	48	35	7	10	11	-3	-10	11
79	42	34	2	3	4	-4	-12	11
79	44	38	7	9	12	-3	-7	7
79	45	30	4	7	8	-1	-4	4
79	45	34	6	9	11	-1	-3	3
79	49	39	8	10	11	-2	-5	6
70	40	40						
70	40	45	3	3	5	-4	-11	9
70	43	45	2	2	3	-3	-9	8
70	44	40	5	6	8	-2	-9	8
70	45	42	8	9	12	-3	-12	11
70	47	49	8	7	10	-3	-8	6
70	48	40	5	6	7	-1	-5	5
70	48	45	7	7	9	-2	-7	7
71	40	41	6	7	11	-3	-10	8
71	41	40	5	6	9	-3	-11	9
71	42	45	7	7	11	-3	-8	7
71	43	45	2	2	3	-4	-11	10
71	46	49	7	6	9	-3	-7	7
71	47	45	3	3	4	-1	-3	3

φ_1	Φ	φ_2	H	K	L	U	V	W
70	40	40						
71	48	42	8	9	11	-1	-4	4
71	49	49	7	6	8	-4	-10	11
72	45	48	9	8	12	-4	-9	9
72	46	45	8	8	11	-3	-8	8
72	47	41	7	8	10	-2	-7	7
73	40	45	3	3	5	-3	-7	6
73	42	49	8	7	12	-6	-12	11
73	45	45	5	5	7	-2	-5	5
73	49	45	4	4	5	-4	-11	12
73	49	45	9	9	11	-3	-8	9
74	41	45	5	5	6	-5	-11	10
74	44	49	8	7	11	-1	-2	2
74	45	45	7	7	10	-3	-7	7
74	46	41	6	7	9	-1	-3	3
74	47	49	8	7	10	-5	-10	11
74	49	41	6	7	8	-3	-10	11
75	40	49	7	6	11	-7	-12	11
75	41	40	5	6	9	-3	-8	7
75	42	41	7	8	12	-4	-10	9
75	43	49	7	6	10	-4	-7	7
75	45	42	8	9	12	-3	-8	8
75	46	49	7	6	9	-6	-11	12
75	48	48	10	9	12	-3	-6	7
75	49	45	4	4	5	-3	-7	8
76	40	45	3	3	5	-5	-10	9
76	40	45	7	7	12	-4	-6	7
76	43	45	2	2	3	-1	-2	2
76	47	45	3	3	4	-5	-11	12
77	42	49	8	7	12	-5	-8	8
77	44	40	5	6	8	-2	-5	5
77	44	41	7	8	11	-3	-7	7
77	47	45	3	3	4	-4	-8	9
77	48	40	5	6	7	-3	-6	9
77	48	45	7	7	9	-3	-6	7
78	47	41	7	8	10	-4	-9	10
78	49	45	4	4	5	-5	-10	12
79	40	49	7	6	11	-5	-7	7
79	41	45	5	5	8	-3	-5	5
79	42	45	7	7	11	-4	-7	7
79	45	45	5	5	7	-5	-9	10
79	46	45	8	8	11	-4	-7	8
70	40	50						
70	41	54	7	5	10	-5	-9	8
70	43	58	8	5	10	-5	-8	8
70	45	56	9	6	11	-5	-9	9
70	46	52	9	7	11	-4	-9	9

φ_1	Φ	φ_2	H	K	L	U	V	W
70	40	50						
70	47	56	9	6	10	-6	-11	12
71	45	53	4	3	5	-1	-2	2
71	49	54	11	8	12	-4	-8	9
71	49	59	5	3	5	-3	-5	6
72	44	54	7	5	9	-4	-7	7
72	47	54	7	5	8	-5	-9	10
73	40	53	4	3	6	-6	-10	9
73	42	56	3	2	4	-2	-3	3
73	44	50	6	5	6	-1	-2	2
73	44	52	9	7	12	-5	-9	9
74	40	59	5	3	7	-4	-5	5
74	41	56	6	5	11	-3	-4	4
74	42	53	6	6	11	-5	-8	8
74	44	59	5	3	6	-6	-8	9
75	41	54	7	5	10	-5	-7	7
75	42	51	5	4	7	-3	-5	5
75	47	51	5	4	6	-4	-7	8
75	48	53	8	6	9	-3	-5	6
76	44	54	7	5	9	-7	-10	11
76	49	59	5	3	5	-4	-5	7
77	40	53	4	3	6	-3	-4	4
77	41	50	6	5	9	-2	-3	3
77	42	56	3	2	4	-8	-10	11
77	43	58	8	5	10	-5	-6	7
77	45	56	9	6	11	-4	-5	6
77	46	55	10	7	11	-5	-7	9
78	42	56	3	2	4	-6	-7	8
78	47	51	5	4	6	-6	-9	11
78	48	50	6	5	7	-5	-6	10
78	49	51	10	8	11	-2	-3	4
79	40	59	5	3	7	-7	-7	8
79	42	51	5	4	7	-8	-11	12
79	42	53	8	6	11	-7	-9	10
79	45	53	4	3	5	-7	-9	11
79	46	52	9	7	11	-3	-4	5
70	40	60						
70	42	63	4	2	5	-3	-4	4
70	42	66	5	2	6	-10	-11	12
70	44	66	10	4	11	-6	-7	8
70	45	63	8	4	9	-8	-11	12
70	46	61	11	6	12	-6	-9	10
70	47	69	8	3	8	-7	-8	10
70	48	63	2	1	2	-7	-10	12
70	48	66	9	4	9	-7	-9	11
71	40	63	6	3	8	-5	-6	6
71	41	69	8	3	10	-10	-10	11

φ_1	Φ	φ_2	H	K	L	U	V	W
70	40	60						
71	42	60	7	4	9	-5	-7	7
71	49	61	9	5	9	-6	-9	11
72	41	61	9	5	12	-7	-9	9
72	42	66	9	4	11	-9	-10	11
72	42	68	5	2	6	-6	-6	7
72	44	69	8	3	9	-9	-9	11
72	46	61	9	5	10	-5	-7	8
72	47	67	7	3	7	-6	-7	9
72	48	63	2	1	2	-3	-4	5
73	40	67	7	3	9	-9	-9	10
73	42	63	4	2	5	-8	-9	10
73	47	69	8	3	8	-8	-8	11
73	48	66	9	4	9	-8	-9	12
73	49	60	7	4	7	-5	-7	9
74	44	63	6	3	7	-9	-10	12
74	44	67	7	3	8	-4	-4	5
74	45	60	7	4	8	-4	-5	6
74	45	63	10	5	11	-7	-8	10
74	47	68	5	2	5	-5	-5	7
74	48	63	2	1	2	-5	-6	8
74	49	61	9	5	9	-7	-9	12
75	42	68	5	2	6	-8	-7	9
75	45	66	11	5	12	-3	-3	4
75	46	61	11	6	12	-6	-7	9
75	47	67	7	3	7	-7	-7	10
75	47	69	8	3	8	-9	-8	12
75	48	63	2	1	2	-7	-8	11
76	40	63	6	3	8	-8	-8	9
76	42	60	7	4	9	-7	-8	9
76	42	63	4	2	5	-5	-5	6
76	43	63	10	5	12	-4	-4	5
76	44	63	6	3	7	-7	-7	9
76	44	69	8	3	9	-6	-5	7
77	42	68	5	2	6	-10	-8	11
77	45	63	8	4	9	-3	-3	4
77	48	63	2	1	2	-2	-2	3
77	49	60	7	4	7	-6	-7	10
78	40	63	6	3	8	-11	-10	12
78	41	61	9	5	12	-6	-6	7
78	44	68	10	4	11	-5	-4	6
78	45	66	9	4	10	-8	-7	10
78	47	67	7	3	7	-8	-7	11
78	47	68	5	2	5	-6	-5	8
79	42	60	7	4	9	-9	-9	11
79	42	63	4	2	5	-7	-6	8
79	45	60	7	4	8	-8	-8	11
79	46	61	9	5	10	-5	-5	7

φ_1	Φ	φ_2	H	K	L	U	V	W
70	40	70						
70	42	72	6	2	7	-7	-7	8
70	42	74	7	2	8	-10	-9	11
70	43	72	9	3	10	-5	-5	6
70	46	76	4	1	4	-9	-8	11
71	47	72	3	1	3	-3	-3	4
71	48	76	12	3	11	-9	-8	12
71	49	77	9	2	8	-6	-5	8
72	42	72	6	2	7	-9	-8	10
72	44	70	11	4	12	-4	-4	5
72	44	73	10	3	11	-7	-6	8
72	46	74	7	2	7	-8	-7	10
72	46	76	4	1	4	-5	-4	6
73	44	75	11	3	12	-9	-7	10
73	46	79	5	1	5	-7	-5	6
73	49	70	11	4	10	-2	-2	3
74	42	72	6	2	7	-11	-9	12
74	46	74	7	2	7	-9	-7	11
74	47	72	3	1	3	-7	-6	9
75	46	79	5	1	5	-8	-5	9
75	48	76	12	3	11	-7	-5	9
75	49	75	11	3	10	-9	-7	12
76	43	72	9	3	10	-8	-6	9
76	46	76	4	1	4	-6	-4	7
76	49	79	10	2	9	-8	-5	10
77	40	76	4	1	5	-12	-7	11
77	41	73	10	3	12	-3	-2	3
77	44	73	10	3	11	-10	-7	11
77	46	74	7	2	7	-10	-7	12
77	46	79	5	1	5	-9	-5	10
77	47	72	3	1	3	-4	-3	5
78	40	79	5	1	6	-12	-6	11
78	41	72	9	3	11	-3	-2	3
79	40	76	4	1	5	-11	-6	10
79	43	79	10	2	11	-2	-1	2
79	46	76	4	1	4	-7	-4	8
79	46	79	5	1	5	-10	-5	11
79	47	72	3	1	3	-9	-6	11
70	40	80						
70	45	81	6	1	6	-8	-6	9
70	45	82	7	1	7	-10	-7	11
70	46	85	11	1	10	-6	-4	7
72	45	81	6	1	6	-9	-6	10
72	45	82	7	1	7	-11	-7	12
72	48	80	11	2	10	-4	-3	5
73	48	85	12	1	11	-7	-4	8

φ_1	Φ	φ_2	H	K	L	U	V	W
70	40	80						
73	49	84	9	1	8	-5	-3	6
74	40	84	10	1	12	-8	-4	7
75	45	81	6	1	6	-10	-6	11
75	48	81	12	2	11	-5	-3	6
75	49	83	8	1	7	-9	-5	11
76	45	81	6	1	6	-11	-6	12
76	48	84	10	1	9	-6	-3	7
78	43	80	11	2	12	-2	-1	2
78	46	85	11	1	10	-7	-3	8
79	48	80	11	2	10	-10	-5	12
79	48	85	12	1	11	-8	-3	9
70	40	90						
70	42	90	9	0	10	-10	-5	9
70	43	90	11	0	12	-12	-6	11
70	47	90	12	0	11	-11	-6	12
70	48	90	10	0	9	-9	-5	10
71	42	90	10	0	11	-11	-5	10
71	45	90	1	0	1	-2	-1	2
71	48	90	11	0	10	-10	-5	11
72	40	90	5	0	6	-12	-5	10
72	41	90	6	0	7	-7	-3	6
72	42	90	8	0	9	-9	-4	8
72	45	90	1	0	1	-11	-5	11
72	48	90	9	0	8	-8	-4	9
72	49	90	7	0	6	-6	-3	7
73	42	90	9	0	10	-10	-4	9
73	43	90	11	0	12	-12	-5	11
73	45	90	1	0	1	-7	-3	7
73	45	90	1	0	1	-9	-4	9
73	47	90	12	0	11	-11	-5	12
73	48	90	10	0	9	-9	-4	10
74	41	90	7	0	8	-8	-3	7
74	45	90	1	0	1	-12	-5	12
74	45	90	1	0	1	-5	-2	5
74	49	90	8	0	7	-7	-3	8
75	42	90	10	0	11	-11	-4	10
75	45	90	1	0	1	-8	-3	8
75	48	90	11	0	10	-10	-4	11
76	40	90	5	0	6	-6	-2	5
76	42	90	8	0	9	-9	-3	8
76	43	90	11	0	12	-12	-4	11
76	45	90	1	0	1	-11	-4	11
76	47	90	12	0	11	-11	-4	12
76	48	90	9	0	8	-8	-3	9
77	42	90	9	0	10	-10	-3	9
77	45	90	1	0	1	-3	-1	3

φ_1	Φ	φ_2	H	K	L	U	V	W
70	40	90						
77	48	90	10	0	9	-9	-3	10
78	41	90	6	0	7	-7	-2	6
78	45	90	1	0	1	-10	-3	10
78	49	90	7	0	6	-6	-2	7
79	40	90	5	0	6	-12	-3	10
79	41	90	7	0	8	-8	-2	7
79	42	90	10	0	11	-11	-3	10
79	45	90	1	0	1	-7	-2	7
79	45	90	1	0	1	-11	-3	11
79	48	90	11	0	10	-10	-3	11
79	49	90	6	0	7	-7	-2	8
70	50	0						
70	50	8	1	7	6	4	-10	11
70	51	9	1	6	5	3	-8	9
70	52	6	1	9	7	4	-9	11
70	54	0	0	11	8	5	-8	11
70	56	0	0	3	2	4	-6	9
70	59	6	1	10	6	4	-7	11
71	51	0	0	11	9	5	-9	11
71	52	0	0	9	7	4	-7	9
71	54	0	0	7	5	3	-5	7
71	54	9	2	12	9	3	-8	10
71	56	0	0	3	2	5	-8	12
71	58	7	1	8	5	1	-2	3
71	59	0	0	5	3	2	-3	5
72	50	0	0	6	5	5	-10	12
72	50	6	1	7	6	3	-9	10
72	51	6	1	10	6	2	-5	6
72	53	0	0	4	3	5	-9	12
72	53	5	1	12	9	3	-7	9
72	55	0	0	10	7	4	-7	10
72	57	9	1	6	4	2	-5	7
72	58	0	0	8	5	3	-5	8
72	58	5	1	11	7	1	-2	3
73	51	0	0	5	4	2	-4	5
73	51	9	1	6	5	2	-7	8
73	53	0	0	4	3	3	-6	8
73	56	6	1	9	6	3	-7	10
73	58	0	0	11	7	4	-7	11
74	51	0	0	11	9	4	-9	11
74	54	0	0	11	8	4	-8	11
74	55	6	1	10	7	3	-8	11
74	55	8	1	7	5	1	-3	4
74	56	0	0	3	2	1	-2	3
75	50	5	1	12	10	2	-6	7
75	50	8	1	7	6	2	-8	9

TAB. VI

FROM EULER ANGLES INTO HKL-UVW

φ_1	Φ	φ_2	H	K	L	U	V	W
70	50	0						
75	52	0	0	9	7	3	-7	9
75	53	0	0	4	3	4	-9	12
75	57	9	1	6	4	2	-7	10
75	59	6	1	10	6	2	-5	8
76	50	0	0	6	5	2	-5	6
76	53	7	1	8	6	2	-7	9
76	54	5	1	11	8	1	-3	4
76	55	0	0	10	7	3	-7	10
76	59	0	0	5	3	3	-6	10
77	51	0	0	5	4	3	-8	10
77	52	6	1	9	7	1	-4	5
77	54	0	0	7	5	2	-5	7
77	58	0	0	11	7	3	-7	11
78	51	0	0	11	9	3	-9	11
78	51	6	1	10	8	2	-9	11
78	51	9	1	6	5	1	-6	7
78	54	0	0	11	8	3	-8	11
78	56	0	0	3	2	3	-8	12
78	58	0	0	6	5	2	-5	8
79	50	0	0	6	5	3	-10	12
79	50	8	1	7	6	1	-7	8
79	51	5	1	11	9	1	-5	6
79	53	0	0	4	3	1	-3	4
70	50	10						
70	52	18	2	6	5	2	-9	10
70	57	13	2	9	6	3	-8	11
71	51	14	3	12	10	2	-8	9
71	52	11	1	5	4	2	-6	7
71	53	13	2	9	7	3	-10	12
71	54	18	3	9	7	1	-5	6
71	56	16	2	7	5	2	-7	9
71	58	18	1	3	2	2	-8	11
72	51	16	2	7	6	2	-10	11
72	56	11	2	10	7	1	-3	4
73	52	18	2	6	5	1	-7	8
73	54	14	1	4	3	1	-4	5
73	56	17	3	10	7	2	-9	12
73	58	18	1	3	2	1	-5	7
74	50	14	2	8	7	2	-11	12
74	50	18	3	9	8	1	-11	12
74	52	15	3	11	9	1	-6	7
74	53	13	2	9	7	2	-9	11
75	51	16	2	7	6	1	-8	9
75	56	16	2	7	5	1	-6	8
75	58	18	1	3	2	1	-7	10
75	59	14	2	8	5	1	-4	6

φ_1	Φ	φ_2	H	K	L	U	V	W
70	50	10						
76	50	14	2	8	7	1	-9	10
76	52	11	1	5	4	1	-5	6
76	56	17	3	10	7	1	-6	11
76	58	15	3	11	7	1	-6	9
77	54	14	1	4	3	1	-7	9
77	55	15	3	11	8	1	-9	12
77	58	10	2	11	7	1	-4	6
78	50	18	3	9	8	0	-8	9
78	52	18	2	6	5	0	-5	6
78	53	13	2	9	7	1	-8	10
78	54	10	2	11	8	1	-6	8
79	52	11	1	5	4	1	-9	11
79	54	18	3	9	7	0	-7	9
79	55	18	4	12	9	0	-3	4
70	50	20						
70	53	22	2	5	4	1	-6	7
70	56	27	2	4	3	1	-6	10
71	51	29	6	11	10	0	-10	11
71	52	23	3	7	6	1	-9	10
71	52	29	5	9	8	0	-6	9
71	55	24	4	9	7	1	-9	11
71	56	24	5	11	8	1	-7	9
72	51	27	6	12	11	0	-11	12
72	54	29	6	11	9	0	-9	11
72	57	22	4	10	7	1	-6	8
73	51	21	3	8	7	1	-10	11
73	51	27	5	10	9	0	-9	10
73	52	27	4	8	7	0	-7	8
73	53	22	2	5	4	1	-10	12
73	53	27	3	6	5	0	-5	6
73	56	29	5	9	7	0	-7	9
73	57	23	3	7	5	1	-9	12
74	50	24	5	11	10	0	-10	11
74	51	24	4	9	8	0	-8	9
74	54	27	5	10	8	0	-4	5
74	56	27	2	4	3	0	-3	4
74	57	29	6	11	8	0	-8	11
75	50	23	5	12	11	0	-11	12
75	52	23	3	7	6	0	-6	7
75	53	24	5	11	9	0	-9	11
75	58	27	5	10	7	0	-7	10
75	59	20	4	11	7	1	-8	12
76	50	22	4	10	9	0	-9	10
76	52	23	5	12	10	0	-5	6
76	55	24	4	9	7	0	-7	9
76	56	24	5	11	8	0	-8	11

φ_1	Φ	φ_2	H	K	L	U	V	W
70	50	20						
76	59	27	3	6	4	0	-2	3
77	51	21	3	8	7	0	-7	8
77	52	27	4	8	7	-1	-10	12
77	53	22	2	5	4	0	-4	5
77	55	23	5	12	9	0	-3	4
77	56	29	5	9	7	-1	-8	11
77	57	23	3	7	5	0	-5	7
77	59	24	4	9	6	0	-2	3
78	52	20	4	11	9	0	-9	11
78	55	21	3	8	6	0	-3	4
78	56	20	4	11	8	0	-8	11
78	57	22	4	10	7	0	-7	10
78	58	23	5	12	8	0	-2	3
79	52	23	3	7	6	-1	-9	11
79	53	27	3	6	5	-1	-7	9
79	54	29	6	11	9	-1	-6	8
79	56	27	2	4	3	-1	-7	10
79	57	29	6	11	8	-1	-6	9
79	59	20	4	11	7	0	-7	11
70	50	30						
70	52	30	7	12	11	0	-11	12
70	52	31	6	10	9	0	-9	10
70	53	32	5	8	7	0	-7	8
70	55	32	7	11	9	0	-9	11
70	55	39	8	10	9	-1	-10	12
70	57	34	6	9	7	0	-7	9
71	53	30	4	7	6	0	-6	7
71	54	30	7	12	10	0	-5	6
71	55	38	7	9	8	-1	-9	11
71	56	31	3	5	4	0	-4	5
71	58	32	5	8	6	0	-3	4
71	58	34	8	12	9	0	-3	4
71	58	39	9	11	9	-1	-9	12
72	54	34	6	9	8	-1	-10	12
72	55	37	6	8	7	-1	-8	10
72	57	30	7	12	9	0	-3	4
72	58	32	7	11	8	0	-6	11
72	58	39	4	5	4	-1	-8	11
73	55	36	5	7	6	-1	-7	9
73	58	30	4	7	5	0	-5	7
73	58	38	7	9	7	-1	-7	10
73	59	31	6	10	7	0	-7	10
74	53	32	5	8	7	-1	-9	11
75	55	37	6	8	7	-2	-9	12
75	56	31	3	5	4	-1	-9	12
75	59	37	3	4	3	-1	-6	9

φ_1	Φ	φ_2	H	K	L	U	V	W
70	50	30						
76	50	39	9	11	12	-2	-6	7
76	53	30	4	7	6	-1	-6	10
76	55	34	4	6	5	-1	-6	6
76	57	36	8	11	9	-1	-5	7
76	58	39	4	5	4	-1	-4	6
77	55	36	5	7	6	-2	-8	11
78	58	30	4	7	5	-1	-8	12
78	58	38	7	9	7	-2	-7	11
79	50	34	2	3	3	-3	-10	12
79	57	34	6	9	7	-1	-4	6
70	50	40						
70	50	45	6	6	7	-3	-11	12
70	52	45	9	9	10	-2	-8	9
70	53	48	10	9	10	-3	-10	12
70	55	45	1	1	1	-2	-9	11
70	57	41	7	8	7	-1	-7	9
71	51	42	10	11	12	-2	-8	9
71	51	45	8	8	9	-2	-7	8
71	53	48	9	6	9	-3	-9	11
71	55	45	1	1	1	-1	-4	5
71	57	45	12	12	11	-2	-9	12
71	59	45	7	7	6	-1	-5	7
72	50	45	5	5	6	-3	-9	10
72	53	49	8	7	8	-3	-8	10
72	55	45	1	1	1	-2	-7	9
72	57	41	6	7	6	-1	-6	8
72	57	45	11	11	10	-2	-6	11
73	51	42	9	10	11	-2	-7	8
73	51	45	7	7	8	-2	-6	7
73	51	48	11	10	12	-2	-5	6
73	52	45	11	11	12	-3	-9	11
73	53	49	7	6	7	-3	-7	9
73	56	42	9	10	9	-2	-9	12
73	58	45	10	10	9	-2	-7	10
73	59	45	6	6	5	-1	-4	6
74	51	48	10	9	11	-4	-9	11
74	52	45	10	10	11	-3	-8	10
74	55	45	1	1	1	-1	-3	4
74	56	42	8	9	8	-2	-8	11
74	57	40	5	6	5	-1	-5	7
74	58	47	12	11	10	-2	-6	9
74	59	42	10	11	9	-1	-4	6
75	50	42	8	9	10	-2	-6	7
75	50	45	6	6	7	-2	-5	6
75	53	48	9	8	9	-4	-9	12
75	55	45	1	1	1	-3	-8	11

φ_1	Φ	φ_2	H	K	L	U	V	W
70	50	40						
75	57	41	7	8	7	-2	-7	10
75	58	45	9	9	8	-2	-6	9
76	50	48	9	8	10	-2	-4	5
76	52	45	9	9	10	-3	-7	9
76	53	49	8	7	8	-4	-8	11
76	55	45	1	1	1	-2	-5	7
76	57	45	12	12	11	-3	-8	12
76	59	42	9	10	8	-2	-7	11
76	59	48	11	10	9	-2	-5	8
77	50	41	7	8	9	-2	-5	6
77	50	49	8	7	9	-4	-7	9
77	55	45	1	1	1	-3	-7	10
77	57	41	6	7	6	-2	-6	9
77	58	45	8	8	7	-2	-5	8
78	50	45	5	5	6	-2	-4	5
78	51	45	8	8	9	-3	-6	8
78	52	45	11	11	12	-4	-8	11
78	53	49	7	6	7	-4	-7	10
78	56	42	8	9	8	-3	-8	12
78	57	45	11	11	10	-3	-7	11
79	55	45	1	1	1	-1	-2	3
79	59	45	6	6	5	-3	-7	12
79	59	48	10	9	8	-2	-4	7
70	50	50						
70	50	56	3	2	3	-3	-6	7
70	53	58	11	7	10	-2	-4	5
70	55	56	6	4	5	-4	-9	12
70	56	59	5	3	4	-4	-8	11
71	50	51	11	9	12	-3	-7	8
71	51	55	10	7	10	-5	-10	12
71	52	51	5	4	5	-2	-5	6
71	54	53	12	9	11	-3	-7	9
72	50	56	3	2	3	-5	-9	11
72	52	52	9	7	9	-4	-9	11
72	54	55	10	7	9	-3	-6	8
72	58	58	8	5	6	-1	-2	3
72	59	53	4	3	3	-3	-8	12
73	50	58	8	5	8	-5	-8	10
73	51	53	4	3	4	-2	-4	5
73	53	56	12	8	11	-5	-9	12
73	53	58	8	5	7	-3	-5	7
73	57	55	10	7	8	-1	-2	3
73	57	56	9	6	7	-1	-2	3
73	59	53	4	3	3	-3	-7	11
74	51	54	7	5	7	-4	-7	9
74	56	53	12	9	10	-1	-2	3

φ_1	Φ	φ_2	H	K	L	U	V	W
70	50	50						
74	57	54	11	8	9	-1	-2	3
75	50	56	3	2	3	-2	-3	4
75	52	50	6	5	6	-3	-6	8
75	52	52	9	7	9	-5	-9	12
75	54	54	11	8	10	-4	-7	10
76	50	58	8	5	8	-6	-8	11
76	51	53	4	3	4	-5	-8	11
76	52	59	10	6	9	-3	-4	6
76	54	56	9	6	8	-4	-6	9
76	59	53	4	3	3	-3	-6	10
77	52	51	5	4	5	-3	-5	7
77	54	53	12	9	11	-5	-8	12
78	51	54	7	5	7	-5	-7	10
78	54	55	10	7	9	-5	-7	11
78	58	51	5	4	4	-4	-7	12
78	59	53	4	3	3	-3	-5	9
79	50	56	3	2	3	-5	-6	9
79	50	58	8	5	8	-7	-8	12
79	53	56	12	8	11	-4	-5	6
79	53	58	8	5	7	-5	-6	10
79	59	59	10	6	7	-4	-5	10
70	50	60						
70	50	66	11	5	10	-5	-7	9
70	52	67	7	3	6	-6	-8	11
70	55	69	8	3	6	-3	-4	6
70	56	63	4	2	3	-4	-7	10
71	52	63	8	4	7	-2	-3	4
71	53	66	11	5	9	-5	-7	10
71	54	61	11	6	9	-3	-5	7
71	56	63	4	2	3	-5	-8	12
71	58	60	7	4	5	-1	-2	3
72	51	61	11	6	10	-2	-3	4
72	51	66	9	4	8	-4	-5	7
72	52	67	12	5	10	-5	-6	9
72	59	66	9	4	6	-2	-3	5
73	53	63	6	3	5	-3	-4	6
73	53	68	5	2	4	-6	-7	11
73	55	66	9	4	7	-4	-5	8
73	55	67	12	5	9	-5	-6	10
73	58	63	10	5	7	-2	-3	5
73	58	67	12	5	8	-3	-4	7
74	51	69	8	3	7	-7	-7	11
74	52	61	9	5	8	-5	-7	10
74	54	60	12	7	10	-4	-6	9
74	57	61	11	6	8	-2	-3	5
75	51	63	12	6	11	-7	-8	12

φ_1	Φ	φ_2	H	K	L	U	V	W
70	50	60						
75	55	69	8	3	6	-6	-6	11
75	57	60	12	7	9	-2	-3	5
76	52	67	7	3	6	-3	-3	5
76	53	68	5	2	4	-4	-4	7
76	54	63	10	5	8	-5	-6	10
77	50	66	11	5	10	-5	-5	8
77	50	68	10	4	9	-8	-7	12
77	53	60	7	4	6	-4	-5	8
77	57	68	10	4	7	-1	-1	2
77	59	63	6	3	4	-5	-6	12
78	51	63	10	5	9	-3	-3	5
78	51	69	8	3	7	-6	-5	9
78	52	60	12	7	11	-6	-7	11
78	52	63	8	4	7	-7	-7	12
78	54	61	11	6	9	-6	-7	12
78	56	66	11	5	8	-1	-1	2
78	57	67	7	3	5	-1	-1	2
79	53	63	6	3	5	-5	-5	9
79	53	68	5	2	4	-6	-5	10
79	56	63	4	2	3	-1	-1	2
70	50	70						
70	52	72	6	2	5	-6	-7	10
70	54	72	9	3	7	-5	-6	9
71	51	74	7	2	6	-2	-2	3
71	54	76	4	1	3	-3	-3	5
71	56	70	11	4	8	-4	-5	6
71	57	77	9	2	6	-6	-6	11
71	59	70	11	4	7	-3	-4	7
72	50	72	9	3	8	-2	-2	3
72	55	75	11	3	8	-4	-4	7
73	52	72	6	2	5	-5	-5	8
73	54	72	9	3	7	-7	-7	12
73	56	74	7	2	5	-5	-5	9
73	56	79	10	2	7	-6	-5	10
73	58	75	11	3	7	-1	-1	2
73	59	76	8	2	5	-1	-1	2
74	52	70	11	4	9	-3	-3	5
74	52	75	11	3	9	-6	-5	9
74	52	79	5	1	4	-7	-5	10
75	53	73	10	3	8	-7	-6	11
75	53	77	9	2	7	-4	-3	6
75	58	72	3	1	2	-1	-1	2
76	57	76	12	3	8	-5	-4	9
76	57	77	9	2	6	-4	-3	7
77	50	76	8	2	7	-3	-2	4
77	54	76	4	1	3	-7	-5	11

φ_1	Φ	φ_2	H	K	L	U	V	W
70	50	70						
77	55	72	12	4	9	-7	-6	12
78	52	72	6	2	5	-4	-3	6
78	52	79	5	1	4	-5	-3	7
78	56	73	10	3	7	-4	-3	7
79	53	73	10	3	8	-6	-4	9
79	56	79	10	2	7	-5	-3	8
70	50	80						
70	55	82	7	1	5	-7	-6	11
70	56	85	12	1	8	-5	-4	8
71	50	82	7	1	6	-7	-5	9
71	51	80	11	2	9	-5	-4	7
72	51	84	10	1	8	-3	-2	4
72	57	81	6	1	4	-7	-6	12
72	59	84	10	1	6	-5	-4	9
73	54	81	12	2	9	-4	-3	6
73	58	83	8	1	5	-4	-3	7
74	51	61	6	1	5	-3	-2	4
74	51	65	11	1	9	-7	-4	9
74	52	84	9	1	7	-8	-5	11
74	53	85	12	1	9	-5	-3	7
74	56	80	11	2	7	-5	-4	9
74	58	65	11	1	7	-3	-2	5
76	50	65	12	1	10	-4	-2	5
76	53	83	8	1	6	-7	-4	10
76	56	64	9	1	6	-5	-3	8
76	56	85	12	1	8	-7	-4	11
76	57	61	6	1	4	-3	-2	5
78	50	82	7	1	6	-4	-2	5
78	55	84	10	1	7	-2	-1	3
78	58	83	8	1	5	-7	-4	12
79	54	85	11	1	8	-7	-3	10
79	55	62	7	1	5	-2	-1	3
79	59	64	10	1	6	-4	-2	7
70	50	90						
70	54	90	11	0	8	-8	-5	11
70	56	90	3	0	2	-6	-4	9
71	51	90	11	0	9	-9	-5	11
71	52	90	9	0	7	-7	-4	9
71	54	90	7	0	5	-5	-3	7
71	56	90	3	0	2	-8	-5	12
71	59	90	5	0	3	-3	-2	5
72	50	90	6	0	5	-10	-5	12
72	53	90	4	0	3	-9	-5	12
72	55	90	10	0	7	-7	-4	10

φ_1	Φ	φ_2	H	K	L	U	V	W
70	50	90						
72	58	90	8	0	5	-5	-3	8
73	51	90	5	0	4	-4	-2	5
73	53	90	4	0	3	-6	-3	8
73	58	90	11	0	7	-7	-4	11
74	51	90	11	0	9	-9	-4	11
74	54	90	11	0	8	-8	-4	11
74	56	90	3	0	2	-2	-1	3
75	52	90	9	0	7	-7	-3	9
75	53	90	4	0	3	-9	-4	12
76	50	90	6	0	5	-5	-2	6
76	55	90	10	0	7	-7	-3	10
76	59	90	5	0	3	-6	-3	10
77	51	90	5	0	4	-8	-3	10
77	54	90	7	0	5	-5	-2	7
77	58	90	11	0	7	-7	-3	11
78	51	90	11	0	9	-9	-3	11
78	54	90	11	0	8	-8	-3	11
78	56	90	3	0	2	-8	-3	12
78	58	90	8	0	5	-5	-2	8
79	50	90	6	0	5	-10	-3	12
79	53	90	4	0	3	-3	-1	4
70	60	0						
70	60	0	0	7	4	3	-4	7
70	60	0	0	12	7	5	-7	12
70	63	0	0	2	1	5	-6	12
70	63	0	0	2	1	4	-5	10
70	68	0	0	5	2	2	-2	5
71	63	0	0	2	1	3	-4	8
71	67	5	1	12	5	3	-4	9
71	69	0	0	8	3	3	-3	8
72	61	0	0	11	6	4	-6	11
72	64	9	1	6	3	3	-6	11
72	66	0	0	11	5	4	-5	11
73	60	9	2	12	7	2	-5	8
73	63	0	0	2	1	2	-3	6
73	66	0	0	9	4	3	-4	9
73	67	0	0	12	5	4	-5	12
73	68	6	1	10	4	2	-3	7
74	60	0	0	12	7	4	-7	12
74	61	0	0	9	5	3	-5	9
74	61	5	1	11	6	2	-4	7
74	68	0	0	5	2	3	-4	10
75	61	6	1	9	5	3	-7	12
75	63	0	0	2	1	3	-5	10
75	67	0	0	7	3	2	-3	7
76	60	0	0	7	4	2	-4	7

φ_1	Φ	φ_2	H	K	L	U	V	W
70	60	0						
76	61	8	1	7	4	1	-3	5
76	66	0	0	11	5	3	-5	11
77	61	0	0	11	6	3	-6	11
77	63	0	0	2	1	1	-2	4
77	67	0	0	12	5	3	-5	12
77	67	8	1	7	3	2	-5	11
77	69	0	0	8	3	2	-3	8
78	60	0	0	12	7	3	-7	12
78	60	5	1	12	7	1	-3	5
79	61	0	0	9	5	2	-5	9
79	66	0	0	9	4	2	-4	9
79	68	0	0	5	2	1	-2	5
79	68	6	1	10	4	2	-5	12
70	60	10						
70	62	10	2	11	6	3	-6	10
71	60	17	3	10	6	2	-6	9
71	68	16	2	7	3	1	-2	4
72	67	13	2	9	4	1	-2	4
72	69	11	1	5	2	3	-5	11
73	62	15	3	11	6	1	-3	5
73	64	14	1	4	2	2	-5	9
73	66	10	2	11	5	1	-2	4
74	62	13	2	9	5	1	-3	5
74	64	14	1	4	2	2	-6	11
74	67	18	3	9	4	1	-3	6
74	69	11	1	5	2	2	-4	9
74	69	17	3	10	4	2	-5	11
75	60	11	1	5	3	2	-7	11
75	61	18	4	12	7	1	-5	8
75	66	15	3	11	5	1	-3	6
77	61	16	2	7	4	1	-6	10
77	62	18	3	9	5	1	-7	12
78	60	11	1	5	3	1	-5	8
78	62	10	2	11	6	1	-4	7
78	69	11	1	5	2	1	-3	7
79	68	14	3	12	5	1	-4	9
79	68	16	2	7	3	1	-5	11
70	60	20						
70	60	21	3	8	5	2	-7	10
70	64	29	5	9	5	1	-5	8
71	61	22	2	5	3	1	-4	6
71	64	29	6	11	6	1	-6	10
71	66	27	1	2	1	1	-4	7
71	69	23	5	12	5	2	-5	10

φ_1	Φ	φ_2	H	K	L	U	V	W
70	60	20						
72	61	27	4	8	5	1	-8	12
72	62	23	3	7	4	1	-5	8
72	66	27	1	2	1	1	-5	9
72	68	23	3	7	3	1	-3	6
73	63	24	4	9	5	1	-6	10
73	66	27	1	2	1	1	-6	11
73	69	29	5	9	4	1	-5	10
74	60	21	3	8	5	1	-6	9
74	60	29	5	9	6	0	-2	3
74	61	22	2	5	3	1	-7	11
74	64	24	5	11	6	1	-7	12
74	68	24	4	9	4	1	-4	8
74	68	29	6	11	5	1	-6	12
75	61	29	6	11	7	0	-7	11
75	67	20	4	11	5	1	-4	8
75	68	24	5	11	5	1	-5	10
76	61	27	4	8	5	0	-5	8
76	64	29	5	9	5	0	-5	9
76	69	23	5	12	5	1	-5	11
77	60	24	5	11	7	0	-7	11
77	62	27	5	10	6	0	-3	5
77	62	27	6	12	7	0	-7	12
77	64	29	6	11	6	0	-6	11
78	66	27	1	2	1	0	-1	2
79	60	21	3	8	5	0	-5	8
79	61	22	2	5	3	0	-3	5
79	62	23	3	7	4	0	-4	7
79	62	23	5	12	7	0	-7	12
79	63	24	4	9	5	0	-5	9
79	64	24	5	11	6	0	-6	11
79	68	29	6	11	5	0	-5	11
79	69	29	5	9	4	0	-4	9
70	60	30						
70	60	36	5	7	5	0	-5	7
70	60	36	8	11	8	0	-6	11
70	62	38	7	9	6	0	-2	3
70	63	30	7	12	7	1	-7	11
70	64	39	9	11	7	0	-7	11
71	60	35	7	10	7	0	-7	10
71	62	37	9	12	8	0	-2	3
71	63	37	6	8	5	0	-5	6
71	65	39	4	5	3	0	-3	5
72	61	34	2	3	2	0	-2	3
72	63	36	8	11	7	0	-7	11
72	65	37	9	12	7	0	-7	12
72	67	39	9	11	6	0	-6	11

φ_1	Φ	φ_2	H	K	L	U	V	W
70	60	30						
73	62	32	7	11	7	0	-7	11
73	64	35	7	10	6	0	-3	5
73	65	36	5	7	4	0	-4	7
73	66	36	7	9	5	0	-5	9
74	60	30	7	12	8	0	-2	3
74	60	36	8	11	6	-1	-8	12
74	62	32	5	8	5	0	-5	8
74	64	34	8	12	7	0	-7	12
74	65	34	6	9	5	0	-5	9
74	66	36	8	11	6	0	-6	11
74	68	37	3	4	2	0	-1	2
74	69	39	8	10	5	0	-1	2
75	61	39	9	11	8	-1	-5	8
75	63	30	7	12	7	0	-7	12
75	63	31	3	5	3	0	-3	5
75	65	32	7	11	6	0	-6	11
75	68	35	7	10	5	0	-1	2
76	60	35	7	10	7	-1	-7	11
76	64	30	4	7	4	0	-4	7
76	67	32	5	8	4	0	-1	2
76	67	34	4	6	3	0	-1	2
77	60	36	5	7	5	-1	-5	8
77	67	30	7	12	6	0	-1	2
77	67	31	6	10	5	0	-1	2
77	69	32	7	11	5	0	-5	11
78	61	34	2	3	2	-1	-6	10
78	62	32	7	11	7	-1	-7	12
78	63	36	8	11	7	-1	-5	9
79	65	34	6	9	5	-1	-6	12
79	65	36	5	7	4	-1	-5	10
79	65	39	4	5	3	-1	-4	8
70	60	40						
70	61	45	5	5	4	-1	-7	10
70	62	45	4	4	3	-1	-8	12
70	65	41	7	8	5	0	-5	8
70	66	42	9	10	6	0	-3	5
70	67	43	11	12	7	0	-7	12
70	69	45	9	9	5	0	-5	9
70	69	45	11	11	6	0	-6	11
71	61	45	9	9	7	-1	-6	9
71	61	49	8	7	6	-1	-4	6
71	63	45	11	11	8	-1	-7	11
71	65	49	8	7	5	-1	-6	10
71	66	40	10	12	7	0	-7	12
71	67	41	6	7	4	0	-4	7
71	67	42	8	9	5	0	-5	9

φ_1	Φ	φ_2	H	K	L	U	V	W
70	60	40						
71	68	42	10	11	6	0	-6	11
72	64	45	10	10	7	-1	-6	10
72	65	45	3	3	2	-1	-7	12
73	62	41	6	7	5	-1	-7	11
73	62	45	4	4	3	-1	-5	8
73	69	40	5	6	3	0	-1	2
73	69	41	7	8	4	0	-1	2
74	60	45	11	11	9	-2	-7	11
74	62	49	7	6	5	-1	-3	5
74	64	47	12	11	8	-1	-4	7
74	65	45	3	3	2	-1	-5	9
74	66	45	11	11	7	-1	-6	11
75	61	47	12	11	9	-1	-3	5
75	62	45	4	4	3	-2	-7	12
75	63	45	7	7	5	-1	-4	7
75	65	41	7	8	5	-1	-6	11
75	67	45	5	5	3	-1	-5	10
75	67	47	12	11	7	-1	-4	8
75	68	45	12	12	7	-1	-6	12
76	61	45	5	5	4	-1	-3	5
76	63	42	9	10	7	-1	-4	7
76	64	48	9	8	6	-2	-6	11
76	66	45	8	8	5	-1	-4	8
76	67	41	6	7	4	-1	-6	12
76	69	49	8	7	4	-1	-4	9
77	62	48	11	10	8	-2	-5	9
77	63	45	11	11	8	-2	-6	11
77	66	40	10	12	7	-1	-5	10
77	69	45	11	11	6	-1	-5	11
78	60	42	8	9	7	-1	-3	5
78	61	45	9	9	7	-2	-5	9
78	65	45	3	3	2	-1	-3	6
78	67	49	7	6	4	-2	-5	11
78	69	45	9	9	5	-1	-4	9
79	61	49	8	7	6	-3	-6	11
79	64	45	10	10	7	-2	-5	10
79	68	42	10	11	6	-1	-4	9
79	68	45	7	7	4	-1	-3	7
70	60	50						
70	61	51	10	8	7	-1	-4	6
70	64	51	11	9	7	-1	-5	8
70	65	56	9	6	5	-2	-7	12
70	68	53	4	3	2	-1	-6	11
70	68	55	10	7	5	-1	-5	9
72	61	56	3	2	2	-2	-5	6
72	63	59	5	3	3	-3	-7	12

φ_1	Φ	φ_2	H	K	L	U	V	W
70	60	50						
72	65	51	5	4	3	-1	-4	7
73	62	52	9	7	6	-1	-3	5
73	65	54	7	5	4	-2	-6	11
73	67	56	6	4	3	-1	-3	6
73	68	53	4	3	2	-1	-4	6
73	69	50	6	5	3	-1	-6	12
73	69	51	10	8	5	-1	-5	10
74	63	59	5	3	3	-3	-6	11
75	61	56	3	2	2	-2	-4	7
75	66	52	9	7	5	-1	-3	6
76	61	51	11	9	8	-3	-7	12
76	63	59	5	3	3	-3	-5	10
76	64	56	12	8	7	-1	-2	4
76	65	51	5	4	3	-2	-5	10
76	66	50	12	10	7	-1	-3	6
76	68	55	10	7	5	-2	-5	11
77	63	53	8	6	5	-1	-2	4
77	64	55	10	7	6	-1	-2	4
77	67	58	8	5	4	-2	-4	9
77	69	50	6	5	3	-1	-3	7
78	66	54	11	8	6	-2	-4	9
78	67	59	10	6	5	-3	-5	12
78	69	51	10	8	5	-2	-5	12
78	69	58	11	7	5	-1	-2	5
79	61	56	3	2	2	-2	-3	6
79	63	50	6	5	4	-1	-2	4
79	63	59	5	3	3	-3	-4	9
79	68	53	4	3	2	-1	-2	5
70	60	60						
70	60	69	8	3	5	-2	-3	5
70	62	63	10	5	6	-3	-6	10
70	64	66	11	5	6	-2	-4	7
70	65	67	12	5	6	-3	-6	11
70	66	63	2	1	1	-2	-5	9
71	62	67	7	3	4	-3	-5	9
71	67	60	12	7	6	-2	-6	11
72	61	63	8	4	5	-4	-7	12
72	65	68	10	4	5	-3	-5	10
72	69	61	9	5	4	-1	-3	6
73	66	63	2	1	1	-1	-2	4
74	61	61	11	6	7	-3	-5	9
74	62	67	12	5	7	-3	-4	6
74	63	66	9	4	5	-2	-3	6
75	60	61	9	5	6	-4	-6	11
75	61	68	5	2	3	-4	-5	10
75	62	63	12	6	7	-2	-3	6

φ_1	Φ	φ_2	H	K	L	U	V	W
70	60	60						
75	65	69	8	3	4	-3	-4	9
75	66	63	2	1	1	-3	-5	11
75	68	67	7	3	3	-3	-5	12
76	61	63	8	4	5	-3	-4	8
77	60	69	8	3	5	-5	-5	11
77	65	68	10	4	5	-4	-5	12
77	66	63	2	1	1	-2	-3	7
78	60	60	12	7	8	-3	-4	8
78	61	68	5	2	3	-3	-3	7
78	66	63	2	1	1	-3	-4	10
78	68	67	7	3	3	-3	-4	11
79	62	67	7	3	4	-2	-2	5
79	65	69	8	3	4	-4	-4	11
79	68	61	11	6	5	-2	-3	8
70	60	70						
70	60	73	10	3	6	-3	-4	7
70	64	76	4	1	2	-3	-4	8
70	64	79	10	2	5	-4	-5	10
70	68	72	12	4	5	-1	-2	4
71	60	79	5	1	3	-1	-1	2
72	65	72	6	2	3	-2	-3	6
73	61	72	12	4	7	-5	-6	12
73	62	77	9	2	5	-5	-5	11
73	64	73	10	3	5	-4	-5	11
73	66	75	11	3	5	-3	-4	9
74	64	79	10	2	5	-5	-5	12
75	61	74	7	2	4	-4	-4	9
75	62	75	11	3	6	-3	-3	7
75	64	76	4	1	2	-2	-2	5
76	67	77	9	2	4	-4	-4	11
76	69	79	5	1	2	-1	-1	3
77	62	72	9	3	5	-5	-5	12
77	68	76	12	3	5	-1	-1	3
78	63	70	11	4	6	-2	-2	5
78	65	72	6	2	3	-3	-3	8
78	68	74	7	2	3	-1	-1	3
79	67	72	9	3	4	-1	-1	3
70	60	80						
70	60	81	12	2	7	-1	-1	2
70	61	82	7	1	4	-1	-1	2
70	64	84	10	1	5	-5	-5	11
70	67	82	7	1	3	-4	-5	11
71	60	85	12	1	7	-6	-5	11
71	64	83	8	1	4	-4	-4	9

φ_1	Φ	φ_2	H	K	L	U	V	W
70	60	80						
71	66	85	11	1	5	-5	-5	12
72	64	81	6	1	3	-3	-3	7
72	66	84	9	1	4	-2	-2	5
73	68	84	10	1	4	-4	-4	11
74	67	82	7	1	3	-3	-3	8
75	64	83	8	1	4	-5	-4	11
76	60	85	12	1	7	-5	-3	9
77	64	81	6	1	3	-4	-3	9
77	66	85	11	1	5	-3	-2	7
78	66	80	11	2	5	-4	-3	10
79	68	84	10	1	4	-3	-2	8
70	60	90						
70	60	90	7	0	4	-4	-3	7
70	60	90	12	0	7	-7	-5	12
70	63	90	2	0	1	-6	-5	12
70	63	90	2	0	1	-5	-4	10
70	66	90	5	0	2	-2	-2	5
71	63	90	2	0	1	-4	-3	8
71	69	90	8	0	3	-3	-3	8
72	61	90	11	0	6	-6	-4	11
72	66	90	11	0	5	-5	-4	11
73	63	90	2	0	1	-3	-2	6
73	66	90	9	0	4	-4	-3	9
73	67	90	12	0	5	-5	-4	12
74	60	90	12	0	7	-7	-4	12
74	61	90	9	0	5	-5	-3	9
74	68	90	5	0	2	-4	-3	10
75	63	90	2	0	1	-5	-3	10
75	67	90	7	0	3	-3	-2	7
76	60	90	7	0	4	-4	-2	7
76	66	90	11	0	5	-5	-3	11
77	61	90	11	0	6	-6	-3	11
77	63	90	2	0	1	-2	-1	4
77	67	90	12	0	5	-5	-3	12
77	69	90	8	0	3	-3	-2	8
78	60	90	12	0	7	-7	-3	12
79	61	90	9	0	5	-5	-2	9
79	66	90	9	0	4	-4	-2	9
79	68	90	5	0	2	-2	-1	5
70	70	0						
70	70	7	1	8	3	4	-5	12
70	72	5	1	12	4	4	-4	11
70	74	6	1	7	2	1	-1	3
70	76	0	0	4	1	3	-2	6

φ_1	Φ	φ_2	H	K	L	U	V	W
70	70	0						
71	70	0	0	11	4	4	-4	11
71	73	6	1	10	3	1	-1	3
71	75	0	0	11	3	4	-3	11
71	76	9	2	12	3	3	-3	10
72	72	0	0	3	1	1	-1	3
72	76	0	0	4	1	4	-3	12
72	77	0	0	9	2	3	-2	9
73	75	5	1	11	3	3	-3	10
73	76	7	1	8	2	2	-2	7
74	72	6	1	9	3	3	-4	11
74	72	9	1	6	2	2	-3	8
74	73	0	0	10	3	3	-3	10
74	76	5	1	12	3	3	-3	11
74	79	0	0	5	1	3	-2	10
75	74	0	0	7	2	2	-2	7
75	75	0	0	11	3	3	-3	11
75	78	6	1	9	2	1	-1	4
76	70	0	0	11	4	3	-4	11
76	76	0	0	4	1	1	-1	4
77	70	7	1	8	3	1	-2	5
77	72	0	0	3	1	3	-4	12
77	72	9	1	6	2	2	-4	11
77	79	6	1	10	2	2	-2	9
78	72	0	0	3	1	2	-3	9
78	76	7	1	8	2	2	-3	11
78	77	0	0	9	2	2	-2	9
79	73	0	0	10	3	2	-3	10
79	79	0	0	5	1	1	-1	5
70	70	10						
70	72	13	2	9	3	3	-4	10
71	72	18	1	3	1	3	-5	12
71	75	16	2	7	2	3	-4	11
71	79	17	3	10	2	2	-2	7
72	74	17	3	10	3	2	-3	8
72	78	13	2	9	2	2	-2	7
74	75	15	3	11	3	2	-3	9
74	79	11	1	5	1	1	-1	4
75	76	14	1	4	1	2	-3	10
76	70	10	2	11	4	1	-2	5
76	70	14	2	8	3	2	-5	12
76	75	16	2	7	2	1	-2	6
77	72	18	1	3	1	1	-3	8
77	79	17	3	10	2	2	-3	12
78	74	11	2	10	3	1	-2	6
79	72	18	1	3	1	1	-4	11
79	74	17	3	10	3	1	-3	9

φ_1	Φ	φ_2	H	K	L	U	V	W
70	70	10						
79	76	14	1	4	1	1	-2	7
70	70	20						
70	79	24	4	9	2	3	-4	12
71	70	27	6	12	5	1	-3	6
71	71	20	4	11	4	2	-4	9
71	75	27	5	10	3	1	-2	5
72	73	27	3	6	2	2	-5	12
72	77	21	3	6	2	2	-3	9
73	75	23	3	7	2	2	-4	11
74	72	24	5	11	4	1	-3	7
74	77	27	2	4	1	1	-2	6
75	71	21	3	8	3	1	-3	7
75	76	20	4	11	3	1	-2	6
75	79	22	2	5	1	2	-3	11
76	70	22	2	5	2	1	-4	9
76	71	27	4	8	3	1	-5	12
76	77	29	6	11	3	1	-3	9
76	79	24	4	9	2	1	-2	7
77	75	23	3	7	2	1	-3	9
78	71	20	4	11	4	1	-4	10
78	77	27	2	4	1	1	-3	10
78	79	22	2	5	1	1	-2	8
78	79	29	5	9	2	1	-3	11
79	74	22	4	10	3	1	-4	12
70	70	30						
70	71	34	8	12	5	1	-4	8
70	76	30	4	7	2	1	-2	5
70	76	35	7	10	3	2	-5	12
70	77	32	7	11	3	1	-2	5
71	71	36	5	7	3	1	-5	10
71	74	34	2	3	1	1	-3	7
71	74	36	8	11	4	1	-4	9
71	78	32	5	8	2	2	-4	11
72	70	30	4	7	3	1	-4	8
72	70	34	6	9	4	1	-6	12
72	78	39	9	11	3	1	-3	8
73	72	32	5	8	3	1	-4	9
73	77	36	5	7	2	1	-3	8
74	71	31	3	5	2	1	-5	11
74	73	32	7	11	4	1	-5	12
74	74	34	2	3	1	1	-4	10
74	76	31	6	10	3	1	-3	6
74	79	37	3	4	1	1	-3	9
75	71	39	9	11	5	0	-5	11

φ_1	Φ	φ_2	H	K	L	U	V	W
70	70	30						
75	76	35	7	10	3	1	-4	11
76	70	36	8	11	5	0	-5	11
76	71	38	7	9	4	0	-4	9
76	78	36	8	11	3	1	-4	12
77	70	34	6	9	4	0	-4	9
77	71	36	5	7	3	0	-3	7
77	72	37	9	12	5	0	-5	12
77	73	39	4	5	2	0	-2	5
77	76	30	4	7	2	1	-4	12
78	71	34	8	12	5	0	-5	12
78	72	35	7	10	4	0	-2	5
78	73	37	6	8	3	0	-3	8
78	74	36	8	11	4	0	-4	11
78	74	39	9	11	4	0	-4	11
79	70	30	4	7	3	0	-3	7
79	70	30	7	12	5	0	-5	12
79	71	31	3	5	2	0	-2	5
79	72	32	5	8	3	0	-3	8
79	73	32	7	11	4	0	-4	11
79	75	37	9	12	4	0	-1	3
79	75	38	7	9	3	0	-1	3
70	70	40						
70	72	41	6	7	3	1	-6	12
70	72	48	9	8	4	0	-1	2
70	72	49	7	6	3	0	-1	2
70	76	45	11	11	4	1	-5	11
70	79	49	8	7	2	1	-4	10
71	71	48	11	10	5	0	-1	2
71	74	41	7	8	3	1	-5	11
71	77	45	3	3	1	1	-5	12
71	78	45	10	10	3	1	-4	10
72	70	43	11	12	6	0	-1	2
72	71	45	2	2	1	0	-1	2
72	73	47	12	11	5	0	-5	11
72	73	48	10	9	4	0	-4	9
72	79	45	11	11	3	1	-4	11
73	70	42	9	10	5	0	-1	2
73	72	45	11	11	5	0	-5	11
73	73	45	9	9	4	0	-4	9
73	74	49	8	7	3	0	-3	7
73	78	41	6	7	2	1	-4	11
74	71	42	10	11	5	0	-5	11
74	72	42	8	9	4	0	-4	9
74	73	45	7	7	3	0	-3	7
74	74	45	12	12	5	0	-5	12
74	75	48	11	10	4	0	-2	5

φ_1	Φ	φ_2	H	K	L	U	V	W
70	70	40						
75	70	48	10	9	5	-1	-5	11
75	72	41	6	7	3	0	-3	7
75	73	43	11	12	5	0	-5	12
75	74	45	5	5	2	0	-2	5
75	76	47	12	11	4	0	-4	11
75	76	46	9	8	3	0	-3	8
76	72	40	10	12	5	0	-5	12
76	73	42	9	10	4	0	-2	5
76	75	45	8	8	3	0	-3	8
76	76	45	11	11	4	0	-4	11
76	77	48	10	9	3	0	-1	3
76	78	49	7	6	2	0	-1	3
77	71	45	2	2	1	-1	-5	12
77	74	41	7	8	3	0	-3	8
77	75	42	10	11	4	0	-4	11
77	77	45	3	3	1	0	-1	3
78	76	40	5	6	2	0	-1	3
78	76	42	8	9	3	0	-1	3
78	76	43	11	12	4	0	-1	3
78	78	45	10	10	3	0	-3	10
78	79	48	11	10	3	0	-3	10
78	79	49	8	7	2	0	-2	7
79	71	45	2	2	1	-1	-4	10
79	74	49	8	7	3	-1	-4	12
79	77	42	9	10	3	0	-3	10
79	79	45	7	7	2	0	-2	7
79	79	45	11	11	3	0	-3	11
70	70	50						
70	71	59	5	3	2	-1	-5	10
70	72	50	12	10	5	0	-1	2
70	73	51	5	4	2	0	-1	2
71	71	56	12	8	5	-1	-6	12
71	75	53	12	9	4	0	-4	9
71	76	55	10	7	3	0	-3	7
71	77	58	11	7	3	0	-3	7
72	72	58	8	5	3	-1	-5	11
72	74	51	11	9	4	0	-4	9
72	75	52	9	7	3	0	-3	7
72	77	54	7	5	2	0	-2	5
72	78	58	8	5	2	0	-2	5
73	76	50	6	5	2	0	-2	5
73	78	56	12	6	3	0	-3	8
74	71	52	9	7	4	-1	-5	11
74	71	54	7	5	3	-1	-4	9
74	71	59	5	3	2	-1	-3	7
74	77	51	10	8	3	0	-3	8

70 70 50

φ_1	Φ	φ_2	H	K	L	U	V	W
74	78	54	11	8	3	0	-3	8
75	79	53	4	3	1	0	-1	3
76	70	54	11	8	5	-1	-3	7
76	74	56	3	2	1	-1	-4	11
76	78	51	11	9	3	0	-1	3
77	70	56	9	6	4	-2	-5	12
77	73	58	11	7	4	-1	-3	8
77	79	50	12	10	3	0	-3	10
78	71	59	5	3	2	-2	-4	11
78	74	56	3	2	1	-1	-3	9
78	75	53	12	9	4	-1	-4	12

70 70 60

φ_1	Φ	φ_2	H	K	L	U	V	W
70	70	68	5	2	2	-2	-5	10
70	71	63	8	4	3	-1	-4	6
70	78	60	12	7	3	0	-3	7
71	70	60	12	7	5	-1	-4	8
71	76	66	11	5	3	-1	-5	12
71	79	61	9	5	2	0	-2	5
72	73	63	6	3	2	-1	-4	9
72	73	66	9	4	3	-1	-3	7
73	70	68	5	2	2	-2	-4	9
73	75	63	10	5	3	-1	-4	10
74	71	63	8	4	3	-2	-5	12
74	74	60	12	7	4	-1	-4	10
74	75	67	7	3	2	-1	-3	8
76	70	63	10	5	4	-1	-2	5
76	73	67	12	5	4	-2	-4	11
76	74	61	9	5	3	-1	-3	8
76	77	63	4	2	1	-1	-4	12
76	77	67	12	5	3	-1	-3	9
77	70	60	7	4	3	-1	-2	5
77	70	68	5	2	2	-2	-3	8
77	74	68	10	4	3	-1	-2	6
78	71	69	8	3	3	-3	-4	12
76	73	63	6	3	2	-1	-2	6
78	77	69	8	3	2	-1	-2	7
78	79	68	5	2	1	-1	-3	11
79	76	66	11	5	3	-1	-2	7
79	78	60	12	7	3	-1	-3	11

70 70 70

φ_1	Φ	φ_2	H	K	L	U	V	W
70	72	72	3	1	1	-2	-5	11
70	72	76	12	3	4	-2	-4	9
70	76	70	11	4	3	-1	-4	9
70	76	76	4	1	1	-1	-3	7

70 70 70

φ_1	Φ	φ_2	H	K	L	U	V	W
70	78	72	9	3	2	-1	-5	12
71	75	74	7	2	2	-2	-5	12
71	79	79	5	1	1	-1	-3	8
72	71	75	11	3	4	-3	-5	12
72	79	73	10	3	2	-1	-4	11
73	72	72	3	1	1	-1	-2	5
73	72	77	9	2	3	-2	-3	8
73	77	72	12	4	3	-1	-3	8
74	75	74	7	2	2	-2	-4	11
75	76	76	4	1	1	-1	-2	6
75	78	72	9	3	2	-1	-3	9
76	72	76	12	3	4	-3	-4	12
76	79	79	5	1	1	-1	-2	7
77	72	72	3	1	1	-2	-3	9
77	75	74	7	2	2	-2	-3	10
78	70	76	8	2	3	-3	-3	10
78	76	76	4	1	1	-2	-3	11
79	71	75	11	3	4	-2	-2	7
79	72	77	9	2	3	-3	-3	11
79	74	79	10	2	3	-1	-1	4
79	78	77	9	2	2	-2	-3	12
79	79	73	10	3	2	-1	-2	8

70 70 80

φ_1	Φ	φ_2	H	K	L	U	V	W
70	72	85	12	1	4	-3	-4	10
70	76	83	8	1	2	-1	-2	5
71	70	80	11	2	4	-2	-3	7
71	75	80	11	2	3	-1	-2	5
72	78	84	9	1	2	-2	-4	11
73	72	81	6	1	2	-3	-4	11
73	76	85	12	1	3	-2	-3	9
73	79	84	10	1	2	-1	-2	6
74	70	85	11	1	4	-1	-1	3
75	70	83	8	1	3	-1	-1	3
76	72	84	9	1	3	-3	-3	10
76	76	81	12	2	3	-2	-3	10
77	72	81	6	1	2	-2	-2	7
77	73	84	10	1	3	-3	-3	11
78	74	82	7	1	2	-1	-1	4
78	75	85	11	1	3	-1	-1	4
79	76	83	6	1	2	-2	-2	9

70 70 90

φ_1	Φ	φ_2	H	K	L	U	V	W
70	76	90	4	0	1	-2	-3	8
71	70	90	11	0	4	-4	-4	11
71	75	90	11	0	3	-3	-4	11

70 70 90

φ_1	Φ	φ_2	H	K	L	U	V	W
72	72	90	3	0	1	-1	-1	3
72	76	90	4	0	1	-3	-4	12
72	77	90	9	0	2	-2	-3	9
74	73	90	10	0	3	-3	-3	10
74	79	90	5	0	1	-2	-3	10
75	74	90	7	0	2	-2	-2	7
75	75	90	11	0	3	-3	-3	11
76	70	90	11	0	4	-4	-3	11
76	76	90	4	0	1	-1	-1	4
77	72	90	3	0	1	-4	-3	12
78	72	90	3	0	1	-3	-2	9
78	77	90	9	0	2	-2	-2	9
79	73	90	10	0	3	-3	-2	10
79	79	90	5	0	1	-1	-1	5

70 80 0

φ_1	Φ	φ_2	H	K	L	U	V	W
70	80	0	0	11	2	4	-2	11
70	81	9	1	6	1	3	-2	9
70	83	0	0	8	1	3	-1	8
70	85	0	0	11	1	4	-1	11
71	83	7	1	8	1	2	-1	6
71	85	5	1	12	1	3	-1	9
72	81	0	0	6	1	2	-1	6
72	84	0	0	9	1	3	-1	9
72	85	0	0	12	1	4	-1	12
73	84	0	0	10	1	3	-1	10
73	84	6	1	9	1	2	-1	7
74	82	0	0	7	1	2	-1	7
74	82	8	1	7	1	3	-2	11
75	80	0	0	11	2	3	-2	11
75	84	6	1	10	1	2	-1	8
75	85	0	0	11	1	3	-1	11
75	85	9	2	12	1	2	-1	8
76	81	0	0	6	1	3	-2	12
76	83	0	0	6	1	2	-1	8
76	85	0	0	12	1	3	-1	12
77	81	9	1	6	1	1	-1	5
77	85	5	1	11	1	2	-1	9
78	80	5	1	11	2	1	-1	5
78	84	0	0	9	1	2	-1	9
78	85	5	1	12	1	2	-1	10
79	81	5	1	12	2	2	-2	11
79	82	8	1	7	1	1	-1	6
79	84	0	0	10	1	2	-1	10

70 80 10

φ_1	Φ	φ_2	H	K	L	U	V	W
70	85	14	3	12	1	2	-1	6
71	83	14	2	8	1	3	-2	10
73	80	15	3	11	2	1	-1	4
73	81	18	2	6	1	1	-1	4
73	85	10	2	11	1	2	-1	7
73	85	17	3	10	1	3	-2	11
74	84	13	2	9	1	3	-2	12
74	85	16	4	12	1	3	-2	12
75	81	14	3	12	2	2	-2	9
76	80	10	2	11	2	2	-2	9
76	82	16	2	7	1	1	-1	5
78	84	18	3	9	1	1	-1	6
79	83	14	2	8	1	1	-1	6

70 80 20

φ_1	Φ	φ_2	H	K	L	U	V	W
70	80	20	4	11	2	2	-2	7
70	81	23	5	12	2	2	-2	7
71	84	29	5	9	1	1	-1	4
71	85	20	4	11	1	3	-2	10
72	83	23	3	7	1	1	-1	4
72	84	27	4	8	1	1	-1	4
73	80	27	5	10	2	2	-3	10
75	84	24	4	9	1	1	-1	5
75	85	27	5	10	1	1	-1	5
75	85	29	6	11	1	1	-1	5
76	82	27	3	6	1	2	-3	12
76	83	21	3	8	1	1	-1	5
77	81	29	6	11	2	1	-2	8
77	86	27	6	12	1	1	-1	6
78	85	22	4	10	1	1	-1	6
78	85	24	5	11	1	1	-1	6
79	82	27	3	6	1	1	-2	9
79	86	23	5	12	1	1	-1	7

70 80 30

φ_1	Φ	φ_2	H	K	L	U	V	W
70	80	31	3	5	1	2	-3	9
70	80	38	7	9	2	2	-4	11
71	85	31	6	10	1	1	-1	4
71	86	32	7	11	1	1	-1	4
71	86	34	8	12	1	1	-1	4
72	80	34	6	9	2	1	-2	6
72	81	39	4	5	1	1	-2	6
72	82	34	4	6	1	2	-3	10
73	83	36	5	7	1	2	-3	11
74	82	30	7	12	2	2	-3	11

φ_1	Φ	φ_2	H	K	L	U	V	W
70	80	30						
74	82	36	8	11	2	1	-2	7
74	84	37	6	8	1	2	-3	12
75	80	31	3	5	1	1	-2	7
75	80	38	7	9	2	1	-3	10
75	86	30	7	12	1	1	-1	5
76	82	34	4	6	1	1	-2	6
77	81	39	4	5	1	1	-3	11
78	82	39	9	11	2	1	-3	12
78	83	36	5	7	1	1	-2	9
79	80	31	3	5	1	1	-3	12
79	84	37	6	8	1	1	-2	10
70	80	40						
70	85	49	8	7	1	1	-2	6
71	80	45	4	4	1	1	-3	8
71	82	42	10	11	2	1	-2	6
71	83	45	6	6	1	1	-2	6
71	85	41	7	8	1	2	-3	10
71	86	45	10	10	1	2	-3	10
72	85	42	8	9	1	2	-3	11
72	86	45	11	11	1	2	-3	11
73	81	45	9	9	2	1	-3	9
73	84	45	7	7	1	1	-2	7
73	85	48	9	8	1	1	-2	7
74	80	45	4	4	1	1	-4	12
74	82	45	5	5	1	1	-3	10
74	83	40	5	6	1	1	-2	7
74	86	42	9	10	1	2	-3	12
74	87	45	12	12	1	2	-3	12
75	84	49	7	6	1	1	-3	11
75	85	45	8	8	1	1	-2	8
75	86	48	10	9	1	1	-2	8
76	83	45	11	11	2	1	-3	11
76	84	41	6	7	1	1	-2	8
77	83	45	6	6	1	1	-3	12
77	85	41	7	8	1	1	-2	9
77	86	45	9	9	1	1	-2	9
77	86	48	11	10	1	1	-2	9
78	85	42	8	9	1	1	-2	10
78	86	45	10	10	1	1	-2	10
78	86	47	12	11	1	1	-2	10
79	80	47	12	11	3	0	-3	11
79	86	42	9	10	1	1	-2	11
79	86	45	11	11	1	1	-2	11

φ_1	Φ	φ_2	H	K	L	U	V	W
70	80	50						
70	82	51	11	9	2	1	-3	8
70	86	51	10	8	1	1	-2	6
70	86	53	12	9	1	1	-2	6
71	81	51	5	4	1	1	-4	11
71	85	56	9	6	1	1	-3	9
72	83	50	6	5	1	1	-3	9
72	84	58	8	5	1	1	-4	12
73	82	53	12	9	2	1	-4	12
73	84	53	8	6	1	1	-3	10
73	86	51	11	9	1	1	-2	7
73	86	58	11	7	1	1	-3	10
74	80	59	5	3	1	0	-1	3
75	80	56	9	6	2	0	-1	3
75	85	55	10	7	1	1	-3	11
75	86	50	12	10	1	1	-2	8
76	85	52	9	7	1	1	-3	12
76	86	56	12	8	1	1	-3	12
77	80	52	9	7	2	0	-2	7
77	81	55	10	7	2	0	-2	7
77	81	58	11	7	2	0	-2	7
78	82	56	6	4	1	0	-1	4
79	81	51	5	4	1	0	-1	4
79	82	54	11	8	2	0	-1	4
70	80	60						
70	81	66	11	5	2	0	-2	5
70	81	67	12	5	2	0	-2	5
70	84	61	9	5	1	1	-4	11
71	80	63	10	5	2	0	-2	5
71	86	60	12	7	1	1	-3	9
72	86	63	12	6	1	1	-4	12
73	83	67	7	3	1	0	-1	3
73	83	69	8	3	1	0	-1	3
74	81	61	11	6	2	0	-1	3
74	82	63	6	3	1	0	-1	3
76	82	60	12	7	2	0	-2	7
77	84	63	8	4	1	0	-1	4
77	84	66	9	4	1	0	-1	4
77	85	68	10	4	1	0	-1	4
78	83	60	7	4	1	0	-1	4
70	80	70						
72	85	73	10	3	1	0	-1	3
72	85	75	11	3	1	0	-1	3
72	85	76	12	3	1	0	-1	3

φ_1	Φ	φ_2	H	K	L	U	V	W
70	80	70						
73	81	76	12	3	2	-1	-4	12
75	80	75	11	3	2	-1	-3	10
77	85	70	11	4	1	0	-1	4
77	85	72	12	4	1	0	-1	4
78	81	72	6	2	1	-1	-3	12
79	81	76	12	3	2	-1	-2	9
70	80	80						
70	81	81	6	1	1	-1	-4	10
71	82	82	7	1	1	-1	-4	11
72	83	83	8	1	1	-1	-4	12
73	81	81	6	1	1	-1	-3	9
74	82	82	7	1	1	-1	-3	10
75	81	85	12	1	2	-1	-2	7
76	83	83	8	1	1	-1	-3	11
77	81	81	6	1	1	-1	-2	8
77	84	84	9	1	1	-1	-3	12
79	82	82	7	1	1	-1	-2	9
70	80	90						
70	80	90	11	0	2	-2	-4	11
70	83	90	8	0	1	-1	-3	8
70	85	90	11	0	1	-1	-4	11
72	81	90	6	0	1	-1	-2	6
72	84	90	9	0	1	-1	-3	9
72	85	90	12	0	1	-1	-4	12
73	84	90	10	0	1	-1	-3	10
74	82	90	7	0	1	-1	-2	7
75	80	90	11	0	2	-2	-3	11
75	85	90	11	0	1	-1	-3	11
76	81	90	6	0	1	-2	-3	12
76	83	90	8	0	1	-1	-2	8
76	85	90	12	0	1	-1	-3	12
78	84	90	9	0	1	-1	-2	9
79	84	90	10	0	1	-1	-2	10
70	90	0						
70	90	0	0	1	0	4	0	11
72	90	0	0	1	0	1	0	3
73	90	0	0	1	0	3	0	10
74	90	0	0	1	0	2	0	7
75	90	0	0	1	0	3	0	11
76	90	0	0	1	0	1	0	4
77	90	0	0	1	0	2	0	9
79	90	0	0	1	0	1	0	5

φ_1	Φ	φ_2	H	K	L	U	V	W
70	90	10						
71	90	14	1	4	0	4	-1	12
72	90	18	1	3	0	3	-1	10
74	90	18	1	3	0	3	-1	11
75	90	18	1	3	0	3	-1	12
70	90	20						
70	90	27	1	2	0	2	-1	6
72	90	27	1	2	0	2	-1	7
74	90	27	1	2	0	2	-1	8
76	90	27	1	2	0	2	-1	9
77	90	27	1	2	0	2	-1	10
79	90	27	1	2	0	2	-1	12
79	90	27	1	2	0	2	-1	11
70	90	30						
70	90	34	2	3	0	3	-2	10
72	90	34	2	3	0	3	-2	11
73	90	34	2	3	0	3	-2	12
70	90	40						
71	90	45	1	1	0	1	-1	4
73	90	45	1	1	0	2	-2	9
74	90	45	1	1	0	1	-1	5
76	90	45	1	1	0	2	-2	11
77	90	45	1	1	0	1	-1	6
79	90	45	1	1	0	1	-1	7
70	90	50						
70	90	56	3	2	0	2	-3	10
72	90	56	3	2	0	2	-3	11
73	90	56	3	2	0	2	-3	12
70	90	60						
70	90	63	2	1	0	1	-2	6
72	90	63	2	1	0	1	-2	7
74	90	63	2	1	0	1	-2	8
76	90	63	2	1	0	1	-2	9
77	90	63	2	1	0	1	-2	10
79	90	63	2	1	0	1	-2	12
79	90	63	2	1	0	1	-2	11

TAB. VI

FROM EULER ANGLES INTO HKL-UVW

φ_1	Φ	φ_2	H	K	L	U	V	W
70	90	70						
71	90	76	4	1	0	1	-4	12
72	90	72	3	1	0	1	-3	10
74	90	72	3	1	0	1	-3	11
75	90	72	3	1	0	1	-3	12
70	90	90						
70	90	90	1	0	0	0	-4	11
72	90	90	1	0	0	0	-1	3
73	90	90	1	0	0	0	-3	10
74	90	90	1	0	0	0	-2	7
75	90	90	1	0	0	0	-3	11
76	90	90	1	0	0	0	-1	4
77	90	90	1	0	0	0	-2	9
79	90	90	1	0	0	0	-1	5
80	0	0						
80	5	0	0	1	11	2	-11	1
81	5	0	0	1	12	2	-12	1
81	9	0	0	1	6	1	-6	1
82	8	0	0	1	7	1	-7	1
83	7	0	0	1	8	1	-8	1
84	6	0	0	1	9	1	-9	1
84	6	0	0	1	10	1	-10	1
85	5	0	0	1	11	1	-11	1
85	5	0	0	1	12	1	-12	1
85	9	0	0	1	6	1	-12	2
80	0	40						
81	7	45	1	1	12	-5	-7	1
84	8	45	1	1	10	-9	-11	2
84	9	45	1	1	9	-4	-5	1
85	7	45	1	1	11	-5	-6	1
80	0	80						
80	0	80	0	0	1	-11	-4	0
80	0	80	0	0	1	-8	-3	0
81	0	81	0	0	1	-3	-1	0
82	0	82	0	0	1	-11	-3	0
82	0	82	0	0	1	-7	-2	0
82	0	82	0	0	1	-10	-3	0
83	0	83	0	0	1	-4	-1	0
84	0	84	0	0	1	-9	-2	0
84	0	84	0	0	1	-5	-1	0

φ_1	Φ	φ_2	H	K	L	U	V	W
80	0	80						
85	0	85	0	0	1	-11	-2	0
85	0	85	0	0	1	-6	-1	0
86	0	86	0	0	1	-8	-1	0
86	0	86	0	0	1	-7	-1	0
87	0	87	0	0	1	-11	-1	0
87	0	87	0	0	1	-9	-1	0
87	0	87	0	0	1	-10	-1	0
88	0	88	0	0	1	-12	-1	0
80	0	90						
80	5	90	1	0	11	-11	-2	1
81	5	90	1	0	12	-12	-2	1
81	9	90	1	0	6	-6	-1	1
82	8	90	1	0	7	-7	-1	1
83	7	90	1	0	8	-8	-1	1
84	6	90	1	0	9	-9	-1	1
84	6	90	1	0	10	-10	-1	1
85	5	90	1	0	11	-11	-1	1
85	5	90	1	0	12	-12	-1	1
85	9	90	1	0	6	-12	-1	2
80	10	0						
80	10	0	0	2	11	2	-11	2
80	15	0	0	3	11	2	-11	3
81	14	0	0	1	4	2	-12	3
81	18	0	0	1	3	1	-6	2
82	16	0	0	2	7	1	-7	2
83	14	0	0	1	4	1	-8	2
84	11	0	0	1	5	1	-10	2
84	13	0	0	2	9	1	-9	2
84	16	0	0	1	3	1	-9	3
85	10	0	0	2	11	1	-11	2
85	14	0	0	1	4	1	-12	3
85	15	0	0	3	11	1	-11	3
85	17	0	0	3	10	1	-10	3
85	18	0	0	1	3	1	-12	4
80	10	10						
80	16	18	1	3	11	-1	-7	2
87	15	18	1	3	12	-3	-11	3
87	19	18	1	3	9	-3	-11	4
88	16	18	1	3	11	-3	-10	3

φ_1	Φ	φ_2	H	K	L	U	V	W
80	10	20						
80	16	27	1	2	8	-2	-7	2
82	18	27	1	2	7	-1	-3	1
85	11	27	1	2	12	-2	-5	1
85	16	27	1	2	8	-4	-10	3
87	11	27	1	2	11	-4	-9	2
88	14	27	1	2	9	-5	-11	3
80	10	30						
83	17	34	2	3	12	-3	-6	2
80	10	40						
81	13	45	1	1	6	-5	-7	2
81	19	45	1	1	4	-5	-7	3
82	11	45	1	1	7	-3	-4	1
83	10	45	1	1	8	-7	-9	2
83	19	45	1	1	4	-7	-9	4
84	13	45	1	1	6	-8	-10	3
84	17	45	2	2	9	-4	-5	2
85	14	45	2	2	11	-5	-6	2
85	16	45	1	1	5	-9	-11	4
85	19	45	1	1	4	-9	-11	5
86	16	45	1	1	5	-7	-6	3
87	11	45	1	1	7	-10	-11	3
80	10	50						
82	17	56	3	2	12	-10	-9	4
83	18	56	3	2	11	-7	-6	3
87	17	56	3	2	12	-8	-6	3
89	18	56	3	2	11	-10	-7	4
80	10	60						
80	11	63	2	1	11	-4	-3	1
81	13	63	2	1	10	-11	-8	3
82	14	63	2	1	9	-10	-7	3
83	11	63	2	1	12	-9	-6	2
83	16	63	2	1	8	-3	-2	1
85	16	63	2	1	7	-6	-5	3
87	14	63	2	1	9	-7	-4	2
88	18	63	2	1	7	-11	-6	4
80	10	70						
84	19	76	4	1	12	-11	-4	4

φ_1	Φ	φ_2	H	K	L	U	V	W
80	10	70						
85	15	72	3	1	12	-7	-3	2
86	19	72	3	1	9	-8	-3	3
80	10	90						
80	10	90	2	0	11	-11	-2	2
80	15	90	3	0	11	-11	-2	3
81	14	90	1	0	4	-12	-2	3
81	18	90	1	0	3	-6	-1	2
82	16	90	2	0	7	-7	-1	2
83	14	90	1	0	4	-8	-1	2
84	11	90	1	0	5	-10	-1	2
84	13	90	2	0	9	-9	-1	2
84	16	90	1	0	3	-9	-1	3
85	10	90	2	0	11	-11	-1	2
85	14	90	1	0	4	-12	-1	3
85	15	90	3	0	11	-11	-1	3
85	17	90	3	0	10	-10	-1	3
85	18	90	1	0	3	-12	-1	4
80	20	0						
80	20	0	0	4	11	2	-11	4
80	27	0	0	1	2	2	-10	5
81	23	0	0	5	12	2	-12	5
81	24	0	0	5	11	2	-11	5
81	29	0	0	6	11	2	-11	6
82	27	0	0	1	2	1	-6	3
82	27	9	1	6	12	0	-2	1
82	29	9	1	6	11	0	-11	6
83	21	0	0	3	8	1	-8	3
83	23	0	0	3	7	1	-7	3
84	24	0	0	4	9	1	-9	4
84	27	0	0	1	2	1	-8	4
84	29	0	0	5	9	1	-9	5
85	20	0	0	4	11	1	-11	4
85	22	0	0	2	5	1	-10	4
85	24	0	0	5	11	1	-11	5
85	27	0	0	1	2	1	-10	5
85	29	0	0	6	11	1	-11	6
86	23	0	0	5	12	1	-12	5
86	27	0	0	1	2	1	-12	6
87	29	9	1	6	11	-1	-9	5
80	20	10						
80	23	11	1	5	12	0	-12	5
80	25	11	1	5	11	0	-11	5

φ_1	Φ	φ_2	H	K	L	U	V	W
80	20	10						
80	27	11	1	5	10	0	-2	1
82	25	14	1	4	9	-1	-11	5
83	22	18	1	3	8	-1	-5	2
83	24	18	1	3	7	-2	-11	5
84	21	14	1	4	11	-1	-8	3
86	22	14	1	4	10	-2	-12	5
87	23	11	1	5	12	-1	-7	3
87	28	18	1	3	6	-3	-11	6
80	20	20						
80	24	27	1	2	5	-3	-11	5
81	29	27	1	2	4	-2	-7	4
83	26	22	2	5	11	-1	-4	2
83	29	27	1	2	4	-4	-12	7
86	20	27	1	2	6	-2	-5	2
86	29	27	1	2	4	-2	-5	3
88	22	27	2	4	11	-4	-9	4
88	26	27	2	4	9	-5	-11	6
80	20	30						
80	22	34	2	3	9	-3	-7	3
81	24	37	3	4	11	-1	-2	1
81	27	34	2	3	7	-4	-9	5
83	20	34	2	3	10	-5	-10	4
83	24	34	2	3	8	-1	-2	1
83	28	31	3	5	11	-3	-7	4
84	27	37	3	4	10	-4	-7	4
84	28	39	4	5	12	-5	-8	5
85	22	34	2	3	9	-6	-11	5
87	20	34	2	3	10	-7	-12	5
87	23	37	3	4	12	-4	-6	3
87	29	37	3	4	9	-2	-3	2
88	26	31	3	5	12	-5	-9	5
88	27	34	2	3	7	-3	-5	3
88	28	39	4	5	12	-9	-12	8
80	20	40						
80	23	45	3	3	10	-4	-6	3
80	25	45	1	1	3	-6	-9	5
80	29	45	2	2	5	-2	-3	2
81	25	45	1	1	3	-5	-7	4
82	22	45	2	2	7	-3	-4	2
83	25	45	1	1	3	-9	-12	7
84	25	45	1	1	3	-4	-5	3
84	28	45	3	3	8	-7	-9	6

φ_1	Φ	φ_2	H	K	L	U	V	W
80	20	40						
85	21	45	3	3	11	-5	-6	3
85	23	45	3	3	10	-9	-11	6
85	27	45	4	4	11	-5	-6	4
85	29	45	2	2	5	-9	-11	8
87	22	45	2	2	7	-10	-11	6
87	25	45	1	1	3	-7	-8	5
87	29	45	2	2	5	-7	-8	6
88	25	45	1	1	3	-10	-11	7
80	20	50						
80	24	53	4	3	11	-9	-10	6
80	24	56	3	2	8	-8	-8	5
80	27	53	4	3	10	-11	-12	8
80	27	56	3	2	7	-7	-7	5
80	28	59	5	3	11	-10	-9	7
82	23	53	4	3	12	-12	-12	7
82	24	56	3	2	8	-10	-9	6
83	24	53	4	3	11	-11	-11	7
83	27	53	4	3	10	-10	-10	7
83	29	53	4	3	9	-9	-9	7
84	24	56	3	2	8	-12	-10	7
84	28	51	5	4	12	-4	-4	3
84	28	59	5	3	11	-9	-7	6
85	27	56	3	2	7	-6	-5	4
85	29	53	4	3	9	-12	-11	9
86	23	53	4	3	12	-9	-8	5
86	27	53	4	3	10	-9	-8	6
87	22	56	3	2	9	-4	-3	2
88	26	59	5	3	12	-12	-6	7
88	27	56	3	2	7	-11	-8	7
89	28	59	5	3	11	-8	-5	5
80	20	60						
80	22	63	4	2	11	-4	-3	2
80	24	63	2	1	5	-9	-7	5
81	20	63	2	1	6	-11	-8	5
81	24	63	2	1	5	-11	-8	6
81	28	68	5	2	10	-8	-5	5
82	26	63	4	2	9	-10	-7	6
83	20	63	2	1	6	-9	-6	4
83	26	68	5	2	11	-9	-5	5
84	29	63	2	1	4	-3	-2	2
86	24	68	5	2	12	-2	-1	1
86	28	68	5	2	10	-10	-5	6
87	20	63	2	1	6	-7	-4	3
87	26	63	4	2	9	-7	-4	4

φ_1	Φ	φ_2	H	K	L	U	V	W
80	20	60						
88	29	63	2	1	4	-11	-6	7
89	28	68	5	2	10	-12	-5	7
80	20	70						
80	22	72	3	1	8	-9	-5	4
81	22	76	4	1	10	-9	-4	4
81	27	76	4	1	8	-9	-4	5
83	21	76	4	1	11	-5	-2	2
83	23	79	5	1	12	-9	-3	4
83	24	72	3	1	7	-2	-1	1
83	28	72	3	1	6	-12	-6	7
85	27	76	4	1	8	-11	-4	6
86	22	72	3	1	8	-7	-3	3
86	28	72	3	1	6	-7	-3	4
86	22	76	4	1	10	-7	-2	3
80	20	80						
84	29	81	6	1	11	-7	-2	4
80	20	90						
80	20	90	4	0	11	-11	-2	4
80	27	90	1	0	2	-10	-2	5
81	23	90	5	0	12	-12	-2	5
81	24	90	5	0	11	-11	-2	5
81	29	90	6	0	11	-11	-2	6
82	27	90	1	0	2	-6	-1	3
83	21	90	3	0	8	-8	-1	3
83	23	90	3	0	7	-7	-1	3
84	24	90	4	0	9	-9	-1	4
84	27	90	1	0	2	-8	-1	4
84	29	90	5	0	9	-9	-1	5
85	20	90	4	0	11	-11	-1	4
85	22	90	2	0	5	-10	-1	4
85	24	90	5	0	11	-11	-1	5
85	27	90	1	0	2	-10	-1	5
85	29	90	6	0	11	-11	-1	6
86	23	90	5	0	12	-12	-1	5
86	27	90	1	0	2	-12	-1	6
80	30	0						
80	31	0	0	3	5	1	-5	3
80	34	0	0	2	3	2	-9	6
80	38	0	0	7	9	2	-9	7
81	32	0	0	7	11	2	-11	7

φ_1	Φ	φ_2	H	K	L	U	V	W
80	30	0						
81	35	0	0	7	10	2	-10	7
81	39	0	0	4	5	1	-5	4
82	30	0	0	7	12	2	-12	7
82	31	9	1	6	10	0	-5	3
82	34	0	0	2	3	1	-6	4
82	34	9	1	6	9	0	-3	2
82	36	0	0	8	11	2	-11	8
82	37	0	0	3	4	2	-12	9
82	37	9	1	6	8	0	-4	3
82	39	0	0	9	11	2	-11	9
83	30	0	0	4	7	1	-7	4
83	31	6	1	7	12	0	-12	7
83	33	8	1	7	11	0	-11	7
83	35	8	1	7	10	0	-10	7
83	36	0	0	5	7	1	-7	5
84	32	0	0	5	8	1	-8	5
84	34	7	1	8	12	0	-3	2
84	36	7	1	8	11	0	-11	8
84	37	0	0	3	4	1	-8	6
84	38	8	1	7	9	0	-9	7
84	39	7	1	8	10	0	-5	4
85	31	0	0	3	5	1	-10	6
85	34	0	0	2	3	1	-9	6
85	35	0	0	7	10	1	-10	7
85	37	6	1	9	12	0	-4	3
85	38	0	0	7	9	1	-9	7
85	39	6	1	9	11	0	-11	9
86	30	0	0	7	12	1	-12	7
86	32	0	0	7	11	1	-11	7
86	34	0	0	2	3	1	-12	8
86	36	0	0	8	11	1	-11	8
86	37	0	0	3	4	1	-12	9
86	39	0	0	4	5	1	-10	8
86	39	0	0	9	11	1	-11	9
80	30	10						
80	30	11	1	5	9	0	-9	5
80	32	18	1	3	5	-1	-8	5
80	33	11	1	5	8	0	-8	5
80	38	13	2	9	12	0	-4	3
80	38	18	1	3	4	-1	-9	7
81	30	18	2	6	11	-1	-7	4
81	31	16	2	7	12	-1	-10	6
81	36	11	1	5	7	0	-7	5
82	30	14	1	4	7	-1	-12	7
82	39	16	2	7	9	-1	-10	8
83	32	18	1	3	5	-2	-11	7

φ1	Φ	φ2	H	K	L	U	V	W
80	**30**	**10**						
84	37	14	2	8	11	-1	-8	6
84	38	18	1	3	4	-1	-5	4
85	33	11	1	5	8	-1	-11	7
85	33	16	2	7	11	-1	-6	4
85	36	11	1	5	7	-1	-11	8
86	39	16	2	7	9	-2	-11	9
87	30	11	1	5	9	-1	-7	4
87	35	18	2	6	9	-3	-11	8
88	30	14	1	4	7	-1	-5	3
88	30	18	2	6	11	-3	-10	6
88	31	16	2	7	12	-2	-8	5
88	34	14	1	4	6	-2	-10	7
88	36	16	2	7	10	-1	-4	3
88	38	18	1	3	4	-3	-11	9
89	39	16	2	7	9	-3	-12	10
80	**30**	**20**						
80	34	22	2	5	8	-1	-6	4
80	37	27	1	2	3	-1	-4	3
81	31	22	2	5	9	-1	-5	3
82	39	24	4	9	12	-3	-12	10
83	33	27	2	4	7	-1	-3	2
83	37	27	1	2	3	-3	-9	7
83	38	22	2	5	7	-2	-9	7
84	35	23	3	7	11	-2	-7	5
86	32	23	3	7	12	-1	-3	2
86	34	22	2	5	8	-3	-10	7
86	38	22	2	5	7	-3	-10	8
87	39	24	4	9	12	-3	-8	7
88	31	22	2	5	9	-4	-11	7
88	31	27	3	6	11	-4	-9	6
68	35	21	3	8	12	-4	-12	9
88	37	27	1	2	3	-5	-11	9
88	39	27	4	8	11	-4	-9	8
89	38	22	2	5	7	-4	-11	9
80	**30**	**30**						
80	33	39	4	5	10	-5	-10	7
80	36	31	3	5	8	-4	-12	9
80	38	32	5	8	12	-4	-11	9
81	39	30	4	7	10	-2	-6	5
81	39	39	4	5	8	-4	-8	7
82	36	34	2	3	5	-4	-9	7
82	39	34	4	6	9	-3	-7	6
63	30	39	4	5	11	-7	-12	6
83	39	39	4	5	8	-7	-12	11

φ1	Φ	φ2	H	K	L	U	V	W
80	**30**	**30**						
84	30	31	3	5	10	-5	-11	7
84	31	34	2	3	6	-3	-6	4
84	32	37	3	4	8	-4	-7	5
84	33	39	4	5	10	-5	-8	6
84	36	34	2	3	5	-5	-10	8
85	38	36	5	7	11	-5	-9	8
86	36	34	2	3	5	-6	-11	9
86	38	32	5	8	12	-4	-6	7
86	39	34	4	6	9	-6	-11	10
87	30	39	4	5	11	-5	-7	5
87	32	37	3	4	8	-8	-12	9
87	34	30	4	7	12	-2	-4	3
87	36	34	2	3	5	-7	-12	10
88	30	31	3	5	10	-5	-9	6
88	31	34	2	3	6	-6	-10	7
88	39	30	4	7	10	-5	-10	9
89	33	31	3	5	9	-7	-12	9
89	39	39	4	5	8	-3	-4	4
80	**30**	**40**						
80	35	45	1	1	2	-7	-11	9
80	38	45	5	5	9	-7	-11	10
81	33	40	5	6	12	-6	-11	8
81	35	45	1	1	2	-4	-6	5
82	31	45	5	5	12	-5	-7	5
82	35	45	1	1	2	-5	-7	6
83	31	45	3	3	7	-3	-4	3
83	35	45	1	1	2	-6	-8	7
83	38	40	5	6	10	-6	-10	9
83	38	49	7	6	12	-6	-7	7
84	35	45	1	1	2	-7	-9	8
84	39	45	4	4	7	-3	-4	4
85	32	45	4	4	9	-4	-5	4
85	33	40	5	6	12	-6	-9	7
85	35	45	1	1	2	-9	-11	10
85	35	45	1	1	2	-8	-10	9
85	38	41	6	7	12	-8	-12	11
85	38	45	5	5	9	-4	-5	5
86	33	45	5	5	11	-5	-6	5
86	35	45	1	1	2	-10	-12	11
86	38	45	6	6	11	-5	-6	6
88	31	45	3	3	7	-10	-11	9
88	39	45	4	4	7	-10	-11	12
89	38	40	5	6	10	-4	-5	5
89	38	41	6	7	12	-5	-6	6

φ1	Φ	φ2	H	K	L	U	V	W
80	**30**	**50**						
80	31	56	3	2	6	-6	-6	5
80	39	51	5	4	8	-4	-5	5
81	32	53	4	3	8	-11	-12	10
81	33	50	6	5	12	-5	-6	5
81	33	56	6	4	11	-11	-11	10
81	36	56	3	2	5	-1	-1	1
81	39	56	6	4	9	-9	-9	10
82	30	59	5	3	10	-12	-10	9
82	36	54	7	5	12	-1	-1	1
82	38	50	6	5	10	-5	-6	6
83	31	56	3	2	6	-10	-9	8
83	32	53	4	3	8	-8	-8	7
63	33	59	5	3	9	-6	-5	5
83	36	53	4	3	7	-1	-1	1
83	38	54	7	5	11	-11	-11	12
84	33	56	6	4	11	-7	-6	6
85	30	51	5	4	11	-11	-11	9
85	33	51	5	4	10	-10	-10	9
85	35	51	5	4	9	-1	-1	1
85	38	58	8	5	12	-5	-4	5
65	39	51	5	4	8	-8	-8	9
86	30	59	5	3	10	-7	-5	5
86	33	50	6	5	12	-12	-12	11
86	35	50	6	5	11	-1	-1	1
86	38	50	6	5	10	-10	-10	11
67	30	51	5	4	11	-9	-8	7
67	31	56	3	2	6	-4	-3	3
88	32	53	4	3	8	-5	-4	4
88	33	59	5	3	9	-9	-6	7
88	38	54	7	5	11	-5	-4	5
68	39	56	6	4	9	-4	-3	4
89	33	50	6	5	12	-7	-6	6
69	33	51	5	4	10	-6	-5	5
89	33	56	6	4	11	-10	-7	8
80	**30**	**60**						
80	33	63	4	2	7	-5	-4	4
60	34	60	7	4	12	-8	-7	7
60	39	66	9	4	12	-4	-3	4
81	31	63	6	3	11	-4	-3	3
61	32	67	7	3	12	-12	-8	9
82	34	63	6	3	10	-11	-8	9
82	37	63	2	1	3	-11	-8	10
62	37	63	2	1	3	-12	-9	11
62	39	63	8	4	11	-4	-3	4
83	37	63	2	1	3	-10	-7	9

φ1	Φ	φ2	H	K	L	U	V	W
80	**30**	**60**						
84	34	60	7	4	12	-12	-9	10
84	35	67	7	3	11	-5	-3	4
84	37	63	2	1	3	-9	-6	8
84	39	60	7	4	10	-4	-3	4
85	33	63	4	2	7	-8	-5	6
86	31	68	5	2	9	-6	-3	4
86	34	68	5	2	8	-4	-2	3
86	37	63	2	1	3	-8	-5	7
87	35	69	8	3	12	-9	-4	7
87	37	63	2	1	3	-7	-4	6
88	33	63	4	2	7	-11	-6	8
88	39	66	9	4	12	-12	-6	11
89	36	68	5	2	7	-12	-5	10
80	**30**	**70**						
80	32	72	3	1	5	-7	-4	5
80	34	76	4	1	6	-4	-2	3
80	35	72	6	2	9	-5	-3	4
80	36	79	5	1	7	-9	-4	7
80	38	72	3	1	4	-10	-6	9
81	30	76	4	1	7	-11	-5	7
81	30	79	5	1	9	-5	-2	3
81	31	74	7	2	12	-6	-3	4
61	36	74	7	2	10	-10	-5	8
82	38	72	3	1	4	-9	-5	8
63	32	72	3	1	5	-10	-5	7
64	30	76	4	1	7	-8	-3	5
64	33	74	7	2	11	-7	-3	5
84	33	79	5	1	8	-3	-1	2
84	37	76	8	2	11	-5	-2	4
84	38	72	3	1	4	-8	-4	7
85	34	76	4	1	6	-11	-4	8
86	38	72	3	1	4	-7	-3	6
87	38	77	9	2	12	-10	-3	8
88	30	79	5	1	9	-12	-3	7
88	34	76	4	1	6	-7	-2	5
88	35	72	6	2	9	-8	-3	6
88	36	79	5	1	7	-4	-1	3
80	**30**	**60**						
80	37	81	6	1	8	-5	-2	4
60	37	64	9	1	12	-9	-3	7
81	31	61	6	1	10	-11	-4	7
81	33	62	7	1	11	-3	-1	2
81	36	82	7	1	9	-11	-4	9
61	39	63	8	1	10	-6	-2	5

TAB. VI FROM EULER ANGLES INTO HKL-UVW

φ_1	Φ	φ_2	H	K	L	U	V	W
80	30	80						
84	34	81	6	1	9	-10	-3	7
84	35	82	7	1	10	-11	-3	8
84	36	83	8	1	11	-4	-1	3
86	31	81	6	1	10	-8	-2	5
87	31	82	7	1	12	-5	-1	3
87	38	82	7	1	9	-5	-1	4
88	37	81	6	1	8	-9	-2	7
88	39	83	8	1	10	-11	-2	9
88	39	84	9	1	11	-6	-1	5
80	30	90						
80	31	90	3	0	5	-5	-1	3
80	34	90	2	0	3	-9	-2	6
80	38	90	7	0	9	-9	-2	7
81	32	90	7	0	11	-11	-2	7
81	35	90	7	0	10	-10	-2	7
81	39	90	4	0	5	-5	-1	4
82	30	90	7	0	12	-12	-2	7
82	34	90	2	0	3	-6	-1	4
82	36	90	8	0	11	-11	-2	8
82	37	90	3	0	4	-12	-2	9
82	39	90	9	0	11	-11	-2	9
83	30	90	4	0	7	-7	-1	4
83	36	90	5	0	7	-7	-1	5
84	32	90	5	0	8	-8	-1	5
84	37	90	3	0	4	-8	-1	6
85	31	90	3	0	5	-10	-1	6
85	34	90	2	0	3	-9	-1	6
85	35	90	7	0	10	-10	-1	7
85	38	90	7	0	9	-9	-1	7
86	30	90	7	0	12	-12	-1	7
86	32	90	7	0	11	-11	-1	7
86	34	90	2	0	3	-12	-1	8
86	36	90	8	0	11	-11	-1	8
86	37	90	3	0	4	-12	-1	9
86	39	90	4	0	5	-10	-1	8
86	39	90	9	0	11	-11	-1	9
80	40	0						
80	43	0	0	11	12	3	-12	11
80	45	0	0	1	1	1	-4	4
80	47	0	0	12	11	3	-11	12
81	42	0	0	8	9	2	-9	8
81	42	6	1	9	10	1	-9	8
81	45	0	0	1	1	2	-9	9
81	48	0	0	9	8	2	-8	9

φ_1	Φ	φ_2	H	K	L	U	V	W
80	40	0						
81	49	7	1	8	7	1	-8	9
82	42	0	0	9	10	2	-10	9
82	42	0	0	10	11	2	-11	10
82	42	6	1	10	11	1	-10	9
82	43	5	1	11	12	1	-11	10
82	45	0	0	1	1	1	-5	5
82	48	0	0	10	9	2	-9	10
82	48	0	0	11	10	2	-10	11
82	48	6	1	10	9	1	-10	11
82	49	6	1	9	8	1	-9	10
83	40	0	0	5	6	1	-6	5
83	41	9	1	6	7	0	-7	6
83	43	0	0	11	12	2	-12	11
83	45	0	0	1	1	2	-11	11
83	45	0	0	1	1	1	-6	6
83	45	9	1	6	6	0	-1	1
83	47	0	0	12	11	2	-11	12
83	48	5	1	11	10	1	-11	12
84	41	0	0	6	7	1	-7	6
84	41	8	1	7	8	0	-6	7
84	45	0	0	1	1	1	-7	7
84	45	8	1	7	7	0	-1	1
84	48	9	2	12	11	0	-11	12
84	49	0	0	7	6	1	-6	7
85	41	0	0	7	8	1	-8	7
85	42	0	0	8	9	1	-9	8
85	42	6	1	9	10	0	-10	9
85	42	7	1	8	9	0	-9	8
85	45	0	0	1	1	1	-8	8
85	45	7	1	8	8	0	-1	1
85	48	0	0	9	8	1	-6	9
85	49	0	0	6	7	1	-7	8
85	49	7	1	8	7	0	-7	8
86	40	0	0	5	6	1	-12	10
86	40	6	1	10	12	0	-6	5
86	42	0	0	9	10	1	-10	9
86	42	0	0	10	11	1	-11	10
86	42	6	1	10	11	0	-11	10
86	43	0	0	11	12	1	-12	11
86	43	5	1	11	12	0	-12	11
86	45	0	0	1	1	1	-9	9
86	45	0	0	1	1	1	-11	11
86	45	0	0	1	1	1	-10	10
86	45	5	1	11	11	0	-1	1
86	45	6	1	9	9	0	-1	1
86	45	6	1	10	10	0	-1	1
86	47	0	0	12	11	1	-11	12
86	48	0	0	10	9	1	-9	10

φ_1	Φ	φ_2	H	K	L	U	V	W
80	40	0						
86	48	0	0	11	10	1	-10	11
86	48	6	1	10	9	0	-9	10
86	49	6	1	9	8	0	-8	9
87	45	0	0	1	1	1	-12	12
87	45	5	1	12	12	0	-1	1
87	48	5	1	11	10	0	-10	11
87	48	5	1	12	11	0	-11	12
88	41	9	1	6	7	-1	-8	7
88	48	9	2	12	11	-1	-9	10
89	41	8	1	7	8	-1	-9	8
89	42	6	1	9	10	-1	-11	10
89	42	6	1	10	11	-1	-12	11
89	42	7	1	8	9	-1	-10	9
80	40	10						
80	40	13	2	9	11	0	-11	9
80	42	14	2	8	9	0	-9	8
80	46	14	1	4	4	0	-1	1
80	49	15	3	11	10	0	-10	11
81	40	11	1	5	6	0	-6	5
81	42	18	2	6	7	-1	-9	8
81	43	13	2	9	10	0	-10	9
81	46	13	2	9	9	0	-1	1
81	48	14	3	12	11	0	-11	12
82	41	18	3	9	11	-1	-7	6
82	42	16	2	7	8	-1	-10	9
82	43	10	2	11	12	0	-12	11
82	43	11	2	10	11	0	-11	10
82	46	11	1	5	5	0	-1	1
82	49	11	2	10	9	0	-9	10
82	49	13	2	9	8	0	-8	9
83	40	14	1	4	5	-1	-11	9
83	42	14	2	8	9	-1	-11	10
83	45	10	2	11	11	0	-1	1
83	48	10	2	11	10	0	-10	11
83	49	18	4	12	11	-1	-7	8
84	40	13	2	9	11	-1	-12	10
84	42	18	2	6	7	-2	-11	10
84	43	13	2	9	10	-1	-12	11
85	42	16	2	7	8	-2	-12	11
85	48	14	3	12	11	-1	-8	9
85	49	15	3	11	10	-1	-7	8
85	49	17	3	10	9	-1	-6	7
86	40	14	1	4	5	-1	-6	5
87	41	17	3	10	12	-2	-9	8
88	40	11	1	5	6	-1	-7	6
88	47	18	1	3	3	-3	-11	12

φ_1	Φ	φ_2	H	K	L	U	V	W
80	40	10						
88	48	10	2	11	10	-1	-8	9
88	49	11	2	10	9	-1	-7	8
88	49	13	2	9	8	-1	-6	7
89	41	18	3	9	11	-3	-10	9
89	47	16	1	3	3	-3	-10	11
89	49	18	4	12	11	-3	-10	12
80	40	20						
80	46	27	1	2	2	-2	-10	11
81	45	24	5	11	12	-1	-5	5
81	45	27	4	8	9	-1	-4	4
81	46	27	1	2	2	-2	-9	10
82	40	27	3	6	8	-2	-7	6
82	41	29	5	9	12	-3	-9	8
82	42	22	2	5	6	-2	-10	9
82	42	27	2	4	5	-3	-11	10
82	46	27	1	2	2	-2	-8	9
83	45	24	4	9	10	-1	-4	4
83	46	27	1	2	2	-2	-7	8
83	49	20	4	11	10	-1	-6	7
84	42	22	2	5	6	-3	-12	11
84	44	22	4	10	11	-1	-4	4
84	44	27	3	6	7	-1	-3	3
85	41	21	3	8	10	-2	-8	7
85	43	29	5	9	11	-2	-5	5
85	46	29	5	9	10	-4	-10	11
85	48	27	1	2	2	-2	-6	7
86	40	23	3	7	9	-3	-9	8
86	40	27	3	6	8	-4	-10	9
86	44	20	4	11	12	-1	-4	4
87	43	27	5	10	12	-2	-5	5
87	44	23	3	7	8	-1	-3	3
87	48	27	1	2	2	-2	-5	6
88	44	21	3	8	9	-1	-3	3
88	45	27	4	8	9	-5	-11	12
88	45	27	5	10	11	-4	-9	10
88	46	27	1	2	2	-4	-9	11
89	46	29	6	11	12	-3	-6	7
80	40	30						
80	40	37	3	4	6	-4	-9	8
80	41	36	5	7	10	-4	-10	9
80	44	36	5	7	9	-2	-5	5
80	47	32	7	11	12	-3	-9	10
81	42	39	4	5	7	-1	-2	2
81	44	31	3	5	6	-1	-3	3

φ_1	Φ	φ_2	H	K	L	U	V	W
80	40	30						
81	47	34	6	9	10	-3	-8	9
82	40	31	3	5	7	-3	-8	7
82	40	37	3	4	6	-6	-12	11
82	42	37	6	8	11	-1	-2	2
82	46	32	5	8	9	-4	-11	12
82	47	36	5	7	8	-3	-7	8
82	47	39	4	5	6	-3	-6	7
82	49	38	7	9	10	-2	-4	5
83	41	32	5	8	11	-5	-12	11
83	43	32	5	8	10	-2	-5	5
83	45	37	3	4	5	-5	-10	11
83	46	34	4	6	7	-4	-9	10
84	45	35	7	10	12	-4	-8	9
84	49	36	8	11	12	-2	-4	5
85	40	31	3	5	7	-5	-11	10
85	42	34	2	3	4	-1	-2	2
85	44	38	7	9	12	-6	-10	11
85	45	37	3	4	5	-4	-7	8
85	46	38	7	9	11	-3	-5	6
85	47	31	6	10	11	-3	-7	8
85	47	34	6	9	10	-5	-10	12
85	47	39	4	5	6	-5	-8	10
86	46	32	5	8	9	-3	-6	7
86	47	36	5	7	8	-5	-9	11
87	41	36	5	7	10	-3	-5	5
87	44	36	5	7	9	-5	-8	9
88	40	37	3	4	6	-2	-3	3
88	42	30	4	7	9	-1	-2	2
88	45	30	4	7	8	-4	-8	9
88	46	34	4	6	7	-3	-5	6
88	48	37	6	8	9	-2	-3	4
88	49	39	8	10	11	-5	-7	10
89	41	32	5	8	11	-3	-5	5
89	43	32	5	8	10	-6	-10	11
89	44	31	3	5	6	-5	-9	10
89	45	35	7	10	12	-6	-9	11
89	48	35	7	10	11	-2	-3	4
89	49	31	3	5	5	-5	-9	12
80	40	40						
80	43	41	6	7	10	-1	-2	2
80	45	48	9	8	12	-4	-6	7
80	47	45	3	3	4	-3	-5	6
80	49	45	9	9	11	-4	-7	9
80	49	49	7	6	8	-2	-3	4
81	40	45	3	3	5	-2	-3	3
81	42	41	7	8	12	-4	-7	7

φ_1	Φ	φ_2	H	K	L	U	V	W
80	40	40						
81	43	45	2	2	3	-7	-11	12
81	43	49	7	6	10	-6	-8	9
81	46	49	7	6	9	-3	-4	5
81	48	42	8	9	11	-5	-9	11
81	49	41	6	7	8	-2	-4	5
82	43	45	2	2	3	-6	-9	10
82	44	40	5	6	8	-6	-11	12
82	45	45	7	7	10	-4	-6	7
82	49	45	4	4	5	-2	-3	4
83	40	45	7	7	12	-5	-7	7
83	41	40	5	6	9	-3	-5	5
83	43	45	2	2	3	-5	-7	8
83	44	41	7	8	11	-5	-8	9
83	47	45	3	3	4	-5	-7	9
83	48	48	10	9	12	-6	-8	11
84	40	49	7	6	11	-8	-9	10
84	43	49	7	6	10	-8	-9	11
84	45	45	5	5	7	-3	-4	5
84	48	42	9	10	12	-2	-3	4
85	40	41	6	7	11	-2	-3	3
85	41	45	5	5	8	-7	-9	10
85	43	45	2	2	3	-4	-5	6
85	45	48	9	8	12	-8	-9	12
85	46	41	6	7	9	-6	-9	11
85	47	45	3	3	4	-7	-9	12
85	48	48	9	8	11	-6	-7	10
86	40	45	3	3	5	-9	-11	12
86	42	41	7	8	12	-8	-11	12
86	42	45	7	7	11	-5	-6	7
86	44	40	5	6	8	-4	-6	7
86	46	45	8	8	11	-5	-6	8
86	48	40	5	6	7	-2	-3	4
86	48	45	7	7	9	-4	-5	7
87	40	45	3	3	5	-7	-8	9
87	42	49	8	7	12	-4	-4	5
87	43	45	2	2	3	-7	-8	10
87	45	42	8	9	12	-3	-4	5
87	47	49	8	7	10	-2	-2	3
87	49	45	4	4	5	-7	-8	12
87	49	45	9	9	11	-5	-6	9
88	44	41	7	8	11	-7	-9	11
88	48	42	8	9	11	-4	-5	7
89	46	41	6	7	9	-5	-6	8
89	49	41	6	7	8	-5	-6	9
80	40	50						
80	44	59	5	3	6	-3	-3	4

φ_1	Φ	φ_2	H	K	L	U	V	W
80	40	50						
80	46	56	6	4	7	-8	-9	12
80	49	52	9	7	10	-6	-8	11
80	49	54	11	8	12	-4	-5	7
81	40	53	4	3	6	-9	-10	11
81	40	59	5	3	7	-10	-9	11
81	42	51	5	4	7	-5	-6	7
81	45	53	4	3	5	-6	-7	9
81	47	58	11	7	12	-2	-2	3
81	49	59	5	3	5	-5	-5	8
82	42	53	8	6	11	-9	-10	12
82	42	56	3	2	4	-4	-4	5
82	46	56	6	4	7	-7	-7	10
82	47	56	9	6	10	-2	-2	3
82	48	50	6	5	7	-4	-5	7
82	48	53	8	6	9	-6	-7	10
83	41	54	7	5	10	-5	-5	6
83	41	58	8	5	11	-8	-7	9
83	44	54	7	5	9	-3	-3	4
84	40	53	4	3	6	-6	-6	7
84	42	56	3	2	4	-10	-9	12
84	45	53	4	3	5	-5	-5	7
84	46	58	8	5	9	-8	-7	11
84	47	54	7	5	8	-2	-2	3
84	49	59	5	3	5	-6	-5	9
85	42	51	5	4	7	-7	-7	9
85	42	56	3	2	4	-6	-5	7
85	43	58	8	5	10	-5	-4	6
85	44	52	9	7	12	-3	-3	4
85	44	59	5	3	6	-9	-7	11
85	45	56	9	6	11	-7	-6	9
85	47	59	10	6	11	-9	-7	12
85	49	52	9	7	10	-5	-5	8
86	40	53	4	3	6	-9	-8	10
86	41	50	6	5	9	-9	-9	11
86	44	50	6	5	8	-6	-6	11
86	46	56	6	4	7	-6	-5	8
86	47	51	5	4	6	-2	-2	3
87	45	53	4	3	5	-9	-8	12
87	48	50	6	5	7	-7	-7	11
87	49	54	11	8	12	-8	-7	12
87	49	59	5	3	5	-7	-5	10
88	40	59	5	3	7	-3	-2	3
88	41	54	7	5	10	-10	-8	11
88	42	51	5	4	7	-9	-6	11
88	42	56	3	2	4	-8	-6	9
88	44	59	5	3	6	-6	-4	7
88	46	52	9	7	11	-8	-7	11
88	46	58	8	5	9	-7	-5	9

φ_1	Φ	φ_2	H	K	L	U	V	W
80	40	50						
89	41	50	6	5	9	-7	-6	8
89	42	56	3	2	4	-10	-7	11
89	44	50	6	5	8	-7	-6	9
89	45	56	9	6	11	-10	-7	12
89	47	59	10	6	11	-8	-5	10
89	49	59	5	3	5	-8	-5	11
80	40	60						
80	41	69	6	3	10	-3	-2	3
80	44	69	8	3	9	-9	-6	10
80	47	67	7	3	7	-9	-7	12
80	49	60	7	4	7	-7	-7	11
81	43	61	9	5	11	-9	-8	11
81	44	63	6	3	7	-5	-4	6
81	44	67	7	3	8	-7	-5	8
81	47	68	5	2	5	-7	-5	9
81	48	63	2	1	2	-7	-6	10
82	40	67	7	3	9	-3	-2	3
82	41	61	9	5	12	-11	-9	12
82	42	63	4	2	5	-9	-7	10
82	46	61	11	6	12	-6	-5	8
82	48	63	2	1	2	-5	-4	7
82	49	60	7	4	7	-8	-7	12
83	42	63	4	2	5	-11	-8	12
83	45	63	10	5	11	-4	-3	5
83	45	66	9	4	10	-6	-4	7
83	47	68	5	2	5	-8	-5	10
83	48	63	2	1	2	-8	-6	11
84	44	67	7	3	8	-10	-6	11
84	45	63	8	4	9	-10	-7	12
85	40	63	6	3	8	-3	-2	3
85	43	61	9	5	11	-7	-5	8
85	43	63	10	5	12	-9	-6	10
85	44	66	10	4	11	-9	-5	10
85	45	60	7	4	8	-4	-3	5
85	47	66	5	2	5	-9	-5	11
85	48	63	2	1	2	-3	-2	4
86	42	68	5	2	6	-2	-1	2
86	44	63	6	3	7	-8	-5	9
87	47	66	5	2	5	-10	-5	12
88	42	66	9	4	11	-2	-1	2
88	45	63	8	4	9	-7	-4	8
88	45	66	9	4	10	-10	-5	11
88	48	63	2	1	2	-7	-4	9
89	41	61	9	5	12	-5	-3	5
89	44	63	6	3	7	-11	-6	12
89	45	66	11	5	12	-8	-4	9

φ_1	Φ	φ_2	H	K	L	U	V	W
80	40	60						
89	46	61	9	5	10	-5	-3	6
80	40	70						
80	40	76	4	1	5	-10	-5	9
80	40	79	5	1	6	-11	-5	10
80	43	72	9	3	10	-11	-7	12
80	43	77	9	2	10	-2	-1	2
80	49	72	12	4	11	-3	-2	4
81	42	76	8	2	9	-2	-1	2
81	44	70	11	4	12	-8	-5	9
81	46	76	4	1	4	-8	-4	9
81	46	79	5	1	5	-11	-5	12
81	47	72	3	1	3	-5	-3	6
81	49	77	9	2	8	-4	-2	5
82	40	76	4	1	5	-9	-4	8
82	40	79	5	1	6	-10	-4	9
82	42	74	7	2	8	-2	-1	2
82	49	75	11	3	10	-4	-2	5
83	46	76	4	1	4	-9	-4	10
83	49	79	10	2	9	-5	-2	6
84	42	72	6	2	7	-2	-1	2
84	47	72	3	1	3	-6	-3	7
85	40	76	4	1	5	-8	-3	7
85	40	79	5	1	6	-9	-3	8
85	46	76	4	1	4	-10	-4	11
85	48	76	12	3	11	-5	-2	6
86	46	76	4	1	4	-11	-4	12
87	47	72	3	1	3	-7	-3	8
87	49	77	9	2	8	-10	-3	12
88	40	79	5	1	6	-8	-2	7
88	44	75	11	3	12	-3	-1	3
89	40	76	4	1	5	-7	-2	6
89	41	73	10	3	12	-12	-4	11
89	44	73	10	3	11	-3	-1	3
89	47	72	3	1	3	-8	-3	9
89	49	73	10	3	9	-9	-3	11
80	40	80						
80	41	81	6	1	7	-12	-5	11
80	49	83	8	1	7	-5	-2	6
82	40	84	10	1	12	-7	-2	6
82	41	81	6	1	7	-11	-4	10
82	41	82	7	1	8	-12	-4	11
82	49	84	9	1	8	-6	-2	7
83	48	84	10	1	9	-7	-2	8
84	48	85	11	1	10	-8	-2	9

φ_1	Φ	φ_2	H	K	L	U	V	W
80	40	80						
85	41	81	6	1	7	-10	-3	9
85	41	82	7	1	8	-11	-3	10
85	42	83	8	1	9	-12	-3	11
85	48	80	11	2	10	-6	-2	7
85	48	85	12	1	11	-9	-2	10
86	48	81	12	2	11	-7	-2	6
88	41	81	6	1	7	-9	-2	6
88	41	82	7	1	8	-10	-2	9
88	42	83	8	1	9	-11	-2	10
88	42	84	9	1	10	-12	-2	11
88	49	83	8	1	7	-6	-1	7
89	48	84	10	1	9	-8	-1	9
89	48	85	11	1	10	-9	-1	10
89	48	85	12	1	11	-10	-1	11
89	49	84	9	1	8	-7	-1	8
80	40	90						
80	43	90	11	0	12	-12	-3	11
80	45	90	1	0	1	-4	-1	4
80	47	90	12	0	11	-11	-3	12
81	42	90	8	0	9	-9	-2	8
81	45	90	1	0	1	-9	-2	9
81	48	90	9	0	8	-8	-2	9
82	42	90	9	0	10	-10	-2	9
82	42	90	10	0	11	-11	-2	10
82	45	90	1	0	1	-5	-1	5
82	48	90	10	0	9	-9	-2	10
82	48	90	11	0	10	-10	-2	11
83	40	90	5	0	6	-6	-1	5
83	43	90	11	0	12	-12	-2	11
83	45	90	1	0	1	-11	-2	11
83	45	90	1	0	1	-6	-1	6
83	47	90	12	0	11	-11	-2	12
84	41	90	6	0	7	-7	-1	6
84	45	90	1	0	1	-7	-1	7
84	49	90	7	0	6	-6	-1	7
85	41	90	7	0	8	-8	-1	7
85	42	90	8	0	9	-9	-1	8
85	45	90	1	0	1	-8	-1	8
85	48	90	9	0	8	-8	-1	9
85	49	90	8	0	7	-7	-1	8
86	40	90	5	0	6	-12	-1	10
86	42	90	9	0	10	-10	-1	9
86	42	90	10	0	11	-11	-1	10
86	43	90	11	0	12	-12	-1	11
86	45	90	1	0	1	-11	-1	11
86	45	90	1	0	1	-10	-1	10

φ_1	Φ	φ_2	H	K	L	U	V	W
80	40	90						
86	45	90	1	0	1	-9	-1	9
86	47	90	12	0	11	-11	-1	12
86	48	90	10	0	9	-9	-1	10
86	48	90	11	0	10	-10	-1	11
87	45	90	1	0	1	-12	-1	12
80	50	0						
80	52	0	0	9	7	2	-7	9
80	55	6	1	10	7	1	-5	7
80	56	0	0	3	2	2	-6	9
80	59	0	0	5	3	1	-3	5
81	51	0	0	5	4	1	-4	5
81	55	0	0	10	7	2	-7	10
81	55	8	1	7	5	1	-8	11
81	58	0	0	11	7	2	-7	11
82	51	0	0	11	9	2	-9	11
82	53	0	0	4	3	2	-9	12
82	54	0	0	11	8	2	-8	11
82	56	0	0	3	2	1	-4	6
82	58	7	1	8	5	1	-7	11
83	50	0	0	6	5	1	-5	6
83	54	0	0	7	5	1	-5	7
84	51	9	1	6	5	0	-5	6
84	53	0	0	4	3	1	-6	8
84	54	9	2	12	9	0	-3	4
84	58	0	0	8	5	1	-5	6
85	50	8	1	7	6	0	-6	7
85	52	0	0	9	7	1	-7	9
85	55	0	0	10	7	1	-7	10
85	55	8	1	7	5	0	-5	7
85	56	0	0	3	2	1	-6	9
85	57	9	1	6	4	0	-2	3
85	59	0	0	5	3	1	-6	10
86	50	0	0	6	5	1	-10	12
86	51	0	0	5	4	1	-8	10
86	51	0	0	11	9	1	-9	11
86	51	6	1	10	8	0	-4	5
86	52	6	1	9	7	0	-7	9
86	53	0	0	4	3	1	-9	12
86	53	7	1	8	6	0	-3	4
86	54	0	0	11	8	1	-8	11
86	56	0	0	3	2	1	-8	12
86	56	6	1	9	6	0	-2	3
86	58	0	0	11	7	1	-7	11
86	58	7	1	8	5	0	-5	8
87	50	5	1	12	10	0	-5	6
87	51	5	1	11	9	0	-9	11

φ_1	Φ	φ_2	H	K	L	U	V	W
80	50	0						
87	53	5	1	12	9	0	-3	4
87	54	5	1	11	8	0	-8	11
87	55	6	1	10	7	0	-7	10
87	56	5	1	12	8	0	-2	3
87	56	5	1	11	7	0	-7	11
87	59	6	1	10	6	0	-3	5
88	51	9	1	6	5	-1	-9	11
80	50	10						
80	51	10	2	11	9	1	-10	12
80	51	16	2	7	6	0	-6	7
80	52	15	3	11	9	0	-9	11
80	53	17	3	10	8	0	-4	5
80	58	18	1	3	2	0	-2	3
81	50	14	2	8	7	0	-7	8
81	51	14	3	12	10	0	-5	6
81	55	15	3	11	8	0	-8	11
81	56	16	2	7	5	0	-5	7
81	56	17	3	10	7	0	-7	10
82	53	13	2	9	7	0	-7	9
82	54	14	1	4	3	0	-3	4
82	57	14	3	12	8	0	-2	3
82	58	15	3	11	7	0	-7	11
83	51	10	2	11	9	0	-9	11
83	51	16	2	7	6	-1	-10	12
83	52	11	1	5	4	0	-4	5
83	52	18	2	6	5	-1	-8	10
83	54	16	3	9	7	-1	-9	12
83	57	13	2	9	6	0	-2	3
83	59	14	2	8	5	0	-5	8
84	54	10	2	11	8	0	-8	11
84	56	11	2	10	7	0	-7	10
84	58	10	2	11	7	0	-7	11
84	56	18	1	3	2	-1	-7	11
85	50	18	3	9	8	-1	-5	6
86	54	14	1	4	3	-1	-8	11
86	55	15	3	11	8	-1	-7	10
86	56	17	3	10	7	-1	-6	9
86	58	18	1	3	2	-1	-5	8
87	53	17	3	10	6	-2	-9	12
88	50	14	2	8	7	-1	-5	6
88	51	10	2	11	9	-1	-8	10
88	52	11	1	5	4	-1	-7	9
88	53	13	2	9	7	-1	-6	8
88	54	14	1	4	3	-1	-5	7
88	59	14	2	6	5	-1	-6	10
89	51	16	2	7	6	-1	-4	5

φ_1	Φ	φ_2	H	K	L	U	V	W
80	50	10						
89	56	16	2	7	5	-1	-4	6
80	50	20						
80	51	27	6	12	11	-1	-5	6
81	50	23	5	12	11	-1	-6	7
81	51	29	6	11	10	-1	-4	5
82	50	24	5	11	10	-1	-5	6
82	51	27	5	10	9	-1	-4	5
82	53	27	3	6	5	-2	-9	12
82	57	22	4	10	7	-1	-8	12
82	57	23	3	7	5	-1	-6	9
83	50	22	4	10	9	-1	-5	6
83	52	20	4	11	9	-1	-7	9
83	52	29	5	9	8	-1	-3	4
83	53	22	2	5	4	-1	-6	8
83	55	24	4	9	7	-1	-5	7
83	56	27	2	4	3	-1	-4	6
84	51	24	4	9	8	-1	-4	5
84	54	27	5	10	8	-2	-7	10
85	52	27	4	8	7	-1	-3	4
65	53	22	2	5	4	-2	-8	11
85	59	20	4	11	7	-1	-6	10
85	59	27	3	6	4	-2	-7	12
86	51	21	3	8	7	-1	-4	5
86	57	23	3	7	5	-2	-7	11
86	58	27	5	10	7	-1	-3	5
87	52	23	3	7	6	-1	-3	4
87	56	20	4	11	8	-1	-4	6
87	56	27	2	4	3	-2	-5	8
88	51	27	6	12	11	-4	-9	12
88	55	24	4	9	7	-3	-8	12
88	59	27	3	6	4	-2	-5	9
89	55	21	3	6	6	-2	-6	9
89	57	29	6	11	8	-2	-4	7
80	50	30						
80	50	34	2	3	3	-3	-9	11
80	52	30	7	12	11	-1	-4	5
80	54	30	7	12	10	-2	-8	11
80	55	34	4	6	5	-2	-7	10
80	56	37	9	12	10	-2	-6	9
80	59	31	6	10	7	-1	-5	8
60	59	37	3	4	3	-1	-3	5
81	53	32	7	11	10	-1	-3	4
81	53	34	8	12	11	-1	-3	4
81	58	32	5	8	6	-2	-7	11

φ_1	Φ	φ_2	H	K	L	U	V	W
80	50	30						
82	50	34	2	3	3	-3	-8	10
82	51	37	3	4	4	-4	-9	12
82	52	31	6	10	9	-1	-3	4
82	55	34	4	6	5	-3	-8	12
82	58	38	7	9	7	-3	-7	12
83	50	34	2	3	3	-3	-7	9
83	50	39	9	11	12	-5	-9	12
83	56	31	3	5	4	-2	-6	9
83	57	35	7	10	6	-2	-5	8
83	58	32	7	11	8	-1	-3	5
84	51	37	3	4	4	-4	-8	11
84	54	36	8	11	10	-1	-2	3
84	54	37	9	12	11	-1	-2	3
84	58	30	4	7	5	-1	-3	5
84	58	34	8	12	9	-3	-7	12
84	58	39	4	5	4	-2	-4	7
85	50	34	2	3	3	-3	-6	8
65	54	35	7	10	9	-1	-2	3
85	57	36	8	11	9	-3	-6	10
85	59	37	3	4	3	-3	-6	11
86	51	37	3	4	4	-4	-7	10
86	52	39	4	5	5	-5	-8	12
86	54	34	6	9	8	-1	-2	3
86	56	31	3	5	4	-3	-7	11
86	56	37	9	12	10	-4	-7	12
87	53	32	5	8	7	-1	-2	3
87	58	32	5	8	6	-2	-4	7
88	50	34	2	3	3	-3	-5	7
88	51	37	3	4	4	-4	-6	9
88	52	39	4	5	5	-5	-7	11
88	53	30	4	7	6	-1	-2	3
88	59	37	3	4	3	-2	-3	6
89	57	34	6	9	7	-3	-5	9
89	57	35	7	10	8	-4	-6	11
89	58	39	4	5	4	-3	-4	8
80	50	40						
80	50	45	6	6	7	-5	-9	12
80	51	42	10	11	12	-3	-6	8
80	52	45	10	10	11	-4	-7	10
80	53	49	8	7	8	-5	-8	12
80	57	40	5	6	5	-2	-5	8
80	57	41	7	8	7	-3	-7	11
80	58	45	10	10	9	-3	-6	10
80	59	45	7	7	6	-2	-4	7
81	51	45	7	7	6	-3	-5	7
81	55	45	1	1	1	-4	-7	11

φ_1	Φ	φ_2	H	K	L	U	V	W
80	50	40						
81	58	43	11	12	10	-2	-4	7
82	53	49	7	6	7	-5	-7	11
82	55	45	1	1	1	-3	-5	8
82	57	41	6	7	6	-3	-6	10
62	57	45	12	12	11	-4	-7	12
82	58	45	9	9	8	-3	-5	9
63	51	42	9	10	11	-3	-5	7
83	51	48	10	9	11	-3	-4	6
83	52	45	9	9	10	-4	-6	9
83	55	45	1	1	1	-2	-3	5
84	50	41	7	8	9	-5	-8	11
64	50	45	5	5	6	-5	-7	10
84	52	45	11	11	12	-5	-7	11
64	57	41	7	8	7	-4	-7	12
84	57	45	11	11	10	-4	-6	11
64	58	47	12	11	10	-3	-4	8
84	59	45	6	6	5	-2	-3	6
85	50	45	6	6	7	-3	-4	6
65	50	48	9	8	10	-6	-7	11
85	53	49	7	6	7	-6	-7	12
65	55	45	1	1	1	-3	-4	7
85	55	45	1	1	1	-5	-7	12
65	57	40	5	6	5	-3	-5	9
86	51	45	8	8	9	-4	-5	8
86	55	45	1	1	1	-4	-5	9
66	57	41	6	7	6	-4	-6	11
86	58	45	8	8	7	-3	-4	8
86	59	42	9	10	8	-2	-3	6
67	50	42	8	9	10	-3	-4	6
87	52	45	10	10	11	-5	-6	10
87	55	45	1	1	1	-5	-6	11
87	57	45	12	12	11	-5	-6	12
87	58	45	10	10	9	-4	-5	10
68	50	49	8	7	9	-3	-3	5
86	51	48	11	10	12	-4	-4	7
89	57	40	5	6	5	-4	-5	10
89	59	46	11	10	9	-3	-3	7
80	50	50						
80	51	53	4	3	4	-3	-4	6
60	52	50	6	5	6	-4	-6	9
80	58	51	5	4	4	-4	-6	11
61	51	54	7	5	7	-6	-7	11
81	54	56	9	6	8	-6	-7	12
61	58	56	12	8	9	-5	-6	12
61	59	53	4	3	3	-3	-4	8
82	52	51	5	4	5	-4	-5	6

φ_1	Φ	φ_2	H	K	L	U	V	W
80	50	50						
82	53	56	11	7	10	-5	-5	9
82	56	59	5	3	4	-1	-1	2
63	50	56	3	2	3	-3	-3	5
83	55	56	6	4	5	-1	-1	2
83	55	58	11	7	9	-1	-1	2
83	58	51	5	4	4	-4	-5	10
83	56	58	11	7	8	-4	-4	9
64	51	54	7	5	7	-7	-7	12
84	52	50	6	5	6	-5	-6	10
84	57	50	6	5	5	-5	-6	12
85	51	53	4	3	4	-4	-4	7
85	52	59	10	6	9	-6	-5	10
85	55	53	8	6	7	-1	-1	2
85	55	54	7	5	6	-1	-1	2
86	50	51	11	9	12	-3	-3	5
86	50	56	3	2	3	-7	-6	11
86	52	51	5	4	5	-5	-5	9
66	53	56	12	8	11	-7	-6	12
86	55	51	10	8	9	-1	-1	2
86	55	52	9	7	8	-1	-1	2
66	58	58	8	5	6	-5	-4	10
66	59	53	4	3	3	-3	-3	7
87	52	50	6	5	6	-6	-6	11
87	55	50	12	10	11	-1	-1	2
87	55	51	11	9	10	-1	-1	2
87	57	50	6	5	5	-5	-5	11
67	57	56	9	6	7	-6	-5	12
67	58	51	5	4	4	-4	-4	9
88	50	56	3	2	3	-4	-3	6
68	58	56	12	8	9	-4	-3	8
69	51	53	4	3	4	-5	-4	8
69	52	51	5	4	5	-6	-5	10
89	59	59	10	6	7	-3	-2	6
80	50	60						
80	52	61	9	5	6	-4	-4	7
60	52	67	12	5	10	-5	-4	8
80	59	63	6	3	4	-4	-4	9
81	50	68	10	4	9	-7	-5	10
61	53	60	7	4	6	-6	-6	11
61	56	61	9	5	7	-1	-1	2
81	58	67	12	5	6	-5	-4	10
82	51	66	9	4	6	-4	-3	6
62	52	63	8	4	7	-5	-4	8
82	53	63	6	3	5	-7	-6	12
62	55	67	12	5	9	-4	-3	7
62	55	69	6	3	6	-3	-2	5

φ_1	Φ	φ_2	H	K	L	U	V	W
80	50	60						
82	58	60	7	4	5	-5	-5	11
83	51	63	12	6	11	-4	-3	6
83	51	69	8	3	7	-5	-3	7
83	52	67	7	3	6	-6	-4	9
83	59	66	9	4	6	-4	-3	8
84	50	67	12	5	11	-8	-5	11
84	55	66	9	4	7	-3	-2	5
84	57	61	11	6	8	-6	-5	12
84	58	63	10	5	7	-5	-4	10
85	52	60	12	7	11	-5	-4	8
86	50	66	11	5	10	-5	-3	7
86	52	61	9	5	8	-7	-5	11
86	54	63	10	5	8	-3	-2	5
86	58	60	7	4	5	-4	-3	8
86	59	63	6	3	4	-3	-2	6
87	50	68	10	4	9	-6	-3	8
87	51	61	11	6	10	-6	-4	9
87	52	63	8	4	7	-8	-5	12
87	54	61	11	6	9	-3	-2	5
88	51	63	10	5	9	-7	-4	10
88	51	66	9	4	8	-8	-4	11
88	51	69	8	3	7	-9	-4	12
88	54	60	12	7	10	-3	-2	5
88	56	63	4	2	3	-7	-4	12
89	53	66	11	5	9	-2	-1	3
89	59	66	9	4	6	-6	-3	11
80	50	70						
80	51	74	7	2	6	-8	-5	11
80	52	79	5	1	4	-8	-4	11
80	56	70	11	4	8	-4	-3	7
80	56	74	7	2	5	-3	-2	5
81	55	75	11	3	8	-5	-3	8
81	57	76	12	3	8	-7	-4	12
81	59	70	11	4	7	-4	-3	8
82	57	77	9	2	6	-6	-3	10
82	58	72	3	1	2	-6	-4	11
83	50	72	9	3	8	-9	-5	12
83	51	74	7	2	6	-6	-3	8
83	52	70	11	4	9	-8	-5	12
83	52	72	6	2	5	-7	-4	10
83	54	76	4	1	3	-2	-1	3
83	55	72	12	4	9	-5	-3	8
84	51	76	12	3	10	-9	-4	12
84	53	77	9	2	7	-5	-2	7
85	54	72	9	3	7	-2	-1	3
86	50	76	8	2	7	-8	-3	10

φ_1	Φ	φ_2	H	K	L	U	V	W
80	50	70						
86	52	79	5	1	4	-3	-1	4
86	56	74	7	2	5	-7	-3	11
86	58	72	3	1	2	-4	-2	7
87	50	72	9	3	8	-7	-3	9
87	53	73	10	3	8	-5	-2	7
87	56	73	10	3	7	-5	-2	8
87	58	72	3	1	2	-7	-3	12
88	52	75	11	3	9	-3	-1	4
88	55	75	11	3	8	-6	-2	9
88	56	79	10	2	7	-4	-1	6
89	51	76	12	3	10	-7	-2	9
89	52	70	11	4	9	-5	-2	7
89	54	76	4	1	3	-7	-2	10
89	55	72	12	4	9	-8	-3	12
89	59	70	11	4	7	-5	-2	9
89	59	76	8	2	5	-7	-2	12
80	50	80						
80	51	80	11	2	9	-6	-3	8
80	53	85	12	1	9	-8	-3	11
80	54	80	11	2	8	-2	-1	3
80	56	84	9	1	6	-7	-3	11
81	51	81	6	1	5	-7	-3	9
81	53	83	8	1	6	-5	-2	7
82	54	81	12	2	9	-7	-3	10
83	50	82	7	1	6	-9	-3	11
83	52	84	9	1	7	-3	-1	4
83	57	81	6	1	4	-5	-2	8
84	51	84	10	1	8	-7	-2	9
84	51	85	11	1	9	-4	-1	5
84	58	83	8	1	5	-3	-1	5
85	50	85	12	1	10	-9	-2	11
85	59	84	10	1	6	-7	-2	12
86	51	80	11	2	9	-7	-2	9
86	53	83	8	1	6	-8	-2	11
86	54	85	11	1	8	-5	-1	7
86	57	81	6	1	4	-7	-2	11
86	58	80	11	2	7	-3	-1	5
87	51	81	6	1	5	-4	-1	5
87	58	85	11	1	7	-5	-1	8
88	50	82	7	1	6	-5	-1	6
80	50	90						
80	52	90	9	0	7	-7	-2	9
80	56	90	3	0	2	-6	-2	9
80	59	90	5	0	3	-3	-1	5

φ_1	Φ	φ_2	H	K	L	U	V	W
80	50	90						
81	51	90	5	0	4	-4	-1	5
81	55	90	10	0	7	-7	-2	10
81	58	90	11	0	7	-7	-2	11
82	51	90	11	0	9	-9	-2	11
82	53	90	4	0	3	-9	-2	12
82	54	90	11	0	8	-8	-2	11
82	56	90	3	0	2	-4	-1	6
83	50	90	6	0	5	-5	-1	6
83	54	90	7	0	5	-5	-1	7
84	53	90	4	0	3	-6	-1	8
84	58	90	8	0	5	-5	-1	8
85	52	90	9	0	7	-7	-1	9
85	55	90	10	0	7	-7	-1	10
85	56	90	3	0	2	-6	-1	9
85	59	90	5	0	3	-6	-1	10
86	50	90	6	0	5	-10	-1	12
86	51	90	5	0	4	-8	-1	10
86	51	90	11	0	9	-9	-1	11
86	53	90	4	0	3	-9	-1	12
86	54	90	11	0	8	-8	-1	11
86	56	90	3	0	2	-6	-1	12
86	58	90	11	0	7	-7	-1	11
80	60	0						
80	60	9	2	12	7	1	-6	10
80	61	6	1	9	5	1	-4	7
80	63	0	0	2	1	2	-5	10
81	61	0	0	11	6	2	-6	11
81	66	0	0	11	5	2	-5	11
81	67	0	0	12	5	2	-5	12
81	67	5	1	12	5	1	-3	7
81	67	8	1	7	3	1	-4	9
82	60	0	0	12	7	2	-7	12
82	61	5	1	11	6	1	-5	9
82	61	8	1	7	4	1	-7	12
82	63	0	0	2	1	1	-3	6
83	60	0	0	7	4	1	-4	7
83	66	6	1	9	4	1	-5	11
83	67	0	0	7	3	1	-3	7
83	69	0	0	8	3	1	-3	8
84	61	0	0	9	5	1	-5	9
84	63	0	0	2	1	1	-4	6
84	66	0	0	9	4	1	-4	9
85	60	9	2	12	7	0	-7	12
85	61	0	0	11	6	1	-6	11
85	63	0	0	2	1	1	-5	10
85	66	0	0	11	5	1	-5	11

φ_1	Φ	φ_2	H	K	L	U	V	W
80	60	0						
85	68	0	0	5	2	1	-4	10
86	60	0	0	12	7	1	-7	12
86	61	8	1	7	4	0	-4	7
86	63	0	0	2	1	1	-6	12
86	64	9	1	6	3	0	-1	2
86	67	0	0	12	5	1	-5	12
86	68	9	2	12	5	0	-5	12
87	61	6	1	9	5	0	-5	9
87	64	6	1	10	5	0	-1	2
87	64	7	1	6	4	0	-1	2
87	66	6	1	9	4	0	-4	9
87	67	6	1	7	3	0	-3	7
88	60	5	1	12	7	0	-7	12
88	61	5	1	11	6	0	-6	11
86	64	5	1	12	6	0	-1	2
86	66	5	1	11	5	0	-5	11
88	67	5	1	12	5	0	-5	12
88	66	6	1	10	4	0	-2	5
80	60	10						
81	60	17	3	10	6	0	-3	5
81	61	18	4	12	7	0	-7	12
81	62	16	3	9	5	0	-5	9
81	69	11	1	5	2	1	-5	12
82	61	16	2	7	4	0	-4	7
82	65	18	2	6	3	0	-1	2
83	60	14	3	12	7	0	-7	12
83	62	15	3	11	6	0	-6	11
83	64	17	3	10	5	0	-1	2
83	67	18	3	9	4	0	-4	9
83	68	18	4	12	5	0	-5	12
84	60	11	1	5	3	0	-3	5
84	62	13	2	9	5	0	-5	9
84	64	14	1	4	2	0	-1	2
84	66	15	3	11	5	0	-5	11
84	68	16	2	7	3	0	-3	7
84	69	17	3	10	4	0	-2	5
85	62	10	2	11	6	0	-6	11
85	64	11	2	10	5	0	-1	2
85	67	13	2	9	4	0	-4	9
85	68	14	3	12	5	0	-5	12
86	66	10	2	11	5	0	-5	11
86	69	11	1	5	2	0	-2	5
87	61	16	2	7	4	-1	-6	11
87	67	18	3	9	4	-1	-5	12
88	60	11	1	5	3	-1	-7	12
89	60	14	3	12	7	-1	-5	9

φ_1	Φ	φ_2	H	K	L	U	V	W
80	60	10						
89	68	16	2	7	3	-1	-4	10
80	60	20						
80	65	22	4	10	5	0	-1	2
80	65	23	5	12	6	0	-1	2
80	68	24	5	11	5	0	-5	11
81	61	27	4	8	5	-1	-7	12
81	63	20	4	11	6	0	-6	11
81	64	29	6	11	6	-1	-6	12
81	65	21	3	8	4	0	-1	2
81	68	23	3	7	3	0	-3	7
81	68	24	4	9	4	0	-4	9
81	69	23	5	12	5	0	-5	12
82	64	29	5	9	5	-1	-5	10
82	67	20	4	11	5	0	-5	11
84	62	23	5	12	7	-1	-6	11
84	64	24	5	11	6	-1	-5	10
84	66	27	1	2	1	-1	-5	11
84	68	29	6	11	5	-1	-4	10
85	60	24	5	11	7	-1	-4	7
85	61	22	2	5	3	-1	-5	9
85	63	24	4	9	5	-1	-4	8
85	66	27	1	2	1	-1	-4	9
85	68	24	5	11	5	-1	-5	12
86	62	27	6	12	7	-1	-3	6
86	68	24	4	9	4	-1	-4	10
86	69	29	5	9	4	-1	-3	8
87	60	21	3	8	5	-1	-4	7
87	63	20	4	11	6	-1	-4	8
87	64	29	5	9	5	-2	-5	11
87	66	27	1	2	1	-1	-3	7
88	62	23	3	7	4	-1	-3	6
88	62	27	5	10	6	-2	-5	10
88	66	27	1	2	1	-2	-5	12
88	68	23	3	7	3	-1	-3	8
89	61	29	6	11	7	-1	-2	4
80	60	30						
80	60	35	7	10	7	-2	-7	12
80	61	34	2	3	2	-1	-4	7
80	62	32	5	8	5	-1	-5	9
80	63	31	3	5	3	-1	-6	11
80	64	39	9	11	7	-1	-3	6
81	64	34	8	12	7	-1	-4	8
82	61	34	2	3	2	-2	-6	11
82	63	37	6	8	5	-1	-3	6

φ_1	Φ	φ_2	H	K	L	U	V	W
80	60	30						
82	67	39	9	11	6	-1	-3	7
83	60	36	5	7	5	-2	-5	9
83	64	30	4	7	4	-1	-4	8
83	65	39	4	5	3	-2	-5	11
84	61	39	8	10	7	-1	-2	4
84	63	31	3	5	3	-1	-3	6
85	62	32	5	8	5	-2	-5	10
85	66	38	7	9	5	-1	-2	5
85	69	32	7	11	5	-1	-3	8
86	64	35	7	10	6	-2	-4	9
86	66	36	8	11	6	-1	-2	5
86	68	37	3	4	2	-2	-4	11
87	61	34	2	3	2	-1	-2	4
87	62	38	7	9	6	-3	-5	11
88	60	36	5	7	5	-3	-5	10
89	63	36	8	11	7	-2	-3	7
89	64	30	4	7	4	-2	-4	9
89	64	34	8	12	7	-3	-5	12
89	68	37	3	4	2	-2	-3	9
80	60	40						
80	61	41	7	8	6	-2	-5	9
80	63	48	10	9	7	-1	-2	4
80	66	45	11	11	7	-2	-5	11
80	68	48	11	10	6	-2	-5	12
81	60	45	11	11	9	-3	-6	11
81	62	45	4	4	3	-1	-2	4
81	67	42	8	9	5	-1	-3	7
81	67	48	9	8	5	-1	-2	5
81	68	45	12	12	7	-2	-5	12
82	62	42	10	11	8	-1	-2	4
82	63	40	5	6	4	-2	-5	10
82	65	45	3	3	2	-2	-4	9
82	69	49	8	7	4	-2	-4	11
83	60	48	9	8	7	-2	-3	6
83	61	45	5	5	4	-3	-5	10
83	62	41	6	7	5	-1	-2	4
83	63	45	11	11	8	-3	-5	11
83	64	48	9	8	6	-2	-3	7
83	67	45	5	5	3	-1	-2	5
83	69	45	11	11	6	-2	-4	11
84	62	49	7	6	5	-3	-4	9
84	65	45	3	3	2	-3	-5	12
84	67	41	6	7	4	-1	-2	5
84	67	43	11	12	7	-1	-2	5
84	67	49	7	6	4	-2	-3	8
85	63	42	9	10	7	-3	-5	11

φ_1	Φ	φ_2	H	K	L	U	V	W
80	60	40						
85	63	45	7	7	5	-2	-3	7
85	64	47	12	11	8	-3	-4	10
85	66	45	8	8	5	-2	-3	8
85	66	48	10	9	6	-3	-4	11
86	60	40	10	12	9	-3	-5	10
86	61	45	9	9	7	-3	-4	9
86	64	45	10	10	7	-3	-4	10
86	69	45	9	9	5	-2	-3	9
87	60	45	11	11	9	-4	-5	11
87	61	43	11	12	9	-3	-4	9
87	62	45	4	4	3	-4	-5	12
87	63	40	5	6	4	-2	-3	7
87	66	40	10	12	7	-2	-3	8
87	66	42	9	10	6	-2	-3	8
87	66	45	11	11	7	-3	-4	11
87	68	45	12	12	7	-3	-4	12
88	61	49	8	7	6	-2	-2	5
88	64	42	8	9	6	-3	-4	10
88	65	49	8	7	5	-1	-1	3
88	67	42	8	9	5	-3	-4	12
89	60	42	8	9	7	-4	-5	11
89	65	48	11	10	7	-1	-1	3
89	68	48	11	10	6	-2	-2	7
80	60	50						
80	61	51	10	8	7	-3	-5	10
81	61	56	3	2	2	-4	-5	11
81	67	56	6	4	3	-2	-3	8
81	67	58	6	5	4	-3	-4	11
82	60	50	12	10	9	-2	-3	6
82	62	53	12	9	8	-3	-4	9
82	64	51	11	9	7	-2	-3	7
83	63	54	11	8	7	-4	-5	12
83	63	59	5	3	3	-3	-3	8
84	61	56	3	2	2	-2	-2	5
84	66	52	9	7	5	-3	-4	11
84	68	53	4	3	2	-3	-4	12
84	69	50	6	5	3	-2	-3	9
85	60	54	7	5	5	-5	-5	12
85	65	56	9	6	5	-1	-1	3
85	65	58	11	7	6	-1	-1	3
86	65	54	7	5	4	-1	-1	3
86	67	56	6	4	3	-3	-3	10
87	61	51	11	9	8	-2	-2	5
87	62	52	9	7	6	-3	-3	6
87	62	58	8	5	5	-5	-4	12
87	65	51	5	4	3	-1	-1	3

φ_1	Φ	φ_2	H	K	L	U	V	W
80	60	50						
87	65	53	12	9	7	-1	-1	3
87	68	53	4	3	2	-2	-2	7
88	60	54	7	5	5	-5	-4	11
88	61	56	3	2	2	-4	-3	9
88	63	50	6	5	4	-4	-4	11
88	67	51	11	9	6	-3	-3	10
88	69	50	6	5	3	-3	-3	11
89	62	53	12	9	8	-5	-4	12
89	63	59	5	3	3	-3	-2	7
89	67	56	6	4	3	-4	-3	12
80	60	60						
80	64	60	7	4	4	-4	-5	12
81	61	63	8	4	5	-5	-5	12
81	62	63	10	5	6	-2	-2	5
81	64	66	11	5	6	-3	-3	8
82	60	61	9	5	6	-3	-3	7
82	61	68	5	2	3	-5	-4	11
82	66	63	2	1	1	-1	-1	3
82	68	67	7	3	3	-3	-3	10
83	60	69	8	3	5	-3	-2	6
83	62	67	12	5	7	-4	-3	9
83	64	60	7	4	4	-4	-4	11
84	62	63	12	6	7	-5	-4	12
84	69	61	9	5	4	-2	-2	7
85	68	66	9	4	4	-4	-3	12
86	60	60	12	7	8	-5	-4	11
86	62	67	7	3	4	-5	-3	11
86	63	66	9	4	5	-3	-2	7
86	66	63	2	1	1	-4	-3	11
86	68	67	7	3	3	-3	-2	9
87	60	66	11	5	7	-5	-3	10
87	64	60	7	4	4	-4	-3	10
87	66	63	2	1	1	-3	-2	8
88	61	68	5	2	3	-2	-1	4
88	66	61	11	6	5	-3	-2	9
89	60	61	9	5	6	-5	-3	10
89	64	61	9	5	5	-5	-3	12
89	64	66	11	5	6	-4	-2	9
89	66	66	9	4	4	-4	-2	11
80	60	70						
80	60	76	12	3	7	-3	-2	6
80	67	70	11	4	5	-1	-1	3
81	62	72	9	3	5	-4	-3	9
81	64	76	4	1	2	-3	-2	7

TAB. VI FROM EULER ANGLES INTO HKL-UVW

φ_1	Φ	φ_2	H	K	L	U	V	W
80	60	70						
82	60	79	5	1	3	-6	-3	11
82	65	72	6	2	3	-4	-3	10
83	62	77	9	2	5	-2	-1	4
83	68	72	12	4	5	-4	-3	12
84	64	76	4	1	2	-4	-2	9
84	67	77	9	2	4	-2	-1	5
84	69	73	10	3	4	-3	-2	9
84	69	79	5	1	2	-4	-2	11
85	60	79	5	1	3	-5	-2	9
85	61	74	7	2	4	-2	-1	4
85	65	72	6	2	3	-5	-3	12
85	66	75	11	3	5	-2	-1	5
86	61	72	12	4	7	-2	-1	4
87	64	76	4	1	2	-5	-2	11
87	69	79	5	1	2	-3	-1	8
89	60	73	10	3	6	-6	-2	11
89	60	79	5	1	3	-4	-1	7
89	62	75	11	3	6	-3	-1	6
80	60	80						
80	64	81	6	1	3	-5	-3	11
80	66	84	9	1	4	-5	-3	12
80	68	81	12	2	5	-3	-2	8
81	67	85	12	1	5	-2	-1	5
82	62	80	11	2	6	-2	-1	4
83	67	82	7	1	3	-2	-1	5
85	60	85	12	1	7	-4	-1	7
85	61	82	7	1	4	-6	-2	11
85	66	84	9	1	4	-3	-1	7
86	66	85	11	1	5	-4	-1	9
87	66	80	11	2	5	-3	-1	7
88	61	82	7	1	4	-5	-1	9
88	61	84	9	1	5	-6	-1	11
88	68	81	12	2	5	-4	-1	10
89	67	82	7	1	3	-5	-1	12
80	60	90						
80	63	90	2	0	1	-5	-2	10
81	61	90	11	0	6	-6	-2	11
81	66	90	11	0	5	-5	-2	11
81	67	90	12	0	5	-5	-2	12
82	60	90	12	0	7	-7	-2	12
82	63	90	2	0	1	-3	-1	6
83	60	90	7	0	4	-4	-1	7
83	67	90	7	0	3	-3	-1	7
83	69	90	8	0	3	-3	-1	8

φ_1	Φ	φ_2	H	K	L	U	V	W
80	60	90						
84	61	90	9	0	5	-5	-1	9
84	63	90	2	0	1	-4	-1	6
84	66	90	9	0	4	-4	-1	9
85	61	90	11	0	6	-6	-1	11
85	63	90	2	0	1	-5	-1	10
85	66	90	11	0	5	-5	-1	11
85	68	90	5	0	2	-4	-1	10
86	60	90	12	0	7	-7	-1	12
86	63	90	2	0	1	-6	-1	12
86	67	90	12	0	5	-5	-1	12
80	70	0						
80	70	0	0	11	4	2	-4	11
80	75	0	0	11	3	2	-3	11
81	72	0	0	3	1	1	-2	6
81	75	5	1	11	3	1	-2	7
81	76	0	0	4	1	2	-3	12
82	70	5	1	11	4	1	-3	8
82	74	0	0	7	2	1	-2	7
82	74	8	1	7	2	1	-3	10
83	76	0	0	4	1	1	-2	8
84	72	0	0	3	1	1	-3	9
84	77	0	0	9	2	1	-2	9
84	79	0	0	5	1	1	-2	10
85	70	0	0	11	4	1	-4	11
85	72	0	0	3	1	1	-4	12
85	73	0	0	10	3	1	-3	10
85	75	0	0	11	3	1	-3	11
85	76	0	0	4	1	1	-3	12
87	72	9	1	6	2	0	-1	3
88	70	5	1	11	4	0	-4	11
88	70	7	1	8	3	0	-3	8
88	72	5	1	12	4	0	-1	3
88	72	6	1	9	3	0	-1	3
88	73	6	1	10	3	0	-3	10
88	74	8	1	7	2	0	-2	7
88	76	7	1	8	2	0	-1	4
88	76	9	2	12	3	0	-1	4
89	75	5	1	11	3	0	-3	11
89	76	5	1	12	3	0	-1	4
89	78	6	1	9	2	0	-2	9
89	79	6	1	10	2	0	-1	5
80	70	10						
80	70	14	2	8	3	1	-4	10
80	75	15	3	11	3	1	-3	10

φ_1	Φ	φ_2	H	K	L	U	V	W
80	70	10						
80	78	13	2	9	2	1	-2	8
81	76	14	1	4	1	1	-3	11
81	78	18	3	9	2	1	-3	12
81	79	11	1	5	1	1	-2	9
84	72	18	1	3	1	0	-1	3
85	70	14	2	8	3	0	-3	8
85	71	15	3	11	4	0	-4	11
85	74	17	3	10	3	0	-3	10
86	70	10	2	11	4	0	-4	11
86	72	13	2	9	3	0	-1	3
86	72	14	3	12	4	0	-1	3
86	75	15	3	11	3	0	-3	11
86	75	16	2	7	2	0	-2	7
86	77	18	4	12	3	0	-1	4
86	78	18	3	9	2	0	-2	9
87	74	11	2	10	3	0	-3	10
87	75	10	2	11	3	0	-3	11
87	76	14	1	4	1	0	-1	4
87	78	13	2	9	2	0	-2	9
87	79	17	3	10	2	0	-1	5
88	79	11	1	5	1	0	-1	5
80	70	20						
80	70	27	5	10	4	0	-2	5
80	70	27	6	12	5	0	-5	12
81	71	27	4	8	3	0	-3	8
81	72	29	6	11	4	0	-4	11
81	74	29	5	9	3	0	-1	3
82	70	22	2	5	2	0	-2	5
82	72	24	5	11	4	0	-4	11
82	73	27	3	6	2	0	-1	3
83	71	20	4	11	4	0	-4	11
83	71	21	3	8	3	0	-3	8
83	73	23	5	12	4	0	-1	3
83	73	24	4	9	3	0	-1	3
83	75	27	5	10	3	0	-3	10
83	77	29	6	11	3	0	-3	11
84	74	22	4	10	3	0	-3	10
84	75	23	3	7	2	0	-2	7
84	76	24	5	11	3	0	-3	11
84	77	27	2	4	1	0	-1	4
84	79	29	5	9	2	0	-2	9
85	76	20	4	11	3	0	-3	11
85	77	21	3	8	2	0	-1	4
85	77	23	5	12	3	0	-1	4
85	79	24	4	9	2	0	-2	9
86	71	27	4	8	3	-1	-4	12

φ_1	Φ	φ_2	H	K	L	U	V	W
80	70	20						
86	79	22	2	5	1	0	-1	5
87	70	22	2	5	2	-1	-4	11
88	71	20	4	11	4	-1	-4	12
89	71	21	3	8	3	-1	-3	9
89	72	29	6	11	4	-1	-2	7
89	75	23	3	7	2	-1	-3	12
80	70	30						
80	74	34	2	3	1	0	-1	3
80	77	39	8	10	3	0	-3	10
80	78	39	9	11	3	0	-3	11
81	74	30	7	12	4	0	-1	3
81	76	31	6	10	3	0	-3	10
81	76	35	7	10	3	0	-3	10
81	77	36	5	7	2	0	-2	7
81	78	36	8	11	3	0	-3	11
82	73	39	4	5	2	-1	-4	12
82	76	30	4	7	2	0	-2	7
82	77	32	7	11	3	0	-3	11
82	78	34	8	12	3	0	-1	4
82	79	37	3	4	1	0	-1	4
83	71	36	5	7	3	-1	-4	11
83	78	30	7	12	3	0	-1	4
83	78	32	5	8	2	0	-1	4
84	70	30	7	12	5	-1	-4	11
84	72	37	9	12	5	-1	-3	9
85	73	37	6	8	3	-1	-3	10
86	71	31	3	5	2	-1	-3	9
86	73	32	7	11	4	-1	-3	10
86	73	39	4	5	2	-1	-2	7
86	74	34	2	3	1	-1	-3	11
87	70	36	8	11	5	-1	-2	6
87	76	31	6	10	3	-1	-3	12
88	70	34	6	9	4	-1	-2	6
88	72	32	5	8	3	-1	-2	7
88	74	34	2	3	1	-1	-2	8
88	76	35	7	10	3	-1	-2	9
88	76	36	8	11	3	-1	-2	10
88	79	37	3	4	1	-1	-2	11
89	70	30	4	7	3	-1	-2	6
89	73	37	6	8	3	-2	-3	12
89	76	30	4	7	2	-1	-2	9
80	70	40						
80	72	45	11	11	5	-1	-4	11
80	74	45	12	12	5	-1	-4	12

φ_1	Φ	φ_2	H	K	L	U	V	W
80	70	40						
80	78	41	6	7	2	0	-2	7
80	79	42	10	11	3	0	-3	11
81	71	45	2	2	1	-1	-3	8
81	72	42	8	9	4	-1	-4	11
81	73	45	9	9	4	-1	-3	9
81	73	47	12	11	5	-1	-3	9
81	79	40	10	12	3	0	-1	4
81	79	41	7	8	2	0	-1	4
82	74	45	5	5	2	-1	-3	10
82	76	48	9	8	3	-1	-3	11
83	72	41	6	7	3	-1	-3	9
83	76	45	11	11	4	-1	-3	11
83	77	45	3	3	1	-1	-3	12
84	71	45	2	2	1	-1	-2	6
84	73	45	7	7	3	-1	-2	7
84	73	48	10	9	4	-1	-2	7
85	73	43	11	12	5	-1	-2	7
85	75	45	8	8	3	-1	-2	8
86	71	45	2	2	1	-2	-3	10
86	72	45	11	11	5	-2	-3	11
86	75	42	10	11	4	-1	-2	8
86	77	45	3	3	1	-1	-2	9
86	78	45	10	10	3	-1	-2	10
86	79	45	11	11	3	-1	-2	11
86	79	49	8	7	2	-1	-2	11
87	74	45	12	12	5	-2	-3	12
87	78	41	6	7	2	-1	-2	10
88	72	41	6	7	3	-2	-3	11
88	73	42	9	10	4	-2	-3	12
89	74	49	8	7	3	-1	-1	5
89	79	48	11	10	3	-1	-1	7

φ_1	Φ	φ_2	H	K	L	U	V	W
80	70	50						
80	72	55	10	7	4	-1	-2	6
80	72	58	8	5	3	-1	-2	6
80	75	52	9	7	3	-1	-3	10
81	72	53	12	9	5	-1	-2	6
81	74	56	3	2	1	-1	-2	7
81	77	54	7	5	2	-1	-3	11
81	78	56	12	8	3	-1	-3	12
82	70	56	9	6	4	-2	-3	9
83	76	55	10	7	3	-1	-2	8
83	78	58	8	5	2	-1	-2	9
84	74	56	3	2	1	-2	-3	12
84	76	50	6	5	2	-1	-2	8
84	78	54	11	8	3	-1	-2	9
84	79	53	4	3	1	-1	-2	10

φ_1	Φ	φ_2	H	K	L	U	V	W
80	70	50						
85	71	59	5	3	2	-1	-1	4
85	73	51	5	4	2	-2	-3	11
86	71	56	12	8	5	-1	-1	4
86	73	58	11	7	4	-2	-2	9
87	71	54	7	5	3	-1	-1	4
87	74	56	3	2	1	-1	-1	5
87	77	58	11	7	3	-1	-1	6
88	71	51	11	9	5	-1	-1	4
88	71	52	9	7	4	-1	-1	4
88	73	51	5	4	2	-2	-2	9
88	74	51	11	9	4	-1	-1	5
88	77	54	7	5	2	-1	-1	6
88	79	53	4	3	1	-1	-1	7
89	76	50	6	5	2	-2	-2	11
89	77	51	10	8	3	-1	-1	6

φ_1	Φ	φ_2	H	K	L	U	V	W
80	70	60						
80	70	63	12	6	5	-3	-4	12
80	73	66	9	4	3	-2	-3	10
80	79	61	9	5	2	-1	-3	12
81	70	60	12	7	5	-2	-3	9
81	71	69	8	3	3	-3	-3	11
81	72	61	11	6	4	-2	-3	10
81	77	63	4	2	1	-1	-2	8
81	79	68	5	2	1	-1	-2	9
82	70	68	5	2	2	-2	-2	7
82	74	61	9	5	3	-2	-3	11
83	72	66	11	5	4	-1	-1	4
84	71	63	8	4	3	-1	-1	4
84	75	67	7	3	2	-1	-1	5
84	77	69	8	3	2	-2	-2	11
85	70	60	7	4	3	-3	-3	11
85	73	63	6	3	2	-2	-2	9
85	75	63	10	5	3	-1	-1	5
86	71	69	8	3	3	-3	-2	10
86	76	60	7	4	2	-2	-2	11
86	77	63	4	2	1	-1	-1	6
86	79	68	5	2	1	-1	-1	7
87	79	61	9	5	2	-1	-1	7
88	70	63	10	5	4	-3	-2	10
88	70	68	5	2	2	-2	-1	6
88	73	63	6	3	2	-3	-2	12
89	70	60	12	7	5	-3	-2	10
89	74	68	10	4	3	-2	-1	6
89	76	66	11	5	3	-2	-1	9
89	79	66	9	4	2	-2	-1	11
89	79	68	5	2	1	-2	-1	12

φ_1	Φ	φ_2	H	K	L	U	V	W
80	70	70						
80	77	72	12	4	3	-2	-3	12
81	72	72	3	1	1	-1	-1	4
82	75	74	7	2	2	-2	-2	9
82	76	76	4	1	1	-1	-1	5
82	78	77	9	2	2	-2	-2	11
83	76	70	11	4	3	-1	-1	5
83	79	79	5	1	1	-1	-1	6
84	78	72	9	3	2	-1	-1	6
85	72	72	3	1	1	-3	-2	11
85	74	73	10	3	3	-3	-2	12
86	70	76	8	2	3	-2	-1	6
87	75	74	7	2	2	-2	-1	8
87	76	76	4	1	1	-2	-1	9
87	78	77	9	2	2	-2	-1	10
87	79	79	5	1	1	-2	-1	11
88	72	72	3	1	1	-2	-1	7
89	71	75	11	3	4	-3	-1	9
89	75	75	11	3	3	-3	-1	12

φ_1	Φ	φ_2	H	K	L	U	V	W
80	70	80						
80	78	84	9	1	2	-1	-1	5
81	79	84	10	1	2	-2	-2	11
82	72	81	6	1	2	-3	-2	10
84	70	80	11	2	4	-2	-1	6
84	73	84	10	1	3	-2	-1	7
86	75	80	11	2	3	-2	-1	8
87	74	82	7	1	2	-3	-1	11
88	70	83	8	1	3	-4	-1	11

φ_1	Φ	φ_2	H	K	L	U	V	W
80	70	90						
80	70	90	11	0	4	-4	-2	11
80	75	90	11	0	3	-3	-2	11
81	72	90	3	0	1	-2	-1	6
81	76	90	4	0	1	-3	-2	12
82	74	90	7	0	2	-2	-1	7
83	76	90	4	0	1	-2	-1	8
84	72	90	3	0	1	-3	-1	9
84	77	90	9	0	2	-2	-1	9
84	79	90	5	0	1	-2	-1	10
85	70	90	11	0	4	-4	-1	11
85	72	90	3	0	1	-4	-1	12
85	73	90	10	0	3	-3	-1	10
85	75	90	11	0	3	-3	-1	11
85	76	90	4	0	1	-3	-1	12

φ_1	Φ	φ_2	H	K	L	U	V	W
80	80	0						
80	80	0	0	11	2	2	-2	11
80	85	0	0	11	1	2	-1	11
81	81	0	0	6	1	1	-1	6
81	83	7	1	8	1	1	-1	7
81	85	0	0	12	1	2	-1	12
82	82	0	0	7	1	1	-1	7
82	84	6	1	9	1	1	-1	8
83	81	9	1	6	1	1	-2	11
83	83	0	0	8	1	1	-1	8
83	84	6	1	10	1	1	-1	9
83	85	9	2	12	1	1	-1	10
84	84	0	0	9	1	1	-1	9
84	84	0	0	10	1	1	-1	10
84	85	5	1	11	1	1	-1	10
84	85	5	1	12	1	1	-1	11
85	80	0	0	11	2	1	-2	11
85	81	0	0	6	1	1	-2	12
85	85	0	0	11	1	1	-1	11
85	85	0	0	12	1	1	-1	12
86	81	9	1	6	1	0	-1	6
89	80	5	1	11	2	0	-2	11
89	81	5	1	12	2	0	-1	6
89	82	8	1	7	1	0	-1	7
89	83	7	1	8	1	0	-1	8
89	84	6	1	9	1	0	-1	9
89	84	6	1	10	1	0	-1	10
89	85	9	2	12	1	0	-1	12

φ_1	Φ	φ_2	H	K	L	U	V	W
80	80	10						
80	84	13	2	9	1	1	-1	7
80	85	17	3	10	1	1	-1	7
81	81	18	2	6	1	1	-2	10
81	85	15	3	11	1	1	-1	8
81	85	18	4	12	1	1	-1	8
82	80	10	2	11	2	1	-2	10
82	84	11	2	10	1	1	-1	8
82	85	14	3	12	1	1	-1	9
83	82	16	2	7	1	1	-2	12
83	85	10	2	11	1	1	-1	9
87	80	15	3	11	2	0	-2	11
87	81	18	2	6	1	0	-1	6
88	80	10	2	11	2	0	-2	11
88	81	14	3	12	2	0	-1	6
88	82	16	2	7	1	0	-1	7
88	83	14	2	8	1	0	-1	8
88	84	18	3	9	1	0	-1	9

φ_1	Φ	φ_2	H	K	L	U	V	W
80	80	10						
88	85	17	3	10	1	0	-1	10
88	85	18	4	12	1	0	-1	12
89	84	11	2	10	1	0	-1	10
89	84	13	2	9	1	0	-1	9
89	85	10	2	11	1	0	-1	11
89	85	14	3	12	1	0	-1	12
89	85	15	3	11	1	0	-1	11
80	80	20						
80	80	20	4	11	2	1	-2	9
80	85	20	4	11	1	1	-1	7
81	83	23	3	7	1	1	-2	11
82	84	27	4	8	1	1	-2	12
85	80	27	5	10	2	0	-1	5
85	81	29	6	11	2	0	-2	11
86	80	20	4	11	2	0	-2	11
86	81	23	5	12	2	0	-1	6
86	81	24	5	11	2	0	-2	11
86	82	27	3	6	1	0	-1	6
87	83	23	3	7	1	0	-1	7
87	84	24	4	9	1	0	-1	9
87	84	27	4	8	1	0	-1	8
87	84	29	5	9	1	0	-1	9
87	85	27	5	10	1	0	-1	10
88	83	21	3	8	1	0	-1	8
88	85	20	4	11	1	0	-1	11
88	85	22	4	10	1	0	-1	10
88	85	24	5	11	1	0	-1	11
88	85	29	6	11	1	0	-1	11
88	86	23	5	12	1	0	-1	12
88	86	27	6	12	1	0	-1	12
80	80	30						
80	83	30	4	7	1	1	-2	10
80	84	32	5	8	1	1	-2	11
80	85	38	7	9	1	1	-2	11
80	86	39	8	10	1	1	-2	12
81	85	34	6	9	1	1	-2	12
82	80	38	7	9	2	0	-2	9
83	80	34	6	9	2	0	-2	9
83	81	39	4	5	1	0	-1	5
83	82	39	9	11	2	0	-2	11
84	80	31	3	5	1	0	-1	5
84	81	32	7	11	2	0	-2	11
84	81	35	7	10	2	0	-1	5
84	82	36	8	11	2	0	-2	11

φ_1	Φ	φ_2	H	K	L	U	V	W
80	80	30						
84	82	37	9	12	2	0	-1	6
85	82	30	7	12	2	0	-1	6
85	82	34	4	6	1	0	-1	6
85	83	36	5	7	1	0	-1	7
86	83	30	4	7	1	0	-1	7
86	84	32	5	8	1	0	-1	8
86	84	37	6	8	1	0	-1	8
86	85	34	6	9	1	0	-1	9
86	85	38	7	9	1	0	-1	9
86	86	39	8	10	1	0	-1	10
87	85	31	6	10	1	0	-1	10
87	85	35	7	10	1	0	-1	10
87	86	32	7	11	1	0	-1	11
87	86	34	8	12	1	0	-1	12
87	86	36	8	11	1	0	-1	11
87	86	37	9	12	1	0	-1	12
87	86	39	9	11	1	0	-1	11
88	86	30	7	12	1	0	-1	12
89	80	34	6	9	2	-1	-2	12
80	80	40						
80	80	45	4	4	1	0	-1	4
80	81	48	9	8	2	0	-1	4
80	86	42	10	11	1	1	-2	12
80	87	45	12	12	1	1	-2	12
81	80	43	11	12	3	0	-1	4
81	81	45	9	9	2	0	-2	9
81	82	48	10	9	2	0	-2	9
82	81	42	8	9	2	0	-2	9
82	82	42	9	10	2	0	-1	5
82	82	45	5	5	1	0	-1	5
82	82	48	11	10	2	0	-1	5
82	83	47	12	11	2	0	-2	11
83	82	42	10	11	2	0	-2	11
83	83	45	6	6	1	0	-1	6
83	83	45	11	11	2	0	-2	11
83	84	49	7	6	1	0	-1	6
84	83	40	5	6	1	0	-1	6
84	83	43	11	12	2	0	-1	6
84	84	45	7	7	1	0	-1	7
84	85	49	8	7	1	0	-1	7
85	84	41	6	7	1	0	-1	7
85	85	41	7	8	1	0	-1	8
85	85	45	8	8	1	0	-1	8
85	85	48	9	8	1	0	-1	8
85	86	48	10	9	1	0	-1	9
86	85	42	8	9	1	0	-1	9

φ_1	Φ	φ_2	H	K	L	U	V	W
80	80	40						
86	86	42	9	10	1	0	-1	10
86	86	45	9	9	1	0	-1	9
86	86	45	10	10	1	0	-1	10
86	86	45	11	11	1	0	-1	11
86	86	47	12	11	1	0	-1	11
86	86	48	11	10	1	0	-1	10
87	80	45	4	4	1	-1	-2	12
87	86	40	10	12	1	0	-1	12
87	86	42	10	11	1	0	-1	11
87	86	43	11	12	1	0	-1	12
87	87	45	12	12	1	0	-1	12
80	80	50						
80	82	51	11	9	2	0	-2	9
80	82	53	12	9	2	0	-2	9
80	84	58	8	5	1	0	-1	5
81	83	50	6	5	1	0	-1	5
81	83	54	7	5	1	0	-1	5
82	84	53	8	6	1	0	-1	6
82	85	56	9	6	1	0	-1	6
82	85	59	10	6	1	0	-1	6
83	85	55	10	7	1	0	-1	7
83	86	58	11	7	1	0	-1	7
84	80	59	5	3	1	-1	-2	11
84	85	52	9	7	1	0	-1	7
84	86	51	10	8	1	0	-1	8
84	86	54	11	8	1	0	-1	8
84	86	56	12	8	1	0	-1	8
85	81	55	10	7	2	-1	-2	12
85	86	51	11	9	1	0	-1	9
85	86	53	12	9	1	0	-1	9
86	86	50	12	10	1	0	-1	10
88	80	59	5	3	1	-1	-1	8
88	81	58	11	7	2	-1	-1	9
88	82	56	6	4	1	-1	-1	10
89	80	52	9	7	2	-1	-1	8
89	81	51	5	4	1	-1	-1	9
89	82	51	11	9	2	-1	-1	10
89	83	50	6	5	1	-1	-1	11
89	83	54	7	5	1	-1	-1	12
80	80	60						
80	84	61	9	5	1	0	-1	5
80	85	63	10	5	1	0	-1	5
80	85	66	11	5	1	0	-1	5
80	86	67	12	5	1	0	-1	5

φ_1	Φ	φ_2	H	K	L	U	V	W
80	80	60						
82	80	63	10	5	2	-1	-2	10
82	85	61	11	6	1	0	-1	6
82	86	63	12	6	1	0	-1	6
83	81	67	12	5	2	-1	-2	11
83	86	60	12	7	1	0	-1	7
84	82	63	6	3	1	-1	-2	12
86	81	66	11	5	2	-1	-1	8
87	82	63	6	3	1	-1	-1	9
87	83	67	7	3	1	-1	-1	10
87	83	69	8	3	1	-1	-1	11
88	83	60	7	4	1	-1	-1	11
88	84	63	8	4	1	-1	-1	12
80	80	70						
82	82	74	7	2	1	-1	-2	11
82	83	76	8	2	1	-1	-2	12
84	80	75	11	3	2	-1	-1	7
86	81	72	6	2	1	-1	-1	8
86	82	74	7	2	1	-1	-1	9
86	83	76	8	2	1	-1	-1	10
86	84	77	9	2	1	-1	-1	11
86	84	79	10	2	1	-1	-1	12
87	84	72	9	3	1	-1	-1	12
80	80	80						
80	83	83	8	1	1	-1	-2	10
80	84	84	9	1	1	-1	-2	11
81	84	84	10	1	1	-1	-2	12
82	80	85	11	1	2	-1	-1	6
83	81	61	6	1	1	-1	-1	7
84	82	82	7	1	1	-1	-1	8
85	83	83	8	1	1	-1	-1	9
85	84	84	9	1	1	-1	-1	10
85	84	84	10	1	1	-1	-1	11
86	85	85	11	1	1	-1	-1	12
87	80	80	11	2	2	-2	-1	12
80	80	90						
80	80	90	11	0	2	-2	-2	11
80	85	90	11	0	1	-1	-2	11
81	81	90	6	0	1	-1	-1	6
81	85	90	12	0	1	-1	-2	12
82	82	90	7	0	1	-1	-1	7
83	83	90	8	0	1	-1	-1	8
84	84	90	9	0	1	-1	-1	9

φ_1	Φ	φ_2	H	K	L	U	V	W
80	80	90						
84	84	90	10	0	1	-1	-1	10
85	80	90	11	0	2	-2	-1	11
85	81	90	6	0	1	-2	-1	12
85	85	90	11	0	1	-1	-1	11
85	85	90	12	0	1	-1	-1	12
80	90	0						
80	90	0	0	1	0	2	0	11
81	90	0	0	1	0	1	0	6
82	90	0	0	1	0	1	0	7
83	90	0	0	1	0	1	0	8
84	90	0	0	1	0	1	0	9
84	90	0	0	1	0	1	0	10
85	90	0	0	1	0	1	0	12
85	90	0	0	1	0	1	0	11
80	90	40						
80	90	45	1	1	0	1	-1	8
81	90	45	1	1	0	1	-1	9
82	90	45	1	1	0	1	-1	10
83	90	45	1	1	0	1	-1	11
83	90	45	1	1	0	1	-1	12
80	90	90						
80	90	90	1	0	0	0	-2	11
81	90	90	1	0	0	0	-1	6
82	90	90	1	0	0	0	-1	7
83	90	90	1	0	0	0	-1	8
84	90	90	1	0	0	0	-1	10
84	90	90	1	0	0	0	-1	9
85	90	90	1	0	0	0	-1	12
85	90	90	1	0	0	0	-1	11
90	0	0						
90	5	0	0	1	11	0	-11	1
90	5	0	0	1	12	0	-12	1
90	6	0	0	1	9	0	-9	1
90	6	0	0	1	10	0	-10	1
90	7	0	0	1	8	0	-8	1
90	8	0	0	1	7	0	-7	1
90	9	0	0	1	6	0	-6	1

φ_1	Φ	φ_2	H	K	L	U	V	W
90	0	40						
90	7	45	1	1	11	-11	-11	2
90	7	45	1	1	12	-6	-6	1
90	8	45	1	1	10	-5	-5	1
90	9	45	1	1	9	-9	-9	2
90	0	90						
90	0	90	0	0	1	-1	0	0
90	5	90	1	0	11	-11	0	1
90	5	90	1	0	12	-12	0	1
90	6	90	1	0	9	-9	0	1
90	6	90	1	0	10	-10	0	1
90	7	90	1	0	8	-8	0	1
90	8	90	1	0	7	-7	0	1
90	9	90	1	0	6	-6	0	1
90	10	0						
90	10	0	0	2	11	0	-11	2
90	11	0	0	1	5	0	-5	1
90	13	0	0	2	9	0	-9	2
90	14	0	0	1	4	0	-4	1
90	15	0	0	3	11	0	-11	3
90	16	0	0	2	7	0	-7	2
90	17	0	0	3	10	0	-10	3
90	18	0	0	1	3	0	-3	1
90	10	10						
90	18	18	1	3	10	-1	-3	1
90	10	20						
90	13	27	1	2	10	-2	-4	1
90	10	40						
90	10	45	1	1	8	-4	-4	1
90	11	45	1	1	7	-7	-7	2
90	13	45	1	1	6	-3	-3	1
90	14	45	2	2	11	-11	-11	4
90	16	45	1	1	5	-5	-5	2
90	17	45	2	2	9	-9	-9	4
90	19	45	1	1	4	-2	-2	1

φ_1	Φ	φ_2	H	K	L	U	V	W
90	10	60						
90	13	63	2	1	10	-4	-2	1
90	10	70						
90	18	72	3	1	10	-3	-1	1
90	10	90						
90	10	90	2	0	11	-11	0	2
90	11	90	1	0	5	-5	0	1
90	13	90	2	0	9	-9	0	2
90	14	90	1	0	4	-4	0	1
90	15	90	3	0	11	-11	0	3
90	16	90	2	0	7	-7	0	2
90	17	90	3	0	10	-10	0	3
90	18	90	1	0	3	-3	0	1
90	20	0						
90	20	0	0	4	11	0	-11	4
90	21	0	0	3	8	0	-8	3
90	22	0	0	2	5	0	-5	2
90	23	0	0	3	7	0	-7	3
90	23	0	0	5	12	0	-12	5
90	24	0	0	4	9	0	-9	4
90	24	0	0	5	11	0	-11	5
90	27	0	0	1	2	0	-2	1
90	29	0	0	5	9	0	-9	5
90	29	0	0	6	11	0	-11	6
90	20	10						
90	22	18	1	3	8	-4	-12	5
90	28	18	1	3	6	-3	-9	5
90	20	20						
90	20	27	1	2	6	-6	-12	5
90	24	27	1	2	5	-1	-2	1
90	29	27	1	2	4	-4	-8	5
90	20	30						
90	27	37	3	4	10	-6	-8	5

φ_1	Φ	φ_2	H	K	L	U	V	W
90	20	40						
90	21	45	3	3	11	-11	-11	6
90	22	45	2	2	7	-7	-7	4
90	23	45	3	3	10	-5	-5	3
90	25	45	1	1	3	-3	-3	2
90	27	45	4	4	11	-11	-11	8
90	28	45	3	3	8	-4	-4	3
90	29	45	2	2	5	-5	-5	4
90	20	50						
90	27	53	4	3	10	-8	-6	5
90	20	60						
90	20	63	2	1	6	-12	-6	5
90	24	63	2	1	5	-2	-1	1
90	29	63	2	1	4	-8	-4	5
90	20	70						
90	22	72	3	1	8	-12	-4	5
90	28	72	3	1	6	-9	-3	5
90	20	90						
90	20	90	4	0	11	-11	0	4
90	21	90	3	0	8	-8	0	3
90	22	90	2	0	5	-5	0	2
90	23	90	3	0	7	-7	0	3
90	23	90	5	0	12	-12	0	5
90	24	90	4	0	9	-9	0	4
90	24	90	5	0	11	-11	0	5
90	27	90	1	0	2	-2	0	1
90	29	90	5	0	9	-9	0	5
90	29	90	6	0	11	-11	0	6
90	30	0						
90	30	0	0	4	7	0	-7	4
90	30	0	0	7	12	0	-12	7
90	31	0	0	3	5	0	-5	3
90	32	0	0	5	8	0	-8	5
90	32	0	0	7	11	0	-11	7
90	34	0	0	2	3	0	-3	2
90	35	0	0	7	10	0	-10	7
90	35	8	1	7	10	-1	-7	5
90	36	0	0	5	7	0	-7	5

TAB. VI

FROM EULER ANGLES INTO HKL-UVW

φ_1	Φ	φ_2	H	K	L	U	V	W
90	30	0						
90	36	0	0	8	11	0	-11	8
90	37	0	0	3	4	0	-4	3
90	38	0	0	7	9	0	-9	7
90	39	0	0	4	5	0	-5	4
90	39	0	0	9	11	0	-11	9
90	30	10						
90	32	18	1	3	5	-1	-3	2
90	38	18	1	3	4	-2	-6	5
90	30	20						
90	34	27	3	6	10	-2	-4	3
90	35	23	3	7	11	-5	-12	9
90	37	27	1	2	3	-3	-6	5
90	30	40						
90	31	45	3	3	7	-7	-7	6
90	31	45	5	5	12	-6	-6	5
90	32	45	4	4	9	-9	-9	8
90	33	45	5	5	11	-11	-11	10
90	35	45	1	1	2	-1	-1	1
90	38	45	5	5	9	-9	-9	10
90	38	45	6	6	11	-11	-11	12
90	39	45	4	4	7	-7	-7	8
90	30	60						
90	34	63	6	3	10	-4	-2	3
90	37	63	2	1	3	-6	-3	5
90	30	70						
90	32	72	3	1	5	-3	-1	2
90	38	72	3	1	4	-6	-2	5
90	30	80						
90	35	82	7	1	10	-7	-1	5
90	30	90						
90	30	90	4	0	7	-7	0	4
90	30	90	7	0	12	-12	0	7
90	31	90	3	0	5	-5	0	3
90	32	90	5	0	8	-8	0	5

φ_1	Φ	φ_2	H	K	L	U	V	W
90	30	90						
90	32	90	7	0	11	-11	0	7
90	34	90	2	0	3	-3	0	2
90	35	90	7	0	10	-10	0	7
90	36	90	5	0	7	-7	0	5
90	36	90	8	0	11	-11	0	8
90	37	90	3	0	4	-4	0	3
90	38	90	7	0	9	-9	0	7
90	39	90	4	0	5	-5	0	4
90	39	90	9	0	11	-11	0	9
90	40	0						
90	40	0	0	5	6	0	-6	5
90	41	0	0	6	7	0	-7	6
90	41	0	0	7	8	0	-8	7
90	42	0	0	8	9	0	-9	8
90	42	0	0	9	10	0	-10	9
90	42	0	0	10	11	0	-11	10
90	43	0	0	11	12	0	-12	11
90	45	0	0	1	1	0	-1	1
90	47	0	0	12	11	0	-11	12
90	48	0	0	9	8	0	-8	9
90	48	0	0	10	9	0	-9	10
90	48	0	0	11	10	0	-10	11
90	49	0	0	7	6	0	-6	7
90	49	0	0	8	7	0	-7	8
90	40	10						
90	43	18	3	9	10	-1	-3	3
90	47	18	1	3	3	-3	-9	10
90	40	20						
90	40	23	3	7	9	-5	-12	11
90	41	21	3	8	10	-4	-11	10
90	42	27	2	4	5	-1	-2	2
90	48	27	1	2	2	-2	-4	5
90	40	30						
90	45	37	3	4	5	-3	-4	5
90	40	40						
90	40	45	3	3	5	-5	-5	6
90	40	45	7	7	12	-6	-6	7
90	41	45	5	5	8	-4	-4	5

φ_1	Φ	φ_2	H	K	L	U	V	W
90	40	40						
90	43	45	2	2	3	-3	-3	4
90	45	45	5	5	7	-7	-7	10
90	45	45	7	7	10	-5	-5	7
90	47	45	3	3	4	-2	-2	3
90	49	45	4	4	5	-5	-5	8
90	40	50						
90	45	53	4	3	5	-4	-3	5
90	40	60						
90	42	63	4	2	5	-2	-1	2
90	48	63	2	1	2	-4	-2	5
90	40	70						
90	43	72	9	3	10	-3	-1	3
90	47	72	3	1	3	-9	-3	10
90	49	70	11	4	10	-8	-3	10
90	40	90						
90	40	90	5	0	6	-6	0	5
90	41	90	6	0	7	-7	0	6
90	41	90	7	0	8	-8	0	7
90	42	90	8	0	9	-9	0	8
90	42	90	9	0	10	-10	0	9
90	42	90	10	0	11	-11	0	10
90	43	90	11	0	12	-12	0	11
90	45	90	1	0	1	-1	0	1
90	47	90	12	0	11	-11	0	12
90	48	90	9	0	8	-8	0	9
90	48	90	10	0	9	-9	0	10
90	48	90	11	0	10	-10	0	11
90	49	90	7	0	6	-6	0	7
90	49	90	8	0	7	-7	0	8
90	50	0						
90	50	0	0	6	5	0	-5	6
90	51	0	0	5	4	0	-4	5
90	51	0	0	11	9	0	-9	11
90	52	0	0	9	7	0	-7	9
90	53	0	0	4	3	0	-3	4
90	54	0	0	7	5	0	-5	7
90	54	0	0	11	8	0	-8	11
90	55	0	0	10	7	0	-7	10

φ_1	Φ	φ_2	H	K	L	U	V	W
90	50	0						
90	55	6	1	7	5	-1	-7	10
90	56	0	0	3	2	0	-2	3
90	56	0	0	8	5	0	-5	8
90	58	0	0	11	7	0	-7	11
90	59	0	0	5	3	0	-3	5
90	50	10						
90	52	18	2	6	5	-1	-3	4
90	58	18	1	3	2	-1	-3	5
90	50	20						
90	52	23	5	12	10	-2	-5	7
90	53	27	3	6	5	-1	-2	3
90	56	27	2	4	3	-3	-6	10
90	50	40						
90	50	45	5	5	6	-3	-3	5
90	50	45	6	6	7	-7	-7	12
90	51	45	7	7	8	-4	-4	7
90	52	45	9	9	10	-5	-5	9
90	52	45	11	11	12	-6	-6	11
90	55	45	1	1	1	-1	-1	2
90	57	41	6	7	6	-5	-6	12
90	57	45	11	11	10	-5	-5	11
90	58	45	9	9	8	-4	-4	9
90	59	45	6	6	5	-5	-5	12
90	59	45	7	7	6	-3	-3	7
90	50	50						
90	52	50	6	5	6	-7	-6	12
90	50	60						
90	50	67	12	5	11	-7	-3	9
90	53	63	6	3	5	-2	-1	3
90	55	67	12	5	9	-7	-3	11
90	56	63	4	2	3	-6	-3	10
90	57	60	12	7	9	-5	-3	9
90	50	70						
90	52	72	6	2	5	-3	-1	4
90	58	72	3	1	2	-3	-1	5

TAB. VI FROM EULER ANGLES INTO HKL-UVW

φ_1	Φ	φ_2	H	K	L	U	V	W
90	50	80						
90	55	82	7	1	5	-7	-1	10
90	50	90						
90	50	90	6	0	5	-5	0	6
90	51	90	5	0	4	-4	0	5
90	51	90	11	0	9	-9	0	11
90	52	90	9	0	7	-7	0	9
90	53	90	4	0	3	-3	0	4
90	54	90	7	0	5	-5	0	7
90	54	90	11	0	8	-8	0	11
90	55	90	10	0	7	-7	0	10
90	56	90	3	0	2	-2	0	3
90	58	90	8	0	5	-5	0	8
90	58	90	11	0	7	-7	0	11
90	59	90	5	0	3	-3	0	5
90	60	0						
90	60	0	0	7	4	0	-4	7
90	60	0	0	12	7	0	-7	12
90	61	0	0	9	5	0	-5	9
90	61	0	0	11	6	0	-6	11
90	63	0	0	2	1	0	-1	2
90	66	0	0	9	4	0	-4	9
90	66	0	0	11	5	0	-5	11
90	67	0	0	7	3	0	-3	7
90	67	0	0	12	5	0	-5	12
90	68	0	0	5	2	0	-2	5
90	69	0	0	8	3	0	-3	8
90	60	10						
90	62	18	3	9	5	-1	-3	6
90	68	18	4	12	5	-1	-3	8
90	60	20						
90	61	27	4	8	5	-1	-2	4
90	62	23	5	12	7	-2	-5	10
90	66	27	1	2	1	-1	-2	5
90	60	30						
90	62	32	5	8	5	-3	-5	11
90	63	37	6	8	5	-3	-4	10

φ_1	Φ	φ_2	H	K	L	U	V	W
90	60	40						
90	61	45	5	5	4	-2	-2	5
90	62	45	4	4	3	-3	-3	8
90	63	45	11	11	8	-4	-4	11
90	65	45	3	3	2	-1	-1	3
90	67	45	5	5	3	-3	-3	10
90	68	45	7	7	4	-2	-2	7
90	69	45	11	11	6	-3	-3	11
90	60	50						
90	63	53	8	6	5	-4	-3	10
90	60	60						
90	61	63	6	4	5	-2	-1	4
90	66	63	2	1	1	-2	-1	5
90	60	70						
90	62	72	9	3	5	-3	-1	6
90	68	72	12	4	5	-3	-1	8
90	60	90						
90	60	90	7	0	4	-4	0	7
90	60	90	12	0	7	-7	0	12
90	61	90	9	0	5	-5	0	9
90	61	90	11	0	6	-6	0	11
90	63	90	2	0	1	-1	0	2
90	66	90	9	0	4	-4	0	9
90	66	90	11	0	5	-5	0	11
90	67	90	7	0	3	-3	0	7
90	67	90	12	0	5	-5	0	12
90	68	90	5	0	2	-2	0	5
90	69	90	8	0	3	-3	0	8
90	70	0						
90	70	0	0	11	4	0	-4	11
90	72	0	0	3	1	0	-1	3
90	73	0	0	10	3	0	-3	10
90	74	0	0	7	2	0	-2	7
90	75	0	0	11	3	0	-3	11
90	76	0	0	4	1	0	-1	4
90	77	0	0	9	2	0	-2	9
90	79	0	0	5	1	0	-1	5

φ_1	Φ	φ_2	H	K	L	U	V	W
90	70	10						
90	72	18	1	3	1	-1	-3	10
90	70	20						
90	70	27	6	12	5	-1	-2	6
90	77	27	2	4	1	-1	-2	10
90	70	30						
90	72	35	7	10	4	-2	-3	11
90	70	40						
90	71	45	2	2	1	-1	-1	4
90	73	45	9	9	4	-2	-2	9
90	74	45	5	5	2	-1	-1	5
90	76	45	11	11	4	-2	-2	11
90	77	45	3	3	1	-1	-1	6
90	79	45	7	7	2	-1	-1	7
90	70	60						
90	70	63	12	6	5	-2	-1	6
90	77	63	4	2	1	-2	-1	10
90	70	70						
90	72	72	3	1	1	-3	-1	10
90	74	73	10	3	3	-3	-1	11
90	70	90						
90	70	90	11	0	4	-4	0	11
90	72	90	3	0	1	-1	0	3
90	73	90	10	0	3	-3	0	10
90	74	90	7	0	2	-2	0	7
90	75	90	11	0	3	-3	0	11
90	76	90	4	0	1	-1	0	4
90	77	90	9	0	2	-2	0	9
90	79	90	5	0	1	-1	0	5
90	80	0						
90	80	0	0	11	2	0	-2	11
90	81	0	0	6	1	0	-1	6
90	82	0	0	7	1	0	-1	7
90	83	0	0	8	1	0	-1	8

φ_1	Φ	φ_2	H	K	L	U	V	W
90	80	0						
90	84	0	0	9	1	0	-1	9
90	84	0	0	10	1	0	-1	10
90	85	0	0	11	1	0	-1	11
90	85	0	0	12	1	0	-1	12
90	85	5	1	11	1	0	-1	11
90	85	5	1	12	1	0	-1	12
90	80	40						
90	80	45	4	4	1	-1	-1	6
90	81	45	9	9	2	-1	-1	9
90	82	45	5	5	1	-1	-1	10
90	83	45	6	6	1	-1	-1	12
90	83	45	11	11	2	-1	-1	11
90	80	90						
90	80	90	11	0	2	-2	0	11
90	81	90	6	0	1	-1	0	6
90	82	90	7	0	1	-1	0	7
90	83	90	8	0	1	-1	0	8
90	84	90	9	0	1	-1	0	9
90	84	90	10	0	1	-1	0	10
90	85	90	11	0	1	-1	0	11
90	85	90	12	0	1	-1	0	12
90	90	0						
90	90	0	0	1	0	0	0	1
90	90	5	1	11	0	0	0	1
90	90	5	1	12	0	0	0	1
90	90	6	1	9	0	0	0	1
90	90	6	1	10	0	0	0	1
90	90	7	1	6	0	0	0	1
90	90	8	1	7	0	0	0	1
90	90	9	1	6	0	0	0	1
90	90	10						
90	90	10	2	11	0	0	0	1
90	90	11	1	5	0	0	0	1
90	90	13	2	9	0	0	0	1
90	90	14	1	4	0	0	0	1
90	90	15	3	11	0	0	0	1
90	90	16	2	7	0	0	0	1
90	90	17	3	10	0	0	0	1
90	90	18	1	3	0	0	0	1

TAB. VI FROM EULER ANGLES INTO HKL-UVW

φ_1 Φ φ_2	H	K	L	U	V	W
90 90 20						
90 90 20	4	11	0	0	0	1
90 90 21	3	8	0	0	0	1
90 90 22	2	5	0	0	0	1
90 90 23	3	7	0	0	0	1
90 90 23	5	12	0	0	0	1
90 90 24	4	9	0	0	0	1
90 90 24	5	11	0	0	0	1
90 90 27	1	2	0	0	0	1
90 90 29	5	9	0	0	0	1
90 90 29	6	11	0	0	0	1
90 90 30						
90 90 30	4	7	0	0	0	1
90 90 30	7	12	0	0	0	1
90 90 31	3	5	0	0	0	1
90 90 32	5	8	0	0	0	1
90 90 32	7	11	0	0	0	1
90 90 34	2	3	0	0	0	1
90 90 35	7	10	0	0	0	1
90 90 36	5	7	0	0	0	1
90 90 36	8	11	0	0	0	1
90 90 37	3	4	0	0	0	1
90 90 38	7	9	0	0	0	1
90 90 39	4	5	0	0	0	1
90 90 39	9	11	0	0	0	1
90 90 40						
90 90 40	5	6	0	0	0	1
90 90 41	6	7	0	0	0	1
90 90 41	7	8	0	0	0	1
90 90 42	8	9	0	0	0	1
90 90 42	9	10	0	0	0	1
90 90 42	10	11	0	0	0	1
90 90 43	11	12	0	0	0	1
90 90 45	1	1	0	0	0	1
90 90 47	12	11	0	0	0	1
90 90 48	9	8	0	0	0	1
90 90 48	10	9	0	0	0	1
90 90 48	11	10	0	0	0	1
90 90 49	7	6	0	0	0	1
90 90 49	8	7	0	0	0	1

φ_1 Φ φ_2	H	K	L	U	V	W
90 90 50						
90 90 51	5	4	0	0	0	1
90 90 51	11	9	0	0	0	1
90 90 52	9	7	0	0	0	1
90 90 53	4	3	0	0	0	1
90 90 54	7	5	0	0	0	1
90 90 54	11	8	0	0	0	1
90 90 55	10	7	0	0	0	1
90 90 56	3	2	0	0	0	1
90 90 58	8	5	0	0	0	1
90 90 58	11	7	0	0	0	1
90 90 59	5	3	0	0	0	1
90 90 60						
90 90 60	7	4	0	0	0	1
90 90 60	12	7	0	0	0	1
90 90 61	9	5	0	0	0	1
90 90 61	11	6	0	0	0	1
90 90 63	2	1	0	0	0	1
90 90 66	9	4	0	0	0	1
90 90 66	11	5	0	0	0	1
90 90 67	7	3	0	0	0	1
90 90 67	12	5	0	0	0	1
90 90 68	5	2	0	0	0	1
90 90 69	8	3	0	0	0	1
90 90 70						
90 90 70	11	4	0	0	0	1
90 90 72	3	1	0	0	0	1
90 90 73	10	3	0	0	0	1
90 90 74	7	2	0	0	0	1
90 90 75	11	3	0	0	0	1
90 90 76	4	1	0	0	0	1
90 90 77	9	2	0	0	0	1
90 90 79	5	1	0	0	0	1
90 90 80						
90 90 80	11	2	0	0	0	1
90 90 81	6	1	0	0	0	1
90 90 82	7	1	0	0	0	1
90 90 83	8	1	0	0	0	1
90 90 84	9	1	0	0	0	1

φ_1 Φ φ_2	H	K	L	U	V	W
90 90 90						
90 90 90	1	0	0	0	0	1

φ_1 Φ φ_2	H	K	L	U	V	W

TABLE VII.

Angles between Sample and Crystal Axes

This table gives the angles between a sample axis XYZ and the main crystal axes. XYZ can mean HKL (i.e. ND), UVW (i.e. RD) or QRS (TD). The crystal axes are numbered in the following way:

$\{100\}$: 1 = (100); 2 = (010); 3 = (001)

$\{111\}$: 1 = (111); 2 = $(\bar{1}\bar{1}1)$; 3 = $(1\bar{1}1)$; 4 = $(\bar{1}11)$

$\{110\}$: 1 = (110); 2 = (101); 3 = $(\bar{1}10)$; 4 = $(\bar{1}01)$; 5 = (011); 6 = $(0\bar{1}1)$.

The table is ordered according to increasing XYZ with $X \leqslant Y \leqslant Z \leqslant 15$.

TAB. VII FROM HKL INTO ALPHA ANGLES

MILL IND X	Y	Z	(100)- α1	α2	α3	(111)- α1	α2	α3	α4	(110)-POLES α1	α2	α3	α4	α5	α6
0	0	1	90	90	0	55	55	55	55	90	45	90	45	45	45
0	1	1	90	45	45	35	90	90	35	60	60	60	60	0	90
0	1	2	90	63	27	39	75	75	39	72	51	72	51	18	72
0	1	3	90	72	18	43	69	69	43	77	48	77	48	27	63
0	1	4	90	76	14	46	65	65	46	80	47	80	47	31	59
0	1	5	90	79	11	47	63	63	47	82	46	82	46	34	56
0	1	6	90	81	9	48	62	62	48	83	46	83	46	36	54
0	1	7	90	82	8	49	61	61	49	84	46	84	46	37	53
0	1	8	90	83	7	50	60	60	50	85	45	85	45	38	52
0	1	9	90	84	6	50	59	59	50	86	45	86	45	39	51
0	1	10	90	84	6	51	59	59	51	86	45	86	45	39	51
0	1	11	90	85	5	51	58	58	51	86	45	86	45	40	50
0	1	12	90	85	5	51	58	58	51	87	45	87	45	40	50
0	1	13	90	86	4	52	58	58	52	87	45	87	45	41	49
0	1	14	90	86	4	52	58	58	52	87	45	87	45	41	49
0	1	15	90	86	4	52	57	57	52	37	45	87	45	41	49
0	2	3	90	56	34	37	81	81	37	67	54	67	54	11	79
0	2	5	90	68	22	41	71	71	41	75	49	75	49	23	67
0	2	7	90	74	16	44	67	67	44	79	47	79	47	29	61
0	2	9	90	77	13	46	64	64	46	81	46	81	46	32	58
0	2	11	90	80	10	48	62	62	48	83	46	83	46	35	55
0	2	13	90	81	9	49	61	61	49	84	46	84	46	36	54
0	2	15	90	82	8	50	60	60	50	85	46	85	46	37	53
0	3	4	90	53	37	36	83	83	36	65	56	65	56	8	82
0	3	5	90	59	31	38	79	79	38	69	53	69	53	14	76
0	3	7	90	67	23	41	72	72	41	74	49	74	49	22	68
0	3	8	90	69	21	42	70	70	42	76	49	76	49	24	66
0	3	10	90	73	17	44	67	67	44	78	47	78	47	28	62
0	3	11	90	75	15	45	66	66	45	79	47	79	47	30	60
0	3	13	90	77	13	46	64	64	46	81	46	81	46	32	58
0	3	14	90	78	12	47	64	64	47	81	46	81	46	33	57
0	4	5	90	51	39	36	85	85	36	64	56	64	56	6	84
0	4	7	90	60	30	38	78	78	38	69	52	69	52	15	75
0	4	9	90	66	24	40	73	73	40	73	50	73	50	21	69
0	4	11	90	70	20	42	70	70	42	76	48	76	48	25	65
0	4	13	90	73	17	44	68	68	44	78	47	78	47	28	62
0	4	15	90	75	15	45	66	66	45	80	47	80	47	30	60
0	5	6	90	50	40	36	86	86	36	63	57	63	57	5	85
0	5	7	90	54	36	36	82	82	36	66	55	66	55	9	81
0	5	8	90	58	32	37	79	79	37	68	53	68	53	13	77
0	5	9	90	61	29	38	77	77	38	70	52	70	52	16	74
0	5	11	90	66	24	40	73	73	40	73	50	73	50	21	69
0	5	12	90	67	23	41	72	72	41	74	49	74	49	22	68
0	5	13	90	69	21	42	71	71	42	75	49	75	49	24	66
0	5	14	90	70	20	42	70	70	42	76	48	76	48	25	65
0	6	7	90	49	41	36	86	86	36	63	58	63	58	4	86
0	6	11	90	61	29	38	77	77	38	70	52	70	52	16	74
0	6	13	90	65	25	40	74	74	40	73	50	73	50	20	70
0	7	8	90	49	41	35	87	87	35	62	58	62	58	4	86

MILL IND X	Y	Z	(100)- α1	α2	α3	(111)- α1	α2	α3	α4	(110)-POLES α1	α2	α3	α4	α5	α6
0	7	9	90	52	38	36	84	84	36	64	56	64	56	7	83
0	7	10	90	55	35	36	82	82	36	66	55	66	55	10	80
0	7	11	90	58	32	37	80	80	37	68	53	68	53	13	77
0	7	12	90	60	30	38	78	78	38	69	52	69	52	15	75
0	7	13	90	62	28	39	76	76	39	70	51	70	51	17	73
0	7	15	90	65	25	40	74	74	40	73	50	73	50	20	70
0	8	9	90	48	42	35	87	87	35	62	58	62	58	3	87
0	8	11	90	54	36	36	83	83	36	65	55	65	55	9	81
0	8	13	90	58	32	37	79	79	37	68	53	68	53	13	77
0	8	15	90	62	28	39	76	76	39	71	51	71	51	17	73
0	9	10	90	48	42	35	88	88	35	62	58	62	58	3	87
0	9	11	90	51	39	36	85	85	36	63	57	63	57	6	84
0	9	13	90	55	35	37	82	82	37	66	54	66	54	10	80
0	9	14	90	57	33	37	80	80	37	68	54	68	54	12	78
0	10	11	90	48	42	35	88	88	35	62	58	62	58	3	87
0	10	13	90	52	38	36	84	84	36	64	56	64	56	7	83
0	11	12	90	47	43	35	88	88	35	61	59	61	59	2	88
0	11	13	90	50	40	36	86	86	36	63	57	63	57	5	85
0	11	14	90	52	38	36	84	84	36	64	56	64	56	7	83
0	11	15	90	54	36	36	83	83	36	65	55	65	55	9	81
0	12	13	90	47	43	35	88	88	35	61	59	61	59	2	88
0	13	14	90	47	43	35	88	88	35	61	59	61	59	2	88
0	13	15	90	49	41	35	87	87	35	62	58	62	58	4	86
0	14	15	90	47	43	35	88	88	35	61	59	61	59	2	88
1	1	1	55	55	55	0	71	71	71	35	35	90	90	35	90
1	1	2	66	66	35	19	90	62	62	55	30	90	73	30	73
1	1	3	72	72	25	29	80	59	59	65	31	90	65	31	65
1	1	4	76	76	19	35	74	57	57	71	34	90	60	34	60
1	1	5	79	79	16	39	71	56	56	74	35	90	57	35	57
1	1	6	81	81	13	41	68	56	56	77	37	90	55	37	55
1	1	7	82	82	11	43	66	56	56	79	38	90	54	38	54
1	1	8	83	83	10	45	65	55	55	80	38	90	52	38	52
1	1	9	84	84	9	46	64	55	55	81	39	90	52	39	52
1	1	10	84	84	8	47	63	55	55	82	40	90	51	40	51
1	1	11	85	85	7	47	62	55	55	83	40	90	50	40	50
1	1	12	85	85	7	48	61	55	55	83	40	90	50	40	50
1	1	13	86	86	6	49	61	55	55	84	41	90	50	41	50
1	1	14	86	86	6	49	61	55	55	84	41	90	49	41	49
1	1	15	66	86	5	49	60	55	55	65	41	90	49	41	49
1	2	2	71	48	48	16	79	79	55	45	45	76	76	19	90
1	2	3	74	58	37	22	90	72	52	55	41	79	68	19	79
1	2	4	77	64	29	28	83	68	51	62	40	81	62	22	72
1	2	5	79	69	24	33	78	65	51	67	39	83	59	25	67
1	2	6	81	72	20	36	74	63	51	71	39	84	56	28	64
1	2	7	82	74	18	38	72	62	51	73	40	84	55	30	61
1	2	8	83	76	16	40	70	61	51	75	40	85	53	32	59
1	2	9	84	78	14	42	68	60	51	77	40	86	52	33	58
1	2	10	84	79	13	43	67	60	52	78	41	86	52	34	56
1	2	11	85	80	11	44	66	59	52	79	41	86	51	35	55

TAB. VII FROM HKL INTO ALPHA ANGLES

MILL IND X	Y	Z	(100)- α1	α2	α3	(111)- α1	α2	α3	α4	(110)-POLES α1	α2	α3	α4	α5	α6
1	2	12	85	81	11	45	65	59	52	80	41	87	50	36	55
1	2	13	86	81	10	46	64	58	52	81	41	87	50	36	54
1	2	14	86	82	9	46	63	58	52	81	42	87	50	37	53
1	2	15	86	82	8	47	63	58	52	82	42	87	49	38	53
1	3	3	77	47	47	22	82	82	49	50	50	71	71	13	90
1	3	4	79	54	38	25	90	77	47	56	46	74	65	14	82
1	3	5	80	60	32	29	84	73	47	61	44	76	61	17	76
1	3	6	82	64	28	32	80	70	47	65	43	78	59	20	72
1	3	7	83	67	24	34	77	68	47	68	43	79	56	23	68
1	3	8	83	70	22	36	74	66	48	71	42	81	55	25	66
1	3	9	84	72	19	38	72	65	48	73	42	81	54	27	64
1	3	10	85	73	18	40	71	64	49	74	42	82	53	29	62
1	3	11	85	75	16	41	69	63	49	76	42	83	52	30	60
1	3	12	85	76	15	42	68	62	49	77	42	83	51	31	59
1	3	13	86	77	14	43	67	62	50	78	42	84	51	32	58
1	3	14	86	78	13	44	66	61	50	79	42	84	50	33	57
1	3	15	86	79	12	44	66	61	50	79	42	85	50	34	56
1	4	4	80	46	46	25	84	84	45	52	52	68	68	10	90
1	4	5	81	52	40	27	90	80	45	57	49	71	64	11	84
1	4	6	82	57	34	29	85	76	44	61	47	73	61	14	79
1	4	7	83	61	30	31	82	73	45	64	46	75	59	17	75
1	4	8	84	64	27	33	79	71	45	67	45	76	57	19	72
1	4	9	84	66	25	35	77	70	46	69	44	78	55	22	69
1	4	10	85	68	22	37	75	68	46	71	44	79	54	24	67
1	4	11	85	70	21	38	73	67	47	72	44	80	53	25	65
1	4	12	85	72	19	39	71	66	47	74	44	80	52	27	64
1	4	13	86	73	18	40	70	65	47	75	43	81	52	28	62
1	4	14	86	74	16	41	69	64	48	76	43	82	51	29	61
1	4	15	86	75	15	42	68	64	48	77	43	82	50	30	60
1	5	5	82	46	46	27	85	85	43	54	54	67	67	8	90
1	5	6	83	51	40	28	90	82	43	57	51	69	63	9	85
1	5	7	83	55	36	30	86	78	43	61	49	71	61	12	81
1	5	8	84	58	33	32	83	76	43	63	48	73	59	14	77
1	5	9	84	61	30	33	80	74	43	66	47	74	57	17	74
1	5	10	85	64	27	35	78	72	44	68	46	75	55	19	72
1	5	11	85	66	25	36	76	71	44	70	46	77	54	21	70
1	5	12	86	67	23	37	75	69	45	71	45	77	53	23	68
1	5	13	86	69	21	38	73	68	45	72	45	78	53	24	66
1	5	14	86	70	20	39	72	67	46	73	45	79	52	26	65
1	5	15	86	72	19	40	71	66	46	74	44	80	51	27	63
1	6	6	83	45	45	29	86	86	42	55	55	66	66	7	90
1	6	7	84	50	41	29	90	83	42	58	52	68	63	8	86
1	6	8	84	53	37	30	87	80	42	60	51	69	60	10	82
1	6	9	85	56	34	32	84	78	42	63	49	71	59	12	79
1	6	10	85	59	31	33	81	76	42	65	48	72	57	15	76
1	6	11	85	61	29	34	79	74	43	67	48	74	56	17	74
1	6	12	86	64	27	35	78	73	43	68	47	75	55	19	72
1	6	13	86	65	25	36	76	71	44	70	46	76	54	21	70
1	6	14	86	67	23	37	75	70	44	71	46	77	53	22	68

MILL IND X	Y	Z	(100)- α1	α2	α3	(111)- α1	α2	α3	α4	(110)-POLES α1	α2	α3	α4	α5	α6
1	6	15	86	58	22	38	73	69	44	72	46	77	52	23	67
1	7	7	84	45	45	29	87	87	41	55	55	65	65	6	90
1	7	8	85	49	41	30	90	84	41	58	53	67	62	7	86
1	7	9	85	52	38	31	87	81	41	60	52	68	60	9	83
1	7	10	85	55	35	32	85	79	41	62	51	70	59	11	80
1	7	11	86	58	33	33	82	77	41	64	50	71	57	13	78
1	7	12	86	60	31	34	80	76	42	66	49	72	56	15	75
1	7	13	86	62	29	35	79	74	42	68	48	73	55	17	73
1	7	14	86	63	27	36	77	73	43	69	47	74	54	19	72
1	7	15	87	65	25	37	76	72	43	70	47	75	53	20	70
1	8	8	85	45	45	30	87	87	40	56	56	64	64	5	90
1	8	9	85	49	42	31	90	85	40	58	54	66	62	6	87
1	8	10	86	51	39	31	87	82	40	60	53	67	60	8	84
1	8	11	86	54	36	32	85	80	40	62	52	69	59	10	81
1	8	12	86	56	34	33	83	78	41	64	51	70	57	12	79
1	8	13	86	58	32	34	81	77	41	65	50	71	56	14	77
1	8	14	86	60	30	35	80	76	41	67	49	72	55	16	75
1	8	15	87	62	28	36	78	74	42	68	48	73	54	17	73
1	9	9	86	45	45	31	87	87	40	56	56	64	64	4	90
1	9	10	86	48	42	31	90	85	40	58	55	65	62	5	87
1	9	11	86	51	39	32	88	83	40	60	53	67	60	7	84
1	9	12	86	53	37	32	86	81	40	62	52	68	59	9	82
1	9	13	86	55	35	33	84	80	40	63	51	69	58	11	80
1	9	14	87	57	33	34	82	78	40	65	50	70	57	13	78
1	9	15	87	59	31	35	81	77	41	66	50	71	56	14	76
1	10	10	86	45	45	31	88	88	39	57	57	63	63	4	90
1	10	11	86	48	42	32	90	86	39	59	55	65	62	5	87
1	10	12	86	50	40	32	88	84	39	60	54	66	60	6	85
1	10	13	87	53	38	33	86	82	39	62	53	67	59	8	83
1	10	14	87	55	36	33	84	80	40	63	52	68	58	10	81
1	10	15	87	56	34	34	83	79	40	64	51	69	57	12	79
1	11	11	86	45	45	32	88	88	39	57	57	63	63	4	90
1	11	12	86	48	43	32	90	86	39	59	56	64	62	4	88
1	11	13	87	50	40	32	88	84	39	60	55	66	60	6	85
1	11	14	87	52	38	33	86	83	39	62	54	67	59	8	83
1	11	15	87	54	36	33	85	81	39	63	53	68	58	9	81
1	12	12	87	45	45	32	88	88	39	57	57	63	63	3	90
1	12	13	87	47	43	32	90	86	39	59	56	64	61	4	88
1	12	14	87	49	41	32	88	85	39	60	55	65	60	5	86
1	12	15	87	51	39	33	87	83	39	61	54	66	59	7	84
1	13	13	87	45	45	32	88	88	38	57	57	63	63	3	90
1	13	14	87	47	43	32	90	87	38	59	56	64	61	4	88
1	13	15	87	49	41	33	88	85	38	60	55	65	60	5	86
1	14	14	87	45	45	32	88	88	38	58	58	62	62	3	90
1	14	15	87	47	43	33	90	87	38	59	57	63	61	3	88
1	15	15	87	45	45	33	88	88	38	58	58	62	62	3	90
2	2	3	61	61	43	11	82	65	65	47	31	90	80	31	80
2	2	5	70	70	29	25	84	60	60	61	30	90	68	30	68
2	2	7	75	75	22	33	77	58	58	68	33	90	62	33	62

TAB. VII — FROM HKL INTO ALPHA ANGLES

MILL IND X	Y	Z	(100)- α1	α2	α3	(111)- α1	α2	α3	α4	(110)-POLES α1	α2	α3	α4	α5	α6
2	2	9	78	78	17	37	72	57	57	73	34	90	58	34	58
2	2	11	80	80	14	40	69	56	56	76	36	90	56	36	56
2	2	13	81	81	12	42	67	56	56	78	37	90	54	37	54
2	2	15	82	82	11	44	65	55	55	79	38	90	53	38	53
2	3	3	65	50	50	10	76	76	61	41	41	81	81	25	90
2	3	4	68	56	42	15	84	71	58	49	38	82	75	23	82
2	3	5	71	61	36	21	90	68	56	55	37	83	70	23	77
2	3	6	73	65	31	25	85	66	55	60	36	84	66	25	72
2	3	7	75	68	27	28	82	64	54	63	36	85	63	26	69
2	3	8	77	70	24	31	79	63	54	66	36	85	61	28	66
2	3	9	78	72	22	34	76	62	53	69	37	86	59	29	64
2	3	10	79	74	20	35	74	61	53	71	37	86	58	30	62
2	3	11	80	75	18	37	73	60	53	72	37	86	57	31	61
2	3	12	81	76	17	38	71	60	53	74	38	87	56	32	59
2	3	13	81	77	16	40	70	59	53	75	38	87	55	33	58
2	3	14	82	78	14	41	69	59	53	76	39	87	54	34	57
2	3	15	83	79	14	42	68	58	53	77	39	87	53	34	57
2	4	5	73	53	42	19	85	75	53	51	42	78	72	18	84
2	4	7	76	61	33	25	86	70	51	59	40	80	65	21	75
2	4	9	79	67	26	30	80	66	51	65	39	82	60	24	69
2	4	11	80	70	22	34	76	64	51	69	39	83	58	27	65
2	4	13	82	73	19	37	73	62	51	72	40	84	56	29	62
2	4	15	83	75	17	39	71	61	51	74	40	85	54	31	60
2	5	5	74	47	47	19	81	81	51	48	48	73	73	16	90
2	5	6	76	52	42	21	86	78	50	52	45	75	69	15	85
2	5	7	77	56	38	24	90	75	49	56	44	76	66	16	81
2	5	8	78	59	34	26	87	73	49	59	43	77	64	18	77
2	5	9	79	62	31	28	84	71	49	62	42	78	62	19	74
2	5	10	80	64	28	30	81	69	49	64	42	79	60	21	72
2	5	11	81	66	26	32	79	68	49	66	41	80	59	23	70
2	5	12	81	68	24	33	77	67	49	68	41	81	57	24	68
2	5	13	82	69	23	35	76	66	49	69	41	81	56	25	66
2	5	14	82	71	21	36	74	65	49	71	41	82	56	26	65
2	5	15	83	72	20	37	73	64	49	72	41	82	55	27	64
2	6	7	78	51	42	23	86	79	48	53	48	73	68	13	86
2	6	9	80	57	35	27	87	75	47	59	45	75	63	15	79
2	6	11	81	62	30	30	82	71	47	64	44	77	60	19	74
2	6	13	82	65	26	33	78	69	47	67	43	79	57	22	70
2	6	15	83	68	23	35	76	67	48	70	42	80	56	24	67
2	7	7	79	46	46	24	83	83	47	51	51	70	70	11	90
2	7	8	79	50	42	25	87	81	46	54	49	71	67	11	86
2	7	9	80	53	39	26	90	78	46	57	48	72	65	12	83
2	7	10	81	56	36	28	87	77	46	59	47	73	63	14	80
2	7	11	81	58	33	29	85	75	46	61	46	74	61	15	78
2	7	12	82	60	31	30	83	73	46	63	45	75	60	17	75
2	7	13	82	62	29	32	81	72	46	65	45	76	59	18	73
2	7	14	83	64	27	33	79	71	46	66	44	77	57	20	72
2	7	15	83	65	26	34	78	70	46	68	44	78	57	21	70
2	8	9	81	49	42	25	87	82	45	55	50	70	66	10	87

MILL IND X	Y	Z	(100)- α1	α2	α3	(111)- α1	α2	α3	α4	(110)-POLES α1	α2	α3	α4	α5	α6
2	8	11	82	54	37	28	88	78	44	59	48	72	62	12	81
2	8	13	83	59	32	30	84	75	45	63	46	74	60	15	77
2	8	15	83	62	29	33	80	72	45	66	45	76	58	18	73
2	9	9	81	46	46	26	85	85	44	53	53	67	67	9	90
2	9	10	82	49	43	27	88	83	44	55	51	69	65	9	87
2	9	11	82	51	40	28	90	81	44	57	50	70	64	10	84
2	9	12	82	54	38	29	88	79	44	59	49	71	62	11	82
2	9	13	83	56	35	30	86	77	44	61	48	72	61	13	80
2	9	14	83	58	33	31	84	76	44	62	48	73	60	14	78
2	9	15	83	59	32	32	82	75	44	64	47	74	59	15	76
2	10	11	82	48	43	28	88	83	43	56	52	68	65	8	87
2	10	13	83	53	38	29	88	80	43	59	50	70	62	10	83
2	10	15	84	57	34	31	85	77	43	62	48	72	60	13	79
2	11	11	83	45	45	28	86	86	43	54	54	66	66	7	90
2	11	12	83	48	43	28	88	84	42	56	53	67	64	7	88
2	11	13	83	50	41	29	90	82	42	58	52	68	63	8	85
2	11	14	84	52	39	30	88	81	42	59	51	69	62	9	83
2	11	15	84	54	37	30	86	79	42	61	50	70	61	11	81
2	12	13	84	48	43	29	88	84	42	56	53	67	64	7	88
2	12	15	84	52	39	30	88	81	42	59	52	69	62	9	84
2	13	13	84	45	45	29	86	86	41	55	55	65	65	6	90
2	13	14	84	47	43	29	88	85	41	56	54	66	64	6	88
2	13	15	84	49	41	30	90	83	41	58	53	67	63	7	86
2	14	15	84	47	43	30	88	85	41	57	54	66	64	6	88
2	15	15	85	45	45	30	87	87	41	56	56	64	64	5	90
3	3	4	59	59	47	8	79	67	67	43	32	90	83	32	83
3	3	5	63	63	40	14	85	64	64	50	30	90	78	30	78
3	3	7	68	68	31	24	86	60	60	59	30	90	70	30	70
3	3	8	71	71	28	27	83	59	59	62	31	90	67	31	67
3	3	10	74	74	23	32	78	58	58	67	32	90	63	32	63
3	3	11	75	75	21	34	76	57	57	69	33	90	61	33	61
3	3	13	77	77	18	37	73	57	57	72	34	90	59	34	59
3	3	14	78	78	17	38	72	56	56	73	35	90	58	35	58
3	4	4	62	51	51	7	74	74	63	39	39	84	84	28	90
3	4	5	65	56	45	12	81	71	61	46	37	84	78	26	84
3	4	6	67	59	40	16	86	68	59	51	35	85	74	25	80
3	4	7	70	62	36	20	90	66	58	55	35	85	71	25	76
3	4	8	71	65	32	23	86	65	57	58	34	86	68	26	73
3	4	9	73	67	29	26	84	63	56	61	34	86	66	27	70
3	4	10	74	69	27	29	81	62	55	64	35	86	64	28	68
3	4	11	76	71	24	31	79	61	55	66	35	87	62	29	66
3	4	12	77	72	23	32	77	61	55	68	35	87	61	30	64
3	4	13	78	73	21	34	76	60	55	69	36	87	59	30	63
3	4	14	78	74	20	35	74	60	54	71	36	87	58	31	62
3	4	15	79	75	18	37	73	59	54	72	36	87	56	32	61
3	5	5	67	49	49	12	77	77	58	43	43	79	79	23	90
3	5	6	69	53	44	15	82	74	56	47	40	80	75	22	85
3	5	7	71	57	40	18	86	72	55	52	39	81	72	21	81
3	5	8	72	60	36	21	90	70	54	55	38	82	69	22	78

TAB. VII FROM HKL INTO ALPHA ANGLES

MILL IND X	Y	Z	(100)- $\alpha 1$	$\alpha 2$	$\alpha 3$	(111)- $\alpha 1$	$\alpha 2$	$\alpha 3$	$\alpha 4$	(110)-POLES $\alpha 1$	$\alpha 2$	$\alpha 3$	$\alpha 4$	$\alpha 5$	$\alpha 6$
3	5	9	74	62	33	24	87	68	54	58	38	82	67	23	75
3	5	10	75	64	30	26	84	66	53	61	37	83	65	24	72
3	5	11	76	66	28	28	82	65	53	63	37	83	63	25	70
3	5	12	77	68	26	30	80	64	53	65	37	84	62	26	68
3	5	13	78	69	24	32	78	64	53	67	37	84	60	27	67
3	5	14	79	71	23	33	77	63	52	68	38	85	59	28	65
3	5	15	79	72	21	34	75	62	52	69	38	85	58	29	64
3	6	7	72	52	44	18	83	76	53	49	43	77	73	19	86
3	6	8	73	55	40	20	87	74	53	52	42	78	70	19	82
3	6	10	76	60	34	24	87	70	51	58	40	80	66	20	76
3	6	11	77	62	31	26	85	69	51	60	40	81	64	21	74
3	6	13	78	66	27	30	81	67	51	64	39	82	61	23	70
3	6	14	79	67	26	31	79	66	51	66	39	82	60	24	69
3	7	7	73	47	47	18	80	80	52	47	47	74	74	17	90
3	7	8	74	51	44	20	84	78	51	50	45	75	71	16	86
3	7	9	75	54	40	21	87	76	50	53	44	76	69	16	83
3	7	10	76	56	37	23	90	74	50	56	43	77	67	17	80
3	7	11	77	58	35	25	88	72	50	58	42	78	65	18	78
3	7	12	78	60	32	27	85	71	49	60	42	79	63	19	76
3	7	13	79	62	30	28	83	70	49	62	41	79	62	20	74
3	7	14	79	64	29	30	82	69	49	64	41	80	61	21	72
3	7	15	80	65	27	31	80	68	49	65	41	80	60	22	70
3	8	8	75	47	47	20	81	81	50	48	48	72	72	15	90
3	8	9	76	50	44	21	85	79	49	51	47	73	70	14	87
3	8	10	77	53	41	23	87	77	49	54	46	74	68	15	84
3	8	11	78	55	38	24	90	76	48	56	45	75	66	15	81
3	8	12	78	57	35	26	88	74	48	58	44	76	64	16	79
3	8	13	79	59	33	27	86	73	48	60	43	77	63	17	77
3	8	14	79	61	31	28	84	72	48	62	43	78	62	18	75
3	8	15	80	62	30	30	82	70	48	63	42	78	61	20	73
3	9	10	77	49	43	23	85	80	48	52	48	72	69	13	87
3	9	11	78	52	41	24	88	79	47	54	47	73	67	13	84
3	9	13	79	56	36	26	88	75	47	58	45	75	64	15	80
3	9	14	80	58	34	27	86	74	47	60	45	75	63	16	78
3	10	10	78	46	46	23	83	83	47	51	51	70	70	12	90
3	10	11	79	49	44	24	86	81	47	53	49	71	68	12	87
3	10	12	79	51	41	25	88	80	46	55	48	72	66	12	85
3	10	13	80	53	39	26	90	78	46	57	47	73	65	13	83
3	10	14	80	55	37	27	88	77	46	58	47	74	64	14	81
3	10	15	81	57	35	28	86	75	46	60	46	74	62	15	79
3	11	11	79	46	46	24	84	84	46	51	51	69	69	11	90
3	11	12	80	48	44	25	86	82	46	53	50	70	67	11	88
3	11	13	80	50	41	26	88	80	45	55	49	71	66	11	85
3	11	14	80	52	39	26	90	79	45	57	48	72	64	12	83
3	11	15	81	54	37	27	88	78	45	58	48	73	63	13	81
3	12	13	80	48	44	26	86	83	45	54	51	69	67	10	88
3	12	14	81	50	41	26	88	81	45	55	50	70	65	10	86
3	13	13	81	46	46	26	85	85	45	53	53	68	68	9	90
3	13	14	81	48	44	26	87	83	44	54	52	69	66	9	88

MILL IND X	Y	Z	(100)- $\alpha 1$	$\alpha 2$	$\alpha 3$	(111)- $\alpha 1$	$\alpha 2$	$\alpha 3$	$\alpha 4$	(110)-POLES $\alpha 1$	$\alpha 2$	$\alpha 3$	$\alpha 4$	$\alpha 5$	$\alpha 6$
3	13	15	81	50	42	27	88	82	44	56	51	69	65	10	86
3	14	14	81	46	46	27	85	85	44	53	53	67	67	9	90
3	14	15	82	48	44	27	87	84	44	55	52	68	66	9	88
4	4	5	58	58	49	6	77	68	68	41	33	90	85	33	85
4	4	7	64	64	39	16	86	63	63	51	30	90	76	30	76
4	4	9	68	68	32	23	87	61	61	58	30	90	71	30	71
4	4	11	71	71	27	28	82	59	59	63	31	90	66	31	66
4	4	13	74	74	24	31	78	58	58	66	32	90	63	32	63
4	4	15	76	76	21	34	75	57	57	69	33	90	61	33	61
4	5	5	61	52	52	6	73	73	65	38	38	85	85	29	90
4	5	6	63	55	47	9	79	71	63	44	36	85	81	28	85
4	5	7	65	58	42	13	83	69	61	48	35	86	77	27	81
4	5	8	67	61	39	17	87	67	60	52	34	86	74	26	78
4	5	9	69	63	35	20	90	65	58	55	34	86	71	26	75
4	5	10	70	65	33	23	87	64	58	58	34	87	69	27	73
4	5	11	72	67	30	25	85	63	57	60	34	87	67	27	71
4	5	12	73	68	28	27	83	62	57	62	34	87	65	28	69
4	5	13	74	70	26	29	81	61	56	64	34	87	64	29	67
4	5	14	75	71	25	30	79	61	56	66	34	87	63	29	66
4	5	15	76	72	23	32	78	60	56	67	35	88	62	30	64
4	6	7	67	53	46	12	80	73	59	45	39	82	78	24	86
4	6	9	70	59	39	18	87	69	57	52	37	83	72	23	79
4	6	11	72	63	33	23	87	67	55	57	36	84	68	24	74
4	6	13	74	66	29	27	83	65	54	62	36	85	65	25	71
4	6	15	76	69	26	30	80	63	54	65	36	85	62	27	68
4	7	7	68	49	49	13	78	78	57	43	43	79	79	22	90
4	7	8	69	52	45	15	81	75	56	47	42	79	76	21	86
4	7	9	71	55	42	17	85	73	55	50	40	80	73	21	83
4	7	10	72	57	39	19	87	72	54	53	40	80	71	21	80
4	7	11	73	59	36	21	90	70	54	55	39	81	69	21	78
4	7	12	74	61	34	23	88	69	53	57	39	82	67	22	76
4	7	13	75	63	32	25	86	68	53	59	38	82	65	22	74
4	7	14	76	64	30	27	84	67	53	61	38	82	64	23	72
4	7	15	76	66	28	28	82	66	52	63	38	83	63	24	71
4	8	9	72	51	45	17	82	77	54	48	44	77	74	19	87
4	8	11	74	56	39	21	88	73	52	53	42	78	70	19	81
4	8	13	75	60	35	24	88	71	52	57	40	80	66	20	77
4	8	15	77	63	31	27	84	69	51	61	40	81	64	21	74
4	9	9	73	48	48	18	80	80	53	46	46	75	75	17	90
4	9	10	73	50	45	19	83	78	52	49	45	75	72	17	87
4	9	11	74	52	42	20	86	76	51	51	44	76	70	17	85
4	9	12	75	55	39	22	88	75	51	54	43	77	69	17	82
4	9	13	76	57	37	23	90	74	50	56	43	77	67	17	80
4	9	14	76	58	35	24	88	72	50	58	42	78	66	18	78
4	9	15	77	60	33	26	86	71	50	59	42	79	64	19	76
4	10	11	75	49	44	20	84	79	50	50	46	74	71	15	87
4	10	13	76	54	40	23	88	76	49	54	45	75	68	16	83
4	10	15	77	57	36	25	88	74	49	58	43	77	65	17	79
4	11	11	76	47	47	21	82	82	50	49	49	72	72	14	90

MILL IND			(100)-			(111)-				(110)-POLES					
X	Y	Z	α1	α2	α3	α1	α2	α3	α4	α1	α2	α3	α4	α5	α6
4	11	12	76	49	44	22	84	80	49	51	48	73	70	14	88
4	11	13	77	51	42	22	86	79	49	53	47	74	69	14	85
4	11	14	77	53	40	23	88	77	48	54	46	74	67	14	83
4	11	15	78	55	38	24	90	76	48	56	45	75	66	15	81
4	12	13	77	49	44	23	85	81	48	51	48	72	69	13	88
4	12	15	78	52	40	24	88	78	47	55	47	73	67	13	84
4	13	13	78	46	46	23	83	83	48	50	50	70	70	12	90
4	13	14	78	48	44	24	85	81	47	52	49	71	69	12	88
4	13	15	79	50	42	24	87	80	47	54	48	72	67	12	86
4	14	15	79	48	44	24	85	82	46	52	50	70	68	11	88
4	15	15	79	46	46	25	84	84	46	52	52	69	69	11	90
5	5	6	57	57	50	5	76	68	68	40	33	90	86	33	86
5	5	7	60	60	45	9	80	66	66	45	31	90	82	31	82
5	5	8	62	62	41	13	84	64	64	49	31	90	79	31	79
5	5	9	64	64	38	17	87	63	63	52	30	90	76	30	76
5	5	11	68	68	33	22	87	61	61	57	30	90	71	30	71
5	5	12	69	69	31	24	85	60	60	59	30	90	69	30	69
5	5	13	70	70	29	26	83	60	60	61	31	90	68	31	68
5	5	14	71	71	27	28	82	59	59	63	31	90	66	31	66
5	6	6	59	52	52	5	73	73	66	38	38	86	86	31	90
5	6	7	62	55	48	8	77	71	64	42	36	86	82	29	86
5	6	8	63	58	44	11	81	69	62	46	35	86	79	28	83
5	6	9	65	60	41	14	84	67	61	49	34	87	76	27	80
5	6	10	67	62	38	17	87	66	60	52	33	87	74	27	77
5	6	11	68	64	35	20	90	65	59	55	33	87	72	27	75
5	6	12	70	65	33	22	88	64	58	57	33	87	70	27	73
5	6	13	71	67	31	24	86	63	58	59	33	87	68	28	71
5	6	14	72	68	29	26	84	62	57	61	33	87	67	28	69
5	6	15	73	69	28	27	82	61	57	63	33	88	65	29	68
5	7	7	63	51	51	8	75	75	62	40	40	83	83	27	90
5	7	8	65	53	47	11	79	73	61	44	39	83	80	25	87
5	7	9	66	56	44	13	82	71	59	47	37	83	77	25	83
5	7	10	68	58	41	16	85	70	58	50	36	84	74	24	81
5	7	11	69	60	38	18	88	68	57	53	36	84	72	24	78
5	7	12	70	62	36	20	90	67	57	55	35	85	70	25	76
5	7	13	71	63	33	22	88	66	56	57	35	85	69	25	74
5	7	14	72	65	32	24	86	65	56	59	35	85	67	25	72
5	7	15	73	66	30	26	84	64	55	61	35	85	66	26	71
5	8	8	66	50	50	11	77	77	59	42	42	80	80	24	90
5	8	9	67	52	46	13	80	75	58	45	41	81	77	23	87
5	8	10	69	54	43	15	83	73	57	48	40	81	75	22	84
5	8	11	70	56	41	17	85	71	56	51	39	82	73	22	82
5	8	12	71	58	38	19	88	70	55	53	38	82	71	22	79
5	8	13	72	60	36	21	90	69	55	55	38	82	69	22	77
5	8	14	73	62	34	23	88	68	54	57	37	83	68	23	75
5	8	15	74	63	32	24	86	67	54	59	37	83	66	23	74
5	9	9	69	49	49	14	78	78	57	44	44	78	78	21	90
5	9	10	70	51	46	15	81	76	56	46	42	79	76	21	87
5	9	11	71	53	43	17	83	74	55	49	41	79	74	20	85

MILL IND			(100)-			(111)-				(110)-POLES					
X	Y	Z	α1	α2	α3	α1	α2	α3	α4	α1	α2	α3	α4	α5	α6
5	9	12	72	55	41	16	86	73	54	51	41	80	72	20	82
5	9	13	72	57	38	20	88	72	54	53	40	80	70	20	80
5	9	14	73	59	36	22	90	71	53	55	39	81	69	21	78
5	9	15	74	60	34	23	88	70	53	57	39	81	67	21	77
5	10	11	71	50	45	17	82	77	54	47	44	77	74	19	87
5	10	12	72	52	43	18	84	76	53	50	43	78	72	18	85
5	10	13	73	54	41	19	86	74	53	52	42	78	71	18	83
5	10	14	74	56	39	21	88	73	52	54	41	79	69	19	81
5	11	11	72	48	48	17	80	80	53	46	46	75	75	18	90
5	11	12	73	50	45	18	82	78	52	48	45	76	73	17	88
5	11	13	74	52	43	19	84	77	52	50	44	76	71	17	85
5	11	14	74	54	41	21	86	76	51	52	43	77	70	17	83
5	11	15	75	55	39	22	88	74	51	54	43	77	68	17	82
5	12	12	74	47	47	19	81	81	52	47	47	74	74	16	90
5	12	13	74	49	45	20	83	79	51	49	46	74	72	16	88
5	12	14	75	51	43	20	85	78	51	51	45	75	71	16	86
5	12	15	75	53	41	21	87	77	50	53	45	76	69	16	84
5	13	13	75	47	47	20	81	81	50	48	48	73	73	15	90
5	13	14	75	49	45	21	83	80	50	50	47	73	71	15	88
5	13	15	76	51	43	21	85	79	50	52	46	74	70	15	86
5	14	14	76	47	47	21	82	82	49	49	49	72	72	14	90
5	14	15	76	48	45	22	84	81	49	50	48	72	70	14	88
6	6	7	57	57	50	4	75	68	68	40	33	90	86	33	86
6	6	11	64	64	38	17	88	63	63	52	30	90	75	30	75
6	6	13	67	67	33	22	88	61	61	57	30	90	71	30	71
6	7	7	59	53	53	4	73	73	66	37	37	86	86	31	90
6	7	8	61	55	49	7	76	71	65	41	36	87	83	30	87
6	7	9	62	57	46	10	80	69	63	44	35	87	81	29	84
6	7	10	64	59	43	13	83	68	62	47	34	87	78	28	81
6	7	11	65	61	40	15	85	66	61	50	33	87	76	28	79
6	7	12	67	62	38	17	88	65	60	53	33	87	74	27	76
6	7	13	68	64	35	20	90	64	60	55	33	87	72	27	75
6	7	14	69	65	33	22	88	63	59	57	32	88	70	28	73
6	7	15	70	67	32	23	86	63	58	59	33	88	69	28	71
6	8	9	64	54	48	9	78	73	62	43	38	84	81	27	87
6	8	11	66	57	42	14	83	70	60	48	36	85	76	25	82
6	8	13	69	61	38	18	88	67	58	53	35	85	72	25	78
6	8	15	71	64	34	22	88	65	57	57	35	86	69	26	74
6	9	10	66	52	47	12	79	74	59	44	40	82	79	24	87
6	9	11	67	54	45	13	81	73	58	47	39	82	77	24	85
6	9	13	69	58	40	17	86	70	57	51	37	83	73	23	80
6	9	14	70	59	38	19	88	69	56	53	37	83	71	23	78
6	10	11	68	51	47	13	80	75	57	45	41	80	77	22	87
6	10	13	70	55	42	17	84	73	56	50	40	81	74	21	83
6	10	15	72	58	38	20	88	70	55	53	39	81	70	22	79
6	11	11	69	49	49	14	78	78	56	44	44	78	78	21	90
6	11	12	70	51	46	15	80	77	56	46	43	78	76	20	88
6	11	13	71	52	44	16	83	75	55	48	42	79	74	20	86
6	11	14	71	54	42	18	85	74	54	50	41	79	72	20	84

TAB. VII FROM HKL INTO ALPHA ANGLES

MILL IND X	Y	Z	(100)- α1	α2	α3	(111)- α1	α2	α3	α4	(110)-POLES α1	α2	α3	α4	α5	α6
6	11	15	72	56	40	19	87	73	54	52	41	80	71	20	82
6	12	13	71	50	46	17	81	78	54	47	44	77	75	19	88
6	13	13	72	48	48	17	80	80	53	46	46	75	75	18	90
6	13	14	73	50	46	18	82	78	53	48	45	76	74	18	88
6	13	15	73	51	44	19	84	77	52	50	44	76	72	17	86
6	14	15	74	49	45	19	82	79	52	49	46	75	73	16	88
7	7	8	57	57	51	4	74	69	69	39	34	90	87	34	87
7	7	9	58	58	48	7	78	67	67	42	32	90	84	32	84
7	7	10	60	60	45	10	81	66	66	45	31	90	81	31	81
7	7	11	62	62	42	13	83	65	65	48	31	90	79	31	79
7	7	12	63	63	40	15	86	64	64	50	30	90	77	30	77
7	7	13	65	65	37	17	88	63	63	53	30	90	75	30	75
7	7	15	67	67	33	21	88	61	61	57	30	90	72	30	72
7	8	8	58	53	53	4	72	72	67	37	37	87	87	32	90
7	8	9	60	55	50	6	76	71	66	40	36	87	84	30	87
7	8	10	61	57	47	9	79	69	64	43	35	87	82	29	84
7	8	11	63	58	44	11	81	68	63	46	34	87	79	29	82
7	8	12	64	60	42	13	84	67	62	49	33	87	77	28	80
7	8	13	65	62	39	16	86	66	61	51	33	88	75	28	78
7	8	14	67	63	37	18	88	65	60	53	32	88	74	28	76
7	8	15	68	64	35	20	90	64	60	55	32	88	72	28	74
7	9	9	61	52	52	6	74	74	64	39	39	84	84	29	90
7	9	10	63	54	49	8	77	72	63	42	38	85	82	28	87
7	9	11	64	55	46	10	80	71	62	44	37	85	80	27	85
7	9	12	65	57	44	12	82	70	61	47	36	85	78	26	83
7	9	13	66	59	41	14	84	68	60	49	35	85	76	26	81
7	9	14	67	60	39	16	86	67	59	51	35	86	74	26	79
7	9	15	68	61	37	18	88	67	59	53	34	86	73	26	77
7	10	10	64	51	51	9	75	75	62	40	40	82	82	26	90
7	10	11	65	53	48	10	78	74	61	43	39	83	80	25	88
7	10	12	66	54	45	12	80	72	60	45	38	83	78	25	85
7	10	13	67	56	43	14	83	71	59	48	38	83	76	24	83
7	10	14	68	57	41	16	85	70	58	50	37	83	75	24	81
7	10	15	69	59	39	17	87	69	57	52	36	84	73	24	79
7	11	11	66	50	50	11	76	76	59	42	42	80	80	24	90
7	11	12	67	52	47	12	79	75	59	44	41	81	78	23	88
7	11	13	68	53	45	14	81	74	58	46	40	81	77	23	86
7	11	14	69	55	43	15	83	72	57	48	39	81	75	22	84
7	11	15	69	56	41	17	85	71	56	50	38	82	73	22	82
7	12	12	68	49	49	13	77	77	58	43	43	79	79	22	90
7	12	13	68	51	47	14	80	76	57	45	42	79	77	22	88
7	12	14	69	53	45	15	82	75	56	47	41	80	75	21	86
7	12	15	70	54	43	16	84	74	56	49	40	80	74	21	84
7	13	13	69	49	49	14	78	78	56	44	44	78	78	21	90
7	13	14	70	50	47	15	80	77	55	46	43	78	76	20	88
7	13	15	71	52	45	16	82	76	55	48	42	78	74	20	86
7	14	15	71	50	46	17	81	78	54	47	44	77	75	19	88
7	15	15	72	48	48	17	80	80	54	46	46	75	75	18	90
8	8	9	56	56	51	3	74	69	69	39	34	90	87	34	87

MILL IND X	Y	Z	(100)- α1	α2	α3	(111)- α1	α2	α3	α4	(110)-POLES α1	α2	α3	α4	α5	α6
8	8	11	60	60	46	9	79	66	66	44	32	90	82	32	82
8	8	13	62	62	41	14	84	64	64	49	30	90	78	30	78
8	8	15	65	65	37	18	88	63	63	53	30	90	75	30	75
8	9	9	58	53	53	3	72	72	67	37	37	87	87	32	90
8	9	10	59	55	50	5	75	71	66	40	36	87	85	31	87
8	9	11	61	57	48	8	78	69	65	43	35	88	83	30	85
8	9	12	62	58	45	10	80	68	64	45	34	88	80	29	83
8	9	13	63	59	43	12	83	67	63	47	33	88	78	29	81
8	9	14	64	61	41	14	85	66	62	49	33	88	77	28	79
8	9	15	65	62	39	16	87	65	61	51	32	88	75	28	77
8	10	11	62	54	49	7	76	72	64	41	37	85	83	28	88
8	10	13	64	57	45	11	81	70	62	46	36	86	79	27	83
8	10	15	66	60	40	15	85	68	60	50	34	86	75	26	80
8	11	11	63	51	51	8	75	75	62	40	40	83	83	27	90
8	11	12	64	53	49	9	77	73	61	42	39	83	81	26	88
8	11	13	65	54	46	11	79	72	61	44	38	84	79	26	86
8	11	14	66	56	44	13	81	71	60	47	37	84	77	25	84
8	11	15	67	57	42	14	83	70	59	48	37	84	76	25	82
8	12	13	66	52	48	11	78	74	60	43	40	82	80	24	88
8	12	15	67	55	44	14	82	72	58	47	39	82	76	23	84
8	13	13	66	50	50	12	77	77	59	42	42	80	80	24	90
8	13	14	67	51	47	13	79	75	58	44	41	80	78	23	88
8	13	15	68	53	46	14	81	74	57	46	41	80	77	22	86
8	14	15	69	51	47	14	79	76	57	45	42	79	77	21	88
8	15	15	69	49	49	15	78	78	56	44	44	77	77	21	90
9	9	10	56	56	52	3	73	69	69	38	34	90	87	34	87
9	9	11	58	58	49	6	76	68	68	41	33	90	85	33	85
9	9	13	60	60	44	10	81	66	66	46	31	90	81	31	81
9	9	14	62	62	42	12	83	65	65	48	31	90	79	31	79
9	10	10	58	53	53	3	72	72	68	37	37	88	88	32	90
9	10	11	59	55	51	5	75	71	67	39	36	88	85	31	88
9	10	12	60	56	48	7	77	69	65	42	35	88	83	30	86
9	10	13	61	58	46	9	79	68	64	44	34	88	81	30	83
9	10	14	62	59	44	11	81	67	64	46	33	88	80	29	82
9	10	15	63	60	42	13	83	66	63	48	33	88	78	29	80
9	11	11	60	52	52	5	73	73	65	38	38	85	85	30	90
9	11	12	61	54	50	7	76	72	64	41	37	86	83	29	88
9	11	13	62	55	48	8	78	71	63	43	36	86	82	28	86
9	11	14	63	57	45	10	80	70	62	45	35	86	80	28	84
9	11	15	64	58	43	12	82	69	62	47	35	86	78	27	82
9	12	13	63	53	49	9	77	73	62	42	38	84	82	27	88
9	12	14	64	54	47	10	79	72	61	44	38	84	80	26	86
9	13	13	64	51	51	9	75	75	61	41	41	82	82	26	90
9	13	14	65	52	48	10	77	74	61	43	40	82	80	25	88
9	13	15	66	53	47	11	79	73	60	44	39	83	79	25	86
9	14	14	66	50	50	11	76	76	60	42	42	81	81	24	90
9	14	15	66	51	48	12	78	75	59	43	41	81	79	24	88
10	10	11	56	56	52	3	73	69	69	38	34	90	88	34	88
10	10	13	59	59	47	7	78	67	67	43	32	90	84	32	84

TAB. VII FROM HKL INTO ALPHA ANGLES

MILL IND X	Y	Z	(100)- $\alpha1$	$\alpha2$	$\alpha3$	(111)- $\alpha1$	$\alpha2$	$\alpha3$	$\alpha4$	(110)-POLES $\alpha1$	$\alpha2$	$\alpha3$	$\alpha4$	$\alpha5$	$\alpha6$
10	11	11	57	54	54	3	72	72	68	37	37	88	88	33	90
10	11	12	58	55	51	4	74	71	67	39	35	88	86	32	88
10	11	13	60	56	49	6	76	69	66	41	35	88	84	31	86
10	11	14	61	57	47	8	79	68	65	43	34	88	82	30	84
10	11	15	62	59	45	10	81	67	64	45	33	88	80	29	82
10	12	13	61	54	50	6	75	72	65	40	37	86	84	30	88
10	12	15	62	56	46	9	79	70	63	44	35	86	81	28	84
10	13	13	61	52	52	7	74	74	64	39	39	84	84	29	90
10	13	14	62	53	50	8	76	73	63	41	38	84	82	28	88
10	13	15	63	54	48	9	78	72	62	43	37	85	81	27	86
10	14	15	64	52	49	9	77	74	61	42	39	83	81	26	88
11	11	12	56	56	52	2	73	69	69	38	34	90	88	34	88
11	11	13	57	57	50	5	75	68	68	40	33	90	86	33	86
11	11	14	58	58	48	7	77	67	67	42	32	90	84	32	84
11	11	15	59	59	46	9	79	66	66	44	32	90	82	32	82
11	12	12	57	54	54	2	72	72	68	36	36	88	88	33	90
11	12	13	58	55	51	4	74	71	67	39	35	88	86	32	88
11	12	14	59	56	49	6	76	70	66	41	35	88	84	31	86
11	12	15	60	57	47	8	78	69	65	43	34	88	83	30	85
11	13	13	59	53	53	4	73	73	66	38	38	86	86	31	90
11	13	14	60	54	51	6	75	72	65	40	37	86	84	30	88
11	13	15	61	55	49	7	77	71	64	42	36	86	83	29	86
11	14	14	61	52	52	6	74	74	64	39	39	85	85	29	90
11	14	15	62	53	50	7	76	73	63	41	38	85	83	28	88
11	15	15	63	51	51	8	75	75	63	40	40	83	83	27	90
12	12	13	56	56	53	2	73	69	69	37	34	90	88	34	88
12	13	13	57	54	54	2	72	72	68	36	36	88	88	33	90
12	13	14	58	55	52	4	74	71	67	38	35	88	86	32	88
12	13	15	59	56	50	5	76	70	67	40	35	88	85	31	87
12	14	15	60	54	51	5	75	72	66	39	37	87	85	30	88
13	13	14	56	56	53	2	73	70	70	37	34	90	88	34	88
13	13	15	57	57	51	4	74	69	69	39	33	90	87	33	87
13	14	14	57	54	54	2	72	72	69	36	36	88	88	33	90
13	14	15	58	55	52	3	73	71	68	38	35	88	87	32	88
13	15	15	58	53	53	4	72	72	67	37	37	87	87	32	90
14	14	15	56	56	53	2	72	70	70	37	34	90	88	34	88
14	15	15	57	54	54	2	71	71	69	36	36	88	88	33	90